Student Solutions Manual

College Algebra

TENTH EDITION

Ron Larson
The Pennsylvania State University, The Behrend College

Prepared by

Ron Larson
The Pennsylvania State University, The Behrend College

Australia • Brazil • Mexico • Singapore • United Kingdom • United States

ISBN: 978-1-337-29150-7

Cengage Learning
20 Channel Center Street
Boston, MA 02210
USA

Cengage Learning is a leading provider of customized learning solutions with office locations around the globe, including Singapore, the United Kingdom, Australia, Mexico, Brazil, and Japan. Locate your local office at: **www.cengage.com/global**.

Cengage Learning products are represented in Canada by Nelson Education, Ltd.

To learn more about Cengage Learning Solutions, visit **www.cengage.com**.

Purchase any of our products at your local college store or at our preferred online store **www.cengagebrain.com**.

Printed in the United States of America
Print Number: 01 Print Year: 2017

CONTENTS

CHAPTER P
Prerequisites

CHAPTER P
Prerequisites

Section P.1 Review of Real Numbers and Their Properties

1. irrational

3. absolute value

5. terms

7. $-9, -\frac{7}{2}, 5, \frac{2}{3}, \sqrt{2}, 0, 1, -4, 2, -11$

(a) Natural numbers: 5, 1, 2

(b) Whole numbers: 0, 5, 1, 2

(c) Integers: $-9, 5, 0, 1, -4, 2, -11$

(d) Rational numbers: $-9, -\frac{7}{2}, 5, \frac{2}{3}, 0, 1, -4, 2, -11$

(e) Irrational numbers: $\sqrt{2}$

9. $2.01, 0.\overline{6}, -13, 0.010110111\ldots, 1, -6$

(a) Natural numbers: 1

(b) Whole numbers: 1

(c) Integers: $-13, 1, -6$

(d) Rational numbers: $2.01, 0.\overline{6}, -13, 1, -6$

(e) Irrational numbers: $0.010110111\ldots$

11. (a)

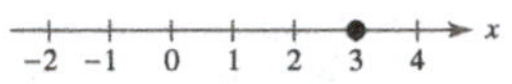

(b) $\frac{7}{2}$

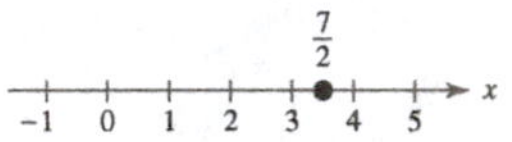

(c) $-\frac{5}{2}$

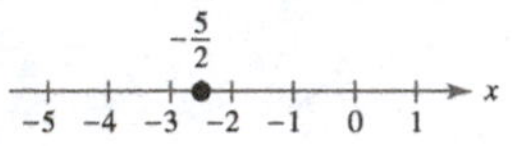

(d) -5.2

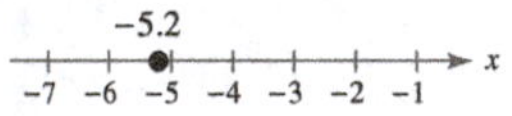

13. $-4 > -8$

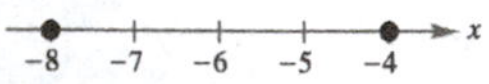

15. $\frac{5}{6} > \frac{2}{3}$

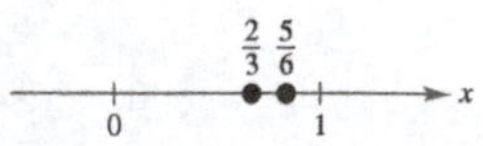

17. (a) The inequality $x \le 5$ denotes the set of all real numbers less than or equal to 5.

(b)

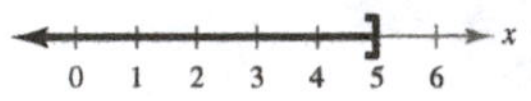

(c) The interval is unbounded.

19. (a) The inequality $-2 < x < 2$ denotes the set of all real numbers greater than -2 and less than 2.

(b)

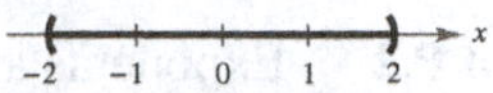

(c) The interval is bounded.

21. (a) The interval $[4, \infty)$ denotes the set of all real numbers greater than or equal to 4.

(b)

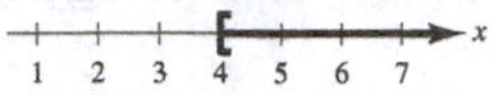

(c) The interval is unbounded.

23. (a) The interval $[-5, 2)$ denotes the set of all real numbers greater than or equal to -5 and less than 2.

(b)

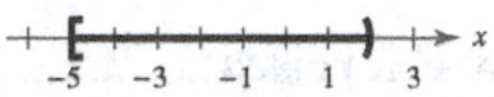

(c) The interval is bounded.

25. $y \ge 0; [0, \infty)$

27. $10 \le t \le 22; [10, 22]$

29. $|-10| = -(-10) = 10$

31. $|3 - 8| = |-5| = -(-5) = 5$

33. $|-1| - |-2| = 1 - 2 = -1$

35. $5|-5| = 5(5) = 25$

37. If $x < -2$, then $x + 2$ is negative.

So, $\dfrac{|x + 2|}{x + 2} = \dfrac{-(x + 2)}{x + 2} = -1.$

39. $|-4| = |4|$ because $|-4| = 4$ and $|4| = 4.$

41. $-|-6| < |-6|$ because $|-6| = 6$ and $-|-6| = -(6) = -6.$

43. $d(126, 75) = |75 - 126| = 51$

45. $d\left(-\frac{5}{2}, 0\right) = \left|0 - \left(-\frac{5}{2}\right)\right| = \frac{5}{2}$

47. $d(x, 5) = |x - 5|$ and $d(x, 5) \le 3$, so $|x - 5| \le 3$.

	Receipts, R	Expenditures, E	$\lvert R - E \rvert$
49.	\$2524.0 billion	\$2982.5 billion	$\lvert 2524.0 - 2982.5 \rvert = \$458.5$ billion
51.	\$2450.0 billion	\$3537.0 billion	$\lvert 2450.0 - 3537.0 \rvert = \$1087.0$ billion

53. $7x + 4$

Terms: $7x, 4$

Coefficient: 7

55. $6x^3 - 5x$

Terms: $6x^3, -5x$

Coefficients: $6, -5$

57. $3\sqrt{3}x^2 + 1$

Terms: $3\sqrt{3}x^2, 1$

Coefficient: $3\sqrt{3}$

59. $4x - 6$

(a) $4(-1) - 6 = -4 - 6 = -10$

(b) $4(0) - 6 = 0 - 6 = -6$

61. $x^2 - 3x + 2$

(a) $(0)^2 - 3(0) + 2 = 2$

(b) $(-1)^2 - 3(-1) + 2 = 1 + 3 + 2 = 6$

63. $\dfrac{x+1}{x-1}$

(a) $\dfrac{1+1}{1-1} = \dfrac{2}{0}$

Division by zero is undefined.

(b) $\dfrac{-1+1}{-1-1} = \dfrac{0}{-2} = 0$

65. $\dfrac{1}{(h+6)}(h+6) = 1, h \neq -6$

Multiplicative Inverse Property

67. $x(3y) = (x \cdot 3)y$ Associative Property of Multiplication

$= (3x)y$ Commutative Property of Multiplication

69. $\dfrac{2x}{3} - \dfrac{x}{4} = \dfrac{8x}{12} - \dfrac{3x}{12} = \dfrac{5x}{12}$

71. $\dfrac{3x}{10} \cdot \dfrac{5}{6} = \dfrac{x}{2} \cdot \dfrac{1}{2} = \dfrac{x}{4}$

73. False. Because zero is nonnegative but not positive, not every nonnegative number is positive.

75. The product of two negative numbers is positive.

77. (a)

n	0.0001	0.01	1	100	10,000
$5/n$	50,000	500	5	0.05	0.0005

(b) (i) As n approaches 0, the value of $5/n$ increases without bound (approaches infinity).

(ii) As n increases without bound (approaches infinity), the value of $5/n$ approaches 0.

Section P.2 Exponents and Radicals

1. exponent; base

3. square root

5. like radicals

7. rationalizing

9. (a) $5 \cdot 5^3 = 5^4 = 625$

(b) $\dfrac{5^2}{5^4} = 5^{-2} = \dfrac{1}{5^2} = \dfrac{1}{25}$

11. (a) $(2^3 \cdot 3^2)^2 = 2^{3 \cdot 2} \cdot 3^{2 \cdot 2}$

$= 2^6 \cdot 3^4 = 64 \cdot 81 = 5184$

(b) $\left(-\dfrac{3}{5}\right)^3\left(\dfrac{5}{3}\right)^2 = (-1)^3\dfrac{3^3}{5^3} \cdot \dfrac{5^2}{3^2} = -1 \cdot 3^{3-2} \cdot 5^{2-3}$

$= -3 \cdot 5^{-1} = -\dfrac{3}{5}$

13. (a) $\dfrac{4 \cdot 3^{-2}}{2^{-2} \cdot 3^{-1}} = 4 \cdot 2^2 \cdot 3^{-2-(-1)} = 4 \cdot 4 \cdot 3^{-1} = \dfrac{16}{3}$

(b) $(-2)^0 = 1$

15. When $x = 2$,

$-3x^3 = -3(2)^3 = -24.$

17. When $x = 10$,

$6x^0 = 6(10)^0 = 6(1) = 6.$

19. When $x = -2$,

$$-3x^4 = -3(-2)^4 = -3(16) = -48.$$

21. (a) $(5z)^3 = 5^3z^3 = 125z^3$

(b) $5x^4(x^2) = 5x^{4+2} = 5x^6$

23. (a) $6y^2(2y^0)^2 = 6y^2(2 \cdot 1)^2 = 6y^2(4) = 24y^2$

(b) $(-z)^3(3z^4) = (-1)^3(z^3)3z^4 = -1 \cdot 3 \cdot z^{3+4} = -3z^7$

25. (a) $\left(\dfrac{4}{y}\right)^3\left(\dfrac{3}{y}\right)^4 = \dfrac{4^3}{y^3} \cdot \dfrac{3^4}{y^4} = \dfrac{64 \cdot 81}{y^{3+4}} = \dfrac{5184}{y^7}$

(b) $\left(\dfrac{b^{-2}}{a^{-2}}\right)\left(\dfrac{b}{a}\right)^2 = \left(\dfrac{a^2}{b^2}\right)\left(\dfrac{b^2}{a^2}\right) = 1,\ a \neq 0,\ b \neq 0$

27. (a) $(x + 5)^0 = 1,\ x \neq -5$

(b) $(2x^2)^{-2} = \dfrac{1}{(2x^2)^2} = \dfrac{1}{4x^4}$

29. (a) $\left(\dfrac{x^{-3}y^4}{5}\right)^{-3} = \left(\dfrac{5x^3}{y^4}\right)^3 = \dfrac{125x^9}{y^{12}}$

(b) $\left(\dfrac{a^{-2}}{b^{-2}}\right)\left(\dfrac{b}{a}\right)^3 = \left(\dfrac{b^2}{a^2}\right)\left(\dfrac{b^3}{a^3}\right) = \dfrac{b^5}{a^5}$

31. $10{,}250.4 = 1.02504 \times 10^4$

33. $3.14 \times 10^{-4} = 0.000314$

35. $9.46 \times 10^{12} = 9{,}460{,}000{,}000{,}000$ kilometers

37. (a) $(2.0 \times 10^9)(3.4 \times 10^{-4}) = 6.8 \times 10^5$

(b) $(1.2 \times 10^7)(5.0 \times 10^{-3}) = 6.0 \times 10^4$

39. (a) $\sqrt{9} = 3$

(b) $\sqrt[3]{\dfrac{27}{8}} = \dfrac{\sqrt[3]{27}}{\sqrt[3]{8}} = \dfrac{3}{2}$

41. (a) $(\sqrt[5]{2})^5 = 2^{5/5} = 2^1 = 2$

(b) $\sqrt[5]{32x^5} = \sqrt[5]{(2x)^5} = 2x$

43. (a) $\sqrt{20} = \sqrt{4 \cdot 5} = \sqrt{4}\sqrt{5} = 2\sqrt{5}$

(b) $\sqrt[3]{128} = \sqrt[3]{64 \cdot 2} = \sqrt[3]{64}\sqrt[3]{2} = 4\sqrt[3]{2}$

45. (a) $\sqrt{72x^3} = \sqrt{36x^2 \cdot 2x} = 6x\sqrt{2x}$

(b) $\sqrt{54xy^4} = \sqrt{6 \cdot 3^2 \cdot x \cdot (y^2)^2} = 3y^2\sqrt{6x}$

47. (a) $\sqrt[3]{16x^5} = \sqrt[3]{(2x)^3 \cdot 2x^2} = 2x\sqrt[3]{2x^2}$

(b)
$$\sqrt{75x^2y^{-4}} = \sqrt{\frac{75x^2}{y^4}} = \frac{\sqrt{25x^2 \cdot 3}}{\sqrt{y^4}} = \frac{\sqrt{(5x)^2 \cdot 3}}{\sqrt{(y^2)^2}} = \frac{5|x|\sqrt{3}}{y^2}$$

49. (a)
$$\begin{aligned} 2\sqrt{20x^2} + 5\sqrt{125x^2} &= 2\sqrt{4x^2 \cdot 5} + 5\sqrt{25x^2 \cdot 5} \\ &= 2\sqrt{(2x)^2 \cdot 5} + 5\sqrt{(5x)^2 \cdot 5} \\ &= 4|x|\sqrt{5} + 25|x|\sqrt{5} \\ &= 29|x|\sqrt{5} \end{aligned}$$

(b)
$$\begin{aligned} 8\sqrt{147x} - 3\sqrt{48x} &= 8\sqrt{49 \cdot 3x} - 3\sqrt{16 \cdot 3x} \\ &= 8\sqrt{7^2 \cdot 3x} - 3\sqrt{4^2 \cdot 3x} \\ &= 56\sqrt{3x} - 12\sqrt{3x} \\ &= 44\sqrt{3x} \end{aligned}$$

51. $\dfrac{1}{\sqrt{3}} = \dfrac{1}{\sqrt{3}} \cdot \dfrac{\sqrt{3}}{\sqrt{3}} = \dfrac{\sqrt{3}}{3}$

53. $\dfrac{5}{\sqrt{14}-2} = \dfrac{5}{\sqrt{14}-2} \cdot \dfrac{\sqrt{14}+2}{\sqrt{14}+2} = \dfrac{5(\sqrt{14}+2)}{(\sqrt{14})^2-(2)^2} = \dfrac{5(\sqrt{14}+2)}{14-4} = \dfrac{5(\sqrt{14}+2)}{10} = \dfrac{\sqrt{14}+2}{2}$

55. $\dfrac{\sqrt{5}+\sqrt{3}}{3} = \dfrac{\sqrt{5}+\sqrt{3}}{3} \cdot \dfrac{\sqrt{5}-\sqrt{3}}{\sqrt{5}-\sqrt{3}} = \dfrac{5-3}{3(\sqrt{5}-\sqrt{3})} = \dfrac{2}{3(\sqrt{5}-\sqrt{3})}$

57. $\sqrt[3]{64} = 64^{1/3}$

59. $3x^{-2/3} = \dfrac{3}{x^{2/3}}$

$= \dfrac{3}{\sqrt[3]{x^2}}, x \neq 0$

61. (a) $32^{-3/5} = \dfrac{1}{32^{3/5}} = \dfrac{1}{(\sqrt[5]{32})^3} = \dfrac{1}{(2)^3} = \dfrac{1}{8}$

(b) $\left(\dfrac{16}{81}\right)^{-3/4} = \left(\dfrac{81}{16}\right)^{3/4} = \left(\sqrt[4]{\dfrac{81}{16}}\right)^3 = \left(\dfrac{3}{2}\right)^3 = \dfrac{27}{8}$

63. (a) $\sqrt[4]{3^2} = 3^{2/4} = 3^{1/2} = \sqrt{3}$

(b) $\sqrt[6]{(x+1)^4} = (x+1)^{4/6} = (x+1)^{2/3} = \sqrt[3]{(x+1)^2}$

65. (a) $\sqrt{\sqrt{32}} = \left(32^{1/2}\right)^{1/2}$

$= 32^{1/4} = \sqrt[4]{32} = \sqrt[4]{16 \cdot 2} = 2\sqrt[4]{2}$

(b) $\sqrt{\sqrt[4]{2x}} = \left((2x)^{1/4}\right)^{1/2} = (2x)^{1/8} = \sqrt[8]{2x}$

67. (a) $(x-1)^{1/3}(x-1)^{2/3} = (x-1)^{3/3} = x-1$

(b) $(x-1)^{1/3}(x-1)^{-4/3} = (x-1)^{-3/3}$

$= (x-1)^{-1}$

$= \dfrac{1}{x-1}$

69. $t = 0.03\left[12^{5/2} - (12-h)^{5/2}\right], 0 \le h \le 12$

h (in centimeters)	t (in seconds)
0	0
1	2.93
2	5.48
3	7.67
4	9.53
5	11.08
6	12.32
7	13.29
8	14.00
9	14.50
10	14.80
11	14.93
12	14.96

71. False. When $x = 0$, the expressions are not equal.

73. False. When a sum is raised to a power, you multiply the sum by itself using the Distributive Property.

$(a+b)^2 = a^2 + 2ab + b^2 \neq a^2 + b^2$

Section P.3 Polynomials and Special Products

1. n; a_n; a_0

3. like terms

5. (a) Standard form: $7x$

(b) Degree: 1
Leading coefficient: 7

(c) Monomial

7. (a) Standard form: $-\frac{1}{2}x^5 + 14x$

(b) Degree: 5
Leading coefficient: $-\frac{1}{2}$

(c) Binomial

9. (a) Standard form: $-4x^5 + 6x^4 + 1$

(b) Degree: 5
Leading coefficient: -4

(c) Trinomial

11. $2x - 3x^3 + 8$ *is* a polynomial.

Standard form: $-3x^3 + 2x + 8$

13. $\dfrac{3x + 4}{x} = 3 + \dfrac{4}{x} = 3 + 4x^{-1}$ is *not* a polynomial because it includes a term with a negative exponent.

15. $y^2 - y^4 + y^3$ *is* a polynomial.

Standard form: $-y^4 + y^3 + y^2$

17. $$\begin{aligned}(6x + 5) - (8x + 15) &= 6x + 5 - 8x - 15\\ &= (6x - 8x) + (5 - 15)\\ &= -2x - 10\end{aligned}$$

19. $$\begin{aligned}(t^3 - 1) + (6t^3 - 5t) &= t^3 - 1 + 6t^3 - 5t\\ &= (t^3 + 6t^3) - 5t - 1\\ &= 7t^3 - 5t - 1\end{aligned}$$

21. $$\begin{aligned}(15x^2 - 6) + (-8.3x^3 - 14.7x^2 - 17) &= 15x^2 - 6 - 8.3x^3 - 14.7x^2 - 17\\ &= -8.3x^3 + (15x^2 - 14.7x^2) + (-6 - 17)\\ &= -8.3x^3 + 0.3x^2 - 23\end{aligned}$$

23. $$\begin{aligned}5z - [3z - (10z + 8)] &= 5z - (3z - 10z - 8)\\ &= 5z - 3z + 10z + 8\\ &= (5z - 3z + 10z) + 8\\ &= 12z + 8\end{aligned}$$

25. $$\begin{aligned}3x(x^2 - 2x + 1) &= 3x(x^2) + 3x(-2x) + 3x(1)\\ &= 3x^3 - 6x^2 + 3x\end{aligned}$$

27. $$\begin{aligned}-5z(3z - 1) &= -5z(3z) + (-5z)(-1)\\ &= -15z^2 + 5z\end{aligned}$$

29. $$\begin{aligned}(1.5t^2 + 5)(-3t) &= (1.5t^2)(-3t) + (5)(-3t)\\ &= -4.5t^3 - 15t\end{aligned}$$

31. $$\begin{aligned}-2x(0.1x + 17) &= (-2x)(0.1x) + (-2x)(17)\\ &= -0.2x^2 - 34x\end{aligned}$$

33. $$\begin{aligned}(x + 7)(x + 5) &= x^2 + 5x + 7x + 35\\ &= x^2 + 12x + 35\end{aligned}$$

35. $$\begin{aligned}(3x - 5)(2x + 1) &= 6x^2 + 3x - 10x - 5\\ &= 6x^2 - 7x - 5\end{aligned}$$

37. $(x^2 - x + 2)(x^2 + x + 1)$

$$\begin{array}{r} x^2 - x + 2\\ \times\ x^2 + x + 1\\ \hline x^4 - x^3 + 2x^2 \\ x^3 - x^2 + 2x \\ x^2 - x + 2\\ \hline x^4 + 0x^3 + 2x^2 + x + 2 \end{array} = x^4 + 2x^2 + x + 2$$

39. $(x + 10)(x - 10) = x^2 - 10^2 = x^2 - 100$

41. $(x + 2y)(x - 2y) = x^2 - (2y)^2 = x^2 - 4y^2$

43. $$\begin{aligned}(2x + 3)^2 &= (2x)^2 + 2(2x)(3) + 3^2\\ &= 4x^2 + 12x + 9\end{aligned}$$

45. $$\begin{aligned}(4x^3 - 3)^2 &= (4x^3)^2 - 2(4x^3)(3) + (3)^2\\ &= 16x^6 - 24x^3 + 9\end{aligned}$$

47. $$\begin{aligned}(x + 3)^3 &= x^3 + 3(x)^2(3) + 3(x)(3)^2 + 3^3\\ &= x^3 + 9x^2 + 27x + 27\end{aligned}$$

49. $$\begin{aligned}(2x - y)^3 &= (2x)^3 - 3(2x)^2y + 3(2x)y^2 - y^3\\ &= 8x^3 - 12x^2y + 6xy^2 - y^3\end{aligned}$$

51. $$\begin{aligned}\left(\tfrac{1}{5}x - 3\right)\left(\tfrac{1}{5}x + 3\right) &= \left(\tfrac{1}{5}x\right)^2 - (3)^2\\ &= \tfrac{1}{25}x^2 - 9\end{aligned}$$

53. $$\begin{aligned}(-6x + 3y)(-6x - 3y) &= (-6x)^2 - (3y)^2\\ &= 36x^2 - 9y^2\end{aligned}$$

55. $$\begin{aligned}\left(\tfrac{1}{4}x - 5\right)^2 &= \left(\tfrac{1}{4}x\right)^2 - 2\left(\tfrac{1}{4}x\right)(5) + (-5)^2\\ &= \tfrac{1}{16}x^2 - \tfrac{5}{2}x + 25\end{aligned}$$

57. $$\begin{aligned}[(x - 3) + y]^2 &= (x - 3)^2 + 2y(x - 3) + y^2\\ &= x^2 - 6x + 9 + 2xy - 6y + y^2\\ &= x^2 + 2xy + y^2 - 6x - 6y + 9\end{aligned}$$

59. $[(m-3)+n][(m-3)-n] = (m-3)^2 - (n)^2$
$= m^2 - 6m + 9 - n^2$
$= m^2 - n^2 - 6m + 9$

61. $(u+2)(u-2)(u^2+4) = (u^2-4)(u^2+4)$
$= u^4 - 16$

63. $(-3x^3 + x^2 + 9) - (4x^2 - 5) = -3x^3 + x^2 + 9 - 4x^2 + 5$
$= -3x^3 + (x^2 - 4x^2) + (9 + 5)$
$= -3x^3 - 3x^2 + 14$

65. $(y^2 + 3y - 5)(y^2 - 6y + 4)$

$$\begin{array}{r} y^2 + 3y - 5 \\ \times\ y^2 - 6y + 4 \\ \hline y^4 + 3y^3 - 5y^2 \qquad\qquad\qquad \\ 6y^3 - 18y^2 + 30y \qquad \\ + 4y^2 + 12y - 20 \\ \hline y^4 - 3y^3 - 19y^2 + 42y - 20 \end{array}$$

67. $(\sqrt{x} + \sqrt{y})(\sqrt{x} - \sqrt{y}) = (\sqrt{x})^2 - (\sqrt{y})^2$
$= x - y$

69. $(x - \sqrt{5})^2 = x^2 - 2(x)(\sqrt{5}) + (\sqrt{5})^2$
$= x^2 - 2\sqrt{5}x + 5$

71. (a) Profit = Revenue − Cost
$P = 135x - (93x + 35{,}000)$
$= 135x - 93x - 35{,}000$
$= 42x - 35{,}000$

(b) For $x = 5000$:
$P = 42(5000) - 35{,}000$
$= 210{,}000 - 35{,}000$
$= 175{,}000$

So, the profit is $175,000.

73. (a) The possible gene combinations of an offspring with albino coloring is $\frac{1}{4}$, or 25%.

(b) $(0.5N + 0.5a)^2 = (0.5N)^2 + 2(0.5N)(0.5a) + (0.5a)^2$
$= 0.25N^2 + 0.5Na + 0.25a^2$

(c) The probability of an offspring with albino coloring is represented by the coefficient of the a^2 term, which is 0.25, or 25%.

75. Area of shaded region = Area of outer rectangle − Area of inner rectangle
$A = 2x(2x + 6) - x(x + 4)$
$= 4x^2 + 12x - x^2 - 4x$
$= 3x^2 + 8x$

77. Area of shaded region = Area of larger square − Area of smaller square
$A = (4x + 2)^2 - (x - 1)^2$
$= 16x^2 + 16x + 4 - (x^2 - 2x + 1)$
$= 15x^2 + 18x + 3$

79. (a) $V = l \cdot w \cdot h$
$= (26 - 2x)(18 - 2x)(x)$
$= 2(13 - x)(2)(9 - x)(x)$
$= 4x(-1)(x - 13)(-1)(x - 9)$
$= 4x(x - 13)(x - 9)$
$= 4x^3 - 88x^2 + 468x$

(b) When $x = 1$: $V = 4(1)^3 - 88(1)^2 + 468(1) = 384 \text{ cm}^3$

When $x = 2$: $V = 4(2)^3 - 88(2)^2 + 468(2) = 616 \text{ cm}^3$

When $x = 3$: $V = 4(3)^3 - 88(3)^2 + 468(3) = 720 \text{ cm}^3$

x (cm)	1	2	3
V (cm^3)	384	616	720

81. (a) Approximations will vary. Actual safe loads for $x = 12$:

$S_6 = \left[0.06(12)^2 - 2.42(12) + 38.71\right]^2 = 335.2561$ (using a calculator)

$S_8 = \left[0.08(12)^2 - 3.30(12) + 51.93\right]^2 = 568.8225$ (using a calculator)

Difference in safe loads $= 568.8225 - 335.2561 = 233.5664$ pounds

(b) The difference in safe loads decreases in magnitude as the span increases.

83. False. $(4x^2 + 1)(3x + 1) = 12x^3 + 4x^2 + 3x + 1$

85. False. $(4x + 3) + (-4x + 6) = 4x + 3 - 4x + 6$
$= 3 + 6$
$= 9$

87. Because $x^m x^n = x^{m+n}$, the degree of the product is $m + n$.

89. The middle term was omitted when squaring the binomial.

$(x - 3)^2 = (x)^2 - 2(x)(3) + (3)^2$
$= x^2 - 6x + 9$
$\neq x^2 + 9$

91. The unknown polynomial may be found by adding $-x^3 + 3x^2 + 2x - 1$ and $5x^2 + 8$:

$(-x^3 + 3x^2 + 2x - 1) + (5x^2 + 8) = -x^3 + (3x^2 + 5x^2) + 2x + (-1 + 8)$
$= -x^3 + 8x^2 + 2x + 7$

Section P.4 Factoring Polynomials

1. factoring

3. perfect square binomial

5. $2x^3 - 6x = 2x(x^2 - 3)$

7. $3x(x - 5) + 8(x - 5) = (x - 5)(3x + 8)$

9. $x^2 - 81 = x^2 - 9^2$
$= (x + 9)(x - 9)$

11. $25y^2 - 4 = (5y)^2 - 2^2 = (5y + 2)(5y - 2)$

13. $64 - 9z^2 = 8^2 - (3z)^2 = (8 + 3z)(8 - 3z)$

15. $(x - 1)^2 - 4 = (x - 1)^2 - (2)^2$
$= \left[(x - 1) + 2\right]\left[(x - 1) - 2\right]$
$= (x + 1)(x - 3)$

17. $81u^4 - 1 = (9u^2 + 1)(9u^2 - 1)$
$= (9u^2 + 1)(3u + 1)(3u - 1)$

19. $x^2 - 4x + 4 = x^2 - 2(2)x + 2^2 = (x - 2)^2$

21. $25z^2 - 30z + 9 = (5z)^2 - 2(5z)(3) + 3^2 = (5z - 3)^2$

23. $4y^2 - 12y + 9 = (2y)^2 - 2(2y)(3) + (3)^2 = (2y - 3)^2$

25. $x^3 - 8 = x^3 - 2^3 = (x - 2)(x^2 + 2x + 4)$

27. $8t^3 - 1 = (2t)^3 - 1^3 = (2t - 1)(4t^2 + 2t + 1)$

29. $27x^3 + 8 = (3x)^3 + 2^3 = (3x + 2)(9x^2 - 6x + 4)$

31. $u^3 + 27v^3 = u^3 + (3v)^3 = (u + 3v)(u^2 - 3uv + 9v^2)$

33. $x^2 + x - 2 = (x + 2)(x - 1)$

35. $s^2 - 5s + 6 = (s - 3)(s - 2)$

37. $3x^2 + 10x - 8 = (3x - 2)(x + 4)$

39. $5x^2 + 31x + 6 = (5x + 1)(x + 6)$

41. $-5y^2 - 8y + 4 = -(5y^2 + 8y - 4)$
$= -(5y - 2)(y + 2)$

43. $x^3 - x^2 + 2x - 2 = x^2(x - 1) + 2(x - 1)$
$= (x - 1)(x^2 + 2)$

45. $2x^3 - x^2 - 6x + 3 = x^2(2x - 1) - 3(2x - 1)$
$= (2x - 1)(x^2 - 3)$

47. $3x^5 + 6x^3 - 2x^2 - 4 = 3x^3(x^2 + 2) - 2(x^2 + 2)$
$= (3x^3 - 2)(x^2 + 2)$

49. $a \cdot c = (2)(9) = 18$. Rewrite the middle term,
$9x = 6x + 3x$, because $(6)(3) = 18$ and $6 + 3 = 9$.
$2x^2 + 9x + 9 = 2x^2 + 6x + 3x + 9$
$= 2x(x + 3) + 3(x + 3)$
$= (x + 3)(2x + 3)$

51. $a \cdot c = (6)(-15) = -90$. Rewrite the middle term,
$-x = -10x + 9x$, because $(-10)(9) = -90$ and
$-10 + 9 = -1$.
$6x^2 - x - 15 = 6x^2 - 10x + 9x - 15$
$= 2x(3x - 5) + 3(3x - 5)$
$= (2x + 3)(3x - 5)$

53. $6x^2 - 54 = 6(x^2 - 9) = 6(x + 3)(x - 3)$

55. $x^3 - x^2 = x^2(x - 1)$

57. $1 - 4x + 4x^2 = (1 - 2x)^2$

59. $2x^2 + 4x - 2x^3 = -2x(-x - 2 + x^2)$
$= -2x(x^2 - x - 2)$
$= -2x(x + 1)(x - 2)$

61. $(x^2 + 3)^2 - 16x^2 = [(x^2 + 3) + 4x][(x^2 + 3) - 4x]$
$= (x^2 + 4x + 3)(x^2 - 4x + 3)$
$= (x + 3)(x + 1)(x - 3)(x - 1)$

63. $2x^3 + x^2 - 8x - 4 = x^2(2x + 1) - 4(2x + 1)$
$= (2x + 1)(x^2 - 4)$
$= (2x + 1)(x + 2)(x - 2)$

65. $2x(3x + 1) + (3x + 1)^2 = (3x + 1)[2x + (3x + 1)]$
$= (3x + 1)(5x + 1)$

67. $2(x - 2)(x + 1)^2 - 3(x - 2)^2(x + 1) = (x - 2)(x + 1)[2(x + 1) - 3(x - 2)]$
$= (x - 2)(x + 1)[2x + 2 - 3x + 6]$
$= (x - 2)(x + 1)(-x + 8)$
$= -(x - 2)(x + 1)(x - 8)$

69. $5(2x + 1)^2(x + 1)^2 + (2x + 1)(x + 1)^3 = (2x + 1)(x + 1)^2[5(2x + 1) + (x + 1)]$
$= (2x + 1)(x + 1)^2(10x + 5 + x + 1)$
$= (2x + 1)(x + 1)^2(11x + 6)$

71. $16x^2 - \frac{1}{9} = (4x)^2 - \left(\frac{1}{3}\right)^2 = \left(4x + \frac{1}{3}\right)\left(4x - \frac{1}{3}\right)$

73. $z^2 + z + \frac{1}{4} = z^2 + 2(z)\left(\frac{1}{2}\right) + \left(\frac{1}{2}\right)^2 = \left(z + \frac{1}{2}\right)^2$

75. $y^3 + \frac{8}{27} = y^3 + \left(\frac{2}{3}\right)^3 = \left(y + \frac{2}{3}\right)\left(y^2 - \frac{2}{3}y + \frac{4}{9}\right)$

77. $x^2 + 3x + 2 = (x + 2)(x + 1)$

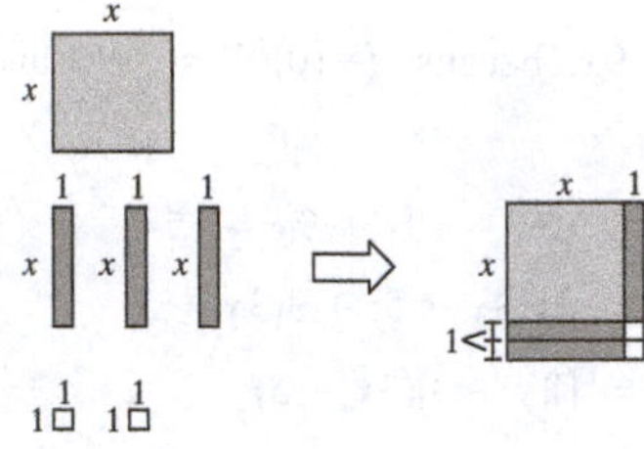

79. $2x^2 + 7x + 3 = (2x + 1)(x + 3)$

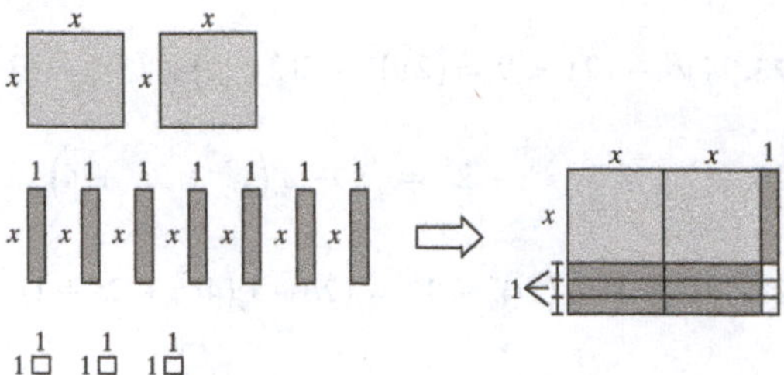

81.
$$\begin{aligned} A &= \pi(r + 2)^2 - \pi r^2 \\ &= \pi\left[(r + 2)^2 - r^2\right] \\ &= \pi\left[r^2 + 4r + 4 - r^2\right] \\ &= \pi(4r + 4) \\ &= 4\pi(r + 1) \end{aligned}$$

83. (a)
$$\begin{aligned} V &= \pi R^2 h - \pi r^2 h \\ &= \pi h(R^2 - r^2) \\ &= \pi h(R + r)(R - r) \end{aligned}$$

(b) Let w = thickness of the shell and let p = average radius of the shell.

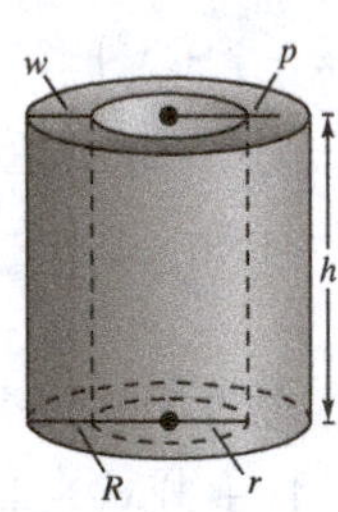

So, $R = p + \frac{1}{2}w$ and $r = p - \frac{1}{2}w$

$$\begin{aligned} V &= \pi h(R + r)(R - r) \\ &= \pi h\left[\left(p + \frac{1}{2}w\right) + \left(p - \frac{1}{2}w\right)\right]\left[\left(p + \frac{1}{2}w\right) - \left(p - \frac{1}{2}w\right)\right] \\ &= \pi h(2p)(w) \\ &= 2\pi pwh \\ &= 2\pi\,(\text{average radius})(\text{thickness of shell})\,h \end{aligned}$$

85. For $x^2 + bx - 15$ to be factorable, b must equal $m + n$ where $mn = -15$.

Factors of -15	Sum of factors
$(15)(-1)$	$15 + (-1) = 14$
$(-15)(1)$	$-15 + 1 = -14$
$(3)(-5)$	$3 + (-5) = -2$
$(-3)(5)$	$-3 + 5 = 2$

The possible b-values are 14, −14, −2, and 2.

87. For $2x^2 + 5x + c$ to be factorable, the factors of $2c$ must add up to 5.

Possible c-values	$2c$	Factors of $2c$ that add up to 5
2	4	$(1)(4) = 4$ and $1 + 4 = 5$
3	6	$(2)(3) = 6$ and $2 + 3 = 5$
−3	−6	$(6)(-1) = -6$ and $6 + (-1) = 5$
−7	−14	$(7)(-2) = -14$ and $7 + (-2) = 5$
−12	−24	$(8)(-3) = -24$ and $8 + (-3) = 5$

These are a few possible c-values. There are *many* correct answers.

If $c = 2$: $2x^2 + 5x + 2 = (2x + 1)(x + 2)$

If $c = 3$: $2x^2 + 5x + 3 = (2x + 3)(x + 1)$

If $c = -3$: $2x^2 + 5x - 3 = (2x - 1)(x + 3)$

If $c = -7$: $2x^2 + 5x - 7 = (2x + 7)(x - 1)$

If $c = -12$: $2x^2 + 5x - 12 = (2x - 3)(x + 4)$

89. True. $a^2 - b^2 = (a + b)(a - b)$

91. 3 should be factored out of both binomials to yield $(3x + 6)(3x - 9) = 3(x + 2)(3)(x - 3) = 9(x + 2)(x - 3)$.

93. $x^{2n} - y^{2n} = (x^n)^2 - (y^n)^2 = (x^n + y^n)(x^n - y^n)$

This is not completely factored unless $n = 1$.

For $n = 2$: $(x^2 + y^2)(x^2 - y^2) = (x^2 + y^2)(x + y)(x - y)$

For $n = 3$: $(x^3 + y^3)(x^3 - y^3) = (x + y)(x^2 - xy + y^2)(x - y)(x^2 + xy + y^2)$

For $n = 4$: $(x^4 + y^4)(x^4 - y^4) = (x^4 + y^4)(x^2 + y^2)(x + y)(x - y)$

95. Answers will vary. *Sample answer:* $x^2 - 3$

97.
$$\begin{aligned} u^6 - v^6 &= (u^3)^2 - (v^3)^2 \\ &= (u^3 + v^3)(u^3 - v^3) \\ &= \left[(u + v)(u^2 - uv + v^2)\right]\left[(u - v)(u^2 + uv + v^2)\right] \\ &= (u + v)(u - v)(u^2 + uv + v^2)(u^2 - uv + v^2) \end{aligned}$$

$x^6 - 1 = (x + 1)(x - 1)(x^2 + x + 1)(x^2 - x + 1)$

$x^6 - 64 = x^6 - 2^6 = (x + 2)(x - 2)(x^2 + 2x + 4)(x^2 - 2x + 4)$

Section P.5 Rational Expressions

1. domain

3. complex

5. The domain of $3x^2 - 4x + 7$ is the set of all real numbers.

7. The domain of $\dfrac{1}{3 - x}$ is the set of all real numbers x such that $x \neq 3$.

9. The domain of $\frac{x+6}{3x+2}$ is the set of all real numbers x such that $x \neq -\frac{2}{3}$.

11. The domain of $\frac{x^2-5x+6}{x^2+6x+8} = \frac{(x-2)(x-3)}{(x+4)(x+2)}$ is the set of all real numbers x such that $x \neq -4, -2$.

13. The domain of $\sqrt{x-7}$ is the set of all real numbers x such that $x \geq 7$.

15. The domain of $\frac{1}{\sqrt{x-3}}$ is the set of all real numbers x such that $x > 3$.

17. $\frac{15x^2}{10x} = \frac{5x(3x)}{5x(2)} = \frac{3x}{2}, x \neq 0$

19. $\frac{x-5}{10-2x} = \frac{x-5}{-2(x-5)} = -\frac{1}{2}, x \neq 5$

21. $\frac{y^2-16}{y+4} = \frac{(y+4)(y-4)}{y+4} = y-4, y \neq -4$

23. $\frac{6y+9y^2}{12y+8} = \frac{3y(3y+2)}{4(3y+2)} = \frac{3y}{4}, y \neq -\frac{2}{3}$

25. $\frac{x^2+4x-5}{x^2+8x+15} = \frac{(x+5)(x-1)}{(x+5)(x+3)} = \frac{x-1}{x+3}, x \neq -5$

27. $\frac{x^2-x-2}{10-3x-x^2} = \frac{x^2-x-2}{-(x^2+3x-10)}$

$$= \frac{(x+1)(x-2)}{-(x+5)(x-2)} = -\frac{x+1}{x+5}, x \neq 2$$

29. $\frac{x^2-16}{x^3+x^2-16x-16} = \frac{x^2-16}{x^2(x+1)-16(x+1)}$

$$= \frac{x^2-16}{(x+1)(x^2-16)}$$

$$= \frac{1}{x+1}, x \neq \pm 4$$

31. $\frac{5x^3}{2x^3+4} = \frac{5x^3}{2(x^3+2)}$

When simplifying fractions, only common factors can be divided out, not terms.

33. $\frac{5}{x-1} \cdot \frac{x-1}{25(x-2)} = \frac{1}{5(x-2)}, x \neq 1$

35. $\frac{x^2-4}{12} \div \frac{2-x}{2x+4} = \frac{x^2-4}{12} \cdot \frac{2x+4}{2-x}$

$$= \frac{(x+2)(x-2)}{12} \cdot \frac{2(x+2)}{-(x-2)}$$

$$= -\frac{(x+2)^2}{6}, x \neq \pm 2$$

37. $\frac{x^2+xy-2y^2}{x^3+x^2y} \cdot \frac{x}{x^2+3xy+2y^2} = \frac{(x+2y)(x-y)}{x^2(x+y)} \cdot \frac{x}{(x+2y)(x+y)} = \frac{x-y}{x(x+y)^2}, x \neq -2y$

39. $\frac{x-1}{x+2} - \frac{x-4}{x+2} = \frac{x-1-(x-4)}{x+2} = \frac{x-1-x+4}{x+2} = \frac{3}{x+2}$

41. $\frac{1}{3x+2} + \frac{x}{x+1} = \frac{(1)(x+1)}{(3x+2)(x+1)} + \frac{x(3x+2)}{(3x+2)(x+1)} = \frac{x+1+3x^2+2x}{(3x+2)(x+1)} = \frac{3x^2+3x+1}{(3x+2)(x+1)}$

43. $\frac{3}{2x+4} - \frac{x}{x+2} = \frac{3}{2(x+2)} - \frac{x}{x+2} = \frac{3}{2(x+2)} - \frac{2x}{2(x+2)} = \frac{3-2x}{2(x+2)}$

45. $-\frac{1}{x} + \frac{2}{x^2+1} + \frac{1}{x^3+x} = \frac{-(x^2+1)}{x(x^2+1)} + \frac{2x}{x(x^2+1)} + \frac{1}{x(x^2+1)}$

$$= \frac{-x^2-1+2x+1}{x(x^2+1)} = \frac{-x^2+2x}{x(x^2+1)} = \frac{-x(x-2)}{x(x^2+1)}$$

$$= -\frac{x-2}{x^2+1} = \frac{2-x}{x^2+1}, x \neq 0$$

47. The minus sign should be distributed to each term in the numerator of the second fraction.

$$\frac{x+4}{x+2} - \frac{3x-8}{x+2} = \frac{(x+4)-(3x-8)}{x+2}$$
$$= \frac{x+4-3x+8}{x+2}$$
$$= \frac{-2x+12}{x+2}$$
$$= \frac{-2(x-6)}{x+2}$$

49. $$\frac{\left(\frac{x}{2}-1\right)}{(x-2)} = \frac{\left(\frac{x}{2}-\frac{2}{2}\right)}{\left(\frac{x-2}{1}\right)} = \frac{x-2}{2}\cdot\frac{1}{x-2} = \frac{1}{2}, x \neq 2$$

51. $$\frac{\left[\frac{x^2}{(x+1)^2}\right]}{\left[\frac{x}{(x+1)^3}\right]} = \frac{x^2}{(x+1)^2}\cdot\frac{(x+1)^3}{x} = x(x+1), x \neq -1, 0$$

53. $$\frac{\left(\sqrt{x}-\frac{1}{2\sqrt{x}}\right)}{\sqrt{x}} = \frac{\left(\sqrt{x}-\frac{1}{2\sqrt{x}}\right)}{\sqrt{x}}\cdot\frac{2\sqrt{x}}{2\sqrt{x}}$$
$$= \frac{2x-1}{2x}, x > 0$$

55. $$x^2(x^2+3)^{-4} + (x^2+3)^3 = (x^2+3)^{-4}\left[x^2 + (x^2+3)^7\right] = \frac{x^2+(x^2+3)^7}{(x^2+3)^4}$$

57. $$2x^2(x-1)^{1/2} - 5(x-1)^{-1/2} = (x-1)^{-1/2}\left[2x^2(x-1)^1 - 5\right] = \frac{2x^3-2x^2-5}{(x-1)^{1/2}}$$

59. $$\frac{3x^{1/3}-x^{-2/3}}{3x^{-2/3}} = \frac{3x^{1/3}-x^{-2/3}}{3x^{-2/3}}\cdot\frac{x^{2/3}}{x^{2/3}}$$
$$= \frac{3x^1-x^0}{3x^0}$$
$$= \frac{3x-1}{3}, x \neq 0$$

61. $$\frac{\left(\frac{1}{x+h}-\frac{1}{x}\right)}{h} = \frac{\left(\frac{1}{x+h}-\frac{1}{x}\right)}{h}\cdot\frac{x(x+h)}{x(x+h)}$$
$$= \frac{x-(x+h)}{hx(x+h)}$$
$$= \frac{-h}{hx(x+h)}$$
$$= -\frac{1}{x(x+h)}, h \neq 0$$

63. $$\frac{\left(\frac{1}{x+h-4}-\frac{1}{x-4}\right)}{h} = \frac{\left(\frac{1}{x+h-4}-\frac{1}{x-4}\right)}{h}\cdot\frac{(x-4)(x+h-4)}{(x-4)(x+h-4)}$$
$$= \frac{(x-4)-(x+h-4)}{h(x-4)(x+h-4)}$$
$$= \frac{-h}{h(x-4)(x+h-4)}$$
$$= -\frac{1}{(x-4)(x+h-4)}, h \neq 0$$

65. $$\frac{\sqrt{x+2}-\sqrt{x}}{2} = \frac{\sqrt{x+2}-\sqrt{x}}{2}\cdot\frac{\sqrt{x+2}+\sqrt{x}}{\sqrt{x+2}+\sqrt{x}}$$

67. $\dfrac{\sqrt{t+3}-\sqrt{3}}{t} = \dfrac{\sqrt{t+3}-\sqrt{3}}{t}\cdot\dfrac{\sqrt{t+3}+\sqrt{3}}{\sqrt{t+3}+\sqrt{3}} = \dfrac{(t+3)-3}{t\left(\sqrt{t+3}+\sqrt{3}\right)} = \dfrac{t}{t\left(\sqrt{t+3}+\sqrt{3}\right)} = \dfrac{1}{\sqrt{t+3}+\sqrt{3}}, t \neq 0$

69.
$$\frac{\sqrt{x+h+1}-\sqrt{x+1}}{h} = \frac{\sqrt{x+h+1}-\sqrt{x+1}}{h}\cdot\frac{\sqrt{x+h+1}+\sqrt{x+1}}{\sqrt{x+h+1}+\sqrt{x+1}}$$
$$= \frac{(x+h+1)-(x+1)}{h\left(\sqrt{x+h+1}+\sqrt{x+1}\right)}$$
$$= \frac{h}{h\left(\sqrt{x+h+1}+\sqrt{x+1}\right)}$$
$$= \frac{1}{\sqrt{x+h+1}+\sqrt{x+1}}, h \neq 0$$

71. $T = 10\left(\dfrac{4t^2+16t+75}{t^2+4t+10}\right)$

(a)

t	0	2	4	6	8	10	12	14	16	18	20	22
T	75°	55.9°	48.3°	45°	43.3°	42.3°	41.7°	41.3°	41.1°	40.9°	40.7°	40.6°

(b) T is approaching 40°.

73. Probability $= \dfrac{\text{Shaded area}}{\text{Total area}} = \dfrac{x(x/2)}{x(2x+1)} = \dfrac{x/2}{2x+1}\cdot\dfrac{2}{2} = \dfrac{x}{2(2x+1)}, x \neq 0$

75. (a)

Year, t	Online Banking	Mobile Banking
11	79.1	17.9
12	80.9	24.0
13	83.1	29.6
14	86.0	34.8

(b) The values from the models are close to the actual data.

(c)
$$\frac{\text{Number of households using mobile banking}}{\text{Number of households using online banking}} = \frac{\left(0.661t^2-47\right)/\left(0.007t^2+1\right)}{(-2.9709t+70.517)/(-0.0474t+1)}$$
$$= \frac{0.661t^2-47}{0.007t^2+1}\cdot\frac{-0.0474t+1}{-2.9709t+70.517}$$
$$= \frac{\left(0.661t^2-47\right)(-0.0474t+1)}{\left(0.007t^2+1\right)(-2.9709t+70.517)}$$
$$= \frac{-0.313t^2+0.661t^2+2.23t-47}{-0.0208t^3+0.494t^2-2.97t+70.5}$$
$$= \frac{0.0313t^3-0.661t^2-2.23t+47}{0.0208t^3-0.494t^2+2.97t-70.5}$$
$$= \frac{0.0313t^3-0.661t^2-2.23t+47}{0.0208t^3-0.494t^2+2.97t-70.5}$$

(d) When $t = 11$, $\dfrac{0.0313(11)^3 - 0.661(11)^2 - 2.23(11) + 47}{0.0208(11)^3 - 0.494(11)^2 + 2.97(11) - 70.5} \approx 0.2267.$

When $t = 12$, $\dfrac{0.0313(12)^3 - 0.661(12)^2 - 2.23(12) + 47}{0.0208(12)^3 - 0.494(12)^2 + 2.97(12) - 70.5} \approx 0.2977.$

When $t = 13$, $\dfrac{0.0313(13)^3 - 0.661(13)^2 - 2.23(13) + 47}{0.0208(13)^3 - 0.494(13)^2 - 2.97(13) - 70.5} \approx 0.3578.$

When $t = 14$, $\dfrac{0.0313(14)^3 - 0.661(14)^2 - 2.23(14) + 47}{0.0208(14)^3 - 0.494(14)^2 + 2.97(14) - 70.5} \approx 0.4061$

Answers will vary.

77. $R_T = \dfrac{1}{\dfrac{1}{R_1} + \dfrac{1}{R_2}} = \dfrac{1}{\dfrac{R_2 + R_1}{R_1 R_2}} = \dfrac{R_1 R_2}{R_1 + R_2}$

79. False. In order for the simplified expression to be equivalent to the original expression, the domain of the simplified expression needs to be restricted. If n is even, $x \neq \pm 1$. If n is odd, $x \neq 1$.

Section P.6 The Rectangular Coordinate System and Graphs

1. Cartesian

3. Distance Formula

5.

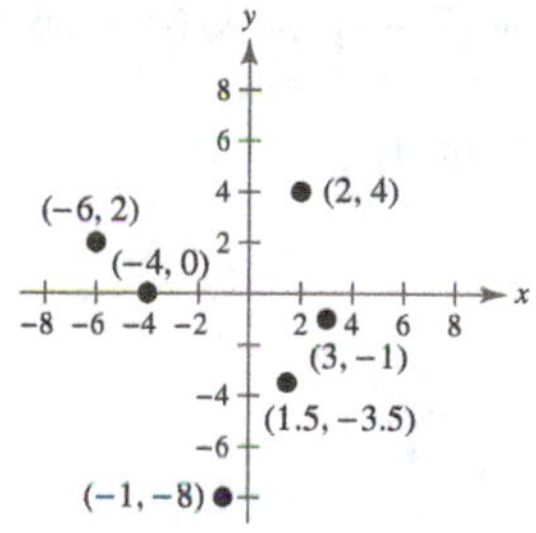

7. $(-3, 4)$

9. $x > 0$ and $y < 0$ in Quadrant IV.

11. $x = -4$ and $y > 0$ in Quadrant II.

13. $x + y = 0, x \neq 0, y \neq 0$ means $x = -y$ or $y = -x$. This occurs in Quadrant II or IV.

15.

Year, x	Number of Stores, y
2007	7276
2008	7720
2009	8416
2010	8970
2011	10,130
2012	10,773
2013	10,942
2014	11,453

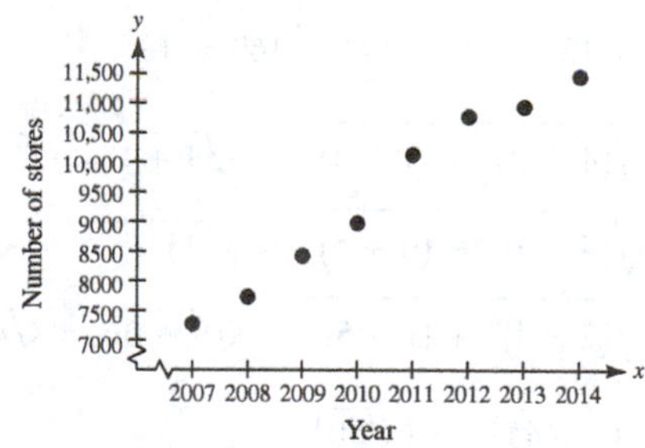

17. $d = \sqrt{(x_2 - x_1)^2 + (y_2 - y_1)^2}$

$= \sqrt{(3 - (-2))^2 + (-6 - 6)^2}$

$= \sqrt{(5)^2 + (-12)^2}$

$= \sqrt{25 + 144}$

$= 13$ units

19. $d = \sqrt{(x_2 - x_1)^2 + (y_2 - y_1)^2}$
$= \sqrt{(-5 - 1)^2 + (-1 - 4)^2}$
$= \sqrt{(-6)^2 + (-5)^2}$
$= \sqrt{36 + 25}$
$= \sqrt{61}$ units

21. $d = \sqrt{(x_2 - x_1)^2 + (y_2 - y_1)^2}$
$= \sqrt{\left(2 - \frac{1}{2}\right)^2 + \left(-1 - \frac{4}{3}\right)^2}$
$= \sqrt{\left(\frac{3}{2}\right)^2 + \left(-\frac{7}{3}\right)^2}$
$= \sqrt{\frac{9}{4} + \frac{49}{9}}$
$= \sqrt{\frac{277}{36}}$
$= \frac{\sqrt{277}}{6}$ units

23. (a) (1, 0), (13, 5)

Distance $= \sqrt{(13 - 1)^2 + (5 - 0)^2}$
$= \sqrt{12^2 + 5^2} = \sqrt{169} = 13$

(13, 5), (13, 0)

Distance $= |5 - 0| = |5| = 5$

(1, 0), (13, 0)

Distance $= |1 - 13| = |-12| = 12$

(b) $5^2 + 12^2 = 25 + 144 = 169 = 13^2$

25. $d_1 = \sqrt{(4 - 2)^2 + (0 - 1)^2} = \sqrt{4 + 1} = \sqrt{5}$
$d_2 = \sqrt{(4 + 1)^2 + (0 + 5)^2} = \sqrt{25 + 25} = \sqrt{50}$
$d_3 = \sqrt{(2 + 1)^2 + (1 + 5)^2} = \sqrt{9 + 36} = \sqrt{45}$
$\left(\sqrt{5}\right)^2 + \left(\sqrt{45}\right)^2 = \left(\sqrt{50}\right)^2$

27. $d_1 = \sqrt{(1 - 3)^2 + (-3 - 2)^2} = \sqrt{4 + 25} = \sqrt{29}$
$d_2 = \sqrt{(3 + 2)^2 + (2 - 4)^2} = \sqrt{25 + 4} = \sqrt{29}$
$d_3 = \sqrt{(1 + 2)^2 + (-3 - 4)^2} = \sqrt{9 + 49} = \sqrt{58}$
$d_1 = d_2$

29. (a)

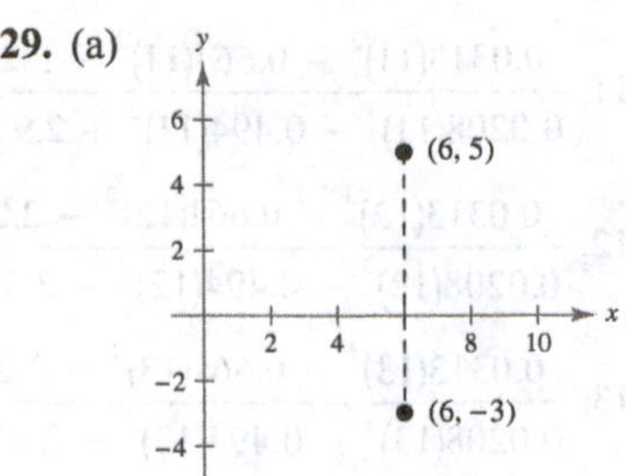

(b) $d = \sqrt{(5 - (-3))^2 + (6 - 6)^2} = \sqrt{64} = 8$

(c) $\left(\frac{6 + 6}{2}, \frac{5 + (-3)}{2}\right) = (6, 1)$

31. (a)

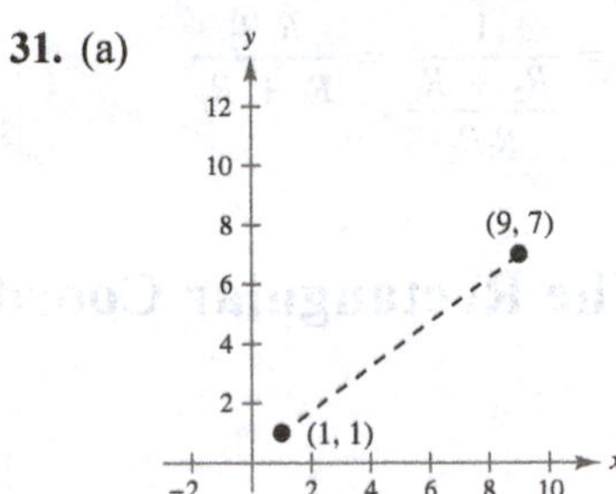

(b) $d = \sqrt{(9 - 1)^2 + (7 - 1)^2} = \sqrt{64 + 36} = 10$

(c) $\left(\frac{9 + 1}{2}, \frac{7 + 1}{2}\right) = (5, 4)$

33. (a)

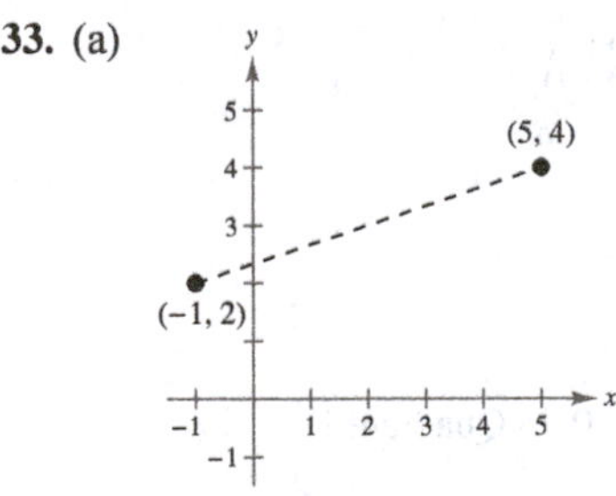

(b) $d = \sqrt{(5 + 1)^2 + (4 - 2)^2}$
$= \sqrt{36 + 4} = 2\sqrt{10}$

(c) $\left(\frac{-1 + 5}{2}, \frac{2 + 4}{2}\right) = (2, 3)$

35. (a)

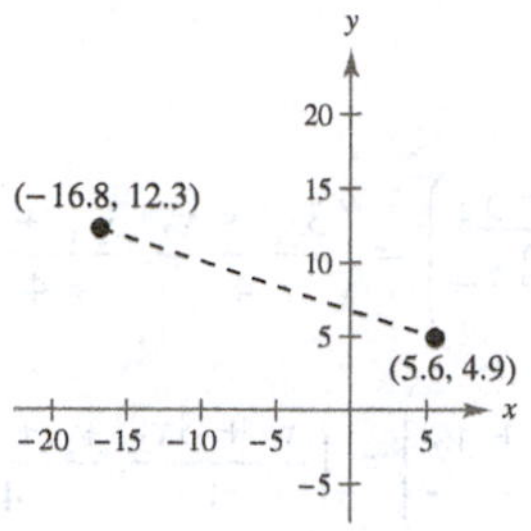

(−16.8, 12.3)

(5.6, 4.9)

(b) $d = \sqrt{(-16.8 - 5.6)^2 + (12.3 - 4.9)^2}$

$= \sqrt{501.76 + 54.76} = \sqrt{556.52}$

(c) $\left(\dfrac{-16.8 + 5.6}{2}, \dfrac{12.3 + 4.9}{2}\right) = (-5.6, 8.6)$

37. $d = \sqrt{120^2 + 150^2} = \sqrt{36{,}900} = 30\sqrt{41} \approx 192.09$

The plane flies about 192 kilometers.

39. $\text{midpoint} = \left(\dfrac{x_1 + x_2}{2}, \dfrac{y_1 + y_2}{2}\right)$

$= \left(\dfrac{2010 + 2014}{2}, \dfrac{35{,}123 + 45{,}998}{2}\right)$

$= (2012, 40{,}560.5)$

In 2012, the sales for the Coca-Cola Company were about $40,560.5 million.

41. $(-2 + 2, -4 + 5) = (0, 1)$

$(2 + 2, -3 + 5) = (4, 2)$

$(-1 + 2, -1 + 5) = (1, 4)$

43. $(-7 + 4, -2 + 8) = (-3, 6)$

$(-2 + 4, 2 + 8) = (2, 10)$

$(-2 + 4, -4 + 8) = (2, 4)$

$(-7 + 4, -4 + 8) = (-3, 4)$

45. (a) The minimum wage had the greatest increase from 2000 to 2010.

(b) Minimum wage in 1985: $3.35

Minimum wage in 2000: $5.15

Percent increase: $\left(\dfrac{5.15 - 3.35}{3.35}\right) \times 100 \approx 53.7\%$

Minimum wage in 2000: $4.25

Minimum wage in 2015: $7.25

Percent increase: $\left(\dfrac{7.25 - 5.15}{5.15}\right) \times 100 \approx 40.8\%$

So, the minimum wage increased 53.7% from 1985 to 2000 and 40.8% from 2000 to 2015.

(c) $\begin{matrix}\text{Minimum wage}\\ \text{in 2030}\end{matrix} = \begin{matrix}\text{Minimum wage}\\ \text{in 2015}\end{matrix} + \begin{pmatrix}\text{Percent}\\ \text{increase}\end{pmatrix}\begin{pmatrix}\text{Minimum wage}\\ \text{in 2015}\end{pmatrix} \approx \$7.25 + 0.408(\$7.25) \approx \$10.21$

So, the minimum wage will be about $10.21 in the year 2030.

(d) Answers will vary. *Sample answer:* Yes, the prediction is reasonable because the percent increase is over an equal time period of 15 years.

47. True. Because $x < 0$ and $y > 0$, $2x < 0$ and $-3y < 0$, which is located in Quadrant III.

49. True. Two sides of the triangle have lengths $\sqrt{149}$ and the third side has a length of $\sqrt{18}$.

51. Answers will vary. *Sample answer:* When the x-values are much larger or smaller than the y-values, different scales for the coordinate axes should be used.

53. Because $x_m = \dfrac{x_1 + x_2}{2}$ and $y_m = \dfrac{y_1 + y_2}{2}$ we have:

$2x_m = x_1 + x_2 \qquad 2y_m = y_1 + y_2$

$2x_m - x_1 = x_2 \qquad 2y_m - y_1 = y_2$

So, $(x_2, y_2) = (2x_m - x_1, 2y_m - y_1)$.

55. The midpoint of the given line segment is $\left(\dfrac{x_1 + x_2}{2}, \dfrac{y_1 + y_2}{2}\right)$.

The midpoint between (x_1, y_1) and $\left(\dfrac{x_1 + x_2}{2}, \dfrac{y_1 + y_2}{2}\right)$ is $\left(\dfrac{x_1 + \dfrac{x_1 + x_2}{2}}{2}, \dfrac{y_1 + \dfrac{y_1 + y_2}{2}}{2}\right) = \left(\dfrac{3x_1 + x_2}{4}, \dfrac{3y_1 + y_2}{4}\right)$.

The midpoint between $\left(\dfrac{x_1 + x_2}{2}, \dfrac{y_1 + y_2}{2}\right)$ and (x_2, y_2) is $\left(\dfrac{\dfrac{x_1 + x_2}{2} + x_2}{2}, \dfrac{\dfrac{y_1 + y_2}{2} + y_2}{2}\right) = \left(\dfrac{x_1 + 3x_2}{4}, \dfrac{y_1 + 3y_2}{4}\right)$.

So, the three points are $\left(\dfrac{3x_1 + x_2}{4}, \dfrac{3y_1 + y_2}{4}\right)$, $\left(\dfrac{x_1 + x_2}{2}, \dfrac{y_1 + y_2}{2}\right)$, and $\left(\dfrac{x_1 + 3x_2}{4}, \dfrac{y_1 + 3y_2}{4}\right)$.

57. Use the Midpoint Formula to prove the diagonals of the parallelogram bisect each other.

$$\left(\frac{b + a}{2}, \frac{c + 0}{2}\right) = \left(\frac{a + b}{2}, \frac{c}{2}\right)$$

$$\left(\frac{a + b + 0}{2}, \frac{c + 0}{2}\right) = \left(\frac{a + b}{2}, \frac{c}{2}\right)$$

59. (a) **First Set**

$d(A, B) = \sqrt{(2 - 2)^2 + (3 - 6)^2} = \sqrt{9} = 3$

$d(B, C) = \sqrt{(2 - 6)^2 + (6 - 3)^2} = \sqrt{16 + 9} = 5$

$d(A, C) = \sqrt{(2 - 6)^2 + (3 - 3)^2} = \sqrt{16} = 4$

Because $3^2 + 4^2 = 5^2$, A, B, and C are the vertices of a right triangle.

Second Set

$d(A, B) = \sqrt{(8 - 5)^2 + (3 - 2)^2} = \sqrt{10}$

$d(B, C) = \sqrt{(5 - 2)^2 + (2 - 1)^2} = \sqrt{10}$

$d(A, C) = \sqrt{(8 - 2)^2 + (3 - 1)^2} = \sqrt{40}$

A, B, and C are the vertices of an isosceles triangle or are collinear: $\sqrt{10} + \sqrt{10} = 2\sqrt{10} = \sqrt{40}$.

(b)

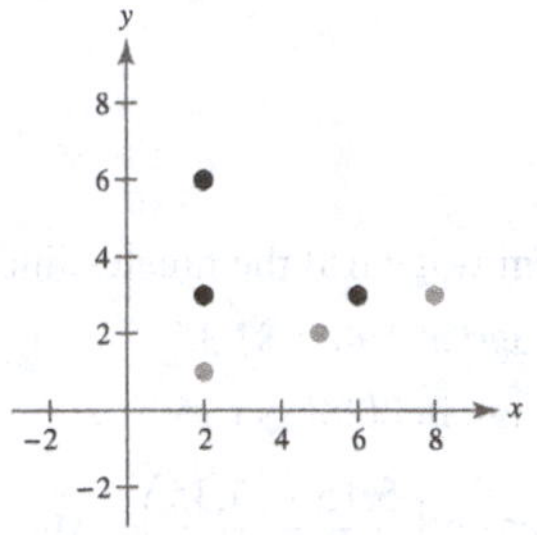

First set: Not collinear

Second set: Collinear.

(c) A set of three points is collinear when the sum of two distances among the points is exactly equal to the third distance.

Review Exercises for Chapter P

1. $\left\{11, -14, -\frac{8}{9}, \frac{5}{2}, \sqrt{6}, 0.4\right\}$

(a) Natural numbers: 11

(b) Whole numbers: 11

(c) Integers: 11, −14

(d) Rational numbers: 11, −14, $-\frac{8}{9}$, $\frac{5}{2}$, 0.4

(e) Irrational numbers: $\sqrt{6}$

3.

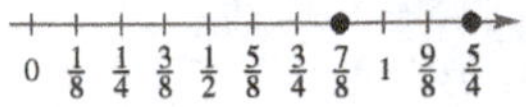

$\dfrac{5}{4} > \dfrac{7}{8}$

5. (a) $x \geq 6$ denotes the set of all real numbers greater than or equal to 6.

(b)

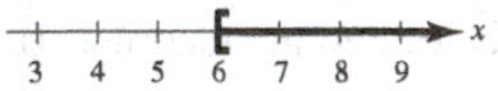

(c) The set is unbounded.

7. (a) $-3 \leq x < 4$ denotes the set of all real numbers greater than or equal to −3 and less than 4.

(b)

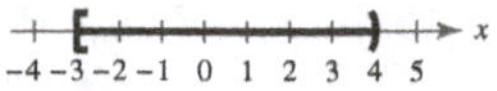

(c) The set is bounded.

9. $d(-74, 48) = |48 - (-74)| = 122$

11. $d(x, 7) = |x - 7|$ and $d(x, 7) \geq 4$, thus $|x - 7| \geq 4$.

13. $12x - 7$

(a) $12(0) - 7 = -7$

(b) $12(-1) - 7 = -19$

15. $-x^2 + x - 1$

(a) $-(1)^2 + 1 - 1 = -1$

(b) $-(-1)^2 + (-1) - 1 = -3$

17. $0 + (a - 5) = a - 5$

Illustrates the Additive Identity Property

19. $2x + (3x - 10) = (2x + 3x) - 10$

Illustrates the Associative Property of Addition

21. $(t^2 + 1) + 3 = 3 + (t^2 + 1)$

Illustrates the Commutative Property of Addition

23. $-6 + 6 = 0$

25. $(-8)(-4) = 32$

27. $\dfrac{x}{5} + \dfrac{7x}{12} = \dfrac{12x}{60} + \dfrac{35x}{60} = \dfrac{47x}{60}$

29. $\dfrac{3x}{10} \cdot \dfrac{5}{3} = \dfrac{x}{2} \cdot \dfrac{1}{1} = \dfrac{x}{2}$

31. (a) $3x^2(4x^3)^3 = 3x^2(64x^9) = 192x^{11}$

(b) $\dfrac{5y^6}{10y} = \dfrac{y^{6-1}}{2} = \dfrac{y^5}{2}, \quad y \neq 0$

33. (a) $(-2z)^3 = -8z^3$

(b) $\dfrac{(8y)^0}{y^2} = \dfrac{1}{y^2}$

35. (a) $\dfrac{a^2}{b^{-2}} = a^2b^2$

(b) $(a^2b^4)(3ab^{-2}) = 3a^{2+1}b^{4-2} = 3a^3b^2$

37. (a) $\dfrac{(5a)^{-2}}{(5a)^2} = (5a)^{-2-2} = (5a)^{-4} = \dfrac{1}{(5a)^4} = \dfrac{1}{625a^4}$

(b) $\dfrac{4(x^{-1})^{-3}}{4^{-2}(x^{-1})^{-1}} = \dfrac{4^{1-(-2)}x^3}{x} = 4^3x^{3-1} = 64x^2$

39. $274,400,000 = 2.744 \times 10^8$

41. $4.84 \times 10^8 = 484,000,000$

43. (a) $\sqrt[3]{27^2} = (\sqrt[3]{27})^2 = (3)^2 = 9$

(b) $\sqrt{49^3} = (\sqrt{49})^3 = (7)^3 = 343$

45. (a) $(\sqrt[3]{216})^3 = (\sqrt[3]{6^3})^3 = (6)^3 = 216$

(b) $\sqrt[4]{32^4} = (\sqrt[4]{32})^4 = 32$

47. (a) $\sqrt{12x^3} + \sqrt{3x} = \sqrt{4x^2 \cdot 3x} + \sqrt{3x}$
$= 2x\sqrt{3x} + \sqrt{3x}$
$= (2x + 1)\sqrt{3x}$

(b) $\sqrt{27x^3} - \sqrt{3x^3} = \sqrt{9x^2 \cdot 3x} - \sqrt{x^2 \cdot 3x}$
$= 3x\sqrt{3x} - x\sqrt{3x}$
$= (3x - x)\sqrt{3x}$
$= 2x\sqrt{3x}$

49. These are not like terms. Radicals cannot be combined by addition or subtraction unless the index and the radicand are the same.

51. $\dfrac{3}{4\sqrt{3}} = \dfrac{3}{4\sqrt{3}} \cdot \dfrac{\sqrt{3}}{\sqrt{3}} = \dfrac{3\sqrt{3}}{4(3)} = \dfrac{\sqrt{3}}{4}$

53. $\dfrac{1}{2 - \sqrt{3}} = \dfrac{1}{2 - \sqrt{3}} \cdot \dfrac{2 + \sqrt{3}}{2 + \sqrt{3}} = \dfrac{2 + \sqrt{3}}{2^2 - (\sqrt{3})^2} = \dfrac{2 + \sqrt{3}}{4 - 3} = \dfrac{2 + \sqrt{3}}{1} = 2 + \sqrt{3}$

55. $\dfrac{\sqrt{7} + 1}{2} = \dfrac{\sqrt{7} + 1}{2} \cdot \dfrac{\sqrt{7} - 1}{\sqrt{7} - 1} = \dfrac{(\sqrt{7})^2 - 1^2}{2(\sqrt{7} - 1)} = \dfrac{7 - 1}{2(\sqrt{7} - 1)} = \dfrac{6}{2(\sqrt{7} - 1)} = \dfrac{3}{\sqrt{7} - 1}$

57. $16^{3/2} = \sqrt{16^3} = (\sqrt{16})^3 = (4)^3 = 64$

59. $(3x^{2/5})(2x^{1/2}) = 6x^{2/5+1/2} = 6x^{9/10}$

61. Standard form: $-11x^2 + 3$

Degree: 2

Leading coefficient: -11

63. Standard form: $-12x^2 - 4$

Degree: 2

Leading coefficient: -12

65. $-(3x^2 + 2x) + (1 - 5x) = -3x^2 - 2x + 1 - 5x$
$= -3x^2 - 7x + 1$

67. $2x(x^2 - 5x + 6) = (2x)(x^2) + (2x)(-5x) + (2x)(6)$
$= 2x^3 - 10x^2 + 12x$

69. $(3x - 6)(5x + 1) = 15x^2 + 3x - 30x - 6$
$= 15x^2 - 27x - 6$

71. $(6x + 5)(6x - 5) = (6x)^2 - 5^2 = 36x^2 - 25$

73. $(2x - 3)^2 = (2x)^2 - 2(2x)(3) + 3^3 = 4x^2 - 12x + 9$

75.

$$\begin{array}{rrrrr} & & x^2 & +\ x & +\ 5 \\ \times & & x^2 & -\ 7x & -\ 2 \\ \hline & & -\ 2x^2 & -\ 2x & -\ 10 \\ & -\ 7x^3 & -\ 7x^2 & -\ 35x & \\ x^4 & +\ x^3 & +\ 5x^2 & & \\ \hline x^4 & -\ 6x^3 & -\ 4x^2 & -\ 37x & -\ 10 \end{array}$$

77. $2500(1 + r)^2 = 2500(r + 1)^2$
$= 2500(r^2 + 2r + 1)$
$= 2500r^2 + 5000r + 2500$

79. Area $= (x + 12)(x + 16)$
$= x^2 + 16x + 12x + 192$
$= x^2 + 28x + 192$ square feet

81. $x^3 - x = x(x^2 - 1) = x(x + 1)(x - 1)$

83. $25x^2 - 49 = (5x)^2 - 7^2 = (5x + 7)(5x - 7)$

85. $x^3 - 64 = x^3 - 4^3 = (x - 4)(x^2 + 4x + 16)$

87. $2x^2 + 21x + 10 = (2x + 1)(x + 10)$

89. $x^3 - x^2 + 2x - 2 = x^2(x - 1) + 2(x - 1)$
$= (x - 1)(x^2 + 2)$

91. The domain of $\dfrac{1}{x + 1}$ is the set of all real numbers x such that $x \neq -1$.

93. The domain of $\sqrt{x + 2}$ is the set of all real numbers x such that $x \geq -2$.

95. $\dfrac{x^2 - 64}{5(3x + 24)} = \dfrac{(x + 8)(x - 8)}{5 \cdot 3(x + 8)} = \dfrac{x - 8}{15}, \quad x \neq -8$

97. $\dfrac{x^2 - 7x + 12}{x^2 + 8x + 16} \cdot \dfrac{x + 4}{x^2 - 9} = \dfrac{(x - 3)(x - 4)}{(x + 4)^2} \cdot \dfrac{x + 4}{(x + 3)(x - 3)} = \dfrac{x - 4}{(x + 4)(x + 3)}, x \neq 3$

99. $\dfrac{2}{x - 3} + \dfrac{6}{3 - x} = \dfrac{2}{x - 3} - \dfrac{6}{x - 3} = -\dfrac{4}{x - 3}$

101. $\dfrac{\left(\dfrac{\frac{3a}{a^2}}{\frac{a^2}{x} - 1}\right)}{\left(\dfrac{a}{x} - 1\right)} = \dfrac{\left(\dfrac{\frac{3a}{a^2 - x}}{x}\right)}{\left(\dfrac{a - x}{x}\right)} = \dfrac{3a}{1} \cdot \dfrac{x}{a^2 - x} \cdot \dfrac{x}{a - x}$

$= \dfrac{3ax^2}{(a^2 - x)(a - x)}, \ x \neq 0$

103. $\dfrac{\left[\dfrac{1}{2(x + h)} - \dfrac{1}{2x}\right]}{h} = \dfrac{\dfrac{x - (x + 6)}{2x(x + h)}}{h}$

$= \dfrac{-h}{2x(x + h)} \cdot \dfrac{1}{h}$

$= \dfrac{-1}{2x(x + h)}, \ h \neq 0$

105.

(−3, 6) (5, 5) (−2, 0) (−1, −7)

107. $x > 0$ and $y = -2$ in Quadrant IV.

109. (a)

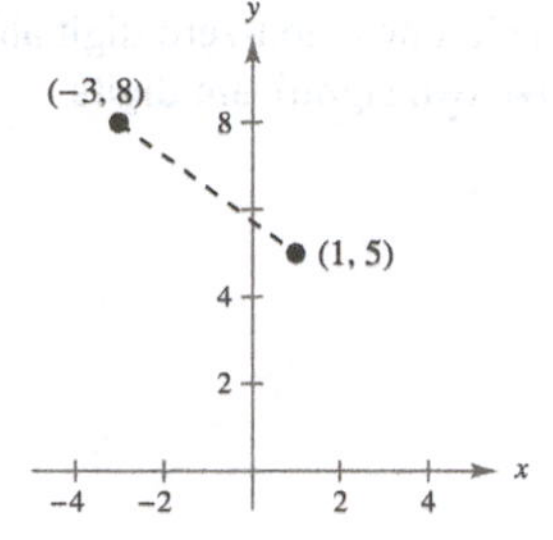

(b) $d = \sqrt{(-3-1)^2 + (8-5)^2} = \sqrt{16+9} = 5$

(c) Midpoint: $\left(\dfrac{-3+1}{2}, \dfrac{8+5}{2}\right) = \left(-1, \dfrac{13}{2}\right)$

111. (a)

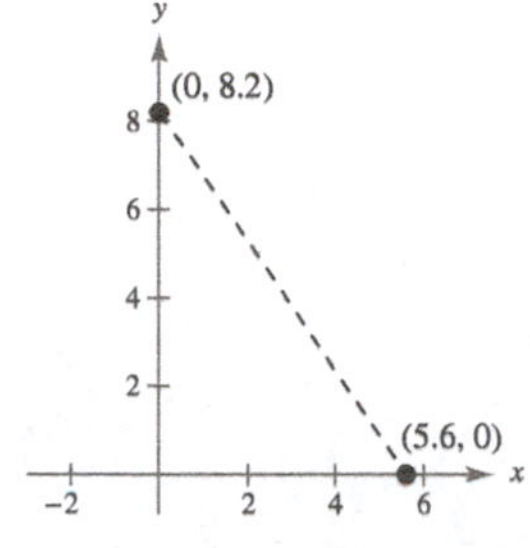

(b) $d = \sqrt{(5.6-0)^2 + (0-8.2)^2}$

$= \sqrt{31.36 + 67.24} = \sqrt{98.6}$

(c) Midpoint: $\left(\dfrac{0+5.6}{2}, \dfrac{8.2+0}{2}\right) = (2.8, 4.1)$

113. $(4-4, 8-8) = (0, 0)$

$(6-4, 8-8) = (2, 0)$

$(4-4, 3-8) = (0, -5)$

$(6-4, 3-8) = (2, -5)$

115. $\text{midpoint} = \left(\dfrac{x_1+x_2}{2}, \dfrac{y_1+y_2}{2}\right)$

$= \left(\dfrac{2013+2015}{2}, \dfrac{6.8+6.1}{2}\right)$

$= (2014, 6.45)$

In 2014, Barnes & Noble had annual sales of \$6.45 billion.

117. False, $(a+b)^2 = a^2 + 2ab + b^2 \neq a^2 + b^2$

There is also a cross-product term when a binomial sum is squared.

Problem Solving for Chapter P

1. (a) Men's

Maximum Volume: $V = \dfrac{4}{3}\pi(65)^3 \approx 1{,}150{,}347 \text{ mm}^3$

Minimum Volume: $V = \dfrac{4}{3}\pi(55)^3 \approx 696{,}910 \text{ mm}^3$

Women's

Maximum Volume: $V = \dfrac{4}{3}\pi(55)^3 \approx 696{,}910 \text{ mm}^3$

Minimum Volume: $V = \dfrac{4}{3}\pi\left(\dfrac{95}{2}\right)^3 \approx 448{,}921 \text{ mm}^3$

(b) Men's

Maximum density: $\dfrac{7.26}{696{,}910} \approx 1.04 \times 10^{-5} \text{ kg/mm}^3$

Minimum density: $\dfrac{7.26}{1{,}150{,}347} \approx 6.31 \times 10^{-6} \text{ kg/mm}^3$

Women's

Maximum density: $\dfrac{4.00}{448{,}921} \approx 8.91 \times 10^{-6} \text{ kg/mm}^3$

Minimum density: $\dfrac{4.00}{696{,}910} \approx 5.74 \times 10^{-6} \text{ kg/mm}^3$

(c) No. The weight would be different. Cork is much lighter than iron so it would have a much smaller density.

3. To say that a number has n significant digits means that the number has n digits with the leftmost non-zero digit and ending with the rightmost non-zero digit. For example; 28,000, 1.400, 0.00079 each have two significant digits.

5. Men: $70 \text{ beats} \times \frac{60}{1} \times \frac{24}{1} \times \frac{365.25}{1} \times \frac{76.4}{1 \text{ lifetime}}$

$$= 2{,}812{,}834{,}080 \frac{\text{beats}}{\text{lifetime}}$$

So, the number of beats in a lifetime for a man is 2,812,834,080.

Women: $70 \text{ beats} \times \frac{60}{1} \times \frac{24}{1} \times \frac{365.25}{1} \times \frac{81.2}{1 \text{ lifetime}}$

$$= 2{,}989{,}556{,}640 \frac{\text{beats}}{\text{lifetime}}$$

So, the number of beats in a lifetime for a woman is 2,989,556,640.

7. $r = 1 - \left(\frac{3225}{12{,}000}\right)^{1/4} \approx 0.280$ or 28%

9. Volume:

$$\begin{aligned} V &= lwh \\ 2x^3 + x^2 - 8x - 4 &= lw(2x + 1) \\ \frac{2x^3 + x^2 - 8x - 4}{2x + 1} &= lw \\ \frac{x^2(2x + 1) - 4(2x + 1)}{2x + 1} &= lw \\ \frac{(x^2 - 4)(2x + 1)}{2x + 1} &= lw \end{aligned}$$

$lw = x^2 - 4 = (x + 2)(x - 2)$

Let $l = x + 2$ and $w = x - 2$.

Surface Area:

$$\begin{aligned} S &= 2lw + 2lh + 2wh \\ &= 2(lw + lh + wh) \\ &= 2\left[(x + 2)(x - 2) + (x + 2)(2x + 1) + (x - 2)(2x + 1)\right] \\ &= 2\left[x^2 - 4 + 2x^2 + 5x + 2 + 2x^2 - 3x - 2\right] \\ &= 2\left[5x^2 + 2x - 4\right] \\ &= 10x^2 + 4x - 8 \end{aligned}$$

When $x = 6$ inches: $S = 10(6)^2 + 4(6) - 8 = 376$ cubic inches.

11. (a) $(1, -2)$ and $(4, 1)$

The points of trisection are:

$$\left(\frac{2(1) + 4}{3}, \frac{2(-2) + 1}{3}\right) = (2, -1)$$

$$\left(\frac{1 + 2(4)}{3}, \frac{-2 + 2(1)}{3}\right) = (3, 0)$$

(b) $(-2, -3)$ and $(0, 0)$

The points of trisection are:

$$\left(\frac{2(-2) + 0}{3}, \frac{2(-3) + 0}{3}\right) = \left(-\frac{4}{3}, -2\right)$$

$$\left(\frac{-2 + 2(0)}{3}, \frac{-3 + 2(0)}{3}\right) = \left(-\frac{2}{3}, -1\right)$$

13. One golf ball: $\frac{1.6 \times 10^7}{1.58 \times 10^8} \approx 0.101$ pound

$0.101(16) = 1.616$ ounces

Practice Test for Chapter P

1. Evaluate $\dfrac{|-42| - 20}{15 - |-4|}$.

2. Simplify $\dfrac{x}{z} - \dfrac{z}{y}$.

3. The distance between x and 7 is no more than 4. Use absolute value notation to describe this expression.

4. Evaluate $10(-x)^3$ for $x = 5$.

5. Simplify $(-4x^3)(-2x^{-5})\left(\frac{1}{16}x\right)$.

6. Change 0.0000412 to scientific notation.

7. Evaluate $125^{2/3}$.

8. Simplify $\sqrt[4]{64x^7y^9}$.

9. Rationalize the denominator and simplify $\dfrac{6}{\sqrt{12}}$.

10. Simplify $3\sqrt{80} - 7\sqrt{500}$.

11. Simplify $(8x^4 - 9x^2 + 2x - 1) - (3x^3 + 5x + 4)$.

12. Multiply $(x - 3)(x^2 + x - 7)$.

13. Multiply $[(x - 2) - y]^2$.

14. Factor $16x^4 - 1$.

15. Factor $6x^2 + 5x - 4$.

16. Factor $x^3 - 64$.

17. Combine and simplify $-\dfrac{3}{x} + \dfrac{x}{x^2 + 2}$.

18. Combine and simplify $\dfrac{x - 3}{4x} \div \dfrac{x^2 - 9}{x^2}$.

19. Simplify $\dfrac{1 - \left(\dfrac{1}{x}\right)}{1 - \dfrac{1}{1 - \left(\dfrac{1}{x}\right)}}$.

20. (a) Plot the points $(-3, 6)$ and $(5, -1)$,

(b) find the distance between the points, and

(c) find the midpoint of the line segment joining the points.

C H A P T E R 1
Equations, Inequalities, and Mathematical Modeling

CHAPTER 1
Equations, Inequalities, and Mathematical Modeling

Section 1.1 Graphs of Equations

1. solution or solution point

3. intercepts

5. circle; (h, k); r

7. (a) $(0, 2)$: $2 \stackrel{?}{=} \sqrt{0 + 4}$

$2 = 2$

Yes, the point *is* on the graph.

(b) $(5, 3)$: $3 \stackrel{?}{=} \sqrt{5 + 4}$

$3 \stackrel{?}{=} \sqrt{9}$

$3 = 3$

Yes, the point *is* on the graph.

9. (a) $(2, 0)$: $(2)^2 - 3(2) + 2 \stackrel{?}{=} 0$

$4 - 6 + 2 \stackrel{?}{=} 0$

$0 = 0$

Yes, the point *is* on the graph.

(b) $(-2, 8)$: $(-2)^2 - 3(-2) + 2 \stackrel{?}{=} 8$

$4 + 6 + 2 \stackrel{?}{=} 8$

$12 \neq 8$

No, the point *is not* on the graph.

11. (a) $(1, 5)$: $5 \stackrel{?}{=} 4 - |1 - 2|$

$5 \stackrel{?}{=} 4 - 1$

$5 \neq 3$

No, the point *is not* on the graph.

(b) $(6, 0)$: $0 \stackrel{?}{=} 4 - |6 - 2|$

$0 \stackrel{?}{=} 4 - 4$

$0 = 0$

Yes, the point *is* on the graph.

13. (a) $(3, -2)$: $(3)^2 + (-2)^2 \stackrel{?}{=} 20$

$9 + 4 \stackrel{?}{=} 20$

$13 \neq 20$

No, the point *is not* on the graph.

(b) $(-4, 2)$: $(-4)^2 + (2)^2 \stackrel{?}{=} 20$

$16 + 4 \stackrel{?}{=} 20$

$20 = 20$

Yes, the point *is* on the graph.

15. $y = -2x + 5$

x	-1	0	1	2	$\frac{5}{2}$
y	7	5	3	1	0
(x, y)	$(-1, 7)$	$(0, 5)$	$(1, 3)$	$(2, 1)$	$\left(\frac{5}{2}, 0\right)$

17. $y = x^2 - 3x$

x	-1	0	1	2	3
y	4	0	-2	-2	0
(x, y)	$(-1, 4)$	$(0, 0)$	$(1, -2)$	$(2, -2)$	$(3, 0)$

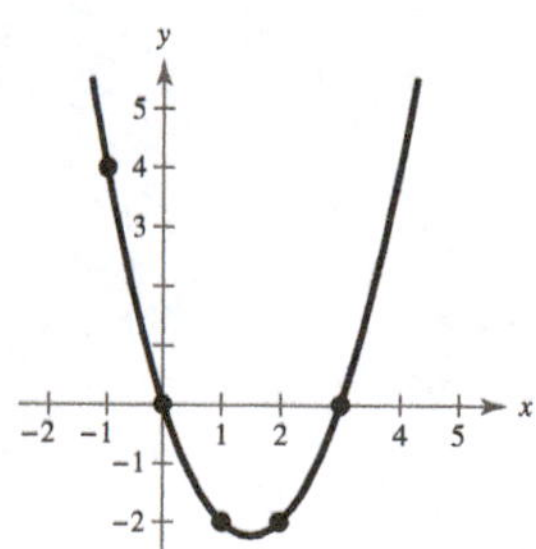

19. x-intercept: $(3, 0)$

y-intercept: $(0, 9)$

21. x-intercept: $(-2, 0)$

y-intercept: $(0, 2)$

23. x-intercept: $(1, 0)$

y-intercept: $(0, 2)$

25. $x^2 - y = 0$

$(-x)^2 - y = 0 \Rightarrow x^2 - y = 0 \Rightarrow$ y-axis symmetry

$x^2 - (-y) = 0 \Rightarrow x^2 + y = 0 \Rightarrow$ No x-axis symmetry

$(-x)^2 - (-y) = 0 \Rightarrow x^2 + y = 0 \Rightarrow$ No origin symmetry

27. $y = x^3$

$y = (-x)^3 \Rightarrow y = -x^3 \Rightarrow$ No y-axis symmetry

$-y = x^3 \Rightarrow y = -x^3 \Rightarrow$ No x-axis symmetry

$-y = (-x)^3 \Rightarrow -y = -x^3 \Rightarrow y = x^3 \Rightarrow$ Origin symmetry

29. $y = \dfrac{x}{x^2 + 1}$

$y = \dfrac{-x}{(-x)^2 + 1} \Rightarrow y = \dfrac{-x}{x^2 + 1} \Rightarrow$ No y-axis symmetry

$-y = \dfrac{x}{x^2 + 1} \Rightarrow y = \dfrac{-x}{x^2 + 1} \Rightarrow$ No x-axis symmetry

$-y = \dfrac{-x}{(-x)^2 + 1} \Rightarrow -y = \dfrac{-x}{x^2 + 1} \Rightarrow y = \dfrac{x}{x^2 + 1} \Rightarrow$ Origin symmetry

31. $xy^2 + 10 = 0$

$(-x)y^2 + 10 = 0 \Rightarrow -xy^2 + 10 = 0 \Rightarrow$ No y-axis symmetry

$x(-y)^2 + 10 = 0 \Rightarrow xy^2 + 10 = 0 \Rightarrow$ x-axis symmetry

$(-x)(-y)^2 + 10 = 0 \Rightarrow -xy^2 + 10 = 0 \Rightarrow$ No origin symmetry

33.

35.

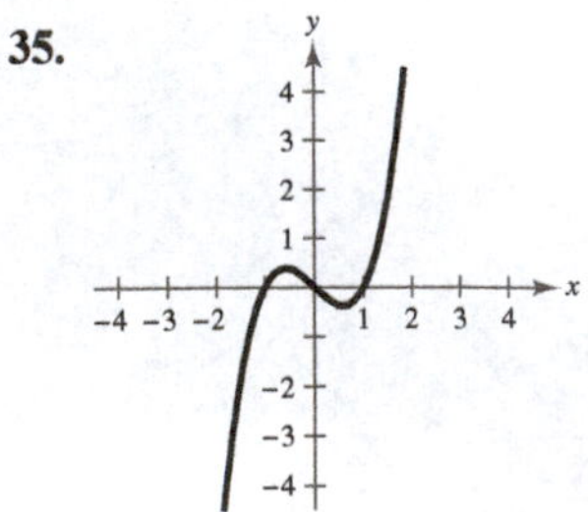

37. $y = -3x + 1$

x-intercept: $\left(\frac{1}{3}, 0\right)$

y-intercept: $(0, 1)$

No symmetry

39. $y = x^2 - 2x$

x-intercepts: $(0, 0), (2, 0)$

y-intercept: $(0, 0)$

No symmetry

x	−1	0	1	2	3
y	3	0	−1	0	3

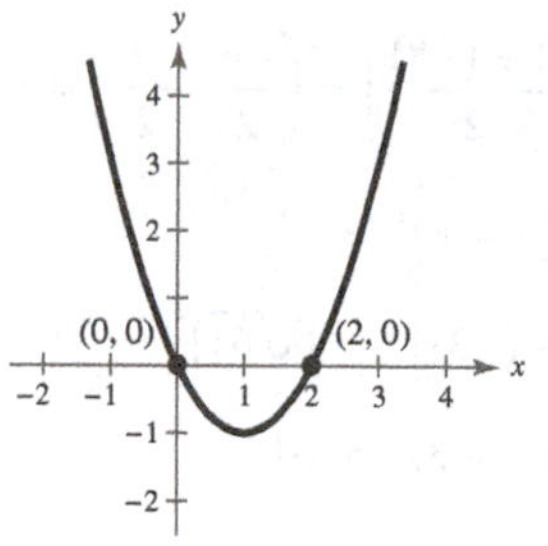

41. $y = x^3 + 3$

x-intercept: $\left(\sqrt[3]{-3}, 0\right)$

y-intercept: $(0, 3)$

No symmetry

x	−2	−1	0	1	2
y	−5	2	3	4	11

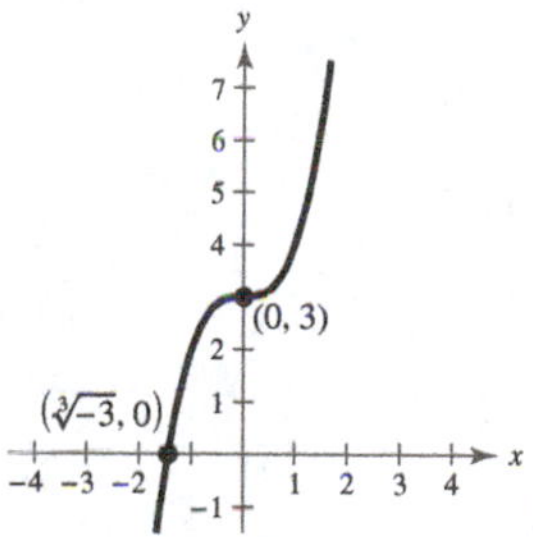

43. $y = \sqrt{x - 3}$

x-intercept: $(3, 0)$

y-intercept: none

No symmetry

x	3	4	7	12
y	0	1	2	3

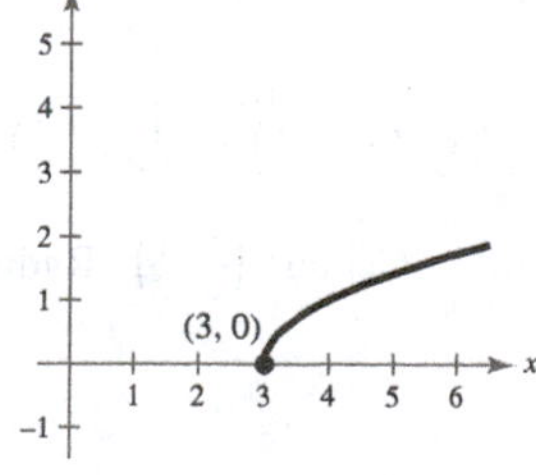

45. $y = |x - 6|$

x-intercept: $(6, 0)$

y-intercept: $(0, 6)$

No symmetry

x	−2	0	2	4	6	8	10
y	8	6	4	2	0	2	4

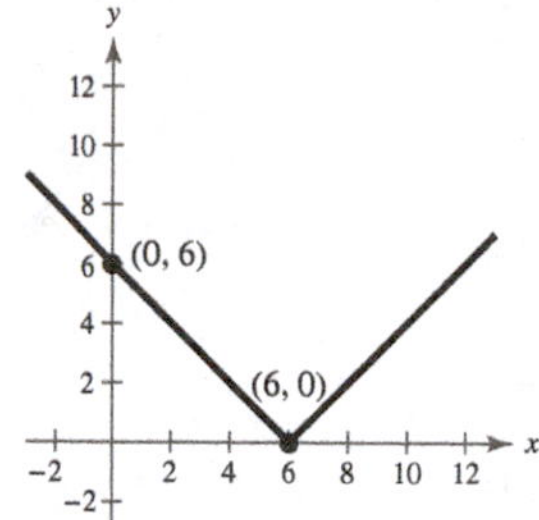

47. $x = y^2 - 1$

x-intercept: $(-1, 0)$

y-intercepts: $(0, -1), (0, 1)$

x-axis symmetry

x	−1	0	3
y	0	±1	±2

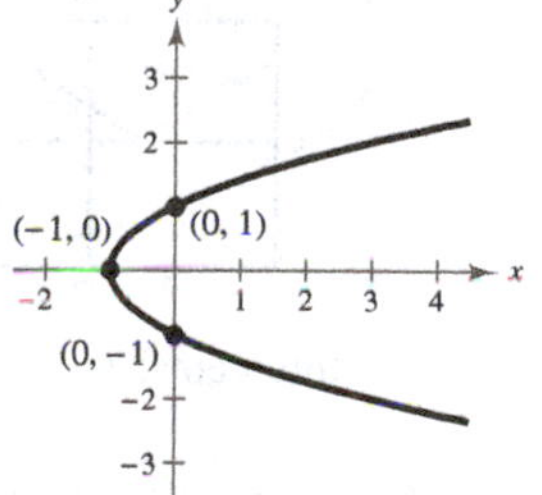

49. $y = 5 - \frac{1}{2}x$

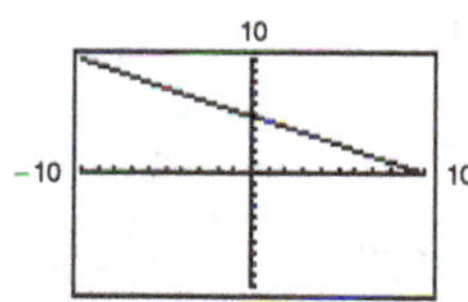

Intercepts: $(10, 0), (0, 5)$

51. $y = x^2 - 4x + 3$

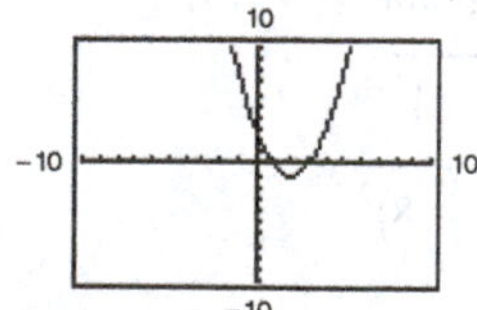

Intercepts: $(3, 0), (1, 0), (0, 3)$

53. $y = \dfrac{2x}{x - 1}$

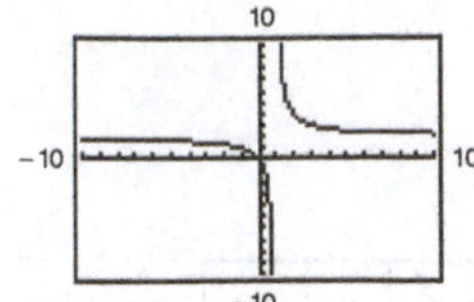

Intercept: $(0, 0)$

55. $y = \sqrt[3]{x + 1}$

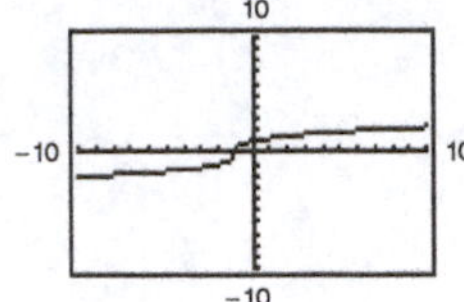

Intercepts: $(-1, 0), (0, 1)$

57. $y = |x + 3|$

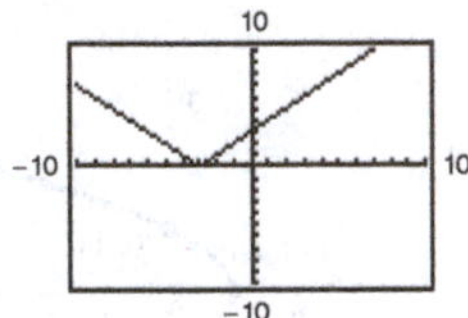

Intercepts: $(-3, 0), (0, 3)$

59. Center: $(0, 0)$; Radius: 3

$$(x - 0)^2 + (y - 0)^2 = 3^2$$
$$x^2 + y^2 = 9$$

61. Center: $(-4, 5)$; Radius: 2

$$(x - h)^2 + (y - k)^2 = r^2$$
$$[x - (-4)]^2 + [y - 5]^2 = 2^2$$
$$(x + 4)^2 + (y - 5)^2 = 4$$

63. Center: $(3, 8)$; Solution point: $(-9, 13)$

$$\begin{aligned} r &= \sqrt{(x - h)^2 + (y - k)^2} \\ &= \sqrt{(-9 - 3)^2 + (13 - 8)^2} \\ &= \sqrt{(-12)^2 + (5)^2} \\ &= \sqrt{144 + 25} \\ &= \sqrt{169} \\ &= 13 \end{aligned}$$

$$(x - h)^2 + (y - k)^2 = r^2$$
$$(x - 3)^2 + (y - 8)^2 = 13^2$$
$$(x - 3)^2 + (y - 8)^2 = 169$$

65. Endpoints of a diameter: $(3, 2), (-9, -8)$

$$\begin{aligned} r &= \frac{1}{2}\sqrt{(-9 - 3)^2 + (-8 - 2)^2} \\ &= \frac{1}{2}\sqrt{(-12)^2 + (-10)^2} \\ &= \frac{1}{2}\sqrt{144 + 100} \\ &= \frac{1}{2}\sqrt{244} = \frac{1}{2}(2)\sqrt{61} = \sqrt{61} \end{aligned}$$

$$(h, k): \left(\frac{3 + (-9)}{2}, \frac{2 + (-8)}{2}\right) = \left(\frac{-6}{2}, \frac{-6}{2}\right) = (-3, -3)$$

$$(x - h)^2 + (y - k)^2 = r^2$$
$$[x - (-3)]^2 + [y - (-3)]^2 = \left(\sqrt{61}\right)^2$$
$$(x + 3)^2 + (y + 3)^2 = 61$$

67. $x^2 + y^2 = 25$

Center: $(0, 0)$, Radius: 5

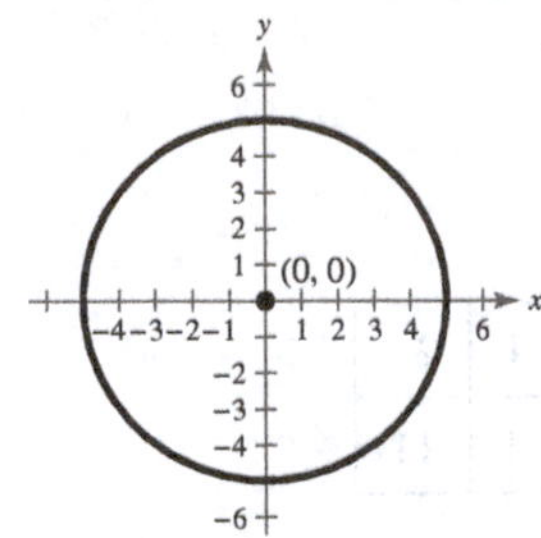

69. $(x - 1)^2 + (y + 3)^2 = 9$

Center: $(1, -3)$, Radius: 3

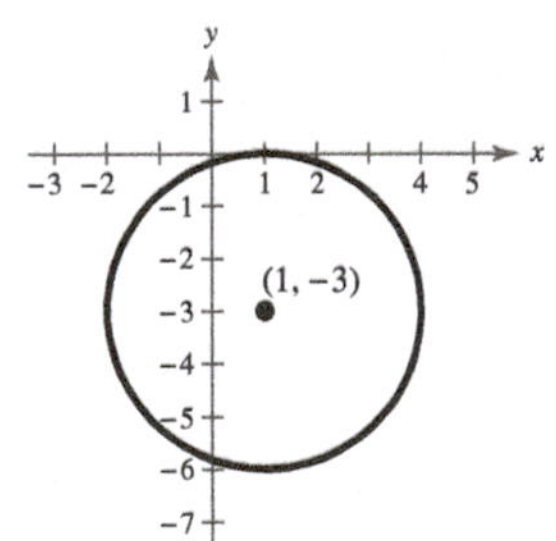

71. $\left(x - \frac{1}{2}\right)^2 + \left(y - \frac{1}{2}\right)^2 = \frac{9}{4}$

Center: $\left(\frac{1}{2}, \frac{1}{2}\right)$, Radius: $\frac{3}{2}$

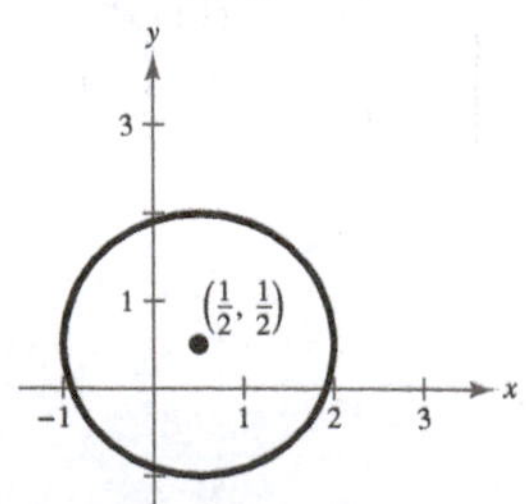

73. $y = 1{,}200{,}000 - 80{,}000t,\ 0 \le t \le 10$

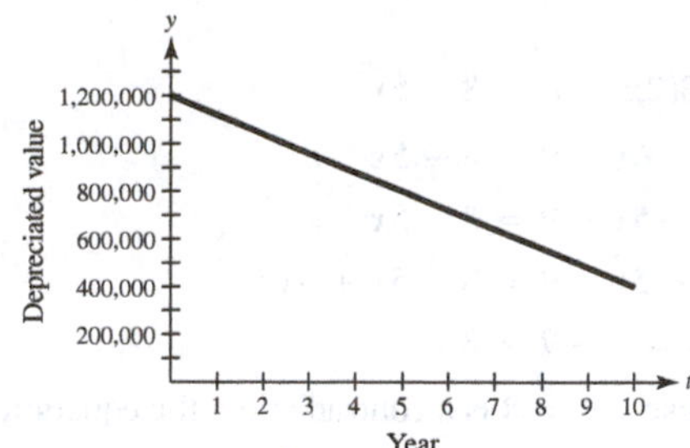

75. (a)

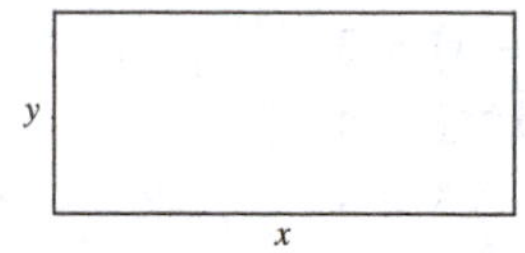

(b) $2x + 2y = \frac{1040}{3}$

$$2y = \frac{1040}{3} - 2x$$

$$y = \frac{520}{3} - x$$

$$A = xy = x\left(\frac{520}{3} - x\right)$$

(c)

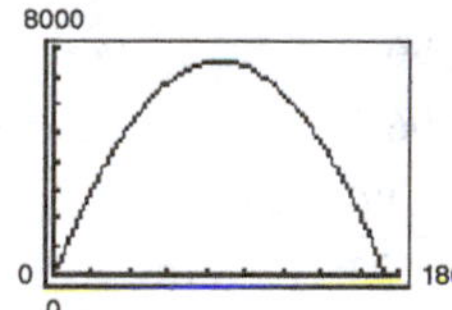

(d) When $x = y = 86\frac{2}{3}$ yards, the area is a maximum of $7511\frac{1}{9}$ square yards.

(e) A regulation NFL playing field is 120 yards long and $53\frac{1}{3}$ yards wide. The actual area is 6400 square yards.

77. (a)

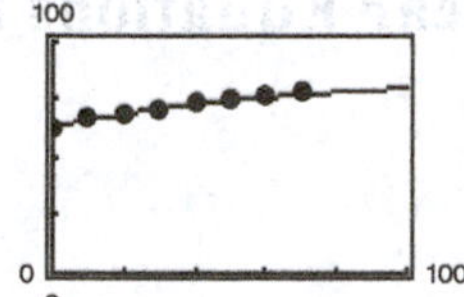

The model fits the data well.

(b) Graphically: The point $(50, 74.7)$ represents a life expectancy of 74.7 years in 1990.

Algebraically: $y = \dfrac{63.6 + 0.97(50)}{1 + 0.01(50)}$

$$= \frac{112.1}{1.5}$$

$$= 74.7$$

So, the life expectancy in 1990 was about 74.7 years.

(c) Graphically: The point $(24.2, 70.1)$ represents a life expectancy of 70.1 years during the year 1964.

Algebraically:

$$y = \frac{63.6 + 0.97t}{1 + 0.01t}$$

$$70.1 = \frac{63.6 + 0.97t}{1 + 0.01t}$$

$$70.1(1 + 0.01t) = 63.6 + 0.97t$$

$$70.1 + 0.701t = 63.6 + 0.97t$$

$$6.5 = 0.269t$$

$$t = 24.2$$

When $y = 70.1$, $t = 24.2$ which represents the year 1964.

(d) $y = \dfrac{63.6 + 0.97(0)}{1 + 0.01(0)} = \dfrac{63.6}{1} = 63.6$

The y-intercept is $(0, 63.6)$. In 1940, the life expectancy of a child (at birth) was 63.6 years.

(e) Answers will vary.

79. False. The line $y = x$ is symmetric with respect to the origin.

81. True. Depending upon the center and radius, the graph of a circle could intersect one, both, or neither axis.

83. $y = ax^2 + bx^3$

(a) $y = a(-x)^2 + b(-x)^3 = ax^2 - bx^3$

To be symmetric with respect to the y-axis; a can be any non-zero real number, b must be zero.

Sample answer: $a = 1, b = 0$

(b) $-y = a(-x)^2 + b(-x)^3$

$$-y = ax^2 - bx^3$$

$$y = -ax^2 + bx^3$$

To be symmetric with respect to the origin; a must be zero, b can be any non-zero real number.

Sample answer: $a = 0, b = 1$

Section 1.2 Linear Equations in One Variable

1. equation

3. $ax + b = 0$

5. rational

7. The equation $3(x - 1) = 3x - 3$ is an *identity* by the Distributive Property. The equation is true for all real values of x.

9. The equation $2(x - 1) = 3x + 1$ is a *conditional equation.* The only value in the domain that satisfies the equation is $x = -3$.

11. The equation $3(x + 2) = 3x + 2$ is a *contradiction.* There are no real values of x for which the equation is true.

13. The equation $2(x + 3) - 5 = 2x + 1$ is an *identity* by simplification. The equation is true for all real values of x.

15.
$$\begin{aligned} x + 11 &= 15 \\ x + 11 - 11 &= 15 - 11 \\ x &= 4 \end{aligned}$$

17.
$$\begin{aligned} 7 - 2x &= 25 \\ 7 - 7 - 2x &= 25 - 7 \\ -2x &= 18 \\ \frac{-2x}{-2} &= \frac{18}{-2} \\ x &= -9 \end{aligned}$$

19.
$$\begin{aligned} 3x - 5 &= 2x + 7 \\ 3x - 2x - 5 &= 2x - 2x + 7 \\ x - 5 &= 7 \\ x - 5 + 5 &= 7 + 5 \\ x &= 12 \end{aligned}$$

21.
$$\begin{aligned} 4y + 2 - 5y &= 7 - 6y \\ 4y - 5y + 2 &= 7 - 6y \\ -y + 2 &= 7 - 6y \\ -y + 6y + 2 &= 7 - 6y + 6y \\ 5y + 2 &= 7 \\ 5y + 2 - 2 &= 7 - 2 \\ 5y &= 5 \\ \frac{5y}{5} &= \frac{5}{5} \\ y &= 1 \end{aligned}$$

23.
$$\begin{aligned} x - 3(2x + 3) &= 8 - 5x \\ x - 6x - 9 &= 8 - 5x \\ -5x - 9 &= 8 - 5x \\ -5x + 5x - 9 &= 8 - 5x + 5x \\ -9 &\neq 8 \end{aligned}$$

Because $-9 = 8$ is a contradiction, the equation has no solution.

25.
$$\begin{aligned} 0.25x + 0.75(10 - x) &= 3 \\ 0.25x + 7.5 - 0.75x &= 3 \\ -0.50x + 7.5 &= 3 \\ -0.50x &= -4.5 \\ x &= 9 \end{aligned}$$

27.
$$\begin{aligned} \frac{3x}{8} - \frac{4x}{3} &= 4 \\ (24)\frac{3x}{8} - (24)\frac{4x}{3} &= (24)4 \\ 9x - 32x &= 96 \\ -23x &= 96 \\ x &= -\frac{96}{23} \end{aligned}$$

29.
$$\begin{aligned} \frac{5x}{4} + \frac{1}{2} &= x - \frac{1}{2} \\ (4)\frac{5x}{4} + (4)\frac{1}{2} &= (4)x - (4)\frac{1}{2} \\ 5x + 2 &= 4x - 2 \\ x &= -4 \end{aligned}$$

31.
$$\begin{aligned} \frac{5x - 4}{5x + 4} &= \frac{2}{3} \\ 3(5x - 4) &= 2(5x + 4) \\ 15x - 12 &= 10x + 8 \\ 5x &= 20 \\ x &= 4 \end{aligned}$$

33.
$$\begin{aligned} 10 - \frac{13}{x} &= 4 + \frac{5}{x} \\ \frac{10x - 13}{x} &= \frac{4x + 5}{x} \\ 10x - 13 &= 4x + 5 \\ 6x &= 18 \\ x &= 3 \end{aligned}$$

35.
$$\begin{aligned} 3 &= 2 + \frac{2}{z+2} \\ 3(z+2) &= \left(2 + \frac{2}{z+2}\right)(z+2) \\ 3z + 6 &= 2z + 4 + 2 \\ z &= 0 \end{aligned}$$

37.
$$\begin{aligned} \frac{x}{x+4} + \frac{4}{x+4} + 2 &= 0 \\ \frac{x+4}{x+4} + 2 &= 0 \\ 1 + 2 &= 0 \\ 3 &\neq 0 \end{aligned}$$

Because $3 = 0$ is a contradiction, the equation has no solution.

39.
$$\begin{aligned} \frac{2}{(x-4)(x-2)} &= \frac{1}{x-4} + \frac{2}{x-2} \quad \text{Multiply each term by } (x-4)(x-2). \\ 2 &= 1(x-2) + 2(x-4) \\ 2 &= x - 2 + 2x - 8 \\ 2 &= 3x - 10 \\ 12 &= 3x \\ 4 &= x \end{aligned}$$

A check reveals that $x = 4$ yields a denominator of zero. So, $x = 4$ is an extraneous solution, and the original equation has no real solution.

41.
$$\begin{aligned} \frac{1}{x-3} + \frac{1}{x+3} &= \frac{10}{x^2-9} \quad \text{Multiply each term by } (x+3)(x-3). \\ \frac{1}{x-3} + \frac{1}{x+3} &= \frac{10}{(x+3)(x-3)} \\ 1(x+3) + 1(x-3) &= 10 \\ 2x &= 10 \\ x &= 5 \end{aligned}$$

43.
$$\begin{aligned} \frac{3}{x^2-3x} + \frac{4}{x} &= \frac{1}{x-3} \\ \frac{3}{x(x-3)} + \frac{4}{x} &= \frac{1}{x-3} \quad \text{Multiply each term by } x(x-3). \\ 3 + 4(x-3) &= x \\ 3 + 4x - 12 &= x \\ 3x &= 9 \\ x &= 3 \end{aligned}$$

A check reveals that $x = 3$ yields a denominator of zero. So, $x = 3$ is an extraneous solution, and the original equation has no solution.

45.
$$\begin{aligned} y &= 12 - 5x & y &= 12 - 5x \\ 0 &= 12 - 5x & y &= 12 - 5(0) \\ 5x &= 12 & y &= 12 \\ x &= \tfrac{12}{5} \end{aligned}$$

The x-intercept is $\left(\frac{12}{5}, 0\right)$ and the y-intercept is $(0, 12)$.

47.
$$\begin{aligned} y &= -3(2x+1) & y &= -3(2x+1) \\ 0 &= -3(2x+1) & y &= -3(2(0)+1) \\ 0 &= 2x + 1 & y &= -3 \\ x &= -\tfrac{1}{2} \end{aligned}$$

The x-intercept is $\left(-\frac{1}{2}, 0\right)$ and the y-intercept is $(0, -3)$.

49.
$$2x + 3y = 10 \qquad 2x + 3y = 10$$
$$2x + 3(0) = 10 \qquad 2(0) + 3y = 10$$
$$2x = 10 \qquad 3y = 10$$
$$x = 5 \qquad y = \tfrac{10}{3}$$

The x-intercept is $(5, 0)$ and the y-intercept is $\left(0, \tfrac{10}{3}\right)$.

51.
$$4y - 0.75x + 1.2 = 0 \qquad 4y - 0.75x + 1.2 = 0$$
$$4(0) - 0.75x + 1.2 = 0 \qquad 4y - 0.75(0) + 1.2 = 0$$
$$-0.75 + 1.2 = 0 \qquad 4y + 1.2 = 0$$
$$x = \frac{1.2}{0.75} = 1.6 \qquad y = \frac{-1.2}{4} = -0.3$$

The x-intercept is $(1.6, 0)$ and the y-intercept is $(0, -0.3)$.

53.
$$\frac{2x}{5} + 8 - 3y = 0 \qquad 2x + 40 - 15y = 0$$
$$2x + 40 - 15y = 0 \qquad 2(0) + 40 - 15y = 0$$
$$2x + 40 - 15(0) = 0 \qquad 40 - 15y = 0$$
$$2x + 40 = 0 \qquad y = \frac{40}{15} = \frac{8}{3}$$
$$x = -20$$

The x-intercept is $(-20, 0)$ and the y-intercept is $\left(0, \tfrac{8}{3}\right)$.

55. $y = 2(x - 1) - 4$

$$0 = 2(x - 1 - 4)$$
$$0 = 2x - 2 - 4$$
$$0 = 2x - 6$$
$$6 = 2x$$
$$3 = x$$
$$x = 3$$

The x-intercept is $x = 3$. The solution of $0 = 2(x - 1) - 4$ and the x-intercept of $y = 2(x - 1) - 4$ are the same. They are both $x = 3$. The x-intercept is $(3, 0)$.

57. $y = 20 - (3x - 10)$

$$0 = 20 - (3x - 10)$$
$$0 = 20 - 3x + 10$$
$$0 = 30 - 3x$$
$$3x = 30$$
$$x = 10$$

The x-intercept is $x = 10$. The solution of $0 = 20 - (3x - 10)$ and the x-intercept of $y = 20 - (3x - 10)$ are the same. They are both $x = 10$. The x-intercept is $(10, 0)$.

59. $y = -38 + 5(9 - x)$

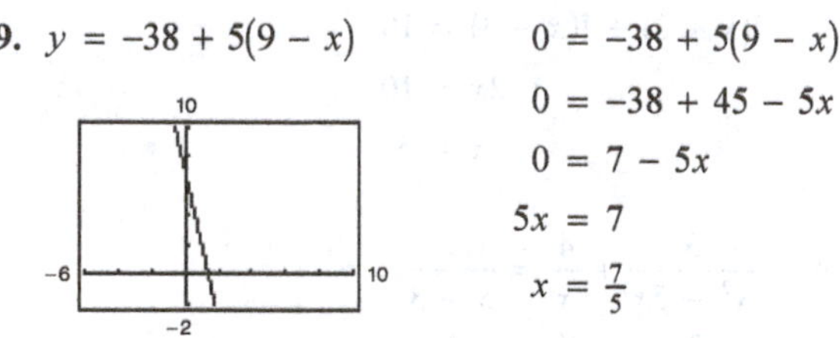

$$0 = -38 + 5(9 - x)$$
$$0 = -38 + 45 - 5x$$
$$0 = 7 - 5x$$
$$5x = 7$$
$$x = \tfrac{7}{5}$$

The x-intercept is at $x = \tfrac{7}{5}$. The solution of $0 = -38 + 5(9 - x)$ and the x-intercept of $y = -38 + 5(9 - x)$ are the same. They are both $x = \tfrac{7}{5}$. The x-intercept is $\left(\tfrac{7}{5}, 0\right)$.

61.
$$0.275x + 0.725(500 - x) = 300$$
$$0.275x + 362.5 - 0.725x = 300$$
$$-0.45x = -62.5$$
$$x = \frac{62.5}{0.45} \approx 138.889$$

63. $\dfrac{2}{7.398} - \dfrac{4.405}{x} = \dfrac{1}{x}$ Multiply both sides by $7.398x$.

$$2x - (4.405)(7.398) = 7.398$$
$$2x = (4.405)(7.398) + 7.398$$
$$2x = (5.405)(7.398)$$
$$x = \frac{(5.405)(7.398)}{2} \approx 19.993$$

65.
$$\begin{aligned}471 &= 2\pi(25) + 2\pi(5h)\\ 471 &= 50\pi + 10\pi h\\ 471 - 50\pi &= 10\pi h\\ h &= \frac{471 - 50\pi}{10\pi} = \frac{471 - 50(3.14)}{10(3.14)} = 10\\ h &= 10 \text{ feet}\end{aligned}$$

67. Let $y = 18$.

$$\begin{aligned}y &= 0.514x - 14.75\\ 18 &= 0.514x - 14.75\\ 32.75 &= 0.514x\\ \frac{32.75}{0.514} &= x\\ 63.7 &= x\end{aligned}$$

So, the height of the female is about 63.7 inches.

69. (a) The y-intercept is about $(0, 295)$.

(b) Let $t = 0$.

$y = 11.09t + 293.4 = 11.09(0) + 293.4 = 293.4$

The y-intercept is $(0, 293.4)$.

In 2000, the population of Raleigh was about 293.4 thousand, or 293,400.

(c) Let $y = 538$.

$$\begin{aligned}y &= 11.09t + 293.4\\ 538 &= 11.09t + 293.4\\ 244.6 &= 11.09t\\ \frac{244.6}{11.09} &= t\\ 22.1 &\approx t\end{aligned}$$

In 2022, the population is expected to reach 538,000. Answers will vary. *Sample answer:* If the population continues to increase at a constant (linear) rate, the answer seems reasonable.

71. Let $c = 10{,}000$.

$$\begin{aligned}c &= 0.37m + 2600\\ 10{,}000 &= 0.37m + 2600\\ 7400 &= 0.37m\\ \frac{7400}{0.37} &= m\\ m &= 20{,}000\end{aligned}$$

So, the number of miles is 20,000.

73. $x(3 - x) = 10$

$3x - x^2 = 10$

False. This is a quadratic equation. The equation cannot be written in the form $ax + b = 0$.

75.
$$\begin{aligned}2(x - 3) + 1 &= 2x - 5\\ 2x - 6 + 1 &= 2x - 5\\ 2x - 5 &= 2x - 5\end{aligned}$$

False. The equation is an identity, so every real number is a solution.

77.
$$\begin{aligned}3(x - 1) - 2 &= 3x - 6\\ 3x - 3 - 2 &= 3x - 6\\ 3x - 5 &= 3x - 6\\ -5 &\neq -6\end{aligned}$$

False. The equation $-5 = -6$ is a contradiction, so the original equation has no solution.

79. (a)

x	−1	0	1	2	3	4
$3.2x - 5.8$	−9	−5.8	−2.6	0.6	3.8	7

(b) Since the sign changes from negative at 1 to positive at 2, the root is somewhere between 1 and 2.

$1 < x < 2$

(c)

x	1.5	1.6	1.7	1.8	1.9	2
$3.2x - 5.8$	−1	−0.68	−0.36	−0.04	0.28	0.6

(d) Since the sign changes from negative at 1.8 to positive at 1.9, the root is somewhere between 1.8 and 1.9.

$1.8 < x < 1.9$

To improve accuracy, evaluate the expression at subintervals within this interval and determine where the sign changes.

(e) $0.3(x - 1.5) - 2 = 0$

x	6	7	8	9	10
$0.3(x - 1.5) - 2$	−0.65	−0.35	−0.05	0.25	0.55

The solution of $0.3(x - 1.5) - 2 = 0$ is in the interval $8 < x < 9$.

x	8.1	8.2	8.3	8.4
$0.3(x - 1.5) - 2$	−0.02	0.01	0.04	0.07

The solution is in the interval $8.1 < x < 8.2$.

x	8.14	8.15	8.16	8.17	8.18
$0.3(x - 1.5) - 2$	−0.008	−0.005	−0.002	0.001	0.004

The solution is in the interval $8.16 < x < 8.17$.

81. (a)

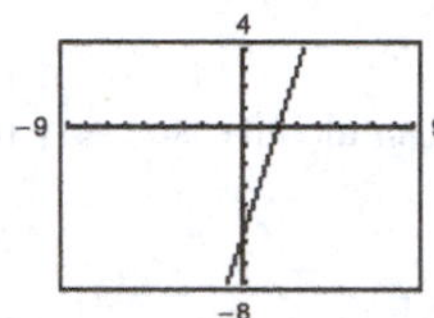

(b) x-intercept: $(2, 0)$

(c) The x-intercept is the solution of the equation $3x - 6 = 0$.

83. (a) To find the x-intercept, let $y = 0$ and solve for x.

$0 = ax + b$

$-b = ax$

$\frac{-b}{a} = x$

The x-intercept is $\left(-\frac{b}{a}, 0\right)$.

(b) To find the y-intercept, let $x = 0$, and solve for y.

$y = a(0) + b$

$y = b$

The y-intercept is $(0, b)$.

(c) x-intercept: $x = \frac{-b}{a} = \frac{-10}{5} = -2$

The x-intercept is $(-2, 0)$.

y-intercept: $y = b = 10$

The y-intercept is $(0, 10)$.

Section 1.3 Modeling with Linear Equations

1. mathematical modeling

3. $y + 2$

The sum of a number and 2

A number increased by 2

5. $\frac{t}{6}$

A number divided by 6

7. $\frac{z - 2}{3}$

A number decreased by 2, then divided by 3

9. $-2(d + 5)$

The product of -2 and a number increased by 5

11. $\frac{3(x - 2)}{x}$

The product of 2 less than a number and 3, then divided by the same number

13. *Verbal Model*: (Sum) = (first number) + (second number)

Labels: Sum = S, first number = n, second number = $n + 1$

Equation: $S = n + (n + 1) = 2n + 1$

15. *Verbal Model:* Product = (first odd integer) · (second odd integer)

Labels: Product = P, first odd integer = $2n - 1$, second odd integer = $2n - 1 + 2 = 2n + 1$

Equation: $P = (2n - 1)(2n + 1) = 4n^2 - 1$

17. *Verbal Model:* (Distance) = (rate) · (time)

Labels: Distance = d, rate = 55 mph, time = t

Equation: $d = 55t$

19. *Verbal Model:* (Amount of acid) = 20% · (amount of solution)

Labels: Amount of acid (in gallons) = A, amount of solution (in gallons) = x

Equation: $A = 0.20x$

21. *Verbal Model:* Perimeter = 2(width) + 2(length)

Labels: Perimeter = P, width = x, length = 2(width) = $2x$

Equation: $P = 2x + 2(2x) = 6x$

23. *Verbal Model:* (Total cost) = (unit cost)(number of units) + (fixed cost)

Labels: Total cost = C, unit cost = \$40, number of units = x, fixed cost = \$2500

Equation: $C = 2500 + 40x$

25. *Verbal Model:* (Discount) = (percent) · (list price)

Equation: $d = 0.30L$

27. *Labels:* N = the number, p = percent % of the number

Verbal Model: $(\text{The number}) = \dfrac{(\text{percent})}{100} \cdot 672$

Equation: $N = \dfrac{p}{100} \cdot 672$

29.

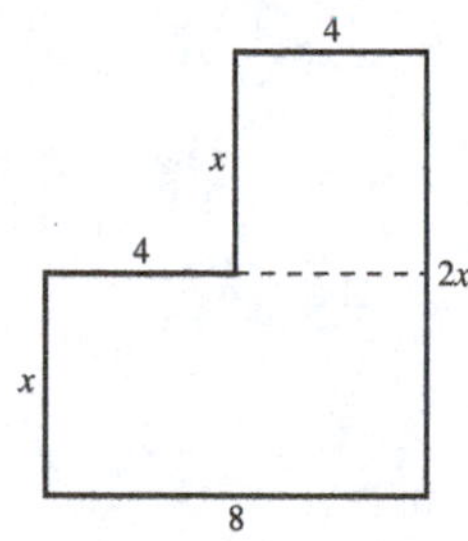

Area = Area of top rectangle + Area of bottom rectangle

$A = 4x + 8x = 12x$

31. *Verbal Model:* Sum = (first number) + (second number)

Labels: Sum = 525, first number = n, second number = $n + 1$

Equation:
$$525 = n + (n + 1)$$
$$525 = 2n + 1$$
$$524 = 2n$$
$$n = 262$$

Answer: First number = n = 262, second number = $n + 1$ = 263

33. *Verbal Model:* Difference = (one number) − (another number)

Labels: Difference = 148, one number = $5x$, another number = x

Equation:
$$148 = 5x - x$$
$$148 = 4x$$
$$x = 37$$
$$5x = 185$$

Answer: The two numbers are 37 and 185.

35. *Verbal Model:* Product = (smaller number) · (larger number) = $(\text{smaller number})^2 - 5$

Labels: Smaller number = n, larger number = $n + 1$

Equation:
$$n(n + 1) = n^2 - 5$$
$$n^2 + n = n^2 - 5$$
$$n = -5$$

Answer: Smaller number = $n = -5$, larger number = $n + 1 = -4$

37. *Verbal Model:* (first paycheck) + (second paycheck) = total

Labels: second paycheck = x, first paycheck = $0.85x$, total = $1125

Equation:
$$0.85x + x = 1125$$
$$1.85x = 1125$$
$$x \approx 608.11$$
$$0.85x \approx 516.89$$

Answer: The first salesperson's weekly paycheck is $516.89 and the second salesperson's weekly paycheck is $608.11.

39. *Verbal Model:* (Loan payments) = (Percent) · (Annual Income)

Labels: Loan payments = 15,680 (dollars)
Percent = 0.32
Annual income = I (dollars)

Equation:
$$15{,}680 = 0.32I$$
$$\frac{15{,}680}{0.32} = \frac{0.32I}{0.32}$$
$$49{,}000 = I$$

Answer: The family's annual income is $49,000.

41. (a)

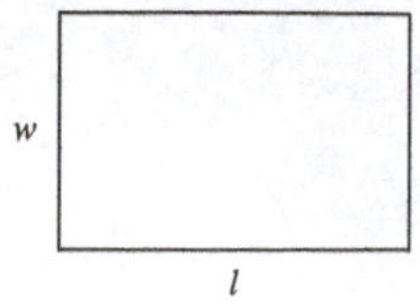

(b) $l = 1.5w$

$$P = 2l + 2w = 2(1.5w) + 2w = 5w$$

(c) $25 = 5w$

$5 = w$

Width: w = 5 meters

Length: $l = 1.5w$ = 7.5 meters

Dimensions: 7.5 meters × 5 meters

43. *Verbal Model:* Average $= \dfrac{(\text{test \#1}) + (\text{test \#2}) + (\text{test \#3}) + (\text{test \#4})}{4}$

Labels: Average = 90, test #1 = 87, test #2 = 92, test #3 = 84, test #4 = x

Equation: $90 = \dfrac{87 + 92 + 84 + x}{4}$

Answer: You must score 97 or better on test #4 to earn an A for the course.

45. $\text{Rate} = \dfrac{\text{distance}}{\text{time}} = \dfrac{50 \text{ kilometers}}{\frac{1}{2} \text{ hour}} = 100 \text{ kilometers/hour}$

$\text{Total time} = \dfrac{\text{total distance}}{\text{rate}} = \dfrac{500 \text{ kilometers}}{100 \text{ kilometers/hour}} = 5 \text{ hours}$

The entire trip takes 5 hours.

47. *Verbal Model:* (Distance) = (rate)(time)

Labels: Distance $= 1.5 \times 10^{11}$ (meters)

Rate $= 3.0 \times 10^{8}$ (meters per second)

Time $= t$

Equation: $1.5 \times 10^{11} = \left(3.0 \times 10^{8}\right)t$

$500 = t$

Light from the sun travels to the Earth in 500 seconds or approximately 8.33 minutes.

49. *Verbal Model:* $\dfrac{(\text{Height of building})}{(\text{Length of building's shadow})} = \dfrac{(\text{Height of post})}{(\text{Length of post's shadow})}$

Labels: Height of building $= x$ (feet)

Length of building's shadow = 105 (feet)

Height of post $= 3 \cdot 12 = 36$ (inches)

Length of post's shadow = 4 (inches)

Equation: $\dfrac{x}{105} = \dfrac{36}{4}$

$x = 945$

One Liberty Place is 945 feet tall.

51. (a)

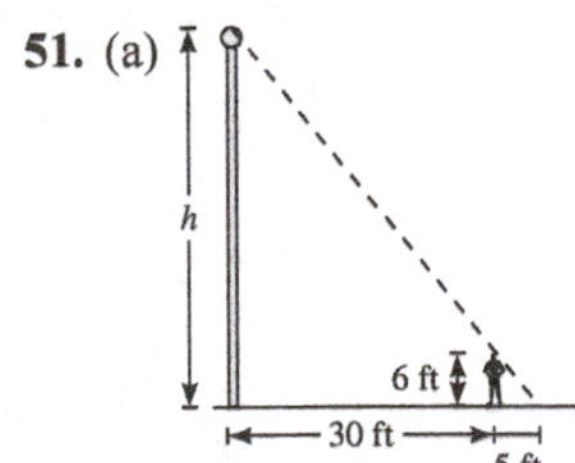

(b) *Verbal Model:* $\dfrac{(\text{height of pole})}{(\text{height of pole's shadow})} = \dfrac{(\text{height of person})}{(\text{height of person's shadow})}$

Labels: Height of pole $= h$, height of pole's shadow $= 30 + 5 = 35$ feet, height of person = 6 feet, height of person's shadow = 5 feet

Equation: $\dfrac{h}{35} = \dfrac{6}{5}$

$h = \dfrac{6}{5} \cdot 35 = 42$

The pole is 42 feet tall.

53. *Verbal Model:* [Interest from $4\frac{1}{2}\%$] + [Interest from 5%] = [Total interest]

Labels:
Amount invested at $4\frac{1}{2}\% = x$ dollars
Amount invested at $5\% = 12{,}000 - x$ dollars
Interest from $4\frac{1}{2}\% = x(0.045)$ dollars
Interest from $5\% = (12{,}000 - x)(0.05)$ dollars
Total annual interest = 580 dollars

Equation:

$$0.045x + 0.05(12{,}000 - x) = 580$$
$$0.045x + 600 - 0.05x = 580$$
$$-0.005x = -20$$
$$x = 4000$$

So, \$4000 was invested at $4\frac{1}{2}\%$ and \$12,000 − \$4000 = \$8000 was invested at 5%.

55. *Verbal Model:* (Profit from dogwood trees) + (profit from red maple trees) = (total profit)

Labels:
Inventory of dogwood trees $= x$, inventory of red maple trees $= 40{,}000 - x$,
profit from dogwood trees $= 0.25x$, profit from red maple trees $= 0.17(40{,}000 - x)$,
total profit $= 0.20(40{,}000) = 8000$

Equation:

$$0.25x + 0.17(40{,}000) = 8000$$
$$0.25x + 6800 - 0.17x = 8000$$
$$0.08x = 1200$$
$$x = 15{,}000$$

The amount invested in dogwood trees was \$15,000 and the amount invested in red maple trees was \$40,000 −\$15,000 = \$25,000.

57. *Verbal Model:* [Amount of gasoline in mixture] + [Amount of gasoline to add] = [Amount of gasoline in final mixture]

Labels:
Amount of gasoline in mixture $= \frac{32}{33}(2)$ (gallons)
Amount of gasoline to add $= x$ (gallons)
Amount of gasoline in final mixture $= \frac{50}{51}(2 + x)$ (gallons)

Equation:

$$\frac{64}{33} + x = \frac{50}{51}(2 + x)$$
$$\frac{64}{33} + x = \frac{100}{51} + \frac{50}{51}x$$
$$3264 + 1683x = 3300 + 1650x$$
$$33x = 36$$
$$x \approx 1.09$$

The forester should add about 1.09 gallons of gasoline to the mixture.

59.
$$A = \frac{1}{2}bh$$
$$2A = bh$$
$$\frac{2A}{b} = h$$

61.
$$S = C + RC$$
$$S = C(1 + R)$$
$$\frac{S}{1 + R} = C$$

63. $A = P + Prt$

$A - P = Prt$

$\dfrac{A - P}{Pt} = r$

65. $C = \frac{5}{9}(F - 32)$

$= \frac{5}{9}(98.6 - 32)$

$= \frac{5}{9}(66.6)$

$= 37^\circ$

The temperature is 37°C.

67. $F = \frac{9}{5}C + 32$

$= \frac{9}{5}(27) + 32$

$= 48.6 + 32$

$= 80.6^\circ F$

The temperature is 80.6°F.

69. $V = \dfrac{4}{3}\pi r^3$

$5.96 = \dfrac{4}{3}\pi r^3$

$17.88 = 4\pi r^3$

$\dfrac{17.88}{4\pi} = r^3$

$r = \sqrt[3]{\dfrac{4.47}{\pi}} \approx 1.12$ inches

71. (a) $W_1x = W_2(L - x)$

$50x = 75(10 - x)$

$50x = 750 - 75x$

$125x = 750$

$x = 6$ feet from 50-pound child

(b) $W_1x = W_2(L - x)$

$W_1 = 200$ pounds

$W_2 = 550$ pounds

$L = 5$ feet

$200x = 550(5 - x)$

$200x = 2750 - 550x$

$750x = 2750$

$x = 3\frac{2}{3}$ feet from the person

73. False, it should be written as $\dfrac{z^3 - 8}{z^2 - 9}$.

75. Area of circle: $A = \pi r^2 = \pi(2)^2 = 4\pi \approx 12.56$ in.^{2}

Area of square: $A = s^2 = (4)^2 = 16$ in.^{2}

True. 12.56 in.$^2 < 16$ in.^{2}, so the area of the circle is less than the area of the square.

Section 1.4 Quadratic Equations and Applications

1. quadratic equation

3. factoring; square roots; completing; square; Quadratic Formula

5. position equation

7. $6x^2 + 3x = 0$

$3x(2x + 1) = 0$

$3x = 0$ or $2x + 1 = 0$

$x = 0$ or $x = -\frac{1}{2}$

9. $3 + 5x - 2x^2 = 0$

$(3 - x)(1 + 2x) = 0$

$3 - x = 0$ or $1 + 2x = 0$

$x = 3$ or $x = -\frac{1}{2}$

11. $x^2 + 10x + 25 = 0$

$(x + 5)(x + 5) = 0$

$x + 5 = 0$

$x = -5$

13. $16x^2 - 9 = 0$

$(4x + 3)(4x + 3) = 0$

$4x + 3 = 0 \Rightarrow x = -\frac{3}{4}$

$4x - 3 = 0 \Rightarrow x = \frac{3}{4}$

15. $2x^2 = 19x + 33$

$2x^2 - 19x - 33 = 0$

$(2x + 3)(x - 11) = 0$

$2x + 3 = 0 \Rightarrow x = -\frac{3}{2}$

$x - 11 = 0 \Rightarrow x = 11$

17. $\frac{3}{4}x^2 + 8x + 20 = 0$

$4\left(\frac{3}{4}x^2 + 8x + 20\right) = 4(0)$

$3x^2 + 32x + 80 = 0$

$(3x + 20)(x + 4) = 0$

$3x + 20 = 0$ or $x + 4 = 0$

$x = -\frac{20}{3}$ or $x = -4$

19. $x^2 = 49$

$x = \pm 7$

21. $x^2 = 19$

$x = \pm\sqrt{19}$

$x \approx \pm 4.36$

23. $3x^2 = 81$

$x^2 = 27$

$x = \pm 3\sqrt{3}$

$\approx \pm 5.20$

25. $(x - 4)^2 = 49$

$x - 4 = \pm 7$

$x = 4 \pm 7$

$x = 11$ or $x = -3$

27. $(x + 2)^2 = 14$

$x + 2 = \pm\sqrt{14}$

$x = -2 \pm \sqrt{14}$

$\approx 1.74, \ -5.74$

29. $(2x - 1)^2 = 18$

$2x - 1 = \pm\sqrt{18}$

$2x = 1 \pm 3\sqrt{2}$

$x = \dfrac{1 \pm 3\sqrt{2}}{2}$

$\approx 2.62, \ -1.62$

31. $(x - 7)^2 = (x + 3)^2$

$x - 7 = \pm(x + 3)$

$x - 7 = x + 3$ or $x - 7 = -x - 3$

$-7 \neq 3$ or $2x = 4$

$x = 2$

The only solution of the equation is $x = 2$.

33. $x^2 + 4x - 32 = 0$

$x^2 + 4x = 32$

$x^2 + 4x + 2^2 = 32 + 2^2$

$(x + 2)^2 = 36$

$x + 2 = \pm 6$

$x = -2 \pm 6$

$x = 4$ or $x = -8$

35. $x^2 + 4x + 2 = 0$

$x^2 + 4x = -2$

$x^2 + 4x + 2^2 = -2 + 2^2$

$(x + 2)^2 = 2$

$x + 2 = \pm\sqrt{2}$

$x = -2 \pm \sqrt{2}$

37. $6x^2 - 12x = -3$

$x^2 - 2x = -\dfrac{1}{2}$

$x^2 - 2x + 1^2 = -\dfrac{1}{2} + 1^2$

$(x - 1)^2 = \dfrac{1}{2}$

$x - 1 = \pm\sqrt{\dfrac{1}{2}}$

$x = 1 \pm \sqrt{\dfrac{1}{2}}$

$x = 1 \pm \dfrac{\sqrt{2}}{2}$

39. $7 + 2x - x^2 = 0$

$-x^2 + 2x + 7 = 0$

$x^2 - 2x - 7 = 0$

$x^2 - 2x = 7$

$x^2 - 2x + (-1)^2 = 7 + (-1)^2$

$(x - 1)^2 = 8$

$x - 1 = \pm 2\sqrt{2}$

$x = 1 \pm 2\sqrt{2}$

41.
$$\begin{aligned}
2x^2 + 5x - 8 &= 0\\
2x^2 + 5x &= 8\\
x^2 + \frac{5}{2}x &= 4\\
x^2 + \frac{5}{2}x + \left(\frac{5}{4}\right)^2 &= 4 + \left(\frac{5}{4}\right)^2\\
\left(x + \frac{5}{4}\right)^2 &= \frac{89}{16}\\
x + \frac{5}{4} &= \pm\frac{\sqrt{89}}{4}\\
x &= -\frac{5}{4} \pm \frac{\sqrt{89}}{4}\\
x &= \frac{-5 \pm \sqrt{89}}{4}
\end{aligned}$$

43.
$$\begin{aligned}
\frac{1}{x^2 - 2x + 5} &= \frac{1}{x^2 - 2x + 1^2 + 5}\\
&= \frac{1}{(x - 1)^2 + 4}
\end{aligned}$$

45.
$$\begin{aligned}
\frac{4}{x^2 + 10x + 74} &= \frac{4}{x^2 + 10x + (5)^2 - (5)^2 + 74}\\
&= \frac{4}{x^2 + 10x + 25 + 49}\\
&= \frac{4}{(x + 5)^2 + 49}
\end{aligned}$$

47.
$$\begin{aligned}
\frac{1}{\sqrt{3 + 2x - x^2}} &= \frac{1}{\sqrt{-1(x^2 - 2x - 3)}}\\
&= \frac{1}{\sqrt{-1\left[x^2 - 2x + (1)^2 - (1)^2 - 3\right]}}\\
&= \frac{1}{\sqrt{-1(x^2 - 2x + 1) + 4}}\\
&= \frac{1}{\sqrt{4 - (x - 1)^2}}
\end{aligned}$$

49.
$$\begin{aligned}
\frac{1}{\sqrt{12 + 4x - x^2}} &= \frac{1}{\sqrt{-1(x^2 - 4x - 12)}} = \frac{1}{\sqrt{-1\left[x^2 - 4x + (2)^2 - (2)^2 - 12\right]}}\\
&= \frac{1}{\sqrt{-1\left[(x^2 - 4x + 4) - 16\right]}} = \frac{1}{\sqrt{16 - (x - 2)^2}}
\end{aligned}$$

51. (a) $y = (x + 3)^2 - 4$

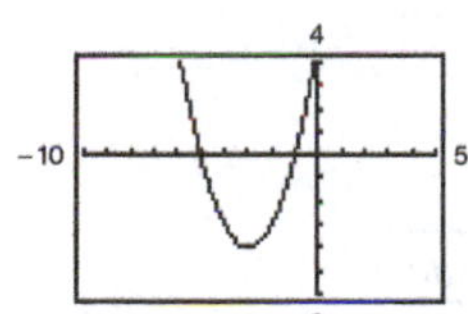

(b) The x-intercepts are $(-1, 0)$ and $(-5, 0)$.

(c)
$$\begin{aligned}
0 &= (x + 3)^2 - 4\\
4 &= (x + 3)^2\\
\pm\sqrt{4} &= x + 3\\
-3 \pm 2 &= x\\
x = -1 \quad &\text{or} \quad x = -5
\end{aligned}$$

(d) The x-intercepts of the graphs are solutions of the equation $0 = (x + 3)^2 - 4$.

53. (a) $y = 1 - (x - 2)^2$

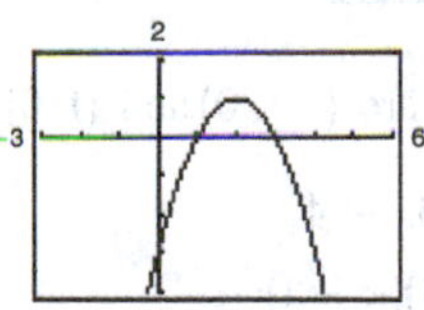

(b) The x-intercepts are $(1, 0)$ and $(3, 0)$.

(c)
$$\begin{aligned}
0 &= 1 - (x - 2)^2\\
(x - 2)^2 &= 1\\
x - 2 &= \pm 1\\
x &= 2 \pm 1\\
x = 3 \quad &\text{or} \quad x = 1
\end{aligned}$$

(d) The x-intercepts of the graphs are solutions of the equation $0 = 1 - (x - 2)^2$.

55. (a) $y = -4x^2 + 4x + 3$

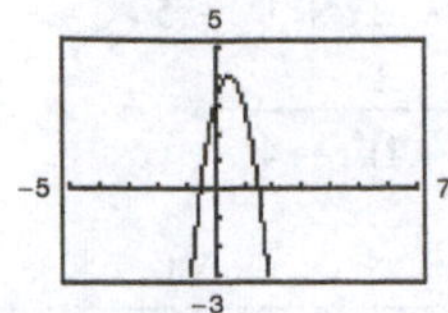

(b) The x-intercepts are $\left(-\frac{1}{2}, 0\right)$ and $\left(\frac{3}{2}, 0\right)$.

(c)
$$\begin{aligned} 0 &= -4x^2 + 4x + 3 \\ 4x^2 - 4x &= 3 \\ 4(x^2 - x) &= 3 \\ x^2 - x &= \tfrac{3}{4} \\ x^2 - x + \left(\tfrac{1}{2}\right)^2 &= \tfrac{3}{4} + \left(\tfrac{1}{2}\right)^2 \\ \left(x - \tfrac{1}{2}\right)^2 &= 1 \\ x - \tfrac{1}{2} &= \pm\sqrt{1} \\ x &= \tfrac{1}{2} \pm 1 \\ x &= \tfrac{3}{2} \text{ or } x = -\tfrac{1}{2} \end{aligned}$$

(d) The x-intercepts of the graphs are solutions of the equation $0 = -4x^2 + 4x + 3$.

57. (a) $y = x^2 + 3x - 4$

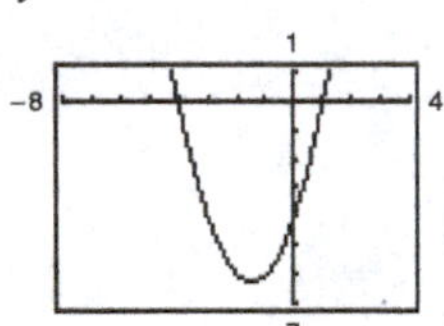

(b) The x-intercepts are $(-4, 0)$ and $(1, 0)$.

(c)
$$\begin{aligned} 0 &= x^2 + 3x - 4 \\ 0 &= (x + 4)(x - 1) \\ x + 4 = 0 \quad &\text{or} \quad x - 1 = 0 \\ x = -4 \quad &\text{or} \quad x = 1 \end{aligned}$$

(d) The x-intercepts of the graphs are solutions of the equation $0 = x^2 + 3x - 4$.

59. $9x^2 + 12x + 4 = 0$

$b^2 - 4ac = (12)^2 - 4(9)(4) = 0$

One repeated real solution

61. $2x^2 - 5x + 5 = 0$

$b^2 - 4ac = (-5)^2 - 4(2)(5) = -15 < 0$

No real solution

63. $2x^2 - z - 1 = 0$

$b^2 - 4ac = (-1)^2 - 4(2)(-1) = 9 > 0$

Two real solutions

65. $\frac{1}{3}x^2 - 5x + 25 = 0$

$b^2 - 4ac = (-5)^2 - 4\left(\frac{1}{3}\right)(25) = -\frac{25}{3} < 0$

No real solution

67. $0.2x^2 + 1.2x - 8 = 0$

$b^2 - 4ac = (1.2)^2 - 4(0.2)(-8) = 7.84 > 0$

Two real solutions

69. $2x^2 + x - 1 = 0$

$$\begin{aligned} x &= \frac{-b \pm \sqrt{b^2 - 4ac}}{2a} \\ &= \frac{-1 \pm \sqrt{1^2 - 4(2)(-1)}}{2(2)} \\ &= \frac{-1 \pm 3}{4} = \frac{1}{2}, -1 \end{aligned}$$

71. $16x^2 + 8x - 3 = 0$

$$\begin{aligned} x &= \frac{-b \pm \sqrt{b^2 - 4ac}}{2a} \\ &= \frac{-8 \pm \sqrt{8^2 - 4(16)(-3)}}{2(16)} \\ &= \frac{-8 \pm 16}{32} = \frac{1}{4}, -\frac{3}{4} \end{aligned}$$

73. $x^2 + 8x - 4 = 0$

$$\begin{aligned} x &= \frac{-b \pm \sqrt{b^2 - 4ac}}{2a} \\ &= \frac{-8 \pm \sqrt{8^2 - 4(1)(-4)}}{2(1)} \\ &= \frac{-8 \pm 4\sqrt{5}}{2} = -4 \pm 2\sqrt{5} \end{aligned}$$

75. $2x^2 - 7x + 1 = 0$

$$x = \frac{-b \pm \sqrt{b^2 - 4ac}}{2a}$$
$$= \frac{-(-7) \pm \sqrt{(-7)^2 - 4(2)(1)}}{2(2)}$$
$$= \frac{7 \pm \sqrt{49 - 8}}{2(2)}$$
$$= \frac{7 \pm \sqrt{41}}{4}$$
$$= \frac{7}{4} \pm \frac{\sqrt{41}}{4}$$

77. $2 + 2x - x^2 = 0$

$-x^2 + 2x + 2 = 0$

$$x = \frac{-b \pm \sqrt{b^2 - 4ac}}{2a}$$
$$= \frac{-2 \pm \sqrt{2^2 - 4(-1)(2)}}{2(-1)}$$
$$= \frac{-2 \pm 2\sqrt{3}}{-2}$$
$$= 1 \pm \sqrt{3}$$

79. $x^2 + 16 = -12x$

$x^2 + 12x + 16 = 0$

$$x = \frac{-b \pm \sqrt{b^2 - 4ac}}{2a}$$
$$= \frac{-(12) \pm \sqrt{(12)^2 - 4(1)(16)}}{2(1)}$$
$$= \frac{-12 \pm \sqrt{144 - 64}}{2(1)}$$
$$= \frac{-12 \pm \sqrt{80}}{2}$$
$$= \frac{-12 \pm 4\sqrt{5}}{2}$$
$$= \frac{4\left(-3 \pm \sqrt{5}\right)}{2}$$
$$= 2\left(-3 \pm \sqrt{5}\right) = -6 \pm 2\sqrt{5}$$

81. $4x^2 + 6x = 8$

$4x^2 + 6x - 8 = 0$

$$x = \frac{-b \pm \sqrt{b^2 - 4ac}}{2a}$$
$$= \frac{-(6) \pm \sqrt{(6)^2 - 4(4)(-8)}}{2(4)}$$
$$= \frac{-6 \pm \sqrt{36 + 128}}{2(4)}$$
$$= \frac{-6 \pm \sqrt{164}}{8}$$
$$= \frac{-6 \pm 2\sqrt{41}}{8}$$
$$= \frac{2\left(-3 \pm \sqrt{41}\right)}{8}$$
$$= \frac{-3 \pm \sqrt{41}}{4}$$
$$= \frac{-3}{4} \pm \frac{\sqrt{41}}{4}$$

83. $28x - 49x^2 = 4$

$-49x^2 + 28x - 4 = 0$

$$x = \frac{-b \pm \sqrt{b^2 - 4ac}}{2a}$$
$$= \frac{-28 \pm \sqrt{28^2 - 4(-49)(-4)}}{2(-49)}$$
$$= \frac{-28 \pm 0}{-98} = \frac{2}{7}$$

85. $8t = 5 + 2t^2$

$-2t^2 + 8t - 5 = 0$

$$t = \frac{-b \pm \sqrt{b^2 - 4ac}}{2a}$$
$$= \frac{-8 \pm \sqrt{8^2 - 4(-2)(-5)}}{2(-2)}$$
$$= \frac{-8 \pm 2\sqrt{6}}{-4} = 2 \pm \frac{\sqrt{6}}{2}$$

87. $(y-5)^2 = 2y$

$y^2 - 12y + 25 = 0$

$y = \dfrac{-b \pm \sqrt{b^2 - 4ac}}{2a}$

$= \dfrac{-(-12) \pm \sqrt{(-12)^2 - 4(1)(25)}}{2(1)}$

$= \dfrac{12 \pm 2\sqrt{11}}{2} = 6 \pm \sqrt{11}$

89. $\frac{1}{2}x^2 + \frac{3}{8}x = 2$

$4x^2 + 3x = 16$

$4x^2 + 3x - 16 = 0$

$x = \dfrac{-b \pm \sqrt{b^2 - 4ac}}{2a}$

$= \dfrac{-3 \pm \sqrt{3^2 - 4(4)(-16)}}{2(4)}$

$= \dfrac{-3 \pm \sqrt{265}}{8} = -\dfrac{3}{8} \pm \dfrac{\sqrt{265}}{8}$

91. $5.1x^2 - 1.7x - 3.2 = 0$

$x = \dfrac{1.7 \pm \sqrt{(-1.7)^2 - 4(5.1)(-3.2)}}{2(5.1)}$

$\approx 0.976, -0.643$

93. $-0.67x^2 + 0.5x + 1.375 = 0$

$x = \dfrac{0.5 \pm \sqrt{(0.5)^2 - 4(-0.67)(1.375)}}{2(5.1)}$

$\approx -1.107, 1.853$

95. $12.67x^2 + 31.55x + 8.09 = 0$

$x = \dfrac{-31.55 \pm \sqrt{(31.55)^2 - 4(12.67)(8.09)}}{2(12.67)}$

$\approx -2.200, -0.290$

97. $x^2 - 2x - 1 = 0$ Complete the square.

$x^2 - 2x = 1$

$x^2 - 2x + 1^2 = 1 + 1^2$

$(x-1)^2 = 2$

$x - 1 = \pm\sqrt{2}$

$x = 1 \pm \sqrt{2}$

99. $(x+2)^2 = 64$ Extract square roots.

$x + 2 = \pm 8$

$x + 2 = 8$ or $x + 2 = -8$

$x = 6$ or $x = -10$

101. $x^2 - x - \frac{11}{4} = 0$ Complete the square.

$x^2 - x = \frac{11}{4}$

$x^2 - x + \left(\frac{1}{2}\right)^2 = \frac{11}{4} + \left(\frac{1}{2}\right)^2$

$\left(x - \frac{1}{2}\right)^2 = \frac{12}{4}$

$x - \frac{1}{2} = \pm\sqrt{\frac{12}{4}}$

$x = \frac{1}{2} \pm \sqrt{3}$

103. $3x + 4 = 2x^2 - 7$ Quadratic Formula

$0 = 2x^2 - 3x - 11$

$x = \dfrac{-(-3) \pm \sqrt{(-3)^2 - 4(2)(-11)}}{2(2)}$

$= \dfrac{3 \pm \sqrt{97}}{4}$

$= \dfrac{3}{4} \pm \dfrac{\sqrt{97}}{4}$

105. (a) $w(w + 14) = 1632$

(b) $w^2 + 14w - 1632 = 0$

$(w + 48)(w - 34) = 0$

$w = -48$ or $w = 34$

Because the width must be greater than zero, $w = 34$ feet and the length is $w + 14 = 48$ feet.

107. $S = x^2 + 4xh$

$108 = x^2 + 4x(3)$

$0 = x^2 + 12x - 108$

$0 = (x + 18)(x - 6)$

$x = -18$ or $x = 6$

Because x must be positive, $x = 6$ inches. The dimensions of the box are 6 inches × 6 inches × 3 inches.

109. (a) Volume is 1024 cubic feet.

$$V = l \cdot w \cdot h = x(x + 1)(4) = 4x^2 + 4x$$

So,

$$4x^2 + 4x = 1024$$
$$4x^2 + 4x - 1024 = 0$$
$$4(x^2 + x - 256) = 0$$
$$x^2 + x - 256 = 0$$
$$x = \frac{-1 \pm \sqrt{1^2 - 4(1) - 256}}{2(1)}$$
$$x \approx 15.51$$
$$x + 1 \approx 16.51$$

So the base of the pool is approximately 15.51 feet × 16.51 feet.

(b) Because 1 cubic foot of water weighs approximately 62.4 pounds, 1024 cubic feet $\cdot \dfrac{62.4 \text{ pounds}}{1 \text{ cubic foot}} = 63{,}897.6$ pounds.

111. (a) $s = -16t^2 + v_0t + s_0$

Since the object was dropped, $v_0 = 0$, and the initial height is $s_0 = 984$. Thus, $s = -16t^2 + 984$.

(b) $s = -16(4)^2 + 984 = 728$ feet

(c)

$$0 = -16t^2 + 984$$
$$16t^2 = 984$$
$$t^2 = \frac{984}{16}$$
$$t = \sqrt{\frac{984}{16}} = \frac{\sqrt{246}}{2} \approx 7.84$$

It will take the coin about 7.84 seconds to strike the ground.

113. (a) $s = -16t^2 + v_0t + s_0$

$s = -16t^2 + 550$

Let $s = 0$ and solve for t.

$$0 = -16t^2 + 550$$
$$16t^2 = 550$$
$$t^2 = \frac{550}{16}$$
$$t = \sqrt{\frac{550}{16}}$$
$$t \approx 5.86$$

The supply package will take about 5.86 seconds to reach the ground.

(b) *Verbal Model:* (Distance) = (Rate) · (Time)

Labels: Distance = d

Rate = 138 miles per hour

Time = $\dfrac{5.86 \text{ seconds}}{3600 \text{ seconds per hour}} \approx 0.0016$ hour

Equation: $d = (138)(0.0016) \approx 0.22$ mile

The supply package will travel about 0.2 mile.

115. (a)

t	8	9	10	11	12	13	14
D	8.98	10.71	12.30	13.75	15.06	16.22	17.24

Sometime during the year 2010, the total public debt reached $13 trillion.

(b) *Algebraically:* $D = -0.071t^2 + 2.94t - 10.0$

$$13 = -0.071t^2 + 2.94t - 10.0$$

$$0 = -0.071t^2 + 2.94t - 23.0$$

Let $a = -0.071$, $b = 2.94$, and $c = 23.0$.

Using the Quadratic Formula,

$$t = \frac{-(2.94) \pm \sqrt{(2.94)^2 - 4 -(0.071)(23.0)}}{2(0.157)}$$

$$= \frac{-2.94 \pm \sqrt{2.1116}}{-0.142}$$

$t \approx 10.47$ and $t \approx 30.94$

Because the domain of the model is $8 \le t \le 14$, $t \approx 10.47$ is the only solution. So, the total public debt reached $13 trillion during 2010.

Graphically: Use a graphing utility to graph $y_1 = -0.071t^2 - 2.94t + 10.0$ and $y_2 = 13$ in the same viewing window. Then use the intersect feature to find that the graphs intersect when $t \approx 10.47$ and $t \approx 30.94$. Choose $t \approx 10.47$ because it is in the domain.

So, the total public debt reached $13 trillion during 2010.

(c) For 2025, let $t = 25$

$$D = -0.071t^2 + 2.94t - 10.0$$

$$= -0.071(25)^2 + 2.94(25) - 10.0$$

$$= \$19.125 \text{ trillion}$$

Using the model, in 2025, the total public debt will be $19.125 trillion.

Answers will vary. *Sample answer*: Yes. In the short time period beyond the interval, $8 \le t \le 14$, the public can be modeled by the equation.

117. $L = -0.270t^2 + 3.59t + 83.1$

$$93 = -0.270t^2 + 3.59t + 83.1$$

$$0 = -0.270t^2 + 3.59t - 9.9$$

$$0 = 0.270t^2 - 3.59t + 9.9$$

Using the Quadratic Formula,

$$t = \frac{-(-3.59) \pm \sqrt{(-3.59)^2 - 4(0.270)(9.9)}}{2(0.270)} = \frac{3.59 \pm \sqrt{2.1961}}{0.54}$$

$t \approx 3.9$ and $t \approx 9.4$

Because the domain of the model is $2 \le t \le 7$, $t \approx 3.9$ is the only solution. The patient's blood oxygen level was 93% at approximately 4:00 P.M.

119. (a) *Model:* $(\text{winch})^2 + (\text{distance to dock})^2 = (\text{length of rope})^2$

Labels: winch $= 15$, distance to dock $= x$, length of rope $= l$

Equation: $15^2 + x^2 = l^2$

(b) When $l = 75$: $15^2 + x^2 = 75^2$

$$x^2 = 5625 - 225 = 5400$$
$$x = \sqrt{5400} = 30\sqrt{6} \approx 73.5$$

The boat is approximately 73.5 feet from the dock when there is 75 feet of rope out.

121. $-3x^2 + x = -5$

$-3x^2 + x + 5 = 0$

$b^2 - 4ac = (1)^2 - 4(-3)(5) = 1 + 60 = 61 > 0.$

True. The quadratic equation has two real solutions.

123. *Sample answer:* $(x - 0)(x - 4) = 0$

$$x(x - 5) = 0$$
$$x^2 - 4x = 0$$

125. One possible equation is:

$(x - 8)(x - 14) = 0$

$x^2 - 22x + 112 = 0$

Any non-zero multiple of this equation would also have these solutions.

127. One possible equation is:

$$\left[x - \left(1 + \sqrt{2}\right)\right]\left[x - \left(1 - \sqrt{2}\right)\right] = 0$$
$$\left[(x - 1) - \sqrt{2}\right]\left[(x - 1) + \sqrt{2}\right] = 0$$
$$(x - 1)^2 - \left(\sqrt{2}\right)^2 = 0$$
$$x^2 - 2x + 1 - 2 = 0$$
$$x^2 - 2x - 1 = 0$$

Any non-zero multiple of this equation would also have these solutions.

129. Yes, the vertex of the parabola would be on the x-axis.

Section 1.5 Complex Numbers

1. real

3. pure imaginary

5. principal square

7. $a + bi = 9 + 8i$

$a = 9$

$b = 8$

9. $(a - 2) + (b + 1)i = 6 + 5i$

$a - 2 = 6 \Rightarrow a = 8$

$b + 1 = 5 \Rightarrow b = 4$

11. $2 + \sqrt{-25} = 2 + 5i$

13. $1 - \sqrt{-12} = 1 - 2\sqrt{3}i$

15. $\sqrt{-40} = 2\sqrt{10}\,i$

17. 23

19. $-6i + i^2 = -6i + (-1) = -1 - 6i$

21. $\sqrt{-0.04} = \sqrt{0.04}\,i = 0.2i$

23. $(5 + i) + (2 + 3i) = 5 + i + 2 + 3i = 7 + 4i$

25. $(9 - i) - (8 - i) = 1$

27. $\left(-2 + \sqrt{-8}\right) + \left(5 - \sqrt{-50}\right) = -2 + 2\sqrt{2}i + 5 - 5\sqrt{2}i$

$$= 3 - 3\sqrt{2}i$$

29. $13i - (14 - 7i) = 13i - 14 + 7i$

$$= -14 + 20i$$

31. $(1 + i)(3 - 2i) = 3 - 2i + 3i - 2i^2$

$$= 3 + i + 2 = 5 + i$$

33. $12i(1 - 9i) = 12i - 108i^2$
$= 12i + 108$
$= 108 + 12i$

35. $(\sqrt{2} + 3i)(\sqrt{2} - 3i) = 2 - 9i^2$
$= 2 + 9 = 11$

37. $(6 + 7i)^2 = 36 + 84i + 49i^2$
$= 36 + 84i - 49$
$= -13 + 84i$

39. The complex conjugate of $9 + 2i$ is $9 - 2i$.

$(9 + 2i)(9 - 2i) = 81 - 4i^2$
$= 81 + 4$
$= 85$

41. The complex conjugate of $-1 - \sqrt{5}i$ is $-1 + \sqrt{5}i$.

$(-1 - \sqrt{5}i)(-1 + \sqrt{5}i) = 1 - 5i^2$
$= 1 + 5 = 6$

43. The complex conjugate of $\sqrt{-20} = 2\sqrt{5}i$ is $-2\sqrt{5}i$.

$(2\sqrt{5}i)(-2\sqrt{5}i) = -20i^2 = 20$

45. The complex conjugate of $\sqrt{6}$ is $\sqrt{6}$.

$(\sqrt{6})(\sqrt{6}) = 6$

47. $\dfrac{2}{4 - 5i} = \dfrac{2}{4 - 5i} \cdot \dfrac{4 + 5i}{4 + 5i}$
$= \dfrac{2(4 + 5i)}{16 + 25} = \dfrac{8 + 10i}{41} = \dfrac{8}{41} + \dfrac{10}{41}i$

49. $\dfrac{5 + i}{5 - i} \cdot \dfrac{(5 + i)}{(5 + i)} = \dfrac{25 + 10i + i^2}{25 - i^2}$
$= \dfrac{24 + 10i}{26} = \dfrac{12}{13} + \dfrac{5}{13}i$

51. $\dfrac{9 - 4i}{i} \cdot \dfrac{-i}{-i} = \dfrac{-9i + 4i^2}{-i^2} = -4 - 9i$

53. $\dfrac{3i}{(4 - 5i)^2} = \dfrac{3i}{16 - 40i + 25i^2} = \dfrac{3i}{-9 - 40i} \cdot \dfrac{-9 + 40i}{-9 + 40i}$
$= \dfrac{-27i + 120i^2}{81 + 1600} = \dfrac{-120 - 27i}{1681}$
$= -\dfrac{120}{1681} - \dfrac{27}{1681}i$

55. $\dfrac{2}{1 + i} - \dfrac{3}{1 - i} = \dfrac{2(1 - i) - 3(1 + i)}{(1 + i)(1 - i)}$
$= \dfrac{2 - 2i - 3 - 3i}{1 + 1}$
$= \dfrac{-1 - 5i}{2}$
$= -\dfrac{1}{2} - \dfrac{5}{2}i$

57. $\dfrac{i}{3 - 2i} + \dfrac{2i}{3 + 8i} = \dfrac{i(3 + 8i) + 2i(3 - 2i)}{(3 - 2i)(3 + 8i)}$
$= \dfrac{3i + 8i^2 + 6i - 4i^2}{9 + 24i - 6i - 16i^2}$
$= \dfrac{4i^2 + 9i}{9 + 18i + 16}$
$= \dfrac{-4 + 9i}{25 + 18i} \cdot \dfrac{25 - 18i}{25 - 18i}$
$= \dfrac{-100 + 72i + 225i - 162i^2}{625 + 324}$
$= \dfrac{62 + 297i}{949} = \dfrac{62}{949} + \dfrac{297}{949}i$

59. $\sqrt{-6} \cdot \sqrt{-2} = (\sqrt{6}i)(\sqrt{2}i) = \sqrt{12}i^2 = (2\sqrt{3})(-1)$
$= -2\sqrt{3}$

61. $(\sqrt{-15})^2 = (\sqrt{15}i)^2 = 15i^2 = -15$

63. $\sqrt{-8} + \sqrt{-50} = \sqrt{8}i + \sqrt{50}i$
$= 2\sqrt{2}i + 5\sqrt{2}i$
$= 7\sqrt{2}i$

65. $(3 + \sqrt{-5})(7 - \sqrt{-10}) = (3 + \sqrt{5}i)(7 - \sqrt{10}i)$
$= 21 - 3\sqrt{10}i + 7\sqrt{5}i - \sqrt{50}i^2$
$= (21 + \sqrt{50}) + (7\sqrt{5} - 3\sqrt{10})i$
$= (21 + 5\sqrt{2}) + (7\sqrt{5} - 3\sqrt{10})i$

67. $x^2 - 2x + 2 = 0;\ a = 1,\ b = -2,\ c = 2$

$$\begin{aligned}x &= \frac{-(-2) \pm \sqrt{(-2)^2 - 4(1)(2)}}{2(1)}\\ &= \frac{2 \pm \sqrt{-4}}{2}\\ &= \frac{2 \pm 2i}{2}\\ &= 1 \pm i\end{aligned}$$

69. $4x^2 + 16x + 17 = 0;\ a = 4,\ b = 16,\ c = 17$

$$\begin{aligned}x &= \frac{-16 \pm \sqrt{(16)^2 - 4(4)(17)}}{2(4)}\\ &= \frac{-16 \pm \sqrt{-16}}{8}\\ &= \frac{-16 \pm 4i}{8}\\ &= -2 \pm \frac{1}{2}i\end{aligned}$$

71. $4x^2 + 16x + 21 = 0;\ a = 4,\ b = 16,\ c = 21$

$$\begin{aligned}x &= \frac{-16 \pm \sqrt{(16)^2 - 4(4)(21)}}{2(4)}\\ &= \frac{-16 \pm \sqrt{-80}}{8}\\ &= \frac{-16 \pm \sqrt{80}\,i}{8}\\ &= \frac{-16 \pm 4\sqrt{5}\,i}{8}\\ &= -2 \pm \frac{\sqrt{5}}{2}i\end{aligned}$$

73. $\frac{3}{2}x^2 - 6x + 9 = 0$ Multiply both sides by 2.

$3x^2 - 12x + 18 = 0;\ a = 3,\ b = -12,\ c = 18$

$$\begin{aligned}x &= \frac{-(-12) \pm \sqrt{(-12)^2 - 4(3)(18)}}{2(3)}\\ &= \frac{12 \pm \sqrt{-72}}{6}\\ &= \frac{12 \pm 6\sqrt{2}i}{6}\\ &= 2 \pm \sqrt{2}i\end{aligned}$$

75. $1.4x^2 - 2x + 10 = 0 \Rightarrow 14x^2 - 20x + 100 = 0;$

$a = 14,\ b = -20,\ c = 100$

$$\begin{aligned}x &= \frac{-(-20) \pm \sqrt{(-20)^2 - 4(14)(100)}}{2(14)}\\ &= \frac{20 \pm \sqrt{-5200}}{28}\\ &= \frac{20 \pm 20\sqrt{13}\,i}{28}\\ &= \frac{20}{28} \pm \frac{20\sqrt{13}\,i}{28}\\ &= \frac{5}{7} \pm \frac{5\sqrt{13}}{7}i\end{aligned}$$

77. $$\begin{aligned}-6i^3 + i^2 &= -6i^2i + i^2\\ &= -6(-1)i + (-1)\\ &= 6i - 1\\ &= -1 + 6i\end{aligned}$$

79. $-14i^5 = -14i^2i^2i = -14(-1)(-1)(i) = -14i$

81. $$\begin{aligned}\left(\sqrt{-72}\right)^3 &= \left(6\sqrt{2}i\right)^3\\ &= 6^3\left(\sqrt{2}\right)^3 i^3\\ &= 216\left(2\sqrt{2}\right)i^2i\\ &= 432\sqrt{2}(-1)i\\ &= -432\sqrt{2}i\end{aligned}$$

83. $\frac{1}{i^3} = \frac{1}{i^2i} = \frac{1}{-i} = \frac{1}{-i}\cdot\frac{i}{i} = \frac{i}{-i^2} = i$

85. $(3i)^4 = 81i^4 = 81i^2i^2 = 81(-1)(-1) = 81$

87. (a) $z_1 = 9 + 16i,\ z_2 = 20 - 10i$

(b) $$\begin{aligned}\frac{1}{z} &= \frac{1}{z_1} + \frac{1}{z_2} = \frac{1}{9 + 16i} + \frac{1}{20 - 10i}\\ &= \frac{20 - 10i + 9 + 16i}{(9 + 16i)(20 - 10i)}\\ &= \frac{29 + 6i}{340 + 230i}\\ z &= \left(\frac{340 + 230i}{29 + 6i}\right)\left(\frac{29 - 6i}{29 - 6i}\right)\\ &= \frac{11{,}240 + 4630i}{877}\\ &= \frac{11{,}240}{877} + \frac{4630}{877}i\end{aligned}$$

89. False.

Sample answer: $(1 + i) + (3 + i) = 4 + 2i$ which is not a real number.

91. True.

$$\begin{aligned} x^4 - x^2 + 14 &= 56 \\ \left(-i\sqrt{6}\right)^4 - \left(-i\sqrt{6}\right)^2 + 14 &\stackrel{?}{=} 56 \\ 36 + 6 + 14 &\stackrel{?}{=} 56 \\ 56 &= 56 \end{aligned}$$

93.
$$\begin{aligned} i &= i \\ i^2 &= -1 \\ i^3 &= -i \\ i^4 &= 1 \\ i^5 &= i^4 i = i \\ i^6 &= i^4 i^2 = -1 \\ i^7 &= i^4 i^3 = -i \\ i^8 &= i^4 i^4 = 1 \\ i^9 &= i^4 i^4 i = i \\ i^{10} &= i^4 i^4 i^2 = -1 \\ i^{11} &= i^4 i^4 i^3 = -i \\ i^{12} &= i^4 i^4 i^4 = 1 \end{aligned}$$

The pattern $i, -1, -i, 1$ repeats. Divide the exponent by 4.

If the remainder is 1, the result is i.

If the remainder is 2, the result is -1.

If the remainder is 3, the result is $-i$.

If the remainder is 0, the result is 1.

95. $\sqrt{-6}\sqrt{-6} = \sqrt{6}i\sqrt{6}i = 6i^2 = -6$

97. $(a_1 + b_1 i) + (a_2 + b_2 i) = (a_1 + a_2) + (b_1 + b_2)i$

The complex conjugate of this sum is $(a_1 + a_2) - (b_1 + b_2)i$.

The sum of the complex conjugates is $(a_1 - b_1 i) + (a_2 - b_2 i) = (a_1 + a_2) - (b_1 + b_2)i$.

So, the complex conjugate of the sum of two complex numbers is the sum of their complex conjugates.

Section 1.6 Other Types of Equations

1. polynomial

3. radical

5.
$$\begin{aligned} 6x^4 - 54x^2 &= 0 \\ 6x^2\left(x^2 - 9\right) &= 0 \\ 6x^2 = 0 &\Rightarrow x = 0 \\ x^2 - 9 = 0 &\Rightarrow x = \pm 3 \end{aligned}$$

7.
$$\begin{aligned} 5x^3 + 3 - x^2 + 45x &= 0 \\ 5x\left(x^2 + 6x + 9\right) &= 0 \\ 5x(x + 3)^2 &= 0 \\ 5x = 0 &\Rightarrow x = 0 \\ x + 3 = 0 &\Rightarrow x = -3 \end{aligned}$$

9.
$$\begin{aligned} x^4 - 81 &= 0 \\ \left(x^2 + 9\right)(x + 3)(x - 3) &= 0 \\ x^2 + 9 = 0 &\Rightarrow x = \pm 3i \\ x + 3 = 0 &\Rightarrow x = -3 \\ x - 3 = 0 &\Rightarrow x = 3 \end{aligned}$$

11.
$$\begin{aligned} x^3 + 512 &= 0 \\ x^3 + 8^3 &= 0 \\ (x + 8)\left(x^2 - 8x + 64\right) &= 0 \\ x + 8 = 0 &\Rightarrow x = -8 \\ x^2 - 8x + 64 = 0 &\Rightarrow x = 4 \pm 4\sqrt{3}i \end{aligned}$$

13. $x^3 - 3x^2 - x + 3 = 0$

$x^2(x - 3) - (x - 3) = 0$

$(x - 3)(x^2 - 1) = 0$

$(x - 3)(x + 3)(x - 1) = 0$

$x - 3 = 0 \Rightarrow x = 3$

$x + 1 = 0 \Rightarrow x = -1$

$x - 1 = 0 \Rightarrow x = 1$

15. $x^4 - x^3 + x - 1 = 0$

$x^3(x - 1) + (x - 1) = 0$

$(x - 1)(x^3 + 1) = 0$

$(x - 1)(x + 1)(x^2 - x + 1) = 0$

$x - 1 = 0 \Rightarrow x = 1$

$x + 1 = 0 \Rightarrow x = -1$

$x^2 - x + 1 = 0 \Rightarrow x = \frac{1}{2} \pm \frac{\sqrt{3}}{2}i$

17. $x^4 - 4x^2 + 3 = 0$

$(x^2)^2 - 4(x^2) + 3 = 0$

Let $u = x^2$.

$u^2 - 4u + 3 = 0$

$(u - 3)(u - 1) = 0$

$u - 3 = 0 \Rightarrow u = 3$

$u - 1 = 0 \Rightarrow u = 1$

$u = 1$ $u = 3$

$x^2 = 1$ $x^2 = 3$

$x = \pm 1$ $x = \pm\sqrt{3}$

19. $4x^4 - 65x^2 + 16 = 0$

$4(x^2) - 65(x^2) + 16 = 0$

Let $u = x^2$.

$4u - 65u + 16 = 0$

$(4u - 1)(u - 16) = 0$

$4u - 1 = 0 \Rightarrow u = \frac{1}{4}$

$u - 16 = 0 \Rightarrow u = 16$

$u = \frac{1}{4}$ $u = 16$

$x^2 = \frac{1}{4}$ $x^2 = 16$

$x = \pm\frac{1}{2}$ $x = \pm 4$

21. $x^6 + 7x^3 - 8 = 0$

$(x^3)^2 + 7(x^3) - 8 = 0$

Let $u = x^3$

$u^2 + 7u - 8 = 0$

$(u + 8)(u - 1) = 0$

$u + 8 = 0$

$x^3 + 8 = 0$

$(x + 2)(x^2 - 2x + 4) = 0$

$x + 2 = 0 \Rightarrow x = -2$

$x^2 - 2x + 4 = 0 \Rightarrow x = 1 \pm \sqrt{3}i$

$u - 1 = 0$

$x^3 - 1 = 0$

$(x - 1)(x^2 + x + 1) = 0$

$x - 1 = 0$

$x = 1$

$x^2 + x + 1 = 0$

$x = -\frac{1}{2} \pm \frac{\sqrt{3}}{2}i$

23. $\frac{1}{x^2} + \frac{8}{x} + 15 = 0$

$\left(\frac{1}{x}\right)^2 + 8\left(\frac{1}{x}\right) + 15 = 0$

Let $u = \frac{1}{x}$.

$u^2 + 8u + 15 = 0$

$(u + 5)(u + 3) = 0$

$u + 5 = 0 \Rightarrow u = -5$

$u + 3 = 0 \Rightarrow u = -3$

$u = -5$ $u = -3$

$\frac{1}{x} = -5$ $\frac{1}{x} = -3$

$x = -\frac{1}{5}$ $x = -\frac{1}{3}$

25. $2\left(\dfrac{x}{x+2}\right)^2 - 3\left(\dfrac{x}{x+2}\right) - 2 = 0$

Let $u = \dfrac{x}{x+2}$.

$2u^2 - 3u - 2 = 0$

$(2u + 1)(u - 2) = 0$

$2u + 1 = 0 \Rightarrow u = -\dfrac{1}{2}$

$u - 2 = 0 \Rightarrow u = 2$

$u = -\dfrac{1}{2}$ $\qquad u = 2$

$\dfrac{x}{x+2} = -\dfrac{1}{2}$ $\qquad \dfrac{x}{x+2} = 2$

$x = -\dfrac{2}{3}$ $\qquad x = -4$

27. $2x + 9\sqrt{x} = 5$

$2x + 9\sqrt{x} - 5 = 0$

$2\left(\sqrt{x}\right)^2 + 9\left(\sqrt{x}\right) - 5 = 0$

Let $u = \sqrt{x}$.

$2u^2 + 9u - 5 = 0$

$(2u - 1)(u + 5) = 0$

$2u - 1 = 0 \Rightarrow u = \frac{1}{2}$

$u + 5 = 0 \Rightarrow u = -5$

$u = \frac{1}{2}$ $\qquad u = -5 \Rightarrow \sqrt{x} \neq -5$

$\sqrt{x} = \frac{1}{2}$ $\qquad \left(\sqrt{x} = -5 \text{ is not a solution.}\right)$

$x = \frac{1}{4}$

29. $9t^{2/3} + 24t^{1/3} + 16 = 0$

$9\left(t^{1/3}\right)^2 + 24\left(t^{1/3}\right) + 16 = 0$

Let $u = t^{1/3}$.

$9u^2 + 24u + 16 = 0$

$(3u + 4)^2 = 0$

$3u + 4 = 0 \Rightarrow u = -\frac{4}{3}$

$u = -\frac{4}{3}$

$t^{1/3} = -\frac{4}{3}$

$t = -\frac{64}{27}$

31. $\sqrt{5x} - 10 = 0$

$\sqrt{5x} = 10$

$\left(\sqrt{5x}\right)^2 = (10)^2$

$5x = 100$

$x = 20$

33. $\sqrt{x+8} - 5 = 0$

$\sqrt{x+8} = 5$

$\left(\sqrt{x+8}\right)^2 = (5)^2$

$x + 8 = 25$

$x = 17$

35. $4 + \sqrt[3]{2x-9} = 0$

$\sqrt[3]{2x-9} = -4$

$\left(\sqrt[3]{2x-9}\right)^3 = (-4)^3$

$2x - 9 = -64$

$2x = -55$

$x = -\dfrac{55}{2}$

37. $\sqrt{x+8} = 2 + x$

$\left(\sqrt{x+8}\right)^2 = (2+x)^2$

$x + 8 = x^2 + 4x + 4$

$0 = x^2 + 3x - 4$

$x^2 + 3x - 4 = 0$

$(x + 4)(x - 1) = 0$

$x + 4 = 0 \Rightarrow x = -4$, extraneous

$x - 1 = 0 \Rightarrow x = 1$

39. $\sqrt{x-3} + 1 = \sqrt{x}$

$\sqrt{x-3} = \sqrt{x} - 1$

$\left(\sqrt{x-3}\right)^2 = \left(\sqrt{x} - 1\right)^2$

$x - 3 = x - 2\sqrt{x} + 1$

$-4 = -2\sqrt{x}$

$2 = \sqrt{x}$

$(2)^2 = \left(\sqrt{x}\right)^2$

$4 = x$

41. $2\sqrt{x+1} - \sqrt{2x+3} = 1$

$$\begin{aligned}
2\sqrt{x+1} &= 1 + \sqrt{2x+3}\\
\left(2\sqrt{x+1}\right)^2 &= \left(1 + \sqrt{2x+3}\right)^2\\
4(x+1) &= 1 + 2\sqrt{2x+3} + 2x + 3\\
2x &= 2\sqrt{2x+3}\\
x &= \sqrt{2x+3}\\
x^2 &= 2x + 3\\
x^2 - 2x - 3 &= 0\\
(x-3)(x+1) &= 0\\
x - 3 = 0 &\Rightarrow x = 3\\
x + 1 = 0 &\Rightarrow x = -1, \text{ extraneous}
\end{aligned}$$

43. $\sqrt{4\sqrt{4x+9}} = \sqrt{8x+2}$

$$\begin{aligned}
\left(\sqrt{4\sqrt{4x+9}}\right)^2 &= \left(\sqrt{8x+2}\right)^2\\
4\sqrt{4x+9} &= 8x + 2\\
2\sqrt{4x+9} &= 4x + 1\\
\left(2\sqrt{4x+9}\right)^2 &= (4x+1)^2\\
4(4x+9) &= 16x^2 + 8x + 1\\
16x + 36 &= 16x^2 + 8x + 1\\
0 &= 16x^2 - 8x - 35\\
0 &= (4x+5)(4x-7)\\
4x + 5 = 0 &\Rightarrow x = -\frac{5}{4}, \text{ extraneous}\\
4x - 7 = 0 &\Rightarrow x = \frac{7}{4}
\end{aligned}$$

45. $(x-5)^{3/2} = 8$

$$\begin{aligned}
(x-5)^3 &= 8^2\\
x - 5 &= \sqrt[3]{64}\\
x &= 5 + 4 = 9
\end{aligned}$$

47. $\left(x^2-5\right)^{3/2} = 27$

$$\begin{aligned}
\left(x^2-5\right)^3 &= 27^2\\
x^2 - 5 &= \sqrt[3]{27^2}\\
x^2 &= 5 + 9\\
x^2 &= 14\\
x &= \pm\sqrt{14}
\end{aligned}$$

49. $3x(x-1)^{1/2} + 2(x-1)^{3/2} = 0$

$$\begin{aligned}
(x-1)^{1/2}\left[3x + 2(x-1)\right] &= 0\\
(x-1)^{1/2}(5x-2) &= 0\\
(x-1)^{1/2} = 0 &\Rightarrow x - 1 = 0 \Rightarrow x = 1\\
5x - 2 = 0 &\Rightarrow x = \tfrac{2}{5}, \text{ extraneous}
\end{aligned}$$

51. $x = \dfrac{3}{x} + \dfrac{1}{2}$

$$\begin{aligned}
(2x)(x) &= (2x)\left(\frac{3}{x}\right) + (2x)\left(\frac{1}{2}\right)\\
2x^2 &= 6 + x\\
2x^2 - x - 6 &= 0\\
(2x+3)(x-2) &= 0\\
2x + 3 = 0 &\Rightarrow x = -\frac{3}{2}\\
x - 2 = 0 &\Rightarrow x = 2
\end{aligned}$$

53. $\dfrac{1}{x} - \dfrac{1}{x+1} = 3$

$$\begin{aligned}
x(x+1)\frac{1}{x} - x(x+1)\frac{1}{x+1} &= x(x+1)(3)\\
x + 1 - x &= 3x(x+1)\\
1 &= 3x^2 + 3x\\
0 &= 3x^2 + 3x - 1\\
x = \frac{-3 \pm \sqrt{(3)^2 - 4(3)(-1)}}{2(3)} &= \frac{-3 \pm \sqrt{21}}{6}
\end{aligned}$$

55. $3 - \dfrac{14}{x} - \dfrac{5}{x^2} = 0$

$$\begin{aligned}
\frac{5}{x^2} + \frac{14}{x} - 3 &= 0\\
5\left(\frac{1}{x}\right)^2 + 14\left(\frac{1}{x}\right) - 3 &= 0
\end{aligned}$$

Let $u = \dfrac{1}{x}$.

$$\begin{aligned}
5u^2 + 14u - 3 &= 0\\
(5u-1)(u+3) &= 0\\
5u - 1 = 0 &\Rightarrow u = \frac{1}{5}\\
u + 3 = 0 &\Rightarrow u = -3
\end{aligned}$$

$u = \dfrac{1}{5}$ $\qquad$ $u = -3$

$\dfrac{1}{x} = \dfrac{1}{5}$ $\qquad$ $\dfrac{1}{x} = -3$

$x = 5$ $\qquad$ $x = -\dfrac{1}{3}$

57.

$$\frac{x}{x^2 - 4} + \frac{1}{x + 2} = 3$$

$$(x + 2)(x - 2)\frac{x}{x^2 - 4} + (x + 2)(x - 2)\frac{1}{x + 2} = 3(x + 2)(x - 2)$$

$$x + x - 2 = 3x^2 - 12$$

$$3x^2 - 2x - 10 = 0$$

$$x = \frac{-(-2) \pm \sqrt{(-2)^2 - 4(3)(-10)}}{2(3)} = \frac{2 \pm \sqrt{124}}{6} = \frac{2 \pm 2\sqrt{31}}{6} = \frac{1 \pm \sqrt{31}}{3}$$

59. $|2x - 5| = 11$

$2x - 5 = 11 \Rightarrow x = 8$

$-(2x - 5) = 11 \Rightarrow x = -3$

61. $|x| = x^2 + x - 24$

First equation:

$x = x^2 + x - 24$

$x^2 - 24 = 0$

$x^2 = 24$

$x = \pm 2\sqrt{6}$

Second equation:

$-x = x^2 + x - 24$

$x^2 + 2x - 24 = 0$

$(x + 6)(x - 4) = 0$

$x + 6 = 0 \Rightarrow x = -6$

$x - 4 = 0 \Rightarrow x = 4$

Only $x = 2\sqrt{6}$ and $x = -6$ are solutions of the original equation. $x = -2\sqrt{6}$ and $x = 4$ are extraneous.

63. $|x + 1| = x^2 - 5$

First equation:

$x + 1 = x^2 - 5$

$x^2 - x - 6 = 0$

$(x - 3)(x + 2) = 0$

$x - 3 = 0 \Rightarrow x = 3$

$x + 2 = 0 \Rightarrow x = -2$

Second equation:

$-(x + 1) = x^2 - 5$

$-x - 1 = x^2 - 5$

$x^2 + x - 4 = 0$

$x = \dfrac{-1 + \sqrt{17}}{2}$

Only $x = 3$ and $x = \dfrac{-1 - \sqrt{17}}{2}$ are solutions of the original equation. $x = -2$ and $x = \dfrac{-1 + \sqrt{17}}{2}$ are extraneous.

65. (a)

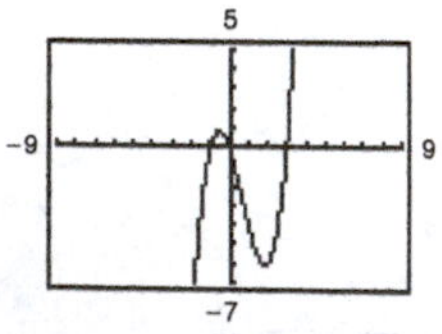

(b) x-intercepts: $(-1, 0)$, $(0, 0)$, $(3, 0)$

(c)

$0 = x^3 - 2x^2 - 3x$

$0 = x(x + 1)(x - 3)$

$x = 0$

$x + 1 = 0 \Rightarrow x = -1$

$x - 3 = 0 \Rightarrow x = 3$

(d) The x-intercepts of the graph are the same as the solutions of the equation.

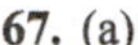

67. (a)

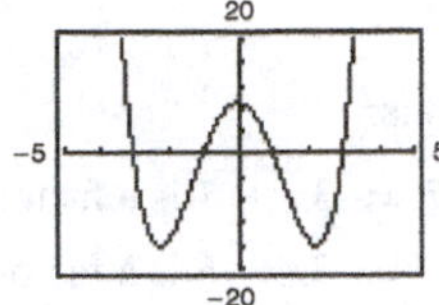

(b) x-intercepts: $(\pm 1, 0), (\pm 3, 0)$

(c) $$0 = x^4 - 10x^2 + 9$$
$$0 = (x^2)^2 - 10(x^2) + 9$$

Let $u = x^2$.
$$0 = (u - 1)(u - 9)$$

$$u - 1 = 0 \qquad u - 9 = 0$$
$$u = 1 \qquad u = 9$$
$$x^2 = 1 \qquad x^2 = 9$$
$$x = \pm 1 \qquad x = \pm 3$$

(d) The x-intercepts of the graph are the same as the solutions of the equation.

69. (a)

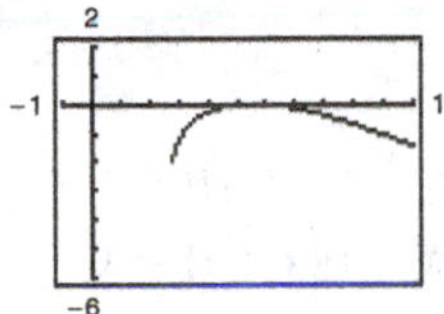

(b) x-intercepts: $(5, 0), (6, 0)$

(c) $$0 = \sqrt{11x - 30} - x$$
$$x = \sqrt{11x - 30}$$
$$x^2 = 11x - 30$$
$$x^2 - 11x + 30 = 0$$
$$(x - 5)(x - 6) = 0$$
$$x - 5 = 0 \Rightarrow x = 5$$
$$x - 6 = 0 \Rightarrow x = 6$$

(d) The x-intercepts of the graph are the same as the solutions of the equation.

71. (a)

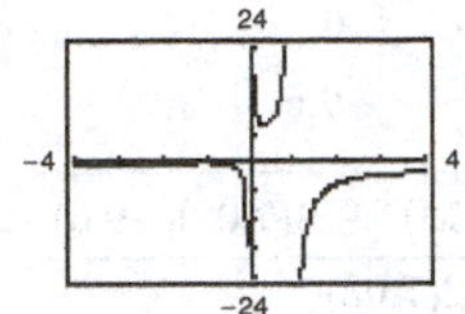

(b) x-intercept: $(-1, 0)$

(c) $$0 = \frac{1}{x} - \frac{4}{x - 1} - 1$$
$$0 = (x - 1) - 4x - x(x - 1)$$
$$0 = x - 1 - 4x - x^2 + x$$
$$0 = -x^2 - 2x - 1$$
$$0 = x^2 + 2x + 1$$
$$0 = (x + 1)^2$$
$$x + 1 = 0 \Rightarrow x = -1$$

(d) The x-intercept of the graph is the same as the solution of the equation.

73. (a)

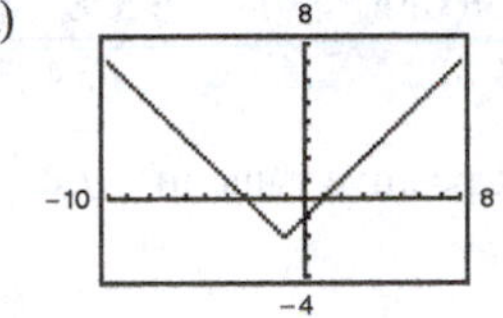

(b) x-intercepts: $(1, 0), (-3, 0)$

(c) $$0 = |x + 1| - 2$$
$$2 = |x + 1|$$
$$x + 1 = 2 \quad \text{or} \quad -(x + 1) = 2$$
$$x = 1 \quad \text{or} \quad -x - 1 = 2$$
$$-x = 3$$
$$x = -3$$

(d) The x-intercepts of the graph are the same as the solutions of the equation.

75. $3.2x^4 - 1.5x^2 - 2.1 = 0$

$$x^2 = \frac{1.5 \pm \sqrt{1.5^2 - 4(3.2)(-2.1)}}{2(3.2)}$$

Using the positive value for x^2, we have

$$x = \pm\sqrt{\frac{1.5 + \sqrt{29.13}}{6.4}} \approx \pm 1.038.$$

77. $7.08x^6 + 4.15x^3 - 9.6 = 0$

$a = 7.8,\ b = 4.15,\ c = -9.6$

$$x^3 = \frac{-4.15 \pm \sqrt{(4.15)^2 - 4(7.08)(-9.6)}}{2(7.08)}$$

$$= \frac{-4.15 \pm \sqrt{2.89.0945}}{14.16}$$

$$x = \sqrt[3]{\frac{-4.15 + \sqrt{289.0945}}{14.16}} \approx 0.968$$

$$x = \sqrt[3]{\frac{-4.15 - \sqrt{289.0945}}{14.16}} \approx -1.143$$

79. $1.8x - 6\sqrt{x} - 5.6 = 0$ Given equation

$1.8\left(\sqrt{x}\right)^2 - 6\sqrt{x} - 5.6 = 0$

Use the Quadratic Formula with $a = 1.8$, $b = -6$, and $c = -5.6$.

$$\sqrt{x} = \frac{6 \pm \sqrt{36 - 4(1.8)(-5.6)}}{2(1.8)} \approx \frac{6 \pm 8.7361}{3.6}$$

Considering only the positive value for $\sqrt{x}$, we have

$\sqrt{x} \approx 4.0934$

$x \approx 16.756.$

81. $4x^{2/3} + 8x^{1/3} + 3.6 = 0$

$a = 4,\ b = 8,\ c = 3.6$

$$x^{1/3} = \frac{-8 \pm \sqrt{8^2 - 4(4)(3.6)}}{2(4)}$$

$$x = \left[\frac{-8 + \sqrt{6.4}}{8}\right]^3 \approx -0.320$$

$$x = \left[\frac{-8 - \sqrt{6.4}}{8}\right]^3 \approx -2.280$$

83. $-4, 7$

Sample answer: $(x - (-4))(x - 7) = 0$

$(x + 4)(x - 7) = 0$

$x^2 - 3x - 28 = 0$

85. $-\frac{7}{3}, \frac{6}{7}$

One possible equation is:

$x = -\frac{7}{3} \Rightarrow 3x = -7 \Rightarrow 3x + 7$ is a factor.

$x = \frac{6}{7} \Rightarrow 7x = 6 \Rightarrow 7x - 6$ is a factor.

$(3x + 7)(7x - 6) = 0$

$21x^2 + 31x - 42 = 0$

Any non-zero multiple of this equation would also have these solutions.

87. $\sqrt{3}, -\sqrt{3}$, and 4

One possible equation is:

$\left(x - \sqrt{3}\right)\left(x - \left(-\sqrt{3}\right)\right)(x - 4) = 0$

$\left(x - \sqrt{3}\right)\left(x + \sqrt{3}\right)(x - 4) = 0$

$\left(x^2 - 3\right)(x - 4) = 0$

$x^3 - 4x^2 - 3x + 12 = 0$

Any non-zero multiple of this equation would also have these solutions.

89. $i, -i$

Sample answer: $(x - i)(x - (-i)) = 0$

$(x - i)(x + i) = 0$

$x^2 - i^2 = 0$

$x^2 + 1 = 0$

91. $-1, 1, i$, and $-i$

One possible equation is:

$\left(x - (-1)\right)(x - 1)(x - i)\left(x - (-i)\right) = 0$

$(x + 1)(x - 1)(x - i)(x + i) = 0$

$\left(x^2 - 1\right)\left(x^2 + 1\right) = 0$

$x^4 - 1 = 0$

Any non-zero multiple of this equation would also have these solutions.

93. *Labels*: Let $x =$ the number of students in the original group. Then, $\frac{1700}{x} =$ the original cost per student.

When six more students join the group, the cost per student becomes $\frac{1700}{x} - 7.50$.

Model: (Cost per student) · (Number of students) = (Total cost)

Equation: $\left(\frac{1700}{x} - 7.5\right)(x + 6) = 1700$

$(3400 - 15x)(x + 6) = 3400x$ Multiply both sides by $2x$ to clear fraction.

$-15x^2 - 90x + 20{,}400 = 0$

$$x = \frac{90 \pm \sqrt{(-90)^2 - 4(-15)(20{,}400)}}{2(-15)} = \frac{90 \pm 1110}{-30}$$

Using the positive value for x, we conclude that the original number was $x = 34$ students.

95. *Model:* $\text{Time} = \frac{\text{Distance}}{\text{Rate}}$

Labels: Let $x =$ average speed of the plane. Then we have a travel time of $t = 145/x$. If the average speed is increased by 40 mph, then

$$t - \frac{12}{60} = \frac{145}{x + 40}$$
$$t = \frac{145}{x + 40} + \frac{1}{5}.$$

Now, we equate these two equations and solve for x.

Equation: $\frac{145}{x} = \frac{145}{x + 40} + \frac{1}{5}$

$$145(5)(x + 40) = 145(5)x + x(x + 40)$$
$$725x + 29{,}000 = 725x + x^2 + 40x$$
$$0 = x^2 + 40x - 29{,}000$$

Using the positive value for x found by the Quadratic Formula, we have $x \approx 151.5$ mph and $x + 40 = 191.5$ mph. The airspeed required to obtain the decrease in travel time is 191.5 miles per hour.

97.
$$A = P\left(1 + \frac{r}{n}\right)^{nt}$$
$$2694.58 = 2500\left(1 + \frac{r}{12}\right)^{(12)(5)}$$
$$\frac{2694.58}{2500} = \left(1 + \frac{r}{12}\right)^{60}$$
$$1.077832 = \left(1 + \frac{r}{12}\right)^{60}$$
$$(1.077832)^{1/60} = 1 + \frac{r}{12}$$
$$\left[(1.0.77832)^{1/60} - 1\right](12) = r$$
$$r \approx 0.015 = 1.5\%$$

99. When $C = 2.5$ we have:

$$2.5 = \sqrt{0.2x + 1}$$
$$6.25 = 0.2x + 1$$
$$5.25 = 0.2x$$
$$x = 26.25 = 26{,}250 \text{ passengers}$$

101. $T = 75.82 - 2.11x + 43.51\sqrt{x}, 5 \le x \le 40$

(a) $212 = 75.82 - 2.11x + 43.51\sqrt{x}$

$0 = -2.11x + 43.51\sqrt{x} - 136.18$

By the Quadratic Formula, we have

$\sqrt{x} \approx 16.77928 \Rightarrow x \approx 281.333$

$\sqrt{x} \approx 3.84787 \Rightarrow x \approx 14.806.$

Since x is restricted to $5 \le x \le 40$, let $x = 14.806$ pounds per square inch.

(b)

103.
$$37.55 = 40 - \sqrt{0.01x + 1}$$
$$\sqrt{0.01x + 1} = 2.45$$
$$0.01x + 1 = 6.0025$$
$$0.01x = 5.0025$$
$$x = 500.25$$

Rounding x to the nearest whole unit yields $x \approx 500$ units.

105.

$$\frac{1}{t} + \frac{1}{t+3} = \frac{1}{y}$$

$$\frac{1}{t} + \frac{1}{t+3} = \frac{1}{2}$$

$$2t(t+3)\frac{1}{t} + 2t(t+3)\frac{1}{t+3} = 2t(t+3)\frac{1}{2}$$

$$2(t+3) + 2t = t(t+3)$$

$$2t + 6 + 2t = t^2 + 3t$$

$$0 = t^2 - t - 6$$

$$0 = (t-3)(t+2)$$

$$t - 3 = 0 \Rightarrow t = 3$$

$$t + 2 = 0 \Rightarrow t = -2$$

Since t represents time, $t = 3$ is the only solution. It takes 3 hours for you working alone to tile the floor.

107.

$$d = \sqrt{\frac{2U}{k}}$$

$$d^2 = \frac{2U}{k}$$

$$d^2k = 2U$$

$$\frac{kd^2}{2} = U$$

109. False. See Example 7 on page 125.

111. The distance between $(3, -5)$ and $(x, 7)$ is 13.

$$\sqrt{(x-3)^2 + (-5-7)^2} = 13$$

$$(x-3)^2 + (-12)^2 = 13^2$$

$$x^2 - 6x + 9 + 144 = 169$$

$$x^2 - 6x - 16 = 0$$

$$(x-8)(x+2) = 0$$

$$x - 8 = 0 \Rightarrow x = 8$$

$$x + 2 = 0 \Rightarrow x = -2$$

Both $(8, 7)$ and $(-2, 7)$ are a distance of 13 from $(3, -5)$.

113. The quadratic equation was not written in general form before the values of a, b, and c were substituted in the Quadratic Formula. As a result, the substitutions in the Quadratic Formula are incorrect.

115. $9 + |9 - a| = b$

$$|9 - a| = b - 9$$

$$9 - a = b - 9 \quad \text{or} \quad 9 - a = -(b - 9)$$

$$-a = b - 18 \qquad 9 - a = -b + 9$$

$$a = 18 - b \qquad -a = -b$$

$$a = b$$

Thus, $a = 18 - b$ *or* $a = b$. From the original equation we know that $b \geq 9$.

Some possibilities are: $b = 9, a = 9$

$b = 10, a = 8$ *or* $a = 10$

$b = 11, a = 7$ *or* $a = 11$

$b = 12, a = 6$ *or* $a = 12$

$b = 13, a = 5$ *or* $a = 13$

$b = 14, a = 4$ *or* $a = 14$

117. $20 + \sqrt{20 - 1} = b$

$$\sqrt{20 - a} = b - 20$$

$$20 - a = b^2 - 40b + 400$$

$$-a = b^2 - 40b + 380$$

$$a = -b^2 + 40b - 380$$

This formula gives the relationship between a and b. From the original equation we know that $a \leq 20$ and $b \geq 20$. Choose a b value, where $b \geq 20$ and then solve for a, keeping in mind that $a \leq 20$.

Some possibilities are: $b = 20, a = 20$

$b = 21, a = 19$

$b = 22, a = 16$

$b = 23, a = 11$

$b = 24, a = 4$

$b = 25, a = -5$

Section 1.7 Linear Inequalities in One Variable

1. solution set

3. double

5. Interval: $[-2, 6)$

Inequality: $-2 \leq x < 6$; The interval is bounded.

7. Interval: $[-1, 5]$

Inequality: $-1 \leq x \leq 5$; The interval is bounded.

9. Interval: $(11, \infty)$

Inequality: $x > 11$; The interval is unbounded.

11. Interval: $(-\infty, -2)$

Inequality: $x < -2$; The interval is unbounded.

13. $4x < 12$

$\frac{1}{4}(4x) < \frac{1}{4}(12)$

$x < 3$

15. $-2x > -3$

$-\frac{1}{2}(-2x) < \left(-\frac{1}{2}\right)(-3)$

$x < \frac{3}{2}$

17. $x - 5 \geq 7$

$x \geq 12$

19. $2x + 7 < 3 + 4x$

$-2x < -4$

$x > 2$

21. $3x - 4 \geq 4 - 5x$

$8x \geq 8$

$x \geq 1$

23. $4 - 2x < 3(3 - x)$

$4 - 2x < 9 - 3x$

$x < 5$

25. $\frac{3}{4}x - 6 \leq x - 7$

$-\frac{1}{4}x \leq -1$

$x \geq 4$

27. $\frac{1}{2}(8x + 1) \geq 3x + \frac{5}{2}$

$4x + \frac{1}{2} \geq 3x + \frac{5}{2}$

$x \geq 2$

29. $3.6x + 11 \geq -3.4$

$3.6x \geq 14.4$

$x \geq -4$

31. $1 < 2x + 3 < 9$

$-2 < 2x < 6$

$-1 < x < 3$

33. $0 < 3(x + 7) \leq 20$

$0 < x + 7 \leq \frac{20}{3}$

$-7 < x \leq -\frac{1}{3}$

35. $-4 < \frac{2x - 3}{3} < 4$

$-12 < 2x - 3 < 12$

$-9 < 2x < 15$

$-\frac{9}{2} < x < \frac{15}{2}$

37. $-1 < \frac{-x - 2}{3} \leq 1$

$-3 < -x - 2 \leq 3$

$-1 < -x \leq 5$

$1 > x \geq -5$

$-5 \leq x < 1$

39. $\frac{3}{4} > x + 1 > \frac{1}{4}$

$-\frac{1}{4} > x > -\frac{3}{4}$

$-\frac{3}{4} < x < -\frac{1}{4}$

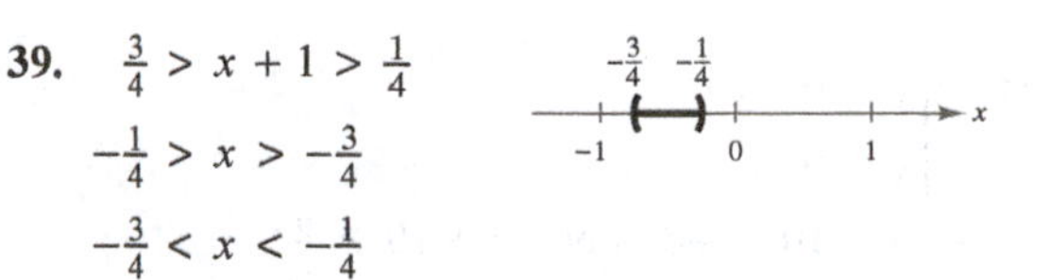

41. $3.2 \leq 0.4x - 1 \leq 4.4$

$4.2 \leq 0.4x \leq 5.4$

$10.5 \leq x \leq 13.5$

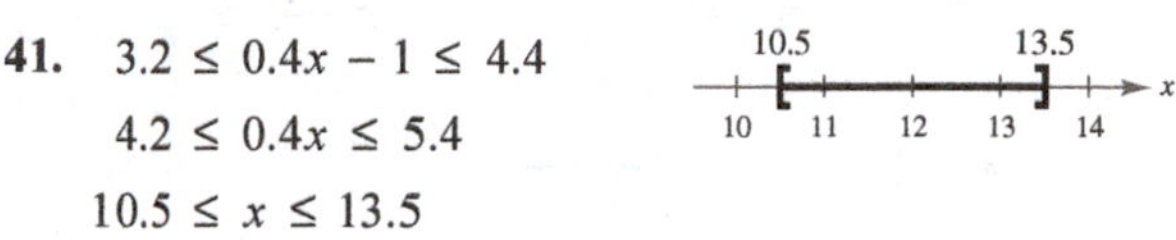

43. $|x| < 5$

$-5 < x < 5$

45. $\left|\frac{x}{2}\right| > 1$

$\frac{x}{2} < -1$ or $\frac{x}{2} > 1$

$x < -2$ $\quad x > 2$

47. $|x - 5| < -1$

No solution. The absolute value of a number cannot be less than a negative number.

49. $|x - 20| \leq 6$

$-6 \leq x - 20 \leq 6$

$14 \leq x \leq 26$

51. $|7 - 2x| \geq 9$

$7 - 2x \leq -9$ or $7 - 2x \geq 9$

$-2x \leq -16$ $\quad -2x \geq 2$

$x \geq 8$ $\quad x \leq -1$

53. $\left|\frac{x - 3}{2}\right| \geq 4$

$\frac{x - 3}{2} \leq -4$ or $\frac{x - 3}{2} \geq 4$

$x - 3 \leq -8$ $\quad x - 3 \geq 8$

$x \leq -5$ $\quad x \geq 11$

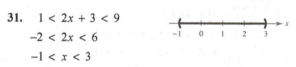

55. $|9 - 2x| - 2 < -1$

$|9 - 2x| < 1$

$-1 < 9 - 2x < 1$

$-10 < -2x < -8$

$5 > x > 4$

$4 < x < 5$

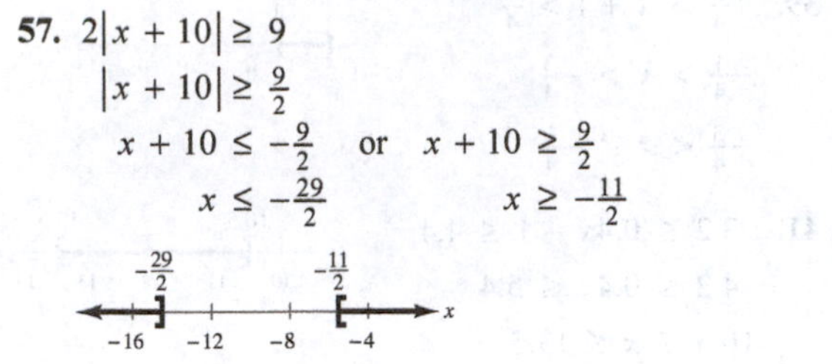

57. $2|x + 10| \ge 9$

$|x + 10| \ge \frac{9}{2}$

$x + 10 \le -\frac{9}{2}$ or $x + 10 \ge \frac{9}{2}$

$x \le -\frac{29}{2}$ $\quad$ $x \ge -\frac{11}{2}$

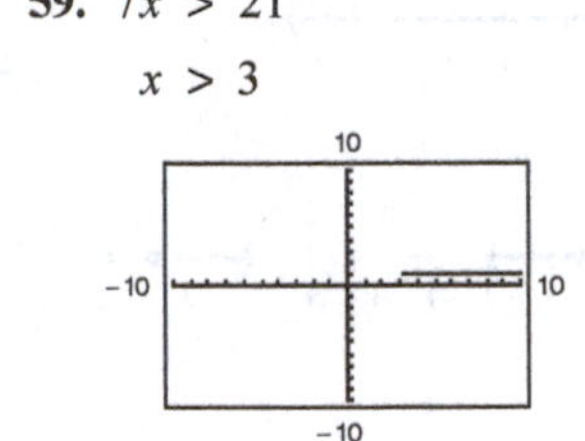

59. $7x > 21$

$x > 3$

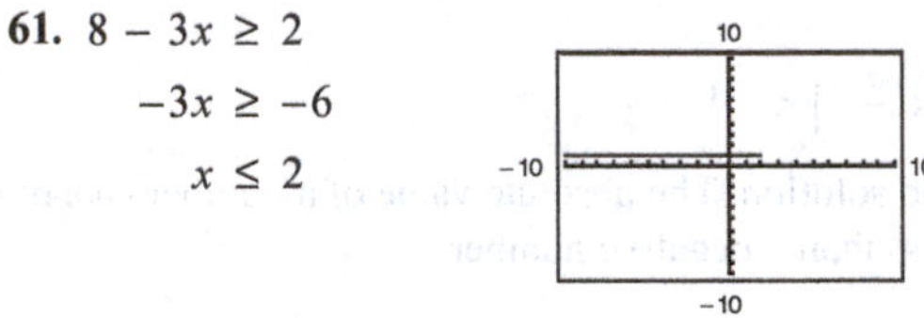

61. $8 - 3x \ge 2$

$-3x \ge -6$

$x \le 2$

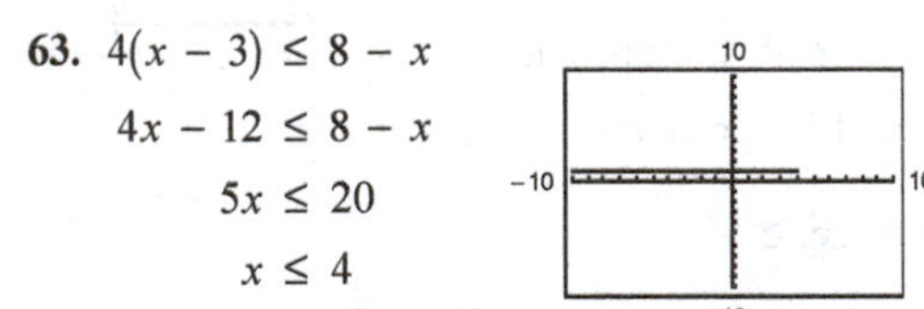

63. $4(x - 3) \le 8 - x$

$4x - 12 \le 8 - x$

$5x \le 20$

$x \le 4$

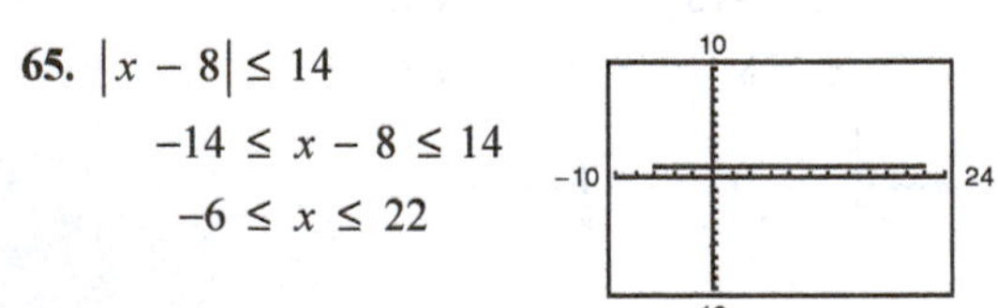

65. $|x - 8| \le 14$

$-14 \le x - 8 \le 14$

$-6 \le x \le 22$

67. $2|x + 7| \ge 13$

$|x + 7| \ge \frac{13}{2}$

$x + 7 \le -\frac{13}{2}$ or $x + 7 \ge \frac{13}{2}$

$x \le -\frac{27}{2}$ $\quad$ $x \ge -\frac{1}{2}$

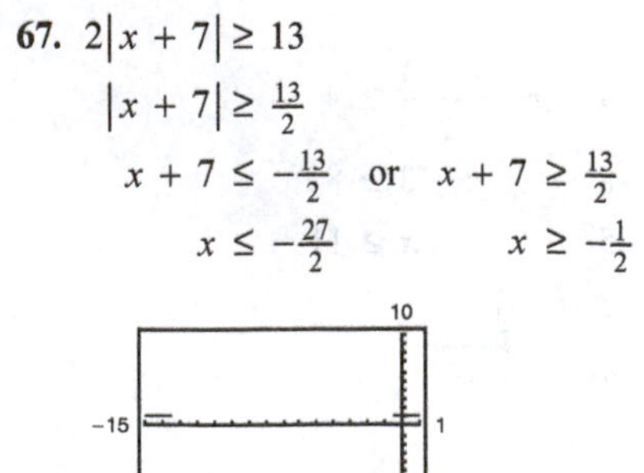

69. $y = 3x - 1$

(a) $y \ge 2$

$3x - 1 \ge 2$

$3x \ge 3$

$x \ge 1$

(b) $y \le 0$

$3x - 1 \le 0$

$3x \le 1$

$x \le \frac{1}{3}$

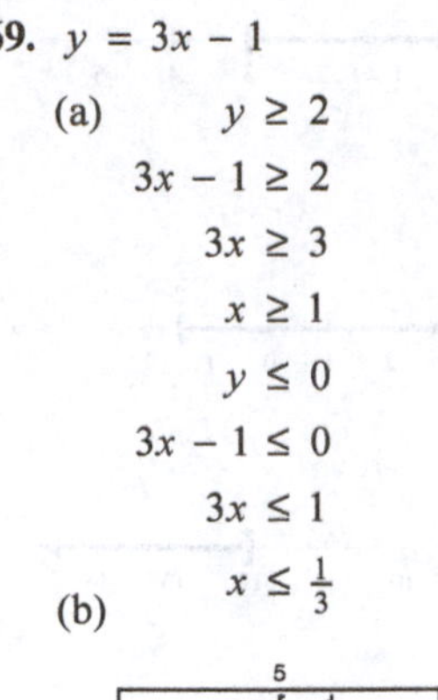

71. $y = -\frac{1}{2}x + 2$

(a) $0 \le y \le 3$

$0 \le -\frac{1}{2}x + 2 \le 3$

$-2 \le -\frac{1}{2}x \le 1$

$4 \ge x \ge -2$

(b) $y \ge 0$

$-\frac{1}{2}x + 2 \ge 0$

$-\frac{1}{2}x \ge -2$

$x \le 4$

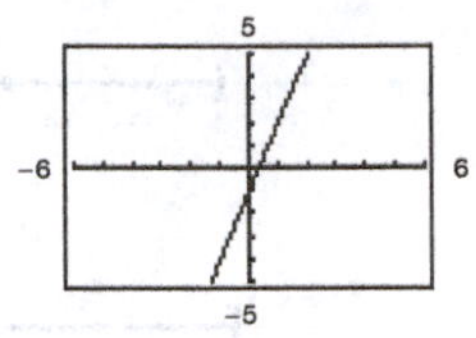

73. $y = |x - 3|$

(a) $y \le 2$

$|x - 3| \le 2$

$-2 \le x - 3 \le 2$

$1 \le x \le 5$

(b) $y \ge 4$

$|x - 3| \ge 4$

$x - 3 \le -4$ or $x - 3 \ge 4$

$x \le -1$ or $x \ge 7$

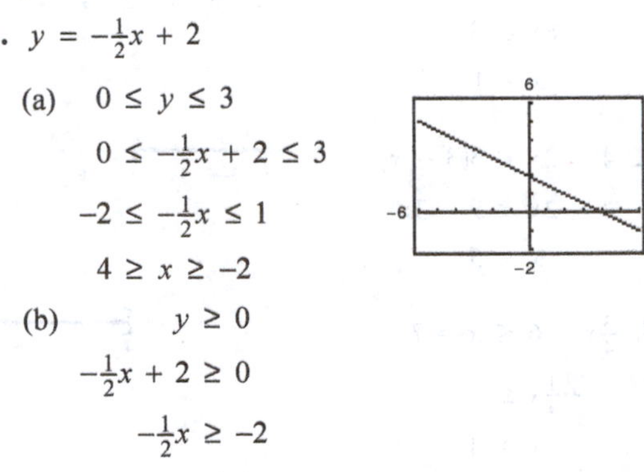

75. All real numbers less than 8 units from 10.

77. The midpoint of the interval $[-3, 3]$ is 0. The interval represents all real numbers x no more than 3 units from 0.

$|x - 0| \le 3$

$|x| \le 3$

79. The graph shows all real numbers at least 3 units from 7.

$|x - 7| \ge 3$

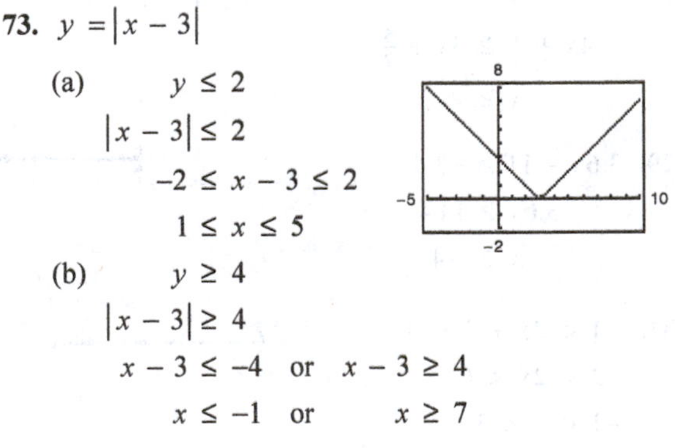

81. All real numbers less than 3 units from 7

$|x - 7| \geq 3$

83. All real numbers less than 4 units from −3

$|x - (-3)| < 4$

$|x + 3| < 4$

85. $\$7.25 \leq P \leq \$7.75$

87. $r \leq 0.08$

89. $r = 220 - A = 220 - 20 = 200$ beats per minute

$0.50(200) \leq r \leq 0.85(200)$

$100 \leq r \leq 170$

The target heart rate is at least 100 beats per minute and at most 170 beats per minute.

91. $9.00 + 0.75x > 13.50$

$0.75x > 4.50$

$x > 6$

You must produce at least 6 units each hour in order to yield a greater hourly wage at the second job.

93. $1000(1 + r(10)) > 2000.00$

$1 + 10r > 2$

$10r > 1$

$r > 0.1$

The rate must be greater than 10%.

95. $R > C$

$115.95x > 95x + 750$

$20.95x > 750$

$x \geq 35.7995$

$x \geq 36$ units

97. Let x = number of dozen doughnuts sold per day.

Revenue: $R = 7.95x$

Cost: $C = 1.45x + 165$

$P = R - C$

$= 7.95x - (1.45x + 165)$

$= 6.50x - 165$

$400 \leq P \leq 1200$

$400 \leq 6.50x - 165 \leq 1200$

$565 \leq 6.50x \leq 1365$

$86.9 \leq x \leq 210$

The daily sales vary between 87 and 210 dozen doughnuts per day.

99. (a)

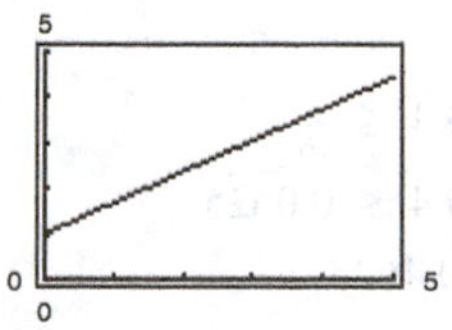

(b) From the graph you see that $y \geq 3$ when $x \geq 2.9$.

(c) Algebraically:

$3 \leq 0.692x + 0.988$

$2.012 \leq 0.692x$

$2.91 \leq x$

$x \geq 2.91$

(d) IQ scores are not a good predictor of GPAs. Other factors include study habits, class attendance, and attitudes.

101. (a) $W = 0.903t + 26.08$

$30 \leq 0.903t + 26.08 \leq 32$

$3.92 \leq 0.903t \leq 6.56$

$4.34 \leq t \leq 5.92$

Between the years 2004 and 2006, the mean hourly wage was between \$30 and \$32.

(b) $0.903t + 26.08 \geq 45$

$0.903t \geq 18.92$

$t \geq 20.95$

The mean hourly wage will exceed \$45 sometime during the year 2020.

103. $\left|\dfrac{t - 15.6}{1.9}\right| < 1$

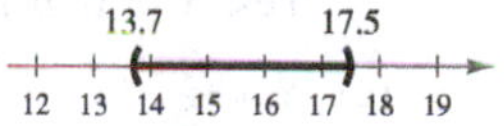

$-1 < \dfrac{t - 15.6}{1.9} < 1$

$-1.9 < t - 15.6 < 1.9$

$13.7 < t < 17.5$

Two-thirds of the workers could perform the task in the time interval between 13.7 minutes and 17.5 minutes.

105. $1 \text{ oz} = \dfrac{1}{16} \text{ lb}$, so $\dfrac{1}{2} \text{ oz} = \dfrac{1}{32} \text{ lb}$.

Because $8.99 \cdot \dfrac{1}{32} = 0.2809375$, you may be undercharged or overcharged by \$0.28.

107. $|s - 10.4| \le \frac{1}{16}$

$-\frac{1}{16} \le s - 10.4 \le \frac{1}{16}$

$-0.0625 \le s - 10.4 \le 0.0625$

$10.3375 \le s \le 10.4625$

Because $A = s^2$,

$(10.3375)^2 \le \text{area} \le (10.4625)^2$

$106.864 \text{ in.}^2 \le \text{area} \le 109.464 \text{ in.}^2$.

109. True. This is the Addition of a Constant Property of Inequalities.

111. False. If $-10 \le x \le 8$, then $10 \ge -x$ and $-x \ge -8$.

113. Answer not unique. Sample answer: $x < x + 1$

115. Answer not unique. *Sample answer:* $|ax - b| \le c$, if $a = 1$, $b = 5$, and $c = 5$, then

$|x - 5| \le 5$

$-5 \le x - 5 \le 5$

$0 \le x \le 10.$

Section 1.8 Other Types of Inequalities

1. positive; negative

3. zeros; undefined values

5. $x^2 - 3 < 0$

(a) $x = 3$

$(3)^2 - 3 \overset{?}{<} 0$

$6 \not< 0$

No, $x = 3$ *is not* a solution.

(b) $x = 0$

$(0)^2 - 3 \overset{?}{<} 0$

$-3 < 0$

Yes, $x = 0$ *is* a solution.

(c) $x = \frac{3}{2}$

$\left(\frac{3}{2}\right)^2 - 3 \overset{?}{<} 0$

$-\frac{3}{4} < 0$

Yes, $x = \frac{3}{2}$ *is* a solution.

(d) $x = -5$

$(-5)^2 - 3 \overset{?}{<} 0$

$22 \not< 0$

No, $x = -5$ *is not* a solution

7. $\dfrac{x + 2}{x - 4} \ge 3$

(a) $x = 5$

$\dfrac{5 + 2}{5 - 4} \overset{?}{\ge} 3$

$7 \ge 3$

Yes, $x = 5$ *is* a solution.

(b) $x = 4$

$\dfrac{4 + 2}{4 - 4} \overset{?}{\ge} 3$

$\dfrac{6}{0}$ is undefined.

No, $x = 4$ *is not* a solution.

(c) $x = -\dfrac{9}{2}$

$\dfrac{-\frac{9}{2} + 2}{-\frac{9}{2} - 4} \overset{?}{\ge} 3$

$\dfrac{5}{17} \not\ge 3$

No, $x = -\dfrac{9}{2}$ *is not* a solution.

(d) $x = \dfrac{9}{2}$

$\dfrac{\frac{9}{2} + 2}{\frac{9}{2} - 4} \overset{?}{\ge} 3$

$13 \ge 3$

Yes, $x = \dfrac{9}{2}$ *is* a solution.

9. $x^2 - 3x - 18 = (x + 3)(x - 6)$

$x + 3 = 0 \Rightarrow x = -3$

$x - 6 = 0 \Rightarrow x = 6$

The key numbers are -3 and 6.

11. $\dfrac{1}{x - 5} + 1 = \dfrac{1 + 1(x - 5)}{x - 5}$

$= \dfrac{x - 4}{x - 5}$

$x - 4 = 0 \Rightarrow x = 4$

$x - 5 = 0 \Rightarrow x = 5$

The key numbers are 4 and 5.

13. $2x^2 + 4x < 0$

$2x(x + 2) < 0$

Key numbers: $x = 0, \ -2$

Test intervals: $(-\infty, -2), (-2, 0), (0, \infty)$

Test: Is $2x(x + 2) < 0$?

Interval	x-Value	Value of $2x(x + 2)$	Conclusion
$(-\infty, -2)$	-3	6	Positive
$(-2, 0)$	-1	-2	Negative
$(0, \infty)$	1	3	Positive

Solution set: $(-2, 0)$

−5 −4 −3 −2 −1 0 1 2 x

15. $x^2 < 9$

$x^2 - 9 < 0$

$(x + 3)(x - 3) < 0$

Key numbers: $x = \pm 3$

Test intervals: $(-\infty, -3), (-3, 3), (3, \infty)$

Test: Is $(x + 3)(x - 3) < 0$?

Interval	x-Value	Value of $x^2 - 9$	Conclusion
$(-\infty, -3)$	-4	7	Positive
$(-3, 3)$	0	-9	Negative
$(3, \infty)$	4	7	Positive

Solution set: $(-3, 3)$

−4 −3 −2 −1 0 1 2 3 4 x

17. $(x + 2)^2 \le 25$

$x^2 + 4x + 4 \le 25$

$x^2 + 4x - 21 \le 0$

$(x + 7)(x - 3) \le 0$

Key numbers: $x = -7, x = 3$

Test intervals: $(-\infty, -7), (-7, 3), (3, \infty)$

Test: Is $(x + 7)(x - 3) \le 0$?

Interval	x-Value	Value of $(x + 7)(x - 3)$	Conclusion
$(-\infty, -7)$	-8	$(-1)(-11) = 11$	Positive
$(-7, 3)$	0	$(7)(-3) = -21$	Negative
$(3, \infty)$	4	$(11)(1) = 11$	Positive

Solution set: $[-7, 3]$

−7 3

−8 −6 −4 −2 0 2 4 6 x

19. $x^2 + 6x + 1 \ge -7$

$x^2 + 6x + 8 \ge 0$

$(x + 2)(x + 4) \ge 0$

Key numbers: $x = -2, x = -4$

Test Intervals: $(-\infty, -4), (-4, -2), (-2, \infty)$

Test: Is $(x + 2)(x + 4) > 0$?

Interval	x-Value	Value of $(x + 2)(x + 4)$	Conclusion
$(-\infty, -4)$	-6	8	Positive
$(-4, -2)$	-3	-1	Negative
$(-2, \infty)$	0	8	Positive

Solution set: $(-\infty, -4] \cup [-2, \infty)$

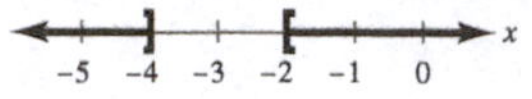

21. $x^2 + x < 6$

$x^2 + x - 6 < 0$

$(x + 3)(x - 2) < 0$

Key numbers: $x = -3, x = 2$

Test intervals: $(-\infty, -3), (-3, 2), (2, \infty)$

Test: Is $(x + 3)(x - 2) < 0$?

Interval	x-Value	Value of $(x + 3)(x - 2)$	Conclusion
$(-\infty, -3)$	-4	$(-1)(-6) = 6$	Positive
$(-3, 2)$	0	$(3)(-2) = -6$	Negative
$(2, \infty)$	3	$(6)(1) = 6$	Positive

Solution set: $(-3, 2)$

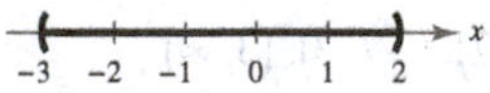

23. $x^2 < 3 - 2x$

$x^2 + 2x - 3 < 0$

$(x + 3)(x - 1) < 0$

Key numbers: $x = -3, x = 1$

Test intervals: $(-\infty, -3), (-3, 1), (1, \infty)$

Test: Is $(x + 3)(x - 1) < 0$?

Interval	x-Value	Value of $(x + 3)(x - 1)$	Conclusion
$(-\infty, -3)$	-4	$(-1)(-5) = 5$	Positive
$(-3, 1)$	0	$(3)(-1) = -3$	Negative
$(1, \infty)$	2	$(5)(1) = 5$	Positive

Solution set: $(-3, 1)$

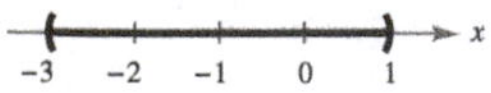

25. $3x^2 - 11x > 20$

$3x^2 - 11x - 20 > 0$

$(3x + 4)(x - 5) > 0$

Key numbers: $x = 5, x = -\frac{4}{3}$

Test intervals: $\left(-\infty, -\frac{4}{5}\right), \left(-\frac{4}{3}, 5\right), (5, \infty)$

Test: Is $(3x + 4)(x - 5) > 0$?

Interval	x-Value	Value of $(3x + 4)(x - 5)$	Conclusion
$\left(-\infty, -\frac{4}{3}\right)$	-3	$(-5)(-8) = 40$	Positive
$\left(-\frac{4}{3}, 5\right)$	0	$(4)(-5) = -20$	Negative
$(5, \infty)$	6	$(22)(1) = 22$	Positive

Solution set: $\left(-\infty, -\frac{4}{3}\right) \cup (5, \infty)$

$-\frac{4}{3}$

−2 −1 0 1 2 3 4 5 6 x

27. $x^3 - 3x^2 - x + 3 > 0$

$x^2(x - 3) - (x - 3) > 0$

$(x - 3)(x^2 - 1) > 0$

$(x - 3)(x + 1)(x - 1) > 0$

Key numbers: $x = -1, x = 1, x = 3$

Test intervals: $(-\infty\ -1), (-1, 1), (1, 3), (3, \infty)$

Test: Is $(x - 3)(x + 1)(x - 1) > 0$?

Interval	x-Value	Value of $(x - 3)(x + 1)(x - 1)$	Conclusion
$(-\infty, -1)$	-2	$(-5)(-1)(-3) = -15$	Negative
$(-1, 1)$	0	$(-3)(1)(-1) = 3$	Positive
$(1, 3)$	2	$(-1)(3)(1) = -3$	Negative
$(3, \infty)$	4	$(1)(5)(3) = 15$	Positive

Solution set: $(-1, 1) \cup (3, \infty)$

−3 −2 −1 0 1 2 3 4 5 6 7 x

29. $-x^3 + 7x^2 + 9x > 63$

$x^3 - 7x^2 - 9x < -63$

$x^3 - 7x^2 - 9x + 63 < 0$

$x^2(x - 7) - 9(x - 7) < 0$

$(x - 7)(x^2 - 9) < 0$

$(x - 7)(x + 3)(x - 3) < 0$

Key numbers: $x = -3, x = 3, x = 7$

Test intervals: $(-\infty, -3), (-3, 3), (3, 7), (7, \infty)$

Test: Is $(x - 7)(x + 3)(x - 3) < 0$?

Interval	x-Value	Value of $(x - 7)(x + 3)(x - 3)$	Conclusion
$(-\infty, -3)$	-4	$(-11)(-1)(-7) = -77$	Negative
$(-3, 3)$	0	$(-7)(3)(-3) = 63$	Positive
$(3, 7)$	4	$(-3)(7)(1) = -21$	Negative
$(7, \infty)$	8	$(1)(11)(5) = 55$	Positive

Solution set: $(-\infty, -3) \cup (3, 7)$

31. $4x^3 - 6x^2 < 0$

$2x^2(2x - 3) < 0$

Key numbers: $x = 0, x = \frac{3}{2}$

Test intervals: $(-\infty, 0) \Rightarrow 2x^2(2x - 3) < 0$

$\left(0, \frac{3}{2}\right) \Rightarrow 2 \Rightarrow 2x^2(2x - 3) < 0$

$\left(\frac{3}{2}, \infty\right) \Rightarrow 2x^2(2x - 3) > 0$

Solution set: $(-\infty, 0) \cup \left(0, \frac{3}{2}\right)$

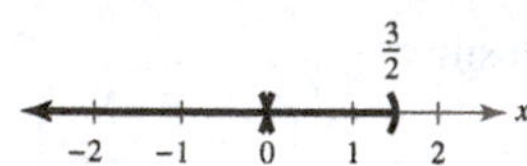

33. $x^3 - 4x \ge 0$

$x(x + 2)(x - 2) \ge 0$

Key numbers: $x = 0, x = \pm 2$

Test intervals: $(-\infty, -2) \Rightarrow x(x + 2)(x - 2) < 0$

$(-2, 0) \Rightarrow x(x + 2)(x - 2) > 0$

$(0, 2) \Rightarrow x(x + 2)(x - 2) < 0$

$(2, \infty) \Rightarrow x(x + 2)(x - 2) > 0$

Solution set: $[-2, 0] \cup [2, \infty)$

35. $(x - 1)^2(x + 2)^3 \ge 0$

Key numbers: $x = 1, x = -2$

Test intervals: $(-\infty, -2) \Rightarrow (x - 1)^2(x + 2)^3 < 0$

$(-2, 1) \Rightarrow (x - 1)^2(x + 2)^3 > 0$

$(1, \infty) \Rightarrow (x - 1)^2(x + 2)^3 > 0$

Solution set: $[-2, \infty)$

37. $4x^2 - 4x + 1 \le 0$

$(2x - 1)^2 \le 0$

Key number: $x = \frac{1}{2}$

Test Interval	x-Value	Polynomial Value	Conclusion
$\left(-\infty, \frac{1}{2}\right)$	$x = 0$	$[2(0) - 1]^2 = 1$	Positive
$\left(\frac{1}{2}, \infty\right)$	$x = 1$	$[2(1) - 1]^2 = 1$	Positive

The solution set consists of the single real number $\frac{1}{2}$.

39. $x^2 - 6x + 12 \le 0$

Using the Quadratic Formula, you can determine that the key numbers are $x = 3 \pm \sqrt{3i}$.

Test Interval	x-Value	Polynomial Value	Conclusion
$(-\infty, \infty)$	$x = 0$	$(0)^2 - 6(0) + 12 = 12$	Positive

The solution set is empty, that is there are no real solutions.

41. $\dfrac{4x - 1}{x} > 0$

Key numbers: $x = 0, x = \dfrac{1}{4}$

Test intervals: $(-\infty, 0), \left(0, \frac{1}{4}\right), \left(\frac{1}{4}, \infty\right)$

Test: Is $\dfrac{4x - 1}{x} > 0$?

Interval	x-Value	Value of $\dfrac{4x - 1}{x}$	Conclusion
$(-\infty, 0)$	-1	$\dfrac{-5}{-1} = 5$	Positive
$\left(0, \dfrac{1}{4}\right)$	$\dfrac{1}{8}$	$\dfrac{-\frac{1}{2}}{\frac{1}{8}} = -4$	Negative
$\left(\dfrac{1}{4}, \infty\right)$	1	$\dfrac{3}{1} = 3$	Positive

Solution set: $(-\infty, 0) \cup \left(\dfrac{1}{4}, \infty\right)$

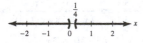

43.
$$\frac{3x+5}{x-1} < 2$$
$$\frac{3x+5}{x-1} - 2 < 0$$
$$\frac{3x+5-2(x-1)}{x-1} < 0$$
$$\frac{x+7}{x-1} < 0$$

Key numbers: $x = -7, x = 1$

Test intervals: $(-\infty, -7), (-7, 1), (1, \infty)$

Test: Is $\frac{x+7}{x-1} < 0$?

Interval	x-Value	Value of $\frac{x+7}{x-1}$	Conclusion
$(-\infty, -7)$	-8	$\frac{-1}{-9} = \frac{1}{9}$	Positive
$(-7, 1)$	0	$\frac{0+7}{0-1} = -7$	Negative
$(1, \infty)$	2	$\frac{2+9}{2-1} = 11$	Positive

Solution set: $(-7, 1)$

45.
$$\frac{2}{x+5} > \frac{1}{x-3}$$
$$\frac{2}{x+5} - \frac{1}{x-3} > 0$$
$$\frac{2(x-3) - 1(x+5)}{(x+5)(x-3)} > 0$$
$$\frac{x-11}{(x+5)(x-3)} > 0$$

Key numbers: $x = -5, x = 3, x = 11$

Test intervals: $(-\infty, -5) \Rightarrow \frac{x-11}{(x+5)(x-3)} < 0$

$(-5, 3) \Rightarrow \frac{x-11}{(x+5)(x-3)} > 0$

$(3, 11) \Rightarrow \frac{x-11}{(x+5)(x-3)} < 0$

$(11, \infty) \Rightarrow \frac{x-11}{(x+5)(x-3)} > 0$

Solution set: $(-5, 3) \cup (11, \infty)$

47.
$$\frac{1}{x-3} \le \frac{9}{4x+3}$$
$$\frac{1}{x-3} - \frac{9}{4x+3} \le 0$$
$$\frac{4x+3-9(x-3)}{(x-3)(4x+3)} \le 0$$
$$\frac{30-5x}{(x-3)(4x+3)} \le 0$$

Key numbers: $x = 3, x = -\frac{3}{4}, x = 6$

Test intervals: $\left(-\infty, -\frac{3}{4}\right) \Rightarrow \frac{30-5x}{(x-3)(4x+3)} > 0$

$\left(-\frac{3}{4}, 3\right) \Rightarrow \frac{30-5x}{(x-3)(4x+3)} < 0$

$(3, 6) \Rightarrow \frac{30-5x}{(x-3)(4x+3)} > 0$

$(6, \infty) \Rightarrow \frac{30-5x}{(x-3)(4x+3)} < 0$

Solution set: $\left(-\frac{3}{4}, 3\right) \cup [6, \infty)$

49. $\dfrac{x^2 + 2x}{x^2 - 9} \le 0$

$\dfrac{x(x + 2)}{(x + 3)(x - 3)} \le 0$

Key numbers: $x = 0, x = -2, x = \pm 3$

Test intervals: $(-\infty, -3) \Rightarrow \dfrac{x(x + 2)}{(x + 3)(x - 3)} > 0$

$(-3, -2) \Rightarrow \dfrac{x(x + 2)}{(x + 3)(x - 3)} < 0$

$(-2, 0) \Rightarrow \dfrac{x(x + 2)}{(x + 3)(x - 3)} > 0$

$(0, 3) \Rightarrow \dfrac{x(x + 2)}{(x + 3)(x - 3)} < 0$

$(3, \infty) \Rightarrow \dfrac{x(x + 2)}{(x + 3)(x - 3)} > 0$

Solution set: $(-3, -2] \cup [0, 3)$

−3 −2 −1 0 1 2 3 x

51. $\dfrac{3}{x - 1} + \dfrac{2x}{x + 1} > -1$

$\dfrac{3(x + 1) + 2x(x - 1) + 1(x + 1)(x - 1)}{(x - 1)(x + 1)} > 0$

$\dfrac{3x^2 + x + 2}{(x - 1)(x + 1)} > 0$

Key numbers: $x = -1, x = 1$

Test intervals: $(-\infty, -1) \Rightarrow \dfrac{3x^2 + x + 2}{(x - 1)(x + 1)} > 0$

$(-1, 1) \Rightarrow \dfrac{3x^2 + x + 2}{(x - 1)(x + 1)} < 0$

$(1, \infty) \Rightarrow \dfrac{3x^2 + x + 2}{(x - 1)(x + 1)} > 0$

Solution set: $(-\infty, -1) \cup (1, \infty)$

−4 −3 −2 −1 0 1 2 3 4 x

53. $y = -x^2 + 2x + 3$

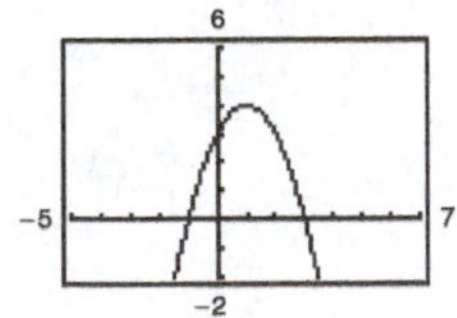

(a) $y \le 0$ when $x \le -1$ or $x \ge 3$.

(b) $y \ge 3$ when $0 \le x \le 2$.

55. $y = \frac{1}{8}x^3 - \frac{1}{2}x$

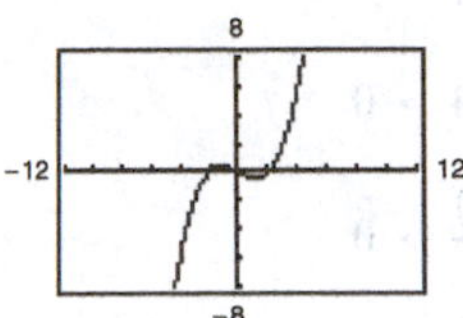

(a) $y \ge 0$ when $-2 \le x \le 0$ or $2 \le x < \infty$.

(b) $y \le 6$ when $x \le 4$.

57. $y = \dfrac{3x}{x - 2}$

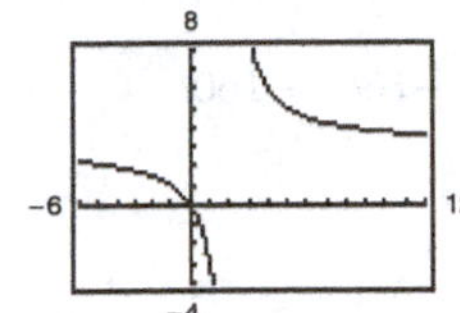

(a) $y \le 0$ when $0 \le x < 2$.

(b) $y \ge 6$ when $2 < x \le 4$.

59. $y = \dfrac{2x^2}{x^2 + 4}$

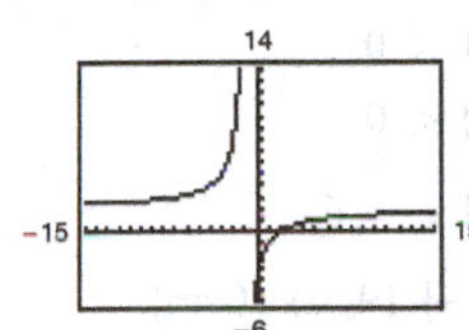

(a) $y \ge 1$ when $x \le -2$ or $x \ge 2$.

This can also be expressed as $|x| \ge 2$.

(b) $y \le 2$ for all real numbers x.

This can also be expressed as $-\infty < x < \infty$.

61. $0.3x^2 + 6.26 < 10.8$

$0.3x^2 + 4.54 < 0$

Key numbers: $x \approx \pm 3.89$

Test intervals: $(-\infty, -3.89), (-3.89, 3.89), (3.89, \infty)$

Solution set: $(-3.89, 3.89)$

63. $-0.5x^2 + 12.5x + 1.6 > 0$

Key numbers: $x \approx -0.13, x \approx 25.13$

Test intervals: $(-\infty, -0.13), (-0.13, 25.13), (25.13, \infty)$

Solution set: $(-0.13, 25.13)$

65. $\dfrac{1}{2.3x - 5.2} > 3.4$

$\dfrac{1}{2.3x - 5.2} - 3.4 > 0$

$\dfrac{1 - 3.4(2.3x - 5.2)}{2.3x - 5.2} > 0$

$\dfrac{-7.82x + 18.68}{2.3x - 5.2} > 0$

Key numbers: $x \approx 2.39, x \approx 2.26$

Test intervals: $(-\infty, 2.26), (2.26, 2.39), (2.39, \infty)$

Solution set: $(2.26, 2.39)$

67. $s = -16t^2 + v_0t + s_0 = -16t^2 + 160t$

(a) $-16t^2 + 160t = 0$

$-16t(t - 10) = 0$

$t = 0, t = 10$

It will be back on the ground in 10 seconds.

(b) $-16t^2 + 160t > 384$

$-16t^2 + 160t - 384 > 0$

$-16(t^2 - 10t + 24) > 0$

$t^2 - 10t + 24 < 0$

$(t - 4)(t - 6) < 0$

Key numbers: $t = 4, t = 6$

Test intervals: $(-\infty, 4), (4, 6), (6, \infty)$

Solution set: 4 seconds $< t <$ 6 seconds

69. $R = x(75 - 0.0005x)$ and $C = 30x + 250{,}000$

$P = R - C$

$= (75x - 0.0005x^2) - (30x + 250{,}000)$

$= -0.0005x^2 + 45x - 250{,}000$

$P \geq 750{,}000$

$-0.0005x^2 + 45x - 250{,}000 \geq 750{,}000$

$-0.0005x^2 + 45x - 1{,}000{,}000 \geq 0$

Key numbers: $x = 40{,}000, x = 50{,}000$

(These were obtained by using the Quadratic Formula.)

Test intervals:

$(0, 40{,}000), (40{,}000, 50{,}000), (50{,}000, \infty)$

The solution set is $[40{,}000, 50{,}000]$ or $40{,}000 \leq x \leq 50{,}000$. The price per unit is

$p = \dfrac{R}{x} = 75 - 0.0005x.$

For $x = 40{,}000$, $p = \$55$. For $x = 50{,}000$, $p = \$50$. So, for $40{,}000 \leq x \leq 50{,}000$, $\$50.00 \leq p \leq \$55.00$.

71. $4 - x^2 \geq 0$

$(2 + x)(2 - x) \geq 0$

Key numbers: $x = \pm 2$

Test intervals: $(-\infty, -2) \Rightarrow 4 - x^2 < 0$

$(-2, 2) \Rightarrow 4 - x^2 > 0$

$(2, \infty) \Rightarrow 4 - x^2 < 0$

Domain: $[-2, 2]$

73. $x^2 - 9x + 20 \geq 0$

$(x - 4)(x - 5) \geq 0$

Key numbers: $x = 4, x = 5$

Test intervals: $(-\infty, 4), (4, 5), (5, \infty)$

Interval	x-Value	Value of $(x - 4)(x - 5)$	Conclusion
$(-\infty, 4)$	0	$(-4)(-5) = 20$	Positive
$(4, 5)$	$\frac{9}{2}$	$\left(\frac{1}{2}\right)\left(-\frac{1}{2}\right) = -\frac{1}{4}$	Negative
$(5, \infty)$	6	$(2)(1) = 2$	Positive

Domain: $(-\infty, 4] \cup [5, \infty)$

75. $\dfrac{x}{x^2 - 2x - 35} \geq 0$

$\dfrac{x}{(x + 5)(x - 7)} \geq 0$

Key numbers: $x = 0, x = -5, x = 7$

Test intervals: $(-\infty, -5) \Rightarrow \dfrac{x}{(x + 5)(x - 7)} < 0$

$(-5, 0) \Rightarrow \dfrac{x}{(x + 5)(x - 7)} > 0$

$(0, 7) \Rightarrow \dfrac{x}{(x + 5)(x - 7)} < 0$

$(7, \infty) \Rightarrow \dfrac{x}{(x + 5)(x - 7)} > 0$

Domain: $(-5, 0] \cup (7, \infty)$

77. (a) and (c)

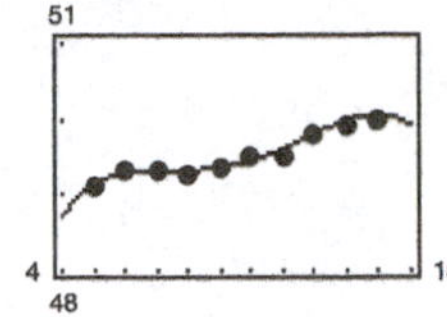

(b) $N = -0.001231t^4 + 0.04723t^3 - 0.6452t^2 + 3.783t + 41.21$

The model fits the data well.

(d) Using the zoom and trace features, the number of students enrolled in elementary and secondary schools fell below 48 million in the year 2017.

(e) No. The model can be used to predict enrollments for years close to those in its domain, $5 \le t \le 14$, but when you project too far into the future, the numbers predicted by the model decrease too rapidly to be considered reasonable.

79. $2L + 2W = 100 \Rightarrow W = 50 - L$

$$LW \ge 500$$
$$L(50 - L) \ge 500$$
$$-L^2 + 50L - 500 \ge 0$$

By the Quadratic Formula you have:

Key numbers: $L = 25 \pm 5\sqrt{5}$

Test: Is $-L^2 + 50L - 500 \ge 0$?

Solution set: $25 - 5\sqrt{5} \le L \le 25 + 5\sqrt{5}$

13.8 meters $\le L \le$ 36.2 meters

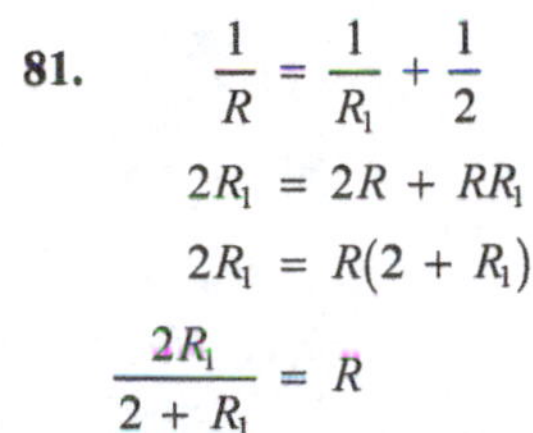

81. $\dfrac{1}{R} = \dfrac{1}{R_1} + \dfrac{1}{2}$

$$2R_1 = 2R + RR_1$$
$$2R_1 = R(2 + R_1)$$
$$\frac{2R_1}{2 + R_1} = R$$

Because $R \ge 1$,

$$\frac{2R_1}{2 + R_1} \ge 1$$
$$\frac{2R_1}{2 + R_1} - 1 \ge 0$$
$$\frac{R_1 - 2}{2 + R_1} \ge 0.$$

Because $R_1 > 0$, the only key number is $R_1 = 2$. The inequality is satisfied when $R_1 \ge 2$ ohms.

83. False.

There are four test intervals. The test intervals are $(-\infty, -3)$, $(-3, 1)$, $(1, 4)$, and $(4, \infty)$.

85.

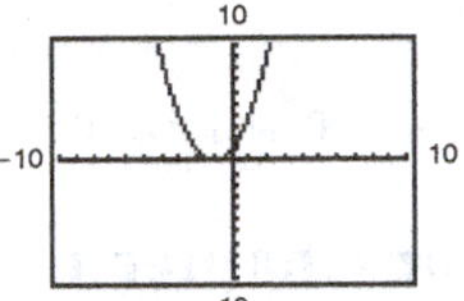

For part (b), the y-values that are less than or equal to 0 occur only at $x = -1$.

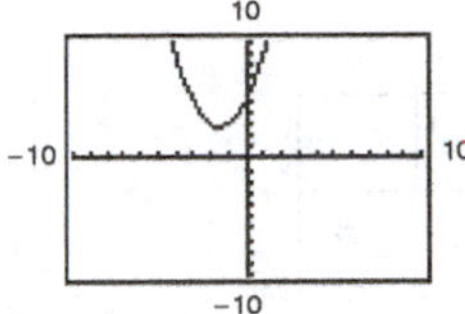

For part (c), there are no y-values that are less than 0.

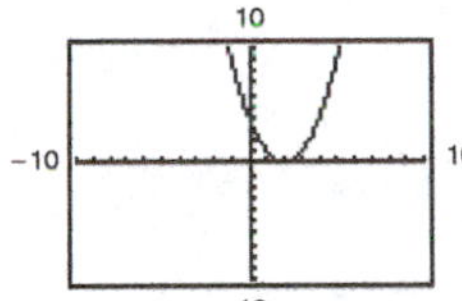

For part (d), the y-values that are greater than 0 occur for all values of x except 2.

87. $x^2 + bx + 9 = 0$

(a) To have at least one real solution, $b^2 - 4ac \ge 0$.

$$b^2 - 4(1)(9) \ge 0$$
$$b^2 - 36 \ge 0$$

Key numbers: $b = -6$, $b = 6$

Test intervals: $(-\infty, -6) \Rightarrow b^2 - 36 > 0$

$(-6, 6) \Rightarrow b^2 - 36 < 0$

$(6, \infty) \Rightarrow b^2 - 36 > 0$

Solution set: $(-\infty, -6] \cup [6, \infty)$

(b) $b^2 - 4ac \ge 0$

Key numbers: $b = -2\sqrt{ac}$, $b = 2\sqrt{ac}$

Similar to part (a), if $a > 0$ and $c > 0$, $b \le -2\sqrt{ac}$ or $b \ge 2\sqrt{ac}$.

89. $3x^2 + bx + 10 = 0$

(a) To have at least one real solution, $b^2 - 4ac \geq 0$.

$b^2 - 4(3)(10) \geq 0$

$b^2 - 120 \geq 0$

Key numbers: $b = -2\sqrt{30}, b = 2\sqrt{30}$

Test intervals: $\left(-\infty, -2\sqrt{30}\right) \Rightarrow b^2 - 120 > 0$

$\left(-2\sqrt{30}, 2\sqrt{30}\right) \Rightarrow b^2 - 120 < 0$

$\left(2\sqrt{30}, \infty\right) \Rightarrow b^2 - 120 > 0$

Solution set: $\left(-\infty, -2\sqrt{30}\right] \cup \left[2\sqrt{30}, \infty\right)$

(b) $b^2 - 4ac \geq 0$

Similar to part (a), if $a > 0$ and $c > 0$, $b \leq -2\sqrt{ac}$ or $b \geq 2\sqrt{ac}$.

Review Exercises for Chapter 1

1. $y = -4x + 1$

x	-2	-1	0	1	2
y	9	5	1	-3	-7

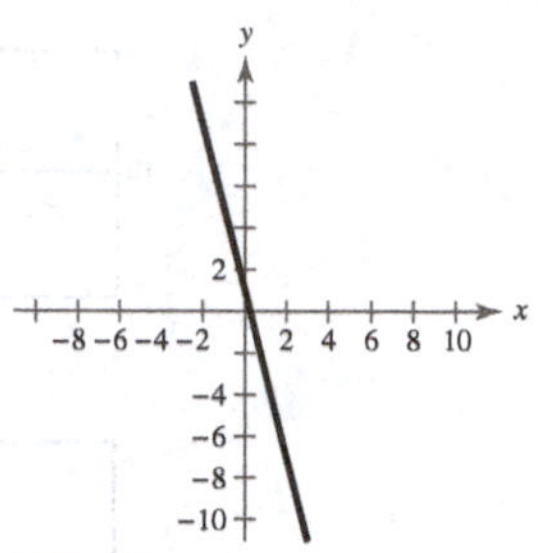

3. x-intercepts: $(1, 0), (5, 0)$

y-intercept: $(0, 5)$

5. $y = -3x + 7$

Intercepts: $\left(\frac{7}{3}, 0\right), (0,7)$

$y = -3(-x) + 7 \Rightarrow y = 3x + 7 \Rightarrow$ No y-axis symmetry

$-y = -3x + 7 \Rightarrow y = 3x - 7 \Rightarrow$ No x-axis symmetry

$-y = -3(-x) + 7 \Rightarrow y = -3x - 7 \Rightarrow$ No origin symmetry

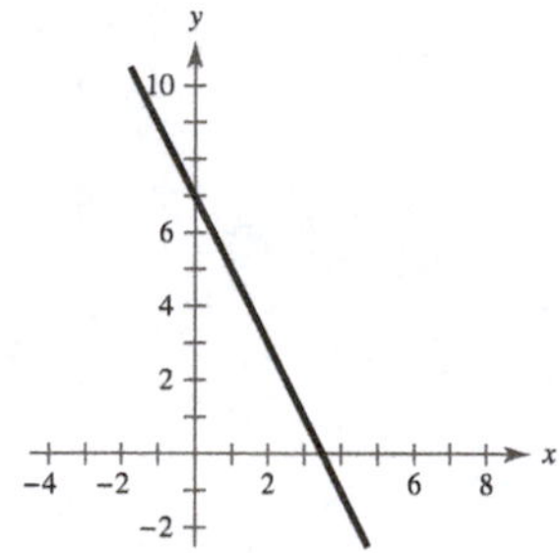

7. $x = y^2 - 5$

Intercepts: $(-5,0), \left(0, \pm\sqrt{5}\right)$

$-x = y^2 - 5 \Rightarrow x = -y^2 + 5 \Rightarrow$ No y-axis symmetry

$x = \left(-y^2\right) - 5 \Rightarrow x = y^2 - 5 \Rightarrow x$-axis symmetry

$-x = (-y)^2 - 5 \Rightarrow x = -y^2 + 5 \Rightarrow$ No origin symmetry

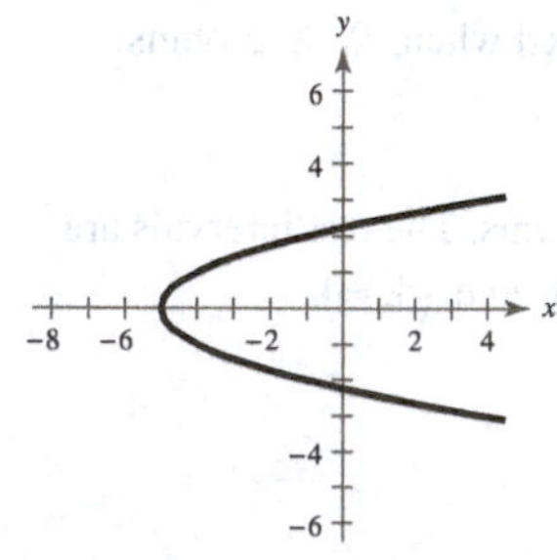

9. $y = -x^4 + 6x^2$

Intercept: $(0, 0)$

$y = -(-x)^4 + 6(-x)^2 \Rightarrow y = -x^4 + 6x^2 \Rightarrow$ y-axis symmetry

$-y = -x^4 + 6x^2 \Rightarrow y = x^4 - 6x^2 \Rightarrow$ No x-axis symmetry

$-y = -(-x)^4 + 6(-x)^2 \Rightarrow y = x^4 - 6x^2 \Rightarrow$ No origin symmetry

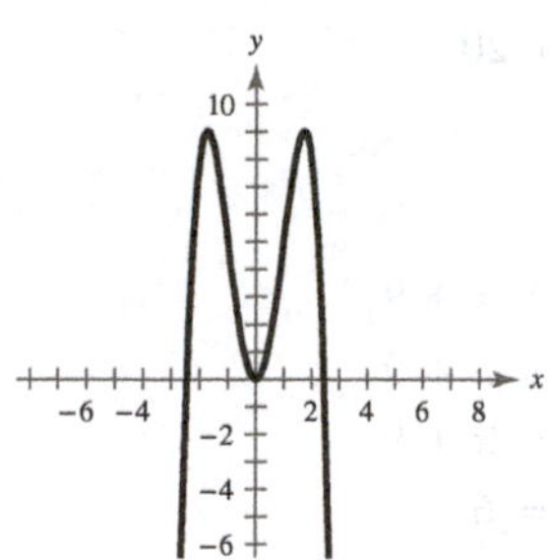

11. $y = \dfrac{3}{x}$

Intercept: None

$y = \dfrac{3}{-x} \Rightarrow y = -\dfrac{3}{x} \Rightarrow$ No y-axis symmetry

$-y = \dfrac{3}{x} \Rightarrow y = -\dfrac{3}{x} \Rightarrow$ No x-axis symmetry

$-y = \dfrac{3}{-x} \Rightarrow y = \dfrac{3}{x} \Rightarrow$ origin symmetry

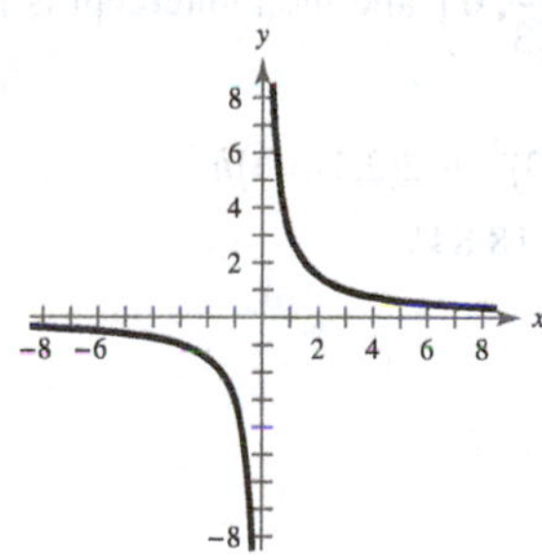

13. $x^2 + y^2 = 9$

Center: $(0, 0)$

Radius: 3

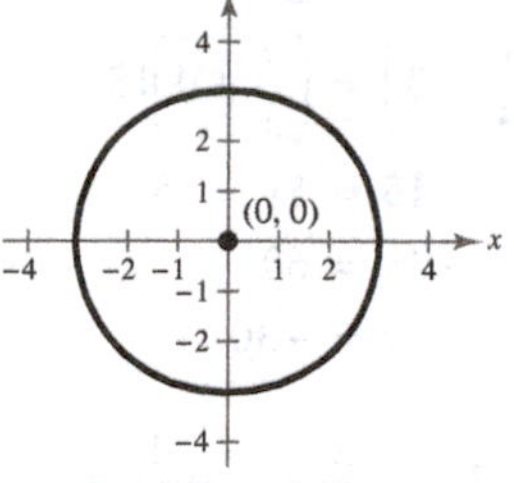

15.
$$(x + 2)^2 + y^2 = 16$$
$$(x - (-2))^2 + (y - 0)^2 = 4^2$$

Center: $(-2, 0)$

Radius: 4

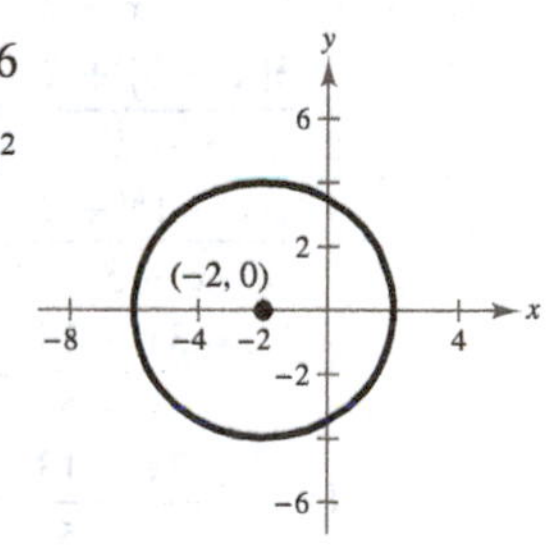

17. Endpoints of a diameter: $(0, 0)$ and $(4, -6)$

Center: $\left(\dfrac{0 + 4}{2}, \dfrac{0 + (-6)}{2}\right) = (2, -3)$

Radius:

$$r = \sqrt{(2 - 0)^2 + (-3 - 0)^2} = \sqrt{4 + 9} = \sqrt{13}$$

Standard form: $(x - 2)^2 + (y - (-3))^2 = \left(\sqrt{13}\right)^2$

$$(x - 2)^2 + (y + 3)^2 = 13$$

19. (a)

(b) $27.215t^2 - 409.06t + 1563.6 = 200$

$27.215t^2 - 409.06t + 1363.6 = 0$

$$t = \frac{-(-409.06) \pm \sqrt{(-409.06)^2 - 4(27.215)(1363.6)}}{2(27.215)}$$

$t = 10.04, 4.99$

Using the zoom and trace features, sales were \$200 million during the year 2010.

21. $2(x - 2) = 2x - 4$

$2x - 4 = 2x - 4$

$0 = 0$ Identity

All real numbers are solutions.

23. $3(x - 2) + 2x = 2(x + 3)$

$3x - 6 + 2x = 2x + 6$

$3x = 12$

$x = 4$

Conditional equation

25. $8x - 5 = 3x + 20$
$5x = 25$
$x = 5$

27. $2(x + 5) - 7 = x + 9$
$2x + 10 - 7 = x + 9$
$2x + 3 = x + 9$
$x = 6$

29. $\frac{x}{5} - 3 = \frac{x}{3} + 1$
$15\left(\frac{x}{5} - 3\right) = \left(\frac{x}{3} + 1\right)15$
$3x - 45 = 5x + 15$
$-2x = 60$
$x = -30$

31. $3 + \frac{2}{x - 5} = \frac{2x}{x - 5}$
$\frac{3(x - 5) + 2}{x - 5} = \frac{2x}{x - 5}$
$\frac{3x - 15 + 2}{x - 5} = \frac{2x}{x - 5}$
$\frac{3x - 13}{x - 5} = \frac{2x}{x - 5}$
$(x - 5)\frac{3x - 13}{x - 5} = \frac{2x}{x - 5}(x - 5)$
$3x - 13 = 2x$
$x = 3$

33. $y = 3x - 1$
x-intercept: $0 = 3x - 1 \Rightarrow x = \frac{1}{3}$
y-intercept: $y = 3(0) - 1 \Rightarrow y = -1$
The x-intercept is $\left(\frac{1}{3}, 0\right)$ and the y-intercept is $(0, -1)$.

35. $y = 2(x - 4)$
x-intercept: $0 = 2(x - 4) \Rightarrow x = 4$
y-intercept: $y = 2(0 - 4) \Rightarrow y = -8$
The x-intercept is $(4, 0)$ and the y-intercept is $(0, -8)$.

37. $y = -\frac{1}{2}x + \frac{2}{3}$
x-intercept: $0 = -\frac{1}{2}x + \frac{2}{3} \Rightarrow x = \frac{2/3}{1/2} = \frac{4}{3}$
y-intercept: $y = -\frac{1}{2}(0) + \frac{2}{3} \Rightarrow y = \frac{2}{3}$
The x-intercept is $\left(\frac{4}{3}, 0\right)$ and the y-intercept is $\left(0, \frac{2}{3}\right)$.

39. $244.92 = 2(3.14)(3)^2 + 2(3.14)(3)h$
$244.92 = 56.52 + 18.84h$
$188.40 = 18.84h$
$10 = h$
The height is 10 inches.

41. *Verbal Model:* Revenue in 2013 + Revenue in 2014 = Total Revenue
Labels: Let x = Revenue in 2014. Then $x = 0.452x$ = Revenue in 2013.
Equation: $x + (x + 0.452x) = 3.75$
$2.452x = 3.75$
$x \approx 1.53$
$x + 0.452x \approx 2.22$

So, Revenue was \$2.22 billion in 2013 and was \$1.53 billion in 2014.

43. Let x = the total investment required.

Each person's share is $\frac{x}{9}$. If three more people invest, each person's share is $\frac{x}{12} + 2500$.

Since this is \$2500 less than the original cost, we have:

$\frac{x}{9} = \frac{x}{12} + 2500$
$\frac{x}{9} - \frac{x}{12} = 2500$
$\frac{8x - 6x}{72} = 2500$
$\frac{2x}{72} = 2500$
$x = 90{,}000$

The total investment to start the business is \$90,000.

45. Let x = the number of liters of pure antifreeze.

$$\begin{aligned} 30\% \text{ of } (10 - x) + 100\% \text{ of } x &= 50\% \text{ of } 10 \\ 0.30(10 - x) + 1.00x &= 0.50(10) \\ 3 - 0.30x + 1.00x &= 5 \\ 0.70x &= 2 \\ x &= \frac{2}{0.70} = \frac{20}{7} = 2.857 \text{ liters} \end{aligned}$$

47.
$$\begin{aligned} V &= \frac{1}{3}\pi r^2 h \\ 3V &= \pi r^2 h \\ \frac{3V}{\pi r^2} &= h \end{aligned}$$

49.
$$\begin{aligned} 15 + x - 2x^2 &= 0 \\ 0 &= 2x^2 - x - 15 \\ 0 &= (2x + 5)(x - 3) \\ 2x + 5 = 0 &\Rightarrow x = -\tfrac{5}{2} \\ x - 3 = 0 &\Rightarrow x = 3 \end{aligned}$$

51.
$$\begin{aligned} 6 &= 3x^2 \\ 2 &= x^2 \\ \pm\sqrt{2} &= x \end{aligned}$$

53.
$$\begin{aligned} (x + 13)^2 &= 25 \\ x + 13 &= \pm 5 \\ x &= -13 \pm 5 \\ x &= -18 \text{ or } x = -8 \end{aligned}$$

55.
$$\begin{aligned} x^2 + 12x + 250 &= 0 \\ x &= \frac{-12 \pm \sqrt{12^2 - 4(1)(25)}}{2(1)} \\ &= \frac{-12 \pm 2\sqrt{11}}{2} \\ &= -6 \pm \sqrt{11} \end{aligned}$$

57.
$$\begin{aligned} -2x^2 - 5x + 27 &= 0 \\ 2x^2 + 5x - 27 &= 0 \end{aligned}$$

59. $M = 500x(20 - x)$

(a) $500x(20 - x) = 0$ when $x = 0$ feet and $x = 20$ feet.

(b)

55,000
0
22
0

(c) The bending moment is greatest when $x = 10$ feet.

61. $4 + \sqrt{-9} = 4 + 3i$

63. $i^2 + 3i = -1 + 3i$

65.
$$\begin{aligned} (6 - 4i) + (-9 + i) &= (6 + (-9)) + (-4i + i) \\ &= -3 - 3i \end{aligned}$$

67.
$$\begin{aligned} -3i(-2 + 5i) &= 6i - 15i^2 \\ &= 6i - 15(-1) \\ &= 15 + 6i \end{aligned}$$

69.
$$\begin{aligned} (1 + 7i)(1 - 7i) &= 1 - 49i^2 \\ &= 1 - 49(-1) \\ &= 1 + 49 \\ &= 50 \end{aligned}$$

71.
$$\begin{aligned} \frac{4}{1 - 2i} &= \frac{4}{1 - 2i} \cdot \frac{1 + 2i}{1 + 2i} \\ &= \frac{4 + 8i}{1 - 4i^2} \\ &= \frac{4 + 8i}{5} \\ &= \frac{4}{5} + \frac{8}{5}i \end{aligned}$$

73.
$$\begin{aligned} \frac{3 + 2i}{5 + i} &= \frac{3 + 2i}{5 + i} \cdot \frac{5 - i}{5 - i} \\ &= \frac{15 - 3i + 10i - 2i^2}{25 - i^2} \\ &= \frac{17 + 7i}{26} \\ &= \frac{17}{26} + \frac{7i}{26} \end{aligned}$$

75. $\dfrac{4}{2-3i}+\dfrac{2}{1+i}=\dfrac{4}{2-3i}\cdot\dfrac{2+3i}{2+3i}+\dfrac{2}{1+i}\cdot\dfrac{1-i}{1-i}$

$$=\frac{8+12i}{4+9}+\frac{2-2i}{1+1}$$
$$=\frac{8}{13}+\frac{12}{13}i+1-i$$
$$=\left(\frac{8}{13}+1\right)+\left(\frac{12}{13}i-i\right)$$
$$=\frac{21}{13}-\frac{1}{13}i$$

77. $x^2-2x+10=0$

$$x=\frac{-b\pm\sqrt{b^2-4ac}}{2a}$$
$$=\frac{-(-2)\pm\sqrt{(-2)^2-4(1)(10)}}{2(1)}$$
$$=\frac{2\pm\sqrt{-36}}{2}$$
$$=\frac{2\pm 6i}{2}$$
$$=1\pm 3i$$

79. $4x^2+4x+7=0$

$$x=\frac{-b\pm\sqrt{b^2-4ac}}{2a}$$
$$=\frac{-4\pm\sqrt{(4)^2-4(4)(7)}}{2(4)}$$
$$=\frac{-4\pm\sqrt{-96}}{8}$$
$$=\frac{-4\pm 4\sqrt{6}i}{8}$$
$$=-\frac{1}{2}\pm\frac{\sqrt{6}}{2}i$$

81. $5x^4-12x^3=0$

$$x^3(5x-12)=0$$
$$x^3=0 \text{ or } 5x-12=0$$
$$x=0 \text{ or } x=\tfrac{12}{5}$$

83. $x^3-7x^2+4x-28=0$

$$x^2(x-7)+4(x-7)=0$$
$$(x-7)(x^2+4)=0$$
$$x-7=0\Rightarrow x=7$$
$$x^2+4=0\Rightarrow x^2=-4$$
$$x=\pm\sqrt{-4}=\pm 2i$$

85. $x^6-7x^3-8=0$

$$(x^3)^2-7(x^3)-8=0$$
$$u^2-7u-8=0$$
$$(u-8)(u+1)=0$$
$$u-8=0$$
$$x^3-8=0$$
$$(x-2)(x^2+2x+4)=0$$
$$x-2=0\Rightarrow x=2$$
$$x^2+2x+4=0\Rightarrow x=-1\pm\sqrt{3}i$$
$$u+1=0$$
$$x^3+1=0$$
$$(x+1)(x^2-x+1)=0$$
$$x+1=0\Rightarrow x=-1$$
$$x^2-x+1=0\Rightarrow x=\frac{1}{2}\pm\frac{\sqrt{3}}{2}i$$

87. $\sqrt{2x+3}+\sqrt{x-2}=2$

$$\left(\sqrt{2x+3}\right)^2=\left(2-\sqrt{x-2}\right)^2$$
$$2x+3=4-4\sqrt{x-2}+x-2$$
$$x+1=-4\sqrt{x-2}$$
$$(x+1)^2=\left(-4\sqrt{x-2}\right)^2$$
$$x^2+2x+1=16(x-2)$$
$$x^2-14x+33=0$$
$$(x-3)(x-11)=0$$

$x=3$, extraneous or $x=11$, extraneous

No solution

89. $(x-1)^{2/3}-25=0$

$$(x-1)^{2/3}=25$$
$$(x-1)^2=25^3$$
$$x-1=\pm\sqrt{25^3}$$
$$x=1\pm 125$$
$$x=126 \text{ or } x=-124$$

91. $\dfrac{5}{x}=1+\dfrac{3}{x+2}$

$$5(x+2)=1(x)(x+2)+3x$$
$$5x+10=x^2+2x+3x$$
$$10=x^2$$
$$\pm\sqrt{10}=x$$

93. $|x - 5| = 10$

$x - 5 = -10$ or $x - 5 = 10$

$x = -5 \qquad x = 15$

95. $|x^2 - 3| = 2x$

$x^2 - 3 = 2x$ or $x^2 - 3 = -2x$

$x^2 - 2x - 3 = 0 \qquad x^2 + 2x - 3 = 0$

$(x - 3)(x + 1) = 0 \qquad (x + 3)(x - 1) = 0$

$x = 3$ or $x = -1 \qquad x = -3$ or $x = 1$

The only solutions of the original equation are $x = 3$ or $x = 1$. ($x = 3$ and $x = -1$ are extraneous.)

97.

$$29.95 = 42 - \sqrt{0.001x + 2}$$
$$-12.05 = -\sqrt{0.001x + 2}$$
$$\sqrt{0.001x + 2} = 12.05$$
$$0.001x + 2 = 145.2025$$
$$0.001x = 143.2025$$
$$x = 143{,}202.5$$
$$\approx 143{,}203 \text{ units}$$

99. Interval: $(-7, 2]$

Inequality: $-7 < x \le 2$

The interval is bounded.

101. Interval: $(-\infty, -10]$

Inequality: $x \le -10$

The interval is unbounded.

103. $3(x + 2) + 7 < 2x - 5$

$3x + 6 + 7 < 2x - 5$

$3x + 13 < 2x - 5$

$x < -18$

$(-\infty\ -18)$

−20 −19 −18 −17 −16 x

105. $4(5 - 2x) \le \frac{1}{2}(8 - x)$

$20 - 8x \le 4 - \frac{1}{2}x$

$-\frac{15}{2}x \le -16$

$x \ge \frac{32}{15}$

$\left[\frac{32}{15}, \infty\right)$

$\frac{32}{15}$ 1 2 3 4 5 x

107. $|x + 6| < 5$

$-5 < x + 6 < 5$

$-11 < x < -1$

$(-11, -1)$

−11 −1 −12 −10 −8 −6 −4 −2 0 x

109. $125.33x > 92x + 1200$

$33.33x > 1200$

$x > 36$ units

So, the smallest value of x for which the product returns a profit is 37 units.

111. $x^2 - 6x - 27 < 0$

$(x + 3)(x - 9) < 0$

Key numbers: $x = -3, x = 9$

Test intervals: $(-\infty, -3), (-3, 9), (9, \infty)$

Test: Is $(x + 3)(x - 9) < 0$?

By testing an x-value in each test interval in the inequality, we see that the solution set is $(-3, 9)$.

−3 9 −4 −2 0 2 4 6 8 10 x

113. $5x^3 - 45x < 0$

$5x(x^2 - 9) < 0$

$5x(x + 3)(x - 3) < 0$

Key numbers: $x = \pm 3, x = 0$

Test intervals: $(-\infty, -3), (-3, 0), (0, 3), (3, \infty)$

Test: Is $5x(x + 3)(x - 3) < 0$?

By testing an x-value in each test interval in the inequality, the solution set is $(-\infty, -3) \cup (0, 3)$.

−5 −4 −3 −2 −1 0 1 2 3 4 5 x

115.

$$\frac{2}{x + 1} \le \frac{3}{x - 1}$$
$$\frac{2(x - 1) - 3(x + 1)}{(x + 1)(x - 1)} \le 0$$
$$\frac{2x - 2 - 3x - 3}{(x + 1)(x - 1)} \le 0$$
$$\frac{-(x + 5)}{(x + 1)(x - 1)} \le 0$$

Key numbers: $x = -5, x = -1, x = 1$

Test intervals: $(-5, -1), (-1, 1), (1, \infty)$

Test: Is $\frac{-(x + 5)}{(x + 1)(x - 1)} \le 0$?

By testing an x-value in each test interval in the inequality, we see that the solution set is $[-5, -1) \cup (1, \infty)$.

−6 −5 −4 −3 −2 −1 0 1 2 x

117. $5000(1 + r)^2 > 5500$

$$(1 + r)^2 > 1.1$$
$$1 + r > 1.0488$$
$$r > 0.0488$$
$$r > 4.9\%$$

119. False.

$$\sqrt{-18}\sqrt{-2} = (\sqrt{18}i)(\sqrt{2}i) = \sqrt{36}i^2 = -6$$
$$\sqrt{(-8)(-2)} = \sqrt{36} = 6$$

121. Rational equations, equations involving radicals, and absolute value equations, may have "solutions" that are extraneous. So checking solutions, in the original equations, is crucial to eliminate these extraneous values.

Problem Solving for Chapter 1

1.

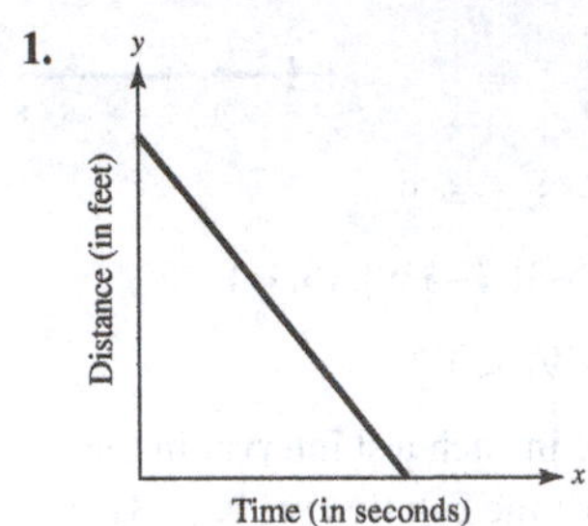

3. (a) $A = \pi ab$

$a + b = 20 \Rightarrow b = 20 - a$, thus:

$A = \pi a(20 - a)$

(b)

a	4	7	10	13	16
A	64π	91π	100π	91π	64π

(c)

$$300 = \pi a(20 - a)$$
$$300 = 20\pi a - \pi a^2$$
$$\pi a^2 - 20\pi a + 300 = 0$$
$$a = \frac{20\pi \pm \sqrt{(-20\pi)^2 - 4\pi(300)}}{2\pi}$$
$$= \frac{20\pi \pm \sqrt{400\pi^2 - 1200\pi}}{2\pi}$$
$$= \frac{20\pi \pm 20\sqrt{\pi(\pi - 3)}}{2\pi}$$
$$= 10 \pm \frac{10}{\pi}\sqrt{\pi(\pi - 3)}$$

$a \approx 12.12$ or $a \approx 7.88$

(d)

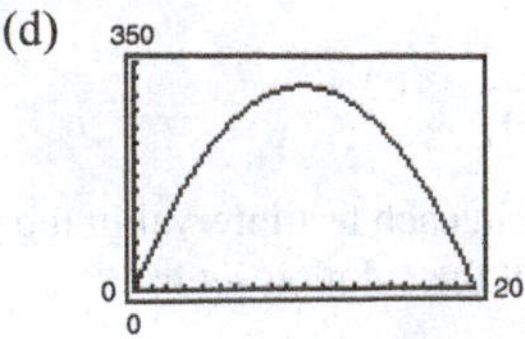

(e) The a-intercepts occur at $a = 0$ and $a = 20$. Both yield an area of 0. When $a = 0$, $b = 20$ and you have a vertical line of length 40. Likewise when $a = 20$, $b = 0$ and you have a horizontal line of length 40. They represent the minimum and maximum values of a.

(f) The maximum value of A is $100\pi \approx 314.159$. This occurs when $a = b = 10$ and the ellipse is actually a circle.

5. $P = 0.00256s^2$

(a)
$$0.00256s^2 = 20$$
$$s^2 = 7812.5$$
$$s \approx 88.4 \text{ miles per hour}$$

(b)
$$0.00256s^2 = 40$$
$$s^2 = 15625$$
$$s = 125 \text{ miles per hour}$$

No, actually it can survive wind blowing at $\sqrt{2}$ times the speed found in part (a).

(c) The wind speed in the formula is squared, so a small increase in wind speed could have potentially serious effects on a building.

7. (a) If $x^2 + 9 = (x + m)(x + n)$ then

$mn = 9$ and $m + n = 0$.

(b) $m + n = 0 \Rightarrow n = -m$

$m(-m) = 9 \Rightarrow -m^2 = 9 \Rightarrow m^2 = -9$

There is no **integer** m such that m^2 equals a negative number. $x^2 + 9$ cannot be factored over the integers.

9. (a) 5, 12, and 13; 8, 15, and 17

7, 24, and 25

(b) $5 \cdot 12 \cdot 13 = 780$ which is divisible by 3, 4, and 5.

$8 \cdot 15 \cdot 17 = 2040$ which is divisible by 3, 4, and 5.

$7 \cdot 24 \cdot 25 = 4200$ which is also divisible by 3, 4, and 5.

(c) Conjecture: If $a^2 + b^2 = c^2$ where a, b, and c are positive integers, then abc is divisible by 60.

11. (a) $S = \dfrac{-b + \sqrt{b^2 - 4ac}}{2a} + \dfrac{-b - \sqrt{b^2 - 4ac}}{2a} = \dfrac{-2a}{2a} = -\dfrac{b}{a}$

(b) $P = \left(\dfrac{-b + \sqrt{b^2 - 4ac}}{2a}\right)\left(\dfrac{-b - \sqrt{b^2 - 4ac}}{2a}\right) = \dfrac{b^2 - (b^2 - 4ac)}{4a^2} = \dfrac{4ac}{4a^2} = \dfrac{c}{a}$

13. (a) $z_m = \dfrac{1}{z} = \dfrac{1}{1+i} = \dfrac{1}{1+i} \cdot \dfrac{1-i}{1-i} = \dfrac{1-i}{2} = \dfrac{1}{2} - \dfrac{1}{2}i$

(b) $z_m = \dfrac{1}{z} = \dfrac{1}{3-i} = \dfrac{1}{3-i} \cdot \dfrac{3+i}{3+i} = \dfrac{3+i}{10} = \dfrac{3}{10} + \dfrac{1}{10}i$

(c) $z_m = \dfrac{1}{z} = \dfrac{1}{-2+8i} = \dfrac{1}{-2+8i} \cdot \dfrac{-2-8i}{-2-8i} = \dfrac{-2-8i}{68} = -\dfrac{1}{34} - \dfrac{2}{17}i$

15. (a) $c = 1$

The terms are: $i, -1 + i, -i, -1 + i, -i, -1 + i, -i, -1 + i, -i, \ldots$

The sequence is bounded so $c = i$ *is* in the Mandelbrot Set.

(b) $c = -2$

The terms are: $1 + i, 1 + 3i, -7 + 7i, 1 - 97i, -9407 - 193li, \ldots$

The sequence is unbounded so $c = 1 + i$ *is not* in the Mandelbrot Set.

(c) $c = -2$

The terms are: $-2, 2, 2, 2, 2, \ldots$

The sequence is bounded so $c = -2$ *is* in the Mandelbrot Set.

Practice Test for Chapter 1

1. Graph $3x - 5y = 15$.

2. Graph $y = \sqrt{9 - x}$.

3. Solve $5x + 4 = 7x - 8$.

4. Solve $\frac{x}{3} - 5 = \frac{x}{5} + 1$.

5. Solve $\frac{3x + 1}{6x - 7} = \frac{2}{5}$.

6. Solve $(x - 3)^2 + 4 = (x + 1)^2$.

7. Solve $A = \frac{1}{2}(a + b)h$ for a.

8. 301 is what percent of 4300?

9. Cindy has \$6.05 in quarters and nickels. How many of each coin does she have if there are 53 coins in all?

10. Ed has \$15,000 invested in two fund paying $9\frac{1}{2}\%$ and 11% simple interest, respectively. How much is invested in each if the yearly interest is \$1582.50?

11. Solve $28 + 5x - 3x^2 = 0$ by factoring.

12. Solve $(x - 2)^2 = 24$ by taking the square root of both sides.

13. Solve $x^2 - 4x - 9 = 0$ by completing the square.

14. Solve $x^2 + 5x - 1 = 0$ by the Quadratic Formula.

15. Solve $3x^2 - 2x + 4 = 0$ by the Quadratic Formula.

16. The perimeter of a rectangle is 1100 feet. Find the dimensions so that the enclosed area will be 60,000 square feet.

17. Find two consecutive even positive integers whose product is 624.

18. Solve $x^3 - 10x^2 + 24x = 0$ by factoring.

19. Solve $\sqrt[3]{6 - x} = 4$.

20. Solve $(x^2 - 8)^{2/5} = 4$.

21. Solve $x^4 - x^2 - 12 = 0$.

22. Solve $4 - 3x > 16$.

23. Solve $\left|\frac{x - 3}{2}\right| < 5$.

24. Solve $\frac{x + 1}{x - 3} < 2$.

25. Solve $|3x - 4| \geq 9$.

CHAPTER 2
Functions and Their Graphs

CHAPTER 2
Functions and Their Graphs

Section 2.1 Linear Equations in Two Variables

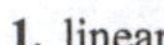

1. linear

3. point-slope

5. perpendicular

7. linear extrapolation

9. (a) $m = \frac{2}{3}$. Because the slope is positive, the line rises. Matches L_2.

   (b) m is undefined. The line is vertical. Matches L_3.

   (c) $m = -2$. The line falls. Matches L_1.

11.

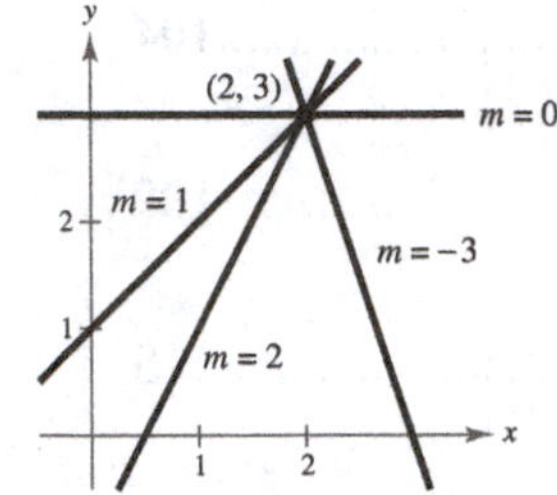

13. Two points on the line: $(0, 0)$ and $(4, 6)$

$$\text{Slope} = \frac{y_2 - y_1}{x_2 - x_1} = \frac{6}{4} = \frac{3}{2}$$

15. $y = 5x + 3$

    Slope: $m = 5$

    y-intercept: $(0, 3)$

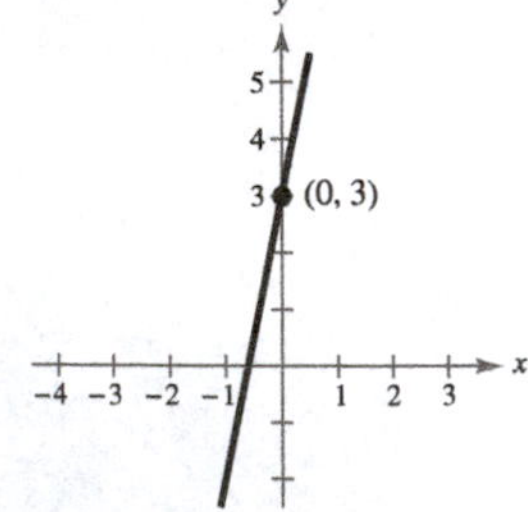

17. $y = -\frac{3}{4}x - 1$

    Slope: $m = -\frac{3}{4}$

    y-intercept: $(0, -1)$

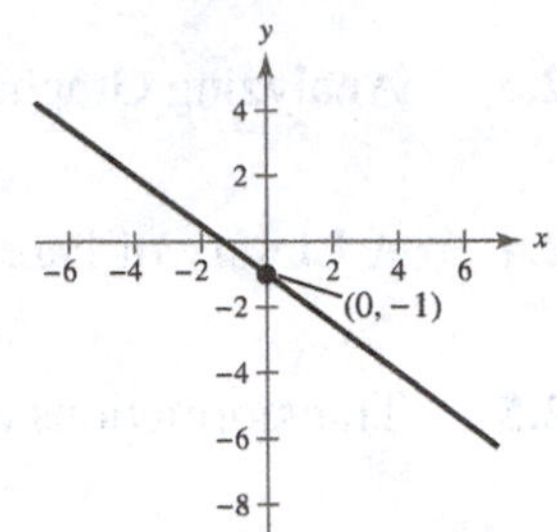

19. $y - 5 = 0$

    $y = 5$

    Slope: $m = 0$

    y-intercept: $(0, 5)$

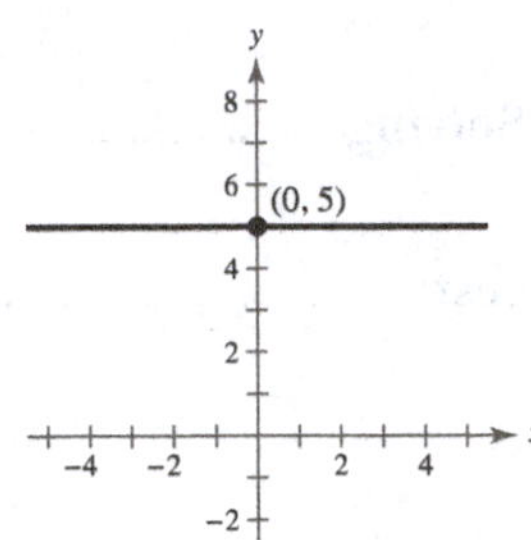

21. $5x - 2 = 0$

    $x = \frac{2}{5}$, vertical line

    Slope: undefined

    y-intercept: none

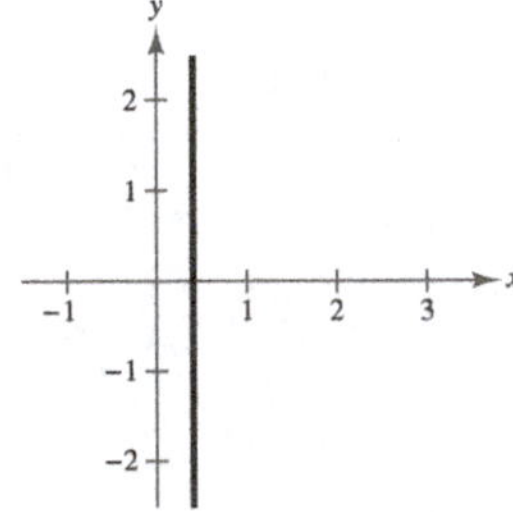

23. $7x - 6y = 30$

$$-6y = -7x + 30$$

$$y = \frac{7}{6}x - 5$$

Slope: $m = \frac{7}{6}$

y-intercept: $(0, -5)$

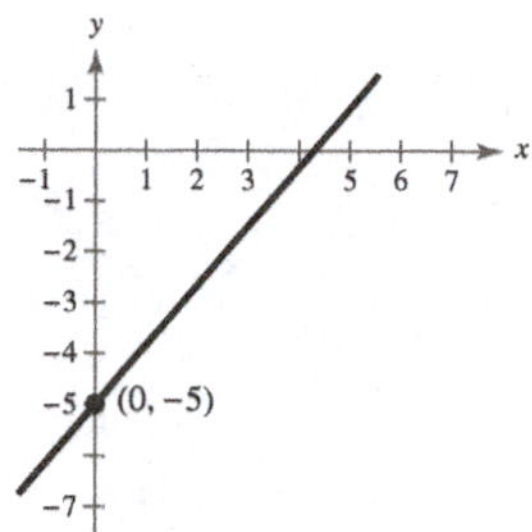

25. $m = \frac{0 - 9}{6 - 0} = \frac{-9}{6} = -\frac{3}{2}$

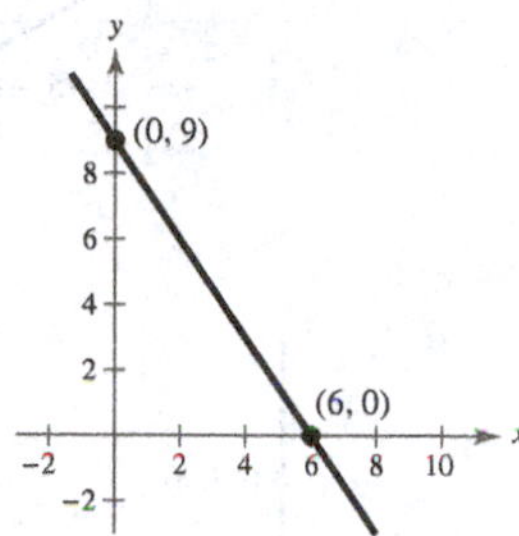

27. $m = \frac{6 - (-2)}{1 - (-3)} = \frac{8}{4} = 2$

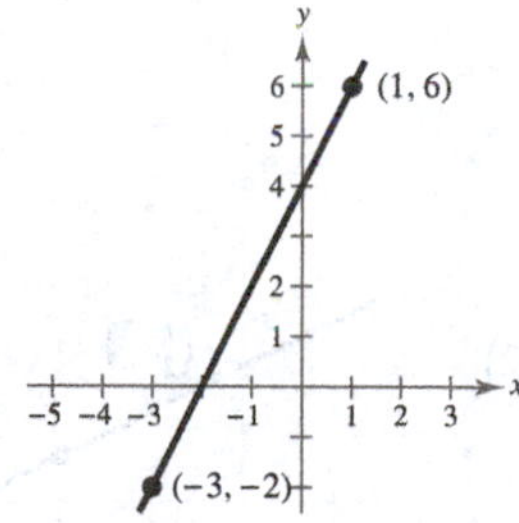

29. $m = \frac{-7 - (-7)}{8 - 5} = \frac{0}{3} = 0$

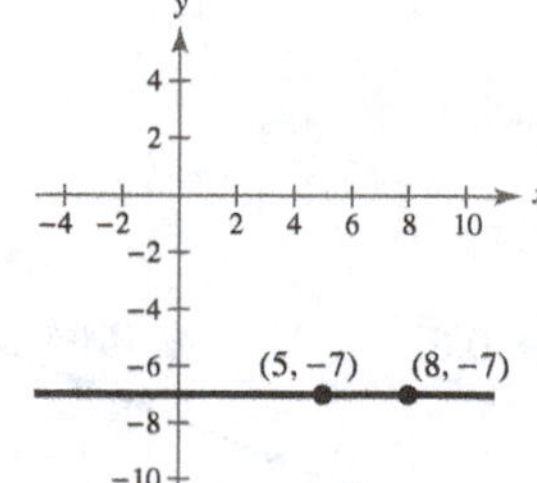

31. $m = \frac{4 - (-1)}{-6 - (-6)} = \frac{5}{0}$

m is undefined.

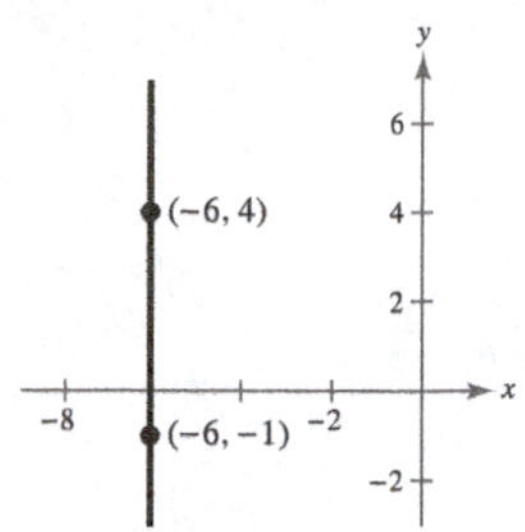

33. $m = \frac{1.6 - 3.1}{-5.2 - 4.8} = \frac{-1.5}{-10} = 0.15$

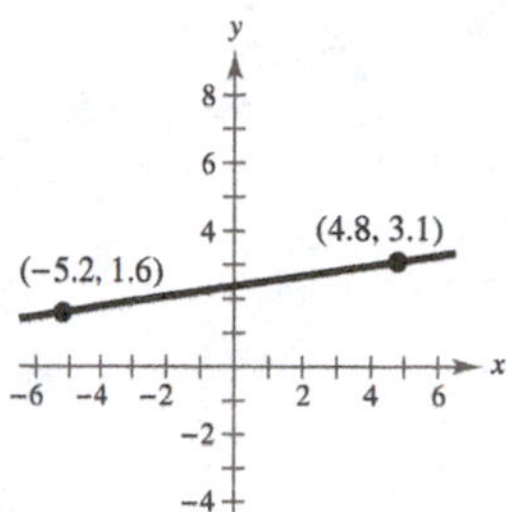

35. Point: $(5, 7)$, Slope: $m = 0$

Because $m = 0$, y does not change. Three other points are $(-1, 7)$, $(0, 7)$, and $(4, 7)$.

37. Point: $(-5, 4)$, Slope: $m = 2$

Because $m = 2 = \frac{2}{1}$, y increases by 2 for every one unit increase in x. Three additional points are $(-4, 6)$, $(-3, 8)$, and $(-2, 10)$.

39. Point: $(4, 5)$, Slope: $m = -\frac{1}{3}$

Because $m = -\frac{1}{3}$, y decreases by 1 unit for every three unit increase in x. Three additional points are $(-2, 7)$, $\left(0, -\frac{19}{4}\right)$, and $(1, 6)$.

41. Point: $(-4, 3)$, Slope is undefined.

Because m is undefined, x does not change. Three points are $(-4, 0)$, $(-4, 5)$, and $(-4, 2)$.

43. Point: $(0, -2)$; $m = 3$

$$y + 2 = 3(x - 0)$$

$$y = 3x - 2$$

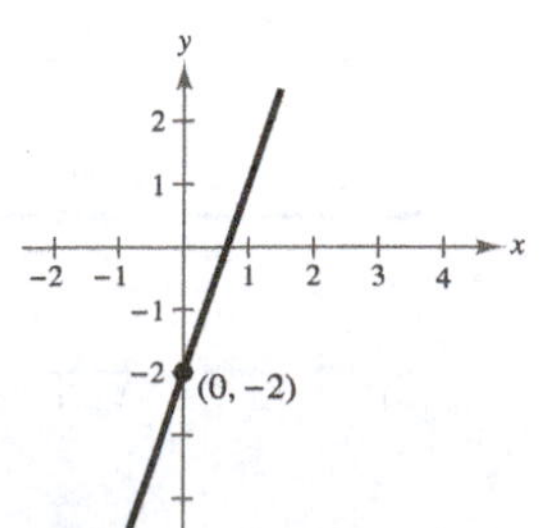

45. Point: $(-3, 6)$; $m = -2$

$$y - 6 = -2(x + 3)$$
$$y = -2x$$

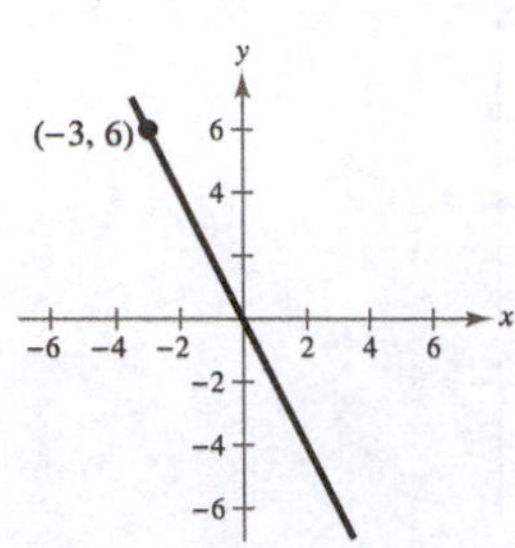

47. Point: $(4, 0)$; $m = -\frac{1}{3}$

$$y - 0 = -\tfrac{1}{3}(x - 4)$$
$$y = -\tfrac{1}{3}x + \tfrac{4}{3}$$

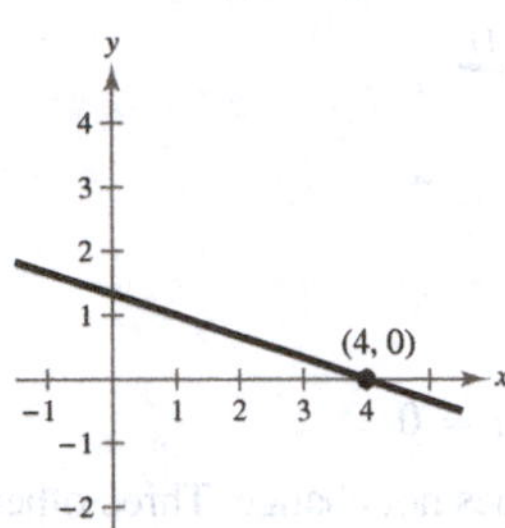

49. Point: $(2, -3)$; $m = -\dfrac{1}{2}$

$$y - (-3) = -\frac{1}{2}(x - 2)$$
$$y + 3 = -\frac{1}{2}x + 1$$
$$y = -\frac{1}{2}x - 2$$

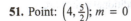

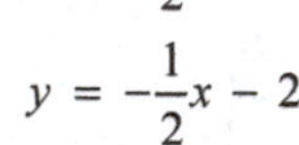

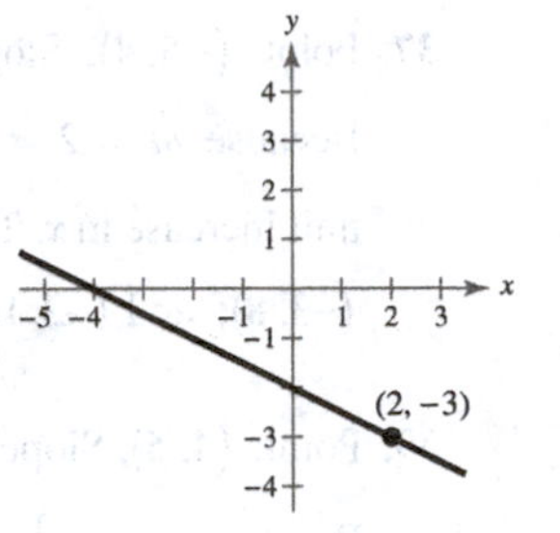

51. Point: $\left(4, \frac{5}{2}\right)$; $m = 0$

$$y - \tfrac{5}{2} = 0(x - 4)$$
$$y - \tfrac{5}{2} = 0$$
$$y = \tfrac{5}{2}$$

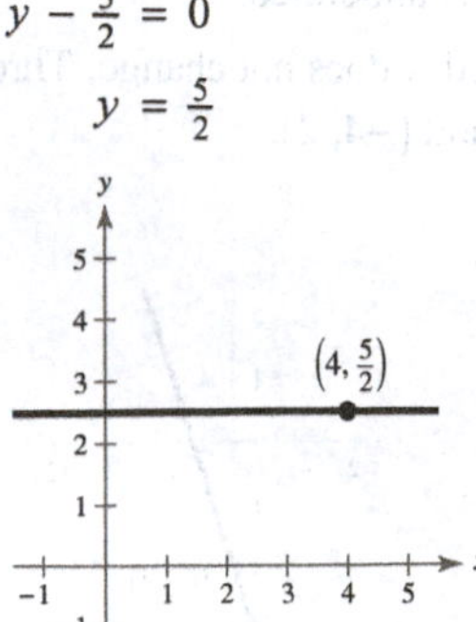

53. Point: $(-5.1, 1.8)$; $m = 5$

$$y - 1.8 = 5(x - (5.1))$$
$$y = 5x + 27.3$$

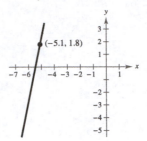

55. $(5, -1)$, $(-5, 5)$

$$y + 1 = \frac{5 + 1}{-5 - 5}(x - 5)$$
$$y = -\frac{3}{5}(x - 5) - 1$$
$$y = -\frac{3}{5}x + 2$$

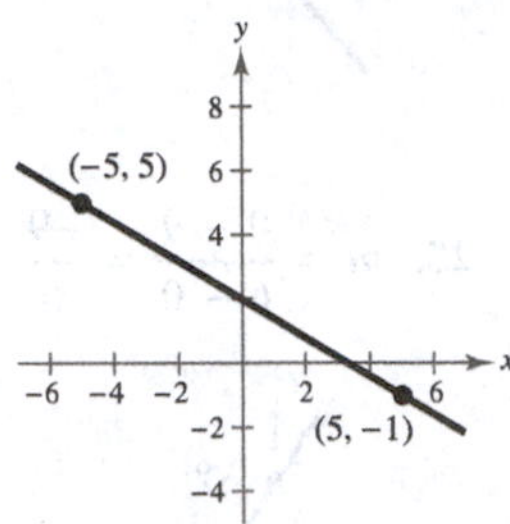

57. $(-7, 2)$, $(-7, 5)$

$$m = \frac{5 - 2}{-7 - (-7)} = \frac{3}{0}$$

m is undefined.

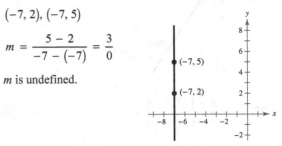

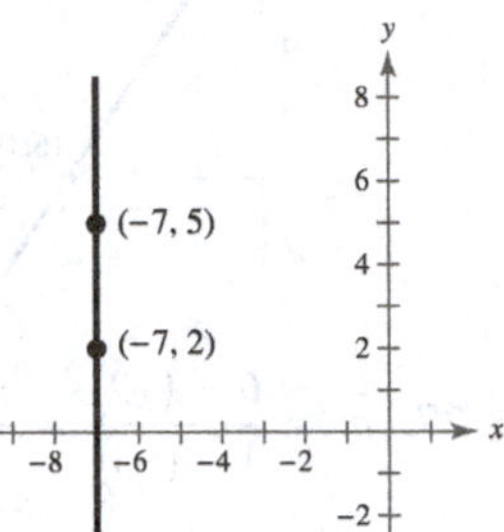

59. $\left(2, \dfrac{1}{2}\right), \left(\dfrac{1}{2}, \dfrac{5}{4}\right)$

$$y - \frac{1}{2} = \frac{\frac{5}{4} - \frac{1}{2}}{\frac{1}{2} - 2}(x - 2)$$
$$y = -\frac{1}{2}(x - 2) + \frac{1}{2}$$
$$y = -\frac{1}{2}x + \frac{3}{2}$$

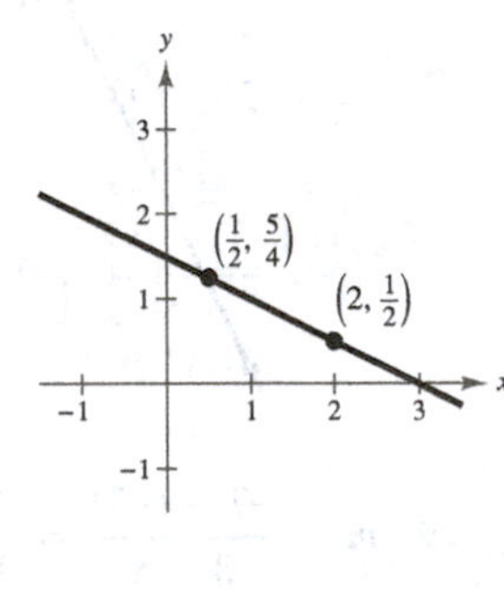

61. $(1, 0.6)$, $(-2, -0.6)$

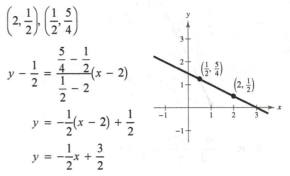

$$y - 0.6 = \frac{-0.6 - 0.6}{-2 - 1}(x - 1)$$
$$y = 0.4(x - 1) + 0.6$$
$$y = 0.4x + 0.2$$

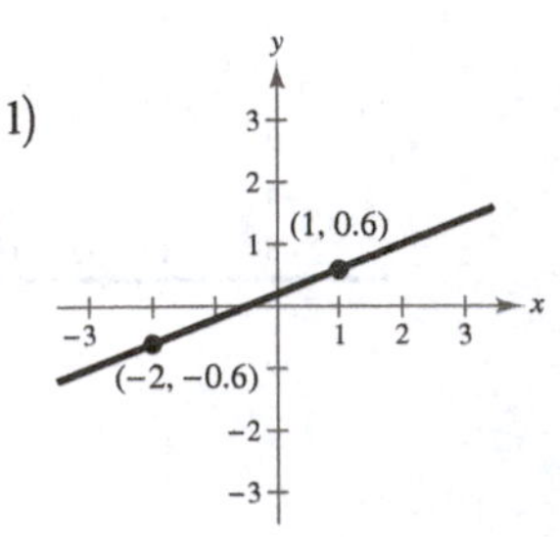

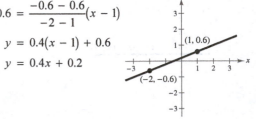

63. $(2, -1), \left(\dfrac{1}{3}, -1\right)$

$$y + 1 = \frac{-1-(-1)}{\frac{1}{3}-2}(x-2)$$

$$y + 1 = 0$$

$$y = -1$$

The line is horizontal.

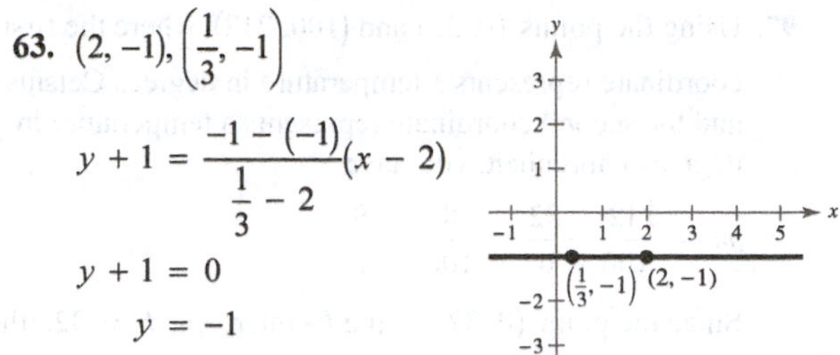

65. $L_1: y = -\frac{2}{3}x - 3$

$$m_1 = -\frac{2}{3}$$

$L_2: y = -\frac{2}{3}x - 1$

$$m_2 = -\frac{2}{3}$$

The slopes are equal, so the lines are parallel.

67. $L_1: y = \frac{1}{2}x - 3$

$$m_1 = \frac{1}{2}$$

$L_2: y = -\frac{1}{2}x + 1$

$$m_2 = -\frac{1}{2}$$

The lines are neither parallel nor perpendicular.

69. $L_1: (0, -1), (5, 9)$

$$m_1 = \frac{9+1}{5-0} = 2$$

$L_2: (0, 3), (4, 1)$

$$m_2 = \frac{1-3}{4-0} = -\frac{1}{2}$$

The slopes are negative reciprocals, so the lines are perpendicular.

71. $L_1: (-6, -3), (2, -3)$

$$m_1 = \frac{-3-(-3)}{2-(-6)} = \frac{0}{8} = 0$$

$L_2: \left(3, -\frac{1}{2}\right), \left(6, -\frac{1}{2}\right)$

$$m_2 = \frac{-\frac{1}{2}-\left(-\frac{1}{2}\right)}{6-3} = \frac{0}{3} = 0$$

L_1 and L_2 are both horizontal lines, so they are parallel.

73. $4x - 2y = 3$

$$y = 2x - \frac{3}{2}$$

Slope: $m = 2$

(a) $(2, 1), m = 2$

$$y - 1 = 2(x - 2)$$

$$y = 2x - 3$$

(b) $(2, 1), m = -\frac{1}{2}$

$$y - 1 = -\frac{1}{2}(x - 2)$$

$$y = -\frac{1}{2}x + 2$$

75. $3x + 4y = 7$

$$y = -\frac{3}{4}x + \frac{7}{4}$$

Slope: $m = -\frac{3}{4}$

(a) $\left(-\frac{2}{3}, \frac{7}{8}\right), m = -\frac{3}{4}$

$$y - \frac{7}{8} = -\frac{3}{4}\left(x - \left(-\frac{2}{3}\right)\right)$$

$$y = -\frac{3}{4}x + \frac{3}{8}$$

(b) $\left(-\frac{2}{3}, \frac{7}{8}\right), m = \frac{4}{3}$

$$y - \frac{7}{8} = \frac{4}{3}\left(x - \left(-\frac{2}{3}\right)\right)$$

$$y = \frac{4}{3}x + \frac{127}{72}$$

77. $y + 5 = 0$

$$y = -5$$

Slope: $m = 0$

(a) $(-2, 4), m = 0$

$$y = 4$$

(b) $(-2, 4)$, m is undefined.

$$x = -2$$

79. $x - y = 4$

$$y = x - 4$$

Slope: $m = 1$

(a) $(2.5, 6.8), m = 1$

$$y - 6.8 = 1(x - 2.5)$$

$$y = x + 4.3$$

(b) $(2.5, 6.8), m = -1$

$$y - 6.8 = (-1)(x - 2.5)$$

$$y = -x + 9.3$$

81. $\frac{x}{3} + \frac{y}{5} = 1$

$(15)\left(\frac{x}{3} + \frac{y}{5}\right) = 1(15)$

$5x + 3y - 15 = 0$

83. $\frac{x}{-1/6} + \frac{y}{-2/3} = 1$

$6x + \frac{3}{2}y = -1$

$12x + 3y + 2 = 0$

85. $\frac{x}{c} + \frac{y}{c} = 1, c \neq 0$

$x + y = c$

$1 + 2 = c$

$3 = c$

$x + y = 3$

$x + y - 3 = 0$

87. (a) $m = 135$. The sales are increasing 135 units per year.

(b) $m = 0$. There is no change in sales during the year.

(c) $m = -40$. The sales are decreasing 40 units per year.

89. $y = \frac{6}{100}x$

$y = \frac{6}{100}(200) = 12$ feet

91. $(16, 3000), m = -150$

$V - 3000 = -150(t - 16)$

$V - 3000 = -150t + 2400$

$V = -150t + 5400, 16 \leq t \leq 21$

93. The C-intercept measures the fixed costs of manufacturing when zero bags are produced.

The slope measures the cost to produce one laptop bag.

95. Using the points $(0, 875)$ and $(5, 0)$, where the first coordinate represents the year t and the second coordinate represents the value V, you have

$m = \frac{0 - 875}{5 - 0} = -175$

$V = -175t + 875, 0 \leq t \leq 5.$

97. Using the points $(0, 32)$ and $(100, 212)$, where the first coordinate represents a temperature in degrees Celsius and the second coordinate represents a temperature in degrees Fahrenheit, you have

$m = \frac{212 - 32}{100 - 0} = \frac{180}{100} = \frac{9}{5}.$

Since the point $(0, 32)$ is the F- intercept, $b = 32$, the equation is $F = 1.8C + 32$ or $C = \frac{5}{9}F - \frac{160}{9}$.

99. (a) Total Cost = cost for fuel and maintainance + cost for operator + purchase cost

$C = 9.5t + 11.5t + 42{,}000$

$C = 21t + 42{,}000$

(b) Revenue = Rate per hour · Hours

$R = 45t$

(c) $P = R - C$

$P = 45t - (21t + 42{,}000)$

$P = 24t - 42{,}000$

(d) Let $P = 0$, and solve for t.

$0 = 24t - 42{,}000$

$42{,}000 = 24t$

$1750 = t$

The equipment must be used 1750 hours to yield a profit of 0 dollars.

101. False. The slope with the greatest magnitude corresponds to the steepest line.

103. Find the slope of the line segments between the points A and B, and B and C.

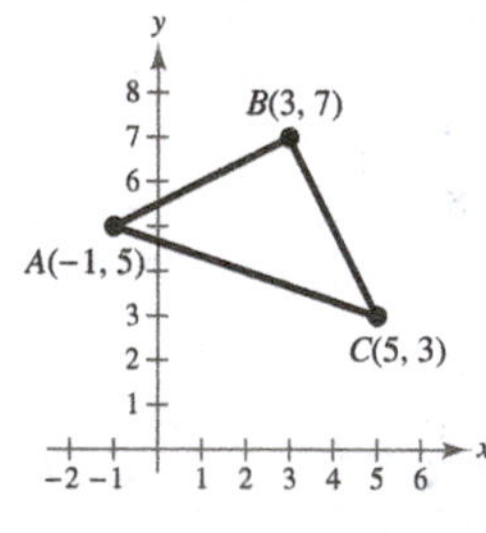

$m_{AB} = \frac{7 - 5}{3 - (-1)} = \frac{2}{4} = \frac{1}{2}$

$m_{BC} = \frac{3 - 7}{5 - 3} = \frac{-4}{2} = -2$

Since the slopes are negative reciprocals, the line segments are perpendicular and therefore intersect to form a right angle. So, the triangle is a right triangle.

105. Since the scales for the y-axis on each graph is unknown, the slopes of the lines cannot be determined.

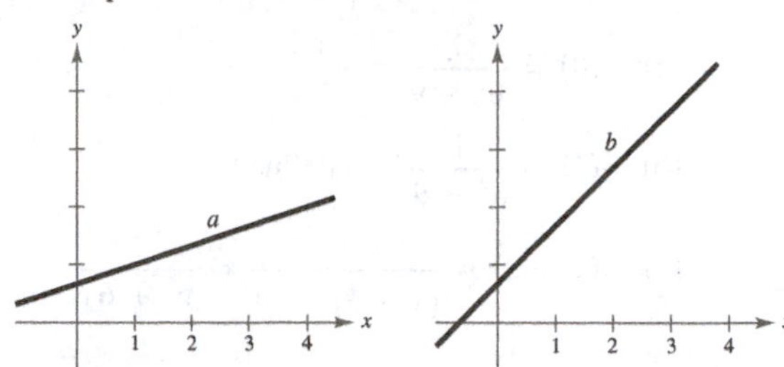

107. No, the slopes of two perpendicular lines have opposite signs. (Assume that neither line is vertical or horizontal.)

109. The line $y = 4x$ rises most quickly.

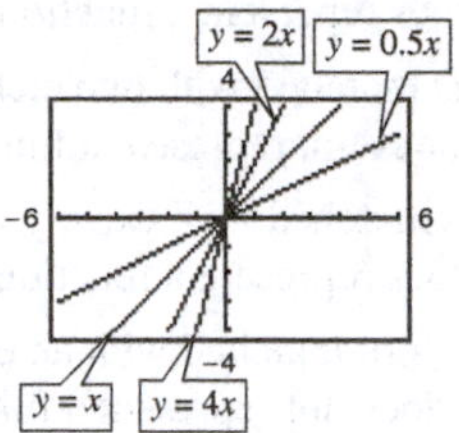

The line $y = -4x$ falls most quickly.

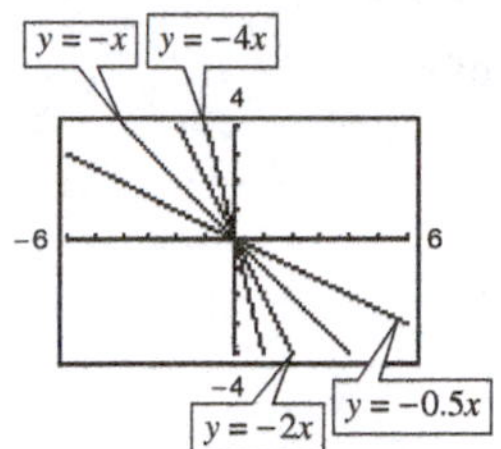

The greater the magnitude of the slope (the absolute value of the slope), the faster the line rises or falls.

111. Set the distance between $(4, -1)$ and (x, y) equal to the distance between $(-2, 3)$ and (x, y).

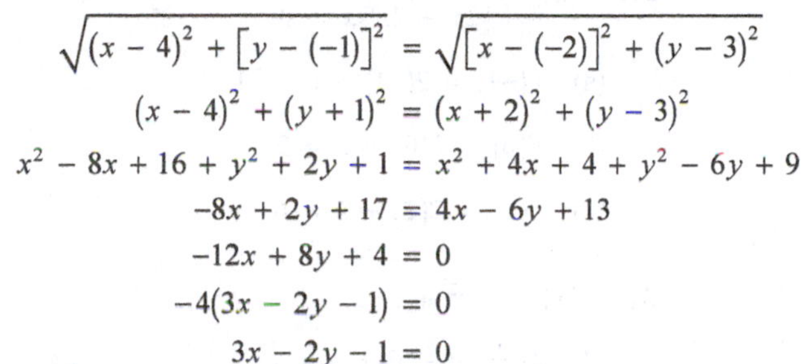

$$\begin{aligned}
\sqrt{(x-4)^2 + [y-(-1)]^2} &= \sqrt{[x-(-2)]^2 + (y-3)^2} \\
(x-4)^2 + (y+1)^2 &= (x+2)^2 + (y-3)^2 \\
x^2 - 8x + 16 + y^2 + 2y + 1 &= x^2 + 4x + 4 + y^2 - 6y + 9 \\
-8x + 2y + 17 &= 4x - 6y + 13 \\
-12x + 8y + 4 &= 0 \\
-4(3x - 2y - 1) &= 0 \\
3x - 2y - 1 &= 0
\end{aligned}$$

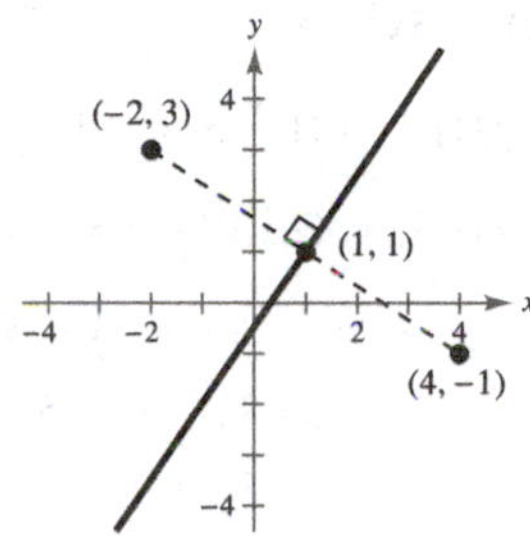

This line is the perpendicular bisector of the line segment connecting $(4, -1)$ and $(-2, 3)$.

113. Set the distance between $\left(3, \frac{5}{2}\right)$ and (x, y) equal to the distance between $(-7, 1)$ and (x, y).

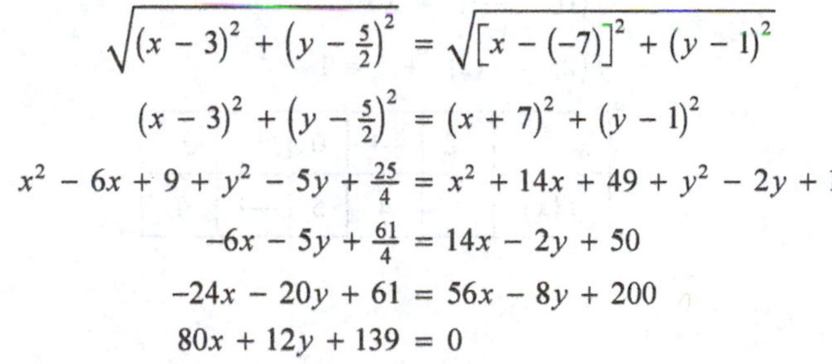

$$\begin{aligned}
\sqrt{(x-3)^2 + \left(y-\tfrac{5}{2}\right)^2} &= \sqrt{[x-(-7)]^2 + (y-1)^2} \\
(x-3)^2 + \left(y-\tfrac{5}{2}\right)^2 &= (x+7)^2 + (y-1)^2 \\
x^2 - 6x + 9 + y^2 - 5y + \tfrac{25}{4} &= x^2 + 14x + 49 + y^2 - 2y + 1 \\
-6x - 5y + \tfrac{61}{4} &= 14x - 2y + 50 \\
-24x - 20y + 61 &= 56x - 8y + 200 \\
80x + 12y + 139 &= 0
\end{aligned}$$

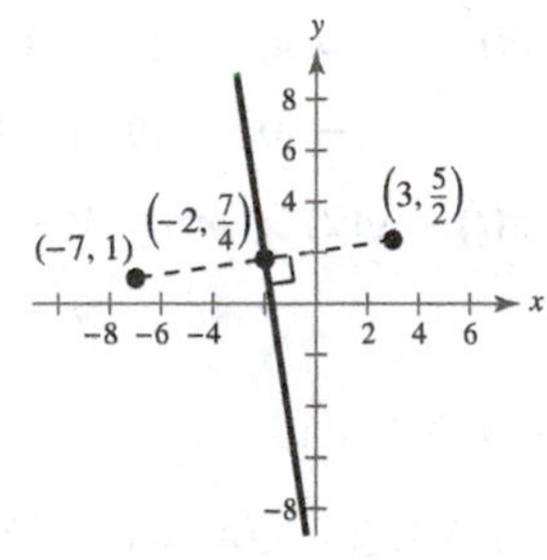

This line is the perpendicular bisector of the line segment connecting $\left(3, \frac{5}{2}\right)$ and $(-7, 1)$.

Section 2.2 Functions

1. domain; range; function

3. implied domain

5. Yes, the relationship is a function. Each domain value is matched with exactly one range value.

7. No, it does not represent a function. The input values of 10 and 7 are each matched with two output values.

9. (a) Each element of A is matched with exactly one element of B, so it does represent a function.

(b) The element 1 in A is matched with two elements, –2 and 1 of B, so it does not represent a function.

(c) Each element of A is matched with exactly one element of B, so it does represent a function.

(d) The element 2 in A is not matched with an element of B, so the relation does not represent a function.

11. $x^2 + y^2 = 4 \Rightarrow y = \pm\sqrt{4 - x^2}$

No, y *is not* a function of x.

13. $y = \sqrt{16 - x^2}$

Yes, y *is* a function of x.

15. $y = 4 - |x|$

Yes, y *is* a function of x.

17. $y = -75$ or $y = -75 + 0x$

Yes, y *is* a function of x.

19. $f(x) = 3x - 5$

(a) $f(1) = 3(1) - 5 = -2$

(b) $f(-3) = 3(-3) - 5 = -14$

(c) $f(x + 2) = 3(x + 2) - 5$
$= 3x + 6 - 5$
$= 3x + 1$

21. $g(t) = 4t^2 - 3t + 5$

(a) $g(2) = 4(2)^2 - 3(2) + 5$
$= 15$

(b) $g(t - 2) = 4(t - 2)^2 - 3(t - 2) + 5$
$= 4t^2 - 19t + 27$

(c) $g(t) - g(2) = 4t^2 - 3t + 5 - 15$
$= 4t^2 - 3t - 10$

23. $f(y) = 3 - \sqrt{y}$

(a) $f(4) = 3 - \sqrt{4} = 1$

(b) $f(0.25) = 3 - \sqrt{0.25} = 2.5$

(c) $f(4x^2) = 3 - \sqrt{4x^2} = 3 - 2|x|$

25. $q(x) = \dfrac{1}{x^2 - 9}$

(a) $q(0) = \dfrac{1}{0^2 - 9} = -\dfrac{1}{9}$

(b) $q(3) = \dfrac{1}{3^2 - 9}$ is undefined.

(c) $q(y + 3) = \dfrac{1}{(y + 3)^2 - 9} = \dfrac{1}{y^2 + 6y}$

27. $f(x) = \dfrac{|x|}{x}$

(a) $f(2) = \dfrac{|2|}{2} = 1$

(b) $f(-2) = \dfrac{|-2|}{-2} = -1$

(c) $f(x - 1) = \dfrac{|x - 1|}{x - 1} = \begin{cases} -1, & \text{if } x < 1 \\ 1, & \text{if } x > 1 \end{cases}$

29. $f(x) = \begin{cases} 2x + 1, & x < 0 \\ 2x + 2, & x \ge 0 \end{cases}$

(a) $f(-1) = 2(-1) + 1 = -1$

(b) $f(0) = 2(0) + 2 = 2$

(c) $f(2) = 2(2) + 2 = 6$

31. $f(x) = -x^2 + 5$

$f(-2) = -(-2)^2 + 5 = 1$

$f(-1) = -(-1)^2 + 5 = 4$

$f(0) = -(0)^2 + 5 = 5$

$f(1) = -(1)^2 + 5 = 4$

$f(2) = -(2)^2 + 5 = 1$

x	–2	–1	0	1	2
$f(x)$	1	4	5	–4	1

33. $f(x) = \begin{cases} -\frac{1}{2}x + 4, & x \le 0 \\ (x-2)^2, & x > 0 \end{cases}$

$f(-2) = -\frac{1}{2}(-2) + 4 = 5$

$f(-1) = -\frac{1}{2}(-1) + 4 = 4\frac{1}{2} = \frac{9}{2}$

$f(0) = -\frac{1}{2}(0) + 4 = 4$

$f(1) = (1-2)^2 = 1$

$f(2) = (2-2)^2 = 0$

x	-2	-1	0	1	2
$f(x)$	5	$\frac{9}{2}$	4	1	0

35. $15 - 3x = 0$

$3x = 15$

$x = 5$

37. $\dfrac{3x-4}{5} = 0$

$3x - 4 = 0$

$x = \dfrac{4}{3}$

39. $f(x) = x^2 - 81$

$x^2 - 81 = 0$

$x^2 = 81$

$x = \pm 9$

41. $x^3 - x = 0$

$x(x^2 - 1) = 0$

$x(x+1)(x-1) = 0$

$x = 0, x = -1$, or $x = 1$

43. $f(x) = g(x)$

$x^2 = x + 2$

$x^2 - x - 2 = 0$

$(x-2)(x+1) = 0$

$x - 2 = 0 \quad x + 1 = 0$

$x = 2 \quad x = -1$

45. $f(x) = g(x)$

$x^4 - 2x^2 = 2x^2$

$x^4 - 4x^2 = 0$

$x^2(x^2 - 4) = 0$

$x^2(x+2)(x-2) = 0$

$x^2 = 0 \Rightarrow x = 0$

$x + 2 = 0 \Rightarrow x = -2$

$x - 2 = 0 \Rightarrow x = 2$

47. $f(x) = 5x^2 + 2x - 1$

Because $f(x)$ is a polynomial, the domain is all real numbers x.

49. $g(y) = \sqrt{y+6}$

Domain: $y + 6 \ge 0$

$y \ge -6$

The domain is all real numbers y such that $y \ge -6$.

51. $g(x) = \dfrac{1}{x} - \dfrac{3}{x+2}$

The domain is all real numbers x except $x = 0, x = -2$.

53. $f(s) = \dfrac{\sqrt{s-1}}{s-4}$

Domain: $s - 1 \ge 0 \Rightarrow s \ge 1$ and $s \ne 4$

The domain consists of all real numbers s, such that $s \ge 1$ and $s \ne 4$.

55. $f(x) = \dfrac{x-4}{\sqrt{x}}$

The domain is all real numbers x such that $x > 0$ or $(0, \infty)$.

57. (a)

Height, x	Volume, V
1	484
2	800
3	972
4	1024
5	980
6	864

The volume is maximum when $x = 4$ and $V = 1024$ cubic centimeters.

(b)

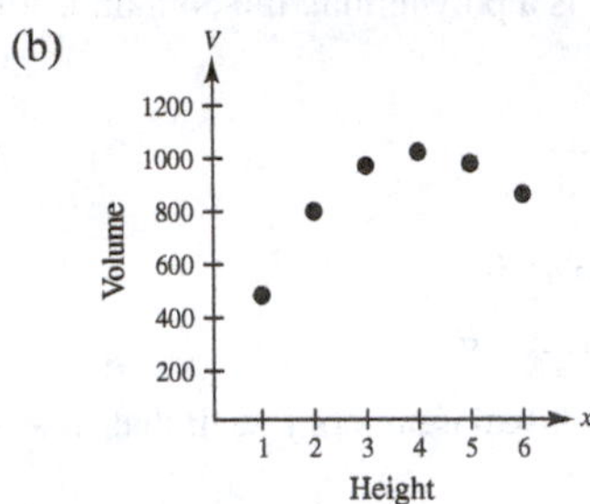

V is a function of x.

(c) $V = x(24 - 2x)^2$

Domain: $0 < x < 12$

59. $A = s^2$ and $P = 4s \Rightarrow \dfrac{P}{4} = s$

$$A = \left(\frac{P}{4}\right)^2 = \frac{P^2}{16}$$

61. $y = -\frac{1}{10}x^2 + 3x + 6$

$y(25) = -\frac{1}{10}(25)^2 + 3(25) + 6 = 18.5$ feet

If the child holds a glove at a height of 5 feet, then the ball *will* be over the child's head because it will be at a height of 18.5 feet.

63. $A = \dfrac{1}{2}bh = \dfrac{1}{2}xy$

Because $(0, y)$, $(2, 1)$, and $(x, 0)$ all lie on the same line, the slopes between any pair are equal.

$$\frac{1 - y}{2 - 0} = \frac{0 - 1}{x - 2}$$

$$\frac{1 - y}{2} = \frac{-1}{x - 2}$$

$$y = \frac{2}{x - 2} + 1$$

$$y = \frac{x}{x - 2}$$

So, $A = \dfrac{1}{2}x\left(\dfrac{x}{x - 2}\right) = \dfrac{x^2}{2(x - 2)}$.

The domain of A includes x-values such that $x^2/[2(x - 2)] > 0$. By solving this inequality, the domain is $x > 2$.

65. For 2008 through 2011, use

$$p(t) = 2.77t + 45.2.$$

2008: $p(8) = 2.77(8) + 45.2 = 67.36\%$

2009: $p(9) = 2.77(9) + 45.2 = 70.13\%$

2010: $p(10) = 2.77(10) + 45.2 = 72.90\%$

2011: $p(11) = 2.77(11) + 45.2 = 75.67\%$

For 2011 through 2014, use

$$p(t) = 1.95t + 55.9.$$

2012: $p(12) = 1.95(12) + 55.9 = 79.30\%$

2013: $p(13) = 1.95(13) + 55.9 = 81.25\%$

2014: $p(14) = 1.95(14) + 55.9 = 83.20\%$

67. (a) Cost = variable costs + fixed costs

$C = 12.30x + 98{,}000$

(b) Revenue = price per unit × number of units

$R = 17.98x$

(c) Profit = Revenue − Cost

$P = 17.98x - (12.30x + 98{,}000)$

$P = 5.68x - 98{,}000$

69. (a)

(b) $(3000)^2 + h^2 = d^2$

$$h = \sqrt{d^2 - (3000)^2}$$

Domain: $d \geq 3000$ (because both $d \geq 0$ and $d^2 - (3000)^2 \geq 0$)

71. (a) $R = n(\text{rate}) = n[8.00 - 0.05(n - 80)], n \geq 80$

$$R = 12.00n - 0.05n^2 = 12n - \frac{n^2}{20} = \frac{240n - n^2}{20}, n \geq 80$$

(b)

n	90	100	110	120	130	140	150
$R(n)$	\$675	\$700	\$715	\$720	\$715	\$700	\$675

The revenue is maximum when 120 people take the trip.

73.

$$f(x) = x^2 - 2x + 4$$
$$f(2 + h) = (2 + h)^2 - 2(2 + h) + 4 = 4 + 4h + h^2 - 4 - 2h + 4 = h^2 + 2h + 4$$
$$f(2) = (2)^2 - 2(2) + 4 = 4$$
$$f(2 + h) - f(2) = h^2 + 2h$$
$$\frac{f(2 + h) - f(2)}{h} = \frac{h^2 + 2h}{h} = h + 2, h \neq 0$$

75.

$$f(x) = x^3 + 3x$$
$$f(x + h) = (x + h)^3 + 3(x + h)$$
$$= x^3 + 3x^2h + 3xh^2 + h^3 + 3x + 3h$$
$$\frac{f(x + h) - f(x)}{h} = \frac{(x^3 + 3x^2h + 3xh^2 + h^3 + 3x + 3h) - (x^3 + 3x)}{h}$$
$$= \frac{h(3x^2 + 3xh + h^2 + 3)}{h}$$
$$= 3x^2 + 3xh + h^2 + 3, h \neq 0$$

77.

$$g(x) = \frac{1}{x^2}$$
$$\frac{g(x) - g(3)}{x - 3} = \frac{\frac{1}{x^2} - \frac{1}{9}}{x - 3} = \frac{9 - x^2}{9x^2(x - 3)} = \frac{-(x + 3)(x - 3)}{9x^2(x - 3)} = -\frac{x + 3}{9x^2}, x \neq 3$$

79. $f(x) = \sqrt{5x}$

$$\frac{f(x) - f(5)}{x - 5} = \frac{\sqrt{5x} - 5}{x - 5}, x \neq 5$$

81. By plotting the points, we have a parabola, so $g(x) = cx^2$. Because $(-4, -32)$ is on the graph, you have $-32 = c(-4)^2 \Rightarrow c = -2$. So, $g(x) = -2x^2$.

83. Because the function is undefined at 0, we have $r(x) = c/x$. Because $(-4, -8)$ is on the graph, you have $-8 = c/-4 \Rightarrow c = 32$. So, $r(x) = 32/x$.

85. False. The equation $y^2 = x^2 + 4$ is a relation between x and y. However, $y = \pm\sqrt{x^2 + 4}$ does not represent a function.

87. False. The range is $[-1, \infty)$.

89. The domain of $f(x) = \sqrt{x - 1}$ includes $x = 1$, $x \geq 1$ and the domain of $g(x) = \frac{1}{\sqrt{x - 1}}$ does not include $x = 1$ because you cannot divide by 0. The domain of $g(x) = \frac{1}{\sqrt{x - 1}}$ is $x > 1$. So, the functions do not have the same domain.

91. No; x is the independent variable, f is the name of the function.

93. (a) Yes. The amount that you pay in sales tax will increase as the price of the item purchased increases.

(b) No. The length of time that you study the night before an exam does not necessarily determine your score on the exam.

Section 2.3 Analyzing Graphs of Functions

1. Vertical Line Test

3. decreasing

5. average rate of change; secant

7. Domain: $(-2, 2]$; Range: $[-1, 8]$

(a) $f(-1) = -1$

(b) $f(0) = 0$

(c) $f(1) = -1$

(d) $f(2) = 8$

9. Domain: $(-\infty, \infty)$; Range: $(-2, \infty)$

(a) $f(2) = 0$

(b) $f(1) = 1$

(c) $f(3) = 2$

(d) $f(-1) = 3$

11. A vertical line intersects the graph at most once, so y *is* a function of x.

13. A vertical line intersects the graph more than once, so y *is not* a function of x.

15.
$$\begin{aligned} f(x) &= 3x + 18 \\ 3x + 18 &= 0 \\ 3x &= -18 \\ x &= -6 \end{aligned}$$

17.
$$\begin{aligned} f(x) &= 2x^2 - 7x - 30 \\ 2x^2 - 7x - 30 &= 0 \\ (2x + 5)(x - 6) &= 0 \end{aligned}$$
$$2x + 5 = 0 \quad \text{or} \quad x - 6 = 0$$
$$x = -\tfrac{5}{2} \qquad\qquad x = 6$$

19.
$$f(x) = \frac{x + 3}{2x^2 - 6}$$
$$\frac{x + 3}{2x^2 - 6} = 0$$
$$x + 3 = 0$$
$$x = -3$$

21.
$$\begin{aligned} f(x) &= \tfrac{1}{3}x^3 - 2x \\ \tfrac{1}{3}x^3 - 2x &= 0 \\ (3)\left(\tfrac{1}{3}x^3 - 2x\right) &= 0(3) \\ x^3 - 6x &= 0 \\ x\left(x^2 - 6\right) &= 0 \end{aligned}$$
$$x = 0 \quad \text{or} \quad x^2 - 6 = 0$$
$$x^2 = 6$$
$$x = \pm\sqrt{6}$$

23.
$$\begin{aligned} f(x) &= x^3 - 4x^2 - 9x + 36 \\ x^3 - 4x^2 - 9x + 36 &= 0 \\ x^2(x - 4) - 9(x - 4) &= 0 \\ (x - 4)\left(x^2 - 9\right) &= 0 \end{aligned}$$
$$x - 4 = 0 \Rightarrow x = 4$$
$$x^2 - 9 = 0 \Rightarrow x = \pm 3$$

25.
$$\begin{aligned} f(x) &= \sqrt{2x} - 1 \\ \sqrt{2x} - 1 &= 0 \\ \sqrt{2x} &= 1 \\ 2x &= 1 \\ x &= \tfrac{1}{2} \end{aligned}$$

27. (a)

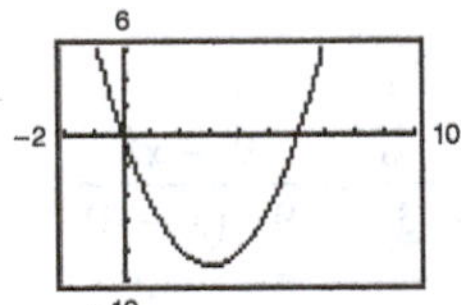

Zeros: $x = 0, 6$

(b)
$$\begin{aligned} f(x) &= x^2 - 6x \\ x^2 - 6x &= 0 \\ x(x - 6) &= 0 \end{aligned}$$
$$x = 0 \Rightarrow x = 0$$
$$x - 6 = 0 \Rightarrow x = 6$$

29. (a)

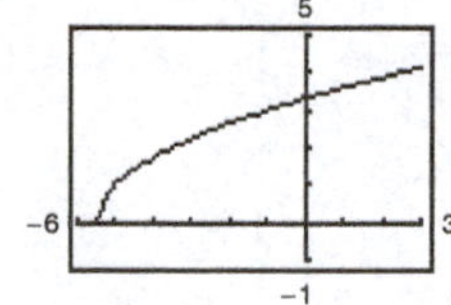

Zero: $x = -5.5$

(b) $$f(x) = \sqrt{2x + 11}$$
$$\sqrt{2x + 11} = 0$$
$$2x + 11 = 0$$
$$x = -\frac{11}{2}$$

31. (a)

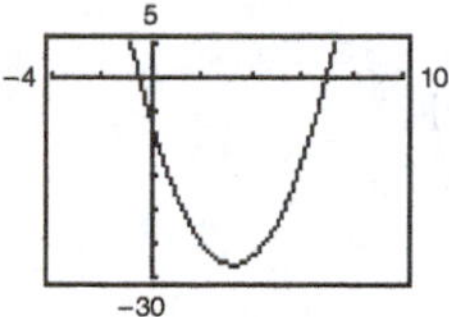

Zero: $x = 0.3333$

(b) $$f(x) = \frac{3x - 1}{x - 6}$$
$$\frac{3x - 1}{x - 6} = 0$$
$$3x - 1 = 0$$
$$x = \frac{1}{3}$$

33. $f(x) = -\frac{1}{2}x^3$

The function is decreasing on $(-\infty, \infty)$.

35. $f(x) = \sqrt{x^2 - 1}$

The function is decreasing on $(-\infty, -1)$ and increasing on $(1, \infty)$.

37. $f(x) = |x + 1| + |x - 1|$

The function is increasing on $(1, \infty)$.

The function is constant on $(-1, 1)$.

The function is decreasing on $(-\infty, -1)$.

39. $f(x) = \begin{cases} 2x + 1, & x \le -1 \\ x^2 - 2, & x > -1 \end{cases}$

The function is decreasing on $(-1, 0)$ and increasing on $(-\infty, -1)$ and $(0, \infty)$.

41. $f(x) = 3$

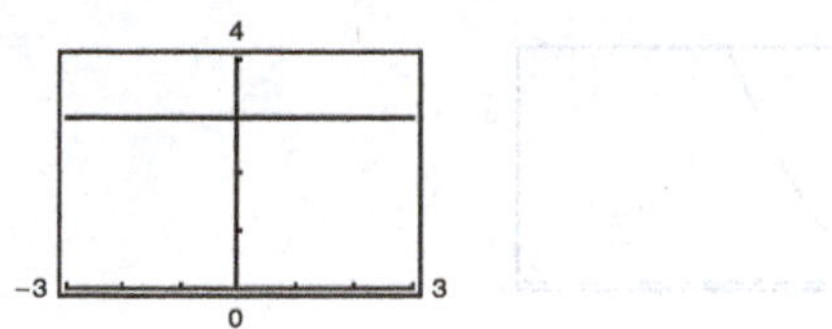

Constant on $(-\infty, \infty)$

x	-2	-1	0	1	2
$f(x)$	3	3	3	3	3

43. $g(x) = \frac{1}{2}x^2 - 3$ b

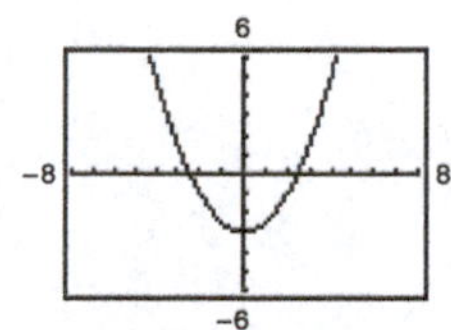

Decreasing on $(-\infty, 0)$.

Increasing on b

x	-2	-1	0	1	2
$g(x)$	-1	$-\frac{5}{2}$	-3	$-\frac{5}{2}$	-1

45. $f(x) = \sqrt{1 - x}$

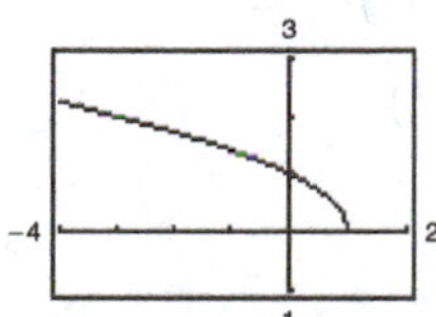

Decreasing on $(-\infty, 1)$

x	-3	-2	-1	0	1
$f(x)$	2	$\sqrt{3}$	$\sqrt{2}$	1	0

47. $f(x) = x^{3/2}$

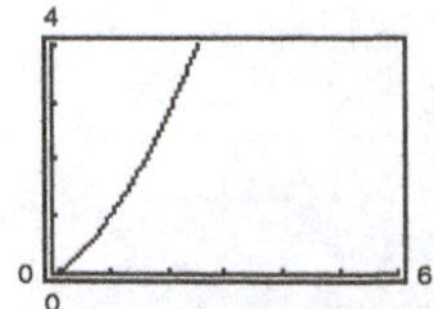

Increasing on $(0, \infty)$

x	0	1	2	3	4
$f(x)$	0	1	2.8	5.2	8

49. $f(x) = x(x + 3)$

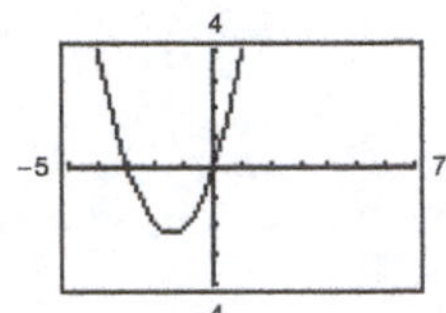

Relative minimum: $(-1.5, -2.25)$

51. $h(x) = x^3 - 6x^2 + 15$

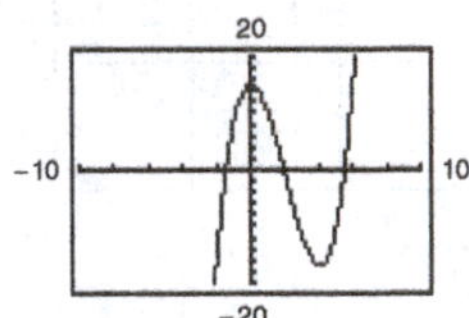

Relative minimum: $(4, -17)$

Relative maximum: $(0, 15)$

53. $h(x) = (x - 1)\sqrt{x}$

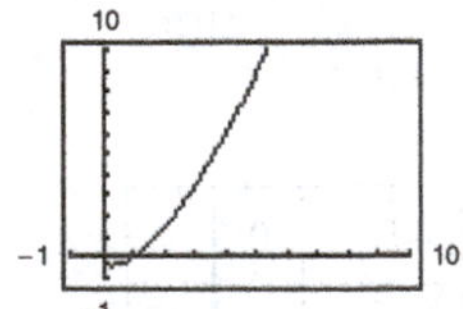

Relative minimum: $(0.33, -0.38)$

55. $f(x) = 4 - x$

$f(x) \geq 0$ on $(-\infty, 4]$

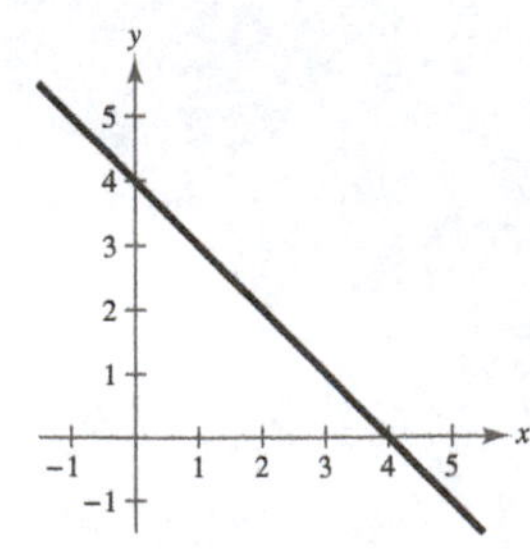

57. $f(x) = 9 - x^2$

$f(x) \geq 0$ on $[-3, 3]$

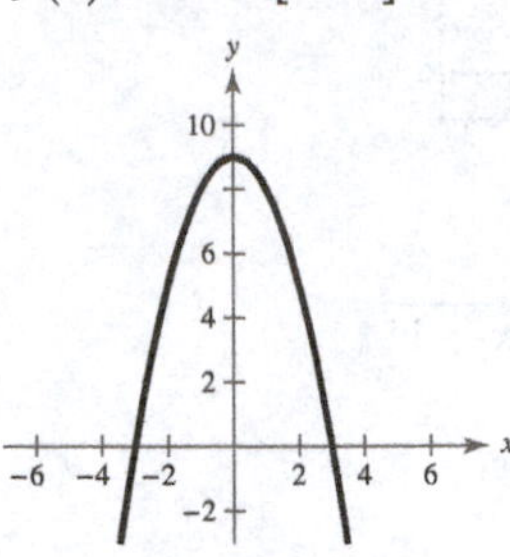

59. $f(x) = \sqrt{x - 1}$

$f(x) \geq 0$ on $[1, \infty)$

$\sqrt{x - 1} \geq 0$

$x - 1 \geq 0$

$x \geq 1$

$[1, \infty)$

61. $f(x) = -2x + 15$

$$\frac{f(3) - f(0)}{3 - 0} = \frac{9 - 15}{3} = -2$$

The average rate of change from $x_1 = 0$ to $x_2 = 3$ is -2.

63. $f(x) = x^3 - 3x^2 - x$

$$\frac{f(2) - f(-1)}{2 - (-1)} = \frac{-6 - (-3)}{3} = \frac{-3}{3} = -1$$

The average rate of change from $x_1 = -1$ to $x_2 = 2$ is -1.

65. (a)

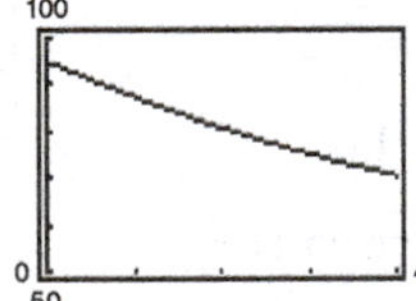

(b) To find the average rate of change of the amount the U.S. Department of Energy spent for research and development from 2010 to 2014, find the average rate of change from $(0, f(0))$ to $(4, f(4))$.

$$\frac{f(4) - f(0)}{4 - 0} = \frac{70.5344 - 95.08}{4} = \frac{-24.5456}{4} = -6.1364$$

The amount the U.S. Department of Energy spent on research and development for defense decreased by about $6.14 billion each year from 2010 to 2014.

67. $s_0 = 6, v_0 = 64$

(a) $s = -16t^2 + 64t + 6$

(b)

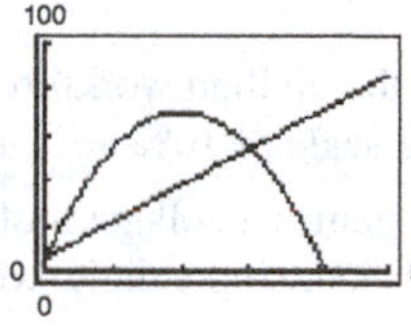

(c) $\frac{s(3) - s(0)}{3 - 0} = \frac{54 - 6}{3} = 16$

(d) The slope of the secant line is positive.

(e) $s(0) = 6, m = 16$

Secant line: $y - 6 = 16(t - 0)$

$y = 16t + 6$

(f)

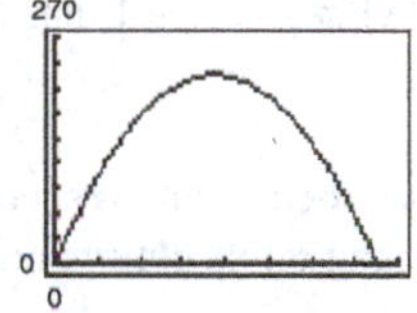

69. $v_0 = 120, s_0 = 0$

(a) $s = -16t^2 + 120t$

(b)

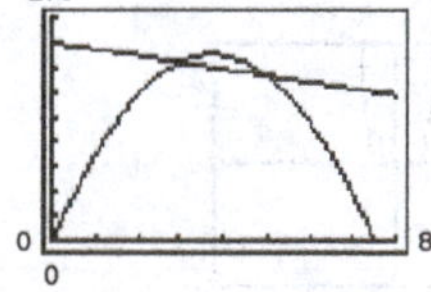

(c) The average rate of change from $t = 3$ to $t = 5$:

$\frac{s(5) - s(3)}{5 - 3} = \frac{200 - 216}{2} = -\frac{16}{2} = -8$ feet per second

(d) The slope of the secant line through $(3, s(3))$ and $(5, s(5))$ is negative.

(e) The equation of the secant line: $m = -8$

Using $(5, s(5)) = (5, 200)$ we have

$y - 200 = -8(t - 5)$

$y = -8t + 240.$

(f)

270
0
0
8

71. $f(x) = x^6 - 2x^2 + 3$

$f(-x) = (-x)^6 - 2(-x)^2 + 3$

$= x^6 - 2x^2 + 3$

$= f(x)$

The function is even. y-axis symmetry.

73. $h(x) = x\sqrt{x + 5}$

$h(-x) = (-x)\sqrt{-x + 5}$

$= -x\sqrt{5 - x}$

$\neq h(x)$

$\neq -h(x)$

The function is neither odd nor even. No symmetry.

75. $f(s) = 4s^{3/2}$

$= 4(-s)^{3/2}$

$\neq f(s)$

$\neq -f(s)$

The function is neither odd nor even. No symmetry.

77.

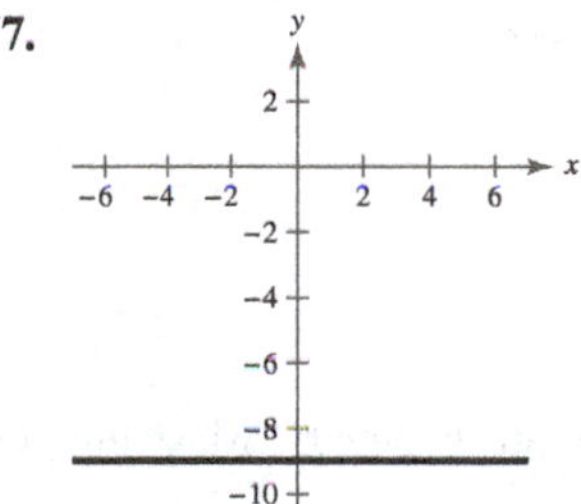

The graph of $f(x) = -9$ is symmetric to the y-axis, which implies $f(x)$ is even.

$f(-x) = -9$

$= f(x)$

The function is even.

79. $f(x) = -|x - 5|$

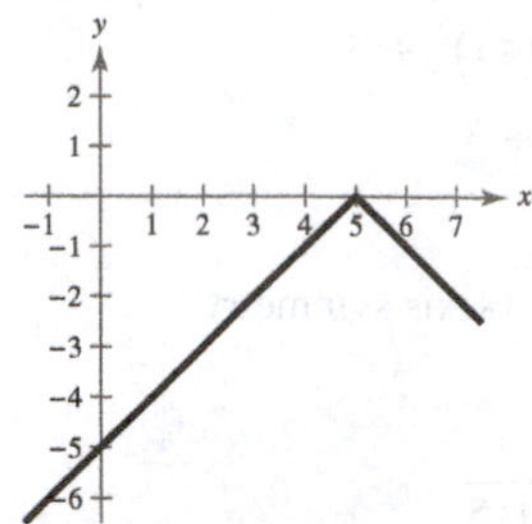

The graph displays no symmetry, which implies $f(x)$ is neither odd nor even.

$$\begin{aligned} f(x) &= -|(-x) - 5| \\ &= -|-x - 5| \\ &\neq f(x) \\ &\neq -f(x) \end{aligned}$$

The function is neither even nor odd.

81. $f(x) = \sqrt[3]{4x}$

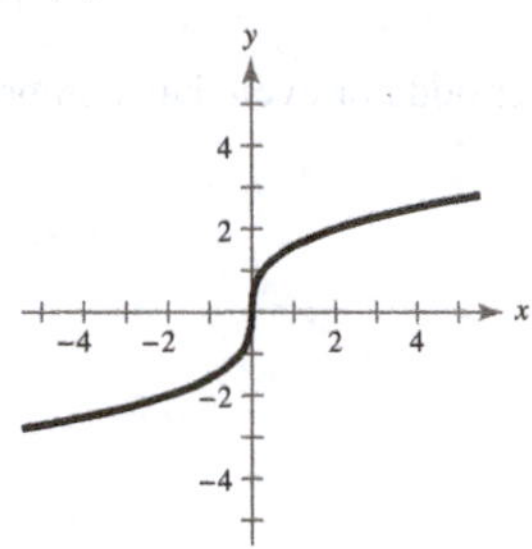

The graph displays origin symmetry, which implies $f(x)$ is odd.

$$f(-x) = \sqrt[3]{4(-x)} = \sqrt[3]{-4xz} = -\sqrt[3]{4x} = -f(x)$$

The function is odd.

83. $h = \text{top} - \text{bottom} = 3 - (4x - x^2) = 3 - 4x + x^2$

85. $L = \text{right} - \text{left} = 2 - \sqrt[3]{2y}$

87. The error is that $-2x^3 - 5 \neq -(2x^3 - 5)$. The correct process is as follows.

$$f(x) = 2x^3 - 5$$

$$f(-x) = 2(-x)^3 - 5 = -2x^3 - 5 = -(2x^3 + 5)$$

$f(-x) \neq -f(x)$ and $f(-x) \neq f(x)$, so the function $f(x) = 2x^3 - 5$ is neither odd nor even.

89. (a) For the average salary of college professors, a scale of \$10,000 would be appropriate.

(b) For the population of the United States, use a scale of 10,000,000.

(c) For the percent of the civilian workforce that is unemployed, use a scale of 10%.

(d) For the number of games a college football team wins in a single season, single digits would be appropriate.

For each of the graphs, using the suggested scale would show yearly changes in the data clearly.

91. False. The function $f(x) = \sqrt{x^2 + 1}$ has a domain of all real numbers.

93. True. A graph that is symmetric with respect to the y-axis cannot be increasing on its entire domain.

95. $\left(-\frac{5}{3}, -7\right)$

(a) If f is even, another point is $\left(\frac{5}{3}, -7\right)$.

(b) If f is odd, another point is $\left(\frac{5}{3}, 7\right)$.

97. (a) $y = x$

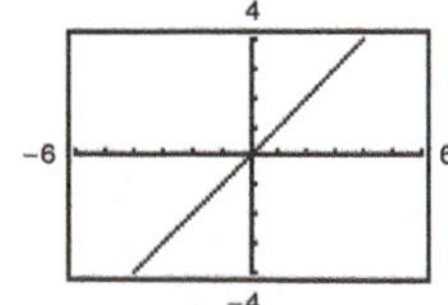

(b) $y = x^2$

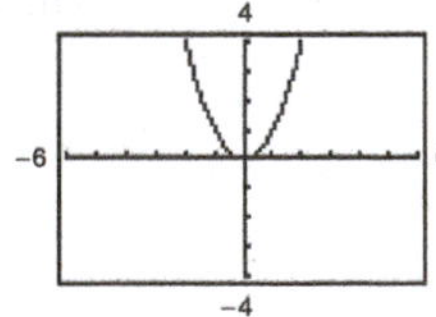

(c) $y = x^3$

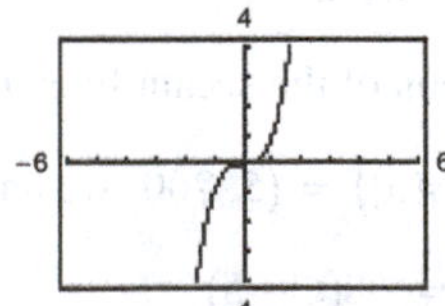

(d) $y = x^4$

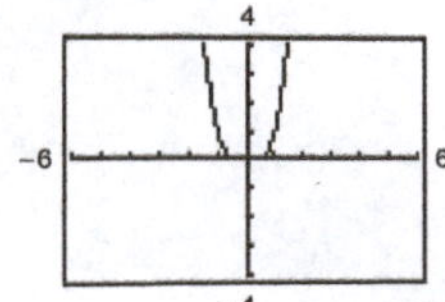

(e) $y = x^5$

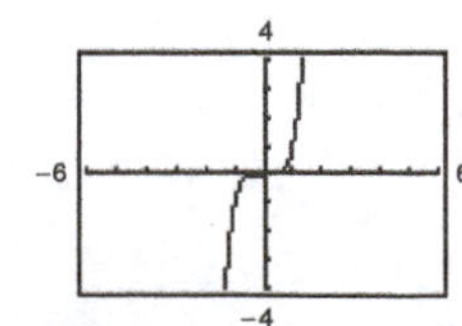

(f) $y = x^6$

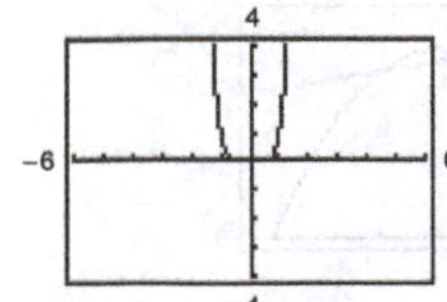

All the graphs pass through the origin. The graphs of the odd powers of x are symmetric with respect to the origin and the graphs of the even powers are symmetric with respect to the y-axis. As the powers increase, the graphs become flatter in the interval $-1 < x < 1$.

99. (a) Even. The graph is a reflection in the x-axis.
(b) Even. The graph is a reflection in the y-axis.
(c) Even. The graph is a vertical translation of f.
(d) Neither. The graph is a horizontal translation of f.

Section 2.4 A Library of Parent Functions

1. Greatest integer function

3. Reciprocal function

5. Square root function

7. Absolute value function

9. Linear function

11. (a) $f(1) = 4, f(0) = 6$

$(1, 4), (0, 6)$

$$m = \frac{6 - 4}{0 - 1} = -2$$

$$y - 6 = -2(x - 0)$$

$$y = -2x + 6$$

$$f(x) = -2x + 6$$

(b)

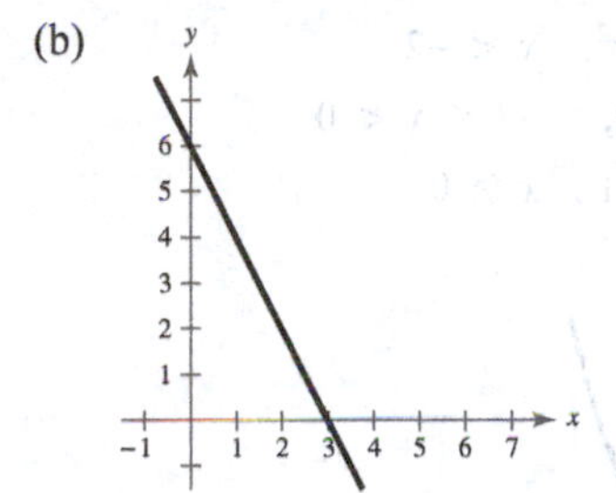

13. (a) $f\left(\frac{1}{2}\right) = -\frac{5}{3}, f(6) = 2$

$\left(\frac{1}{2}, -\frac{5}{3}\right), (6, 2)$

$$m = \frac{2 - \left(-\frac{5}{3}\right)}{6 - \left(\frac{1}{2}\right)}$$

$$= \frac{\frac{11}{3}}{\frac{11}{2}} = \left(\frac{11}{3}\right) \cdot \left(\frac{2}{11}\right) = \frac{2}{3}$$

(b)

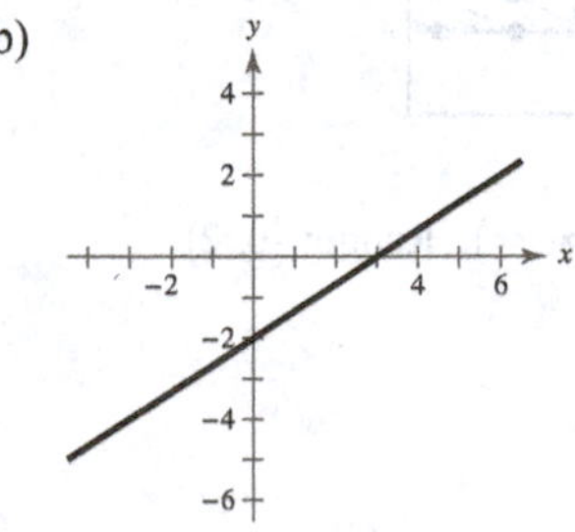

15. $f(x) = 2.5x - 4.25$

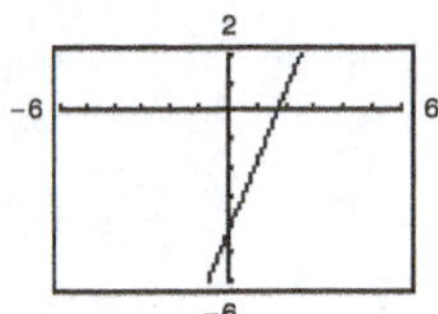

17. $g(x) = x^2 + 3$

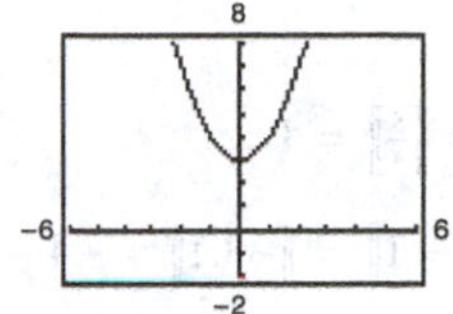

19. $f(x) = x^3 - 1$

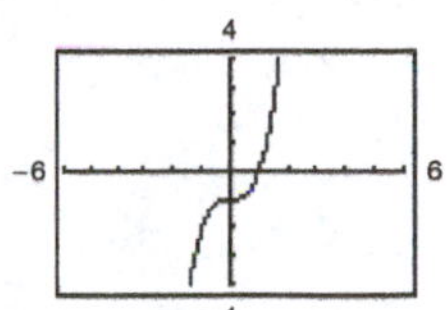

21. $f(x) = \sqrt{x} + 4$

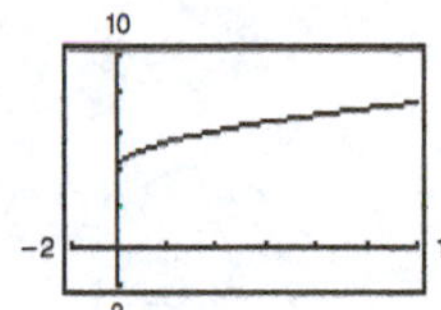

23. $f(x) = \dfrac{1}{x - 2}$

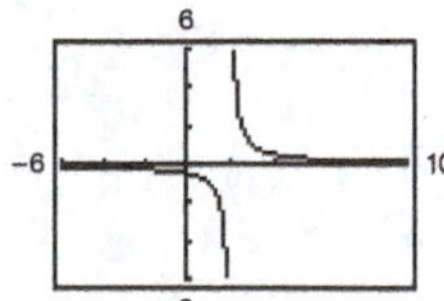

25. $g(x) = |x| - 5$

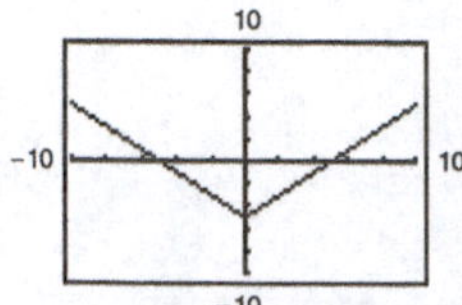

27. $f(x) = [\![x]\!]$

(a) $f(2.1) = 2$

(b) $f(2.9) = 2$

(c) $f(-3.1) = -4$

(d) $f\left(\frac{7}{2}\right) = 3$

29. $k(x) = [\![2x + 1]\!]$

(a) $k\left(\frac{1}{3}\right) = [\![2\left(\frac{1}{3}\right) + 1]\!] = [\![\frac{5}{3}]\!] = 1$

(b) $k(-2.1) = [\![2(-2.1) + 1]\!] = [\![-3.1]\!] = -4$

(c) $k(1.1) = [\![2(1.1) + 1]\!] = [\![3.2]\!] = 3$

(d) $k\left(\frac{2}{3}\right) = [\![2\left(\frac{2}{3}\right) + 1]\!] = [\![\frac{7}{3}]\!] = 2$

31. $g(x) = -[\![x]\!]$

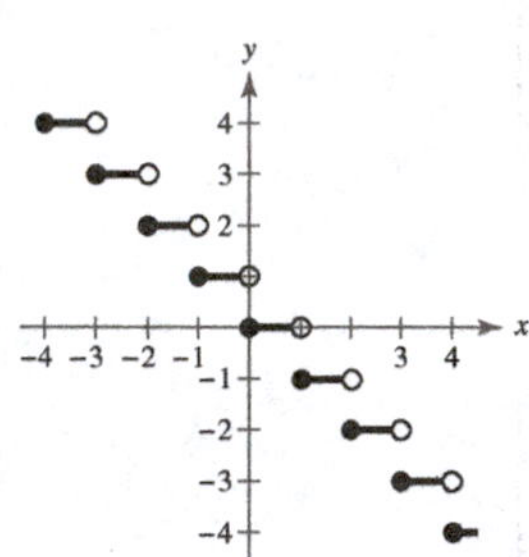

33. $g(x) = [\![x]\!] - 1$

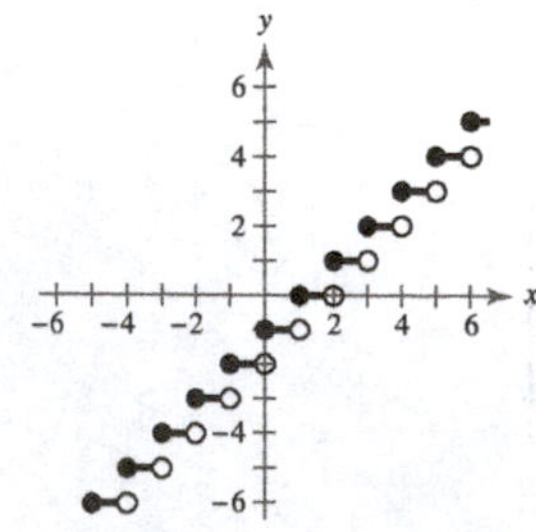

35. $g(x) = \begin{cases} x + 6, & x \le -4 \\ \frac{1}{2}x - 4, & x > -4 \end{cases}$

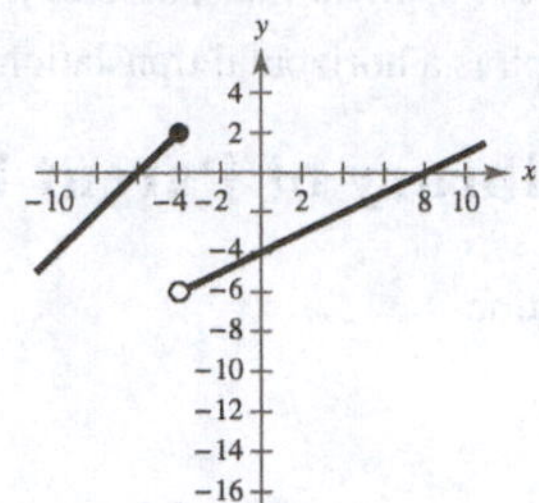

37. $f(x) = \begin{cases} 1 - (x - 1)^2, & x \le 2 \\ \sqrt{x - 2}, & x > 2 \end{cases}$

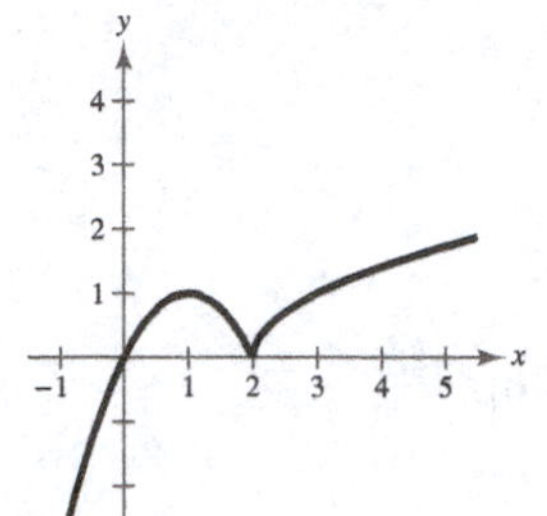

39. $h(x) = \begin{cases} 4 - x^2, & x < -2 \\ 3 + x, & -2 \le x < 0 \\ x^2 + 1, & x \ge 0 \end{cases}$

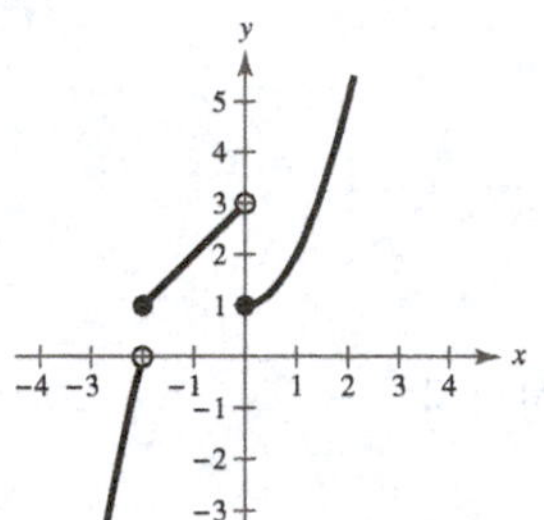

41. $s(x) = 2\left(\frac{1}{4}x - [\![\frac{1}{4}x]\!]\right)$

(a)

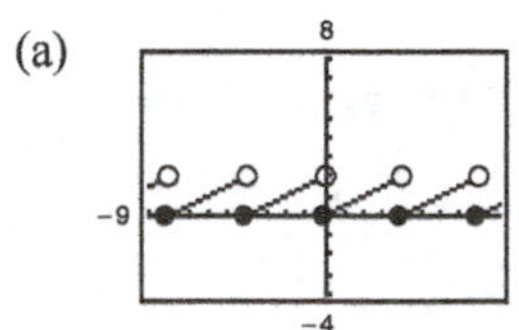

(b) Domain: $(-\infty, \infty)$; Range: $[0, 2)$

43. (a) $W(30) = 14(30) = 420$

$W(40) = 14(40) = 560$

$W(45) = 21(45 - 40) + 560 = 665$

$W(50) = 21(50 - 40) + 560 = 770$

(b) $W(h) = \begin{cases} 14h, & 0 < h \le 36 \\ 21(h - 36) + 504, & h > 36 \end{cases}$

(c) $W(h) = \begin{cases} 16h, & 0 < h \le 40 \\ 24(h - 40) + 640, & h > 40 \end{cases}$

45. Answers will vary. *Sample answer:*

Interval	*Input Pipe*	*Drain Pipe 1*	*Drain Pipe 2*
[0, 5]	Open	Closed	Closed
[5, 10]	Open	Open	Closed
[10, 20]	Closed	Closed	Closed
[20, 30]	Closed	Closed	Open
[30, 40]	Open	Open	Open
[40, 45]	Open	Closed	Open
[45, 50]	Open	Open	Open
[50, 60]	Open	Open	Closed

47. For the first two hours, the slope is 1. For the next six hours, the slope is 2. For the final hour, the slope is $\frac{1}{2}$.

$f(t) = \begin{cases} t, & 0 \le t \le 2 \\ 2t - 2, & 2 < t \le 8 \\ \frac{1}{2}t + 10, & 8 < t \le 9 \end{cases}$

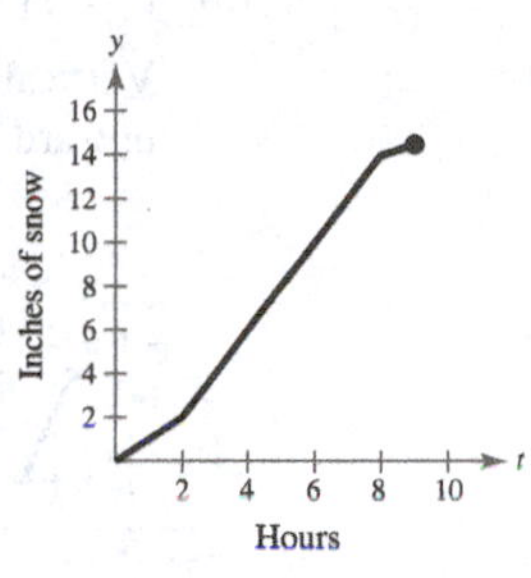

To find $f(t) = 2t - 2$, use $m = 2$ and $(2, 2)$.

$y - 2 = 2(t - 2) \Rightarrow y = 2t - 2$

To find $f(t) = \frac{1}{2}t + 10$, use $m = \frac{1}{2}$ and $(8, 14)$.

$y - 14 = \frac{1}{2}(t - 8) \Rightarrow y = \frac{1}{2}t + 10$

Total accumulation $= 14.5$ inches

49. False. A piecewise-defined function is a function that is defined by two or more equations over a specified domain. That domain may or may not include x- and y-intercepts.

Section 2.5 Transformations of Functions

1. rigid

3. vertical stretch; vertical shrink

5. (a) $f(x) = |x| + c$ Vertical shifts

$c = -2$: $f(x) = |x| - 2$ 2 units down

$c = -1$: $f(x) = |x| - 1$ 1 unit down

$c = 1$: $f(x) = |x| + 1$ 1 unit up

$c = 2$: $f(x) = |x| + 2$ 2 units up

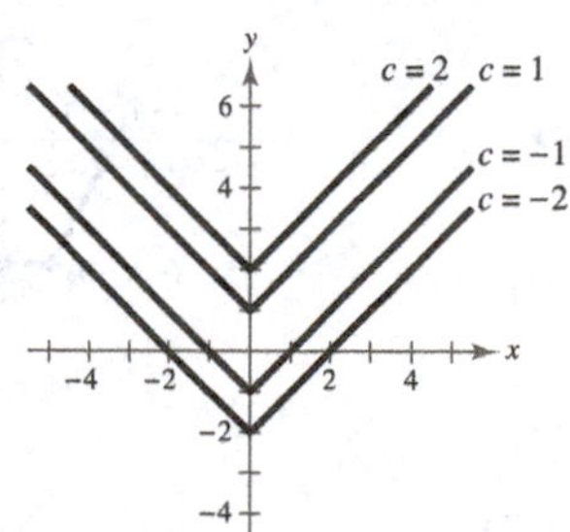

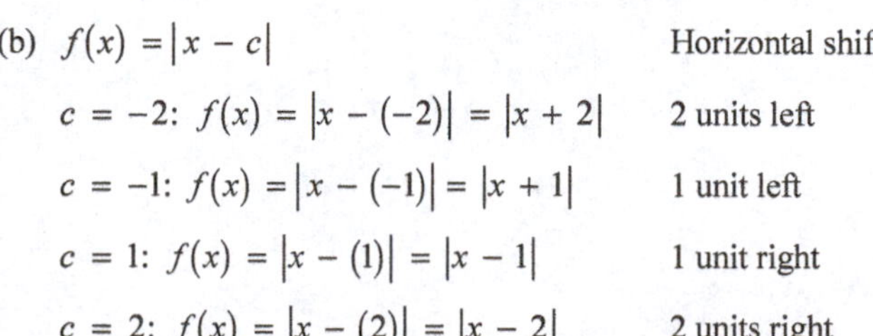

(b) $f(x) = |x - c|$ Horizontal shifts

$c = -2$: $f(x) = |x - (-2)| = |x + 2|$ 2 units left

$c = -1$: $f(x) = |x - (-1)| = |x + 1|$ 1 unit left

$c = 1$: $f(x) = |x - (1)| = |x - 1|$ 1 unit right

$c = 2$: $f(x) = |x - (2)| = |x - 2|$ 2 units right

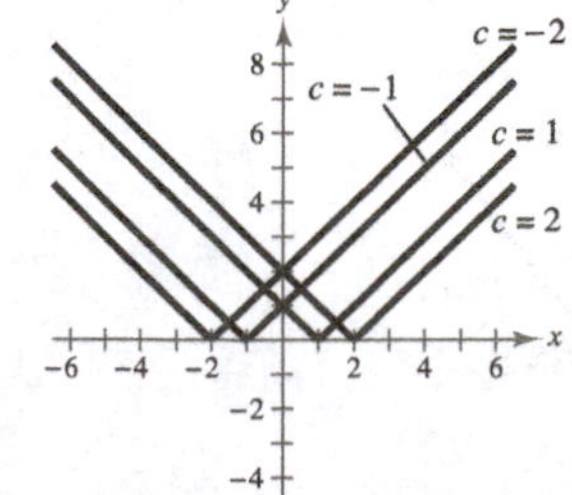

7. (a) $f(x) = [\![x]\!] + c$ Vertical shifts

$c = -4$: $f(x) = [\![x]\!] - 4$ 4 units down

$c = -1$: $f(x) = [\![x]\!] - 1$ 1 unit down

$c = 2$: $f(x) = [\![x]\!] + 2$ 2 units up

$c = 5$: $f(x) = [\![x]\!] + 5$ 5 units up

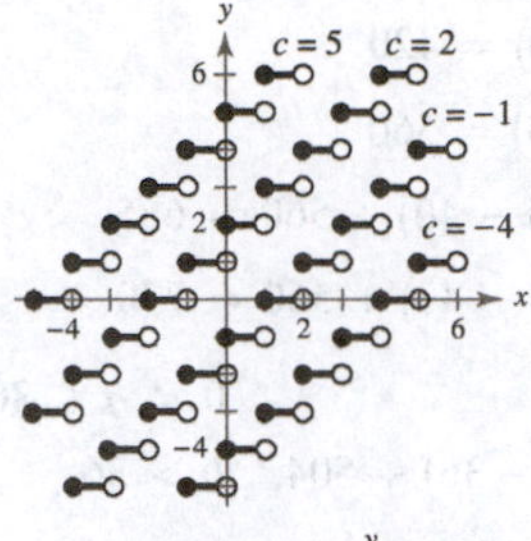

(b) $f(x) = [\![x + c]\!]$ Horizontal shifts

$c = -4$: $f(x) = [\![x - (-4)]\!] = [\![x + 4]\!]$ 4 units left

$c = -1$: $f(x) = [\![x - (-1)]\!] = [\![x + 1]\!]$ 1 unit left

$c = 2$: $f(x) = [\![x - (2)]\!] = [\![x - 2]\!]$ 2 units right

$c = 5$: $f(x) = [\![x - (5)]\!] = [\![x - 5]\!]$ 5 units right

9. (a) $y = f(-x)$

Reflection in the y-axis

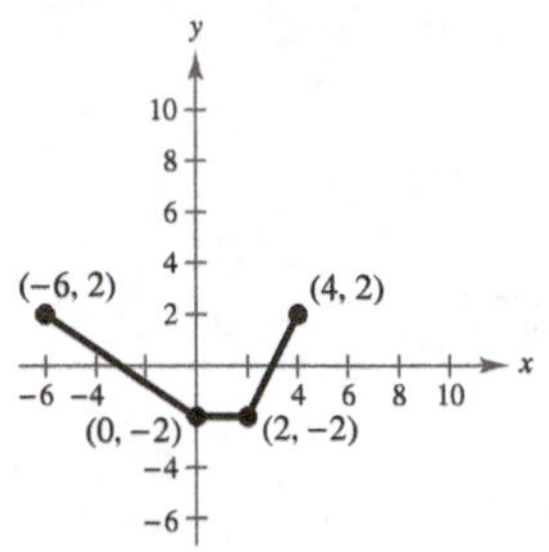

(b) $y = f(x) + 4$

Vertical shift 4 units upward

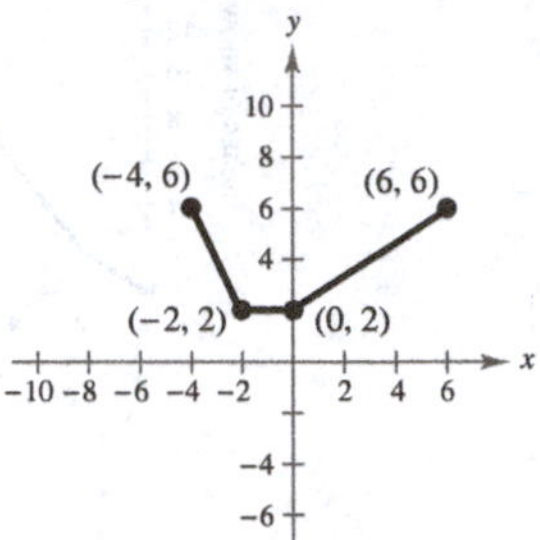

(c) $y = 2f(x)$

Vertical stretch (each y-value is multiplied by 2)

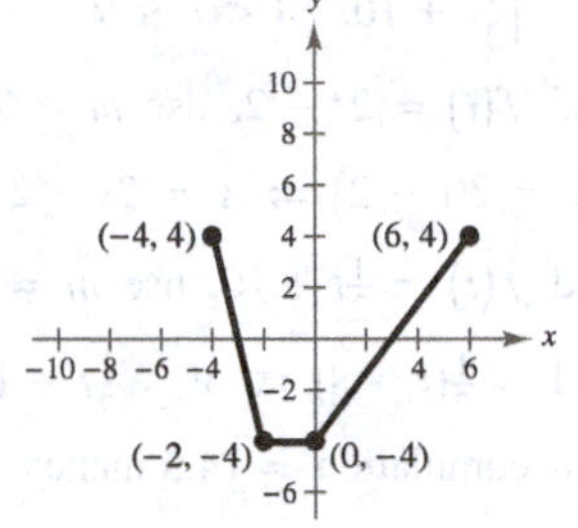

(d) $y = -f(x - 4)$

Reflection in the x-axis and a horizontal shift 4 units to the right

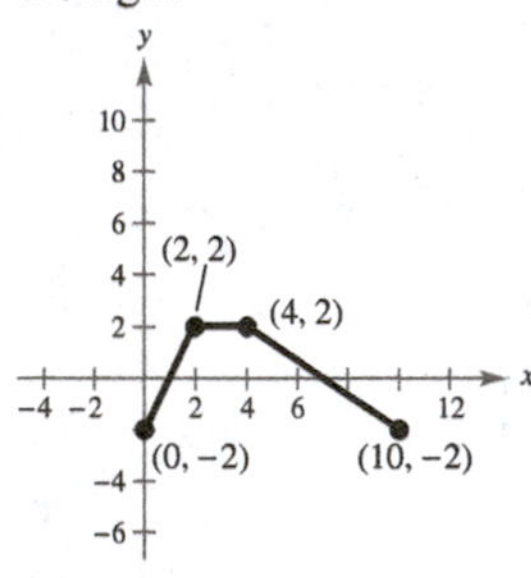

(e) $y = f(x) - 3$

Vertical shift 3 units downward

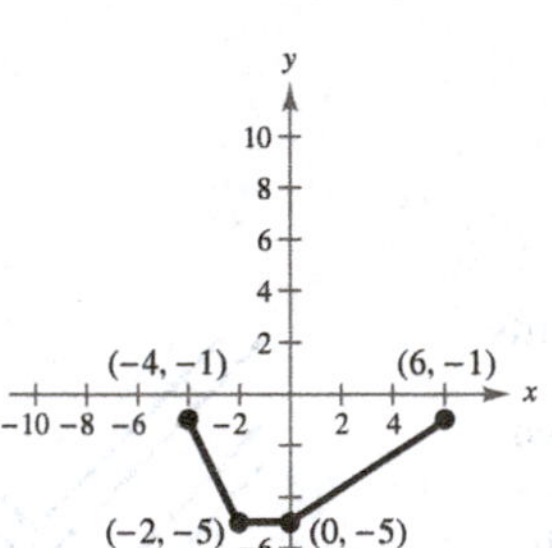
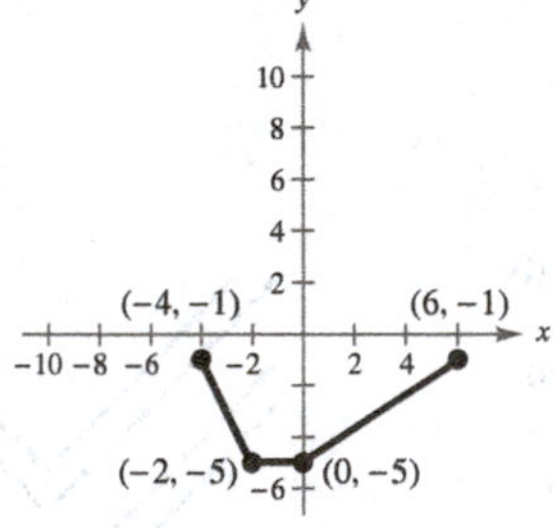

(f) $y = -f(x) - 1$

Reflection in the x-axis and a vertical shift 1 unit downward

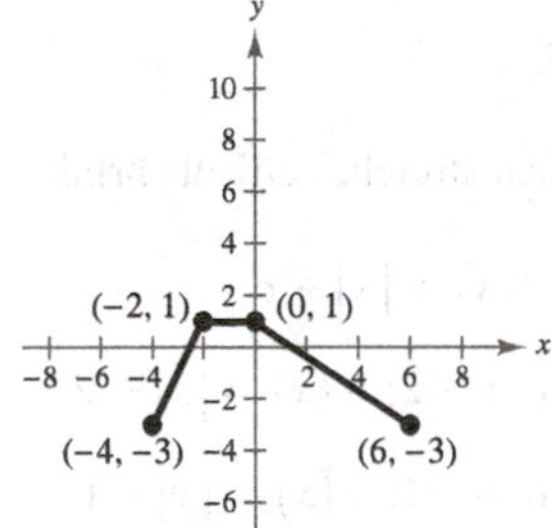

(g) $y = f(2x)$

Horizontal shrink (each x-value is divided by 2)

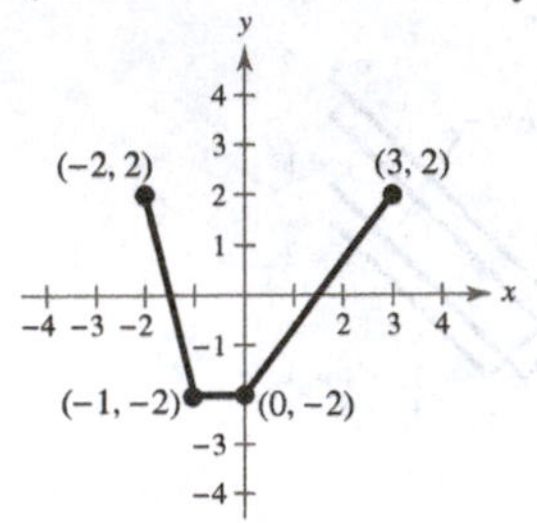

11. Parent function: $f(x) = x^2$

(a) Vertical shift 1 unit downward

$g(x) = x^2 - 1$

(b) Reflection in the x-axis, horizontal shift 1 unit to the left, and a vertical shift 1 unit upward

$g(x) = -(x + 1)^2 + 1$

13. Parent function: $f(x) = |x|$

(a) Reflection in the x-axis and a horizontal shift 3 units to the left

$g(x) = -|x + 3|$

(b) Horizontal shift 2 units to the right and a vertical shift 4 units downward

$g(x) = |x - 2| - 4$

15. Parent function: $f(x) = x^3$

Horizontal shift 2 units to the right

$y = (x - 2)^3$

17. Parent function: $f(x) = x^2$

Reflection in the x-axis

$y = -x^2$

19. Parent function: $f(x) = \sqrt{x}$

Reflection in the x-axis and a vertical shift 1 unit upward

$y = -\sqrt{x} + 1$

21. $g(x) = x^2 + 6$

(a) Parent function: $f(x) = x^2$

(b) A vertical shift 6 units upward

(c)

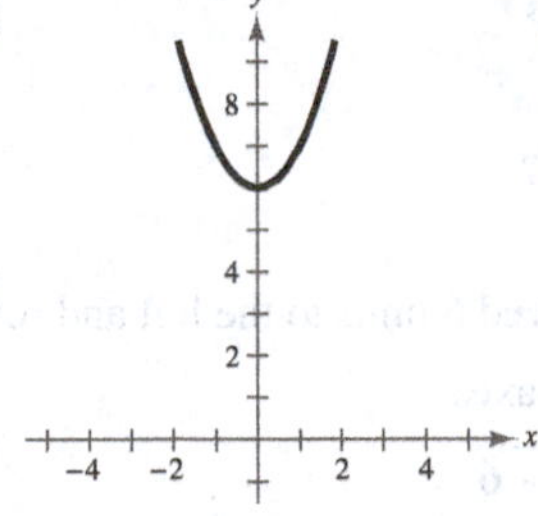

(d) $g(x) = f(x) + 6$

23. $g(x) = -(x - 2)^3$

(a) Parent function: $f(x) = x^3$

(b) Horizontal shift of 2 units to the right and a reflection in the x-axis

(c)

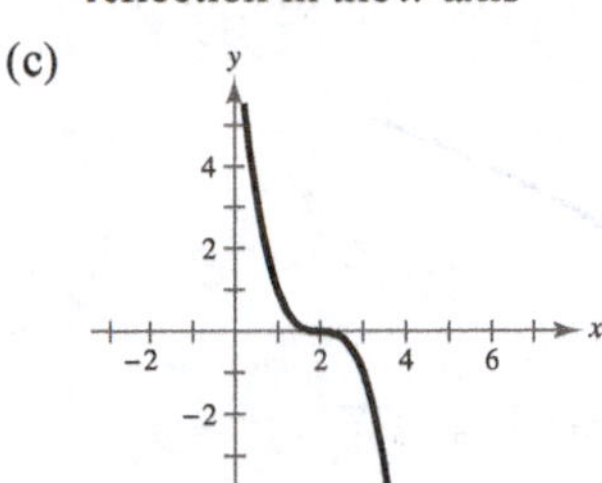

(d) $g(x) = -f(x - 2)$

25. $g(x) = -3 - (x + 1)^2$

(a) Parent function: $f(x) = x^2$

(b) Reflection in the x-axis, a vertical shift 3 units downward and a horizontal shift 1 unit left

(c)

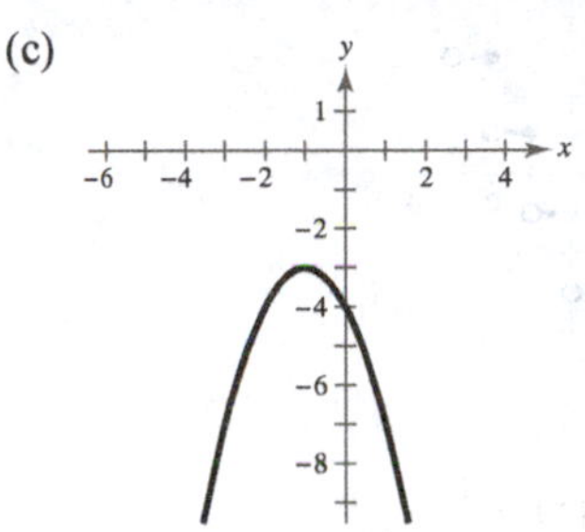

(d) $g(x) = -f(x + 1) - 3$

27. $g(x) = |x - 1| + 2$

(a) Parent function: $f(x) = |x|$

(b) A horizontal shift 1 unit right and a vertical shift 2 units upward

(c)

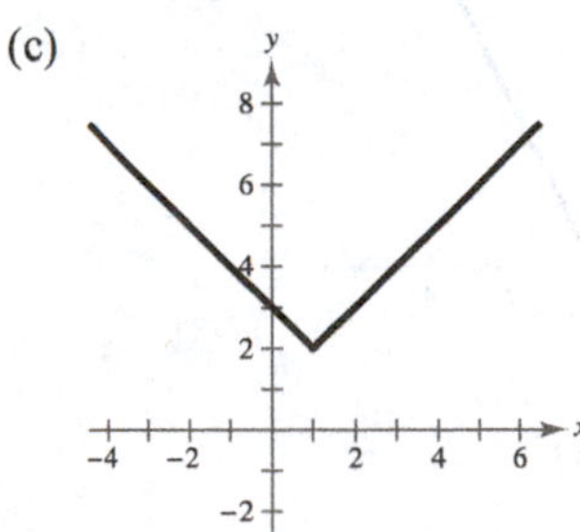

(d) $g(x) = f(x - 1) + 2$

29. $g(x) = 2\sqrt{x}$

(a) Parent function: $f(x) = \sqrt{x}$

(b) A vertical stretch (each y value is multiplied by 2)

(c)

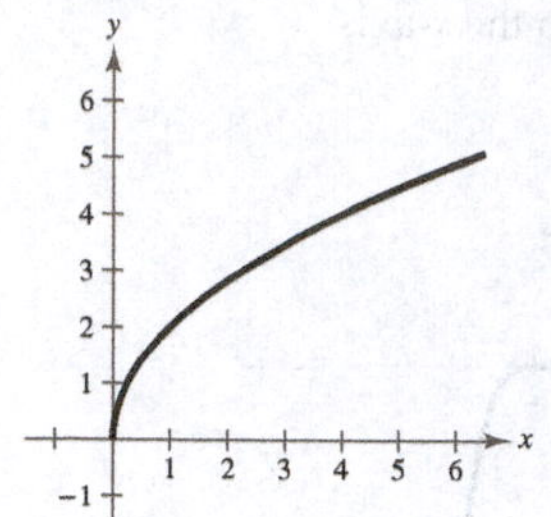

(d) $g(x) = 2f(x)$

31. $g(x) = 2[\![x]\!] - 1$

(a) Parent function: $f(x) = [\![x]\!]$

(b) A vertical shift of 1 unit downward and a vertical stretch (each y value is multiplied by 2)

(c)

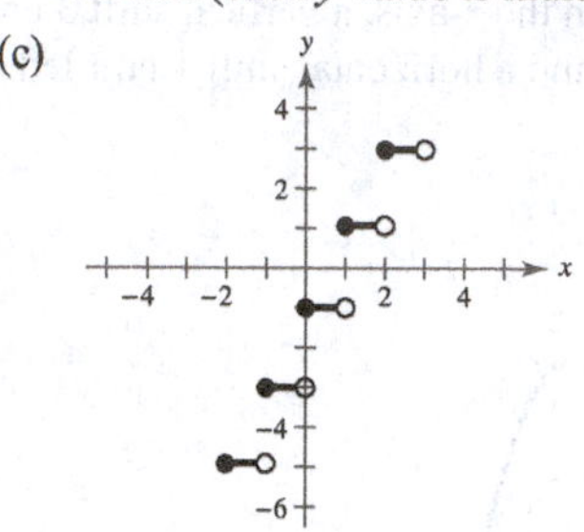

(d) $g(x) = 2f(x) - 1$

33. $g(x) = |2x|$

(a) Parent function: $f(x) = |x|$

(b) A horizontal shrink

(c)

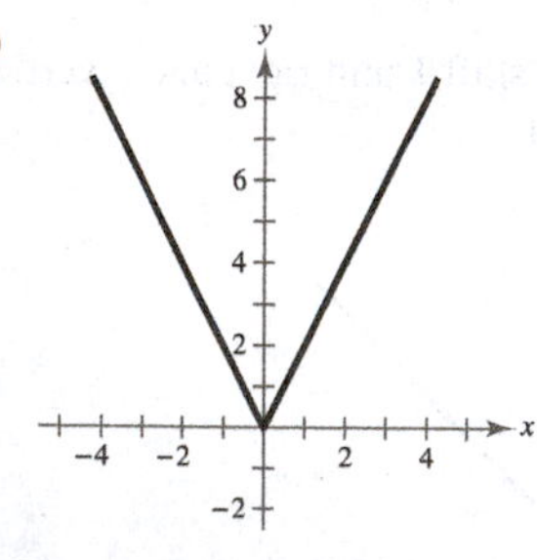

(d) $g(x) = f(2x)$

35. $g(x) = -2x^2 + 1$

(a) Parent function: $f(x) = x^2$

(b) A vertical stretch, reflection in the x-axis and a vertical shift 1 unit upward

(c)

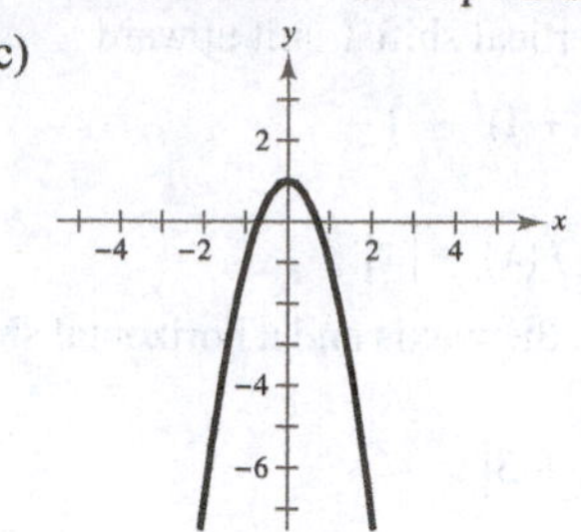

(d) $g(x) = -2f(x) + 1$

37. $g(x) = 3|x - 1| + 2$

(a) Parent function: $f(x) = |x|$

(b) A horizontal shift of 1 unit to the right, a vertical stretch, and a vertical shift 2 units upward

(c)

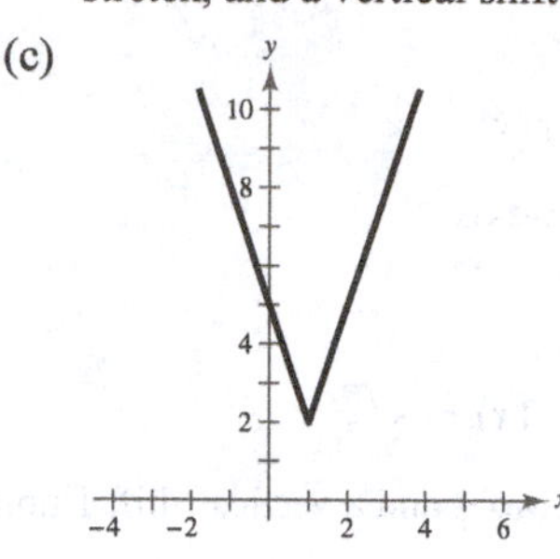

(d) $g(x) = 3f(x - 1) + 2$

39. $g(x) = (x - 3)^2 - 7$

41. $f(x) = x^3$ moved 13 units to the right

$g(x) = (x - 13)^3$

43. $g(x) = -|x| + 12$

45. $f(x) = \sqrt{x}$ moved 6 units to the left and reflected in both the x- and y-axes

$g(x) = -\sqrt{-x + 6}$

47. $f(x) = x^2$

(a) Reflection in the x-axis and a vertical stretch (each y-value is multiplied by 3)

$g(x) = -3x^2$

(b) Vertical shift 3 units upward and a vertical stretch (each y-value is multiplied by 4)

$g(x) = 4x^2 + 3$

49. $f(x) = |x|$

(a) Reflection in the x-axis and a vertical shrink $\left(\text{each } y\text{-value is multiplied by } \frac{1}{2}\right)$

$g(x) = -\frac{1}{2}|x|$

(b) Vertical stretch (each y-value is multiplied by 3) and a vertical shift 3 units downward

$g(x) = 3|x| - 3$

51. Parent function: $f(x) = x^3$

Vertical stretch (each y-value is multiplied by 2)

$g(x) = 2x^3$

53. Parent function: $f(x) = x^2$

Reflection in the x-axis, vertical shrink $\left(\text{each } y\text{-value is multiplied by } \frac{1}{2}\right)$

$g(x) = -\frac{1}{2}x^2$

55. Parent function: $f(x) = \sqrt{x}$

Reflection in the y-axis, vertical shrink $\left(\text{each } y\text{-value is multiplied by } \frac{1}{2}\right)$

$g(x) = \frac{1}{2}\sqrt{-x}$

57. Parent function: $f(x) = x^3$

Reflection in the x-axis, horizontal shift 2 units to the right and a vertical shift 2 units upward

$g(x) = -(x - 2)^3 + 2$

59. Parent function: $f(x) = \sqrt{x}$

Reflection in the x-axis and a vertical shift 3 units downward

$g(x) = -\sqrt{x} - 3$

61. (a)

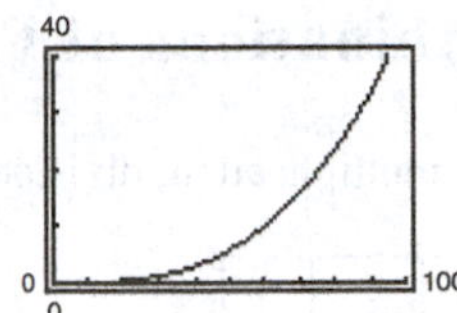

(b) $H(x) = 0.00004636x^3$

$$H\left(\frac{x}{1.6}\right) = 0.00004636\left(\frac{x}{1.6}\right)^3$$

$$= 0.00004636\left(\frac{x^3}{4.096}\right)$$

$$= 0.0000113184x^3 = 0.00001132x^3$$

The graph of $H\left(\frac{x}{1.6}\right)$ is a horizontal stretch of the graph of $H(x)$.

63. False. $y = f(-x)$ is a reflection in the y-axis.

65. True. Because $|x| = |-x|$, the graphs of $f(x) = |x| + 6$ and $f(x) = |-x| + 6$ are identical.

67. $y = f(x + 2) - 1$

Horizontal shift 2 units to the left and a vertical shift 1 unit downward

$(0, 1) \rightarrow (0 - 2, 1 - 1) = (-2, 0)$

$(1, 2) \rightarrow (1 - 2, 2 - 1) = (-1, 1)$

$(2, 3) \rightarrow (2 - 2, 3 - 1) = (0, 2)$

69. Since the graph of $g(x)$ is a horizontal shift one unit to the right of $f(x) = x^3$, the equation should be $g(x) = (x - 1)^3$ and not $g(x) = (x + 1)^3$.

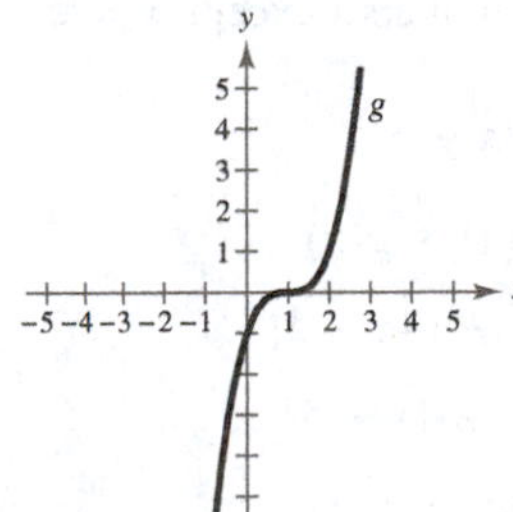

71. (a) The profits were only $\frac{3}{4}$ as large as expected:

$g(t) = \frac{3}{4}f(t)$

(b) The profits were \$10,000 greater than predicted:

$g(t) = f(t) + 10{,}000$

(c) There was a two-year delay: $g(t) = f(t - 2)$

Section 2.6 Combinations of Functions: Composite Functions

1. addition; subtraction; multiplication; division

3.

x	0	1	2	3
f	2	3	1	2
g	-1	0	$\frac{1}{2}$	0
$f + g$	1	3	$\frac{3}{2}$	2

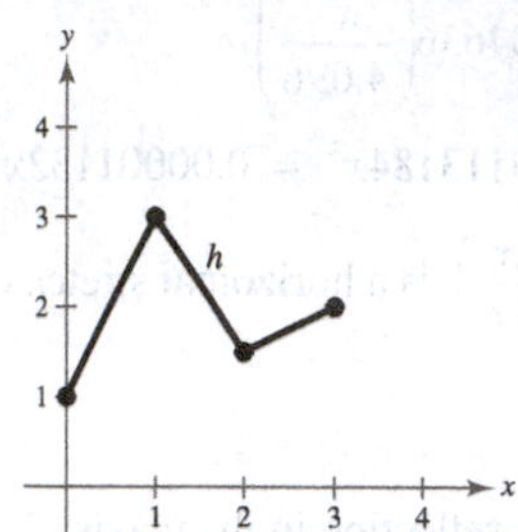

5. $f(x) = x + 2, g(x) = x - 2$

(a) $(f + g)(x) = f(x) + g(x)$
$= (x + 2) + (x - 2)$
$= 2x$

(b) $(f - g)(x) = f(x) - g(x)$
$= (x + 2) - (x - 2)$
$= 4$

(c) $(fg)(x) = f(x) \cdot g(x)$
$= (x + 2)(x - 2)$
$= x^2 - 4$

(d) $\left(\frac{f}{x}\right)(x) = \frac{f(x)}{g(x)} = \frac{x + 2}{x - 2}$

Domain: all real numbers x except $x = 2$

7. $f(x) = x^2, g(x) = 4x - 5$

(a) $(f + g)(x) = f(x) + g(x)$
$= x^2 + (4x - 5)$
$= x^2 + 4x - 5$

(b) $(f - g)(x) = f(x) - g(x)$
$= x^2 - (4x - 5)$
$= x^2 - 4x + 5$

(c) $(fg)(x) = f(x) \cdot g(x)$
$= x^2(4x - 5)$
$= 4x^3 - 5x^2$

(d) $\left(\frac{f}{g}\right)(x) = \frac{f(x)}{g(x)}$
$= \frac{x^2}{4x - 5}$

Domain: all real numbers x except $x = \frac{5}{4}$

9. $f(x) = x^2 + 6, g(x) = \sqrt{1 - x}$

(a) $(f + g)(x) = f(x) + g(x) = x^2 + 6 + \sqrt{1 - x}$

(b) $(f - g)(x) = f(x) - g(x) = x^2 + 6 - \sqrt{1 - x}$

(c) $(fg)(x) = f(x) \cdot g(x) = (x^2 + 6)\sqrt{1 - x}$

(d) $\left(\frac{f}{g}\right)(x) = \frac{f(x)}{g(x)} = \frac{x^2 + 6}{\sqrt{1 - x}} = \frac{(x^2 + 6)\sqrt{1 - x}}{1 - x}$

Domain: $x < 1$

11. $f(x) = \frac{x}{x + 1}, g(x) = x^3$

(a) $(f + g)(x) = \frac{x}{x + 1} + x^3 = \frac{x + x^4 + x^3}{x + 1}$

(b) $(f - g)(x) = \frac{x}{x + 1} - x^3 = \frac{x - x^4 - x^3}{x + 1}$

(c) $(fg)(x) = \frac{x}{x + 1} \cdot x^3 = \frac{x^4}{x + 1}$

(d) $\left(\frac{f}{g}\right)(x) = \frac{x}{x + 1} \div x^3 = \frac{x}{x + 1} \cdot \frac{1}{x^3} = \frac{1}{x^2(x + 1)}$

Domain: all real numbers x except $x = 0$ and $x = -1$

For Exercises 13–23, $f(x) = x + 3$ and $g(x) = x^2 - 2$.

13. $(f + g)(2) = f(2) + g(2)$
$= (2 + 3) + (2^2 - 2)$
$= 7$

15. $(f - g)(0) = f(0) - g(0)$
$= (0 + 3) - ((0)^2 - 2)$
$= 5$

17. $(f - g)(3t) = f(3t) - g(3t)$
$= ((3t) + 3) - ((3t)^2 - 2)$
$= 3t + 3 - (9t^2 - 2)$
$= -9t^2 + 3t + 5$

19. $(fg)(6) = f(6)g(6)$
$= ((6) + 3)((6)^2 - 2)$
$= (9)(34)$
$= 306$

21. $(f/g)(5) = f(5) / g(5)$
$= ((5) + 3) / ((5)^2 - 2)$
$= \frac{8}{23}$

23. $(f/g)(-1) - g(3) = f(-1) / g(-1) - g(3)$
$= ((-1) + 3) / ((-1)^2 - 2) - ((3)^2 - 2)$
$= (2/-1) - 7$
$= -2 - 7 = -9$

25. $f(x) = 3x, g(x) = -\frac{x^3}{10}$

$(f + g)(x) = 3x - \frac{x^3}{10}$

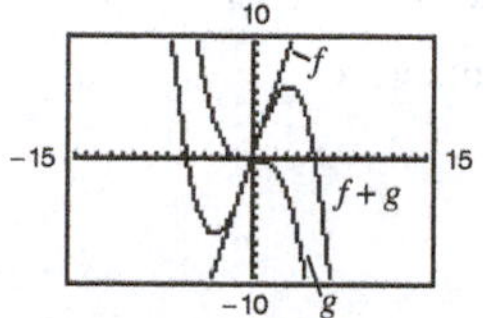

For $0 \le x \le 2$, $f(x)$ contributes most to the magnitude.

For $x > 6$, $g(x)$ contributes most to the magnitude.

27. $f(x) = 3x + 2, g(x) = -\sqrt{x + 5}$

$(f + g)x = 3x - \sqrt{x + 5} + 2$

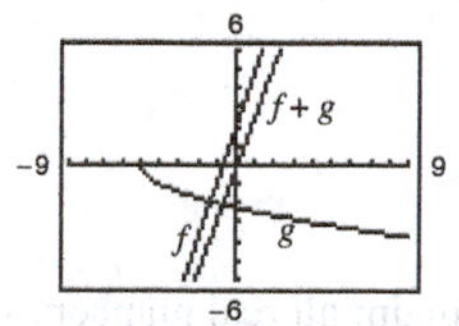

For $0 \le x \le 2$, $f(x)$ contributes most to the magnitude.

For $x > 6$, $f(x)$ contributes most to the magnitude.

29. $f(x) = x + 8, g(x) = x - 3$

(a) $(f \circ g)(x) = f(g(x)) = f(x - 3) = (x - 3) + 8 = x + 5$

(b) $(g \circ f)(x) = g(f(x)) = g(x + 8) = (x + 8) - 3 = x + 5$

(c) $(g \circ g)(x) = g(g(x)) = g(x - 3) = (x - 3) - 3 = x - 6$

31. $f(x) = x^2, g(x) = x - 1$

(a) $(f \circ g)(x) = f(g(x)) = f(x - 1) = (x - 1)^2$

(b) $(g \circ f)(x) = g(f(x)) = g(x^2) = x^2 - 1$

(c) $(g \circ g)(x) = g(g(x)) = g(x - 1) = x - 2$

33. $f(x) = \sqrt[3]{x - 1}, g(x) = x^3 + 1$

(a) $(f \circ g)(x) = f(g(x))$
$= f(x^3 + 1)$
$= \sqrt[3]{(x^3 + 1) - 1}$
$= \sqrt[3]{x^3} = x$

(b) $(g \circ f)(x) = g(f(x))$
$= g(\sqrt[3]{x - 1})$
$= (\sqrt[3]{x - 1})^3 + 1$
$= (x - 1) + 1 = x$

(c) $(g \circ g)(x) = g(g(x))$
$= g(x^3 + 1)$
$= (x^3 + 1)^3 + 1$
$= x^9 + 3x^6 + 3x^3 + 2$

35. $f(x) = \sqrt{x+4}$ Domain: $x \geq -4$

$g(x) = x^2$ Domain: all real numbers x

(a) $(f \circ g)(x) = f(g(x)) = f(x^2) = \sqrt{x^2+4}$

Domain: all real numbers x

(b) $(g \circ f)(x) = g(f(x))$

$= g(\sqrt{x+4}) = (\sqrt{x+4})^2 = x + 4$

Domain: $x \geq -4$

37. $f(x) = x^3$ Domain: all real numbers x

$g(x) = x^{2/3}$ Domain: all real numbers x

(a) $(f \circ g)(x) = f(g(x)) = f(x^{2/3}) = (x^{2/3})^3 = x^2$

Domain: all real numbers x.

(b) $(g \circ f)(x) = g(f(x)) = g(x^3) = (x^3)^{2/3} = x^2$

Domain: all real numbers x.

39. $f(x) = |x|$ Domain: all real numbers x

$g(x) = x + 6$ Domain: all real numbers x

(a) $(f \circ g)(x) = f(g(x)) = f(x+6) = |x+6|$

Domain: all real numbers x

(b) $(g \circ f)(x) = g(f(x)) = g(|x|) = |x| + 6$

Domain: all real numbers x

41. $f(x) = \dfrac{1}{x}$ Domain: all real numbers x except $x = 0$

$g(x) = x + 3$ Domain: all real numbers x

(a) $(f \circ g)(x) = f(g(x)) = f(x+3) = \dfrac{1}{x+3}$

Domain: all real numbers x except $x = -3$

(b) $(g \circ f)(x) = g(f(x)) = g\left(\dfrac{1}{x}\right) = \dfrac{1}{x} + 3$

Domain: all real numbers x except $x = 0$

43. $f(x) = \frac{1}{2}x$, $g(x) = x - 4$

(a)

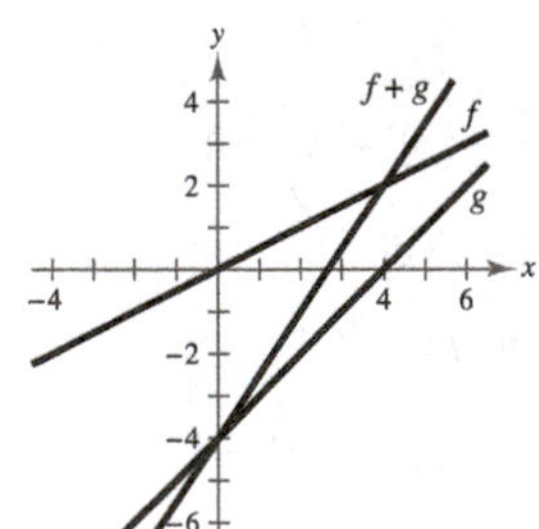

(b)

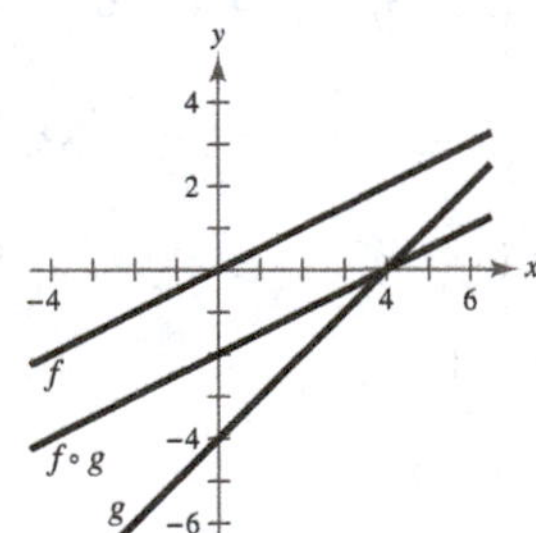

45. (a) $(f+g)(3) = f(3) + g(3) = 2 + 1 = 3$

(b) $\left(\dfrac{f}{g}\right)(2) = \dfrac{f(2)}{g(2)} = \dfrac{0}{2} = 0$

47. (a) $(f \circ g)(2) = f(g(2)) = f(2) = 0$

(b) $(g \circ f)(2) = g(f(2)) = g(0) = 4$

49. $h(x) = (2x^2 + 1)^2$

One possibility: Let $f(x) = x^2$ and $g(x) = 2x + 1$, then $(f \circ g)(x) = h(x)$.

51. $h(x) = \sqrt[3]{x^2 - 4}$

One possibility: Let $f(x) = \sqrt[3]{x}$ and $g(x) = x^2 - 4$, then $(f \circ g)(x) = h(x)$.

53. $h(x) = \dfrac{1}{x + 2}$

One possibility: Let $f(x) = 1/x$ and $g(x) = x + 2$, then $(f \circ g)(x) = h(x)$.

55. $h(x) = \dfrac{-x^2 + 3}{4 - x^2}$

One possibility: Let $f(x) = \dfrac{x + 3}{4 + x}$ and $g(x) = -x^2$, then $(f \circ g)(x) = h(x)$.

57. (a) $T(x) = R(x) + B(x) = \frac{3}{4}x + \frac{1}{15}x^2$

(b)

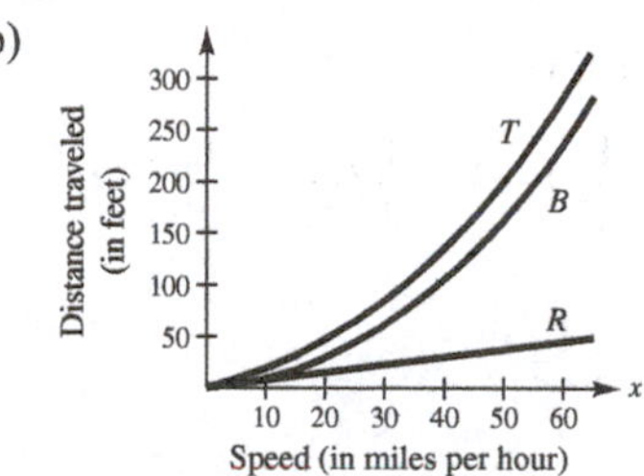

(c) $B(x)$; As x increases, $B(x)$ increases at a faster rate.

59. (a) $c(t) = \dfrac{b(t) - d(t)}{p(t)} \times 100$

(b) $c(16)$ represents the percent change in the population due to births and deaths in the year 2016.

61. (a) $r(x) = \dfrac{x}{2}$

(b) $A(r) = \pi r^2$

(c) $(A \circ r)(x) = A(r(x)) = A\left(\dfrac{x}{2}\right) = \pi\left(\dfrac{x}{2}\right)^2$

$(A \circ r)(x)$ represents the area of the circular base of the tank on the square foundation with side length x.

63. (a) $f(g(x)) = f(0.03x) = 0.03x - 500{,}000$

(b) $g(f(x)) = g(x - 500{,}000) = 0.03(x - 500{,}000)$

$g(f(x))$ represents your bonus of 3% of an amount over \$500,000.

65. False. $(f \circ g)(x) = 6x + 1$ and $(g \circ f)(x) = 6x + 6$

67. Let O = oldest sibling, M = middle sibling, Y = youngest sibling.

Then the ages of each sibling can be found using the equations:

$O = 2M$

$M = \frac{1}{2}Y + 6$

(a) $O(M(Y)) = 2\left(\frac{1}{2}(Y) + 6\right) = 12 + Y$; Answers will vary.

(b) Oldest sibling is 16: $O = 16$

Middle sibling: $O = 2M$

$16 = 2M$

$M = 8$ years old

Youngest sibling: $M = \frac{1}{2}Y + 6$

$8 = \frac{1}{2}Y + 6$

$2 = \frac{1}{2}Y$

$Y = 4$ years old

69. Let $f(x)$ and $g(x)$ be two odd functions and define $h(x) = f(x)g(x)$. Then

$$\begin{aligned} h(-x) &= f(-x)g(-x) \\ &= [-f(x)][-g(x)] \quad \text{because } f \text{ and } g \text{ are odd} \\ &= f(x)g(x) \\ &= h(x). \end{aligned}$$

So, $h(x)$ is even.

Let $f(x)$ and $g(x)$ be two even functions and define $h(x) = f(x)g(x)$. Then

$$\begin{aligned} h(-x) &= f(-x)g(-x) \\ &= f(x)g(x) \quad \text{because } f \text{ and } g \text{ are even} \\ &= h(x). \end{aligned}$$

So, $h(x)$ is even.

71. (a) Answer not unique. *Sample answer:* $f(x) = x + 3$, $g(x) = x + 2$

$(f \circ g)(x) = f(g(x)) = (x + 2) + 3 = x + 5$

$(g \circ f)(x) = g(f(x)) = (x + 3) + 2 = x + 5$

(b) Answer not unique. *Sample answer:* $f(x) = x^2$, $g(x) = x^3$

$(f \circ g)(x) = f(g(x)) = (x^3)^2 = x^6$

$(g \circ f)(x) = g(f(x)) = (x^2)^3 = x^6$

73. (a) $g(x) = \frac{1}{2}[f(x) + f(-x)]$

To determine if $g(x)$ is even, show $g(-x) = g(x)$.

$$g(-x) = \frac{1}{2}[f(-x) + f(-(-x))] = \frac{1}{2}[f(-x) + f(x)] = \frac{1}{2}[f(x) + f(-x)] = g(x) \checkmark$$

$$h(x) = \frac{1}{2}[f(x) - f(-x)]$$

To determine if $h(x)$ is odd show $h(-x) = -h(x)$.

$$h(-x) = \frac{1}{2}[f(-x) - f(-(-x))] = \frac{1}{2}[f(-x) - f(x)] = -\frac{1}{2}[f(x) - f(-x)] = -h(x) \checkmark$$

(b) Let $f(x) = a$ function

$f(x) =$ even function + odd function.

Using the result from part (a) $g(x)$ is an even function and $h(x)$ is an odd function.

$$f(x) = g(x) + h(x) = \frac{1}{2}[f(x) + f(-x)] + \frac{1}{2}[f(x) - f(-x)] = \frac{1}{2}f(x) + \frac{1}{2}f(-x) + \frac{1}{2}f(x) - \frac{1}{2}f(-x) = f(x) \checkmark$$

(c) $f(x) = x^2 - 2x + 1$

$f(x) = g(x) + h(x)$

$$g(x) = \frac{1}{2}[f(x) + f(-x)] = \frac{1}{2}\left[x^2 - 2x + 1 + (-x)^2 - 2(-x) + 1\right]$$
$$= \frac{1}{2}\left[x^2 - 2x + 1 + x^2 + 2x + 1\right] = \frac{1}{2}\left[2x^2 + 2\right] = x^2 + 1$$

$$h(x) = \frac{1}{2}[f(x) - f(-x)] = \frac{1}{2}\left[x^2 - 2x + 1 - \left((-x)^2 - 2(-x) + 1\right)\right]$$
$$= \frac{1}{2}\left[x^2 - 2x + 1 - x^2 - 2x - 1\right] = \frac{1}{2}[-4x] = -2x$$

$$f(x) = \left(x^2 + 1\right) + (-2x)$$

$$k(x) = \frac{1}{x + 1}$$

$$k(x) = g(x) + h(x)$$

$$g(x) = \frac{1}{2}[k(x) + k(-x)] = \frac{1}{2}\left[\frac{1}{x + 1} + \frac{1}{-x + 1}\right]$$
$$= \frac{1}{2}\left[\frac{1 - x + x + 1}{(x + 1)(1 - x)}\right] = \frac{1}{2}\left[\frac{2}{(x + 1)(1 - x)}\right]$$
$$= \frac{1}{(x + 1)(1 - x)} = \frac{-1}{(x + 1)(x - 1)}$$

$$h(x) = \frac{1}{2}[k(x) - k(-x)] = \frac{1}{2}\left[\frac{1}{x + 1} - \frac{1}{1 - x}\right]$$
$$= \frac{1}{2}\left[\frac{1 - x - (x + 1)}{(x + 1)(1 - x)}\right] = \frac{1}{2}\left[\frac{-2x}{(x + 1)(1 - x)}\right]$$
$$= \frac{-x}{(x + 1)(1 - x)} = \frac{x}{(x + 1)(x - 1)}$$

$$k(x) = \left(\frac{-1}{(x + 1)(x - 1)}\right) + \left(\frac{x}{(x + 1)(x - 1)}\right)$$

Section 2.7 Inverse Functions

1. inverse

3. range; domain

5. one-to-one

7. $f(x) = 6x$

$$f^{-1}(x) = \frac{x}{6} = \frac{1}{6}x$$

$$f(f^{-1}(x)) = f\left(\frac{x}{6}\right) = 6\left(\frac{x}{6}\right) = x$$

$$f^{-1}(f(x)) = f^{-1}(6x) = \frac{6x}{6} = x$$

9. $f(x) = 3x + 1$

$$f^{-1}(x) = \frac{x-1}{3}$$

$$f(f^{-1}(x)) = f\left(\frac{x-1}{3}\right) = 3\left(\frac{x-1}{3}\right) + 1 = x$$

$$f^{-1}(f(x)) = f^{-1}(3x+1) = \frac{(3x+1)-1}{3} = x$$

11. $f(x) = x^2 - 4, x \geq 0$

$$f^{-1}(x) = \sqrt{x+4}$$

$$f(f^{-1}(x)) = f(\sqrt{x+4}) = (\sqrt{x+4})^2 - 4 = (x+4) - 4 = x$$

$$f^{-1}(f(x)) = f^{-1}(x^2 - 4) = \sqrt{(x^2-4)+4} = \sqrt{x^2} = x$$

13. $f(x) = x^3 + 1$

$$f^{-1}(x) = \sqrt[3]{x-1}$$

$$f(f^{-1}(x)) = f(\sqrt[3]{x-1}) = (\sqrt[3]{x-1})^3 + 1 = (x-1) + 1 = x$$

$$f^{-1}(f(x)) = f^{-1}(x^3+1) = \sqrt[3]{(x^3+1)-1} = \sqrt[3]{x^3} = x$$

15. $(f \circ g)(x) = f(g(x)) = f(4x+9) = \dfrac{4x+9-9}{4} = \dfrac{4x}{4} = x$

$$(g \circ f)(x) = g(f(x)) = g\left(\frac{x-9}{4}\right) = 4\left(\frac{x-9}{4}\right) + 9 = x - 9 + 9 = x$$

17. $f(g(x)) = f(\sqrt[3]{4x}) = \dfrac{(\sqrt[3]{4x})^3}{4} = \dfrac{4x}{4} = x$

$$g(f(x)) = g\left(\frac{x^3}{4}\right) = \sqrt[3]{4\left(\frac{x^3}{4}\right)} = \sqrt[3]{x^3} = x$$

19.

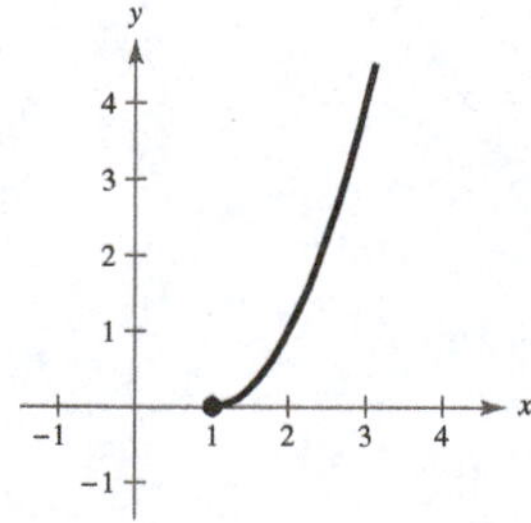

21. $f(x) = x - 5, g(x) = x + 5$

(a) $f(g(x)) = f(x+5) = (x+5) - 5 = x$

$g(f(x)) = g(x-5) = (x-5) + 5 = x$

(b)

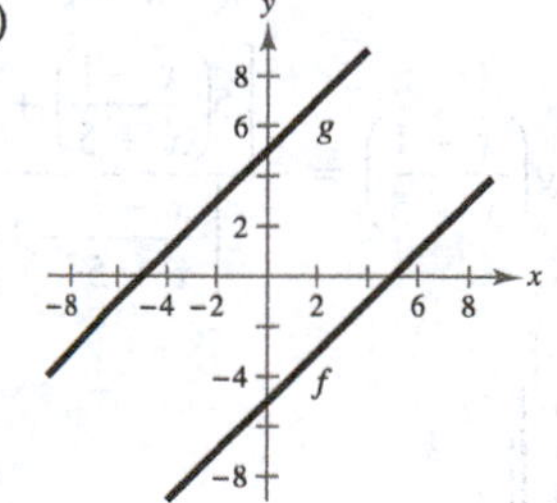

23. $f(x) = 7x + 1,\ g(x) = \dfrac{x-1}{7}$

(a) $f(g(x)) = f\left(\dfrac{x-1}{7}\right) = 7\left(\dfrac{x-1}{7}\right) + 1 = x$

$g(f(x)) = g(7x+1) = \dfrac{(7x+1)-1}{7} = x$

(b)

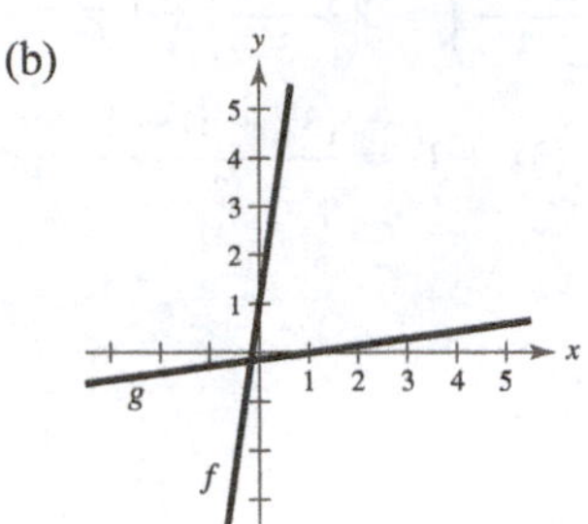

25. $f(x) = x^3,\ g(x) = \sqrt[3]{x}$

(a) $f(g(x)) = f\left(\sqrt[3]{x}\right) = \left(\sqrt[3]{x}\right)^3 = x$

$g(f(x)) = g(x^3) = \sqrt[3]{(x^3)} = x$

(b)

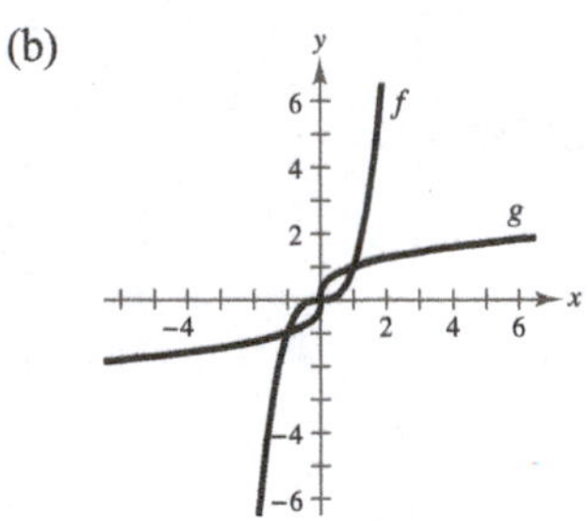

27. $f(x) = \sqrt{x+5},\ g(x) = x^2 - 5,\ x \ge 0$

(a) $f(g(x)) = f(x^2 - 5),\ x \ge 0$

$= \sqrt{(x^2 - 5) + 5} = x$

$g(f(x)) = g\left(\sqrt{x+5}\right)$

$= \left(\sqrt{x+5}\right)^2 - 5 = x$

(b)

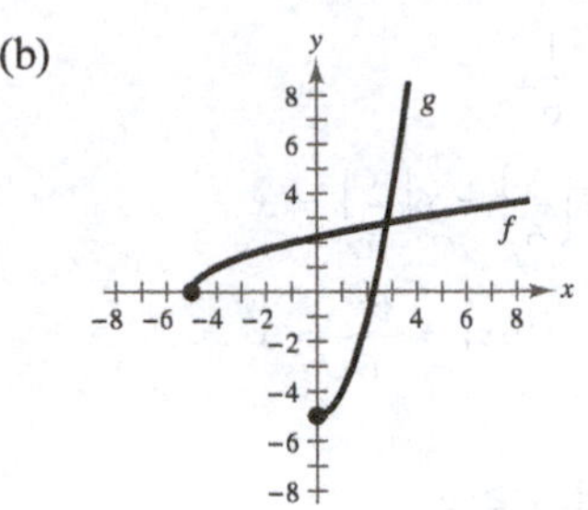

29. $f(x) = \dfrac{1}{x},\ g(x) = \dfrac{1}{x}$

(a) $f(g(x)) = f\left(\dfrac{1}{x}\right) = \dfrac{1}{1/x} = 1 \div \dfrac{1}{x} = 1 \cdot \dfrac{x}{1} = x$

$g(f(x)) = g\left(\dfrac{1}{x}\right) = \dfrac{1}{1/x} = 1 \div \dfrac{1}{x} = 1 \cdot \dfrac{x}{1} = x$

(b)

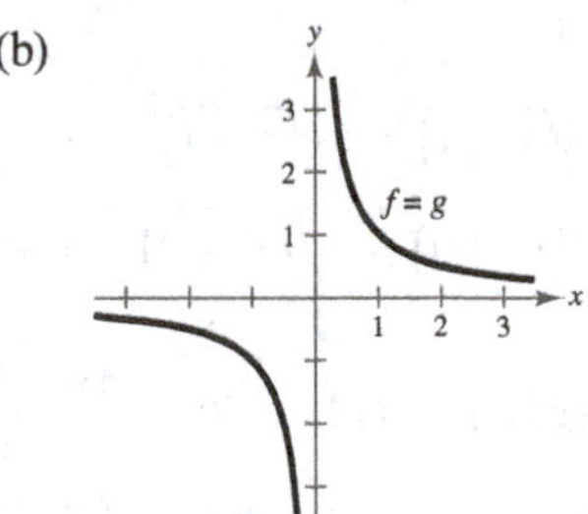

31. $f(x) = \dfrac{x-1}{x+5},\ g(x) = -\dfrac{5x+1}{x-1}$

(a) $f(g(x)) = f\left(-\dfrac{5x+1}{x-1}\right) = \dfrac{\left(-\dfrac{5x+1}{x-1} - 1\right)}{\left(-\dfrac{5x+1}{x-1} + 5\right)} \cdot \dfrac{x-1}{x-1} = \dfrac{-(5x+1)-(x-1)}{-(5x+1)+5(x-1)} = \dfrac{-6x}{-6} = x$

$g(f(x)) = g\left(\dfrac{x-1}{x+5}\right) = -\dfrac{\left[5\left(\dfrac{x-1}{x+5}\right) + 1\right]}{\left[\dfrac{x-1}{x+5} - 1\right]} \cdot \dfrac{x+5}{x+5} = -\dfrac{5(x-1)+(x+5)}{(x-1)-(x+5)} = -\dfrac{6x}{-6} = x$

(b)

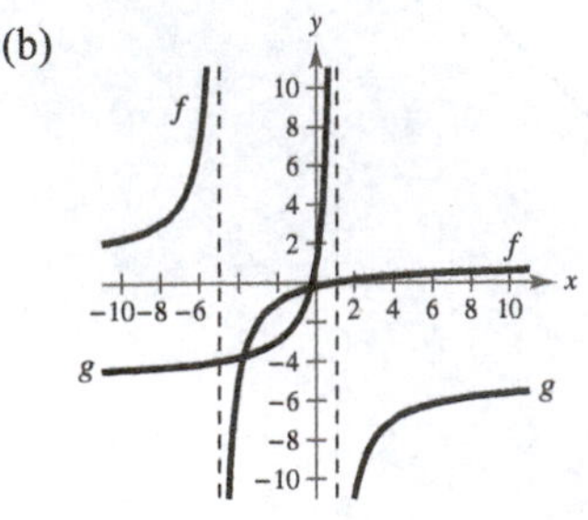

33. No, $\{(-2, -1), (1, 0), (2, 1), (1, 2), (-2, 3), (-6, 4)\}$ does not represent a function. -2 and 1 are paired with two different values.

35.

x	3	5	7	9	11	13
$f^{-1}(x)$	-1	0	1	2	3	4

37. Yes, because no horizontal line crosses the graph of f at more than one point, f *has* an inverse.

39. No, because some horizontal lines cross the graph of f twice, f *does not* have an inverse.

41. $g(x) = (x + 3)^2 + 2$

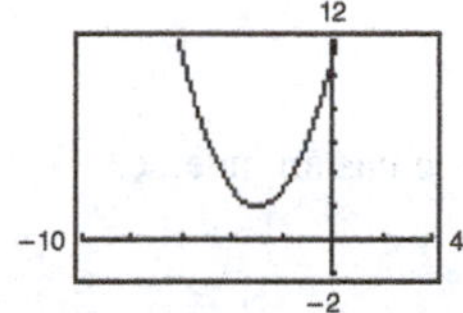

g does not pass the Horizontal Line Test, so g *does not* have an inverse.

43. $f(x) = x\sqrt{9 - x^2}$

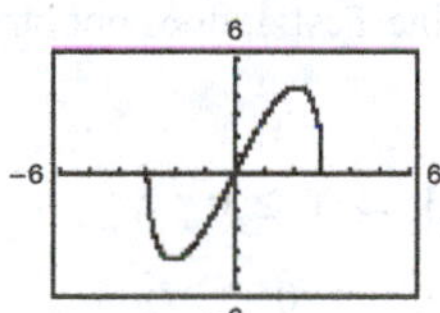

f does not pass the Horizontal Line Test, so f *does not* have an inverse.

45. (a)

$$f(x) = x^5 - 2$$
$$y = x^5 - 2$$
$$x = y^5 - 2$$
$$y = \sqrt[5]{x + 2}$$
$$f^{-1}(x) = \sqrt[5]{x + 2}$$

(b)

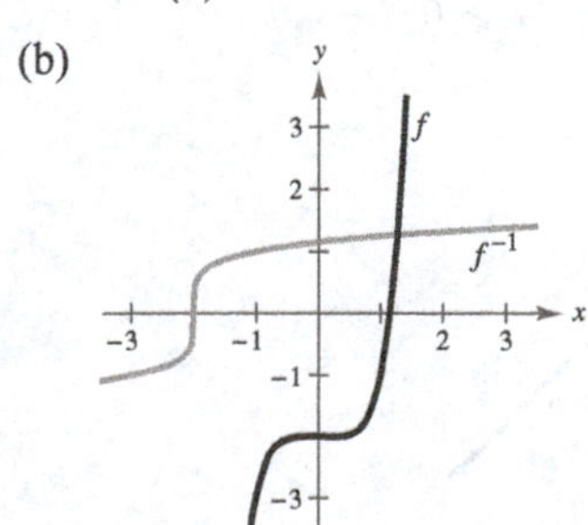

(c) The graph of f^{-1} is the reflection of the graph of f in the line $y = x$.

(d) The domains and ranges of f and f^{-1} are all real numbers.

47. (a)

$$f(x) = \sqrt{4 - x^2}, 0 \le x \le 2$$
$$y = \sqrt{4 - x^2}$$
$$x = \sqrt{4 - y^2}$$
$$x^2 = 4 - y^2$$
$$y^2 = 4 - x^2$$
$$y = \sqrt{4 - x^2}$$
$$f^{-1}(x) = \sqrt{4 - x^2}, 0 \le x \le 2$$

(b)

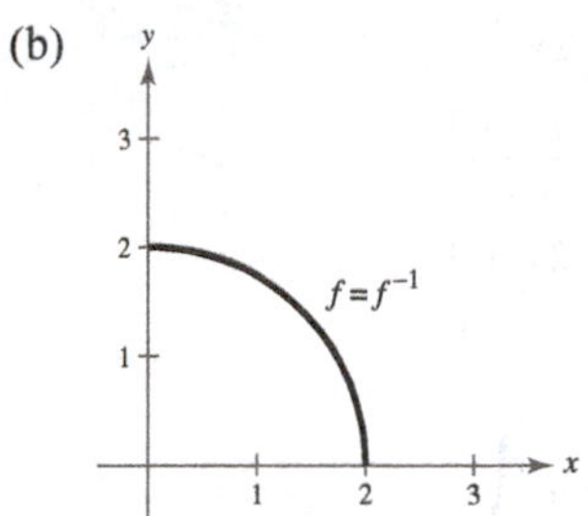

(c) The graph of f^{-1} is the same as the graph of f.

(d) The domains and ranges of f and f^{-1} are all real numbers x such that $0 \le x \le 2$.

49. (a)

$$f(x) = \frac{4}{x}$$
$$y = \frac{4}{x}$$
$$x = \frac{4}{y}$$
$$xy = 4$$
$$y = \frac{4}{x}$$
$$f^{-1}(x) = \frac{4}{x}$$

(b)

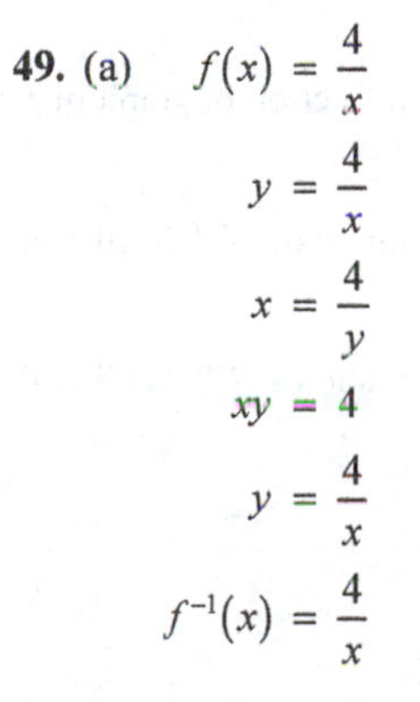

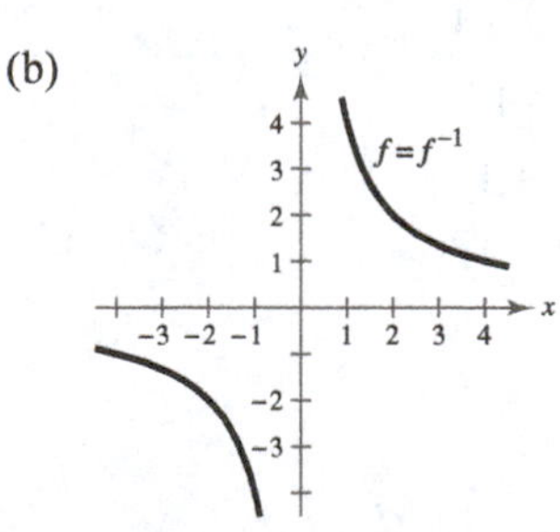

(c) The graph of f^{-1} is the same as the graph of f.

(d) The domains and ranges of f and f^{-1} are all real numbers except for 0.

51. (a)
$$f(x) = \frac{x+1}{x-2}$$
$$y = \frac{x+1}{x-2}$$
$$x = \frac{y+1}{y-2}$$
$$x(y-2) = y+1$$
$$xy - 2x = y+1$$
$$xy - y = 2x+1$$
$$y(x-1) = 2x+1$$
$$y = \frac{2x+1}{x-1}$$
$$f^{-1}(x) = \frac{2x+1}{x-1}$$

(b)

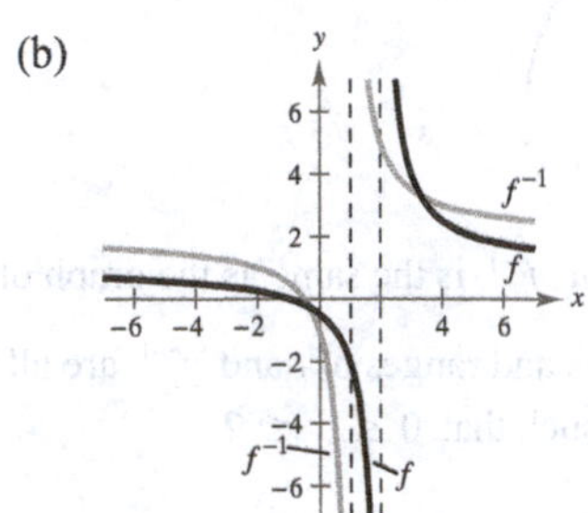

(c) The graph of f^{-1} is the reflection of graph of *f* in the line $y = x$.

(d) The domain of *f* and the range of f^{-1} is all real numbers except 2.

The range of *f* and the domain of f^{-1} is all real numbers except 1.

53. (a)
$$f(x) = \sqrt[3]{x-1}$$
$$y = \sqrt[3]{x-1}$$
$$x = \sqrt[3]{y-1}$$
$$x^3 = y-1$$
$$y = x^3 + 1$$
$$f^{-1}(x) = x^3 + 1$$

(b)

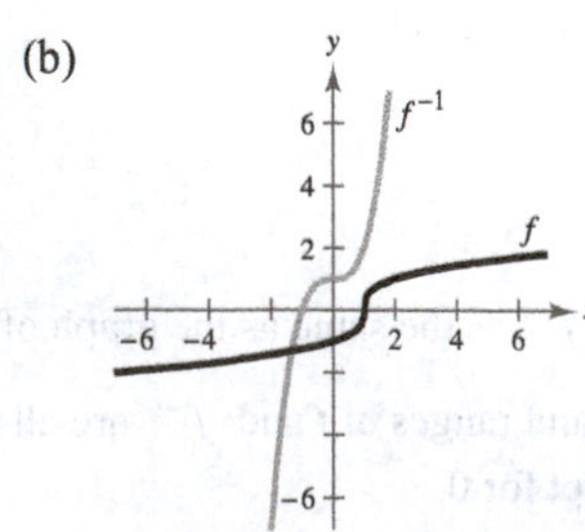

(c) The graph of f^{-1} is the reflection of the graph of *f* in the line $y = x$.

(d) The domains and ranges of *f* and f^{-1} are all real numbers.

55.
$$f(x) = x^4$$
$$y = x^4$$
$$x = y^4$$
$$y = \pm\sqrt[4]{x}$$

This does not represent *y* as a function of *x*. *f* does not have an inverse.

57.
$$g(x) = \frac{x+1}{6}$$
$$y = \frac{x+1}{6}$$
$$x = \frac{y+1}{6}$$
$$6x = y+1$$
$$y = 6x - 1$$

This is a function of *x*, so *g* has an inverse.

$$g^{-1}(x) = 6x - 1$$

59.
$$p(x) = -4$$
$$y = -4$$

Because $y = -4$ for all *x*, the graph is a horizontal line and fails the Horizontal Line Test. *p* does not have an inverse.

61.
$$f(x) = (x+3)^2,\ x \ge -3 \Rightarrow y \ge 0$$
$$y = (x+3)^2,\ x \ge -3,\ y \ge 0$$
$$x = (y+3)^2,\ y \ge -3,\ x \ge 0$$
$$\sqrt{x} = y+3,\ y \ge -3,\ x \ge 0$$
$$y = \sqrt{x} - 3,\ x \ge 0,\ y \ge -3$$

This is a function of *x*, so *f* has an inverse.

$$f^{-1}(x) = \sqrt{x} - 3,\ x \ge 0$$

63.
$$f(x) = \begin{cases} x+3, & x < 0 \\ 6-x, & x \ge 0 \end{cases}$$

This graph fails the Horizontal Line Test, so *f* does not have an inverse.

65. $h(x) = |x + 1| - 1$

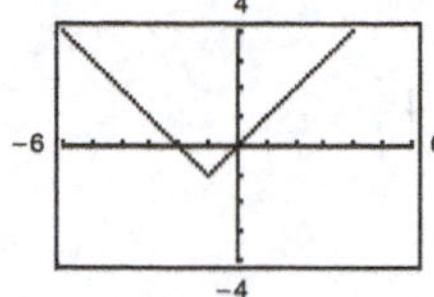

The graph fails the Horizontal Line Test, so h does not have an inverse.

67. $f(x) = \sqrt{2x + 3} \Rightarrow x \ge -\frac{3}{2}, y \ge 0$

$$y = \sqrt{2x + 3}, x \ge -\frac{3}{2}, y \ge 0$$
$$x = \sqrt{2y + 3}, y \ge -\frac{3}{2}, x \ge 0$$
$$x^2 = 2y + 3, x \ge 0, y \ge -\frac{3}{2}$$
$$y = \frac{x^2 - 3}{2}, x \ge 0, y \ge -\frac{3}{2}$$

This is a function of x, so f has an inverse.

$$f^{-1}(x) = \frac{x^2 - 3}{2}, x \ge 0$$

69.
$$f(x) = \frac{6x + 4}{4x + 5}$$
$$y = \frac{6x + 4}{4x + 5}$$
$$x = \frac{6y + 4}{4y + 5}$$
$$x(4y + 5) = 6y + 4$$
$$4xy + 5x = 6y + 4$$
$$4xy - 6y = -5x + 4$$
$$y(4x - 6) = -5x + 4$$
$$y = \frac{-5x + 4}{4x - 6}$$
$$= \frac{5x - 4}{6 - 4x}$$

This is a function of x, so f has an inverse.

$$f^{-1}(x) = \frac{5x - 4}{6 - 4x}$$

71. $f(x) = |x + 2|$

domain of $f: x \ge -2$, range of $f: y \ge 0$

$$f(x) = |x + 2|$$
$$y = |x + 2|$$
$$x = y + 2$$
$$x - 2 = y$$

So, $f^{-1}(x) = x - 2$.

domain of $f^{-1}: x \ge 0$, range of $f^{-1}: y \ge -2$

73. $f(x) = (x + 6)^2$

domain of $f: x \ge -6$, range of $f: y \ge 0$

$$f(x) = (x + 6)^2$$
$$y = (x + 6)^2$$
$$x = (y + 6)^2$$
$$\sqrt{x} = y + 6$$
$$\sqrt{x} - 6 = y$$

So, $f^{-1}(x) = \sqrt{x} - 6$.

domain of $f^{-1}: x \ge 0$, range of $f^{-1}: y \ge -6$

75. $f(x) = -2x^2 + 5$

domain of $f: x \ge 0$, range of $f: y \le 5$

$$f(x) = -2x^2 + 5$$
$$y = -2x^2 + 5$$
$$x = -2y^2 + 5$$
$$x - 5 = -2y^2$$
$$5 - x = 2y^2$$
$$\sqrt{\frac{5 - x}{2}} = y$$
$$\frac{\sqrt{5 - x}}{\sqrt{2}} \cdot \frac{\sqrt{2}}{\sqrt{2}} = y$$
$$\frac{\sqrt{2(5 - x)}}{2} = y$$

So, $f^{-1}(x) = \frac{\sqrt{-2(x - 5)}}{2}$.

domain of $f^{-1}(x): x \le 5$, range of $f^{-1}(x): y \ge 0$

77. $f(x) = |x - 4| + 1$

domain of $f: x \geq 4$, range of $f: y \geq 1$

$$\begin{aligned} f(x) &= |x - 4| + 1 \\ y &= x - 3 \\ x &= y - 3 \\ x + 3 &= y \end{aligned}$$

So, $f^{-1}(x) = x + 3$.

domain of $f^{-1}: x \geq 1$, range of $f^{-1}: y \geq 4$

In Exercises 79–83, $f(x) = \frac{1}{8}x - 3$, $f^{-1}(x) = 8(x + 3)$, $g(x) = x^3$, $g^{-1}(x) = \sqrt[3]{x}$.

79. $$\begin{aligned} \left(f^{-1} \circ g^{-1}\right)(1) &= f^{-1}\left(g^{-1}(1)\right) \\ &= f^{-1}\left(\sqrt[3]{1}\right) \\ &= 8\left(\sqrt[3]{1} + 3\right) = 32 \end{aligned}$$

81. $$\begin{aligned} \left(f^{-1} \circ f^{-1}\right)(4) &= f^{-1}\left(f^{-1}(4)\right) \\ &= f^{-1}(8[4 + 3]) \\ &= 8\left[8(4 + 3) + 3\right] \\ &= 8\left[8(7) + 3\right] \\ &= 8(59) = 472 \end{aligned}$$

83. $$\begin{aligned} (f \circ g)(x) &= f(g(x)) = f\left(x^3\right) = \tfrac{1}{8}x^3 - 3 \\ y &= \tfrac{1}{8}x^3 - 3 \\ x &= \tfrac{1}{8}y^3 - 3 \\ x + 3 &= \tfrac{1}{8}y^3 \\ 8(x + 3) &= y^3 \\ \sqrt[3]{8(x + 3)} &= y \\ (f \circ g)^{-1}(x) &= 2\sqrt[3]{x + 3} \end{aligned}$$

In Exercises 85 and 87, $f(x) = x + 4$, $f^{-1}(x) = x - 4$, $g(x) = 2x - 5$, $g^{-1}(x) = \dfrac{x + 5}{2}$.

85. $$\begin{aligned} \left(g^{-1} \circ f^{-1}\right)(x) &= g^{-1}\left(f^{-1}(x)\right) \\ &= g^{-1}(x - 4) \\ &= \frac{(x - 4) + 5}{2} \\ &= \frac{x + 1}{2} \end{aligned}$$

87. $$\begin{aligned} (f \circ g)(x) &= f(g(x)) \\ &= f(2x - 5) \\ &= (2x - 5) + 4 \\ &= 2x - 1 \end{aligned}$$

$$(f \circ g)^{-1}(x) = \frac{x + 1}{2}$$

Note: Comparing Exercises 85 and 87, $(f \circ g)^{-1}(x) = \left(g^{-1} \circ f^{-1}\right)(x)$.

89. (a) $$\begin{aligned} y &= 10 + 0.75x \\ x &= 10 + 0.75y \\ x - 10 &= 0.75y \\ \frac{x - 10}{0.75} &= y \end{aligned}$$

So, $f^{-1}(x) = \dfrac{x - 10}{0.75}$.

x = hourly wage, y = number of units produced

(b) $y = \dfrac{24.25 - 10}{0.75} = 19$

So, 19 units are produced.

91. False. $f(x) = x^2$ is even and does not have an inverse.

93.

x	1	3	4	6
f	1	2	6	7

x	1	2	6	7
$f^{-1}(x)$	1	3	4	6

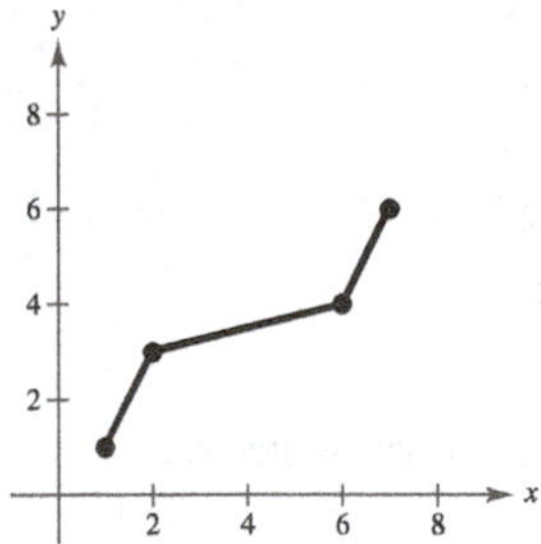

95. Let $(f \circ g)(x) = y$. Then $x = (f \circ g)^{-1}(y)$. Also,

$$\begin{aligned} (f \circ g)(x) = y \Rightarrow f(g(x)) &= y \\ g(x) &= f^{-1}(y) \\ x &= g^{-1}\left(f^{-1}(y)\right) \\ x &= \left(g^{-1} \circ f^{-1}\right)(y). \end{aligned}$$

Because f and g are both one-to-one functions, $(f \circ g)^{-1} = g^{-1} \circ f^{-1}$.

97. If $f(x) = k(2 - x - x^3)$ has an inverse and $f^{-1}(3) = -2$, then $f(-2) = 3$. So,

$$f(-2) = k\left(2 - (-2) - (-2)^3\right) = 3$$
$$k(2 + 2 + 8) = 3$$
$$12k = 3$$
$$k = \tfrac{3}{12} = \tfrac{1}{4}.$$

So, $k = \frac{1}{4}$.

99.

There is an inverse function $f^{-1}(x) = \sqrt{x - 1}$ because the domain of f is equal to the range of f^{-1} and the range of f is equal to the domain of f^{-1}.

101. This situation could be represented by a one-to-one function if the runner does not stop to rest. The inverse function would represent the time in hours for a given number of miles completed.

Review Exercises for Chapter 2

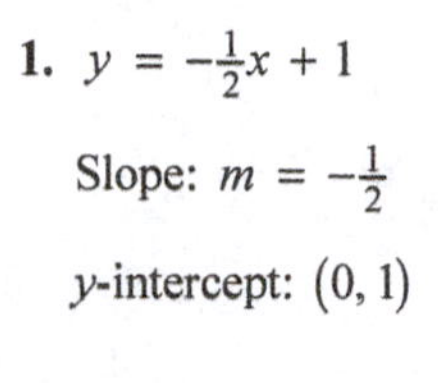

1. $y = -\frac{1}{2}x + 1$

Slope: $m = -\frac{1}{2}$

y-intercept: $(0, 1)$

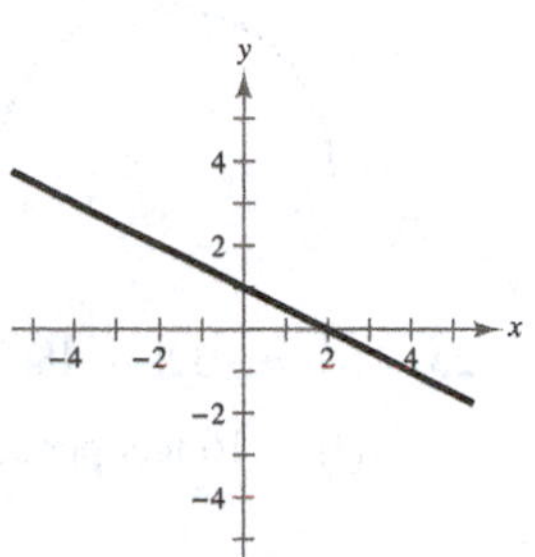

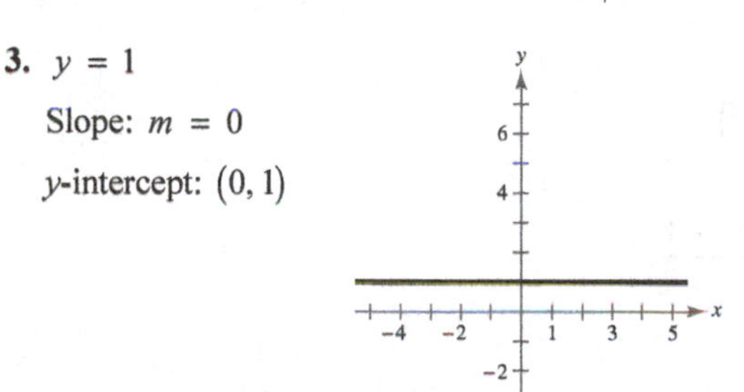

3. $y = 1$

Slope: $m = 0$

y-intercept: $(0, 1)$

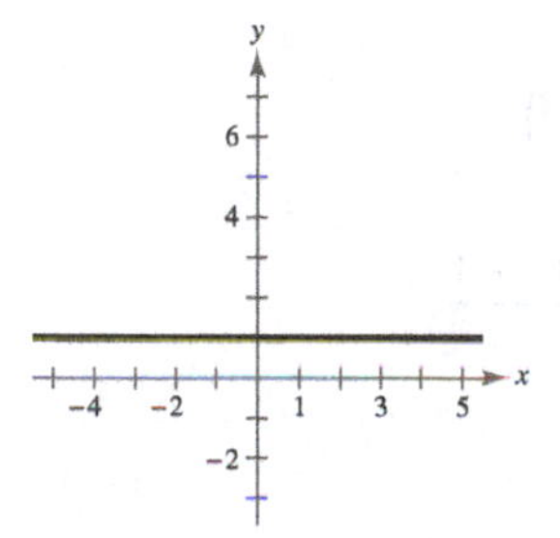

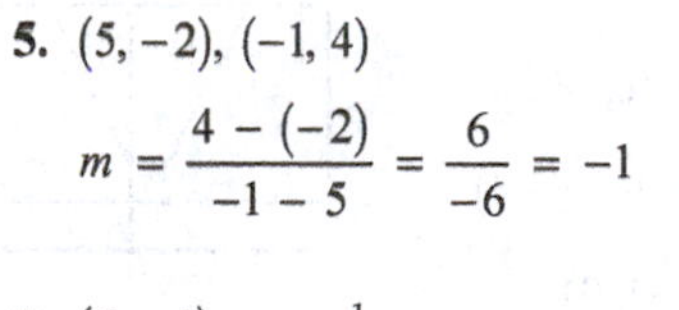

5. $(5, -2), (-1, 4)$

$$m = \frac{4 - (-2)}{-1 - 5} = \frac{6}{-6} = -1$$

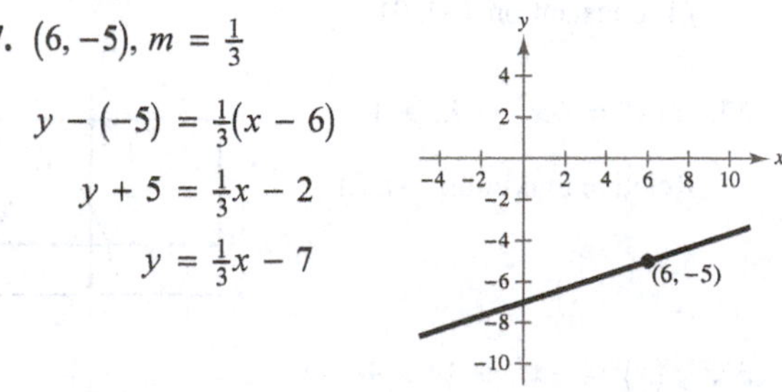

7. $(6, -5), m = \frac{1}{3}$

$$y - (-5) = \tfrac{1}{3}(x - 6)$$
$$y + 5 = \tfrac{1}{3}x - 2$$
$$y = \tfrac{1}{3}x - 7$$

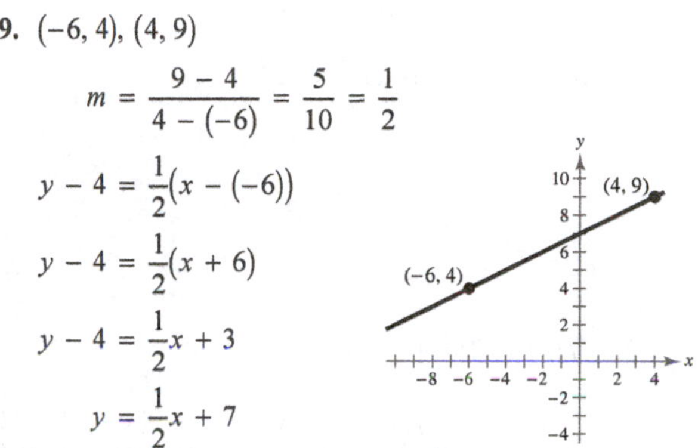

9. $(-6, 4), (4, 9)$

$$m = \frac{9 - 4}{4 - (-6)} = \frac{5}{10} = \frac{1}{2}$$
$$y - 4 = \frac{1}{2}(x - (-6))$$
$$y - 4 = \frac{1}{2}(x + 6)$$
$$y - 4 = \frac{1}{2}x + 3$$
$$y = \frac{1}{2}x + 7$$

11. Point: $(3, -2)$

$$5x - 4y = 8$$
$$y = \tfrac{5}{4}x - 2$$

(a) Parallel slope: $m = \frac{5}{4}$

$$y - (-2) = \tfrac{5}{4}(x - 3)$$
$$y + 2 = \tfrac{5}{4}x - \tfrac{15}{4}$$
$$y = \tfrac{5}{4}x - \tfrac{23}{4}$$

(b) Perpendicular slope: $m = -\frac{4}{5}$

$$y - (-2) = -\tfrac{4}{5}(x - 3)$$
$$y + 2 = -\tfrac{4}{5}x + \tfrac{12}{5}$$
$$y = -\tfrac{4}{5}x + \tfrac{2}{5}$$

13. *Verbal Model:* Sale price = (List price) − (Discount)

Labels: Sale price $= S$
List price $= L$
Discount = 20% of $L = 0.2L$

Equation: $S = L - 0.2L$
$S = 0.8L$

15. $16x - y^4 = 0$

$y^4 = 16x$

$y = \pm 2\sqrt[4]{x}$

No, y is not a function of x. Some x-values correspond to two y-values.

17. $y = \sqrt{1 - x}$

Yes, the equation represents y as a function of x. Each x-value, $x \le 1$, corresponds to only one y-value.

19. $g(x) = x^{4/3}$

(a) $g(8) = 8^{4/3} = 2^4 = 16$

(b) $g(t + 1) = (t + 1)^{4/3}$

(c) $(-27)^{4/3} = (-3)^4 = 81$

(d) $g(-x) = (-x)^{4/3} = x^{4/3}$

21. $f(x) = \sqrt{25 - x^2}$

Domain: $25 - x^2 \ge 0$

$(5 + x)(5 - x) \ge 0$

Critical numbers: $x = \pm 5$

Test intervals: $(-\infty, -5), (-5, 5), (5, \infty)$

Test: Is $25 - x^2 \ge 0$?

Solution set: $-5 \le x \le 5$

Domain: all real numbers x such that $-5 \le x \le 5$, or $[-5, 5]$

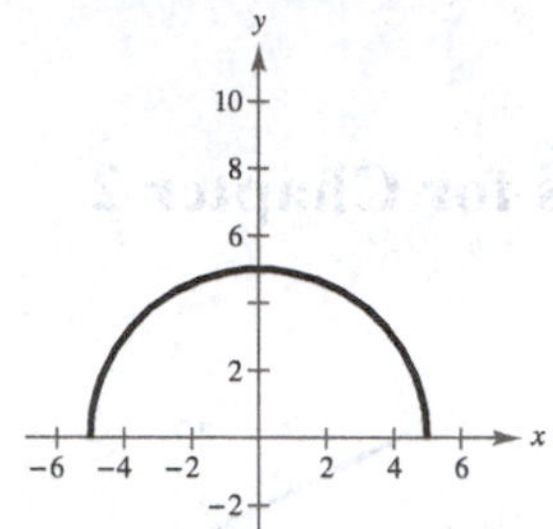

23. $v(t) = -32t + 48$

$v(1) = 16$ feet per second

25. $f(x) = 2x^2 + 3x - 1$

$$\frac{f(x + h) - f(x)}{h} = \frac{\left[2(x + h)^2 + 3(x + h) - 1\right] - \left(2x^2 + 3x - 1\right)}{h}$$

$$= \frac{2x^2 + 4xh + 2h^2 + 3x + 3h - 1 - 2x^2 - 3x + 1}{h}$$

$$= \frac{h(4x + 2h + 3)}{h} = 4x + 2h + 3, \quad h \ne 0$$

27. $y = (x - 3)^2$

A vertical line intersects the graph no more than once, so y *is* a function of x.

29. $f(x) = 5x^2 + 4x - 1$

$5x^2 + 4x - 1 = 0$

$(5x - 1)(x + 1) = 0$

$5x - 1 = 0 \Rightarrow x = \frac{1}{5}$

$x + 1 = 0 \Rightarrow x = -1$

31. $f(x) = \sqrt{2x + 1}$

$\sqrt{2x + 1} = 0$

$2x + 1 = 0$

$2x = -1$

$x = -\frac{1}{2}$

33. $f(x) = |x| + |x + 1|$

f is increasing on $(0, \infty)$.

f is decreasing on $(-\infty, -1)$.

f is constant on $(-1, 0)$.

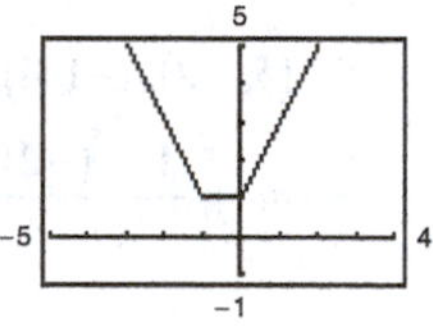

35. $f(x) = -x^2 + 2x + 1$

Relative maximum: $(1, 2)$

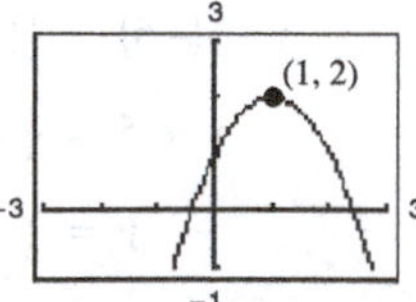

37. $f(x) = -x^2 + 8x - 4$

$$\frac{f(4) - f(0)}{4 - 0} = \frac{12 - (-4)}{4} = 4$$

The average rate of change of f from $x_1 = 0$ to $x_2 = 4$ is 4.

39. $f(x) = x^5 + 4x - 7$

$$\begin{aligned} f(-x) &= (-x)^5 + 4(-x) - 7 \\ &= -x^5 - 4x - 7 \\ &\neq f(x) \neq -f(x) \end{aligned}$$

The function is neither even nor odd, so the graph has no symmetry.

41. $f(x) = 2x\sqrt{x^2 + 3}$

$$f(-x) = 2(-x)\sqrt{(-x)^2 + 3} = -2x\sqrt{x^2 + 3} = -f(x)$$

The function is odd, so the graph has origin symmetry.

43. (a) $f(2) = -6, f(-1) = 3$

Points: $(2, -6), (-1, 3)$

$$m = \frac{3 - (-6)}{-1 - 2} = \frac{9}{-3} = -3$$

$$\begin{aligned} y - (-6) &= -3(x - 2) \\ y + 6 &= -3x + 6 \\ y &= -3x \\ f(x) &= -3x \end{aligned}$$

(b)

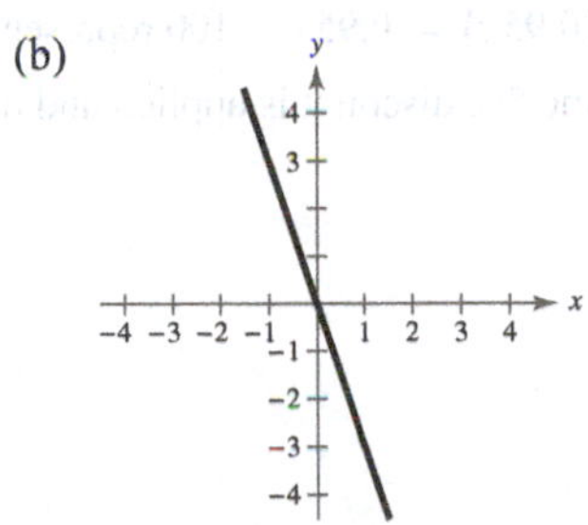

45. $g(x) = [\![x]\!] - 2$

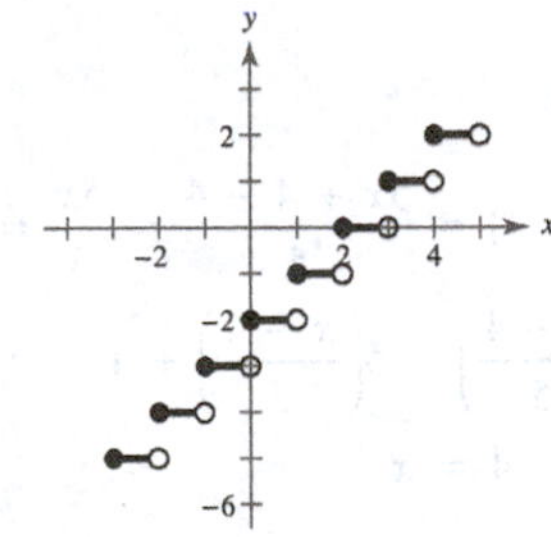

47. $f(x) = \begin{cases} 5x - 3, & x \geq -1 \\ -4x + 5, & x < -1 \end{cases}$

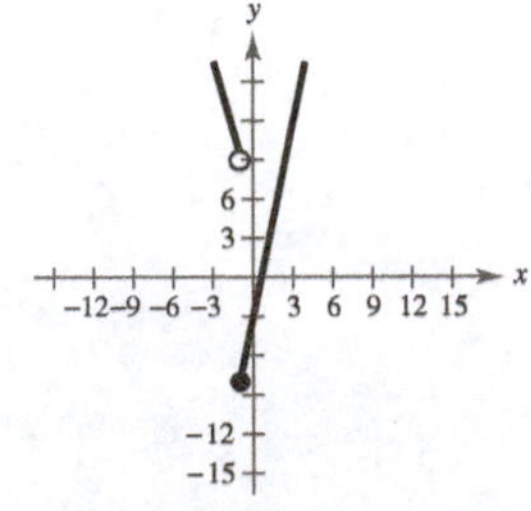

49. (a) $f(x) = x^2$

(b) $h(x) = x^2 - 9$

Vertical shift 9 units downward

(c)

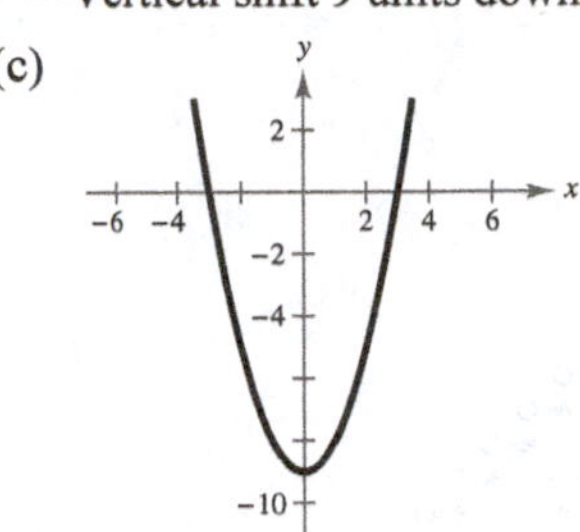

(d) $h(x) = f(x) - 9$

51. (a) $f(x) = \sqrt{x}$

(b) $h(x) = -\sqrt{x} + 4$

Reflection in the x-axis and a vertical shift 4 units upward

(c)

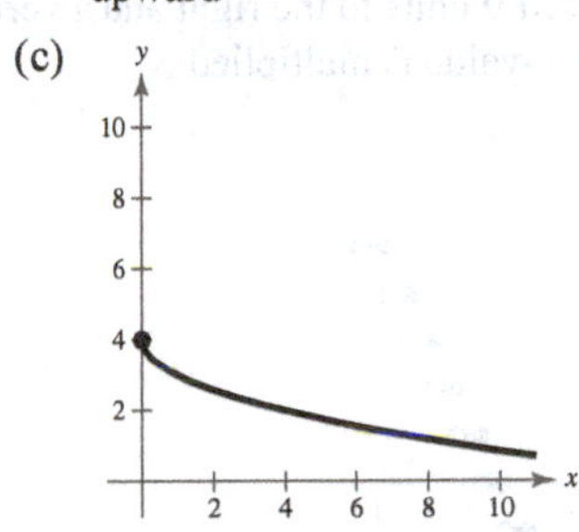

(d) $h(x) = -f(x) + 4$

53. (a) $f(x) = x^2$

(b) $h(x) = -(x + 2)^2 + 3$

Reflection in the x-axis, a horizontal shift 2 units to the left, and a vertical shift 3 units upward

(c)

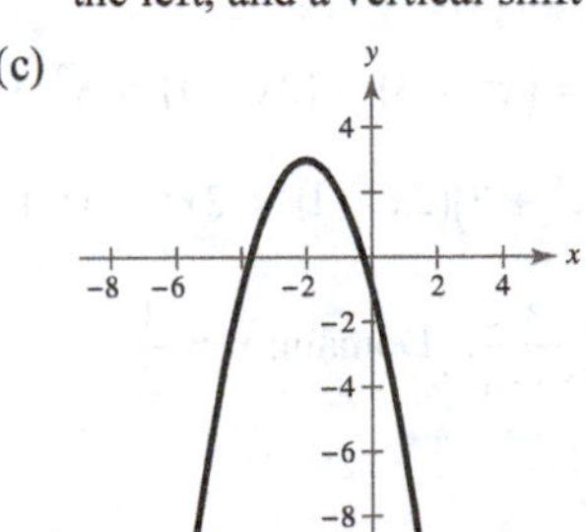

(d) $h(x) = -f(x + 2) + 3$

55. (a) $f(x) = [\![x]\!]$

(b) $h(x) = -[\![x]\!] + 6$

Reflection in the x-axis and a vertical shift 6 units upward

(c)

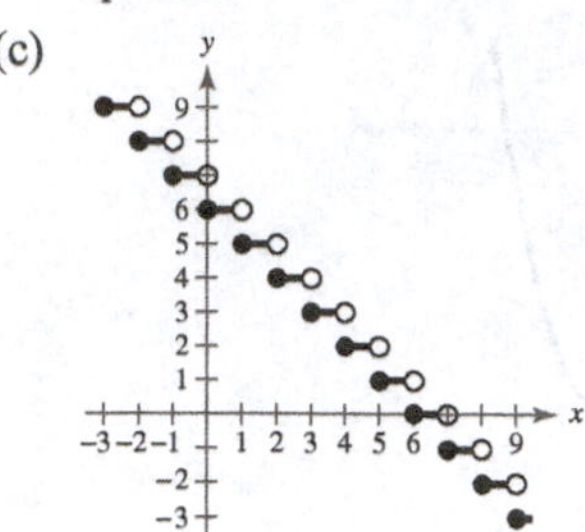

(d) $h(x) = -f(x) + 6$

57. (a) $f(x) = [\![x]\!]$

(b) $h(x) = 5[\![x - 9]\!]$

Horizontal shift 9 units to the right and a vertical stretch (each y-value is multiplied by 5)

(c)

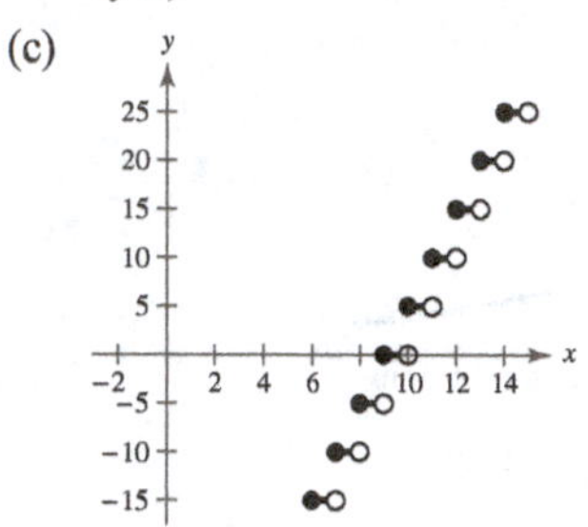

(d) $h(x) = 5f(x - 9)$

59. $f(x) = x^2 + 3$, $g(x) = 2x - 1$

(a) $(f + g)(x) = (x^2 + 3) + (2x - 1) = x^2 + 2x + 2$

(b) $(f - g)(x) = (x^2 + 3) - (2x - 1) = x^2 - 2x + 4$

(c) $(fg)(x) = (x^2 + 3)(2x - 1) = 2x^3 - x^2 + 6x - 3$

(d) $\left(\dfrac{f}{g}\right)(x) = \dfrac{x^2 + 3}{2x - 1}$, Domain: $x \neq \dfrac{1}{2}$

61. $f(x) = \frac{1}{3}x - 3$, $g(x) = 3x + 1$

The domains of f and g are all real numbers.

(a)
$$\begin{aligned}(f \circ g)(x) &= f(g(x))\\ &= f(3x + 1)\\ &= \tfrac{1}{3}(3x + 1) - 3\\ &= x + \tfrac{1}{3} - 3\\ &= x - \tfrac{8}{3}\end{aligned}$$

Domain: all real numbers

(b)
$$\begin{aligned}(g \circ f)(x) &= g(f(x))\\ &= g\left(\tfrac{1}{3}x - 3\right)\\ &= 3\left(\tfrac{1}{3}x - 3\right) + 1\\ &= x - 9 + 1\\ &= x - 8\end{aligned}$$

Domain: all real numbers

In Exercise 63 use the following functions.
$f(x) = x - 100$, $g(x) = 0.95x$

63. $(f \circ g)(x) = f(0.95x) = 0.95x - 100$ represents the sale price if first the 5% discount is applied and then the \$100 rebate.

65.
$$\begin{aligned}f(x) &= \frac{x - 4}{5}\\ y &= \frac{x - 4}{5}\\ x &= \frac{y - 4}{5}\\ 5x &= y - 4\\ y &= 5x + 4\end{aligned}$$

So, $f^{-1}(x) = 5x + 4$.

$$f(f^{-1}(x)) = f(5x + 4) = \frac{5x + 4 - 4}{5} = \frac{5x}{5} = x$$

$$\begin{aligned}f^{-1}(f(x)) &= f^{-1}\left(\frac{x - 4}{5}\right) = 5\left(\frac{x - 4}{5}\right) + 4\\ &= x - 4 + 4 = x\end{aligned}$$

67. $f(x) = (x - 1)^2$

No, the function does not have an inverse because the horizontal line test fails.

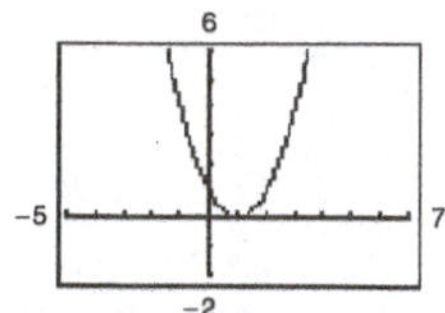

69. (a)

$$\begin{aligned} f(x) &= \tfrac{1}{2}x - 3 \\ y &= \tfrac{1}{2}x - 3 \\ x &= \tfrac{1}{2}y - 3 \\ x + 3 &= \tfrac{1}{2}y \\ 2(x + 3) &= y \\ f^{-1}(x) &= 2x + 6 \end{aligned}$$

(b)

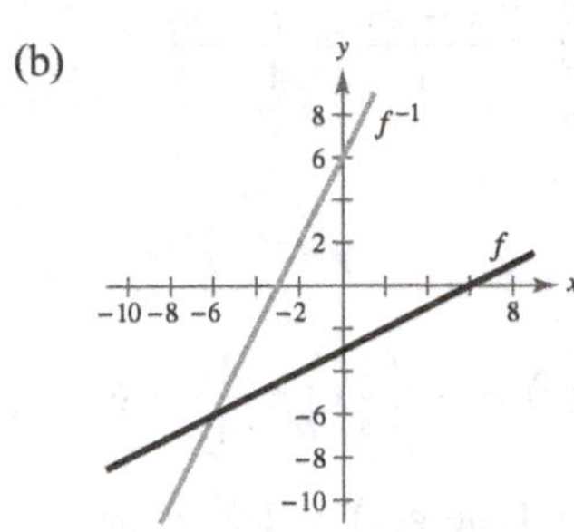

(c) The graph of f^{-1} is the reflection of the graph of f in the line $y = x$.

(d) The domains and ranges of f and f^{-1} are the set of all real numbers.

71. $f(x) = 2(x - 4)^2$ is increasing on $(4, \infty)$.

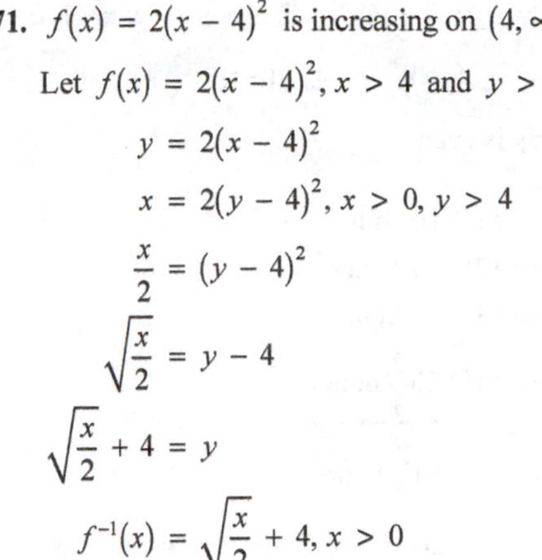

Let $f(x) = 2(x - 4)^2, x > 4$ and $y > 0$.

$$\begin{aligned} y &= 2(x - 4)^2 \\ x &= 2(y - 4)^2, x > 0, y > 4 \\ \frac{x}{2} &= (y - 4)^2 \\ \sqrt{\frac{x}{2}} &= y - 4 \\ \sqrt{\frac{x}{2}} + 4 &= y \\ f^{-1}(x) &= \sqrt{\frac{x}{2}} + 4, x > 0 \end{aligned}$$

73. False. The graph is reflected in the x-axis, shifted 9 units to the left, then shifted 13 units downward.

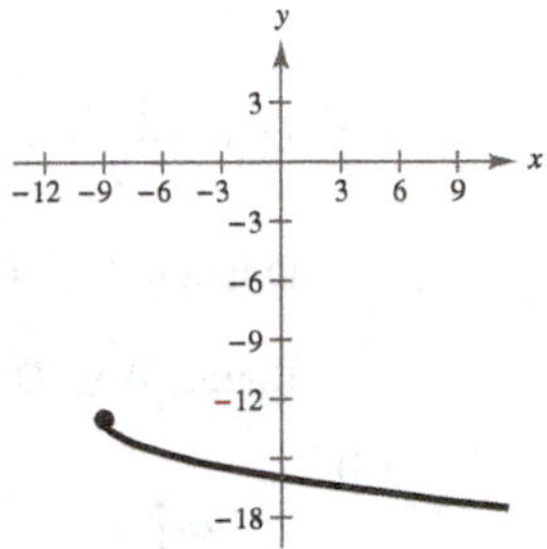

Problem Solving for Chapter 2

1. (a) $W_1 = 0.07S + 2000$

(b) $W_2 = 0.05S + 2300$

(c)

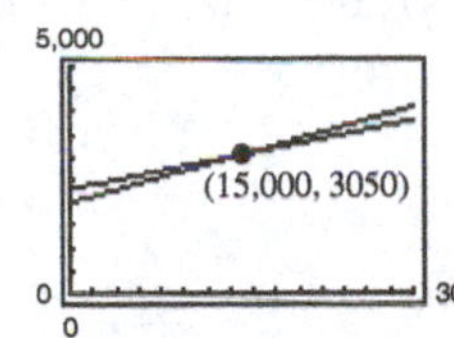

Point of intersection: (15,000, 3050)

Both jobs pay the same, \$3050, if you sell \$15,000 per month.

(d) No. If you think you can sell \$20,000 per month, keep your current job with the higher commission rate. For sales over \$15,000 it pays more than the other job.

3. (a) Let $f(x)$ and $g(x)$ be two even functions.

Then define $h(x) = f(x) \pm g(x)$.

$$\begin{aligned} h(-x) &= f(-x) \pm g(-x) \\ &= f(x) \pm g(x) \text{ because } f \text{ and } g \text{ are even} \\ &= h(x) \end{aligned}$$

So, $h(x)$ is also even.

(b) Let $f(x)$ and $g(x)$ be two odd functions

Then define $h(x) = f(x) \pm g(x)$.

$$\begin{aligned} h(-x) &= f(-x) \pm g(-x) \\ &= -f(x) \pm g(x) \text{ because } f \text{ and } g \text{ are odd} \\ &= -h(x) \end{aligned}$$

So, $h(x)$ is also odd. $\left(\text{If } f(x) \neq g(x)\right)$

(c) Let $f(x)$ be odd and $g(x)$ be even. Then define $h(x) = f(x) \pm g(x)$.

$$\begin{aligned} h(-x) &= f(-x) \pm g(-x) \\ &= -f(x) \pm g(x) \text{ because } f \text{ is odd and } g \text{ is even} \\ &\neq h(x) \neq -h(x) \end{aligned}$$

So, $h(x)$ is neither odd nor even.

5. $f(x) = a_{2n}x^{2n} + a_{2n-2}x^{2n-2} + \cdots + a_2x^2 + a_0$

$f(-x) = a_{2n}(-x)^{2n} + a_{2n-2}(-x)^{2n-2} + \cdots + a_2(-x)^2 + a_0 = a_{2n}x^{2n} + a_{2n-2}x^{2n-2} + \cdots + a_2x^2 + a_0 = f(x)$

So, $f(x)$ is even.

7. (a) April 11: 10 hours

April 12: 24 hours

April 13: 24 hours

April 14: $23\frac{2}{3}$ hours

Total: $81\frac{2}{3}$ hours

(b) $\text{Speed} = \dfrac{\text{distance}}{\text{time}} = \dfrac{2100}{81\frac{2}{3}} = \dfrac{180}{7} = 25\frac{5}{7}$ mph

(c) $D = -\dfrac{180}{7}t + 3400$

Domain: $0 \le t \le \dfrac{1190}{9}$

Range: $0 \le D \le 3400$

(d)

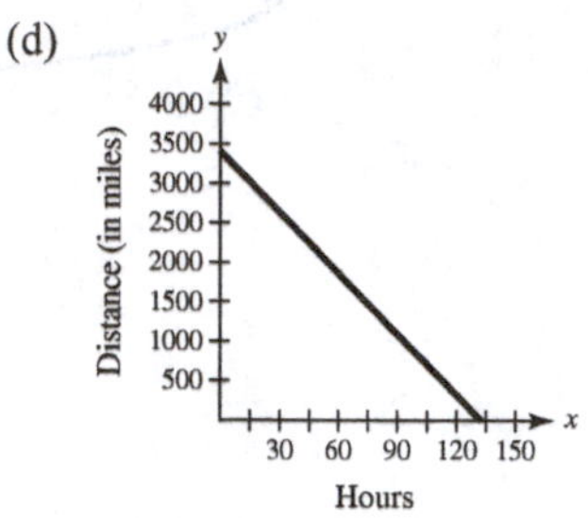

9. (a)–(d) Use $f(x) = 4x$ and $g(x) = x + 6$.

(a) $(f \circ g)(x) = f(x + 6) = 4(x + 6) = 4x + 24$

(b) $(f \circ g)^{-1}(x) = \dfrac{x - 24}{4} = \dfrac{1}{4}x - 6$

(c) $f^{-1}(x) = \dfrac{1}{4}x$

$g^{-1}(x) = x - 6$

(d) $(g^{-1} \circ f^{-1})(x) = g^{-1}\left(\dfrac{1}{4}x\right) = \dfrac{1}{4}x - 6$

(e) $f(x) = x^3 + 1$ and $g(x) = 2x$

$(f \circ g)(x) = f(2x) = (2x)^3 + 1 = 8x^3 + 1$

$(f \circ g)^{-1}(x) = \sqrt[3]{\dfrac{x - 1}{8}} = \dfrac{1}{2}\sqrt[3]{x - 1}$

$f^{-1}(x) = \sqrt[3]{x - 1}$

$g^{-1}(x) = \dfrac{1}{2}x$

$(g^{-1} \circ f^{-1})(x) = g^{-1}\left(\sqrt[3]{x - 1}\right) = \dfrac{1}{2}\sqrt[3]{x - 1}$

(f) Answers will vary.

(g) Conjecture: $(f \circ g)^{-1}(x) = (g^{-1} \circ f^{-1})(x)$

11. $H(x) = \begin{cases} 1, & x \ge 0 \\ 0, & x < 0 \end{cases}$

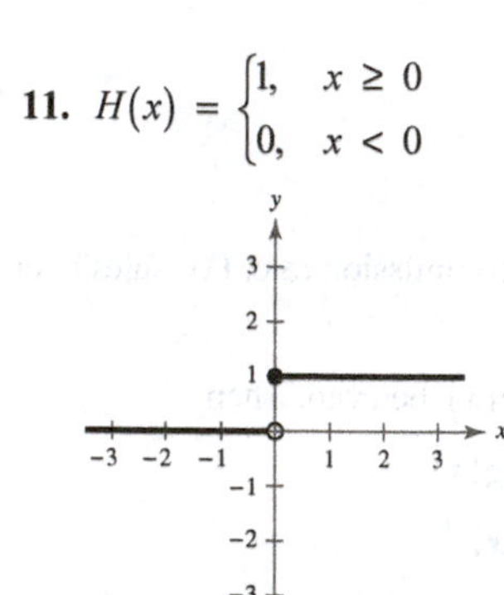

(a) $H(x) - 2$

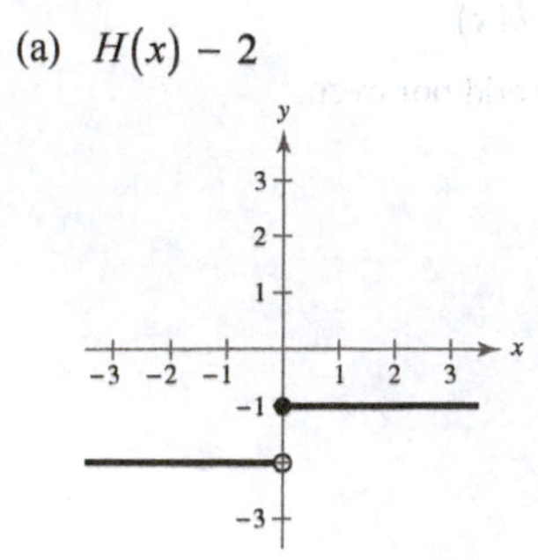

(b) $H(x - 2)$

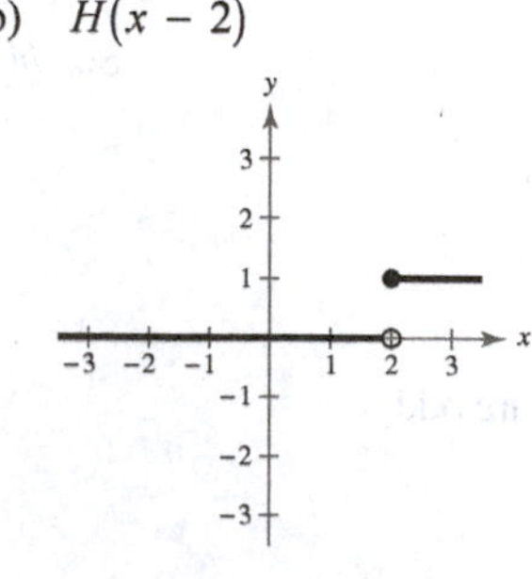

(c) $-H(x)$

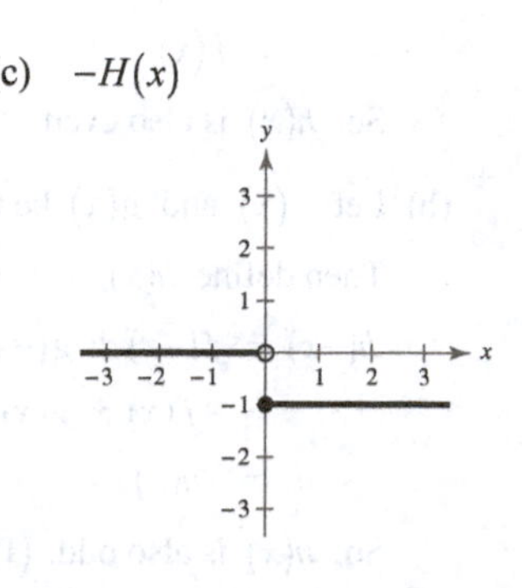

(d) $H(-x)$

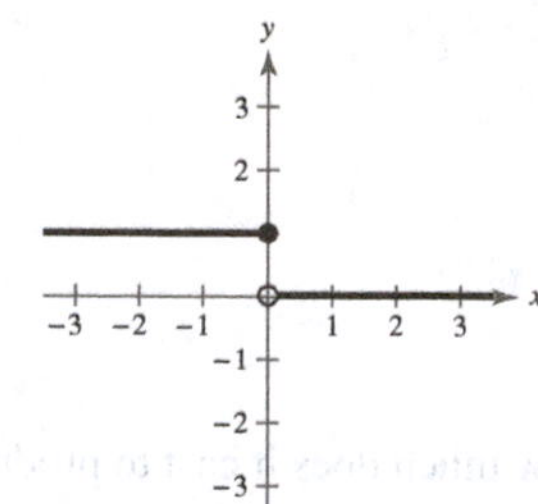

(e) $\frac{1}{2}H(x)$

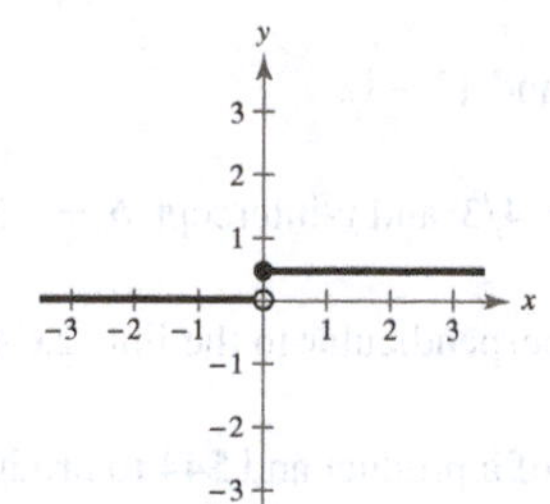

(f) $-H(x-2)+2$

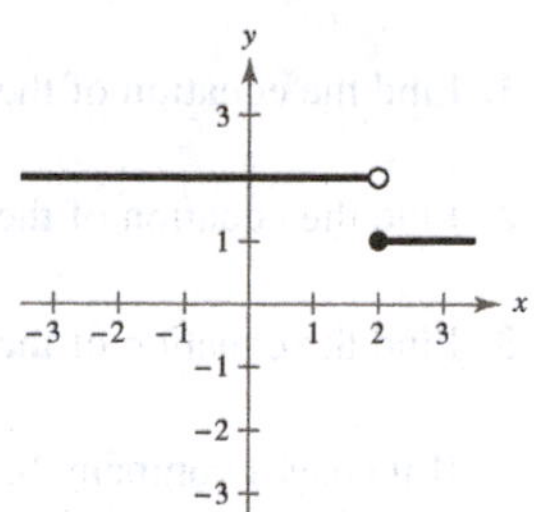

13. $(f \circ (g \circ h))(x) = f((g \circ h)(x)) = f(g(h(x))) = (f \circ g \circ h)(x)$

$((f \circ g) \circ h)(x) = (f \circ g)(h(x)) = f(g(h(x))) = (f \circ g \circ h)(x)$

15.

x	$f(x)$	$f^{-1}(x)$
−4	—	2
−3	4	1
−2	1	0
−1	0	—
0	−2	−1
1	−3	−2
2	−4	—
3	—	—
4	—	−3

(a)

x	$f(f^{-1}(x))$
−4	$f(f^{-1}(-4)) = f(2) = -4$
−2	$f(f^{-1}(-2)) = f(0) = -2$
0	$f(f^{-1}(0)) = f(-1) = 0$
4	$f(f^{-1}(4)) = f(-3) = 4$

(b)

x	$(f + f^{-1})(x)$
−3	$f(-3) + f^{-1}(-3) = 4 + 1 = 5$
−2	$f(-2) + f^{-1}(-2) = 1 + 0 = 1$
0	$f(0) + f^{-1}(0) = -2 + (-1) = -3$
1	$f(1) + f^{-1}(1) = -3 + (-2) = -5$

(c)

x	$(f \cdot f^{-1})(x)$
−3	$f(-3)f^{-1}(-3) = (4)(1) = 4$
−2	$f(-2)f^{-1}(-2) = (1)(0) = 0$
0	$f(0)f^{-1}(0) = (-2)(-1) = 2$
1	$f(1)f^{-1}(1) = (-3)(-2) = 6$

(d)

x	$\lvert f^{-1}(x)\rvert$
−4	$\lvert f^{-1}(-4)\rvert = \lvert 2\rvert = 2$
−3	$\lvert f^{-1}(-3)\rvert = \lvert 1\rvert = 1$
0	$\lvert f^{-1}(0)\rvert = \lvert -1\rvert = 1$
4	$\lvert f^{-1}(4)\rvert = \lvert -3\rvert = 3$

Practice Test for Chapter 2

1. Find the equation of the line through $(2, 4)$ and $(3, -1)$.

2. Find the equation of the line with slope $m = 4/3$ and y-intercept $b = -3$.

3. Find the equation of the line through $(4, 1)$ perpendicular to the line $2x + 3y = 0$.

4. If it costs a company \$32 to produce 5 units of a product and \$44 to produce 9 units, how much does it cost to produce 20 units? (Assume that the cost function is linear.)

5. Given $f(x) = x^2 - 2x + 1$, find $f(x - 3)$.

6. Given $f(x) = 4x - 11$, find $\dfrac{f(x) - f(3)}{x - 3}$.

7. Find the domain and range of $f(x) = \sqrt{36 - x^2}$.

8. Which equations determine y as a function of x?
   (a) $6x - 5y + 4 = 0$
   (b) $x^2 + y^2 = 9$
   (c) $y^3 = x^2 + 6$

9. Sketch the graph of $f(x) = x^2 - 5$.

10. Sketch the graph of $f(x) = |x + 3|$.

11. Sketch the graph of $f(x) = \begin{cases} 2x + 1, & \text{if } x \ge 0, \\ x^2 - x, & \text{if } x < 0. \end{cases}$

12. Use the graph of $f(x) = |x|$ to graph the following:
    (a) $f(x + 2)$
    (b) $-f(x) + 2$

13. Given $f(x) = 3x + 7$ and $g(x) = 2x^2 - 5$, find the following:
    (a) $(g - f)(x)$
    (b) $(fg)(x)$

14. Given $f(x) = x^2 - 2x + 16$ and $g(x) = 2x + 3$, find $f(g(x))$.

15. Given $f(x) = x^3 + 7$, find $f^{-1}(x)$.

16. Which of the following functions have inverses?
    (a) $f(x) = |x - 6|$
    (b) $f(x) = ax + b, a \ne 0$
    (c) $f(x) = x^3 - 19$

17. Given $f(x) = \sqrt{\dfrac{3 - x}{x}}, 0 < x \le 3,$ find $f^{-1}(x)$.

Exercises 18–20, true or false?

18. $y = 3x + 7$ and $y = \frac{1}{3}x - 4$ are perpendicular.

19. $(f \circ g)^{-1} = g^{-1} \circ f^{-1}$

20. If a function has an inverse, then it must pass both the Vertical Line Test and the Horizontal Line Test.

CHAPTER 3
Polynomial Functions

CHAPTER 3
Polynomial Functions

Section 3.1 Quadratic Functions and Models

1. polynomial

3. quadratic; parabola

5. $f(x) = x^2 - 2$ opens upward and has vertex $(0, -2)$. Matches graph (b).

6. $f(x) = (x + 1)^2 - 2$ opens upward and has vertex $(-1, -2)$. Matches graph (a).

7. $f(x) = -(x - 4)^2$ opens downward and has vertex $(4, 0)$. Matches graph (c).

8. $f(x) = 4 - (x - 2)^2 = -(x - 2)^2 + 4$ opens downward and has vertex $(2, 4)$. Matches graph (d).

9. (a) $y = \frac{1}{2}x^2$

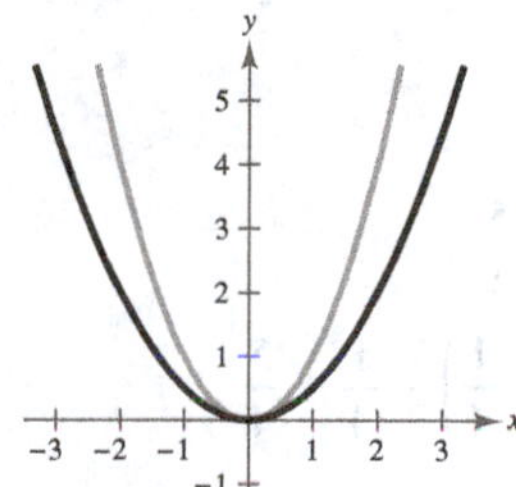

Vertical shrink

(b) $y = -\frac{1}{8}x^2$

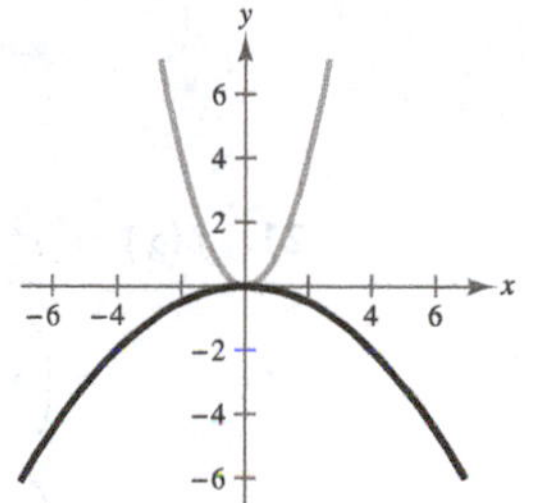

Vertical shrink and a reflection in the x-axis

(c) $y = \frac{3}{2}x^2$

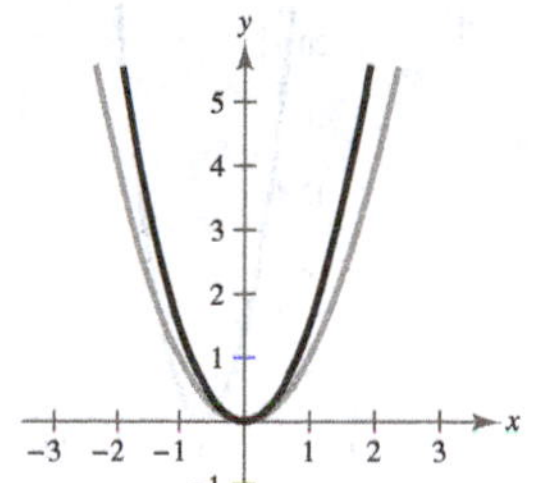

Vertical stretch

(d) $y = -3x^2$

Vertical stretch and a reflection in the x-axis

11. (a) $y = (x - 1)^2$

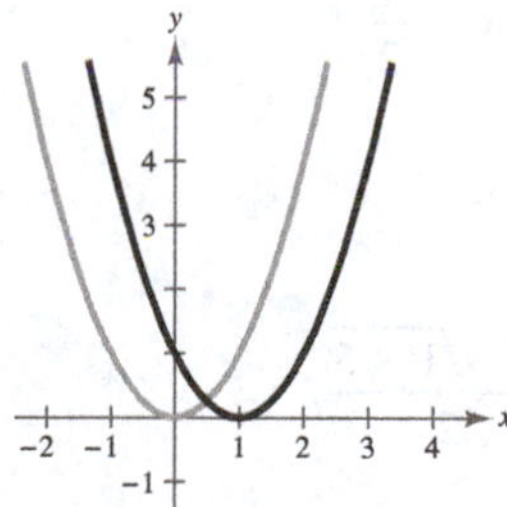

Horizontal shift one unit to the right

(b) $y = (3x)^2 + 1$

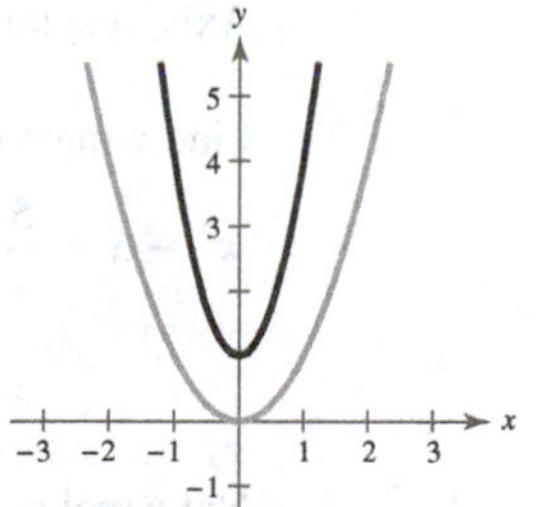

Horizontal shrink and a vertical shift one unit upward

(c) $y = \left(\frac{1}{3}x\right)^2 - 3$

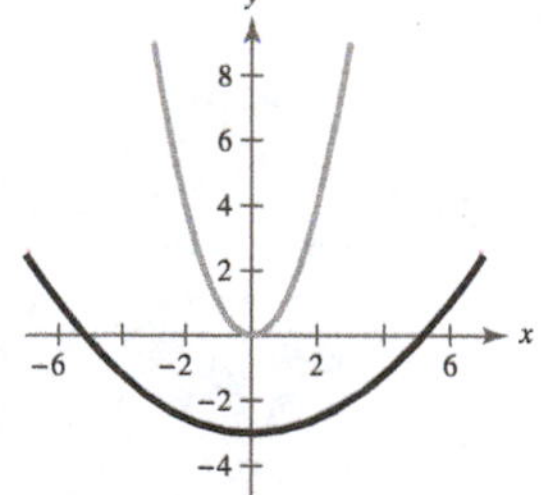

Horizontal stretch and a vertical shift three units downward

(d) $y = (x + 3)^2$

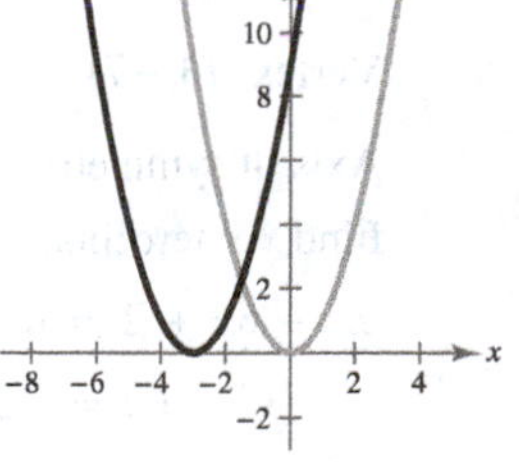

Horizontal shift three units to the left

13. $f(x) = x^2 - 6x$

$= (x^2 - 6x + 9) - 9$

$= (x - 3)^2 - 9$

Vertex: $(3, -9)$

Axis of symmetry: $x = 3$

Find x-intercepts:

$x^2 - 6x = 0$

$x(x - 6) = 0$

$x = 0$

$x - 6 = 0 \Rightarrow x = 6$

x-intercepts: $(0, 0), (6, 0)$

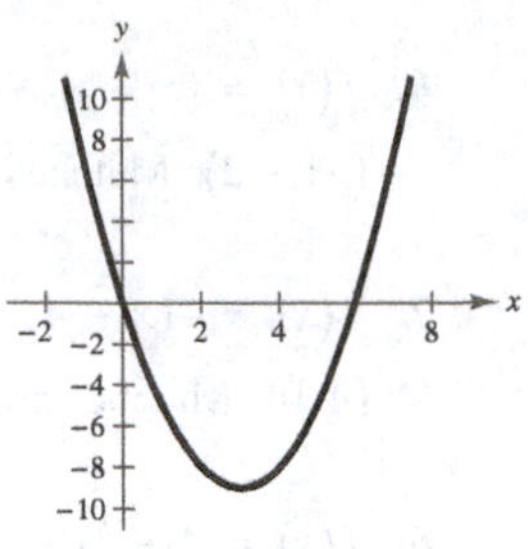

15. $h(x) = x^2 - 8x + 16 = (x - 4)^2$

Vertex: $(4, 0)$

Axis of symmetry: $x = 4$

Find x-intercepts:

$(x - 4)^2 = 0$

$x - 4 = 0$

$x = 4$

x-intercept: $(4, 0)$

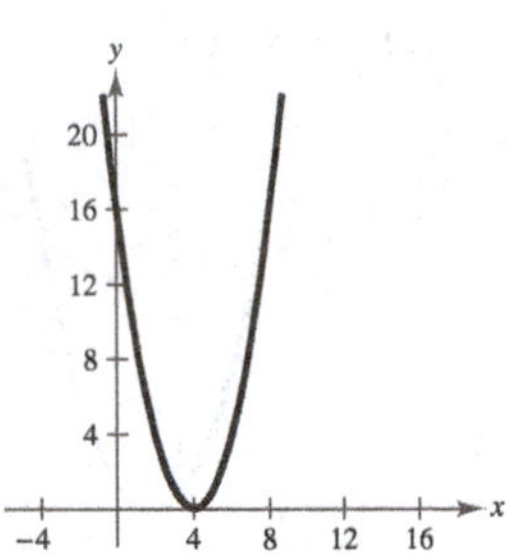

17. $f(x) = x^2 - 6x + 2$

$= (x^2 - 6x + 9) - 9 + 2$

$= (x^2 - 6x + 9) - 7$

$= (x - 3)^2 - 7$

Vertex: $(3, -7)$

Axis of symmetry: $x = 3$

Find x-intercepts:

$x^2 - 6x + 2 = 0$

$x^2 - 6x = -2$

$x^2 - 6x + 9 = -2 + 9$

$(x - 3)^2 = 7$

$x = 3 \pm \sqrt{7}$

x-intercepts: $(3 \pm 7, 0)$

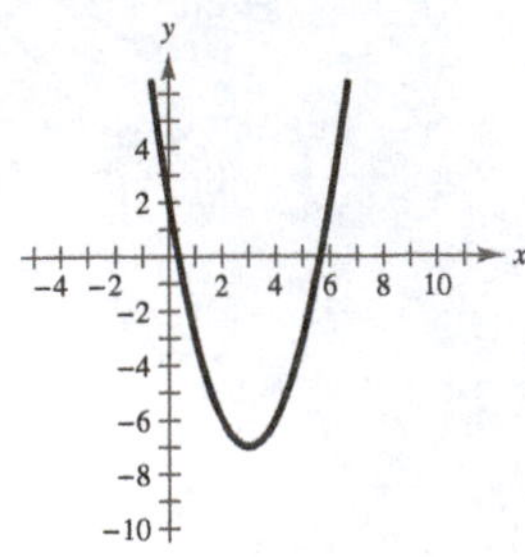

19. $f(x) = x^2 - 8x + 21$

$= (x^2 - 8x + 16) - 16 + 21$

$= (x - 4)^2 + 5$

Vertex: $(4, 5)$

Axis of symmetry: $x = 4$

Find x-intercepts:

$x^2 - 8x + 21 = 0$

$x^2 - 8x = -21$

$x^2 - 8x + 16 = -21 + 16$

$(x - 4)^2 = -5$

$x - 4 = \pm\sqrt{-5}$

$x = 4 \pm \sqrt{5}i$

Not a real number

No x-intercepts

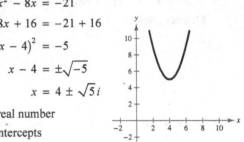

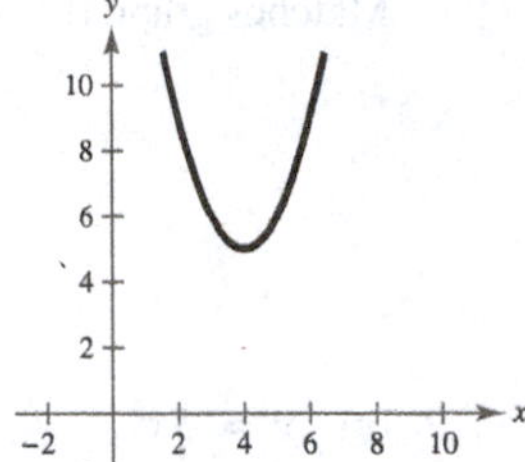

21. $f(x) = x^2 - x + \frac{5}{4}$

$= \left(x^2 - x + \frac{1}{4}\right) - \frac{1}{4} + \frac{5}{4}$

$= \left(x - \frac{1}{2}\right)^2 + 1$

Vertex: $\left(\frac{1}{2}, 1\right)$

Axis of symmetry: $x = \frac{1}{2}$

Find x-intercepts:

$x^2 - x + \frac{5}{4} = 0$

$x = \frac{1 \pm \sqrt{1 - 5}}{2}$

Not a real number

No x-intercepts

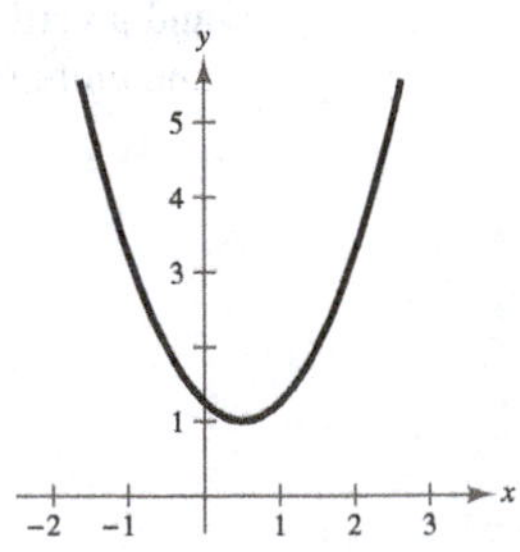

23. $f(x) = -x^2 + 2x + 5$

$= -(x^2 - 2x + 1) - (-1) + 5$

$= -(x - 1)^2 + 6$

Vertex: $(1, 6)$

Axis of symmetry: $x = 1$

Find x-intercepts:

$-x^2 + 2x + 5 = 0$

$x^2 - 2x - 5 = 0$

$$x = \frac{2 \pm \sqrt{4 + 20}}{2}$$

$= 1 \pm \sqrt{6}$

x-intercepts: $(1 - \sqrt{6}, 0), (1 + \sqrt{6}, 0)$

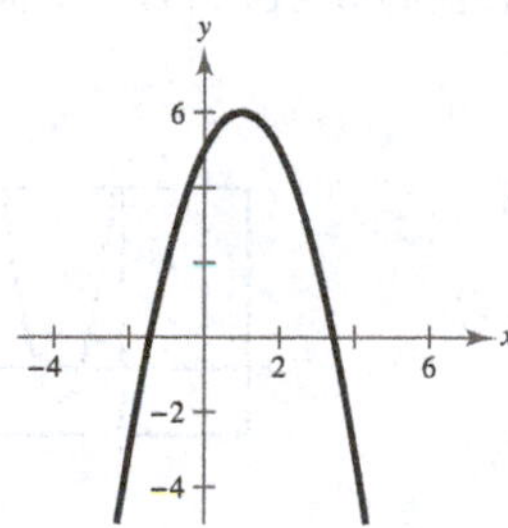

25. $h(x) = 4x^2 - 4x + 21$

$$= 4\left(x^2 - x + \frac{1}{4}\right) - 4\left(\frac{1}{4}\right) + 21$$

$$= 4\left(x - \frac{1}{2}\right)^2 + 20$$

Vertex: $\left(\frac{1}{2}, 20\right)$

Axis of symmetry: $x = \frac{1}{2}$

Find x-intercepts:

$4x^2 - 4x + 21 = 0$

$$x = \frac{4 \pm \sqrt{16 - 336}}{2(4)}$$

Not a real number

No x-intercepts

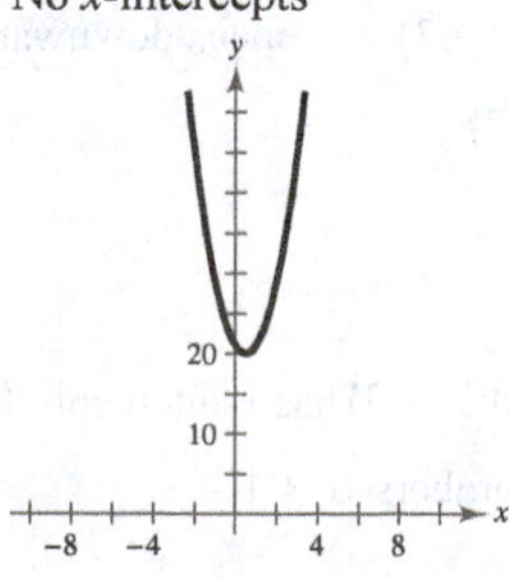

27. $f(x) = -(x^2 + 2x - 3) = -(x + 1)^2 + 4$

Vertex: $(-1, 4)$

Axis of symmetry: $x = -1$

x-intercepts: $(-3, 0), (1, 0)$

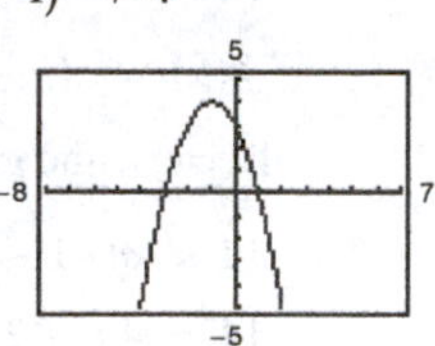

29. $g(x) = x^2 + 8x + 11 = (x + 4)^2 - 5$

Vertex: $(-4, -5)$

Axis of symmetry: $x = -4$

x-intercepts: $(-4 \pm \sqrt{5}, 0)$

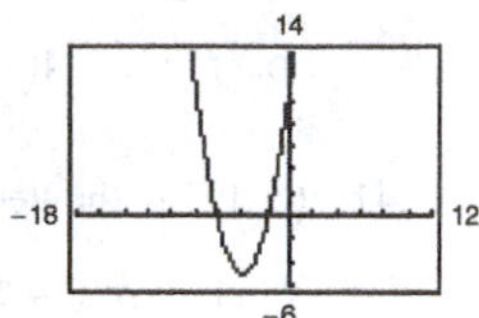

31. $f(x) = -2x^2 + 12x - 18$

$= -2(x^2 - 6x + 9 - 9) - 18$

$= -2(x^2 - 6x + 9) + 18 - 18$

$= -2(x^2 - 6x + 9)$

$= -2(x - 3)^2$

Vertex: $(3, 0)$

Axis of symmetry: $x = 3$

x-intercept: $(3, 0)$

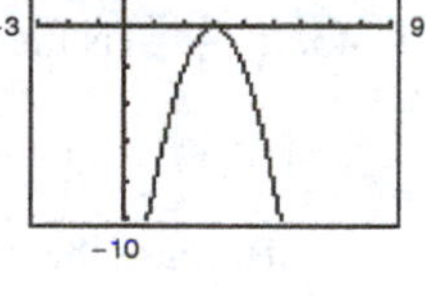

33. $g(x) = \frac{1}{2}(x^2 + 4x - 2) = \frac{1}{2}(x + 2)^2 - 3$

Vertex: $(-2, -3)$

Axis of symmetry: $x = -2$

x-intercepts: $(-2 \pm \sqrt{6}, 0)$

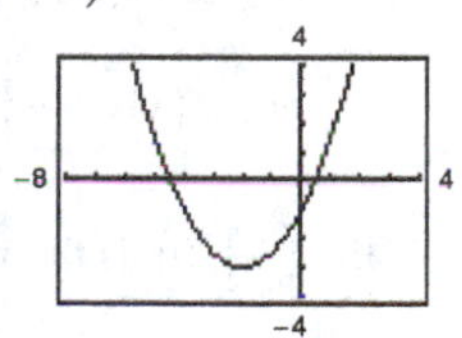

35. $(-2, -1)$ is the vertex.

$f(x) = a(x + 2)^2 - 1$

Because the graph passes through $(0, 3)$,

$3 = a(0 + 2)^2 - 1$

$3 = 4a - 1$

$4 = 4a$

$1 = a.$

So, $y = (x + 2)^2 - 1.$

37. $(-2, 5)$ is the vertex.

$f(x) = a(x + 2)^2 + 5$

Because the graph passes through $(0, 9)$,

$9 = a(0 + 2)^2 + 5$

$4 = 4a$

$1 = a.$

So, $f(x) = 1(x + 2)^2 + 5 = (x + 2)^2 + 5.$

39. $(1, -2)$ is the vertex.

$f(x) = a(x - 1)^2 - 2$

Because the graph passes through $(-1, 14)$,

$14 = a(-1 - 1)^2 - 2$

$14 = 4a - 2$

$16 = 4a$

$4 = a.$

So, $f(x) = 4(x - 1)^2 - 2.$

41. $(5, 12)$ is the vertex.

$f(x) = a(x - 5)^2 + 12$

Because the graph passes through $(7, 15)$,

$15 = a(7 - 5)^2 + 12$

$3 = 4a \Rightarrow a = \frac{3}{4}.$

So, $f(x) = \frac{3}{4}(x - 5)^2 + 12.$

43. $\left(-\frac{1}{4}, \frac{3}{2}\right)$ is the vertex.

$f(x) = a\left(x + \frac{1}{4}\right)^2 + \frac{3}{2}$

Because the graph passes through $(-2, 0)$,

$0 = a\left(-2 + \frac{1}{4}\right)^2 + \frac{3}{2}$

$-\frac{3}{2} = \frac{49}{16}a \Rightarrow a = -\frac{24}{49}.$

So, $f(x) = -\frac{24}{49}\left(x + \frac{1}{4}\right)^2 + \frac{3}{2}.$

45. $\left(-\frac{5}{2}, 0\right)$ is the vertex.

$f(x) = a\left(x + \frac{5}{2}\right)^2$

Because the graph passes through $\left(-\frac{7}{2}, -\frac{16}{3}\right)$,

$-\frac{16}{3} = a\left(-\frac{7}{2} + \frac{5}{2}\right)^2$

$-\frac{16}{3} = a.$

So, $f(x) = -\frac{16}{3}\left(x + \frac{5}{2}\right)^2.$

47. $y = x^2 - 2x - 3$

$0 = x^2 - 2x - 3$

$0 = (x - 3)(x + 1)$

$x = 3$ or $x = -1$

x-intercepts: $(3, 0), (-1, 0)$

49. $y = 2x^2 + 5x - 3$

$0 = 2x^2 + 5x - 3$

$0 = (2x - 1)(x + 3)$

$2x - 1 = 0 \Rightarrow x = \frac{1}{2}$

$x + 3 = 0 \Rightarrow x = -3$

x-intercepts: $\left(\frac{1}{2}, 0\right), (-3, 0)$

51. $f(x) = x^2 - 4x$

x-intercepts: $(0, 0), (4, 0)$

$0 = x^2 - 4x$

$0 = x(x - 4)$

$x = 0$ or $x = 4$

The x-intercepts and the solutions of $f(x) = 0$ are the same.

53. $f(x) = x^2 - 9x + 18$

x-intercepts: $(3, 0), (6, 0)$

$0 = x^2 - 9x + 18$

$0 = (x - 3)(x - 6)$

$x = 3$ or $x = 6$

The x-intercepts and the solutions of $f(x) = 0$ are the same.

55. $f(x) = 2x^2 - 7x - 30$

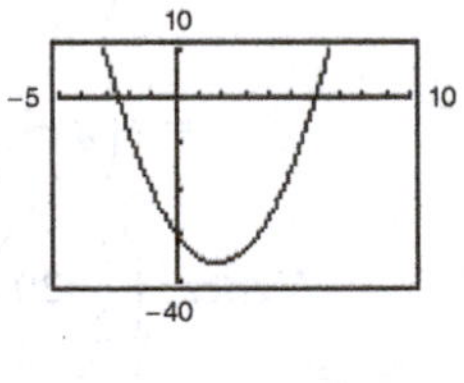

x-intercepts: $\left(-\frac{5}{2}, 0\right), (6, 0)$

$0 = 2x^2 - 7x - 30$

$0 = (2x + 5)(x - 6)$

$x = -\frac{5}{2}$ or $x = 6$

The x-intercepts and the solutions of $f(x) = 0$ are the same.

57. $f(x) = [x - (-3)](x - 3)$ opens upward

$= (x + 3)(x - 3)$

$= x^2 - 9$

$g(x) = -[x - (-3)](x - 3)$ opens downward

$= -(x + 3)(x - 3)$

$= -(x^2 - 9)$

$= -x^2 + 9$

Note: $f(x) = a(x + 3)(x - 3)$ has x-intercepts $(-3, 0)$ and $(3, 0)$ for all real numbers $a \neq 0$.

59. $f(x) = [x - (-1)](x - 4)$ opens upward

$= (x + 1)(x - 4)$

$= x^2 - 3x - 4$

$g(x) = -[x - (-1)](x - 4)$ opens downward

$= -(x + 1)(x - 4)$

$= -(x^2 - 3x - 4)$

$= -x^2 + 3x + 4$

Note: $f(x) = a(x + 1)(x - 4)$ has x-intercepts $(-1, 0)$ and $(4, 0)$ for all real numbers $a \neq 0$.

61. $f(x) = [x - (-3)]\left[x - \left(-\frac{1}{2}\right)\right](2)$ opens upward

$= (x + 3)\left(x + \frac{1}{2}\right)(2)$

$= (x + 3)(2x + 1)$

$= 2x^2 + 7x + 3$

$g(x) = -(2x^2 + 7x + 3)$ opens downward

$= -2x^2 - 7x - 3$

Note: $f(x) = a(x + 3)(2x + 1)$ has x-intercepts $(-3, 0)$ and $\left(-\frac{1}{2}, 0\right)$ for all real numbers $a \neq 0$.

63. Let x = the first number and y = the second number. Then the sum is

$x + y = 110 \Rightarrow y = 110 - x.$

The product is $P(x) = xy = x(110 - x) = 110x - x^2.$

$P(x) = -x^2 + 110x$

$= -(x^2 - 110x + 3025 - 3025)$

$= -\left[(x - 55)^2 - 3025\right]$

$= -(x - 55)^2 + 3025$

The maximum value of the product occurs at the vertex of $P(x)$ and is 3025. This happens when $x = y = 55$.

65. Let x = the first number and y = the second number. Then the sum is

$x + 2y = 24 \Rightarrow y = \dfrac{24 - x}{2}.$

The product is $P(x) = xy = x\left(\dfrac{24 - x}{2}\right).$

$P(x) = \frac{1}{2}(-x^2 + 24x)$

$= -\frac{1}{2}(x^2 - 24x + 144 - 144)$

$= -\frac{1}{2}\left[(x - 12)^2 - 144\right] = -\frac{1}{2}(x - 12)^2 + 72$

The maximum value of the product occurs at the vertex of $P(x)$ and is 72. This happens when $x = 12$ and $y = (24 - 12)/2 = 6$. So, the numbers are 12 and 6.

67. $y = -\frac{4}{9}x^2 + \frac{24}{9}x + 12$

The vertex occurs at $-\dfrac{b}{2a} = \dfrac{-24/9}{2(-4/9)} = 3.$

The maximum height is

$y(3) = -\frac{4}{9}(3)^2 + \frac{24}{9}(3) + 12 = 16$ feet.

69. $C = 800 - 10x + 0.25x^2 = 0.25x^2 - 10x + 800$

The vertex occurs at $x = -\dfrac{b}{2a} = -\dfrac{-10}{2(0.25)} = 20.$

The cost is minimum when $x = 20$ fixtures.

71. $R(p) = -25p^2 + 1200p$

(a) $R(20) = \$14,000$ thousand $= \$14,000,000$

$R(25) = \$14,375$ thousand $= \$14,375,000$

$R(30) = \$13,500$ thousand $= \$13,500,000$

(b) The revenue is a maximum at the vertex.

$-\dfrac{b}{2a} = \dfrac{-1200}{2(-25)} = 24$

$R(24) = 14,400$

The unit price that will yield a maximum revenue of $14,400,000 is $24.

73. (a) [Diagram: rectangle of height y divided into two parts, each of width x]

$$4x + 3y = 200 \Rightarrow y = \frac{1}{3}(200 - 4x) = \frac{4}{3}(50 - x)$$

$$A = 2xy = 2x\left[\frac{4}{3}(50 - x)\right] = \frac{8}{3}x(50 - x) = \frac{8x(50 - x)}{3}$$

(b) To find the dimensions that produce a maximum enclosed area, you can find the vertex.

To do this, either write the quadratic function in standard form or use $x = -\frac{b}{2a}$, so the coordinates of the vertex are $\left(-\frac{b}{2a}, f\left(-\frac{b}{2a}\right)\right)$.

$$A = \frac{8}{3}x(50 - x) = -\frac{8}{3}(x^2 - 50x) = -\frac{8}{3}(x^2 - 50x + 625 - 625) = -\frac{8}{3}\left[(x - 25)^2 - 625\right] = -\frac{8}{3}(x - 25)^2 + \frac{5000}{3}$$

So the vertex is $\left(25, \frac{5000}{3}\right)$ from the standard form, or is $x = -\frac{b}{2a} = -\frac{\frac{400}{3}}{2\left(-\frac{8}{3}\right)} = \frac{400}{16} = 25$ and $A(25) = \frac{5000}{3}$.

When $x = 25$ feet and $y = \frac{(200 - 4(25))}{3} = \frac{100}{3}$ feet.

The dimensions are $2x = 50$ feet by $33\frac{1}{3}$ feet, and the maximum enclosed area is $\frac{5000}{3} \approx 1666.67$ square feet.

75. True. The equation $-12x^2 - 1 = 0$ has no real solution, so the graph has no x-intercepts.

77. $f(x) = -x^2 + bx - 75$, maximum value: 25

The maximum value, 25, is the y-coordinate of the vertex.

Find the x-coordinate of the vertex:

$$x = -\frac{b}{2a} = -\frac{b}{2(-1)} = \frac{b}{2}$$

$$f(x) = -x^2 + bx - 75$$

$$f\left(\frac{b}{2}\right) = -\left(\frac{b}{2}\right)^2 + b\left(\frac{b}{2}\right) - 75$$

$$25 = -\frac{b^2}{4} + \frac{b^2}{2} - 75$$

$$100 = \frac{b^2}{4}$$

$$400 = b^2$$

$$\pm 20 = b$$

79. $f(x) = ax^2 + bx + c$

$$= a\left(x^2 + \frac{b}{a}x\right) + c$$

$$= a\left(x^2 + \frac{b}{a}x + \frac{b^2}{4a^2} - \frac{b^2}{4a^2}\right) + c$$

$$= a\left(x + \frac{b}{2a}\right)^2 - \frac{b^2}{4a} + c$$

$$= a\left(x + \frac{b}{2a}\right)^2 + \frac{4ac - b^2}{4a}$$

$$f\left(-\frac{b}{2a}\right) = a\left(\frac{b^2}{4a^2}\right) + b\left(-\frac{b}{2a}\right) + c$$

$$= \frac{b^2}{4a} - \frac{b^2}{2a} + c$$

$$= \frac{b^2 - 2b^2 + 4ac}{4a}$$

$$= \frac{4ac - b^2}{4a}$$

So, the vertex occurs at

$$\left(-\frac{b}{2a}, \frac{4ac - b^2}{4a}\right) = \left(-\frac{b}{2a}, f\left(-\frac{b}{2a}\right)\right).$$

81. If $f(x) = ax^2 + bx + c$ has two real zeros, then by the Quadratic Formula they are

$$x = \frac{-b \pm \sqrt{b^2 - 4ac}}{2a}.$$

The average of the zeros of f is

$$\frac{\dfrac{-b - \sqrt{b^2 - 4ac}}{2a} + \dfrac{-b + \sqrt{b^2 - 4ac}}{2a}}{2} = \frac{\dfrac{-2b}{2a}}{2} = -\frac{b}{2a}.$$

This is the x-coordinate of the vertex of the graph.

Section 3.2 Polynomial Functions of Higher Degree

1. continuous

3. n; $n - 1$

5. touches; crosses

7. standard

9. $f(x) = -2x^2 - 5x$ is a parabola with x-intercepts $(0, 0)$ and $\left(-\frac{5}{2}, 0\right)$ and opens downward. Matches graph (c).

10. $f(x) = 2x^3 - 3x + 1$ has intercepts $(0, 1)$, $(1, 0)$, $\left(-\frac{1}{2} - \frac{1}{2}\sqrt{3}, 0\right)$ and $\left(-\frac{1}{2} + \frac{1}{2}\sqrt{3}, 0\right)$. Matches graph (f).

11. $f(x) = -\frac{1}{4}x^4 + 3x^2$ has intercepts $(0, 0)$ and $\left(\pm 2\sqrt{3}, 0\right)$. Matches graph (a).

12. $f(x) = -\frac{1}{3}x^3 + x^2 - \frac{4}{3}$ has y-intercept $\left(0, -\frac{4}{3}\right)$. Matches graph (e).

13. $f(x) = x^4 + 2x^3$ has intercepts $(0, 0)$ and $(-2, 0)$. Matches graph (d).

14. $f(x) = \frac{1}{5}x^5 - 2x^3 + \frac{9}{5}x$ has intercepts $(0, 0)$, $(1, 0)$, $(-1, 0)$, $(3, 0)$, $(-3, 0)$. Matches graph (b).

15. $y = x^3$

(a) $f(x) = (x - 4)^3$

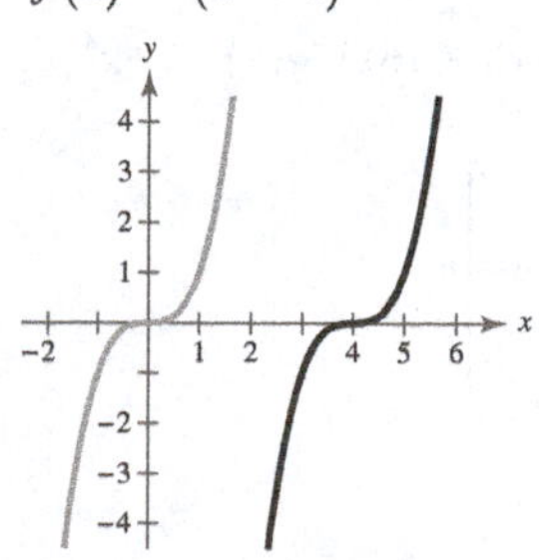

Horizontal shift four units to the right

(b) $f(x) = x^3 - 4$

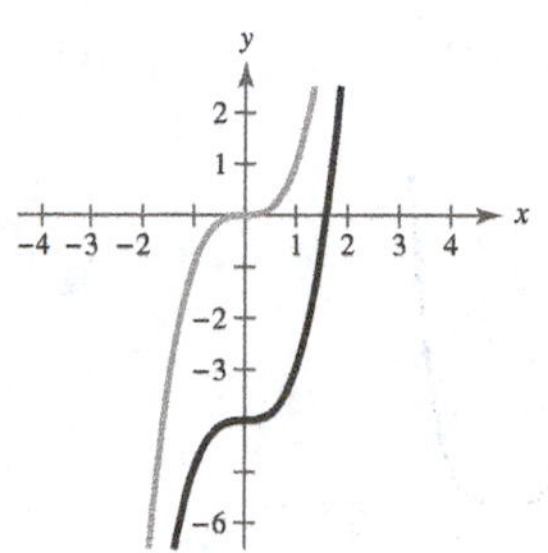

Vertical shift four units downward

(c) $f(x) = -\frac{1}{4}x^3$

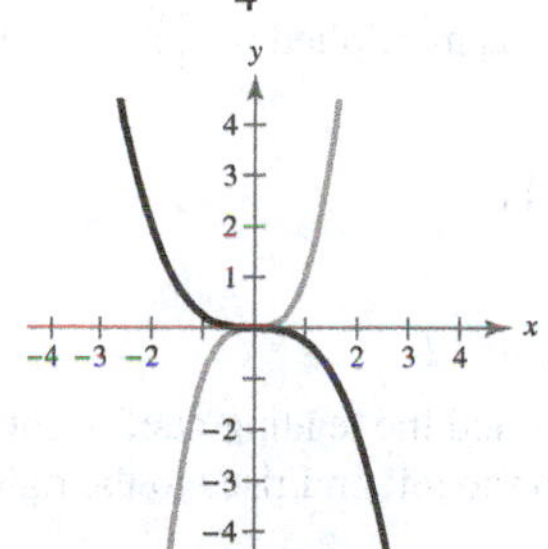

Reflection in the x-axis and a vertical shrink (each y-value is multiplied by $\frac{1}{4}$)

(d) $f(x) = (x - 4)^3 - 4$

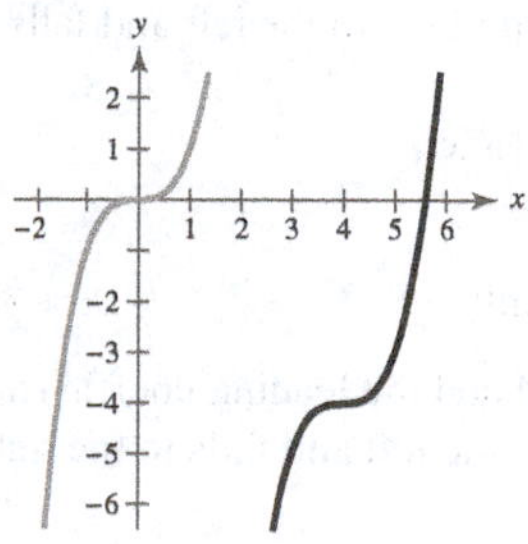

Horizontal shift four units to the right and vertical shift four units downward

17. $y = x^4$

(a) $f(x) = (x + 3)^4$

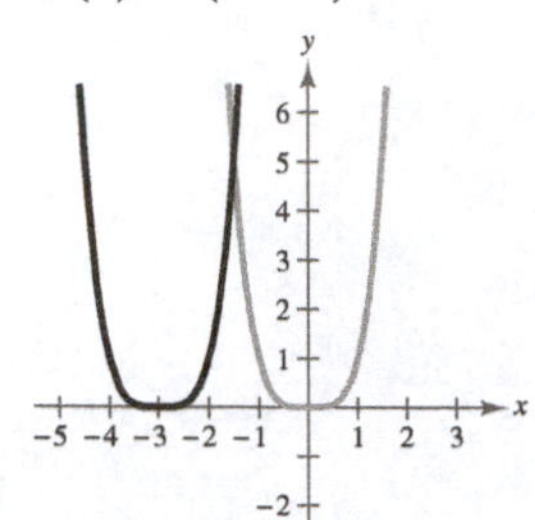

Horizontal shift three units to the left

(b) $f(x) = x^4 - 3$

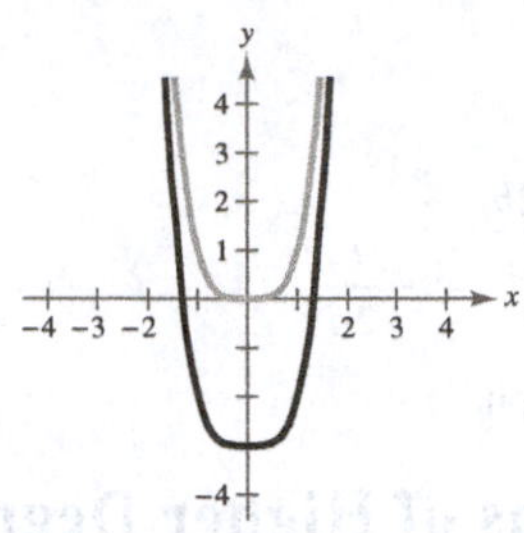

Vertical shift three units downward

(c) $f(x) = 4 - x^4$

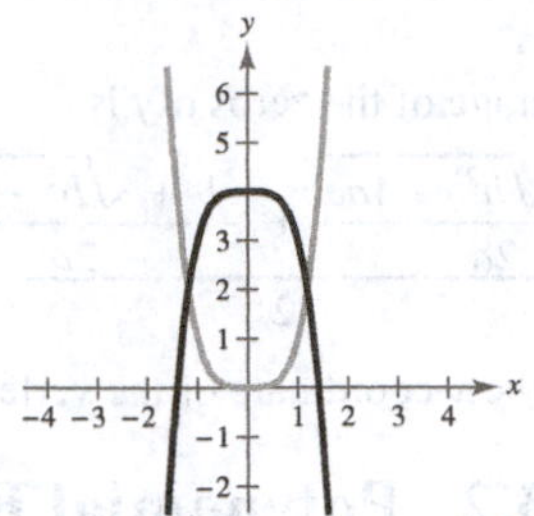

Reflection in the x-axis and then a vertical shift four units upward

(d) $f(x) = \frac{1}{2}(x - 1)^4$

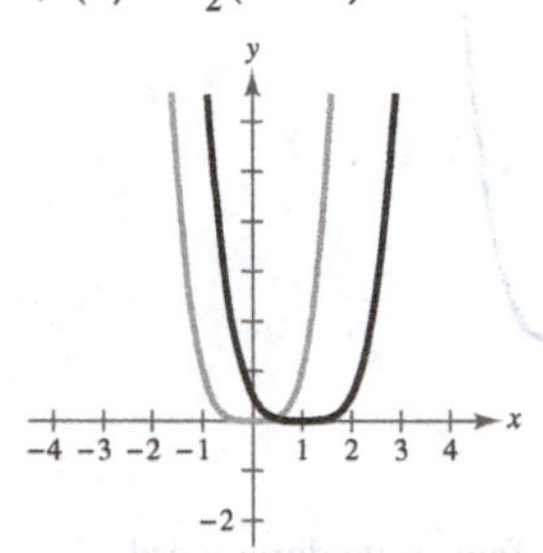

Horizontal shift one unit to the right and a vertical shrink (each y-value is multiplied by $\frac{1}{2}$)

(e) $f(x) = (2x)^4 + 1$

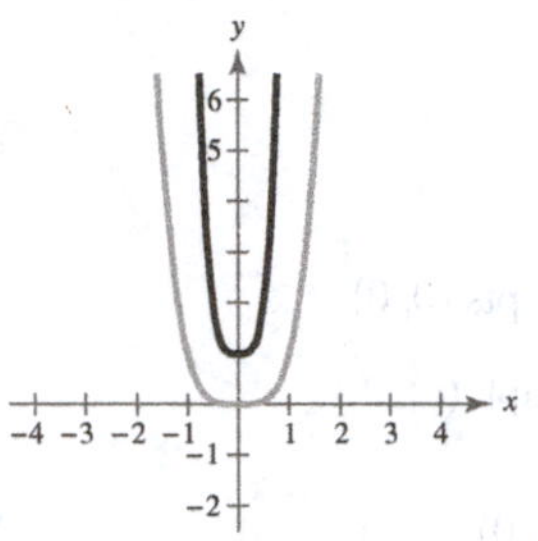

Vertical shift one unit upward and a horizontal shrink (each y-value is multiplied by 16)

(f) $f(x) = \left(\frac{1}{2}x\right)^4 - 2$

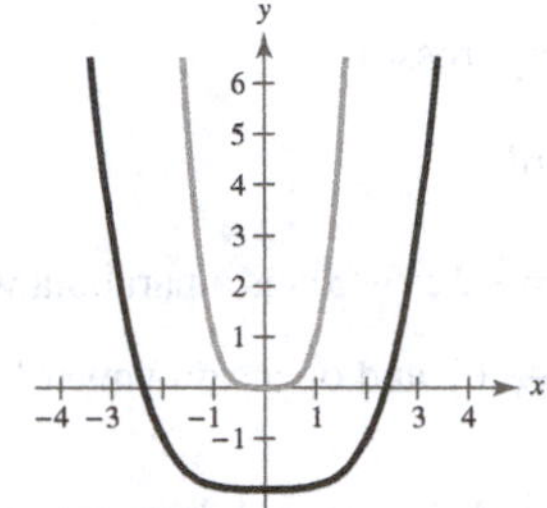

Vertical shift two units downward and a horizontal stretch (each y-value is multipied by $\frac{1}{16}$)

19. $f(x) = 12x^3 + 4x$

Degree: 3

Leading coefficient: 12

The degree is odd and the leading coefficient is positive. The graph falls to the left and rises to the right.

21. $g(x) = 5 - \frac{7}{2}x - 3x^2$

Degree: 2

Leading coefficient: -3

The degree is even and the leading coefficient is negative. The graph falls to the left and falls to the right.

23. $g(x) = 6x - x^3 + x^2$

Degree: 3

Leading coefficient: -1

The degree is odd and the leading coefficient is negative. The graph rises to the left and falls to the right.

25. $f(x) = 9.8x^6 - 1.2x^3$

Degree: 6

Leading coefficient: 9.8

The degree is even and the leading coefficient is positive.

The graph rises to the left and rises to the right.

27. $f(s) = -\frac{7}{8}(s^3 + 5s^2 - 7s + 1)$

Degree: 3

Leading coefficient: $-\frac{7}{8}$

The degree is odd and the leading coefficient is negative.

The graph rises to the left and falls to the right.

29. $f(x) = 3x^3 - 9x + 1$; $g(x) = 3x^3$

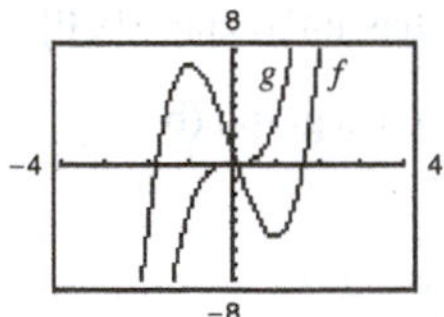

31. $f(x) = -(x^4 - 4x^3 + 16x)$; $g(x) = -x^4$

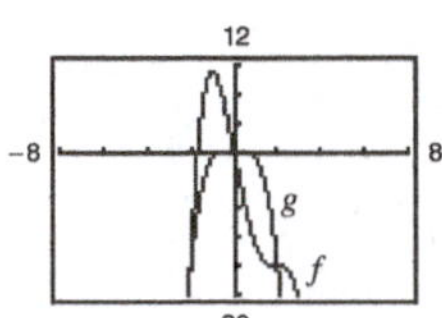

33. $f(x) = x^2 - 36$

(a) $0 = x^2 - 36$

$0 = (x + 6)(x - 6)$

$x + 6 = 0 \qquad x - 6 = 0$

$x = -6 \qquad x = 6$

Zeros: ± 6

(b) Each zero has a multiplicity of one (odd multiplicity).

(c) Turning points: 1 (the vertex of the parabola)

(d)

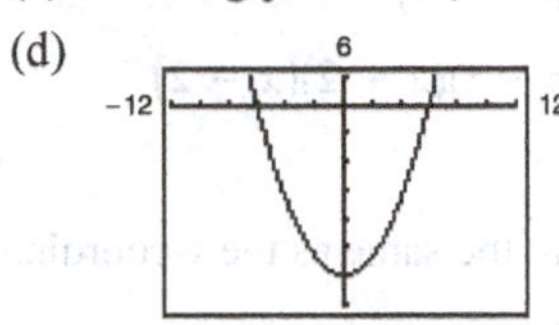

35. $h(t) = t^2 - 6t + 9$

(a) $0 = t^2 - 6t + 9 = (t - 3)^2$

Zero: $t = 3$

(b) $t = 3$ has a multiplicity of 2 (even multiplicity).

(c) Turning points: 1 (the vertex of the parabola)

(d)

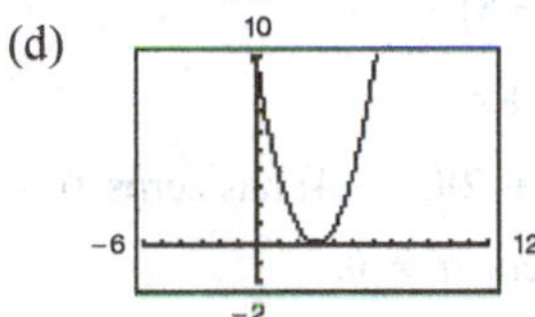

37. $f(x) = \frac{1}{3}x^2 + \frac{1}{3}x - \frac{2}{3}$

(a) $0 = \frac{1}{3}x^2 + \frac{1}{3}x - \frac{2}{3}$

$= \frac{1}{3}(x^2 + x - 2)$

$= \frac{1}{3}(x + 2)(x - 1)$

Zeros: $x = -2, x = 1$

(b) Each zero has a multiplicity of 1 (odd multiplicity).

(c) Turning points: 1 (the vertex of the parabola)

(d)

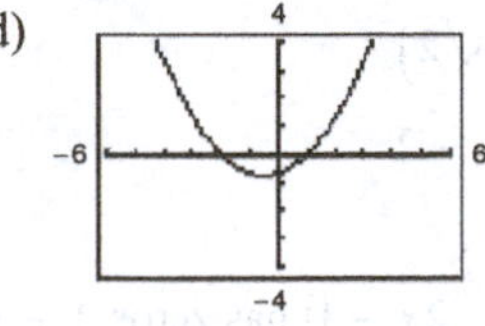

39. $g(x) = 5x(x^2 - 2x - 1)$

(a) $0 = 5x(x^2 - 2x - 1)$

$0 = x(x^2 - 2x - 1)$

For $x^2 - 2x - 1 = 0, a = 1, b = -2, c = -1$.

$$x = \frac{-(-2) \pm \sqrt{(-2)^2 - 4(1)(-1)}}{2(1)}$$

$$= \frac{2 \pm \sqrt{8}}{2}$$

$$= 1 \pm \sqrt{2}$$

Zeros: $x = 0, x = 1 \pm \sqrt{2}$

(b) Each zero has a multiplicity of 1 (odd multiplicity).

(c) Turning points: 2

(d)

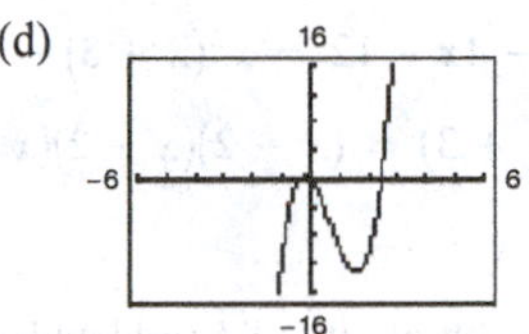

41. $f(x) = 3x^3 - 12x^2 + 3x$

(a) $0 = 3x^3 - 12x^2 + 3x = 3x(x^2 - 4x + 1)$

Zeros: $x = 0, x = 2 \pm \sqrt{3}$ (by the Quadratic Formula)

(b) Each zero has a multiplicity of 1 (odd multiplicity).

(c) Turning points: 2

(d)

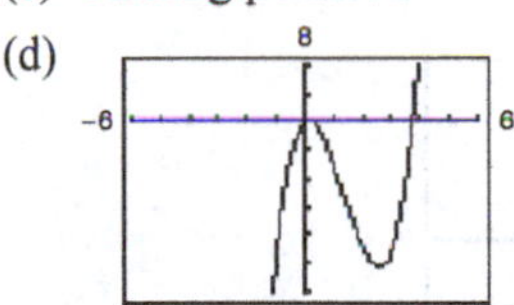

43. $g(t) = t^5 - 6t^3 + 9t$

(a) $0 = t^5 - 6t^3 + 9t = t(t^4 - 6t^2 + 9) = t(t^2 - 3)^2$

$= t(t + \sqrt{3})^2(t - \sqrt{3})^2$

Zeros: $t = 0, t = \pm\sqrt{3}$

(b) $t = 0$ has a multiplicity of 1 (odd multiplicity).

$t = \pm\sqrt{3}$ each have a multiplicity of 2 (even multiplicity).

(c) Turning points: 4

(d)

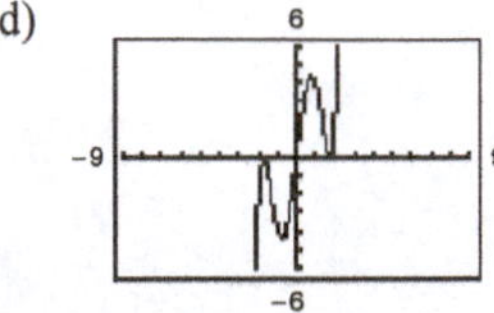

45. $f(x) = 3x^4 + 9x^2 + 6$

(a) $0 = 3x^4 + 9x^2 + 6$

$0 = 3(x^4 + 3x^2 + 2)$

$0 = 3(x^2 + 1)(x^2 + 2)$

No real zeros

(b) No multiplicity

(c) Turning points: 1

(d)

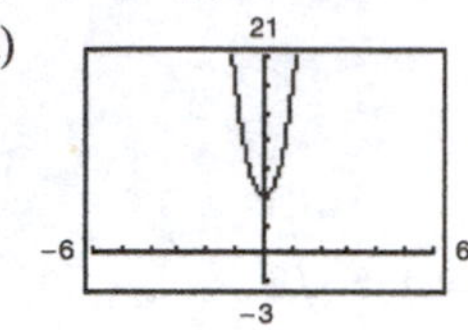

47. $g(x) = x^3 + 3x^2 - 4x - 12$

(a) $0 = x^3 + 3x^2 - 4x - 12 = x^2(x + 3) - 4(x + 3)$

$= (x^2 - 4)(x + 3) = (x - 2)(x + 2)(x + 3)$

Zeros: $x = \pm 2, x = -3$

(b) Each zero has a multiplicity of 1 (odd multiplicity).

(c) Turning points: 2

(d)

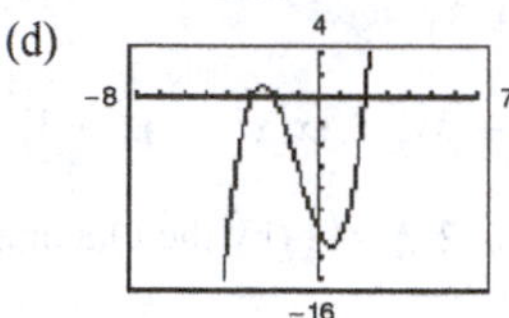

49. $y = 4x^3 - 20x^2 + 25x$

(a)

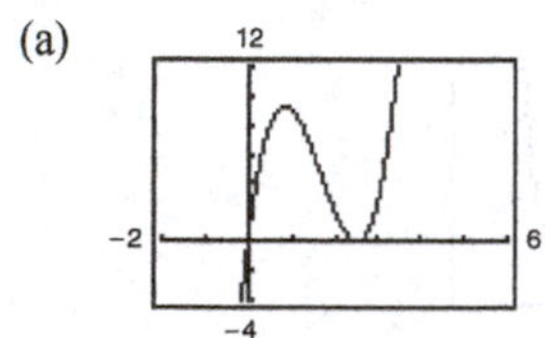

(b) x-intercepts: $(0, 0), \left(\frac{5}{2}, 0\right)$

(c) $0 = 4x^3 - 20x^2 + 25x$

$0 = x(4x^2 - 20x + 25)$

$0 = x(2x - 5)^2$

$x = 0, \frac{5}{2}$

(d) The solutions are the same as the x-coordinates of the x-intercepts.

51. $y = x^5 - 5x^3 + 4x$

(a)

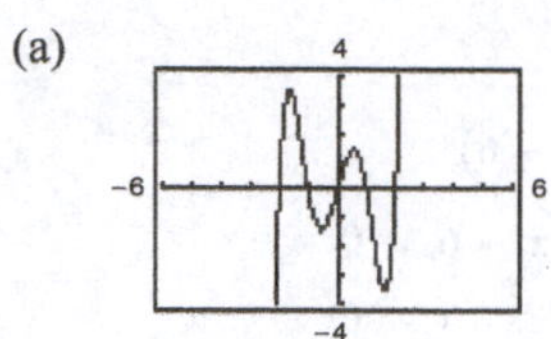

(b) x-intercepts: $(0, 0), (\pm 1, 0), (\pm 2, 0)$

(c) $0 = x^5 - 5x^3 + 4x$

$0 = x(x^2 - 1)(x^2 - 4)$

$0 = x(x + 1)(x - 1)(x + 2)(x - 2)$

$x = 0, \pm 1, \pm 2$

(d) The solutions are the same as the x-coordinates of the x-intercepts.

53. $f(x) = (x - 0)(x - 7)$

$= x^2 - 7x$

Note: $f(x) = ax(x - 7)$ has zeros 0 and 7 for all real numbers $a \neq 0$.

55. $f(x) = (x - 0)(x + 2)(x + 4)$

$= x(x^2 + 6x + 8)$

$= x^3 + 6x^2 + 8x$

Note: $f(x) = ax(x + 2)(x + 4)$ has zeros $0, -2,$ and -4 for all real numbers $a \neq 0$.

57. $f(x) = (x - 4)(x + 3)(x - 3)(x - 0)$

$= (x - 4)(x^2 - 9)x$

$= x^4 - 4x^3 - 9x^2 + 36x$

Note: $f(x) = a(x^4 - 4x^3 - 9x^2 + 36x)$ has zeros 4, −3, 3, and 0 for all real numbers $a \neq 0$.

59. $f(x) = \left[x - (1 + \sqrt{2})\right]\left[x - (1 - \sqrt{2})\right]$

$= \left[(x - 1) - \sqrt{2}\right]\left[(x - 1) + \sqrt{2}\right]$

$= (x - 1)^2 - (\sqrt{2})^2$

$= x^2 - 2x + 1 - 2$

$= x^2 - 2x - 1$

Note: $f(x) = a(x^2 - 2x - 1)$ has zeros $1 + \sqrt{2}$ and $1 - \sqrt{2}$ for all real numbers $a \neq 0$.

61. $f(x) = (x - 2)\left[x - \left(2 + \sqrt{5}\right)\right]\left[x - \left(2 - \sqrt{5}\right)\right]$
$= (x - 2)\left[(x - 2) - \sqrt{5}\right]\left[(x - 2) + \sqrt{5}\right]$
$= (x - 2)\left[(x - 2)^2 - 5\right]$
$= (x - 2)\left[x^2 - 4x + 4 - 5\right]$
$= (x - 2)\left(x^2 - 4x - 1\right)$
$= x^3 - 6x^2 + 7x + 2$

Note: $f(x) = a\left(x^3 - 6x^2 + 7x + 2\right)$ has zeros 2, $2 + \sqrt{5}$, and $2 - \sqrt{5}$ for all real numbers $a \neq 0$.

63. $f(x) = (x + 3)(x + 3) = x^2 + 6x + 9$

Note: $f(x) = a\left(x^2 + 6x + 9\right)$, $a \neq 0$, has degree 2 and zero $x = -3$.

65. $f(x) = (x - 0)(x + 5)(x - 1)$
$= x\left(x^2 + 4x - 5\right)$
$= x^3 + 4x^2 - 5x$

Note: $f(x) = ax\left(x^2 + 4x - 5\right)$, $a \neq 0$, has degree 3 and zeros $x = 0, -5,$ and 1.

67. $f(x) = \left(x - (-5)\right)^2(x - 1)(x - 2) = x^4 + 7x^3 - 3x^2 - 55x + 50$
or $f(x) = \left(x - (-5)\right)(x - 1)^2(x - 2) = x^4 + x^3 - 15x^2 + 23x - 10$
or $f(x) = \left(x - (-5)\right)(x - 1)(x - 2)^2 = x^4 - 17x^2 + 36x - 20$

Note: Any nonzero scalar multiple of these functions would also have degree 4 and zeros $x = -5, 1,$ and 2.

69. $f(x) = (x - 0)(x - 0)(x - 0)\left(x - \sqrt{3}\right)\left(x - \left(-\sqrt{3}\right)\right)$
$= x^3\left(x - \sqrt{3}\right)\left(x + \sqrt{3}\right)$
$= x^3\left(x^2 - 3\right)$
$= x^5 - 3x^3$

Note: $f(x) = a\left(x^5 - 3x^3\right)$, $a \neq 0$, has degree 5 and zeros $x = 0, \sqrt{3}$, and $-\sqrt{3}$.

71. $f(t) = \frac{1}{4}\left(t^2 - 2t + 15\right) = \frac{1}{4}(t - 1)^2 + \frac{7}{2}$

(a) Rises to the left; rises to the right

(b) No real zeros (no x-intercepts)

(c)

t	−1	0	1	2	3
$f(t)$	4.5	3.75	3.5	3.75	4.5

(d) The graph is a parabola with vertex $\left(1, \frac{7}{2}\right)$.

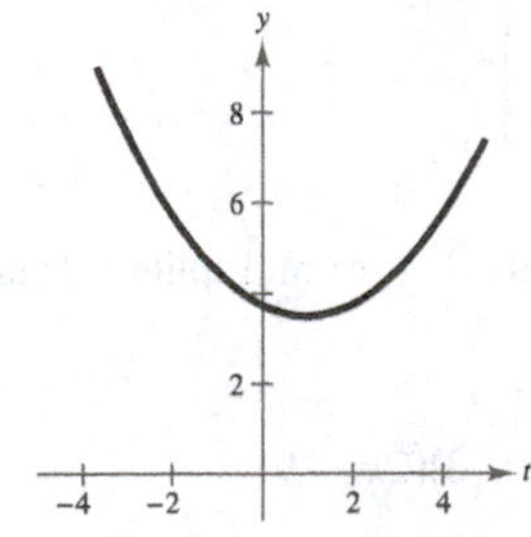

73. $f(x) = x^3 - 25x = x(x + 5)(x - 5)$

(a) Falls to the left; rises to the right

(b) Zeros: 0, −5, 5

(c)

x	−2	−1	0	1	2
$f(x)$	42	24	0	−24	−42

(d)

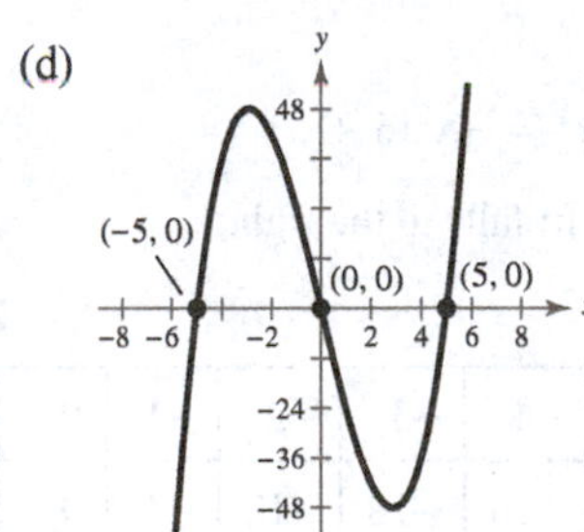

75. $f(x) = -8 + \frac{1}{2}x^4 = \frac{1}{2}(x^4 - 16)$
$= \frac{1}{2}(x^2 + 4)(x - 2)(x + 2)$

(a) Rises to the left; rises to the right

(b) Zeros $x = \pm 2$:

(c)

x	-2	-1	0	1	2
$f(x)$	0	$-\frac{15}{2}$	-8	$-\frac{15}{2}$	0

(d)

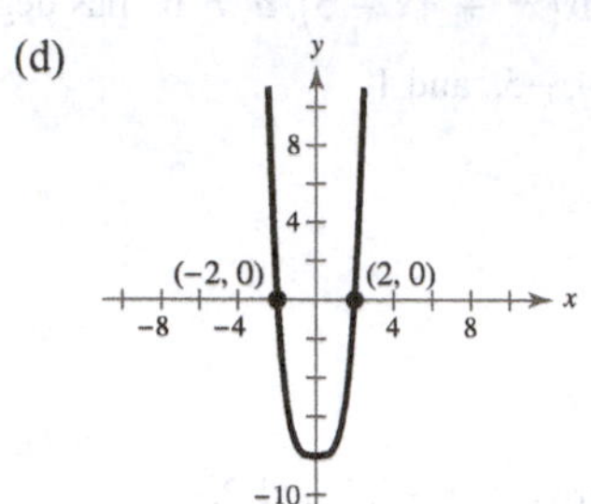

77. $f(x) = 3x^3 - 15x^2 + 18x = 3x(x - 2)(x - 3)$

(a) Falls to the left; rises to the right

(b) Zeros: 0, 2, 3

(c)

x	0	1	2	2.5	3	3.5
$f(x)$	0	6	0	-1.875	0	7.875

(d)

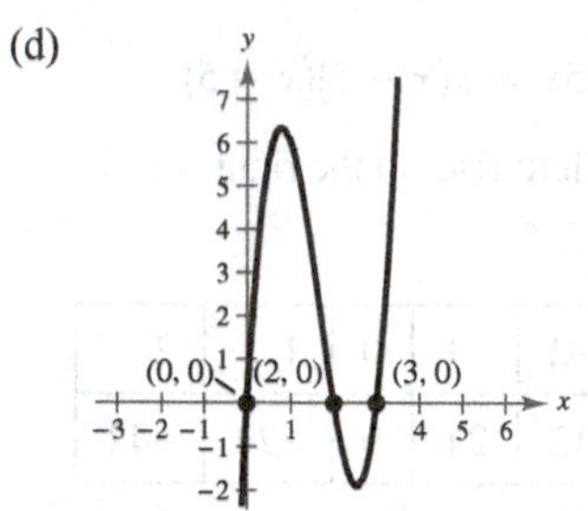

79. $f(x) = -5x^2 - x^3 = -x^2(5 + x)$

(a) Rises to the left; falls to the right

(b) Zeros: 0, −5

(c)

x	-5	-4	-3	-2	-1	0	1
$f(x)$	0	-16	-18	-12	-4	0	-6

(d)

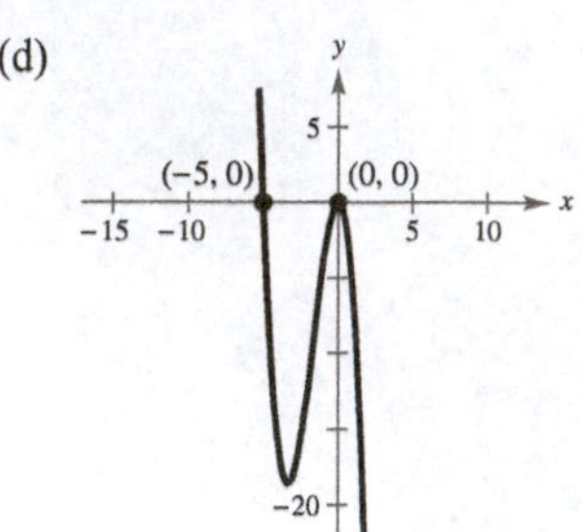

81. $f(x) = 9x^2(x + 2)^2$

(a) Falls to the left, rises to the right

(b) Zeros: $x = 0, -2$

(c)

x	-3	-2	-1	0	1
$f(x)$	81	0	9	0	81

(d)

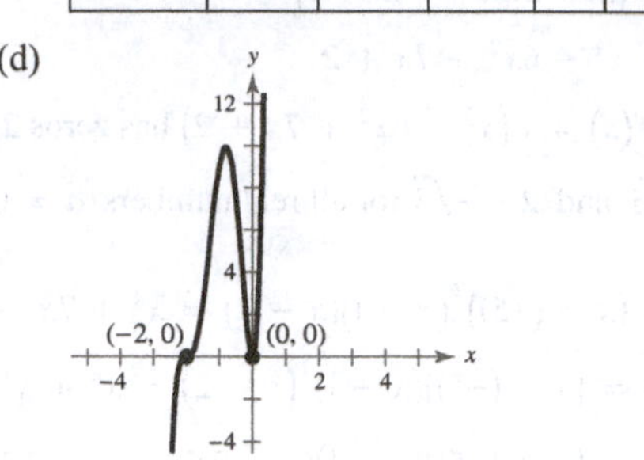

83. $g(t) = -\frac{1}{4}(t - 2)^2(t + 2)^2$

(a) Falls to the left; falls to the right

(b) Zeros: 2, −2

(c)

t	-3	-2	-1	0	1	2	3
$g(t)$	$-\frac{25}{4}$	0	$-\frac{9}{4}$	-4	$-\frac{9}{4}$	0	$-\frac{25}{4}$

(d)

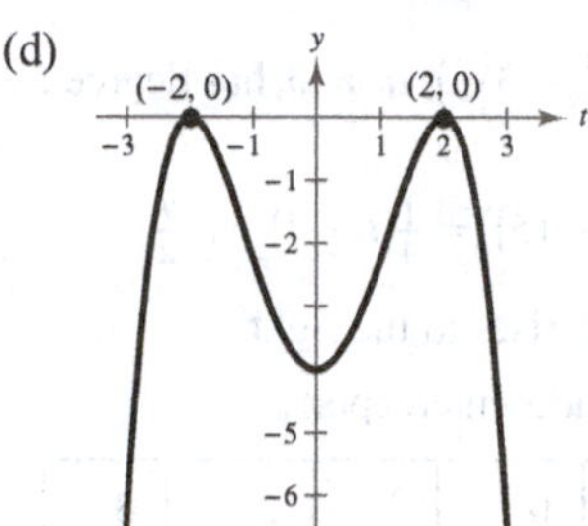

85. $f(x) = x^3 - 16x = x(x - 4)(x + 4)$

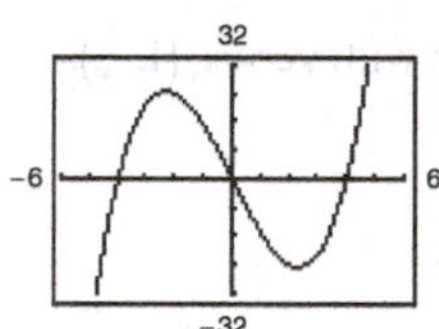

Zeros: 0 of multiplicity 1; 4 of multiplicity 1; and −4 of multiplicity 1

87. $g(x) = \frac{1}{5}(x + 1)^2(x - 3)(2x - 9)$

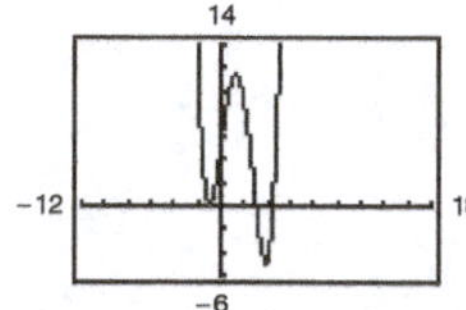

Zeros: −1 of multiplicity 2; 3 of multiplicity 1; $\frac{9}{2}$ of multiplicity 1

89. $f(x) = x^3 - 3x^2 + 3$

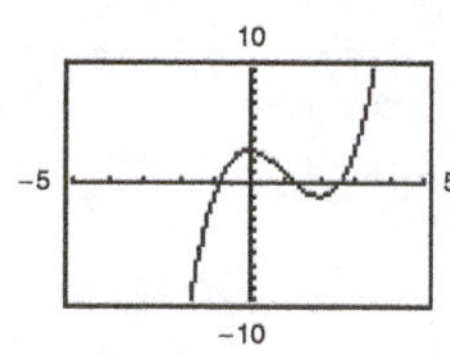

The function has three zeros. They are in the intervals $[-1, 0]$, $[1, 2]$, and $[2, 3]$. They are $x \approx -0.879, 1.347, 2.532$.

x	y
−3	−51
−2	−17
−1	−1
0	3
1	1
2	−1
3	3
4	19

91. $g(x) = 3x^4 + 4x^3 - 3$

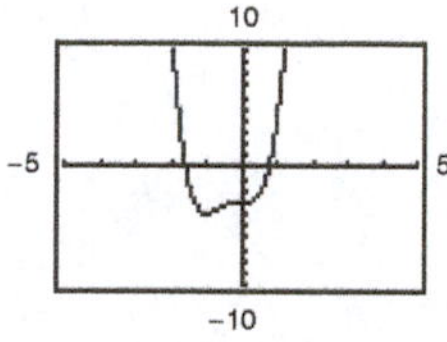

The function has two zeros. They are in the intervals $[-2, -1]$ and $[0, 1]$. They are $x \approx -1.585, 0.779$.

x	y
−4	509
−3	132
−2	13
−1	−4
0	−3
1	4
2	77
3	348

93. (a) Volume $= l \cdot w \cdot h$

height $= x$

length $=$ width $= 36 - 2x$

So, $V(x) = (36 - 2x)(36 - 2x)(x) = x(36 - 2x)^2$.

(b) Domain: $0 < x < 18$

The length and width must be positive.

(c)

Box Height	Box Width	Box Volume, V
1	$36 - 2(1)$	$1[36 - 2(1)]^2 = 1156$
2	$36 - 2(2)$	$2[36 - 2(2)]^2 = 2048$
3	$36 - 2(3)$	$3[36 - 2(3)]^2 = 2700$
4	$36 - 2(4)$	$4[36 - 2(4)]^2 = 3136$
5	$36 - 2(5)$	$5[36 - 2(5)]^2 = 3380$
6	$36 - 2(6)$	$6[36 - 2(6)]^2 = 3456$
7	$36 - 2(7)$	$7[36 - 2(7)]^2 = 3388$

The volume is a maximum of 3456 cubic inches when the height is 6 inches and the length and width are each 24 inches. So the dimensions are $6 \times 24 \times 24$ inches.

(d)

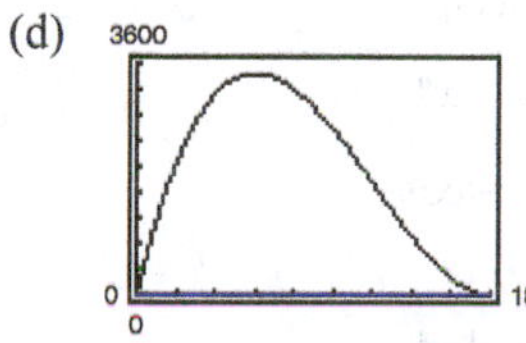

The maximum point on the graph occurs at $x = 6$. This agrees with the maximum found in part (c).

95. (a)

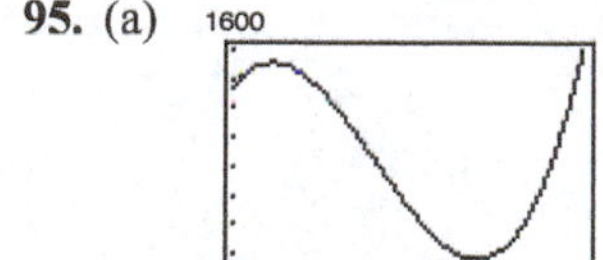

Using trace and zoom features, the relative maximum is approximately $(4.44, 1512.60)$ and the relative minimum is approximately $(11.97, 189.37)$.

(b) The revenue is increasing on $(3, 4.44)$ and $(11.97, 16)$ and decreasing on $(4.44, 11.97)$

(c) Answers will vary. *Sample answer:* The revenue for the software company was increasing from 2003 to midway through 2004 when it reached a maximum of approximately \$1.5 trillion. Then from 2004 to 2012 the revenue was decreasing. It decreased to \$189 million. From 2012 to 2016 the revenue has been increasing.

97. $R = \dfrac{1}{100{,}000}\left(-x^3 + 600x^2\right)$

The point of diminishing returns (where the graph changes from curving upward to curving downward) occurs when $x = 200$. The point is $(200, 160)$ which corresponds to spending \$2,000,000 on advertising to obtain a revenue of \$160 million.

99. True. A polynomial function only falls to the right when the leading coefficient is negative.

101. False. The range of an even function cannot be $(-\infty, \infty)$. An even function's graph will fall to the left and right or rise to the left and right.

103. Answers will vary. *Sample answers:*

$a_4 < 0$ $\qquad$ $a_4 > 0$

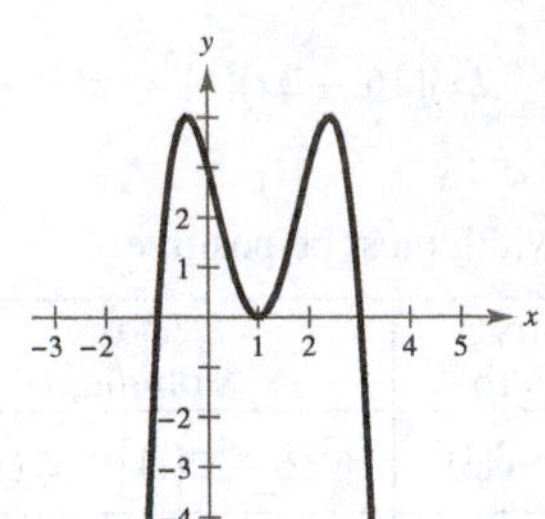

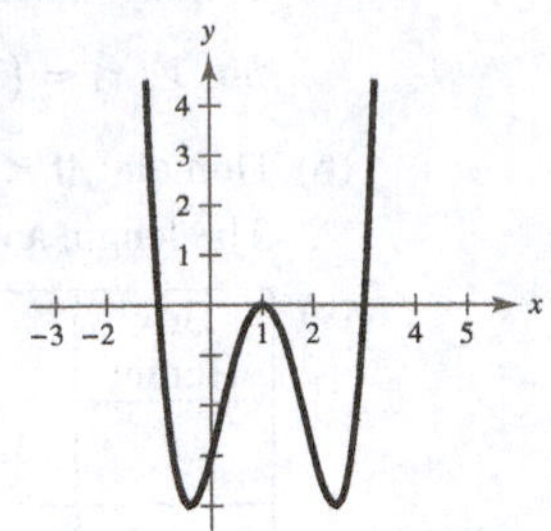

105. $f(x) = x^4$; $f(x)$ is even.

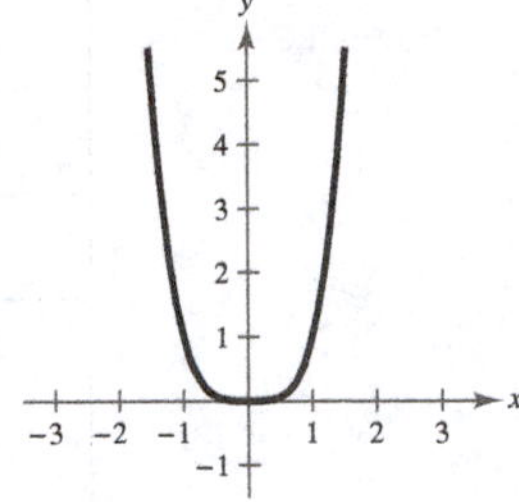

(a) $g(x) = f(x) + 2$

Vertical shift two units upward

$g(-x) = f(-x) + 2$

$= f(x) + 2$

$= g(x)$

Even

(b) $g(x) = f(x + 2)$

Horizontal shift two units to the left

Neither odd nor even

(c) $g(x) = f(-x) = (-x)^4 = x^4$

Reflection in the y-axis. The graph looks the same.

Even

(d) $g(x) = -f(x) = -x^4$

Reflection in the x-axis

Even

(e) $g(x) = f\left(\frac{1}{2}x\right) = \frac{1}{16}x^4$

Horizontal stretch

Even

(f) $g(x) = \frac{1}{2}f(x) = \frac{1}{2}x^4$

Vertical shrink

Even

(g) $g(x) = f\left(x^{3/4}\right) = \left(x^{3/4}\right)^4 = x^3, x \geq 0$

Neither odd nor even

(h) $g(x) = (f \circ f)(x) = f(f(x)) = f\left(x^4\right) = \left(x^4\right)^4 = x^{16}$

Even

107.

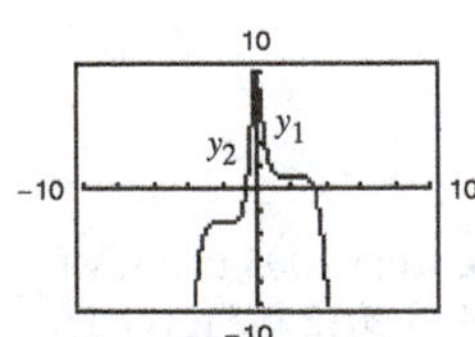

(a) $y_1 = -\frac{1}{3}(x - 2)^5 + 1$ is decreasing and $y_2 = \frac{3}{5}(x + 2)^5 - 3$ is increasing.

(b) It is possible for $g(x) = a(x - h)^5 + k$ to be strictly increasing if $a > 0$ and strictly decreasing if $a < 0$.

(c) f cannot be written in the form $f(x) = a(x - h)^5 + k$ because f is not strictly increasing or strictly decreasing.

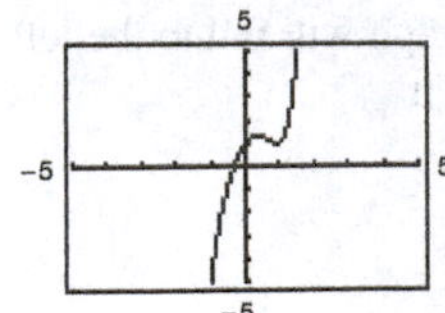

Section 3.3 Polynomial and Synthetic Division

1. $f(x)$ is the dividend; $d(x)$ is the divisor; $q(x)$ is the quotient; $r(x)$ is the remainder

3. improper

5. Factor

7. $y_1 = \dfrac{x^2}{x+2}$ and $y_2 = x - 2 + \dfrac{4}{x+2}$

$$\begin{array}{r}
x - 2 \\
x + 2\overline{\big)x^2 + 0x + 0} \\
\underline{x^2 + 2x} \\
-2x + 0 \\
\underline{-2x - 4} \\
4
\end{array}$$

So, $\dfrac{x^2}{x+2} = x - 2 + \dfrac{4}{x+2}$ and $y_1 = y_2$.

9. $y_1 = \dfrac{x^2 + 2x - 1}{x+3}$, $y_2 = x - 1 + \dfrac{2}{x+3}$

(a) and (b)

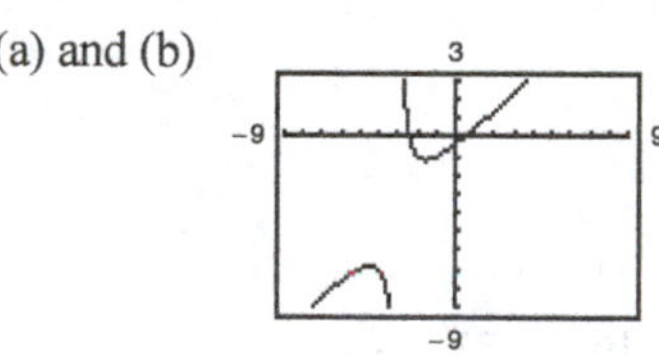

(c)
$$\begin{array}{r}
x - 1 \\
x + 3\overline{\big)x^2 + 2x - 1} \\
\underline{x^2 + 3x} \\
-x - 1 \\
\underline{-x - 3} \\
2
\end{array}$$

So, $\dfrac{x^2 + 2x - 1}{x+3} = x - 1 + \dfrac{2}{x+3}$ and $y_1 = y_2$.

11.
$$\begin{array}{r}
2x + 4 \\
x + 3\overline{\big)2x^2 + 10x + 12} \\
\underline{2x^2 + 6x} \\
4x + 12 \\
\underline{4x + 12} \\
0
\end{array}$$

$\dfrac{2x^2 + 10x + 12}{x+3} = 2x + 4, x \neq 3$

13.
$$\begin{array}{r}
x^2 - 3x + 1 \\
4x + 5\overline{\big)4x^3 - 7x^2 - 11x + 5} \\
\underline{4x^3 + 5x^2} \\
-12x^2 - 11x \\
\underline{-12x^2 - 15x} \\
4x + 5 \\
\underline{4x + 5} \\
0
\end{array}$$

$\dfrac{4x^3 - 7x^2 - 11x + 5}{4x+5} = x^2 - 3x + 1, x \neq -\dfrac{5}{4}$

15.
$$\begin{array}{r}
x^3 + 3x^2 - 1 \\
x + 2\overline{\big)x^4 + 5x^3 + 6x^2 - x - 2} \\
\underline{x^4 + 2x^3} \\
3x^3 + 6x^2 \\
\underline{3x^3 + 6x^2} \\
-x - 2 \\
\underline{-x - 2} \\
0
\end{array}$$

$\dfrac{x^4 + 5x^3 + 6x^2 - x - 2}{x+2} = x^3 + 3x^2 - 1, x \neq -2$

17.
$$\begin{array}{r}
6 \\
x + 1\overline{\big)6x + 5} \\
\underline{6x + 6} \\
-1
\end{array}$$

$\dfrac{6x+5}{x+1} = 6 - \dfrac{1}{x+1}$

19.
$$\begin{array}{r}
x \\
x^2 + 0x + 1\overline{\big)x^3 + 0x^2 + 0x - 9} \\
\underline{x^3 + 0x^2 + x} \\
-x - 9
\end{array}$$

$\dfrac{x^3 - 9}{x^2 + 1} = x - \dfrac{x + 9}{x^2 + 1}$

21.
$$\begin{array}{r}
2x - 8 \\
x^2 + 0x + 1\overline{\big)2x^3 - 8x^2 + 3x - 9} \\
\underline{2x^3 + 0x^2 + 2x} \\
-8x^2 + x - 9 \\
\underline{-8x^2 - 0x - 8} \\
x - 1
\end{array}$$

$\dfrac{2x^3 - 8x^2 + 3x - 9}{x^2 + 1} = 2x - 8 + \dfrac{x - 1}{x^2 + 1}$

23.
$$\begin{array}{r}
x + 3 \\
x^3 - 3x^2 + 3x - 1\overline{\big)x^4 + 0x^3 + 0x^2 + 0x + 0} \\
\underline{x^4 - 3x^3 + 3x^2 - x} \\
3x^3 - 3x^2 + x + 0 \\
\underline{3x^3 - 9x^2 + 9x - 3} \\
6x^2 - 8x + 3
\end{array}$$

$\dfrac{x^4}{(x-1)^3} = x + 3 + \dfrac{6x^2 - 8x + 3}{(x-1)^3}$

25.
$$\begin{array}{r|rrrr} 4 & 2 & -10 & 14 & -24\\ & & 8 & -8 & 24\\ \hline & 2 & -2 & 6 & 0\end{array}$$

$$\frac{2x^3 - 10x^2 + 14x - 24}{x - 4} = 2x^2 - 2x + 6, x \neq 4$$

27.
$$\begin{array}{r|rrrr} 3 & 6 & 7 & -1 & 26\\ & & 18 & 75 & 222\\ \hline & 6 & 25 & 74 & 248\end{array}$$

$$\frac{6x^3 + 7x^2 - x + 26}{x - 3} = 6x^2 + 25x + 74 + \frac{248}{x - 3}$$

29.
$$\begin{array}{r|rrrr} -2 & 4 & 8 & -9 & -18\\ & & -8 & 0 & 18\\ \hline & 4 & 0 & -9 & 0\end{array}$$

$$\frac{4x^3 + 8x^2 - 9x - 18}{x + 2} = 4x^2 - 9, x \neq -2$$

31.
$$\begin{array}{r|rrrr} -10 & -1 & 0 & 75 & -250\\ & & 10 & -100 & 250\\ \hline & -1 & 10 & -25 & 0\end{array}$$

$$\frac{-x^3 + 75x - 250}{x + 10} = -x^2 + 10x - 25, x \neq -10$$

33.
$$\begin{array}{r|rrrr} 4 & 1 & -3 & 0 & 5\\ & & 4 & 4 & 16\\ \hline & 1 & 1 & 4 & 21\end{array}$$

$$\frac{x^3 - 3x^2 + 5}{x - 4} = x^2 + x + 4 + \frac{21}{x - 4}$$

35.
$$\begin{array}{r|rrrrr} 6 & 10 & -50 & 0 & 0 & -800\\ & & 60 & 60 & 360 & 2160\\ \hline & 10 & 10 & 60 & 360 & 1360\end{array}$$

$$\frac{10x^4 - 50x^3 - 800}{x - 6} = 10x^3 + 10x^2 + 60x + 360 + \frac{1360}{x - 6}$$

37.
$$\begin{array}{r|rrrr} -8 & 1 & 0 & 0 & 512\\ & & -8 & 64 & -512\\ \hline & 1 & -8 & 64 & 0\end{array}$$

$$\frac{x^3 + 512}{x + 8} = x^2 - 8x + 64, x \neq -8$$

39.
$$\begin{array}{r|rrrrr} 2 & -3 & 0 & 0 & 0 & 0\\ & & -6 & -12 & -24 & -48\\ \hline & -3 & -6 & -12 & -24 & -48\end{array}$$

$$\frac{-3x^4}{x - 2} = -3x^3 - 6x^2 - 12x - 24 - \frac{48}{x - 2}$$

41.
$$\begin{array}{r|rrrrr} 6 & -1 & 0 & 0 & 180 & 0\\ & & -6 & -36 & -216 & -216\\ \hline & -1 & -6 & -36 & -36 & -216\end{array}$$

$$\frac{180x - x^4}{x - 6} = -x^3 - 6x^2 - 36x - 36 - \frac{216}{x - 6}$$

43.
$$\begin{array}{r|rrrr} -\frac{1}{2} & 4 & 16 & -23 & -15\\ & & -2 & -7 & 15\\ \hline & 4 & 14 & -30 & 0\end{array}$$

$$\frac{4x^3 + 16x^2 - 23x - 15}{x + \frac{1}{2}} = 4x^2 + 14x - 30, x \neq -\frac{1}{2}$$

45. $f(x) = x^3 - x^2 - 10x + 7, k = 3$

$$\begin{array}{r|rrrr} 3 & 1 & -1 & -10 & 7\\ & & 3 & 6 & -12\\ \hline & 1 & 2 & -4 & -5\end{array}$$

$$f(x) = (x - 3)(x^2 + 2x - 4) - 5$$
$$f(3) = 3^3 - 3^2 - 10(3) + 7 = -5$$

47. $f(x) = 15x^4 + 10x^3 - 6x^2 + 14, k = -\frac{2}{3}$

$$\begin{array}{r|rrrrr} -\frac{2}{3} & 15 & 10 & -6 & 0 & 14\\ & & -10 & 0 & 4 & -\frac{8}{3}\\ \hline & 15 & 0 & -6 & 4 & \frac{34}{3}\end{array}$$

$$f(x) = \left(x + \tfrac{2}{3}\right)\left(15x^3 - 6x + 4\right) + \tfrac{34}{3}$$
$$f\left(-\tfrac{2}{3}\right) = 15\left(-\tfrac{2}{3}\right)^4 + 10\left(-\tfrac{2}{3}\right)^3 - 6\left(-\tfrac{2}{3}\right)^2 + 14 = \tfrac{34}{3}$$

49. $f(x) = -4x^3 + 6x^2 + 12x + 4, k = 1 - \sqrt{3}$

$$\begin{array}{r|rrrr} 1-\sqrt{3} & -4 & 6 & 12 & 4 \\ & & -4+4\sqrt{3} & -10+2\sqrt{3} & -4 \\ \hline & -4 & 2+4\sqrt{3} & 2+2\sqrt{3} & 0 \end{array}$$

$$f(x) = \left(x - 1 + \sqrt{3}\right)\left[-4x^2 + \left(2 + 4\sqrt{3}\right)x + \left(2 + 2\sqrt{3}\right)\right]$$

$$f\left(1 - \sqrt{3}\right) = -4\left(1 - \sqrt{3}\right)^3 + 6\left(1 - \sqrt{3}\right)^2 + 12\left(1 - \sqrt{3}\right) + 4 = 0$$

51. $f(x) = 2x^3 - 7x + 3$

(a) Using the Remainder Theorem:

$f(1) = 2(1)^3 - 7(1) + 3 = -2$

Using synthetic division:

$$\begin{array}{r|rrrr} 1 & 2 & 0 & -7 & 3 \\ & & 2 & 2 & -5 \\ \hline & 2 & 2 & -5 & -2 \end{array}$$

Verify using long division:

$$\begin{array}{r} 2x^2 + 2x - 5 \\ x - 1\overline{)2x^3 + 0x^2 - 7x + 3} \\ \underline{2x^3 - 2x^2} \\ 2x^2 - 7x \\ \underline{2x^2 - 2x} \\ -5x + 3 \\ \underline{-5x + 5} \\ -2 \end{array}$$

(b) Using the Remainder Theorem:

$f(-2) = 2(-2)^3 - 7(-2) + 3 = 1$

Using synthetic division:

$$\begin{array}{r|rrrr} -2 & 2 & 0 & -7 & 3 \\ & & -4 & 8 & -2 \\ \hline & 2 & -4 & 1 & 1 \end{array}$$

Verify using long division:

$$\begin{array}{r} 2x^2 - 4x + 1 \\ x + 2\overline{)2x^3 + 0x^2 - 7x + 3} \\ \underline{2x^3 + 4x^2} \\ -4x^2 - 7x \\ \underline{-4x^2 - 8x} \\ x + 3 \\ \underline{x + 2} \\ 1 \end{array}$$

(c) Using the Remainder Theorem:

$f(3) = 2(3)^3 - 7(3) + 3 = 36$

Using synthetic division:

$$\begin{array}{r|rrrr} 3 & 2 & 0 & -7 & 3 \\ & & 6 & 18 & 33 \\ \hline & 2 & 6 & 11 & 36 \end{array}$$

Verify using long division:

$$\begin{array}{r} 2x^2 + 6x + 11 \\ x - 3\overline{)2x^3 + 0x^2 - 7x + 3} \\ \underline{2x^3 - 6x^2} \\ 6x^2 - 7x + 3 \\ \underline{6x^2 - 18x} \\ 11x + 3 \\ \underline{11x - 33} \\ 36 \end{array}$$

(d) Using the Remainder Theorem:

$f(2) = 2(2)^3 - 7(2) + 3 = 5$

Using synthetic division:

$$\begin{array}{r|rrrr} 2 & 2 & 0 & -7 & 3 \\ & & 4 & 8 & 2 \\ \hline & 2 & 4 & 1 & 5 \end{array}$$

Verify using long division:

$$\begin{array}{r} 2x^2 + 4x + 1 \\ x - 2\overline{)2x^3 + 0x^2 - 7x + 3} \\ \underline{2x^3 - 4x^2} \\ 4x^2 - 7x \\ \underline{4x^2 - 8x} \\ x + 3 \\ \underline{x - 2} \\ 5 \end{array}$$

53. $h(x) = x^3 - 5x^2 - 7x + 4$

(a) Using the Remainder Theorem:

$h(3) = (3)^3 - 5(3)^2 - 7(3) + 4 = -35$

Using synthetic division:

$$\begin{array}{r|rrrr} 3 & 1 & -5 & -7 & 4 \\ & & 3 & -6 & -39 \\ \hline & 1 & -2 & -13 & -35 \end{array}$$

Verify using long division:

$$\begin{array}{r} x^2 - 2x - 13 \\ x - 3\overline{)x^3 - 5x^2 - 7x + 4} \\ \underline{x^3 - 3x^2} \qquad\qquad \\ -2x^2 - 7x \qquad \\ \underline{-2x^2 + 6x} \qquad \\ -13x + 4 \\ \underline{-13x + 39} \\ -35 \end{array}$$

(b) Using the Remainder Theorem:

$h\left(\frac{1}{2}\right) = \left(\frac{1}{2}\right)^3 - 5\left(\frac{1}{2}\right)^2 - 7\left(\frac{1}{2}\right) + 4 = -\frac{5}{8}$

Using synthetic division:

$$\begin{array}{r|rrrr} \frac{1}{2} & 1 & -5 & -7 & 4 \\ & & \frac{1}{2} & -\frac{9}{4} & -\frac{37}{8} \\ \hline & 1 & -\frac{9}{2} & -\frac{37}{4} & -\frac{5}{8} \end{array}$$

Verify using long division:

$$\begin{array}{r} x^2 - \frac{9}{2}x - \frac{37}{4} \\ x - \frac{1}{2}\overline{)x^3 - 5x^2 - 7x + 4} \\ \underline{x^3 - \frac{1}{2}x^2} \qquad\qquad \\ -\frac{9}{2}x^2 - 7x + 4 \\ \underline{-\frac{9}{2}x^2 + \frac{9}{4}x} \qquad \\ -\frac{37}{4}x + 4 \\ \underline{-\frac{37}{4}x + \frac{37}{8}} \\ -\frac{5}{8} \end{array}$$

(c) Using the Remainder Theorem:

$h(-2) = (-2)^3 - 5(-2)^2 - 7(-2) + 4 = -10$

Using synthetic division:

$$\begin{array}{r|rrrr} -2 & 1 & -5 & -7 & 4 \\ & & -2 & 14 & -14 \\ \hline & 1 & -7 & 7 & -10 \end{array}$$

Verify using long division:

$$\begin{array}{r} x^2 - 7x + 7 \\ x + 2\overline{)x^3 - 5x^2 - 7x + 4} \\ \underline{x^3 + 2x^2} \qquad\qquad \\ -7x^2 - 7x \qquad \\ \underline{-7x^2 - 14x} \qquad \\ 7x + 4 \\ \underline{7x + 14} \\ -10 \end{array}$$

(d) Using the Remainder Theorem:

$h(-5) = (-5)^3 - 5(-5)^2 - 7(-5) + 4 = -211$

Using synthetic division:

$$\begin{array}{r|rrrr} -5 & 1 & -5 & -7 & 4 \\ & & -5 & 50 & -215 \\ \hline & 1 & -10 & 43 & -211 \end{array}$$

Verify using long division:

$$\begin{array}{r} x^2 - 10x + 43 \\ x + 5\overline{)x^3 - 5x^2 - 7x + 4} \\ \underline{x^3 + 5x^2} \qquad\qquad \\ -10x^2 - 7x \qquad \\ \underline{-10x^2 - 50x} \qquad \\ 43x + 4 \\ \underline{43x + 215} \\ -211 \end{array}$$

55.

$$\begin{array}{r|rrrr} -3 & 1 & 6 & 11 & 6 \\ & & -3 & -9 & -6 \\ \hline & 1 & 3 & 2 & 0 \end{array}$$

$x^3 + 6x^2 + 11x + 6 = (x + 3)(x^2 + 3x + 2) = (x + 3)(x + 2)(x + 1)$

Zeros: $-3, -2, -1$

57.

$$\begin{array}{r|rrrr} \frac{1}{2} & 2 & -15 & 27 & -10 \\ & & 1 & -7 & 10 \\ \hline & 2 & -14 & 20 & 0 \end{array}$$

$2x^3 - 15x^2 + 27x - 10 = \left(x - \frac{1}{2}\right)(2x^2 - 14x + 20) = (2x - 1)(x - 2)(x - 5)$

Zeros: $\frac{1}{2}, 2, 5$

59. $\begin{array}{r|rrrr} \sqrt{3} & 1 & 2 & -3 & -6 \\ & & \sqrt{3} & 3+2\sqrt{3} & 6 \\ \hline & 1 & 2+\sqrt{3} & 2\sqrt{3} & 0 \end{array}$

$\begin{array}{r|rrr} -\sqrt{3} & 1 & 2+\sqrt{3} & 2\sqrt{3} \\ & & -\sqrt{3} & -2\sqrt{3} \\ \hline & 1 & 2 & 0 \end{array}$

$x^3 + 2x^2 - 3x - 6 = \left(x - \sqrt{3}\right)\left(x + \sqrt{3}\right)(x + 2)$

Zeros: $-\sqrt{3}, \sqrt{3}, -2$

61. $\begin{array}{r|rrrr} 1+\sqrt{3} & 1 & -3 & 0 & 2 \\ & & 1+\sqrt{3} & 1-\sqrt{3} & -2 \\ \hline & 1 & -2+\sqrt{3} & 1-\sqrt{3} & 0 \end{array}$

$\begin{array}{r|rrr} 1-\sqrt{3} & 1 & -2+\sqrt{3} & 1-\sqrt{3} \\ & & 1-\sqrt{3} & -1+\sqrt{3} \\ \hline & 1 & -1 & 0 \end{array}$

$$\begin{aligned} x^3 - 3x^2 + 2 &= \left[x - \left(1+\sqrt{3}\right)\right]\left[x - \left(1-\sqrt{3}\right)\right](x-1) \\ &= (x-1)\left(x - 1 - \sqrt{3}\right)\left(x - 1 + \sqrt{3}\right) \end{aligned}$$

Zeros: $1, 1 - \sqrt{3}, 1 + \sqrt{3}$

63. $f(x) = 2x^3 + x^2 - 5x + 2$; Factors: $(x + 2), (x - 1)$

(a) $\begin{array}{r|rrrr} -2 & 2 & 1 & -5 & 2 \\ & & -4 & 6 & -2 \\ \hline & 2 & -3 & 1 & 0 \end{array}$

$\begin{array}{r|rrr} 1 & 2 & -3 & 1 \\ & & 2 & -1 \\ \hline & 2 & -1 & 0 \end{array}$

Both are factors of $f(x)$ because the remainders are zero.

(b) The remaining factor of $f(x)$ is $(2x - 1)$.

(c) $f(x) = (2x - 1)(x + 2)(x - 1)$

(d) Zeros: $\frac{1}{2}, -2, 1$

(e)

65. $f(x) = x^4 - 8x^3 + 9x^2 + 38x - 40$;

Factors: $(x - 5), (x + 2)$

(a) $\begin{array}{r|rrrrr} 5 & 1 & -8 & 9 & 38 & -40 \\ & & 5 & -15 & -30 & 40 \\ \hline & 1 & -3 & -6 & 8 & 0 \end{array}$

$\begin{array}{r|rrrr} -2 & 1 & -3 & -6 & 8 \\ & & -2 & 10 & -8 \\ \hline & 1 & -5 & 4 & 0 \end{array}$

Both are factors of $f(x)$ because the remainders are zero.

(b) $x^2 - 5x + 4 = (x - 1)(x - 4)$

The remaining factors are $(x - 1)$ and $(x - 2)$.

(c) $f(x) = (x - 5)(x + 2)(x - 1)(x - 4)$

(d) Zeros: $-2, 1, 4, 5$

(e) 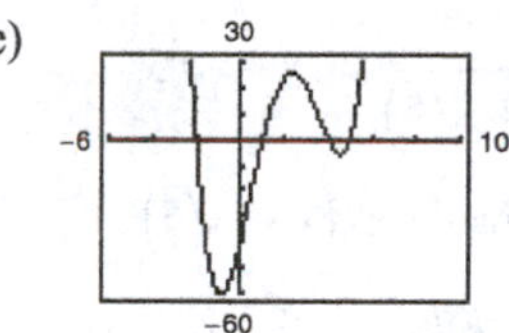

67. $f(x) = 6x^3 + 41x^2 - 9x - 14$;

Factors: $(2x + 1), (3x - 2)$

(a) $\begin{array}{r|rrrr} -\frac{1}{2} & 6 & 41 & -9 & -14 \\ & & -3 & -19 & 14 \\ \hline & 6 & 38 & -28 & 0 \end{array}$

$\begin{array}{r|rrr} \frac{2}{3} & 6 & 38 & -28 \\ & & 4 & 28 \\ \hline & 6 & 42 & 0 \end{array}$

Both are factors of $f(x)$ because the remainders are zero.

(b) $6x + 42 = 6(x + 7)$

This shows that $\dfrac{f(x)}{\left(x + \frac{1}{2}\right)\left(x - \frac{2}{3}\right)} = 6(x + 7)$,

so $\dfrac{f(x)}{(2x + 1)(3x - 2)} = x + 7$.

The remaining factor is $(x + 7)$.

(c) $f(x) = (x + 7)(2x + 1)(3x - 2)$

(d) Zeros: $-7, -\frac{1}{2}, \frac{2}{3}$

(e) 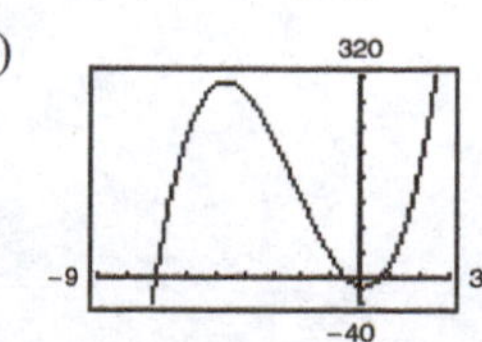

69. $f(x) = 2x^3 - x^2 - 10x + 5;$

Factors: $(2x - 1), \left(x + \sqrt{5}\right)$

(a)
$$\begin{array}{r|rrrr} \frac{1}{2} & 2 & -1 & -10 & 5 \\ & & 1 & 0 & -5 \\ \hline & 2 & 0 & -10 & 0 \end{array}$$

$$\begin{array}{r|rrr} -\sqrt{5} & 2 & 0 & -10 \\ & & -2\sqrt{5} & 10 \\ \hline & 2 & -2\sqrt{5} & 0 \end{array}$$

Both are factors of $f(x)$ because the remainders are zero.

(b) $2x - 2\sqrt{5} = 2\left(x - \sqrt{5}\right)$

This shows that $\dfrac{f(x)}{\left(x - \frac{1}{2}\right)\left(x + \sqrt{5}\right)} = 2\left(x - \sqrt{5}\right),$

so $\dfrac{f(x)}{(2x - 1)\left(x + \sqrt{5}\right)} = x - \sqrt{5}.$

The remaining factor is $\left(x - \sqrt{5}\right)$.

(c) $f(x) = \left(x + \sqrt{5}\right)\left(x - \sqrt{5}\right)(2x - 1)$

(d) Zeros: $-\sqrt{5}, \sqrt{5}, \dfrac{1}{2}$

(e)

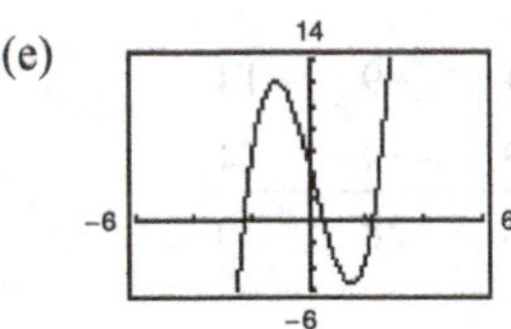

71. $f(x) = x^3 - 2x^2 - 5x + 10$

(a) The zeros of f are $x = 2$ and $x \approx \pm 2.236$.

(b) An exact zero is $x = 2$.

(c)
$$\begin{array}{r|rrrr} 2 & 1 & -2 & -5 & 10 \\ & & 2 & 0 & -10 \\ \hline & 1 & 0 & -5 & 0 \end{array}$$

$$\begin{aligned} f(x) &= (x - 2)\left(x^2 - 5\right) \\ &= (x - 2)\left(x - \sqrt{5}\right)\left(x + \sqrt{5}\right) \end{aligned}$$

73. $h(t) = t^3 - 2t^2 - 7t + 2$

(a) The zeros of h are $t = -2, t \approx 3.732, t \approx 0.268$.

(b) An exact zero is $t = -2$.

(c)
$$\begin{array}{r|rrrr} -2 & 1 & -2 & -7 & 2 \\ & & -2 & 8 & -2 \\ \hline & 1 & -4 & 1 & 0 \end{array}$$

$$h(t) = (t + 2)\left(t^2 - 4t + 1\right)$$

By the Quadratic Formula, the zeros of $t^2 - 4t + 1$ are $2 \pm \sqrt{3}$. Thus,

$$h(t) = (t + 2)\left[t - \left(2 + \sqrt{3}\right)\right]\left[t - \left(2 - \sqrt{3}\right)\right].$$

75. $h(x) = x^5 - 7x^4 + 10x^3 + 14x^2 - 24x$

(a) The zeros of h are $x = 0, x = 3, x = 4,$ $x \approx 1.414, x \approx -1.414$.

(b) An exact zero is $x = 4$.

(c)
$$\begin{array}{r|rrrrr} 4 & 1 & -7 & 10 & 14 & -24 \\ & & 4 & -12 & -8 & 24 \\ \hline & 1 & -3 & -2 & 6 & 0 \end{array}$$

$$\begin{aligned} h(x) &= (x - 4)\left(x^4 - 3x^3 - 2x^2 + 6x\right) \\ &= x(x - 4)(x - 3)\left(x + \sqrt{2}\right)\left(x - \sqrt{2}\right) \end{aligned}$$

77. $\dfrac{x^3 + x^2 - 64x - 64}{x + 8}$

$$\begin{array}{r|rrrr} -8 & 1 & 1 & -64 & -64 \\ & & -8 & 56 & 64 \\ \hline & 1 & -7 & -8 & 0 \end{array}$$

$$\frac{x^3 + x^2 - 64x - 64}{x + 8} = x^2 - 7x - 8, \; x \neq -8$$

79. $\dfrac{x^4 + 6x^3 + 11x^2 + 6x}{x^2 + 3x + 2} = \dfrac{x^4 + 6x^3 + 11x^2 + 6x}{(x + 1)(x + 2)}$

$$\begin{array}{r|rrrrr} -1 & 1 & 6 & 11 & 6 & 0 \\ & & -1 & -5 & -6 & 0 \\ \hline & 1 & 5 & 6 & 0 & 0 \end{array}$$

$$\begin{array}{r|rrrr} -2 & 1 & 5 & 6 & 0 \\ & & -2 & -6 & 0 \\ \hline & 1 & 3 & 0 & 0 \end{array}$$

$$\frac{x^4 + 6x^3 + 11x^2 + 6x}{(x + 1)(x + 2)} = x^2 + 3x, \; x \neq -2, -1$$

81. (a)

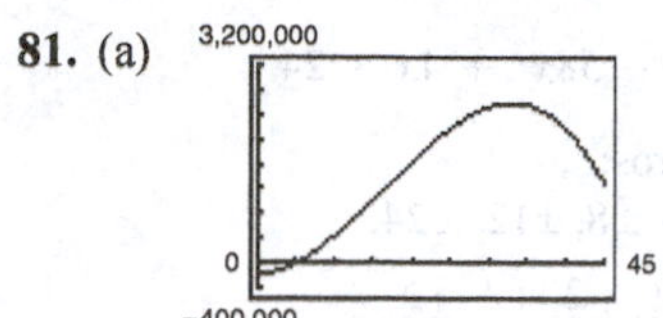

(b) Using the trace and zoom features, when $x = 25$, an advertising expense of about \$250,000 would produce the same profit of \$2,174,375.

(c) $x = 25$

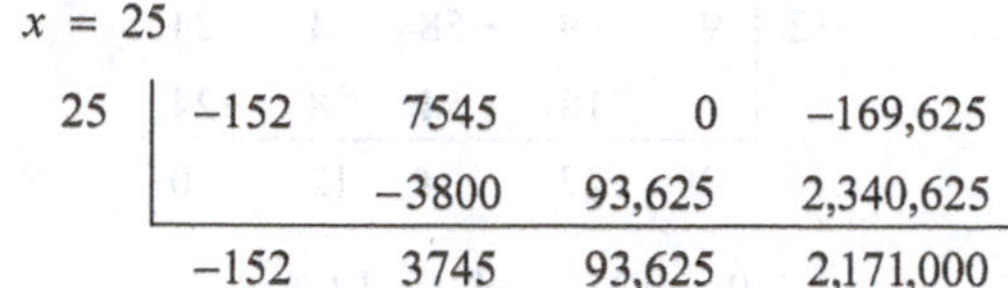

$$\begin{array}{r|rrrr} 25 & -152 & 7545 & 0 & -169{,}625 \\ & & -3800 & 93{,}625 & 2{,}340{,}625 \\ \hline & -152 & 3745 & 93{,}625 & 2{,}171{,}000 \end{array}$$

So, an advertising expense of \$250,000 yields a profit of \$2,171,000, which is close to \$2,174,375.

83. False. If $(7x + 4)$ is a factor of f, then $-\frac{4}{7}$ is a zero of f.

85. True. The degree of the numerator is greater than the degree of the denominator.

87.

$$\begin{array}{r} x^{2n} + 6x^n + 9 \\ x^n + 3 \overline{\big) x^{3n} + 9x^{2n} + 27x^n + 27} \\ \underline{x^{3n} + 3x^{2n}} \qquad\qquad\quad \\ 6x^{2n} + 27x^n \qquad \\ \underline{6x^{2n} + 18x^n} \qquad \\ 9x^n + 27 \\ \underline{9x^n + 27} \\ 0 \end{array}$$

$$\frac{x^{3n} + 9x^{2n} + 27x^n + 27}{x^n + 3} = x^{2n} + 6x^n + 9,\ x^n \neq -3$$

89. To divide $x^2 + 3x - 5$ by $x + 1$ using synthetic division, the value of k is $k = -1$ not $k = 1$ as shown.

$$\begin{array}{r|rrr} -1 & 1 & 3 & -5 \\ & & -1 & -2 \\ \hline & 1 & 2 & \textcircled{-7} \end{array} \leftarrow \text{Remainder: } -7$$

91.

$$\begin{array}{r|rrrr} 5 & 1 & 4 & -3 & c \\ & & 5 & 45 & 210 \\ \hline & 1 & 9 & 42 & c + 210 \end{array}$$

To divide evenly, $c + 210$ must equal zero. So, c must equal -210.

93. If $x - 4$ is a factor of $f(x) = x^3 - kx^2 + 2kx - 8$, then $f(4) = 0$.

$$\begin{aligned} f(4) &= (4)^3 - k(4)^2 + 2k(4) - 8 \\ 0 &= 64 - 16k + 8k - 8 \\ -56 &= -8k \\ 7 &= k \end{aligned}$$

Section 3.4 Zeros of Polynomial Functions

1. Fundamental Theorem of Algebra

3. Rational Zero

5. linear; quadratic; quadratic

7. Descartes's Rule of Signs

9. f is a 3rd degree polynomial function, so there are three zeros.

11. f is a 5th degree polynomial function, so there are five zeros.

13. f is a 2nd degree polynomial function, so there are two zeros.

15. $f(x) = x^3 + 2x^2 - x - 2$

Possible rational zeros: $\pm 1, \pm 2$

Zeros shown on graph: $-2, -1, 1, 2$

17. $f(x) = 2x^4 - 17x^3 + 35x^2 + 9x - 45$

Possible rational zeros: $\pm 1, \pm 3, \pm 5, \pm 9, \pm 15, \pm 45, \pm\frac{1}{2}, \pm\frac{3}{2}, \pm\frac{5}{2}, \pm\frac{9}{2}, \pm\frac{15}{2}, \pm\frac{45}{2}$

Zeros shown on graph: $-1, \frac{3}{2}, 3, 5$

19. $f(x) = x^3 - 7x - 6$

Possible rational zeros: $\pm1, \pm2, \pm3, \pm6$

$$\begin{array}{r|rrrr} 3 & 1 & 0 & -7 & -6 \\ & & 3 & 9 & 6 \\ \hline & 1 & 3 & 2 & 0 \end{array}$$

$$\begin{aligned} f(x) &= (x - 3)(x^2 + 3x + 2) \\ &= (x - 3)(x + 2)(x + 1) \end{aligned}$$

So, the rational zeros are $-2, -1$, and 3.

21. $g(t) = t^3 - 4t^2 + 4$

Possible rational zeros: $\pm1, \pm 2, \pm 4$

After testing all six possible rational zeros by synthetic division, you can conclude there are no rational zeros.

23. $h(t) = t^3 + 8t^2 + 13t + 6$

Possible rational zeros: $\pm1, \pm2, \pm3, \pm6$

$$\begin{array}{r|rrrr} -6 & 1 & 8 & 13 & 6 \\ & & -6 & -12 & -6 \\ \hline & 1 & 2 & 1 & 0 \end{array}$$

$$\begin{aligned} t^3 + 8t^2 + 13t + 6 &= (t + 6)(t^2 + 2t + 1) \\ &= (t + 6)(t + 1)(t + 1) \end{aligned}$$

So, the rational zeros are -1 and -6.

25. $C(x) = 2x^3 + 3x^2 - 1$

Possible rational zeros: $\pm1, \pm\frac{1}{2}$

$$\begin{array}{r|rrrr} -1 & 2 & 3 & 0 & -1 \\ & & -2 & -1 & 1 \\ \hline & 2 & 1 & -1 & 0 \end{array}$$

$$\begin{aligned} 2x^3 + 3x^2 - 1 &= (x + 1)(2x^2 + x - 1) \\ &= (x + 1)(x + 1)(2x - 1) \\ &= (x + 1)^2(2x - 1) \end{aligned}$$

So, the rational zeros are -1 and $\frac{1}{2}$.

27. $f(x) = 9x^4 - 9x^3 - 58x^2 + 4x + 24$

Possible rational zeros:
$\pm1, \pm2, \pm3, \pm4, \pm6, \pm8, \pm12, \pm24,$
$\pm\frac{1}{3}, \pm\frac{2}{3}, \pm\frac{4}{3}, \pm\frac{8}{3}, \pm\frac{1}{9}, \pm\frac{2}{9}, \pm\frac{4}{9}, \pm\frac{8}{9}$

$$\begin{array}{r|rrrrr} -2 & 9 & -9 & -58 & 4 & 24 \\ & & -18 & 54 & 8 & -24 \\ \hline & 9 & -27 & -4 & 12 & 0 \end{array}$$

$$\begin{array}{r|rrrr} 3 & 9 & -27 & -4 & 12 \\ & & 27 & 0 & -12 \\ \hline & 9 & 0 & -4 & 0 \end{array}$$

$$\begin{aligned} f(x) &= (x + 2)(x - 3)(9x^2 - 4) \\ &= (x + 2)(x - 3)(3x - 2)(3x + 2) \end{aligned}$$

So, the rational zeros are $-2, 3, \frac{2}{3}$, and $-\frac{2}{3}$.

29. $-5x^3 + 11x^2 - 4x - 2 = 0$

Possible rational zeros: $\frac{\pm1, \pm2}{\pm1, \pm5} = \pm\frac{1}{5}, \pm\frac{2}{5}, \pm 1, \pm 2$

$$\begin{array}{r|rrrr} 1 & -5 & 11 & -4 & -2 \\ & & -5 & 6 & 2 \\ \hline & -5 & 6 & 2 & 0 \end{array}$$

$$\begin{aligned} (x - 1)(-5x^2 + 6x + 2) &= 0 \\ -5x^2 + 6x + 2 &= 0 \\ 5x^2 - 6x - 2 &= 0 \end{aligned}$$

$$x = \frac{-b \pm \sqrt{b^2 - 4ac}}{2a}$$

$$x = \frac{-(-6) \pm \sqrt{(-6)^2 - 4(5)(-2)}}{2(5)}$$

$$x = \frac{6 \pm \sqrt{76}}{10}$$

$$x = \frac{2(3 \pm \sqrt{19})}{10}$$

$$x = \frac{3 \pm \sqrt{19}}{5}$$

So, the real zeros are $x = 1, x = \frac{3}{5} \pm \frac{\sqrt{19}}{5}$.

31. $x^4 + 6x^3 + 3x^2 - 16x + 6 = 0$

Possible rational zeros: $\pm1, \pm2, \pm3, \pm6$

$$\begin{array}{r|rrrrr} 1 & 1 & 6 & 3 & -16 & 6 \\ & & 1 & 7 & 10 & -6 \\ \hline & 1 & 7 & 10 & -6 & 0 \end{array}$$

$$\begin{array}{r|rrrr} -3 & 1 & 7 & 10 & -6 \\ & & -3 & -12 & 6 \\ \hline & 1 & 4 & -2 & 0 \end{array}$$

$$\begin{aligned} (x - 1)(x + 3)(x^2 + 4x - 2) &= 0 \\ x^2 + 4x - 2 &= 0 \\ x^2 + 4x + 4 &= 2 + 4 \\ (x + 2)^2 &= 6 \\ x + 2 &= \pm\sqrt{6} \\ x &= -2 \pm \sqrt{6} \end{aligned}$$

So the real zeros are $x = 1, -3, -2 \pm 2\sqrt{6}$.

33. $f(x) = x^3 + x^2 - 4x - 4$

(a) Possible rational zeros: $\pm1, \pm2, \pm4$

(b)

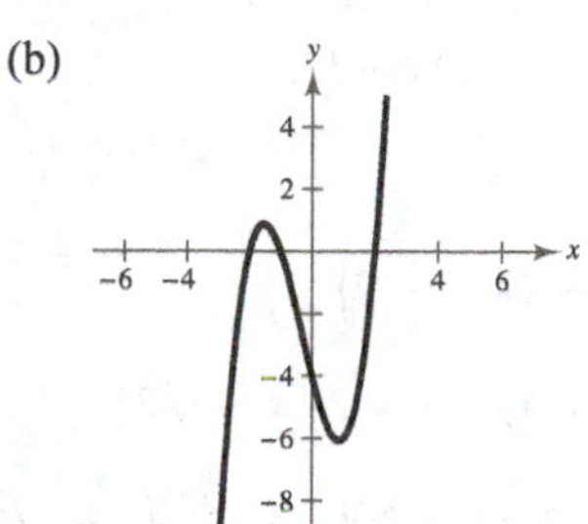

(c) Real zeros: $-2, -1, 2$

35. $f(x) = -4x^3 + 15x^2 - 8x - 3$

(a) Possible rational zeros: $\pm1, \pm3, \pm\frac{1}{2}, \pm\frac{3}{2}, \pm\frac{1}{4}, \pm\frac{3}{4}$

(b)

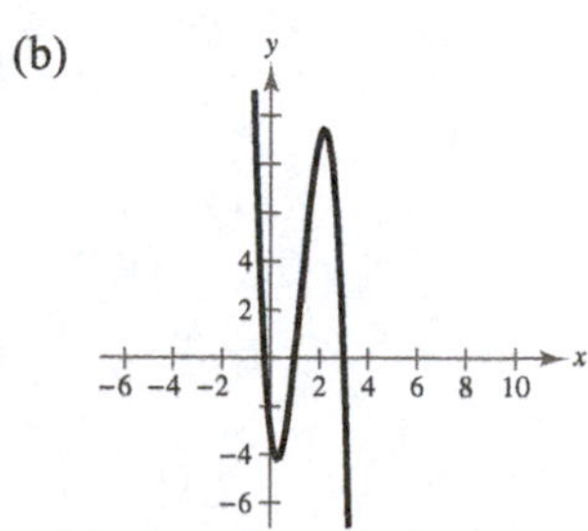

(c) Real zeros: $-\frac{1}{4}, 1, 3$

37. $f(x) = -2x^4 + 13x^3 - 21x^2 + 2x + 8$

(a) Possible rational zeros: $\pm1, \pm2, \pm4, \pm8, \pm\frac{1}{2}$

(b)

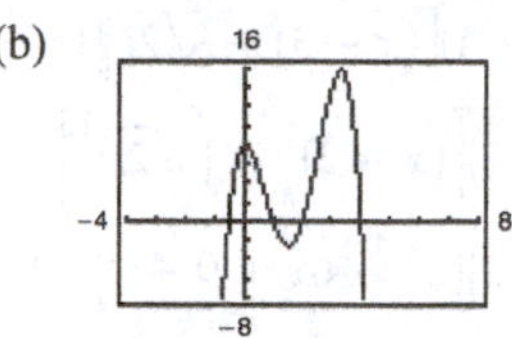

(c) Real zeros: $-\frac{1}{2}, 1, 2, 4$

39. $f(x) = 32x^3 - 52x^2 + 17x + 3$

(a) Possible rational zeros: $\pm1, \pm3, \pm\frac{1}{2}, \pm\frac{3}{2}, \pm\frac{1}{4}, \pm\frac{3}{4},$
$\pm\frac{1}{8}, \pm\frac{3}{8}, \pm\frac{1}{16}, \pm\frac{3}{16}, \pm\frac{1}{32}, \pm\frac{3}{32}$

(b)

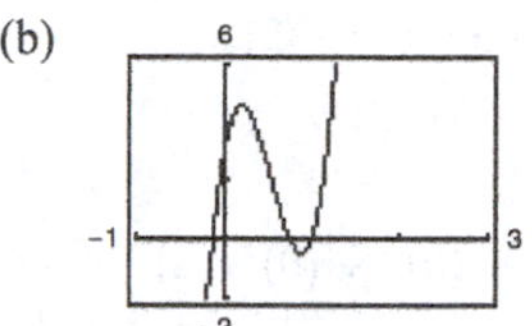

(c) Real zeros: $-\frac{1}{8}, \frac{3}{4}, 1$

41.
$$\begin{aligned} f(x) &= (x - 1)(x - 5i)(x + 5i) \\ &= (x - 1)(x^2 + 25) \\ &= x^3 - x^2 + 25x - 25 \end{aligned}$$

Note: $f(x) = a(x^3 - x^2 + 25x - 25)$, where a is any nonzero real number, has the zeros 1 and $\pm5i$.

43. If $1 + i$ is a zero, so is its conjugate, $1 - i$.

$$\begin{aligned} f(x) &= (x - 2)(x - 2)(x - (1 + i))(x - (1 - i)) \\ &= (x^2 - 4x + 4)(x^2 - 2x + 2) \\ &= x^4 - 6x^3 + 14x^2 - 16x + 8 \end{aligned}$$

Note: $f(x) = a(x^4 - 6x^3 + 14x^2 - 16x + 8)$, where a is any nonzero real number, has the zeros 2, 2 and $1 \pm i$.

45. If $3 + \sqrt{2}i$ is a zero, so is its conjugate, $3 - \sqrt{2}i$.

$$\begin{aligned} f(x) &= (3x - 2)(x + 1)\left[x - \left(3 + \sqrt{2}i\right)\right]\left[x - \left(3 - \sqrt{2}i\right)\right] \\ &= (3x - 2)(x + 1)\left[(x - 3) - \sqrt{2}i\right]\left[(x - 3) + \sqrt{2}i\right] \\ &= \left(3x^2 + x - 2\right)\left[(x - 3)^2 - \left(\sqrt{2}i\right)^2\right] \\ &= \left(3x^2 + x - 2\right)\left(x^2 - 6x + 9 + 2\right) \\ &= \left(3x^2 + x - 2\right)\left(x^2 - 6x + 11\right) \\ &= 3x^4 - 17x^3 + 25x^2 + 23x - 22 \end{aligned}$$

Note: $f(x) = a\left(3x^4 - 17x^3 + 25x^2 + 23x - 22\right)$, where a is any nonzero real number, has the zeros $\frac{2}{3}$, -1, and $3 \pm \sqrt{2}i$.

47.
$$\begin{aligned} f(x) &= a(x + 2)(x - 1)(x - i)(x + i) \\ &= a\left(x^2 + x - 2\right)\left(x^2 + 1\right) \\ &= a\left(x^4 + x^3 - x^2 + x - 2\right) \end{aligned}$$

Since $f(0) = -4$

$$\begin{aligned} -4 &= a\left((0)^4 + (0)^3 - (0)^2 + (0) - 2\right) \\ -4 &= -2a \\ a &= 2 \end{aligned}$$

So, $f(x) = 2\left(x^4 + x^3 - x^2 + x - 2\right)$
$= 2x^4 + 2x^3 - 2x^2 + 2x - 4.$

49.
$$\begin{aligned} f(x) &= a(x + 3)\left(x - \left(1 + \sqrt{3}i\right)\right)\left(x - \left(1 - \sqrt{3}i\right)\right) \\ &= a(x + 3)\left(x^2 - 2x + 4\right) \\ &= a\left(x^3 + x^2 - 2x + 12\right) \end{aligned}$$

Since $f(-2) = 12$

$$\begin{aligned} 12 &= a\left((-2)^3 + (-2)^2 - 2(-2) + 12\right) \\ 12 &= 12a \\ a &= 1 \end{aligned}$$

So, $f(x) = (1)\left(x^4 - x^3 - 2x - 4\right)$
$= x^3 + x^2 - 2x + 12.$

51. $f(x) = x^4 + 2x^2 - 8$

(a) $f(x) = \left(x^2 + 4\right)\left(x^2 - 2\right)$

(b) $f(x) = \left(x^2 + 4\right)\left(x + \sqrt{2}\right)\left(x - \sqrt{2}\right)$

(c) $f(x) = (x + 2i)(x - 2i)\left(x + \sqrt{2}\right)\left(x - \sqrt{2}\right)$

53. $f(x) = x^4 - 2x^3 - 3x^2 + 12x - 18$

$$\begin{array}{r} x^2 - 2x + 3 \\ x^2 - 6\overline{\big)\,x^4 - 2x^3 - 3x^2 + 12x - 18} \\ \underline{x^4 \qquad\quad - 6x^2 \qquad\qquad\quad} \\ -2x^3 + 3x^2 + 12x \qquad \\ \underline{-2x^3 \qquad\quad + 12x \qquad} \\ 3x^2 \qquad - 18 \\ \underline{3x^2 \qquad - 18} \\ 0 \end{array}$$

(a) $f(x) = \left(x^2 - 6\right)\left(x^2 - 2x + 3\right)$

(b) $f(x) = \left(x + \sqrt{6}\right)\left(x - \sqrt{6}\right)\left(x^2 - 2x + 3\right)$

(c) $f(x) = \left(x + \sqrt{6}\right)\left(x - \sqrt{6}\right)\left(x - 1 - \sqrt{2}i\right)\left(x - 1 + \sqrt{2}i\right)$

Note: Use the Quadratic Formula for (c).

55. $f(x) = x^3 - x^2 + 4x - 4$

Because $2i$ is a zero, so is $-2i$.

$$\begin{array}{r|rrrr} 2i & 1 & -1 & 4 & -4 \\ & & 2i & -4-2i & 4 \\ \hline & 1 & 2i-1 & -2i & 0 \end{array}$$

$$\begin{array}{r|rrr} -2i & 1 & 2i-1 & -2i \\ & & -2i & 2i \\ \hline & 1 & -1 & 0 \end{array}$$

$f(x) = (x - 2i)(x + 2i)(x - 1)$

The zeros of $f(x)$ are $x = 1, \pm 2i$.

Alternate Solution:

Because $x = \pm 2i$ are zeros of $f(x)$,

$(x + 2i)(x - 2i) = x^2 + 4$ is a factor of $f(x)$.

By long division, you have:

$$\begin{array}{r} x - 1 \\ x^2 + 0x + 4 \overline{)\, x^3 - x^2 + 4x - 4} \\ \underline{x^3 + 0x^2 + 4x} \\ -x^2 + 0x - 4 \\ \underline{-x^2 + 0x - 4} \\ 0 \end{array}$$

$f(x) = (x^2 + 4)(x - 1)$

The zeros of $f(x)$ are $x = 1, \pm 2i$.

57. $f(x) = x^3 - 8x^2 + 25x - 26$

Because $3 + 2i$ is a zero, so is $3 - 2i$.

$$\begin{array}{r|rrrr} 3+2i & 1 & -8 & 25 & -26 \\ & & 3+2i & -19-4i & 26 \\ \hline & 1 & -5+2i & 6-4i & 0 \end{array}$$

$$\begin{array}{r|rrr} 3-2i & 1 & -5+2i & 6-4i \\ & & 3-2i & -6+4i \\ \hline & 1 & -2 & 0 \end{array}$$

$f(x) = (x - (3 + 2i))(x - (3 - 2i))(x - 2)$

The zeros of $f(x)$ are $x = 3 \pm 2i, 2$.

Alternate Solution:

Because $x = 3 \pm 2i$ are zeros of $f(x)$, $(x - (3 + 2i))(x - (3 - 2i)) = x^2 - 6x + 13$ is a factor of $f(x.)$

By long division, you have:

$$\begin{array}{r} x - 2 \\ x^2 - 6x + 13 \overline{)\, x^3 - 8x^2 + 25x - 26} \\ \underline{x^3 - 6x^2 + 13x} \\ -2x^2 + 12x - 26 \\ \underline{-2x^2 + 12x^2 - 26} \\ 0 \end{array}$$

$f(x) = (x^2 - 6x + 13)(x - 2)$

The zeros of $f(x)$ are $x = 3 \pm 2i, 2$.

59. $f(x) = x^4 - 6x^3 + 14x^2 - 18x + 9$

Because $1 - \sqrt{2}\,i$ is a zero, so is $1 + \sqrt{2}\,i$, and

$$\begin{aligned} \left[x - \left(1 - \sqrt{2}i\right)\right]\left[x - \left(1 + \sqrt{2}i\right)\right] &= \left[(x-1) - \sqrt{2}i\right]\left[(x-1) - \sqrt{2}i\right]\left[(x-1) + \sqrt{2}i\right] \\ &= (x-1)^2 - \left(\sqrt{2}i\right)^2 \\ &= x^2 - 2x + 1 - 2i^2 \\ &= x^2 - 2x + 3 \end{aligned}$$

is a factor of $f(x)$. By long division, you have:

$$\begin{array}{r} x^2 - 4x + 3 \\ x^2 - 2x + 3 \overline{)\, x^4 - 6x^3 + 14x^2 - 18x + 9} \\ \underline{x^4 - 2x^3 + 3x^2} \\ -4x^3 + 11x^2 - 18x + 9 \\ \underline{-4x^3 + 8x^2 - 12x} \\ 3x^2 - 6x + 9 \\ \underline{3x^2 - 6x + 9} \\ 0 \end{array}$$

$$\begin{aligned} f(x) &= (x^2 - 2x + 3)(x^2 - 4x + 3) \\ &= (x^2 - 2x + 3)(x - 1)(x - 3) \end{aligned}$$

The zeros of $f(x)$ are $x = 1 \pm \sqrt{2}i, 1, 3$.

61. $f(x) = x^2 + 36$
$= (x + 6i)(x - 6i)$

The zeros of $f(x)$ are $x = \pm 6i$.

63. $h(x) = x^2 - 2x + 17$

By the Quadratic Formula, the zeros of $f(x)$ are

$$x = \frac{2 \pm \sqrt{4 - 68}}{2} = \frac{2 \pm \sqrt{-64}}{2} = 1 \pm 4i.$$

$f(x) = (x - (1 + 4i))(x - (1 - 4i))$
$= (x - 1 - 4i)(x - 1 + 4i)$

65. $f(x) = x^4 - 16$
$= (x^2 - 4)(x^2 + 4)$
$= (x - 2)(x + 2)(x - 2i)(x + 2i)$

Zeros: $\pm 2, \pm 2i$

67. $f(z) = z^2 - 2z + 2$

By the Quadratic Formula, the zeros of $f(z)$ are

$$z = \frac{2 \pm \sqrt{4 - 8}}{2} = 1 \pm i.$$

$f(z) = [z - (1 + i)][z - (1 - i)]$
$= (z - 1 - i)(z - 1 + i)$

69. $g(x) = x^3 - 3x^2 + x + 5$

Possible rational zeros: $\pm 1, \pm 5$

−1	1	−3	1	5
		−1	4	−5
	1	−4	5	0

By the Quadratic Formula, the zeros of $x^2 - 4x + 5$ are: $x = \dfrac{4 \pm \sqrt{16 - 20}}{2} = 2 \pm i$

Zeros: $-1, 2 \pm i$

$g(x) = (x + 1)(x - 2 - i)(x - 2 + i)$

71. $g(x) = x^4 - 4x^3 + 8x^2 - 16x + 16$

Possible rational zeros: $\pm 1, \pm 2, \pm 4, \pm 8, \pm 16$

2	1	−4	8	−16	16
		2	−4	8	−16
	1	−2	4	−8	0

2	1	−2	4	−8
		2	0	8
	1	0	4	0

$g(x) = (x - 2)(x - 2)(x^2 + 4)$
$= (x - 2)^2(x + 2i)(x - 2i)$

Zeros: $2, \pm 2i$

73. $f(x) = x^3 + 24x^2 + 214x + 740$

Possible rational zeros: $\pm 1, \pm 2, \pm 4, \pm 5, \pm 10, \pm 20, \pm 37, \pm 74, \pm 148, \pm 185, \pm 370, \pm 740$

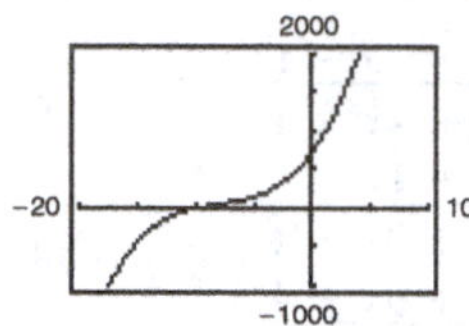

Based on the graph, try $x = -10$.

−10	1	24	214	740
		−10	−140	−740
	1	14	74	0

By the Quadratic Formula, the zeros of $x^2 + 14x + 74$ are $x = \dfrac{-14 \pm \sqrt{196 - 296}}{2} = -7 \pm 5i.$

The zeros of $f(x)$ are $x = -10$ and $x = -7 \pm 5i$.

75. $f(x) = 16x^3 - 20x^2 - 4x + 15$

Possible rational zeros:

$\pm 1, \pm 3, \pm 5, \pm 15, \pm\frac{1}{2}, \pm\frac{3}{2}, \pm\frac{5}{2}, \pm\frac{15}{2}, \pm\frac{1}{4}, \pm\frac{3}{4},$

$\pm\frac{5}{4}, \pm\frac{15}{4}, \pm\frac{1}{8}, \pm\frac{3}{8}, \pm\frac{5}{8}, \pm\frac{15}{8}, \pm\frac{1}{16}, \pm\frac{3}{16}, \pm\frac{5}{16}, \pm\frac{15}{16}$

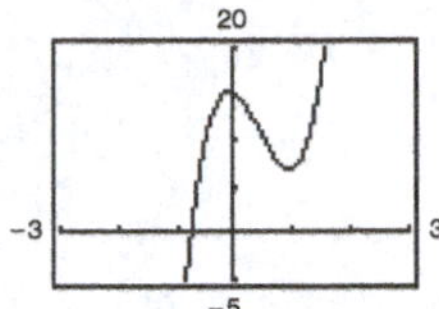

Based on the graph, try $x = -\frac{3}{4}$.

$$\begin{array}{r|rrrr} -\frac{3}{4} & 16 & -20 & -4 & 15 \\ & & -12 & 24 & -15 \\ \hline & 16 & -32 & 20 & 0 \end{array}$$

By the Quadratic Formula, the zeros of $16x^2 - 32x + 20 = 4(4x^2 - 8x + 5)$ are

$$x = \frac{8 \pm \sqrt{64 - 80}}{8} = 1 \pm \frac{1}{2}i.$$

The zeros of $f(x)$ are $x = -\frac{3}{4}$ and $x = 1 \pm \frac{1}{2}i$.

77. $f(x) = 2x^4 + 5x^3 + 4x^2 + 5x + 2$

Possible rational zeros: $\pm 1, \pm 2, \pm\frac{1}{2}$

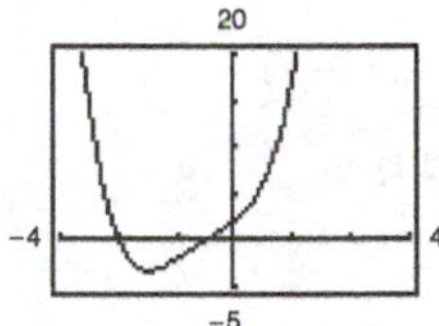

Based on the graph, try $x = -2$ and $x = -\frac{1}{2}$.

$$\begin{array}{r|rrrrr} -2 & 2 & 5 & 4 & 5 & 2 \\ & & -4 & -2 & -4 & -2 \\ \hline & 2 & 1 & 2 & 1 & 0 \end{array}$$

$$\begin{array}{r|rrrr} -\frac{1}{2} & 2 & 1 & 2 & 1 \\ & & -1 & 0 & -1 \\ \hline & 2 & 0 & 2 & 0 \end{array}$$

The zeros of $2x^2 + 2 = 2(x^2 + 1)$ are $x = \pm i$.

The zeros of $f(x)$ are $x = -2$, $x = -\frac{1}{2}$, and $x = \pm i$.

79. $g(x) = 2x^3 - 3x^2 - 3$

Sign variations: 1, positive zeros: 1

$g(-x) = -2x^3 - 3x^2 - 3$

Sign variations: 0, negative zeros: 0

81. $h(x) = 2x^3 + 3x^2 + 1$

Sign variations: 0, positive zeros: 0

$h(-x) = -2x^3 + 3x^2 + 1$

Sign variations: 1, negative zeros: 1

83. $g(x) = 6x^4 + 2x^3 - 3x^2 + 2$

Sign variations: 2, positive zeros: 2 or 0

$g(-x) = 6x^4 - 2x^3 - 3x^2 + 2$

Sign variations: 2, negative zeros: 2 or 0

85. $f(x) = 5x^3 + x^2 - x + 5$

Sign variations: 2, positive zeros: 2 or 0

$f(-x) = -5x^3 + x^2 + x + 5$

Sign variations: 1, negative zeros: 1.

87. $f(x) = x^3 + 3x^2 - 2x + 1$

(a)
$$\begin{array}{r|rrrr} 1 & 1 & 3 & -2 & 1 \\ & & 1 & 4 & 2 \\ \hline & 1 & 4 & 2 & 3 \end{array}$$

1 is an upper bound.

(b)
$$\begin{array}{r|rrrr} -4 & 1 & 3 & -2 & 1 \\ & & -4 & 4 & -8 \\ \hline & 1 & -1 & 2 & -7 \end{array}$$

-4 is a lower bound.

89. $f(x) = x^4 - 4x^3 + 16x - 16$

(a)
$$\begin{array}{r|rrrrr} 5 & 1 & -4 & 0 & 16 & -16 \\ & & 5 & 5 & 25 & 205 \\ \hline & 1 & 1 & 5 & 41 & 189 \end{array}$$

5 is an upper bound.

(b)
$$\begin{array}{r|rrrrr} -3 & 1 & -4 & 0 & 16 & -16 \\ & & -3 & 21 & -63 & 141 \\ \hline & 1 & -7 & 21 & -47 & 125 \end{array}$$

-3 is a lower bound.

91. $f(x) = 16x^3 - 12x^2 - 4x + 3$

Possible rational zeros: $\dfrac{\pm1, \pm3}{\pm, \pm2, \pm4, \pm8, \pm16} = \pm\frac{1}{16}, \pm\frac{1}{8}, \pm\frac{3}{16}, \pm\frac{1}{4}, \pm\frac{3}{8}, \pm\frac{3}{4}, \pm\frac{1}{2}, \pm1, \pm\frac{3}{2}, \pm3$

However, the function factors by grouping.

$$16x^2 - 12x^2 - 4x + 3 = 0$$
$$4x^2(4x - 3) - (4x - 3) = 0$$
$$(4x - 3)(4x^2 - 1) = 0$$
$$(4x - 3)(2x - 1)(2x + 1) = 0$$

$4x - 3 = 0 \to x = \frac{3}{4}$

$2x - 1 = 0 \to x = \frac{1}{2}$

$2x + 1 = 0 \to x = -\frac{1}{2}$

So, the zeros are $x = \frac{3}{4}, \pm\frac{1}{2}$.

93. $f(y) = 4y^3 + 3y^2 + 8y + 6$

Possible rational zeros: $\pm1, \pm2, \pm3, \pm6, \pm\frac{1}{2}, \pm\frac{3}{2}, \pm\frac{1}{4}, \pm\frac{3}{4}$

$$\begin{array}{r|rrrr} -\frac{3}{4} & 4 & 3 & 8 & 6 \\ & & -3 & 0 & -6 \\ \hline & 4 & 0 & 8 & 0 \end{array}$$

$$\begin{aligned} 4y^3 + 3y^2 + 8y + 6 &= \left(y + \tfrac{3}{4}\right)\left(4y^2 + 8\right) \\ &= \left(y + \tfrac{3}{4}\right)4\left(y^2 + 2\right) \\ &= (4y + 3)\left(y^2 + 2\right) \end{aligned}$$

So, the only real zero is $-\frac{3}{4}$.

95. $$\begin{aligned} P(x) &= x^4 - \tfrac{25}{4}x^2 + 9 \\ &= \tfrac{1}{4}\left(4x^4 - 25x^2 + 36\right) \\ &= \tfrac{1}{4}\left(4x^2 - 9\right)\left(x^2 - 4\right) \\ &= \tfrac{1}{4}(2x + 3)(2x - 3)(x + 2)(x - 2) \end{aligned}$$

The rational zeros are $\pm\frac{3}{2}$ and ±2.

97. $$\begin{aligned} f(x) &= x^3 - \tfrac{1}{4}x^2 - x + \tfrac{1}{4} \\ &= \tfrac{1}{4}\left(4x^3 - x^2 - 4x + 1\right) \\ &= \tfrac{1}{4}\left[x^2(4x - 1) - 1(4x - 1)\right] \\ &= \tfrac{1}{4}(4x - 1)\left(x^2 - 1\right) \\ &= \tfrac{1}{4}(4x - 1)(x + 1)(x - 1) \end{aligned}$$

The rational zeros are $\frac{1}{4}$ and ±1.

99. $f(x) = x^3 - 1 = (x - 1)\left(x^2 + x + 1\right)$

Rational zeros: $1\ (x = 1)$

Irrational zeros: 0

Matches (d).

100. $$\begin{aligned} f(x) &= x^3 - 2 \\ &= \left(x - \sqrt[3]{2}\right)\left(x^2 + \sqrt[3]{2}x + \sqrt[3]{4}\right) \end{aligned}$$

Rational zeros: 0

Irrational zeros: $1\left(x = \sqrt[3]{2}\right)$

Matches (a).

101. $f(x) = x^3 - x = x(x + 1)(x - 1)$

Rational zeros: $3\ (x = 0, \pm1)$

Irrational zeros: 0

Matches (b).

102. $$\begin{aligned} f(x) &= x^3 - 2x \\ &= x\left(x^2 - 2\right) \\ &= x\left(x + \sqrt{2}\right)\left(x - \sqrt{2}\right) \end{aligned}$$

Rational zeros: $1\ (x = 0)$

Irrational zeros: $2\left(x = \pm\sqrt{2}\right)$

Matches (c).

103. (a)

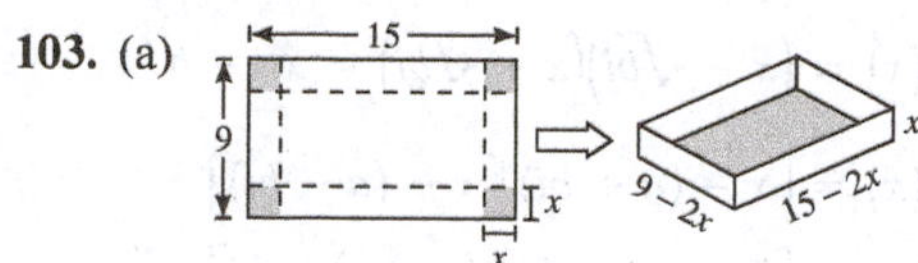

(b) $V = l \cdot w \cdot h = (15 - 2x)(9 - 2x)x$

$= x(9 - 2x)(15 - 2x)$

Because length, width, and height must be positive, you have $0 < x < \frac{9}{2}$ for the domain.

(c)

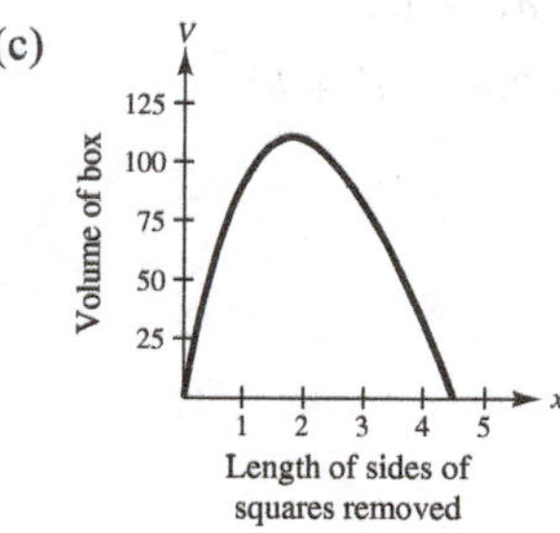

The volume is maximum when $x \approx 1.82$.

The dimensions are:

$r = ak^3 + bk^2 + ck + d, f(k) = r.$

1.82 cm × 5.36 cm × 11.36 cm

(d) $56 = x(9 - 2x)(15 - 2x)$

$56 = 135x - 48x^2 + 4x^3$

$0 = 4x^3 - 48x^2 + 135x - 56$

The zeros of this polynomial are $\frac{1}{2}, \frac{7}{2},$ and 8.

x cannot equal 8 because it is not in the domain of V. [The length cannot equal −1 and the width cannot equal −7. The product of $(8)(-1)(-7) = 56$ so it showed up as an extraneous solution.]

So, the volume is 56 cubic centimeters when $x = \frac{1}{2}$ centimeter or $x = \frac{7}{2}$ centimeters.

105. (a) Current bin: $V = 2 \times 3 \times 4 = 24$ cubic feet

New bin: $V = 5(24) = 120$ cubic feet

$V(x) = (2 + x)(3 + x)(4 + x) = 120$

(b) $x^3 + 9x^2 + 26x + 24 = 120$

$x^3 + 9x^2 + 26x - 96 = 0$

The only real zero of this polynomial is $x = 2$. All the dimensions should be increased by 2 feet, so the new bin will have dimensions of 4 feet by 5 feet by 6 feet.

107. False. The most complex zeros it can have is two, and the Linear Factorization Theorem guarantees that there are three linear factors, so one zero must be real.

109. $g(x) = -f(x)$. This function would have the same zeros as $f(x)$, so $r_1, r_2,$ and r_3 are also zeros of $g(x)$.

111. $g(x) = f(x - 5)$. The graph of $g(x)$ is a horizontal shift of the graph of $f(x)$ five units of the right, so the zeros of $g(x)$ are $5 + r_1, 5 + r_2,$ and $5 + r_3$.

113. $g(x) = 3 + f(x)$. Because $g(x)$ is a vertical shift of the graph of $f(x)$, the zeros of $g(x)$ cannot be determined.

115. Zeros: $-2, \frac{1}{2}, 3$

$f(x) = -(x + 2)(2x - 1)(x - 3)$

$= -2x^3 + 3x^2 + 11x - 6$

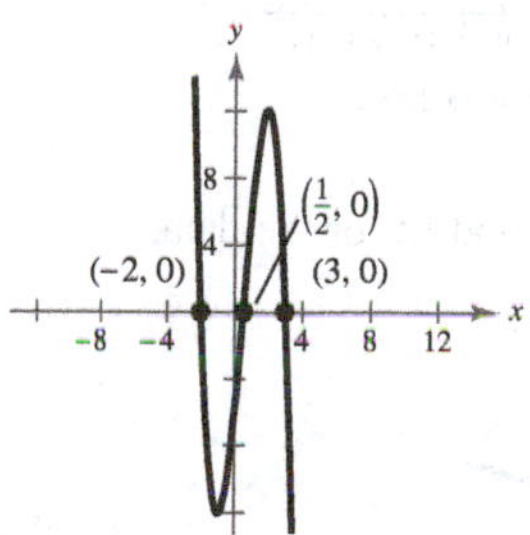

Any nonzero scalar multiple of f would have the same three zeros. Let $g(x) = af(x), a > 0$. There are infinitely many possible functions for f.

117. Because $1 + i$ is a zero of f, so is $1 - i$. From the graph, 1 is also a zero.

$f(x) = (x - (1 + i))(x - (1 - i))(x - 1)$

$= (x^2 - 2x + 2)(x - 1)$

$= x^3 - 3x^2 + 4x - 2$

119.

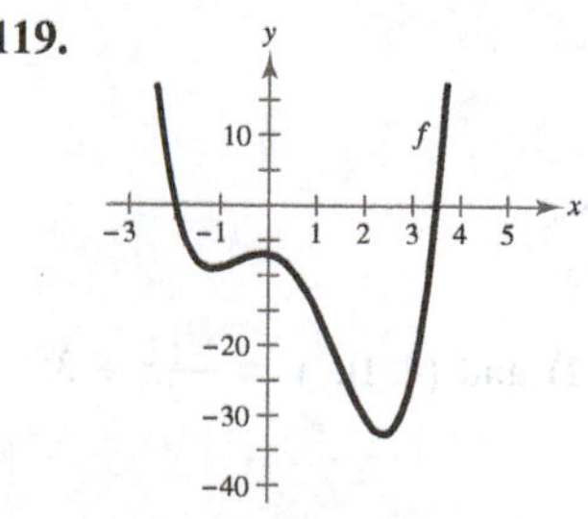

If $x = i$ is a zero, then $x = -i$ is also a zero. So the function is $f(x) = (x - 2)(x - 3.5)(x - i)(x + i)$.

121. Because $f(i) = f(2i) = 0$, then i and $2i$ are zeros of f. Because i and $2i$ are zeros of f, so are $-i$ and $-2i$.

$$\begin{aligned} f(x) &= (x - i)(x + i)(x - 2i)(x + 2i) \\ &= (x^2 + 1)(x^2 + 4) \\ &= x^4 + 5x^2 + 4 \end{aligned}$$

123. (a) $f(x) = \left(x - \sqrt{b}i\right)\left(x + \sqrt{b}i\right) = x^2 + b$

(b) $$\begin{aligned} f(x) &= \left[x - (a + bi)\right]\left[x - (a - bi)\right] \\ &= \left[(x - a) - bi\right]\left[(x - a) + bi\right] \\ &= (x - a)^2 - (bi)^2 \\ &= x^2 - 2ax + a^2 + b^2 \end{aligned}$$

Section 3.5 Mathematical Modeling and Variation

1. variation; regression

3. least squares regression

5. directly proportional

7. combined

9. (a)

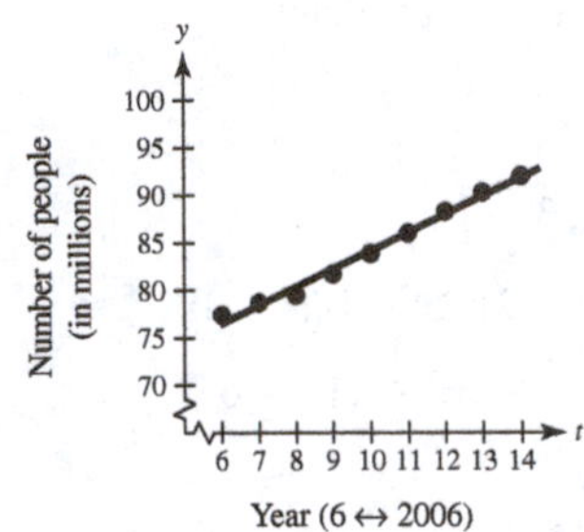

(b) The model is a good fit for the data.

11.

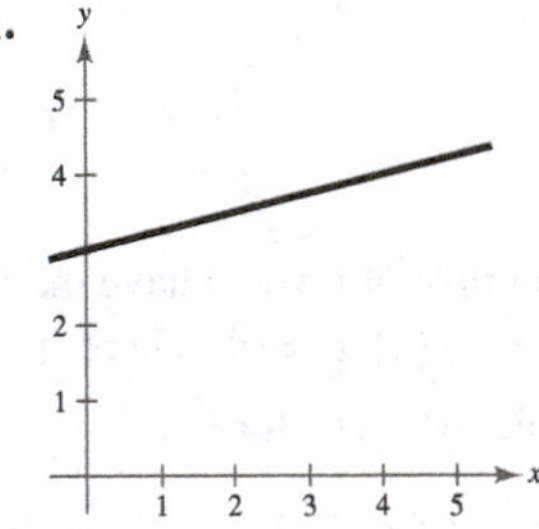

Using the point $(0, 3)$ and $(4, 4)$, $y = \frac{1}{4}x + 3$.

13.

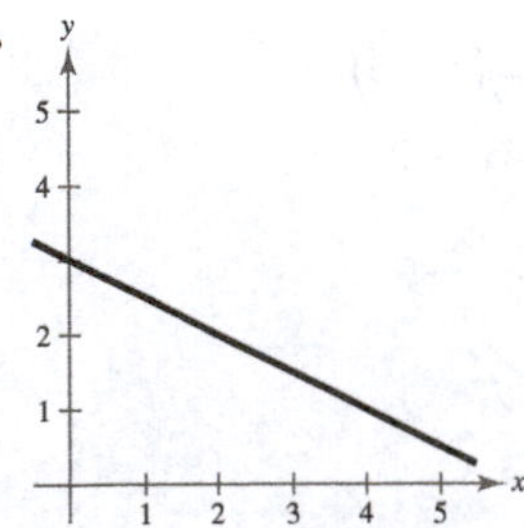

Using the points $(2, 2)$ and $(4, 1)$, $y = -\frac{1}{2}x + 3$.

15.

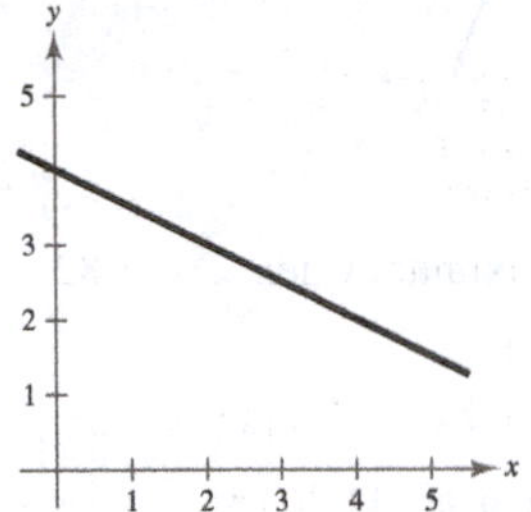

The line appears to pass through $(2, 3)$ and $(4, 2)$, so its equation is $y = -\frac{1}{2}x + 4$.

17. (a) and (b)

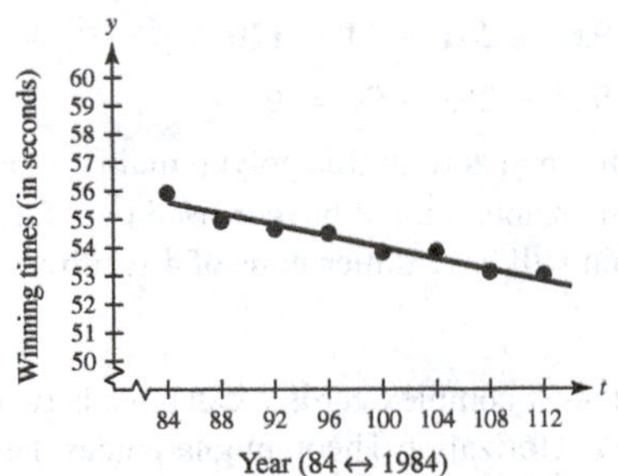

(b) Let 1984 correspond to $t = 84$, using the points $(84, 55.92)$ and $(112, 53.00)$:

$$\begin{aligned} m &= \frac{53.00 - 55.92}{112 - 84} \\ &= -0.1 \\ y - 53.00 &= -0.1(t - 112) \\ y - 53 &= -0.1t + 11.2 \\ y &= -0.1t + 64 \end{aligned}$$

(c) $y = -0.1t + 64.2$

(d) The models are similar.

19. $y = kx$

$14 = k(2)$

$7 = k$

$y = 7x$

21. $y = kx$

$1 = k(5)$

$\frac{1}{5} = k$

$y = \frac{1}{5}x$

23. $y = kx$

$8\pi = k(4)$

$2\pi = k$

$y = 2\pi x$

25. $k = 1$

x	2	4	6	8	10
$y = kx^2$	4	16	36	64	100

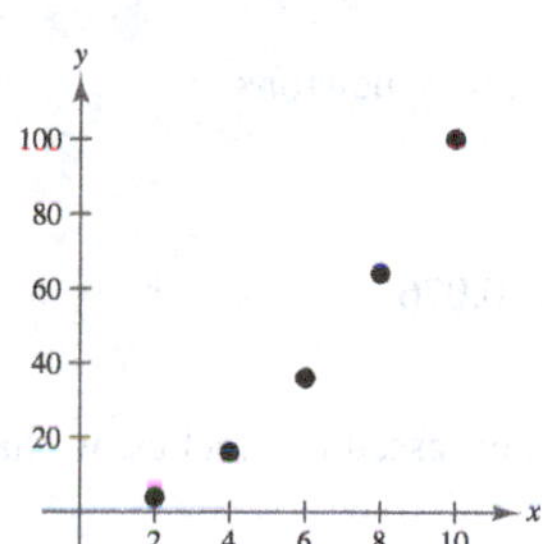

27. $k = \frac{1}{2}$

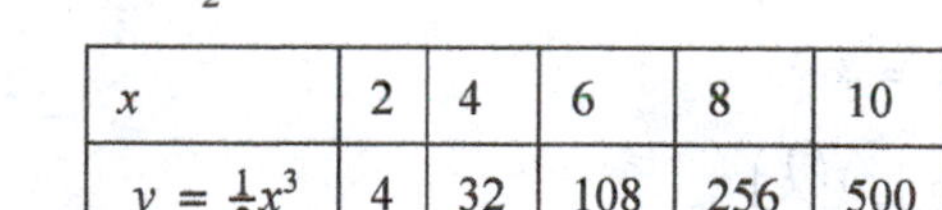

x	2	4	6	8	10
$y = \frac{1}{2}x^3$	4	32	108	256	500

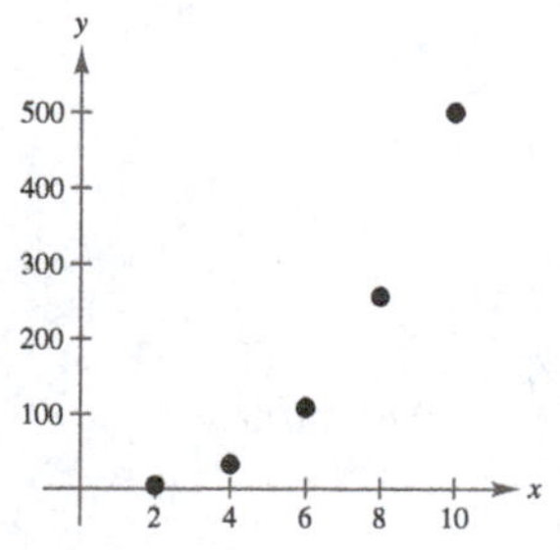

29. $k = 2, n = 1$

x	2	4	6	8	10
$y = \frac{2}{x}$	1	$\frac{1}{2}$	$\frac{1}{3}$	$\frac{1}{4}$	$\frac{1}{5}$

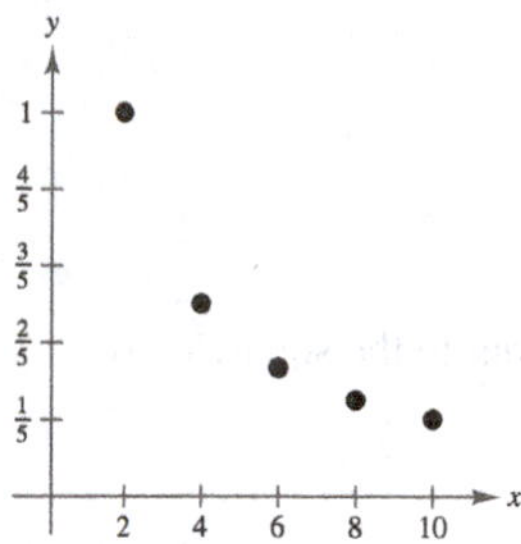

31. $k = 10$

x	2	4	6	8	10
$y = \frac{k}{x^2}$	$\frac{5}{2}$	$\frac{5}{8}$	$\frac{5}{18}$	$\frac{5}{32}$	$\frac{1}{10}$

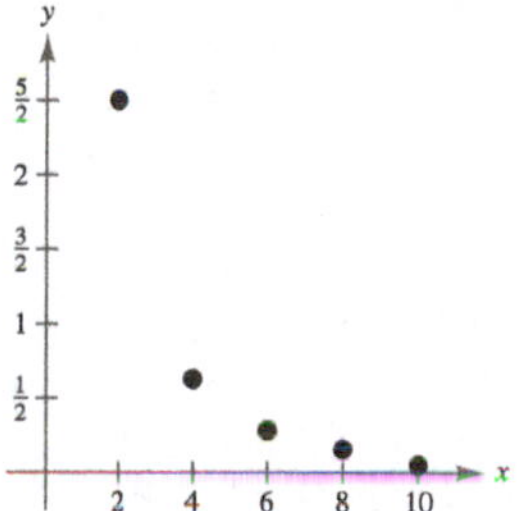

33. The graph appears to represent $y = 4/x$, so y varies inversely as x.

35. $y = \frac{k}{x}$

$1 = \frac{k}{5}$

$5 = k$

$y = \frac{5}{x}$

This equation checks with the other points given in the table.

37. $y = kx$

$-7 = k(10)$

$-\frac{7}{10} = k$

$y = -\frac{7}{10}x$

This equation checks with the other points given in the table.

39. $A = kr^2$

41. $y = \dfrac{k}{x^2}$

43. $F = \dfrac{kg}{r^2}$

45. $R = k(T - T_e)$

47. $P = kVI$

49. y is directly proportional to the square of x.

51. $A = \frac{1}{2}bh$

The area of a triangle is jointly proportional to its base and height.

53.
$$\begin{aligned} y &= kx \\ 54 &= k(3) \\ 18 &= k \\ y &= 18x \end{aligned}$$

55.
$$\begin{aligned} y &= \frac{k}{x} \\ 3 &= \frac{k}{25} \\ 75 &= k \\ y &= \frac{75}{x} \end{aligned}$$

57.
$$\begin{aligned} z &= kxy \\ 64 &= k(4)(8) \\ 2 &= k \\ z &= 2xy \end{aligned}$$

59.
$$\begin{aligned} P &= \frac{kx}{y^2} \\ \frac{28}{3} &= \frac{k(42)}{9^2} \\ \frac{28}{3} \cdot \frac{81}{42} &= k \\ \frac{2 \cdot 27}{3} &= k \\ 18 &= k \\ P &= \frac{18x}{y^2} \end{aligned}$$

61.
$$\begin{aligned} I &= kP \\ 113.75 &= k(3250) \\ 0.035 &= k \\ I &= 0.035P \end{aligned}$$

63.
$$\begin{aligned} y &= kx \\ 33 &= k(13) \\ \tfrac{33}{13} &= k \\ y &= \tfrac{33}{13}x \end{aligned}$$

When $x = 10$ inches, $y \approx 25.4$ centimeters.

When $x = 20$ inches, $y \approx 50.8$ centimeters.

65.
$$\begin{aligned} d &= kF \\ 0.12 &= k(220) \\ \tfrac{3}{5500} &= k \\ d &= \tfrac{3}{5500}F \\ 0.16 &= \tfrac{3}{5500}F \\ \tfrac{880}{3} &= F \end{aligned}$$

The required force is $293\frac{1}{3}$ newtons.

67.
$$\begin{aligned} d &= kF \\ 1.9 &= k(25) \Rightarrow k = 0.076 \\ d &= 0.076F \end{aligned}$$

When the distance compressed is 3 inches, we have

$$\begin{aligned} 3 &= 0.076F \\ F &\approx 39.47. \end{aligned}$$

No child over 39.47 pounds should use the toy.

69.
$$\begin{aligned} d &= kv^2 \\ 0.02 &= k\left(\frac{1}{4}\right)^2 \\ k &= 0.32 \\ d &= 0.32v^2 \\ 0.12 &= 0.32v^2 \\ v^2 &= \frac{0.12}{0.32} = \frac{3}{8} \\ v &= \frac{\sqrt{3}}{2\sqrt{2}} = \frac{\sqrt{6}}{4} \approx 0.61 \text{ mi/hr} \end{aligned}$$

71. (a)

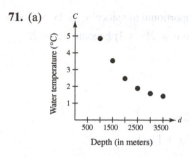

(b) The data shows an inverse variation model fits.

(c) $4.85 = \frac{k_1}{1000}$ $\quad 3.525 = \frac{k_2}{1500}$ $\quad 2.468 = \frac{k_3}{2000}$ $\quad 1.888 = \frac{k_4}{2500}$ $\quad 1.583 = \frac{k_5}{3000}$ $\quad 1.422 = \frac{k_6}{3500}$

$4850.0 = k_1$ $\quad 5287.5 = k_2$ $\quad 4936.0 = k_3$ $\quad 4720.0 = k_5$ $\quad 4749.0 = k_5$ $\quad 4977.0 = k_6$

Mean: $k = \frac{4850 + 5287.5 + 4936 + 4720 + 4977}{6} \approx 4919.92$, Model: $C = \frac{4919.92}{d}$

(d) $C = \frac{4919.92}{d}$

$3 = \frac{4919.92}{d}$

$d = \frac{4919.92}{3} \approx 1639.97$ meters

The temperature is $3°$ C at approximately 1640 meters.

73. $f_0 = k\frac{\sqrt{T_0}}{l_0\sqrt{p}}$ where f_0 = original frequency, T_0 = original tension, l_0 = original length of string, and p = mass density

$f_0 = 100$

(a) The frequency of a string with four times the tension would double the frequency in order to maintain the direct proportion.

Let $f_{\text{new}} = \frac{k \cdot \sqrt{T_{\text{new}}}}{l_{\text{new}} \cdot \sqrt{p}}$ and $T_{\text{new}} = 4T_0$, $l_{\text{new}} = l_0$

$$f_{\text{new}} = \frac{k \cdot \sqrt{4T_0}}{l_0 \cdot \sqrt{p}} = 2 \cdot \frac{k \cdot \sqrt{T_0}}{l_0 \cdot \sqrt{p}} = 2 \cdot f_0 = 2 \cdot 100 = 200Hz$$

(b) The frequency of a string with two times the length would half the frequency in order to maintain the inverse proportion.

Let $f_{\text{new}} = \frac{k \cdot \sqrt{T_{\text{new}}}}{l_{\text{new}} \cdot \sqrt{p}}$ and $l_{\text{new}} = 2l_0$, $\text{T}_{\text{new}} = T_0$

$$f_{\text{new}} = \frac{k \cdot \sqrt{T_0}}{2l_0 \cdot \sqrt{p}} = \frac{1}{2} \cdot \frac{k \cdot \sqrt{T_0}}{l_0 \cdot \sqrt{p}} = \frac{1}{2}f_0 = \frac{1}{2} \cdot 100 = 50Hz$$

(c) The frequency of a string with four times the tension would double the frequency, and two times the length would half the frequency in order to maintain the inverse proportion. Therefore the frequency would remain the same,

Let $f_{\text{new}} = \frac{k \cdot \sqrt{T_{\text{new}}}}{l_{\text{new}} \cdot \sqrt{p}}$ and $T_{\text{new}} = 4T_0$, $l_{\text{new}} = 2l_0$

$$f_{\text{new}} = \frac{k \cdot \sqrt{4T_0}}{2l_0 \cdot \sqrt{p}} = \frac{2 \cdot k\sqrt{T_0}}{2 \cdot l_0 \cdot \sqrt{p}} = \frac{2}{2}f_0 = 1 \cdot 100 = 100Hz$$

75. True. If $y = k_1x$ and $x = k_2z$, then $y = k_1(k_2z) = (k_1k_2)z$.

77. π is a constant and not a variable, so S does not vary with π. In the formula $S = 4\pi r^2$, the surface area, S is directly proportional to the square of the radius, r.

79. y is directly proportional to t since $y = 2x + 2$ and $t = x + 1$, then $y = 2(x + 1)$ becomes $y = 2t$.

Review Exercises for Chapter 3

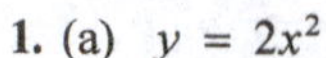

1. (a) $y = 2x^2$

Vertical stretch

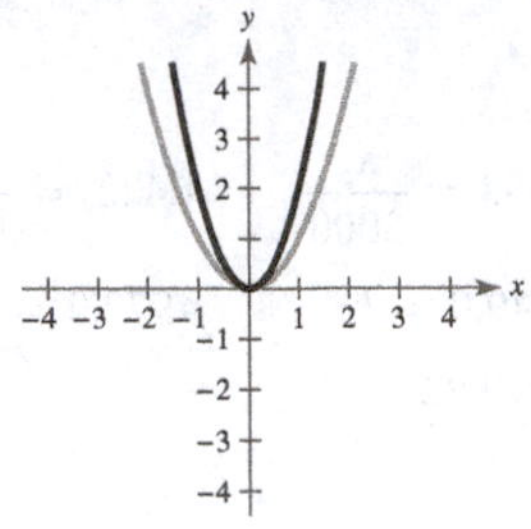

(b) $y = -2x^2$

Vertical stretch and a reflection in the x-axis

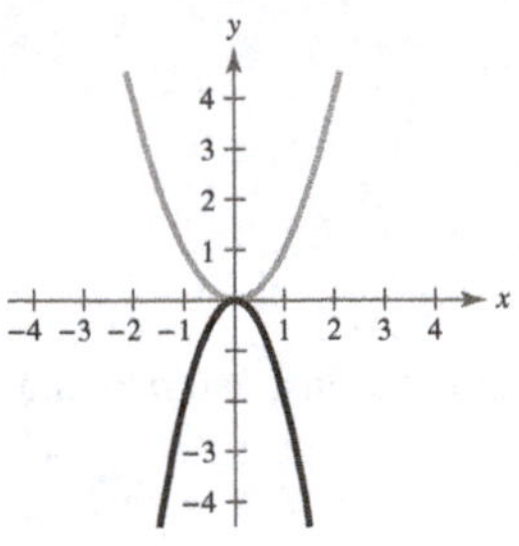

(c) $y = x^2 + 2$

Vertical shift two units upward

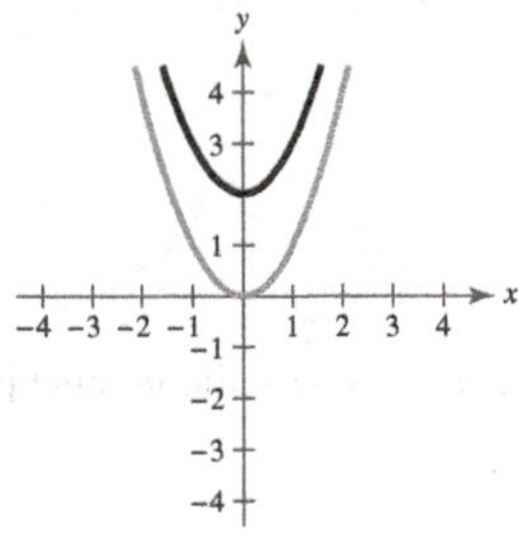

(d) $y = (x + 2)^2$

Horizontal shift two units to the left

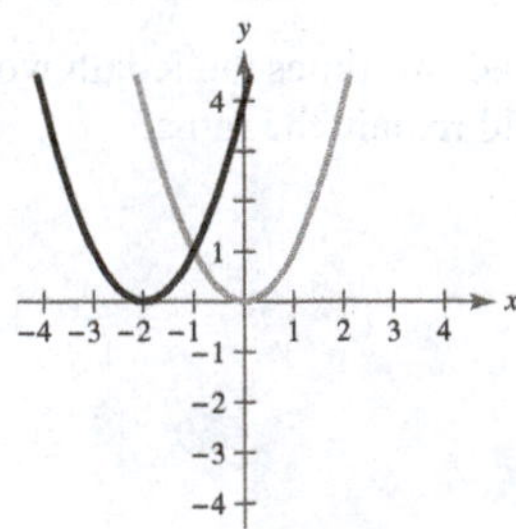

3.
$$\begin{aligned} g(x) &= x^2 - 2x \\ &= x^2 - 2x + 1 - 1 \\ &= (x - 1)^2 - 1 \end{aligned}$$

Vertex: $(1, -1)$

Axis of symmetry: $x = 1$

$0 = x^2 - 2x = x(x - 2)$

x-intercepts: $(0, 0), (2, 0)$

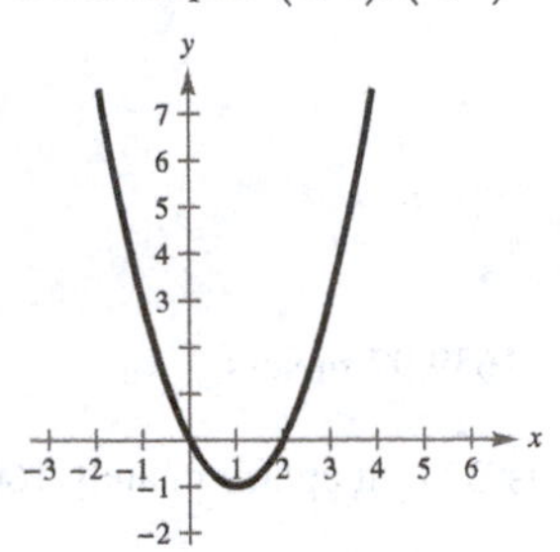

5.
$$\begin{aligned} f(x) &= x^2 - 6x + 1 \\ &= x^2 - 6x + 9 - 9 + 1 \\ &= (x - 3)^2 - 8 \end{aligned}$$

Vertex: $(3, -8)$

Axis of symmetry: $x = 3$

$0 = x^2 - 6x + 1$

$$\begin{aligned} x &= \frac{-(-6) \pm \sqrt{(-6)^2 - 4(1)(1)}}{2(1)} \\ &= \frac{6 \pm \sqrt{32}}{2} = 3 \pm 2\sqrt{2} \end{aligned}$$

x-intercepts: $\left(3 \pm 2\sqrt{2}, 0\right)$

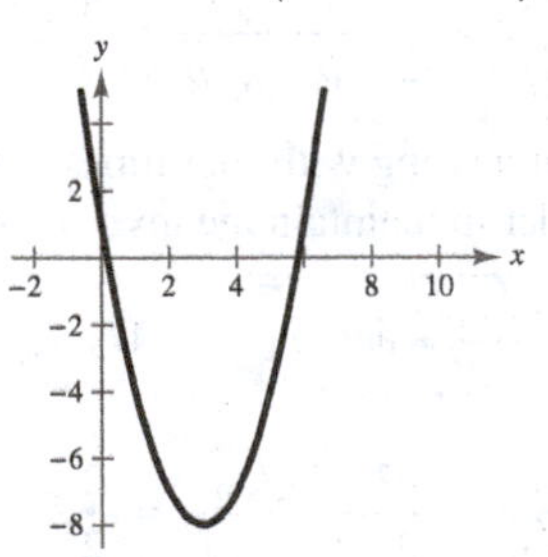

7. $f(x) = x^2 + 8x + 10$

$= x^2 + 8x + 16 - 16 + 10$

$= (x + 4)^2 - 6$

Vertex: $(-4, -6)$

Axis of symmetry: $x = -4$

$0 = (x + 4)^2 - 6$

$(x + 4)^2 = 6$

$x + 4 = \pm\sqrt{6}$

$x = -4 \pm \sqrt{6}$

x-intercepts: $\left(-4 \pm \sqrt{6}, 0\right)$

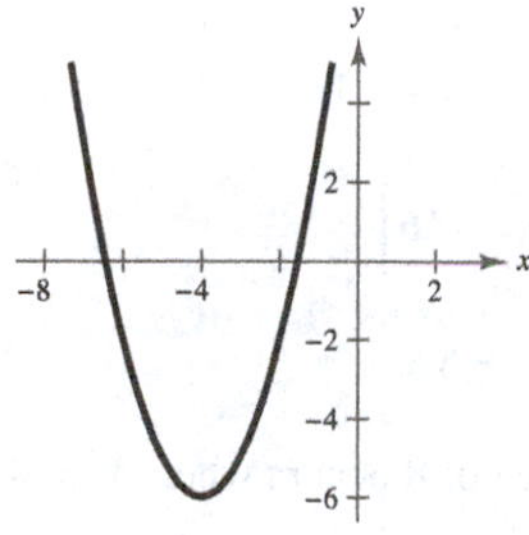

9. $h(x) = 3 + 4x - x^2$

$= -(x^2 - 4x - 3)$

$= -(x^2 - 4x + 4 - 4 - 3)$

$= -\left[(x - 2)^2 - 7\right]$

$= -(x - 2)^2 + 7$

Vertex: $(2, 7)$

Axis of symmetry: $x = 2$

$0 = 3 + 4x - x^2$

$0 = x^2 - 4x - 3$

$x = \dfrac{-(-4) \pm \sqrt{(-4)^2 - 4(1)(-3)}}{2(1)}$

$= \dfrac{4 \pm \sqrt{28}}{2} = 2 \pm \sqrt{7}$

x-intercepts: $\left(2 \pm \sqrt{7}, 0\right)$

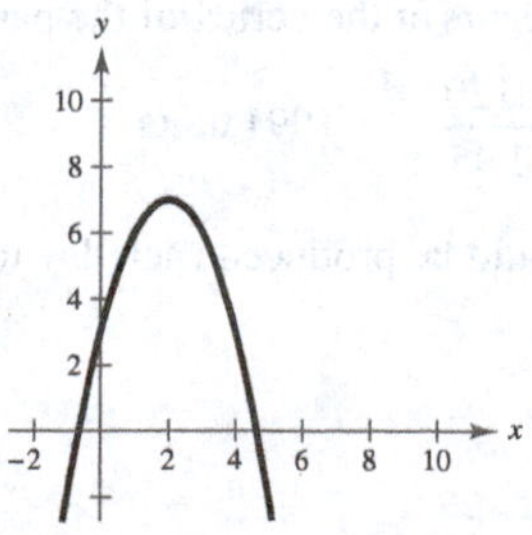

11. $h(x) = 4x^2 + 4x + 13$

$= 4(x^2 + x) + 13$

$= 4\left(x^2 + x + \frac{1}{4} - \frac{1}{4}\right) + 13$

$= 4\left(x^2 + x + \frac{1}{4}\right) - 1 + 13$

$= 4\left(x + \frac{1}{2}\right)^2 + 12$

Vertex: $\left(-\frac{1}{2}, 12\right)$

Axis of symmetry: $x = -\frac{1}{2}$

$0 = 4\left(x + \frac{1}{2}\right)^2 + 12$

$\left(x + \frac{1}{2}\right)^2 = -3$

No real zeros

x-intercepts: none

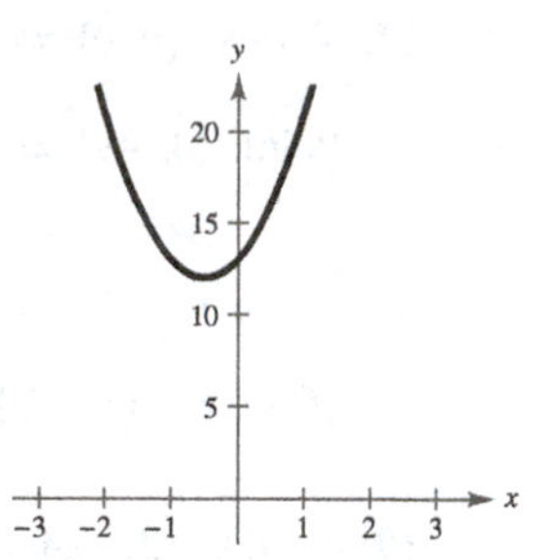

13. $f(x) = \dfrac{1}{3}(x^2 + 5x - 4)$

$= \dfrac{1}{3}\left(x^2 + 5x + \dfrac{25}{4} - \dfrac{25}{4} - 4\right)$

$= \dfrac{1}{3}\left[\left(x + \dfrac{5}{2}\right)^2 - \dfrac{41}{4}\right]$

$= \dfrac{1}{3}\left(x + \dfrac{5}{2}\right)^2 - \dfrac{41}{12}$

Vertex: $\left(-\dfrac{5}{2}, -\dfrac{41}{12}\right)$

Axis of symmetry: $x = -\dfrac{5}{2}$

$0 = x^2 + 5x - 4$

$x = \dfrac{-5 \pm \sqrt{5^2 - 4(1)(-4)}}{2(1)} = \dfrac{-5 \pm \sqrt{41}}{2}$

x-intercepts: $\left(\dfrac{-5 \pm \sqrt{41}}{2}, 0\right)$

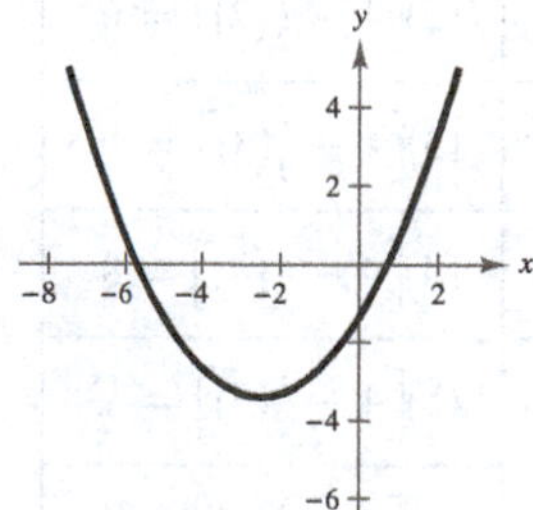

15. Vertex: $(4, 1) \Rightarrow f(x) = a(x - 4)^2 + 1$

Point: $(2, -1) \Rightarrow -1 = a(2 - 4)^2 + 1$

$$-2 = 4a$$
$$-\tfrac{1}{2} = a$$

$f(x) = -\tfrac{1}{2}(x - 4)^2 + 1$

17. Vertex: $(6, 0) \Rightarrow f(x) = a(x - 6)^2$

Point: $(3, -9) \Rightarrow -9 = a(3 - 6)^2$

$$-9 = 9a$$
$$-1 = a$$

$f(x) = -(x - 6)^2$

19. Vertex: $\left(2, -\tfrac{5}{2}\right) \Rightarrow f(x) = a(x - 2)^2 - \tfrac{5}{2}$

Point: $\left(4, \tfrac{1}{2}\right) \Rightarrow \tfrac{1}{2} = a(4 - 2)^2 - \tfrac{5}{2}$

$$\tfrac{1}{2} = 4a - \tfrac{5}{2}$$
$$3 = 4a$$
$$\tfrac{3}{4} = a$$

$f(x) = \tfrac{3}{4}(x - 2)^2 - \tfrac{5}{2}$

21. (a) $A = xy = x\left(\dfrac{8 - x}{2}\right)$ because

$x + 2y - 8 = 0 \Rightarrow y = \dfrac{8 - x}{2}$.

(b) The figure is in the first quadrant and x and y must be positive, so the domain of

$A = x\left(\dfrac{8 - x}{2}\right)$ is $0 < x < 8$.

(c)

x	y	Area
1	$4 - \frac{1}{2}(1)$	$(1)\left[4 - \frac{1}{2}(1)\right] = \frac{7}{2}$
2	$4 - \frac{1}{2}(2)$	$(2)\left[4 - \frac{1}{2}(2)\right] = 6$
3	$4 - \frac{1}{2}(3)$	$(3)\left[4 - \frac{1}{2}(3)\right] = \frac{15}{2}$
4	$4 - \frac{1}{2}(4)$	$(4)\left[4 - \frac{1}{2}(4)\right] = 8$
5	$4 - \frac{1}{2}(5)$	$(5)\left[4 - \frac{1}{2}(5)\right] = \frac{15}{2}$
6	$4 - \frac{1}{2}(6)$	$(6)\left[4 - \frac{1}{2}(6)\right] = 6$

Using the table, when $x = 4$ and $y = 2$, the area A is 8 square units.

(d)

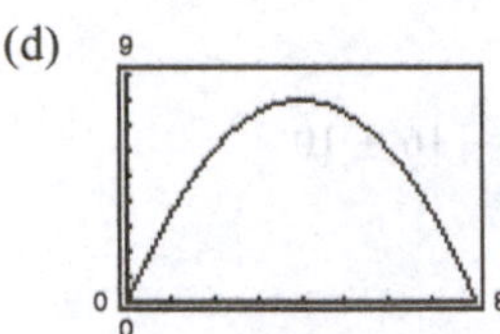

The maximum area of 8 occurs at the vertex when $x = 4$ and $y = \dfrac{8 - 4}{2} = 2$.

(e)
$$\begin{aligned} A &= x\left(\frac{8 - x}{2}\right) \\ &= \frac{1}{2}(8x - x^2) \\ &= -\frac{1}{2}(x^2 - 8x) \\ &= -\frac{1}{2}(x^2 - 8x + 16 - 16) \\ &= -\frac{1}{2}\left[(x - 4)^2 - 16\right] \\ &= -\frac{1}{2}(x - 4)^2 + 8 \end{aligned}$$

The maximum area of 8 occurs when $x = 4$ and $y = \dfrac{8 - 4}{2} = 2$.

23. $R = -10p^2 + 800p$

(a) $R(20) = \$12{,}000$

$R(25) = \$13{,}750$

$R(30) = \$15{,}000$

(b) The maximum revenue occurs at the vertex of the parabola.

$$-\frac{b}{2a} = \frac{-800}{2(-10)} = \$40$$

$R(40) = \$16{,}000$

The revenue is maximum when the price is \$40 per unit.

The maximum revenue is \$16,000.

Any price greater or less than \$40 per unit will not yield as much revenue.

25. $C = 70{,}000 - 120x + 0.055x^2$

The minimum cost occurs at the vertex of the parabola.

Vertex: $-\dfrac{b}{2a} = -\dfrac{-120}{2(0.055)} \approx 1091$ units

About 1091 units should be produced each day to yield a minimum cost.

27. $y = x^3$, $f(x) = -(x - 2)^3$

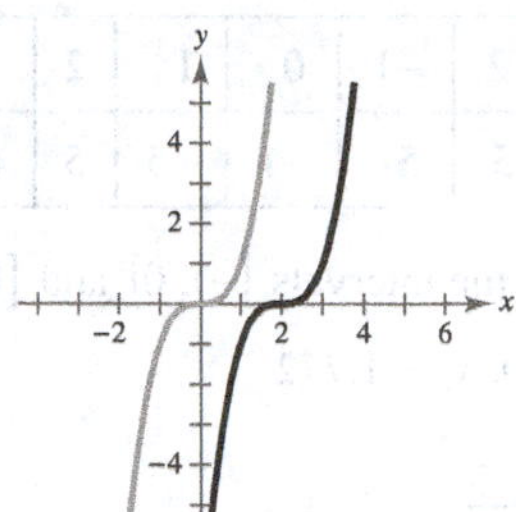

Transformation: A horizontal shift two units to the right

29. $y = x^4$, $f(x) = 6 - x^4$

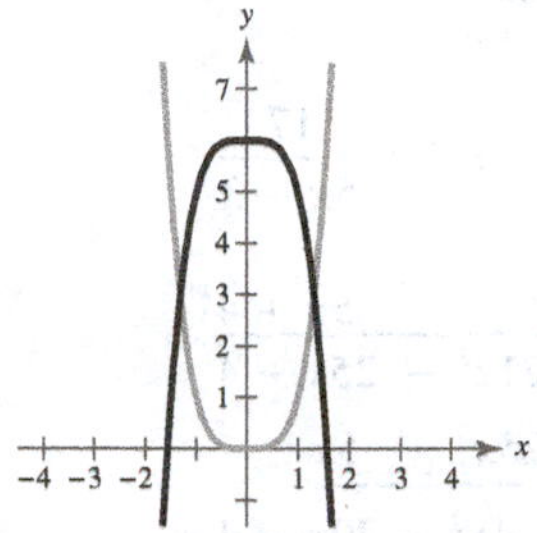

Transformation: Reflection in the x-axis and a vertical shift six units upward

31. $y = x^5, f(x) = (x - 5)^5 + 1$

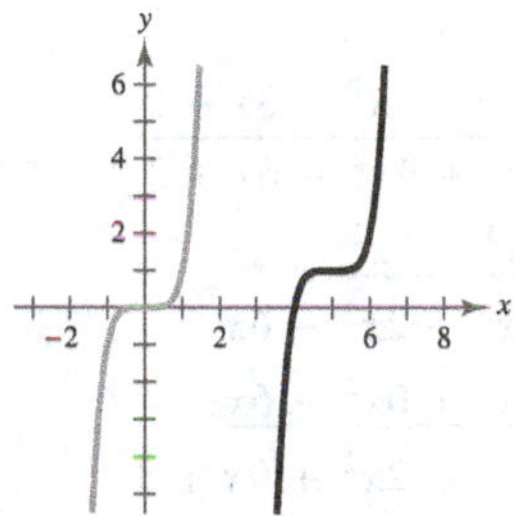

Transformation: Horizontal shift five units to the right and a vertical shift one unit upward

33. $f(x) = -2x^2 - 5x + 12$

The degree is even and the leading coefficient is negative. The graph falls to the left and falls to the right.

35. $g(x) = -3x^3 - 8x^4 + x^5$

The degree is odd and the leading coefficient is positive. The graph falls to the left and rises to the right.

37. (a) $f(x) = 3x^2 + 20x - 32$

$$0 = 3x^2 + 20x - 32$$
$$0 = (3x - 4)(x + 8)$$

Zeros: $x = \frac{4}{3}$ and $x = -8$,

(b) both of multiplicity 1 (odd multiplicity)

(c) Turning points: 1

(d)

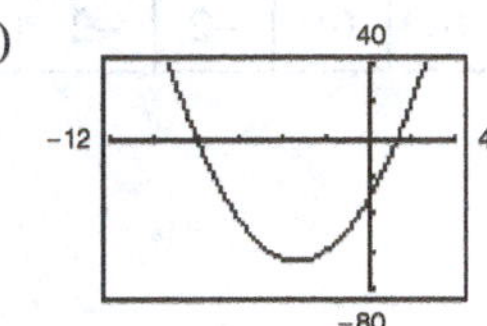

39. (a) $f(t) = t^3 - 3t$

$$0 = t^3 - 3t$$
$$0 = t(t^2 - 3)$$

Zeros: $t = 0, t = \pm\sqrt{3}$

(b) All of the zeros have a multiplicity of 1 (odd multiplicity)

(c) Turning points: 2

(d)

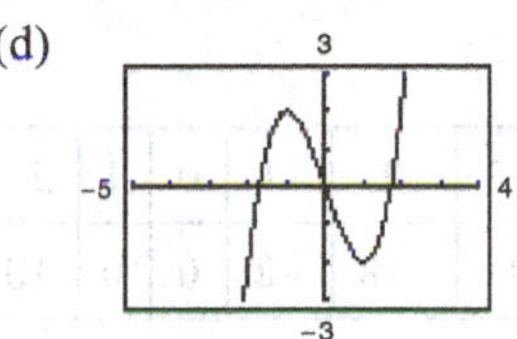

41. (a)

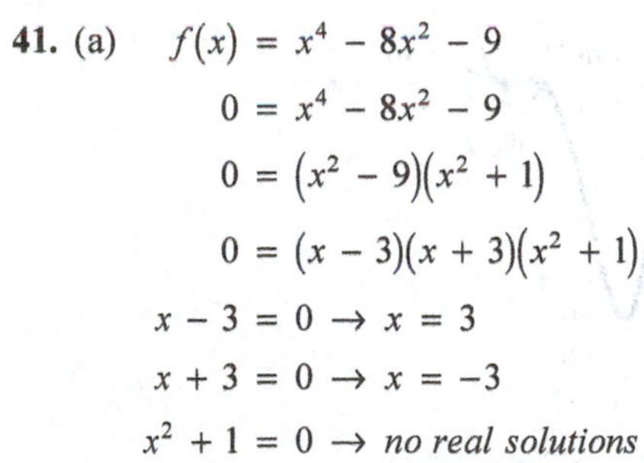

$$f(x) = x^4 - 8x^2 - 9$$
$$0 = x^4 - 8x^2 - 9$$
$$0 = (x^2 - 9)(x^2 + 1)$$
$$0 = (x - 3)(x + 3)(x^2 + 1)$$

$x - 3 = 0 \rightarrow x = 3$

$x + 3 = 0 \rightarrow x = -3$

$x^2 + 1 = 0 \rightarrow$ *no real solutions*

Zeros: $x = \pm 3$

(b) $x = 3$ of multiplicity 1 (odd multiplicity)

$x = -3$ of multiplicity 1 (odd multiplicity)

(c) Turning points: 3

(d)

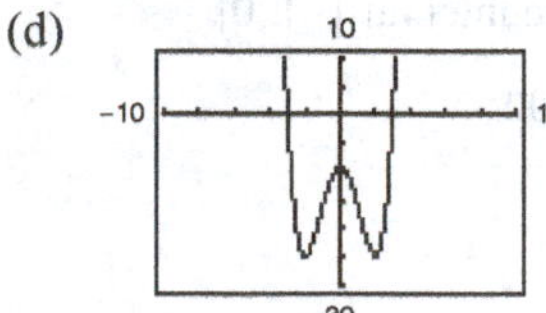

43. $f(x) = -x^3 + x^2 - 2$

(a) The degree is odd and the leading coefficient is negative. The graph rises to the left and falls to the right.

(b) Zero: $x = -1$

(c)

x	-3	-2	-1	0	1	2
$f(x)$	34	10	0	-2	-2	-6

(d)

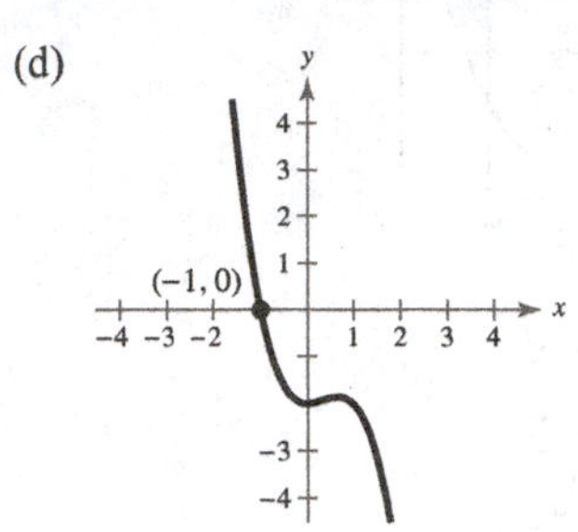

45. $f(x) = x(x^3 + x^2 - 5x + 3)$

(a) The degree is even and the leading coefficient is positive. The graph rises to the left and rises to the right.

(b) Zeros: $x = 0, 1, -3$

(c)

x	-4	-3	-2	-1	0	1	2	3
$f(x)$	100	0	-18	-8	0	0	10	72

(d)

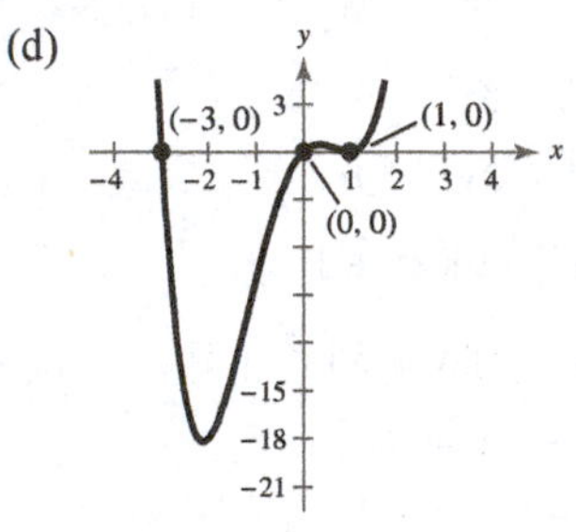

47. (a) $f(x) = 3x^3 - x^2 + 3$

x	-3	-2	-1	0	1	2	3
$f(x)$	-87	-25	-1	3	5	23	75

The zero is in the interval $[-1, 0]$.

(b) Zero: $x \approx -0.900$

49. (a) $f(x) = x^4 - 5x - 1$

x	-3	-2	-1	0	1	2	3
$f(x)$	95	25	5	-1	-5	5	65

There are zeros in the intervals $[-1, 0]$ and $[1, 2]$.

(b) Zeros: $x \approx -0.200$, $x \approx 1.772$

51.

$$\begin{array}{r}
6x + 3 \\
5x - 3\overline{)30x^2 - 3x + 8} \\
\underline{30x^2 - 18x} \\
15x + 8 \\
\underline{15x - 9} \\
17
\end{array}$$

$$\frac{30x^2 - 3x + 8}{5x - 3} = 6x + 3 + \frac{17}{5x - 3}$$

53.

$$\begin{array}{r}
5x + 4 \\
x^2 - 5x - 1\overline{)5x^3 - 21x^2 - 25x - 4} \\
\underline{5x^3 - 25x^2 - 5x} \\
4x^2 - 20x - 4 \\
\underline{4x^2 - 20x - 4} \\
0
\end{array}$$

$$\frac{5x^3 - 21x^2 - 25x - 4}{x^2 - 5x - 1} = 5x + 4,\ x \neq \frac{5}{2} \pm \frac{\sqrt{29}}{2}$$

55.

$$\begin{array}{r}
x^2 - 3x + 2 \\
x^2 + 0x + 2\overline{)x^4 - 3x^3 + 4x^2 - 6x + 3} \\
\underline{x^4 + 0x^3 + 2x^2} \\
-3x^3 + 2x^2 - 6x \\
\underline{-3x^3 + 0x^2 - 6x} \\
2x^2 + 0x + 3 \\
\underline{2x^2 + 0x + 4} \\
-1
\end{array}$$

$$\frac{x^4 - 3x^3 + 4x^2 - 6x + 3}{x^2 + 2} = x^2 - 3x + 2 - \frac{1}{x^2 + 2}$$

57.

$$\begin{array}{r|rrrr}
8 & 2 & -25 & 66 & 48 \\
 & & 16 & -72 & -48 \\
\hline
 & 2 & -9 & -6 & 0
\end{array}$$

$$\frac{2x^3 - 25x^2 + 66x + 48}{x - 8} = 2x^2 - 9x - 6, \quad x \neq 8$$

59.

$$\begin{array}{r|rrrrr}
-3 & 1 & 0 & -2 & 9 & 0 \\
 & & -3 & 9 & -21 & 36 \\
\hline
 & 1 & -3 & 7 & -12 & 36
\end{array}$$

$$\frac{x^4 - 2x^2 + 9x}{x + 3} = x^3 - 3x^2 + 7x - 12 + \frac{36}{x + 3}$$

61. $f(x) = x^4 + 10x^3 - 24x^2 + 20x + 44$

(a) Remainder Theorem:

$$f(-3) = (-3)^4 + 10(-3)^3 - 24(-3)^2 + 20(-3) + 44$$
$$= -421$$

Synthetic Division:

$$\begin{array}{r|rrrrr} -3 & 1 & 10 & -24 & 20 & 44 \\ & & -3 & -21 & 135 & -465 \\ \hline & 1 & 7 & -45 & 155 & -421 \end{array}$$

So, $f(-3) = -421$.

(b) Remainder Theorem:

$$f(-1) = (-1)^4 + 10(-1)^3 - 24(-1)^2 + 20(-1) + 44$$
$$= -9$$

Synthetic Division:

$$\begin{array}{r|rrrrr} -1 & 1 & 10 & -24 & 20 & 44 \\ & & -1 & -9 & 33 & -53 \\ \hline & 1 & 9 & -33 & 53 & -9 \end{array}$$

So, $f(-1) = -9$.

63. $f(x) = 20x^4 + 9x^3 - 14x^2 - 3x$

(a) $$\begin{array}{r|rrrrr} -1 & 20 & 9 & -14 & -3 & 0 \\ & & -20 & 11 & 3 & 0 \\ \hline & 20 & -11 & -3 & 0 & 0 \end{array}$$

Yes, $x = -1$ is a zero of f.

(b) $$\begin{array}{r|rrrrr} \frac{3}{4} & 20 & 9 & -14 & -3 & 0 \\ & & 15 & 18 & 3 & 0 \\ \hline & 20 & 24 & 4 & 0 & 0 \end{array}$$

Yes, $x = \frac{3}{4}$ is a zero of f.

(c) $$\begin{array}{r|rrrrr} 0 & 20 & 9 & -14 & -3 & 0 \\ & & 0 & 0 & 0 & 0 \\ \hline & 20 & 9 & -14 & -3 & 0 \end{array}$$

Yes, $x = 0$ is a zero of f.

(d) $$\begin{array}{r|rrrrr} 1 & 20 & 9 & -14 & -3 & 0 \\ & & 20 & 29 & 15 & 12 \\ \hline & 20 & 29 & 15 & 12 & 12 \end{array}$$

No, $x = 1$ is not a zero of f.

65. $f(x) = x^3 + 4x^2 - 25x - 28$; Factor: $(x - 4)$

(a) $$\begin{array}{r|rrrr} 4 & 1 & 4 & -25 & -28 \\ & & 4 & 32 & 28 \\ \hline & 1 & 8 & 7 & 0 \end{array}$$

Yes, $(x - 4)$ is a factor of $f(x)$.

(b) $x^2 + 8x + 7 = (x + 7)(x + 1)$

The remaining factors of f are $(x + 7)$ and $(x + 1)$.

(c) $f(x) = x^3 + 4x^2 - 25x - 28$

$= (x + 7)(x + 1)(x - 4)$

(d) Zeros: $-7, -1, 4$

(e) 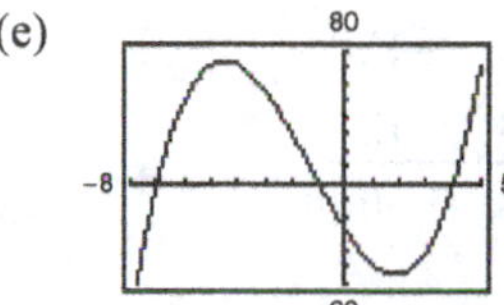

67. $f(x) = x^4 - 4x^3 - 7x^2 + 22x + 24$

Factors: $(x + 2), (x - 3)$

(a) $$\begin{array}{r|rrrrr} -2 & 1 & -4 & -7 & 22 & 24 \\ & & -2 & 12 & -10 & -24 \\ \hline & 1 & -6 & 5 & 12 & 0 \end{array}$$

$$\begin{array}{r|rrrr} 3 & 1 & -6 & 5 & 12 \\ & & 3 & -9 & -12 \\ \hline & 1 & -3 & -4 & 0 \end{array}$$

Yes, $(x + 2)$ and $(x - 3)$ are both factors of $f(x)$.

(b) $x^2 - 3x - 4 = (x + 1)(x - 4)$

The remaining factors are $(x + 1)$ and $(x - 4)$.

(c) $f(x) = (x + 1)(x - 4)(x + 2)(x - 3)$

(d) Zeros: $-2, -1, 3, 4$

(e) 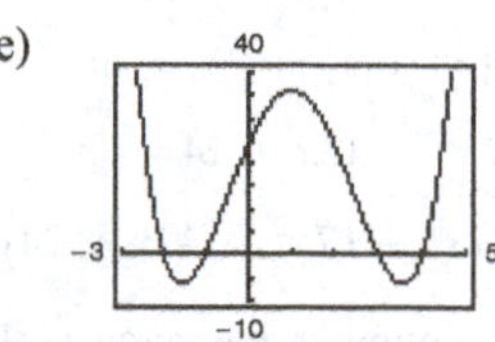

69. Since $f(x) = x - 6$ is a 1st degree polynomial function, it has one zero.

71. Since $h(t) = t^2 - t^5$ is a 5th degree polynomial function, it has five zeros.

73. Since $f(x) = (x - 8)^3 = x^3 - 24x^2 + 19x^2 - 512$ is a 3rd degree polynomial function, it has three zeros.

75. $f(x) = x^3 + 3x^2 - 28x - 60$

Possible rational zeros: $\pm1, \pm2, \pm3, \pm4, \pm5, \pm6, \pm10, \pm12, \pm15, \pm20, \pm30, \pm60$

$$\begin{array}{r|rrrr} -2 & 1 & 3 & -28 & -60 \\ & & -2 & -2 & 60 \\ \hline & 1 & 1 & -30 & 0 \end{array}$$

$$\begin{aligned} x^3 + 3x^2 - 28x - 60 &= (x + 2)(x^2 + x - 30) \\ &= (x + 2)(x + 6)(x - 5) \end{aligned}$$

The zeros of $f(x)$ are $x = -2$, $x = -6$, and $x = 5$.

77. $f(x) = 3x^3 + 8x^2 - 4x - 16$

Possible rational zeros: $\dfrac{\pm1, \pm2, \pm4, \pm8, \pm16}{\pm1, \pm3} = \pm\frac{1}{3}, \pm\frac{2}{3}, \pm1, \pm\frac{4}{3}, \pm2, \pm\frac{8}{3}, \pm4, \pm\frac{16}{3}, \pm8, \pm16$

$$\begin{array}{r|rrrr} -2 & 3 & 8 & -4 & -16 \\ & & -6 & -4 & 16 \\ \hline & 3 & 2 & -8 & 0 \end{array}$$

$3x^3 + 8x^2 - 4x - 16 = (x + 2)(3x^2 + 2x - 8) = (x + 2)(3x - 4)(x + 2)$

The zeros of $f(x)$ are $x = \frac{4}{3}$, $x = -2$.

79. $f(x) = x^4 + x^3 - 11x^2 + x - 12$

Possible rational zeros: $\pm1, \pm2, \pm3, \pm4, \pm6, \pm12$

$$\begin{array}{r|rrrrr} 3 & 1 & 1 & -11 & 1 & -12 \\ & & 3 & 12 & 3 & 12 \\ \hline & 1 & 4 & 1 & 4 & 0 \end{array}$$

$$\begin{array}{r|rrrr} -4 & 1 & 4 & 1 & 4 \\ & & -4 & 0 & -4 \\ \hline & 1 & 0 & 1 & 0 \end{array}$$

$x^4 + x^3 - 11x^2 + x - 12 = (x - 3)(x + 4)(x^2 + 1)$

The zeros of $f(x)$ are $x = 3$ and $x = -4$.

81. Because $\sqrt{3}i$ is a zero, so is $-\sqrt{3}i$.

Multiply by 3 to clear the fraction.

$$\begin{aligned} f(x) &= 3\left(x - \tfrac{2}{3}\right)(x - 4)\left(x - \sqrt{3}i\right)\left(x + \sqrt{3}i\right) \\ &= (3x - 2)(x - 4)(x^2 + 3) \\ &= (3x^2 - 14x + 8)(x^2 + 3) \\ &= 3x^4 - 14x^3 + 17x^2 - 42x + 24 \end{aligned}$$

Note: $f(x) = a(3x^4 - 14x^3 + 17x^2 - 42x + 24)$, where a is any real nonzero number, has zeros $\frac{2}{3}$, 4, and $\pm\sqrt{3}i$.

83. $h(x) = -x^3 + 2x^2 - 16x + 32$

Because $-4i$ is a zero, so is $4i$.

$$\begin{array}{r|rrrr} -4i & -1 & 2 & -16 & 32 \\ & & 4i & 16 - 8i & -32 \\ \hline & -1 & 2 + 4i & -8i & 0 \end{array}$$

$$\begin{array}{r|rrr} 4i & -1 & 2 + 4i & -8i \\ & & -4i & 8i \\ \hline & -1 & 2 & 0 \end{array}$$

$h(x) = (x + 4i)(x - 4i)(-x + 2)$

Zeros: $x = \pm4i, 2$

85.
$$\begin{aligned} f(x) &= x^3 + 4x^2 - 5x \\ &= x(x^2 + 4x - 5) \\ &= x(x + 5)(x - 1) \end{aligned}$$

Zeros: $x = 0, -5, 1$

87. $g(x) = x^4 + 4x^3 - 3x^2 + 40x + 208$, Zero: $x = -4$

$$\begin{array}{r|rrrrr} -4 & 1 & 4 & -3 & 40 & 208 \\ & & -4 & 0 & 12 & -208 \\ \hline & 1 & 0 & -3 & 52 & 0 \end{array}$$

$$\begin{array}{r|rrrr} -4 & 1 & 0 & -3 & 52 \\ & & -4 & 16 & -52 \\ \hline & 1 & -4 & 13 & 0 \end{array}$$

$g(x) = (x + 4)^2(x^2 - 4x + 13)$

By the Quadratic Formula, the zeros of $x^2 - 4x + 13$ are $x = 2 \pm 3i$. The zeros of $g(x)$ are $x = -4$ and $x = 2 \pm 3i$.

$$\begin{aligned} g(x) &= (x + 4)^2[x - (2 + 3i)][x - (2 - 3i)] \\ &= (x + 4)^2(x - 2 - 3i)(x - 2 + 3i) \end{aligned}$$

89. $f(x) = x^3 - 16x^2 + x - 16$

Possible rational zeros: $\pm1, \pm2, \pm4, \pm8, \pm16$

$$\begin{array}{r|rrrr} 16 & 1 & -16 & 1 & -16 \\ & & 16 & 0 & 16 \\ \hline & 1 & 0 & 1 & 0 \end{array}$$

$$x^3 - 16x^2 + x - 16 = (x - 16)(x^2 + 1)$$

$$\begin{aligned} x^2 + 1 &= 0 \\ x^2 &= -1 \\ x &= \pm\sqrt{-1} \\ x &= \pm i \end{aligned}$$

The zero of $f(x)$ are $x = 16$ and $x = \pm i$.

91. $g(x) = x^4 - 3x^3 - 14x^2 - 12x - 72$

Possible rational zeros: $\pm1, \pm2, \pm3, \pm4, \pm6, \pm8, \pm9, \pm12, \pm18, \pm24, \pm36, \pm72$

$$\begin{array}{r|rrrrr} -3 & 1 & -3 & -14 & -12 & -72 \\ & & -3 & 18 & -12 & 72 \\ \hline & 1 & -6 & 4 & -24 & 0 \end{array}$$

$$\begin{array}{r|rrrr} 6 & 1 & -6 & 4 & -24 \\ & & 6 & 0 & 24 \\ \hline & 1 & 0 & 4 & 0 \end{array}$$

$$x^4 - 3x^3 - 14x^2 - 12x - 72 = (x + 3)(x - 6)(x^2 + 4)$$

$$\begin{aligned} x^2 + 4 &= 0 \\ x^2 &= -4 \\ x &= \pm\sqrt{-4} \\ x &= \pm 2i \end{aligned}$$

The zeros of $g(x)$ are $x = -3$, $x = 6$, and $x = \pm 2i$.

93. $g(x) = 5x^3 + 3x^2 - 6x + 9$

$g(x)$ has two variations in sign, so g has either two or no positive real zeros.

$g(-x) = -5x^3 + 3x^2 + 6x + 9$

$g(-x)$ has one variation in sign, so g has one negative real zero.

95. $f(x) = 4x^3 - 3x^2 + 4x - 3$

(a) $\begin{array}{r|rrrr} 1 & 4 & -3 & 4 & -3 \\ & & 4 & 1 & 5 \\ \hline & 4 & 1 & 5 & 2 \end{array}$

Because the last row has all positive entries, $x = 1$ is an upper bound.

(b) $\begin{array}{r|rrrr} -\frac{1}{4} & 4 & -3 & 4 & -3 \\ & & -1 & 1 & -\frac{5}{4} \\ \hline & 4 & -4 & 5 & -\frac{17}{4} \end{array}$

Because the last row entries alternate in sign, $x = -\frac{1}{4}$ is a lower bound.

97. The volume of a right circular cylinder is given by: $V = \pi r^2 h.$

So, $36\pi = \pi r^2 h$ and because $h = r + 9$,

$$36\pi = \pi r^2(r + 9)$$
$$36 = r^3 + 9r^2$$
$$0 = r^3 + 9r^2 - 36$$

Using the list of possible rational zeros and synthetic division, you can determine the only real solution is irrational. Therefore, use a graphing utility to approximate the solution. So, the radius $r \approx 1.82$ inches and the height $h \approx 1.82 + 9 = 10.82$ inches.

99.

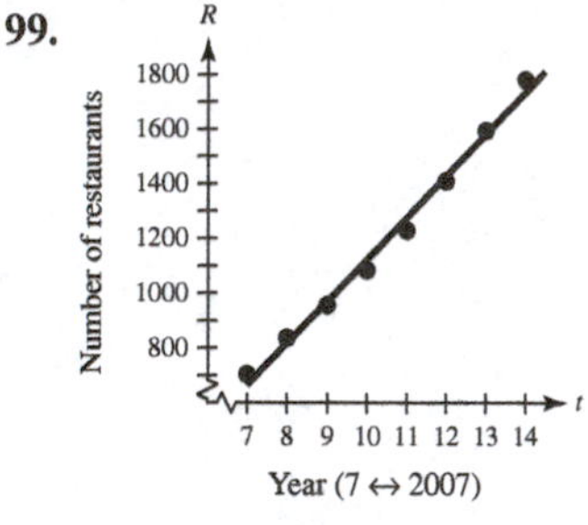

The data fits the model well.

101.
$$y = kx$$
$$20 = 12.5k$$
$$\frac{20}{12.5} = k$$
$$k = \frac{8}{5}$$
$$y = \frac{8}{5}x \quad \text{or} \quad y = 1.6x$$

In 5 miles: $y = 1.6(5) = 8.0$ kilometers

In 25 miles: $y = 1.6(25) = 40.0$ kilometers

103. $F = ks^2$

If speed is doubled,

$$F = k(2s)^2$$
$$F = 4ks^2.$$

So, the force will be changed by a factor of 4.

105.
$$T = \frac{k}{r}$$
$$3 = \frac{k}{65}$$
$$k = 3(65) = 195$$
$$T = \frac{195}{r}$$

When $r = 80$ mph,

$$T = \frac{195}{80} = 2.4375 \text{ hours}$$
$$\approx 2 \text{ hours, 26 minutes.}$$

107. True. The graph of $f(x) = 2 + x - x^2 + x^3 - x^4 + x^5 + x^6 - x^7$ rises to the left and falls to the right because it is an odd degree and the leading coefficient is negative.

109. True. If y is directly proportional to x, then $y = kx$, so $x = \left(\frac{1}{k}\right)y$. Therefore, x is directly proportional to y.

111. Answers will vary. *Sample answer:*

Polynomials of degree $n > 0$ with real coefficients can be written as the product of linear and quadratic factors with real coefficients, where the quadratic factors have no real zeros.

Setting the factors equal to zero and solving for the variable can find the zeros of a polynomial function.

To solve an equation is to find all the values of the variable for which the equation is true.

Problem Solving for Chapter 3

1. (a) (i) $g(x) = x^2 - 4x - 12$

$0 = (x - 6)(x + 2)$

Zeros: 6, −2

(ii) $g(x) = x^2 + 5x$

$0 = x(x + 5)$

Zeros: 0, −5

(iii) $g(x) = x^2 + 3x - 10$

$0 = (x + 5)(x - 2)$

Zeros: −5, 2

(iv) $g(x) = x^2 - 4x + 4$

$0 = (x - 2)(x - 2)$

Zeros: 2 (repeated zero)

(v) $g(x) = x^2 - 2x - 6$

$0 = x^2 - 2x - 6$

By the Quadratic Formula, $x = 1 \pm \sqrt{7}$.

Zeros: $1 \pm \sqrt{7}$

(vi) $g(x) = x^2 + 3x + 4$

$0 = x^2 + 3x + 4$

By the Quadratic Formula, $x = \dfrac{-3 \pm \sqrt{7}i}{2}$.

Zeros: $\dfrac{-3 \pm \sqrt{7}i}{2}$

(b) (i) $f(x) = (x - 2)(x^2 - 4x - 12)$

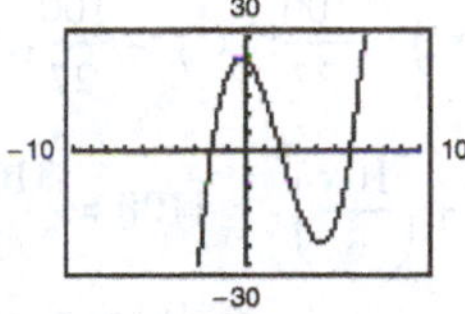

(ii) $f(x) = (x - 2)(x^2 + 5x)$

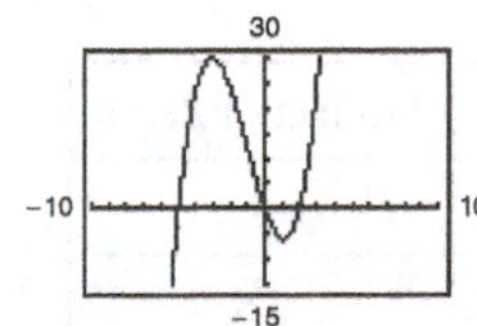

(iii) $f(x) = (x - 2)(x^2 + 3x - 10)$

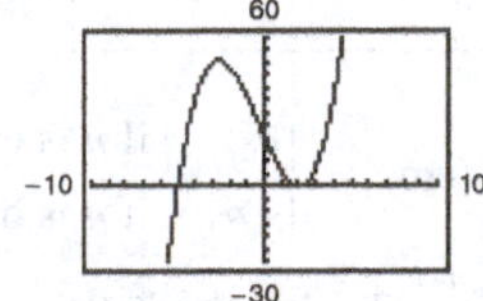

(iv) $f(x) = (x - 2)(x^2 - 4x + 4)$

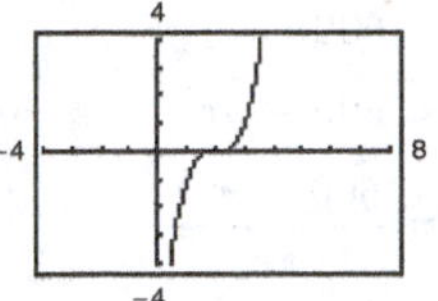

(v) $f(x) = (x - 2)(x^2 - 2x - 6)$

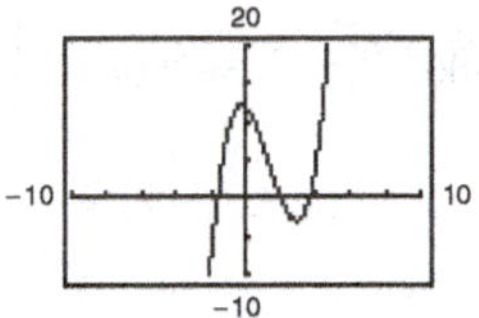

(vi) $f(x) = (x - 2)(x^2 + 3x + 4)$

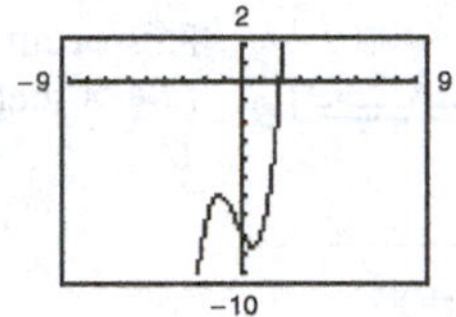

$x = 2$ is an x-intercept of $f(x)$ in all six graphs. All the graphs, except (iii), cross the x-axis at $x = 2$.

(c) (i) other x-intercepts: $(-2, 0)$, $(6, 0)$

(ii) other x-intercepts: $(-5, 0)$, $(0, 0)$

(iii) other x-intercepts: $(-5, 0)$

(iv) other x-intercepts: No additional x-intercepts

(v) other x-intercepts: $(-1.6, 0)$, $(3.6, 0)$

(vi) other x-intercepts: No additional x-intercepts

(d) When the function has two real zeros, the results are the same. When the function has one real zero, the graph touches the x-axis at the zero. When there are no real zeros, there is no x-intercept.

3. 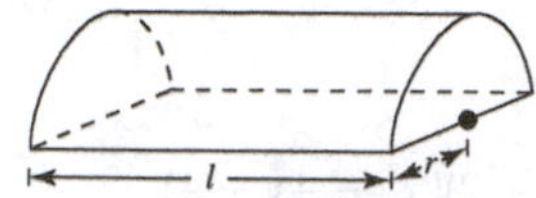

(a) $\pi r^2 + \pi rl = 600$

$\pi rl = 600 - \pi r^2$

$l = \dfrac{600 - \pi r^2}{\pi r}$

(b) $V = \dfrac{1}{2}\pi r^2 l - \dfrac{1}{2}\pi r^2\left(\dfrac{600 - \pi r^2}{\pi r}\right)$

$= \dfrac{1}{2}r\left(600 - \pi r^2\right) = 300r - \dfrac{1}{2}\pi r^3$

$= -\dfrac{1}{2}\pi r^3 + 300r$

(c) 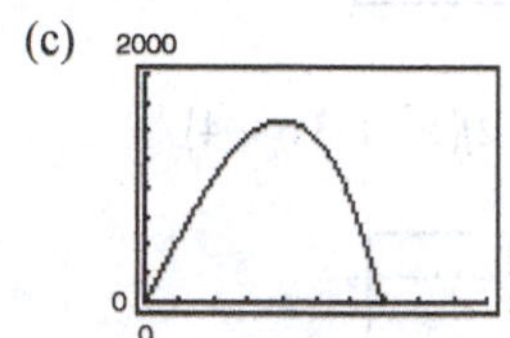

The maximum volume is $V \approx 1595.8$ cubic feet. This occurs when $r \approx 8$ feet and $l \approx 16$ feet.

5. (a)

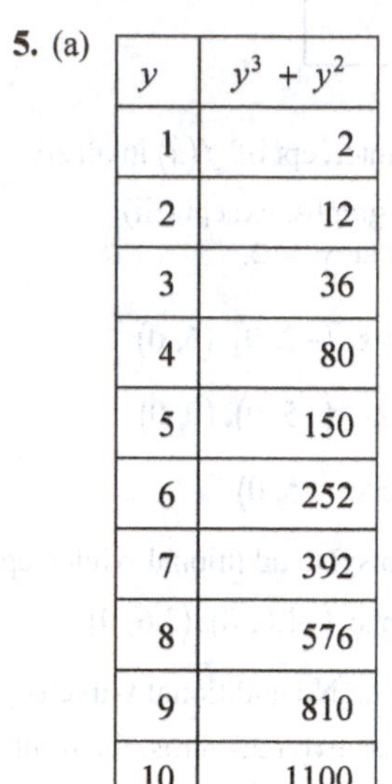

y	$y^3 + y^2$
1	2
2	12
3	36
4	80
5	150
6	252
7	392
8	576
9	810
10	1100

(b) (i) $x^3 + x^2 = 252 \Rightarrow x = 6$

(ii) $x^3 + 2x^2 = 288;\ a = 1, b = 2 \Rightarrow \dfrac{a^2}{b^3} = \dfrac{1}{8}$

$\dfrac{1}{8}x^3 + \dfrac{1}{8}\left(2x^2\right) = \dfrac{1}{8}(288)$

$\left(\dfrac{x}{2}\right)^3 + \left(\dfrac{x}{2}\right)^2 = 36 \Rightarrow \dfrac{x}{2} = 3 \Rightarrow x = 6$

(iii) $3x^3 + x^2 = 90;\ a = 3, b = 1 \Rightarrow \dfrac{a^2}{b^3} = 9$

$9\left(3x^3\right) + 9x^2 = 9(90)$

$(3x)^3 + (3x)^2 = 810 \Rightarrow 3x = 9 \Rightarrow x = 3$

(iv) $2x^3 + 5x^2 = 2500;\ a = 2, b = 5 \Rightarrow \dfrac{a^2}{b^3} = \dfrac{4}{125}$

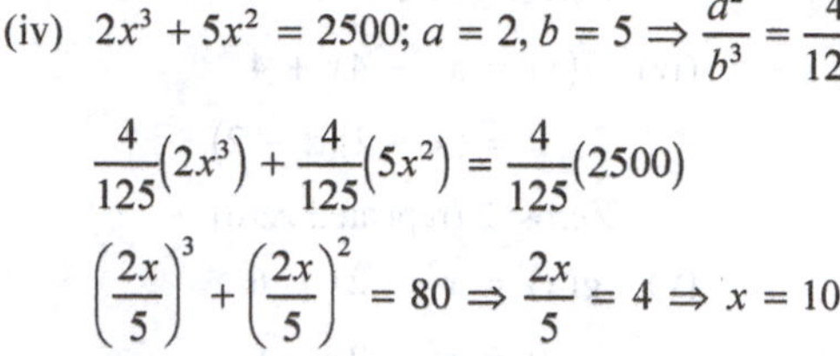

$\dfrac{4}{125}\left(2x^3\right) + \dfrac{4}{125}\left(5x^2\right) = \dfrac{4}{125}(2500)$

$\left(\dfrac{2x}{5}\right)^3 + \left(\dfrac{2x}{5}\right)^2 = 80 \Rightarrow \dfrac{2x}{5} = 4 \Rightarrow x = 10$

(v) $7x^3 + 6x^2 = 1728;$

$a = 7, b = 6 \Rightarrow \dfrac{a^2}{b^3} = \dfrac{49}{216}$

$\dfrac{49}{216}\left(7x^3\right) + \dfrac{49}{216}\left(6x^2\right) = \dfrac{49}{216}(1728)$

$\left(\dfrac{7x}{6}\right)^3 + \left(\dfrac{7x}{6}\right)^2 = 392 \Rightarrow \dfrac{7x}{6} = 7 \Rightarrow x = 6$

(vi) $10x^3 + 3x^2 = 297;$

$a = 10, b = 3 \Rightarrow \dfrac{a^2}{b^3} = \dfrac{100}{27}$

$\dfrac{100}{27}\left(10x^3\right) + \dfrac{100}{27}\left(3x^2\right) = \dfrac{100}{27}(297)$

$\left(\dfrac{10x}{3}\right)^3 + \left(\dfrac{10x}{3}\right)^2 = 1100 \Rightarrow \dfrac{10x}{3}$

$= 10 \Rightarrow x = 3$

(c) Answers will vary.

7. (a)

Function	Zeros	Sum of Zeros	Product of Zeros
$f_1(x) = x^2 - 5x + 6$	2, 3	5	6
$f_2(x) = x^3 - 7x + 6$	$-3, 1, 2$	0	-6
$f_3(x) = x^4 + 2x^3 + x^2 + 8x - 12$	$-3, 1, \pm 2i$	-2	-12
$f^4(x) = x^5 - 3x^4 - 9x^3 + 25x^2 - 6x$	$-3, 0, 2, 2 \pm \sqrt{3}$	3	0

(b) Conjecture: Sum of Zeros $= -a_{n-1}$

if $f(x) = a_nx^n + a_{n-1}x^{n-1} + \cdots + a_1x + a_0$

(c) Conjecture: Product of Zeros $= \begin{cases} a_0, & \text{if } n \text{ is even} \\ -a_0, & \text{if } n \text{ is odd} \end{cases}$

if $f(x) = a_nx^n + a_{n-1}x^{n-1} + \cdots + a_1x + a_0$

9. (a) $y = ax^2 + bx + c$

$$(0, -4): -4 = a(0)^2 + b(0) + c$$
$$-4 = c$$
$$(4, 0): 0 = a(4)^2 + b(4) - 4$$
$$0 = 16a + 4b - 4 = 4(4a + b - 1)$$
$$0 = 4a + b - 1 \quad \text{or} \quad b = 1 - 4a$$
$$(1, 0): 0 = a(1)^2 + b(1) - 4$$
$$4 = a + b$$
$$4 = a + (1 - 4a)$$
$$4 = 1 - 3a$$
$$3 = -3a$$
$$a = -1$$
$$b = 1 - 4(-1) = 5$$
$$y = -x^2 + 5x - 4$$

(b) Enter the data points $(0, -4)$, $(1, 0)$, $(2, 2)$, $(4, 0)$, $(6, -10)$ and use the regression feature to obtain $y = -x^2 + 5x - 4$.

11. (a) $x + 2y = 100$

$$y = \frac{100 - x}{2}$$
$$A(x) = xy = x\left(\frac{100 - x}{2}\right) = \frac{100x - x^2}{2}$$
$$= 50x - \frac{x^2}{2}$$

Domain: $0 < x < 100$

(b)

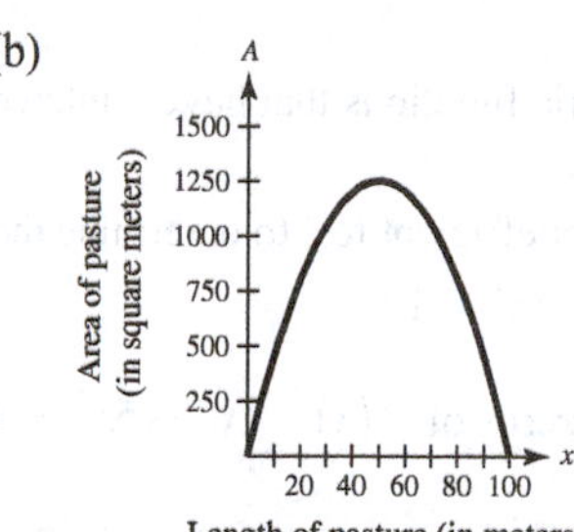

The vertex occurs when $x \approx 50$,

so $y = \dfrac{100 - 50}{2} = 25$.

The dimensions are $x = 50$, $y = 25$

(c) $A = 50x - \dfrac{x^2}{2} = -\dfrac{1}{2}x^2 + 50x$

$$= -\frac{1}{2}(x^2 - 100x + 2500) + 1250$$
$$= -\frac{1}{2}(x - 50)^2 + 1250$$

The vertex is $(h, k) = (50, 1250)$, so the maximum area of 1250 square meters occurs when $x = 50$ meters and $y = \dfrac{100 - 50}{2} = 25$ meters.

13. $V = l \cdot w \cdot h = x^2(x + 3)$

$$x^2(x + 3) = 20$$
$$x^3 + 3x^2 - 20 = 0$$

Possible rational zeros: $\pm1, \pm2, \pm4, \pm5, \pm10, \pm20$

$$\begin{array}{r|rrrr} 2 & 1 & 3 & 0 & -20 \\ & & 2 & 10 & 20 \\ \hline & 1 & 5 & 10 & 0 \end{array}$$

$$(x - 2)(x^2 + 5x + 10) = 0$$
$$x = 2 \text{ or } x = \frac{-5 \pm \sqrt{15}i}{2}$$

Choosing the real positive value for x we have: $x = 2$ and $x + 3 = 5$.

The dimensions of the mold are 2 inches × 2 inches × 5 inches.

Practice Test for Chapter 3

1. Sketch the graph of $f(x) = x^2 - 6x + 5$ and identify the vertex and the intercepts.

2. Find the number of units x that produce a minimum cost C if
$C = 0.01x^2 - 90x + 15{,}000$.

3. Find the quadratic function that has a maximum at $(1, 7)$ and passes through the point $(2, 5)$.

4. Find two quadratic functions that have x-intercepts $(2, 0)$ and $\left(\frac{4}{3}, 0\right)$.

5. Use the leading coefficient test to determine the right and left end behavior of the graph of the polynomial function $f(x) = -3x^5 + 2x^3 - 17$.

6. Find all the real zeros of $f(x) = x^5 - 5x^3 + 4x$.

7. Find a polynomial function with 0, 3, and -2 as zeros.

8. Sketch $f(x) = x^3 - 12x$.

9. Divide $3x^4 - 7x^2 + 2x - 10$ by $x - 3$ using long division.

10. Divide $x^3 - 11$ by $x^2 + 2x - 1$.

11. Use synthetic division to divide $3x^5 + 13x^4 + 12x - 1$ by $x + 5$.

12. Use synthetic division to find $f(-6)$ given $f(x) = 7x^3 + 40x^2 - 12x + 15$.

13. Find the real zeros of $f(x) = x^3 - 19x - 30$.

14. Find the real zeros of $f(x) = x^4 + x^3 - 8x^2 - 9x - 9$.

15. List all possible rational zeros of the function $f(x) = 6x^3 - 5x^2 + 4x - 15$.

16. Find the rational zeros of the polynomial $f(x) = x^3 - \frac{20}{3}x^2 + 9x - \frac{10}{3}$.

17. Write $f(x) = x^4 + x^3 + 5x - 10$ as a product of linear factors.

18. Find a polynomial with real coefficients that has $2, 3 + i,$ and $3 - 2i$ as zeros.

19. Use synthetic division to show that $3i$ is a zero of $f(x) = x^3 + 4x^2 + 9x + 36$.

20. Find a mathematical model for the statement, "z varies directly as the square of x and inversely as the square of x and inversely as the square root of y."

CHAPTER 4
Rational Functions and Conics

CHAPTER 4
Rational Functions and Conics

Section 4.1 Rational Functions and Asymptotes

1. rational functions

3. horizontal asymptote

5. Because the denominator is zero when $x - 1 = 0$, the domain of f is all real numbers except $x = 1$.

x	0	0.5	0.9	0.99	$\to 1$
$f(x)$	-1	-2	-10	-100	$\to -\infty$

x	$1 \leftarrow$	1.01	1.1	1.5	2
$f(x)$	$\infty \leftarrow$	100	10	2	1

As x approaches 1 from the left, $f(x)$ decreases without bound towards $-\infty$. As x approaches 1 from the right, $f(x)$ increases without bound towards $+\infty$.

7. Because the denominator is zero when $x + 2 = 0$, the domain of f is all real numbers except $x = -2$.

x	-3	-2.5	-2.1	-2.01	$\to -2$
$f(x)$	15	25	105	1005	$\to \infty$

x	$-2 \leftarrow$	-1.99	-1.9	-1.5	-1
$f(x)$	$-\infty \leftarrow$	-955	-95	-15	-5

As x approaches -2 from the left, $f(x)$ increases without bound (∞). As x approaches -2 from the right, $f(x)$ decreases without bound $(-\infty)$.

9. Because the denominator is zero when $x^2 - 1 = 0$, the domain of f is all real numbers except $x = -1$ and $x = 1$.

x	-2	-1.5	-1.1	-1.01	$\to -1$
$f(x)$	4	5.4	17.3	152.3	$\to \infty$

x	$-1 \leftarrow$	-0.99	-0.9	-0.5	0
$f(x)$	$-\infty \leftarrow$	-147.8	-12.8	-1	0

As x approaches -1 from the left, $f(x)$ increases without bound (∞). As x approaches -1 from the right, $f(x)$ decreases without bound $(-\infty)$.

x	0	0.5	0.9	0.99	$\to 1$
$f(x)$	0	-1	-12.8	-147.8	$\to -\infty$

x	$1 \leftarrow$	1.01	1.1	1.5	2
$f(x)$	$\infty \leftarrow$	152.3	17.3	5.4	4

As x approaches 1 from the left, $f(x)$ decreases without bound $(-\infty)$. As x approaches 1 from the right, $f(x)$ increases without bound (∞).

11. Because the denominator is zero when $x^2 - 2x + 1 = (x - 1)^2 = 0$, the domain of f is all real numbers except $x = 1$.

x	0	0.5	0.9	0.99	$\to 1$
$f(x)$	2	15	551	59,501	$\to \infty$

x	$1 \leftarrow$	1.01	1.1	1.5	2
$f(x)$	$\infty \leftarrow$	60,501	651	35	12

As x approaches 1 from the left, $f(x)$ increases without bound (∞). As x approaches 1 from the right, $f(x)$ increases without bound (∞).

13. $f(x) = \frac{4}{x^2}$

Domain: all real numbers except $x = 0$

Vertical asymptote: $x = 0$

Horizontal asymptote: $y = 0$

$\left[\text{Degree of } N(x) < \text{degree of } D(x)\right]$

15. $f(x) = \frac{5 + x}{5 - x} = \frac{x + 5}{-x + 5}$

Domain: all real numbers except $x = 5$

Vertical asymptote: $x = 5$

Horizontal asymptote: $y = -1$

$\left[\text{Degree of } N(x) = \text{degree of } D(x)\right]$

17. $f(x) = \frac{x^3}{x^2 - 1}$

Domain: all real numbers except $x = \pm 1$

Vertical asymptotes: $x = \pm 1$

Horizontal asymptote: None

$\left[\text{Degree of } N(x) > \text{degree of } D(x)\right]$

19. $f(x) = \frac{-4x^2 + 1}{x^2 + x + 3}$

Domain: All real numbers. The denominator has no real zeros.

[Use the Quadratic Formula on the denominator.]

Vertical asymptote: None

Horizontal asymptote: $y = -4$

$\left[\text{Degree of } N(x) = \text{degree of } D(x)\right]$

21. $f(x) = \frac{x - 4}{x^2 - 16} = \frac{1}{x + 4}, x \neq 4$

Horizontal asymptote: $y = 0$

$\left(\text{Degree of } N(x) < \text{degree of } D(x)\right)$

Vertical asymptote: $x = -4$

(Because $x - 4$ is a common factor of $N(x)$ and $D(x)$, $x = 4$ is not a vertical asymptote of $f(x)$.)

23. $f(x) = \frac{x^2 - 1}{x^2 - x - 6} = \frac{(x + 1)(x - 1)}{(x + 2)(x - 3)}, [x \neq -2, 3]$

Horizontal asymptote: $y = 1$

$\left(\text{Degree of } N(x) = \text{degree of } D(x)\right)$

Vertical asymptote: $x = -2$ and $x = 3$

25. $f(x) = \frac{x^2 - 3x - 4}{2x^2 + x - 1}$

$= \frac{(x + 1)(x - 4)}{(2x - 1)(x + 1)}$

$= \frac{x - 4}{2x - 1}, x \neq -1$

Horizontal asymptote: $y = \frac{1}{2}$

$\left(\text{Degree of } N(x) = \text{degree of } D(x)\right)$

Vertical asymptote: $x = \frac{1}{2}$

(Because $x + 1$ is a common factor of $N(x)$ and $D(x)$, $x = -1$ is not a vertical asymptote of $f(x)$.)

27. $f(x) = \frac{6x^2 + 5x - 6}{3x^2 - 8x + 4}$

$= \frac{(3x - 2)(2x + 3)}{(3x - 2)(x - 2)}$

$= \frac{2x + 3}{x - 2}, x \neq \frac{2}{3}$

Horizontal asymptote:

$y = 2$ $\left(\text{Degree of } N(x) = \text{degree of } D(x)\right)$

Vertical asymptote: $x = 2$

(Because $3x - 2$ is a common factor of $N(x)$ and $D(x)$, $x = \frac{2}{3}$ is not a vertical asymptote of $f(x)$.)

29. Because the function has a vertical asymptote at $x = -2$ and a horizontal asymptote at $y = 0$,

$f(x) = \frac{4}{x + 2}$ matches graph (f).

30. Because the function has a vertical asymptote at $x = 2$ and a horizontal asymptote at $y = 0$,

$f(x) = \frac{5}{x - 2}$ matches graph (e).

31. Because the function has a vertical asymptote at $x = 2$ and a horizontal asymptote at $y = -2$,

$f(x) = -\frac{2x - 1}{x - 2}$ matches graph (a).

32. Because the function has a vertical asymptote at $x = 2$ and a horizontal asymptote at $y = -1/2$,

$f(x) = -\frac{x - 1}{2x - 4}$ matches graph (g).

33. Because the function has vertical asymptotes at $x = \pm 2$ and a horizontal asymptote at $y = 2$,

$f(x) = \dfrac{2x^2}{x^2 - 4}$ matches graph (c).

34. Because the function has vertical asymptotes at $x = \pm 2$ and a horizontal asymptote at $y = 0$,

$f(x) = \dfrac{-2x}{x^2 - 4}$ matches graph (b).

35. Because the function has a vertical asymptote at $x = -2$ and no horizontal asymptote,

$f(x) = \dfrac{x^3}{4(x + 2)^2}$ matches graph (d).

36. Because the function has a vertical asymptote at $x = -2$ and a horizontal asymptote at $y = 0$,

$f(x) = \dfrac{3x}{(x + 2)^2}$ matches graph (h).

37. $f(x) = \dfrac{x^2 - 4}{x + 2}$, $g(x) = x - 2$

$f(x) = \dfrac{(x + 2)(x - 2)}{x + 2}$

(a) Domain of f: all real numbers except $x = -2$

Domain of g: all real numbers

(b) $f(x) = x - 2$

Because $x + 2$ is a common factor of both the numerator and the denominator of $f(x)$, $x = -2$ is not a vertical asymptote of f. f has no vertical asymptotes.

(c)

x	-4	-3	-2.5	-2	-1.5	-1	0
$f(x)$	-6	-5	-4.5	Undef.	-3.5	-3	-2
$g(x)$	-6	-5	-4.5	-4	-3.5	-3	-2

(d) f and g differ only at $x = -2$, where f is undefined and g is defined.

39. $f(x) = \dfrac{2x - 1}{2x^2 - x}$, $g(x) = \dfrac{1}{x}$

$f(x) = \dfrac{2x - 1}{x(2x - 1)}$

(a) Domain of f: all real numbers except $x = 0$ and $x = \dfrac{1}{2}$

Domain of g: all real numbers except $x = 0$

(b) $f(x) = \dfrac{1}{x}$

Because $2x - 1$ is a common factor of both the numerator and the denominator of f, $x = 0.5$ is not a vertical asymptote of f. The only vertical asymptote is $x = 0$.

(c)

x	-1	-0.5	0	0.5	2	3	4
$f(x)$	-1	-2	Undef.	Undef.	$\frac{1}{2}$	$\frac{1}{3}$	$\frac{1}{4}$
$g(x)$	-1	-2	Undef.	2	$\frac{1}{2}$	$\frac{1}{3}$	$\frac{1}{4}$

(d) f and g differ only at $x = 0.5$, where f is undefined and g is defined.

41. $C = \dfrac{25{,}000p}{100 - p}$, $0 \le p < 100$

(a)

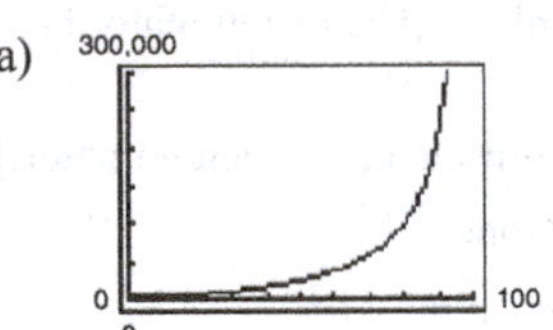

(b) $C = \dfrac{25{,}000(15)}{100 - 15} \approx \$4411.76$

$C = \dfrac{25{,}000(50)}{100 - 50} = \$25{,}000$

$C = \dfrac{25{,}000(90)}{100 - 90} = \$225{,}000$

(c) $C \to \infty$ as $x \to 100$. No, it would not be possible to supply bins to 100% of the residents because the model is undefined for $p = 100$.

43. $N = \dfrac{20(5 + 3t)}{1 + 0.04t}, t \geq 0$

(a) 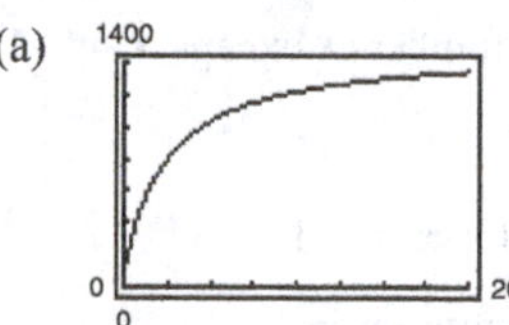

(b) $N(5) \approx 333$ deer

$N(10) = 500$ deer

$N(25) = 800$ deer

(c) The herd is limited by the horizontal asymptote:

$N = \dfrac{60}{0.04} = 1500$ deer

45. $P = \dfrac{0.5 + 0.9(n - 1)}{1 + 0.9(n - 1)}, n > 0$

(a)

n	1	2	3	4	5
P	0.50	0.74	0.82	0.86	0.89

n	6	7	8	9	10
P	0.91	0.92	0.93	0.94	0.95

P approaches 1 as n increases.

(b) $P = \dfrac{0.9n - 0.4}{0.9n + 0.1}$

The percentage of correct responses is limited by the horizontal asymptote:

$P = \dfrac{0.9}{0.9} = 1 = 100\%$

47. False. Polynomial functions do not have vertical asymptotes.

49. True. A rational function has at most one horizontal asymptote. It never has two horizontal asymptotes.

51. $f(x) = 4 - \dfrac{1}{x}$

(a) As $x \to \pm\infty$, $f(x) \to 4$.

(b) As $x \to \infty$, $f(x) \to 4$ but is less than 4.

(c) As $x \to -\infty$, $f(x) \to 4$ but is greater than 4.

53. $f(x) = \dfrac{2x - 1}{x^2 + 1}$

(a) As $x \to \pm\infty$, $f(x) \to 0$.

(b) As $x \to \infty$, $f(x) \to 0$ but is greater than 0.

(c) As $x \to -\infty$, $f(x) \to 0$ but is less than 0.

55. Vertical asymptote: None $\Rightarrow$ The denominator is not zero for any value of x (unless the numerator is also zero there).

Horizontal asymptote: $y = 2 \Rightarrow$ The degree of the numerator equals the degree of the denominator.

$f(x) = \dfrac{2x^2}{x^2 + 1}$ is one possible function. There are many correct answers.

57. If $(x - c)$ is a factor of both the numerator and denominator of the function, then f is not defined at $x = c$ and therefore not a zero of f.

Section 4.2 Graphs of Rational Functions

1. slant asymptote

3. $g(x) = \dfrac{2}{x} + 4$

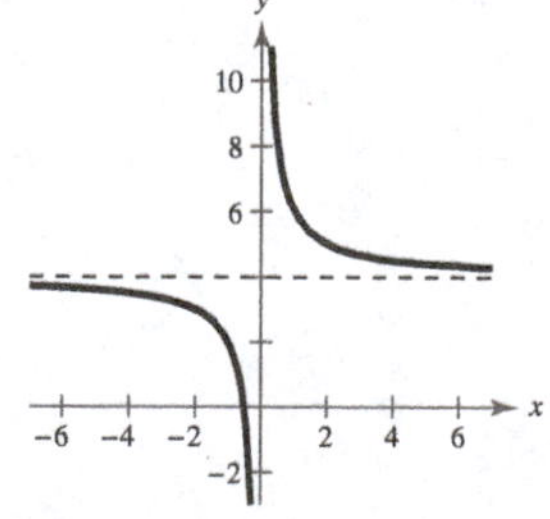

Vertical shift four units upward

5. $g(x) = -\dfrac{2}{x}$

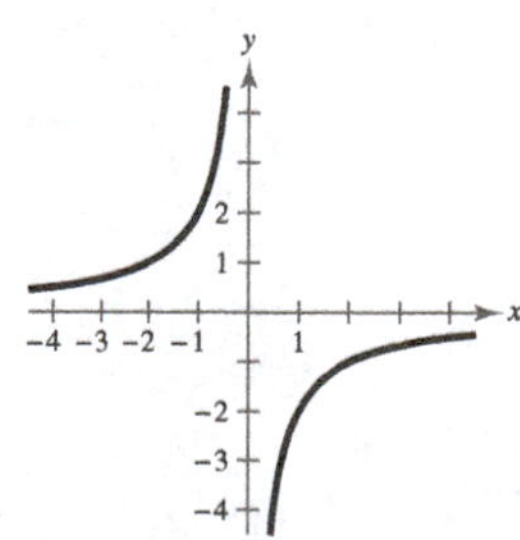

Reflection in the x-axis

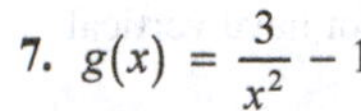

7. $g(x) = \dfrac{3}{x^2} - 1$

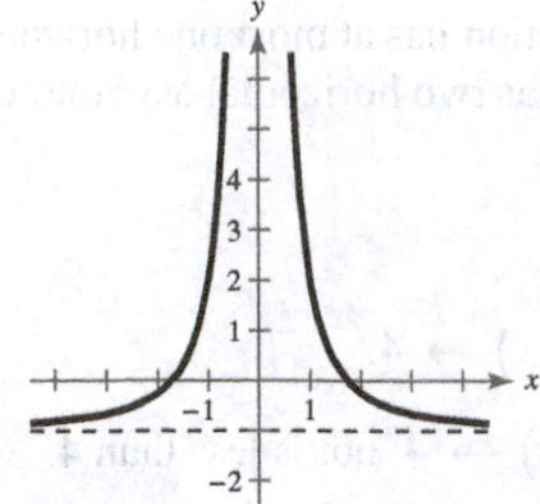

Vertical shift one unit downward

9. $g(x) = \dfrac{3}{(x-1)^2}$

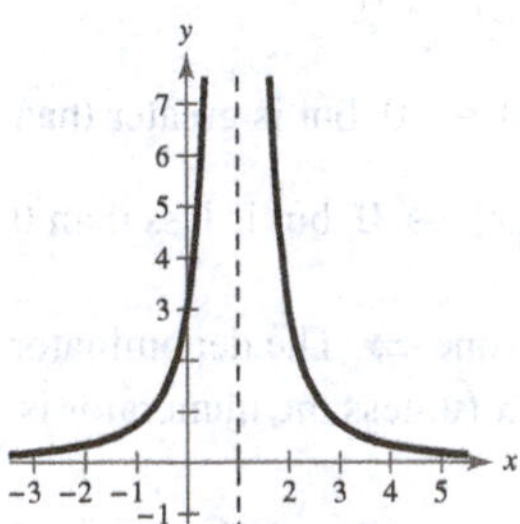

Horizontal shift one unit to the right

11. $g(x) = \dfrac{4}{(x+2)^3}$

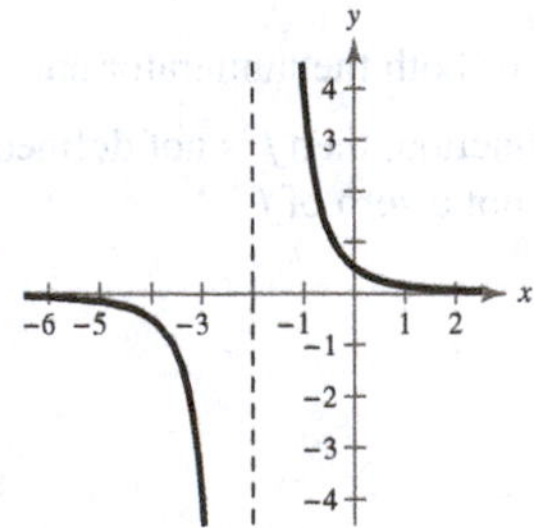

Horizontal shift two units to the left

13. $g(x) = -\dfrac{4}{x^3}$

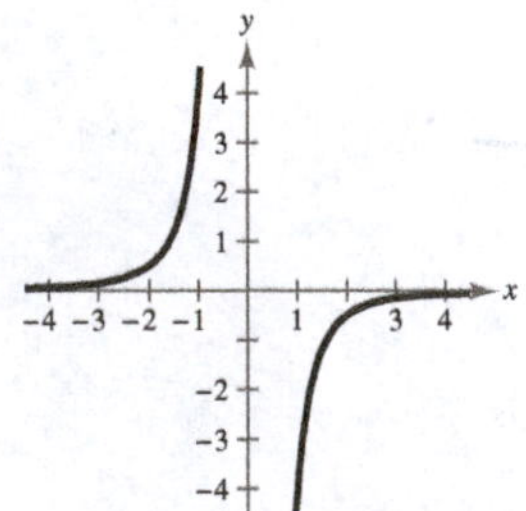

Reflection in the x-axis

15. $f(x) = \dfrac{1}{x+1}$

(a) Domain: all real numbers x except $x = -1$

(b) y-intercept: $(0, 1)$

(c) Vertical asymptote: $x = -1$

Horizontal asymptote: $y = 0$

(d)

x	-4	-3	0	1	2	3
$f(x)$	$-\frac{1}{3}$	$-\frac{1}{2}$	1	$\frac{1}{2}$	$\frac{1}{3}$	$\frac{1}{4}$

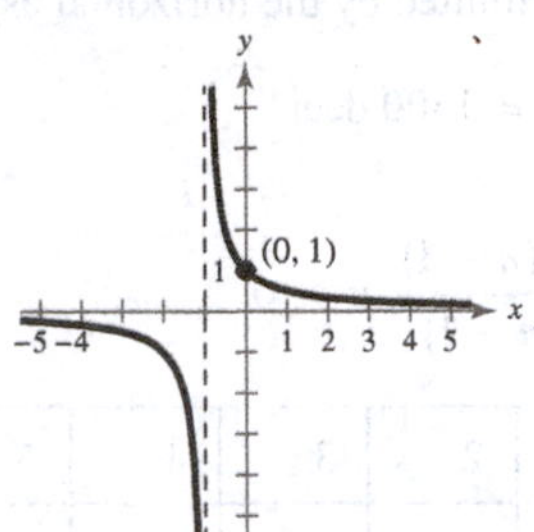

17. $h(x) = \dfrac{-1}{x+4}$

(a) Domain: all real numbers x except $x = -4$

(b) y-intercept: $\left(0, -\frac{1}{4}\right)$

(c) Vertical asymptote: $x = -4$

Horizontal asymptote: $y = 0$

(d)

x	-6	-5	-3	-2	-1	0
$h(x)$	$\frac{1}{2}$	1	-1	$-\frac{1}{2}$	$-\frac{1}{3}$	$-\frac{1}{4}$

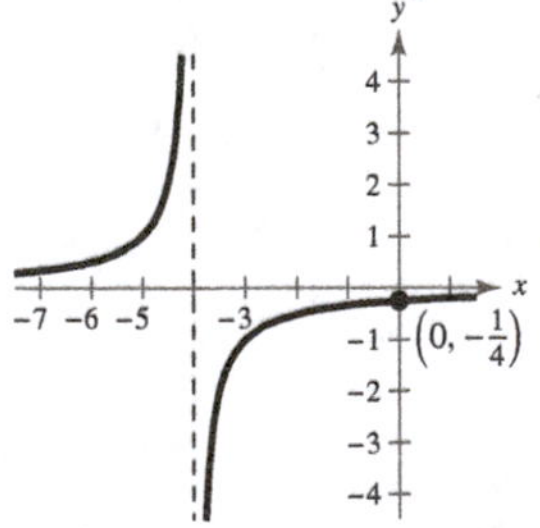

19. $C(x) = \dfrac{2x + 3}{x + 2}$

(a) Domain: all real numbers x except $x = -2$

(b) x-intercept: $\left(-\dfrac{3}{2}, 0\right)$

y-intercept: $\left(0, \dfrac{3}{2}\right)$

(c) Vertical asymptote: $x = -2$

Horizontal asymptote: $y = 2$

(d)

x	-4	-3	-1	0	1	2
$C(x)$	$\dfrac{5}{2}$	3	1	$\dfrac{3}{2}$	$\dfrac{5}{3}$	$\dfrac{7}{4}$

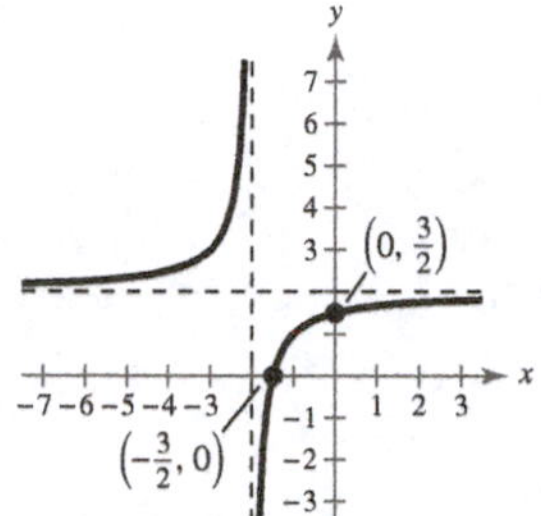

21. $g(x) = \dfrac{1}{x + 2} + 2 = \dfrac{2x + 5}{x + 2}$

(a) Domain: all real numbers x except $x = -2$

(b) x-intercept: $\left(-\dfrac{5}{2}, 0\right)$

y-intercept: $\left(0, \dfrac{5}{2}\right)$

(c) Vertical asymptote: $x = -2$

Horizontal asymptote: $y = 2$

(d)

x	-4	-3	-1	0	1
$g(x)$	$\dfrac{3}{2}$	1	3	$\dfrac{5}{2}$	$\dfrac{7}{3}$

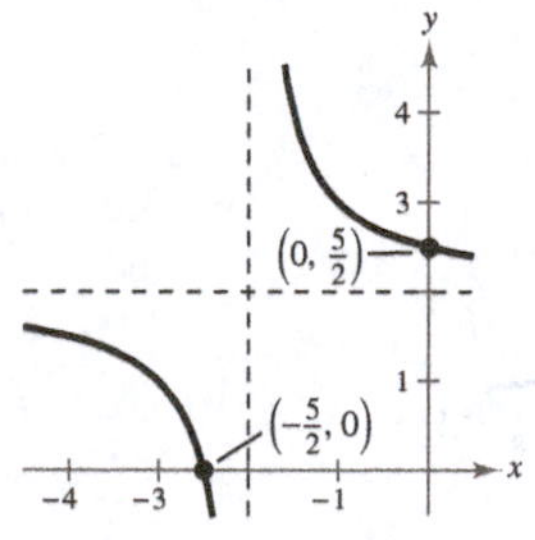

23. $f(x) = \dfrac{x^2}{x^2 + 9}$

(a) Domain: all real numbers x

(b) Intercept: $(0, 0)$

(c) Horizontal asymptote: $y = 1$

(d)

x	± 1	± 2	± 3
$f(x)$	$\dfrac{1}{10}$	$\dfrac{4}{13}$	$\dfrac{1}{2}$

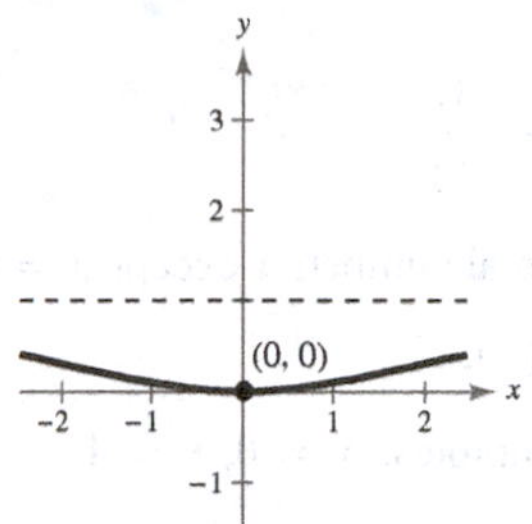

25. $h(x) = \dfrac{x^2}{x^2 - 9}$

(a) Domain: all real numbers x except $x = \pm 3$

(b) Intercept: $(0, 0)$

(c) Vertical asymptotes: $x = \pm 3$

Horizontal asymptote: $y = 1$

(d)

x	± 5	± 4	± 2	± 1	0
$h(x)$	$\dfrac{25}{16}$	$\dfrac{16}{7}$	$-\dfrac{4}{5}$	$-\dfrac{1}{8}$	0

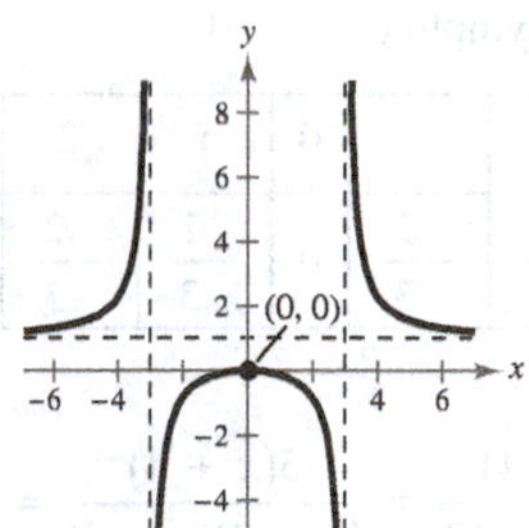

27. $g(s) = \dfrac{4s}{s^2 + 4}$

(a) Domain: all real numbers s

(b) Intercept: $(0, 0)$

(c) Horizontal asymptote: $y = 0$

(d)

s	-2	-1	0	1	2
$g(s)$	-1	$-\frac{4}{5}$	0	$\frac{4}{5}$	1

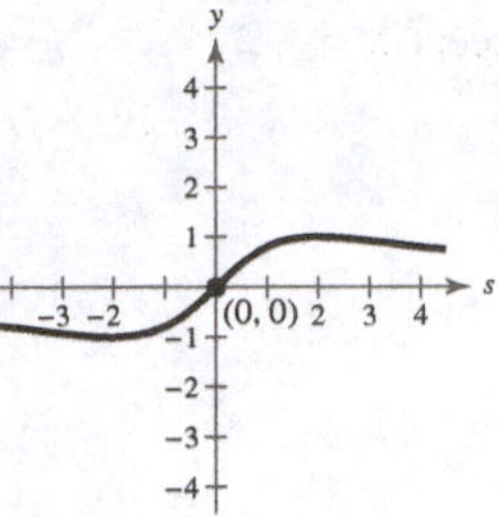

29. $g(x) = \dfrac{4(x + 1)}{x(x - 4)}$

(a) Domain: all real numbers x except $x = 0$ and $x = 4$

(b) x-intercept: $(-1, 0)$

(c) Vertical asymptotes: $x = 0, x = 4$

Horizontal asymptote: $y = 0$

(d)

x	-2	-1	1	2	3	5	6
$g(x)$	$-\frac{1}{3}$	0	$-\frac{8}{3}$	-3	$-\frac{16}{3}$	$\frac{24}{5}$	$\frac{7}{3}$

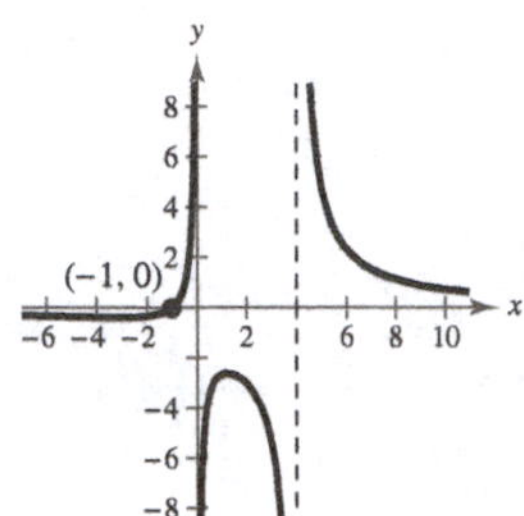

31. $f(x) = \dfrac{2x}{x^2 - 3x - 4} = \dfrac{2x}{(x - 4)(x + 1)}$

(a) Domain: all real numbers x except $x = 4$ and $x = -1$

(b) Intercept: $(0, 0)$

(c) Vertical asymptotes: $x = 4, x = -1$

Horizontal asymptote: $y = 0$

(d)

x	-3	-2	0	1	2	3	5
$f(x)$	$-\frac{3}{7}$	$-\frac{2}{3}$	0	$-\frac{1}{3}$	$-\frac{2}{3}$	$-\frac{3}{2}$	$\frac{5}{3}$

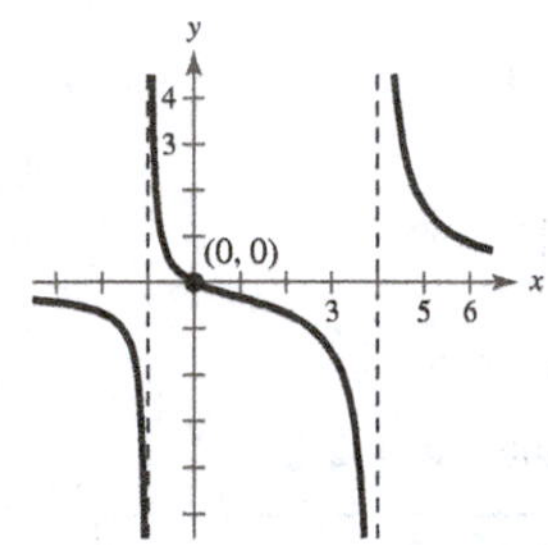

33. $f(x) = \dfrac{5(x + 4)}{x^2 + x - 12} = \dfrac{5(x + 4)}{(x + 4)(x - 3)} = \dfrac{5}{x - 3}, \quad x \neq -4$

(a) Domain: all real numbers x except $x = -4$ or $x = 3$

(b) y-intercept: $\left(0, -\frac{5}{3}\right)$

(c) Vertical asymptote: $x = 3$

Horizontal asymptote: $y = 0$

(d)

x	-2	0	2	5	7
$f(x)$	-1	$-\frac{5}{3}$	-5	$\frac{5}{2}$	$\frac{5}{4}$

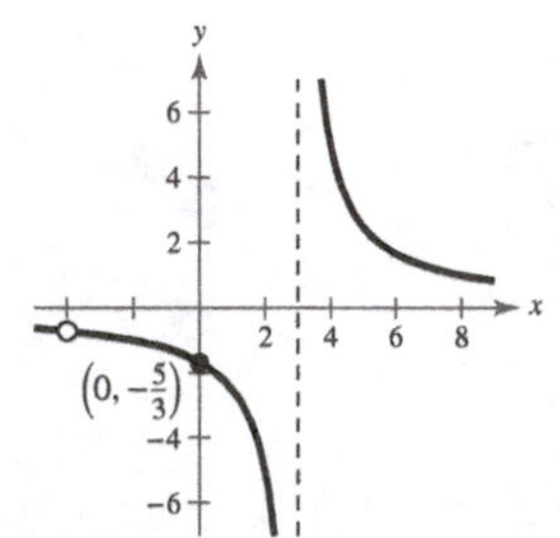

35. $f(t) = \dfrac{t^2 - 1}{t - 1} = \dfrac{(t + 1)(t - 1)}{t - 1} = t + 1, \quad t \neq 1$

(a) Domain: all real numbers t except $t = 1$

(b) t-intercept: $(-1, 0)$

y-intercept: $(0, 1)$

(c) No asymptotes

(d)

t	-3	-2	-1	0	1	2
$f(t)$	-2	-1	0	1	Undef.	3

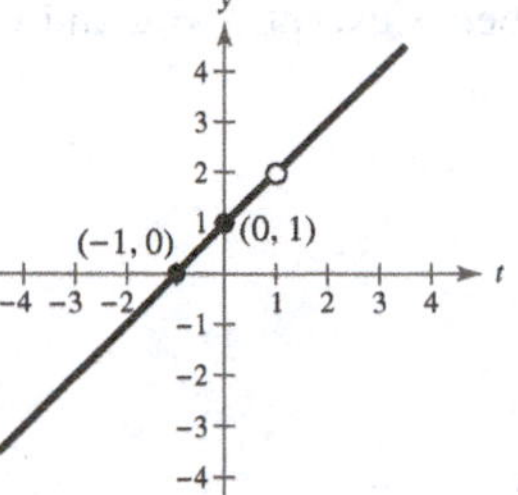

37. $h(x) = \dfrac{x^2 - 5x + 4}{x^2 - 4} = \dfrac{(x - 1)(x - 4)}{(x + 2)(x - 2)}$

(a) Domain: all real numbers x except $x = \pm 2$

(b) x-intercepts: $(1, 0), (4, 0)$

y-intercept: $(0, -1)$

(c) Vertical asymptotes: $x = -2, x = 2$

Horizontal asymptote: $y = 1$

(d)

x	-4	-3	-1	0	1	3	4
$h(x)$	$\frac{10}{3}$	$\frac{28}{5}$	$-\frac{10}{3}$	-1	0	$-\frac{2}{5}$	0

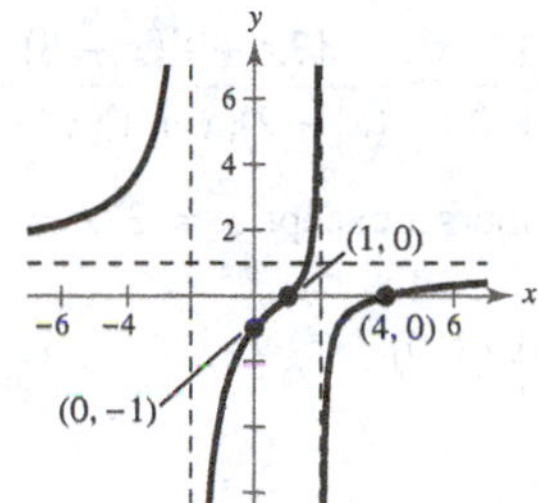

39. $f(x) = \dfrac{x^2 + 4}{x^2 + 3x - 4} = \dfrac{x^2 + 4}{(x + 4)(x - 1)}$

(a) Domain: all real numbers x except $x = -4$ and $x = 1$

(b) no x-intercepts

y-intercept: $(0, -1)$

(c) Vertical asymptotes: $x = -4$ and $x = 1$

Horizontal asymptote: $y = 1$

(d)

x	-5	-3	-2	-1	2	3
$f(x)$	$\frac{29}{6}$	$-\frac{13}{4}$	$-\frac{4}{3}$	$-\frac{5}{6}$	$\frac{4}{3}$	$\frac{13}{14}$

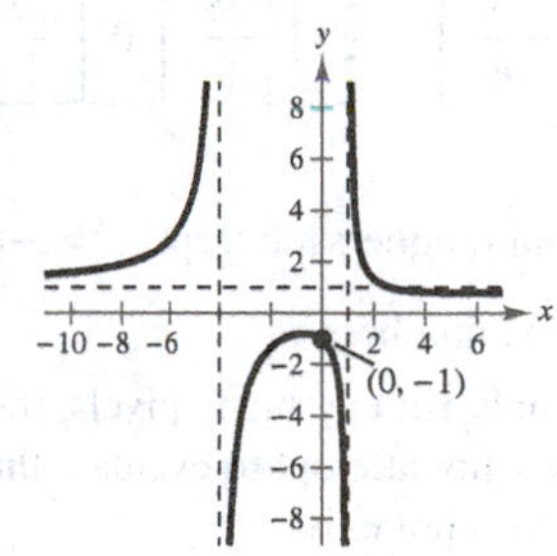

41. $f(x) = \frac{2x^2 - 5x + 2}{2x^2 - x - 6} = \frac{(2x - 1)(x - 2)}{(2x + 3)(x - 2)} = \frac{2x - 1}{2x + 3}, \quad x \neq 2$

(a) Domain: all real numbers x except $x = 2$ and $x = -\frac{3}{2}$

(b) x-intercept: $\left(\frac{1}{2}, 0\right)$

y-intercept: $\left(0, -\frac{1}{3}\right)$

(c) Vertical asymptote: $x = -\frac{3}{2}$

Horizontal asymptote: $y = 1$

(d)

x	-3	-2	-1	0	1
$f(x)$	$\frac{7}{3}$	5	-3	$-\frac{1}{3}$	$\frac{1}{5}$

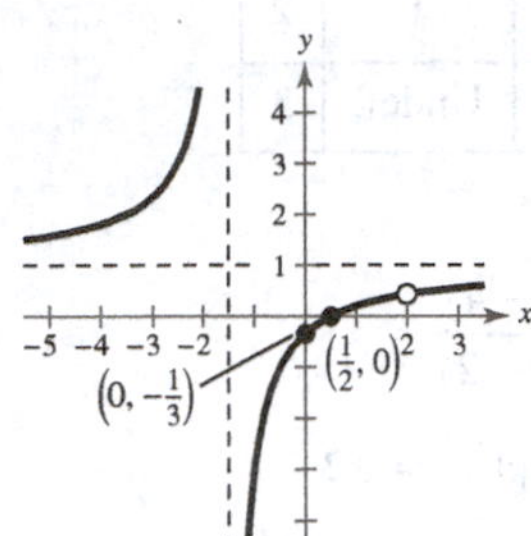

43. $f(x) = \frac{2x^2 - 5x - 3}{x^3 - 2x^2 - x + 2} = \frac{(2x + 1)(x - 3)}{(x - 2)(x + 1)(x - 1)}$

(a) Domain: all real numbers x except $x = 2, x = \pm 1$

(b) x-intercepts: $\left(-\frac{1}{2}, 0\right), (3, 0)$

y-intercept: $\left(0, -\frac{3}{2}\right)$

(c) Vertical asymptotes: $x = 2, x = -1,$ and $x = 1$

Horizontal asymptote: $y = 0$

(d)

x	-3	-2	0	$\frac{3}{2}$	3	4
$f(x)$	$-\frac{3}{4}$	$-\frac{5}{4}$	$-\frac{3}{2}$	$\frac{48}{5}$	0	$\frac{3}{10}$

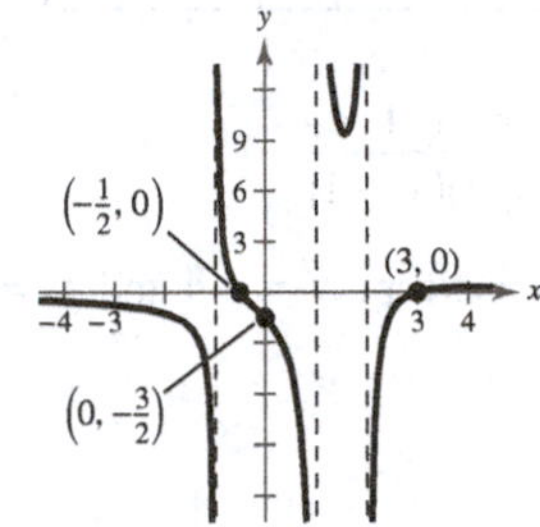

45. (a) Domain of f: all real numbers x except $x = -1$

Domain of g: all real numbers x

(b)

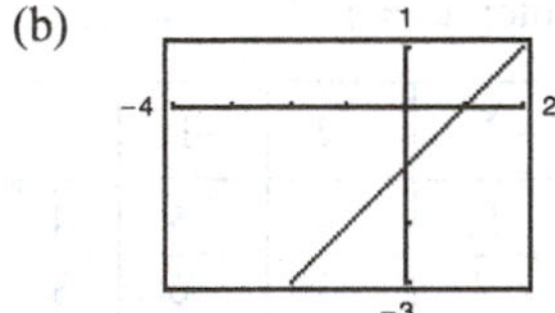

(c) Because there are only finitely many pixels, the graphing utility may not attempt to evaluate the function where it does not exist.

47. $h(x) = \frac{x^2 - 4}{x} = x - \frac{4}{x}$

(a) Domain: all real numbers x except $x = 0$

(b) x-intercepts: $(-2, 0), (2, 0)$

(c) Vertical asymptote: $x = 0$

Slant asymptote: $y = x$

(d)

x	-4	-3	-1	1	3	4
$h(x)$	-3	$\frac{-5}{3}$	3	-3	$\frac{5}{3}$	3

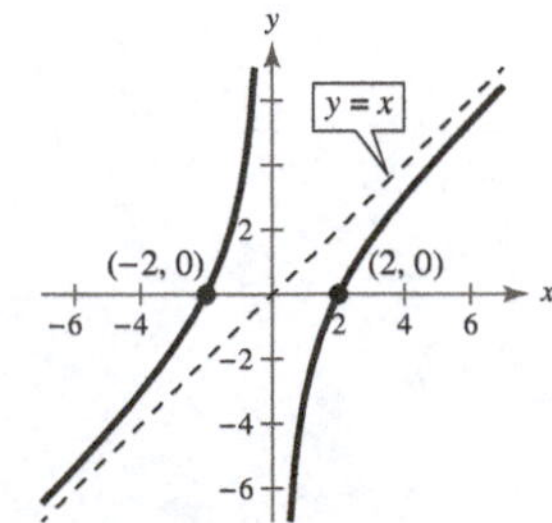

49. $f(x) = \dfrac{2x^2 + 1}{x} = 2x + \dfrac{1}{x}$

(a) Domain: all real numbers x except $x = 0$

(b) No intercepts

(c) Vertical asymptote: $x = 0$

Slant asymptote: $y = 2x$

(d)

x	-4	-2	2	4	6
$f(x)$	$-\frac{33}{4}$	$-\frac{9}{2}$	$\frac{9}{2}$	$\frac{33}{4}$	$\frac{73}{6}$

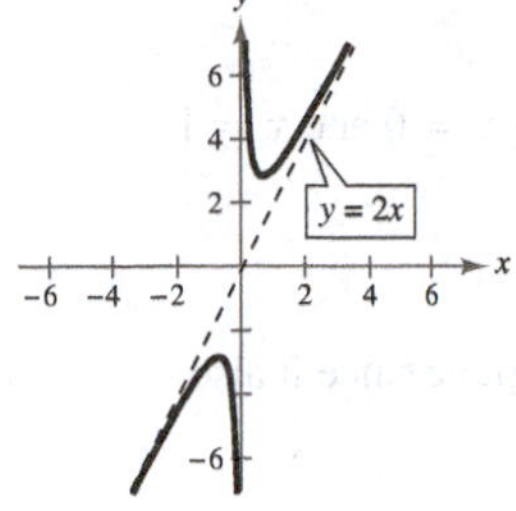

51. $g(x) = \dfrac{x^2 + 1}{x} = x + \dfrac{1}{x}$

(a) Domain: all real numbers x except $x = 0$

(b) No intercepts

(c) Vertical asymptote: $x = 0$

Slant asymptote: $y = x$

(d)

x	-4	-2	2	4	6
$g(x)$	$-\frac{17}{4}$	$-\frac{5}{2}$	$\frac{5}{2}$	$\frac{17}{4}$	$\frac{37}{6}$

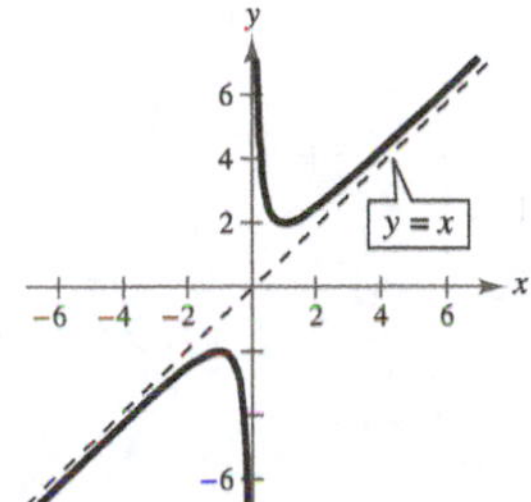

53. $f(t) = \dfrac{t^2 + 1}{t + 5} = -t + 5 - \dfrac{26}{t + 5}$

(a) Domain: all real numbers t except $t = -5$

(b) Intercept: $\left(0, -\dfrac{1}{5}\right)$

(c) Vertical asymptote: $t = -5$

Slant asymptote: $y = -t + 5$

(d)

t	-7	-6	-4	-3	0
$f(t)$	25	37	-17	-5	$-\frac{1}{5}$

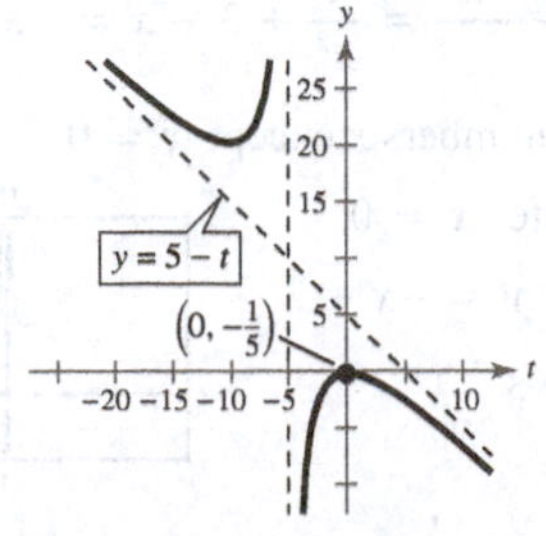

55. $f(x) = \dfrac{x^3}{x^2 - 4} = x + \dfrac{4x}{x^2 - 4}$

(a) Domain: all real numbers x except $x = \pm 2$

(b) Intercept: $(0, 0)$

(c) Vertical asymptotes: $x = \pm 2$

Slant asymptote: $y = x$

(d)

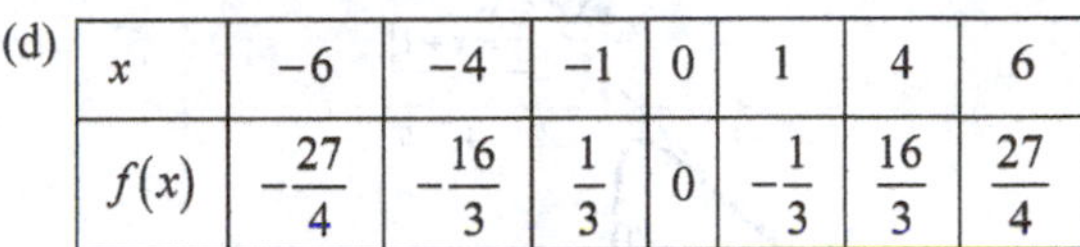

x	-6	-4	-1	0	1	4	6
$f(x)$	$-\frac{27}{4}$	$-\frac{16}{3}$	$\frac{1}{3}$	0	$-\frac{1}{3}$	$\frac{16}{3}$	$\frac{27}{4}$

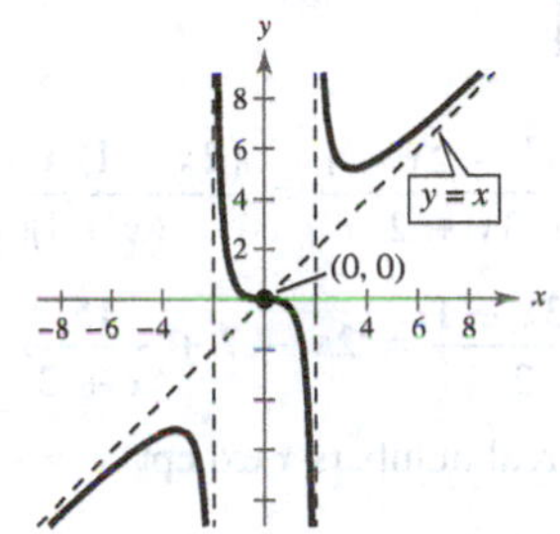

57. $f(x) = \dfrac{x^3 - 1}{x^2 - x} = \dfrac{(x - 1)(x^2 + x + 1)}{x(x - 1)} = \dfrac{x^2 + x + 1}{x}$

$= x + 1 + \dfrac{1}{x}, \quad x \neq 1$

(a) Domain: all real numbers x except $x = 0$ and $x = 1$

(b) Intercepts: none

(c) Vertical asymptote: $x = 0$

Note: $x = 1$ is not a vertical asymptote since it also makes the numerator zero.

Slant asymptote: $y = x + 1$

(d)

x	-2	-1	$\frac{1}{2}$	2
$f(x)$	$-\frac{3}{2}$	-1	$\frac{7}{2}$	$\frac{7}{2}$

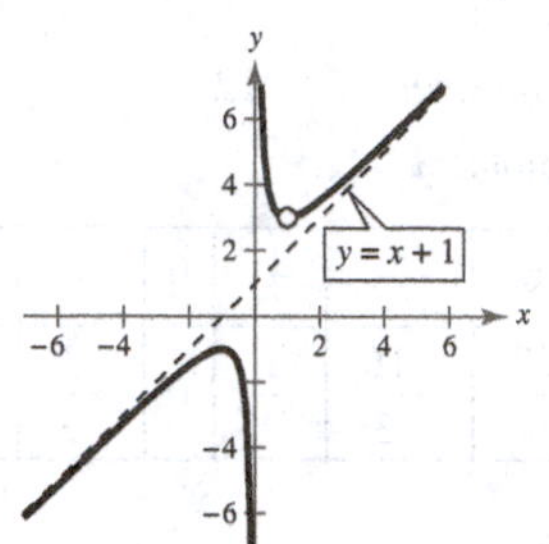

59. $f(x) = \dfrac{x^2 - x + 1}{x - 1} = x + \dfrac{1}{x - 1}$

(a) Domain: all real numbers x except $x = 1$

(b) y-intercept: $(0, -1)$

(c) Vertical asymptote: $x = 1$

Slant asymptote: $y = x$

(d)

x	-4	-2	0	2	4
$f(x)$	$-\frac{21}{5}$	$-\frac{7}{3}$	-1	3	$\frac{13}{3}$

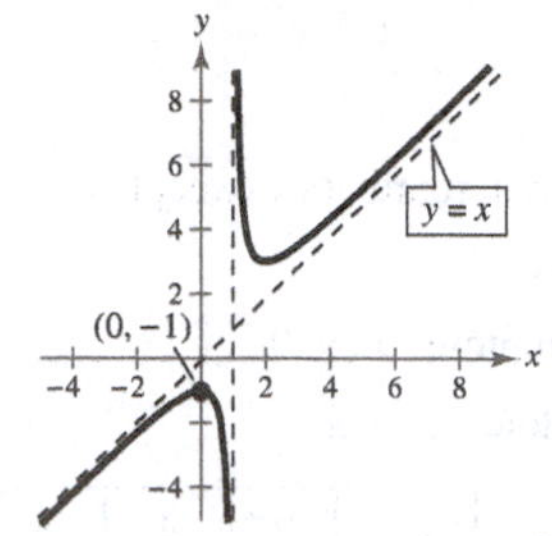

61. $f(x) = \dfrac{2x^3 - x^2 - 2x + 1}{x^2 + 3x + 2} = \dfrac{(2x - 1)(x + 1)(x - 1)}{(x + 1)(x + 2)} = \dfrac{(2x - 1)(x - 1)}{x + 2}, \quad x \neq -1$

$= \dfrac{2x^2 - 3x + 1}{x + 2} = 2x - 7 + \dfrac{15}{x + 2}, \quad x \neq -1$

(a) Domain: all real numbers x except $x = -1$ and $x = -2$

(b) y-intercept: $\left(0, \frac{1}{2}\right)$

x-intercepts: $\left(\frac{1}{2}, 0\right), (1, 0)$

(c) Vertical asymptote: $x = -2$

Slant asymptote: $y = 2x - 7$

(d)

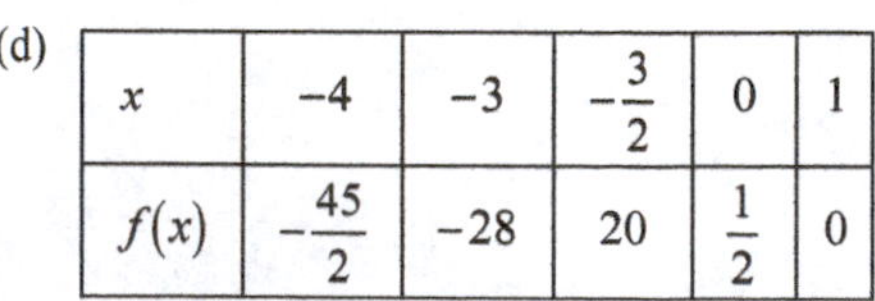

x	-4	-3	$-\frac{3}{2}$	0	1
$f(x)$	$-\frac{45}{2}$	-28	20	$\frac{1}{2}$	0

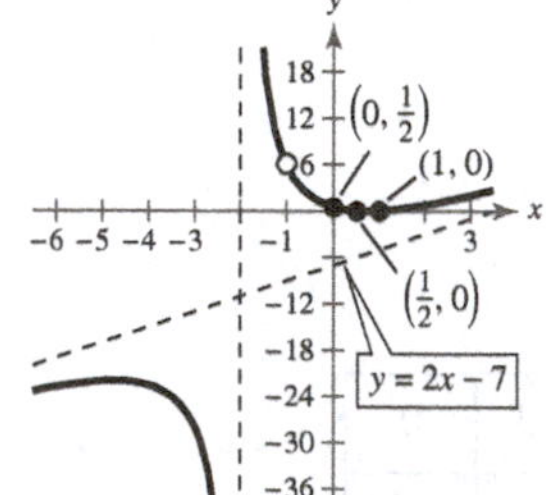

63. $f(x) = \dfrac{x^2 + 2x - 8}{x + 2} = x - \dfrac{8}{x + 2}$

Domain: all real numbers x except $x = -2$

Vertical asymptote: $x = -2$

Slant asymptote: $y = x$

Line: $y = x$

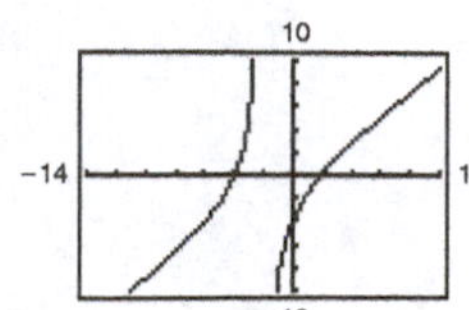

65. $g(x) = \dfrac{1 + 3x^2 - x^3}{x^2} = \dfrac{1}{x^2} + 3 - x = -x + 3 + \dfrac{1}{x^2}$

Domain: all real numbers x except $x = 0$

Vertical asymptote: $x = 0$

Slant asymptote: $y = -x + 3$

Line: $y = -x + 3$

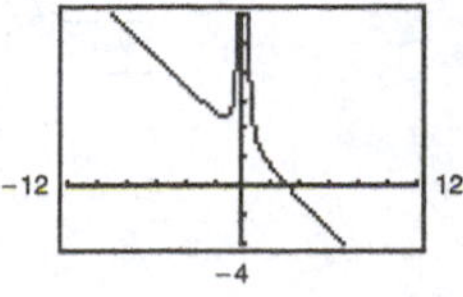

67. $y = \frac{x+1}{x-3}$

(a) x-intercept: $(-1, 0)$

(b) $0 = \frac{x+1}{x-3}$
$0 = x + 1$
$-1 = x$

69. $y = \frac{3x}{x+1} = 3 - \frac{3}{x+1}$

(a) x-intercept: $(0, 0)$

(b) $0 = \frac{3x}{x+1}$
$0 = 3x$
$x = 0$

71. $y = \frac{1}{x} - x$

(a) x-intercepts: $(-1, 0), (1, 0)$

(b) $0 = \frac{1}{x} - x$
$x = \frac{1}{x}$
$x^2 = 1$
$x = \pm 1$

73. $y = x + 2 + \frac{1}{x}$

(a) x-intercepts: $(-1, 0)$

(b) $0 = x + 2 + \frac{1}{x}$
$0 = \frac{x^2 + 2x + 1}{x}$
$0 = x^2 + 2x + 1$
$0 = (x+1)^2$
$0 = x + 1$
$x = -1$

75. $y = \frac{1}{x+5} + \frac{4}{x}$

(a)

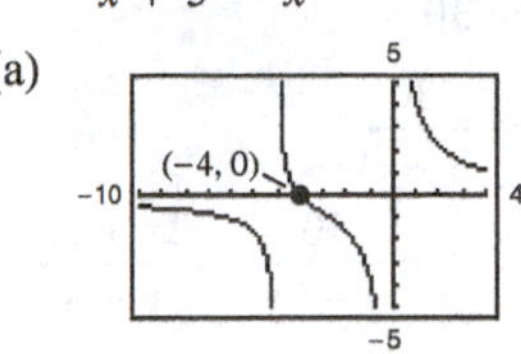

x-intercept: $(-4, 0)$

(b) $0 = \frac{1}{x+5} + \frac{4}{x}$
$-\frac{4}{x} = \frac{1}{x+5}$
$-4(x+5) = x$
$-4x - 20 = x$
$-5x = 20$
$x = -4$

77. $y = x - \frac{6}{x-1}$

(a)

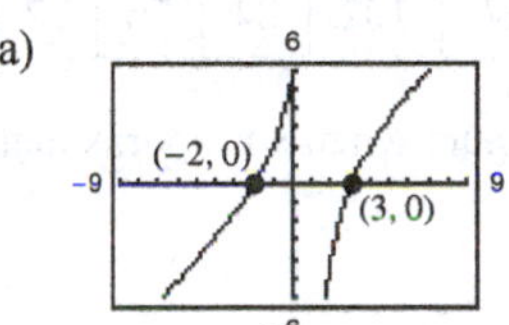

x-intercepts: $(-2, 0), (3, 0)$

(b) $0 = x - \frac{6}{x-1}$
$\frac{6}{x-1} = x$
$6 = x(x-1)$
$0 = x^2 - x - 6$
$0 = (x+2)(x-3)$
$x = -2, \ x = 3$

79. (a) Area $= xy = 600$
$y = \frac{600}{x}$

(b) Domain: $(0, \infty)$

(c)

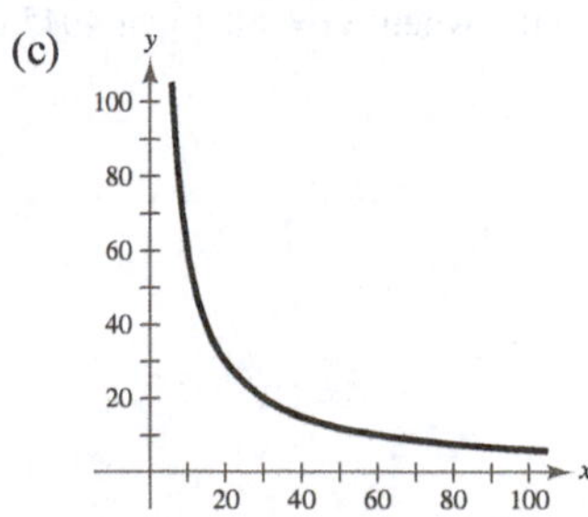

For $x = 35$, $y = \frac{600}{35} = \frac{120}{7} = 17\frac{1}{7}$ meters.

81. (a) $A = xy$ and

$$(x - 4)(y - 2) = 30$$

$$y - 2 = \frac{30}{x - 4}$$

$$y = 2 + \frac{30}{x - 4} = \frac{2x + 22}{x - 4}$$

Thus, $A = xy = x\left(\frac{2x + 22}{x - 4}\right) = \frac{2x(x + 11)}{x - 4}$.

(b) Domain: Since the margins on the left and right are each 2 inches, $x > 4$. In interval notation, the domain is $(4, \infty)$.

(c)

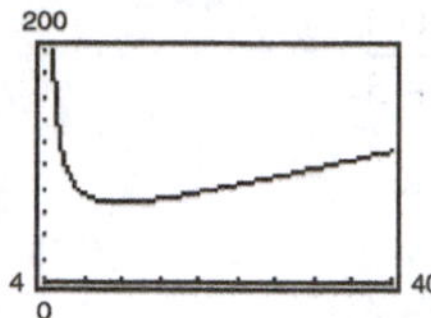

The area is minimum when $x \approx 11.75$ inches and $y \approx 5.87$ inches.

x	5	6	7	8	9	10	11	12	13	14	15
y_1 (Area)	160	102	84	76	72	70	69.143	69	69.333	70	70.909

The area is minimum when x is approximately 12.

83. $f(x) = \frac{3(x + 1)}{x^2 + x + 1}$

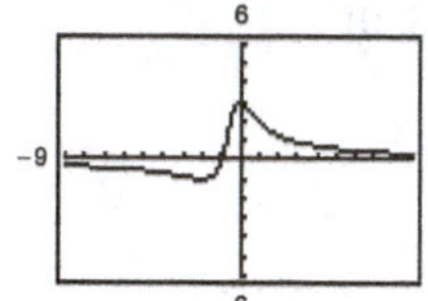

Relative minimum: $(-2, -1)$

Relative maximum: $(0, 3)$

85.

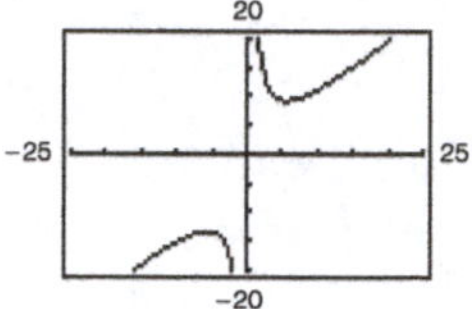

Relative minimum: $(5.657, 9.314)$

Relative maximum: $(-5.657, -13.314)$

87. $C = 100\left(\frac{200}{x^2} + \frac{x}{x + 30}\right), x \geq 1$

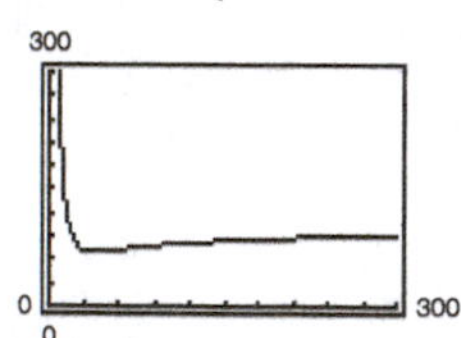

The minimum occurs when $x \approx 40.45$, or 4045 components.

89. (a) Let t_1 = time from Akron to Columbus and t_2 = time from Columbus back to Akron.

$$xt_1 = 100 \Rightarrow t_1 = \frac{100}{x}$$

$$yt_2 = 100 \Rightarrow t_2 = \frac{100}{y}$$

$$50(t_1 + t_2) = 200$$

$$t_1 + t_2 = 4$$

$$\frac{100}{x} + \frac{100}{y} = 4$$

$$100y + 100x = 4xy$$

$$25y + 25x = xy$$

$$25x = xy - 25y$$

$$25x = y(x - 25)$$

Thus, $y = \dfrac{25x}{x - 25}$.

(b) Vertical asymptote: $x = 25$

Horizontal asymptote: $y = 25$

(c)

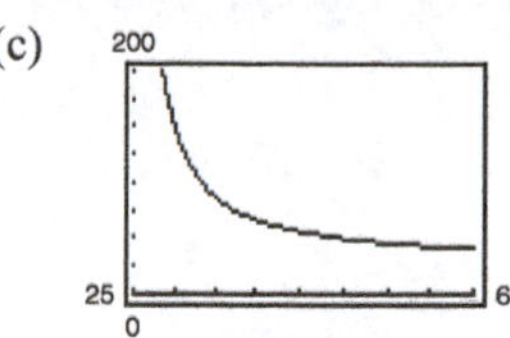

(d)

x	30	35	40	45	50	55	60
y	150	87.5	66.7	56.3	50	45.8	42.9

(e) *Sample answer:* No. You might expect the average speed for the round trip to be the average of the average speeds for the two parts of the trip.

(f) No. At 20 miles per hour you would use more time in one direction than is required for the round trip at an average speed of 50 miles per hour.

91. False. There are two distinct branches of the graph.

93. True. The degree of the numerator is one more than the degree of the denominator.

95. $h(x) = \dfrac{6 - 2x}{3 - x} = \dfrac{2(3 - x)}{3 - x}$

Since $h(x)$ is not reduced and $3 - x$ is a factor of both the numerator and the denominator, $x = 3$ is not a vertical asymptote. Instead, the graph has a hole at $x = 3$.

97. Answers will vary.

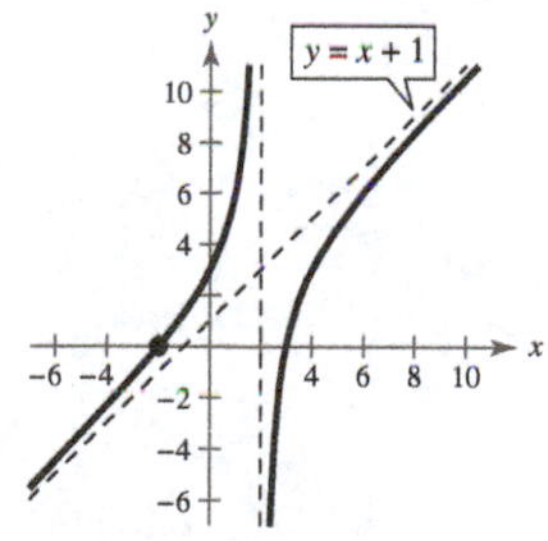

Sample answer:

$$y = x + 1 + \frac{a}{x - 2}$$ This has a slant asymptote of $x + 1$ and a vertical asymptote of $x = 2$.

$$0 = -2 + 1 + \frac{a}{-2 - 2}$$ Since $x = -2$ is a zero, $(-2, 0)$ is on the graph. Use this point to solve for a.

$$1 = \frac{a}{-4}$$

$$-4 = a$$

Thus, $y = x + 1 - \dfrac{4}{x - 2} = \dfrac{(x + 1)(x - 2) - 4}{x - 2} = \dfrac{x^2 - x - 6}{x - 2}$.

$$f(x) = \frac{x^2 - x - 6}{x - 2}$$

99. No, given $f(x) = \dfrac{a_n x^n + \cdots + a_0}{b_m x^m + \cdots b_0}$, if $n > m$, there is no horizontal asymptote and n must be greater than m for a slant asymptote to occur.

Section 4.3 Conics

1. conic or conic section

3. parabola; directrix; focus

5. ellipse; foci

7. hyperbola; foci

9. $x^2 = -2y$

Parabola opening downward

Matches (b).

10. $y^2 = 2x$

Parabola opening to the right

Matches (c).

11. $\frac{x^2}{9} + y^2 = 1$

Ellipse with horizontal major axis

Matches (f).

12. $x^2 - \frac{y^2}{9} = 1$

Hyperbola with horizontal transverse axis

Matches (d)

13. $y^2 - 9x^2 = 9$

$\frac{y^2}{9} - \frac{x^2}{1} = 1$

Hyperbola with vertical transverse axis

Matches (a).

14. $x^2 + y^2 = 16$

Circle with radius 4

Matches (e).

15. $y = \frac{1}{2}x^2$

$x^2 = 2y = 4\left(\frac{1}{2}\right)y;\ p = \frac{1}{2}$

Focus: $\left(0, \frac{1}{2}\right)$

Directrix: $y = -\frac{1}{2}$

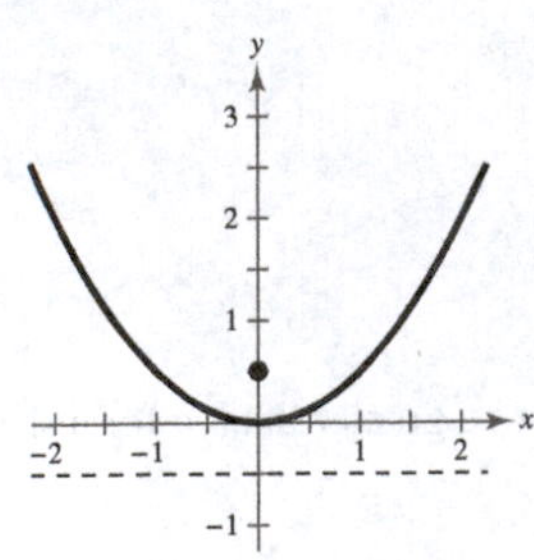

17. $y^2 = -6x$

$y^2 = 4\left(-\frac{3}{2}\right)x;\ p = -\frac{3}{2}$

Focus: $\left(-\frac{3}{2}, 0\right)$

Directrix: $x = \frac{3}{2}$

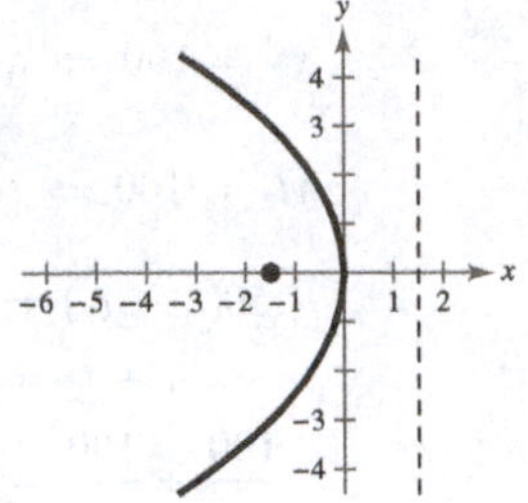

19. $x^2 + 12y = 0$

$x^2 = 4(-3)y;\ p = -3$

Focus: $(0, -3)$

Directrix: $y = 3$

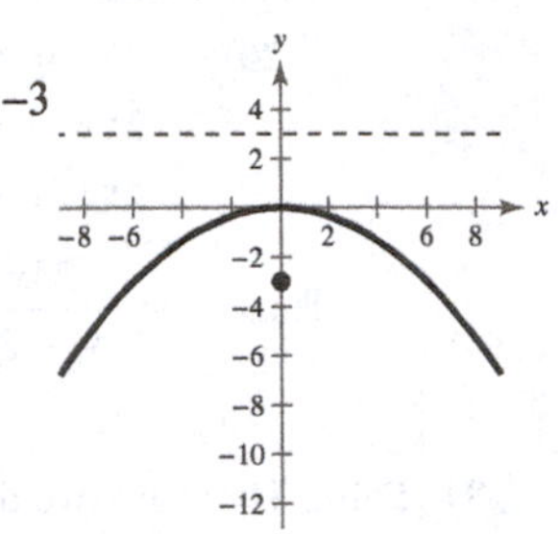

21. Focus: $(3, 0)$

$y^2 = 4(3)x$

$y^2 = 12x$

23. Directrix: $y = 2$

$x^2 = 4(-2)y$

$x^2 = -8y$

25. $y^2 = 4px$

$6^2 = 4p(-4)$

$36 = -16p$

$-\frac{9}{4} = p$

$y^2 = 4\left(-\frac{9}{4}\right)x$

$y^2 = -9x$

27. $x^2 = 4py$

$3^2 = 4p(6)$

$9 = 24p$

$\frac{3}{8} = p$

$x^2 = 4\left(\frac{3}{8}\right)y$

$x^2 = \frac{3}{2}y$

Focus: $\left(0, \frac{3}{8}\right)$

29. $y^2 = 4px,\ p = 1.5$

$y^2 = 4(1.5)x$

$y^2 = 6x$

31. (a)

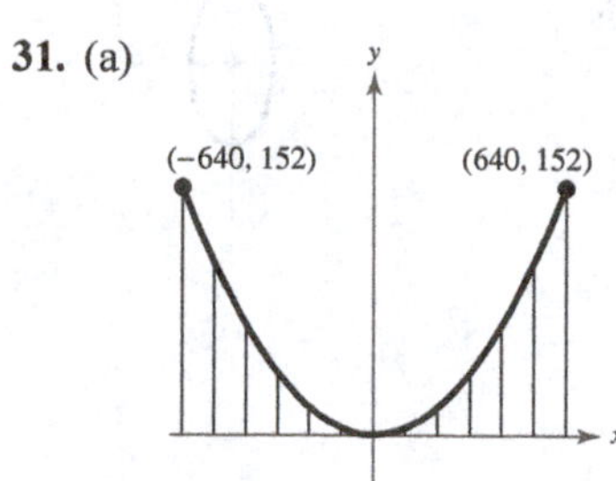

(b) $x^2 = 4py$

$(640)^2 = 4p(152) \Rightarrow p = \dfrac{409{,}600}{608} = \dfrac{12{,}800}{19}$

$x^2 = 4\left(\dfrac{12{,}800}{19}\right)y$

$y = \dfrac{19x^2}{51{,}200}$

33. Vertices: $(0, \pm 2) \Rightarrow a = 2$

Minor axis of length $2 \Rightarrow b = 1$

Vertical major axis

$\dfrac{x^2}{b^2} + \dfrac{y^2}{a^2} = 1$

$\dfrac{x^2}{1} + \dfrac{y^2}{4} = 1$

35. Vertices: $(\pm 5, 0) \Rightarrow a = 5$

Foci: $(\pm 2, 0) \Rightarrow c = 2$

$b = \sqrt{5^2 - 2^2} = \sqrt{21}$

Horizontal major axis

$\dfrac{x^2}{a^2} + \dfrac{y^2}{b^2} = 1$

$\dfrac{x^2}{25} + \dfrac{y^2}{21} = 1$

37. Foci: $(\pm 5, 0) \Rightarrow c = 5$

Major axis of length $14 \Rightarrow a = 7$

$b = \sqrt{7^2 - 5^2} = \sqrt{24}$

Horizontal major axis

$\dfrac{x^2}{a^2} + \dfrac{y^2}{b^2} = 1$

$\dfrac{x^2}{49} + \dfrac{y^2}{24} = 1$

39. Vertices: $(\pm 9, 0) \Rightarrow a = 9$

Minor axis of length $6 \Rightarrow b = 3$

Horizontal major axis

$\dfrac{x^2}{a^2} + \dfrac{y^2}{b^2} = 1$

$\dfrac{x^2}{9^2} + \dfrac{y^2}{3^2} = 1$

$\dfrac{x^2}{81} + \dfrac{y^2}{9} = 1$

41. Major axis vertical $\Rightarrow \dfrac{x^2}{b^2} + \dfrac{y^2}{a^2} = 1$

Passes through $(0, 14) \Rightarrow \dfrac{0^2}{b^2} + \dfrac{14^2}{a^2} = 1 \Rightarrow 196 = a^2$

Passes through $(-7, 0) \Rightarrow \dfrac{(-7)^2}{b^2} + \dfrac{0^2}{a^2} = 1 \Rightarrow 49 = b^2$

$\dfrac{x^2}{49} + \dfrac{y^2}{196} = 1$

43. $\dfrac{x^2}{25} + \dfrac{y^2}{16} = 1$

Horizontal major axis

$a = 5, b = 4$

Center: $(0, 0)$

Vertices: $(\pm 5, 0)$

$5^2 = 4^2 + c^2$ ×

$9 = c^2$

$c = 3$

$e = \dfrac{c}{a}$

$e = 3/5$

45. $\dfrac{x^2}{25/9} + \dfrac{y^2}{16/9} = 1$

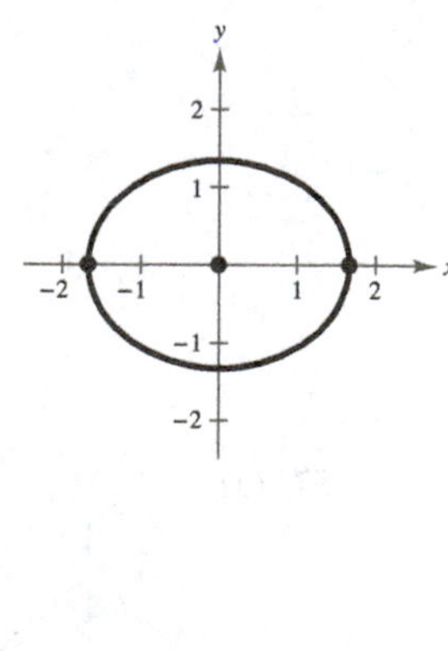

Horizontal major axis

$a = \dfrac{5}{3}, b = \dfrac{4}{3}$

Center: $(0, 0)$

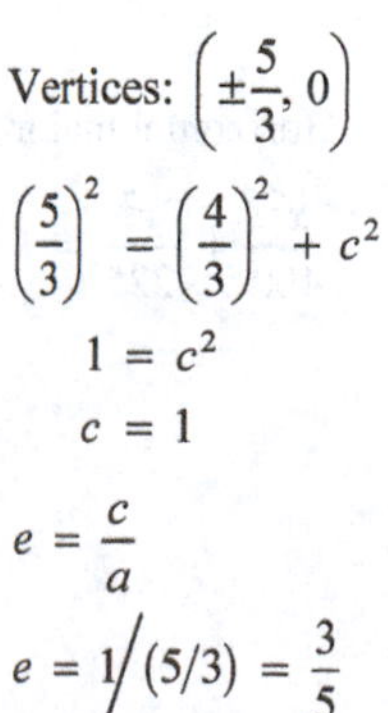

Vertices: $\left(\pm\dfrac{5}{3}, 0\right)$

$\left(\dfrac{5}{3}\right)^2 = \left(\dfrac{4}{3}\right)^2 + c^2$

$1 = c^2$

$c = 1$

$e = \dfrac{c}{a}$

$e = 1/(5/3) = \dfrac{3}{5}$

47. $\dfrac{x^2}{36} + \dfrac{y^2}{7} = 1$

Horizontal major axis

$a = 6, b = \sqrt{7}$

Center: $(0, 0)$

Vertices: $(\pm 6, 0)$

$6^2 = \left(\sqrt{7}\right)^2 + c^2$

$29 = c^2$

$c = \sqrt{29}$

$e = \dfrac{c}{a}$

$e = \dfrac{\sqrt{29}}{6}$

49. $9x^2 + 5y^2 = 45$

$\dfrac{x^2}{5} + \dfrac{y^2}{9} = 1$

Vertical major axis

$a = 3, b = \sqrt{5}$

Center: $(0, 0)$

Vertices: $(0, \pm 3)$

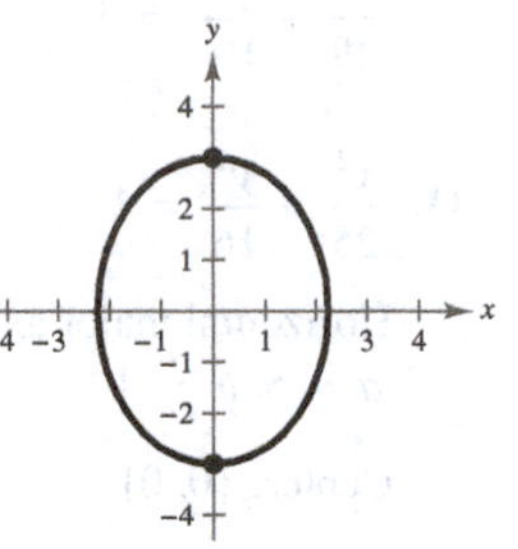

$3^2 = \left(\sqrt{5}\right)^2 + c^2$

$4 = c^2$

$c = 2$

$e = \dfrac{c}{a}$

$e = \dfrac{2}{3}$

51. $4x^2 + y^2 = 1$

$\dfrac{x^2}{1/4} + y^2 = 1$

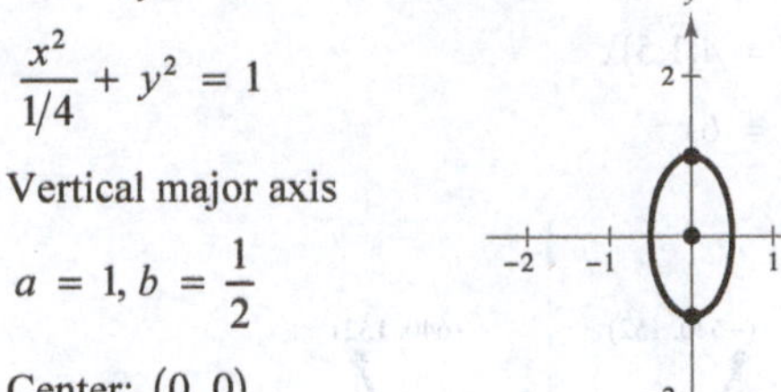

Vertical major axis

$a = 1, b = \dfrac{1}{2}$

Center: $(0, 0)$

Vertices: $(0, \pm 1)$

$1^2 = \left(\dfrac{1}{2}\right)^2 + c^2$

$\dfrac{3}{4} = c^2$

$c = \dfrac{\sqrt{3}}{2}$

$e = \dfrac{c}{a}$

$e = \dfrac{\frac{\sqrt{3}}{2}}{1} = \dfrac{\sqrt{3}}{2}$

53. Vertices: $(\pm 5, 0) \Rightarrow a = 5$

Major axis horizontal $\Rightarrow \dfrac{x^2}{a^2} + \dfrac{y^2}{b^2} = 1$

$e = \dfrac{c}{a} = \dfrac{4}{5} \Rightarrow c = 4$

$a^2 = b^2 + c^2$

$25 = b^2 + 16$

$9 = b^2$

$\dfrac{x^2}{25} + \dfrac{y^2}{9} = 1$

55. The length of string needed is $2a$ or $2(3) = 6$ feet. The positions of the tacks are $(\pm c, 0)$ where c is given by

$c^2 = a^2 - b^2$

$= 3^2 - 2^2 = 9 - 4 = 5 \Rightarrow c = \pm\sqrt{5}.$

Positions: $\left(\pm\sqrt{5}, 0\right)$

57. (a)

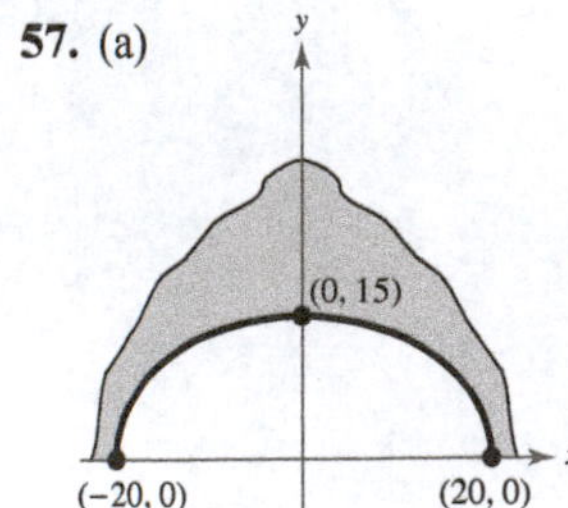

(b) Horizontal major axis $a = 20, b = 15$

$\dfrac{x^2}{400} + \dfrac{y^2}{225} = 1 \Rightarrow y^2 = 225\left(1 - \dfrac{x^2}{400}\right)$

$y = \pm\sqrt{\dfrac{225}{400}\left(400 - x^2\right)}$

$y = \dfrac{3}{4}\sqrt{400 - x^2}$ Top half of ellipse

(c) When $x = 5$:

$y = \dfrac{3}{4}\sqrt{400 - 5^2}$

$y \approx 14.52$ feet

Yes, the truck can clear the tunnel with 0.52 foot clearance.

59. $\dfrac{x^2}{4} + \dfrac{y^2}{1} = 1$

$a = 2, b = 1, c = \sqrt{3}$

Points on the ellipse: $(\pm 2, 0), (0, \pm 1)$

Length of each latus rectum: $\dfrac{2b^2}{a} = \dfrac{2(1)}{2} = 1$

Additional points: $\left(\sqrt{3}, \pm\dfrac{1}{2}\right), \left(-\sqrt{3}, \pm\dfrac{1}{2}\right)$

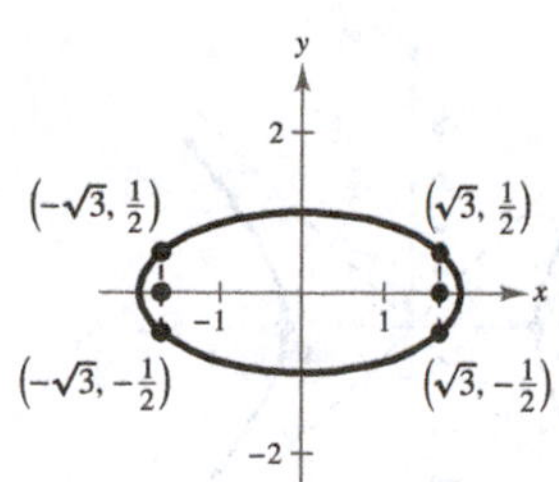

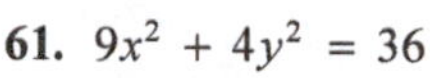

61. $9x^2 + 4y^2 = 36$

$\dfrac{x^2}{4} + \dfrac{y^2}{9} = 1$

$a = 3, b = 2, c = \sqrt{5}$

Points on the ellipse: $(\pm 2, 0), (0, \pm 3)$

Length of each latus rectum: $\dfrac{2b^2}{a} = \dfrac{2(2)^2}{3} = \dfrac{8}{3}$

Additional points: $\left(\pm\dfrac{4}{3}, -\sqrt{5}\right), \left(\pm\dfrac{4}{3}, \sqrt{5}\right)$

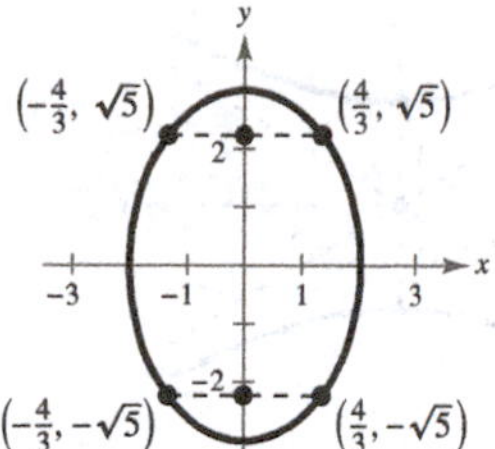

63. Vertices: $(0, \pm 2) \Rightarrow a = 2$

Foci: $(0, \pm 6) \Rightarrow c = 6$

Vertical transverse axis

$b^2 = c^2 - a^2 = 32$

$\dfrac{y^2}{a^2} - \dfrac{x^2}{b^2} = 1$

$\dfrac{y^2}{4} - \dfrac{x^2}{32} = 1$

65. Vertices: $(\pm 1, 0) \Rightarrow a = 1$

Asymptotes: $y = \pm 3x$

Horizontal transverse axis

$3 = \dfrac{b}{a} = \dfrac{b}{1} \Rightarrow b = 3$

$\dfrac{x^2}{a^2} - \dfrac{y^2}{b^2} = 1$

$\dfrac{x^2}{1} - \dfrac{y^2}{9} = 1$

67. Foci: $(\pm 10, 0) \Rightarrow c = 10$

Asymptotes: $y = \pm\dfrac{3}{4}x$

Horizontal transverse axis

$\dfrac{3}{4} = \dfrac{b}{a} \Rightarrow a = \dfrac{4}{3}b$

$\dfrac{16}{9}b^2 + b^2 = (10)^2$

$b^2 = 36 \Rightarrow a^2 = 64$

$\dfrac{x^2}{64} - \dfrac{y^2}{36} = 1$

69. Vertices: $(0, \pm 3) \Rightarrow a = 3$

Vertical transverse axis

$\dfrac{y^2}{9} - \dfrac{x^2}{b^2} = 1$

Point on the graph: $(-2, 5)$

$\dfrac{5^2}{9} - \dfrac{(-2)^2}{b^2} = 1$

$b^2 = \dfrac{9}{4}$

$\dfrac{y^2}{9} - \dfrac{x^2}{9/4} = 1$

71. $\frac{x^2}{25} - \frac{y^2}{25} = 1$

$a = 5, b = 5$

Center: $(0, 0)$

Vertices: $(\pm 5, 0)$

Asymptotes: $y = \pm x$

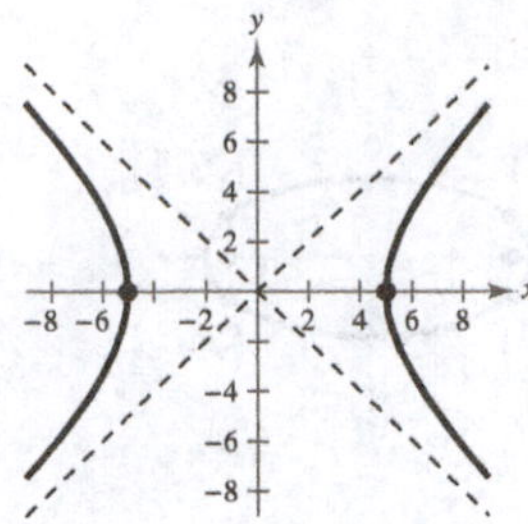

73. $\frac{1}{36}y^2 - \frac{1}{100}x^2 = 1$

$\frac{y^2}{36} - \frac{x^2}{100} = 1$

$a = 6, b = 10$

Center: $(0, 0)$

Vertices: $(0, \pm 6)$

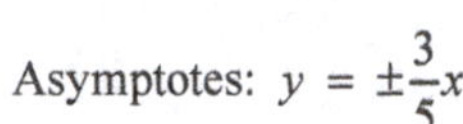

Asymptotes: $y = \pm\frac{3}{5}x$

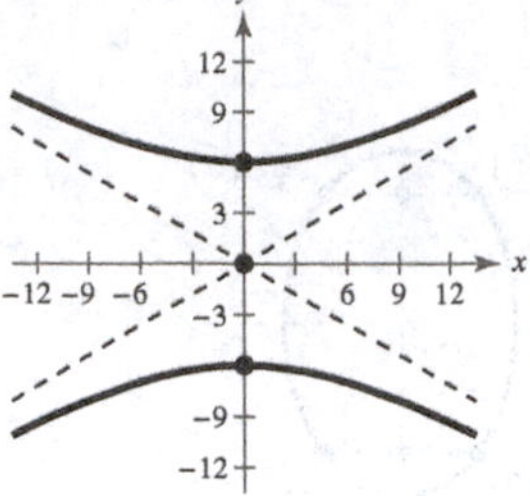

75. $\frac{y^2}{1} - \frac{x^2}{4} = 1$

$a = 1, b = 2$

Center: $(0, 0)$

Vertices: $(0, \pm 1)$

Asymptotes: $y = \pm\frac{1}{2}x$

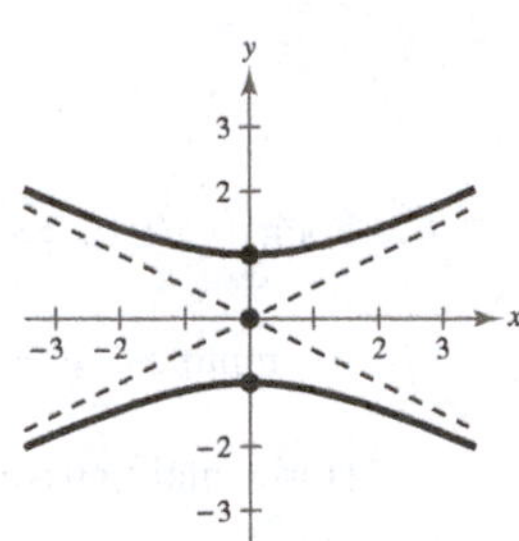

77. $25y^2 - 9x^2 = 225$

$\frac{y^2}{9} - \frac{x^2}{25} = 1$

$a = 3, b = 5$

Center: $(0, 0)$

Vertices: $(0, \pm 3)$

Asymptotes: $y = \pm\frac{3}{5}x$

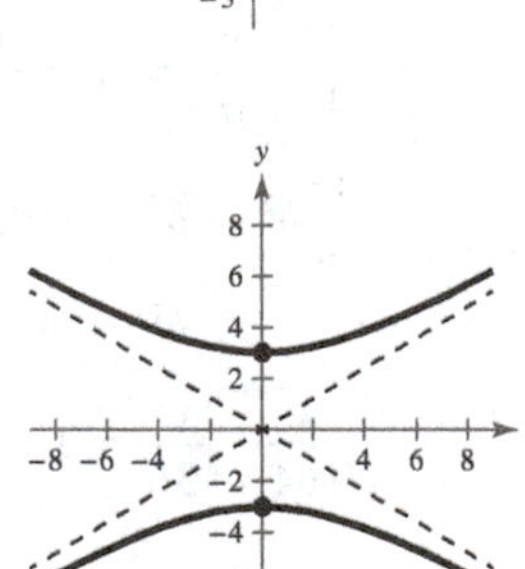

79. $9x^2 - y^2 = 1$

$\frac{x^2}{1/9} - \frac{y^2}{1} = 0$

$a = \frac{1}{3}, b = 1$

Center: $(0, 0)$

Vertices: $\left(\pm\frac{1}{3}, 0\right)$

Asymptotes: $y = \pm\frac{1}{\frac{1}{3}} = \pm 3x$

81. (a) Vertices: $(\pm 1, 0) \Rightarrow a = 1$

Horizontal transverse axis

$\frac{x^2}{a^2} - \frac{y^2}{b^2} = 1$

Point on the graph: $(2, 13)$

$$\frac{2^2}{1^2} - \frac{13^2}{b^2} = 1$$

$$4 - \frac{169}{b^2} = 1$$

$$3b^2 = 169$$

$$b^2 = \frac{169}{3}$$

Thus, we have $\frac{x^2}{1} - \frac{y^2}{\frac{169}{3}} = 1.$

(b) When $y = 5$:

$$x^2 = 1 + \frac{3(5^2)}{169}$$

$$x \approx \sqrt{1 + \frac{75}{169}} \approx 1.2016$$

Width: $2x \approx 2.403$ feet

83. $\frac{x^2}{100} - \frac{y^2}{4} = 1$

The shortest horizontal distance would be the distance between the center and a vertex of the hyperbola, that is, 10 miles.

85. False. The equation represents a hyperbola.

$\frac{x^2}{144} - \frac{y^2}{144} = 1$

87. False. If the graph crossed the directrix, there would exist points nearer the directrix than the focus.

89. (a) $a + b = 20 \Rightarrow b = 20 - a$

$A = \pi ab = \pi a(20 - a)$

(b) $264 = \pi a(20 - a)$

$\pi a^2 - 20\pi a + 264 = 0$

$a \approx 14$ or $a \approx 6$ by the Quadratic Formula

$b = 6$ $\quad b = 14$

Since $a > b$ we choose $a = 14$ and $b = 6$.

$\frac{x^2}{14^2} + \frac{y^2}{6^2} = 1, \frac{x^2}{196} + \frac{y^2}{36} = 1$

(c)

a	8	9	10	11	12	13
A	301.6	311.0	314.2	311.0	301.6	285.9

Conjecture: Area is maximum when $a = b = 10$ and the shape is a circle.

(d) The area is maximum when $a = b = 10$ and the shape is a circle.

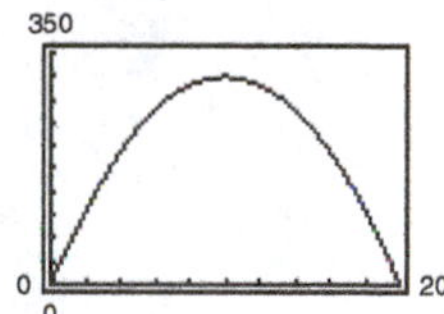

91. *Sample answer.* To graph $\frac{x^2}{25} + \frac{y^2}{16} = 1$, you must solve the equation for y and graph the two resulting functions.

$$\frac{x^2}{25} + \frac{y^2}{16} = 1$$

$$16x^2 + 25y^2 = 400$$

$$25y^2 = 400 - 16x^2$$

$$y^2 = \frac{400 - 16x^2}{25}$$

$$y = \pm\sqrt{\frac{400 - 16x^2}{25}}$$

$$y = \pm\tfrac{4}{5}\sqrt{25 - x^2}$$

$y_1 = \frac{4}{5}\sqrt{25 - x^2}$ is the top half of the ellipse and the bottom half of the ellipse is $y_2 = -\frac{4}{5}\sqrt{25 - x^2}$.

93. No. The y-term has an exponent of 4, not 2. Therefore it is not a second degree equation and not a hyperbola.

95. (a) Left half of ellipse (portion to the left of the y-axis)

(b) Top half of ellipse (portion to the right of the y-axis)

97. *Sample answer:* The smaller the distance between the thumbtacks, the more circular the ellipse becomes. The larger the distance between the thumbtacks, the more long and narrow the ellipse becomes.

99. Let (x, y) be such that the difference of the distances from $(c, 0)$ and $(-c, 0)$ is $2a$. (We are only deriving the form where the transverse axis is horizontal.)

$$2a = \left|\sqrt{(x + c)^2 + y^2} - \sqrt{(x - c)^2 + y^2}\right|$$

$$\pm 2a = \sqrt{(x + c)^2 + y^2} - \sqrt{(x - c)^2 + y^2}$$

$$\pm 2a + \sqrt{(x - c)^2 + y^2} = \sqrt{(x + c)^2 + y^2}$$

$$4a^2 \pm 4a\sqrt{(x - c)^2 + y^2} + (x - c)^2 + y^2 = (x + c)^2 + y^2$$

$$\pm 4a\sqrt{(x - c)^2 + y^2} = 4cx - 4a^2$$

$$\pm a\sqrt{(x - c)^2 + y^2} = cx - a^2$$

$$a^2(x^2 - 2cx + c^2 + y^2) = c^2x^2 - 2a^2cx + a^4$$

$$a^2(c^2 - a^2) = (c^2 - a^2)x^2 - a^2y^2$$

Let $b^2 = c^2 - a^2$. Then, $a^2b^2 = b^2x^2 - a^2y^2 \Rightarrow 1 = \frac{x^2}{a^2} - \frac{y^2}{b^2}$.

Section 4.4 Translations of Conics

1. Hyperbola with horizontal transverse axis $\Rightarrow \dfrac{(x-h)^2}{a^2} - \dfrac{(y-k)^2}{b^2} = 1$

Matches (b).

2. Ellipse with vertical major axis $\Rightarrow \dfrac{(x-h)^2}{b^2} + \dfrac{(y-k)^2}{a^2} = 1$

Matches (d).

3. Parabola with vertical axis $\Rightarrow (x-h)^2 = 4p(y-k)$

Matches (e).

4. Hyperbola with vertical transverse axis $\Rightarrow \dfrac{(y-k)^2}{a^2} - \dfrac{(x-h)^2}{b^2} = 1$

Matches (c).

5. Ellipse with horizontal major axis $\Rightarrow \dfrac{(x-h)^2}{a^2} + \dfrac{(y-k)^2}{b^2} = 1$

Matches (a).

6. Parabola with horizontal axis $\Rightarrow (y-k)^2 = 4p(x-h)$

Matches (f).

7. The graph of $(x+2)^2 + (y-1)^2 = 4$ is a circle.

The graph has been shifted two units to the left and one unit upward from standard position.

9. The graph of $\dfrac{(y+3)^2}{4} - (x-1)^2 = 1$ is a hyperbola.

The graph has been shifted one unit to the right and three units downward from standard position.

11. The graph of $(x+1)^2 = 4(-1)(y-2)$ is a parabola.

The graph has been shifted one unit to the left and two units upward from standard position.

13. The graph of $\dfrac{(x+4)^2}{9} + \dfrac{(y+2)^2}{16} = 1$ is an ellipse.

The graph has been shifted four units to the left and two units downward from standard position.

15. $x^2 + y^2 = 49$

Center: $(0, 0)$

Radius: 7

17. $(x-4)^2 + (y-5)^2 = 36$

Center: $(4, 5)$

Radius: 6

19. $(x-1)^2 + y^2 = 10$

Center: $(1, 0)$

Radius: $\sqrt{10}$

21.

$$\begin{aligned} x^2 + y^2 - 8y &= 0 \\ (x^2) + (y^2 - 8y) &= 0 \\ (x-0)^2 + (y^2 - 8y + 16) &= 0 + 16 \\ (x-0)^2 + (y-4)^2 &= 16 \\ x^2 + (y-4)^2 &= 16 \end{aligned}$$

Center: $(0, 4)$

Radius: 4

23.

$$\begin{aligned} x^2 + y^2 - 2x + 6y + 9 &= 0 \\ (x^2 - 2x) + (y^2 + 6y) &= -9 \\ (x^2 - 2x + 1) + (y^2 + 6y + 9) &= -9 + 1 + 9 \\ (x-1)^2 + (y+3)^2 &= 1 \end{aligned}$$

Center: $(1, -3)$

Radius: 1

25. $4x^2 + 4y^2 + 12x - 24y + 41 = 0$

$$x^2 + y^2 + 3x - 6y + \frac{41}{4} = 0$$

$$(x^2 + 3x) + (y^2 - 6y) = -\frac{41}{4}$$

$$\left(x^2 + 3x + \frac{9}{4}\right) + \left(y^2 - 6y + 9\right) = -\frac{41}{4} + \frac{9}{4} + 9$$

$$\left(x + \frac{3}{2}\right)^2 + (y - 3)^2 = 1$$

Center: $\left(-\frac{3}{2}, 3\right)$

Radius: 1

27. $(x - 1)^2 + 8(y + 2) = 0$

$$(x - 1)^2 = 4(-2)(y + 2);\ p = -2$$

Vertex: $(1, -2)$

Focus: $(1, -4)$

Directrix: $y = 0$

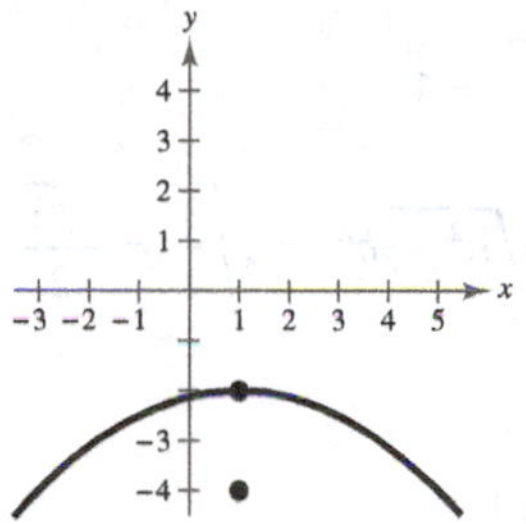

29. $\left(y + \frac{1}{2}\right)^2 = 2(x - 5)$

$\left(y + \frac{1}{2}\right)^2 = 4\left(\frac{1}{2}\right)(x - 5);\ p = \frac{1}{2}$

Vertex: $\left(5, -\frac{1}{2}\right)$

Focus: $\left(\frac{11}{2}, -\frac{1}{2}\right)$

Directrix: $x = \frac{9}{2}$

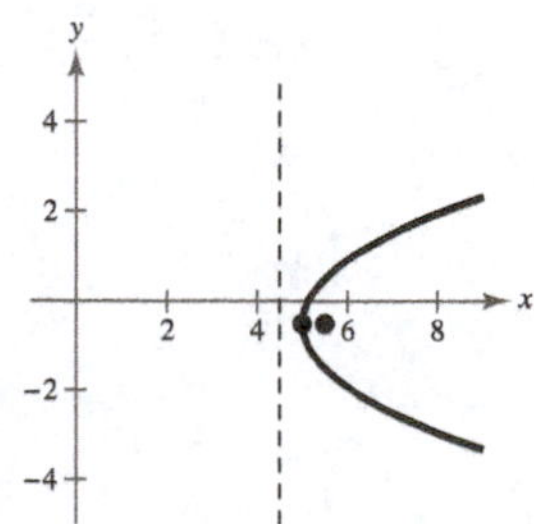

31. $y = \dfrac{x^2 - 2x + 5}{4}$

$$4y = x^2 - 2x + 5$$

$$x^2 - 2x = 4y - 5$$

$$x^2 - 2x + 1 = 4y - 5 + 1$$

$$(x - 1)^2 = 4y - 4$$

$$(x - 1)^2 = 4(1)(y - 1);\ p = 1$$

Vertex: $(1, 1)$

Focus: $(1, 2)$

Directrix: $y = 0$

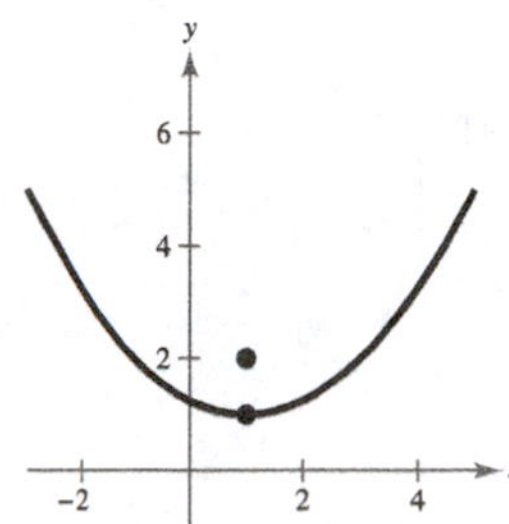

33. $8x + y^2 + 1 = 2y - 4$

$$y^2 - 2y = -8x - 5$$

$$y^2 - 2y + 1 = -8x - 5 + 1$$

$$(y - 1)^2 = -8x - 4$$

$$(y - 1)^2 = -8\left(x + \frac{1}{2}\right)$$

$$(y - 1)^2 = 4(-2)\left(x + \frac{1}{2}\right);\ p = -2$$

Vertex: $\left(-\frac{1}{2}, 1\right)$

Focus: $\left(-\frac{5}{2}, 1\right)$

Directrix: $x = \frac{3}{2}$

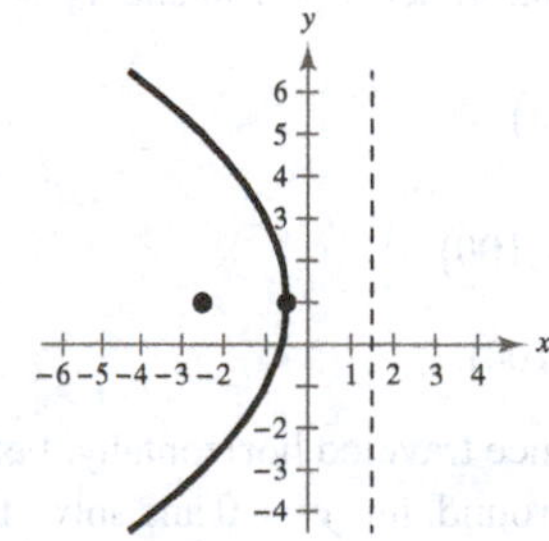

35. Vertex: $(3, 2)$

Focus: $(1, 2)$

Horizontal axis

$$p = 1 - 3 = -2$$
$$(y - 2)^2 = 4(-2)(x - 3)$$
$$(y - 2)^2 = -8(x - 3)$$

37. Vertex: $(0, 4)$

Directrix: $y = 2$

Vertical axis

$$p = 4 - 2 = 2$$
$$(x - 0)^2 = 4(2)(y - 4)$$
$$x^2 = 8(y - 4)$$

39. Focus: $(4, 4)$

Directrix: $x = -4$

Horizontal axis

Vertex: $(0, 4)$

$$p = 4 - 0 = 4$$
$$(y - 4)^2 = 4(4)(x - 0)$$
$$(y - 4)^2 = 16x$$

41. (a) $V = 17{,}500\sqrt{2}$ mi/h

$\approx 24{,}750$ mi/h

(b) $p = -4100, (h, k) = (0, 4100)$

$$(x - 0)^2 = 4(-4100)(y - 4100)$$
$$x^2 = -16{,}400(y - 4100)$$

43. (a) To write the equation, let $s = 100$ and $v_0 = 28$.

$$x^2 = -\frac{v^2}{16}(y - s)$$
$$x^2 = -\frac{28^2}{16}(y - 100)$$
$$x^2 = -49(y - 100)$$

(b) To find the distance traveled horizontally, before the ball strikes the ground, let $y = 0$ and solve for x.

$$x^2 = -49(0 - 100)$$
$$x^2 = 4900$$
$$x = \sqrt{4900}$$
$$= 70$$

The ball travels 70 feet horizontally before striking the ground.

45. $\dfrac{(x - 1)^2}{9} + \dfrac{(y - 5)^2}{25} = 1$

$a = 5, b = 3, c = \sqrt{a^2 - b^2} = 4$

Center: $(1, 5)$

Foci: $(1, 1), (1, 9)$

Vertices: $(1, 0), (1, 10)$

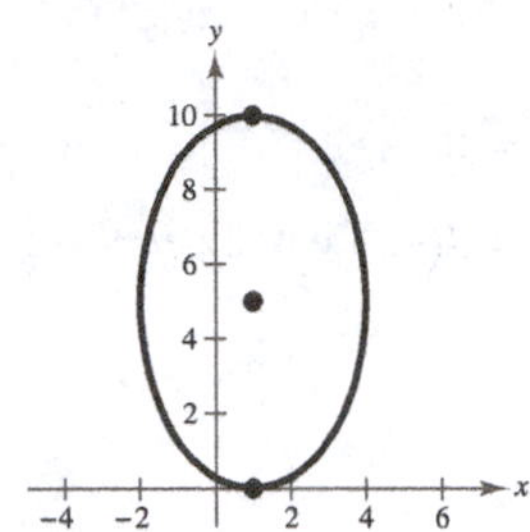

47. $\dfrac{(x + 2)^2}{1} + \dfrac{(y + 4)^2}{1/4} = 1$

$a = 1, b = \dfrac{1}{2}, c = \sqrt{a^2 - b^2} = \sqrt{\dfrac{3}{4}} = \dfrac{\sqrt{3}}{2}$

Center: $(-2, -4)$

Foci: $\left(-2 \pm \dfrac{\sqrt{3}}{2}, -4\right)$

Vertices: $(-3, -4), (-1, -4)$

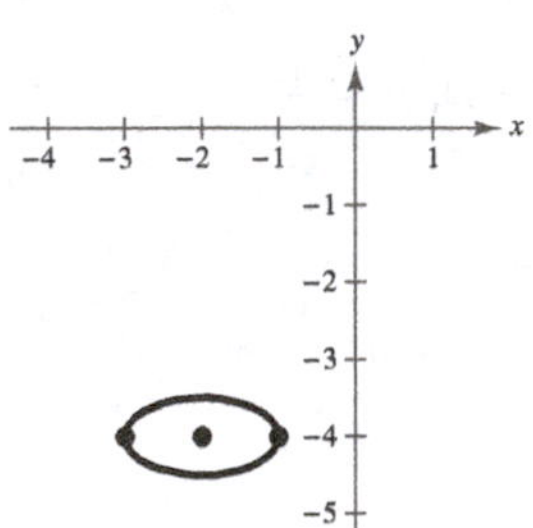

49.
$$9x^2 + 25y^2 - 36x - 50y + 52 = 0$$
$$9(x^2 - 4x) + 25(y^2 - 2y) = -52$$
$$9(x^2 - 4x + 4) + 25(y^2 - 2y + 1) = -52 + 9(4) + 25(1)$$
$$9(x - 2)^2 + 25(y - 1)^2 = 9$$
$$(x - 2)^2 + \frac{(y - 1)^2}{9/25} = 1$$
$$a = 1, b = \frac{3}{5}, c = \sqrt{1 - \frac{9}{25}} = \sqrt{\frac{16}{25}} = \frac{4}{5}$$

Horizontal major axis

Center: $(2, 1)$

Foci: $\left(\frac{14}{5}, 1\right), \left(\frac{6}{5}, 1\right)$

Vertices: $(1, 1), (3, 1)$

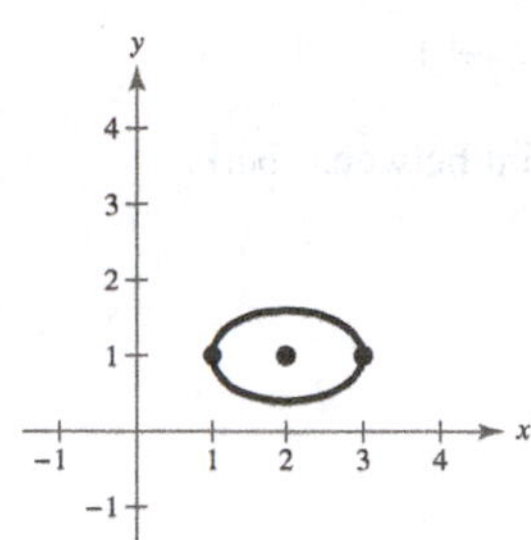

51. $25x^2 + 4y^2 + 50x - 75 = 0$
$$25(x^2 + 2x) + 4(y^2) = 75$$
$$25(x^2 + 2x + 1) + 4(y^2) = 75 + 25(1)$$
$$25(x + 1)^2 + 4(y - 0)^2 = 100$$
$$\frac{(x + 1)^2}{4} + \frac{(y - 0)^2}{25} = 1$$
$$\frac{(x + 1)^2}{4} + \frac{y^2}{25} = 1$$
$$a = 5, b = 2, c = \sqrt{5^2 - 4^2} = \sqrt{21}$$

Vertical major axis

Center: $(-1, 0)$

Foci: $\left(-1, \pm\sqrt{21}\right)$

Vertices: $(-1, \pm 5)$

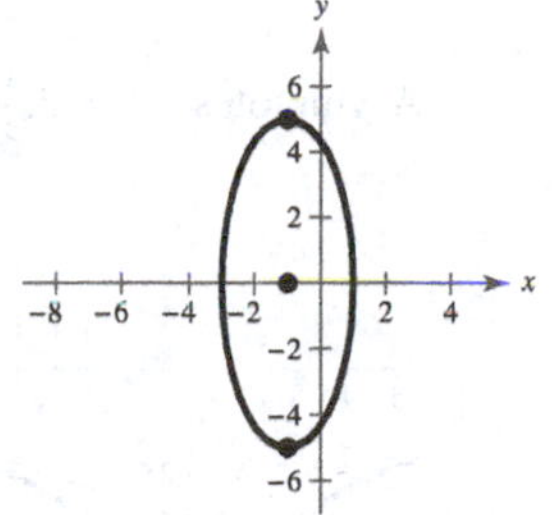

53. Vertices: $(3, 3), (3, -3)$

Minor axis of length $2 \Rightarrow b = 1$

Center: $(3, 0) \Rightarrow a = 3$

Vertical major axis

$$\frac{(x - 3)^2}{1} + \frac{y^2}{9} = 1$$

55. Foci: $(0, 0), (4, 0) \Rightarrow c = 2$

Major axis of length $8 \Rightarrow a = 4$

$b^2 = a^2 - c^2 = 12$

Center: $(2, 0)$

Horizontal major axis

$$\frac{(x - 2)^2}{16} + \frac{y^2}{12} = 1$$

57. Center: $(1, 4)$

Vertices: $(1, 0), (1, 8) \Rightarrow a = 4$

$a = 2c \rightarrow 4 = 2c \rightarrow c = 2$

$b^2 = a^2 - c^2 = 4^2 - 2^2 = 12$

Vertical major axis

$$\frac{(x - 1)^2}{12} + \frac{(y - 4)^2}{16} = 1$$

59. Vertices: $(-3, 0), (7, 0) \Rightarrow a = 5$

Foci: $(0, 0), (4, 0) \Rightarrow c = 2$

Center: $(2, 0)$

Horizontal major axis

$a^2 = b^2 + c^2 \Rightarrow b^2 = 5^2 - 2^2 = 21$

$$\frac{(x - 2)^2}{25} + \frac{y^2}{21} = 1$$

61. Foci: $(-3, -3), (-3, -1)$; $c = 1$

Center: $(-3, -2)$ (midpoint between foci)

$$e = \frac{c}{a} = \frac{1}{3}$$

$$c = 1 \rightarrow a = 3$$

$$b^2 = 3^2 - 1^2 = 8$$

$$\frac{(x+3)^2}{8} + \frac{(y+2)^2}{9} = 1$$

63. Vertices: $(2, -1), (2, 3)$; $a = 2$

Center: $(2, 1)$ (midpoint between vertices)

$$e = \frac{c}{a} = \frac{\sqrt{3}}{2}$$

$$a = 2c = \sqrt{3}, b^2 = 2^2 - \left(\sqrt{3}\right)^2 = 1$$

$$\frac{(x-2)^2}{1} + \frac{(y-1)^2}{4} = 1$$

$$(x-2)^2 + \frac{(y-1)^2}{4} = 1$$

65. $a = 3.67 \times 10^9$

$$e = \frac{c}{a} = 0.249$$

$$c = 913{,}830{,}000$$

Smallest distance: $a - c = 2{,}756{,}170{,}000$ miles

Greatest distance: $a + c = 4{,}583{,}830{,}000$ miles

67. $\dfrac{(x-2)^2}{16} - \dfrac{(y+1)^2}{9} = 1$

$a = 4, b = 3, c = \sqrt{a^2 + b^2} = 5$

Center: $(2, -1)$

Horizontal transverse axis

Vertices: $(6, -1), (-2, -1)$

Foci: $(7, -1), (-3, -1)$

Asymptotes: $y = \pm\frac{3}{4}(x - 2) - 1$

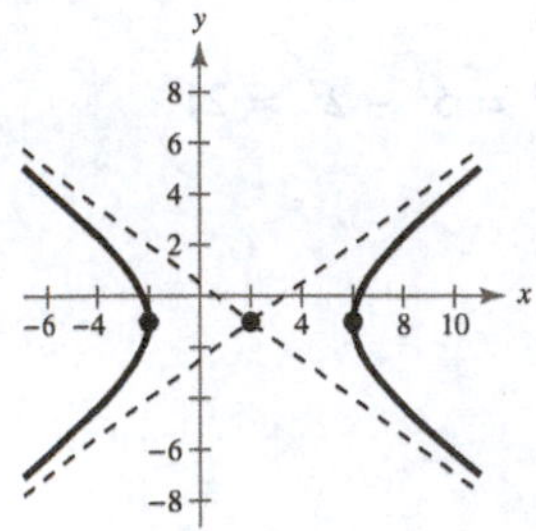

69. $(y + 6)^2 - (x - 2)^2 = 1$

$a = 1, b = 1, c = \sqrt{a^2 + b^2} = \sqrt{2}$

Center: $(2, -6)$

Vertical transverse axis

Vertices: $(2, -7), (2, -5)$

Foci: $\left(2, -6 \pm \sqrt{2}\right)$

Asymptotes:

$y = \pm(x - 2) - 6$

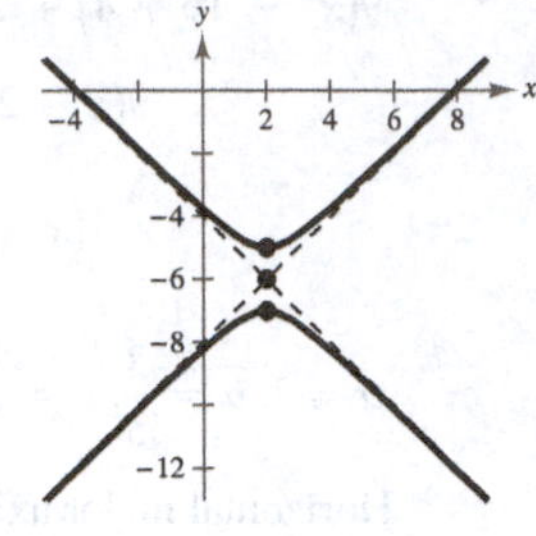

71.

$$x^2 - 9y^2 + 2x - 54y - 85 = 0$$

$$\left(x^2 + 2x\right) - 9\left(y^2 + 6y\right) = 85$$

$$\left(x^2 + 2x + 1\right) - 9\left(y^2 + 6y + 9\right) = 85 + 1 - 9(9)$$

$$(x+1)^2 - 9(y+3)^2 = 5$$

$$\frac{(x+1)^2}{5} - \frac{(y+3)^2}{5/9} = 1$$

$$a = \sqrt{5}, b = \frac{\sqrt{5}}{3}, c = \sqrt{a^2 + b^2} = \frac{5\sqrt{2}}{3}$$

Center: $(-1, -3)$

Horizontal transverse axis

Vertices: $\left(-1 \pm \sqrt{5}, -3\right)$

Foci: $\left(-1 \pm \dfrac{5\sqrt{2}}{3}, -3\right)$

Asymptotes: $y = \pm\frac{1}{3}(x + 1) - 3$

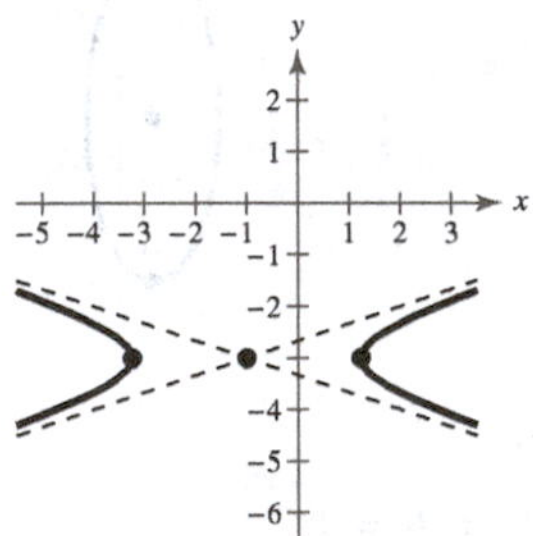

73.
$$9y^2 - x^2 - 36y - 6x + 18 = 0$$
$$9(y^2 - 4y) - (x^2 + 6x) = -18$$
$$9(y^2 - 4y + 4) - (x^2 + 6x + 9) = -18 + 9(4) - 9$$
$$9(y - 2)^2 - (x + 3)^2 = 9$$
$$(y - 2)^2 - \frac{(x + 3)^2}{9} = 1$$

$a = 1, b = 3, c = \sqrt{a^2 + b^2} = \sqrt{10}$

Center: $(-3, 2)$

Vertical transverse axis

Vertices: $(-3, 1), (-3, 3)$

Foci: $\left(-3, 2 \pm \sqrt{10}\right)$

Asymptotes:

$y = \pm\frac{1}{3}(x + 3) + 2$

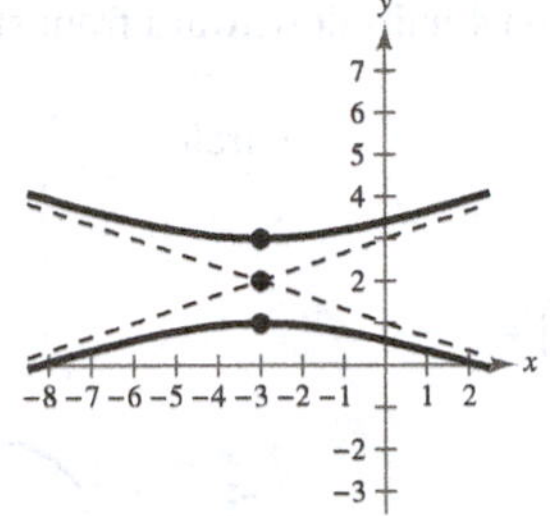

75. Vertices: $(0, 0), (0, 2)$

Foci: $(0, -1), (0, 3)$

Center: $(0, 1)$

Vertical transverse axis

$a = 1, c = 2, b^2 = c^2 - a^2 = 3$

$$\frac{(y - 1)^2}{1} - \frac{x^2}{3} = 1$$

77. Foci: $(-8, 5), (0, 5); c = 4$

Center: $(-4, 5)$; midpoint between foci

Transverse axis of length 4, so $a = 2$

$c = 4, a = 2, b^2 = c^2 - a^2 = 4^2 - 2^2 = 12$

$$\frac{(x + 4)^2}{4} - \frac{(y - 5)^2}{12} = 1$$

79. Vertices: $(2, 3), (2, -3)$

Passes through the point $(0, 5)$

Center: $(2, 0)$

Vertical transverse axis

$a = 3$

$$\frac{y^2}{9} - \frac{(x - 2)^2}{b^2} = 1$$
$$\frac{5^2}{9} - \frac{(0 - 2)^2}{b^2} = 1$$
$$b^2 = \frac{9}{4}$$
$$\frac{y^2}{9} - \frac{4(x - 2)^2}{9} = 1$$

81. Vertices: $(0, 2), (6, 2)$

Asymptotes: $y = \frac{2}{3}x, y = 4 - \frac{2}{3}x$

Center: $(3, 2)$

Horizontal transverse axis

$a = 3, b = 2$

$$\frac{(x - 3)^2}{9} - \frac{(y - 2)^2}{4} = 1$$

83.
$$y^2 - x^2 + 4y = 0$$
$$(y^2 + 4y + 4) - x^2 = 4$$
$$(y + 2)^2 - x^2 = 4$$
$$\frac{(y + 2)^2}{4} - \frac{x^2}{4} = 1$$

Hyperbola

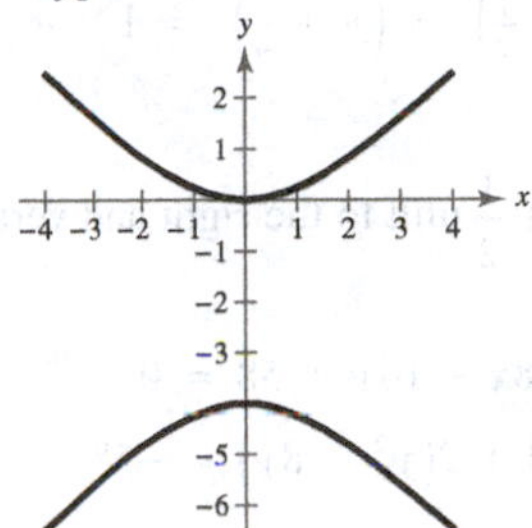

Vertical translation 2 units downward from standard position

85. $16y^2 + 128x + 8y - 7 = 0$

$$16\left(y^2 + \tfrac{1}{2}y\right) = -128x + 7$$
$$16\left(y^2 + \tfrac{1}{2}y + \tfrac{1}{16}\right) = -128x + 7 + 16\left(\tfrac{1}{16}\right)$$
$$16\left(y + \tfrac{1}{4}\right)^2 = -128x + 8$$
$$\left(y + \tfrac{1}{4}\right)^2 = -8x + \tfrac{1}{2}$$
$$\left(y + \tfrac{1}{4}\right)^2 = -8\left(x - \tfrac{1}{16}\right)$$

Parabola

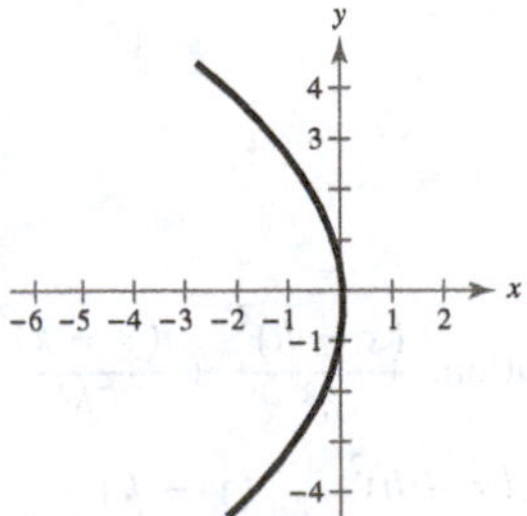

Horizontal translation 1/16 unit to the right and a vertical translation 1/4 unit down from standard position

87.
$$\begin{aligned}
9x^2 + 16y^2 + 36x + 128y + 148 &= 0\\
9(x^2 + 4x) + 16(y^2 + 8y) &= -148\\
9(x^2 + 4x + 4) + 16(y^2 + 8y + 16) &= -148 + 9(4) + 16(16)\\
9(x + 2)^2 + 16(y + 4)^2 &= 144\\
\frac{(x + 2)^2}{16} + \frac{(y + 4)^2}{9} &= 1
\end{aligned}$$

Ellipse

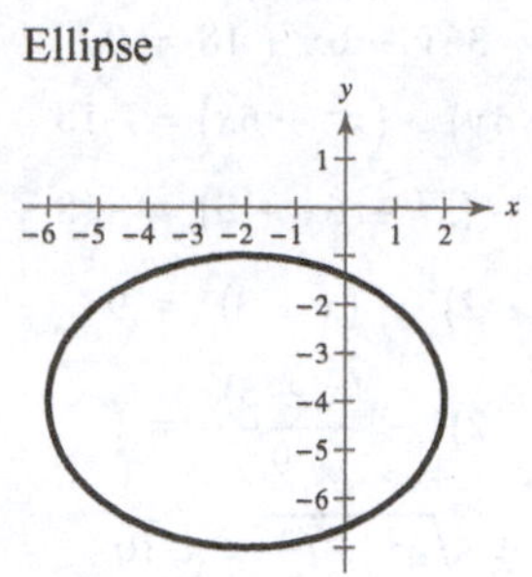

Horizontal translation 2 units to the left and vertical translation 4 units downward from standard position

89.
$$\begin{aligned}
16x^2 + 16y^2 - 16x + 24y - 3 &= 0\\
16(x^2 - x) + 16\left(y^2 + \tfrac{3}{2}y\right) &= 3\\
16\left(x^2 - x + \tfrac{1}{4}\right) + 16\left(y^2 + \tfrac{3}{2}y + \tfrac{9}{16}\right) &= 3 + 16\left(\tfrac{1}{4}\right) + 16\left(\tfrac{9}{16}\right)\\
16\left(x - \tfrac{1}{2}\right)^2 + 16\left(y + \tfrac{3}{4}\right)^2 &= 16\\
\left(x - \tfrac{1}{2}\right)^2 + \left(y + \tfrac{3}{4}\right)^2 &= 1
\end{aligned}$$

Circle

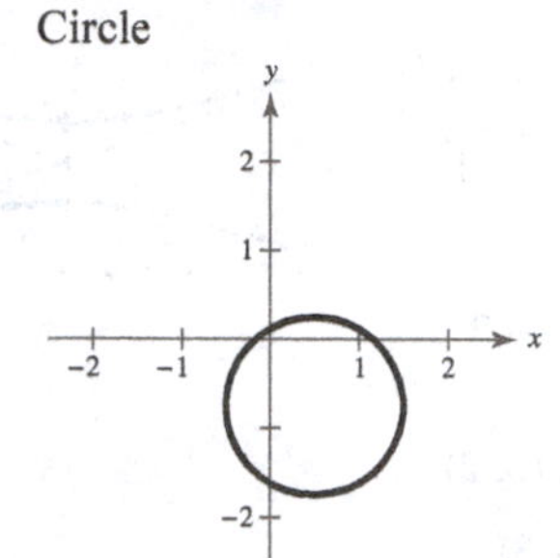

Horizontal translation $\frac{1}{2}$ unit to the right and vertical translation $\frac{3}{4}$ unit downward from standard position

91.
$$\begin{aligned}
3x^2 + 2y^2 - 18x - 16y + 58 &= 0\\
3(x^2 - 6x) + 2(y^2 - 8y) &= -58\\
3(x^2 - 6x + 9) + 2(y^2 - 8y + 16) &= -58 + 27 + 32\\
3(x - 3)^2 + 2(y - 4)^2 &= 1\\
\frac{(x - 3)^2}{1/3} + \frac{(y - 4)^2}{1/2} &= 1
\end{aligned}$$

True. The equation in standard form is $\dfrac{(x - 3)^2}{1/3} + \dfrac{(y - 4)^2}{1/2} = 1$.

93. False. A hyperbola cannot have vertices at $(-9, 4)$ and $(-3, 4)$ while having foci at $(-6, 9)$ and $(-6, -1)$ because the vertices and foci must be collinear, that is they must lie on the same horizontal or vertical line.

95. (a) For an ellipse $e = c/a$ and $c^2 = a^2 - b^2$.

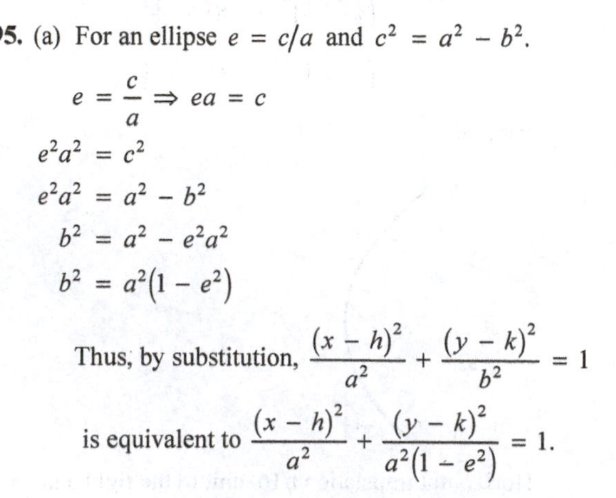

$$\begin{aligned}
e &= \frac{c}{a} \Rightarrow ea = c\\
e^2a^2 &= c^2\\
e^2a^2 &= a^2 - b^2\\
b^2 &= a^2 - e^2a^2\\
b^2 &= a^2(1 - e^2)
\end{aligned}$$

Thus, by substitution, $\dfrac{(x - h)^2}{a^2} + \dfrac{(y - k)^2}{b^2} = 1$

is equivalent to $\dfrac{(x - h)^2}{a^2} + \dfrac{(y - k)^2}{a^2(1 - e^2)} = 1$.

(b)

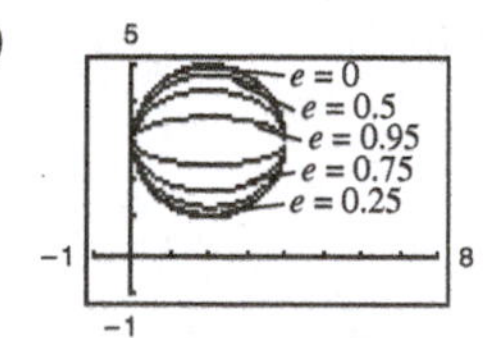

As e approaches 0, the ellipse becomes a circle.

Review Exercises for Chapter 4

1. Because the denominator is zero when $x + 10 = 0$, the domain of f is all real numbers except $x = -10$.

x	-11	-10.5	-10.1	-10.01	-10.001	$\to -10$
$f(x)$	33	63	303	3003	30,003	$\to \infty$

x	$-10 \leftarrow$	-9.999	-9.99	-9.9	-9.5	-9
$f(x)$	$-\infty \leftarrow$	$-29{,}997$	-2997	-297	-57	-27

As x approaches -10 from the left, $f(x)$ increases without bound.

As x approaches -10 from the right, $f(x)$ decreases without bound.

3. Because the denominator is zero when $x^2 - 10x + 24 = (x - 4)(x - 6) = 0$, the domain of f is all real numbers except $x = 4$ and $x = 6$.

x	3	3.5	3.9	3.99	3.999	$\to 4$
$f(x)$	2.667	6.4	38.095	398.010	3998.001	$\to \infty$

x	$4 \leftarrow$	4.001	4.01	4.1	4.5	5
$f(x)$	$-\infty \leftarrow$	-4002.001	-402.010	-42.105	-10.67	-8

As x approaches 4 from the left, $f(x)$ increases without bound.

As x approaches 4 from the right, $f(x)$ decreases without bound.

x	5	5.5	5.9	5.99	5.999	$\to 6$
$f(x)$	-8	-10.67	-42.015	-402.010	-4002.001	$\to -\infty$

x	$6 \leftarrow$	6.001	6.01	6.1	6.5	7
$f(x)$	$\infty \leftarrow$	3998.001	398.010	38.095	6.4	2.667

As x approaches 6 from the left, $f(x)$ decreases without bound.

As x approaches 6 from the right, $f(x)$ increases without bound.

5. $f(x) = \dfrac{6x^2}{x + 3}$

Vertical asymptote: $x = -3$

Horizontal asymptote: none

7. $g(x) = \dfrac{x^2}{x^2 - 4}$

Vertical asymptotes: $x = -2, x = 2$

Horizontal asymptote: $y = 1$

9. $h(x) = \dfrac{5x + 20}{x^2 - 2x - 24}$

$= \dfrac{5(x + 4)}{(x - 6)(x + 4)} = \dfrac{5}{x - 6}, \quad x \neq -4$

Vertical asymptote: $x = 6$

Horizontal asymptote: $y = 0$

11. $\overline{C} = \dfrac{C}{x} = \dfrac{0.5x + 500}{x}, \quad 0 < x$

Horizontal asymptote: $\overline{C} = \dfrac{0.5}{1} = 0.5$

As x increases, the average cost per unit approaches the horizontal asymptote, $\overline{C} = 0.5 = \$0.50$.

13. $f(x) = \dfrac{-3}{2x^2}$

(a) Domain: all real numbers x except $x = 0$

(b) No intercepts

(c) Vertical asymptote: $x = 0$

Horizontal asymptote: $y = 0$

(d)

x	-3	-2	-1	1	2	3
$f(x)$	$-\frac{1}{6}$	$-\frac{3}{8}$	$-\frac{3}{2}$	$-\frac{3}{2}$	$-\frac{3}{8}$	$-\frac{1}{6}$

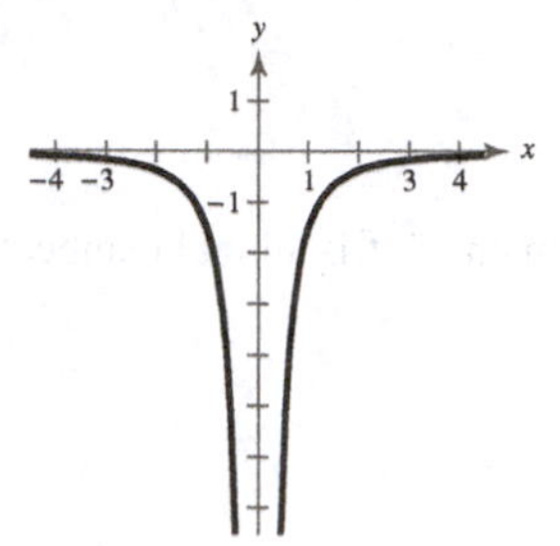

15. $g(x) = \dfrac{2 + x}{1 - x} = -\dfrac{x + 2}{x - 1}$

(a) Domain: all real numbers x except $x = 1$

(b) x-intercept: $(-2, 0)$

y-intercept: $(0, 2)$

(c) Vertical asymptote: $x = 1$

Horizontal asymptote: $y = -1$

(d)

x	-1	0	2	3
$g(x)$	$\frac{1}{2}$	2	-4	$-\frac{5}{2}$

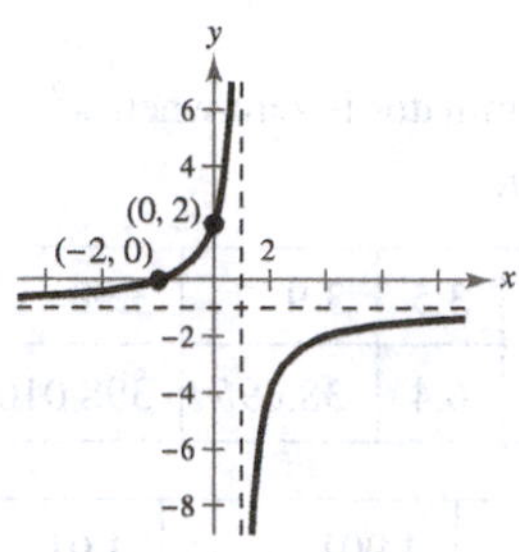

17. $p(x) = \dfrac{5x^2}{4x^2 + 1}$

(a) Domain: all real numbers x

(b) Intercept: $(0, 0)$

(c) Horizontal asymptote: $y = \dfrac{5}{4}$

(d)

x	-3	-2	-1	0	1	2	3
$p(x)$	$\frac{45}{37}$	$\frac{20}{17}$	1	0	1	$\frac{20}{17}$	$\frac{45}{37}$

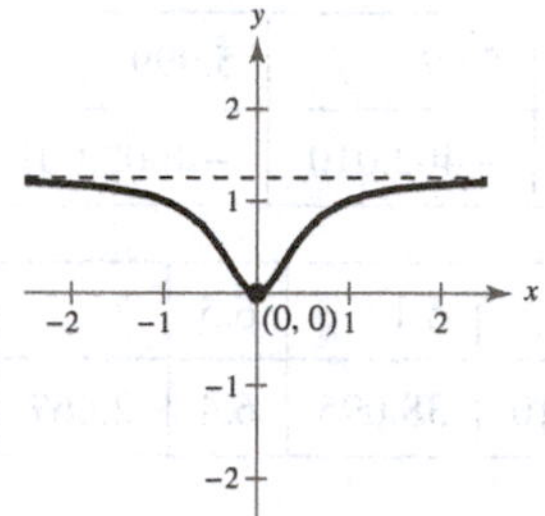

19. $f(x) = \dfrac{x}{x^2 - 16}$

(a) Domain: all real numbers x except $x \neq \pm 4$

(b) Intercept: $(0, 0)$

(c) Vertical asymptotes: $x = \pm 4$

Horizontal asymptote: $y = 0$

(d)

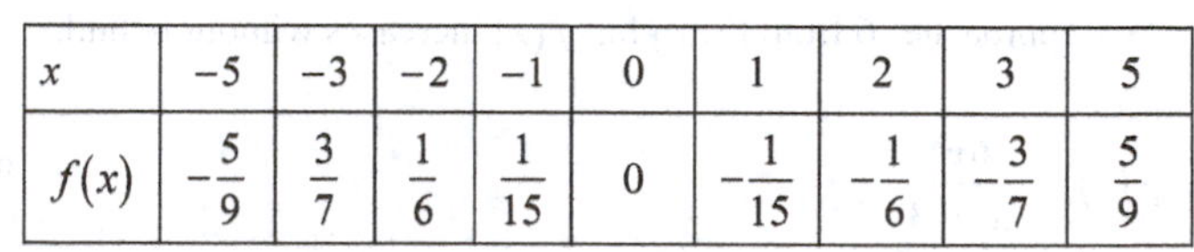

x	-5	-3	-2	-1	0	1	2	3	5
$f(x)$	$-\frac{5}{9}$	$\frac{3}{7}$	$\frac{1}{6}$	$\frac{1}{15}$	0	$-\frac{1}{15}$	$-\frac{1}{6}$	$-\frac{3}{7}$	$\frac{5}{9}$

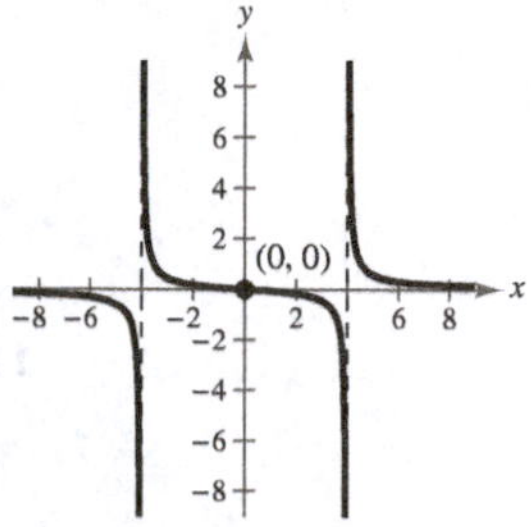

21. $f(x) = \dfrac{-6x^2}{x^2 + 1}$

(a) Domain: all real numbers x

(b) Intercept: $(0, 0)$

(c) Horizontal asymptote: $y = -6$

(d)

x	± 3	± 2	± 1	0
$f(x)$	$-\frac{27}{5}$	$-\frac{24}{5}$	-3	0

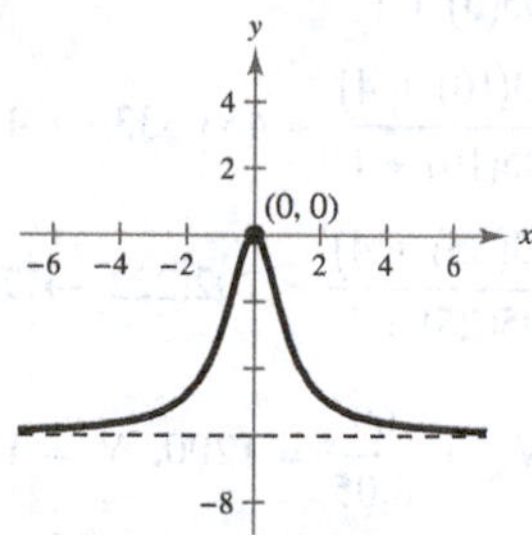

23. $f(x) = \dfrac{6x^2 - 11x + 3}{3x^2 - x}$

$= \dfrac{(3x - 1)(2x - 3)}{x(3x - 1)} = \dfrac{2x - 3}{x}, \; x \neq \dfrac{1}{3}$

(a) Domain: all real numbers x except $x = 0$ and $x = \dfrac{1}{3}$

(b) x-intercept: $\left(\dfrac{3}{2}, 0\right)$

(c) Vertical asymptote: $x = 0$
Horizontal asymptote: $y = 2$

(d)

x	-2	-1	1	2	3	4
$f(x)$	$\frac{7}{2}$	5	-1	$\frac{1}{2}$	1	$\frac{5}{4}$

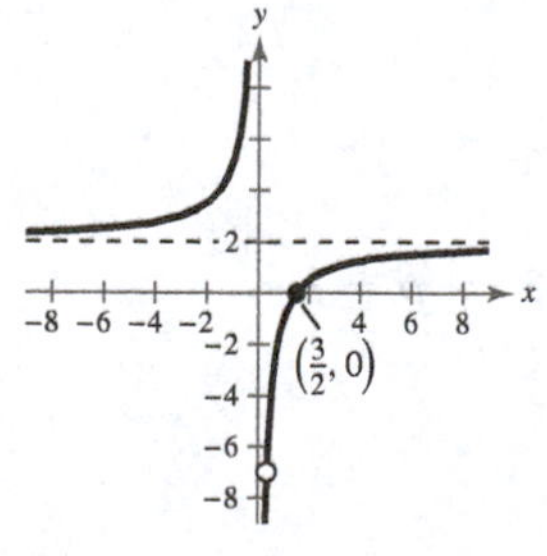

25. $f(x) = \dfrac{2x^3}{x^2 + 1} = 2x - \dfrac{2x}{x^2 + 1}$

(a) Domain: all real numbers x

(b) Intercept: $(0, 0)$

(c) Slant asymptote: $y = 2x$

(d)

x	-2	-1	0	1	2
$f(x)$	$-\frac{16}{5}$	-1	0	1	$\frac{16}{5}$

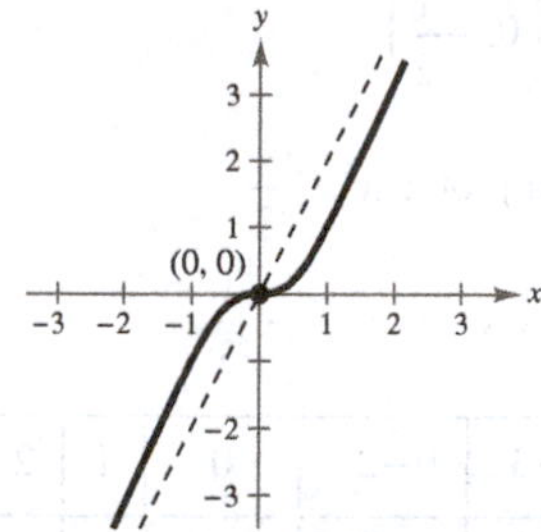

27. $f(x) = \dfrac{x^2 + 3x - 10}{x + 2} = x + 1 - \dfrac{12}{x + 2}$

(a) Domain: all real numbers x except $x = -2$

(b) y-intercept: $(0, -5)$
x-intercepts: $(2, 0), (-5, 0)$

(c) Vertical asymptote: $x = -2$
Slant asymptote: $y = x + 1$

(d)

x	-6	-4	-3	-1	0	1	2	4
$f(x)$	-2	3	10	-12	-5	-2	0	3

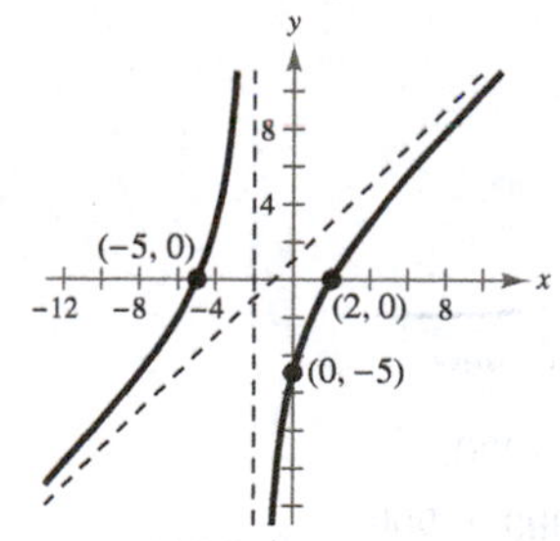

29. $f(x) = \dfrac{3x^3 - 2x^2 - 3x + 2}{3x^2 - x - 4} = \dfrac{(3x - 2)(x + 1)(x - 1)}{(3x - 4)(x + 1)}$

$= \dfrac{(3x - 2)(x - 1)}{3x - 4} = x - \dfrac{1}{3} + \dfrac{2/3}{3x - 4}, \quad x \neq -1$

(a) Domain: all real numbers x except

$x = -1$ and $x = \dfrac{4}{3}$

(b) x-intercepts: $(1, 0), \left(\dfrac{2}{3}, 0\right)$

y-intercept: $\left(0, -\dfrac{1}{2}\right)$

(c) Vertical asymptote: $x = \dfrac{4}{3}$

Slant asymptote: $y = x - \dfrac{1}{3}$

(d)

x	-3	-2	0	1	2	3
$f(x)$	$-\dfrac{44}{13}$	$-\dfrac{12}{5}$	$-\dfrac{1}{2}$	0	2	$\dfrac{14}{5}$

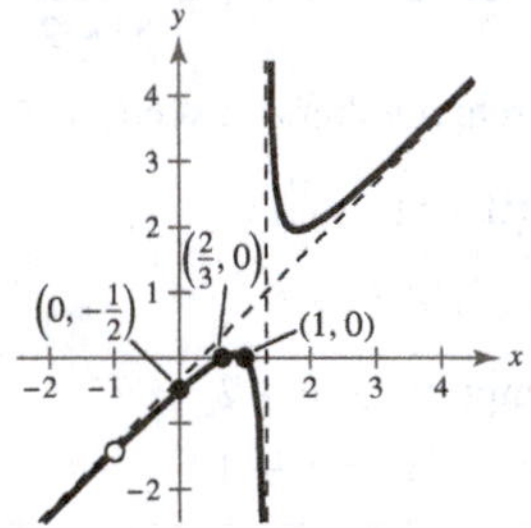

31. $\overline{C} = \dfrac{100{,}000 + 0.9x}{x}, \quad x > 0$

(a)

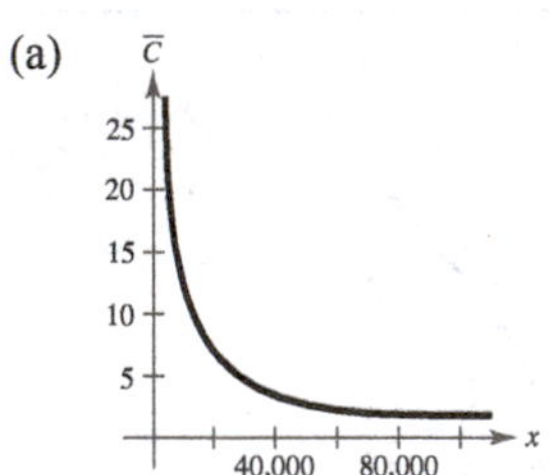

(b) When $x = 1000$,

$C = \dfrac{100{,}000 + 900}{1000} = \$100.90.$

When $x = 10{,}000$,

$C = \dfrac{100{,}000 + 9000}{10{,}000} = \$10.90.$

When $x = 100{,}000$,

$C = \dfrac{100{,}000 + 90{,}000}{100{,}000} = \$1.90.$

(c) The average cost per unit is always greater than \$0.90 because \$0.90 is the horizontal asymptote of the function.

33. (a)

(b) $N(5) = \dfrac{20(3(5) + 4)}{0.05(5) + 1} = 304.0 \rightarrow 304{,}000$ fish

$N(10) = \dfrac{20(3(10) + 4)}{0.05(10) + 1} \approx 453.333 \rightarrow 453{,}333$ fish

$N(25) = \dfrac{20(3(25) + 4)}{0.05(25) + 1} \approx 702.222 \rightarrow 702{,}222$ fish

(c) As $t \rightarrow \infty$, $N \rightarrow \dfrac{60}{0.05} = 1200$. $N = 1200$ is the horizontal asymptote of the graph of the function. Therefore as time passes, the number of fish goes to about 1,200,000 fish.

35. $y^2 = -10x$ is a parabola.

37. $\dfrac{y^2}{12} - \dfrac{x^2}{9} = 1$ is a hyperbola.

39. $4x^2 + 18y^2 = 36$ is an ellipse.

41. $3x^2 + 3y^2 = 75$ is a circle.

43. Vertex: $(0, 0)$

Focus: $(-6, 0) \Rightarrow p = -6$

Horizontal axis

$y^2 = 4px$

$y^2 = 4(-6)x$

$y^2 = -24x$

45. Vertex: $(0, 0)$

Directrix: $y = -3 \Rightarrow p = 3$

Vertical axis

$x^2 = 4py$

$x^2 = 4(3)y$

$x^2 = 12y$

47. Vertex: $(0, 0)$

Point: $(3, 6)$

Horizontal axis

$$y^2 = 4px$$
$$6^2 = 4p(3) \Rightarrow p = 3$$
$$y^2 = 4(3)x$$
$$y^2 = 12x$$

49. $y = \dfrac{x^2}{200}, \ -100 \le x \le 100$

Vertex: $(0, 0)$

$$x^2 = 200y$$
$$x^2 = 4(50)y$$
$$p = 50$$

Focus: $(0, 50)$

51. Vertices: $(0, 6), (0, -6) \Rightarrow a = 6$

Ends of minor axis: $(5, 0), (-5, 0) \Rightarrow b = 5$

Center: $(0, 0)$

Vertical major axis

$$\frac{(x - 0)^2}{5^2} + \frac{(y - 0)^2}{6^2} = 1$$
$$\frac{x^2}{25} + \frac{y^2}{36} = 1$$

53. Vertices: $(0, \pm 7)$; $a = 7$

Foci: $(0, \pm 6)$; $c = 6$

$b^2 = a^2 - c^2 = 49 - 36 = 13$

Vertical major axis

$$\frac{x^2}{13} + \frac{y^2}{49} = 1$$

55. Foci: $(\pm 14, 0) \Rightarrow c = 14$

Center: $(0, 0)$

Minor axis of length $10 \Rightarrow b = 5$

$$c^2 = a^2 - b^2$$
$$196 = a^2 - 25$$
$$a^2 = 221$$
$$\frac{x^2}{221} + \frac{y^2}{25} = 1$$

57. $a = 5, b = 4, c = \sqrt{a^2 - b^2} = 3$

The foci should be placed 3 feet on either side of the center and have the same height as the pillars.

59. Vertices: $(0, \pm 1)$

Foci: $(0, \pm 5)$

Vertical transverse axis

Center: $(0, 0)$

$$a = 1, c = 5$$
$$b = \sqrt{25 - 1} = \sqrt{24}$$
$$\frac{y^2}{a^2} - \frac{x^2}{b^2} = 1$$
$$\frac{y^2}{1} - \frac{x^2}{24} = 1$$

61. Center: $(0, 0)$

Horizontal transverse axis

Vertices: $(\pm 1, 0) \Rightarrow a = 1$

Asymptotes: $y = \pm 2x \Rightarrow \dfrac{b}{a} = 2 \Rightarrow b = 2$

$$\frac{x^2}{1} - \frac{y^2}{4} = 1$$

63. Vertex: $(4, 2)$

Focus: $(4, 0)$

Vertical axis, $p = -2$

$$(x - 4)^2 = 4(-2)(y - 2)$$
$$(x - 4)^2 = -8(y - 2)$$

65. Vertex: $(8, -8)$

Directrix: $x = 1$

Horizontal axis

$$p = 8 - 1 = 7$$
$$(y + 8)^2 = 4(7)(x - 8)$$
$$(y + 8)^2 = 28(x - 8)$$

67. Vertices: $(0, 2), (4, 2) \Rightarrow a = 2$

Minor axis of length $2 \Rightarrow b = 1$

Center: $\left(\frac{0+4}{2}, \frac{2+2}{2}\right) = (2, 2)$

Horizontal major axis

$$\frac{(x-2)^2}{2^2} + \frac{(y-2)^2}{1} = 1$$

$$\frac{(x-2)^2}{4} + (y-2)^2 = 1$$

69. Vertices: $(2, -2), (2, 8) \Rightarrow a = 5$

Foci: $(2, 0), (2, 6) \Rightarrow c = 3$

Center: $\left(\frac{2+2}{2}, \frac{-2+8}{2}\right) = (2, 3)$

$$c^2 = a^2 - b^2 \Rightarrow b^2 = a^2 - c^2$$
$$b^2 = 25 - 9$$
$$b^2 = 16$$

Vertical major axis

$$\frac{(x-2)^2}{4^2} + \frac{(y-3)^2}{5^2} = 1$$

$$\frac{(x-2)^2}{16} + \frac{(y-3)^2}{25} = 1$$

71. Foci: $(0, -4), (0, 0) \Rightarrow c = 2$

Eccentricity: $e = \frac{2}{3}$

$$\frac{2}{3} = \frac{c}{a}$$
$$\frac{2}{3} = \frac{2}{a}$$
$$a = 3$$

Center: $\left(\frac{0+0}{2}, \frac{-4+0}{2}\right) = (0, -2)$

$$c^2 = a^2 - b^2 \Rightarrow b^2 = a^2 - c^2$$
$$b^2 = 9 - 4$$
$$b^2 = 5$$

Vertical major axis

$$\frac{x^2}{5} + \frac{(y+2)^2}{9} = 1$$

73. Vertices: $(-10, 3), (6, 3) \Rightarrow a = 8$

Foci: $(-12, 3), (8, 3) \Rightarrow c = 10 \Rightarrow b$

$$= \sqrt{c^2 - a^2}$$
$$= \sqrt{100 - 64} = 6$$

Horizontal transverse axis

Center: $(-2, 3)$

$$\frac{(x+2)^2}{64} - \frac{(y-3)^2}{36} = 1$$

75. Vertices: $(3, -4), (3, 4) \Rightarrow a = 4$

Vertical transverse axis

Center: $\left(\frac{3+3}{2}, \frac{-4+4}{2}\right) = (3, 0)$

$$\frac{y^2}{16} - \frac{(x-3)^2}{b^2} = 1$$

The hyperbola passes through $(4, 6)$, so let $x = 4$ and $y = 6$

$$\frac{6^2}{16} - \frac{(4-3)^2}{b^2} = 1$$
$$\frac{36}{16} - \frac{1}{b^2} = 1$$
$$36b^2 - 16 = 16b^2$$
$$20b^2 = 16$$
$$b^2 = \frac{4}{5}$$

$$\frac{y^2}{16} - \frac{(x-3)^2}{\frac{4}{5}} = 1$$

77. $x^2 - 6x + 2y + 9 = 0$

$$(x-3)^2 = -2y$$
$$(x-3)^2 = 4\left(-\frac{1}{2}\right)y \Rightarrow p = -\frac{1}{2}$$

Parabola

Vertex: $(3, 0)$

Focus: $\left(3, -\frac{1}{2}\right)$

The graph of $x^2 = -2y$ has been shifted to the right three units.

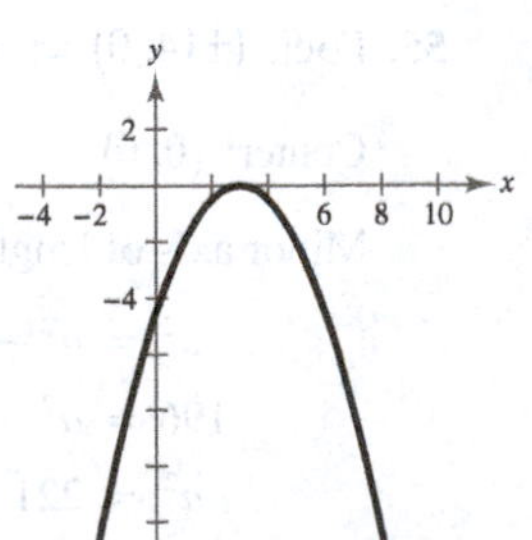

79.
$$x^2 + y^2 - 2x - 4y + 1 = 0$$
$$(x^2 - 2x) + (y^2 - 4y) = -1$$
$$(x^2 - 2x + 1) + (y^2 - 4y + 4) = -1 + 1 + 4$$
$$(x - 1)^2 + (y - 2)^2 = 4$$

Circle

Center: $(1, 2)$

Radius: 2

The graph is the circle $x^2 + y^2 = 1$ translated 1 unit to the right and two units upward.

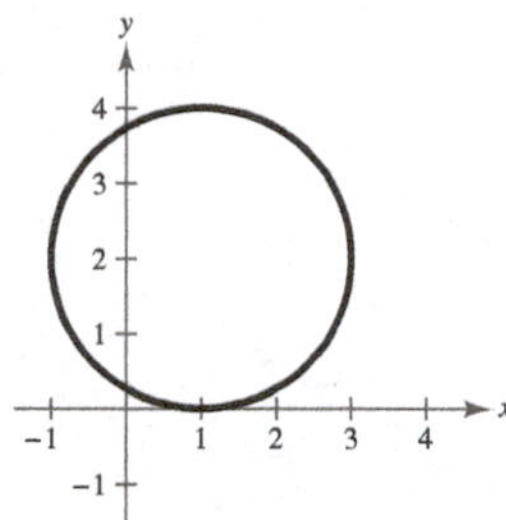

81.
$$x^2 + 9y^2 + 10x - 18y + 25 = 0$$
$$(x^2 + 10x) + 9(y^2 - 2y) = -25$$
$$(x^2 + 10x + 25) + 9(y^2 - 2y + 1) = -25 + 25 + 9$$
$$(x + 5)^2 + 9(y - 1)^2 = 9$$
$$\frac{(x + 5)^2}{9} + (y - 1)^2 = 1$$

Ellipse

Center: $(-5, 1)$

Vertices: $(-8, 1), (-2, 1)$

The graph of $\frac{x^2}{9} + y^2 = 1$ has been shifted left five units and upward one unit.

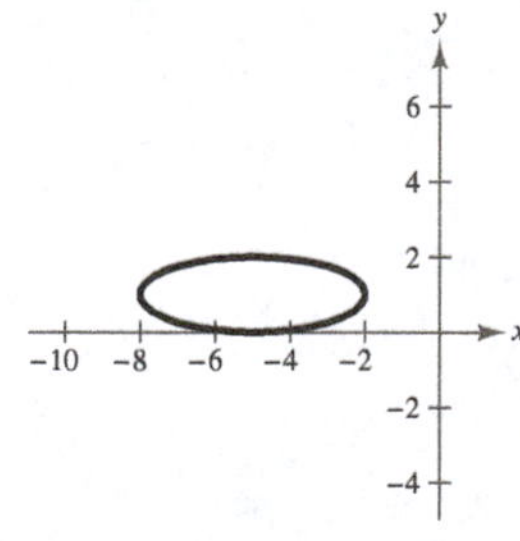

83.
$$9x^2 - y^2 - 72x + 8y + 119 = 0$$
$$9(x^2 - 8x) - (y^2 - 8y) = -119$$
$$9(x^2 - 8x + 16) - (y^2 - 8y + 16) = -119 + 144 - 16$$
$$9(x - 4)^2 - (y - 4)^2 = 9$$
$$\frac{(x - 4)^2}{1} - \frac{(y - 4)^2}{9} = 1$$

Hyperbola

Center: $(4, 4)$

Vertices: $(3, 4), (5, 4)$

The graph of $x^2 - \frac{y^2}{9} = 1$ has been shifted right four units and upward four units from standard position.

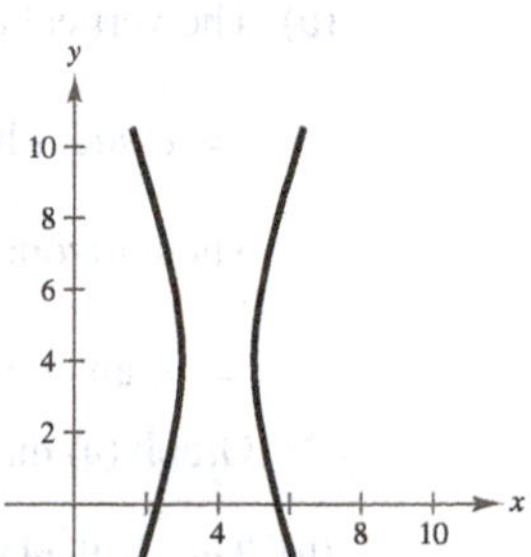

85. $x^2 = 4p(y - 12)$

$(\pm 4)^2 = 4p(10 - 12)$

$16 = -8p$

$-2 = p$

$x^2 = 4(-2)(y - 12)$

$x^2 = -8(y - 12)$

When $y = 0$, we have:

$x^2 = 96$

$x = \pm\sqrt{96} = \pm 4\sqrt{6}$

At ground level, the width is $2x = 8\sqrt{6} \approx 19.6$ meters.

87. $x^2 + y^2 - 200x - 52{,}500 = 0$

(a) The conic is a circle because x^2 and y^2 have identical coefficients.

(b) $x^2 - 200x + 10{,}000 + y^2 = 52{,}500 + 10{,}000$

$$(x - 100)^2 + y^2 = 62{,}500$$

Center: $(100, 0)$

Radius: 250

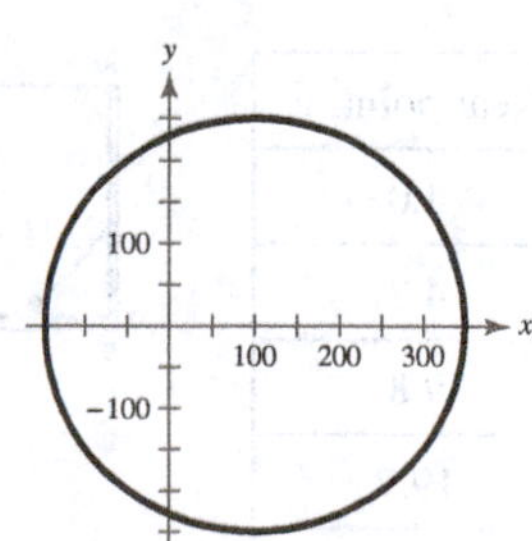

89. True. It could be a degenerate conic.

Problem Solving for Chapter 4

1. $f(x) = \dfrac{ax + b}{cx + d}$

Vertical asymptote: $x = -\dfrac{d}{c}$

Horizontal asymptote: $y = \dfrac{a}{c}$

(a) The vertical asymptote is to the right of the y-axis, so $-\dfrac{d}{c} > 0$.

$\Rightarrow$ c and d have opposite signs.

The horizontal asymptote is under the x-axis, so $\dfrac{a}{c} < 0$.

$\Rightarrow$ a and c have opposite signs.

Graph (a) matches (iii).

(b) The vertical asymptote is to the left of the y-axis, so $-\dfrac{d}{c} < 0$.

$\Rightarrow$ c and d have the same sign.

The horizontal asymptote is under the x-axis, so $\dfrac{a}{c} < 0$.

$\Rightarrow$ a and c have opposite signs.

Graph (b) matches (ii).

(c) The vertical asymptote is to the left of the y-axis, so $-\dfrac{d}{c} < 0$.

$\Rightarrow$ c and d have the same sign.

The horizontal asymptote is above the x-axis, so $\dfrac{a}{c} > 0$.

$\Rightarrow$ a and c have the same sign.

Graph (c) matches (iv).

(d) The vertical asymptote is to the right of the y-axis, so $-\dfrac{d}{c} > 0$.

$\Rightarrow$ c and d have opposite signs.

The horizontal asymptote is above the x-axis, so $\dfrac{a}{c} > 0$.

$\Rightarrow$ a and c have the same sign.

Graph (d) matches (i).

3. (a)

Age, x	Near point, y
16	3.0
32	4.7
44	9.8
50	19.7
60	39.4

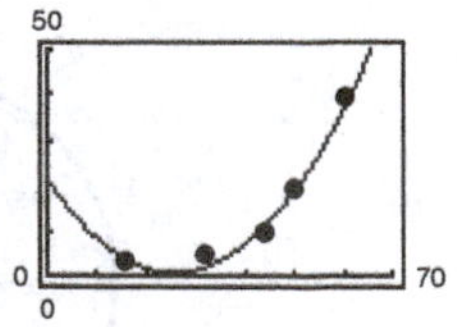

50

0 70

0

$y \approx 0.031x^2 - 1.59x + 21.0$

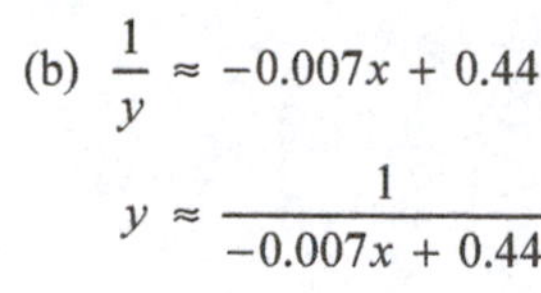

(b) $\frac{1}{y} \approx -0.007x + 0.44$

$y \approx \frac{1}{-0.007x + 0.44}$

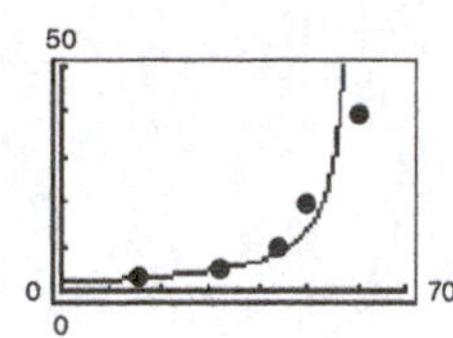

(c)

Age, x	Near point, y	Quadratic Model	Rational Model
16	3.0	3.66	3.05
32	4.7	2.32	4.63
44	9.8	11.83	7.58
50	19.7	19.97	11.11
60	39.4	38.54	50.00

The models are fairly good fits to the data. The quadratic model seems to be a better fit for older ages and the rational model a better fit for younger ages.

(d) For $x = 25$, the quadratic model yields $y \approx 0.625$ inch and the rational model yields $y \approx 3.774$ inches.

(e) The reciprocal model cannot be used to predict the near point for a person who is 70 years old because it results in a negative value $(y \approx -20)$. The quadratic model yields $y \approx 63.37$ inches.

5. A hyperbola is the set of all points (x, y) in a plane, the difference of whose distances from two distinct fixed points (foci) is a positive constant.

Using the vertex $(a, 0)$, the distance to $(-c, 0)$ is $a + c$, and the distance to $(c, 0)$ is $(c - a)$.

$d_2 - d_1 = (a + c) - (c - a) = 2a$

By definition, $|d_2 - d_1|$ is constant for any point (x, y) on the hyperbola. Thus, $|d_2 - d_1| = 2a$.

7. (a)

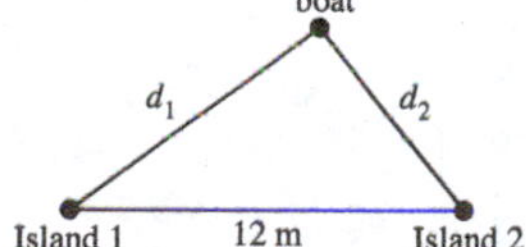

Since $d_1 + d_2 \le 20$, by definition, the outer bound that the boat can travel is an ellipse. The islands are the foci.

(b)

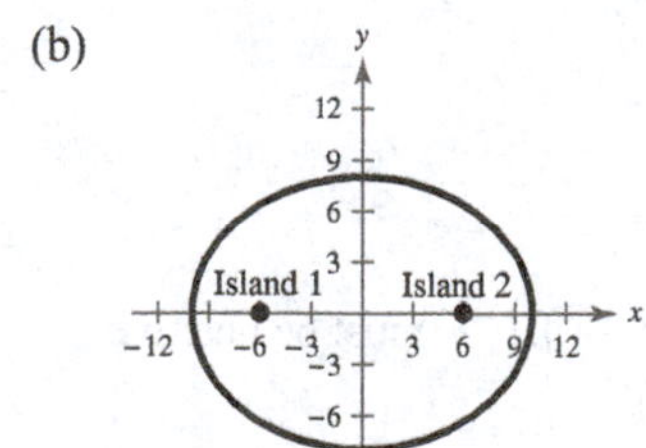

Island 1 is located at $(-6, 0)$ and Island 2 is located at $(6, 0)$.

(c) $d_1 + d_2 = 2a = 20 \Rightarrow a = 10$

The boat traveled 20 miles. The vertex is $(10, 0)$.

(d) $c = 6, a = 10 \Rightarrow b^2 = a^2 - c^2 = 64$

$\frac{x^2}{100} + \frac{y^2}{64} = 1$

9. $x^2 = 4py$

(a)

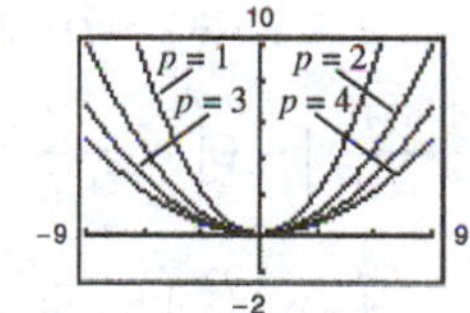

As p increases, the graph flattens near $(0, 0)$. It produces a vertical shrink and the parabola becomes wider.

(b) $p = 1$: $x^2 = 4y$ Focus: $(0, 1)$

$p = 2$: $x^2 = 8y$ Focus: $(0, 2)$

$p = 3$: $x^2 = 12y$ Focus: $(0, 3)$

$p = 4$: $x^2 = 16y$ Focus: $(0, 4)$

(c) $p = 1$ Chord length $= 4$

$p = 2$ Chord length $= 8$

$p = 3$ Chord length $= 12$

$p = 4$ Chord length $= 16$

In general, the chord length equals $4|p|$.

(d) To graph a parabola, first plot the vertex and focus. At the focus, two additional points can be plotted by moving $2|p|$ units to the right and $2|p|$ units to the left if the directrix is horizontal. If the directrix is vertical, move $2|p|$ units up and $2|p|$ units down from the focus to plot two additional points.

11. $Ax^2 + Cy^2 + Dx + Ey + F = 0$

Assume that the conic is *not* degenerate.

(a) $A = C, A \neq 0$

$$Ax^2 + Ay^2 + Dx + Ey + F = 0$$

$$x^2 + y^2 + \frac{D}{A}x + \frac{E}{A}y + \frac{F}{A} = 0$$

$$\left(x^2 + \frac{D}{A}x + \frac{D^2}{4A^2}\right) + \left(y^2 + \frac{E}{A}y + \frac{E^2}{4A^2}\right) = -\frac{F}{A} + \frac{D^2}{4A^2} + \frac{E^2}{4A^2}$$

$$\left(x + \frac{D}{2A}\right)^2 + \left(y + \frac{E}{2A}\right)^2 = \frac{D^2 + E^2 - 4AF}{4A^2}$$

This is a circle with center $\left(-\frac{D}{2A}, -\frac{E}{2A}\right)$ and radius $\frac{\sqrt{D^2 + E^2 - 4AF}}{2|A|}$.

(b) $A = 0$ or $C = 0$ (but not both). Let $C = 0$.

$$Ax^2 + Dx + Ey + F = 0$$

$$x^2 + \frac{D}{A}x = -\frac{E}{A}y - \frac{F}{A}$$

$$x^2 + \frac{D}{A}x + \frac{D^2}{4A^2} = -\frac{E}{A}y - \frac{F}{A} + \frac{D^2}{4A^2}$$

$$\left(x + \frac{D}{2A}\right)^2 = -\frac{E}{A}\left(y + \frac{F}{E} - \frac{D^2}{4AE}\right)$$

This is a parabola with vertex $\left(-\frac{D}{2A}, \frac{D^2 - 4AF}{4AE}\right)$. $A = 0$ yields a similar result.

(c) $AC > 0 \Rightarrow A$ and C are either both positive or are both negative (if that is the case, move the terms to the other side of the equation so that they are both positive).

$$Ax^2 + Cy^2 + Dx + Ey + F = 0$$

$$A\left(x^2 + \frac{D}{A}x + \frac{D^2}{4A^2}\right) + C\left(y^2 + \frac{E}{C}y + \frac{E^2}{4C^2}\right) = -F + \frac{D^2}{4A} + \frac{E^2}{4C}$$

$$A\left(x + \frac{D}{2A}\right)^2 + C\left(y + \frac{E}{2C}\right)^2 = \frac{CD^2 + AE^2 - 4ACF}{4AC}$$

$$\frac{\left(x + \frac{D}{2A}\right)^2}{\frac{CD^2 + AE^2 - 4ACF}{4A^2C}} + \frac{\left(y + \frac{E}{2C}\right)^2}{\frac{CD^2 + AE^2 - 4ACF}{4AC^2}} = 1$$

Since A and C are both positive, $4A^2C$ and $4AC^2$ are both positive. $CD^2 + AE^2 - 4ACF$ must be positive or the conic is degenerate. Thus, we have an ellipse with center $(-D/2A, -E/2C)$.

(d) $AC < 0 \Rightarrow A$ and C have opposite signs. Let's assume that A is positive and C is negative. (If A is negative and C is positive, move the terms to the other side of the equation.) From part (c) we have

$$\frac{\left(x + \frac{D}{2A}\right)^2}{\frac{CD^2 + AE^2 - 4ACF}{4A^2C}} + \frac{\left(y + \frac{E}{2C}\right)^2}{\frac{CD^2 + AE^2 - 4ACF}{4AC^2}} = 1.$$

Since $A > 0$ and $C < 0$, the first denominator is positive if $CD^2 + AE^2 - 4ACF < 0$ and is negative if $CD^2 + AE^2 - 4ACF > 0$, since $4A^2C$ is negative. The second denominator would have the *opposite* sign since $4AC^2 > 0$. Thus, we have a hyperbola with center $(-D/2A, -E/2C)$.

Practice Test for Chapter 4

1. Sketch the graph of $f(x) = \dfrac{x - 1}{2x}$ and label all intercepts and asymptotes.

2. Sketch the graph of $f(x) = \dfrac{3x^2 - 4}{x}$ and label all intercepts and asymptotes.

3. Find all the asymptotes of $f(x) = \dfrac{8x^2 - 9}{x^2 + 1}$.

4. Find all the asymptotes of $f(x) = \dfrac{4x^2 - 2x + 7}{x - 1}$.

5. Sketch the graph of $f(x) = \dfrac{x - 5}{(x - 5)^2}$.

6. Find the vertex, focus, and directrix of the parabola $x^2 = 20y$.

7. Find the equation of the parabola with vertex $(0, 0)$ and focus $(7, 0)$.

8. Find the center, foci, and vertices of the ellipse $\dfrac{x^2}{144} + \dfrac{y^2}{25} = 1$.

9. Find the equation of the ellipse with foci $(\pm 4, 0)$ and minor axis of length 6.

10. Find the center, vertices, foci, and asymptotes of the hyperbola $\dfrac{y^2}{144} - \dfrac{x^2}{169} = 1$.

11. Find the equation of the hyperbola with vertices $(\pm 4, 0)$ and asymptotes of $y = \pm\frac{1}{2}x$.

12. Find the equation of the parabola with vertex $(6, -1)$ and focus $(6, 3)$.

13. Find the center, foci, and vertices of the ellipse $16x^2 + 9y^2 - 96x + 36y + 36 = 0$.

14. Find the equation of the ellipse with vertices $(-1, 1)$ and $(7, 1)$ and minor axis of length 2.

15. Find the center, vertices, foci, and asymptotes of the hyperbola $4(x + 3)^2 - 9(y - 1)^2 = 1$.

16. Find the equation of the hyperbola with vertices $(3, 4)$ and $(3, -4)$ and foci $(3, 7)$ and $(3, -7)$.

CHAPTER 5
Exponential and Logarithmic Functions

CHAPTER 5
Exponential and Logarithmic Functions

Section 5.1 Exponential Functions and Their Graphs

1. algebraic

3. One-to-One

5. $A = P\left(1 + \dfrac{r}{n}\right)^{nt}$

7. $f(1.4) = (0.9)^{1.4} \approx 0.863$

9. $f\left(\frac{2}{5}\right) = 3^{2/5} \approx 1.552$

11. $f(-1.5) = 5000\left(2^{-1.5}\right)$
    ≈ 1767.767

13. $f(x) = 2^x$
    Increasing
    Asymptote: $y = 0$
    Intercept: $(0, 1)$
    Matches graph (d).

14. $f(x) = 2^x + 1$
    Increasing
    Asymptote: $y = 1$
    Intercept: $(0, 2)$
    Matches graph (c).

15. $f(x) = 2^{-x}$
    Decreasing
    Asymptote: $y = 0$
    Intercept: $(0, 1)$
    Matches graph (a).

16. $f(x) = 2^{x-2}$
    Increasing
    Asymptote: $y = 0$
    Intercept: $\left(0, \frac{1}{4}\right)$
    Matches graph (b).

17. $f(x) = 7^x$

x	-2	-1	0	1	2
$f(x)$	0.020	0.143	1	7	49

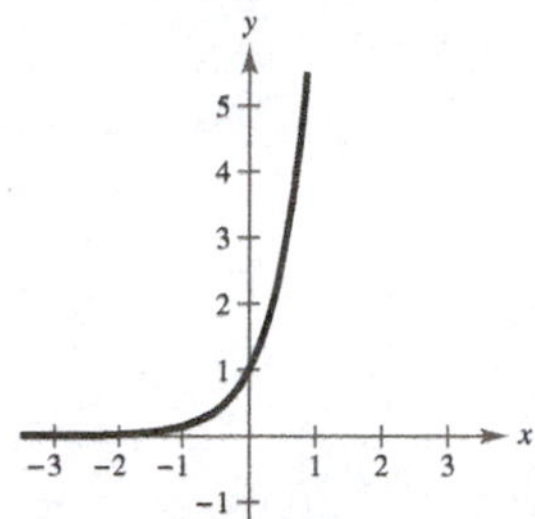

19. $f(x) = \left(\frac{1}{4}\right)^{-x}$

x	-2	-1	0	1	2
$f(x)$	0.063	0.25	1	4	16

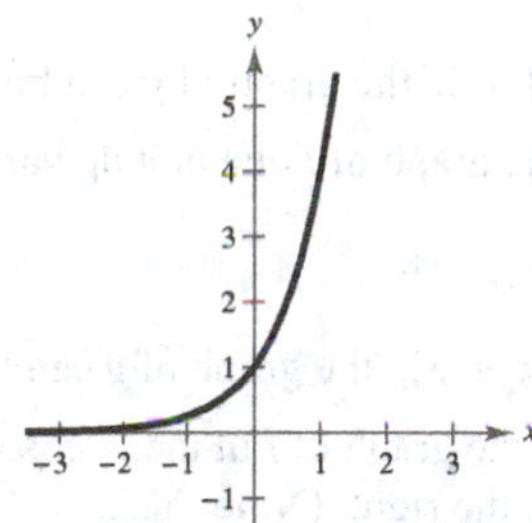

21. $f(x) = 4^{x-1}$

x	-2	-1	0	1	2
$f(x)$	0.016	0.063	0.25	1	4

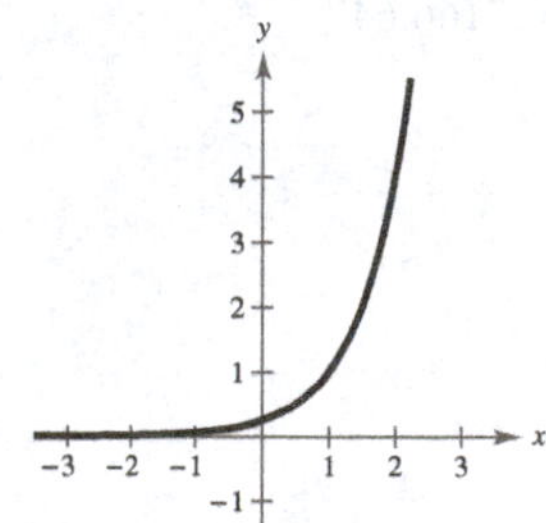

23. $f(x) = 2^{x+1} + 3$

x	-3	-2	-1	0	1
$f(x)$	3.25	3.5	4	5	7

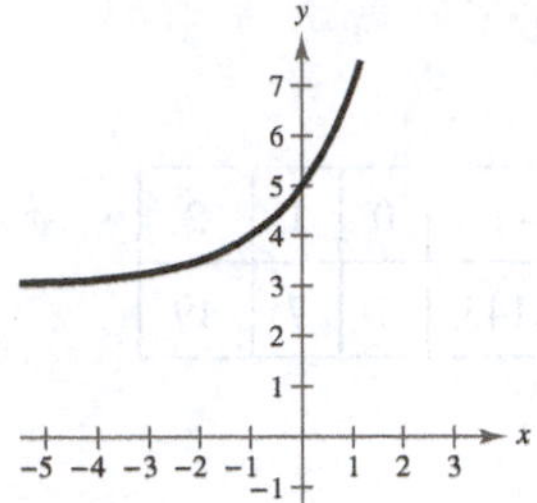

25.
$$3^{x+1} = 27$$
$$3^{x+1} = 3^3$$
$$x + 1 = 3$$
$$x = 2$$

27.
$$\left(\tfrac{1}{2}\right)^x = 32$$
$$\left(\tfrac{1}{2}\right)^x = \left(\tfrac{1}{2}\right)^{-5}$$
$$x = -5$$

29. $f(x) = 3^x$, $g(x) = 3^x + 1$

Because $g(x) = f(x) + 1$, the graph of g can be obtained by shifting the graph of f one unit upward.

31. $f(x) = 10^x$, $g(x) = 10^{-x+3}$

Because $g(x) = f(-x + 3)$, the graph of g can be obtained by reflecting the graph of f in the y-axis and shifting f three units to the right. (**Note:** This is equivalent to shifting f three units to the left and then reflecting the graph in the y-axis.)

33.
$$f(x) = e^x$$
$$f(1.9) = e^{1.9} \approx 6.686$$

35. $f(6) = 5000e^{0.06(6)} \approx 7166.647$

37. $f(x) = 3e^{x+4}$

x	-8	-7	-6	-5	-4
$f(x)$	0.055	0.149	0.406	1.104	3

Asymptote: $y = 0$

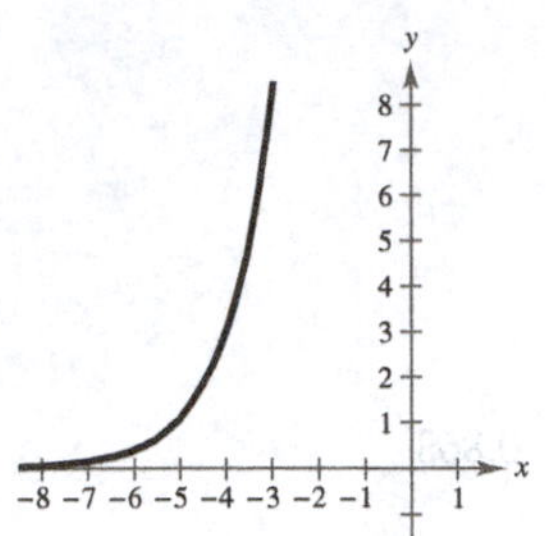

39. $f(x) = 2e^{x-2} + 4$

x	-2	-1	0	1	2
$f(x)$	4.037	4.100	4.271	4.736	6

Asymptote: $y = 4$

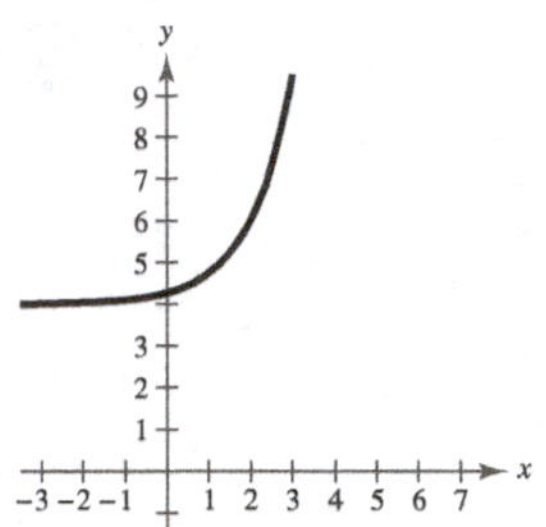

41. $s(t) = 2e^{0.5t}$

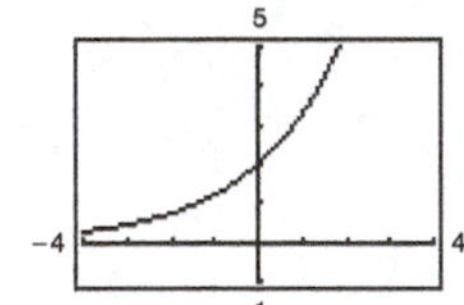

43. $g(x) = 1 + e^{-x}$

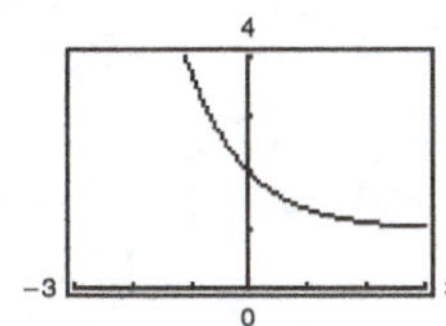

45. $e^{3x+2} = e^3$

$3x + 2 = 3$

$3x = 1$

$x = \frac{1}{3}$

47. $e^{x^2-3} = e^{2x}$

$x^2 - 3 = 2x$

$x^2 - 2x - 3 = 0$

$(x - 3)(x + 1) = 0$

$x = 3 \quad \text{or} \quad x = -1$

49. $P = \$1500, r = 2\%, t = 10$ years

Compounded n times per year: $A = P\left(1 + \frac{r}{n}\right)^{nt} = 1500\left(1 + \frac{0.02}{n}\right)^{10n}$

Compounded continuously: $A = Pe^{rt} = 1500e^{0.02(10)}$

n	1	2	4	12	365	Continuous
A	$1828.49	$1830.29	$1831.19	$1831.80	$1832.09	$1832.10

51. $P = \$2500, r = 4\%, t = 20$ years

Compounded n times per year: $A = P\left(1 + \frac{r}{n}\right)^{nt} = 2500\left(1 + \frac{0.04}{n}\right)^{20n}$

Compounded continuously: $A = Pe^{rt} = 2500e^{0.04(20)}$

n	1	2	4	12	365	Continuous
A	$5477.81	$5520.10	$5541.79	$5556.46	$5563.61	$5563.85

53. $A = Pe^{rt} = 12{,}000e^{0.04t}$

t	10	20	30	40	50
A	$17,901.90	$26,706.49	$39,841.40	$59,436.39	$88,668.67

55. $A = Pe^{rt} = 12{,}000e^{0.065t}$

t	10	20	30	40	50
A	$22,986.49	$44,031.56	$84,344.25	$161,564.86	$309,484.08

57. $A = 30{,}000e^{(0.05)(25)} \approx \$104{,}710.29$

59. $C(t) = 29.88(1.04)^t$

Ten years from today, $t = 10$: $C(10) = 29.88(1.04)^{10} \approx \$44.23$

61. (a)

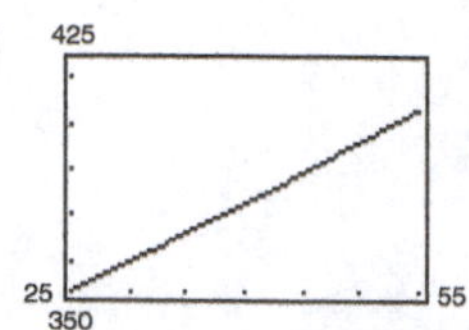

(b)

t	25	26	27	28
P (in millions)	350.281	352.107	353.943	355.788

t	29	30	31	32
P (in millions)	357.643	359.508	361.382	363.266

t	33	34	35	36
P (in millions)	365.160	367.064	368.977	370.901

t	37	38	39	40
P (in millions)	372.835	374.779	376.732	378.697

t	41	42	43	44
P (in millions)	380.671	382.656	384.651	386.656

t	45	46	47	48
P (in millions)	388.672	390.698	392.735	394.783

t	49	50	51	52
P (in millions)	396.841	398.910	400.989	403.080

t	53	54	55
P (in millions)	405.182	407.294	409.417

(c) Using the model and extending the table beyond the year 2055, the population will exceed 430 million in 2064.

t	55	56	57	58	59	60	61	62	63	64	65
P (in millions)	409.417	411.552	413.698	415.854	418.022	420.202	422.393	424.595	426.808	429.034	431.270

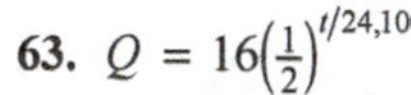

63. $Q = 16\left(\frac{1}{2}\right)^{t/24,100}$

(a) $Q(0) = 16$ grams

(b) $Q(75{,}000) \approx 1.85$ grams

(c)

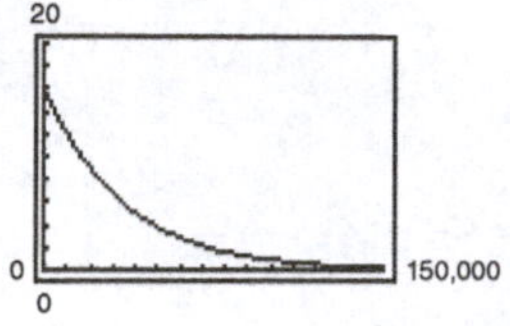

65. (a) $V(t) = 49{,}810\left(\frac{7}{8}\right)^t$ where t is the number of years since it was purchased.

(b) $V(4) = 49{,}810\left(\frac{7}{8}\right)^4 \approx 29{,}197.71$

After 4 years, the value of the van is about $29,198.

67. True. The line $y = -2$ is a horizontal asymptote for the graph of $f(x) = 10^x - 2$. As $x \to -\infty$, $f(x) \to -2$ but never reaches -2.

69. $f(x) = 3^{x-2} = 3^x 3^{-2} = 3^x\left(\frac{1}{3^2}\right) = \frac{1}{9}(3^x) = h(x)$

So, $f(x) \neq g(x)$, but $f(x) = h(x)$.

71. $f(x) = 16(4^{-x})$ and $f(x) = 16(4^{-x})$

$$= 4^2(4^{-x}) \qquad = 16(2^2)^{-x}$$

$$= 4^{2-x} \qquad = 16(2^{-2x})$$

$$= \left(\tfrac{1}{4}\right)^{-(2-x)} \qquad = h(x)$$

$$= \left(\tfrac{1}{4}\right)^{x-2}$$

$$= g(x)$$

So, $f(x) = g(x) = h(x)$.

73. $y = 3^x$ and $y = 4^x$

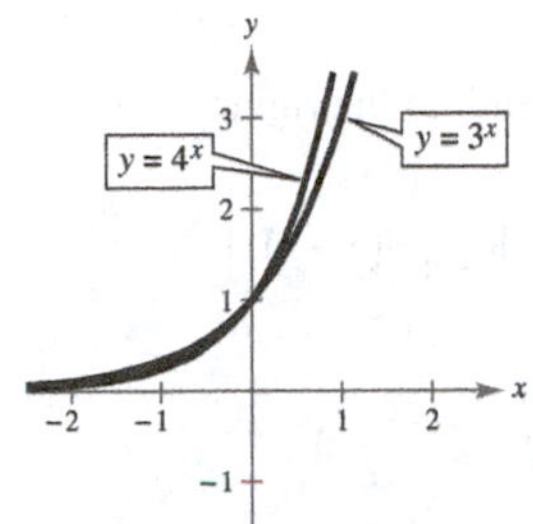

x	-2	-1	0	1	2
3^x	$\frac{1}{9}$	$\frac{1}{3}$	1	3	9
4^x	$\frac{1}{16}$	$\frac{1}{4}$	1	4	16

(a) $4^x < 3^x$ when $x < 0$.

(b) $4^x > 3^x$ when $x > 0$.

75. 

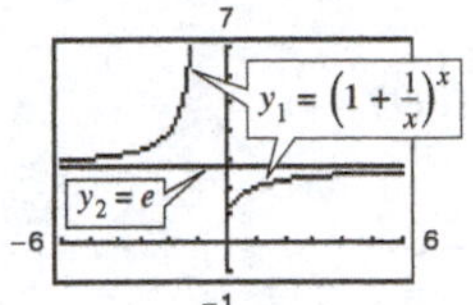

As x increases, the graph of y_1 approaches e, which is y_2.

77. (a)

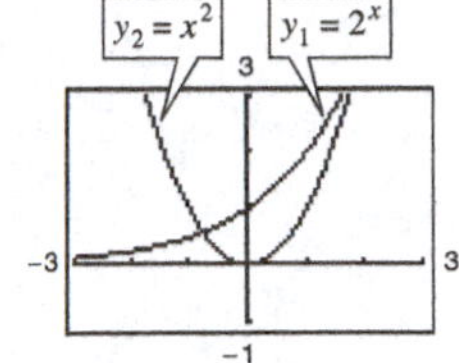

At $x = 2$, both functions have a value of 4. The function y_1 increases for all values of x. The function y_2 is symmetric with respect to the y-axis.

(b)

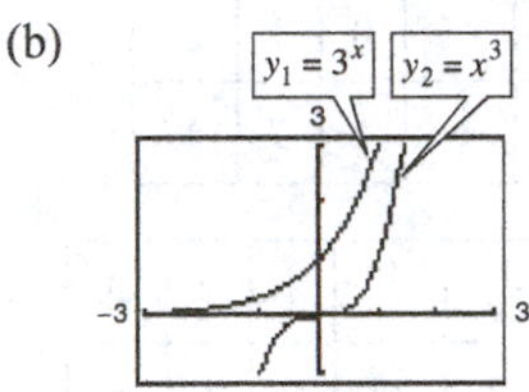

Both functions are increasing for all values of x. For $x > 0$, both functions have a similar shape. The function y_2 is symmetric with respect to the origin.

In both viewing windows, the constant raised to a variable power increases more rapidly than the variable raised to a constant power.

79. The functions (c) $h(x) = 3^x$ and (d) $k(x) = 2^{-x}$ are exponential.

Section 5.2 Logarithmic Functions and Their Graphs

1. logarithmic

3. natural; e

5. $x = y$

7. $\log_4 16 = 2 \Rightarrow 4^2 = 16$

9. $\log_{12} 12 = 1 \Rightarrow 12^1 = 12$

11. $5^3 = 125 \Rightarrow \log_5 125 = 3$

13. $4^{-3} = \frac{1}{64} \Rightarrow \log_4 \frac{1}{64} = -3$

15. $f(x) = \log_2 x$

$f(64) = \log_2 64 = 6$ because $2^6 = 64$

17. $f(x) = \log_8 x$

$f(1) = \log_8 1 = 0$ because $8^0 = 1$

19. $g(x) = \log_a x$

$g(a^2) = \log_a a^{-2} = -2$

21. $f(x) = \log x$

$f\left(\frac{7}{8}\right) = \log\left(\frac{7}{8}\right) \approx -0.058$

23. $f(x) = \log x$

$f(12.5) = \log 12.5 \approx 1.097$

25. $\log_8 8 = 1$ because $8^1 = 8$

27. $\log_{7.5} 1 = 0$ because $7.5^0 = 1$

29. $\log_5(x + 1) = \log_5 6$

$x + 1 = 6$

$x = 5$

31. $\log 11 = \log(x^2 + 7)$

$11 = x^2 + 7$

$x^2 = 4$

$x = \pm 2$

33.

x	-2	-1	0	1	2
$f(x) = 7^x$	$\frac{1}{49}$	$\frac{1}{7}$	1	7	49

x	$\frac{1}{49}$	$\frac{1}{7}$	1	7	49
$g(x) = \log_7 x$	-2	-1	0	1	2

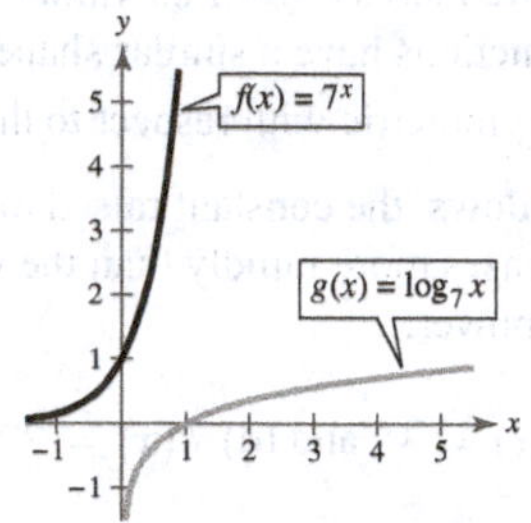

35.

x	-2	-1	0	1	2
$f(x) = 6^x$	$\frac{1}{36}$	$\frac{1}{6}$	1	6	36

x	$\frac{1}{36}$	$\frac{1}{6}$	1	6	36
$g(x) = \log_6 x$	-2	-1	0	1	2

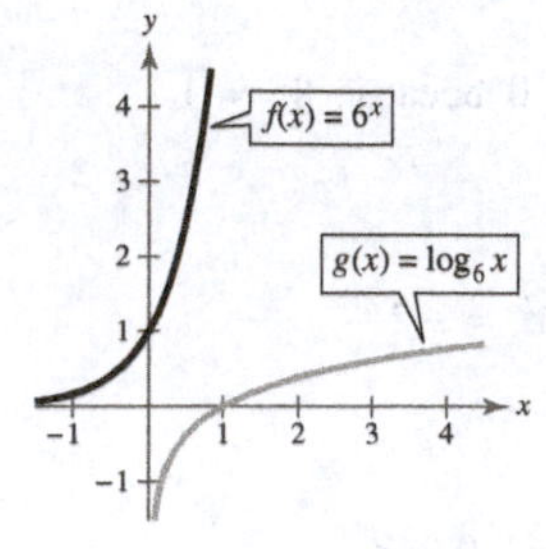

37. $f(x) = \log_3 x + 2$

Asymptote: $x = 0$

Point on graph: $(1, 2)$

Matches graph (a).

The graph of $f(x)$ is obtained from $g(x)$ by shifting the graph two units upward.

38. $f(x) = \log_3(x - 1)$

Asymptote: $x = 1$

Point on graph: $(2, 0)$

Matches graph (d).

$f(x)$ shifts $g(x)$ one unit to the right.

39. $f(x) = \log_3(1 - x) = \log_3[-(x - 1)]$

Asymptote: $x = 1$

Point on graph: $(0, 0)$

Matches graph (b).

The graph of $f(x)$ is obtained by reflecting the graph of $g(x)$ in the y-axis and shifting the graph one unit to the right.

40. $f(x) = -\log_3 x$

Asymptote: $x = 0$

Point on graph: $(1, 0)$

Matches graph (c).

$f(x)$ reflects $g(x)$ in the x-axis.

41. $f(x) = \log_4 x$

Domain: $(0, \infty)$

x-intercept: $(1, 0)$

Vertical asymptote: $x = 0$

$y = \log_4 x \Rightarrow 4^y = x$

x	$\frac{1}{4}$	1	4	2
$f(x)$	-1	0	1	$\frac{1}{2}$

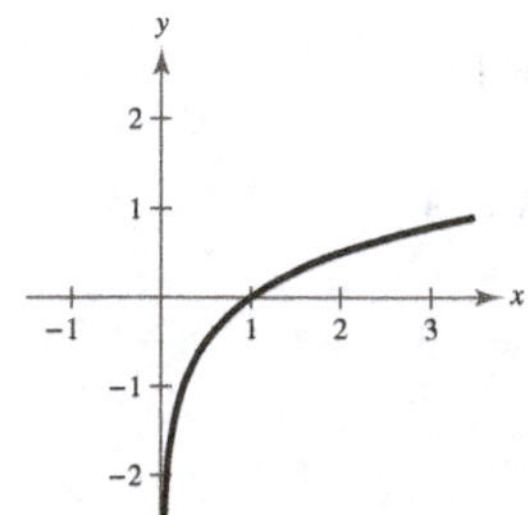

43. $y = \log_3 x + 1$

Domain: $(0, \infty)$

x-intercept:

$\log_3 x + 1 = 0$

$\log_3 x = -1$

$3^{-1} = x$

$\frac{1}{3} = x$

The x-intercept is $\left(\frac{1}{3}, 0\right)$.

Vertical asymptote: $x = 0$

$y = \log_3 x + 1$

$\log_3 x = y - 1 \Rightarrow 3^{y-1} = x$

x	$\frac{1}{9}$	$\frac{1}{3}$	0	3	9
y	−1	0	1	2	3

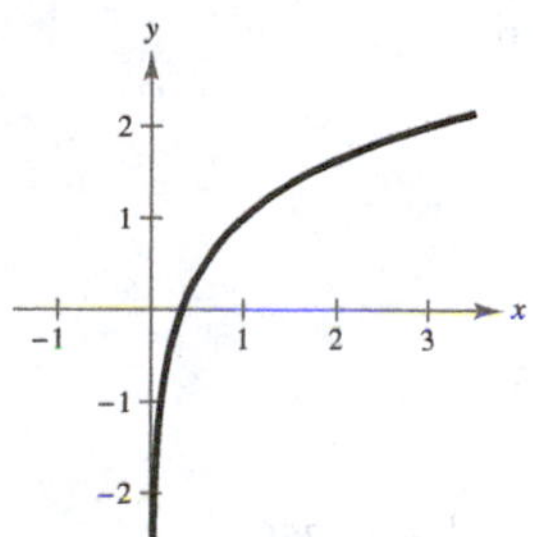

45. $f(x) = -\log_6(x + 2)$

Domain: $x + 2 > 0 \Rightarrow x > -2$

The domain is $(-2, \infty)$.

x-intercept:

$0 = -\log_6(x + 2)$

$0 = \log_6(x + 2)$

$6^0 = x + 2$

$1 = x + 2$

$-1 = x$

The x-intercept is $(-1, 0)$.

Vertical asymptote: $x + 2 = 0 \Rightarrow x = -2$

$y = -\log_6(x + 2)$

$-y = \log_6(x + 2)$

$6^{-y} - 2 = x$

x	4	−1	$-1\frac{5}{6}$	$-1\frac{35}{36}$
$f(x)$	−1	0	1	2

47. $y = \log\left(\frac{x}{7}\right)$

Domain: $\frac{x}{7} > 0 \Rightarrow x > 0$

The domain is $(0, \infty)$.

x-intercept: $\log\left(\frac{x}{7}\right) = 0$

$\frac{x}{7} = 10^0$

$\frac{x}{7} = 1$

$x = 7$

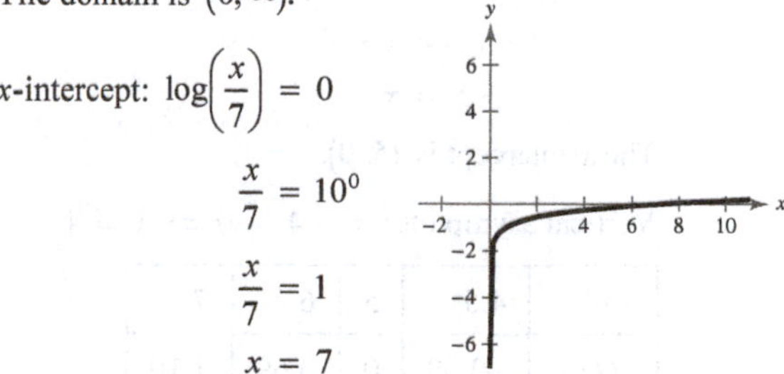

The x-intercept is $(7, 0)$.

Vertical asymptote: $\frac{x}{7} = 0 \Rightarrow x = 0$

The vertical asymptote is the y-axis.

x	1	2	3	4	5
y	−0.85	−0.54	−0.37	−0.24	−0.15

x	6	7	8
y	−0.069	0	0.06

49. $\ln \frac{1}{2} = -0.693\ldots \Rightarrow e^{-0.693\ldots} = \frac{1}{2}$

51. $\ln 250 = 5.521\ldots \Rightarrow e^{5.521\ldots} = 250$

53. $e^2 = 7.3890\ldots \Rightarrow \ln 7.3890\ldots = 2$

55. $e^{-4x} = \frac{1}{2} \Rightarrow \ln \frac{1}{2} = -4x$

57. $f(x) = \ln x$

$f(18.42) = \ln 18.42 \approx 2.913$

59. $g(x) = 8 \ln x$

$g\left(\sqrt{5}\right) = 8 \ln \sqrt{5} \approx 6.438$

61. $e^{\ln 4} = 4$

63. $2.5 \ln 1 = 2.5(0) = 0$

65. $\ln e^{\ln e} = \ln e^1 = 1$

67. $f(x) = \ln(x - 4)$

Domain: $x - 4 > 0 \Rightarrow x > 4$

The domain is $(4, \infty)$.

x-intercept: $0 = \ln(x - 4)$

$e^0 = x - 4$

$5 = x$

The x-intercept is $(5, 0)$.

Vertical asymptote: $x - 4 = 0 \Rightarrow x = 4$

x	4.5	5	6	7
$f(x)$	–0.69	0	0.69	1.10

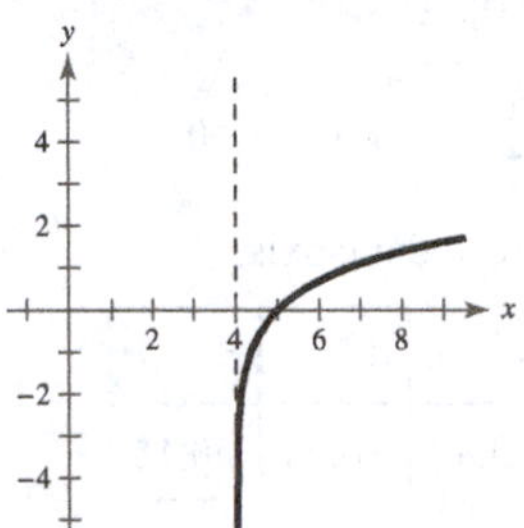

69. $g(x) = \ln(-x)$

Domain: $-x > 0 \Rightarrow x < 0$

The domain is $(-\infty, 0)$.

x-intercept:

$0 = \ln(-x)$

$e^0 = -x$

$-1 = x$

The x-intercept is $(-1, 0)$.

Vertical asymptote: $-x = 0 \Rightarrow x = 0$

x	–0.5	–1	–2	–3
$g(x)$	–0.69	0	0.69	1.10

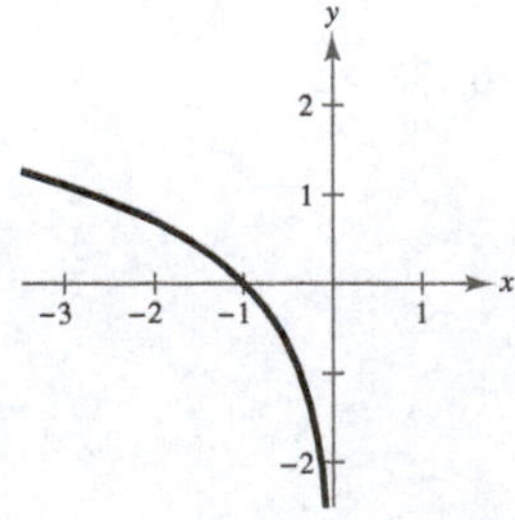

71. $f(x) = \ln(x - 1)$

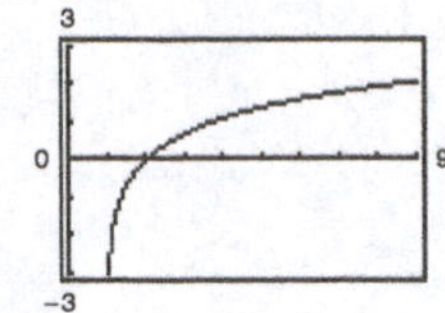

73. $f(x) = -\ln x + 8$

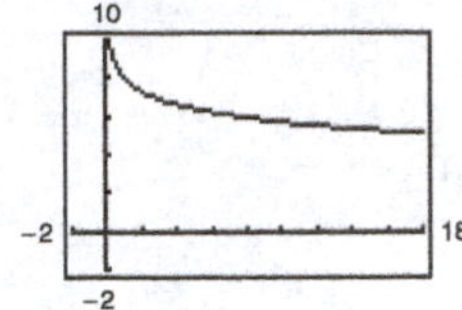

75. $\ln(x + 4) = \ln 12$

$x + 4 = 12$

$x = 8$

77. $\ln(x^2 - x) = \ln 6$

$x^2 - x = 6$

$x^2 - x - 6 = 0$

$(x - 3)(x + 2) = 0$

$x = -2$ or $x = 3$

79. $t = 16.625 \ln\left(\dfrac{x}{x - 750}\right), x > 750$

(a) When $x = \$897.72$:

$$t = 16.625 \ln\left(\frac{897.72}{897.72 - 750}\right) \approx 30 \text{ years}$$

When $x = \$1659.24$:

$$t = 16.625 \ln\left(\frac{1659.24}{1659.24 - 750}\right) \approx 10 \text{ years}$$

(b) Total amounts:

$(897.72)(12)(30) = \$323{,}179.20 \approx \$323{,}179$

$(1659.24)(12)(10) = \$199{,}108.80 \approx \$199{,}109$

Interest charges:

$323{,}179.20 - 150{,}000 = \$173{,}179.20 \approx \$173{,}179$

$199{,}108.80 - 150{,}000 = \$49{,}108.80 \approx \$49{,}109$

(c) The vertical asymptote is $x = 750$. The closer the payment is to \$750 per month, the longer the length of the mortgage will be. Also, the monthly payment must be greater than \$750.

81. $t = \dfrac{\ln 2}{r}$

(a)

r	0.005	0.010	0.015	0.020	0.025	0.030
t	138.6	69.3	46.2	34.7	27.7	23.1

(b)

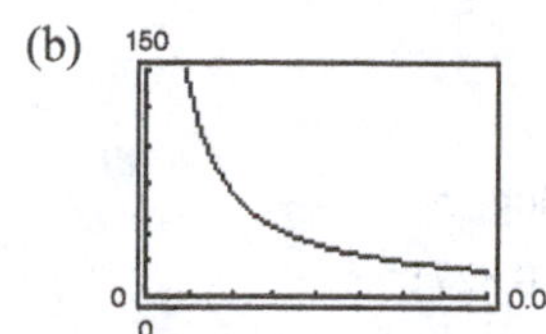

83. $f(t) = 80 - 17\log(t + 1),\ 0 \le t \le 12$

(a)

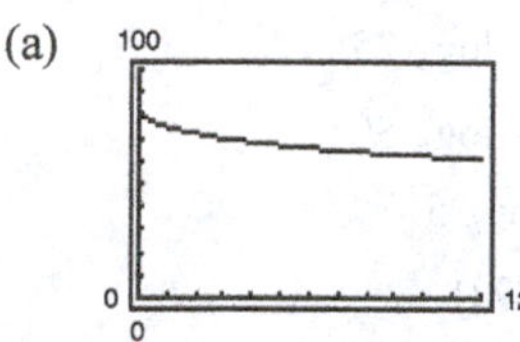

(b) $f(0) = 80 - 17\log 1 = 80.0$

(c) $f(4) = 80 - 17\log 5 \approx 68.1$

(d) $f(10) = 80 - 17\log 11 \approx 62.3$

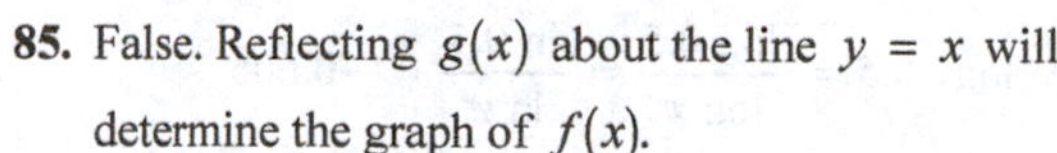

85. False. Reflecting $g(x)$ about the line $y = x$ will determine the graph of $f(x)$.

87. (a) $f(x) = \ln x,\ g(x) = \sqrt{x}$

The natural log function grows at a slower rate than the square root function.

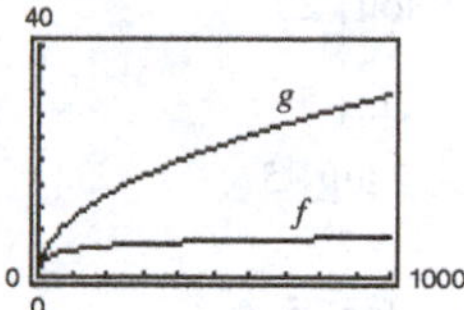

(b) $f(x) = \ln x,\ g(x) = \sqrt[4]{x}$

The natural log function grows at a slower rate than the fourth root function.

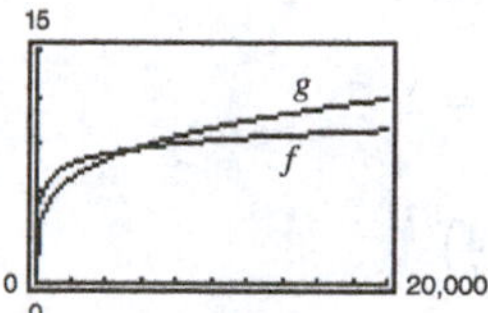

89.

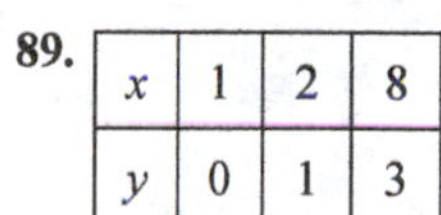

x	1	2	8
y	0	1	3

y is not an exponential function of x, but it is a logarithmic function of x, $y = \log_2 x$.

91. $f(x) = \dfrac{\ln x}{x}$

(a)

x	1	5	10	10^2	10^4	10^6
$f(x)$	0	0.322	0.230	0.046	0.00092	0.0000138

(b) As $x \to \infty$, $f(x) \to 0$.

(c)

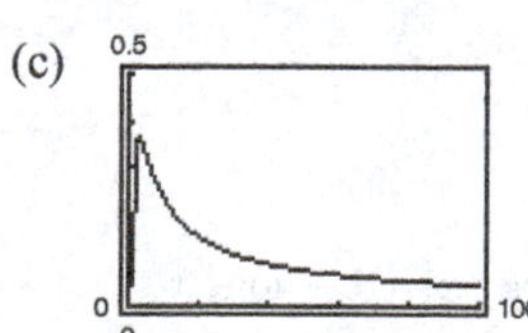

Section 5.3 Properties of Logarithms

1. change-of-base

3. $\dfrac{1}{\log_b a}$

5. (a) $\log_5 16 = \dfrac{\log 16}{\log 5}$

(b) $\log_5 16 = \dfrac{\ln 16}{\ln 5}$

7. (a) $\log_x \dfrac{3}{10} = \dfrac{\log(3/10)}{\log x}$

(b) $\log_x \dfrac{3}{10} = \dfrac{\ln(3/10)}{\ln x}$

9. $\log_3 17 = \dfrac{\log 17}{\log 3} = \dfrac{\ln 17}{\ln 3} \approx 2.579$

11. $\log_\pi 0.5 = \dfrac{\log 0.5}{\log \pi} = \dfrac{\ln 0.5}{\ln \pi} \approx -0.606$

13. $\log_3 35 = \log_3 (5 \cdot 7)$
$= \log_3 5 + \log_3 7$

15. $\log_3 \left(\frac{7}{25}\right) = \log_3 7 - \log_3 25$
$= \log_3 7 - \log_3 5^2$
$= \log_3 7 - 2 \log_3 5$

17. $\log_3 \left(\frac{21}{5}\right) = \log_3 21 - \log_3 5$
$= \log_3 (3 \cdot 7) - \log_3 5$
$= \log_3 3 + \log_3 7 - \log_3 5$
$= 1 + \log_3 7 - \log_3 5$

19. $\log_3 9 = 2 \log_3 3 = 2$

21. $\log_6 \sqrt[3]{\frac{1}{6}} = \log_6 \left(\frac{1}{6}\right)^{1/3}$
$= \frac{1}{3} \log_6 \left(\frac{1}{6}\right)$
$= \frac{1}{3} \log_6 6^{-1}$
$= \frac{1}{3}(-1)$
$= -\frac{1}{3}$

23. $\log_2(-2)$ is undefined. -2 is not in the domain of $\log_2 x$.

25. $\ln \sqrt[4]{e^3} = \ln e^{3/4}$
$= \frac{3}{4} \ln e$
$= \frac{3}{4}(1)$
$= \frac{3}{4}$

27. $\ln e^2 + \ln e^5 = 2 + 5 = 7$

29. $\log_5 75 - \log_5 3 = \log_5 \frac{75}{3}$
$= \log_5 25$
$= \log_5 5^2$
$= 2 \log_5 5$
$= 2$

31. $\log_4 8 = \log_4 (4 \cdot 2)$
$= \log_4 4 + \log_4 2$
$= \log_4 4 + \log_4 4^{1/2}$
$= 1 + \frac{1}{2}$
$= \frac{3}{2}$

33. $\log_b 10 = \log_b 2.5$
$= \log_b 2 + \log_b 5$
$\approx 0.3562 + 0.8271$
$= 1.1833$

35. $\log_b 0.04 = \log_b \frac{4}{100} = \log_b \frac{1}{25}$
$= \log_b 1 - \log_b 25$
$= \log_b 1 - \log_b 5^2$
$= 0 - 2 \log_b 5$
$\approx -2(0.8271)$
$= -1.6542$

37. $\log_b 45 = \log_b 9.5$
$= \log_b 9 + \log_b 5$
$= \log_b 3^2 + \log_b 5$
$= 2\log_b 3 + \log_b 5$
$\approx 2(0.5646) + 0.8271$
$= 1.9563$

39. $\log_b (2b)^{-2} = -2 \log_b 2b$
$= -2(\log_b 2 + \log_b b)$
$\approx -2(0.3562 + 1)$
$= -2.7124$

41. $\ln 7x = \ln 7 + \ln x$

43. $\log_8 x^4 = 4 \log_8 x$

45. $\log_5 \dfrac{5}{x} = \log_5 5 - \log_5 x = 1 - \log_5 x$

47. $\ln \sqrt{z} = \ln z^{1/2} = \frac{1}{2} \ln z$

49. $\ln xyz^2 = \ln x + \ln y + \ln z^2 = \ln x + \ln y + 2 \ln z$

51. $\ln z(z - 1)^2 = \ln z + \ln(z - 1)^2$
$= \ln z + 2 \ln(z - 1), z > 1$

53. $\log_2\left(\dfrac{\sqrt{a^2-4}}{7}\right) = \log_2\sqrt{a^2-4} - \log_2 7$

$= \log_2\left(a^2-4\right)^{1/2} - \log_2 7$

$= \frac{1}{2}\log_2\left(a^2-4\right) - \log_2 7$

$= \frac{1}{2}\log_2\left[(a-2)(a+2)\right] - \log_2 7$

$= \frac{1}{2}\left[\log_2(a-2) + \log_2(a+2)\right] - \log_2 7$

$= \frac{1}{2}\log_2(a-2) + \frac{1}{2}\log_2(a+2) - \log_2 7$

55. $\log_5\left(\dfrac{x^2}{y^2z^3}\right) = \log_5 x^2 - \log_5 y^2z^3$

$= \log_5 x^2 - \left(\log_5 y^2 + \log_5 z^3\right)$

$= 2\log_5 x - 2\log_5 y - 3\log_5 z$

57. $\ln\sqrt[3]{\dfrac{yz}{x^2}} = \ln\left(\dfrac{yz}{x^2}\right)^{1/3}$

$= \dfrac{1}{3}\ln\left(\dfrac{yz}{x^2}\right)$

$= \dfrac{1}{3}\left[\ln(yz) - \ln x^2\right]$

$= \dfrac{1}{3}\left[\ln(yz) - 2\ln x\right]$

$= \dfrac{1}{3}\left[\ln y + \ln z - 2\ln x\right]$

$= \dfrac{1}{3}\ln y + \dfrac{1}{3}\ln z - \dfrac{2}{3}\ln x$

59. $\ln\sqrt[4]{x^3\left(x^2+3\right)} = \frac{1}{4}\ln x^3\left(x^2+3\right)$

$= \frac{1}{4}\left[\ln x^3 + \ln\left(x^2+3\right)\right]$

$= \frac{1}{4}\left[3\ln x + \ln\left(x^2+3\right)\right]$

$= \frac{3}{4}\ln x + \frac{1}{4}\ln\left(x^2+3\right)$

61. $\ln 3 + \ln x = \ln(3x)$

63. $\frac{2}{3}\log_7(z-2) = \log_7(z-2)^{2/3}$

65. $\log_3 5x - 4\log_3 x = \log_3 5x - \log_3 x^4$

$= \log_3\left(\dfrac{5x}{x^4}\right)$

$= \log_3\left(\dfrac{5}{x^3}\right)$

67. $\log x + 2\log(x+1) = \log x + \log(x+1)^2$

$= \log\left[x(x+1)^2\right]$

69. $\log x - 2\log y + 3\log z = \log x - \log y^2 + \log z^3$

$= \log\dfrac{x}{y^2} + \log z^3$

$= \log\dfrac{xz^3}{y^2}$

71. $\ln x - \left[\ln(x+1) + \ln(x-1)\right] = \ln x - \ln(x+1)(x-1) = \ln\dfrac{x}{(x+1)(x-1)}$

73. $\frac{1}{2}\left[2\ln(x+3) + \ln x - \ln(x^2-1)\right] = \frac{1}{2}\left[\ln(x+3)^2 + \ln x - \ln(x^2-1)\right]$

$$= \frac{1}{2}\left[\ln\left[x(x+3)^2\right] - \ln(x^2-1)\right]$$

$$= \frac{1}{2}\left[\ln\left(\frac{x(x+3)^2}{x^2-1}\right)\right]$$

$$= \frac{1}{2}\ln\left[\frac{x(x+3)^2}{x^2-1}\right]$$

$$= \ln\sqrt{\frac{x(x+3)^2}{x^2-1}}$$

75. $\frac{1}{3}\left[\log_8 y + 2\log_8(y+4)\right] - \log_8(y-1) = \frac{1}{3}\left[\log_8 y + \log_8(y+4)^2\right] - \log_8(y-1)$

$$= \frac{1}{3}\log_8 y(y+4)^2 - \log_8(y-1)$$

$$= \log_8 \sqrt[3]{y(y+4)^2} - \log_8(y-1)$$

$$= \log_8\left(\frac{\sqrt[3]{y(y+4)^2}}{y-1}\right)$$

77. $\log_2 \frac{32}{4} = \log_2 32 - \log_2 4 \neq \frac{\log_2 32}{\log_2 4}$

The second and third expressions are equal by Property 2.

79. $\beta = 10\log\left(\frac{I}{10^{-12}}\right) = 10\left[\log I - \log 10^{-12}\right] = 10\left[\log I + 12\right] = 120 + 10\log I$

When $I = 10^{-6}$: $\beta = 120 + 10\log 10^{-6} = 120 + 10(-6) = 60$ decibels

81. $\beta = 10\log\left(\frac{I}{10^{-12}}\right)$

Difference $= 10\log\left(\frac{10^{-4}}{10^{-12}}\right) - 10\log\left(\frac{10^{-11}}{10^{-12}}\right) = 10\left[\log 10^8 - \log 10\right] = 10(8-1) = 10(7) = 70$ dB

83.

x	1	2	3	4	5	6
y	1.000	1.189	1.316	1.414	1.495	1.565
$\ln x$	0	0.693	1.099	1.386	1.609	1.792
$\ln y$	0	0.173	0.275	0.346	0.402	0.448

The slope of the line is $\frac{1}{4}$. So, $\ln y = \frac{1}{4}\ln x$

85.

x	1	2	3	4	5	6
y	2.500	2.102	1.900	1.768	1.672	1.597
$\ln x$	0	0.693	1.099	1.386	1.609	1.792
$\ln y$	0.916	0.743	0.642	0.570	0.514	0.468

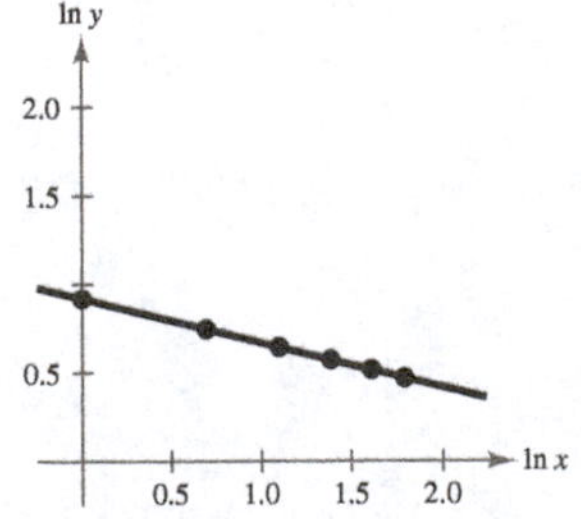

The slope of the line is $-\frac{1}{4}$. So, $\ln y = -\frac{1}{4}\ln x + \ln\frac{5}{2}$.

87.

Weight, x	25	35	50	75	500	1000
Galloping Speed, y	191.5	182.7	173.8	164.2	125.9	114.2
$\ln x$	3.219	3.555	3.912	4.317	6.215	6.908
$\ln y$	5.255	5.208	5.158	5.101	4.835	4.738

$\ln y = -0.14 \ln x + 5.7$

89. (a)

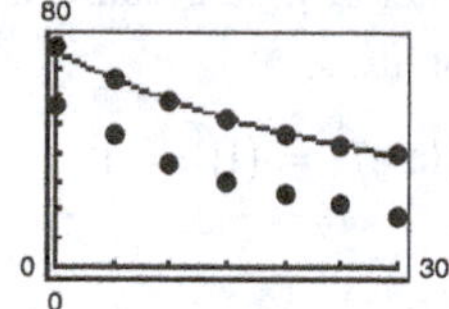

(b) $T - 21 = 54.4(0.964)^t$

$T = 54.4(0.964)^t + 21$

See graph in (a).

(c)

t (in minutes)	T (°C)	$T - 21$ (°C)	$\ln(T - 21)$	$1/(T - 21)$
0	78	57	4.043	0.0175
5	66	45	3.807	0.0222
10	57.5	36.5	3.597	0.0274
15	51.2	30.2	3.408	0.0331
20	46.3	25.3	3.231	0.0395
25	42.5	21.5	3.068	0.0465
30	39.6	18.6	2.923	0.0538

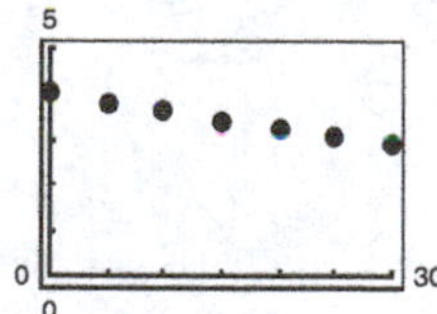

$\ln(T - 21) = -0.037t + 4$

$T = e^{-0.037t+3.997} + 21$

This graph is identical to T in (b).

(d) $\dfrac{1}{T - 21} = 0.0012t + 0.016$

$T = \dfrac{1}{0.001t + 0.016} + 21$

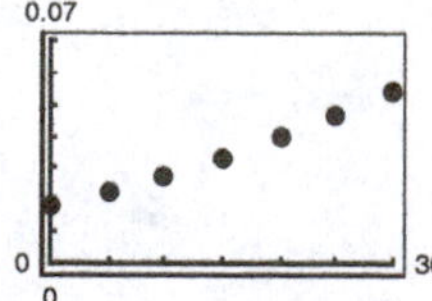

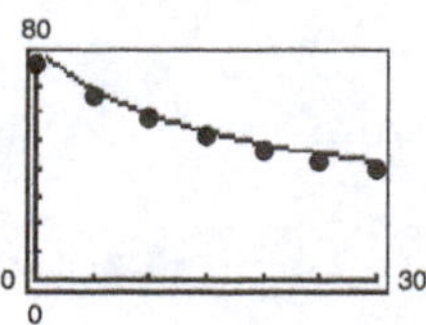

91. $f(x) = \ln x$

False, $f(0) \neq 0$ because 0 is not in the domain of $f(x)$.

$f(1) = \ln 1 = 0$

93. False.

$f(x) - f(2) = \ln x - \ln 2 = \ln \frac{x}{2} \neq \ln(x - 2)$

95. False.

$f(u) = 2f(v) \Rightarrow \ln u = 2 \ln v \Rightarrow \ln u = \ln v^2 \Rightarrow u = v^2$

97. $f(x) = \log_2 x = \dfrac{\log x}{\log 2} = \dfrac{\ln x}{\ln 2}$

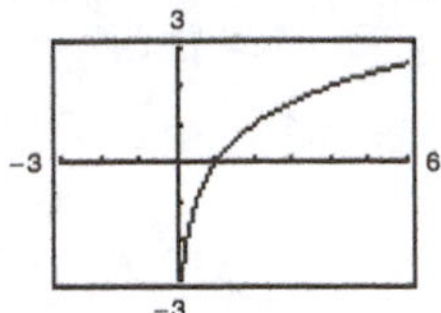

99. $f(x) = \log_{1/4} x = \dfrac{\log x}{\log(1/4)} = \dfrac{\ln x}{\ln(1/4)}$

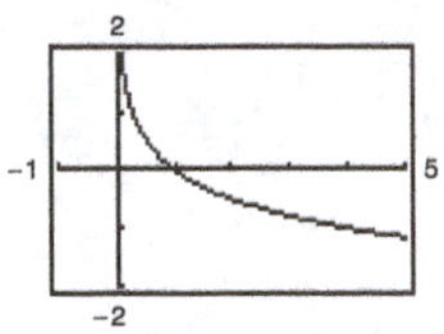

101. The power property cannot be used because $\ln e$ is raised to the second power, not just e.

A correct statement is $(\ln e)^2 = (1)^2 = 1$.

103.

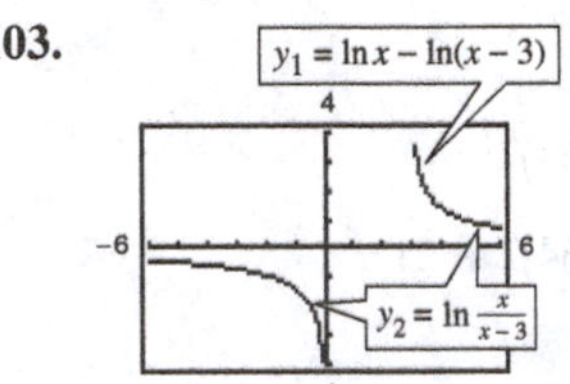

The graphing utility does not show the functions with the same domain. The domain of $y_1 = \ln x - \ln(x - 3)$ is $(3, \infty)$ and the domain of $y_2 = \ln \dfrac{x}{x - 3}$ is $(-\infty, 0) \cup (3, \infty)$.

105. $\ln 2 \approx 0.6931, \ln 3 \approx 1.0986, \ln 5 \approx 1.6094$

$\ln 1 = 0$

$\ln 2 \approx 0.6931$

$\ln 3 \approx 1.0986$

$\ln 4 = \ln(2 \cdot 2) = \ln 2 + \ln 2 \approx 0.6931 + 0.6931 = 1.3862$

$\ln 5 \approx 1.6094$

$\ln 6 = \ln(2 \cdot 3) = \ln 2 + \ln 3 \approx 0.6931 + 1.0986 = 1.7917$

$\ln 8 = \ln 2^3 = 3 \ln 2 \approx 3(0.6931) = 2.0793$

$\ln 9 = \ln 3^2 = 2 \ln 3 \approx 2(1.0986) = 2.1972$

$\ln 10 = \ln(5 \cdot 2) = \ln 5 + \ln 2 \approx 1.6094 + 0.6931 = 2.3025$

$\ln 12 = \ln(2^2 \cdot 3) = \ln 2^2 + \ln 3 = 2 \ln 2 + \ln 3 \approx 2(0.6931) + 1.0986 = 2.4848$

$\ln 15 = \ln(5 \cdot 3) = \ln 5 + \ln 3 \approx 1.6094 + 1.0986 = 2.7080$

$\ln 16 = \ln 2^4 = 4 \ln 2 \approx 4(0.6931) = 2.7724$

$\ln 18 = \ln(3^2 \cdot 2) = \ln 3^2 + \ln 2 = 2 \ln 3 + \ln 2 \approx 2(1.0986) + 0.6931 = 2.8903$

$\ln 20 = \ln(5 \cdot 2^2) = \ln 5 + \ln 2^2 = \ln 5 + 2 \ln 2 \approx 1.6094 + 2(0.6931) = 2.9956$

Section 5.4 Exponential and Logarithmic Equations

1. (a) $x = y$

(b) $x = y$

(c) x

(d) x

3. $4^{2x-7} = 64$

(a) $x = 5$

$4^{2(5)-7} = 4^3 = 64$

Yes, $x = 5$ *is* a solution.

(b) $x = 2$

$4^{2(2)-7} = 4^{-3} = \frac{1}{64} \neq 64$

No, $x = 2$ *is not* a solution.

(c) $x = \frac{1}{2}(\log_4 64 + 7)$

$4^{2(1/2(\log_4 64+7))-7} = 64$

$4^{(\log_4 64+7)-7} = 64$

$4^{(3+7)-7} = 64$

$4^3 = 64$

Yes, $x = \frac{1}{2}(\log_4 64 + 7)$ *is* a solution.

5. $\log_2(x + 3) = 10$

(a) $x = 1021$

$\log_2(1021 + 3) = \log_2(1024)$

Because $2^{10} = 1024$, $x = 1021$ *is* a solution.

(b) $x = 17$

$\log_2(17 + 3) = \log_2(20)$

Because $2^{10} \neq 20$, $x = 17$ *is not* a solution.

(c) $x = 10^2 - 3 = 97$

$\log_2(97 + 3) = \log_2(100)$

Because $2^{10} \neq 100$, $10^2 - 3$ *is not* a solution.

7. $4^x = 16$

$4^x = 4^2$

$x = 2$

9. $\ln x - \ln 2 = 0$

$\ln x = \ln 2$

$x = 2$

11. $e^x = 2$

$\ln e^x = \ln 2$

$x = \ln 2$

$x \approx 0.693$

13. $\ln x = -1$

$e^{\ln x} = e^{-1}$

$x = e^{-1}$

$x \approx 0.368$

15. $\log_4 x = 3$

$4^{\log_4 x} = 4^3$

$x = 4^3$

$x = 64$

17. $f(x) = g(x)$

$2^x = 8$

$2^x = 2^3$

$x = 3$

Point of intersection: $(3, 8)$

19. $e^x = e^{x^2-2}$

$x = x^2 - 2$

$0 = x^2 - x - 2$

$0 = (x + 1)(x - 2)$

$x = -1, x = 2$

21. $4(3^x) = 20$

$3^x = 5$

$\log_3 3^x = \log_3 5$

$x = \log_3 5 = \frac{\log 5}{\log 3}$ or $\frac{\ln 5}{\ln 3}$

$x \approx 1.465$

23. $e^x - 8 = 31$

$e^x = 39$

$\ln e^x = \ln 39$

$x = \ln 39 \approx 3.664$

25. $3^{2x} = 80$

$\ln 3^{2x} = \ln 80$

$2x \ln 3 = \ln 80$

$x = \frac{\ln 80}{2 \ln 3} \approx 1.994$

27.
$$\begin{aligned} 3^{2-x} &= 400 \\ \ln 3^{2-x} &= \ln 400 \\ (2-x)\ln 3 &= \ln 400 \\ 2\ln 3 - x\ln 3 &= \ln 400 \\ -x\ln 3 &= \ln 400 - 2\ln 3 \\ x\ln 3 &= 2\ln 3 - \ln 400 \\ x &= \frac{2\ln 3 - \ln 400}{\ln 3} \\ x &= 2 - \frac{\ln 400}{\ln 3} \approx -3.454 \end{aligned}$$

29.
$$\begin{aligned} 8\left(10^{3x}\right) &= 12 \\ 10^{3x} &= \tfrac{12}{8} \\ \log 10^{3x} &= \log\left(\tfrac{3}{2}\right) \\ 3x &= \log\left(\tfrac{3}{2}\right) \\ x &= \tfrac{1}{3}\log\left(\tfrac{3}{2}\right) \\ x &\approx 0.059 \end{aligned}$$

31.
$$\begin{aligned} e^{3x} &= 12 \\ 3x &= \ln 12 \\ x &= \frac{\ln 12}{3} \approx 0.828 \end{aligned}$$

33.
$$\begin{aligned} 7 - 2e^x &= 5 \\ -2e^x &= -2 \\ e^x &= 1 \\ x &= \ln 1 = 0 \end{aligned}$$

35.
$$\begin{aligned} 6\left(2^{3x-1}\right) - 7 &= 9 \\ 6\left(2^{3x-1}\right) &= 16 \\ 2^{3x-1} &= \frac{8}{3} \\ \log_2 2^{3x-1} &= \log_2\left(\frac{8}{3}\right) \\ 3x - 1 &= \log_2\left(\frac{8}{3}\right) = \frac{\log(8/3)}{\log 2} \text{ or } \frac{\ln(8/3)}{\ln 2} \\ x &= \frac{1}{3}\left[\frac{\log(8/3)}{\log 2} + 1\right] \approx 0.805 \end{aligned}$$

37.
$$\begin{aligned} 3^x &= 2^{x-1} \\ \ln 3^x &= \ln 2^{x-1} \\ x\ln 3 &= (x-1)\ln 2 \\ x\ln 3 &= x\ln 2 - \ln 2 \\ x\ln 3 - x\ln 2 &= -\ln 2 \\ x(\ln 3 - \ln 2) &= -\ln 2 \\ x &= \frac{\ln 2}{\ln 2 - \ln 3} \approx -1.710 \end{aligned}$$

39.
$$\begin{aligned} 4^x &= 5^{x^2} \\ \ln 4^x &= \ln 5^{x^2} \\ x\ln 4 &= x^2\ln 5 \\ x^2\ln 5 - x\ln 4 &= 0 \\ x(x\ln 5 - \ln 4) &= 0 \\ x &= 0 \\ x\ln 5 - \ln 4 = 0 \Rightarrow x &= \frac{\ln 4}{\ln 5} \approx 0.861 \end{aligned}$$

41.
$$\begin{aligned} e^{2x} - 4e^x - 5 &= 0 \\ \left(e^x + 1\right)\left(e^x - 5\right) &= 0 \end{aligned}$$
$e^x = -1$ or $e^x = 5$

(No solution) $x = \ln 5 \approx 1.609$

43.
$$\begin{aligned} \frac{1}{1 - e^x} &= 5 \\ 1 &= 5\left(1 - e^x\right) \\ \frac{1}{5} &= 1 - e^x \\ \frac{1}{5} - 1 &= -e^x \\ -\frac{4}{5} &= -e^x \\ \frac{4}{5} &= e^x \\ \ln\frac{4}{5} &= \ln e^x \\ \ln\frac{4}{5} &= x \\ x &\approx -0.223 \end{aligned}$$

45.
$$\begin{aligned} \left(1 + \frac{0.065}{365}\right)^{365t} &= 4 \\ \ln\left(1 + \frac{0.065}{365}\right)^{365t} &= \ln 4 \\ 365t\ln\left(1 + \frac{0.065}{365}\right) &= \ln 4 \\ t &= \frac{\ln 4}{365\ln\left(1 + \frac{0.065}{365}\right)} \approx 21.330 \end{aligned}$$

47.
$$\begin{aligned} \ln x &= -3 \\ x &= e^{-3} \approx 0.050 \end{aligned}$$

49.
$$\begin{aligned} 2.1 &= \ln 6x \\ e^{2.1} &= 6x \\ \frac{e^{2.1}}{6} &= x \\ 1.361 &\approx x \end{aligned}$$

51. $3 - 4 \ln x = 11$

$$-4 \ln x = 8$$
$$\ln x = -2$$
$$x = e^{-2} = \frac{1}{e^2} \approx 0.135$$

53. $6 \log_3(0.5x) = 11$

$$\log_3(0.5x) = \tfrac{11}{6}$$
$$3^{\log_3(0.5x)} = 3^{11/6}$$
$$0.5x = 3^{11/6}$$
$$x = 2\left(3^{11/6}\right) \approx 14.988$$

55. $\ln x - \ln(x + 1) = 2$

$$\ln\left(\frac{x}{x+1}\right) = 2$$
$$\frac{x}{x+1} = e^2$$
$$x = e^2(x + 1)$$
$$x = e^2x + e^2$$
$$x - e^2x = e^2$$
$$x\left(1 - e^2\right) = e^2$$
$$x = \frac{e^2}{1 - e^2} \approx -1.157$$

This negative value is extraneous. The equation has no solution.

57. $\ln(x + 5) = \ln(x - 1) - \ln(x + 1)$

$$\ln(x + 5) = \ln\left(\frac{x-1}{x+1}\right)$$
$$x + 5 = \frac{x-1}{x+1}$$
$$(x + 5)(x + 1) = x - 1$$
$$x^2 + 6x + 5 = x - 1$$
$$x^2 + 5x + 6 = 0$$
$$(x + 2)(x + 3) = 0$$
$$x = -2 \quad \text{or} \quad x = -3$$

Both of these solutions are extraneous, so the equation has no solution.

59. $\log(3x + 4) = \log(x - 10)$

$$3x + 4 = x - 10$$
$$2x = -14$$
$$x = -7$$

The negative value is extraneous.

The equation has no solution.

61. $\log_4 x - \log_4(x - 1) = \frac{1}{2}$

$$\log_4\left(\frac{x}{x-1}\right) = \frac{1}{2}$$
$$4^{\log_4[x/(x-1)]} = 4^{1/2}$$
$$\frac{x}{x-1} = 4^{1/2}$$
$$x = 2(x - 1)$$
$$x = 2x - 2$$
$$-x = -2$$
$$x = 2$$

63. $f(x) = 5^x - 212$

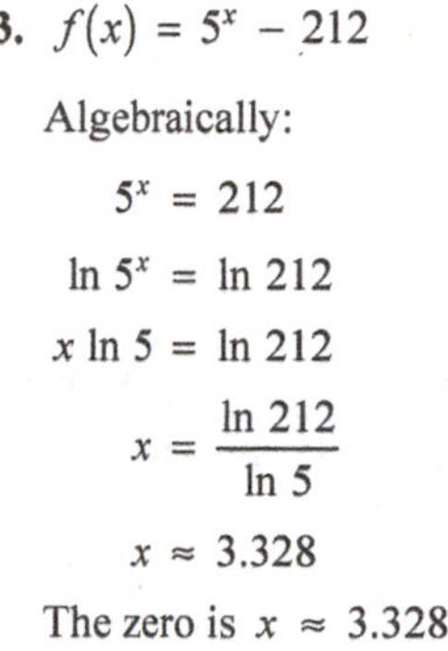

Algebraically:

$$5^x = 212$$
$$\ln 5^x = \ln 212$$
$$x \ln 5 = \ln 212$$
$$x = \frac{\ln 212}{\ln 5}$$
$$x \approx 3.328$$

The zero is $x \approx 3.328$.

65. $g(x) = 8e^{-2x/3} - 11$

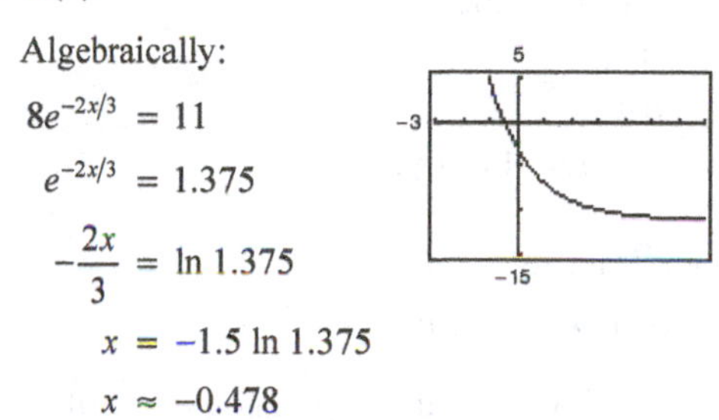

Algebraically:

$$8e^{-2x/3} = 11$$
$$e^{-2x/3} = 1.375$$
$$-\frac{2x}{3} = \ln 1.375$$
$$x = -1.5 \ln 1.375$$
$$x \approx -0.478$$

The zero is $x \approx -0.478$.

67. $y_1 = 3$

$y_2 = \ln x$

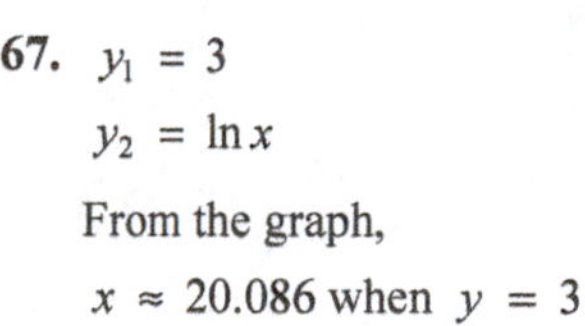

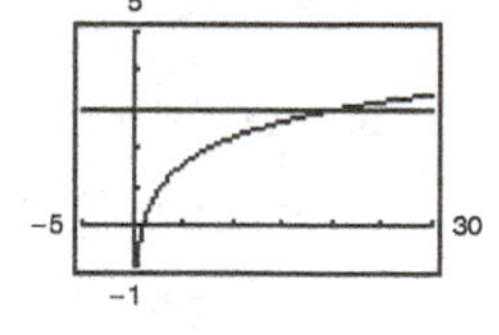

From the graph,

$x \approx 20.086$ when $y = 3$.

Algebraically:

$$3 - \ln x = 0$$
$$\ln x = 3$$
$$x = e^3 \approx 20.086$$

69. $y_1 = 2\ln(x + 3)$

$y_2 = 3$

From the graph, $x \approx 1.482$ when $y = 3$.

Algebraically:

$$2\ln(x + 3) = 3$$
$$\ln(x + 3) = \frac{3}{2}$$
$$x + 3 = e^{3/2}$$
$$x = e^{3/2} - 3 \approx 1.482$$

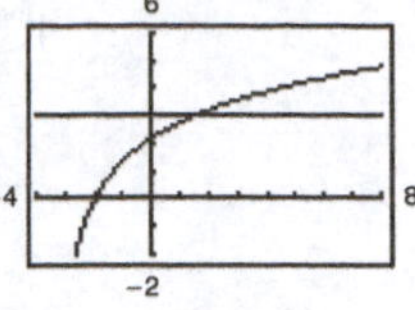

71. (a)

$$r = 0.025$$
$$A = Pe^{rt}$$
$$5000 = 2500e^{0.025t}$$
$$2 = e^{0.025t}$$
$$\ln 2 = 0.025t$$
$$\frac{\ln 2}{0.025} = t$$
$$t \approx 27.73 \text{ years}$$

(b)

$$r = 0.025$$
$$A = Pe^{rt}$$
$$7500 = 2500e^{0.025t}$$
$$3 = e^{0.025t}$$
$$\ln 3 = 0.025t$$
$$\frac{\ln 3}{0.025} = t$$
$$t \approx 43.94 \text{ years}$$

73. $2x^2e^{2x} + 2xe^{2x} = 0$

$$(2x^2 + 2x)e^{2x} = 0$$
$$2x^2 + 2x = 0 \quad (\text{because } e^{2x} \neq 0)$$
$$2x(x + 1) = 0$$
$$x = 0, -1$$

75. $-xe^{-x} + e^{-x} = 0$

$$(-x + 1)e^{-x} = 0$$
$$-x + 1 = 0 \quad (\text{because } e^{-x} \neq 0)$$
$$x = 1$$

77. $\frac{1 + \ln x}{2} = 0$

$$1 + \ln x = 0$$
$$\ln x = -1$$
$$x = e^{-1} = \frac{1}{e} \approx 0.368$$

79. $2x\ln x + x = 0$

$$x(2\ln x + 1) = 0$$
$$2\ln x + 1 = 0 \quad (\text{because } x > 0)$$
$$\ln x = -\frac{1}{2}$$
$$x = e^{-1/2} \approx 0.607$$

81. (a)

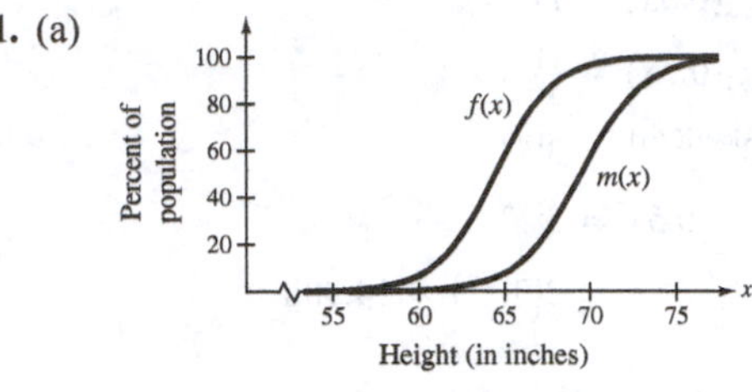

From the graph you see horizontal asymptotes at $y = 0$ and $y = 100$.

These represent the lower and upper percent bounds; the range falls between 0% and 100%.

(b) Males:

$$50 = \frac{100}{1 + e^{-0.5536(x-69.51)}}$$
$$1 + e^{-0.5536(x-69.51)} = 2$$
$$e^{-0.5536(x-69.51)} = 1$$
$$-0.5536(x - 69.51) = \ln 1$$
$$-0.5536(x - 69.51) = 0$$
$$x = 69.51$$

The average height of an American male is 69.51 inches.

Females:

$$50 = \frac{100}{1 + e^{-0.5834(x-64.49)}}$$
$$1 + e^{-0.5834(x-64.49)} = 2$$
$$e^{-0.5834(x-64.49)} = 1$$
$$-0.5834(x - 64.49) = \ln 1$$
$$-0.5834(x - 64.49) = 0$$
$$x = 64.49$$

The average height of an American female is 64.49 inches.

83. $N = 5.5 \cdot 10^{0.23x}$

When $N = 78$:

$$78 = 5.5 \cdot 10^{0.23x}$$
$$\frac{78}{5.5} = 10^{0.23x}$$
$$\log_{10}\frac{78}{5.5} = 0.23x$$
$$x = \frac{\log_{10}(78/5.5)}{0.23} \approx 5.008 \text{ years}$$

The beaver population will reach 78 in about 5 years.

85. $P = 75 \ln t + 540$

Let $P = 720$

$720 = 75 \ln t + 540$

$180 = 75 \ln t$

$\dfrac{180}{75} = \ln t$

$\ln t = 2.4$

$t = e^{2.4} \approx 11.02$ or 2011

87. $T = 20 + 60e^{-0.06m}$

Let $T = 70$

$70 = 20 + 60e^{-0.06m}$

$50 = 60e^{-0.06m}$

$\frac{5}{6} = e^{-0.06m}$

$\ln \frac{5}{6} = -0.06m$

$m = -\frac{1}{0.06} \ln \frac{5}{6}$

$m \approx 3.039$ minutes

89. $\log_a(uv) = \log_a u + \log_a v$

True by Property 1 in Section 5.3.

91. $\log_a(u - v) = \log_a u - \log_a v$

False.

$1.95 = \log(100 - 10)$

$\neq \log 100 - \log 10 = 1$

93. Yes, a logarithmic equation can have more than one extraneous solution. See Exercise 57.

95. $A = Pe^{rt}$

(a) $A = (2P)e^{rt} = 2(Pe^{rt})$ This doubles your money.

(b) $A = Pe^{(2r)t} = Pe^{rt}e^{rt} = e^{rt}(Pe^{rt})$

(c) $A = Pe^{r(2t)} = Pe^{rt}e^{rt} = e^{rt}(Pe^{rt})$

Doubling the interest rate yields the same result as doubling the number of years.

If $2 > e^{rt}$ (i.e., $rt < \ln 2$), then doubling your investment would yield the most money. If $rt > \ln 2$, then doubling either the interest rate or the number of years would yield more money.

97. (a) $P = 1000, r = 0.07$, compounded annually, $n = 1$

Effective yield: $A = P\left(1 + \dfrac{r}{n}\right)^{nt} = 1000\left(1 + \dfrac{0.07}{1}\right)^{1} = \$1070$

$\dfrac{1070 - 1000}{1000} = 7\%$

The effective yield is 7%.

Balance after 5 years: $A = P\left(1 + \dfrac{r}{n}\right)^{nt} = 1000\left(1 + \dfrac{0.07}{1}\right)^{1(5)} \approx \$1402.55$

(b) $P = 1000, r = 0.07$, compounded continuously

Effective yield: $A = Pe^{rt} = 1000e^{0.07(1)} \approx \$1072.51$

$\dfrac{1072.51 - 1000}{1000} = 7.25\%$

The effective yield is about 7.25%.

Balance after 5 years: $A = Pe^{rt} = 1000e^{0.07(5)} \approx \$1419.07$

(c) $P = 1000, r = 0.07$, compounded quarterly, $n = 4$

Effective yield: $A = P\left(1 + \dfrac{r}{n}\right)^{nt} = 1000\left(1 + \dfrac{0.07}{4}\right)^{4(1)} \approx \$1071.86$

$\dfrac{1071.86 - 1000}{1000} = 7.19\%$

The effective yield is about 7.19%.

Balance after 5 years: $A = P\left(1 + \dfrac{r}{n}\right)^{nt} = 1000\left(1 + \dfrac{0.07}{4}\right)^{4(5)} \approx \$1414.78$

(d) $P = 1000, r = 0.0725,$ compounded quarterly, $n = 4$

Effective yield: $A = P\left(1 + \frac{r}{n}\right)^{nt} = 1000\left(1 + \frac{0.0725}{4}\right)^{4(1)} \approx \$1074.50$

$\frac{1074.50 - 1000}{1000} \approx 7.45\%$

The effective yield is about 7.45%.

Balance after 5 years: $A = P\left(1 + \frac{r}{n}\right)^{nt} = 1000\left(1 + \frac{0.0725}{4}\right)^{4(5)} \approx \$1432.26$

Savings plan (d) has the greatest effective yield and the highest balance after 5 years.

Section 5.5 Exponential and Logarithmic Models

1. $y = ae^{bx}; y = ae^{-bx}$

3. normally distributed

5. (a) $A = Pe^{rt}$

$\frac{A}{e^{rt}} = P$

(b) $A = Pe^{rt}$

$\frac{A}{P} = e^{rt}$

$\ln\frac{A}{P} = \ln e^{rt}$

$\ln\frac{A}{P} = rt$

$\frac{\ln(A/P)}{r} = t$

7. Because $A = 1000e^{0.035t}$, the time to double is given by $2000 = 1000e^{0.035t}$ and you have

$2 = e^{0.035t}$

$\ln 2 = \ln e^{0.035t}$

$\ln 2 = 0.035t$

$t = \frac{\ln 2}{0.035} \approx 19.8$ years.

Amount after 10 years: $A = 1000e^{0.35} \approx \$1419.07$

9. Because $A = 750e^{rt}$ and $A = 1500$ when $t = 7.75$, you have

$1500 = 750e^{7.75r}$

$2 = e^{7.75r}$

$\ln 2 = \ln e^{7.75r}$

$\ln 2 = 7.75r$

$r = \frac{\ln 2}{7.75} \approx 0.089438 = 8.9438\%.$

Amount after 10 years: $A = 750e^{0.089438(10)} \approx \$1834.37$

11. Because $A = Pe^{0.045t}$ and $A = 10{,}000.00$ when $t = 10$, you have

$10{,}000.00 = Pe^{0.045(10)}$

$\frac{10{,}000.00}{e^{0.045(10)}} = P \approx \$6376.28.$

The time to double is given by

$t = \frac{\ln 2}{0.045} \approx 15.40$ years.

13. $A = 500{,}000, r = 0.05, n = 12, t = 10$

$A = P\left(1 + \frac{r}{n}\right)^{nt}$

$500{,}000 = P\left(1 + \frac{0.05}{12}\right)^{12(10)}$

$P = \frac{500{,}000}{\left(1 + \frac{0.05}{12}\right)^{12(10)}}$

$\approx \$303{,}580.52$

15. $P = 1000, r = 0.1, A = 2000$

$$A = P\left(1 + \frac{r}{n}\right)^{nt}$$

$$2000 = 1000\left(1 + \frac{0.1}{n}\right)^{nt}$$

$$2 = \left(1 + \frac{0.1}{n}\right)^{nt}$$

(a) $n = 1$

$$(1 + 0.1)^t = 2$$
$$(1.1)^t = 2$$
$$\ln(1.1)^t = \ln 2$$
$$t \ln 1.1 = \ln 2$$
$$t = \frac{\ln 2}{\ln 1.1} \approx 7.27 \text{ years}$$

(b) $n = 12$

$$\left(1 + \frac{0.1}{12}\right)^{12t} = 2$$
$$\ln\left(\frac{12.1}{12}\right)^{12t} = \ln 2$$
$$12t \ln\left(\frac{12.1}{12}\right) = \ln 2$$
$$12t = \frac{\ln 2}{\ln(12.1/12)}$$
$$t = \frac{\ln 2}{12 \ln(12.1/12)} \approx 6.96 \text{ years}$$

(c) $n = 365$

$$\left(1 + \frac{0.1}{365}\right)^{365t} = 2$$
$$\ln\left(\frac{365.1}{365}\right)^{365t} = \ln 2$$
$$365t \ln\left(\frac{365.1}{365}\right) = \ln 2$$
$$365t = \frac{\ln 2}{\ln(365.1/365)}$$
$$t = \frac{\ln 2}{365 \ln(365.1/365)} \approx 6.93 \text{ years}$$

(d) Compounded continuously

$$A = Pe^{rt}$$
$$2000 = 1000e^{0.1t}$$
$$2 = e^{0.1t}$$
$$\ln 2 = \ln e^{0.1t}$$
$$0.1t = \ln 2$$
$$t = \frac{\ln 2}{0.1} \approx 6.93 \text{ years}$$

17. (a)

$$3P = Pe^{rt}$$
$$3 = e^{rt}$$
$$\ln 3 = rt$$
$$\frac{\ln 3}{r} = t$$

r	2%	4%	6%	8%	10%	12%
$t = \frac{\ln 3}{r}$ (years)	54.93	27.47	18.31	13.73	10.99	9.16

(b)

$$3P = P(1 + r)^t$$
$$3 = (1 + r)^t$$
$$\ln 3 = \ln (1 + r)^t$$
$$\frac{\ln 3}{\ln (1 + r)} = t$$

r	2%	4%	6%	8%	10%	12%
$t = \frac{\ln 3}{\ln (1 + r)}$ (years)	55.48	28.01	18.85	14.27	11.53	9.69

19. Continuous compounding results in faster growth.

$A = 1 + 0.075[\![t]\!]$ and $A = e^{0.07t}$

21. $a = 10, y = \frac{1}{2}(10) = 5, t = 1599$

$$y = ae^{-bt}$$
$$5 = 10e^{-b(1599)}$$
$$0.5 = e^{-1599b}$$
$$\ln 0.5 = \ln e^{-1599b}$$
$$\ln 0.5 = -1599b$$
$$b = -\frac{\ln 0.5}{1599}$$

Given an initial quantity of 10 grams, after 1000 years, you have

$$y = 10e^{-[-(\ln 0.5)/1599](1000)} \approx 6.48 \text{ grams.}$$

23. $y = 2, a = 2(2) = 4, t = 5715$

$$y = ae^{-bt}$$
$$2 = 4e^{-b(5715)}$$
$$0.5 = e^{-5715b}$$
$$\ln 0.5 = \ln e^{-5715b}$$
$$\ln 0.5 = -5715b$$
$$b = -\frac{\ln 0.5}{5715}$$

Given 2 grams after 1000 years, the initial amount is

$$2 = ae^{-[-(\ln 0.5)/5715](1000)}$$
$$a \approx 2.26 \text{ grams.}$$

25.
$$y = ae^{bx}$$
$$1 = ae^{b(0)} \Rightarrow 1 = a$$
$$10 = e^{b(3)}$$
$$\ln 10 = 3b$$
$$\frac{\ln 10}{3} = b \Rightarrow b \approx 0.7675$$

So, $y = e^{0.7675x}$.

27.
$$y = ae^{bx}$$
$$5 = ae^{b(0)} \Rightarrow 5 = a$$
$$1 = 5e^{b(4)}$$
$$\frac{1}{5} = e^{4b}$$
$$\ln\left(\frac{1}{5}\right) = 4b$$
$$\frac{\ln(1/5)}{4} = b \Rightarrow b \approx -0.4024$$

So, $y = 5e^{-0.4024x}$.

29. (a) $P = 76.6e^{0.0313t}$

Year	1980	1990	2000	2010
P	104.752	143.251	195.899	267.896
Population	104,752	143,251	195,899	267,896

(b) Let $P = 360$, and solve for t.

$$360 = 76.6e^{0.0313t}$$
$$\frac{360}{76.6} = e^{0.0313t}$$
$$\ln\left(\frac{360}{76.6}\right) = 0.0313t$$
$$\frac{1}{0.0313}\ln\left(\frac{360}{76.6}\right) = t$$
$$49.4 \approx t$$

According to the model, the population will reach 360,000 in 2019.

(c) No; As t increases, the population increases rapidly.

31. $y = 4080e^{kt}$

When $t = 3, y = 10{,}000$:

$$10{,}000 = 4080e^{k(3)}$$
$$\frac{10{,}000}{4080} = e^{3k}$$
$$\ln\left(\frac{10{,}000}{4080}\right) = 3k$$
$$k = \frac{\ln(10{,}000/4080)}{3} \approx 0.2988$$

When $t = 24$: $y = 4080e^{0.2988(24)} \approx 5{,}309{,}734$ hits

33. $y = ae^{bt}$

When $t = 3$, $y = 100$:

$$100 = ae^{3b}$$
$$\frac{100}{e^{3b}} = a$$

When $t = 5$, $y = 400$:

$$400 = ae^{5b}$$

Substitute $\frac{100}{e^{3b}}$ for a in the equation on the right.

$$400 = \frac{100}{e^{3b}}e^{5b}$$
$$400 = 100e^{2b}$$
$$4 = e^{2b}$$
$$\ln 4 = 2b$$
$$\ln 2^2 = 2b$$
$$2 \ln 2 = 2b$$
$$\ln 2 = b$$
$$a = \frac{100}{e^{3b}} = \frac{100}{e^{3 \ln 2}} = \frac{100}{e^{\ln 2^3}} = \frac{100}{2^3} = \frac{100}{8} = 12.5$$
$$y = 12.5e^{(\ln 2)t}$$

After 6 hours, there are $y = 12.5e^{(\ln 2)(6)} = 800$ bacteria.

35. $(0, 575), (2, 275)$

(a) $m = \dfrac{275 - 575}{2 - 0} = -150$

$V = -150t + 575$

(b) Since $V = 575$, when

$t = 0,\ 575 = ae^{(b)(0)} \rightarrow a = 575$

Then $275 = 575e^{k(2)}$

$\ln\left(\dfrac{275}{575}\right) = 2k \Rightarrow k \approx -0.3688$

$V = 575e^{-0.3688t}$

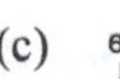
(c)

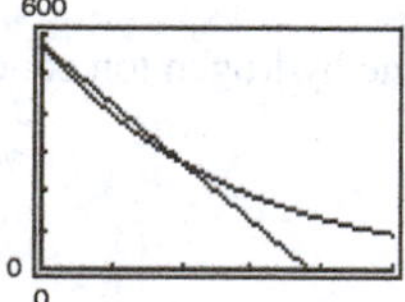

The exponential model depreciates faster in the first two years.

(d)

t	1	3
$V = -150t + 575$	\$425	\$125
$V = 575e^{-0.3688t}$	\$397.65	\$190.18

(e) Answers will vary. Sample Answer: The slope of the linear model means that the laptop depreciates \$150 per year, then loses all value late in the third year. The exponential model depreciates faster in the first three years but maintains value longer.

37. $R = \dfrac{1}{10^{12}}e^{-t/8223}$

$$R = \frac{1}{8^{14}}$$
$$\frac{1}{10^{12}}e^{-t/8223} = \frac{1}{8^{14}}$$
$$e^{-t/8223} = \frac{10^{12}}{8^{14}}$$
$$-\frac{t}{8223} = \ln\left(\frac{10^{12}}{8^{14}}\right)$$
$$t = -8223 \ln\left(\frac{10^{12}}{8^{14}}\right) \approx 12{,}180 \text{ years old}$$

39. $y = 0.0266e^{-(x-100)^2/450},\ 70 \le x \le 116$

(a)

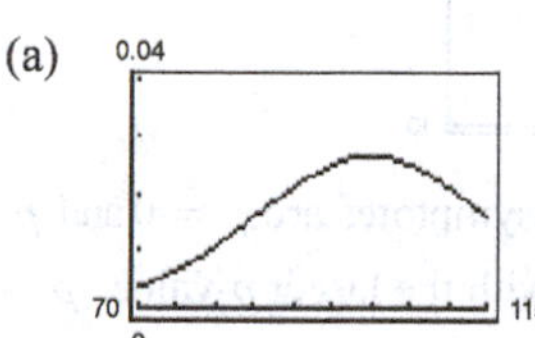

(b) The average IQ score of an adult student is 100.

41. (a) 1998: $t = 18$, $y = \dfrac{320{,}110}{1 + 374e^{-0.252(18)}} \approx 63{,}992$ sites

2003: $t = 23$, $y = \dfrac{320{,}110}{1 + 374e^{-0.252(23)}}$

$\approx 149{,}805$ sites

2006: $t = 26$, $y = \dfrac{320{,}110}{1 + 374e^{-0.252(26)}}$

$\approx 208{,}705$ sites

(b)

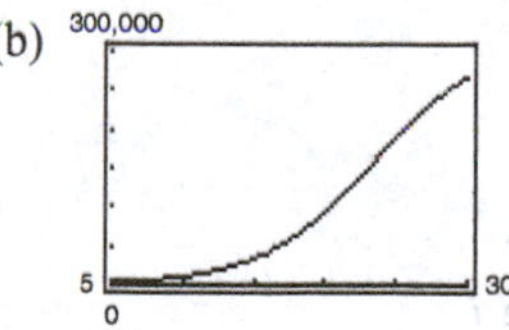

(c) When $y = 270{,}000$, $t \approx 30.2$. So, the number of cell sites will reach 270,000 in the year 2010.

(d) Let $y = 270{,}000$ and solve for t.

$$270{,}000 = \frac{320{,}110}{1 + 374e^{-0.252t}}$$
$$1 + 374e^{-0.252t} = \frac{320{,}110}{270{,}000}$$
$$374e^{-0.252t} = 0.1855926$$
$$e^{-0.252t} \approx 0.000496237$$
$$-0.252t \approx \ln(0.000496237)$$
$$t \approx 30.2$$

The number of cell sites will reach 270,000 during the year 2010.

43. $p(t) = \dfrac{1000}{1 + 9e^{-0.1656t}}$

(a) $p(5) = \dfrac{1000}{1 + 9e^{-0.1656(5)}} \approx 203$ animals

(b)
$$500 = \frac{1000}{1 + 9e^{-0.1656t}}$$
$$1 + 9e^{-0.1656t} = 2$$
$$9e^{-0.1656t} = 1$$
$$e^{-0.1656t} = \frac{1}{9}$$
$$t = \frac{\ln(1/9)}{0.1656} \approx 13 \text{ months}$$

(c)

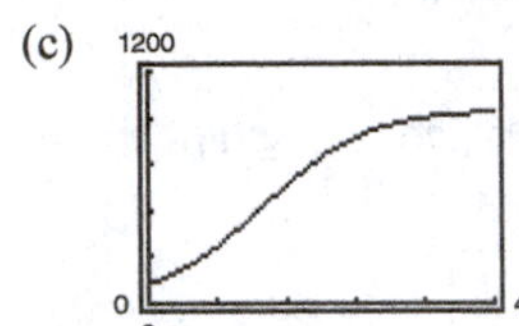

The horizontal asymptotes are $p = 0$ and $p = 1000$. The asymptote with the larger p-value, $p = 1000$, indicates that the population size will approach 1000 as time increases.

45. $R = \log \dfrac{I}{I_0} = \log I$ because $I_0 = 1$.

(a)
$$R = 7.6$$
$$7.6 = \log I$$
$$10^{7.6} = 10^{\log I}$$
$$39{,}810{,}717 \approx I$$

(b)
$$R = 5.6$$
$$5.6 = \log I$$
$$10^{5.6} = 10^{\log I}$$
$$10^{5.6} = I$$
$$398{,}107 \approx I$$

(c)
$$R = 6.6$$
$$6.6 = \log I$$
$$10^{6.6} = 10^{\log I}$$
$$3{,}981{,}072 \approx I$$

47. $\beta = 10 \log \dfrac{I}{I_0}$ where $I_0 = 10^{-12}$ watt/m^2.

(a) $\beta = 10 \log \dfrac{10^{-10}}{10^{-12}} = 10 \log 10^2 = 20$ decibels

(b) $\beta = 10 \log \dfrac{10^{-5}}{10^{-12}} = 10 \log 10^7 = 70$ decibels

(c) $\beta = 10 \log \dfrac{10^{-8}}{10^{-12}} = 10 \log 10^4 = 40$ decibels

(d) $\beta = 10 \log \dfrac{10^{-3}}{10^{-12}} = 10 \log 10^9 = 90$ decibels

49.
$$\beta = 10 \log \frac{I}{I_0}$$
$$\frac{\beta}{10} = \log \frac{I}{I_0}$$
$$10^{\beta/10} = 10^{\log I/I_0}$$
$$10^{\beta/10} = \frac{I}{I_0}$$
$$I = I_0 10^{\beta/10}$$
$$\% \text{ decrease} = \frac{I_0 10^{9.3} - I_0 10^{8.0}}{I_0 10^{9.3}} \times 100 \approx 95\%$$

51. $\text{pH} = -\log[H^+]$

$-\log(2.3 \times 10^{-5}) \approx 4.64$

53.
$$5.8 = -\log[H^+]$$
$$-5.8 = \log[H^+]$$
$$10^{-5.8} = 10^{\log[H^+]}$$
$$10^{-5.8} = [H^+]$$
$$[H^+] \approx 1.58 \times 10^{-6} \text{ moles per liter}$$

55.
$$2.9 = -\log[H^+]$$
$$-2.9 = \log[H^+]$$
$$[H^+] = 10^{-2.9} \text{ for the apple juice}$$
$$8.0 = -\log[H^+]$$
$$-8.0 = \log[H^+]$$
$$[H^+] = 10^{-8} \text{ for the drinking water}$$

$\dfrac{10^{-2.9}}{10^{-8}} = 10^{5.1}$ times the hydrogen ion concentration of drinking water

57. $t = -10 \ln \dfrac{T - 70}{98.6 - 70}$

At 9:00 A.M. you have:

$t = -10 \ln \dfrac{85.7 - 70}{98.6 - 70} \approx 6$ hours

From this you can conclude that the person died at 3:00 A.M.

59. $u = 120{,}000\left[\dfrac{0.075t}{1 - \left(\dfrac{1}{1 + 0.075/12}\right)^{12t}} - 1\right]$

(a)

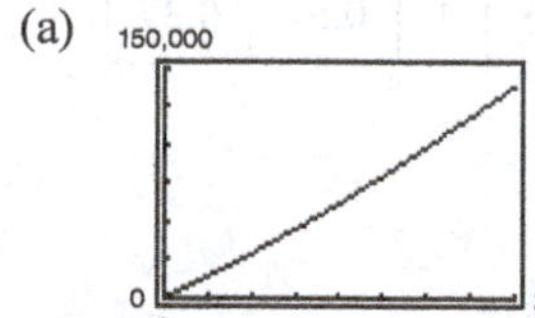

(b) From the graph, $u = \$120{,}000$ when $t \approx 21$ years. It would take approximately 37.6 years to pay \$240,000 in interest. Yes, it is possible to pay twice as much in interest charges as the size of the mortgage. It is especially likely when the interest rates are higher.

61. False. The domain can be the set of real numbers for a logistic growth function.

63. False. The graph of $f(x)$ is the graph of $g(x)$ shifted upward five units.

65. Answers will vary.

Review Exercises for Chapter 5

1. $f(x) = 0.3^x$

$f(1.5) = 0.3^{1.5} \approx 0.164$

3. $f(x) = 2^x$

$f\left(\frac{2}{3}\right) = 2^{2/3} \approx 1.587$

5. $f(x) = 7(0.2^x)$

$f\left(-\sqrt{11}\right) = 7\left(0.2^{-\sqrt{11}}\right) \approx 1456.529$

7. $f(x) = 4^{-x} + 4$

Horizontal asymptote: $y = 4$

x	−1	0	1	2	3
$f(x)$	8	5	4.25	4.063	4.016

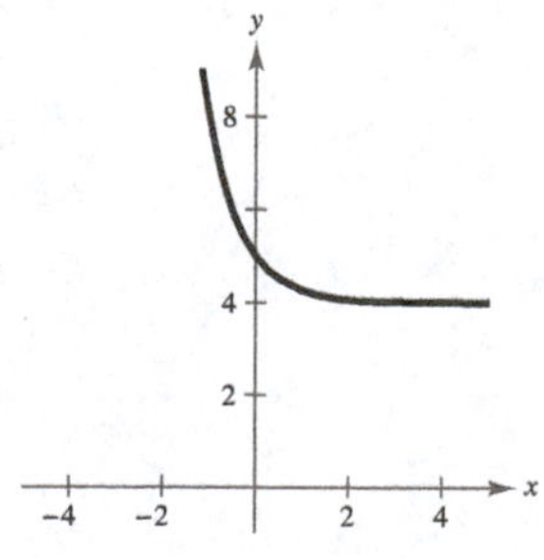

9. $f(x) = 5^{x-2} + 4$

Horizontal asymptote: $y = 4$

x	−1	0	1	2	3
$f(x)$	4.008	4.04	4.2	5	9

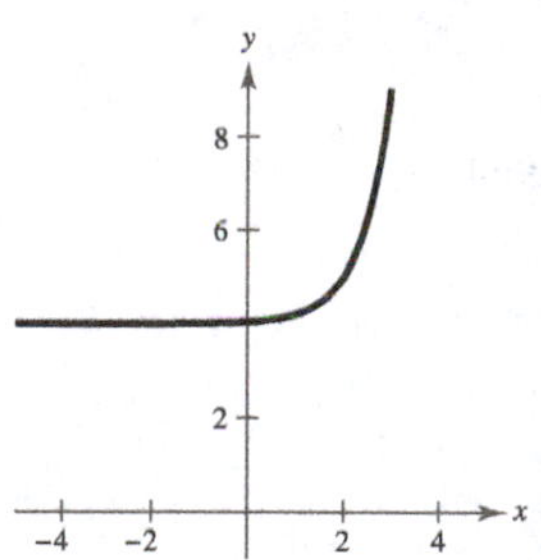

11. $f(x) = \left(\frac{1}{2}\right)^{-x} + 3 = 2^x + 3$

Horizontal asymptote: $y = 3$

x	−2	−1	0	1	2
$f(x)$	3.25	3.5	4	5	7

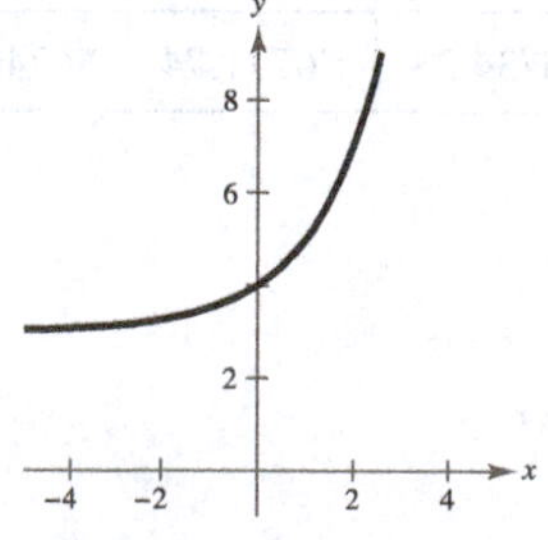

13. $\left(\frac{1}{3}\right)^{x-3} = 9$

$\left(\frac{1}{3}\right)^{x-3} = 3^2$

$\left(\frac{1}{3}\right)^{x-3} = \left(\frac{1}{3}\right)^{-2}$

$x - 3 = -2$

$x = 1$

15. $e^{3x-5} = e^7$

$3x - 5 = 7$

$3x = 12$

$x = 4$

17. $f(x) = 5^x, g(x) = 5^x + 1$

Because $g(x) = f(x) + 1$, the graph of g can be obtained by shifting the graph of f one unit upward.

19. $f(x) = 3^x, g(x) = 1 - 3^x$

Because $g(x) = 1 - f(x)$, the graph of g can be obtained by reflecting the graph of f in the x-axis and shifting the graph one unit upward. (**Note:** This is equivalent to shifting the graph of f one unit upward and then reflecting the graph in the x-axis.)

21. $f(x) = e^x$

$f(3.4) = e^{3.4} \approx 29.964$

23. $f(x) = e^x$

$f\left(\frac{3}{5}\right) = e^{3/5} \approx 1.822$

25. $h(x) = e^{-x/2}$

x	-2	-1	0	1	2
$h(x)$	2.72	1.65	1	0.61	0.37

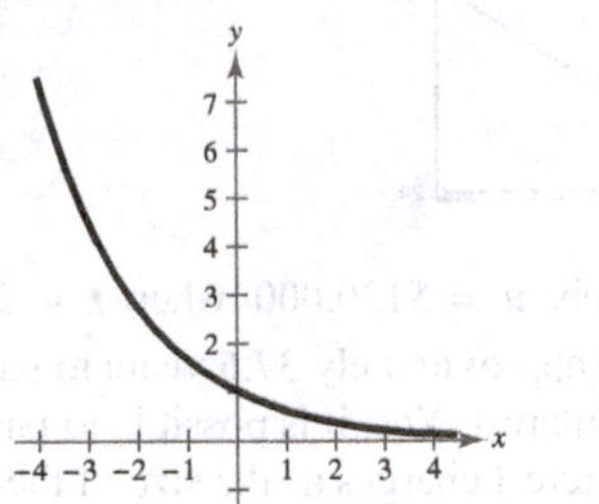

27. $f(x) = e^{x+2}$

x	-3	-2	-1	0	1
$f(x)$	0.37	1	2.72	7.39	20.09

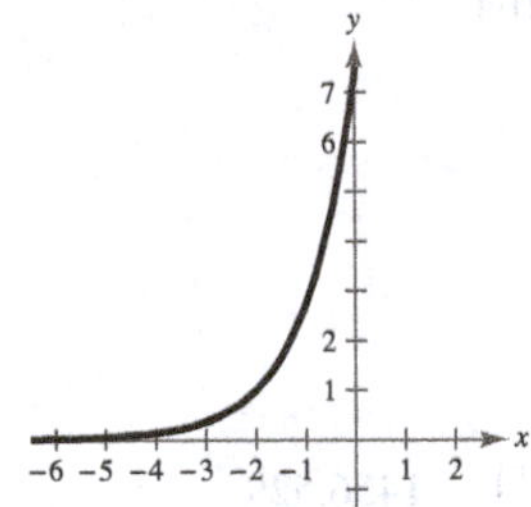

29. $F(t) = 1 - e^{-t/3}$

(a) $F(1) \approx 0.283$

(b) $F(2) \approx 0.487$

(c) $F(5) \approx 0.811$

31. $P = \$5000, r = 3\%, t = 10$ years

Compounded n times per year: $A = P\left(1 + \frac{r}{n}\right)^{nt} = 5000\left(1 + \frac{0.03}{n}\right)^{10n}$

Compounded continuously: $A = Pe^{rt} = 5000e^{0.03(10)}$

n	1	2	4	12	365	Continuous
A	\$6719.58	\$6734.28	\$6741.74	\$6746.77	\$6749.21	\$6749.29

33. $3^3 = 27$

$\log_3 27 = 3$

35. $e^{0.8} = 2.2255\ldots$

$\ln 2.2255\ldots = 0.8$

37. $f(x) = \log x$

$f(1000) = \log 1000 = \log 10^3 = 3$

39. $g(x) = \log_2 x$

$g\left(\frac{1}{4}\right) = \log_2 \frac{1}{4} = \log_2 2^{-2} = -2$

41. $\log_4(x + 7) = \log_4 14$

$x + 7 = 14$

$x = 7$

43. $\ln(x + 9) = \ln 4$

$x + 9 = 4$

$x = -5$

45. $g(x) = \log_7 x \Rightarrow x = 7^y$

Domain: $(0, \infty)$

x-intercept: $(1, 0)$

Vertical asymptote: $x = 0$

x	$\frac{1}{7}$	1	7	49
$g(x)$	-1	0	1	2

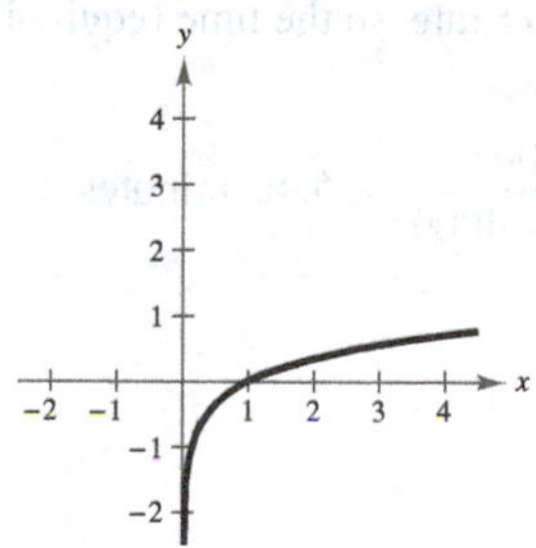

47. $f(x) = 4 - \log(x + 5)$

Domain: $(-5, \infty)$

Because

$4 - \log(x + 5) = 0 \Rightarrow \log(x + 5) = 4$

$x + 5 = 10^4$

$x = 10^4 - 5$

$= 9995.$

x-intercept: $(9995, 0)$

Vertical asymptote: $x = -5$

x	-4	-3	-2	-1	0	1
$f(x)$	4	3.70	3.52	3.40	3.30	3.22

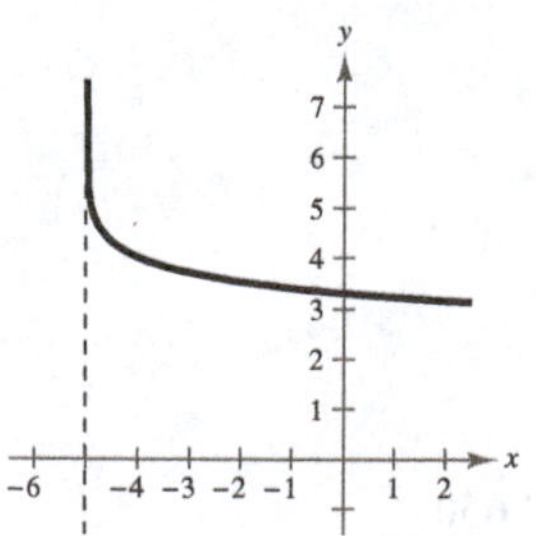

49. $f(22.6) = \ln 22.6 \approx 3.118$

51. $f\left(\sqrt{e}\right) = \frac{1}{2}\ln\sqrt{e} = 0.25$

53. $f(x) = \ln x + 6 = 6 + \ln x$

Domain: $(0, \infty)$

$\ln x + 6 = 0$

$\ln x = -6$

$x = e^{-6}$

x-intercept: $\left(e^{-6}, 0\right)$

Vertical asymptote: $x = 0$

x	$\frac{1}{4}$	$\frac{1}{2}$	1	2	3
$f(x)$	4.613	5.037	6	6.693	7.098

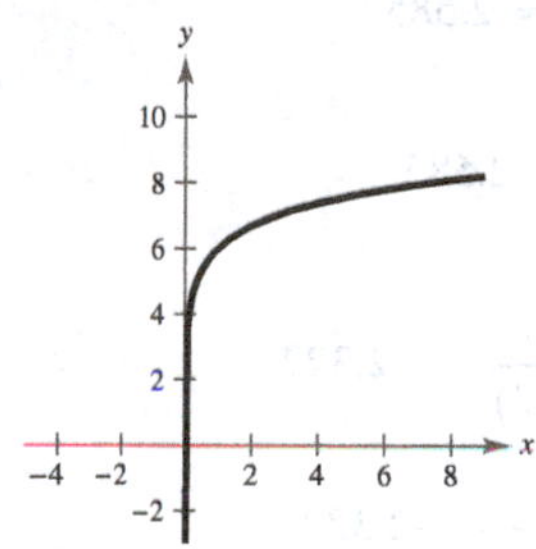

55. $f(x) = \ln(x - 6)$

Domain: $(6, \infty)$

$\ln(x - 6) = 0$

$x - 6 = e^0$

$x - 6 = 1$

$x = 7$

x-intercept: $(7, 0)$

Vertical asymptote: $x = 6$

x	6.5	7	8	9	10
$f(x)$	-0.693	0	0.693	1.099	1.386

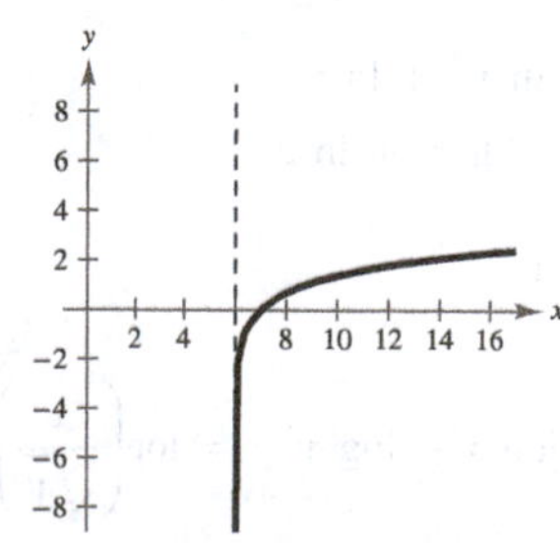

57. $M = m - 5\log\left(\dfrac{d}{10}\right)$

Let $m = 2.08$ and $M = 1.3$ and solve for d.

$$1.3 = 2.08 - 5\log\left(\frac{d}{10}\right)$$
$$-0.78 = -5\log\left(\frac{d}{10}\right)$$
$$0.156 = \log\left(\frac{d}{10}\right)$$
$$10^{0.156} = 10^{\log(d/10)}$$
$$10^{0.156} = \frac{d}{10}$$
$$10 \cdot 10^{0.156} = d$$
$$d = 10^{1.156} \approx 14.32 \text{ parsecs}$$

59. (a) $\log_2 6 = \dfrac{\log 6}{\log 2} \approx 2.585$

(b) $\log_2 6 = \dfrac{\ln 6}{\ln 2} \approx 2.585$

61. (a) $\log_{1/2} 5 = \dfrac{\log 5}{\log(1/2)} \approx -2.322$

(b) $\log_{1/2} 5 = \dfrac{\ln 5}{\ln(1/2)} \approx -2.322$

63. $\log_2 \frac{5}{3} = \log_2 5 - \log_2 3$

65. $\log_2 \frac{9}{5} = \log_2 9 - \log_2 5$
$$= \log_2 3^2 - \log_2 9$$
$$= 2\log_2 3 - \log_2 5$$

67. $\log 7x^2 = \log 7 + \log x^2 = \log 7 + 2\log x$

69. $\log_3 \dfrac{9}{\sqrt{x}} = \log_3 9 - \log_3 \sqrt{x}$
$$= \log_3 3^2 - \log_3 x^{1/2}$$
$$= 2 - \frac{1}{2}\log_3 x$$

71. $\ln x^2y^2z = \ln x^2 + \ln y^2 + \ln z$
$$= 2\ln x + 2\ln y + \ln z$$

73. $\ln 7 + \ln x = \ln(7x)$

75. $\log x - \dfrac{1}{2}\log y = \log x - \log y^{1/2} = \log\left(\dfrac{x}{\sqrt{y}}\right)$

77. $\dfrac{1}{2}\log_3 x - 2\log_3(y + 8) = \log_3 x^{1/2} - \log_3(y + 8)^2$
$$= \log_3 \sqrt{x} - \log_3(y + 8)^2$$
$$= \log_3 \frac{\sqrt{x}}{(y + 8)^2}$$

79. $t = 50\log \dfrac{18{,}000}{18{,}000 - h}$

(a) Domain: $0 \le h < 18{,}000$

(b)

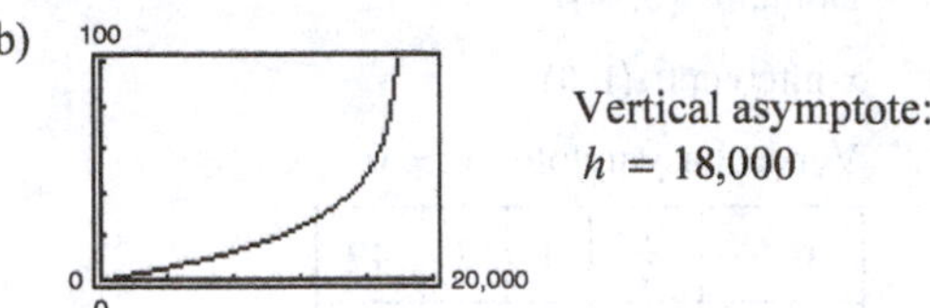

Vertical asymptote: $h = 18{,}000$

(c) As the plane approaches its absolute ceiling, it climbs at a slower rate, so the time required increases.

(d) $50\log \dfrac{18{,}000}{18{,}000 - 4000} \approx 5.46$ minutes

81. $5^x = 125$
$$5^x = 5^3$$
$$x = 3$$

83. $e^x = 3$
$$x = \ln 3 \approx 1.099$$

85. $\ln x = 4$
$$x = e^4 \approx 54.598$$

87. $e^{4x} = e^{x^2+3}$
$$4x = x^2 + 3$$
$$0 = x^2 - 4x + 3$$
$$0 = (x - 1)(x - 3)$$
$$x = 1, x = 3$$

89. $2^x - 3 = 29$
$$2^x = 32$$
$$2^x = 2^5$$
$$x = 5$$

91. $\ln 3x = 8.2$
$$e^{\ln 3x} = e^{8.2}$$
$$3x = e^{8.2}$$
$$x = \frac{e^{8.2}}{3} \approx 1213.650$$

93. $\ln x + \ln(x - 3) = 1$

$$\ln[x(x - 3)] = 1$$
$$\ln(x^2 - 3x) = 1$$
$$e^{\ln(x^2 - 3x)} = e^1$$
$$x^2 - 3x - e = 0$$
$$x = \frac{-b \pm \sqrt{b^2 - 4ac}}{2a}$$
$$x = \frac{-(-3) \pm \sqrt{(-3)^2 - 4(1)(-e)}}{2(1)}$$
$$x = \frac{3 \pm \sqrt{9 + 4e}}{2}$$
$$x = \frac{3 + \sqrt{9 + 4e}}{2} \approx 3.729$$

$x = \dfrac{3 - \sqrt{9 + 4e}}{2}$ is extraneous since the domain of the $\ln x$ term is $x > 0$.

95. $\log_8(x - 1) = \log_8(x - 2) - \log_8(x + 2)$

$$\log_8(x - 1) = \log_8\left(\frac{x - 2}{x + 2}\right)$$
$$x - 1 = \frac{x - 2}{x + 2}$$
$$(x - 1)(x + 2) = x - 2$$
$$x^2 + x - 2 = x - 2$$
$$x^2 = 0$$
$$x = 0$$

Because $x = 0$ is not in the domain of $\log_8(x - 1)$ or of $\log_8(x - 2)$, it is an extraneous solution. The equation has no solution.

97. $\log(1 - x) = -1$

$$1 - x = 10^{-1}$$
$$1 - \tfrac{1}{10} = x$$
$$x = 0.900$$

99. $25e^{-0.3x} = 12$

Graph $y_1 = 25e^{-0.3x}$ and $y_2 = 12$.

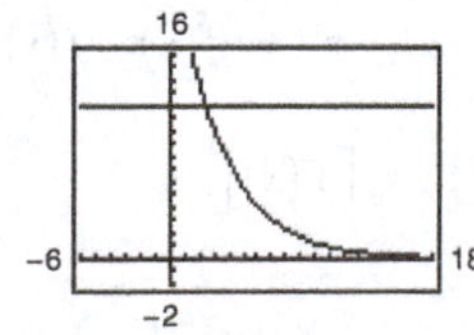

The graphs intersect at $x \approx 2.447$.

101. $2\ln(x + 3) - 3 = 0$

Graph $y_1 = 2\ln(x + 3) - 3$.

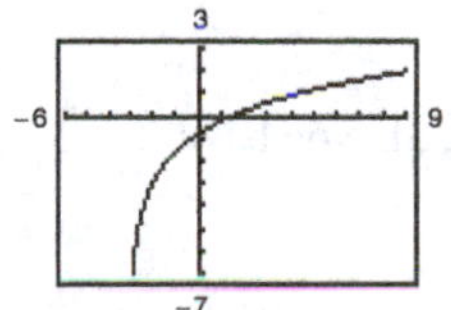

The x-intercept is at $x \approx 1.482$.

103. $P = 8500$, $A = 3(8500) = 25{,}500$, $r = 1.5\%$

$$A = Pe^{rt}$$
$$25{,}500 = 8500e^{0.015t}$$
$$3 = e^{0.015t}$$
$$\ln 3 = 0.015t$$
$$t = \frac{\ln 3}{0.015} \approx 73.2 \text{ years}$$

105. $y = 3e^{-2x/3}$

Exponential decay model

Matches graph (e).

106. $y = 4e^{2x/3}$

Exponential growth model

Matches graph (b).

107. $y = \ln(x + 3)$

Logarithmic model

Vertical asymptote: $x = -3$

Graph includes $(-2, 0)$

Matches graph (f).

108. $y = 7 - \log(x + 3)$

Logarithmic model

Vertical asymptote: $x = -3$

Matches graph (d).

109. $y = 2e^{-(x+4)^2/3}$

Gaussian model

Matches graph (a).

110. $y = \dfrac{6}{1 + 2e^{-2x}}$

Logistics growth model

Matches graph (c).

111. $y = ae^{bx}$

Using the point $(0, 2)$, you have

$$2 = ae^{b(0)}$$
$$2 = ae^0$$
$$2 = a(1)$$
$$2 = a$$

Then, using the point $(4, 3)$, you have

$$3 = 2e^{b(4)}$$
$$3 = 2e^{4b}$$
$$\tfrac{3}{2} = e^{4b}$$
$$\ln\tfrac{3}{2} = 4b$$
$$\tfrac{1}{4}\ln\left(\tfrac{3}{2}\right) = b$$

So, $y = 2e^{\frac{1}{4}\ln\left(\frac{3}{2}\right)x}$

or

$y = 2e^{0.1014x}$

113. $y = 0.0499e^{-(x-71)^2/128}, \quad 40 \le x \le 100$

Graph $y_1 = 0.0499e^{-(x-71)^2/128}$.

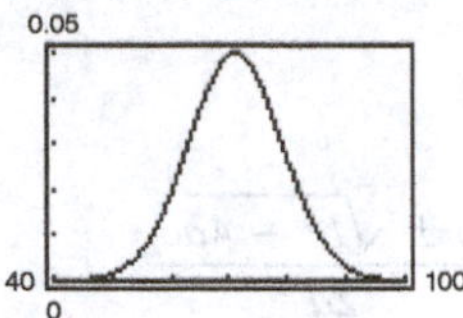

The average test score is 71.

115.
$$\beta = 10\log\left(\frac{I}{10^{-12}}\right)$$
$$\frac{\beta}{10} = \log\left(\frac{I}{10^{-12}}\right)$$
$$10^{\beta/10} = \frac{I}{10^{-12}}$$
$$I = 10^{\beta/10-12}$$

(a) $\beta = 60$

$I = 10^{60/10-12} = 10^{-6}$ watt/m^2

(b) $\beta = 135$

$I = 10^{135/10-12} = 10^{1.5} = 10\sqrt{10}$ watts/m^2

(c) $\beta = 1$

$$I = 10^{1/10-12}$$
$$= 10^{\frac{1}{10}} \times 10^{-12}$$
$$\approx 1.259 \times 10^{-12} \text{ watt/m}^2$$

117. True. By the inverse properties, $\log_b b^{2x} = 2x$.

Problem Solving for Chapter 5

1. $y = a^x$

$y_1 = 0.5^x$

$y_2 = 1.2^x$

$y_3 = 2.0^x$

$y_4 = x$

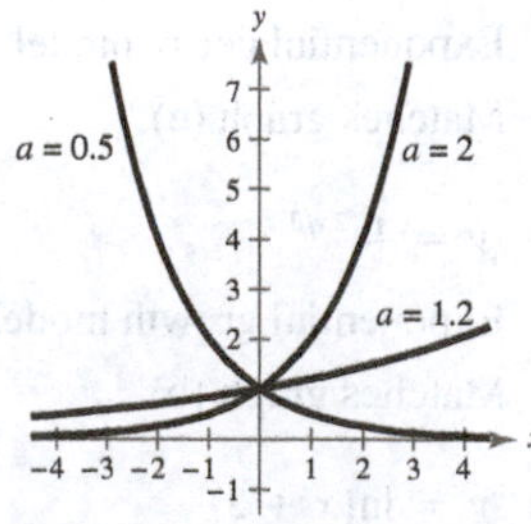

The curves $y = 0.5^x$ and $y = 1.2^x$ cross the line $y = x$. From checking the graphs it appears that $y = x$ will cross $y = a^x$ for $0 \le a \le 1.44$.

3. The exponential function, $y = e^x$, increases at a faster rate than the polynomial $y = x^n$.

5. (a) $f(u + v) = a^{u+v} = a^u \cdot a^v = f(u) \cdot f(v)$

(b) $f(2x) = a^{2x} = \left(a^x\right)^2 = \left[f(x)\right]^2$

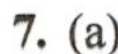

7. (a)

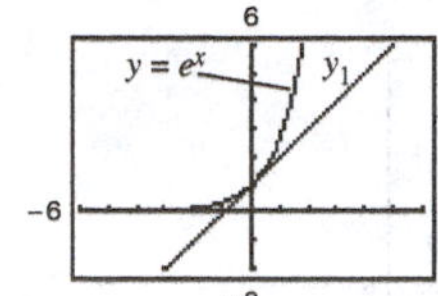

(b)

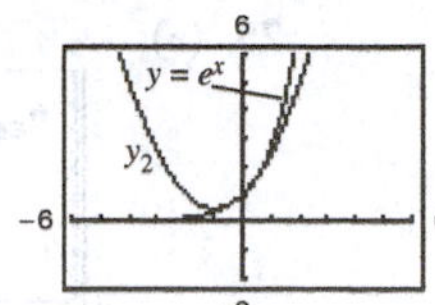

(c)

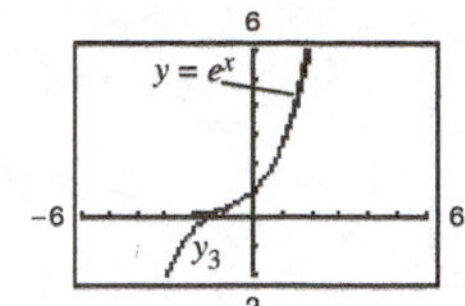

9.

$$f(x) = e^x - e^{-x}$$
$$y = e^x - e^{-x}$$
$$x = e^y - e^{-y}$$
$$x = \frac{e^{2y} - 1}{e^y}$$
$$xe^y = e^{2y} - 1$$
$$e^{2y} - xe^y - 1 = 0$$
$$e^y = \frac{x \pm \sqrt{x^2 + 4}}{2} \quad \text{Quadratic Formula}$$

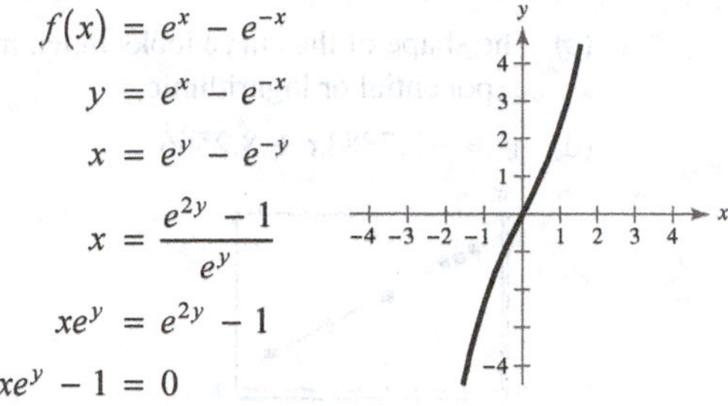

Choosing the positive quantity for e^y you have

$y = \ln\left(\frac{x + \sqrt{x^2 + 4}}{2}\right)$. So,

$f^{-1}(x) = \ln\left(\frac{x + \sqrt{x^2 + 4}}{2}\right)$.

11. Answer (c). $y = 6\left(1 - e^{-x^2/2}\right)$

The graph passes through (0, 0) and neither (a) nor (b) pass through the origin. Also, the graph has y-axis symmetry and a horizontal asymptote at $y = 6$.

13. $y_1 = c_1\left(\frac{1}{2}\right)^{t/k_1}$ and $y_2 = c_2\left(\frac{1}{2}\right)^{t/k_2}$

$$c_1\left(\frac{1}{2}\right)^{t/k_1} = c_2\left(\frac{1}{2}\right)^{t/k_2}$$
$$\frac{c_1}{c_2} = \left(\frac{1}{2}\right)^{(t/k_2 - t/k_1)}$$
$$\ln\left(\frac{c_1}{c_2}\right) = \left(\frac{t}{k_2} - \frac{t}{k_1}\right)\ln\left(\frac{1}{2}\right)$$
$$\ln c_1 - \ln c_2 = t\left(\frac{1}{k_2} - \frac{1}{k_1}\right)\ln\left(\frac{1}{2}\right)$$
$$t = \frac{\ln c_1 - \ln c_2}{\left[(1/k_2) - (1/k_1)\right]\ln(1/2)}$$

15. (a) $y_1 \approx 252{,}606(1.0310)^t$

(b) $y_2 \approx 400.88t^2 - 1464.6t + 291{,}782$

(c)

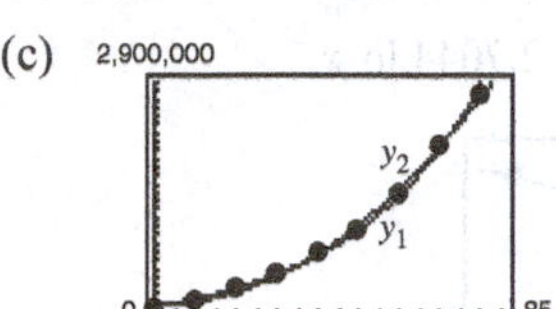

(d) The exponential model is a better fit for the data, but neither would be reliable to predict the population of the United States in 2020. The exponential model approaches infinity rapidly.

17.

$$(\ln x)^2 = \ln x^2$$
$$(\ln x)^2 - 2\ln x = 0$$
$$\ln x(\ln x - 2) = 0$$
$$\ln x = 0 \text{ or } \ln x = 2$$
$$x = 1 \text{ or } x = e^2$$

19. $y_4 = (x - 1) - \frac{1}{2}(x - 1)^2 + \frac{1}{3}(x - 1)^3 - \frac{1}{4}(x - 1)^4$

The pattern implies that

$\ln x = (x - 1) - \frac{1}{2}(x - 1)^2 + \frac{1}{3}(x - 1)^3 - \frac{1}{4}(x - 1)^4 + \ldots.$

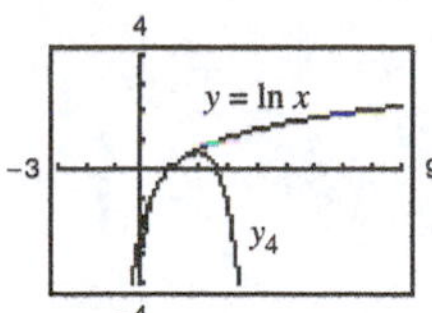

21. $y = 80.4 - 11\ln x$

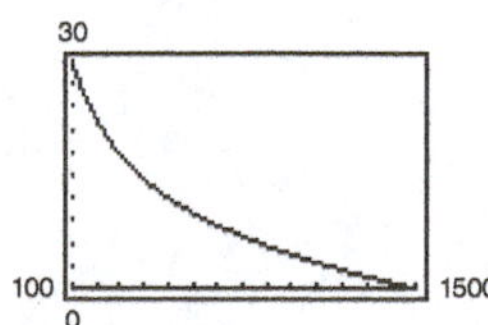

$y(300) = 80.4 - 11\ln 300 \approx 17.7 \text{ ft}^3/\text{min}$

23. (a)

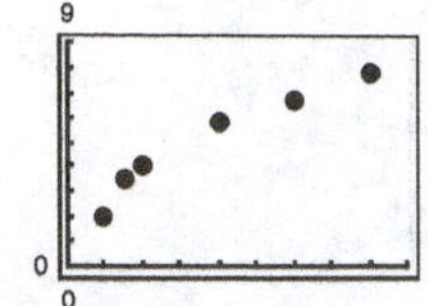

(b) The data could best be modeled by a logarithmic model.

(c) The shape of the curve looks much more logarithmic than linear or exponential.

(d) $y \approx 2.1518 + 2.7044 \ln x$

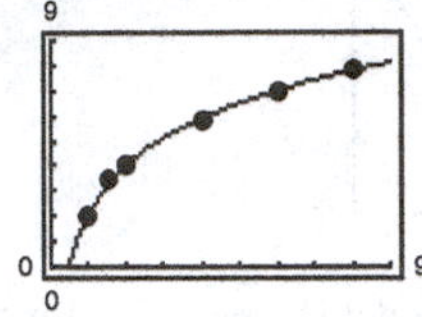

(e) The model is a good fit to the actual data.

25. (a)

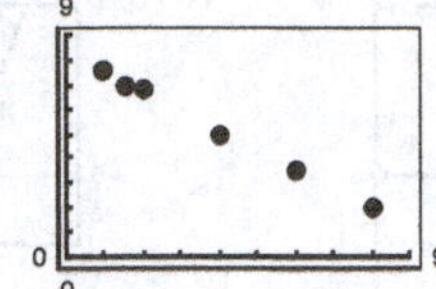

(b) The data could best be modeled by a linear model.

(c) The shape of the curve looks much more linear than exponential or logarithmic.

(d) $y \approx -0.7884x + 8.2566$

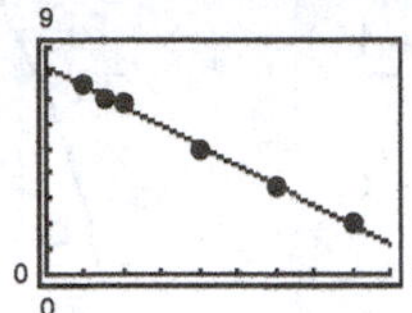

(e) The model is a good fit to the actual data.

Practice Test for Chapter 5

1. Solve for x: $x^{3/5} = 8$.

2. Solve for x: $3^{x-1} = \frac{1}{81}$.

3. Graph $f(x) = 2^{-x}$.

4. Graph $g(x) = e^x + 1$.

5. If \$5000 is invested at 9% interest, find the amount after three years if the interest is compounded
   (a) monthly.
   (b) quarterly.
   (c) continuously.

6. Write the equation in logarithmic form: $7^{-2} = \frac{1}{49}$.

7. Solve for x: $x - 4 = \log_2 \frac{1}{64}$.

8. Given $\log_b 2 = 0.3562$ and $\log_b 5 = 0.8271$, evaluate $\log_b \sqrt[4]{8/25}$.

9. Write $5 \ln x - \frac{1}{2} \ln y + 6 \ln z$ as a single logarithm.

10. Using your calculator and the change of base formula, evaluate $\log_9 28$.

11. Use your calculator to solve for N: $\log_{10} N = 0.6646$

12. Graph $y = \log_4 x$.

13. Determine the domain of $f(x) = \log_3(x^2 - 9)$.

14. Graph $y = \ln(x - 2)$.

15. True or false: $\dfrac{\ln x}{\ln y} = \ln(x - y)$

16. Solve for x: $5^x = 41$

17. Solve for x: $x - x^2 = \log_5 \frac{1}{25}$

18. Solve for x: $\log_2 x + \log_2(x - 3) = 2$

19. Solve for x: $\dfrac{e^x + e^{-x}}{3} = 4$

20. Six thousand dollars is deposited into a fund at an annual interest rate of 13%. Find the time required for the investment to double if the interest is compounded continuously.

CHAPTER 6
Systems of Equations and Inequalities

CHAPTER 6
Systems of Equations and Inequalities

Section 6.1 Linear and Nonlinear Systems of Equations

1. solution

3. points; intersection

5. $\begin{cases} 2x - y = 4 \\ 8x + y = -9 \end{cases}$

(a) $(0, -4)$

$8(0) - 4 \neq -9$

$(0, -4)$ *is not* a solution.

(b) $(3, -1)$

$2(3) - (-1) \neq 4$

$(3, -1)$ *is not* a solution.

(c) $\left(\frac{3}{2}, -1\right)$

$8\left(\frac{3}{2}\right) - 1 \neq -9$

$\left(\frac{3}{2}, -1\right)$ *is not* a solution.

(d) $\left(-\frac{1}{2}, -5\right)$

$2\left(-\frac{1}{2}\right) + 5 \stackrel{?}{=} 4$

$-1 + 5 = 4$

$8\left(-\frac{1}{2}\right) - 5 \stackrel{?}{=} -9$

$-4 - 5 = -9$

$\left(-\frac{1}{2}, -5\right)$ *is* a solution.

7. $\begin{cases} 2x + y = 6 & \text{Equation 1} \\ -x + y = 0 & \text{Equation 2} \end{cases}$

Solve for y in Equation 1: $y = 6 - 2x$

Substitute for y in Equation 2: $-x + (6 - 2x) = 0$

Solve for x: $-3x + 6 = 0 \Rightarrow x = 2$

Back-substitute $x = 2$: $y = 6 - 2(2) = 2$

Solution: $(2, 2)$

9. $\begin{cases} x - y = -4 & \text{Equation 1} \\ x^2 - y = -2 & \text{Equation 2} \end{cases}$

Solve for y in Equation 1: $y = x + 4$

Substitute for y in Equation 2: $x^2 - (x + 4) = -2$

Solve for x:

$x^2 - x - 2 = 0 \Rightarrow (x + 1)(x - 2) = 0 \Rightarrow x = -1, 2$

Back-substitute $x = -1$: $y = -1 + 4 = 3$

Back-substitute $x = 2$: $y = 2 + 4 = 6$

Solutions: $(-1, 3), (2, 6)$

11. $\begin{cases} x^2 + y = 0 & \text{Equation 1} \\ x^2 - 4x - y = 0 & \text{Equation 2} \end{cases}$

Solve for y in Equation 1: $y = -x^2$

Substitute for y in Equation 2: $x^2 - 4x - (-x^2) = 0$

Solve for x:

$2x^2 - 4x = 0 \Rightarrow 2x(x - 2) = 0 \Rightarrow x = 0, 2$

Back-substitute $x = 0$: $y = -0^2 = 0$

Back-substitute $x = 2$: $y = -2^2 = -4$

Solutions: $(0, 0), (2, -4)$

13. $\begin{cases} y = x^3 - 3x^2 + 1 & \text{Equation 1} \\ y = x^2 - 3x + 1 & \text{Equation 2} \end{cases}$

Substitute for y in Equation 2:

$x^3 - 3x^2 + 1 = x^2 - 3x + 1$

$x^3 - 4x^2 + 3x = 0$

$x(x - 1)(x - 3) = 0 \Rightarrow x = 0, 1, 3$

Back-substitute $x = 0$: $y = 0^3 - 3(0)^2 + 1 = 1$

Back-substitute $x = 1$: $y = 1^3 - 3(1)^2 + 1 = -1$

Back-substitute $x = 3$: $y = 3^3 - 3(3)^2 + 1 = 1$

Solutions: $(0, 1), (1, -1), (3, 1)$

15. $\begin{cases} x - y = 2 & \text{Equation 1} \\ 6x - 5y = 16 & \text{Equation 2} \end{cases}$

Solve for x in Equation 1: $x = y + 2$

Substitute for x in Equation 2:
$6(y + 2) - 5y = 16 \Rightarrow 6y + 12 - 5y = 16 \Rightarrow y = 4$

Back-substitute $y = 4$: $x - 4 = 2 \Rightarrow x = 6$

Solution: $(6, 4)$

17. $\begin{cases} 2x - y + 2 = 0 & \text{Equation 1} \\ 4x + y - 5 = 0 & \text{Equation 2} \end{cases}$

Solve for y in Equation 1: $y = 2x + 2$

Substitute for y in Equation 2: $4x + (2x + 2) - 5 = 0$

Solve for x: $6x - 3 = 0 \Rightarrow x = \frac{1}{2}$

Back-substitute $x = \frac{1}{2}$: $y = 2x + 2 = 2\left(\frac{1}{2}\right) + 2 = 3$

Solution: $\left(\frac{1}{2}, 3\right)$

19. $\begin{cases} 1.5x + 0.8y = 2.3 & \text{Equation 1} \\ 0.3x - 0.2y = 0.1 & \text{Equation 2} \end{cases}$

Multiply the equations by 10.

$15x + 8y = 23$ Revised Equation 1

$3x - 2y = 1$ Revised Equation 2

Solve for y in revised Equation 2: $y = \frac{3}{2}x - \frac{1}{2}$

Substitute for y in revised Equation 1:
$15x + 8\left(\frac{3}{2}x - \frac{1}{2}\right) = 23$

Solve for x:
$15x + 12x - 4 = 23 \Rightarrow 27x = 27 \Rightarrow x = 1$

Back-substitute $x = 1$: $y = \frac{3}{2}(1) - \frac{1}{2} = 1$

Solution: $(1, 1)$

21. $\begin{cases} \frac{1}{5}x + \frac{1}{2}y = 8 & \text{Equation 1} \\ x + y = 20 & \text{Equation 2} \end{cases}$

Solve for x in Equation 2: $x = 20 - y$

Substitute for x in Equation 1: $\frac{1}{5}(20 - y) + \frac{1}{2}y = 8$

Solve for y: $4 + \frac{3}{10}y = 8 \Rightarrow y = \frac{40}{3}$

Back-substitute $y = \frac{40}{3}$: $x = 20 - y = 20 - \frac{40}{3} = \frac{20}{3}$

Solution: $\left(\frac{20}{3}, \frac{40}{3}\right)$

23. $\begin{cases} 6x + 5y = -3 & \text{Equation 1} \\ -x - \frac{5}{6}y = -7 & \text{Equation 2} \end{cases}$

Solve for x in Equation 2: $x = 7 - \frac{5}{6}y$

Substitute for x in Equation 1: $6\left(7 - \frac{5}{6}y\right) + 5y = -3$

Solve for y: $42 - 5y + 5y = -3 \Rightarrow 42 = -3$ (False)

No solution

25. $\begin{cases} x + y = 12{,}000 \\ 0.02x + 0.06y = 500 \end{cases}$

Solve for y in Equation 1: $y = 12{,}000 - x$

Substitute for y in Equation 2:
$0.02x + 0.06(12{,}000 - x) = 500$

Solve for x: $0.02x + 720 - 0.06x = 500$

$-0.04x = -220$

$x = 5500$

Back-substitute $x = 5500$:
$y = 12{,}000 - 5500 = 6500$

So, \$5500 is invested at 2% and \$6500 is invested at 6%.

27. $\begin{cases} x + y = 12{,}000 \\ 0.028x + 0.038y = 396 \end{cases}$

Solve for y in Equation 1: $y = 12{,}000 - x$

Substitute for y in Equation 2:
$0.028x + 0.038(12{,}000 - x) = 396$

Solve for x: $0.028x + 456 - 0.038x = 396$

$-0.01x = -60$

$x = 6000$

Back-substitute $x = 6000$:
$y = 12{,}000 - 6000 = 6000$

So, \$6000 is invested at 2.8% and \$6000 is invested at 3.8%.

29. $\begin{cases} x^2 - y = 0 & \text{Equation 1} \\ 2x + y = 0 & \text{Equation 2} \end{cases}$

Solve for y in Equation 2: $y = -2x$

Substitute for y in Equation 1: $x^2 - (-2x) = 0$

Solve for x:
$x^2 + 2x = 0 \Rightarrow x(x + 2) = 0 \Rightarrow x = 0, -2$

Back-substitute $x = 0$: $y = -2(0) = 0$

Back-substitute $x = -2$: $y = -2(-2) = 4$

Solutions: $(0, 0), (-2, 4)$

31. $\begin{cases} x - y = -1 & \text{Equation 1} \\ x^2 - y = -4 & \text{Equation 2} \end{cases}$

Solve for y in Equation 1: $y = x + 1$

Substitute for y in Equation 2: $x^2 - (x + 1) = -4$

Solve for x: $x^2 - x - 1 = -4 \Rightarrow x^2 - x + 3 = 0$

The Quadratic Formula yields no real solutions.

33. $\begin{cases} -x + 2y = -2 \\ 3x + y = 20 \end{cases}$

Point of intersection: $(6, 2)$

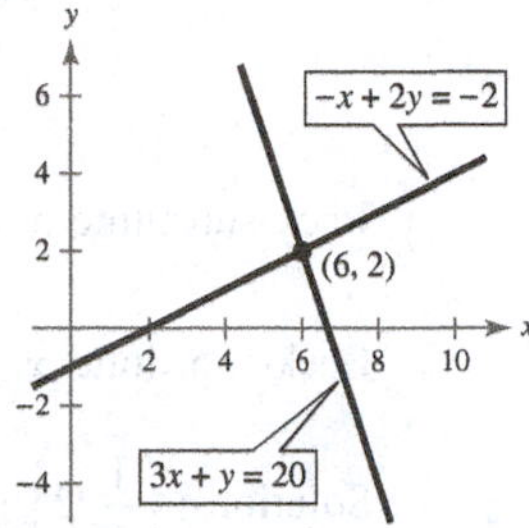

35. $\begin{cases} x - 3y = -3 \\ 5x + 3y = -6 \end{cases}$

Point of intersection: $\left(-\frac{3}{2}, \frac{1}{2}\right)$

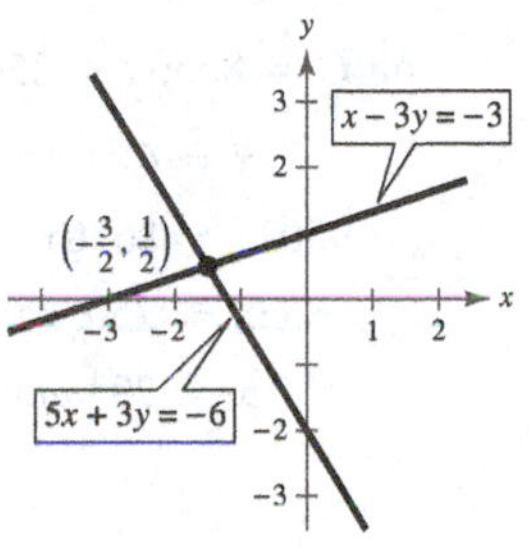

37. $\begin{cases} x + y = 4 \\ x^2 + y^2 - 4x = 0 \end{cases}$

Points of intersection: $(2, 2)$, $(4, 0)$

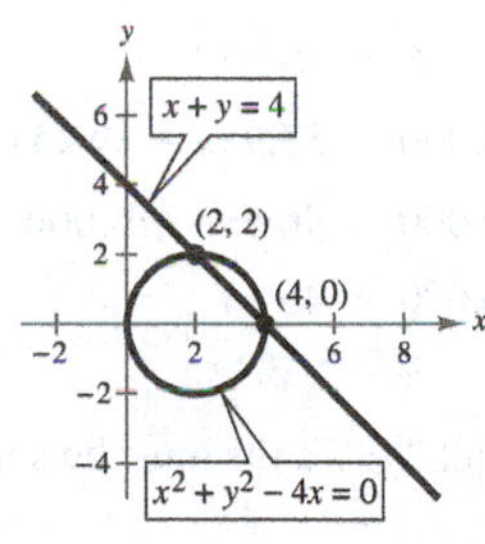

39. $\begin{cases} 3x - 2y = 0 \\ x^2 - y^2 = 4 \end{cases}$

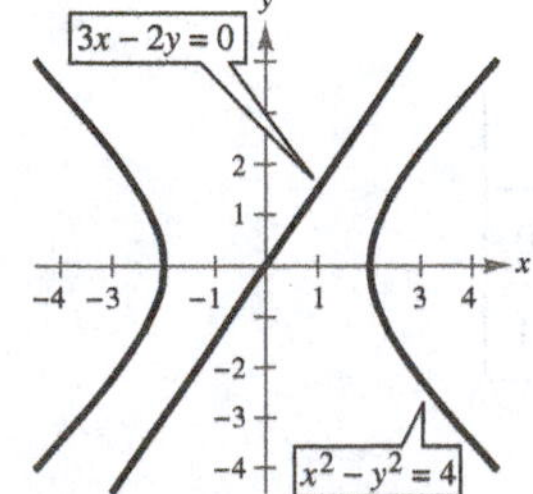

No points of intersection $\Rightarrow$ No solution

41. $\begin{cases} x^2 + y^2 = 25 \\ 3x^2 - 16y = 0 \end{cases}$

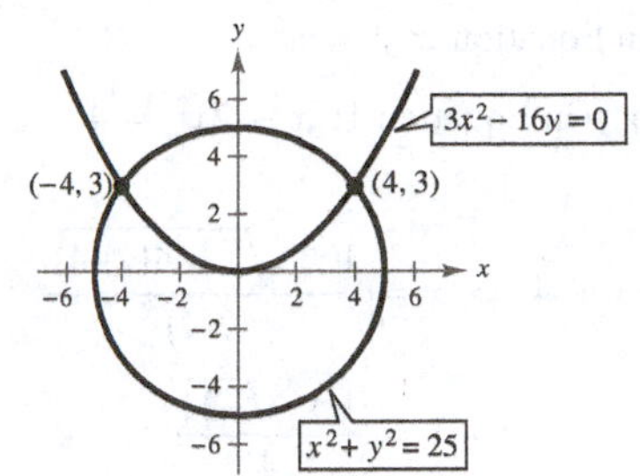

Points of intersection: $(-4, 3)$, $(4, 3)$

43. $\begin{cases} y = e^x \\ x - y + 1 = 0 \Rightarrow y = x + 1 \end{cases}$

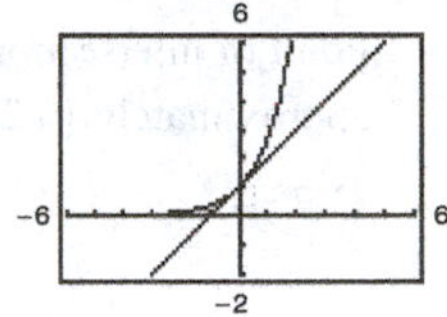

Point of intersection: $(0, 1)$

45. $\begin{cases} y = -2 + \ln(x - 1) \\ 3y + 2x = 9 \Rightarrow y = -\frac{2}{3}x + 3 \end{cases}$

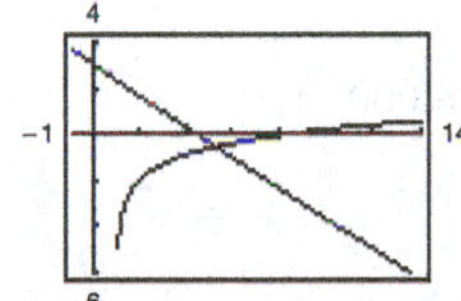

Point of intersection: $(5.31, -0.54)$

47. $\begin{cases} y = 2x & \text{Equation 1} \\ y = x^2 + 1 & \text{Equation 2} \end{cases}$

Substitute for y in Equation 2: $2x = x^2 + 1$

Solve for x: $x^2 - 2x + 1 = (x - 1)^2 = 0 \Rightarrow x = 1$

Back-substitute $x = 1$ in Equation 1: $y = 2x = 2$

Solution: $(1, 2)$

49. $\begin{cases} x - 2y = 4 & \text{Equation 1} \\ x^2 - y = 0 & \text{Equation 2} \end{cases}$

Solve for y in Equation 2: $y = x^2$

Substitute for y in Equation 1: $x - 2x^2 = 4$

Solve for x:

$$0 = 2x^2 - x + 4 \Rightarrow x = \frac{1 \pm \sqrt{1 - 4(2)(4)}}{2(2)}$$

$$\Rightarrow x = \frac{1 \pm \sqrt{-31}}{4}$$

The discriminant in the Quadratic Formula is negative.

No real solution

51. $\begin{cases} y - e^{-x} = 1 \Rightarrow y = e^{-x} + 1 \\ y - \ln x = 3 \Rightarrow y = \ln x + 3 \end{cases}$

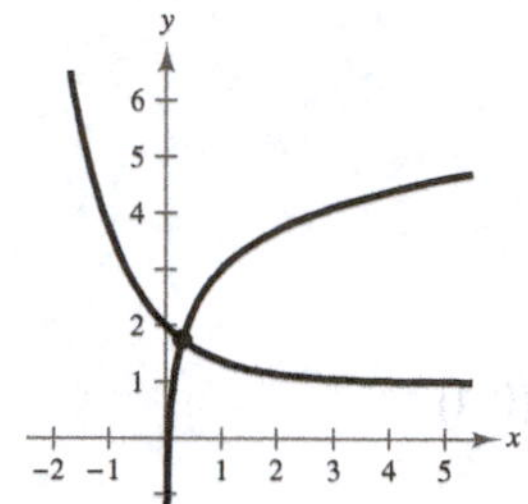

Point of intersection: approximately $(0.287), (1.751)$

53. $\begin{cases} xy - 1 = 0 & \text{Equation 1} \\ 2x - 4y + 7 = 0 & \text{Equation 2} \end{cases}$

Solve for y in Equation 1: $y = \dfrac{1}{x}$

Substitute for y in Equation 2: $2x - 4\left(\dfrac{1}{x}\right) + 7 = 0$

Solve for x:

$$2x^2 - 4 + 7x = 0 \Rightarrow (2x - 1)(x + 4) = 0$$

$$\Rightarrow x = \frac{1}{2}, -4$$

Back-substitute $x = \dfrac{1}{2}$: $y = \dfrac{1}{1/2} = 2$

Back-substitute $x = -4$: $y = \dfrac{1}{-4} = -\dfrac{1}{4}$

Solutions: $\left(\dfrac{1}{2}, 2\right), \left(-4, -\dfrac{1}{4}\right)$

55. $C = 8650x + 250{,}000$, $R = 9502x$

$$R = C$$
$$9502x = 8650x + 250{,}000$$
$$852x = 250{,}000$$
$$x \approx 293 \text{ units}$$

57. $C = 9.45x + 16{,}000$; $R = 55.95x$

(a)
$$R = C$$
$$55.95x = 9.45x + 16{,}000$$
$$46.5x = 16{,}000$$
$$x \approx 344$$

About 344 units must be sold to break even.

(b)
$$P = R - C$$
$$100{,}000 = 55.95x - (9.45x + 16{,}000)$$
$$100{,}000 = 46.5x - 16{,}000$$
$$116{,}000 = 46.5x$$
$$x \approx 2495$$

About 2495 units must be sold to earn a $100,000 profit.

59. $\begin{cases} R = 360 - 24x & \text{Equation 1} \\ R = 24 + 18x & \text{Equation 2} \end{cases}$

(a) Substitute for R in Equation 2: $360 - 24x = 24 + 18x$

Solve for x: $336 = 42x \Rightarrow x = 8$ weeks

(b)

Weeks, x	1	2	3	4	5	6	7	8	9	10
$R = 360 - 24x$	336	312	288	264	240	216	192	168	144	120
$R = 24 + 18x$	42	60	78	96	114	132	150	168	186	204

The rentals are equal when $x = 8$ weeks.

61. The error was when the second equation was solved for y.

$$x^2 + 2x - y = 3$$
$$2x - y = 2 \Rightarrow y = 2x - 2$$
$$x^2 + 2x - (2x - 2) = 3$$
$$x^2 + 2x - 2x + 2 = 3$$
$$x^2 + 2 = 3$$
$$x^2 = 1$$
$$x = \pm 1$$

When $x = 1, y = 2(1) - 2 = 0$

When $x = -1, y = 2(-1) - 2 = -4$

Solutions: $(1, 0), (-1, -4)$.

63. $2l + 2w = 56 \quad \Rightarrow \quad l + w = 28$

$$l = w + 4 \Rightarrow (w + 4) + w = 28$$
$$2w + 4 = 28$$
$$2w = 24$$
$$w = 12 \text{ meters}$$

$l = w + 4 = 12 + 4 = 16$ meters

Dimensions: 12 meters × 16 meters

65. $44 = 2l + 2w$

$$22 = l + w \Rightarrow l = 22 - w$$
$$A = lw$$
$$120 = lw$$
$$120 = (22 - w)w$$
$$120 = 22w - w^2$$
$$w^2 - 22w + 120 = 0$$
$$(w - 10)(w - 12) = 0$$
$$w = 10, \quad w = 12$$

When $w = 10, l = 22 - 10 = 12$.

When $w = 12, l = 22 - 12 = 10$.

Dimensions: 10 kilometers × 12 kilometers

67. False. To solve a system of equations by substitution, you can solve for either variable in one of the two equations and then back-substitute.

69. *Sample answer*: If the result is a contradictory equation such as $0 = N$, then you know there are no solutions. When solving a system of equations that is a nonlinear system, there may be an equation with imaginary or extraneous solutions.

71. Answers will vary.

Section 6.2 Two-Variable Linear Systems

1. elimination

3. consistent; inconsistent

5. $\begin{cases} 2x + y = 7 & \text{Equation 1} \\ x - y = -4 & \text{Equation 2} \end{cases}$

Add to eliminate y:

$$\begin{aligned} 2x + y &= 7 \\ x - y &= -4 \\ \hline 3x \quad &= 3 \Rightarrow x = 1 \end{aligned}$$

Substitute $x = 1$ in Equation 2: $1 - y = -4 \Rightarrow y = 5$

Solution: $(1, 5)$

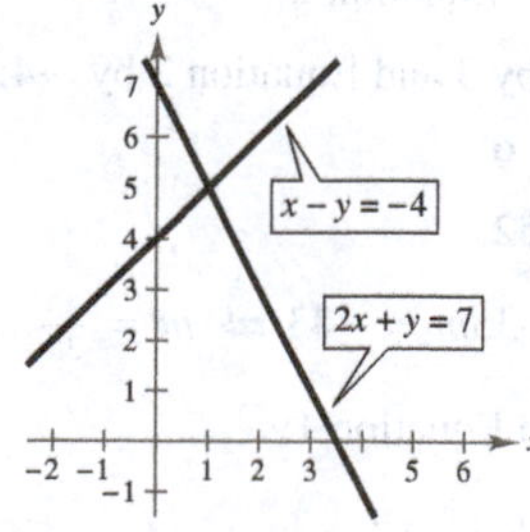

7. $\begin{cases} x + y = 0 & \text{Equation 1} \\ 3x + 2y = 1 & \text{Equation 2} \end{cases}$

Multiply Equation 1 by -2: $-2x - 2y = 0$

Add this to Equation 2 to eliminate y:

$$\begin{aligned} -2x - 2y &= 0 \\ 3x + 2y &= 1 \\ \hline x \quad &= 1 \end{aligned}$$

Substitute $x = 1$ in Equation 1: $1 + y = 0 \Rightarrow y = -1$

Solution: $(1, -1)$

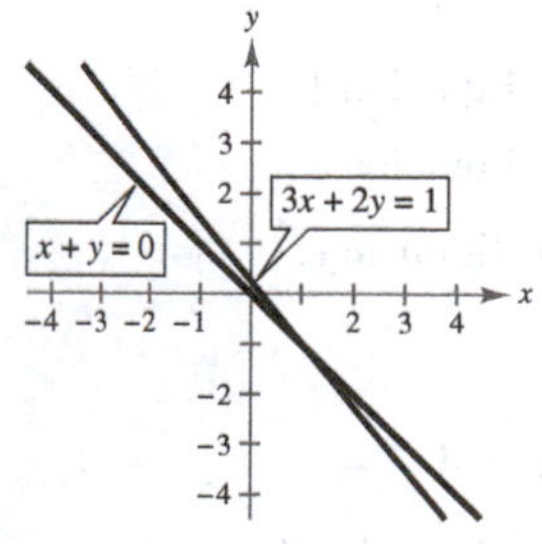

9. $\begin{cases} x - y = 2 & \text{Equation 1} \\ -2x + 2y = 5 & \text{Equation 2} \end{cases}$

Multiply Equation 1 by 2: $2x - 2y = 4$

Add this to Equation 2: $\begin{array}{r} 2x - 2y = 4 \\ -2x + 2y = 5 \\ \hline 0 = 9 \end{array}$

There are no solutions.

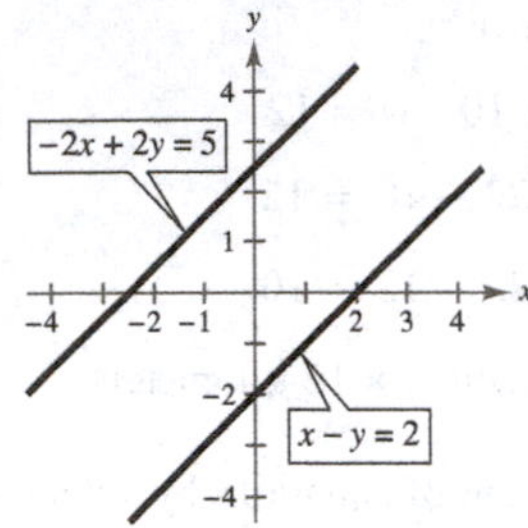

11. $\begin{cases} 3x - 2y = 5 & \text{Equation 1} \\ -6x + 4y = -10 & \text{Equation 2} \end{cases}$

Multiply Equation 1 by 2: $6x - 4y = 10$

Add this to Equation 2: $\begin{array}{r} 6x - 4y = 10 \\ -6x + 4y = -10 \\ \hline 0 = 0 \end{array}$

The equations are dependent. There are infinitely many solutions.

Let $x = a$, then $y = \dfrac{3a - 5}{2} = \dfrac{3}{2}a - \dfrac{5}{2}$.

Solution: $\left(a, \dfrac{3}{2}a - \dfrac{5}{2}\right)$, where a is any real number.

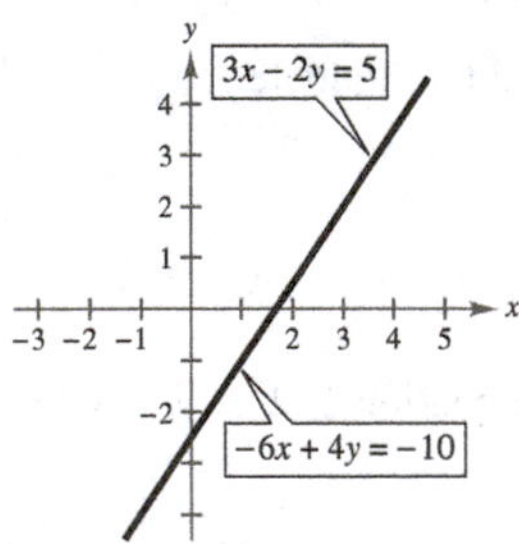

13. $\begin{cases} x + 2y = 6 & \text{Equation 1} \\ x - 2y = 2 & \text{Equation 2} \end{cases}$

Add the equations to eliminate y:

$$\begin{array}{rl} x + 2y &= 6 \\ x - 2y &= 2 \\ \hline 2x &= 8 \Rightarrow x = 4 \end{array}$$

Substitute $x = 4$ into Equation 1:

$4 + 2y = 6 \Rightarrow y = 1$

Solution: $(4, 1)$

15. $\begin{cases} 5x + 3y = 6 & \text{Equation 1} \\ 3x - y = 5 & \text{Equation 2} \end{cases}$

Multiply Equation 2 by 3: $9x - 3y = 15$

Add this to Equation 1 to eliminate y:

$$\begin{array}{rl} 5x + 3y &= 6 \\ 9x - 3y &= 15 \\ \hline 14x &= 21 \Rightarrow x = \frac{3}{2} \end{array}$$

Substitute $x = \frac{3}{2}$ into Equation 1:

$5\left(\frac{3}{2}\right) + 3y = 6 \Rightarrow y = -\frac{1}{2}$

Solution: $\left(\frac{3}{2}, -\frac{1}{2}\right)$

17. $\begin{cases} 2u + 3v = -1 & \text{Equation 1} \\ 7u + 15v = 4 & \text{Equation 2} \end{cases}$

Multiply Equation 1 by -5 and add to Equation 2.

$\begin{cases} -10u - 15v = 5 \\ 7u + 15v = 4 \end{cases}$

Solve for u: $-3u = 9 \Rightarrow u = -3$

Substitute $u = -3$ in Equation 1:

$2(-3) + 3v = -1 \Rightarrow v = \dfrac{5}{3}$

Solution: $\left(-3, \dfrac{5}{3}\right)$

19. $\begin{cases} 3x + 2y = 10 & \text{Equation 1} \\ 2x + 5y = 3 & \text{Equation 2} \end{cases}$

Multiply Equation 1 by 2 and Equation 2 by -3:

$\begin{cases} 6x + 4y = 20 \\ -6x - 15y = -9 \end{cases}$

Add to eliminate x: $-11y = 11 \Rightarrow y = -1$

Substitute $y = -1$ in Equation 1:

$3x - 2 = 10 \Rightarrow x = 4$

Solution: $(4, -1)$

21. $\begin{cases} 4b + 3m = 3 & \text{Equation 1} \\ 3b + 11m = 13 & \text{Equation 2} \end{cases}$

Multiply Equation 1 by 3 and Equation 2 by -4:

$\begin{cases} 12b + 9m = 9 \\ -12b - 44m = -52 \end{cases}$

Add to eliminate b: $-35m = -43 \Rightarrow m = \frac{43}{35}$

Substitute $m = \frac{43}{35}$ in Equation 1:

$4b + 3\left(\frac{43}{35}\right) = 3 \Rightarrow b = -\frac{6}{35}$

Solution: $\left(-\frac{6}{35}, \frac{43}{35}\right)$

23. $\begin{cases} 0.2x - 0.5y = -27.8 & \text{Equation 1} \\ 0.3x + 0.4y = 68.7 & \text{Equation 2} \end{cases}$

Multiply Equation 1 by 4 and Equation 2 by 5:

$\begin{cases} 0.8x - 2y = -111.2 \\ 1.5x + 2y = 343.5 \end{cases}$

Add these to eliminate y: $\begin{array}{rcl} 0.8x - 2y &=& -111.2 \\ 1.5x + 2y &=& 343.5 \\ \hline 2.3x &=& 232.3 \\ x &=& 101 \end{array}$

Substitute $x = 101$ in Equation 1:

$0.2(101) - 0.5y = -27.8 \Rightarrow y = 96$

Solution: $(101, 96)$

25. $\begin{cases} 3x + 2y = 4 & \text{Equation 1} \\ 9x + 6y = 3 & \text{Equation 2} \end{cases}$

Multiply Equation 1 by -3 and add to Equation 2.

$\begin{cases} -9x - 6y = -12 \\ 9x + 6y = 3 \end{cases}$

Add: $\begin{array}{rcl} -9x - 6y &=& -12 \\ 9x + 6y &=& 3 \\ \hline 0 &\neq& -9 \end{array}$

No solution.

27. $\begin{cases} -5x + 6y = -3 & \text{Equation 1} \\ 20x - 24y = 12 & \text{Equation 2} \end{cases}$

Multiply Equation 1 by 4:

$\begin{cases} -20x + 24y = -12 \\ 20x - 24y = 12 \end{cases}$

Add these two together: $0 = 0$

The equations are dependent. There are infinitely many solutions.

Let $x = a$, then

$-5a + 6y = -3 \Rightarrow y = \dfrac{5a - 3}{6} = \dfrac{5}{6}a - \dfrac{1}{2}.$

Solution: $\left(a, \dfrac{5}{6}a - \dfrac{1}{2}\right)$, where a is any real number

29. $\begin{cases} \dfrac{x + 3}{4} + \dfrac{y - 1}{3} = 1 & \text{Equation 1} \\ 2x - y = 12 & \text{Equation 2} \end{cases}$

Multiply Equation 1 by 12 and Equation 2 by 4:

$\begin{cases} 3x + 4y = 7 \\ 8x - 4y = 48 \end{cases}$

Add to eliminate y: $11x = 55 \Rightarrow x = 5$

Substitute $x = 5$ into Equation 2:

$2(5) - y = 12 \Rightarrow y = -2$

Solution: $(5, -2)$

31. $\begin{cases} -7x + 6y = -4 \\ 14x - 12y = 8 \end{cases}$

Multiply Equation 1 by 2:

$\begin{cases} -14x + 12y = -8 \\ 14x - 12y = 8 \end{cases}$

Add this to Equation 2: $0 = 0$

The original equations are dependent.

Matches graph (a).

Number of solutions: Infinite

Consistent

33. $\begin{cases} 7x - 6y = -6 \\ -7x + 6y = -4 \end{cases}$

Add the equations: $0 = -10$

Inconsistent

Matches graph (d).

Number of solutions: None

Inconsistent

35. $\begin{cases} 3x - 5y = 7 & \text{Equation 1} \\ 2x + y = 9 & \text{Equation 2} \end{cases}$

Multiply Equation 2 by 5:

$10x + 5y = 45$

Add this to Equation 1:

$13x = 52 \Rightarrow x = 4$

Back-substitute $x = 4$ into Equation 2:

$2(4) + y = 9 \Rightarrow y = 1$

Solution: $(4, 1)$

37. $\begin{cases} -2x + 8y = 20 & \text{Equation 1} \\ y = x - 5 & \text{Equation 2} \end{cases}$

Substitute Equation 2 into Equation 1:

$$\begin{aligned} -2x + 8(x - 5) &= 20 \\ -2x + 8x - 40 &= 20 \\ 6x &= 60 \\ x &= 10 \end{aligned}$$

Back-substitute $x = 10$ into Equation 2:

$y = 10 - 5 = 5$

Solution: $(10, 5)$

39. $\begin{cases} y = -2x - 17 & \text{Equation 1} \\ y = 2 - 3x & \text{Equation 2} \end{cases}$

Use substitution because both equations are solved for y, set them equal to one another and solve for x.

$-2x - 17 = 2 - 3x$

$x = 19$

Back-substitute $x = 19$ into Equation 1:

$y = -2(19) - 17 = -55$

Solution: $(19, -15)$

41. Let and r_1 = the air speed of the plane and r_2 = the wind air speed.

$\begin{cases} 3(r_1 - r_2) = 1800 & \text{Equation 1} \\ 2.5(r_1 + r_2) = 1800 & \text{Equation 2} \end{cases}$

$$\begin{aligned} r_1 - r_2 &= 600 \\ r_1 + r_2 &= 720 \\ \hline 2r_1 &= 1320 \\ r_1 &= 660 \end{aligned}$$

Back substitute into Equation 2 $660 + r_2 = 720$

$r_2 = 60$

The air speed of the plane is 660 miles per hour and the speed of the wind is 60 miles per hour.

43. Let x = the number of calories in a cheeseburger.

Let y = the number of calories in a small order of french fries.

$\begin{cases} 2x + y = 1420 & \text{Equation 1} \\ 3x + 2y = 2290 & \text{Equation 2} \end{cases}$

Multiply Equation 1 by -2 and add this to Equation 2.

$$\begin{aligned} -4x - 2y &= -2840 \\ 3x + 2y &= 2290 \\ \hline -x &= -550 \\ x &= 550 \text{ calories} \end{aligned}$$

Back-substitute $x = 550$ into Equation 2:

$3(550) + 2y = 2290$

$2y = 640$

$y = 320$ calories

The cheeseburger contains 550 calories and the fries contain 320 calories.

45. $500 - 0.4x = 380 + 0.1x$

$120 = 0.5x$

$x = 240$ units

$p = \$404$

Equilibrium point: $(240, 404)$

47. $140 - 0.00002x = 80 + 0.00001x$

$60 = 0.00003x$

$x = 2{,}000{,}000$ units

$p = \$100.00$

Equilibrium point: $(2{,}000{,}000, 100)$

49. (a) Let x = the number of liters at 25%.

Let y = the number of liters at 50%.

$\begin{cases} 0.25x + 0.50y = 12 \\ x + y = 30 \end{cases}$

(b)

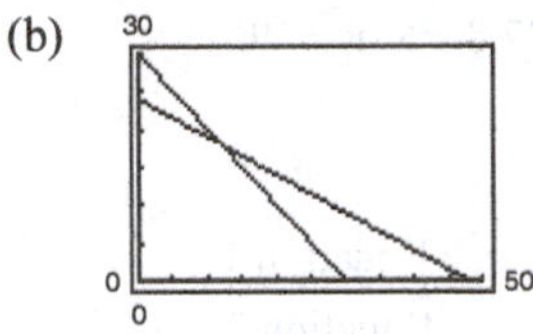

As the amount of 25% solution increases, the amount of 50% solution decreases.

(c) $\begin{cases} 0.25x + 0.50y = 12 & \text{Equation 1} \\ x + y = 30 & \text{Equation 2} \end{cases}$

Solve Equation 2 for y: $y = 30 - x$

Substitute this into Equation 1 to eliminate y:

$0.25x + 0.50(30 - x) = 12$

$0.25x + 15 - 0.50x = 12$

$-0.25x = -3$

$x = 12$ liters

Back-substitute $x = 12$ into Equation 2:

$12 + y = 30 \Rightarrow y = 18$ liters

The final mixture should contain 12 liters of the 25% solution and 18 liters of the 50% solution.

51. Let x = the amount of money invested at 3.5%.

Let y = the amount of money invested at 5%.

$\begin{cases} x + y = 24{,}000 & \text{Equation 1} \\ 0.035x + 0.05y = 930 & \text{Equation 2} \end{cases}$

Solve Equation 1 for x: $x = 24{,}000 - y$

Substitute this into Equation 2 to eliminate x:

$0.035(24{,}000 - y) + 0.05y = 930$

$840 + 0.015y = 930$

$y = \$6000$

Back-substitute $y = 6000$ into Equation 1:

$x + 6000 = 24{,}000$

$x = \$18{,}000$

$18,000 should be invested in the 3.5% bond.

53. (a) Pharmacy A: Pharmacy B:

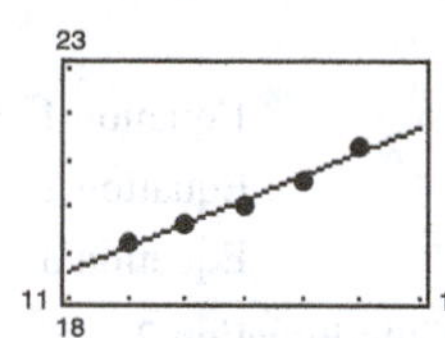
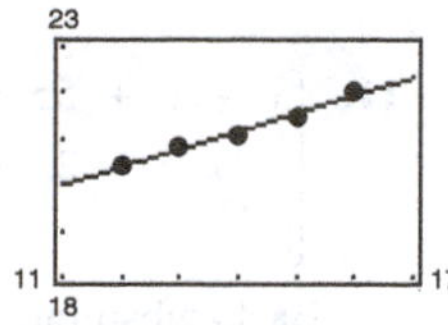

$P_A = 0.52t + 12.9$ $\qquad P_B = 0.39t + 15.7$

(b) To find when the prescriptions filled are equal, solve the system of equations consisting of $P = 0.52t + 12.9$ and $P = 0.39t + 15.7$.

$$\begin{cases} P = 0.52t + 12.9 \\ P = 0.39t + 15.7 \end{cases}$$

$$0.52t + 12.9 = 0.39t + 15.7$$
$$0.13t = 2.8$$
$$t = \frac{2.8}{013} \approx 21.5$$

The number of prescriptions filled at pharmacy A will exceed the number of prescriptions filled at pharmacy B during the year 2021.

55. $\begin{cases} 5b + 10a = 20.2 \Rightarrow & b + 2a = 4.04 \\ 10b + 30a = 50.1 \Rightarrow & -b - 3a = -5.01 \end{cases}$

$$-a = -0.97$$
$$a = 0.97$$

$$b + 2a = 4.04$$
$$b + 2(0.97) = 4.04$$
$$b = 2.1$$

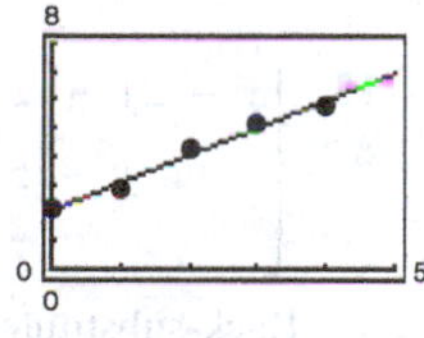

Least squares regression line:

$y = 0.97x + 2.1$

57. (a) $\begin{cases} 4b + 7.0a = 174 \Rightarrow & 28b + 49.0a = 1218 \\ 7b + 13.5a = 322 \Rightarrow & -28b - 54.0a = -1288 \end{cases}$

$$-5a = -70$$
$$a = 14$$

$$4b + 7.0a = 174$$
$$4b + 7.0(14) = 174$$
$$4b = 76$$
$$b = 19$$

Least squares regression line: $y = 14x + 19$

(b) Substitute $x = 1.6$ into $y = 14x + 19$.

$y = 14(1.6) + 19 = 41.4$

The wheat yield is about 41.4 bushels per acre.

59. False. Two lines that coincide have infinitely many points of intersection.

61. $\begin{cases} 4x - 8y = -3 & \text{Equation 1} \\ 2x + ky = 16 & \text{Equation 2} \end{cases}$

Multiply Equation 2 by –2: $-4x - 2ky = -32$

Add this to Equation 1: $\quad 4x - 8y = -3$

$$-4x - 2ky = -32$$
$$-8y - 2ky = -35$$

The system is inconsistent if $-8y - 2ky = 0$.

This occurs when $k = -4$.

63. No, it is not possible for a consistent system of linear equations to have exactly two solutions. Either the lines will intersect once or they will coincide and then the system would have infinite solutions.

65. $\begin{cases} 3x + 2y = 4 \\ 5x - 2y = 12 \end{cases}$

$$2y = -3x + 4$$
$$y = -\tfrac{3}{2}x + 2$$

$$5x - 2\left(-\tfrac{3}{2}x + 2\right) = 12$$
$$5x + 3x - 4 = 12$$
$$8x = 16$$
$$x = 2$$

Back substitute $x = 2$: $3(2) + 2y = 4$

$$6 + 2y = 4$$
$$2y = -2$$
$$y = -1$$

Solution: $(2, -1)$

Answers will vary: *Sample answer*: If the equations can be added or subtracted without having to multiply by any coefficient, elimination of variable may be preferred. If one or both of the equations is already solved for one of the variables, the method of substitution may be more efficient.

67. $\begin{cases} 100y - x = 200 & \text{Equation 1} \\ 99y - x = -198 & \text{Equation 2} \end{cases}$

Subtract Equation 2 from Equation 1 to eliminate x:

$$100y - x = 200$$
$$-99y + x = 198$$
$$y = 398$$

Substitute $y = 398$ into Equation 1:

$100(398) - x = 200 \Rightarrow x = 39{,}600$

Solution: $(39, 600, 398)$

The lines are not parallel. The scale on the axes must be changed to see the point of intersection.

Section 6.3 Multivariable Linear Systems

1. row-echelon

3. Gaussian

5. nonsquare

7. $\begin{cases} 6x - y + z = -1 \\ 4x - 3z = -19 \\ 2y + 5z = 25 \end{cases}$

(a) $(0, 3, 1)$

$6(0) - (3) + (1) \neq 1$

$(0, 3, 1)$ *is not* a solution.

(b) $(-3, 0, 5)$

$6(-3) - 0 + 5 \neq -1$

$(-3, 0, 5)$ *is not* a solution

(c) $(0, -1, 4)$

$4(0) - 3(4) \neq -19$

$(0, -1, 4)$ *is not* a solution.

(d) $(-1, 0, 5)$

$6(-1) - 0 + 5 = -1$

$4(-1) - 3(5) = -19$

$2(0) + 5(5) = 25$

$(-1, 0, 5)$ *is* a solution.

9. $\begin{cases} 4x + y - z = 0 \\ -8x - 6y + z = -\frac{7}{4} \\ 3x - y = -\frac{9}{4} \end{cases}$

(a) $4\left(\frac{1}{2}\right) + \left(-\frac{3}{4}\right) - \left(-\frac{7}{4}\right) \neq 0$

$\left(\frac{1}{2}, -\frac{3}{4}, -\frac{7}{4}\right)$ *is not* a solution.

(b) $4\left(\frac{3}{2}\right) + \left(-\frac{2}{5}\right) - \left(\frac{3}{5}\right) \neq 0$

$\left(\frac{3}{2}, -\frac{2}{5}, \frac{3}{5}\right)$ *is not* a solution.

(c) $4\left(-\frac{1}{2}\right) + \left(\frac{3}{4}\right) - \left(-\frac{5}{4}\right) = 0$

$-8\left(-\frac{1}{2}\right) - 6\left(\frac{3}{4}\right) + \left(-\frac{5}{4}\right) = -\frac{7}{4}$

$3\left(-\frac{1}{2}\right) - \left(\frac{3}{4}\right) = -\frac{9}{4}$

$\left(-\frac{1}{2}, \frac{3}{4}, -\frac{5}{4}\right)$ *is* a solution.

(d) $4\left(-\frac{1}{2}\right) + \left(\frac{1}{6}\right) - \left(-\frac{3}{4}\right) \neq 0$

$\left(-\frac{1}{2}, \frac{1}{6}, -\frac{3}{4}\right)$ *is not* a solution.

11. $\begin{cases} x - y + 5z = 37 & \text{Equation 1} \\ y + 2z = 6 & \text{Equation 2} \\ z = 8 & \text{Equation 3} \end{cases}$

Back-substitute $z = 8$ into Equation 2:

$y + 2(8) = 6$

$y = -10$

Back-substitute $y = -10$ and $z = 8$ into Equation 1:

$x - (-10) + 5(8) = 37$

$x + 10 + 40 = 37$

$x = -13$

Solution: $(-13, -10, 8)$

13. $\begin{cases} x + y - 3z = 7 & \text{Equation 1} \\ y + z = 12 & \text{Equation 2} \\ z = 2 & \text{Equation 3} \end{cases}$

Back-substitute $z = 2$ into Equation 2:

$y + 2 = 12 \Rightarrow y = 10$

Back-substitute $y = 10$ and $z = 2$ into Equation 1:

$x + (10) - 3(2) = 7$

$x + 4 = 7$

$x = 3$

Solution: $(3, 10, 2)$

15. $\begin{cases} x - 2y + z = -\frac{1}{4} & \text{Equation 1} \\ y - z = -4 & \text{Equation 2} \\ z = 11 & \text{Equation 3} \end{cases}$

Back-substitute $z = 11$ into Equation 2:

$y - 11 = -4$

$y = 7$

Back-substitute $y = 7$ and $z = 11$ into Equation 1:

$x - 2(7) + (11) = -\frac{1}{4}$

$x - 3 = -\frac{1}{4}$

$x = \frac{11}{4}$

Solution: $\left(\frac{11}{4}, 7, 11\right)$

17. $\begin{cases} x - 2y + 3z = 5 & \text{Equation 1} \\ -x + 3y - 5z = 4 & \text{Equation 2} \\ 2x - 3z = 0 & \text{Equation 3} \end{cases}$

Add Equation 1 to Equation 2:

$\begin{cases} x - 2y + 3z = 5 \\ y - 2z = 9 \\ 2x - 3z = 0 \end{cases}$

This is the first step in putting the system in row-echelon form.

19. $\begin{cases} x + y = 0 \\ -2x + 3y = 10 \end{cases}$

$\begin{cases} x + y = 0 \\ \quad 5y = 10 \end{cases}$ 2 Eq.1 + Eq.2

$\begin{cases} x + y = 0 \\ \quad y = 2 \end{cases}$ $\frac{1}{5}$ Eq.2

$x + (2) = 0$

$x = -2$

Solution: $(-2, 2)$

21. $\begin{cases} x - 2y = -2 \\ 3x - y = 9 \end{cases}$

$\begin{cases} x - 2y = -2 \\ \quad 5y = 15 \end{cases}$ (-3)Eq.1 + Eq.2

$\begin{cases} x - 2y = -2 \\ \quad y = 3 \end{cases}$ $\frac{1}{5}$ Eq.2

$x - 2(3) = -2$

$x = 4$

Solution: $(4, 3)$

23. $\begin{cases} x + y + z = 7 & \text{Equation 1} \\ 2x - y + z = 9 & \text{Equation 2} \\ 3x \quad - z = 10 & \text{Equation 3} \end{cases}$

$\begin{cases} x + y + z = 7 \\ 3x \quad + 2z = 16 & \text{Eq.2 + Eq.1} \\ 3x \quad - z = 10 \end{cases}$

$\begin{cases} x + y + z = 7 \\ 3x \quad + 2z = 16 \\ 9x \quad = 36 & \text{Eq.2 + 2Eq.3} \end{cases}$

$\begin{cases} x + y + z = 7 \\ 3x \quad + 2z = 16 \\ x \quad = 4 & \frac{1}{4}\text{ Eq.3} \end{cases}$

$3(4) + 2z = 16$

$2z = 4$

$z = 2$

$4 + y + 2 = 7$

$y = 1$

Solution: $(4, 1, 2)$

25. $\begin{cases} 2x + 4y - z = 7 \\ 2x - 4y + 2z = -6 \\ x + 4y + z = 0 \end{cases}$

$\begin{cases} x + 4y + z = 0 \\ 2x - 4y + 2z = -6 \\ 2x + 4y - z = 7 \end{cases}$ Interchange equations.

$\begin{cases} x + 4y + z = 0 \\ \quad -12y \quad = -6 & (-2)\text{Eq. 1 + Eq. 2} \\ \quad -4y - 3z = 7 & (-2)\text{Eq. 1 + Eq. 3} \end{cases}$

$\begin{cases} x + 4y + z = 0 \\ \quad y \quad = \frac{1}{2} & -\frac{1}{12}\text{Eq. 2} \\ \quad -4y - 3z = 7 \end{cases}$

$y = \frac{1}{2}$

$-4\left(\frac{1}{2}\right) - 3z = 7 \Rightarrow z = -3$

$x + 4\left(\frac{1}{2}\right) + (-3) = 0$

$x = 1$

Solution: $\left(1, \frac{1}{2}, -3\right)$

27. $\begin{cases} x - 2y + 2z = -9 \\ 2x + y - z = 7 \\ 3x - y + z = 5 \end{cases}$ Interchange equations.

$\begin{cases} x - 2y + 2z = -9 \\ \quad 5y - 5z = 25 & -2\text{Eq.1 + Eq.2} \\ \quad 5y - 5z = 32 & -3\text{Eq.1 + Eq.3} \end{cases}$

$\begin{cases} x - 2y + 2z = -9 \\ \quad 5y - 5z = 25 \\ \quad 0 = 7 & -\text{Eq.2 + Eq.3} \end{cases}$

Inconsistent, no solution

29. $\begin{cases} 3x - 5y + 5z = 1 & \text{Equation 1} \\ 2x - 2y + 3z = 0 & \text{Equation 2} \\ 7x - y + 3z = 0 & \text{Equation 3} \end{cases}$

$\begin{cases} x - 3y + 2z = 1 & \text{Eq. 1} - \text{Eq. 2} \\ 2x - 2y + 3z = 0 \\ 7x - y + 3z = 0 \end{cases}$

$\begin{cases} x - 3y + 2z = 1 \\ -4y + z = 2 & 2\text{Eq. 1} - \text{Eq.2} \\ 7x - y + 3z = 0 \end{cases}$

$\begin{cases} x - 3y + 2z = 1 \\ -4y + z = 2 \\ -20y + 11z = 7 & 7\text{Eq. 1} - \text{Eq. 3} \end{cases}$

$\begin{cases} x - 3y + 2z = 1 \\ -4y + z = 2 \\ 6z = -3 & -5\text{Eq. 2} + \text{Eq. 3} \end{cases}$

$6z = -3 \Rightarrow z = -\frac{1}{2}$

$-4y + \left(-\frac{1}{2}\right) = 2 \Rightarrow -4y = \frac{5}{2} \Rightarrow y = -\frac{5}{8}$

$x - 3\left(-\frac{5}{8}\right) + 2\left(-\frac{1}{2}\right) = 1 \Rightarrow x + \frac{7}{8} = 1 \Rightarrow x = \frac{1}{8}$

Solution: $\left(\frac{1}{8}, -\frac{5}{8}, -\frac{1}{2}\right)$

31. $\begin{cases} 2x + 3y = 0 & \text{Equation 1} \\ 4x + 3y - z = 0 & \text{Equation 2} \\ 8x + 3y + 3z = 0 & \text{Equation 3} \end{cases}$

$\begin{cases} 2x + 3y = 0 \\ -3y - z = 0 & -2\text{Eq.1} + \text{Eq.2} \\ -9y + 3z = 0 & -4\text{Eq.1} + \text{Eq.3} \end{cases}$

$\begin{cases} 2x + 3y = 0 \\ -3y - z = 0 \\ 6z = 0 & -3\text{Eq.2} + \text{Eq.3} \end{cases}$

$6z = 0 \Rightarrow z = 0$

$-3y - 0 = 0 \Rightarrow y = 0$

$2x + 3(0) = 0 \Rightarrow x = 0$

Solution: $(0, 0, 0)$

33. $\begin{cases} x + 4z = 1 & \text{Equation 1} \\ x + y + 10z = 10 & \text{Equation 2} \\ 2x - y + 2z = -5 & \text{Equation 3} \end{cases}$

$\begin{cases} x + 4z = 1 \\ y + 6z = 9 & -\text{Eq.1} + \text{Eq.2} \\ -y - 6z = -7 & -2\text{Eq.1} + \text{Eq.3} \end{cases}$

$\begin{cases} x + 4z = 1 \\ y + 6z = 9 \\ 0 = 2 & \text{Eq.2} + \text{Eq.3} \end{cases}$

No solution, inconsistent

35. $\begin{cases} 3x - 3y + 6z = 6 & \text{Equation 1} \\ x + 2y - z = 5 & \text{Equation 2} \\ 5x - 8y + 13z = 7 & \text{Equation 3} \end{cases}$

$\begin{cases} x - y + 2z = 2 & \frac{1}{3}\text{Eq.1} \\ x + 2y - z = 5 \\ 5x - 8y + 13z = 7 \end{cases}$

$\begin{cases} x - y + 2z = 2 \\ 3y - 3z = 3 & -\text{Eq.1} + \text{Eq.2} \\ -3y + 3z = -3 & -5\text{Eq.1} + \text{Eq.3} \end{cases}$

$\begin{cases} x - y + 2z = 2 \\ y - z = 1 & \frac{1}{3}\text{Eq.2} \\ 0 = 0 & \text{Eq.2} + \text{Eq.3} \end{cases}$

$\begin{cases} x + z = 3 & \text{Eq.2} + \text{Eq.1} \\ y - z = 1 \end{cases}$

Let $z = a$, then: $y = a + 1$

$x = -a + 3$

Solution: $(-a + 3, a + 1, a)$

37. $\begin{cases} x + 2y - 7z = -4 & \text{Equation 1} \\ 2x + y + z = 13 & \text{Equation 2} \\ 3x + 9y - 36z = -33 & \text{Equation 3} \end{cases}$

$\begin{cases} x + 2y - 7z = -4 \\ -3y + 15z = 21 & -2\text{Eq.1} + \text{Eq.2} \\ 3y - 15z = -21 & -3\text{Eq.1} + \text{Eq.3} \end{cases}$

$\begin{cases} x + 2y - 7z = -4 \\ -3y + 15z = 21 \\ 0 = 0 & \text{Eq.2} + \text{Eq.3} \end{cases}$

$\begin{cases} x + 2y - 7z = -4 \\ y - 5z = -7 & -\frac{1}{3}\text{Eq.2} \end{cases}$

$\begin{cases} x + 3z = 10 & -2\text{Eq.2} + \text{Eq.1} \\ y - 5z = -7 \end{cases}$

Let $z = a$, then: $y = 5a - 7$

$x = -3a + 10$

Solution: $(-3a + 10, 5a - 7, a)$

39. $\begin{cases} x \qquad\qquad + 3w = 4 & \text{Equation 1} \\ 2y - z - w = 0 & \text{Equation 2} \\ 3y \qquad - 2w = 1 & \text{Equation 3} \\ 2x - y + 4z \qquad = 5 & \text{Equation 4} \end{cases}$

$\begin{cases} x \qquad\qquad + 3w = 4 \\ 2y - z - w = 0 \\ 3y \qquad - 2w = 1 \\ -y + 4z - 6w = -3 & -2\text{Eq.1} + \text{Eq.4} \end{cases}$

$\begin{cases} x \qquad\qquad + 3w = 4 \\ y - 4z + 6w = 3 \\ 2y - z - w = 0 \\ 3y \qquad - 2w = 1 \end{cases}$ $-$Eq.4 and interchange the equations.

$\begin{cases} x \qquad\qquad + 3w = 4 \\ y - 4z + 6w = 3 \\ 7z - 13w = -6 & -\text{Eq.2} + \text{Eq.3} \\ 12z - 20w = -8 & -3\text{Eq.2} + \text{Eq.4} \end{cases}$

$\begin{cases} x \qquad\qquad + 3w = 4 \\ y - 4z + 6w = 3 \\ z - 3w = -2 & -\frac{1}{2}\text{Eq.4} + \text{Eq.3} \\ 12z - 20w = -8 \end{cases}$

$\begin{cases} x \qquad\qquad + 3w = 4 \\ y - 4z + 6w = 3 \\ z - 3w = -2 \\ 16w = 16 & -12\text{Eq.3} + \text{Eq.4} \end{cases}$

$16w = 16 \Rightarrow w = 1$

$z - 3(1) = -2 \Rightarrow z = 1$

$y - 4(1) + 6(1) = 3 \Rightarrow y = 1$

$x + 3(1) = 4 \Rightarrow x = 1$

Solution: $(1, 1, 1, 1)$

41. $\begin{cases} x - 2y + 5z = 2 \\ 4x \qquad\quad - z = 0 \end{cases}$

Let $z = a$, then: $x = \frac{1}{4}a$.

$\frac{1}{4}a - 2y + 5a = 2$

$a - 8y + 20a = 8$

$-8y = -21a + 8$

$y = \frac{21}{8}a - 1$

Answer: $\left(\frac{1}{4}a, \frac{21}{8}a - 1, a\right)$

To avoid fractions, we could go back and let $z = 8a$, then $4x - 8a = 0 \Rightarrow x = 2a$.

$2a - 2y + 5(8a) = 2$

$-2y + 42a = 2$

$y = 21a - 1$

Solution: $(2a, 21a - 1, 8a)$

43. $\begin{cases} 2x - 3y + z = -2 & \text{Equation 1} \\ -4x + 9y \qquad = 7 & \text{Equation 2} \end{cases}$

$\begin{cases} 2x - 3y + z = -2 \\ 3y + 2z = 3 & 2\text{Eq.1} + \text{Eq.2} \end{cases}$

$\begin{cases} 2x \qquad + 3z = 1 & \text{Eq.2} + \text{Eq.1} \\ 3y + 2z = 3 \end{cases}$

Let $z = a$, then:

$y = -\frac{2}{3}a + 1$

$x = -\frac{3}{2}a + \frac{1}{2}$

Solution: $\left(-\frac{3}{2}a + \frac{1}{2}, -\frac{2}{3}a + 1, a\right)$

45. $s = \frac{1}{2}at^2 + v_0t + s_0$

$(1, 128), (2, 80), (3, 0)$

$128 = \frac{1}{2}a + v_0 + s_0 \Rightarrow a + 2v_0 + 2s_0 = 256$

$80 = 2a + 2v_0 + s_0 \Rightarrow 2a + 2v_0 + s_0 = 80$

$0 = \frac{9}{2}a + 3v_0 + s_0 \Rightarrow 9a + 6v_0 + 2s_0 = 0$

Solving this system yields $a = -32, v_0 = 0, s_0 = 144$.

So, $s = \frac{1}{2}(-32)t^2 + (0)t + 144 = -16t^2 + 144$.

47. $y = ax^2 + bx + c$ passing through $(0, 0), (2, -2), (4, 0)$

$(0, 0)$: $0 = c$

$(2, -2)$: $-2 = 4a + 2b + c \Rightarrow -1 = 2a + b$

$(4, 0)$: $0 = 16a + 4b + c \Rightarrow 0 = 4a + b$

Solution: $a = \frac{1}{2}, b = -2, c = 0$

The equation of the parabola is $y = \frac{1}{2}x^2 - 2x$.

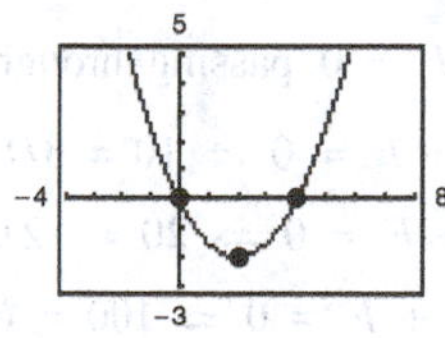

49. $y = ax^2 + bx + c$ passing through $(2, 0), (3, -1), (4, 0)$

$(2, 0)$: $0 = 4a + 2b + c$

$(3, -1)$: $-1 = 9a + 3b + c$

$(4, 0)$: $0 = 16a + 4b + c$

$$\begin{cases} 0 = 4a + 2b + c \\ -1 = 5a + b & -\text{Eq.1} + \text{Eq.2} \\ 0 = 12a + 2b & -\text{Eq.1} + \text{Eq.3} \end{cases}$$

$$\begin{cases} 0 = 4a + 2b + c \\ -1 = 5a + b \\ 2 = 2a & -2\text{Eq.2} + \text{Eq.3} \end{cases}$$

Solution:

$a = 1, b = -6, c = 8$

The equation of the parabola is $y = x^2 - 6x + 8$.

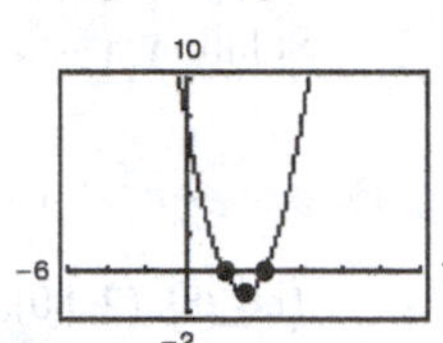

51. $y = ax^2 + bx + c$ passing through $\left(\frac{1}{2}, 1\right), (1, 3), (2, 13)$

$\left(\frac{1}{2}, 1\right)$: $1 = a\left(\frac{1}{2}\right)^2 + b\left(\frac{1}{2}\right) + c$

$(1, 3)$: $3 = a(1)^2 + b(1) + c$

$(2, 13)$: $13 = a(2)^2 + b(2) + c$

$$\begin{cases} a + 2b + 4c = 4 \\ a + b + c = 3 \\ 4a + 2b + c = 13 \end{cases}$$

Solution: $a = 4, b = -2, c = 1$

The equation of the parabola is $y = 4x^2 - 2x + 1$.

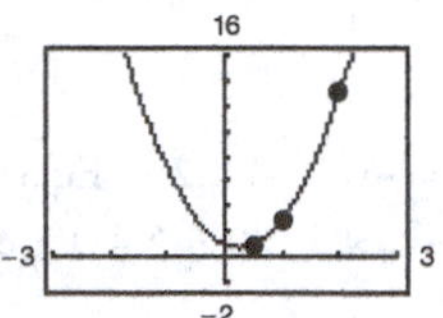

53. $x^2 + y^2 + Dx + Ey + F = 0$ passing through $(0, 0), (5, 5), (10, 0)$

$(0, 0)$: $0^2 + 0^2 + D(0) + E(0) + F = 0 \Rightarrow F = 0$

$(5, 5)$: $5^2 + 5^2 + D(5) + E(5) + F = 0 \Rightarrow 5D + 5E + F = -50$

$(10, 0)$: $10^2 + 0^2 + D(10) + E(0) + F = 0 \Rightarrow 10D + F = -100$

Solution: $D = -10, E = 0, F = 0$

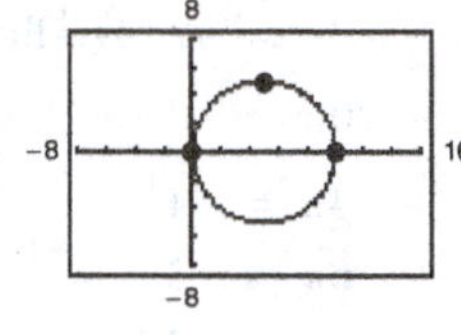

The equation of the circle is $x^2 + y^2 - 10x = 0$. To graph, complete the square first, then solve for y.

$$\begin{aligned} (x^2 - 10x + 25) + y^2 &= 25 \\ (x - 5)^2 + y^2 &= 25 \\ y^2 &= 25 - (x - 5)^2 \\ y &= \pm\sqrt{25 - (x - 5)^2} \end{aligned}$$

Let $y_1 = \sqrt{25 - (x - 5)^2}$ and $y_2 = -\sqrt{25 - (x - 5)^2}$.

55. $x^2 + y^2 + Dx + Ey + F = 0$ passing through $(-3, -1), (2, 4), (-6, 8)$

$(-3, -1)$: $10 - 3D - E + F = 0 \Rightarrow 10 = 3D + E - F$

$(2, 4)$: $20 + 2D + 4E + F = 0 \Rightarrow 20 = -2D - 4E - F$

$(-6, 8)$: $100 - 6D + 8E + F = 0 \Rightarrow 100 = 6D - 8E - F$

Solution: $D = 6, E = -8, F = 0$

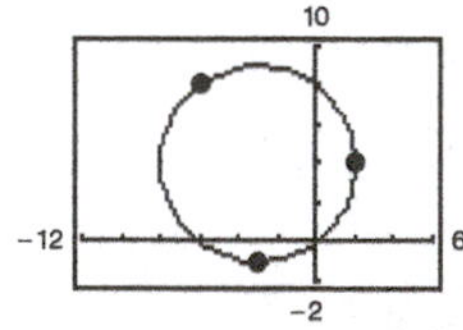

The equation of the circle is $x^2 + y^2 + 6x - 8y = 0$. To graph, complete the squares first, then solve for y.

$$\begin{aligned} (x^2 + 6x + 9) + (y^2 - 8y + 16) &= 0 + 9 + 16 \\ (x + 3)^2 + (y - 4)^2 &= 25 \\ (y - 4)^2 &= 25 - (x + 3)^2 \\ y - 4 &= \pm\sqrt{25 - (x + 3)^2} \\ y &= 4 \pm \sqrt{25 - (x + 3)^2} \end{aligned}$$

Let $y_1 = 4 + \sqrt{25 - (x + 3)^2}$ and $y_2 = 4 - \sqrt{25 - (x + 3)^2}$.

57. The leading coefficient of the third equation is not 1, so the system is not in row-echelon form.

$$\begin{cases} x - 2y + 3x = 12 \\ y + 3z = 5 \\ z = 2 \end{cases}$$

59. Let x = amount at 8%.

Let y = amount at 9%.

Let z = amount at 10%.

$$\begin{cases} x + y + z = 775{,}000 \\ 0.08x + 0.09y + 0.10z = 67{,}500 \\ x = 4z \end{cases}$$

$$\begin{cases} y + 5z = 775{,}000 \\ 0.09y + 0.42z = 67{,}500 \end{cases}$$

$z = 75{,}000$

$y = 775{,}000 - 5z = 400{,}000$

$x = 4z = 300{,}000$

\$300,000 was borrowed at 8%.

\$400,000 was borrowed at 9%.

\$75,000 was borrowed at 10%.

61. $$\begin{cases} x + y + z = 180 \\ 2x + 7 + z = 180 \\ y + 2x - 7 = 180 \end{cases}$$

$$\begin{cases} x + y + z = 180 \\ 2x + z = 173 \\ 2x + y = 187 \end{cases}$$

$$\begin{cases} -x + y = 7 & -\text{Eq.2} + \text{Eq.1} \\ 2x + z = 173 \\ 2x + y = 187 \end{cases}$$

$$\begin{cases} -x + y = 7 \\ 2x + z = 173 \\ 3x = 180 & -\text{Eq.1} + \text{Eq.3} \end{cases}$$

$x = 60°$

$2(60) + z = 173 \Rightarrow z = 53°$

$-60 + y = 7 \Rightarrow y = 67°$

63. Let x = the longest side (hypotenuse).

Let y = leg.

Let z = shortest leg.

$$\begin{cases} x + y + z = 180 \\ x = 2z - 9 \\ y + z = 30 + x \end{cases}$$

$$\begin{cases} x + y + z = 180 \\ x - 2z = -9 \\ -x + y + z = 30 \end{cases}$$

$$\begin{cases} y + 3z = 189 & -\text{Eq.2} + \text{Eq.1} \\ x - 2z = -9 \\ y - z = 21 & \text{Eq.2} + \text{Eq.3} \end{cases}$$

$$\begin{cases} 4z = 168 & -\text{Eq.3} + \text{Eq.1} \\ x - 2z = -9 \\ y - z = 21 \end{cases}$$

$z = 42$

$x - 2(42) = -9 \Rightarrow x = 75$

$y - 42 = 21 \Rightarrow y = 63$

So, the longest side measures 75 feet, the shortest side measures 42 feet, and the third side measures 63 feet.

65. $$\begin{cases} I_1 - I_2 + I_3 = 0 & \text{Equation 1} \\ 3I_1 + 2I_2 = 7 & \text{Equation 2} \\ 2I_2 + 4I_3 = 8 & \text{Equation 3} \end{cases}$$

$$\begin{cases} I_1 - I_2 + I_3 = 0 \\ 5I_2 - 3I_3 = 7 & (-3)\text{Eq.1} + \text{Eq.2} \\ 2I_2 + 4I_3 = 8 \end{cases}$$

$$\begin{cases} I_1 - I_2 + I_3 = 0 \\ 10I_2 - 6I_3 = 14 & 2\text{Eq.2} \\ 10I_2 + 20I_3 = 40 & 5\text{Eq.3} \end{cases}$$

$$\begin{cases} I_1 - I_2 + I_3 = 0 \\ 10I_2 - 6I_3 = 14 \\ 26I_3 = 26 & (-1)\text{Eq.2} + \text{Eq.3} \end{cases}$$

$26I_3 = 26 \Rightarrow I_3 = 1$

$10I_2 - 6(1) = 14 \Rightarrow I_2 = 2$

$I_1 - 2 + 1 = 0 \Rightarrow I_1 = 1$

Solution: $I_1 = 1, I_2 = 2, I_3 = 1$

67. $\begin{cases} 4c + 9b + 29a = 20 \\ 9c + 29b + 99a = 70 \\ 29c + 99b + 353a = 254 \end{cases}$

$\begin{cases} 9c + 29b + 99a = 70 \\ 4c + 9b + 29a = 20 \\ 29c + 99b + 353a = 254 \end{cases}$ Interchange equations.

$\begin{cases} c + 11b + 41a = 30 & -2\text{Eq.}2 + \text{Eq.}1 \\ -35b - 135a = -100 & -4\text{Eq.}1 + \text{Eq.}2 \\ -220b - 836a = -616 & -29\text{Eq.}1 + \text{Eq.}3 \end{cases}$

$\begin{cases} c + 11b + 41a = 30 \\ 1540b + 5940a = 4400 & -44\text{Eq.}2 \\ -1540b - 5852a = -4312 & 7\text{Eq.}3 \end{cases}$

$\begin{cases} c + 11b + 41a = 30 \\ 1540b + 5940a = 4400 \\ 88a = 88 & \text{Eq.}2 + \text{Eq.}3 \end{cases}$

$88a = 88 \Rightarrow a = 1$

$1540b + 5940(1) = 4400 \Rightarrow b = -1$

$c + 11(-1) + 41(1) = 30 \Rightarrow c = 0$

Least squares regression parabola: $y = x^2 - x$

69. (a) $\begin{cases} 5c + 250b + 13{,}500a = 923 \\ 250c + 13{,}500b + 775{,}000a = 52{,}170 \\ 13{,}500c + 775{,}000b + 46{,}590{,}000a = 3{,}101{,}300 \end{cases}$

$\begin{cases} 5c + 250b + 13{,}500a = 923 \\ 1000b + 100{,}000a = 6020 & (-50)\text{Eq. } 1 + \text{Eq. } 2 \\ 100{,}000b + 10{,}140{,}000a = 609{,}200 & (-2700)\text{Eq. } 1 + (3)\text{Eq. } 3 \end{cases}$

$\begin{cases} 5c + 250b + 13{,}500a = 923 \\ 1000b + 100{,}000a = 6020 \\ 140{,}000a = 7200 & (-100)\text{Eq. } 2 + \text{Eq. } 3 \end{cases}$

$140{,}000a = 7200 \Rightarrow a \approx 0.0514$

$1000b + 100{,}000(0.0514) = 6020 \Rightarrow b \approx 0.8771$

$5c + 250(0.8771) + 13{,}500(0.0514) = 923 \Rightarrow c \approx 1.8857$

Least-squares regression parabola: $y = 0.0514x^2 + 0.8771x + 1.8857$

(b) 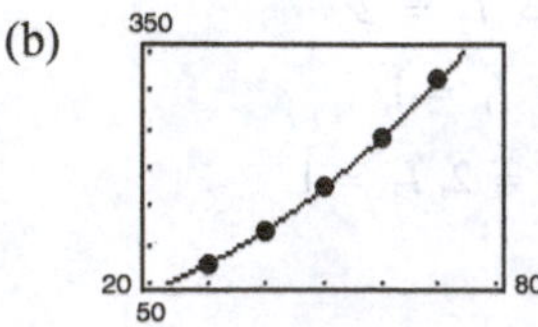

The model fits the data well.

(c) When $x = 75$, $y = 0.0514(75)^2 + 0.8771(75) + 1.8857 \approx 356$ feet.

71. $\begin{cases} 2x - 2x\lambda = 0 \Rightarrow 2x(1 - \lambda) = 0 \Rightarrow \lambda = 1 \text{ or } x = 0 \\ -2y + \lambda = 0 \\ y - x^2 = 0 \end{cases}$

If $\lambda = 1$: $2y = \lambda \Rightarrow y = \frac{1}{2}$

$x^2 = y \Rightarrow x = \pm\sqrt{\frac{1}{2}} = \pm\frac{\sqrt{2}}{2}$

If $x = 0$: $x^2 = y \Rightarrow y = 0$

$2y = \lambda \Rightarrow \lambda = 0$

Solution: $x = \pm\frac{\sqrt{2}}{2}$ or $x = 0$

$y = \frac{1}{2}$ $\qquad y = 0$

$\lambda = 1$ $\qquad \lambda = 0$

73. False. For example, refer to Example 6 on page 449,

$\begin{cases} x - 2y + z = 2 \\ 2x - y - z = 1 \end{cases}$

has the solution set of all ordered triples of the form $(a, a - 1, a)$ where a is a real number. Therefore, it is not an unique solution.

75. No, they are not equivalent. There are two arithmetic errors. The constant in the second equation should be -11 and the coefficient of z in the third equation should be 2.

77. Answers will vary. *Sample answer:*

$(2, 0, -1)$ is a solution to $\begin{cases} x + y + z = 1 \\ 3x + y + z = 5 \\ -x + 2y + 3z = -5 \end{cases}$

79. Answers will vary. *Sample answer:*

$\left(\frac{1}{2}, -3, 0\right)$ is a solution to $\begin{cases} 2x + y + z = -2 \\ 4x + y + z = -1 \\ -2x + 2y + 3z = -7 \end{cases}$

Section 6.4 Partial Fractions

1. partial fraction decomposition

3. partial fraction

5. $\frac{3x - 1}{x(x - 4)} = \frac{A}{x} + \frac{B}{x - 4}$

Matches (b).

6. $\frac{3x - 1}{x^2(x - 4)} = \frac{A}{x} + \frac{B}{x^2} + \frac{C}{x - 4}$

Matches (c).

7. $\frac{3x - 1}{x(x^2 + 4)} = \frac{A}{x} + \frac{Bx + C}{x^2 + 4}$

Matches (d).

8. $\frac{3x - 1}{x(x^2 - 4)} = \frac{3x - 1}{x(x - 2)(x + 2)}$

$= \frac{A}{x} + \frac{B}{x - 2} + \frac{C}{x + 2}$

Matches (a).

9. $\frac{3}{x^2 - 2x} = \frac{3}{x(x - 2)} = \frac{A}{x} + \frac{B}{x - 2}$

11. $\frac{6x + 5}{(x + 2)^4} = \frac{6x + 5}{(x + 2)(x + 2)(x + 2)(x + 2)}$

$= \frac{A}{x + 2} + \frac{B}{(x + 2)^2} + \frac{C}{(x + 2)^3} + \frac{D}{(x + 2)^4}$

13. $\frac{2x - 3}{x^3 + 10x} = \frac{2x - 3}{x(x^2 + 10)} = \frac{A}{x} + \frac{Bx + C}{x^2 + 10}$

15. $\frac{8x}{x^2(x^2 + 3)^2} = \frac{A}{x} + \frac{B}{x^2} + \frac{Cx + D}{x^2 + 3} + \frac{Ex + F}{(x^2 + 3)^2}$

17. $\frac{1}{x^2 + x} = \frac{A}{x} + \frac{B}{x + 1}$

$1 = A(x + 1) + Bx$

Let $x = 0$: $1 = A$

Let $x = -1$: $1 = -B \Rightarrow B = -1$

$\frac{1}{x^2 + x} = \frac{1}{x} - \frac{1}{x + 1}$

19. $\dfrac{3}{x^2+x-2} = \dfrac{A}{x-1} + \dfrac{B}{x+2}$

$3 = A(x+2) + B(x-1)$

Let $x = 1$: $3 = 3A \Rightarrow A = 1$

Let $x = -2$: $3 = -3B \Rightarrow B = -1$

$\dfrac{3}{x^2+x-2} = \dfrac{1}{x-1} - \dfrac{1}{x+2}$

21. $\dfrac{1}{x^2-1} = \dfrac{A}{x+1} + \dfrac{B}{x-1}$

$1 = A(x-1) + B(x+1)$

Let $x = -1$: $1 = -2A \Rightarrow A = -\dfrac{1}{2}$

Let $x = 1$: $1 = 2B \Rightarrow B = \dfrac{1}{2}$

$\dfrac{1}{x^2-1} = \dfrac{1/2}{x-1} - \dfrac{1/2}{x+1} = \dfrac{1}{2}\left(\dfrac{1}{x-1} - \dfrac{1}{x+1}\right)$

23. $\dfrac{x^2+12x+12}{x^3-4x} = \dfrac{A}{x} + \dfrac{B}{x+2} + \dfrac{C}{x-2}$

$x^2 + 12x + 12 = A(x+2)(x-2) + Bx(x-2) + Cx(x+2)$

Let $x = 0$: $12 = -4A \Rightarrow A = -3$

Let $x = -2$: $-8 = 8B \Rightarrow B = -1$

Let $x = 2$: $40 = 8C \Rightarrow C = 5$

$\dfrac{x^2+12x+12}{x^3-4x} = -\dfrac{3}{x} - \dfrac{1}{x+2} + \dfrac{5}{x-2}$

25. $\dfrac{3x}{(x-3)^2} = \dfrac{A}{x-3} + \dfrac{B}{(x-3)^2}$

$3x = A(x-3) + B$

Let $x = 3$: $9 = B$

Let $x = 0$: $0 = -3A + B$

$0 = -3A + 9$

$3 = A$

$\dfrac{3x}{(x-3)^2} = \dfrac{3}{x-3} + \dfrac{9}{(x-3)^2}$

27. $\dfrac{4x^2+2x-1}{x^2(x+1)} = \dfrac{A}{x} + \dfrac{B}{x^2} + \dfrac{C}{x+1}$

$4x^2 + 2x - 1 = Ax(x+1) + B(x+1) + Cx^2$

Let $x = 0$: $-1 = B$

Let $x = -1$: $1 = C$

Let $x = 1$: $5 = 2A + 2B + C$

$5 = 2A - 2 + 1$

$6 = 2A$

$3 = A$

$\dfrac{4x^2+2x-1}{x^2(x+1)} = \dfrac{3}{x} - \dfrac{1}{x^2} + \dfrac{1}{x+1}$

29. $\dfrac{x^2+2x+3}{x^3+x} = \dfrac{A}{x} + \dfrac{Bx+C}{x^2+1}$

$x^2 + 2x + 3 = A(x^2+1) + (Bx+C)(x)$

$x^2 + 2x + 3 = x^2(A+B) + Cx + A$

Equating coefficients of like terms gives

$A + B = 1$, $C = 2$, and $A = 3$.

So, $A = 3$, $B = -2$, and $C = 2$.

$\dfrac{x^2+2x+3}{x^3+x} = \dfrac{3}{x} - \dfrac{2x-2}{x^2+1}$

31. $\dfrac{x}{x^3 - x^2 - 2x + 2} = \dfrac{x}{(x-1)(x^2-2)} = \dfrac{A}{x-1} + \dfrac{Bx + C}{x^2 - 2}$

$$\begin{aligned} x &= A(x^2 - 2) + (Bx + C)(x - 1) \\ &= Ax^2 - 2A + Bx^2 - Bx + Cx - C \\ &= (A + B)x^2 + (C - B)x - (2A + C) \end{aligned}$$

Equating coefficients of like terms gives $0 = A + B, 1 = C - B,$ and $0 = 2A + C$. So, $A = -1, B = 1,$ and $C = 2$.

$$\frac{x}{x^3 - x^2 - 2x + 2} = -\frac{1}{x-1} + \frac{x+2}{x^2-2}$$

33. $\dfrac{x}{16x^4 - 1} = \dfrac{x}{(4x^2-1)(4x^2+1)} = \dfrac{x}{(2x+1)(2x-1)(4x^2+1)} = \dfrac{A}{2x+1} + \dfrac{B}{2x-1} + \dfrac{Cx + D}{4x^2+1}$

$$\begin{aligned} x &= A(2x-1)(4x^2+1) + B(2x+1)(4x^2+1) + (Cx + D)(2x+1)(2x-1) \\ &= A(8x^3 - 4x^2 + 2x - 1) + B(8x^3 + 4x^2 + 2x + 1) + (Cx + D)(4x^2 - 1) \\ &= 8Ax^3 - 4Ax^2 + 2Ax - A + 8Bx^3 + 4Bx^2 + 2Bx + B + 4Cx^3 + 4Dx^2 - Cx - D \\ &= (8A + 8B + 4C)x^3 + (-4A + 4B + 4D)x^2 + (2A + 2B - C)x + (-A + B - D) \end{aligned}$$

Equating coefficients of like terms gives $0 = 8A + 8B + 4C, 0 = -4A + 4B + 4D, 1 = 2A + 2B - C,$ and $0 = -A + B - D$.

Using the first and third equations, $2A + 2B + C = 0$ and $2A + 2B - C = 1$; by subtraction, $2C = -1$, so $C = -\frac{1}{2}$.

Using the second and fourth equations, $-A + B + D = 0$ and $-A + B - D = 0$; by subtraction $2D = 0$, so $D = 0$.

Substituting $-\frac{1}{2}$ for C and 0 for D in the first and second equations, $8A + 8B = 2$ and $-4A + 4B = 0$, so $A = \frac{1}{8}$ and $B = \frac{1}{8}$.

$$\frac{x}{16x^4 - 1} = \frac{\frac{1}{8}}{2x+1} + \frac{\frac{1}{8}}{2x-1} + \frac{\left(-\frac{1}{2}\right)x}{4x^2+1} = \frac{1}{8(2x+1)} + \frac{1}{8(2x-1)} - \frac{x}{2(4x^2+1)} = \frac{1}{8}\left(\frac{1}{2x+1} + \frac{1}{2x-1} - \frac{4x}{4x^2+1}\right)$$

35. $\dfrac{x^2 + 5}{(x+1)(x^2 - 2x + 3)} = \dfrac{A}{x+1} + \dfrac{Bx + C}{x^2 - 2x + 3}$

$$\begin{aligned} x^2 + 5 &= A(x^2 - 2x + 3) + (Bx + C)(x + 1) = Ax^2 - 2Ax + 3A + Bx^2 + Bx + Cx + C \\ &= (A + B)x^2 + (-2A + B + C)x + (3A + C) \end{aligned}$$

Equating coefficients of like terms gives $1 = A + B, 0 = -2A + B + C,$ and $5 = 3A + C$.

Subtracting both sides of the second equation from the first gives $1 = 3A - C$; combining this with the third equation gives $A = 1$ and $C = 2$. Because $A + B = 1, B = 0$.

$$\frac{x^2 + 5}{(x+1)(x^2 - 2x + 3)} = \frac{1}{x+1} + \frac{2}{x^2 - 2x + 3}$$

37. $\dfrac{2x^2 + x + 8}{\left(x^2 + 4\right)^2} = \dfrac{Ax + B}{x^2 + 4} + \dfrac{Cx + D}{\left(x^2 + 4\right)^2}$

$2x^2 + x + 8 = (Ax + B)(x^2 + 4) + Cx + D$

$2x^2 + x + 8 = Ax^3 + Bx^2 + (4A + C)x + (4B + D)$

Equating coefficients of like terms gives

$0 = A$

$2 = B$

$1 = 4A + C \Rightarrow C = 1$

$8 = 4B + D \Rightarrow D = 0$

$\dfrac{2x^2 + x + 8}{\left(x^2 + 4\right)^2} = \dfrac{2}{x^2 + 4} + \dfrac{x}{\left(x^2 + 4\right)^2}$

39. $\dfrac{5x^2 - 2}{\left(x^2 + 3\right)^3} = \dfrac{Ax + B}{x^2 + 3} + \dfrac{Cx + D}{\left(x^2 + 3\right)^2} + \dfrac{Ex + F}{\left(x^2 + 3\right)^3}$

$$\begin{aligned}5x^2 - 2 &= (Ax + B)\left(x^2 + 3\right)^2 + (Cx + D)\left(x^2 + 3\right) + Ex + F\\ &= Ax^5 + 6Ax^3 + 9Ax + Bx^4 + 6Bx^2 + 9B + Cx^3 + 3Cx + Dx^2 + 3D + F + Ex\\ &= Ax^5 + Bx^4 + (6A + C)x^3 + (6B + D)x^2 + (9A + 3C + E)x + (9B + 3D + F)\end{aligned}$$

Equating coefficients of like terms gives

$0 = A$

$0 = B$

$0 = 6A + C \Rightarrow C = 0$

$5 = 6B + D \Rightarrow D = 5$

$0 = 9A + 3C + E \Rightarrow E = 0$

$-2 = 9B + 3D + F \Rightarrow F = -17$

$\dfrac{5x^2 - 2}{\left(x^2 + 3\right)^3} = \dfrac{5}{\left(x^2 + 3\right)^2} - \dfrac{17}{\left(x^2 + 3\right)^3}$

41. $\dfrac{8x - 12}{x^2\left(x^2 + 2\right)^2} = \dfrac{A}{x} + \dfrac{B}{x^2} + \dfrac{Cx + D}{x^2 + 2} + \dfrac{Ey + F}{\left(x^2 + 2\right)^2}$

$$\begin{aligned}8x - 12 &= Ax\left(x^2 + 2\right)^2 + B\left(x^2 + 2\right)^2 + (Cx + D)x^2\left(x^2 + 2\right) + (Ex + F)x^2\\ &= Ax^5 + 4Ax^3 + 4Ax + Bx^4 + 4Bx^2 + 4B + Cx^5 + 2Cx^3 + Dx^4 + 2Dx^2 + Ex^3 + Fx^2\\ &= (A + C)x^5 + (B + D)x^4 + (4A + 2C + E)x^3 + (4B + 2D + F)x^2 + 4Ax + 4B\end{aligned}$$

Equating coefficients of like terms gives

$A + C = 0,$

$B + D = 0,$

$4A + 2C + E = 0,$

$4B + 2D + F = 0,$

$4A = 8,$ and

$4B = -12.$

So, $A = 2, B = -3, C = -2, D = 3, E = -4,$ and $F = 6$.

$\dfrac{8x - 12}{x^2\left(x^2 + 2\right)^2} = \dfrac{2}{x} + \dfrac{-3}{x^2} + \dfrac{-2x + 3}{x^2 + 2} + \dfrac{-4x + 6}{\left(x^2 + 2\right)^2}$

43. $\dfrac{x^2 - x}{x^2 + x + 1} = 1 + \dfrac{-2x - 1}{x^2 + x + 1} = 1 - \dfrac{2x + 1}{x^2 + x + 1}$

45. $\dfrac{2x^3 - x^2 + x + 5}{x^2 + 3x + 2} = 2x - 7 + \dfrac{18x + 19}{(x + 1)(x + 2)}$

$$\frac{18x + 19}{(x + 1)(x + 2)} = \frac{A}{x + 1} + \frac{B}{x + 2}$$

$$18x + 19 = A(x + 2) + B(x + 1)$$

Let $x = -1$: $1 = A$

Let $x = -2$: $-17 = -B \Rightarrow B = 17$

$$\frac{2x^3 - x^2 + x + 5}{x^2 + 3x + 2} = 2x - 7 + \frac{1}{x + 1} + \frac{17}{x + 2}$$

47. $\dfrac{x^4}{(x - 1)^3} = \dfrac{x^4}{x^3 - 3x^2 + 3x - 1}$

$$= x + 3 + \frac{6x^2 - 8x + 3}{(x - 1)^3}$$

$$\frac{6x^2 - 8x + 3}{(x - 1)^3} = \frac{A}{x - 1} + \frac{B}{(x - 1)^2} + \frac{C}{(x - 1)^3}$$

$$6x^2 - 8x + 3 = A(x - 1)^2 + B(x - 1) + C$$

Let $x = 1$: $1 = C$

Let $x = 0$: $3 = A - B + 1$ ⎫ $A - B = 2$
Let $x = 2$: $11 = A + B + 1$ ⎭ $A + B = 10$

So, $A = 6$ and $B = 4$.

$$\frac{x^4}{(x - 1)^3} = x + 3 + \frac{6}{x - 1} + \frac{4}{(x - 1)^2} + \frac{1}{(x - 1)^3}$$

49. $\dfrac{x^4 + 2x^3 + 4x^2 + 8x + 2}{x^3 + 2x^2 + x} = x + \dfrac{3x^2 + 8x + 2}{x^3 + 2x^2 + x}$

$$= x + \frac{3x^2 + 8x + 2}{x(x + 1)^2}$$

$$\frac{3x^2 + 8x + 2}{x(x + 1)^2} = \frac{A}{x} + \frac{B}{x + 1} + \frac{C}{(x + 1)^2}$$

$$3x^2 + 8x + 2 = A(x + 1)^2 + B(x)(x + 1) + C(x)$$

$$3x^2 + 8x + 2 = Ax^2 + 2Ax + A + Bx^2 + Bx + Cx$$

$$3x^2 + 8x + 2 = (A + B)x^2 + (2A + B + C)x + A$$

Equating coefficients of like terms gives
$A + B = 3$, $2A + B + C = 8$, and $A = 2$.

So, $A = 2$, $B = 1$, and $C = 3$.

$$\frac{x^4 + 2x^3 + 4x^2 + 8x + 2}{x^3 + 2x^2 + x} = x + \frac{2}{x} + \frac{1}{x + 1} + \frac{3}{(x + 1)^2}$$

51. $\dfrac{5 - x}{2x^2 + x - 1} = \dfrac{A}{2x - 1} + \dfrac{B}{x + 1}$

$$-x + 5 = A(x + 1) + B(2x - 1)$$

Let $x = \frac{1}{2}$: $\frac{9}{2} = \frac{3}{2}A \Rightarrow A = 3$

Let $x = -1$: $6 = -3B \Rightarrow B = -2$

$$\frac{5 - x}{2x^2 + x - 1} = \frac{3}{2x - 1} - \frac{2}{x + 1}$$

53. $\dfrac{3x^2 - 7x - 2}{x^3 - x} = \dfrac{A}{x} + \dfrac{B}{x + 1} + \dfrac{C}{x - 1}$

$$3x^2 - 7x - 2 = A(x^2 - 1) + Bx(x - 1) + Cx(x + 1)$$

Let $x = 0$: $-2 = -A \Rightarrow A = 2$

Let $x = -1$: $8 = 2B \Rightarrow B = 4$

Let $x = 1$: $-6 = 2C \Rightarrow C = -3$

$$\frac{3x^2 - 7x - 2}{x^3 - x} = \frac{2}{x} + \frac{4}{x + 1} - \frac{3}{x - 1}$$

55. $\dfrac{x^2 + x + 2}{(x^2 + 2)^2} = \dfrac{Ax + B}{x^2 + 2} + \dfrac{Cx + D}{(x^2 + 2)^2}$

$$x^2 + x + 2 = (Ax + B)(x^2 + 2) + Cx + D$$

$$x^2 + x + 2 = Ax^3 + Bx^2 + (2A + C)x + (2B + D)$$

Equating coefficients of like terms gives

$0 = A$

$1 = B$

$1 = 2A + C \Rightarrow C = 1$

$2 = 2B + D \Rightarrow D = 0$

$$\frac{x^2 + x + 2}{(x^2 + 2)^2} = \frac{1}{x^2 + 2} + \frac{x}{(x^2 + 2)^2}$$

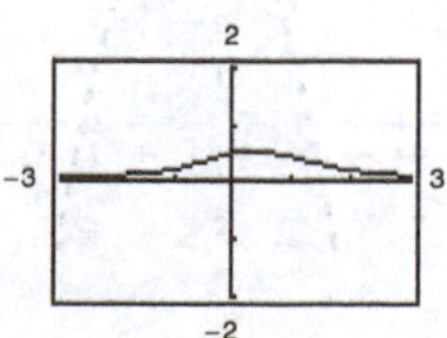

57. $\dfrac{2x^3 - 4x^2 - 15x + 5}{x^2 - 2x - 8} = 2x + \dfrac{x + 5}{(x + 2)(x - 4)}$

$$\frac{x + 5}{(x + 2)(x - 4)} = \frac{A}{x + 2} + \frac{B}{x - 4}$$

$$x + 5 = A(x - 4) + B(x + 2)$$

Let $x = -2$: $3 = -6A \Rightarrow A = -\dfrac{1}{2}$

Let $x = 4$: $9 = 6B \Rightarrow B = \dfrac{3}{2}$

$$\frac{2x^3 - 4x^2 - 15x + 5}{x^2 - 2x - 8} = 2x + \frac{1}{2}\left(\frac{3}{x - 4} - \frac{1}{x + 2}\right)$$

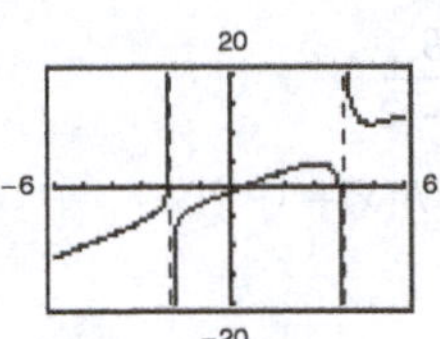

59. $C = \dfrac{120p}{10{,}000 - p^2} = \dfrac{120p}{(100 + p)(100 - p)} = \dfrac{A}{100 + p} + \dfrac{B}{100 - p}$

$$120p = A(100 - p) + B(100 + p)$$

Let $p = 100$: $200B = 12{,}000$

$B = 60$

Let $p = -100$: $200A = -12{,}000$

$A = -60$

$$C = \frac{120p}{10{,}000 - p^2} = -\frac{60}{100 + p} + \frac{60}{100 - p}$$

Let $y_1 = \dfrac{120p}{10{,}000 - p^2}$ and $y_2 = -\dfrac{60}{100 + p} + \dfrac{60}{100 - p}$.

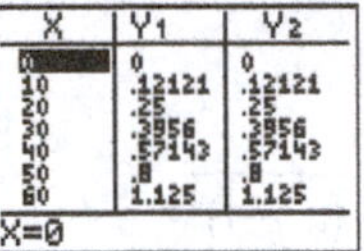

X	Y1	Y2
0	0	0
10	.12121	.12121
20	.25	.25
30	.3956	.3956
40	.57143	.57143
50	.8	.8
60	1.125	1.125

X=0

61. False. The partial fraction decomposition is

$$\frac{A}{x + 10} + \frac{B}{x - 10} + \frac{C}{(x - 10)^2}.$$

63. False. The degrees could be equal. For example,

Exercise 55: $\dfrac{x^2 + x + 2}{(x^2 + 2)^2} = \dfrac{1}{x^2 + 2} + \dfrac{x}{(x^2 + 2)^2}$.

65. The expression is improper, $\dfrac{x^2 + 1}{x(x - 1)} = \dfrac{x^2 + 1}{x^2 - x}$ so first divide the denominator into the numerator to obtain

$$\frac{x^2 + 1}{x^2 - x} = 1 + \frac{x + 1}{x^2 - x} = 1 + \frac{x + 1}{x(x - 1)}.$$

Then find the partial fraction decomposition of

$$\frac{x + 1}{x(x - 1)} = \frac{A}{x} + \frac{B}{x - 1}$$

Section 6.5 Systems of Inequalities

1. solution

3. solution

5. $y < 5 - x^2$

Using a dashed line, graph $y = 5 - x^2$, and shade the region inside the parabola.

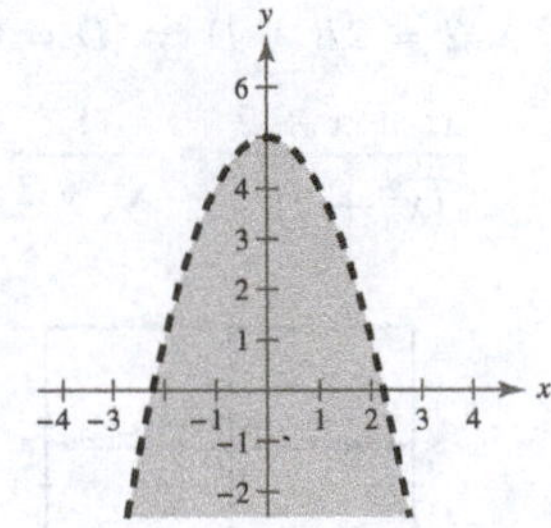

7. $x \ge 6$

Using a solid line, graph the vertical line $x = 6$, and shade to the right of this line.

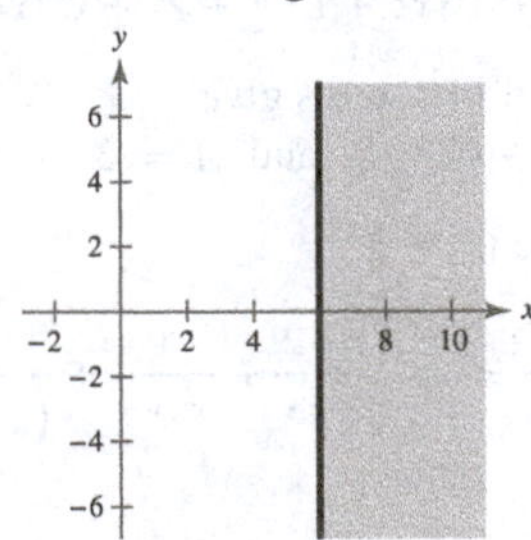

9. $y > -7$

Using a dashed line, graph the horizontal line $y = -7$, and shade above the line.

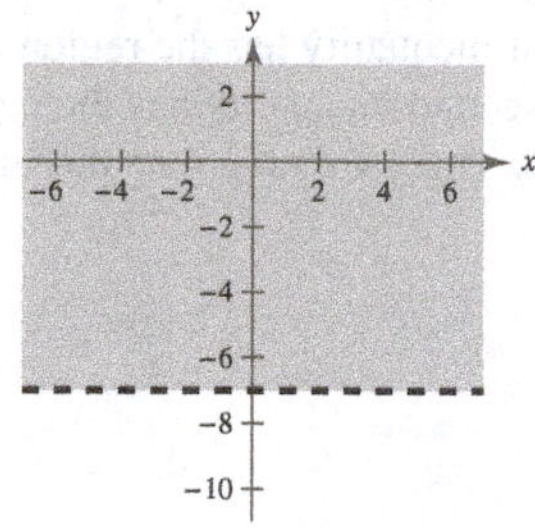

11. $y < 2 - x$

Using a dashed line, graph $y = 2 - x$, and then shade below the line. (Use $(0, 0)$ as a test point.)

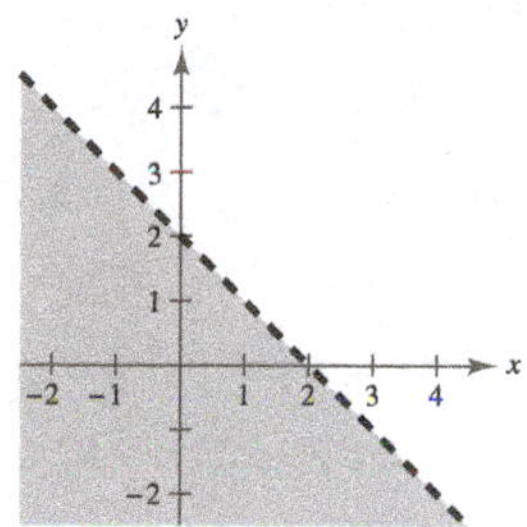

13. $2y - x \geq 4$

Using a solid line, graph $2y - x = 4$, and then shade above the line. (Use $(0, 0)$ as a test point.)

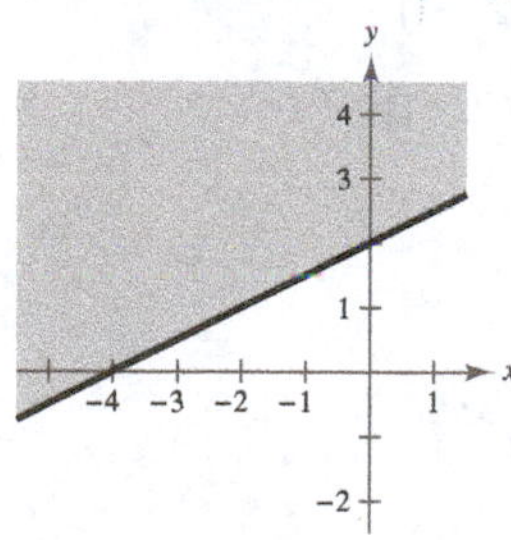

15. $x^2 + (y - 3)^2 < 4$

Using a dashed line, sketch the circle $x^2 + (y - 3)^2 = 4$.

Center: $(0, 3)$

Radius: 2

Test point: $(0, 0)$

Shade the inside of the circle.

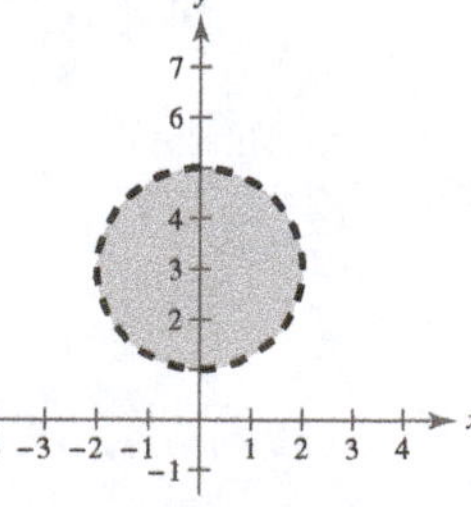

17. $y > -\dfrac{2}{x^2 + 1}$

Using a solid line, graph $y = -\dfrac{2}{x^2 + 1}$. Use $(0, 0)$ as a test point. Then shade above the curve.

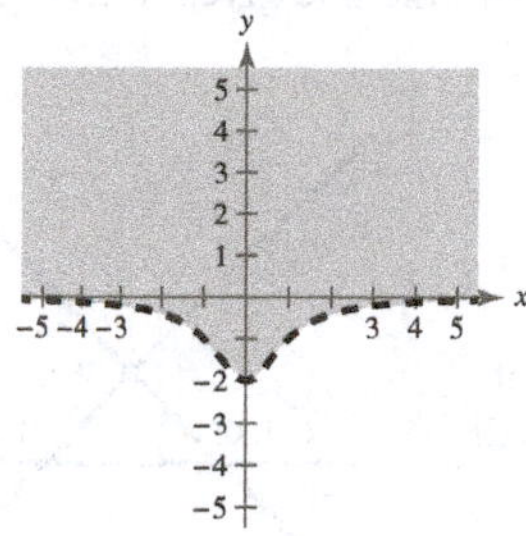

19. $y \geq -\ln(x - 1)$

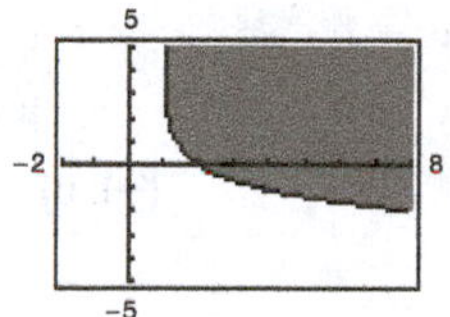

21. $y < 2^x$

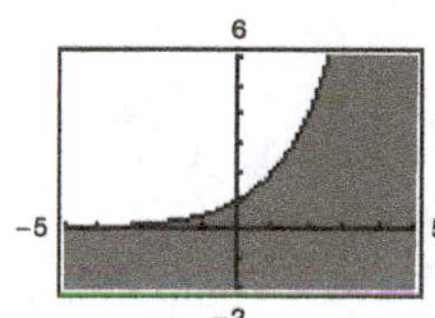

23. $y \leq 2 - \frac{1}{5}x$

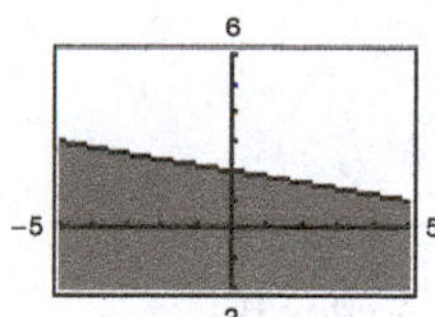

25. $\frac{2}{3}y + 2x^2 - 5 \geq 0$

$$\frac{2}{3}y \geq 5 - 2x^2$$

$$y \geq \frac{3}{2}(5 - 2x^2)$$

$$y \geq \frac{15}{2} - 3x^2$$

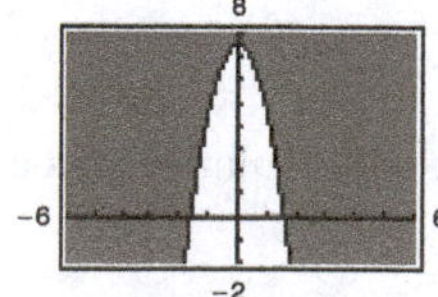

27. The line through $(-5, 0)$ and $(-1, 0)$ is $y = 5x + 5$.
The shaded region below the line gives $y < 5x + 5$.

29. The line through $(0, 2)$ and $(3, 0)$ is $y = -\frac{2}{3}x + 2$.
The shaded region above the line gives $y \geq -\frac{2}{3}x + 2$.

31. $\begin{cases} x + y \leq 1 \\ -x + y \leq 1 \\ y \geq 0 \end{cases}$

First, find the points of intersection of each pair of equations.

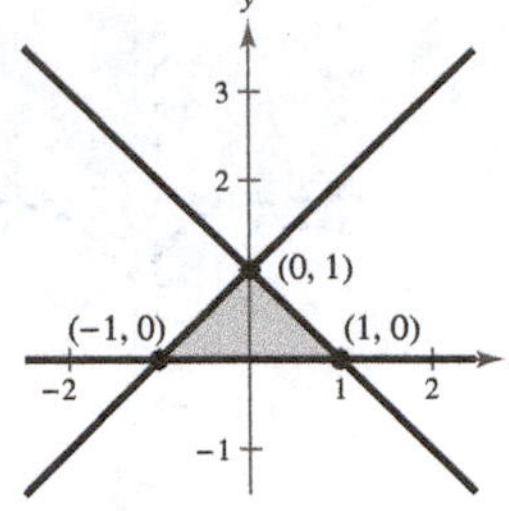

Vertex A	Vertex B	Vertex C
$x + y = 1$	$x + y = 1$	$-x + y = 1$
$-x + y = 1$	$y = 0$	$y = 0$
$(0, 1)$	$(1, 0)$	$(-1, 0)$

33. $\begin{cases} -3x + 2y < 6 \\ x - 4y > -2 \\ 2x + y < 3 \end{cases}$

First, find the points of intersection of each pair of equations.

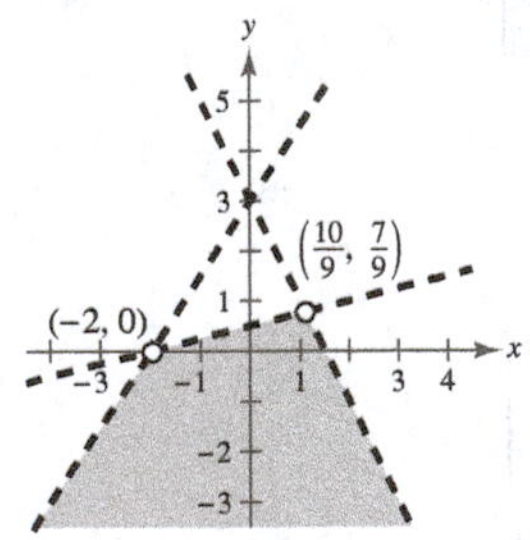

Vertex A

$-3x + 2y = 6$
$x - 4y = -2$
$(-2, 0)$

Vertex B

$-3x + 2y = 6$
$2x + y = 3$
$(0, 3)$

Vertex C

$x - 4y = -2$
$2x + y = 3$
$\left(\frac{10}{9}, \frac{7}{9}\right)$

Note that B is not a vertex of the solution region.

35. $\begin{cases} 2x + y > 2 \\ 6x + 3y < 2 \end{cases}$

The graphs of $2x + y = 2$ and $6x + 3y = 2$ are parallel lines. The first inequality has the region above the line shaded. The second inequality has the region below the line shaded. There are no points that satisfy both inequalities.

No solution

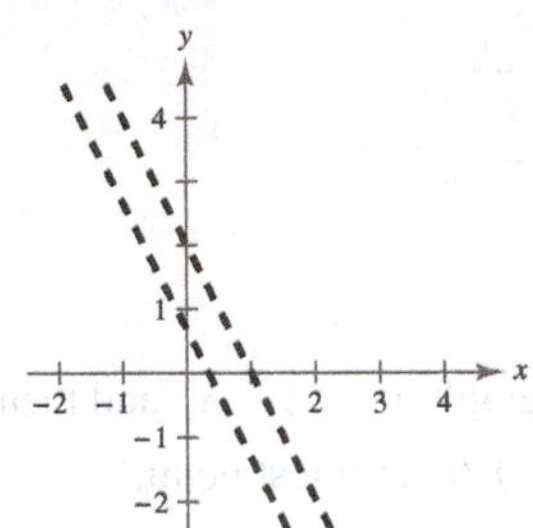

37. $\begin{cases} 2x - 3y > 7 \\ 5x + y < 9 \end{cases}$

$$2x - 3y = 7$$
$$5x + y = 9 \Rightarrow y = -5x + 9$$
$$2x - 3(-5x + 9) = 7$$
$$2x + 15x - 27 = 7$$
$$17x = 34$$
$$x = 2$$
$$y = -5(2) + 9 = -1$$
$$(2, -1)$$

Point of intersection: $(2, -1)$

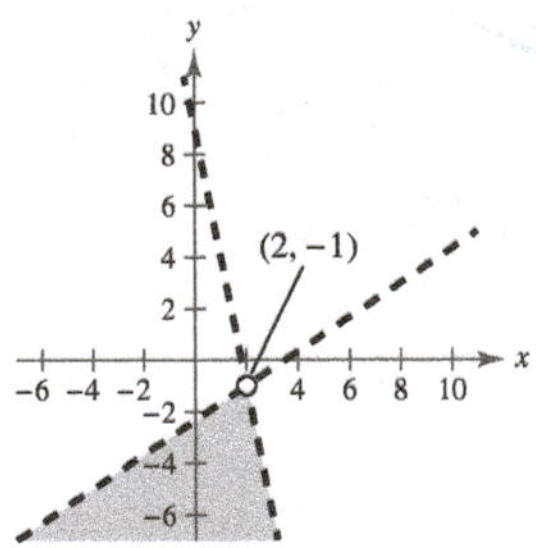

39. $\begin{cases} x^2 + y \le 7 \\ x \ge -2 \\ y \ge 0 \end{cases}$

First, find the points of intersection of each pair of equations.

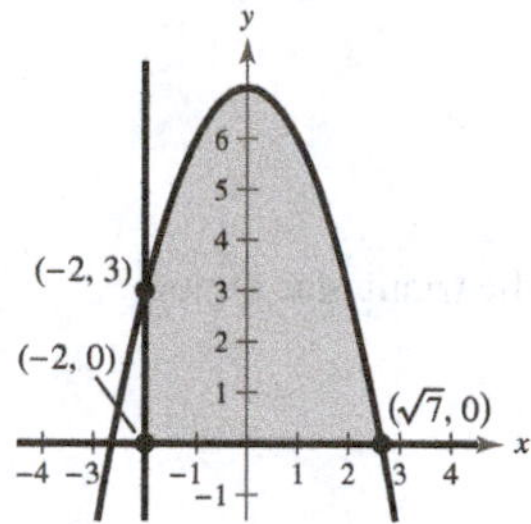

Vertex A

$x^2 + y = 7, x = -2$

$4 + y = 7$

$y = 3$

$(-2, 3)$

Vertex B

$x^2 + y = 7, y = 0$

$x^2 = 7$

$x = \sqrt{7}$

$(\sqrt{7}, 0)$

Vertex C

$x = -2, y = 0$

$(-2, 0)$

41. $\begin{cases} x - y^2 > 0 \\ x - y > 2 \end{cases}$

Points of intersection:

$y^2 = y + 2$

$y^2 - y - 2 = 0$

$(y + 1)(y - 2) = 0$

$y = -1, 2$

$(1, -1), (4, 2)$

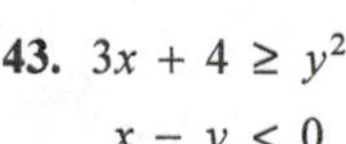

43. $3x + 4 \ge y^2$

$x - y < 0$

Points of intersection:

$x - y = 0 \Rightarrow y = x$

$3y + 4 = y^2$

$0 = y^2 - 3y - 4$

$0 = (y - 4)(y + 1)$

$y = 4$ or $y = -1$

$x = 4 \quad x = -1$

$(4, 4)$ and $(-1, -1)$

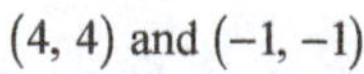

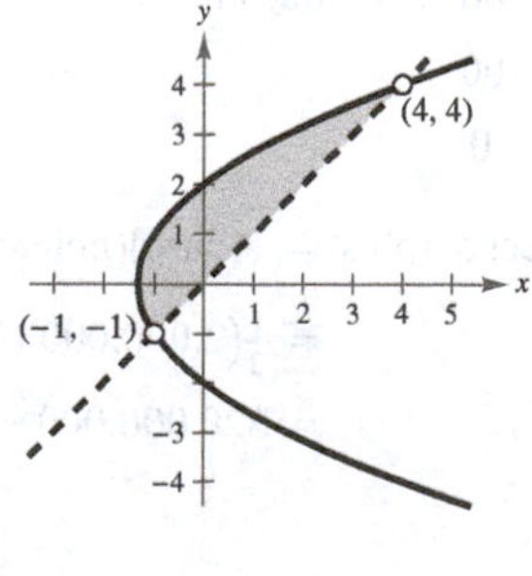

45. $\begin{cases} y \le \sqrt{3x} + 1 \\ y \ge x^2 + 1 \end{cases}$

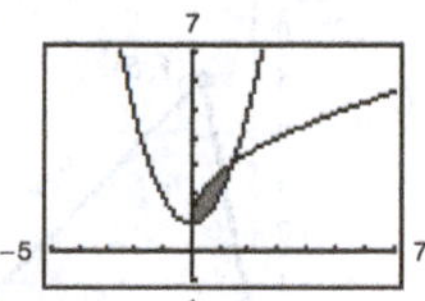

47. $\begin{cases} y < -x^2 + 2x + 3 \\ y > x^2 - 4x + 3 \end{cases}$

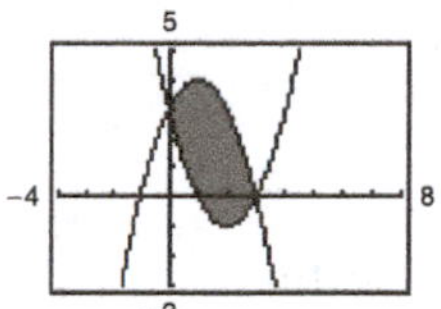

49. $\begin{cases} x^2y \ge 1 \Rightarrow y \ge \dfrac{1}{x^2} \\ 0 < x \le 4 \\ y \le 4 \end{cases}$

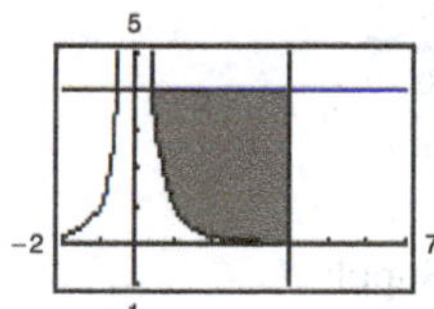

51. Line through points $(6, 0)$ and $(0, 6)$: $y = 6 - x$

$\begin{cases} x \ge 0 \\ y \ge 0 \\ y \le 6 - x \end{cases}$

53. $(8, 0), (0, 8)$

$\begin{cases} x \ge 0 \\ y \ge 0 \\ x^2 + y^2 < 64 \end{cases}$

55. Rectangular region with vertices at $(4, 3), (9, 3), (9, 9), (4, 9)$

$\begin{cases} x \ge 4 \\ x \le 9 \\ y \ge 3 \\ y \le 9 \end{cases}$

This system may be written as:

$\begin{cases} 4 \le x \le 9 \\ 3 \le y \le 9 \end{cases}$

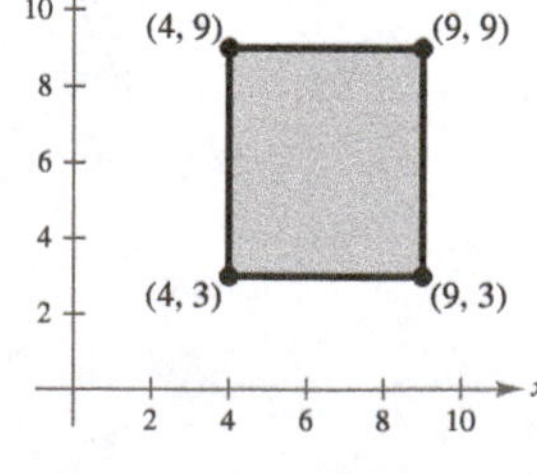

57. Triangle with vertices at $(0, 0), (6, 0), (1, 5)$

$(0, 0), (6, 0)$: $y = 0$

$(0, 0), (1, 5)$: $y = 5x$

$(6, 0), (1, 5)$: $y = -x + 6$

$$\begin{cases} y \geq 0 \\ y \leq 5x \\ y \leq -x + 6 \end{cases}$$

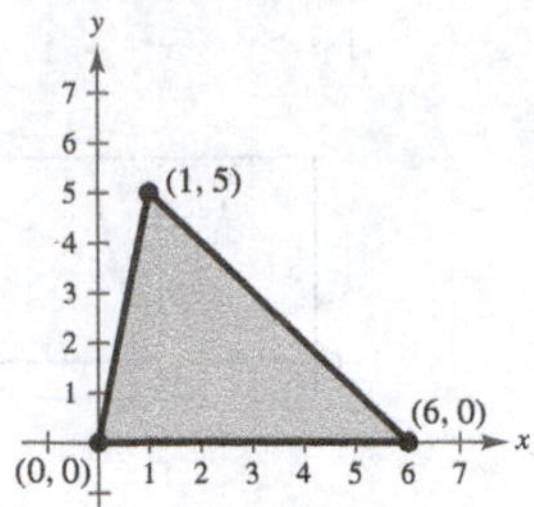

59. (a) Demand = Supply

$$50 - 0.5x = 0.125x$$
$$50 = 0.625x$$
$$80 = x$$
$$10 = p$$

Point of equilibrium: $(80, 10)$

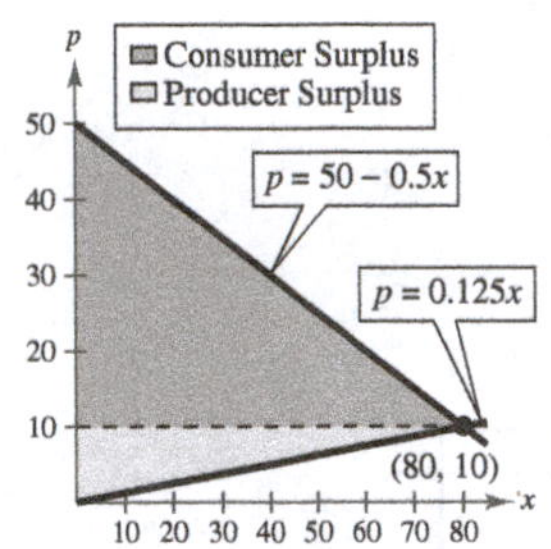

(b) The consumer surplus is the area of the triangular region defined by

$$\begin{cases} p \leq 50 - 0.5x \\ p \geq 10 \\ x \geq 0. \end{cases}$$

Consumer surplus $= \frac{1}{2}(\text{base})(\text{height}) = \frac{1}{2}(80)(40) = \$1600$

The producer surplus is the area of the triangular region defined by

$$\begin{cases} p \geq 0.125x \\ p \leq 10 \\ x \geq 0. \end{cases}$$

Producer surplus $= \frac{1}{2}(\text{base})(\text{height}) = \frac{1}{2}(80)(10) = \$400$

61. (a) Demand = Supply

$$140 - 0.00002x = 80 + 0.00001x$$
$$60 = 0.00003x$$
$$2{,}000{,}000 = x$$
$$100 = p$$

Point of equilibrium: $(2{,}000{,}000, 100)$

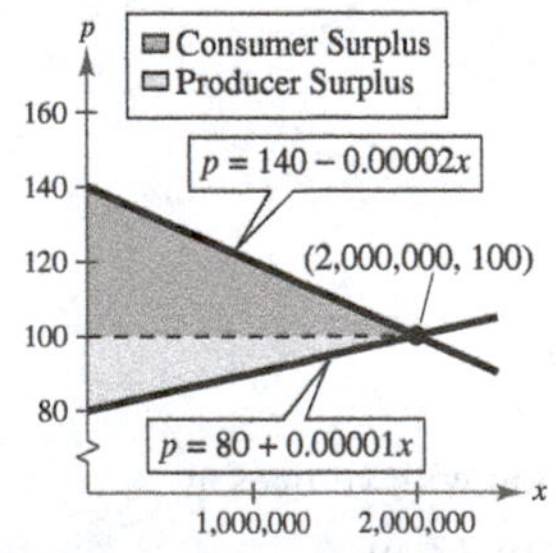

(b) The consumer surplus is the area of the triangular region defined by

$$\begin{cases} p \leq 140 - 0.00002x \\ p \geq 100 \\ x \geq 0. \end{cases}$$

$$\begin{aligned} \text{Consumer surplus} &= \tfrac{1}{2}(\text{base})(\text{height}) \\ &= \tfrac{1}{2}(2{,}000{,}000)(40) \\ &= \$40{,}000{,}000 \end{aligned}$$

The producer surplus is the area of the triangular region defined by

$$\begin{cases} p \geq 80 + 0.00001x \\ p \leq 100 \\ x \geq 0. \end{cases}$$

$$\begin{aligned} \text{Producer surplus} &= \tfrac{1}{2}(\text{base})(\text{height}) \\ &= \tfrac{1}{2}(2{,}000{,}000)(20) \\ &= \$20{,}000{,}000 \end{aligned}$$

63. x = amount in smaller account

y = amount in larger account

Account constraints:

$$\begin{cases} x + y \le 20{,}000 \\ y \ge 2x \\ x \ge 5{,}000 \\ y \ge 5{,}000 \end{cases}$$

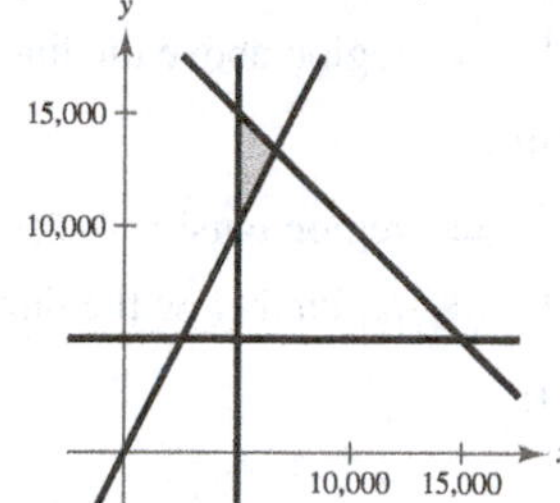

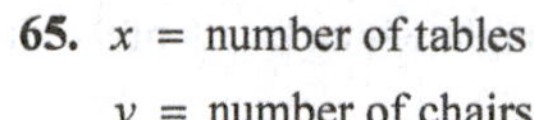

65. x = number of tables

y = number of chairs

$$\begin{cases} x + \frac{3}{2}y \le 12 & \text{Assembly center} \\ \frac{4}{3}x + \frac{3}{2}y \le 15 & \text{Finishing center} \\ x \ge 0 \\ y \ge 0 \end{cases}$$

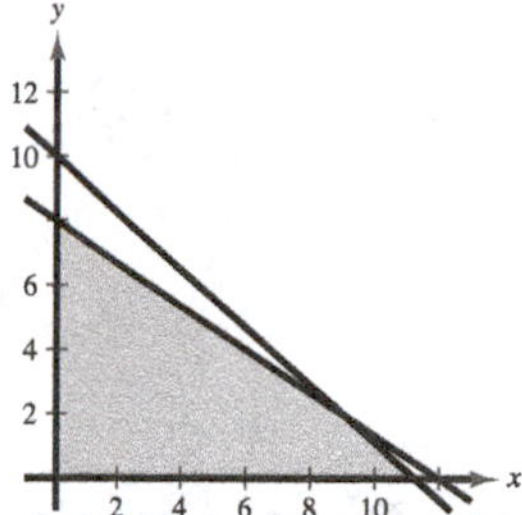

67. (a) x = number of ounces of food X

y = number of ounces of food Y

$$\begin{cases} 180x + 100y \ge 1000 \text{ (calcium)} \\ 6x + y \ge 18 \text{ (iron)} \\ 220x + 40y \ge 400 \text{ (magnesium)} \\ x \ge 0 \\ y \ge 0 \end{cases}$$

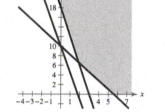

(b) Answers will vary. Some possible solutions which would satisfy the minimum daily requirements for calcium, iron, and magnesium:

$(5, 10) \Rightarrow$ 5 ounces of food X and 10 ounces of food Y

$(4, 12) \Rightarrow$ 4 ounces of food X and 12 ounces of food Y

Either of these will satisfy the minimum daily requirements of the dietician's special dietary diet plan.

69. (a) Let x = number of bags of gravel

Let y = number of bags of stone.

The delivery requirements are:

$$\begin{cases} x \ge 50 \\ y \ge 40 \\ 55x + 70y \le 7500 \end{cases}$$

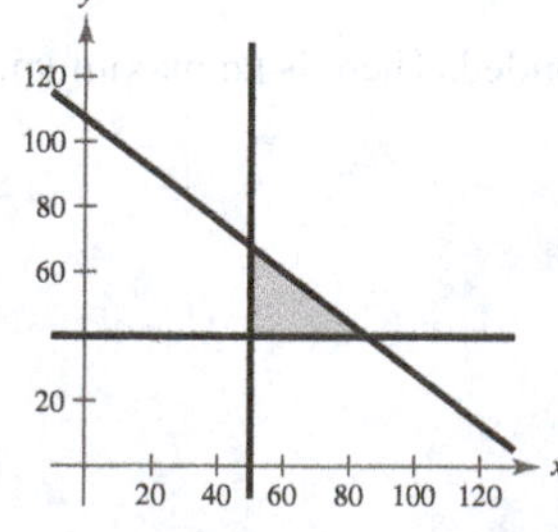

(b) The points $(60, 60)$ and $(70, 52)$ lie in the solution region.

These values would represent the number of bags of each type of fill while maintaining the maximum weight capacity of the truck. The first $(60, 60)$ is to ship 60 bags of gravel and 60 bags of stone. The second $(70, 52)$ is to is to ship 70 bags of gravel and 52 bags of stone.

71. True. The figure is a rectangle with a length of 9 units and a width of 11 units.

73. Test a point on each side of the line $y = -x + 3$. Because the origin $(0, 0)$ satisfies the inequality, the solution set of the inequality lies below the dashed line.

75. (a) $\begin{cases} x^2 + y^2 \le 16 & \Rightarrow \text{region inside the circle} \\ x + y \ge 4 & \Rightarrow \text{region above the line} \end{cases}$

Matches graph (iv).

(b) $\begin{cases} x^2 + y^2 \le 16 & \Rightarrow \text{region inside the circle} \\ x + y \le 4 & \Rightarrow \text{region below the line} \end{cases}$

Matches graph (ii).

(c) $\begin{cases} x^2 + y^2 \ge 16 & \Rightarrow \text{region outside the circle} \\ x + y \ge 4 & \Rightarrow \text{region above the line} \end{cases}$

Matches graph (iii).

(d) $\begin{cases} x^2 + y^2 \ge 16 & \Rightarrow \text{region outside the circle} \\ x + y \le 4 & \Rightarrow \text{region below the line} \end{cases}$

Matches graph (i).

Section 6.6 Linear Programming

1. optimization

3. objective

5. inside; on

7. $z = 4x + 3y$

At $(0, 5)$: $z = 4(0) + 3(5) = 15$

At $(0, 0)$: $z = 4(0) + 3(0) = 0$

At $(5, 0)$: $z = 4(5) + 3(0) = 20$

The minimum value is 0 at $(0, 0)$.

The maximum value is 20 at $(5, 0)$.

9. $z = 2x + 5y$

At $(1, 0)$: $z = 2(1) + 5(0) = 2$

At $(4, 0)$: $z = 2(4) + 5(0) = 8$

At $(3, 4)$: $z = 2(3) + 5(4) = 26$

At $(0, 5)$: $z = 2(0) + 5(5) = 25$

The minimum value is 2 at $(1, 0)$.

The maximum value is 26 at $(3, 4)$.

11. $z = 10x + 7y$

At $(0, 20)$: $z = 10(0) + 7(20) = 140$

At $(30, 45)$: $z = 10(30) + 7(45) = 615$

At $(60, 20)$: $z = 10(60) + 7(20) = 740$

At $(60, 0)$: $z = 10(60) + 7(0) = 600$

At $(0, 45)$: $z = 10(0) + 7(45) = 315$

The minimum value is 140 at $(0, 20)$.

The maximum value is 740 at $(60, 20)$.

13. $z = 3x + 2y$

At $(3, 0)$: $z = 3(3) + 2(0) = 9$

The minimum value is 9 at $(3, 0)$.

The maximum value is 24 at any point on the line $3x + 2y = 24$, that is any point on the line segment between $(0, 12)$ and $(8, 0)$.

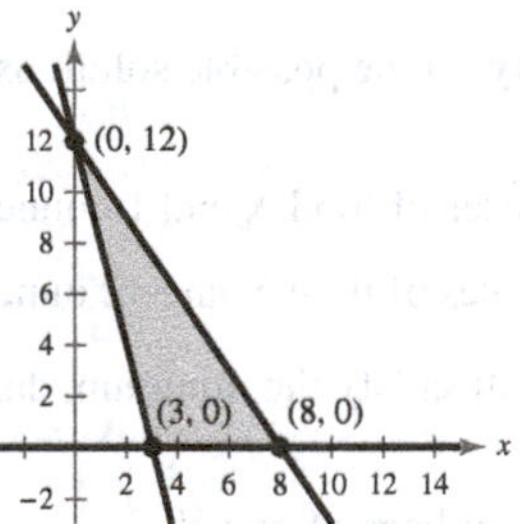

15. $z = 4x + 5y$

At $(10, 0)$: $z = 4(10) + 5(0) = 40$

At $(5, 3)$: $z = 4(5) + 5(3) = 35$

At $(0, 8)$: $z = 4(0) + 5(8) = 40$

The minimum value is 35 at $(5, 3)$.

The region is unbounded. There is no maximum.

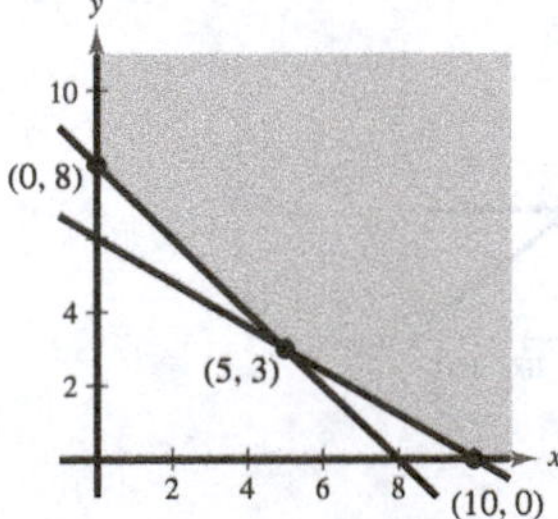

17. $z = 3x + y$

At $(16, 0)$: $z = 3(16) + 0 = 48$

At $(60, 0)$: $z = 3(60) + 0 = 180$

At $(7.2, 13.2)$: $z = 3(7.2) + 13.2 = 34.8$

The minimum value is 34.8 at $(7.2, 13.2)$.

The maximum value is 180 at $(60, 0)$.

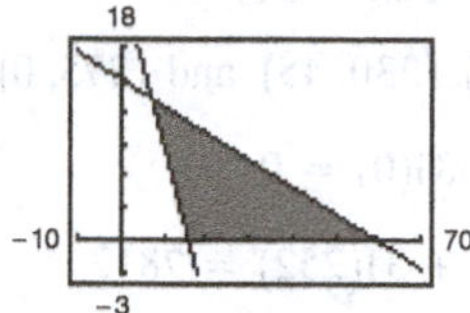

19. $z = x$

At $(60, 0)$: $z = 60$

At $(7.2, 13.2)$: $z = 7.2$

At $(16, 0)$: $z = 16$

The minimum value is 7.2 at $(7.2, 13.2)$.

The maximum value is 60 at $(60, 0)$.

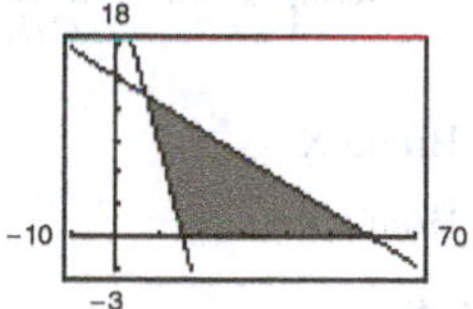

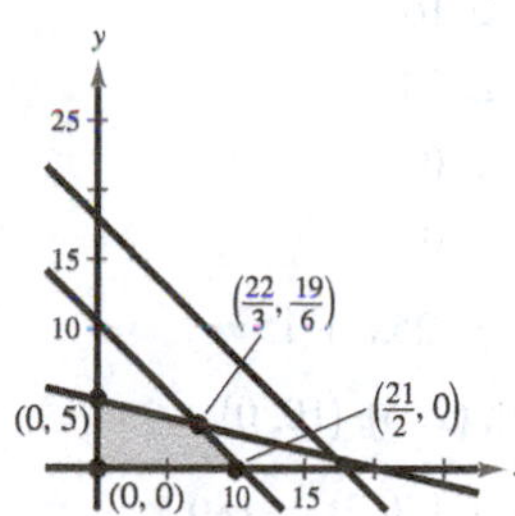

Figure for Exercises 21–23

21. $z = x + 5y$

At $(0, 5)$: $z = 0 + 5(5) = 25$

At $\left(\frac{22}{3}, \frac{19}{6}\right)$: $z = \frac{22}{3} + 5\left(\frac{19}{6}\right) = \frac{139}{6}$

At $\left(\frac{21}{2}, 0\right)$: $z = \frac{21}{2} + 5(0) = \frac{21}{2}$

At $(0, 0)$: $z = 0 + 5(0) = 0$

The minimum value is 0 at $(0, 0)$.

The maximum value is 25 at $(0, 5)$.

23. $z = 4x + 5y$

At $(0, 5)$: $z = 4(0) + 5(5) = 25$

At $\left(\frac{22}{3}, \frac{19}{6}\right)$: $z = 4\left(\frac{22}{3}\right) + 5\left(\frac{19}{6}\right) = \frac{271}{6}$

At $\left(\frac{21}{2}, 0\right)$: $z = 4\left(\frac{21}{2}\right) + 5(0) = 42$

At $(0, 0)$: $z = 4(0) + 5(0) = 0$

The minimum value is 0 at $(0, 0)$.

The maximum value is $\frac{271}{6}$ at $\left(\frac{22}{3}, \frac{19}{6}\right)$.

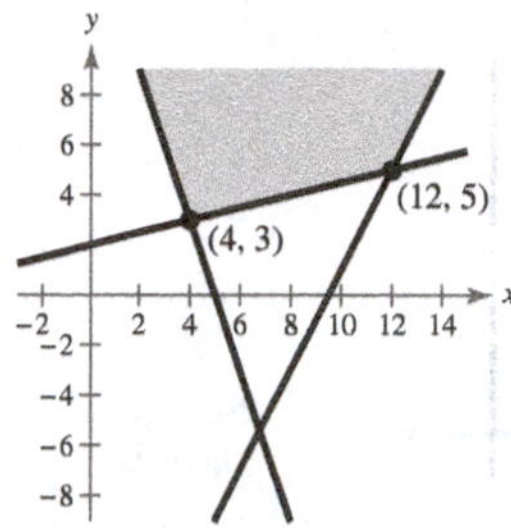

Figure for Exercises 25–27

25. $z = x + 2y$

At $(4, 3)$: $z = 4 + 2(3) = 10$

At $(12, 5)$: $z = 12 + 2(5) = 22$

The minimum value is 10 at $(4, 3)$.

There is no maximum value, the region is unbounded.

27. $z = x - y$

At $(4, 3)$: $z = 4 - 3 = 1$

At $(12, 5)$: $z = 12 - 5 = 7$

There is no minimum value.

The maximum value is 7 at $(12, 5)$.

29. Objective function: $z = 2.5x + y$

Constraints:
$x \geq 0, y \geq 0, 3x + 5y \leq 15, 5x + 2y \leq 10$

At $(0, 0)$: $z = 0$

At $(2, 0)$: $z = 5$

At $\left(\frac{20}{19}, \frac{45}{19}\right)$: $z = \frac{95}{19} = 5$

At $(0, 3)$: $z = 3$

The minimum value is 0 at $(0, 0)$.

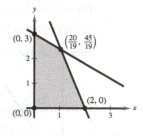

The maximum value of 5 occurs at any point on the line segment connecting $(2, 0)$ and $\left(\frac{20}{19}, \frac{45}{19}\right)$.

31. Objective function: $z = -x + 2y$

Constraints: $x \geq 0, y \geq 0, x \leq 10, x + y \leq 7$

At $(0, 0)$: $z = -0 + 2(0) = 0$

At $(0, 7)$: $z = -0 + 2(7) = 14$

At $(7, 0)$: $z = -7 + 2(0) = -7$

The constraint $x \leq 10$ is extraneous.

The minimum value is -7 at $(7, 0)$.

The maximum value is 14 at $(0, 7)$.

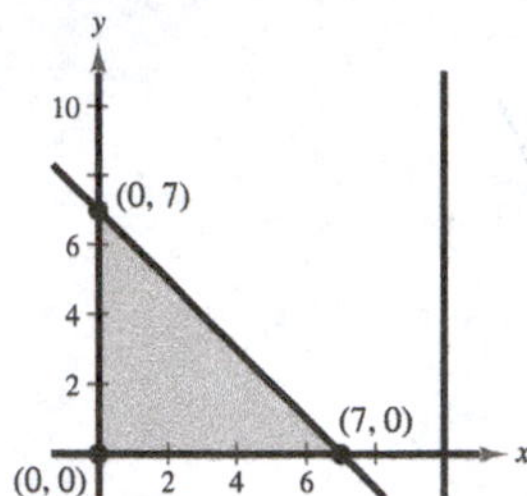

33. Objective function: $z = 3x + 4y$

Constraints: $x \geq 0, y \geq 0, x + y \leq 1, 2x + y \geq 4$

The feasible set is empty.

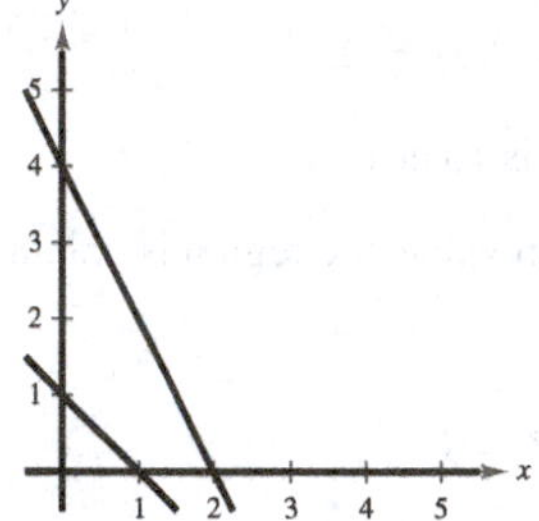

35. Objective function: $z = x + y$

Constraints: $x \geq 9, 0 \leq y \leq 7, -x + 3y \leq -6$

At $(9, 0)$: $z = 9 + 0 = 9$

At $(9, 1)$: $z = 9 + 1 = 10$

At $(27, 7)$: $z = 27 + 7 = 34$

The solution region is unbounded.

The minimum value is 9 at $(9, 0)$.

There is no maximum value.

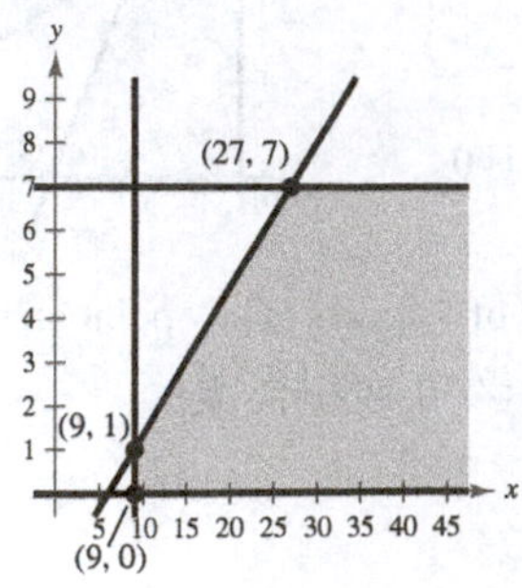

37. x = number of \$225 models

y = number of \$250 models

Constraints:

$$225x + 250y \leq 63{,}000$$
$$x + y \leq 275$$
$$x \geq 0$$
$$y \geq 0$$

Objective function: $P = 30x + 31y$

Vertices: $(0, 0)$, $(0, 252)$, $(230, 45)$ and $(275, 0)$

At $(0, 0)$: $P = 30(0) + 31(0) = 0$

At $(0, 252)$: $P = 30(0) + 31(252) = 7812$

At $(230, 45)$: $P = 30(230) + 31(45) = 8295$

At $(275, 0)$: $P = 30(275) + 31(0) = 8250$

An optimal profit of \$8295 occurs when 230 units of the \$225 model and 45 units of the \$250 model are stocked in inventory.

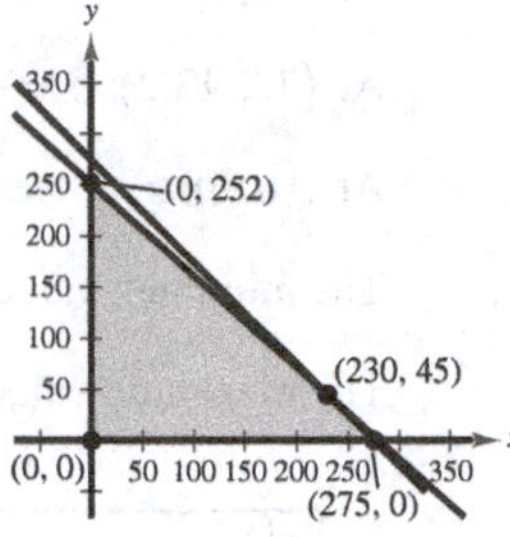

39. x = number of bags of Brand X

y = number of bags of Brand Y

Constraints:

$$3x + 9y \geq 30$$
$$3x + 2y \geq 16$$
$$7x + 2y \geq 24$$
$$x \geq 0$$
$$y \geq 0$$

Objective function: $C = 25x + 15y$

Vertices: $(0, 12)$, $(4, 2)$, $(2, 5)$, $(10, 0)$

At $(0, 12)$: $C = 25(0) + 15(12) = 180$

At $(4, 2)$: $C = 25(4) + 15(2) = 130$

At $(2, 5)$: $C = 25(2) + 15(5) = 125$

At $(10, 0)$: $C = 25(10) + 15(0) = 250$

To minimize cost, use two bags of Brand X and five bags of Brand Y for a minimal cost of \$125.

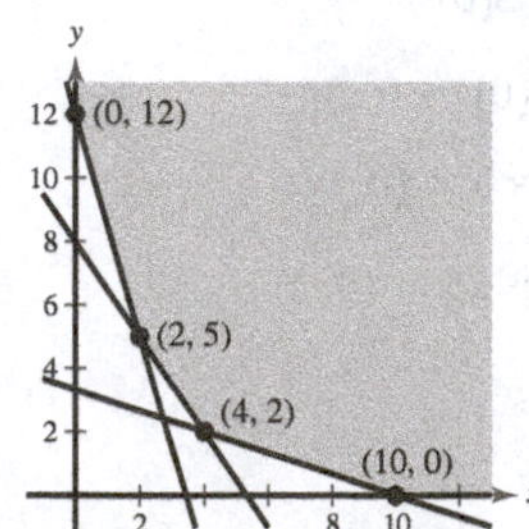

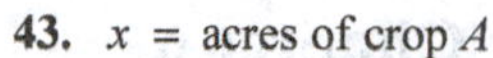

41. x = number of audits

y = number of tax returns

Constraints:

$60x + 10y \le 780$

$16x + 4y \le 272$

$x \ge 0$

$y \ge 0$

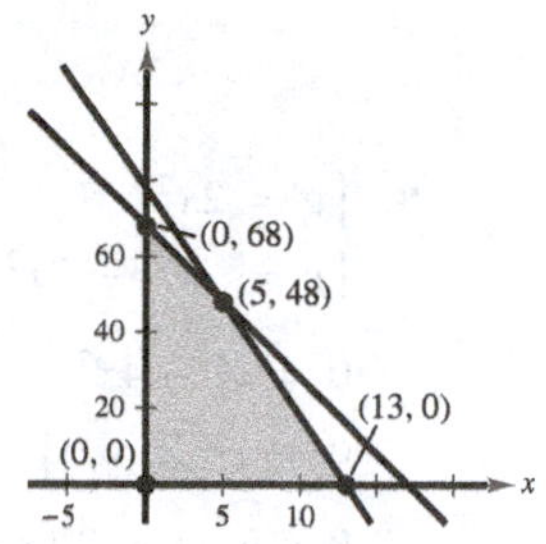

Objective function:

$R = 1600x + 250y$

Vertices: $(0, 0), (13, 0), (5, 48), (0, 68)$

At $(0, 0)$: $R = 1600(0) + 250(0) = 0$

At $(13, 0)$: $R = 1600(13) + 250(0) = 20{,}800$

At $(5, 48)$: $R = 1600(5) + 250(48) = 20{,}000$

At $(0, 68)$: $R = 1600(0) + 250(68) = 17{,}000$

A maximum revenue of $20,800 occurs when the firm conducts 13 audits and 0 tax returns.

43. x = acres of crop A

y = acres of crop B

Constraints: $x + y \le 150, x + 2y \le 240,$

$0.3x + 0.1y \le 30$

Objective function: $z = 300x + 500y$

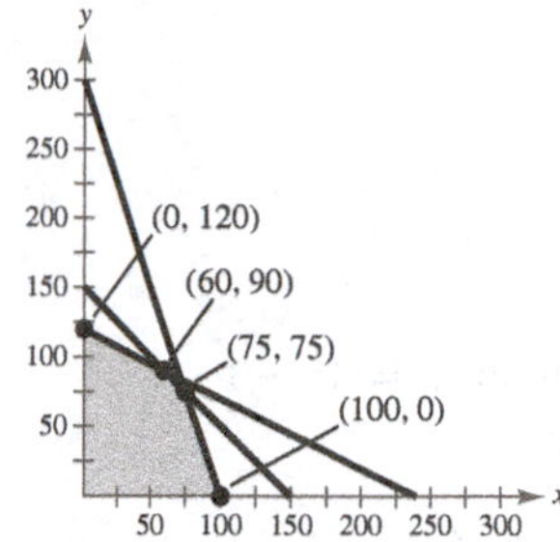

At $(0, 0)$: $z = 300(0) + 500(0) = 0$

At $(0, 20)$: $z = 300(0) + 500(120) = 60{,}000$

At $(60, 90)$: $z = 300(60) + 500(90) = 63{,}000$

At $(75, 75)$: $z = 300(75) + 500(75) = 60{,}000$

At $(100, 0)$: $z = 300(100) + 500(0) = 30{,}000$

So, 60 acres of crop A and 90 acres of crop B yield 63,000 bushels.

45. x = number of TV ads

y = number of newspaper ads

Constraints: $100{,}000x + 20{,}000y \le 1{,}000{,}000$

$100{,}000x \le 800{,}000$

$x \ge 0$

$y \ge 0$

Objective function: $A = 20x + 5y$ (A in millions)

Vertices: $(0, 0), (0, 50), (8, 10), (8, 0)$

At $(0, 0)$: $A = 20(0) + 5(0) = 0$

At $(0, 50)$: $A = 20(0) + 5(50) = 250$ million

At $(8, 10)$: $A = 20(8) + 5(10) = 210$ million

At $(8, 0)$: $A = 20(8) + 5(0) = 160$ million

The company should spend $0 on television ads and $1,000,000 on newspaper ads. The optimal total audience is 250 million people.

47. True. The objective function has a maximum value at any point on the line segment connecting the two vertices. Both of these points are on the line $y = -x + 11$ and lie between $(4, 7)$ and $(8, 3)$.

49. False. In Exercise 27 the constraint region lies in the first quadrant and is unbounded, but the objective function has a maximum value. It will depend upon the objective function. For example, if the objection function is $z = x - y$, as y values increase, the objective function approaches very large negative values. Therefore, there would have existed a maximum for small values of y.

51. If a linear programming problem has an objective function $z = 3x + 5y$ and an infinite number of optimal solutions then the slope of the line connecting two points is $m = -\frac{3}{5}$, that is $z = 3x + 5y \Rightarrow y = -\frac{3}{5}x - \frac{1}{5}z.$

Review Exercises for Chapter 6

1. $\begin{cases} x + y = 2 \\ x - y = 0 \Rightarrow x = y \end{cases}$

$x + x = 2$

$2x = 2$

$x = 1$

$y = 1$

Solution: $(1, 1)$

3. $\begin{cases} 4x - y - 1 = 0 \Rightarrow y = 4x - 1 \\ 8x + y - 17 = 0 \end{cases}$

$8x + (4x - 1) - 17 = 0$

$12x = 18$

$x = \frac{3}{2}$

$4\left(\frac{3}{2}\right) - y - 1 = 0$

$-y + 5 = 0$

$y = 5$

Solution: $\left(\frac{3}{2}, 5\right)$

5. $\begin{cases} 0.5x + y = 0.75 \Rightarrow y = 0.75 - 0.5x \\ 1.25x - 4.5y = -2.5 \end{cases}$

$1.25x - 4.5(0.75 - 0.5x) = -2.5$

$1.25x - 3.375 + 2.25x = -2.5$

$3.50x = 0.875$

$x = 0.25$

$y = 0.625$

Solution: $(0.25, 0.625)$

7. $\begin{cases} x^2 - y^2 = 9 \\ x - y = 1 \Rightarrow x = y + 1 \end{cases}$

$(y + 1)^2 - y^2 = 9$

$2y + 1 = 9$

$y = 4$

$x = 5$

Solution: $(5, 4)$

9. $\begin{cases} y = 2x^2 \\ y = x^4 - 2x^2 \Rightarrow 2x^2 = x^4 - 2x^2 \end{cases}$

$0 = x^4 - 4x^2$

$0 = x^2(x^2 - 4)$

$0 = x^2(x + 2)(x - 2) \Rightarrow x = 0, -2, 2$

$x = 0: y = 2(0)^2 = 0$

$x = -2: y = 2(-2)^2 = 8$

$x = 2: y = 2(2)^2 = 8$

Solutions: $(0, 0), (-2, 8), (2, 8)$

11. $\begin{cases} 2x - y = 10 \\ x + 5y = -6 \end{cases}$

Point of intersection: $(4, -2)$

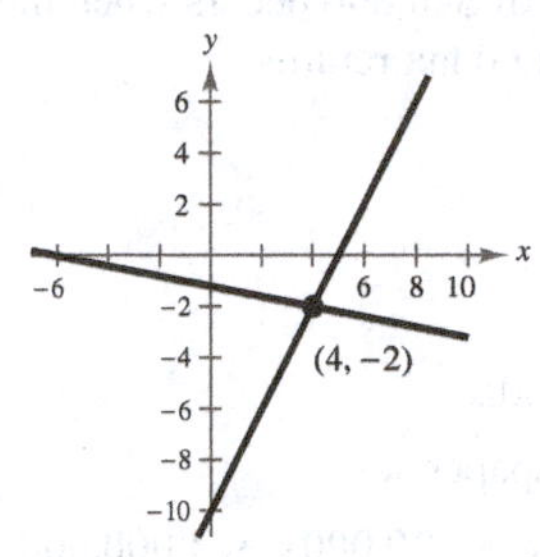

13. $\begin{cases} y = 2x^2 - 4x + 1 \\ y = x^2 - 4x + 3 \end{cases}$

Points of intersection: $(1.41, -0.66), (-1.41, 10.66)$

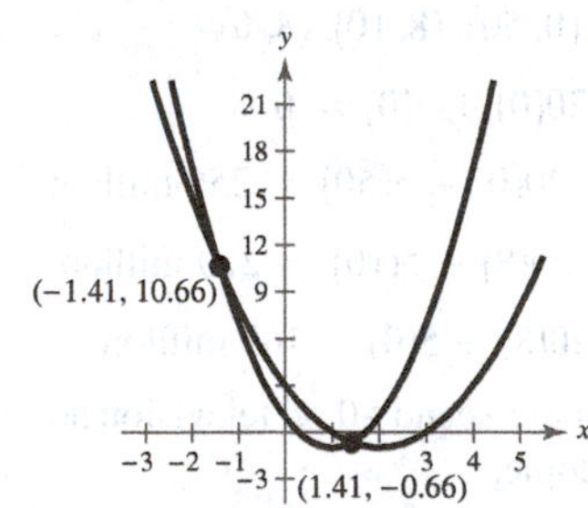

15. $\begin{cases} y = -2e^{-x} \\ 2e^x + y = 0 \Rightarrow y = -2e^x \end{cases}$

Point of intersection: $(0, -2)$

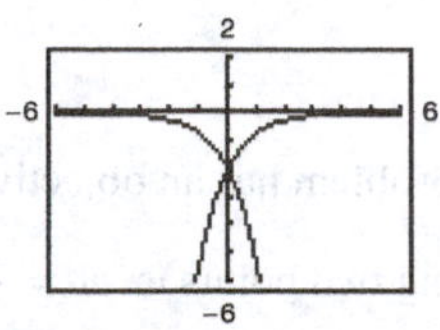

17. $\begin{cases} y = 2 + \log x \\ y = \frac{3}{4}x + 5 \end{cases}$

No Solution

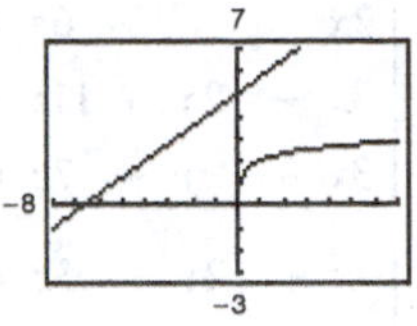

19. 

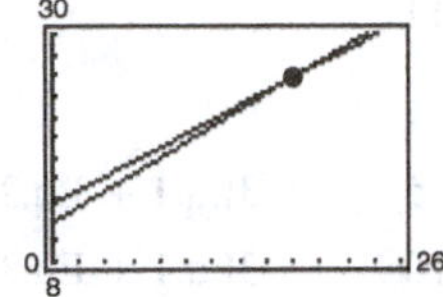

$$0.68a + 13.5 > 0.78a + 11.7$$
$$1.8 > 0.1a$$
$$18 > a$$

The BMI for males exceeds the BMI for females after age 18.

21. $\begin{cases} 2l + 2w = 68 \\ w = \frac{8}{9}l \end{cases}$

$$2l + 2\left(\tfrac{8}{9}\right)l = 68$$
$$\tfrac{34}{9}l = 68$$
$$l = 18$$
$$w = \tfrac{8}{9}l = 16$$

The width of the rectangle is 16 feet, and the length is 18 feet.

23. $\begin{cases} 2x - y = 2 \Rightarrow 16x - 8y = 16 \\ 6x + 8y = 39 \Rightarrow \ \ 6x + 8y = 39 \end{cases}$

$$22x = 55$$
$$x = \tfrac{55}{22} = \tfrac{5}{2}$$

Back-substitute $x = \frac{5}{2}$ into Equation 1.

$$2\left(\tfrac{5}{2}\right) - y = 2$$
$$y = 3$$

Solution: $\left(\frac{5}{2}, 3\right)$

25. $\begin{cases} 3x - 2y = 0 \\ 3x + 2y = 0 \end{cases}$

Add the equations $6x = 0 \Rightarrow x = 0$.

Back substitute into Equation 1.

$$3(0) - 2y = 0$$
$$2y = 0$$
$$y = 0$$

Solution: $(0, 0)$

27. $\begin{cases} 1.25x - 2y = 3.5 \Rightarrow \ \ 5x - 8y = \ \ 14 \\ 5x - 8y = 14 \Rightarrow -5x + 8y = -14 \end{cases}$

$$0 = 0$$

There are infinitely many solutions.

Let $y = a$, then $5x - 8a = 14 \Rightarrow x = \frac{8}{5}a + \frac{14}{5}$.

Solution: $\left(\frac{8}{5}a + \frac{14}{5}, a\right)$ where a is any real number.

29. $\begin{cases} x + 5y = 4 \Rightarrow \ \ x + 5y = \ \ 4 \\ x - 3y = 6 \Rightarrow -x + 3y = -6 \end{cases}$

$$8y = -2 \Rightarrow y = -\tfrac{1}{4}$$

Matches graph (d). The system has one solution and is consistent.

31. $\begin{cases} 3x - y = 7 \Rightarrow \ \ 6x - 2y = 14 \\ -6x + 2y = 8 \Rightarrow -6x + 2y = \ \ 8 \end{cases}$

$$0 \neq 22$$

Matches graph (b). The system has no solution and is inconsistent.

33. $22 + 0.00001x = 43 - 0.0002x$

$$0.00021x = 21$$
$$x = 100{,}000,\ p = 2^3$$

Point of Equilibrium: $(100{,}000,\ 23)$

35. $\begin{cases} x - 4y + 3z = \ \ 3 \\ \quad -y + \ \ z = -1 \\ \qquad\quad\ \ z = -5 \end{cases}$

$$-y + (-5) = -1 \Rightarrow y = -4$$
$$x - 4(-4) + 3(-5) = 3 \Rightarrow x = 2$$

Solution: $(2, -4, -5)$

37. $\begin{cases} 4x - 3y - 2z = -65 \\ \quad\ \ 8y - 7z = -14 \\ \qquad\quad\ \ z = \ \ 10 \end{cases}$

$$8y - 7(10) = -14 \Rightarrow y = 7$$
$$4x - 3(7) - 2(10) = -65 \Rightarrow x = -6$$

Solution: $(-6, 7, 10)$

39. $\begin{cases} x + 2y + 6z = 4 & \text{Equation 1} \\ -3x + 2y - z = -4 & \text{Equation 2} \\ 4x + 2z = 16 & \text{Equation 3} \end{cases}$

$\begin{cases} x + 2y + 6z = 4 \\ 8y + 17z = 8 & 3\text{Eq.}1 + \text{Eq.}2 \\ -8y - 22z = 0 & -4\text{Eq.}1 + \text{Eq.}3 \end{cases}$

$\begin{cases} x + 2y + 6z = 4 \\ 8y + 17z = 8 \\ -5z = 8 & \text{Eq.}2 + \text{Eq.}3 \end{cases}$

$\begin{cases} x + 2y + 6z = 4 \\ 8y + 17z = 8 \\ z = -\frac{8}{5} & -\frac{1}{5}\text{Eq.}3 \end{cases}$

$8y + 17\left(-\frac{8}{5}\right) = 8 \Rightarrow y = \frac{22}{5}$

$x + 2\left(\frac{22}{5}\right) + 6\left(-\frac{8}{5}\right) = 4 \Rightarrow x = \frac{24}{5}$

Solution: $\left(\frac{24}{5}, \frac{22}{5}, -\frac{8}{5}\right)$

41. $\begin{cases} 2x + 6z = -9 & \text{Equation 1} \\ 3x - 2y + 11z = -16 & \text{Equation 2} \\ 3x - y + 7z = -11 & \text{Equation 3} \end{cases}$

$\begin{cases} -x + 2y - 5z = 7 & (-1)\text{Eq.}2 + \text{Eq.}1 \\ 3x - 2y + 11z = -16 \\ 3x - y + 7z = -11 \end{cases}$

$\begin{cases} -x + 2y - 5z = 7 \\ 4y - 4z = 5 & 3\text{Eq.}1 + \text{Eq.}2 \\ 5y - 8z = 10 & 3\text{Eq.}1 + \text{Eq.}3 \end{cases}$

$\begin{cases} -x + 2y - 5z = 7 \\ 4y - 4z = 5 \\ -3y = 0 & (-2)\text{Eq.}2 + \text{Eq.}3 \end{cases}$

$\begin{cases} -x + 2y - 5z = 7 \\ y - z = \frac{5}{4} & \left(\frac{1}{4}\right)\text{Eq.}2 \\ y = 0 & \left(-\frac{1}{3}\right)\text{Eq.}3 \end{cases}$

$0 - z = \frac{5}{4} \Rightarrow z = -\frac{5}{4}$

$-x + 2(0) - 5\left(-\frac{5}{4}\right) = 7 \Rightarrow x = -\frac{3}{4}$

Solution: $\left(-\frac{3}{4}, 0, -\frac{5}{4}\right)$

43. $\begin{cases} 5x - 12y + 7z = 16 \Rightarrow \\ 3x - 7y + 4z = 9 \Rightarrow \end{cases} \begin{cases} 15x - 36y + 21z = 48 \\ -15x + 35y - 20z = -45 \end{cases}$

$-y + z = 3$

Let $z = a$. Then $y = a - 3$ and $5x - 12(a - 3) + 7a = 16 \Rightarrow x = a - 4$.

Solution: $(a - 4, a - 3, a)$ where a is any real number.

45. $y = ax^2 + bx + c$ through $(0, -5)$, $(1, -2)$, and $(2, 5)$.

$(0, -5)$: $-5 = c \Rightarrow c = -5$

$(1, -2)$: $-2 = a + b + c \Rightarrow \begin{cases} a + b = 3 \\ 2a + b = 5 \end{cases}$

$(2, 5)$: $5 = 4a + 2b + c \Rightarrow$

$\begin{cases} 2a + b = 5 \\ -a - b = -3 \end{cases}$

$a = 2$

$b = 1$

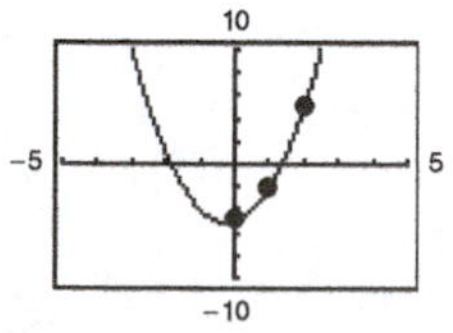

The equation of the parabola is $y = 2x^2 + x - 5$.

47. $x^2 + y^2 + Dx + Ey + F = 0$ through $(-1, -2)$, $(5, -2)$, and $(2, 1)$.

$$\begin{aligned} (-1,-2)&: \quad 5 - D - 2E + F = 0 \Rightarrow \\ (5,-2)&: \quad 29 + 5D - 2E + F = 0 \Rightarrow \\ (2,1)&: \quad 5 + 2D + E + F = 0 \Rightarrow \end{aligned} \begin{cases} D + 2E - F = 5 \\ 5D - 2E + F = -29 \\ 2D + E + F = -5 \end{cases}$$

From the first two equations

$$6D = -24$$
$$D = -4.$$

Substituting $D = -4$ into the second and third equations yields:

$$\begin{aligned} -20 - 2E + F = -29 &\Rightarrow \begin{cases} -2E + F = -9 \\ -E - F = -3 \end{cases} \\ -8 + E + F = -5 &\Rightarrow \end{aligned}$$

$$-3E = -12$$
$$E = 4$$
$$F = -1$$

The equation of the circle is $x^2 + y^2 - 4x + 4y - 1 = 0$.

To verify the result using a graphing utility, solve the equation for y.

$$\left(x^2 - 4x + 4\right) + \left(y^2 + 4y + 4\right) = 1 + 4 + 4$$
$$(x-2)^2 + (y+2)^2 = 9$$
$$(y+2)^2 = 9 - (x-2)^2$$
$$y = -2 \pm \sqrt{9 - (x-2)^2}$$

Let $y_1 = -2 + \sqrt{9 - (x-2)^2}$ and $y_2 = -2 - \sqrt{9 - (x-2)^2}$.

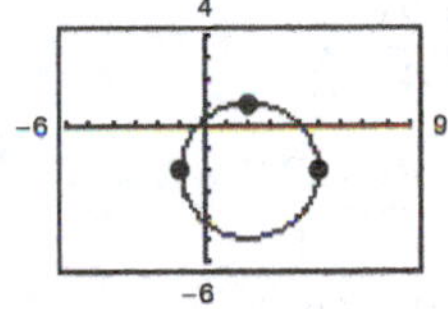

49. From the following chart we obtain our system of equations.

	A	B	C
Mixture X	$\frac{1}{5}$	$\frac{2}{5}$	$\frac{2}{5}$
Mixture Y	0	0	1
Mixture Z	$\frac{1}{3}$	$\frac{1}{3}$	$\frac{1}{3}$
Desired Mixture	$\frac{6}{27}$	$\frac{8}{27}$	$\frac{13}{27}$

$$\left.\begin{aligned} \tfrac{1}{5}x + \tfrac{1}{3}z &= \tfrac{6}{27} \\ \tfrac{2}{5}x + \tfrac{1}{3}z &= \tfrac{8}{27} \end{aligned}\right\} x = \tfrac{10}{27},\ z = \tfrac{12}{27}$$

$$\tfrac{2}{5}x + y + \tfrac{1}{3}z = \tfrac{13}{27} \Rightarrow y = \tfrac{5}{27}$$

To obtain the desired mixture, use 10 gallons of spray X, 5 gallons of spray Y, and 12 gallons of spray Z.

51. Let x = amount invested at 7%

y = amount invested at 9%

z = amount invested at 11%.

$y = x - 3000$ and

$z = x - 5000 \Rightarrow y + z = 2x - 8000$

$$\begin{cases} x + y + z = 40{,}000 \\ 0.07x + 0.09y + 0.11z = 3500 \\ y + z = 2x - 8000 \end{cases}$$

$$\begin{aligned} x + (2x - 8000) = 40{,}000 &\Rightarrow x = 16{,}000 \\ y = 16{,}000 - 3000 &\Rightarrow y = 13{,}000 \\ z = 16{,}000 - 5000 &\Rightarrow z = 11{,}000 \end{aligned}$$

So, \$16,000 was invested at 7%, \$13,000 at 9%, and \$11,000 at 11%.

53. $s = \frac{1}{2}at^2 + v_0t + s_0$

When $t = 1$: $s = 134$: $\frac{1}{2}a(1)^2 + v_0(1) + s_0 = 134 \Rightarrow a + 2v_0 + 2s_0 = 268$

When $t = 2$: $s = 86$: $\frac{1}{2}a(2)^2 + v_0(2) + s_0 = 86 \Rightarrow 2a + 2v_0 + s_0 = 86$

When $t = 3$: $s = 6$: $\frac{1}{2}a(3)^2 + v_0(3) + s_0 = 6 \Rightarrow 9a + 6v_0 + 2s_0 = 12$

$$\begin{cases} a + 2v_0 + 2s_0 = 268 \\ 2a + 2v_0 + s_0 = 86 \\ 9a + 6v_0 + 2s_0 = 12 \end{cases}$$

$$\begin{cases} a + 2v_0 + 2s_0 = 268 & \\ -2v_0 - 3s_0 = -450 & (-2)\text{Eq.}1 + \text{Eq.}2 \\ -12v_0 - 16s_0 = -2400 & (-9)\text{Eq.}1 + \text{Eq.}3 \end{cases}$$

$$\begin{cases} a + 2v_0 + 2s_0 = 268 & \\ -2v_0 - 3s_0 = -450 & \\ 3v_0 + 4s_0 = 600 & \left(-\frac{1}{4}\right)\text{Eq.}3 \end{cases}$$

$$\begin{cases} a + 2v_0 + 2s_0 = 268 & \\ -2v_0 - 3s_0 = -450 & \\ -s_0 = -150 & 3\text{Eq.}2 + 2\text{Eq.}3 \end{cases}$$

$$-s_0 = -150 \Rightarrow s_0 = 150$$
$$-2v_0 - 3(150) = -450 \Rightarrow v_0 = 0$$
$$a + 2(0) + 2(150) = 268 \Rightarrow a = -32$$

The position equation is $s = \frac{1}{2}(-32)t^2 + (0)t + 150$, or $s = -16t^2 + 150$.

55. $\dfrac{3}{x^2 + 20x} = \dfrac{3}{x(x + 20)} = \dfrac{A}{x} + \dfrac{B}{x + 20}$

57. $\dfrac{3x - 4}{x^3 - 5x^2} = \dfrac{3x - 4}{x^2(x - 5)} = \dfrac{A}{x} + \dfrac{B}{x^2} + \dfrac{C}{x - 5}$

59. $\dfrac{4 - x}{x^2 + 6x + 8} = \dfrac{A}{x + 2} + \dfrac{B}{x + 4}$

$$4 - x = A(x + 4) + B(x + 2)$$

Let $x = -2$: $6 = 2A \Rightarrow A = 3$

Let $x = -4$: $8 = -2B \Rightarrow B = -4$

$$\frac{4 - x}{x^2 + 6x + 8} = \frac{3}{x + 2} - \frac{4}{x + 4}$$

61. $\dfrac{x^2}{x^2 + 2x - 15} = 1 - \dfrac{2x - 15}{x^2 + 2x - 15}$

$$\frac{-2x + 15}{(x + 5)(x - 3)} = \frac{A}{x + 5} + \frac{B}{x - 3}$$

$$-2x + 15 = A(x - 3) + B(x + 5)$$

Let $x = -5$: $25 = -8A \Rightarrow A = -\dfrac{25}{8}$

Let $x = 3$: $9 = 8B \Rightarrow B = \dfrac{9}{8}$

$$\frac{x^2}{x^2 + 2x - 15} = 1 - \frac{25}{8(x + 5)} + \frac{9}{8(x - 3)}$$

63. $\dfrac{x^2 + 2x}{x^3 - x^2 + x - 1} = \dfrac{x^2 + 2x}{(x - 1)(x^2 + 1)}$

$$= \frac{A}{x - 1} + \frac{Bx + C}{x^2 + 1}$$

$$x^2 + 2x = A(x^2 + 1) + (Bx + C)(x - 1)$$
$$= Ax^2 + A + Bx^2 - Bx + Cx - C$$
$$= (A + B)x^2 + (-B + C)x + (A - C)$$

Equating coefficients of like terms gives $1 = A + B$, $2 = -B + C$, and $0 = A - C$. Adding both sides of all three equations gives $3 = 2A$. So, $A = \frac{3}{2}$, $B = -\frac{1}{2}$, and $C = \frac{3}{2}$.

$$\frac{x^2 + 2x}{x^3 - x^2 + x - 1} = \frac{\frac{3}{2}}{x - 1} + \frac{-\frac{1}{2}x + \frac{3}{2}}{x^2 + 1}$$
$$= \frac{1}{2}\left(\frac{3}{x - 1} - \frac{x - 3}{x^2 + 1}\right)$$

65. $\dfrac{3x^2 + 4x}{(x^2 + 1)^2} = \dfrac{Ax + B}{x^2 + 1} + \dfrac{Cx + D}{(x^2 + 1)^2}$

$3x^2 + 4x = (Ax + B)(x^2 + 1) + Cx + D$

$= Ax^3 + Bx^2 + (A + C)x + (B + D)$

Equating coefficients of like terms gives

$0 = A$

$3 = B$

$4 = 0 + C \Rightarrow C = 4$

$0 = B + D \Rightarrow D = -3$

$\dfrac{3x^2 + 4x}{(x^2 + 1)^2} = \dfrac{3}{x^2 + 1} + \dfrac{4x - 3}{(x^2 + 1)^2}$

67. $y \geq 5$

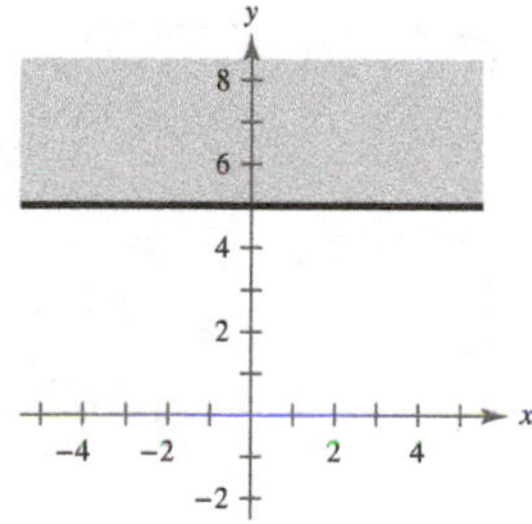

69. $y \leq 5 - 2x$

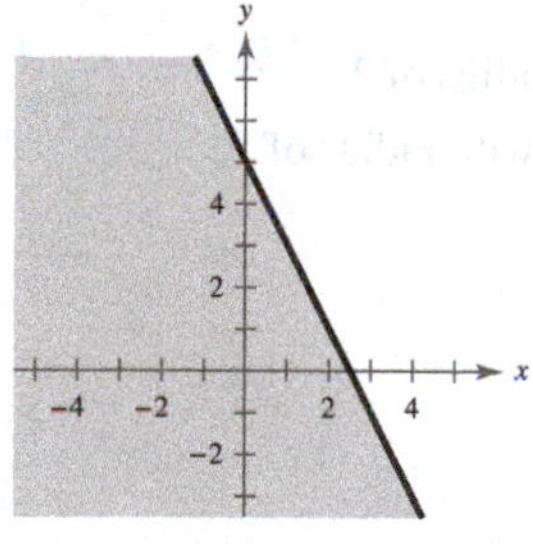

71. $(x - 1)^2 + (y - 3)^2 < 16$

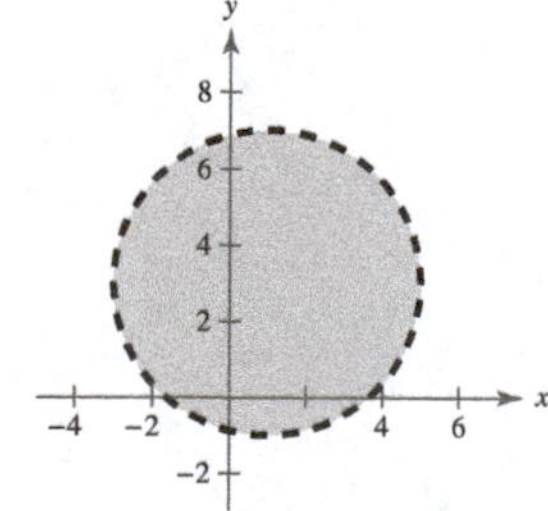

73. $\begin{cases} x + 2y \leq 2 \\ -x + 2y \leq 2 \\ y \geq 0 \end{cases}$

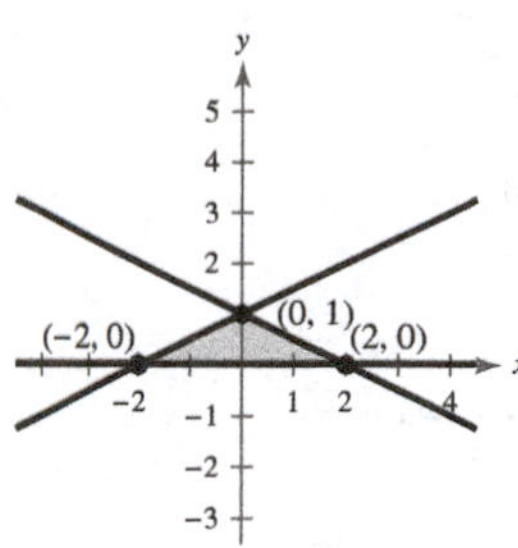

Vertex A

$\begin{cases} x + 2y = 2 \\ -x + 2y = 2 \end{cases}$

$4y = 4 \Rightarrow y = 1$

$x + 2(1) = 2 \Rightarrow x = 0$

$(0, 1)$

Vertex B

$\begin{cases} x + 2y = 2 \\ y = 0 \end{cases}$

$x + 2(0) = 2$

$x = 2$

$(2, 0)$

Vertex C

$\begin{cases} -x + 2y = 2 \\ y = 0 \end{cases}$

$-x + 2(0) = 2$

$x = -2$

$(-2, 0)$

75. $\begin{cases} 2x - y < -1 \\ -3x + 2y > 4 \\ y > 0 \end{cases}$

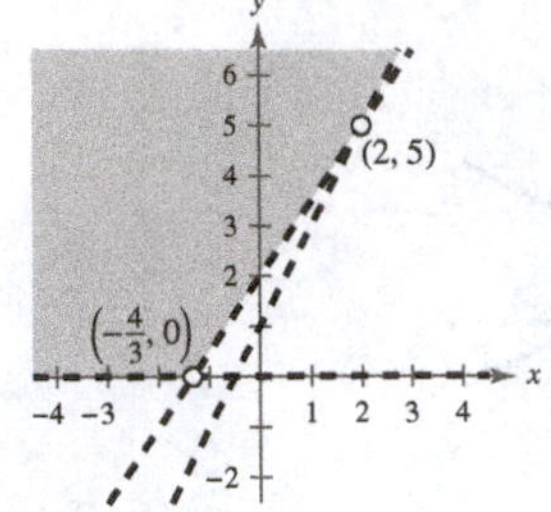

Vertex A

$\begin{cases} 2x - y = -1 \Rightarrow 4x - 2y = -2 \\ -3x + 2y = 4 \Rightarrow -3x + 2y = 4 \end{cases}$

$x = 2$

$2(2) - y = -1 \Rightarrow -y = -5 \Rightarrow y = 5$

$(2, 5)$

Vertex B

$\begin{cases} -3x + 2y = 4 \\ y = 0 \end{cases}$

$-3x + 2(0) = 4 \Rightarrow x = -\frac{4}{3}$

$\left(-\frac{4}{3}, 0\right)$

77. $\begin{cases} y < x + 1 \\ y > x^2 - 1 \end{cases}$

Vertices:

$x + 1 = x^2 - 1$

$0 = x^2 - x - 2 = (x + 1)(x - 2)$

$x = -1$ or $x = 2$

$y = 0 \quad y = 3$

$(-1, 0) \quad (2, 3)$

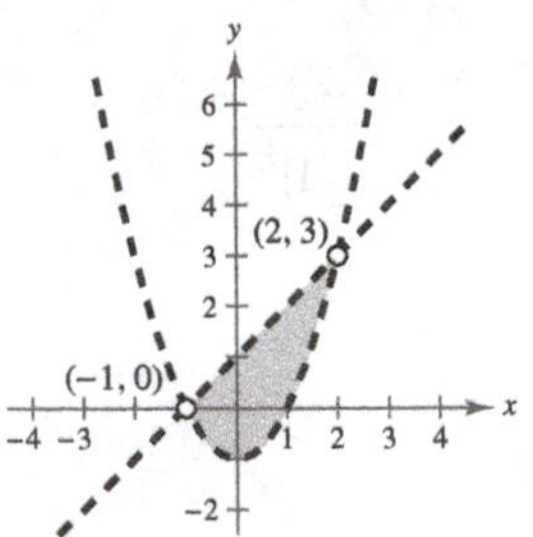

79. $\begin{cases} x^2 + y^2 > 4 \Rightarrow y^2 > 4 - x^2: \text{ The region outside the circle centered at } (0, 0) \text{ with radius of } 2. \\ x^2 + y^2 \le 9 \Rightarrow y^2 \le 9 - x^2: \text{ The region inside and on the circle centered at } (0, 0) \text{ with radius of } 3. \end{cases}$

Vertices: $4 - x^2 = 9 - x^2$

$0 \ne 5$

The circles do not intersect, so there are no vertices.

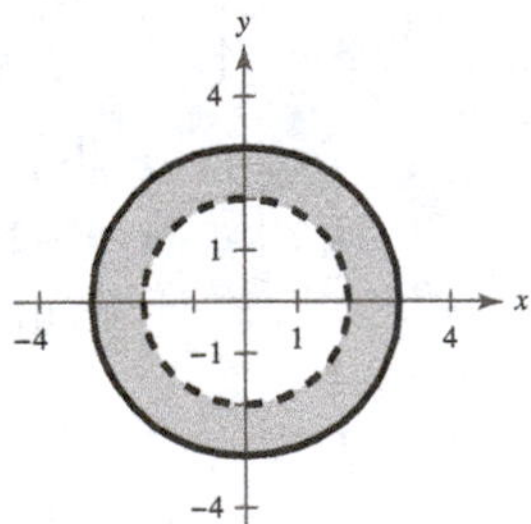

81. Rectangular region with vertices at: $(3, 1), (7, 1), (7, 10)$, and $(3, 10)$

$\begin{cases} x \ge 3 \\ x \le 7 \\ y \ge 1 \\ y \le 10 \end{cases}$

This system may be written as:

$\begin{cases} 3 \le x \le 7 \\ 1 \le y \le 10 \end{cases}$

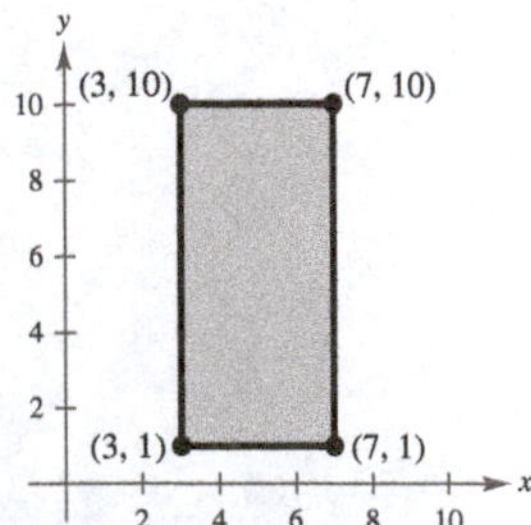

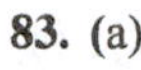

83. (a)

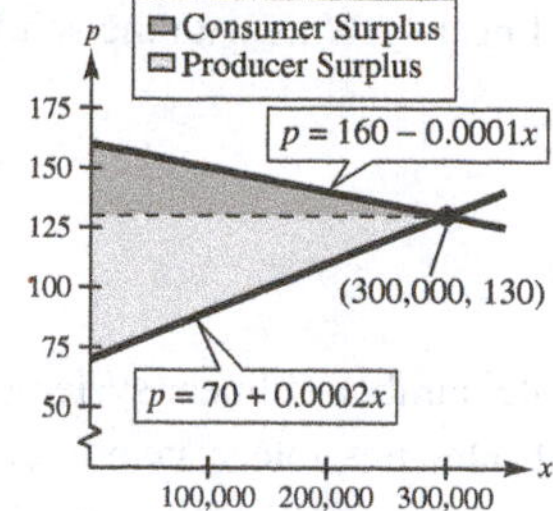

$$160 - 0.0001x = 70 + 0.0002x$$
$$90 = 0.0003x$$
$$x = 300{,}000 \text{ units}$$
$$p = \$130$$

Point of equilibrium: $(300{,}000, 130)$

(b) Consumer surplus: $\frac{1}{2}(300{,}000)(30) = \$4{,}500{,}000$

Producer surplus: $\frac{1}{2}(300{,}000)(60) = \$9{,}000{,}000$

85. x = number of units of Product I

y = number of units of Product II

$$\begin{cases} 20x + 30y \le 24{,}000 \\ 12x + 8y \le 12{,}400 \\ x \ge 0 \\ y \ge 0 \end{cases}$$

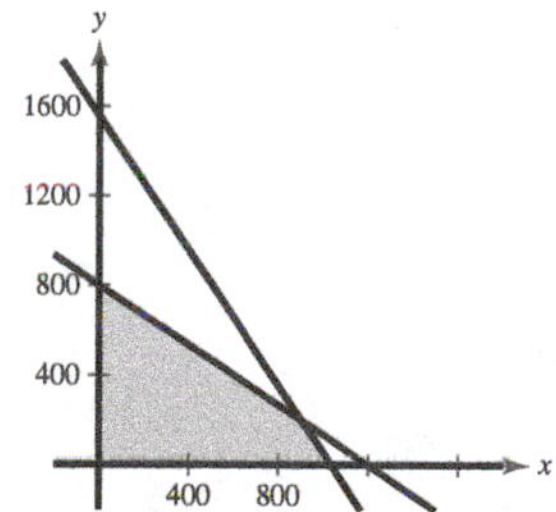

87. Objective function: $z = 3x + 4y$

Constraints: $\begin{cases} x \ge 0 \\ y \ge 0 \\ 2x + 5y \le 50 \\ 4x + y \le 28 \end{cases}$

At $(0, 0)$: $z = 0$

At $(0, 10)$: $z = 40$

At $(5, 8)$: $z = 47$

At $(7, 0)$: $z = 21$

The minimum value is 0 at $(0, 0)$.

The maximum value is 47 at $(5, 8)$.

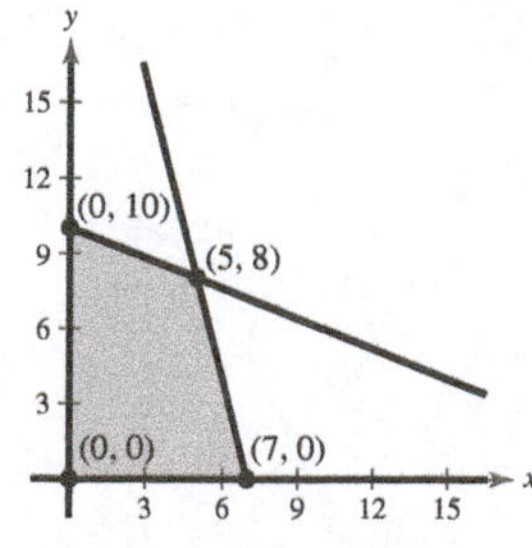

89. Objective function: $z = 1.75x + 2.25y$

Constraints: $\begin{cases} x \ge 0 \\ y \ge 0 \\ 2x + y \ge 25 \\ 3x + 2y \ge 45 \end{cases}$

At $(0, 25)$: $z = 56.25$

At $(5, 15)$: $z = 42.5$

At $(15, 0)$: $z = 26.25$

The minimum value is 26.25 at $(15, 0)$.

Because the region is unbounded, there is no maximum value.

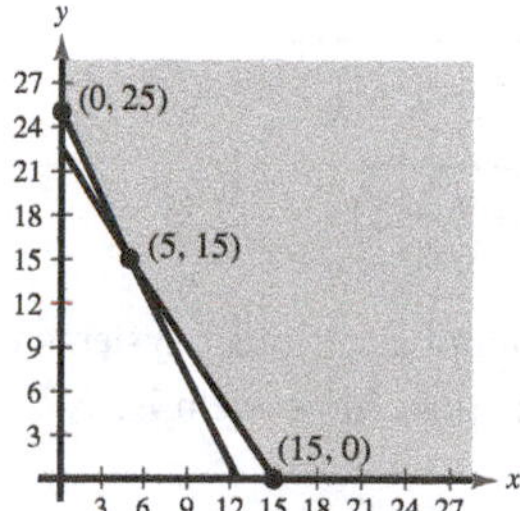

91. x = number of haircuts

y = number of permanents

Objective function: Optimize $R = 25x + 70y$ subject to the following constraints:

$$\begin{cases} x \ge 0 \\ y \ge 0 \\ \left(\frac{20}{60}\right)x + \left(\frac{70}{60}\right)y \le 24 \Rightarrow 2x + 7y \le 144 \end{cases}$$

At $(0, 0)$: $R = 0$

At $(72, 0)$: $R = 1800$

At $\left(0, \frac{144}{7}\right)$: $R = 1440$

The revenue is optimal if the student does 72 haircuts and no permanents. The maximum revenue is $1800.

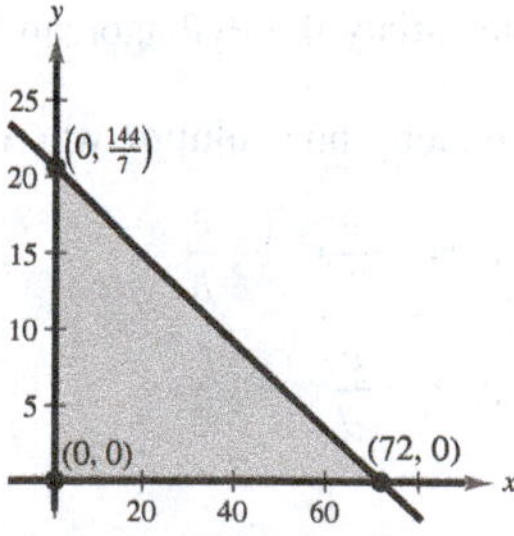

93. True. Because $y = 5$ and $y = -2$ are horizontal lines, exactly one pair of opposite sides are parallel. The non-parallel sides of the trapezoid are equal in length. Therefore, the trapezoid is isosceles as shown below.

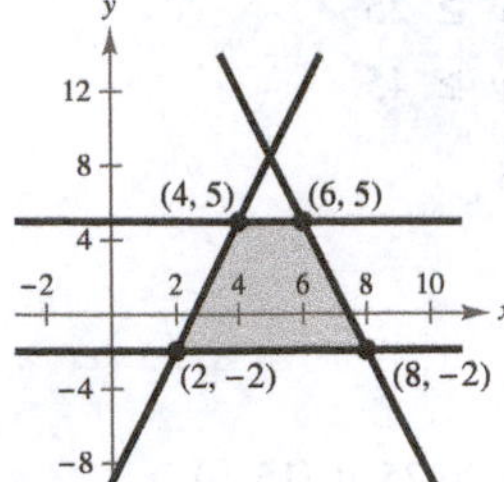

The distance from $(-4, 5)$ to $(2, -2)$ is equal to the distance from $(6, 5)$ to $(8, -2)$.

$$d_1 = \sqrt{(4-2)^2 + [5-(-2)]^2} = \sqrt{53}$$

$$d_2 = \sqrt{(8-6)^2 + (-2-5)^2} = \sqrt{53}$$

95. There are an infinite number of linear systems with the solution $(-8, 10)$. One possible system is:

$$\begin{cases} 4x + y = -22 \\ \frac{1}{2}x + y = \quad 6 \end{cases}$$

97. There are infinite linear systems with the solution $\left(\frac{4}{3}, 3\right)$.

One possible system is:

$$\begin{cases} 3x + \quad y = 7 \\ -6x + 3y = 1 \end{cases}$$

99. There are an infinite number of linear systems with the solution $(4, -1, 3)$. One possible system is as follows:

$$\begin{cases} x + y + z = 6 \\ x + y - z = 0 \\ x - y - z = 2 \end{cases}$$

101. There are an infinite number of linear systems with the solution $\left(5, \frac{3}{2}, 2\right)$. One possible system is:

$$\begin{cases} 2x + 2y - 3z = \quad 7 \\ \quad x - 2y + \quad z = \quad 4 \\ -x + 4y - \quad z = -1 \end{cases}$$

103. A system of linear equations is inconsistent if it has no solution.

Problem Solving for Chapter 6

1. The longest side of the triangle is a diameter of the circle and has a length of 20.

The lines $y = \frac{1}{2}x + 5$ and $y = -2x + 20$ intersect at the point $(6, 8)$.

The distance between $(-10, 0)$ and $(6, 8)$ is:

$$d_1 = \sqrt{(6-(-10))^2 + (8-0)^2} = \sqrt{320} = 8\sqrt{5}$$

The distance between $(6, 8)$ and $(10, 0)$ is:

$$d_2 = \sqrt{(10-6)^2 + (0-8)^2} = \sqrt{80} = 4\sqrt{5}$$

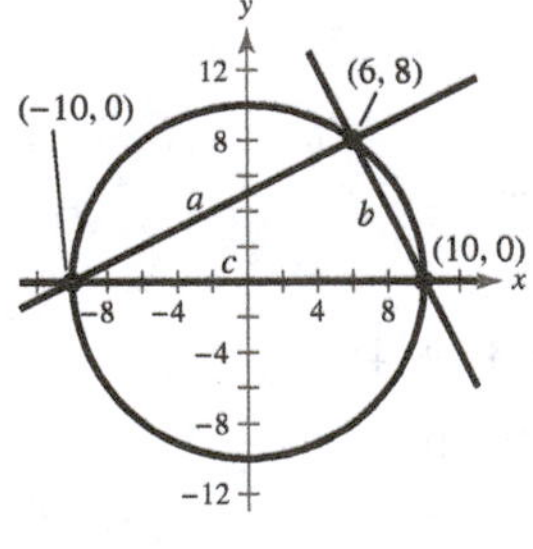

Because $\left(\sqrt{320}\right)^2 + \left(\sqrt{80}\right)^2 = (20)^2$

$$400 = 400,$$

the sides of the triangle satisfy the Pythagorean Theorem. So, the triangle is a right triangle.

3. The system will have exactly one solution when the slopes of the line are *not* equal.

$$\begin{cases} ax + by = e \Rightarrow y = -\dfrac{a}{b}x + \dfrac{e}{b} \\ cx + dy = f \Rightarrow y = -\dfrac{c}{d}x + \dfrac{f}{d} \end{cases}$$

$$-\frac{a}{b} \neq -\frac{c}{d}$$

$$\frac{a}{b} \neq \frac{c}{d}$$

$$ad \neq bc$$

5. (a) $\begin{cases} x - 4y = -3 & \text{Eq. 1} \\ 5x - 6y = 13 & \text{Eq. 2} \end{cases}$

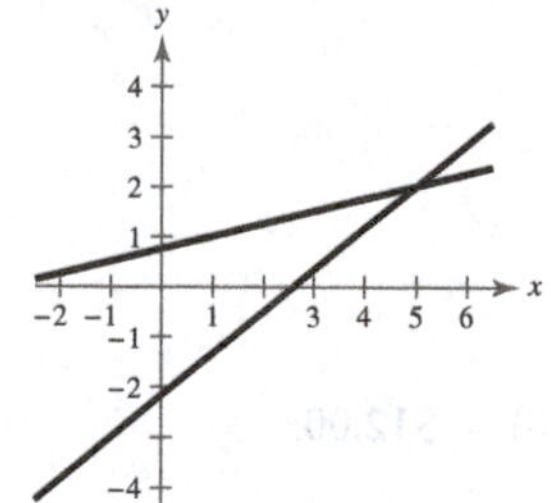

$\begin{cases} x - 4y = -3 & \\ 14y = 28 & -5\text{Eq.1} + \text{Eq.2} \end{cases}$

$\begin{cases} x - 4y = -3 & \\ y = 2 & \frac{1}{14}\text{Eq.2} \end{cases}$

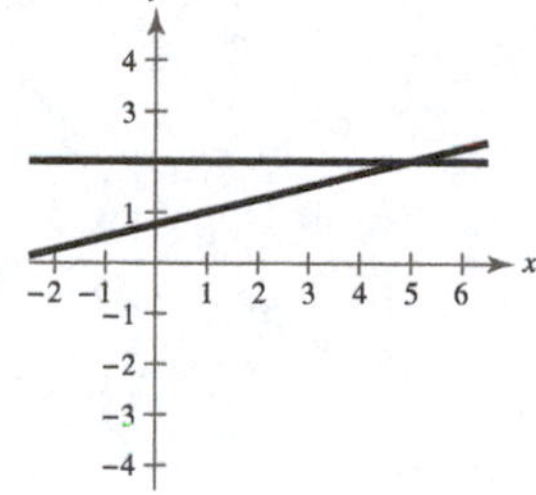

$\begin{cases} x = 5 & 4\text{Eq.2} + \text{Eq.1} \\ y = 2 & \end{cases}$

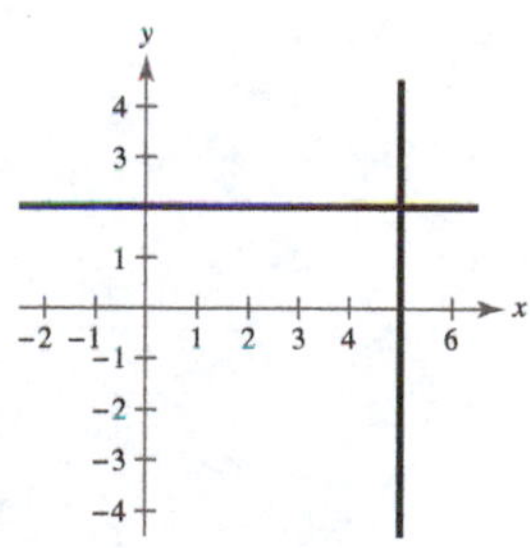

Solution: $(5, 2)$

(b) $\begin{cases} 2x - 3y = 7 & \text{Eq. 1} \\ -4x + 6y = -14 & \text{Eq. 2} \end{cases}$

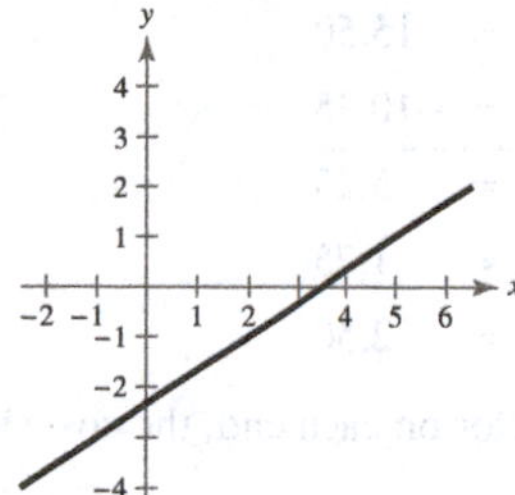

$\begin{cases} 2x - 3y = 7 & \\ 0 = 0 & 2\text{Eq.1} + \text{Eq.2} \end{cases}$

The lines coincide. Infinite solutions.

Let $y = a$, then $2x - 3a = 7 \Rightarrow x = \frac{3}{2}a + \frac{7}{2}$

Solution: $\left(\frac{3}{2}a + \frac{7}{2}, a\right)$

The solution(s) remain the same at each step of the process.

7. The point where the two sections meet is at a depth of 10.1 feet. The distance between $(0, -10.1)$ and $(252.5, 0)$ is:

$d = \sqrt{(252.5 - 0)^2 + (0 - (-10.1))^2} = \sqrt{63{,}858.26}$

$d \approx 252.7$

Each section is approximately 252.7 feet long.

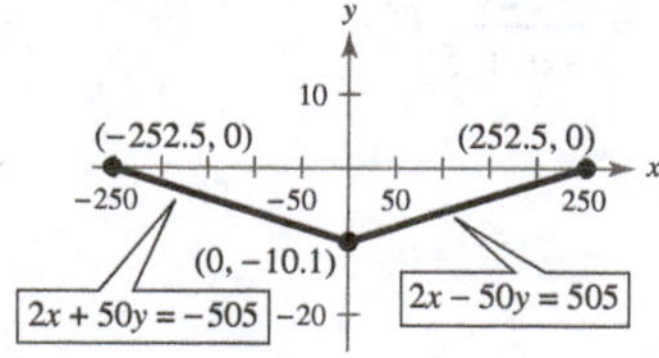

9. Let x = cost of the cable, per foot.

Let y = cost of a connector.

$$\begin{cases} 6x + 2y = 15.50 \Rightarrow & 6x + 2y = 15.50 \\ 3x + 2y = 10.25 \Rightarrow & -3x - 2y = -10.25 \end{cases}$$

$$3x = 5.25$$

$$x = 1.75$$

$$y = 2.50$$

For a four-foot cable with a connector on each end, the cost should be $4(1.75) + 2(2.50) = \$12.00$.

11. Let $X = \frac{1}{x}$, $Y = \frac{1}{y}$, and $Z = \frac{1}{z}$.

(a)
$$\begin{cases} \frac{12}{x} - \frac{12}{y} = 7 \Rightarrow 12X - 12Y = 7 \Rightarrow 12X - 12Y = 7 \\ \frac{3}{x} - \frac{4}{y} = 0 \Rightarrow 3X + 4Y = 0 \Rightarrow 9X + 12Y = 0 \\ 21X = 7 \\ X = \frac{1}{3} \\ Y = -\frac{1}{4} \end{cases}$$

So, $\frac{1}{x} = \frac{1}{3} \Rightarrow x = 3$ and $\frac{1}{y} = -\frac{1}{4} \Rightarrow y = -4$.

Solution: $(3, -4)$

(b)
$$\begin{cases} \frac{2}{x} + \frac{1}{y} - \frac{3}{z} = 4 \Rightarrow 2X + Y - 3Z = 4 & \text{Eq.1} \\ \frac{4}{x} + \frac{2}{z} = 10 \Rightarrow 4X + 2Z = 10 & \text{Eq.2} \\ -\frac{2}{x} + \frac{3}{y} - \frac{13}{z} = -8 \Rightarrow -2X + 3Y - 13Z = -8 & \text{Eq.3} \end{cases}$$

$$\begin{cases} 2X + Y - 3Z = 4 & \\ -2Y + 8Z = 2 & -2\text{Eq.1} + \text{Eq.2} \\ 4Y - 16Z = -4 & \text{Eq.1} + \text{Eq.3} \end{cases}$$

$$\begin{cases} 2X + Y - 3Z = 4 & \\ -2Y + 8Z = 2 & \\ 0 = 0 & 2\text{Eq.2} + \text{Eq.3} \end{cases}$$

The system has infinite solutions.

Let $Z = a$, then $Y = 4a - 1$ and $X = \frac{-a + 5}{2}$.

Then $\frac{1}{z} = a \Rightarrow z = \frac{1}{a}, \frac{1}{y} = 4a - 1 \Rightarrow y = \frac{1}{4a - 1}$, and $\frac{1}{x} = \frac{-a + 5}{2} \Rightarrow x = \frac{2}{-a + 5}$.

Solution: $\left(\frac{2}{-a + 5}, \frac{1}{4a - 1}, \frac{1}{a}\right), a \neq 5, \frac{1}{4}, 0$

13. Solution: $(1, -1, 2)$

$$\begin{cases} 4x - 2y + 5z = 16 & \text{Equation 1} \\ x + y = 0 & \text{Equation 2} \\ -x - 3y + 2z = 6 & \text{Equation 3} \end{cases}$$

(a) $\begin{cases} 4x - 2y + 5z = 16 \\ x + y = 0 \end{cases}$

$\begin{cases} x + y = 0 \\ 4x - 2y + 5z = 16 \end{cases}$ Interchange the equations.

$\begin{cases} x + y = 0 \\ -6y + 5z = 16 \end{cases}$ $-4\text{Eq.}1 + \text{Eq.}2$

Let $z = a$, then $y = \dfrac{5a - 16}{6}$ and $x = \dfrac{-5a + 16}{6}$.

Solution: $\left(\dfrac{-5a + 16}{6}, \dfrac{5a - 16}{6}, a\right)$

When $a = 2$, we have the original solution.

(b) $\begin{cases} 4x - 2y + 5z = 16 \\ -x - 3y + 2z = 6 \end{cases}$

$\begin{cases} -x - 2y + 2z = 6 \\ 4x - 3y + 5z = 16 \end{cases}$ Interchange the equations.

$\begin{cases} -x - 3y + 2z = 6 \\ -14y + 13z = 40 \end{cases}$ $4\text{Eq.}1 + \text{Eq.}2$

Let $z = a$, then $y = \dfrac{13a - 40}{14}$ and $x = \dfrac{-11a + 36}{14}$.

Solution: $\left(\dfrac{-11a + 36}{14}, \dfrac{13a - 40}{14}, a\right)$

When $a = 2$, we have the original solution.

(c) $\begin{cases} x + y = 0 \\ -x - 3y + 2z = 6 \end{cases}$

$\begin{cases} x + y = 0 \\ -2y + 2z = 6 \end{cases}$ $\text{Eq.}1 + \text{Eq.}2$

Let $z = a$, then $y = a - 3$ and $x = -a + 3$.

Solution: $(-a + 3, a - 3, a)$

When $a = 2$, we have the original solution.

(d) Each of these systems has infinite solutions.

15. t = amount of terrestrial vegetation in kilograms
a = amount of aquatic vegetation in kilograms

$$\begin{cases} a + t \le 32 \\ 0.15a \ge 1.9 \\ 193a + 772t \ge 11{,}000 \end{cases}$$

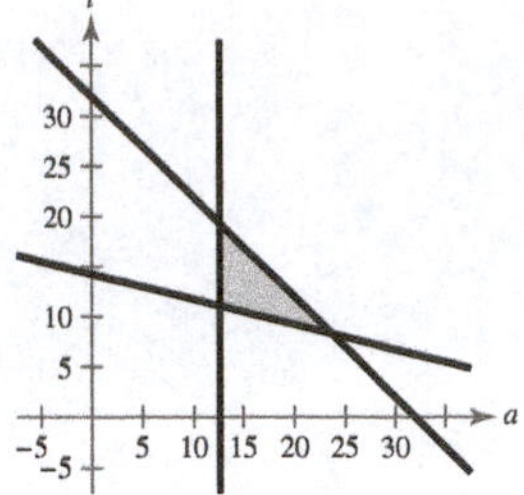

17. x = milligrams of HDL cholestrol
y = milligrams of LDL/VLDL cholestrol

(a) $\begin{cases} 0 < y < 130 \\ x \ge 60 \\ x + y \le 200 \end{cases}$

(b)

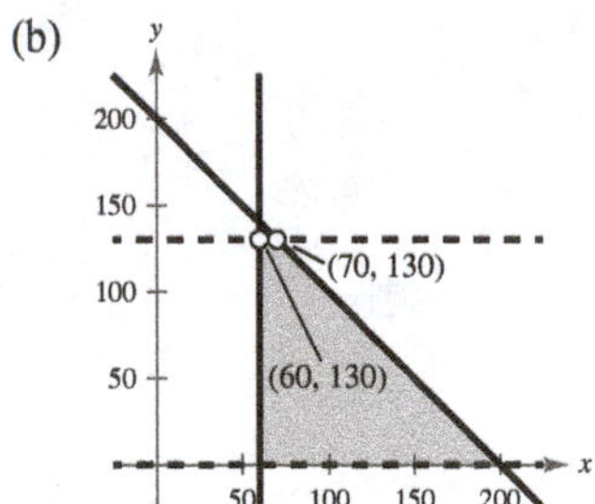

(c) $y = 120$ is in the region because $0 < y < 130$.
$x = 90$ is in the region because $60 \le x \le 200$.
$x + y = 210$ is not the region because $x + y \le 200$.

(d) *Sample answer*: If the LDL/VLDL reading is 135 and the HDL reading is 65, then $x \ge 60$ and $x + y \le 200$, but $y \not< 130$.

(e) $\dfrac{x + y}{x} < 4$
$x + y < 4x$
$y < 3x$

Sample answer: The point $(75, 90)$ is in the region, and $\dfrac{165}{75} = 2.2 < 4$.

Practice Test for Chapter 6

For Exercises 1–3, solve the given system by the method of substitution.

1. $\begin{cases} x + y = 1 \\ 3x - y = 15 \end{cases}$

2. $\begin{cases} x - 3y = -3 \\ x^2 + 6y = 5 \end{cases}$

3. $\begin{cases} x + y + z = 6 \\ 2x - y + 3z = 0 \\ 5x + 2y - z = -3 \end{cases}$

4. Find the two numbers whose sum is 110 and product is 2800.

5. Find the dimensions of a rectangle if its perimeter is 170 feet and its area is 1500 square feet.

For Exercises 6–8, solve the linear system by elimination.

6. $\begin{cases} 2x + 15y = 4 \\ x - 3y = 23 \end{cases}$

7. $\begin{cases} x + y = 2 \\ 38x - 19y = 7 \end{cases}$

8. $\begin{cases} 0.4x + 0.5y = 0.112 \\ 0.3x - 0.7y = -0.131 \end{cases}$

9. Herbert invests \$17,000 in two funds that pay 11% and 13% simple interest, respectively. If he receives \$2080 in yearly interest, how much is invested in each fund?

10. Find the least squares regression line for the points $(4, 3)$, $(1, 1)$, $(-1, -2)$, and $(-2, -1)$.

For Exercises 11–12, solve the system of equations.

11. $\begin{cases} x + y = -2 \\ 2x - y + z = 11 \\ 4y - 3z = -20 \end{cases}$

12. $\begin{cases} 3x + 2y - z = 5 \\ 6x - y + 5z = 2 \end{cases}$

13. Find the equation of the parabola $y = ax^2 + bx + c$ passing through the points $(0, -1)$, $(1, 4)$ and $(2, 13)$.

For Exercises 14–15, write the partial fraction decomposition of the rational functions.

14. $\dfrac{10x - 17}{x^2 - 7x - 8}$

15. $\dfrac{x^2 + 4}{x^4 + x^2}$

16. Graph $x^2 + y^2 \geq 9$.

17. Graph the solution of the system.

$$\begin{cases} x + y \leq 6 \\ x \geq 2 \\ y \geq 0 \end{cases}$$

18. Derive a set of inequalities to describe the triangle with vertices $(0, 0)$, $(0, 7)$, and $(2, 3)$.

19. Find the maximum value of the objective function, $z = 30x + 26y$, subject to the following constraints.

$$\begin{cases} x \geq 0 \\ y \geq 0 \\ 2x + 3y \leq 21 \\ 5x + 3y \leq 30 \end{cases}$$

20. Graph the system of inequalities.

$$\begin{cases} x^2 + y^2 \leq 4 \\ (x - 2)^2 + y^2 \geq 4 \end{cases}$$

For Exercises 21–22, write the partial fraction decomposition for the rational expression.

21. $\dfrac{1 - 2x}{x^2 + x}$

22. $\dfrac{6x - 17}{(x - 3)^2}$

CHAPTER 7
Matrices and Determinants

CHAPTER 7
Matrices and Determinants

Section 7.1 Matrices and Systems of Equations

1. square

3. augmented

5. row-equivalent

7. Because the matrix has one row and two columns, its dimension is 1×2.

9. Because the matrix has three rows and one column, its dimension is 3×1.

11. Because the matrix has two rows and two columns, its dimension is 2×2.

13. Because the matrix has three rows and three columns, its dimension is 3×3.

15. $\begin{cases} 2x - y = 7 \\ x + y = 2 \end{cases}$

$$\left[\begin{array}{cc|c} 2 & -1 & 7 \\ 1 & 1 & 2 \end{array}\right]$$

17. $\begin{cases} x - y + 2z = 2 \\ 4x - 3y + z = -1 \\ 2x + y = 0 \end{cases}$

$$\left[\begin{array}{ccc|c} 1 & -1 & 2 & 2 \\ 4 & -3 & 1 & -1 \\ 2 & 1 & 0 & 0 \end{array}\right]$$

19. $\begin{cases} 3x - 5y + 2z = 12 \\ 12x - 7z = 10 \end{cases}$

$$\left[\begin{array}{ccc|c} 3 & -5 & 2 & 12 \\ 12 & 0 & -7 & 10 \end{array}\right]$$

21. $\left[\begin{array}{cc|c} 1 & 1 & 3 \\ 5 & -3 & -1 \end{array}\right]$

$$\begin{cases} x + y = 3 \\ 5x - 3y = -1 \end{cases}$$

23. $\left[\begin{array}{ccc|c} 2 & 0 & 5 & -12 \\ 0 & 1 & -2 & 7 \\ 6 & 3 & 0 & 2 \end{array}\right]$

$$\begin{cases} 2x + 5z = -12 \\ y - 2z = 7 \\ 6x + 3y = 2 \end{cases}$$

25. $\left[\begin{array}{cccc|c} 9 & 12 & 3 & 0 & 0 \\ -2 & 18 & 5 & 2 & 10 \\ 1 & 7 & -8 & 0 & -4 \\ 3 & 0 & 2 & 0 & -10 \end{array}\right]$

$$\begin{cases} 9x + 12y + 3z = 0 \\ -2x + 18y + 5z + 2w = 10 \\ x + 7y - 8z = -4 \\ 3x + 2z = -10 \end{cases}$$

27. $\begin{bmatrix} -2 & 5 & 1 \\ 3 & -1 & -8 \end{bmatrix} \rightarrow \begin{bmatrix} 13 & 0 & -39 \\ 3 & -1 & -8 \end{bmatrix}$

Add 5 times Row 2 to Row 1.

29. $\begin{bmatrix} 0 & -1 & -5 & 5 \\ -1 & 3 & -7 & 6 \\ 4 & -5 & 1 & 3 \end{bmatrix} \rightarrow \begin{bmatrix} -1 & 3 & -7 & 6 \\ 0 & -1 & -5 & 5 \\ 0 & 7 & -27 & 27 \end{bmatrix}$

Interchange Row 1 and Row 2. Then add 4 times the new Row 1 to Row 3.

31. $\begin{bmatrix} 3 & 6 & 8 \\ 4 & -3 & 6 \end{bmatrix}$

$$\tfrac{1}{3}R_1 \rightarrow \begin{bmatrix} 1 & \boxed{2} & \frac{8}{3} \\ 4 & -3 & 6 \end{bmatrix}$$

33. $\begin{bmatrix} 1 & 1 & 1 \\ 5 & -2 & 4 \end{bmatrix}$

$$-5R_1 + R_2 \rightarrow \begin{bmatrix} 1 & 1 & 1 \\ 0 & \boxed{-7} & -1 \end{bmatrix}$$

35. $\begin{bmatrix} 1 & 5 & 4 & -1 \\ 0 & 1 & -2 & 2 \\ 0 & 0 & 1 & -7 \end{bmatrix}$

$-5R_2 + R_1 \rightarrow \begin{bmatrix} 1 & 0 & \boxed{14} & \boxed{-11} \\ 0 & 1 & -2 & 2 \\ 0 & 0 & 1 & -7 \end{bmatrix}$

37. $\begin{bmatrix} 1 & 1 & 4 & -1 \\ 3 & 8 & 10 & 3 \\ -2 & 1 & 12 & 6 \end{bmatrix}$

$\begin{matrix} -3R_1 + R_2 \rightarrow \\ 2R_1 + R_3 \rightarrow \end{matrix} \begin{bmatrix} 1 & 1 & 4 & -1 \\ 0 & 5 & \boxed{-2} & \boxed{6} \\ 0 & 3 & \boxed{20} & \boxed{4} \end{bmatrix}$

$\frac{1}{5}R_2 \rightarrow \begin{bmatrix} 1 & 1 & 4 & -1 \\ 0 & 1 & -\frac{2}{5} & \frac{6}{5} \\ 0 & 3 & \boxed{20} & \boxed{4} \end{bmatrix}$

39. (a) $\begin{bmatrix} -3 & 4 & \vdots & 22 \\ 6 & -4 & \vdots & -28 \end{bmatrix}$

(i) $R_1 + R_2 \rightarrow \begin{bmatrix} 3 & 0 & \vdots & -6 \\ 6 & -4 & \vdots & -28 \end{bmatrix}$

(ii) $-2R_1 + R_2 \rightarrow \begin{bmatrix} 3 & 0 & \vdots & -6 \\ 0 & -4 & \vdots & -16 \end{bmatrix}$

(iii) $-\frac{1}{4}R_2 \rightarrow \begin{bmatrix} 3 & 0 & \vdots & -6 \\ 0 & 1 & \vdots & 4 \end{bmatrix}$

(iv) $\frac{1}{3}R_1 \rightarrow \begin{bmatrix} 1 & 0 & \vdots & -2 \\ 0 & 1 & \vdots & 4 \end{bmatrix}$

The solution is $x = -2$ and $y = 4$.

(b) $\begin{cases} -3x + 4y = 22 \\ 6x - 4y = -28 \end{cases}$

$$3x = -6$$
$$x = -2$$

Back-substitute $x = -2$ into $-3x + 4y = 22$.

$$-3(-2) + 4y = 22$$
$$4y = 16$$
$$y = 4$$

The solution is $x = -2$ and $y = 4$.

(c) Answers vary. *Sample answer:* In this case, solving the system of linear equations using the elimination method was more efficient.

41. $\begin{bmatrix} 1 & 0 & 0 & 0 \\ 0 & 1 & 1 & 5 \\ 0 & 0 & 0 & 0 \end{bmatrix}$

This matrix is in reduced row-echelon form.

43. $\begin{bmatrix} 1 & 0 & 0 & 1 \\ 0 & 1 & 0 & -1 \\ 0 & 0 & 0 & 2 \end{bmatrix}$

This matrix is not in row-echelon form.

45. $\begin{bmatrix} 1 & 1 & 0 & 5 \\ -2 & -1 & 2 & -10 \\ 3 & 6 & 7 & 14 \end{bmatrix}$

$\begin{matrix} 2R_1 + R_2 \rightarrow \\ -3R_1 + R_3 \rightarrow \end{matrix} \begin{bmatrix} 1 & 1 & 0 & 5 \\ 0 & 1 & 2 & 0 \\ 0 & 3 & 7 & -1 \end{bmatrix}$

$-3R_2 + R_3 \rightarrow \begin{bmatrix} 1 & 1 & 0 & 5 \\ 0 & 1 & 2 & 0 \\ 0 & 0 & 1 & -1 \end{bmatrix}$

47. $\begin{bmatrix} 1 & -1 & -1 & 1 \\ 5 & -4 & 1 & 8 \\ -6 & 8 & 18 & 0 \end{bmatrix}$

$\begin{matrix} -5R_1 + R_2 \rightarrow \\ 6R_1 + R_3 \rightarrow \end{matrix} \begin{bmatrix} 1 & -1 & -1 & 1 \\ 0 & 1 & 6 & 3 \\ 0 & 2 & 12 & 6 \end{bmatrix}$

$-2R_2 + R_3 \rightarrow \begin{bmatrix} 1 & -1 & -1 & 1 \\ 0 & 1 & 6 & 3 \\ 0 & 0 & 0 & 0 \end{bmatrix}$

49. Use the reduced row-echelon form feature of a graphing utility.

$$\begin{bmatrix} 3 & 3 & 3 \\ -1 & 0 & -4 \\ 2 & 4 & -2 \end{bmatrix} \Rightarrow \begin{bmatrix} 1 & 0 & 0 \\ 0 & 1 & 0 \\ 0 & 0 & 1 \end{bmatrix}$$

51. Use the reduced row-echelon form feature of a graphing utility.

$$\begin{bmatrix} 1 & 2 & 3 & -5 \\ 1 & 2 & 4 & -9 \\ -2 & -4 & -4 & 3 \\ 4 & 8 & 11 & -14 \end{bmatrix} \Rightarrow \begin{bmatrix} 1 & 2 & 0 & 0 \\ 0 & 0 & 1 & 0 \\ 0 & 0 & 0 & 1 \\ 0 & 0 & 0 & 0 \end{bmatrix}$$

53. Use the reduced row-echelon form feature of a graphing utility.

$$\begin{bmatrix} -3 & 5 & 1 & 12 \\ 1 & -1 & 1 & 4 \end{bmatrix} \Rightarrow \begin{bmatrix} 1 & 0 & 3 & 16 \\ 0 & 1 & 2 & 12 \end{bmatrix}$$

55. $\begin{cases} x - 2y = 4 \\ y = -1 \end{cases}$

$x - 2(-1) = 4$

$x = 2$

Solution: $(2, -1)$

57. $\begin{cases} x - y + 2z = 4 \\ y - z = 2 \\ z = -2 \end{cases}$

$y - (-2) = 2$

$y = 0$

$x - 0 + 2(-2) = 4$

$x = 8$

Solution: $(8, 0, -2)$

59. $\begin{cases} x + 2y = 7 \\ -x + y = 8 \end{cases}$

$$\begin{bmatrix} 1 & 2 & \vdots & 7 \\ -1 & 1 & \vdots & 8 \end{bmatrix}$$

$$R_1 + R_2 \rightarrow \begin{bmatrix} 1 & 2 & \vdots & 7 \\ 0 & 3 & \vdots & 15 \end{bmatrix}$$

$$\tfrac{1}{3}R_2 \rightarrow \begin{bmatrix} 1 & 2 & \vdots & 7 \\ 0 & 1 & \vdots & 5 \end{bmatrix}$$

$\begin{cases} x + 2y = 7 \\ y = 5 \end{cases}$

$x + 2(5) = 7 \Rightarrow x = -3$

Solution: $(-3, 5)$

61. $\begin{cases} 3x - 2y = -27 \\ x + 3y = 13 \end{cases}$

$$\begin{bmatrix} 3 & -2 & \vdots & -27 \\ 1 & 3 & \vdots & 13 \end{bmatrix}$$

$$\begin{matrix} R_1 \\ R_2 \end{matrix} \begin{bmatrix} 1 & 3 & \vdots & 13 \\ 3 & -2 & \vdots & -27 \end{bmatrix}$$

$$-3R_1 + R_2 \rightarrow \begin{bmatrix} 1 & 3 & \vdots & 13 \\ 0 & -11 & \vdots & -66 \end{bmatrix}$$

$$-\tfrac{1}{11}R_2 \rightarrow \begin{bmatrix} 1 & 3 & \vdots & 13 \\ 0 & 1 & \vdots & 6 \end{bmatrix}$$

$\begin{cases} x + 3y = 13 \\ y = 6 \end{cases}$

$y = 6$

$x + 3(6) = 13 \Rightarrow x = -5$

Solution: $(-5, 6)$

63. $\begin{cases} x + 2y - 3z = -28 \\ 4y + 2z = 0 \\ -x + y - z = -5 \end{cases}$

$$\begin{bmatrix} 1 & 2 & -3 & \vdots & -28 \\ 0 & 4 & 2 & \vdots & 0 \\ -1 & 1 & -1 & \vdots & -5 \end{bmatrix}$$

$$\begin{matrix} \\ \frac{1}{4}R_2 \rightarrow \\ R_1 + R_3 \rightarrow \end{matrix} \begin{bmatrix} 1 & 2 & -3 & \vdots & -28 \\ 0 & 1 & \frac{1}{2} & \vdots & 0 \\ 0 & 3 & -4 & \vdots & -33 \end{bmatrix}$$

$$-3R_2 + R_3 \rightarrow \begin{bmatrix} 1 & 2 & -3 & \vdots & -28 \\ 0 & 1 & \frac{1}{2} & \vdots & 0 \\ 0 & 0 & -\frac{11}{2} & \vdots & -33 \end{bmatrix}$$

$$-\frac{2}{11}R_3 \rightarrow \begin{bmatrix} 1 & 2 & -3 & \vdots & -28 \\ 0 & 1 & \frac{1}{2} & \vdots & 0 \\ 0 & 0 & 1 & \vdots & 6 \end{bmatrix}$$

$\begin{cases} x + 2y - 3z = -28 \\ y + \frac{1}{2}z = 0 \\ z = 6 \end{cases}$

$z = 6$

$y + \frac{1}{2}(6) = 0 \Rightarrow y = -3$

$x + 2(-3) - 3(6) = -28 \Rightarrow x = -4$

Solution: $(-4, -3, 6)$

65. $\begin{cases} -3x + 2y = -22 \\ 3x + 4y = 4 \\ 4x - 8y = 32 \end{cases}$

$$\begin{bmatrix} -3 & 2 & \vdots & -22 \\ 3 & 4 & \vdots & 4 \\ 4 & -8 & \vdots & 32 \end{bmatrix}$$

$$R_1 + R_3 \rightarrow \begin{bmatrix} 1 & -6 & \vdots & 10 \\ 3 & 4 & \vdots & 4 \\ 4 & -8 & \vdots & 32 \end{bmatrix}$$

$$\begin{matrix} \\ -3R_1 + R_2 \rightarrow \\ -4R_1 + R_3 \rightarrow \end{matrix} \begin{bmatrix} 1 & -6 & \vdots & 10 \\ 0 & 22 & \vdots & -26 \\ 0 & 16 & \vdots & -8 \end{bmatrix}$$

$$\begin{matrix} \\ \frac{1}{22}R_2 \rightarrow \\ \frac{1}{16}R_3 \rightarrow \end{matrix} \begin{bmatrix} 1 & -1 & \vdots & 22 \\ 0 & 1 & \vdots & -\frac{13}{10} \\ 0 & 1 & \vdots & -\frac{1}{2} \end{bmatrix}$$

$$-R_2 + R_3 \rightarrow \begin{bmatrix} 1 & -1 & \vdots & 22 \\ 0 & 1 & \vdots & -\frac{13}{10} \\ 0 & 0 & \vdots & \frac{9}{5} \end{bmatrix}$$

The system is inconsistent and there is no solution.

67. Use the reduced row-echelon form feature of a graphing utility.

$$\begin{cases} 3x + 2y - z + w = 0 \\ x - y + 4z + 2w = 25 \\ -2x + y + 2z - w = 2 \\ x + y + z + w = 6 \end{cases}$$

$$\left[\begin{array}{cccc:c} 3 & 2 & -1 & 1 & 0 \\ 1 & -1 & 4 & 2 & 25 \\ -2 & 1 & 2 & -1 & 2 \\ 1 & 1 & 1 & 1 & 6 \end{array}\right] \Rightarrow \left[\begin{array}{cccc:c} 1 & 0 & 0 & 0 & 3 \\ 0 & 1 & 0 & 0 & -2 \\ 0 & 0 & 1 & 0 & 5 \\ 0 & 0 & 0 & 1 & 0 \end{array}\right]$$

$x = 3$

$y = -2$

$z = 5$

$w = 0$

Solution: $(3, -2, 5, 0)$

69. $\left[\begin{array}{cc:c} 1 & 0 & 3 \\ 0 & 1 & -4 \end{array}\right]$

$x = 3$

$y = -4$

Solution: $(3, -4)$

71. $\begin{cases} -2x + 6y = -22 \\ x + 2y = -9 \end{cases}$

$$\left[\begin{array}{cc:c} -2 & 6 & -22 \\ 1 & 2 & -9 \end{array}\right]$$

$$\begin{matrix} R_1 \\ R_2 \end{matrix} \left[\begin{array}{cc:c} 1 & 2 & -9 \\ -2 & 6 & -22 \end{array}\right]$$

$$2R_1 + R_2 \rightarrow \left[\begin{array}{cc:c} 1 & 2 & -9 \\ 0 & 10 & -40 \end{array}\right]$$

$$\tfrac{1}{10}R_2 \rightarrow \left[\begin{array}{cc:c} 1 & 2 & -9 \\ 0 & 1 & -4 \end{array}\right]$$

$\begin{cases} x + 2y = -9 \\ y = -4 \end{cases}$

$y = -4$

$x + 2(-4) = -9 \Rightarrow x = -1$

Solution: $(-1, -4)$

73. $\begin{cases} x + 2y + z = 8 \\ 3x + 7y + 6z = 26 \end{cases}$

$$\left[\begin{array}{ccc:c} 1 & 2 & 1 & 8 \\ 3 & 7 & 6 & 26 \end{array}\right]$$

$$-3R_1 + R_2 \rightarrow \left[\begin{array}{ccc:c} 1 & 2 & 1 & 8 \\ 0 & 1 & 3 & 2 \end{array}\right]$$

$\begin{cases} x + 2y + z = 8 \\ y + 3z = 2 \end{cases}$

Let $z = a$.

$y + 3a = 2 \Rightarrow y = -3a + 2$

$x + 2(-3a + 2) + a = 8 \Rightarrow x = 5a + 4$

Solution: $(5a + 4, -3a + 2, a)$ where a is a real number

75. $\begin{cases} x - 3z = -2 \\ 3x + y - 2z = 5 \\ 2x + 2y + z = 4 \end{cases}$

$$\left[\begin{array}{ccc:c} 1 & 0 & -3 & -2 \\ 3 & 1 & -2 & 5 \\ 2 & 2 & 1 & 4 \end{array}\right]$$

$$\begin{matrix} \\ -3R_1 + R_2 \rightarrow \\ -2R_1 + R_3 \rightarrow \end{matrix} \left[\begin{array}{ccc:c} 1 & 0 & -3 & -2 \\ 0 & 1 & 7 & 11 \\ 0 & 2 & 7 & 8 \end{array}\right]$$

$$-2R_2 + R_3 \rightarrow \left[\begin{array}{ccc:c} 1 & 0 & -3 & -2 \\ 0 & 1 & 7 & 11 \\ 0 & 0 & -7 & -14 \end{array}\right]$$

$$-\tfrac{1}{7}R_3 \rightarrow \left[\begin{array}{ccc:c} 1 & 0 & -3 & -2 \\ 0 & 1 & 7 & 11 \\ 0 & 0 & 1 & 2 \end{array}\right]$$

$\begin{cases} x - 3z = -2 \\ y + 7z = 11 \\ z = 2 \end{cases}$

$z = 2$

$y + 7(2) = 11 \Rightarrow y = -3$

$x - 3(2) = -2 \Rightarrow x = 4$

Solution: $(4, -3, 2)$

77. $\begin{cases} -x + y - z = -14 \\ 2x - y + z = 21 \\ 3x + 2y + z = 19 \end{cases}$

$$\begin{bmatrix} -1 & 1 & -1 & \vdots & -14 \\ 2 & -1 & 1 & \vdots & 21 \\ 3 & 2 & 1 & \vdots & 19 \end{bmatrix}$$

$$\begin{matrix} -R_1 \to \\ \\ \\ \end{matrix} \begin{bmatrix} 1 & -1 & 1 & \vdots & 14 \\ 2 & -1 & 1 & \vdots & 21 \\ 3 & 2 & 1 & \vdots & 19 \end{bmatrix}$$

$$\begin{matrix} \\ -2R_1 + R_2 \to \\ -3R_1 + R_3 \to \end{matrix} \begin{bmatrix} 1 & -1 & 1 & \vdots & 14 \\ 0 & 1 & -1 & \vdots & -7 \\ 0 & 5 & -2 & \vdots & -23 \end{bmatrix}$$

$$\begin{matrix} \\ \\ -5R_2 + R_3 \to \end{matrix} \begin{bmatrix} 1 & -1 & 1 & \vdots & 14 \\ 0 & 1 & -1 & \vdots & -7 \\ 0 & 0 & 3 & \vdots & 12 \end{bmatrix}$$

$$\begin{matrix} \\ \\ \frac{1}{3}R_3 \to \end{matrix} \begin{bmatrix} 1 & -1 & 1 & \vdots & 14 \\ 0 & 1 & -1 & \vdots & -7 \\ 0 & 0 & 1 & \vdots & 4 \end{bmatrix}$$

$\begin{cases} x - y + z = 14 \\ y - z = -7 \\ z = 4 \end{cases}$

$z = 4$

$y - 4 = -7 \Rightarrow y = -3$

$x - (-3) + 4 = 14 \Rightarrow x = 7$

Solution: $(7, -3, 4)$

79. Use the reduced row-echelon form feature of a graphic utility.

$\begin{cases} 3x + 3y + 12z = 6 \\ x + y + 4z = 2 \\ 2x + 5y + 20z = 10 \\ -x + 2y + 8z = 4 \end{cases}$

$$\begin{bmatrix} 3 & 3 & 12 & \vdots & 6 \\ 1 & 1 & 4 & \vdots & 2 \\ 2 & 5 & 20 & \vdots & 10 \\ -1 & 2 & 8 & \vdots & 4 \end{bmatrix} \Rightarrow \begin{bmatrix} 1 & 0 & 0 & \vdots & 0 \\ 0 & 1 & 4 & \vdots & 2 \\ 0 & 0 & 0 & \vdots & 0 \\ 0 & 0 & 0 & \vdots & 0 \end{bmatrix} \Rightarrow \begin{cases} x = 0 \\ y + 4z = 2 \end{cases}$$

Let $z = a$.

$y = 2 - 4a$

$x = 0$

Solution: $(0, 2 - 4a, a)$ where a is any real number

81. Use the reduced row-echelon form feature of a graphing utility.

$\begin{cases} 2x + y - z + 2w = -6 \\ 3x + 4y + w = 1 \\ x + 5y + 2z + 6w = -3 \\ 5x + 2y - z - w = 3 \end{cases}$

$$\begin{bmatrix} 2 & 1 & -1 & 2 & \vdots & -6 \\ 3 & 4 & 0 & 1 & \vdots & 1 \\ 1 & 5 & 2 & 6 & \vdots & -3 \\ 5 & 2 & -1 & -1 & \vdots & 3 \end{bmatrix} \Rightarrow \begin{bmatrix} 1 & 0 & 0 & 0 & \vdots & 1 \\ 0 & 1 & 0 & 0 & \vdots & 0 \\ 0 & 0 & 1 & 0 & \vdots & 4 \\ 0 & 0 & 0 & 1 & \vdots & -2 \end{bmatrix}$$

$x = 1$

$y = 0$

$z = 4$

$w = -2$

Solution: $(1, 0, 4, -2)$

83. Use the reduced row-echelon form feature of a graphing utility.

$\begin{cases} x + y + z + w = 0 \\ 2x + 3y + z - 2w = 0 \\ 3x + 5y + z = 0 \end{cases}$

$$\begin{bmatrix} 1 & 1 & 1 & 1 & \vdots & 0 \\ 2 & 3 & 1 & -2 & \vdots & 0 \\ 3 & 5 & 1 & 0 & \vdots & 0 \end{bmatrix} \Rightarrow \begin{bmatrix} 1 & 0 & 2 & 0 & \vdots & 0 \\ 0 & 1 & -1 & 0 & \vdots & 0 \\ 0 & 0 & 0 & 1 & \vdots & 0 \end{bmatrix}$$

$\begin{cases} x + 2z = 0 \\ y - z = 0 \\ w = 0 \end{cases}$

Let $z = a$. Then $x = -2a$ and $y = a$.

Solution: $(-2a, a, a, 0)$ where a is a real number

85. The dimension of the matrix is 4×1.

87. $\begin{cases} a + b + c = 1 \\ 4a + 2b + c = -1 \\ 9a + 3b + c = -5 \end{cases}$

$$\begin{bmatrix} 1 & 1 & 1 & \vdots & 1 \\ 4 & 2 & 1 & \vdots & -1 \\ 9 & 3 & 1 & \vdots & -5 \end{bmatrix} \Rightarrow \begin{bmatrix} 1 & 0 & 0 & \vdots & -1 \\ 0 & 1 & 0 & \vdots & 1 \\ 0 & 0 & 1 & \vdots & 1 \end{bmatrix}$$

$a = 1$

$b = 1$

$c = 1$

So, $f(x) = -x^2 + x + 1$.

89. $\begin{cases} 4a - 2b + c = -15 \\ a - b + c = 7 \\ a + b + c = -3 \end{cases}$

$$\begin{bmatrix} 4 & -2 & 1 & \vdots & -15 \\ 4 & -1 & 1 & \vdots & 7 \\ 1 & 1 & 1 & \vdots & -3 \end{bmatrix} \Rightarrow \begin{bmatrix} 1 & 0 & 0 & \vdots & -9 \\ 0 & 1 & 0 & \vdots & -5 \\ 0 & 0 & 1 & \vdots & 11 \end{bmatrix}$$

$a = -9$

$b = -5$

$c = 11$

So, $f(x) = -9x^2 - 5x + 11$.

91. $\begin{cases} a + b + c = 8 \\ 4a + 2b + c = 13 \\ 9a + 3b + c = 20 \end{cases}$

$$\begin{bmatrix} 1 & 1 & 1 & \vdots & 8 \\ 4 & 2 & 1 & \vdots & 13 \\ 9 & 3 & 1 & \vdots & 20 \end{bmatrix} \Rightarrow \begin{bmatrix} 1 & 0 & 0 & \vdots & 1 \\ 0 & 1 & 0 & \vdots & 2 \\ 0 & 0 & 1 & \vdots & 5 \end{bmatrix}$$

$a = 1$

$b = 2$

$c = 5$

So, $f(x) = x^2 + 2x + 5$.

93. $\begin{cases} 12b + 66a = 831 \\ 66b + 506a = 5643 \end{cases}$

$$\begin{bmatrix} 12 & 66 & \vdots & 831 \\ 66 & 506 & \vdots & 5643 \end{bmatrix} \Rightarrow \begin{bmatrix} 1 & 0 & \vdots & 28 \\ 0 & 1 & \vdots & 7.5 \end{bmatrix}$$

$y = 7.5t + 28$

In 2020, the number of new cases of the waterborne disease will be $y = 7.5(15) + 28 = 140.5 \approx 141$ cases.

Because the data values increase in a linear pattern, this prediction is reasonable.

95. x = amount at 8%

y = amount at 9%

z = amount at 12%

$\begin{cases} x + y + z = 2,000,000 \\ 0.08x + 0.09y + 0.12z = 186,000 \\ x - 2z = 0 \end{cases}$

$$\begin{bmatrix} 1 & 1 & 1 & \vdots & 2,000,000 \\ 0.08 & 0.09 & 0.12 & \vdots & 186,000 \\ 1 & 0 & -2 & \vdots & 0 \end{bmatrix}$$

$$\begin{matrix} \\ -0.08R_1 + R_2 \\ -R_1 + R_3 \end{matrix} \begin{bmatrix} 1 & 1 & 1 & \vdots & 2,000,000 \\ 0 & 0.01 & 0.04 & \vdots & 26,000 \\ 0 & -1 & -3 & \vdots & -2,000,000 \end{bmatrix}$$

$$\begin{matrix} \\ 100R_2 \rightarrow \\ \\ \end{matrix} \begin{bmatrix} 1 & 1 & 1 & \vdots & 2,000,000 \\ 0 & 1 & 4 & \vdots & 2,600,000 \\ 0 & -1 & -3 & \vdots & -2,000,000 \end{bmatrix}$$

$$\begin{matrix} \\ \\ R_2 + R_3 \rightarrow \end{matrix} \begin{bmatrix} 1 & 1 & 1 & \vdots & 2,000,000 \\ 0 & 1 & 4 & \vdots & 2,600,000 \\ 0 & 0 & 1 & \vdots & 600,000 \end{bmatrix}$$

The matrix is now in row-echelon form, and the corresponding system is shown.

$\begin{cases} x + y + z = 2,000,000 \\ y + 4z = 2,600,000 \\ z = 600,000 \end{cases}$

Using back-substitution, you can determine the solution.

$$y + 4(600,000) = 2,600,000$$
$$y = 200,000$$
$$x + 200,000 + 600,000 = 2,000,000$$
$$x = 1,200,000$$

So, the zoo borrowed 1,200,000 at 8%, \$200,000 at 9%, and \$600,000 at 12%.

97. False. It is a 2×4 matrix.

99. They are the same.

Section 7.2 Operations with Matrices

1. equal

3. zero; O

5. $\begin{bmatrix} x & -2 \\ 7 & 23 \end{bmatrix} = \begin{bmatrix} -4 & -2 \\ 7 & y \end{bmatrix}$

$x = -4$

$y = 23$

7. $\begin{bmatrix} 16 & 4 & x & 4 \\ 0 & 2 & 4 & 0 \end{bmatrix} = \begin{bmatrix} 16 & 4 & 2x + 1 & 4 \\ 0 & 2 & 3y - 5 & 0 \end{bmatrix}$

$x = 2x + 1 \Rightarrow x = -1$

$4 = 3y - 5 \Rightarrow y = 3$

9. (a) $A + B = \begin{bmatrix} 1 & -1 \\ 2 & -1 \end{bmatrix} + \begin{bmatrix} 2 & -1 \\ -1 & 8 \end{bmatrix} = \begin{bmatrix} 1+2 & -1-1 \\ 2-1 & -1+8 \end{bmatrix} = \begin{bmatrix} 3 & -2 \\ 1 & 7 \end{bmatrix}$

(b) $A - B = \begin{bmatrix} 1 & -1 \\ 2 & -1 \end{bmatrix} - \begin{bmatrix} 2 & -1 \\ -1 & 8 \end{bmatrix} = \begin{bmatrix} 1-2 & -1+1 \\ 2+1 & -1-8 \end{bmatrix} = \begin{bmatrix} -1 & 0 \\ 3 & -9 \end{bmatrix}$

(c) $3A = 3\begin{bmatrix} 1 & -1 \\ 2 & -1 \end{bmatrix} = \begin{bmatrix} 3(1) & 3(-1) \\ 3(2) & 3(-1) \end{bmatrix} = \begin{bmatrix} 3 & -3 \\ 6 & -3 \end{bmatrix}$

(d) $3A - 2B = \begin{bmatrix} 3 & -3 \\ 6 & -3 \end{bmatrix} - 2\begin{bmatrix} 2 & -1 \\ -1 & 8 \end{bmatrix} = \begin{bmatrix} 3 & -3 \\ 6 & -3 \end{bmatrix} + \begin{bmatrix} -4 & 2 \\ 2 & -16 \end{bmatrix} = \begin{bmatrix} -1 & -1 \\ 8 & -19 \end{bmatrix}$

11. $A = \begin{bmatrix} 6 & 0 & 3 \\ -1 & -4 & 0 \end{bmatrix}, B = \begin{bmatrix} 8 & -1 \\ 4 & -3 \end{bmatrix}$

(a) $A + B$ is not possible. A and B do not have the same order.

(b) $A - B$ is not possible. A and B do not have the same order.

(c) $3A = \begin{bmatrix} 18 & 0 & 9 \\ -3 & -12 & 0 \end{bmatrix}$

(d) $3A - 2B$ is not possible. A and B do not have the same order.

13. $A = \begin{bmatrix} 8 & -1 \\ 2 & 3 \\ -4 & 5 \end{bmatrix}, B = \begin{bmatrix} 1 & 6 \\ -1 & -5 \\ 1 & 10 \end{bmatrix}$

(a) $A + B = \begin{bmatrix} 8 & -1 \\ 2 & 3 \\ -4 & 5 \end{bmatrix} + \begin{bmatrix} 1 & 6 \\ -1 & -5 \\ 1 & 10 \end{bmatrix} = \begin{bmatrix} 8+1 & -1+6 \\ 2-1 & 3-5 \\ -4+1 & 5+10 \end{bmatrix} = \begin{bmatrix} 9 & 5 \\ 1 & -2 \\ -3 & 15 \end{bmatrix}$

(b) $A - B = \begin{bmatrix} 8 & -1 \\ 2 & 3 \\ -4 & 5 \end{bmatrix} - \begin{bmatrix} 1 & 6 \\ -1 & -5 \\ 1 & 10 \end{bmatrix} = \begin{bmatrix} 8-1 & -1-6 \\ 2-(-1) & 3-(-5) \\ -4-1 & 5-10 \end{bmatrix} = \begin{bmatrix} 7 & -7 \\ 3 & 8 \\ -5 & -5 \end{bmatrix}$

(c) $3A = 3\begin{bmatrix} 8 & -1 \\ 2 & 3 \\ -4 & 5 \end{bmatrix} = \begin{bmatrix} 3(8) & 3(-1) \\ 3(2) & 3(3) \\ 3(-4) & 3(5) \end{bmatrix} = \begin{bmatrix} 24 & -3 \\ 6 & 9 \\ -12 & 15 \end{bmatrix}$

(d) $3A - 2B = \begin{bmatrix} 24 & -3 \\ 6 & 9 \\ -12 & 15 \end{bmatrix} - 2\begin{bmatrix} 1 & 6 \\ -1 & -5 \\ 1 & 10 \end{bmatrix} = \begin{bmatrix} 24-2 & -3-12 \\ 6+2 & 9+10 \\ -12-2 & 15-20 \end{bmatrix} = \begin{bmatrix} 22 & -15 \\ 8 & 19 \\ -14 & -5 \end{bmatrix}$

15. $A = \begin{bmatrix} 4 & 5 & -1 & 3 & 4 \\ 1 & 2 & -2 & -1 & 0 \end{bmatrix}, B = \begin{bmatrix} 1 & 0 & -1 & 1 & 0 \\ -6 & 8 & 2 & -3 & -7 \end{bmatrix}$

(a) $A + B = \begin{bmatrix} 4 & 5 & -1 & 3 & 4 \\ 1 & 2 & -2 & -1 & 0 \end{bmatrix} + \begin{bmatrix} 1 & 0 & -1 & 1 & 0 \\ -6 & 8 & 2 & -3 & -7 \end{bmatrix} = \begin{bmatrix} 4+1 & 5+0 & -1-1 & 3+1 & 4+0 \\ 1-6 & 2+8 & -2+2 & -1-3 & 0-7 \end{bmatrix}$

$= \begin{bmatrix} 5 & 5 & -2 & 4 & 4 \\ -5 & 10 & 0 & -4 & -7 \end{bmatrix}$

(b) $A - B = \begin{bmatrix} 4 & 5 & -1 & 3 & 4 \\ 1 & 2 & -2 & -1 & 0 \end{bmatrix} - \begin{bmatrix} 1 & 0 & -1 & 1 & 0 \\ -6 & 8 & 2 & -3 & -7 \end{bmatrix} = \begin{bmatrix} 4-1 & 5-0 & -1-(-1) & 3-1 & 4-0 \\ 1-(-6) & 2-8 & -2-2 & -1-(-3) & 0-(-7) \end{bmatrix}$

$= \begin{bmatrix} 3 & 5 & 0 & 2 & 4 \\ 7 & -6 & -4 & 2 & 7 \end{bmatrix}$

(c) $3A = 3\begin{bmatrix} 4 & 5 & -1 & 3 & 4 \\ 1 & 2 & -2 & -1 & 0 \end{bmatrix} = \begin{bmatrix} 3(4) & 3(5) & 3(-1) & 3(3) & 3(4) \\ 3(1) & 3(2) & 3(-2) & 3(-1) & 3(0) \end{bmatrix} = \begin{bmatrix} 12 & 15 & -3 & 9 & 12 \\ 3 & 6 & -6 & -3 & 0 \end{bmatrix}$

(d) $3A - 2B = \begin{bmatrix} 12 & 15 & -3 & 9 & 12 \\ 3 & 6 & -6 & -3 & 0 \end{bmatrix} - 2\begin{bmatrix} 1 & 0 & -1 & 1 & 0 \\ -6 & 8 & 2 & -3 & -7 \end{bmatrix} = \begin{bmatrix} 12-2 & 15+0 & -3+2 & 9-2 & 12-0 \\ 3+12 & 6-16 & -6-4 & -3+6 & 0+14 \end{bmatrix}$

$= \begin{bmatrix} 10 & 15 & -1 & 7 & 12 \\ 15 & -10 & -10 & 3 & 14 \end{bmatrix}$

17. $\begin{bmatrix} -5 & 0 \\ 3 & -6 \end{bmatrix} + \begin{bmatrix} 7 & 1 \\ -2 & -1 \end{bmatrix} + \begin{bmatrix} -10 & -8 \\ 14 & 6 \end{bmatrix} = \begin{bmatrix} -5+7+(-10) & 0+1+(-8) \\ 3+(-2)+14 & -6+(-1)+6 \end{bmatrix} = \begin{bmatrix} -8 & -7 \\ 15 & -1 \end{bmatrix}$

19. $4\left(\begin{bmatrix} -4 & 0 & 1 \\ 0 & 2 & 3 \end{bmatrix} - \begin{bmatrix} 2 & 1 & -2 \\ 3 & -6 & 0 \end{bmatrix}\right) = 4\begin{bmatrix} -6 & -1 & 3 \\ -3 & 8 & 3 \end{bmatrix} = \begin{bmatrix} -24 & -4 & 12 \\ -12 & 32 & 12 \end{bmatrix}$

21. $-3\left(\begin{bmatrix} 0 & -3 \\ 7 & 2 \end{bmatrix} + \begin{bmatrix} -6 & 3 \\ 8 & 1 \end{bmatrix}\right) - 2\begin{bmatrix} 4 & -4 \\ 7 & -9 \end{bmatrix} = -3\begin{bmatrix} -6 & 0 \\ 15 & 3 \end{bmatrix} - \begin{bmatrix} 8 & -8 \\ 14 & -18 \end{bmatrix} = \begin{bmatrix} 18 & 0 \\ -45 & -9 \end{bmatrix} - \begin{bmatrix} 8 & -8 \\ 14 & -18 \end{bmatrix} = \begin{bmatrix} 10 & 8 \\ -59 & 9 \end{bmatrix}$

23. $\frac{11}{25}\begin{bmatrix} 2 & 5 \\ -1 & -4 \end{bmatrix} + 6\begin{bmatrix} -3 & 0 \\ 2 & 2 \end{bmatrix} = \begin{bmatrix} -17.12 & 2.2 \\ 11.56 & 10.24 \end{bmatrix}$

25. $-2\begin{bmatrix} 1.23 & 4.19 & -3.85 \\ 7.21 & -2.60 & 6.54 \end{bmatrix} - \begin{bmatrix} 8.35 & -3.02 & 7.30 \\ -0.38 & -5.49 & 1.68 \end{bmatrix} = \begin{bmatrix} -10.81 & -5.39 & 0.4 \\ -14.04 & 10.69 & -14.76 \end{bmatrix}$

In Exercises 27-33, $A = \begin{bmatrix} -2 & 1 & 3 \\ -1 & 0 & 4 \end{bmatrix}$ **and** $B = \begin{bmatrix} 0 & 2 & -4 \\ 3 & 0 & 1 \end{bmatrix}$

27. $X = 2A + 2B = 2\begin{bmatrix} -2 & 1 & 3 \\ -1 & 0 & 4 \end{bmatrix} + 2\begin{bmatrix} 0 & 2 & -4 \\ 3 & 0 & 1 \end{bmatrix} = \begin{bmatrix} -4 & 2 & 6 \\ -2 & 0 & 8 \end{bmatrix} + \begin{bmatrix} 0 & 4 & -8 \\ 6 & 0 & 2 \end{bmatrix} = \begin{bmatrix} -4 & 6 & -2 \\ 4 & 0 & 10 \end{bmatrix}$

29. $2X = 2A - B$

$X = \frac{1}{2}(2A - B)$

$= \frac{1}{2}\left(2\begin{bmatrix} -2 & 1 & 3 \\ -1 & 0 & 4 \end{bmatrix} - \begin{bmatrix} 0 & 2 & -4 \\ 3 & 0 & 1 \end{bmatrix}\right)$

$= \frac{1}{2}\left(\begin{bmatrix} -4 & 2 & 6 \\ -2 & 0 & 8 \end{bmatrix} - \begin{bmatrix} 0 & 2 & -4 \\ 3 & 0 & 1 \end{bmatrix}\right)$

$= \frac{1}{2}\begin{bmatrix} -4 & 0 & 10 \\ -5 & 0 & 7 \end{bmatrix} = \begin{bmatrix} -2 & 0 & 5 \\ -\frac{5}{2} & 0 & \frac{7}{2} \end{bmatrix}$

31. $2X + 3A = B$

$2X = B - 3A$

$X = \frac{1}{2}(B - 3A)$

$= \frac{1}{2}\left(\begin{bmatrix} 0 & 2 & -4 \\ 3 & 0 & 1 \end{bmatrix} - 3\begin{bmatrix} -2 & 1 & 3 \\ -1 & 0 & 4 \end{bmatrix}\right)$

$= \frac{1}{2}\left(\begin{bmatrix} 0 & 2 & -4 \\ 3 & 0 & 1 \end{bmatrix} - \begin{bmatrix} -6 & 3 & 9 \\ -3 & 0 & 12 \end{bmatrix}\right)$

$= \frac{1}{2}\begin{bmatrix} 6 & -1 & -13 \\ 6 & 0 & -11 \end{bmatrix} = \begin{bmatrix} 3 & -\frac{1}{2} & -\frac{13}{2} \\ 3 & 0 & -\frac{11}{2} \end{bmatrix}$

33. $4B = -2X - 2A$

$2X = -2A - 4B$

$X = \frac{1}{2}(-2A - 4B)$

$= \frac{1}{2}\left(-2\begin{bmatrix} -2 & 1 & 3 \\ -1 & 0 & 4 \end{bmatrix} - 4\begin{bmatrix} 0 & 2 & -4 \\ 3 & 0 & 1 \end{bmatrix}\right)$

$= \frac{1}{2}\left(\begin{bmatrix} 4 & -2 & -6 \\ 2 & 0 & -8 \end{bmatrix} - \begin{bmatrix} 0 & 8 & -16 \\ 12 & 0 & 4 \end{bmatrix}\right)$

$= \frac{1}{2}\begin{bmatrix} 4 & -10 & 10 \\ -10 & 0 & -12 \end{bmatrix}$

$= \begin{bmatrix} 2 & -5 & 5 \\ -5 & 0 & -6 \end{bmatrix}$

35. A is 3×2, B is $2 \times 2 \Rightarrow AB$ is 3×2.

$$A = \begin{bmatrix} -1 & 6 \\ -4 & 5 \\ 0 & 3 \end{bmatrix}, B = \begin{bmatrix} 2 & 3 \\ 0 & 9 \end{bmatrix} \qquad AB = \begin{bmatrix} -1 & 6 \\ -4 & 5 \\ 0 & 3 \end{bmatrix}\begin{bmatrix} 2 & 3 \\ 0 & 9 \end{bmatrix} = \begin{bmatrix} (-1)(2) + (6)(0) & (-1)(3) + (6)(9) \\ (-4)(2) + (5)(0) & (-4)(3) + (5)(9) \\ (0)(2) + (3)(0) & (0)(3) + (3)(9) \end{bmatrix} = \begin{bmatrix} -2 & 51 \\ -8 & 33 \\ 0 & 27 \end{bmatrix}$$

37. A is 3×2 and B is 3×3. AB is not possible.

39. A is 3×3, B is $3 \times 3 \Rightarrow AB$ is 3×3.

$$AB = \begin{bmatrix} 5 & 0 & 0 \\ 0 & -8 & 0 \\ 0 & 0 & 7 \end{bmatrix}\begin{bmatrix} \frac{1}{5} & 0 & 0 \\ 0 & -\frac{1}{8} & 0 \\ 0 & 0 & \frac{1}{2} \end{bmatrix} = \begin{bmatrix} 1 & 0 & 0 \\ 0 & 1 & 0 \\ 0 & 0 & \frac{7}{2} \end{bmatrix}$$

41. $A = \begin{bmatrix} 7 & 5 & -4 \\ -2 & 5 & 1 \\ 10 & -4 & -7 \end{bmatrix}, B = \begin{bmatrix} 2 & -2 & 3 \\ 8 & 1 & 4 \\ -4 & 2 & -8 \end{bmatrix}$

$$AB = \begin{bmatrix} 7 & 5 & -4 \\ -2 & 5 & 1 \\ 10 & -4 & -7 \end{bmatrix}\begin{bmatrix} 2 & -2 & 3 \\ 8 & 1 & 4 \\ -4 & 2 & -8 \end{bmatrix} = \begin{bmatrix} 70 & -17 & 73 \\ 32 & 11 & 6 \\ 16 & -38 & 70 \end{bmatrix}$$

43. A is 3×2, B is 5×3. AB is not possible.

45. $$\begin{bmatrix} -3 & 8 & -6 & 8 \\ -12 & 15 & 9 & 6 \\ 5 & -1 & 1 & 5 \end{bmatrix}\begin{bmatrix} 3 & 1 & 6 \\ 24 & 15 & 14 \\ 16 & 10 & 21 \\ 8 & -4 & 10 \end{bmatrix} = \begin{bmatrix} 151 & 25 & 48 \\ 516 & 279 & 387 \\ 47 & -20 & 87 \end{bmatrix}$$

47. (a) $$AB = \begin{bmatrix} 1 & 2 \\ 4 & 2 \end{bmatrix}\begin{bmatrix} 2 & -1 \\ -1 & 8 \end{bmatrix} = \begin{bmatrix} (1)(2) + (2)(-1) & (1)(-1) + (2)(8) \\ (4)(2) + (2)(-1) & (4)(-1) + (2)(8) \end{bmatrix} = \begin{bmatrix} 0 & 15 \\ 6 & 12 \end{bmatrix}$$

(b) $$BA = \begin{bmatrix} 2 & -1 \\ -1 & 8 \end{bmatrix}\begin{bmatrix} 1 & 2 \\ 4 & 2 \end{bmatrix} = \begin{bmatrix} (2)(1) + (-1)(4) & (2)(2) + (-1)(2) \\ (-1)(1) + (8)(4) & (-1)(2) + (8)(2) \end{bmatrix} = \begin{bmatrix} -2 & 2 \\ 31 & 14 \end{bmatrix}$$

(c) $$A^2 = \begin{bmatrix} 1 & 2 \\ 4 & 2 \end{bmatrix}\begin{bmatrix} 1 & 2 \\ 4 & 2 \end{bmatrix} = \begin{bmatrix} (1)(1) + (2)(4) & (1)(2) + (2)(2) \\ (4)(1) + (2)(4) & (4)(2) + (2)(2) \end{bmatrix} = \begin{bmatrix} 9 & 6 \\ 12 & 12 \end{bmatrix}$$

49. (a) $$AB = \begin{bmatrix} 5 & -9 & 0 \\ 3 & 0 & -8 \\ -1 & 4 & 11 \end{bmatrix}\begin{bmatrix} 1 & 0 & 0 \\ 0 & 1 & 0 \\ 0 & 0 & 1 \end{bmatrix} = \begin{bmatrix} (5)(1) + (-9)(0) + (0)(0) & (5)(0) + (-9)(1) + (0)(0) & (5)(0) + (-9)(0) + (0)(1) \\ (3)(1) + (0)(0) + (-8)(0) & (3)(0) + (0)(1) + (-8)(0) & (3)(0) + (0)(0) + (-8)(1) \\ (-1)(1) + (4)(0) + (11)(0) & (-1)(0) + (4)(1) + (11)(0) & (-1)(0) + (4)(0) + (11)(1) \end{bmatrix}$$

$$= \begin{bmatrix} 5 & -9 & 0 \\ 3 & 0 & -8 \\ -1 & 4 & 11 \end{bmatrix}$$

(b) $$BA = \begin{bmatrix} 1 & 0 & 0 \\ 0 & 1 & 0 \\ 0 & 0 & 1 \end{bmatrix}\begin{bmatrix} 5 & -9 & 0 \\ 3 & 0 & -8 \\ -1 & 4 & 11 \end{bmatrix} = \begin{bmatrix} (1)(5) + (0)(3) + (0)(-1) & (1)(-9) + (0)(0) + (0)(4) & (1)(0) + (0)(-8) + (0)(11) \\ (0)(5) + (1)(3) + (0)(-1) & (0)(-9) + (1)(0) + (0)(4) & (0)(0) + (1)(-8) + (0)(11) \\ (0)(5) + (0)(3) + (1)(-1) & (0)(-9) + (0)(0) + (1)(4) & (0)(0) + (0)(-8) + (1)(11) \end{bmatrix}$$

$$= \begin{bmatrix} 5 & -9 & 0 \\ 3 & 0 & -8 \\ -1 & 4 & 11 \end{bmatrix}$$

(c) $$AA = \begin{bmatrix} 5 & -9 & 0 \\ 3 & 0 & -8 \\ -1 & 4 & 11 \end{bmatrix}\begin{bmatrix} 5 & -9 & 0 \\ 3 & 0 & -8 \\ -1 & 4 & 11 \end{bmatrix} = \begin{bmatrix} (5)(5) + (-9)(3) + (0)(-1) & (5)(-9) + (-9)(0) + (0)(4) & (5)(0) + (-9)(-8) + (0)(11) \\ (3)(5) + (0)(3) + (-8)(-1) & (3)(-9) + (0)(0) + (-8)(4) & (3)(0) + (0)(-8) + (-8)(11) \\ (-1)(5) + (4)(3) + (11)(-1) & (-1)(-9) + (4)(0) + (11)(4) & (-1)(0) + (4)(-8) + (11)(11) \end{bmatrix}$$

$$= \begin{bmatrix} -2 & -45 & 72 \\ 23 & -59 & -88 \\ -4 & 53 & 89 \end{bmatrix}$$

51. (a) $AB = \begin{bmatrix} -4 & -1 \\ 2 & 12 \end{bmatrix}\begin{bmatrix} -6 \\ 5 \end{bmatrix} = \begin{bmatrix} (-4)(-6) + (-1)(5) \\ (2)(-6) + (12)(5) \end{bmatrix} = \begin{bmatrix} 19 \\ 48 \end{bmatrix}$

(b) BA is not possible, B is 2×1 and A is 2×2.

(c) $A^2 = \begin{bmatrix} -4 & -1 \\ 2 & 12 \end{bmatrix}\begin{bmatrix} -4 & -1 \\ 2 & 12 \end{bmatrix} = \begin{bmatrix} (-4)(-4) + (-1)(2) & (-4)(-1) + (-1)(12) \\ (2)(-4) + (12)(2) & (2)(-1) + (12)(12) \end{bmatrix} = \begin{bmatrix} 14 & -8 \\ 16 & 142 \end{bmatrix}$

53. (a) $AB = \begin{bmatrix} 7 \\ 8 \\ -1 \end{bmatrix}\begin{bmatrix} 1 & 1 & 2 \end{bmatrix} = \begin{bmatrix} 7(1) & 7(1) & 7(2) \\ 8(1) & 8(1) & 8(2) \\ -1(1) & -1(1) & -1(2) \end{bmatrix} = \begin{bmatrix} 7 & 7 & 14 \\ 8 & 8 & 16 \\ -1 & -1 & -2 \end{bmatrix}$

(b) $BA = \begin{bmatrix} 1 & 1 & 2 \end{bmatrix}\begin{bmatrix} 7 \\ 8 \\ -1 \end{bmatrix} = \begin{bmatrix} (1)(7) + (1)(8) + (2)(-1) \end{bmatrix} = \begin{bmatrix} 13 \end{bmatrix}$

(c) A^2 is not possible.

55. $\begin{bmatrix} 3 & 1 \\ 0 & -2 \end{bmatrix}\begin{bmatrix} 1 & 0 \\ -2 & 2 \end{bmatrix}\begin{bmatrix} 1 & 0 \\ 2 & 4 \end{bmatrix} = \begin{bmatrix} 1 & 2 \\ 4 & -4 \end{bmatrix}\begin{bmatrix} 1 & 0 \\ 2 & 4 \end{bmatrix} = \begin{bmatrix} 5 & 8 \\ -4 & -16 \end{bmatrix}$

57. $\begin{bmatrix} 0 & 2 & -2 \\ 4 & 1 & 2 \end{bmatrix}\left(\begin{bmatrix} 4 & 0 \\ 0 & -1 \\ -1 & 2 \end{bmatrix} + \begin{bmatrix} -2 & 3 \\ -3 & 5 \\ 0 & -3 \end{bmatrix}\right) = \begin{bmatrix} 0 & 2 & -2 \\ 4 & 1 & 2 \end{bmatrix}\begin{bmatrix} 2 & 3 \\ -3 & 4 \\ -1 & -1 \end{bmatrix} = \begin{bmatrix} -4 & 10 \\ 3 & 14 \end{bmatrix}$

59. $\begin{bmatrix} 3 \\ -1 \\ 5 \\ 7 \end{bmatrix}\left(\begin{bmatrix} 5 & -6 \end{bmatrix} + \begin{bmatrix} 7 & -1 \end{bmatrix} + \begin{bmatrix} -8 & 9 \end{bmatrix}\right) = \begin{bmatrix} 3 \\ -1 \\ 5 \\ 7 \end{bmatrix}\begin{bmatrix} 4 & 2 \end{bmatrix} = \begin{bmatrix} 3(4) & 3(2) \\ (-1)(4) & (-1)(2) \\ 5(4) & 5(2) \\ 7(4) & 7(2) \end{bmatrix} = \begin{bmatrix} 12 & 6 \\ -4 & -2 \\ 20 & 10 \\ 28 & 14 \end{bmatrix}$

61. (a) $\begin{bmatrix} 2 & 3 \\ 1 & 4 \end{bmatrix}\begin{bmatrix} x_1 \\ x_2 \end{bmatrix} = \begin{bmatrix} 5 \\ 10 \end{bmatrix}$

(b)

$$\begin{array}{r} R_2 \\ R_1 \end{array}\begin{bmatrix} 1 & 4 & \vdots & 10 \\ 2 & 3 & \vdots & 5 \end{bmatrix}$$

$$\begin{array}{r} \\ -2R_1 + R_2 \rightarrow \end{array}\begin{bmatrix} 1 & 4 & \vdots & 10 \\ 0 & -5 & \vdots & -15 \end{bmatrix}$$

$$\begin{array}{r} \\ -\frac{1}{5}R_2 \rightarrow \end{array}\begin{bmatrix} 1 & 4 & \vdots & 10 \\ 0 & 1 & \vdots & 3 \end{bmatrix}$$

$$\begin{array}{r} -4R_2 + R_1 \rightarrow \\ -\frac{1}{5}R_2 \rightarrow \end{array}\begin{bmatrix} 1 & 0 & \vdots & -2 \\ 0 & 1 & \vdots & 3 \end{bmatrix}$$

$$X = \begin{bmatrix} -2 \\ 3 \end{bmatrix}$$

63. (a) $\begin{bmatrix} 1 & -2 & 3 \\ -1 & 3 & -1 \\ 2 & -5 & 5 \end{bmatrix}\begin{bmatrix} x_1 \\ x_2 \\ x_3 \end{bmatrix} = \begin{bmatrix} 9 \\ -6 \\ 17 \end{bmatrix}$

(b)

$$\begin{bmatrix} 1 & -2 & 3 & \vdots & 9 \\ -1 & 3 & -1 & \vdots & -6 \\ 2 & -5 & 5 & \vdots & 17 \end{bmatrix}$$

$$\begin{array}{r} \\ R_1 + R_2 \rightarrow \\ -2R_2 + R_3 \rightarrow \end{array}\begin{bmatrix} 1 & -2 & 3 & \vdots & 9 \\ 0 & 1 & 2 & \vdots & 3 \\ 0 & -1 & -1 & \vdots & -1 \end{bmatrix}$$

$$\begin{array}{r} 2R_2 + R_1 \rightarrow \\ \\ R_2 + R_3 \rightarrow \end{array}\begin{bmatrix} 1 & 0 & 7 & \vdots & 15 \\ 0 & 1 & 2 & \vdots & 3 \\ 0 & 0 & 1 & \vdots & 2 \end{bmatrix}$$

$$\begin{array}{r} -7R_3 + R_1 \rightarrow \\ -2R_3 + R_2 \rightarrow \\ \\ \end{array}\begin{bmatrix} 1 & 0 & 0 & \vdots & 1 \\ 0 & 1 & 0 & \vdots & -1 \\ 0 & 0 & 1 & \vdots & 2 \end{bmatrix}$$

$$X = \begin{bmatrix} 1 \\ -1 \\ 2 \end{bmatrix}$$

65. (a) $\begin{bmatrix} 1 & -5 & 2 \\ -3 & 1 & -1 \\ 0 & -2 & 5 \end{bmatrix}\begin{bmatrix} x_1 \\ x_2 \\ x_3 \end{bmatrix} = \begin{bmatrix} -20 \\ 8 \\ -16 \end{bmatrix}$

(b) $\begin{bmatrix} 1 & -5 & 2 & \vdots & -20 \\ -3 & 1 & -1 & \vdots & 8 \\ 0 & -2 & 5 & \vdots & -16 \end{bmatrix}$

$3R_1 + R_2 \rightarrow \begin{bmatrix} 1 & -5 & 2 & \vdots & -20 \\ 0 & -14 & 5 & \vdots & -52 \\ 0 & -2 & 5 & \vdots & -16 \end{bmatrix}$

$-R_3 + R_2 \rightarrow \begin{bmatrix} 1 & -5 & 2 & \vdots & -20 \\ 0 & -12 & 0 & \vdots & -36 \\ 0 & -2 & 5 & \vdots & -16 \end{bmatrix}$

$-\frac{1}{12}R_2 \rightarrow \begin{bmatrix} 1 & -5 & 2 & \vdots & -20 \\ 0 & 1 & 0 & \vdots & 3 \\ 0 & -2 & 5 & \vdots & -16 \end{bmatrix}$

$\begin{matrix} 5R_2 + R_1 \rightarrow \\ \\ 2R_2 + R_3 \rightarrow \end{matrix} \begin{bmatrix} 1 & 0 & 2 & \vdots & -5 \\ 0 & 1 & 0 & \vdots & 3 \\ 0 & 0 & 5 & \vdots & -10 \end{bmatrix}$

$\begin{matrix} \\ \\ \frac{1}{5}R_3 \rightarrow \end{matrix} \begin{bmatrix} 1 & 0 & 2 & \vdots & -5 \\ 0 & 1 & 0 & \vdots & 3 \\ 0 & 0 & 1 & \vdots & -2 \end{bmatrix}$

$-2R_3 + R_1 \rightarrow \begin{bmatrix} 1 & 0 & 0 & \vdots & -1 \\ 0 & 1 & 0 & \vdots & 3 \\ 0 & 0 & 1 & \vdots & -2 \end{bmatrix}$

$X = \begin{bmatrix} -1 \\ 3 \\ -2 \end{bmatrix}$

67. $1.10\begin{bmatrix} 100 & 90 & 70 & 30 \\ 40 & 20 & 60 & 60 \end{bmatrix} = \begin{bmatrix} 110 & 99 & 77 & 33 \\ 44 & 22 & 66 & 66 \end{bmatrix}$

69. $BA = \begin{bmatrix} 3.50 & 6.00 \end{bmatrix}\begin{bmatrix} 125 & 100 & 75 \\ 100 & 175 & 125 \end{bmatrix}$

$= \begin{bmatrix} \$1037.50 & \$1400 & \$1012.50 \end{bmatrix}$

The entries represent the profits from both crops at each of the three outlets.

71. $ST = \begin{bmatrix} 1.0 & 0.5 & 0.2 \\ 1.6 & 1.0 & 0.2 \\ 2.5 & 2.0 & 1.4 \end{bmatrix}\begin{bmatrix} 15 & 13 \\ 12 & 11 \\ 11 & 10 \end{bmatrix} = \begin{bmatrix} \$23.20 & \$20.50 \\ \$38.20 & \$33.80 \\ \$76.90 & \$68.50 \end{bmatrix}$

The entries represent the labor costs at each plant for each size of boat.

73. $P^2 = \begin{bmatrix} 0.6 & 0.1 & 0.1 \\ 0.2 & 0.7 & 0.1 \\ 0.2 & 0.2 & 0.8 \end{bmatrix}\begin{bmatrix} 0.6 & 0.1 & 0.1 \\ 0.2 & 0.7 & 0.1 \\ 0.2 & 0.2 & 0.8 \end{bmatrix}$

$= \begin{bmatrix} 0.40 & 0.15 & 0.15 \\ 0.28 & 0.53 & 0.17 \\ 0.32 & 0.32 & 0.68 \end{bmatrix}$

The P^2 matrix gives the proportion of the voting population that changed parties or remained loyal to their parties from the first election to the third.

75. True.

The sum of two matrices of different orders is undefined.

77. Answers will vary. *Sample answer:*

$(A + B)^2 = \left(\begin{bmatrix} 2 & -1 \\ 1 & 3 \end{bmatrix} + \begin{bmatrix} -1 & 1 \\ 0 & -2 \end{bmatrix}\right)^2 = \begin{bmatrix} 1 & 0 \\ 2 & 1 \end{bmatrix} \neq$

$A^2 + 2AB + B^2 = \left(\begin{bmatrix} 2 & -1 \\ 1 & 3 \end{bmatrix}\right)^2 + 2\begin{bmatrix} 2 & -1 \\ 1 & 3 \end{bmatrix}\begin{bmatrix} -1 & 1 \\ 0 & -2 \end{bmatrix} + \left(\begin{bmatrix} -1 & 1 \\ 0 & -2 \end{bmatrix}\right)^2 = \begin{bmatrix} 0 & 0 \\ 3 & 2 \end{bmatrix}$

79. Answers will vary. *Sample answer:*

$(A + B)(A - B) = \left(\begin{bmatrix} 2 & -1 \\ 1 & 3 \end{bmatrix} + \begin{bmatrix} -1 & 1 \\ 0 & -2 \end{bmatrix}\right)\left(\begin{bmatrix} 2 & -1 \\ 1 & 3 \end{bmatrix} - \begin{bmatrix} -1 & 1 \\ 0 & -2 \end{bmatrix}\right) = \begin{bmatrix} 3 & -2 \\ 4 & 3 \end{bmatrix} \neq$

$A^2 - B^2 = \left(\begin{bmatrix} 2 & -1 \\ 1 & 3 \end{bmatrix}\right)^2 - \left(\begin{bmatrix} -1 & 1 \\ 0 & -2 \end{bmatrix}\right)^2 = \begin{bmatrix} 2 & -2 \\ 5 & 4 \end{bmatrix}$

81. $AC = \begin{bmatrix} 0 & 1 \\ 0 & 1 \end{bmatrix}\begin{bmatrix} 2 & 3 \\ 2 & 3 \end{bmatrix} = \begin{bmatrix} 2 & 3 \\ 2 & 3 \end{bmatrix}$

$BC = \begin{bmatrix} 1 & 0 \\ 1 & 0 \end{bmatrix}\begin{bmatrix} 2 & 3 \\ 2 & 3 \end{bmatrix} = \begin{bmatrix} 2 & 3 \\ 2 & 3 \end{bmatrix}$

So, $AC = BC$ even though $A \neq B$.

83. Answer will vary. *Sample answer*:

$$A = \begin{bmatrix} 1 & 0 \\ 0 & 1 \end{bmatrix}, B = \begin{bmatrix} 0 & 1 \\ 1 & 0 \end{bmatrix}$$

$$AB = \begin{bmatrix} 1 & 0 \\ 0 & 1 \end{bmatrix}\begin{bmatrix} 0 & 1 \\ 1 & 0 \end{bmatrix} = \begin{bmatrix} (1)(0) + (0)(1) & (1)(1) + (0)(0) \\ (0)(0) + (1)(1) & (0)(1) + (1)(0) \end{bmatrix} = \begin{bmatrix} 0 & 1 \\ 1 & 0 \end{bmatrix}$$

$$BA = \begin{bmatrix} 0 & 1 \\ 1 & 0 \end{bmatrix}\begin{bmatrix} 1 & 0 \\ 0 & 1 \end{bmatrix} = \begin{bmatrix} (0)(1) + (1)(0) & (0)(0) + (1)(1) \\ (1)(1) + (0)(0) & (1)(0) + (0)(1) \end{bmatrix} = \begin{bmatrix} 0 & 1 \\ 1 & 0 \end{bmatrix}$$

So, $AB = BA$.

85. The product of two diagonal matrices of the same order is a diagonal matrix whose entries are the products of the corresponding diagonal entries of A and B.

Section 7.3 The Inverse of a Square Matrix

1. inverse

3. determinant

5. $$AB = \begin{bmatrix} 2 & 1 \\ 5 & 3 \end{bmatrix}\begin{bmatrix} 3 & -1 \\ -5 & 2 \end{bmatrix} = \begin{bmatrix} 6 - 5 & -2 + 2 \\ 15 - 15 & -5 + 6 \end{bmatrix} = \begin{bmatrix} 1 & 0 \\ 0 & 1 \end{bmatrix}$$

$$BA = \begin{bmatrix} 3 & -1 \\ -5 & 2 \end{bmatrix}\begin{bmatrix} 2 & 1 \\ 5 & 3 \end{bmatrix} = \begin{bmatrix} 6 - 5 & 3 - 3 \\ -10 + 10 & -5 + 6 \end{bmatrix} = \begin{bmatrix} 1 & 0 \\ 0 & 1 \end{bmatrix}$$

7. $$AB = \frac{1}{10}\begin{bmatrix} 3 & 2 \\ 1 & 4 \end{bmatrix}\begin{bmatrix} 4 & -2 \\ -1 & 3 \end{bmatrix} = \frac{1}{10}\begin{bmatrix} 12 - 2 & -6 + 6 \\ 4 - 4 & -2 + 12 \end{bmatrix} = \frac{1}{10}\begin{bmatrix} 10 & 0 \\ 0 & 10 \end{bmatrix} = \begin{bmatrix} 1 & 0 \\ 0 & 1 \end{bmatrix}$$

$$BA = \frac{1}{10}\begin{bmatrix} 4 & -2 \\ -1 & 3 \end{bmatrix}\begin{bmatrix} 3 & 2 \\ 1 & 4 \end{bmatrix} = \frac{1}{10}\begin{bmatrix} 12 - 2 & 8 - 8 \\ -3 + 3 & -2 + 12 \end{bmatrix} = \frac{1}{10}\begin{bmatrix} 10 & 0 \\ 0 & 10 \end{bmatrix} = \begin{bmatrix} 1 & 0 \\ 0 & 1 \end{bmatrix}$$

9. $$AB = \begin{bmatrix} 2 & -17 & 11 \\ -1 & 11 & -7 \\ 0 & 3 & -2 \end{bmatrix}\begin{bmatrix} 1 & 1 & 2 \\ 2 & 4 & -3 \\ 3 & 6 & -5 \end{bmatrix} = \begin{bmatrix} 2 - 34 + 33 & 2 - 68 + 66 & 4 + 51 - 55 \\ -1 + 22 - 21 & -1 + 44 - 42 & -2 - 33 + 35 \\ 6 - 6 & 12 - 12 & -9 + 10 \end{bmatrix} = \begin{bmatrix} 1 & 0 & 0 \\ 0 & 1 & 0 \\ 0 & 0 & 1 \end{bmatrix}$$

$$BA = \begin{bmatrix} 1 & 1 & 2 \\ 2 & 4 & -3 \\ 3 & 6 & -5 \end{bmatrix}\begin{bmatrix} 2 & -17 & 11 \\ -1 & 11 & -7 \\ 0 & 3 & -2 \end{bmatrix} = \begin{bmatrix} 2 - 1 & -17 + 11 + 6 & 11 - 7 - 4 \\ 4 - 4 & -34 + 44 - 9 & 22 - 28 + 6 \\ 6 - 6 & -51 + 66 - 15 & 33 - 42 + 10 \end{bmatrix} = \begin{bmatrix} 1 & 0 & 0 \\ 0 & 1 & 0 \\ 0 & 0 & 1 \end{bmatrix}$$

11. $$AB = \frac{1}{3}\begin{bmatrix} 2 & 0 & 2 & 1 \\ 3 & 0 & 0 & 1 \\ -1 & 1 & -2 & 1 \\ 3 & -1 & 1 & 0 \end{bmatrix}\begin{bmatrix} -1 & 3 & -2 & -2 \\ -2 & 9 & -7 & -10 \\ 1 & 0 & -1 & -1 \\ 3 & -6 & 6 & 6 \end{bmatrix}$$

$$= \frac{1}{3}\begin{bmatrix} -2 + 0 + 2 + 3 & 6 + 0 + 0 - 6 & -4 + 0 - 2 + 6 & -4 + 0 - 2 + 6 \\ -3 + 0 + 0 + 3 & 9 + 0 + 0 - 6 & -6 + 0 + 0 + 6 & -6 + 0 + 0 + 6 \\ 1 - 2 - 2 + 3 & -3 + 9 + 0 - 6 & 2 - 7 + 2 + 6 & 2 - 10 + 2 + 6 \\ -3 + 2 + 1 + 0 & 9 - 9 + 0 + 0 & -6 + 7 - 1 + 0 & -6 + 10 - 1 + 0 \end{bmatrix} = \begin{bmatrix} 1 & 0 & 0 & 0 \\ 0 & 1 & 0 & 0 \\ 0 & 0 & 1 & 0 \\ 0 & 0 & 0 & 1 \end{bmatrix}$$

$$BA = \frac{1}{3}\begin{bmatrix} -1 & 3 & -2 & -2 \\ -2 & 9 & -7 & -10 \\ 1 & 0 & -1 & -1 \\ 3 & -6 & 6 & 6 \end{bmatrix}\begin{bmatrix} 2 & 0 & 2 & 1 \\ 3 & 0 & 0 & 1 \\ -1 & 1 & -2 & 1 \\ 3 & -1 & 1 & 0 \end{bmatrix}$$

$$= \frac{1}{3}\begin{bmatrix} -2 + 9 + 2 - 6 & 0 + 0 - 2 + 2 & -2 + 0 + 4 - 2 & -1 + 3 - 2 + 0 \\ -4 + 27 + 7 - 30 & 0 + 0 - 7 + 10 & -4 + 0 + 14 - 10 & -2 + 9 - 7 + 0 \\ 2 + 0 + 1 - 3 & 0 + 0 - 1 + 1 & 2 + 0 + 2 - 1 & 1 + 0 - 1 + 0 \\ 6 - 18 - 6 + 18 & 0 + 0 + 6 - 6 & 6 + 0 - 12 + 6 & 3 - 6 + 6 + 0 \end{bmatrix} = \begin{bmatrix} 1 & 0 & 0 & 0 \\ 0 & 1 & 0 & 0 \\ 0 & 0 & 1 & 0 \\ 0 & 0 & 0 & 1 \end{bmatrix}$$

13. $[A \;\vdots\; I] = \left[\begin{array}{cc|cc} 2 & 1 & 1 & 0 \\ 5 & 3 & 0 & 1 \end{array}\right]$

$$\begin{array}{r} -5R_1 + 2R_2 \to \end{array} \left[\begin{array}{cc|cc} 2 & 1 & 1 & 0 \\ 0 & 1 & -5 & 2 \end{array}\right]$$

$$\begin{array}{r} -R_2 + R_1 \to \end{array} \left[\begin{array}{cc|cc} 2 & 0 & 6 & -2 \\ 0 & 1 & -5 & 2 \end{array}\right]$$

$$\begin{array}{r} \tfrac{1}{2}R_1 \to \end{array} \left[\begin{array}{cc|cc} 1 & 0 & 3 & -1 \\ 0 & 1 & -5 & 2 \end{array}\right] = [I \;\vdots\; A^{-1}]$$

$$A^{-1} = \begin{bmatrix} 3 & -1 \\ -5 & 2 \end{bmatrix}$$

15. $[A \;\vdots\; I] = \left[\begin{array}{cc|cc} 1 & -2 & 1 & 0 \\ 2 & -3 & 0 & 1 \end{array}\right]$

$$-2R_1 + R_2 \to \left[\begin{array}{cc|cc} 1 & -2 & 1 & 0 \\ 0 & 1 & -2 & 1 \end{array}\right]$$

$$2R_2 + R_1 \to \left[\begin{array}{cc|cc} 1 & 0 & -3 & 2 \\ 0 & 1 & -2 & 1 \end{array}\right] = [I \;\vdots\; A^{-1}]$$

$$A^{-1} = \begin{bmatrix} -3 & 2 \\ -2 & 1 \end{bmatrix}$$

17. $[A \;\vdots\; I] = \left[\begin{array}{cc|cc} 3 & 1 & 1 & 0 \\ 4 & 2 & 0 & 1 \end{array}\right]$

$$\tfrac{1}{2}R_2 \to \left[\begin{array}{cc|cc} 3 & 1 & 1 & 0 \\ 2 & 1 & 0 & \tfrac{1}{2} \end{array}\right]$$

$$-R_2 + R_1 \to \left[\begin{array}{cc|cc} 1 & 0 & 1 & -\tfrac{1}{2} \\ 2 & 1 & 0 & \tfrac{1}{2} \end{array}\right]$$

$$-2R_1 + R_2 \to \left[\begin{array}{cc|cc} 1 & 0 & 1 & -\tfrac{1}{2} \\ 0 & 1 & -2 & \tfrac{3}{2} \end{array}\right] = [I \;\vdots\; A^{-1}]$$

$$A^{-1} = \begin{bmatrix} 1 & -\tfrac{1}{2} \\ -2 & \tfrac{3}{2} \end{bmatrix}$$

19. $[A \;\vdots\; I] = \left[\begin{array}{ccc|ccc} 1 & 1 & 1 & 1 & 0 & 0 \\ 3 & 5 & 4 & 0 & 1 & 0 \\ 3 & 6 & 5 & 0 & 0 & 1 \end{array}\right]$

$$\begin{array}{r} \\ -3R_1 + R_2 \to \\ -3R_1 + R_3 \to \end{array} \left[\begin{array}{ccc|ccc} 1 & 1 & 1 & 1 & 0 & 0 \\ 0 & 2 & 1 & -3 & 1 & 0 \\ 0 & 3 & 2 & -3 & 0 & 1 \end{array}\right]$$

$$\begin{array}{r} \\ \tfrac{1}{2}R_2 \to \\ \\ \end{array} \left[\begin{array}{ccc|ccc} 1 & 1 & 1 & 1 & 0 & 0 \\ 0 & 1 & \tfrac{1}{2} & -\tfrac{3}{2} & \tfrac{1}{2} & 0 \\ 0 & 3 & 2 & -3 & 0 & 1 \end{array}\right]$$

$$\begin{array}{r} -R_2 + R_1 \to \\ \\ -3R_2 + R_3 \to \end{array} \left[\begin{array}{ccc|ccc} 1 & 0 & \tfrac{1}{2} & \tfrac{5}{2} & -\tfrac{1}{2} & 0 \\ 0 & 1 & \tfrac{1}{2} & -\tfrac{3}{2} & \tfrac{1}{2} & 0 \\ 0 & 0 & \tfrac{1}{2} & \tfrac{3}{2} & -\tfrac{3}{2} & 1 \end{array}\right]$$

$$\begin{array}{r} -R_3 + R_1 \to \\ -R_3 + R_2 \to \\ \\ \end{array} \left[\begin{array}{ccc|ccc} 1 & 0 & 0 & 1 & 1 & -1 \\ 0 & 1 & 0 & -3 & 2 & -1 \\ 0 & 0 & \tfrac{1}{2} & \tfrac{3}{2} & -\tfrac{3}{2} & 1 \end{array}\right]$$

$$\begin{array}{r} \\ \\ 2R_3 \to \end{array} \left[\begin{array}{ccc|ccc} 1 & 0 & 0 & 1 & 1 & -1 \\ 0 & 1 & 0 & -3 & 2 & -1 \\ 0 & 0 & 1 & 3 & -3 & 2 \end{array}\right] = [I \;\vdots\; A^{-1}]$$

$$A^{-1} = \begin{bmatrix} 1 & 1 & -1 \\ -3 & 2 & -1 \\ 3 & -3 & 2 \end{bmatrix}$$

21. $[A \;\vdots\; I] = \left[\begin{array}{ccc|ccc} -5 & 0 & 0 & 1 & 0 & 0 \\ 2 & 0 & 0 & 0 & 1 & 0 \\ -1 & 5 & 7 & 0 & 0 & 1 \end{array}\right] \begin{array}{r} \\ \\ R_2 + 2R_3 \to \end{array} \left[\begin{array}{ccc|ccc} -5 & 0 & 0 & 1 & 0 & 0 \\ 2 & 0 & 0 & 0 & 1 & 0 \\ 0 & 10 & 14 & 0 & 1 & 2 \end{array}\right] \begin{array}{r} \\ 2R_1 + 5R_2 \to \\ \\ \end{array} \left[\begin{array}{ccc|ccc} -5 & 0 & 0 & 1 & 0 & 0 \\ 0 & 0 & 0 & 2 & 5 & 0 \\ 0 & 10 & 14 & 0 & 1 & 2 \end{array}\right]$

Because the first three entries of row 2 are all zeros, the inverse of A does not exist.

23. $[A \;\vdots\; I] = \left[\begin{array}{cccc|cccc} -8 & 0 & 0 & 0 & 1 & 0 & 0 & 0 \\ 0 & 1 & 0 & 0 & 0 & 1 & 0 & 0 \\ 0 & 0 & 4 & 0 & 0 & 0 & 1 & 0 \\ 0 & 0 & 0 & -5 & 0 & 0 & 0 & 1 \end{array}\right] \begin{array}{r} -\tfrac{1}{8}R_1 \to \\ \\ \tfrac{1}{4}R_3 \to \\ -\tfrac{1}{5}R_4 \to \end{array} \left[\begin{array}{cccc|cccc} 1 & 0 & 0 & 0 & -\tfrac{1}{8} & 0 & 0 & 0 \\ 0 & 1 & 0 & 0 & 0 & 1 & 0 & 0 \\ 0 & 0 & 1 & 0 & 0 & 0 & \tfrac{1}{4} & 0 \\ 0 & 0 & 0 & 1 & 0 & 0 & 0 & -\tfrac{1}{5} \end{array}\right] = [I \;\vdots\; A^{-1}]$

$$A^{-1} = \begin{bmatrix} -\tfrac{1}{8} & 0 & 0 & 0 \\ 0 & 1 & 0 & 0 \\ 0 & 0 & \tfrac{1}{4} & 0 \\ 0 & 0 & 0 & -\tfrac{1}{5} \end{bmatrix}$$

25. $A = \begin{bmatrix} 1 & 2 & -1 \\ 3 & 7 & -10 \\ -5 & -7 & -15 \end{bmatrix}$

$A^{-1} = \begin{bmatrix} -175 & 37 & -13 \\ 95 & -20 & 7 \\ 14 & -3 & 1 \end{bmatrix}$

27. $A = \begin{bmatrix} -\frac{1}{2} & \frac{3}{4} & \frac{1}{4} \\ 1 & 0 & -\frac{3}{2} \\ 0 & -1 & \frac{1}{2} \end{bmatrix}$

$A^{-1} = \begin{bmatrix} -12 & -5 & -9 \\ -4 & -2 & -4 \\ -8 & -4 & -6 \end{bmatrix}$

29. $A = \begin{bmatrix} 0.1 & 0.2 & 0.3 \\ -0.3 & 0.2 & 0.2 \\ 0.5 & 0.4 & 0.4 \end{bmatrix}$

$A^{-1} = \begin{bmatrix} 0 & -1.\overline{81} & 0.\overline{90} \\ -10 & 5 & 5 \\ 10 & -2.\overline{72} & -3.\overline{63} \end{bmatrix}$

31. $A = \begin{bmatrix} -1 & 0 & 1 & 0 \\ 0 & 2 & 0 & -1 \\ 2 & 0 & -1 & 0 \\ 0 & -1 & 0 & 1 \end{bmatrix}$

$A^{-1} = \begin{bmatrix} 1 & 0 & 1 & 0 \\ 0 & 1 & 0 & 1 \\ 2 & 0 & 1 & 0 \\ 0 & 1 & 0 & 2 \end{bmatrix}$

33. $A = \begin{bmatrix} a & b \\ c & d \end{bmatrix}, A^{-1} = \dfrac{1}{ad - bc}\begin{bmatrix} d & -b \\ -c & a \end{bmatrix}$

$A = \begin{bmatrix} 2 & 3 \\ -1 & 5 \end{bmatrix}$

$ad - bc = (2)(5) - (3)(-1) = 13$

$A^{-1} = \frac{1}{13}\begin{bmatrix} 5 & -3 \\ 1 & 2 \end{bmatrix} = \begin{bmatrix} \frac{5}{13} & -\frac{3}{13} \\ \frac{1}{13} & \frac{2}{13} \end{bmatrix}$

35. $A = \begin{bmatrix} -4 & -6 \\ 2 & 3 \end{bmatrix}$

$ad - bc = (-4)(3) - (-2)(-6) = 0$

Because $ad - bc = 0$, A^{-1} does not exist.

37. $A = \begin{bmatrix} 0.5 & 0.3 \\ 1.5 & 0.6 \end{bmatrix}$

$ad - bc = (0.5)(0.6) - (0.3)(1.5) = 0.3 - 0.45 = -0.15$

$A^{-1} = -\frac{1}{0.15}\begin{bmatrix} 0.6 & -0.3 \\ -1.5 & 0.5 \end{bmatrix} = \begin{bmatrix} -4 & 2 \\ 10 & -\frac{10}{3} \end{bmatrix}$

39. $\begin{bmatrix} x \\ y \end{bmatrix} = \begin{bmatrix} -3 & 2 \\ -2 & 1 \end{bmatrix}\begin{bmatrix} 5 \\ 10 \end{bmatrix} = \begin{bmatrix} 5 \\ 0 \end{bmatrix}$

Solution: (5, 0)

41. $\begin{bmatrix} x \\ y \end{bmatrix} = \begin{bmatrix} -3 & 2 \\ -2 & 1 \end{bmatrix}\begin{bmatrix} 4 \\ 2 \end{bmatrix} = \begin{bmatrix} -8 \\ -6 \end{bmatrix}$

Solution: (−8, −6)

43. $\begin{bmatrix} x \\ y \\ z \end{bmatrix} = \begin{bmatrix} 1 & 1 & -1 \\ -3 & 2 & -1 \\ 3 & -3 & 2 \end{bmatrix}\begin{bmatrix} 0 \\ 5 \\ 2 \end{bmatrix} = \begin{bmatrix} 3 \\ 8 \\ -11 \end{bmatrix}$

Solution: (3, 8, −11)

45. $\begin{bmatrix} x_1 \\ x_2 \\ x_3 \\ x_4 \end{bmatrix} = \begin{bmatrix} -24 & 7 & 1 & -2 \\ -10 & 3 & 0 & -1 \\ -29 & 7 & 3 & -2 \\ 12 & -3 & -1 & 1 \end{bmatrix}\begin{bmatrix} 0 \\ 1 \\ -1 \\ 2 \end{bmatrix} = \begin{bmatrix} 2 \\ 1 \\ 0 \\ 0 \end{bmatrix}$

Solution: (2, 1, 0, 0)

47. $A = \begin{bmatrix} 5 & 4 \\ 2 & 5 \end{bmatrix}$

$A^{-1} = \dfrac{1}{25 - 8}\begin{bmatrix} 5 & -4 \\ -2 & 5 \end{bmatrix}$

$\begin{bmatrix} x \\ y \end{bmatrix} = -\frac{1}{17}\begin{bmatrix} 5 & -4 \\ -2 & 5 \end{bmatrix}\begin{bmatrix} -1 \\ 3 \end{bmatrix} = -\frac{1}{17}\begin{bmatrix} -17 \\ 17 \end{bmatrix} = \begin{bmatrix} -1 \\ 1 \end{bmatrix}$

Solution: (−1, 1)

49. $A = \begin{bmatrix} -0.4 & 0.8 \\ 2 & -4 \end{bmatrix}$

$A^{-1} = \dfrac{1}{1.6 - 1.6}\begin{bmatrix} -4 & -0.8 \\ -2 & -0.4 \end{bmatrix}$

A^{-1} does not exist.

This implies that there is no unique solution; that is, either the system is inconsistent *or* there are infinitely many solutions.

Find the reduced row-echelon form of the matrix corresponding to the system.

$\begin{bmatrix} -0.4 & 0.8 & \vdots & 1.6 \\ 2 & -4 & \vdots & 5 \end{bmatrix}$

$-2.5R_1 \rightarrow \begin{bmatrix} 1 & -2 & \vdots & -4 \\ 2 & -4 & \vdots & 5 \end{bmatrix}$

$-2R_1 + R_2 \rightarrow \begin{bmatrix} 1 & -2 & \vdots & -4 \\ 0 & 0 & \vdots & 13 \end{bmatrix}$

The given system is inconsistent and there is no solution

51. $A = \begin{bmatrix} -\frac{1}{4} & \frac{3}{8} \\ \frac{3}{2} & \frac{3}{4} \end{bmatrix}$

$$A^{-1} = \frac{1}{-\frac{3}{16} - \frac{9}{16}} \begin{bmatrix} \frac{3}{4} & -\frac{3}{8} \\ -\frac{3}{2} & -\frac{1}{4} \end{bmatrix} = -\frac{4}{3} \begin{bmatrix} \frac{3}{4} & -\frac{3}{8} \\ -\frac{3}{2} & -\frac{1}{4} \end{bmatrix} = \begin{bmatrix} -1 & \frac{1}{2} \\ 2 & \frac{1}{3} \end{bmatrix}$$

$$\begin{bmatrix} x \\ y \end{bmatrix} = \begin{bmatrix} -1 & \frac{1}{2} \\ 2 & \frac{1}{3} \end{bmatrix} \begin{bmatrix} -2 \\ -12 \end{bmatrix} = \begin{bmatrix} -4 \\ -8 \end{bmatrix}$$

Solution: $(-4, -8)$

53. $A = \begin{bmatrix} 4 & -1 & 1 \\ 2 & 2 & 3 \\ 5 & -2 & 6 \end{bmatrix}$

Find A^{-1}.

$$[A \;\vdots\; I] = \left[\begin{array}{ccc:ccc} 4 & -1 & 1 & 1 & 0 & 0 \\ 2 & 2 & 3 & 0 & 1 & 0 \\ 5 & -2 & 6 & 0 & 0 & 1 \end{array}\right]$$

$$\begin{matrix} R_1 \\ \\ R_3 \end{matrix} \left[\begin{array}{ccc:ccc} 5 & -2 & 6 & 0 & 0 & 1 \\ 2 & 2 & 3 & 0 & 1 & 0 \\ 4 & -1 & 1 & 1 & 0 & 0 \end{array}\right]$$

$$\begin{matrix} -R_3 + R_1 \rightarrow \\ \\ \\ \end{matrix} \left[\begin{array}{ccc:ccc} 1 & -1 & 5 & -1 & 0 & 1 \\ 2 & 2 & 3 & 0 & 1 & 0 \\ 4 & -1 & 1 & 1 & 0 & 0 \end{array}\right]$$

$$\begin{matrix} \\ -2R_1 + R_2 \rightarrow \\ -4R_1 + R_3 \rightarrow \end{matrix} \left[\begin{array}{ccc:ccc} 1 & -1 & 5 & -1 & 0 & 1 \\ 0 & 4 & -7 & 2 & 1 & -2 \\ 0 & 3 & -19 & 5 & 0 & -4 \end{array}\right]$$

$$\begin{matrix} \\ -R_3 + R_2 \rightarrow \\ \\ \end{matrix} \left[\begin{array}{ccc:ccc} 1 & -1 & 5 & -1 & 0 & 1 \\ 0 & 1 & 12 & -3 & 1 & 2 \\ 0 & 3 & -19 & 5 & 0 & -4 \end{array}\right]$$

$$\begin{matrix} R_2 + R_1 \rightarrow \\ \\ -3R_2 + R_3 \rightarrow \end{matrix} \left[\begin{array}{ccc:ccc} 1 & 0 & 17 & -4 & 1 & 3 \\ 0 & 1 & 12 & -3 & 1 & 2 \\ 0 & 0 & -55 & 14 & -3 & -10 \end{array}\right]$$

$$\begin{matrix} \\ \\ -\frac{1}{55}R_3 \rightarrow \end{matrix} \left[\begin{array}{ccc:ccc} 1 & 0 & 17 & -4 & 1 & 3 \\ 0 & 1 & 12 & -3 & 1 & 2 \\ 0 & 0 & 1 & -\frac{14}{55} & \frac{3}{55} & \frac{2}{11} \end{array}\right]$$

$$\begin{matrix} -17R_3 + R_1 \rightarrow \\ -12R_3 + R_2 \rightarrow \\ \\ \end{matrix} \left[\begin{array}{ccc:ccc} 1 & 0 & 0 & \frac{18}{55} & \frac{4}{55} & -\frac{1}{11} \\ 0 & 1 & 0 & \frac{3}{55} & \frac{19}{55} & -\frac{2}{11} \\ 0 & 0 & 1 & -\frac{14}{55} & \frac{3}{55} & \frac{2}{11} \end{array}\right] = [I \;\vdots\; A^{-1}]$$

$$A^{-1} = \frac{1}{55}\begin{bmatrix} 18 & 4 & -5 \\ 3 & 19 & -10 \\ -14 & 3 & 10 \end{bmatrix}$$

$$\begin{bmatrix} x \\ y \\ z \end{bmatrix} = \frac{1}{55}\begin{bmatrix} 18 & 4 & -5 \\ 3 & 19 & -10 \\ -14 & 3 & 10 \end{bmatrix} \begin{bmatrix} -5 \\ 10 \\ 1 \end{bmatrix} = \frac{1}{55}\begin{bmatrix} -55 \\ 165 \\ 110 \end{bmatrix} = \begin{bmatrix} -1 \\ 3 \\ 2 \end{bmatrix}$$

Solution: $(-1, 3, 2)$

55. $\begin{cases} 5x - 3y + 2z = 2 \\ 2x + 2y - 3z = 3 \\ x - 7y + 7z = -4 \end{cases}$

Using a graphing utility $(0.8125, 0.6875, 0) = \left(\frac{13}{16}, \frac{11}{16}, 0\right)$

57. $A = \begin{bmatrix} 1 & 1 & 1 \\ 0.045 & 0.05 & 0.09 \\ 0 & 2 & -1 \end{bmatrix}$

$$[A \;\vdots\; I] = \left[\begin{array}{ccc:ccc} 1 & 1 & 1 & 1 & 0 & 0 \\ 0.045 & 0.05 & 0.09 & 0 & 1 & 0 \\ 0 & 2 & -1 & 0 & 0 & 1 \end{array}\right]$$

$$200R_2 \rightarrow \left[\begin{array}{ccc:ccc} 1 & 1 & 1 & 1 & 0 & 0 \\ 9 & 10 & 18 & 0 & 200 & 0 \\ 0 & 2 & -1 & 0 & 0 & 1 \end{array}\right]$$

$$-9R_1 + R_2 \rightarrow \left[\begin{array}{ccc:ccc} 1 & 1 & 1 & 1 & 0 & 0 \\ 0 & 1 & 9 & -9 & 200 & 0 \\ 0 & 2 & -1 & 0 & 0 & 1 \end{array}\right]$$

$$\begin{array}{r} -R_2 + R_1 \rightarrow \\ \\ -2R_2 + R_3 \rightarrow \end{array} \left[\begin{array}{ccc:ccc} 1 & 0 & -8 & 10 & -200 & 0 \\ 0 & 1 & 9 & -9 & 200 & 0 \\ 0 & 0 & -19 & 18 & -400 & 1 \end{array}\right]$$

$$-\tfrac{1}{19}R_3 \rightarrow \left[\begin{array}{ccc:ccc} 1 & 0 & -8 & 10 & -200 & 0 \\ 0 & 1 & 9 & -9 & 200 & 0 \\ 0 & 0 & 1 & -\frac{18}{19} & \frac{400}{19} & -\frac{1}{19} \end{array}\right]$$

$$\begin{array}{r} 8R_3 + R_1 \rightarrow \\ -9R_3 + R_2 \rightarrow \\ \\ \end{array} \left[\begin{array}{ccc:ccc} 1 & 0 & 0 & \frac{46}{19} & -\frac{600}{19} & -\frac{8}{19} \\ 0 & 1 & 0 & -\frac{9}{19} & \frac{200}{19} & \frac{9}{19} \\ 0 & 0 & 1 & -\frac{18}{19} & \frac{400}{19} & -\frac{1}{19} \end{array}\right] = [I \;\vdots\; A^{-1}]$$

$$X = A^{-1}B = \frac{1}{19}\begin{bmatrix} 46 & -600 & -8 \\ -9 & 200 & 9 \\ -18 & 400 & -1 \end{bmatrix}\begin{bmatrix} 10{,}000 \\ 650 \\ 0 \end{bmatrix} \approx \begin{bmatrix} 3684.21 \\ 2105.26 \\ 4210.53 \end{bmatrix}$$

Solution: \$3684.21 in AAA-rated bonds, \$2105.26 in A-rated bonds, \$4210.53 in B-rated bonds

59. $\begin{cases} 2I_1 + 4I_3 = 15 \\ I_2 + 4I_3 = 17 \\ I_1 + I_2 - I_3 = 0 \end{cases}$

$$\underset{A}{\begin{bmatrix} 2 & 0 & 4 \\ 0 & 1 & 4 \\ 1 & 1 & -1 \end{bmatrix}}\underset{X}{\begin{bmatrix} I_1 \\ I_2 \\ I_3 \end{bmatrix}} = \underset{B}{\begin{bmatrix} 15 \\ 17 \\ 0 \end{bmatrix}}$$

$$X = A^{-1}B = \begin{bmatrix} \frac{5}{14} & -\frac{2}{7} & \frac{2}{7} \\ -\frac{2}{7} & \frac{3}{7} & \frac{4}{7} \\ \frac{1}{14} & \frac{1}{7} & -\frac{1}{7} \end{bmatrix}\begin{bmatrix} 15 \\ 17 \\ 0 \end{bmatrix} = \begin{bmatrix} \frac{1}{2} \\ 3 \\ \frac{7}{2} \end{bmatrix}$$

So, $I_1 = 0.5$ ampere, $I_2 = 3.0$ ampere, and $I_3 = 3.5$ ampere.

61. $\begin{cases} 2I_1 + 4I_3 = 28 \\ I_2 + 4I_3 = 21 \\ I_1 + I_2 - I_3 = 0 \end{cases}$

$$\underset{A}{\begin{bmatrix} 2 & 0 & 4 \\ 0 & 1 & 4 \\ 1 & 1 & -1 \end{bmatrix}}\underset{X}{\begin{bmatrix} I_1 \\ I_2 \\ I_3 \end{bmatrix}} = \underset{B}{\begin{bmatrix} 28 \\ 21 \\ 0 \end{bmatrix}}$$

$$X = A^{-1}B = \begin{bmatrix} \frac{5}{14} & -\frac{2}{7} & \frac{2}{7} \\ -\frac{2}{7} & \frac{3}{7} & \frac{4}{7} \\ \frac{1}{14} & \frac{1}{7} & -\frac{1}{7} \end{bmatrix}\begin{bmatrix} 28 \\ 21 \\ 0 \end{bmatrix} = \begin{bmatrix} 4 \\ 1 \\ 5 \end{bmatrix}$$

So, $I_1 = 4$ ampere, $I_2 = 1$ ampere, and $I_3 = 5$ ampere.

In Exercise 63, use the following:

Let x = bags of potting soil for seedlings,

y = bags of potting soil for general potting, and

z = bags of potting soil for hardwood plants.

$$AX = B = \begin{bmatrix} 2 & 1 & 2 \\ 1 & 2 & 1 \\ 1 & 1 & 2 \end{bmatrix}\begin{bmatrix} x \\ y \\ z \end{bmatrix} = \begin{bmatrix} \text{Sand} \\ \text{Loam} \\ \text{Peat Moss} \end{bmatrix}$$

$$A^{-1} = \begin{bmatrix} 1 & 0 & -1 \\ 0 & 1 & -1 \\ -\frac{1}{2} & -\frac{1}{2} & \frac{3}{2} \end{bmatrix}$$

63. $A^{-1}\begin{bmatrix} 500 \\ 500 \\ 400 \end{bmatrix} = \begin{bmatrix} 100 \\ 100 \\ 100 \end{bmatrix}$

Solution:

$x = 100$ bags of potting soil for seedlings,

$y = 100$ bags of potting soil for general potting,

$z = 100$ bags of potting soil for hardwood plants.

65. Let r = number of roses, l = number of lilies, and i = number of irises.

(a) $\begin{cases} r + l + i = 120 \\ 2.5r + 4l + 2i = 300 \\ -r + 2l + 2i = 0 \end{cases}$

$$\begin{bmatrix} 1 & 1 & 1 \\ 2.5 & 4 & 2 \\ -1 & 2 & 2 \end{bmatrix}\begin{bmatrix} r \\ l \\ i \end{bmatrix} = \begin{bmatrix} 120 \\ 300 \\ 0 \end{bmatrix}$$

$A \quad X \quad = \quad B$

(b) $A^{-1} = \begin{bmatrix} \frac{2}{3} & 0 & -\frac{1}{3} \\ -\frac{7}{6} & \frac{1}{2} & \frac{1}{12} \\ \frac{3}{2} & -\frac{1}{2} & -\frac{1}{4} \end{bmatrix}$

$$X = A^{-1}B = \begin{bmatrix} \frac{2}{3} & 0 & -\frac{1}{3} \\ -\frac{7}{6} & \frac{1}{2} & \frac{1}{12} \\ \frac{3}{2} & -\frac{1}{2} & \frac{1}{4} \end{bmatrix}\begin{bmatrix} 120 \\ 300 \\ 0 \end{bmatrix} = \begin{bmatrix} 80 \\ 10 \\ 30 \end{bmatrix}$$

So, 80 roses, 10 lilies, and 30 irises will create 40 centerpieces.

67. True. If B is the inverse of A, then $AB = I = BA$.

69. If the determinant of a 2×2 matrix is not equal to 0, then the inverse exists.

To find the inverse, take 1 divided by the determinant and multiply it by the matrix which has a diagonal from top left to bottom right that has the terms from the original matrix flipped and the other diagonal is the negative of the terms from the original matrix.

71. If A^{-1} does not exist, it is singular.

$$\begin{bmatrix} 4 & 3 \\ -2 & k \end{bmatrix}, (4)(k) - (-2)(3) = 0 \Rightarrow 4k + 6 = 0$$

$$\Rightarrow k = -\tfrac{3}{2}.$$

When $k \neq -\frac{3}{2}$, A^{-1} exists because the determinant does not equal zero.

73. (a) Given $A = \begin{bmatrix} a_{11} & 0 \\ 0 & a_{22} \end{bmatrix}$, $A^{-1} = \begin{bmatrix} \frac{1}{a_{11}} & 0 \\ 0 & \frac{1}{a_{22}} \end{bmatrix}$.

Given $A = \begin{bmatrix} a_{11} & 0 & 0 \\ 0 & a_{22} & 0 \\ 0 & 0 & a_{33} \end{bmatrix}$, $A^{-1} = \begin{bmatrix} \frac{1}{a_{11}} & 0 & 0 \\ 0 & \frac{1}{a_{22}} & 0 \\ 0 & 0 & \frac{1}{a_{33}} \end{bmatrix}$.

(b) In general, the inverse of a matrix in the form of A is

$$\begin{bmatrix} \frac{1}{a_{11}} & 0 & 0 & \cdots & 0 \\ 0 & \frac{1}{a_{22}} & 0 & \cdots & 0 \\ 0 & 0 & \frac{1}{a_{33}} & \cdots & 0 \\ \vdots & \vdots & \vdots & \cdots & \vdots \\ 0 & 0 & 0 & \cdots & \frac{1}{a_{nn}} \end{bmatrix}.$$

75. Answers will vary. *Sample answer.* $A = \begin{bmatrix} a & b \\ c & d \end{bmatrix}$, $A^{-1} = \frac{1}{ad - bc}\begin{bmatrix} d & -b \\ -c & a \end{bmatrix}$

$$A \cdot A^{-1} = \begin{bmatrix} a & b \\ c & d \end{bmatrix}\left(\frac{1}{ad - bc}\begin{bmatrix} d & -b \\ -c & a \end{bmatrix}\right) = \frac{1}{ad - bc}\begin{bmatrix} ad - bc & -ab + ab \\ cd - cd & -bc + ad \end{bmatrix} = \frac{1}{ad - bc}\begin{bmatrix} ad - bc & 0 \\ 0 & ad - bc \end{bmatrix} = \begin{bmatrix} 1 & 0 \\ 0 & 1 \end{bmatrix} = I$$

$$A^{-1} \cdot A = \left(\frac{1}{ad - bc}\begin{bmatrix} d & -b \\ -c & a \end{bmatrix}\right)\begin{bmatrix} a & b \\ c & d \end{bmatrix} = \frac{1}{ad - bc}\begin{bmatrix} ad - bc & -ab + ab \\ cd - cd & -bc + ad \end{bmatrix} = \frac{1}{ad - bc}\begin{bmatrix} ad - bc & 0 \\ 0 & ad - bc \end{bmatrix} = \begin{bmatrix} 1 & 0 \\ 0 & 1 \end{bmatrix} = I$$

Section 7.4 The Determinant of a Square Matrix

1. determinant

3. cofactor

5. 4

7. $\begin{vmatrix} 8 & 4 \\ 2 & 3 \end{vmatrix} = (8)(3) - (4)(2) = 16$

9. $\begin{vmatrix} 6 & -3 \\ -5 & 2 \end{vmatrix} = (6)(2) - (-3)(-5) = -3$

11. $\begin{vmatrix} -7 & 0 \\ 3 & 0 \end{vmatrix} = -7(0) - 0(3) = 0$

13. $\begin{vmatrix} 2 & 6 \\ 0 & 3 \end{vmatrix} = 2(3) - 6(0) = 6$

15. $\begin{vmatrix} -3 & -2 \\ -6 & -4 \end{vmatrix} = (-3)(-4) - (-2)(-6) = 12 - 12 = 0$

17. $\begin{vmatrix} -2 & -7 \\ -3 & 1 \end{vmatrix} = (-2)(1) - (-3)(-7) = -23$

19. $\begin{vmatrix} -7 & 6 \\ \frac{1}{2} & 3 \end{vmatrix} = (-7)(3) - (6)\left(\frac{1}{2}\right) = -24$

21. $\begin{vmatrix} -\frac{1}{2} & \frac{1}{3} \\ -6 & \frac{1}{3} \end{vmatrix} = -\frac{1}{2}\left(\frac{1}{3}\right) - \frac{1}{3}(-6) = -\frac{1}{2} + 2 = \frac{11}{6}$

23. $\begin{vmatrix} 3 & 4 \\ -2 & 1 \end{vmatrix} = 11$

25. $\begin{vmatrix} 19 & 20 \\ 43 & -56 \end{vmatrix} = -1924$

27. $\begin{vmatrix} \frac{1}{10} & \frac{1}{5} \\ -\frac{3}{10} & \frac{1}{5} \end{vmatrix} = 0.08$

29. $\begin{bmatrix} 4 & 5 \\ 3 & -6 \end{bmatrix}$

(a) $M_{11} = -6$
$M_{12} = 3$
$M_{21} = 5$
$M_{22} = 4$

(b) $C_{11} = M_{11} = -6$
$C_{12} = -M_{12} = -3$
$C_{21} = -M_{21} = -5$
$C_{22} = M_{22} = 4$

31. $\begin{bmatrix} 4 & 0 & 2 \\ -3 & 2 & 1 \\ 1 & -1 & 1 \end{bmatrix}$

(a) $M_{11} = \begin{vmatrix} 2 & 1 \\ -1 & 1 \end{vmatrix} = 2 - (-1) = 3$

$M_{12} = \begin{vmatrix} -3 & 1 \\ 1 & 1 \end{vmatrix} = -3 - 1 = -4$

$M_{13} = \begin{vmatrix} -3 & 2 \\ 1 & -1 \end{vmatrix} = 3 - 2 = 1$

$M_{21} = \begin{vmatrix} 0 & 2 \\ -1 & 1 \end{vmatrix} = 0 - (-2) = 2$

$M_{22} = \begin{vmatrix} 4 & 2 \\ 1 & 1 \end{vmatrix} = 4 - 2 = 2$

$M_{23} = \begin{vmatrix} 4 & 0 \\ 1 & -1 \end{vmatrix} = -4 - 0 = -4$

$M_{31} = \begin{vmatrix} 0 & 2 \\ 2 & 1 \end{vmatrix} = 0 - 4 = -4$

$M_{32} = \begin{vmatrix} 4 & 2 \\ -3 & 1 \end{vmatrix} = 4 - (-6) = 10$

$M_{33} = \begin{vmatrix} 4 & 0 \\ -3 & 2 \end{vmatrix} = 8 - 0 = 8$

(b) $C_{11} = (-1)^2 M_{11} = 3$
$C_{12} = (-1)^3 M_{12} = 4$
$C_{13} = (-1)^4 M_{13} = 1$
$C_{21} = (-1)^3 M_{21} = -2$
$C_{22} = (-1)^4 M_{22} = 2$
$C_{23} = (-1)^5 M_{23} = 4$
$C_{31} = (-1)^4 M_{31} = -4$
$C_{32} = (-1)^5 M_{32} = -10$
$C_{33} = (-1)^6 M_{33} = 8$

33. $\begin{bmatrix} -4 & 6 & 3 \\ 7 & -2 & 8 \\ 1 & 0 & -5 \end{bmatrix}$

(a) $M_{11} = \begin{vmatrix} -2 & 8 \\ 0 & -5 \end{vmatrix} = (-2)(-5) - (8)(0) = 10$

$M_{12} = \begin{vmatrix} 7 & 8 \\ 1 & -5 \end{vmatrix} = (7)(-5) - (8)(1) = -43$

$M_{13} = \begin{vmatrix} 7 & -2 \\ 1 & 0 \end{vmatrix} = (7)(0) - (-2)(1) = 2$

$M_{21} = \begin{vmatrix} 6 & 3 \\ 0 & -5 \end{vmatrix} = (6)(-5) - (3)(0) = -30$

$M_{22} = \begin{vmatrix} -4 & 3 \\ 1 & -5 \end{vmatrix} = (-4)(-5) - (3)(1) = 17$

$M_{23} = \begin{vmatrix} -4 & 6 \\ 1 & 0 \end{vmatrix} = (-4)(0) - (6)(1) = -6$

$M_{31} = \begin{vmatrix} 6 & 3 \\ -2 & 8 \end{vmatrix} = (6)(8) - (3)(-2) = 54$

$M_{32} = \begin{vmatrix} -4 & 3 \\ 7 & 8 \end{vmatrix} = (-4)(8) - (3)(7) = -53$

$M_{33} = \begin{vmatrix} -4 & 6 \\ 7 & -2 \end{vmatrix} = (-4)(-2) - (6)(7) = -34$

(b) $C_{11} = (-1)^2 M_{11} = 10$

$C_{12} = (-1)^3 M_{12} = 43$

$C_{13} = (-1)^4 M_{13} = 2$

$C_{21} = (-1)^3 M_{21} = 30$

$C_{22} = (-1)^4 M_{22} = 17$

$C_{23} = (-1)^5 M_{23} = 6$

$C_{31} = (-1)^4 M_{31} = 54$

$C_{32} = (-1)^5 M_{32} = 53$

$C_{33} = (-1)^6 M_{33} = -34$

35. (a) $\begin{vmatrix} 2 & 5 \\ 6 & -3 \end{vmatrix} = 2(-3) - 5(6) = -36$

(b) $\begin{vmatrix} 2 & 5 \\ 6 & -3 \end{vmatrix} = 2(-3) - 6(5) = -36$

37. (a) $\begin{vmatrix} 5 & 0 & -3 \\ 0 & 12 & 4 \\ 1 & 6 & 3 \end{vmatrix} = 0\begin{vmatrix} 0 & -3 \\ 6 & 3 \end{vmatrix} + 12\begin{vmatrix} 5 & -3 \\ 1 & 3 \end{vmatrix} - 4\begin{vmatrix} 5 & 0 \\ 1 & 6 \end{vmatrix}$

$= 0(18) + 12(18) - 4(30)$

$= 96$

(b) $\begin{vmatrix} 5 & 0 & -3 \\ 0 & 12 & 4 \\ 1 & 6 & 3 \end{vmatrix} = 0\begin{vmatrix} 0 & 4 \\ 1 & 3 \end{vmatrix} + 12\begin{vmatrix} 5 & -3 \\ 1 & 3 \end{vmatrix} - 6\begin{vmatrix} 5 & -3 \\ 0 & 4 \end{vmatrix}$

$= 0(-4) + 12(18) - 6(20)$

$= 96$

39. (a) $\begin{vmatrix} -3 & 2 & 1 \\ 4 & 5 & 6 \\ 2 & -3 & 1 \end{vmatrix} = -3\begin{vmatrix} 5 & 6 \\ -3 & 1 \end{vmatrix} - 2\begin{vmatrix} 4 & 6 \\ 2 & 1 \end{vmatrix} + \begin{vmatrix} 4 & 5 \\ 2 & -3 \end{vmatrix}$

$= -3(23) - 2(-8) - 22$

$= -75$

(b) $\begin{vmatrix} -3 & 2 & 1 \\ 4 & 5 & 6 \\ 2 & -3 & 1 \end{vmatrix} = -2\begin{vmatrix} 4 & 6 \\ 2 & 1 \end{vmatrix} + 5\begin{vmatrix} -3 & 1 \\ 2 & 1 \end{vmatrix} + 3\begin{vmatrix} -3 & 1 \\ 4 & 6 \end{vmatrix}$

$= -2(-8) + 5(-5) + 3(-22)$

$= -75$

41. (a) $\begin{vmatrix} 6 & 0 & -3 & 5 \\ 4 & 0 & 6 & -8 \\ -1 & 0 & 7 & 4 \\ 8 & 0 & 0 & 2 \end{vmatrix} = -8\begin{vmatrix} 0 & -3 & 5 \\ 0 & 6 & -8 \\ 0 & 7 & 4 \end{vmatrix} + 0\begin{vmatrix} 6 & -3 & 5 \\ 4 & 6 & -8 \\ -1 & 7 & 4 \end{vmatrix} - 0\begin{vmatrix} 6 & 0 & 5 \\ 4 & 0 & -8 \\ -1 & 0 & 4 \end{vmatrix} + 2\begin{vmatrix} 6 & 0 & -3 \\ 4 & 0 & 6 \\ -1 & 0 & 0 \end{vmatrix}$

$= -8(0) + 2(0) = 0$

(b) $\begin{vmatrix} 6 & 0 & -3 & 5 \\ 4 & 0 & 6 & -8 \\ -1 & 0 & 7 & 4 \\ 8 & 0 & 0 & 2 \end{vmatrix} = -0\begin{vmatrix} 4 & 6 & -8 \\ -1 & 7 & 4 \\ 8 & 0 & 2 \end{vmatrix} + 0\begin{vmatrix} 6 & -3 & 5 \\ -1 & 7 & 4 \\ 8 & 0 & 2 \end{vmatrix} - 0\begin{vmatrix} 6 & -3 & 5 \\ 4 & 6 & -8 \\ 8 & 0 & 2 \end{vmatrix} + 0\begin{vmatrix} 6 & -3 & 5 \\ 4 & 6 & -8 \\ -1 & 7 & 4 \end{vmatrix}$

$= 0$

43. (a) $\begin{vmatrix} -2 & 4 & 7 & 1 \\ 3 & 0 & 0 & 0 \\ 8 & 5 & 10 & 5 \\ 6 & 0 & 5 & 0 \end{vmatrix} = -3\begin{vmatrix} 4 & 7 & 1 \\ 5 & 10 & 5 \\ 0 & 5 & 0 \end{vmatrix} + 0\begin{vmatrix} -2 & 7 & 1 \\ 8 & 10 & 5 \\ 6 & 5 & 0 \end{vmatrix} - 0\begin{vmatrix} -2 & 4 & 1 \\ 8 & 5 & 5 \\ 6 & 0 & 0 \end{vmatrix} + 0\begin{vmatrix} -2 & 4 & 7 \\ 8 & 5 & 10 \\ 6 & 0 & 5 \end{vmatrix} = -3(-75) = 225$

(b) $\begin{vmatrix} -2 & 4 & 7 & 1 \\ 3 & 0 & 0 & 0 \\ 8 & 5 & 10 & 5 \\ 6 & 0 & 5 & 0 \end{vmatrix} = -1\begin{vmatrix} 3 & 0 & 0 \\ 8 & 5 & 10 \\ 6 & 0 & 5 \end{vmatrix} + 0\begin{vmatrix} -2 & 4 & 7 \\ 8 & 5 & 10 \\ 6 & 0 & 5 \end{vmatrix} - 5\begin{vmatrix} -2 & 4 & 7 \\ 3 & 0 & 0 \\ 6 & 0 & 5 \end{vmatrix} + 0\begin{vmatrix} -2 & 4 & 7 \\ 3 & 0 & 0 \\ 8 & 5 & 10 \end{vmatrix} = (-1)(75) - 5(-60) = 225$

45. Expand along Column 1.

$$\begin{vmatrix} -1 & 2 & -5 \\ 0 & 3 & 4 \\ 0 & 0 & 3 \end{vmatrix} = -1\begin{vmatrix} 3 & 4 \\ 0 & 3 \end{vmatrix} - 0\begin{vmatrix} 2 & -5 \\ 0 & 3 \end{vmatrix} + 0\begin{vmatrix} 2 & -5 \\ 3 & 4 \end{vmatrix}$$
$$= -1(9) - 0(6) + 0(23) = -9$$

47. Expand along Row 2.

$$\begin{vmatrix} 6 & 3 & -7 \\ 0 & 0 & 0 \\ 4 & -6 & 3 \end{vmatrix} = 0\begin{vmatrix} 3 & -7 \\ -6 & 3 \end{vmatrix} - 0\begin{vmatrix} 6 & -7 \\ 4 & 3 \end{vmatrix} + 0\begin{vmatrix} 6 & 3 \\ 4 & -6 \end{vmatrix} = 0$$

49. Expand along Column 1.

$$\begin{vmatrix} 2 & -1 & 0 \\ 4 & 2 & 1 \\ 4 & 2 & 1 \end{vmatrix} = 2\begin{vmatrix} 2 & 1 \\ 2 & 1 \end{vmatrix} - 4\begin{vmatrix} -1 & 0 \\ 2 & 1 \end{vmatrix} + 4\begin{vmatrix} -1 & 0 \\ 2 & 1 \end{vmatrix}$$
$$= 2(0) - 4(-1) + 4(-1) = 0$$

51. Expand along Column 3.

$$\begin{vmatrix} 1 & 4 & -2 \\ 3 & 2 & 0 \\ -1 & 4 & 3 \end{vmatrix} = -2\begin{vmatrix} 3 & 2 \\ -1 & 4 \end{vmatrix} + 3\begin{vmatrix} 1 & 4 \\ 3 & 2 \end{vmatrix}$$
$$= -2(14) + 3(-10) = -58$$

53. Expand along Column 3.

$$\begin{vmatrix} 2 & 6 & 0 & 2 \\ 2 & 7 & 3 & 6 \\ 1 & 0 & 0 & 1 \\ 3 & 7 & 0 & 7 \end{vmatrix} = 0\begin{vmatrix} 2 & 7 & 6 \\ 1 & 0 & 1 \\ 3 & 7 & 7 \end{vmatrix} - 3\begin{vmatrix} 2 & 6 & 2 \\ 1 & 0 & 1 \\ 3 & 7 & 7 \end{vmatrix} + 0\begin{vmatrix} 2 & 6 & 2 \\ 2 & 7 & 6 \\ 3 & 7 & 7 \end{vmatrix} - 0\begin{vmatrix} 2 & 6 & 2 \\ 2 & 7 & 6 \\ 1 & 0 & 1 \end{vmatrix} = -3(-24) = 72$$

55. Expand along Column 1.

$$\begin{vmatrix} 5 & 3 & 0 & 6 \\ 4 & 6 & 4 & 12 \\ 0 & 2 & -3 & 4 \\ 0 & 1 & -2 & 2 \end{vmatrix} = 5\begin{vmatrix} 6 & 4 & 12 \\ 2 & -3 & 4 \\ 1 & -2 & 2 \end{vmatrix} - 4\begin{vmatrix} 3 & 0 & 6 \\ 2 & -3 & 4 \\ 1 & -2 & 2 \end{vmatrix} = 5(0) - 4(0) = 0$$

57. Expand along Column 2, then along Column 4.

$$\begin{vmatrix} 3 & 2 & 4 & -1 & 5 \\ -2 & 0 & 1 & 3 & 2 \\ 1 & 0 & 0 & 4 & 0 \\ 6 & 0 & 2 & -1 & 0 \\ 3 & 0 & 5 & 1 & 0 \end{vmatrix} = -2\begin{vmatrix} -2 & 1 & 3 & 2 \\ 1 & 0 & 4 & 0 \\ 6 & 2 & -1 & 0 \\ 3 & 5 & 1 & 0 \end{vmatrix} = (-2)(-2)\begin{vmatrix} 1 & 0 & 4 \\ 6 & 2 & -1 \\ 3 & 5 & 1 \end{vmatrix} = 4(103) = 412$$

59. $\begin{vmatrix} 3 & 8 & -7 \\ 0 & -5 & 4 \\ 8 & 1 & 6 \end{vmatrix} = -126$

61. $\begin{vmatrix} 1 & -1 & 8 & 4 \\ 2 & 6 & 0 & -4 \\ 2 & 0 & 2 & 6 \\ 0 & 2 & 8 & 0 \end{vmatrix} = -336$

63. (a) $\begin{vmatrix} -1 & 0 \\ 0 & 3 \end{vmatrix} = -3$

(b) $\begin{vmatrix} 2 & 0 \\ 0 & -1 \end{vmatrix} = -2$

(c) $\begin{bmatrix} -1 & 0 \\ 0 & 3 \end{bmatrix}\begin{bmatrix} 2 & 0 \\ 0 & -1 \end{bmatrix} = \begin{bmatrix} -2 & 0 \\ 0 & -3 \end{bmatrix}$

(d) $\begin{vmatrix} -2 & 0 \\ 0 & -3 \end{vmatrix} = 6$

65. (a) $\begin{vmatrix} 4 & 0 \\ 3 & -2 \end{vmatrix} = -8$

(b) $\begin{vmatrix} -1 & 1 \\ -2 & 2 \end{vmatrix} = 0$

(c) $\begin{bmatrix} 4 & 0 \\ 3 & -2 \end{bmatrix}\begin{bmatrix} -1 & 1 \\ -2 & 2 \end{bmatrix} = \begin{bmatrix} -4 & 4 \\ 1 & -1 \end{bmatrix}$

(d) $\begin{vmatrix} -4 & 4 \\ 1 & -1 \end{vmatrix} = 0$

67. (a) $\begin{vmatrix} -1 & 2 & 1 \\ 1 & 0 & 1 \\ 0 & 1 & 0 \end{vmatrix} = 2$

(b) $\begin{vmatrix} -1 & 0 & 0 \\ 0 & 2 & 0 \\ 0 & 0 & 3 \end{vmatrix} = -6$

(c) $\begin{bmatrix} -1 & 2 & 1 \\ 1 & 0 & 1 \\ 0 & 1 & 0 \end{bmatrix}\begin{bmatrix} -1 & 0 & 0 \\ 0 & 2 & 0 \\ 0 & 0 & 3 \end{bmatrix} = \begin{bmatrix} 1 & 4 & 3 \\ -1 & 0 & 3 \\ 0 & 2 & 0 \end{bmatrix}$

(d) $\begin{vmatrix} 1 & 4 & 3 \\ -1 & 0 & 3 \\ 0 & 2 & 0 \end{vmatrix} = -12$

69. Answers will vary. *Sample answer*:

$$|A| = \begin{vmatrix} 3 & 2 \\ 3 & 3 \end{vmatrix} = 9 - 6 = 3$$

71. Answers will vary. *Sample answer*:

$$|A| = \begin{vmatrix} 4 & 2 & -1 \\ 2 & 1 & 0 \\ 1 & 1 & 3 \end{vmatrix} = -2\begin{vmatrix} 2 & -1 \\ 1 & 3 \end{vmatrix} + 1\begin{vmatrix} 4 & -1 \\ 1 & 3 \end{vmatrix}$$
$$= -2(7) + 13 = -1$$

73. Answers will vary. *Sample answer*:

$$|A| = \begin{vmatrix} 2 & 3 \\ 8 & 12 \end{vmatrix} = 24 - 24 = 0$$

75. $\begin{vmatrix} w & x \\ y & z \end{vmatrix} = wz - xy$

$-\begin{vmatrix} y & z \\ w & x \end{vmatrix} = -(xy - wz) = wz - xy$

So, $\begin{vmatrix} w & x \\ y & z \end{vmatrix} = -\begin{vmatrix} y & z \\ w & x \end{vmatrix}$.

77. $\begin{vmatrix} w & x \\ y & z \end{vmatrix} = wz - xy$

$\begin{vmatrix} w & x + cw \\ y & z + cy \end{vmatrix} = w(z + cy) - y(x + cw) = wz - xy$

So, $\begin{vmatrix} w & x \\ y & z \end{vmatrix} = \begin{vmatrix} w & x + cw \\ y & z + cy \end{vmatrix}$.

79.
$$\begin{vmatrix} 1 & x & x^2 \\ 1 & y & y^2 \\ 1 & z & z^2 \end{vmatrix} = \begin{vmatrix} y & y^2 \\ z & z^2 \end{vmatrix} - \begin{vmatrix} x & x^2 \\ z & z^2 \end{vmatrix} + \begin{vmatrix} x & x^2 \\ y & y^2 \end{vmatrix}$$
$$= (yz^2 - y^2z) - (xz^2 - x^2z) + (xy^2 - x^2y)$$
$$= yz^2 - xz^2 - y^2z + x^2z + xy(y - x)$$
$$= z^2(y - x) - z(y^2 - x^2) + xy(y - x)$$
$$= z^2(y - x) - z(y - x)(y + x) + xy(y - x)$$
$$= (y - x)[z^2 - z(y + x) + xy]$$
$$= (y - x)[z^2 - zy - zx + xy]$$
$$= (y - x)[z^2 - zx - zy + xy]$$
$$= (y - x)[z(z - x) - y(z - x)]$$
$$= (y - x)(z - x)(z - y)$$

81. $\begin{vmatrix} x & 2 \\ 1 & x \end{vmatrix} = 2$

$$x^2 - 2 = 2$$
$$x^2 = 4$$
$$x = \pm 2$$

83. $\begin{vmatrix} x + 1 & 2 \\ -1 & x \end{vmatrix} = 4$

$$(x + 1)(x) - (2)(-1) = 4$$
$$x^2 + x - 2 = 0$$
$$(x + 2)(x - 1) = 0$$
$$x = -2 \text{ or } x = 1$$

85. $\begin{vmatrix} x+3 & 2 \\ 1 & x+2 \end{vmatrix} = 0$

$(x+3)(x+2) - 2 = 0$

$x^2 + 5x + 4 = 0$

$(x+1)(x+4) = 0$

$x = -1 \text{ or } x = -4$

87. $\begin{vmatrix} 4u & -1 \\ -1 & 2v \end{vmatrix} = 8uv - 1$

89. $\begin{vmatrix} e^{2x} & e^{3x} \\ 2e^{2x} & 3e^{3x} \end{vmatrix} = 3e^{5x} - 2e^{5x} = e^{5x}$

91. $\begin{vmatrix} x & \ln x \\ 1 & \frac{1}{x} \end{vmatrix} = 1 - \ln x$

93. True. If an entire row is zero, then each cofactor in the expansion is multiplied by zero.

95. *Sample answer:* Let $A = \begin{bmatrix} 1 & 3 \\ -2 & 4 \end{bmatrix}$ and $B = \begin{bmatrix} -4 & 0 \\ 3 & 5 \end{bmatrix}$.

$|A| = \begin{vmatrix} 1 & 3 \\ -2 & 4 \end{vmatrix} = 10, \ |B| = \begin{vmatrix} -4 & 0 \\ 3 & 5 \end{vmatrix} = -20, \ |A| + |B| = -10$

$A + B = \begin{bmatrix} -3 & 3 \\ 1 & 9 \end{bmatrix}, \ |A+B| = \begin{vmatrix} -3 & 3 \\ 1 & 9 \end{vmatrix} = -30$

So, $|A + B| \neq |A| + |B|$.

97. The signs of the cofactors should be $-, +, -$.

$$\begin{vmatrix} 1 & 1 & 4 \\ 3 & 2 & 0 \\ 2 & 1 & 3 \end{vmatrix} = 3(-1)\begin{vmatrix} 1 & 4 \\ 1 & 3 \end{vmatrix} + 2(1)\begin{vmatrix} 1 & 4 \\ 2 & 3 \end{vmatrix} + (0)(-1)\begin{vmatrix} 1 & 1 \\ 2 & 1 \end{vmatrix} = 3(1) + 2(-5) + 0 = -7.$$

99. (a) $\begin{vmatrix} 1 & 3 & 4 \\ -7 & 2 & -5 \\ 6 & 1 & 2 \end{vmatrix} = -115$

$-\begin{vmatrix} 1 & 4 & 3 \\ -7 & -5 & 2 \\ 6 & 2 & 1 \end{vmatrix} = -115$

Column 2 and Column 3 were interchanged.

(b) $\begin{vmatrix} 1 & 3 & 4 \\ -2 & 2 & 0 \\ 1 & 6 & 2 \end{vmatrix} = -40$

$-\begin{vmatrix} 1 & 6 & 2 \\ -2 & 2 & 0 \\ 1 & 3 & 4 \end{vmatrix} = -40$

Row 1 and Row 3 were interchanged.

101. (a) $A = \begin{bmatrix} 1 & 2 \\ 2 & -3 \end{bmatrix}, B = \begin{bmatrix} 5 & 10 \\ 2 & -3 \end{bmatrix}$

$|B| = \begin{vmatrix} 5 & 10 \\ 2 & -3 \end{vmatrix} = -35$

$5|A| = 5\begin{vmatrix} 1 & 2 \\ 2 & -3 \end{vmatrix} = -35$

Row 1 was multiplied by 5.

$|B| = 5|A|$

(b) $A = \begin{bmatrix} 1 & 2 & -1 \\ 3 & -3 & 2 \\ 7 & 1 & 3 \end{bmatrix}, B = \begin{bmatrix} 1 & 8 & -3 \\ 3 & -12 & 6 \\ 7 & 4 & 9 \end{bmatrix}$

$|B| = \begin{bmatrix} 1 & 8 & -3 \\ 3 & -12 & 6 \\ 7 & 4 & 9 \end{bmatrix} = -300$

$12|A| = 12\begin{vmatrix} 1 & 2 & -1 \\ 3 & -3 & 2 \\ 7 & 1 & 3 \end{vmatrix} = -300$

Column 2 was multiplied by 4 and Column 3 was multiplied by 3.

$|B| = (4)(3)|A| = 12|A|$

103. (a) $\begin{vmatrix} 7 & 0 \\ 0 & 4 \end{vmatrix} = 7(4) - 0 = 28$

(b) $\begin{vmatrix} -1 & 0 & 0 \\ 0 & 5 & 0 \\ 0 & 0 & 2 \end{vmatrix} = (-1)\begin{vmatrix} 5 & 0 \\ 0 & 2 \end{vmatrix} - 0\begin{vmatrix} 0 & 0 \\ 0 & 2 \end{vmatrix} + 0\begin{vmatrix} 0 & 5 \\ 0 & 0 \end{vmatrix}$

$= (-1)(10) = -10$

(c) $\begin{vmatrix} 2 & 0 & 0 & 0 \\ 0 & -2 & 0 & 0 \\ 0 & 0 & 1 & 0 \\ 0 & 0 & 0 & 3 \end{vmatrix} = (-2)\begin{vmatrix} -2 & 0 & 0 \\ 0 & 1 & 0 \\ 0 & 0 & 3 \end{vmatrix} - 0\begin{vmatrix} 0 & 0 & 0 \\ 0 & 1 & 0 \\ 0 & 0 & 3 \end{vmatrix} + 0\begin{vmatrix} 0 & -2 & 0 \\ 0 & 0 & 0 \\ 0 & 0 & 3 \end{vmatrix} - 0\begin{vmatrix} 0 & -2 & 0 \\ 0 & 0 & 1 \\ 0 & 0 & 0 \end{vmatrix}$

$= (-2)\left((2)\begin{vmatrix} 1 & 0 \\ 0 & 3 \end{vmatrix} - 0\begin{vmatrix} 0 & 0 \\ 0 & 3 \end{vmatrix} + 0\begin{vmatrix} 0 & 1 \\ 0 & 0 \end{vmatrix}\right)$

$= (-2)(2)(3) = -12$

The determinant of a diagonal matrix is the product of the entries on the main diagonal.

Section 7.5 Applications of Matrices and Determinants

1. Cramer's Rule

3. $A = \pm\frac{1}{2}\begin{vmatrix} x_1 & y_1 & 1 \\ x_2 & y_2 & 1 \\ x_3 & y_3 & 1 \end{vmatrix}$

5. uncoded; coded

7. $\begin{cases} -5x + 9y = -14 \\ 3x - 7y = 10 \end{cases}$

$x = \dfrac{\begin{vmatrix} -14 & 9 \\ 10 & -7 \end{vmatrix}}{\begin{vmatrix} -5 & 9 \\ 3 & -7 \end{vmatrix}} = \dfrac{8}{8} = 1$

$y = \dfrac{\begin{vmatrix} -5 & -14 \\ 3 & 10 \end{vmatrix}}{\begin{vmatrix} -5 & 9 \\ 3 & -7 \end{vmatrix}} = \dfrac{-8}{8} = -1$

Solution: $(1, -1)$

9. $\begin{cases} 3x + 2y = -2 \\ 6x + 4y = 4 \end{cases}$

Because $\begin{vmatrix} 3 & 2 \\ 6 & 4 \end{vmatrix} = 0$, Cramer's Rule does not apply.

The system is inconsistent in this case and has no solution.

11. $\begin{cases} 4x - y + z = -5 \\ 2x + 2y + 3z = 10, \\ 5x - 2y + 6z = 1 \end{cases}$ $D = \begin{vmatrix} 4 & -1 & 1 \\ 2 & 2 & 3 \\ 5 & -1 & 6 \end{vmatrix} = 55$

$x = \dfrac{\begin{vmatrix} -5 & -1 & 1 \\ 10 & 2 & 3 \\ 1 & -2 & 6 \end{vmatrix}}{55} = \dfrac{-55}{55} = -1$

$y = \dfrac{\begin{vmatrix} 4 & -5 & 1 \\ 2 & 10 & 3 \\ 5 & 1 & 6 \end{vmatrix}}{55} = \dfrac{165}{55} = 3$

$z = \dfrac{\begin{vmatrix} 4 & -1 & -5 \\ 2 & 2 & 10 \\ 5 & -2 & 1 \end{vmatrix}}{55} = \dfrac{110}{55} = 2$

Solution: $(-1, 3, 2)$

13. $\begin{cases} x + 2y + 3z = -3 \\ -2x + y - z = 6, \\ 3x - 3y + 2z = -11 \end{cases}$ $D = \begin{vmatrix} 1 & 2 & 3 \\ -2 & 1 & -1 \\ 3 & -3 & 2 \end{vmatrix} = 10$

$$x = \frac{\begin{vmatrix} -3 & 2 & 3 \\ 6 & 1 & -1 \\ -11 & -3 & 2 \end{vmatrix}}{10} = \frac{-20}{10} = -2$$

$$y = \frac{\begin{vmatrix} 1 & -3 & 3 \\ -2 & 6 & -1 \\ 3 & -11 & 2 \end{vmatrix}}{10} = \frac{10}{10} = 1$$

$$z = \frac{\begin{vmatrix} 1 & 2 & -3 \\ -2 & 1 & 6 \\ 3 & -3 & -11 \end{vmatrix}}{10} = \frac{-10}{10} = -1$$

Solution: $(-2, 1, -1)$

15. Vertices: $(0, 0)$ $(3, 1)$, $(1, 5)$

$$\text{Area} = \frac{1}{2}\begin{vmatrix} 0 & 0 & 1 \\ 3 & 1 & 1 \\ 1 & 5 & 1 \end{vmatrix} = \frac{1}{2}\begin{vmatrix} 3 & 1 \\ 1 & 5 \end{vmatrix} = 7 \text{ square units}$$

17. Vertices: $(-2, -3)$, $(2, -3)$, $(0, 4)$

$$\text{Area} = \frac{1}{2}\begin{vmatrix} -2 & -3 & 1 \\ 2 & -3 & 1 \\ 0 & 4 & 1 \end{vmatrix} = \frac{1}{2}\left(-2\begin{vmatrix} -3 & 1 \\ 4 & 1 \end{vmatrix} - 2\begin{vmatrix} -3 & 1 \\ 4 & 1 \end{vmatrix}\right)$$

$$= \frac{1}{2}(14 + 14) = 14 \text{ square units}$$

19. $4 = \pm\frac{1}{2}\begin{vmatrix} -5 & 1 & 1 \\ 0 & 2 & 1 \\ -2 & y & 1 \end{vmatrix}$

$$\pm 8 = -5\begin{vmatrix} 2 & 1 \\ y & 1 \end{vmatrix} - 2\begin{vmatrix} 1 & 1 \\ 2 & 1 \end{vmatrix}$$

$$\pm 8 = -5(2 - y) - 2(-1)$$

$$\pm 8 = 5y - 8$$

$$y = \frac{8 \pm 8}{5}$$

$$y = \frac{16}{5} \text{ or } y = 0$$

21. Vertices: $(0, 25)$, $(10, 0)$, $(28, 5)$

$$\text{Area} = \frac{1}{2}\begin{vmatrix} 0 & 25 & 1 \\ 10 & 0 & 1 \\ 28 & 5 & 1 \end{vmatrix} = 250 \text{ square miles}$$

23. Points: $(2, -6)$, $(0, -2)$, $(3, -8)$

$$\begin{vmatrix} 2 & -6 & 1 \\ 0 & -2 & 1 \\ 3 & -8 & 1 \end{vmatrix} = 2\begin{vmatrix} -2 & 1 \\ -8 & 1 \end{vmatrix} + 3\begin{vmatrix} -6 & 1 \\ -2 & 1 \end{vmatrix}$$

$$= 2(6) + 3(-4)$$

$$= 0$$

The points are collinear.

25. Points: $\left(2, -\frac{1}{2}\right)$, $(-4, 4)$, $(6, -3)$

$$\begin{vmatrix} 2 & -\frac{1}{2} & 1 \\ -4 & 4 & 1 \\ 6 & -3 & 1 \end{vmatrix} = \begin{vmatrix} -4 & 4 \\ 6 & -3 \end{vmatrix} - \begin{vmatrix} 2 & -\frac{1}{2} \\ 6 & -3 \end{vmatrix} + \begin{vmatrix} 2 & -\frac{1}{2} \\ -4 & 4 \end{vmatrix}$$

$$= -12 + 3 + 6$$

$$= -3 \neq 0$$

The points are not collinear.

27. Points: $(0, 2)$, $(1, 2.4)$, $(-1, 1.6)$

$$\begin{vmatrix} 0 & 2 & 1 \\ 1 & 2.4 & 1 \\ -1 & 1.6 & 1 \end{vmatrix} = -2\begin{vmatrix} 1 & 1 \\ -1 & 1 \end{vmatrix} + \begin{vmatrix} 1 & 2.4 \\ -1 & 1.6 \end{vmatrix} = -2(2) + 4 = 0$$

The points are collinear.

29. $\begin{vmatrix} 2 & -5 & 1 \\ 4 & y & 1 \\ 5 & -2 & 1 \end{vmatrix} = 0$

$$2\begin{vmatrix} y & 1 \\ -2 & 1 \end{vmatrix} + 5\begin{vmatrix} 4 & 1 \\ 5 & 1 \end{vmatrix} + \begin{vmatrix} 4 & y \\ 5 & -2 \end{vmatrix} = 0$$

$$2(y + 2) + 5(-1) + (-8 - 5y) = 0$$

$$-3y - 9 = 0$$

$$y = -3$$

31. Points: $(0, 0)$, $(5, 3)$

Equation:

$$\begin{vmatrix} x & y & 1 \\ 0 & 0 & 1 \\ 5 & 3 & 1 \end{vmatrix} = -\begin{vmatrix} x & y \\ 5 & 3 \end{vmatrix} = 5y - 3x = 0 \Rightarrow 3x - 5y = 0$$

33. Points: $(-4, 3), (2, 1)$

Equation: $\begin{vmatrix} x & y & 1 \\ -4 & 3 & 1 \\ 2 & 1 & 1 \end{vmatrix} = x\begin{vmatrix} 3 & 1 \\ 1 & 1 \end{vmatrix} - y\begin{vmatrix} -4 & 1 \\ 2 & 1 \end{vmatrix} + \begin{vmatrix} -4 & 3 \\ 2 & 1 \end{vmatrix} = 2x + 6y - 10 = 0 \Rightarrow x + 3y - 5 = 0$

35. Points: $\left(-\frac{1}{2}, 3\right), \left(\frac{5}{2}, 1\right)$

Equation: $\begin{vmatrix} x & y & 1 \\ -\frac{1}{2} & 3 & 1 \\ \frac{5}{2} & 1 & 1 \end{vmatrix} = x\begin{vmatrix} 3 & 1 \\ 1 & 1 \end{vmatrix} - y\begin{vmatrix} -\frac{1}{2} & 1 \\ \frac{5}{2} & 1 \end{vmatrix} + \begin{vmatrix} -\frac{1}{2} & 3 \\ \frac{5}{2} & 1 \end{vmatrix} = 2x + 3y - 8 = 0$

37. A horizontal stretch, $k = 2$, of the square with vertices $(0, 0), (0, 3), (3, 0)$ and $(3, 3)$.

$\begin{bmatrix} 2 & 0 \\ 0 & 1 \end{bmatrix}\begin{bmatrix} 0 \\ 0 \end{bmatrix} = \begin{bmatrix} 0 \\ 0 \end{bmatrix}, \begin{bmatrix} 2 & 0 \\ 0 & 1 \end{bmatrix}\begin{bmatrix} 0 \\ 3 \end{bmatrix} = \begin{bmatrix} 0 \\ 3 \end{bmatrix}, \begin{bmatrix} 2 & 0 \\ 0 & 1 \end{bmatrix}\begin{bmatrix} 3 \\ 0 \end{bmatrix} = \begin{bmatrix} 6 \\ 0 \end{bmatrix}$

and $\begin{bmatrix} 2 & 0 \\ 0 & 1 \end{bmatrix}\begin{bmatrix} 3 \\ 3 \end{bmatrix} = \begin{bmatrix} 6 \\ 3 \end{bmatrix}$.

New Vertices: $(0, 0), (0, 3), (6, 0)$ and $(6, 3)$

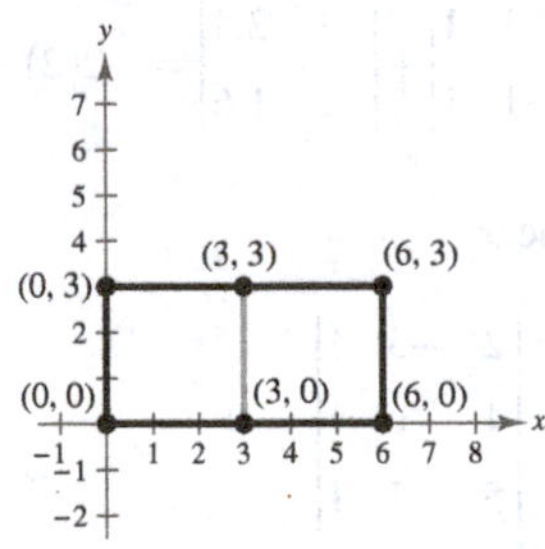

39. A reflection in the y-axis of the square with vertices $(4, 3), (5, 3), (4, 4)$ and $(5, 4)$.

$\begin{bmatrix} -1 & 0 \\ 0 & 1 \end{bmatrix}\begin{bmatrix} 4 \\ 3 \end{bmatrix} = \begin{bmatrix} -4 \\ 3 \end{bmatrix}, \begin{bmatrix} -1 & 0 \\ 0 & 1 \end{bmatrix}\begin{bmatrix} 5 \\ 3 \end{bmatrix} = \begin{bmatrix} -5 \\ 3 \end{bmatrix},$

$\begin{bmatrix} -1 & 0 \\ 0 & 1 \end{bmatrix}\begin{bmatrix} 4 \\ 4 \end{bmatrix} = \begin{bmatrix} -4 \\ 4 \end{bmatrix}$ and $\begin{bmatrix} -1 & 0 \\ 0 & 1 \end{bmatrix}\begin{bmatrix} 5 \\ 4 \end{bmatrix} = \begin{bmatrix} -5 \\ 4 \end{bmatrix}$.

New Vertices: $(-4, 3), (-5, 3), (-4, 4)$ and $(-5, 4)$

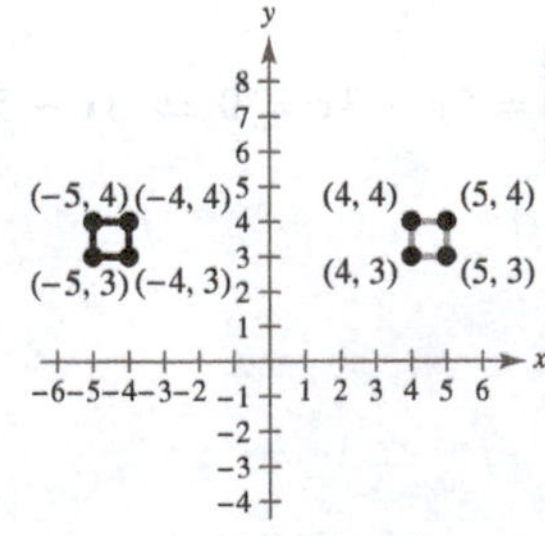

41. The area of the parallelogram with vertices: $(0, 0), (1, 0), (2, 2)$ and $(3, 2) \Rightarrow a = 1, b = 2, c = 2$ and $d = 2$.

$A = \begin{bmatrix} 1 & 0 \\ 2 & 2 \end{bmatrix}$

$\text{Area} = |\det(A)| = |2 - 0| = 2$ square units.

43. The area of the parallelogram with vertices: $(0, 0), (-2, 0), (3, 5)$ and $(1, 5) \Rightarrow a = -2, b = 0, c = 3$ and $d = 5$.

$A = \begin{bmatrix} -2 & 0 \\ 3 & 5 \end{bmatrix}$

$\text{Area} = |\det(A)| = |-10 - 0| = 10$ square units.

45. (a) Uncoded: C O M E _ H O M E _

[3 15] [13 5] [0 8] [15 13] [5 0]

S O O N

[19 15] [15 14]

(b) $[3\ \ 15]\begin{bmatrix} 1 & 2 \\ 3 & 5 \end{bmatrix} = [48\ \ 81]$

$[13\ \ 5]\begin{bmatrix} 1 & 2 \\ 3 & 5 \end{bmatrix} = [28\ \ 51]$

$[0\ \ 8]\begin{bmatrix} 1 & 2 \\ 3 & 5 \end{bmatrix} = [24\ \ 40]$

$[15\ \ 13]\begin{bmatrix} 1 & 2 \\ 3 & 5 \end{bmatrix} = [54\ \ 95]$

$[5\ \ 0]\begin{bmatrix} 1 & 2 \\ 3 & 5 \end{bmatrix} = [5\ \ 10]$

$[19\ \ 15]\begin{bmatrix} 1 & 2 \\ 3 & 5 \end{bmatrix} = [64\ \ 113]$

$[15\ \ 14]\begin{bmatrix} 1 & 2 \\ 3 & 5 \end{bmatrix} = [57\ \ 100]$

Encoded: 48 81 28 51 24 40 54 95 5 10 64 113 57 100

47. (a) Uncoded: C A L L _ M E _ T O M O R R O W _ _

[3 1 12] [12 0 13] [5 0 20] [15 13 15] [18 18 15] [23 0 0]

(b) $[3 \quad 1 \quad 12]\begin{bmatrix} 1 & -1 & 0 \\ 1 & 0 & -1 \\ -6 & 2 & 3 \end{bmatrix} = [-68 \quad 21 \quad 35]$

$[12 \quad 0 \quad 13]\begin{bmatrix} 1 & -1 & 0 \\ 1 & 0 & -1 \\ -6 & 2 & 3 \end{bmatrix} = [-66 \quad 14 \quad 39]$

$[5 \quad 0 \quad 20]\begin{bmatrix} 1 & -1 & 0 \\ 1 & 0 & -1 \\ -6 & 2 & 3 \end{bmatrix} = [-115 \quad 35 \quad 60]$

$[15 \quad 13 \quad 15]\begin{bmatrix} 1 & -1 & 0 \\ 1 & 0 & -1 \\ -6 & 2 & 3 \end{bmatrix} = [-62 \quad 15 \quad 32]$

$[18 \quad 18 \quad 15]\begin{bmatrix} 1 & -1 & 0 \\ 1 & 0 & -1 \\ -6 & 2 & 3 \end{bmatrix} = [-54 \quad 12 \quad 27]$

$[23 \quad 0 \quad 0]\begin{bmatrix} 1 & -1 & 0 \\ 1 & 0 & -1 \\ -6 & 2 & 3 \end{bmatrix} = [23 \quad -23 \quad 0]$

Encoded: −68 21 35 −66 14 39 −115 35 60
−62 15 32 −54 12 27 23 −23 0

In Exercises 49–51, use the matrix $A = \begin{bmatrix} 1 & 2 & 2 \\ 3 & 7 & 9 \\ -1 & -4 & -7 \end{bmatrix}$.

49. L A N D I N G _ S U C C E S S F U L

[12 1 14] [4 9 14] [7 0 19] [21 3 3] [5 19 19] [6 21 12]

$[12 \quad 1 \quad 14]\begin{bmatrix} 1 & 2 & 2 \\ 3 & 7 & 9 \\ -1 & -4 & -7 \end{bmatrix} = [1 \quad -25 \quad -65]$

$[4 \quad 9 \quad 14]\begin{bmatrix} 1 & 2 & 2 \\ 3 & 7 & 9 \\ -1 & -4 & -7 \end{bmatrix} = [17 \quad 15 \quad -9]$

$[7 \quad 0 \quad 19]\begin{bmatrix} 1 & 2 & 2 \\ 3 & 7 & 9 \\ -1 & -4 & -7 \end{bmatrix} = [-12 \quad -62 \quad -119]$

$[21 \quad 3 \quad 3]\begin{bmatrix} 1 & 2 & 2 \\ 3 & 7 & 9 \\ -1 & -4 & -7 \end{bmatrix} = [27 \quad 51 \quad 48]$

$[5 \quad 19 \quad 19]\begin{bmatrix} 1 & 2 & 2 \\ 3 & 7 & 9 \\ -1 & -4 & -7 \end{bmatrix} = [43 \quad 67 \quad 48]$

$[6 \quad 21 \quad 12]\begin{bmatrix} 1 & 2 & 2 \\ 3 & 7 & 9 \\ -1 & -4 & -7 \end{bmatrix} = [57 \quad 111 \quad 117]$

Cryptogram: 1 −25 −65 17 15 −9 −12 −62 −119 27 51 48 43 67 48 57 111 117

51. H A P P Y _ B I R T H D A Y _

[8 1 16][16 25 0][2 9 18][20 8 4][1 25 0]

$$\begin{bmatrix} 8 & 1 & 16 \end{bmatrix}\begin{bmatrix} 1 & 2 & 2 \\ 3 & 7 & 9 \\ -1 & -4 & -7 \end{bmatrix} = \begin{bmatrix} -5 & -41 & -87 \end{bmatrix}$$

$$\begin{bmatrix} 16 & 25 & 0 \end{bmatrix}\begin{bmatrix} 1 & 2 & 2 \\ 3 & 7 & 9 \\ -1 & -4 & -7 \end{bmatrix} = \begin{bmatrix} 91 & 207 & 257 \end{bmatrix}$$

$$\begin{bmatrix} 2 & 9 & 18 \end{bmatrix}\begin{bmatrix} 1 & 2 & 2 \\ 3 & 7 & 9 \\ -1 & -4 & -7 \end{bmatrix} = \begin{bmatrix} 11 & -5 & -41 \end{bmatrix}$$

$$\begin{bmatrix} 20 & 8 & 4 \end{bmatrix}\begin{bmatrix} 1 & 2 & 2 \\ 3 & 7 & 9 \\ -1 & -4 & -7 \end{bmatrix} = \begin{bmatrix} 40 & 80 & 84 \end{bmatrix}$$

$$\begin{bmatrix} 1 & 25 & 0 \end{bmatrix}\begin{bmatrix} 1 & 2 & 2 \\ 3 & 7 & 9 \\ -1 & -4 & -7 \end{bmatrix} = \begin{bmatrix} 76 & 177 & 227 \end{bmatrix}$$

Cryptogram: −5 −41 −87 91 207 257 11 −5 −41 40 80 84 76 177 227

53. $$A^{-1} = \begin{bmatrix} 1 & 2 \\ 3 & 5 \end{bmatrix}^{-1} = \begin{bmatrix} -5 & 2 \\ 3 & -1 \end{bmatrix}$$

$$\begin{bmatrix} 11 & 21 \\ 64 & 112 \\ 25 & 50 \\ 29 & 53 \\ 23 & 46 \\ 40 & 75 \\ 55 & 92 \end{bmatrix}\begin{bmatrix} -5 & 2 \\ 3 & -1 \end{bmatrix} = \begin{bmatrix} 8 & 1 \\ 16 & 16 \\ 25 & 0 \\ 14 & 5 \\ 23 & 0 \\ 25 & 5 \\ 1 & 18 \end{bmatrix} \begin{matrix} \text{H} & \text{A} \\ \text{P} & \text{P} \\ \text{Y} & \_ \\ \text{N} & \text{E} \\ \text{W} & \_ \\ \text{Y} & \text{E} \\ \text{A} & \text{R} \end{matrix}$$

Message: HAPPY NEW YEAR

55. $$A^{-1} = \begin{bmatrix} 1 & -1 & 0 \\ 1 & 0 & -1 \\ -6 & 2 & 3 \end{bmatrix}^{-1} = \begin{bmatrix} -2 & -3 & -1 \\ -3 & -3 & -1 \\ -2 & -4 & -1 \end{bmatrix}$$

$$\begin{bmatrix} 9 & -1 & -9 \\ 38 & -19 & -19 \\ 28 & -9 & -19 \\ -80 & 25 & 41 \\ -64 & 21 & 31 \\ 9 & -5 & -4 \end{bmatrix}\begin{bmatrix} -2 & -3 & -1 \\ -3 & -3 & -1 \\ -2 & -4 & -1 \end{bmatrix} = \begin{bmatrix} 3 & 12 & 1 \\ 19 & 19 & 0 \\ 9 & 19 & 0 \\ 3 & 1 & 14 \\ 3 & 5 & 12 \\ 5 & 4 & 0 \end{bmatrix} \begin{matrix} \text{C} & \text{L} & \text{A} \\ \text{S} & \text{S} & \_ \\ \text{I} & \text{S} & \_ \\ \text{C} & \text{A} & \text{N} \\ \text{C} & \text{E} & \text{L} \\ \text{E} & \text{D} & \_ \end{matrix}$$

Message: CLASS IS CANCELED

57. $A^{-1} = \begin{bmatrix} 1 & 2 & 2 \\ 3 & 7 & 9 \\ -1 & -4 & -7 \end{bmatrix}^{-1} = \begin{bmatrix} -13 & 6 & 4 \\ 12 & -5 & -3 \\ -5 & 2 & 1 \end{bmatrix}$

$$\begin{bmatrix} 20 & 17 & -15 \\ -12 & -56 & -104 \\ 1 & -25 & -65 \\ 62 & 143 & 181 \end{bmatrix} \begin{bmatrix} -13 & 6 & 4 \\ 12 & -5 & -3 \\ -5 & 2 & 1 \end{bmatrix} = \begin{bmatrix} 19 & 5 & 14 \\ 4 & 0 & 16 \\ 12 & 1 & 14 \\ 5 & 19 & 0 \end{bmatrix} \begin{matrix} S & E & N \\ D & \_ & P \\ L & A & N \\ E & S & \_ \end{matrix}$$

Message: SEND PLANES

59. Let A be the 2×2 matrix needed to decode the message.

$$\begin{bmatrix} -18 & -18 \\ 1 & 16 \end{bmatrix} A = \begin{bmatrix} 0 & 18 \\ 15 & 14 \end{bmatrix} \begin{matrix} \_ & R \\ O & N \end{matrix}$$

$$A = \begin{bmatrix} -18 & -18 \\ 1 & 16 \end{bmatrix}^{-1} \begin{bmatrix} 0 & 18 \\ 15 & 14 \end{bmatrix} = \begin{bmatrix} -\frac{8}{135} & -\frac{1}{15} \\ \frac{1}{270} & \frac{1}{15} \end{bmatrix} \begin{bmatrix} 0 & 18 \\ 15 & 14 \end{bmatrix} = \begin{bmatrix} -1 & -2 \\ 1 & 1 \end{bmatrix}$$

$$\begin{bmatrix} 8 & 21 \\ -15 & -10 \\ -13 & -13 \\ 5 & 10 \\ 5 & 25 \\ 5 & 19 \\ -1 & 6 \\ 20 & 40 \\ -18 & -18 \\ 1 & 16 \end{bmatrix} \begin{bmatrix} -1 & -2 \\ 1 & 1 \end{bmatrix} = \begin{bmatrix} 13 & 5 \\ 5 & 20 \\ 0 & 13 \\ 5 & 0 \\ 20 & 15 \\ 14 & 9 \\ 7 & 8 \\ 20 & 0 \\ 0 & 18 \\ 15 & 14 \end{bmatrix} \begin{matrix} M & E \\ E & T \\ \_ & M \\ E & \_ \\ T & O \\ N & I \\ G & H \\ T & \_ \\ \_ & R \\ O & N \end{matrix}$$

Message: MEET ME TONIGHT RON

61. $D = \begin{vmatrix} 4 & 0 & 8 \\ 0 & 2 & 8 \\ 1 & 1 & -1 \end{vmatrix} = -56$

$$I_1 = \frac{\begin{vmatrix} 2 & 0 & 8 \\ 6 & 2 & 8 \\ 0 & 1 & -1 \end{vmatrix}}{-56} = -\frac{28}{56} = -\frac{1}{2}$$

$$I_2 = \frac{\begin{vmatrix} 4 & 2 & 8 \\ 0 & 6 & 8 \\ 1 & 0 & -1 \end{vmatrix}}{-56} = \frac{-56}{-56} = 1$$

$$I_3 = \frac{\begin{vmatrix} 4 & 0 & 2 \\ 0 & 2 & 6 \\ 1 & 1 & 0 \end{vmatrix}}{-56} = \frac{-28}{-56} = \frac{1}{2}$$

So, the solution is $I_1 = -0.5$ ampere, $I_2 = 1$ ampere, and $I_3 = 0.5$ ampere.

63. False. In Cramer's Rule, the denominator is the determinant of the coefficient matrix.

65. If the determinant of the coefficient matrix is zero, the system has either no solution or infinitely many solutions.

67. $\text{Area} = \frac{1}{2}\begin{vmatrix} 3 & -1 & 1 \\ 7 & -1 & 1 \\ 7 & 5 & 1 \end{vmatrix}$

$= \frac{1}{2}\left(3\begin{vmatrix} -1 & 1 \\ 5 & 1 \end{vmatrix} + 1\begin{vmatrix} 7 & 1 \\ 7 & 1 \end{vmatrix} + 1\begin{vmatrix} 7 & -1 \\ 7 & 5 \end{vmatrix}\right)$

$= \frac{1}{2}(-18 + 0 + 42)$

$= 12$ square units

$\text{Area} = \frac{1}{2}(\text{base})(\text{height}) = \frac{1}{2}(7 - 3)(5 - (-1)) = \frac{1}{2}(4)(6) = 12$ square units

Review Exercises for Chapter 7

1. $\begin{bmatrix} -1 & 3 \end{bmatrix}$

Order: 1×2

3. $\begin{bmatrix} 2 & 1 & 0 & 4 & -1 \\ 6 & 2 & 1 & 8 & 0 \end{bmatrix}$

Order: 2×5

5. $\begin{cases} 3x - 10y = 15 \\ 5x + 4y = 22 \end{cases}$

$\left[\begin{array}{cc|c} 3 & -10 & 15 \\ 5 & 4 & 22 \end{array}\right]$

7. $\left[\begin{array}{ccc|c} 1 & 0 & 2 & -8 \\ 2 & -2 & 3 & 12 \\ 4 & 7 & 1 & 3 \end{array}\right]$

$\begin{cases} x + 2z = -8 \\ 2x - 2y + 3z = 12 \\ 4x + 7y + z = 3 \end{cases}$

9. $\begin{bmatrix} 0 & 1 & 1 \\ 1 & 2 & 3 \\ 2 & 2 & 2 \end{bmatrix}$

$\begin{matrix} R_1 \\ R_2 \\ \\ \end{matrix}\begin{bmatrix} 1 & 2 & 3 \\ 0 & 1 & 1 \\ 2 & 2 & 2 \end{bmatrix}$

$-2R_1 + R_3 \rightarrow \begin{bmatrix} 1 & 2 & 3 \\ 0 & 1 & 1 \\ 0 & -2 & -4 \end{bmatrix}$

$2R_2 + R_3 \rightarrow \begin{bmatrix} 1 & 2 & 3 \\ 0 & 1 & 1 \\ 0 & 0 & -2 \end{bmatrix}$

$-\frac{1}{2}R_3 \rightarrow \begin{bmatrix} 1 & 2 & 3 \\ 0 & 1 & 1 \\ 0 & 0 & 1 \end{bmatrix}$

11. $\left[\begin{array}{ccc|c} 1 & 2 & 3 & 9 \\ 0 & 1 & -2 & 2 \\ 0 & 0 & 1 & -1 \end{array}\right] \Rightarrow \begin{cases} x + 2y + 3z = 9 \\ y - 2z = 2 \\ z = -1 \end{cases}$

$y - 2(-1) = 2 \Rightarrow y = 0$

$x + 2(0) + 3(-1) = 9 \Rightarrow x = 12$

Solution: $(12, 0, -1)$

13. $\left[\begin{array}{ccc|c} 1 & 3 & 4 & 1 \\ 0 & 1 & 2 & 3 \\ 0 & 0 & 1 & 4 \end{array}\right] \Rightarrow \begin{cases} x + 3y + 4z = 1 \\ y + 2z = 3 \\ z = 4 \end{cases}$

$y + 2(4) = 3 \Rightarrow y = -5$

$x + 3(-5) + 4(4) = 1 \Rightarrow x = 0$

Solution: $(0, -5, 4)$

15. $\left[\begin{array}{cc|c} 5 & 4 & 2 \\ -1 & 1 & -22 \end{array}\right]$

$4R_2 + R_1 \rightarrow \left[\begin{array}{cc|c} 1 & 8 & -86 \\ -1 & 1 & -22 \end{array}\right]$

$R_1 + R_2 \rightarrow \left[\begin{array}{cc|c} 1 & 8 & -86 \\ 0 & 9 & -108 \end{array}\right]$

$\frac{1}{9}R_2 \rightarrow \left[\begin{array}{cc|c} 1 & 8 & -86 \\ 0 & 1 & -12 \end{array}\right]$

$\begin{cases} x + 8y = -86 \\ y = -12 \end{cases}$

$y = -12$

$x + 8(-12) = -86 \Rightarrow x = 10$

Solution: $(10, -12)$

17. $\begin{bmatrix} 0.3 & -0.1 & \vdots & -0.13 \\ 0.2 & -0.3 & \vdots & -0.25 \end{bmatrix}$

$\begin{matrix} 10R_1 \rightarrow \\ 10R_2 \rightarrow \end{matrix} \begin{bmatrix} 3 & -1 & \vdots & -1.3 \\ 2 & -3 & \vdots & -2.5 \end{bmatrix}$

$-R_2 + R_1 \rightarrow \begin{bmatrix} 1 & 2 & \vdots & 1.2 \\ 2 & -3 & \vdots & -2.5 \end{bmatrix}$

$-2R_1 + R_2 \rightarrow \begin{bmatrix} 1 & 2 & \vdots & 1.2 \\ 0 & -7 & \vdots & -4.9 \end{bmatrix}$

$-\frac{1}{7}R_2 \rightarrow \begin{bmatrix} 1 & 2 & \vdots & 1.2 \\ 0 & 1 & \vdots & 0.7 \end{bmatrix}$

$\begin{cases} x + 2y = 1.2 \\ \quad\ \ y = 0.7 \end{cases}$

$y = 0.7$

$x + 2(0.7) = 1.2 \Rightarrow x = -0.2$

Solution: $(-0.2, 0.7) = \left(-\frac{1}{5}, \frac{7}{10}\right)$

19. $\begin{cases} -x + 2y = 3 \\ 2x - 4y = 6 \end{cases}$

$\begin{bmatrix} -1 & 2 & \vdots & 3 \\ 2 & -4 & \vdots & 6 \end{bmatrix}$

$2R_1 + R_2 \rightarrow \begin{bmatrix} -1 & 2 & \vdots & 3 \\ 0 & 0 & \vdots & 12 \end{bmatrix}$

Because the last row consists of all zeros except for the last entry, the system is inconsistent and there is no solution.

21. $\begin{cases} x - 2y + z = 7 \\ 2x + y - 2z = -4 \\ -x + 3y + 2z = -3 \end{cases}$

$\begin{bmatrix} 1 & -2 & 1 & \vdots & 7 \\ 2 & 1 & -2 & \vdots & -4 \\ -1 & 3 & 2 & \vdots & -3 \end{bmatrix}$

$\begin{matrix} \\ -2R_1 + R_2 \rightarrow \\ R_1 + R_3 \rightarrow \end{matrix} \begin{bmatrix} 1 & -2 & 1 & \vdots & 7 \\ 0 & 5 & -4 & \vdots & -18 \\ 0 & 1 & 3 & \vdots & 4 \end{bmatrix}$

$R_2 + (-5)R_3 \rightarrow \begin{bmatrix} 1 & -2 & 1 & \vdots & 7 \\ 0 & 0 & -19 & \vdots & -38 \\ 0 & 1 & 3 & \vdots & 4 \end{bmatrix}$

$-19z = -38$

$z = 2$

$y + 3(2) = 4 \Rightarrow y = -2$

$x - 2(-2) + 2 = 7 \Rightarrow x = 1$

Solution: $(1, -2, 2)$

23. $\begin{bmatrix} 2 & 1 & 2 & \vdots & 4 \\ 2 & 2 & 0 & \vdots & 5 \\ 2 & -1 & 6 & \vdots & 2 \end{bmatrix}$

$\begin{matrix} \\ -R_1 + R_2 \rightarrow \\ -R_1 + R_3 \rightarrow \end{matrix} \begin{bmatrix} 2 & 1 & 2 & \vdots & 4 \\ 0 & 1 & -2 & \vdots & 1 \\ 0 & -2 & 4 & \vdots & -2 \end{bmatrix}$

$\begin{matrix} -R_2 + R_1 \rightarrow \\ \\ 2R_2 + R_3 \rightarrow \end{matrix} \begin{bmatrix} 2 & 0 & 4 & \vdots & 3 \\ 0 & 1 & -2 & \vdots & 1 \\ 0 & 0 & 0 & \vdots & 0 \end{bmatrix}$

$\begin{matrix} \frac{1}{2}R_1 \rightarrow \\ \\ \\ \end{matrix} \begin{bmatrix} 1 & 0 & 2 & \vdots & \frac{3}{2} \\ 0 & 1 & -2 & \vdots & 1 \\ 0 & 0 & 0 & \vdots & 0 \end{bmatrix}$

Let $z = a$, then:

$y - 2a = 1 \Rightarrow y = 2a + 1$

$x + 2a = \frac{3}{2} \Rightarrow x = -2a + \frac{3}{2}$

Solution: $\left(-2a + \frac{3}{2}, 2a + 1, a\right)$ where a is any real number

25. $\begin{bmatrix} 2 & 3 & 1 & \vdots & 10 \\ 2 & -3 & -3 & \vdots & 22 \\ 4 & -2 & 3 & \vdots & -2 \end{bmatrix}$

$\begin{matrix} \\ -R_1 + R_2 \rightarrow \\ -2R_1 + R_3 \rightarrow \end{matrix} \begin{bmatrix} 2 & 3 & 1 & \vdots & 10 \\ 0 & -6 & -4 & \vdots & 12 \\ 0 & -8 & 1 & \vdots & -22 \end{bmatrix}$

$\begin{matrix} \frac{1}{2}R_1 \rightarrow \\ -\frac{1}{6}R_2 \rightarrow \\ \\ \end{matrix} \begin{bmatrix} 1 & \frac{3}{2} & \frac{1}{2} & \vdots & 5 \\ 0 & 1 & \frac{2}{3} & \vdots & -2 \\ 0 & -8 & 1 & \vdots & -22 \end{bmatrix}$

$\begin{matrix} \\ \\ 8R_2 + R_3 \rightarrow \end{matrix} \begin{bmatrix} 1 & \frac{3}{2} & \frac{1}{2} & \vdots & 5 \\ 0 & 1 & \frac{2}{3} & \vdots & -2 \\ 0 & 0 & \frac{19}{3} & \vdots & -38 \end{bmatrix}$

$\begin{matrix} \\ \\ \frac{3}{19}R_3 \rightarrow \end{matrix} \begin{bmatrix} 1 & \frac{3}{2} & \frac{1}{2} & \vdots & 5 \\ 0 & 1 & \frac{2}{3} & \vdots & -2 \\ 0 & 0 & 1 & \vdots & -6 \end{bmatrix}$

$z = -6$

$y + \frac{2}{3}(-6) = -2 \Rightarrow y = 2$

$x + \frac{3}{2}(2) + \frac{1}{2}(-6) = 5 \Rightarrow x = 5$

Solution: $(5, 2, -6)$

27. $\begin{cases} x + 2y - z = 3 \\ x - y - z = -3 \\ 2x + y + 3z = 10 \end{cases}$

$$\begin{bmatrix} 1 & 2 & -1 & \vdots & 3 \\ 1 & -1 & -1 & \vdots & -3 \\ 2 & 1 & 3 & \vdots & 10 \end{bmatrix}$$

$$\begin{matrix} \\ -R_1 + R_2 \rightarrow \\ -2R_1 + R_3 \rightarrow \end{matrix} \begin{bmatrix} 1 & 2 & -1 & \vdots & 3 \\ 0 & -3 & 0 & \vdots & -6 \\ 0 & 3 & 5 & \vdots & 16 \end{bmatrix}$$

$$\begin{matrix} \\ \\ R_2 + R_3 \rightarrow \end{matrix} \begin{bmatrix} 1 & 2 & -1 & \vdots & 3 \\ 0 & -3 & 0 & \vdots & -6 \\ 0 & 0 & 5 & \vdots & 10 \end{bmatrix}$$

$$\begin{matrix} 3R_1 + 2R_2 \rightarrow \\ \\ \\ \end{matrix} \begin{bmatrix} 3 & 0 & -3 & \vdots & -3 \\ 0 & -3 & 0 & \vdots & -6 \\ 0 & 0 & 5 & \vdots & 10 \end{bmatrix}$$

$$\begin{matrix} 5R_1 + 3R_3 \rightarrow \\ \\ \\ \end{matrix} \begin{bmatrix} 15 & 0 & 0 & \vdots & 15 \\ 0 & -3 & 0 & \vdots & -6 \\ 0 & 0 & 5 & \vdots & 10 \end{bmatrix}$$

$$\begin{matrix} \frac{1}{15}R_1 \rightarrow \\ \frac{1}{3}R_2 \rightarrow \\ \frac{1}{5}R_3 \rightarrow \end{matrix} \begin{bmatrix} 1 & 0 & 0 & \vdots & 1 \\ 0 & 1 & 0 & \vdots & 2 \\ 0 & 0 & 1 & \vdots & 2 \end{bmatrix}$$

$x = 1$

$y = 2$

$z = 2$

Solution: (1, 2, 2)

29. $$\begin{bmatrix} -1 & 1 & 2 & \vdots & 1 \\ 2 & 3 & 1 & \vdots & -2 \\ 5 & 4 & 2 & \vdots & 4 \end{bmatrix}$$

$$\begin{matrix} -R_1 \rightarrow \\ \\ \\ \end{matrix} \begin{bmatrix} 1 & -1 & -2 & \vdots & -1 \\ 2 & 3 & 1 & \vdots & -2 \\ 5 & 4 & 2 & \vdots & 4 \end{bmatrix}$$

$$\begin{matrix} \\ -2R_1 + R_2 \rightarrow \\ -5R_1 + R_3 \rightarrow \end{matrix} \begin{bmatrix} 1 & -1 & -2 & \vdots & -1 \\ 0 & 5 & 5 & \vdots & 0 \\ 0 & 9 & 12 & \vdots & 9 \end{bmatrix}$$

$$\begin{matrix} \\ \frac{1}{5}R_2 \rightarrow \\ \\ \end{matrix} \begin{bmatrix} 1 & -1 & -2 & \vdots & -1 \\ 0 & 1 & 1 & \vdots & 0 \\ 0 & 9 & 12 & \vdots & 9 \end{bmatrix}$$

$$\begin{matrix} R_2 + R_1 \rightarrow \\ \\ -9R_2 + R_3 \rightarrow \end{matrix} \begin{bmatrix} 1 & 0 & -1 & \vdots & -1 \\ 0 & 1 & 1 & \vdots & 0 \\ 0 & 0 & 3 & \vdots & 9 \end{bmatrix}$$

$$\begin{matrix} \\ \\ \frac{1}{3}R_3 \rightarrow \end{matrix} \begin{bmatrix} 1 & 0 & -1 & \vdots & -1 \\ 0 & 1 & 1 & \vdots & 0 \\ 0 & 0 & 1 & \vdots & 3 \end{bmatrix}$$

$$\begin{matrix} R_3 + R_1 \rightarrow \\ -R_3 + R_2 \rightarrow \\ \\ \end{matrix} \begin{bmatrix} 1 & 0 & 0 & \vdots & 2 \\ 0 & 1 & 0 & \vdots & -3 \\ 0 & 0 & 1 & \vdots & 3 \end{bmatrix}$$

$x = 2, y = -3, z = 3$

Solution: (2, −3, 3)

31. Use the reduced row-echelon form feature of a graphing utility.

$$\begin{bmatrix} 3 & -1 & 5 & -2 & \vdots & -44 \\ 1 & 6 & 4 & -1 & \vdots & 1 \\ 5 & -1 & 1 & 3 & \vdots & -15 \\ 0 & 4 & -1 & -8 & \vdots & 58 \end{bmatrix} \Rightarrow \begin{bmatrix} 1 & 0 & 0 & 0 & \vdots & 2 \\ 0 & 1 & 0 & 0 & \vdots & 6 \\ 0 & 0 & 1 & 0 & \vdots & -10 \\ 0 & 0 & 0 & 1 & \vdots & -3 \end{bmatrix}$$

$x = 2, y = 6, z = -10, w = -3$

Solution: (2, 6, −10, −3)

33. $\begin{bmatrix} -1 & x \\ y & 9 \end{bmatrix} = \begin{bmatrix} -1 & 12 \\ 11 & 9 \end{bmatrix} \Rightarrow x = 12$ and $y = 11$

35. $$\begin{bmatrix} x + 3 & -4 & 44 \\ 0 & -3 & 2 \\ -2 & y + 5 & 6 \end{bmatrix} = \begin{bmatrix} 5x - 1 & -4 & 44 \\ 0 & -3 & 2 \\ -2 & 16 & 6 \end{bmatrix}$$

$$\left.\begin{aligned} x + 3 &= 5x - 1 \\ 4 &= 4x \\ y + 5 &= 16 \\ y &= 11 \end{aligned}\right\} x = 1 \text{ and } y = 11$$

37. (a) $A + B = \begin{bmatrix} 2 & -2 \\ 3 & 5 \end{bmatrix} + \begin{bmatrix} -3 & 10 \\ 12 & 8 \end{bmatrix} = \begin{bmatrix} -1 & 8 \\ 15 & 13 \end{bmatrix}$

(b) $A - B = \begin{bmatrix} 2 & -2 \\ 3 & 5 \end{bmatrix} - \begin{bmatrix} -3 & 10 \\ 12 & 8 \end{bmatrix} = \begin{bmatrix} 5 & -12 \\ -9 & -3 \end{bmatrix}$

(c) $4A = 4\begin{bmatrix} 2 & -2 \\ 3 & 5 \end{bmatrix} = \begin{bmatrix} 8 & -8 \\ 12 & 20 \end{bmatrix}$

(d) $2A + 2B = 2\begin{bmatrix} 2 & -2 \\ 3 & 5 \end{bmatrix} + 2\begin{bmatrix} -3 & 10 \\ 12 & 8 \end{bmatrix} = \begin{bmatrix} 4 & -4 \\ 6 & 10 \end{bmatrix} + \begin{bmatrix} -6 & 20 \\ 24 & 16 \end{bmatrix} = \begin{bmatrix} -2 & 16 \\ 30 & 26 \end{bmatrix}$

39. $A = \begin{bmatrix} 4 & 3 \\ -6 & 1 \\ 10 & 1 \end{bmatrix}, B = \begin{bmatrix} 3 & 11 \\ 15 & 25 \\ 20 & 29 \end{bmatrix}$

(a) $A + B = \begin{bmatrix} 4 & 3 \\ -6 & 1 \\ 10 & 1 \end{bmatrix} + \begin{bmatrix} 3 & 11 \\ 15 & 25 \\ 20 & 29 \end{bmatrix} = \begin{bmatrix} 4+3 & 3+11 \\ -6+15 & 1+25 \\ 10+20 & 1+29 \end{bmatrix} = \begin{bmatrix} 7 & 14 \\ 9 & 26 \\ 30 & 30 \end{bmatrix}$

(b) $A - B = \begin{bmatrix} 4 & 3 \\ -6 & 1 \\ 10 & 1 \end{bmatrix} - \begin{bmatrix} 3 & 11 \\ 15 & 25 \\ 20 & 29 \end{bmatrix} = \begin{bmatrix} 4-3 & 3-11 \\ -6-15 & 1-25 \\ 10-20 & 1-29 \end{bmatrix} = \begin{bmatrix} 1 & -8 \\ -21 & -24 \\ -10 & -28 \end{bmatrix}$

(c) $4A = 4\begin{bmatrix} 4 & 3 \\ -6 & 1 \\ 10 & 1 \end{bmatrix} = \begin{bmatrix} 4(4) & 4(3) \\ 4(-6) & 4(1) \\ 4(10) & 4(1) \end{bmatrix} = \begin{bmatrix} 16 & 12 \\ -24 & 4 \\ 40 & 4 \end{bmatrix}$

(d) $2A + 2B = 2\begin{bmatrix} 4 & 3 \\ -6 & 1 \\ 10 & 1 \end{bmatrix} + 2\begin{bmatrix} 3 & 11 \\ 15 & 25 \\ 20 & 29 \end{bmatrix} = \begin{bmatrix} 8 & 6 \\ -12 & 2 \\ 20 & 2 \end{bmatrix} + \begin{bmatrix} 6 & 22 \\ 30 & 50 \\ 40 & 58 \end{bmatrix} = \begin{bmatrix} 14 & 28 \\ 18 & 52 \\ 60 & 60 \end{bmatrix}$

41. (a) $A + B = \begin{bmatrix} -2 & 3 & -2 \\ 3 & -2 & 3 \\ -2 & 3 & -2 \end{bmatrix} + \begin{bmatrix} 3 & -2 & 3 \\ -2 & 3 & -2 \\ 3 & -2 & 3 \end{bmatrix} = \begin{bmatrix} -2+3 & 3+(-2) & -2+3 \\ 3+(-2) & -2+3 & 3+(-2) \\ -2+3 & 3+(-2) & -2+3 \end{bmatrix} = \begin{bmatrix} 1 & 1 & 1 \\ 1 & 1 & 1 \\ 1 & 1 & 1 \end{bmatrix}$

(b) $A - B = \begin{bmatrix} -2 & 3 & -2 \\ 3 & -2 & 3 \\ -2 & 3 & -2 \end{bmatrix} - \begin{bmatrix} 3 & -2 & 3 \\ -2 & 3 & -2 \\ 3 & -2 & 3 \end{bmatrix} = \begin{bmatrix} -2-3 & 3-(-2) & -2-3 \\ 3-(-2) & -2-3 & 3-(-2) \\ -2-3 & 3-(-2) & -2-3 \end{bmatrix} = \begin{bmatrix} -5 & 5 & -5 \\ 5 & -5 & 5 \\ -5 & 5 & -5 \end{bmatrix}$

(c) $4A = 4\begin{bmatrix} -2 & 3 & -2 \\ 3 & -2 & 3 \\ -2 & 3 & -2 \end{bmatrix} = \begin{bmatrix} 4(-2) & 4(3) & 4(-2) \\ 4(3) & 4(-2) & 4(3) \\ 4(-2) & 4(3) & 4(-2) \end{bmatrix} = \begin{bmatrix} -8 & 12 & -8 \\ 12 & -8 & 12 \\ -8 & 12 & -8 \end{bmatrix}$

(d) $2A + 2B = 2\begin{bmatrix} -2 & 3 & -2 \\ 3 & -2 & 3 \\ -2 & 3 & -2 \end{bmatrix} + 2\begin{bmatrix} 3 & -2 & 3 \\ -2 & 3 & -2 \\ 3 & -2 & 3 \end{bmatrix} = \begin{bmatrix} -4 & 6 & -4 \\ 6 & -4 & 6 \\ -4 & 6 & -4 \end{bmatrix} + \begin{bmatrix} 6 & -4 & 6 \\ -4 & 6 & -4 \\ 6 & -4 & 6 \end{bmatrix} = \begin{bmatrix} 2 & 2 & 2 \\ 2 & 2 & 2 \\ 2 & 2 & 2 \end{bmatrix}$

43. $\begin{bmatrix} 7 & 3 \\ -1 & 5 \end{bmatrix} + \begin{bmatrix} 10 & -20 \\ 14 & -3 \end{bmatrix} + \begin{bmatrix} 5 & 0 \\ 1 & 9 \end{bmatrix} = \begin{bmatrix} 7+10+5 & 3-20+0 \\ -1+14+1 & 5-3+9 \end{bmatrix} = \begin{bmatrix} 22 & -17 \\ 14 & 11 \end{bmatrix}$

45. $-2\left(\begin{bmatrix} 1 & 2 \\ 5 & -4 \\ 6 & 0 \end{bmatrix} + \begin{bmatrix} 7 & 1 \\ 1 & 2 \\ 1 & 4 \end{bmatrix}\right) = -2\begin{bmatrix} 8 & 3 \\ 6 & -2 \\ 7 & 4 \end{bmatrix} = \begin{bmatrix} -16 & -6 \\ -12 & 4 \\ -14 & -8 \end{bmatrix}$

47. $X = 2A - 3B = 2\begin{bmatrix} -4 & 0 \\ 1 & -5 \\ -3 & 2 \end{bmatrix} - 3\begin{bmatrix} 1 & 2 \\ -2 & 1 \\ 4 & 4 \end{bmatrix} = \begin{bmatrix} -8 & 0 \\ 2 & -10 \\ -6 & 4 \end{bmatrix} + \begin{bmatrix} -3 & -6 \\ 6 & -3 \\ -12 & -12 \end{bmatrix} = \begin{bmatrix} -11 & -6 \\ 8 & -13 \\ -18 & -8 \end{bmatrix}$

49. $X = \frac{1}{3}[B - 2A] = \frac{1}{3}\left(\begin{bmatrix} 1 & 2 \\ -2 & 1 \\ 4 & 4 \end{bmatrix} - 2\begin{bmatrix} -4 & 0 \\ 1 & -5 \\ -3 & 2 \end{bmatrix}\right) = \frac{1}{3}\begin{bmatrix} 9 & 2 \\ -4 & 11 \\ 10 & 0 \end{bmatrix} = \begin{bmatrix} 3 & \frac{2}{3} \\ -\frac{4}{3} & \frac{11}{3} \\ \frac{10}{3} & 0 \end{bmatrix}$

51. A and B are both 2×2, so AB exists and has dimensions 2×2.

$$AB = \begin{bmatrix} 2 & -2 \\ 3 & 5 \end{bmatrix}\begin{bmatrix} -3 & 10 \\ 12 & 8 \end{bmatrix} = \begin{bmatrix} 2(-3) + (-2)(12) & 2(10) + (-2)(8) \\ 3(-3) + 5(12) & 3(10) + 5(8) \end{bmatrix} = \begin{bmatrix} -30 & 4 \\ 51 & 70 \end{bmatrix}$$

53. Because A is 3×2 and B is 2×2, AB exists and has dimensions 3×2.

$$AB = \begin{bmatrix} 5 & 4 \\ -7 & 2 \\ 11 & 2 \end{bmatrix}\begin{bmatrix} 4 & 12 \\ 20 & 40 \end{bmatrix} = \begin{bmatrix} 5(4) + 4(20) & 5(12) + 4(40) \\ -7(4) + 2(20) & -7(12) + 2(40) \\ 11(4) + 2(20) & 11(12) + 2(40) \end{bmatrix} = \begin{bmatrix} 100 & 220 \\ 12 & -4 \\ 84 & 212 \end{bmatrix}$$

55. $\begin{bmatrix} 4 & 1 \\ 11 & -7 \\ 12 & 3 \end{bmatrix}\begin{bmatrix} 3 & -5 & 6 \\ 2 & -2 & -2 \end{bmatrix} = \begin{bmatrix} 14 & -22 & 22 \\ 19 & -41 & 80 \\ 42 & -66 & 66 \end{bmatrix}$

57. Not possible. The number of columns of the first matrix does not equal the number of rows of the second matrix.

59. (a) $AB = \begin{bmatrix} 1 & 3 \\ 4 & 1 \end{bmatrix}\begin{bmatrix} 5 & -1 \\ -2 & 0 \end{bmatrix} = \begin{bmatrix} (1)(5) + (3)(-2) & (1)(-1) + (3)(0) \\ (4)(5) + (1)(-2) & (4)(-1) + (1)(0) \end{bmatrix} = \begin{bmatrix} -1 & -1 \\ 18 & -4 \end{bmatrix}$

(b) $BA = \begin{bmatrix} 5 & -1 \\ -2 & 0 \end{bmatrix}\begin{bmatrix} 1 & 3 \\ 4 & 1 \end{bmatrix} = \begin{bmatrix} (5)(1) + (-1)(4) & (5)(3) + (-1)(1) \\ (-2)(1) + (0)(4) & (-2)(3) + (0)(1) \end{bmatrix} = \begin{bmatrix} 1 & 14 \\ -2 & -6 \end{bmatrix}$

(c) $A^2 = \begin{bmatrix} 1 & 3 \\ 4 & 1 \end{bmatrix}\begin{bmatrix} 1 & 3 \\ 4 & 1 \end{bmatrix} = \begin{bmatrix} (1)(1) + (3)(4) & (1)(3) + (3)(1) \\ (4)(1) + (1)(4) & (4)(3) + (1)(1) \end{bmatrix} = \begin{bmatrix} 13 & 6 \\ 8 & 13 \end{bmatrix}$

In Exercise 61, v $= \langle 2, 5 \rangle = \begin{bmatrix} 2 \\ 5 \end{bmatrix}$

61. $0.95A = 0.95\begin{bmatrix} 80 & 120 & 140 \\ 40 & 100 & 80 \end{bmatrix} = \begin{bmatrix} 76 & 114 & 133 \\ 38 & 95 & 76 \end{bmatrix}$

63. $AB = \begin{bmatrix} -4 & -1 \\ 7 & 2 \end{bmatrix}\begin{bmatrix} -2 & -1 \\ 7 & 4 \end{bmatrix} = \begin{bmatrix} -4(-2) + (-1)(7) & -4(-1) + (-1)(4) \\ 7(-2) + 2(7) & 7(-1) + 2(4) \end{bmatrix} = \begin{bmatrix} 1 & 0 \\ 0 & 1 \end{bmatrix} = I$

$BA = \begin{bmatrix} -2 & -1 \\ 7 & 4 \end{bmatrix}\begin{bmatrix} -4 & -1 \\ 7 & 2 \end{bmatrix} = \begin{bmatrix} -2(-4) + (-1)(7) & -2(-1) + (-1)(2) \\ 7(-4) + 4(7) & 7(-1) + 4(2) \end{bmatrix} = \begin{bmatrix} 1 & 0 \\ 0 & 1 \end{bmatrix} = I$

65. $AB = \begin{bmatrix} 1 & 1 & 0 \\ 1 & 0 & 1 \\ 6 & 2 & 3 \end{bmatrix}\begin{bmatrix} -2 & -3 & 1 \\ 3 & 3 & -1 \\ 2 & 4 & -1 \end{bmatrix} = \begin{bmatrix} 1(-2)+1(3)+0(2) & 1(-3)+1(3)+0(4) & 1(1)+1(-1)+0(-1) \\ 1(-2)+0(3)+1(2) & 1(-3)+0(3)+1(4) & 1(1)+0(-1)+1(-1) \\ 6(-2)+2(3)+3(2) & 6(-3)+2(3)+3(4) & 6(1)+2(-1)+3(-1) \end{bmatrix} = \begin{bmatrix} 1 & 0 & 0 \\ 0 & 1 & 0 \\ 0 & 0 & 1 \end{bmatrix} = I$

$BA = \begin{bmatrix} -2 & -3 & 1 \\ 3 & 3 & -1 \\ 2 & 4 & -1 \end{bmatrix}\begin{bmatrix} 1 & 1 & 0 \\ 1 & 0 & 1 \\ 6 & 2 & 3 \end{bmatrix} = \begin{bmatrix} -2(1)+(-3)(1)+1(6) & -2(1)+(-3)(0)+1(2) & -2(0)+(-3)(1)+1(3) \\ 3(1)+3(1)+(-1)(6) & 3(1)+3(0)+(-1)(2) & 3(0)+3(1)+(-1)(3) \\ 2(1)+4(1)+(-1)(6) & 2(1)+4(0)+(-1)(2) & 2(0)+4(1)+(-1)(3) \end{bmatrix} = \begin{bmatrix} 1 & 0 & 0 \\ 0 & 1 & 0 \\ 0 & 0 & 1 \end{bmatrix} = I$

67. $[A \ \vdots \ I] = \left[\begin{array}{cc:cc} -6 & 5 & 1 & 0 \\ -5 & 4 & 0 & 1 \end{array}\right]$

$-\frac{1}{6}R_1 \rightarrow \left[\begin{array}{cc:cc} 1 & -\frac{5}{6} & -\frac{1}{6} & 0 \\ -5 & 4 & 0 & 1 \end{array}\right]$

$5R_1 + R_2 \rightarrow \left[\begin{array}{cc:cc} 1 & -\frac{5}{6} & -\frac{1}{6} & 0 \\ 0 & -\frac{1}{6} & -\frac{5}{6} & 1 \end{array}\right]$

$-6R_2 \rightarrow \left[\begin{array}{cc:cc} 1 & -\frac{5}{6} & -\frac{1}{6} & 0 \\ 0 & 1 & 5 & -6 \end{array}\right]$

$\frac{5}{6}R_2 + R_1 \rightarrow \left[\begin{array}{cc:cc} 1 & 0 & 4 & -5 \\ 0 & 1 & 5 & -6 \end{array}\right] = [I \ \vdots \ A^{-1}]$

$A^{-1} = \begin{bmatrix} 4 & -5 \\ 5 & -6 \end{bmatrix}$

69. $[A \ \vdots \ I] = \left[\begin{array}{ccc:ccc} 2 & 0 & 3 & 1 & 0 & 0 \\ -1 & 1 & 1 & 0 & 1 & 0 \\ 2 & -2 & 1 & 0 & 0 & 1 \end{array}\right]$

$2R_2 + R_3 \rightarrow \left[\begin{array}{ccc:ccc} 2 & 0 & 3 & 1 & 0 & 0 \\ -1 & 1 & 1 & 0 & 1 & 0 \\ 0 & 0 & 3 & 0 & 2 & 1 \end{array}\right]$

$-R_3 + R_1 \rightarrow \left[\begin{array}{ccc:ccc} 2 & 0 & 0 & 1 & -2 & -1 \\ -1 & 1 & 1 & 0 & 1 & 0 \\ 0 & 0 & 3 & 0 & 2 & 1 \end{array}\right]$

$\begin{array}{l} \frac{1}{2}R_1 \rightarrow \\ \\ \frac{1}{3}R_3 \rightarrow \end{array} \left[\begin{array}{ccc:ccc} 1 & 0 & 0 & \frac{1}{2} & -1 & -\frac{1}{2} \\ -1 & 1 & 1 & 0 & 1 & 0 \\ 0 & 0 & 1 & 0 & \frac{2}{3} & \frac{1}{3} \end{array}\right]$

$R_1 + R_2 \rightarrow \left[\begin{array}{ccc:ccc} 1 & 0 & 0 & \frac{1}{2} & -1 & -\frac{1}{2} \\ 0 & 1 & 1 & \frac{1}{2} & 0 & -\frac{1}{2} \\ 0 & 0 & 1 & 0 & \frac{2}{3} & \frac{1}{3} \end{array}\right]$

$-R_3 + R_2 \rightarrow \left[\begin{array}{ccc:ccc} 1 & 0 & 0 & \frac{1}{2} & -1 & -\frac{1}{2} \\ 0 & 1 & 0 & \frac{1}{2} & -\frac{2}{3} & -\frac{5}{6} \\ 0 & 0 & 1 & 0 & \frac{2}{3} & \frac{1}{3} \end{array}\right] = [I \ \vdots \ A^{-1}]$

$A^{-1} = \begin{bmatrix} \frac{1}{2} & -1 & -\frac{1}{2} \\ \frac{1}{2} & -\frac{2}{3} & -\frac{5}{6} \\ 0 & \frac{2}{3} & \frac{1}{3} \end{bmatrix}$

71. $\begin{bmatrix} -1 & -2 & -2 \\ 3 & 7 & 9 \\ 1 & 4 & 7 \end{bmatrix}^{-1} = \begin{bmatrix} 13 & 6 & -4 \\ -12 & -5 & 3 \\ 5 & 2 & -1 \end{bmatrix}$

73. $A = \begin{bmatrix} -7 & 2 \\ -8 & 2 \end{bmatrix}$

$A^{-1} = \dfrac{1}{-7(2) - 2(-8)}\begin{bmatrix} 2 & -2 \\ 8 & -7 \end{bmatrix} = \dfrac{1}{2}\begin{bmatrix} 2 & -2 \\ 8 & -7 \end{bmatrix} = \begin{bmatrix} 1 & -1 \\ 4 & -\frac{7}{2} \end{bmatrix}$

75. $A = \begin{bmatrix} -12 & 6 \\ 10 & -5 \end{bmatrix}$

$ad - bc = (-12)(-5) - (6)(10) = 0$

A^{-1} does not exist.

77. $\begin{cases} -x + 4y = 8 \\ 2x - 7y = -5 \end{cases}$

$\begin{bmatrix} x \\ y \end{bmatrix} = \begin{bmatrix} -1 & 4 \\ 2 & -7 \end{bmatrix}^{-1}\begin{bmatrix} 8 \\ -5 \end{bmatrix} = \begin{bmatrix} 7 & 4 \\ 2 & 1 \end{bmatrix}\begin{bmatrix} 8 \\ -5 \end{bmatrix}$

$= \begin{bmatrix} 7(8) + 4(-5) \\ 2(8) + 1(-5) \end{bmatrix} = \begin{bmatrix} 36 \\ 11 \end{bmatrix}$

Solution: $(36, 11)$

79. $\begin{cases} -3x + 10y = 8 \\ 5x - 17y = -13 \end{cases}$

$\begin{bmatrix} x \\ y \end{bmatrix} = \begin{bmatrix} -3 & 10 \\ 5 & -17 \end{bmatrix}^{-1}\begin{bmatrix} 8 \\ -13 \end{bmatrix} = \begin{bmatrix} -17 & -10 \\ -5 & -3 \end{bmatrix}\begin{bmatrix} 8 \\ -13 \end{bmatrix}$

$= \begin{bmatrix} -17(8) + (-10)(-13) \\ -5(8) + (-3)(-13) \end{bmatrix} = \begin{bmatrix} -6 \\ -1 \end{bmatrix}$

Solution: $(-6, -1)$

81. $\begin{cases} \frac{1}{2}x + \frac{1}{3}y = 2 \\ -3x + 2y = 0 \end{cases}$

$\begin{bmatrix} x \\ y \end{bmatrix} = \begin{bmatrix} \frac{1}{2} & \frac{1}{3} \\ -3 & 2 \end{bmatrix}^{-1}\begin{bmatrix} 2 \\ 0 \end{bmatrix} = \begin{bmatrix} 1 & -\frac{1}{6} \\ \frac{3}{2} & \frac{1}{4} \end{bmatrix}\begin{bmatrix} 2 \\ 0 \end{bmatrix} = \begin{bmatrix} 2 \\ 3 \end{bmatrix}$

Solution: $(2, 3)$

83. $\begin{cases} 0.3x + 0.7y = 10.2 \\ 0.4x + 0.6y = 7.6 \end{cases}$

$$\begin{bmatrix} x \\ y \end{bmatrix} = \begin{bmatrix} 0.3 & 0.7 \\ 0.4 & 0.6 \end{bmatrix}^{-1} \begin{bmatrix} 10.2 \\ 7.6 \end{bmatrix} = \begin{bmatrix} -6 & 7 \\ 4 & -3 \end{bmatrix} \begin{bmatrix} 10.2 \\ 7.6 \end{bmatrix} = \begin{bmatrix} -8 \\ 18 \end{bmatrix}$$

Solution: $(-8, 18)$

85. $\begin{cases} 3x + 2y - z = 6 \\ x - y + 2z = -1 \\ 5x + y + z = 7 \end{cases}$

$$\begin{bmatrix} x \\ y \\ z \end{bmatrix} = \begin{bmatrix} 3 & 2 & -1 \\ 1 & -1 & 2 \\ 5 & 1 & 1 \end{bmatrix}^{-1} \begin{bmatrix} 6 \\ -1 \\ 7 \end{bmatrix} = \begin{bmatrix} -1 & -1 & 1 \\ 3 & \frac{8}{3} & -\frac{7}{3} \\ 2 & \frac{7}{3} & -\frac{5}{3} \end{bmatrix} \begin{bmatrix} 6 \\ -1 \\ 7 \end{bmatrix}$$

$$= \begin{bmatrix} -1(6) - 1(-1) + 1(7) \\ 3(6) + \frac{8}{3}(-1) - \frac{7}{3}(7) \\ 2(6) + \frac{7}{3}(-1) - \frac{5}{3}(7) \end{bmatrix} = \begin{bmatrix} 2 \\ -1 \\ -2 \end{bmatrix}$$

Solution: $(2, -1, -2)$

87. $\begin{cases} x + 2y = -1 \\ 3x + 4y = -5 \end{cases}$

$$\begin{bmatrix} x \\ y \end{bmatrix} = \begin{bmatrix} 1 & 2 \\ 3 & 4 \end{bmatrix}^{-1} \begin{bmatrix} -1 \\ -5 \end{bmatrix} = \begin{bmatrix} -2 & 1 \\ \frac{3}{2} & -\frac{1}{2} \end{bmatrix} \begin{bmatrix} -1 \\ -5 \end{bmatrix} = \begin{bmatrix} -3 \\ 1 \end{bmatrix}$$

Solution: $(-3, 1)$

89. $\begin{cases} \frac{6}{5}x - \frac{4}{7}y = \frac{6}{5} \\ -\frac{12}{5}x + \frac{12}{7}y = -\frac{17}{5} \end{cases}$

$$\begin{bmatrix} x \\ y \end{bmatrix} = \begin{bmatrix} \frac{6}{5} & -\frac{4}{7} \\ -\frac{12}{5} & \frac{12}{7} \end{bmatrix}^{-1} \begin{bmatrix} \frac{6}{5} \\ -\frac{17}{5} \end{bmatrix} = \begin{bmatrix} \frac{5}{2} & \frac{5}{6} \\ \frac{7}{2} & \frac{7}{4} \end{bmatrix} \begin{bmatrix} \frac{6}{5} \\ -\frac{17}{5} \end{bmatrix} = \begin{bmatrix} \frac{1}{6} \\ -\frac{7}{4} \end{bmatrix}$$

Solution: $\left(\frac{1}{6}, -\frac{7}{4}\right)$

91. $A = \begin{bmatrix} 2 & 5 \\ -4 & 3 \end{bmatrix} : \begin{vmatrix} 2 & 5 \\ -4 & 3 \end{vmatrix} = (2)(3) - (-4)(5) = 26$

93. $A = \begin{bmatrix} 10 & -2 \\ 18 & 8 \end{bmatrix} : \begin{vmatrix} 10 & -2 \\ 18 & 8 \end{vmatrix} = (10)(8) - (18)(-2) = 116$

95. $\begin{bmatrix} 2 & -1 \\ 7 & 4 \end{bmatrix}$

(a) $M_{11} = 4$
$M_{12} = 7$
$M_{21} = -1$
$M_{22} = 2$

(b) $C_{11} = M_{11} = 4$
$C_{12} = -M_{12} = -7$
$C_{21} = -M_{21} = 1$
$C_{22} = M_{22} = 2$

97. $\begin{bmatrix} 3 & 2 & -1 \\ -2 & 5 & 0 \\ 1 & 8 & 6 \end{bmatrix}$

(a) $M_{11} = \begin{vmatrix} 5 & 0 \\ 8 & 6 \end{vmatrix} = 30$

$M_{12} = \begin{vmatrix} -2 & 0 \\ 1 & 6 \end{vmatrix} = -12$

$M_{13} = \begin{vmatrix} -2 & 5 \\ 1 & 8 \end{vmatrix} = -21$

$M_{21} = \begin{vmatrix} 2 & -1 \\ 8 & 6 \end{vmatrix} = 20$

$M_{22} = \begin{vmatrix} 3 & -1 \\ 1 & 6 \end{vmatrix} = 19$

$M_{23} = \begin{vmatrix} 3 & 2 \\ 1 & 8 \end{vmatrix} = 22$

$M_{31} = \begin{vmatrix} 2 & -1 \\ 5 & 0 \end{vmatrix} = 5$

$M_{32} = \begin{vmatrix} 3 & -1 \\ -2 & 0 \end{vmatrix} = -2$

$M_{33} = \begin{vmatrix} 3 & 2 \\ -2 & 5 \end{vmatrix} = 19$

(b) $C_{11} = M_{11} = 30$
$C_{12} = -M_{12} = 12$
$C_{13} = M_{13} = -21$
$C_{21} = -M_{21} = -20$
$C_{22} = M_{22} = 19$
$C_{23} = -M_{23} = -22$
$C_{31} = M_{31} = 5$
$C_{32} = -M_{32} = 2$
$C_{33} = M_{33} = 19$

99. Expand using Row 1.

$$\begin{vmatrix} -2 & 0 & 0 \\ 2 & -1 & 0 \\ -1 & 1 & -3 \end{vmatrix} = -2\begin{vmatrix} -1 & 0 \\ 1 & -3 \end{vmatrix} - 0\begin{vmatrix} 2 & 0 \\ -1 & 3 \end{vmatrix} + 0\begin{vmatrix} 2 & -1 \\ -1 & 1 \end{vmatrix}$$

$= -2(3) - 0(6) + 0(1)$

$= -6$

101. Expand using Row 3.

$$\begin{vmatrix} 4 & 1 & -1 \\ 2 & 3 & 2 \\ 1 & -1 & 0 \end{vmatrix} = 1\begin{vmatrix} 1 & -1 \\ 3 & 2 \end{vmatrix} + 1\begin{vmatrix} 4 & -1 \\ 2 & 2 \end{vmatrix} + 0\begin{vmatrix} 4 & 1 \\ 2 & 3 \end{vmatrix}$$

$= 1(5) + 1(10) + 0(10)$

$= 15$

103. Expand using Column 2.

$$\begin{vmatrix} -2 & 4 & 1 \\ -6 & 0 & 2 \\ 5 & 3 & 4 \end{vmatrix} = -4\begin{vmatrix} -6 & 2 \\ 5 & 4 \end{vmatrix} - 3\begin{vmatrix} -2 & 1 \\ -6 & 2 \end{vmatrix}$$

$$= -4(-34) - 3(2) = 130$$

105. $\begin{cases} 5x - 2y = 6 \\ -11x + 3y = -23 \end{cases}$

$$x = \frac{\begin{vmatrix} 6 & -2 \\ -23 & 3 \end{vmatrix}}{\begin{vmatrix} 5 & -2 \\ -11 & 3 \end{vmatrix}} = \frac{-28}{-7} = 4$$

$$y = \frac{\begin{vmatrix} 5 & 6 \\ -11 & -23 \end{vmatrix}}{\begin{vmatrix} 5 & -2 \\ -11 & 3 \end{vmatrix}} = \frac{-49}{-7} = 7$$

Solution: $(4, 7)$

107. $\begin{cases} -2x + 3y - 5z = -11 \\ 4x - y + z = -3 \\ -x - 4y + 6z = 15 \end{cases}$

$$D = \begin{vmatrix} -2 & 3 & -5 \\ 4 & -1 & 1 \\ -1 & -4 & 6 \end{vmatrix} = -2(-1)^2\begin{vmatrix} -1 & 1 \\ -4 & 6 \end{vmatrix} + 4(-1)^3\begin{vmatrix} 3 & -5 \\ -4 & 6 \end{vmatrix} - 1(-1)^4\begin{vmatrix} 3 & -5 \\ -1 & 1 \end{vmatrix} = -2(-2) - 4(-2) - (-2) = 14$$

$$x = \frac{\begin{vmatrix} -11 & 3 & -5 \\ -3 & -1 & 1 \\ 15 & -4 & 6 \end{vmatrix}}{14} = \frac{-11(-1)^2\begin{vmatrix} -1 & 1 \\ -4 & 6 \end{vmatrix} - 3(-1)^3\begin{vmatrix} 3 & -5 \\ -4 & 6 \end{vmatrix} + 15(-1)^4\begin{vmatrix} 3 & -5 \\ -1 & 1 \end{vmatrix}}{14} = \frac{-11(-2) + 3(-2) + 15(-2)}{14} = \frac{-14}{14} = -1$$

$$y = \frac{\begin{vmatrix} -2 & -11 & -5 \\ 4 & -3 & 1 \\ -1 & 15 & 6 \end{vmatrix}}{14} = \frac{-2(-1)^2\begin{vmatrix} -3 & 1 \\ 15 & 6 \end{vmatrix} + 4(-1)^3\begin{vmatrix} -11 & -5 \\ 15 & 6 \end{vmatrix} - 1(-1)^4\begin{vmatrix} -11 & -5 \\ -3 & 1 \end{vmatrix}}{14} = \frac{-2(-33) - 4(9) - 1(-26)}{14} = \frac{56}{14} = 4$$

$$z = \frac{\begin{vmatrix} -2 & 3 & -11 \\ 4 & -1 & -3 \\ -1 & -4 & 15 \end{vmatrix}}{14} = \frac{-2(-1)^2\begin{vmatrix} -1 & -3 \\ -4 & 15 \end{vmatrix} + 4(-1)^3\begin{vmatrix} 3 & -11 \\ -4 & 15 \end{vmatrix} - 1(-1)^4\begin{vmatrix} 3 & -11 \\ -1 & -3 \end{vmatrix}}{14} = \frac{-2(-27) - 4(1) - 1(-20)}{14} = \frac{70}{14} = 5$$

Solution: $(-1, 4, 5)$

109. $(1, 0), (5, 0), (5, 8)$

$$\text{Area} = \frac{1}{2}\begin{vmatrix} 1 & 0 & 1 \\ 5 & 0 & 1 \\ 5 & 8 & 1 \end{vmatrix} = \frac{1}{2}\left(1\begin{vmatrix} 0 & 1 \\ 8 & 1 \end{vmatrix} + 1\begin{vmatrix} 5 & 0 \\ 5 & 8 \end{vmatrix}\right) = \frac{1}{2}(-8 + 40) = \frac{1}{2}(32) = 16 \text{ square units}$$

111. $(-1, 7), (3, -9), (-3, 15)$

$$\begin{vmatrix} -1 & 7 & 1 \\ 3 & -9 & 1 \\ -3 & 15 & 1 \end{vmatrix} = \begin{vmatrix} 3 & -9 \\ -3 & 15 \end{vmatrix} - \begin{vmatrix} -1 & 7 \\ -3 & 15 \end{vmatrix} + \begin{vmatrix} -1 & 7 \\ 3 & -9 \end{vmatrix} = 18 - 6 - 12 = 0$$

The points are collinear.

113. $(-4, 0), (4, 4)$

$$\begin{vmatrix} x & y & 1 \\ -4 & 0 & 1 \\ 4 & 4 & 1 \end{vmatrix} = 0$$

$$1\begin{vmatrix} -4 & 0 \\ 4 & 4 \end{vmatrix} - 1\begin{vmatrix} x & y \\ 4 & 4 \end{vmatrix} + 1\begin{vmatrix} x & y \\ -4 & 0 \end{vmatrix} = 0$$

$$-16 - (4x - 4y) + 4y = 0$$
$$-4x + 8y - 16 = 0$$
$$x - 2y + 4 = 0$$

115. $\left(-\frac{5}{2}, 3\right), \left(\frac{7}{2}, 1\right)$

$$\begin{vmatrix} x & y & 1 \\ -\frac{5}{2} & 3 & 1 \\ \frac{7}{2} & 1 & 1 \end{vmatrix} = 0$$

$$1\begin{vmatrix} -\frac{5}{2} & 3 \\ \frac{7}{2} & 1 \end{vmatrix} - 1\begin{vmatrix} x & y \\ \frac{7}{2} & 1 \end{vmatrix} + 1\begin{vmatrix} x & y \\ -\frac{5}{2} & 3 \end{vmatrix} = 0$$

$$-13 - \left(x - \frac{7}{2}y\right) + \left(3x + \frac{5}{2}y\right) = 0$$
$$2x + 6y - 13 = 0$$

117. The area of the parallelogram with vertices: $(0, 0)$, $(2, 0)$, $(1, 4)$ and $(3, 4) \Rightarrow a = 2, b = 0, c = 1$ and $d = 4$.

$$A = \left\| \begin{matrix} 2 & 0 \\ 1 & 4 \end{matrix} \right\| = |8 - 0| = 8 \text{ square units.}$$

119. $A^{-1} = \begin{bmatrix} -1 & 2 & -3 \\ 2 & 1 & 0 \\ 4 & -2 & 5 \end{bmatrix}$

$$\begin{bmatrix} -5 & 11 & -2 \end{bmatrix}\begin{bmatrix} -1 & 2 & -3 \\ 2 & 1 & 0 \\ 4 & -2 & 5 \end{bmatrix} = \begin{bmatrix} 19 & 5 & 5 \end{bmatrix} \quad \text{S E E}$$

$$\begin{bmatrix} 370 & -265 & 225 \end{bmatrix}\begin{bmatrix} -1 & 2 & -3 \\ 2 & 1 & 0 \\ 4 & -2 & 5 \end{bmatrix} = \begin{bmatrix} 0 & 25 & 15 \end{bmatrix} \quad \text{\_ Y O}$$

$$\begin{bmatrix} -57 & 48 & -33 \end{bmatrix}\begin{bmatrix} -1 & 2 & -3 \\ 2 & 1 & 0 \\ 4 & -2 & 5 \end{bmatrix} = \begin{bmatrix} 21 & 0 & 6 \end{bmatrix} \quad \text{U \_ F}$$

$$\begin{bmatrix} 32 & -15 & 20 \end{bmatrix}\begin{bmatrix} -1 & 2 & -3 \\ 2 & 1 & 0 \\ 4 & -2 & 5 \end{bmatrix} = \begin{bmatrix} 18 & 9 & 4 \end{bmatrix} \quad \text{R I D}$$

$$\begin{bmatrix} 245 & -171 & 147 \end{bmatrix}\begin{bmatrix} -1 & 2 & -3 \\ 2 & 1 & 0 \\ 4 & -2 & 5 \end{bmatrix} = \begin{bmatrix} 1 & 25 & 0 \end{bmatrix} \quad \text{A Y \_}$$

Message: SEE YOU FRIDAY

121. False. The matrix must be square.

123. If A is a square matrix, the cofactor C_{ij} of the entry a_{ij} is $(-1)^{i+j}M_{ij}$, where M_{ij} is the determinant obtained by deleting the ith row and jth column of A. The determinant of A is the sum of the entries of any row or column of A multiplied by their respective cofactors.

Problem Solving for Chapter 7

1. $A = \begin{bmatrix} 0 & -1 \\ 1 & 0 \end{bmatrix} \quad T = \begin{bmatrix} 1 & 2 & 3 \\ 1 & 4 & 2 \end{bmatrix}$

(a) $AT = \begin{bmatrix} -1 & -4 & -2 \\ 1 & 2 & 3 \end{bmatrix} \quad AAT = \begin{bmatrix} -1 & -2 & -3 \\ -1 & -4 & -2 \end{bmatrix}$

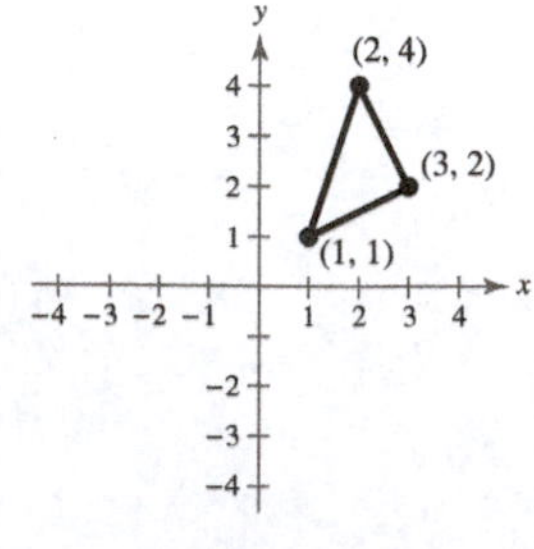

Original Triangle

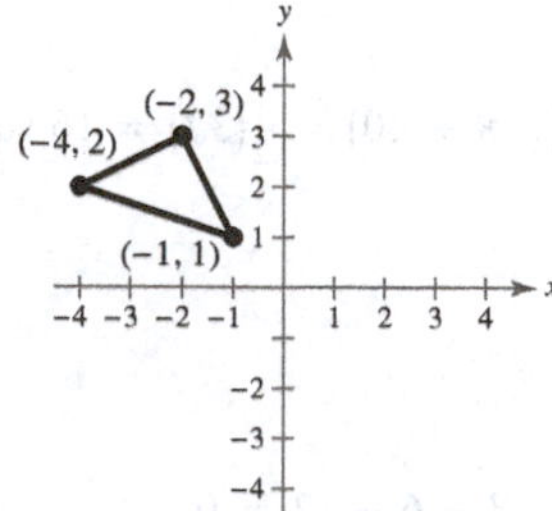

AT Triangle

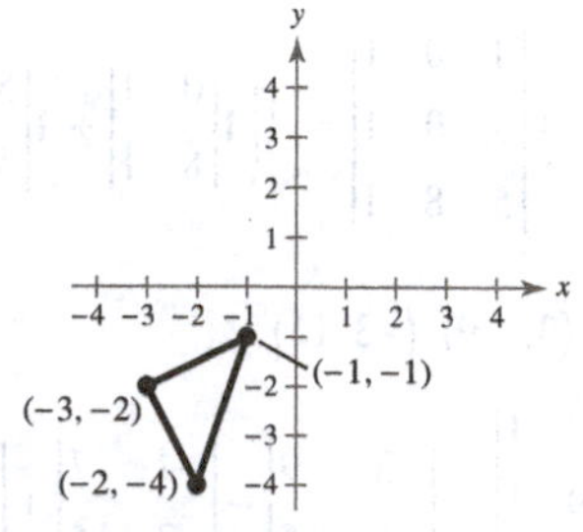

AAT Triangle

The transformation A interchanges the x and y coordinates and then takes the negative of the x coordinate. A represents a counterclockwise rotation by 90°.

(b) AAT is rotated clockwise 90° to obtain AT. AT is then rotated clockwise 90° to obtain T.

3. (a) $A^2 = \begin{bmatrix} 1 & 0 \\ 0 & 0 \end{bmatrix}\begin{bmatrix} 1 & 0 \\ 0 & 0 \end{bmatrix} = \begin{bmatrix} 1 & 0 \\ 0 & 0 \end{bmatrix} = A$

A *is* idempotent.

(b) $A^2 = \begin{bmatrix} 0 & 1 \\ 1 & 0 \end{bmatrix}\begin{bmatrix} 0 & 1 \\ 1 & 0 \end{bmatrix} = \begin{bmatrix} 1 & 0 \\ 0 & 1 \end{bmatrix} \neq A$

A is *not* idempotent.

(c) $A^2 = \begin{bmatrix} 2 & 3 \\ -1 & -2 \end{bmatrix}\begin{bmatrix} 2 & 3 \\ -1 & -2 \end{bmatrix} = \begin{bmatrix} 1 & 0 \\ 0 & 1 \end{bmatrix} \neq A$

A is *not* idempotent.

(d) $A^2 = \begin{bmatrix} 2 & 3 \\ 1 & 2 \end{bmatrix}\begin{bmatrix} 2 & 3 \\ 1 & 2 \end{bmatrix} = \begin{bmatrix} 7 & 12 \\ 4 & 7 \end{bmatrix} \neq A$

A is *not* idempotent.

(e) $A^2 = \begin{bmatrix} 0 & 0 & 1 \\ 0 & 1 & 0 \\ 1 & 0 & 0 \end{bmatrix}\begin{bmatrix} 0 & 0 & 1 \\ 0 & 1 & 0 \\ 1 & 0 & 0 \end{bmatrix} = \begin{bmatrix} 1 & 0 & 0 \\ 0 & 1 & 0 \\ 0 & 0 & 1 \end{bmatrix} \neq A$

A is *not* idempotent.

(f) $A^2 = \begin{bmatrix} 0 & 1 & 0 \\ 1 & 0 & 0 \\ 0 & 0 & 1 \end{bmatrix}\begin{bmatrix} 0 & 1 & 0 \\ 1 & 0 & 0 \\ 0 & 0 & 1 \end{bmatrix} = \begin{bmatrix} 1 & 0 & 0 \\ 0 & 1 & 0 \\ 0 & 0 & 1 \end{bmatrix} \neq A$

A is *not* idempotent.

5. $A = \begin{bmatrix} 1 & 2 \\ -2 & 1 \end{bmatrix}$

(a) $A^2 - 2A + 5I = \begin{bmatrix} 1 & 2 \\ -2 & 1 \end{bmatrix}\begin{bmatrix} 1 & 2 \\ -2 & 1 \end{bmatrix} - 2\begin{bmatrix} 1 & 2 \\ -2 & 1 \end{bmatrix} + 5\begin{bmatrix} 1 & 0 \\ 0 & 1 \end{bmatrix}$

$= \begin{bmatrix} -3 & 4 \\ -4 & -3 \end{bmatrix} + \begin{bmatrix} -2 & -4 \\ 4 & -2 \end{bmatrix} + \begin{bmatrix} 5 & 0 \\ 0 & 5 \end{bmatrix}$

$= \begin{bmatrix} 0 & 0 \\ 0 & 0 \end{bmatrix} = 0$

(b) $A^{-1} = \dfrac{1}{(1) - (-4)}\begin{bmatrix} 1 & -2 \\ 2 & 1 \end{bmatrix} = \dfrac{1}{5}\begin{bmatrix} 1 & -2 \\ 2 & 1 \end{bmatrix}$

$\dfrac{1}{5}(2I - A) = \dfrac{1}{5}\left[\begin{bmatrix} 2 & 0 \\ 0 & 2 \end{bmatrix} - \begin{bmatrix} 1 & 2 \\ -2 & 1 \end{bmatrix}\right] = \dfrac{1}{5}\begin{bmatrix} 1 & -2 \\ 2 & 1 \end{bmatrix}$

So, $A^{-1} = \dfrac{1}{5}(2I - A)$.

(c) $A^2 - 2A + 5I = 0$

$A^2 - 2A = -5I$

$(A - 2I)A = -5I$

$-\dfrac{1}{5}(A - 2I)A = I$

$\dfrac{1}{5}(2I - A)A = I$

So, $A^{-1} = \dfrac{1}{5}(2I - A)$.

7. $A = \begin{bmatrix} -1 & 1 & -2 \\ 2 & 0 & 1 \end{bmatrix}$, $B = \begin{bmatrix} -3 & 0 \\ 1 & 2 \\ 1 & -1 \end{bmatrix}$

$A^T = \begin{bmatrix} -1 & 2 \\ 1 & 0 \\ -2 & 1 \end{bmatrix}$, $B^T = \begin{bmatrix} -3 & 1 & 1 \\ 0 & 2 & -1 \end{bmatrix}$

$AB = \begin{bmatrix} 2 & 4 \\ -5 & -1 \end{bmatrix}$, $(AB)^T = \begin{bmatrix} 2 & -5 \\ 4 & -1 \end{bmatrix}$

$B^TA^T = \begin{bmatrix} -3 & 1 & 1 \\ 0 & 2 & -1 \end{bmatrix}\begin{bmatrix} -1 & 2 \\ 1 & 0 \\ -2 & 1 \end{bmatrix} = \begin{bmatrix} 2 & -5 \\ 4 & -1 \end{bmatrix}$

So, $(AB)^T = B^TA^T$.

9. If $A = \begin{bmatrix} 4 & x \\ -2 & -3 \end{bmatrix}$ is singular then

$ad - bc = -12 + 2x = 0$.

So, $x = 6$.

11. $(a - b)(b - c)(c - a)(a + b + c) = -a^3b + a^3c + ab^3 - ac^3 - b^3c + bc^3$

$$\begin{vmatrix} 1 & 1 & 1 \\ a & b & c \\ a^3 & b^3 & c^3 \end{vmatrix} = \begin{vmatrix} b & c \\ b^3 & c^3 \end{vmatrix} - \begin{vmatrix} a & c \\ a^3 & c^3 \end{vmatrix} + \begin{vmatrix} a & b \\ a^3 & b^3 \end{vmatrix} = bc^3 - b^3c - ac^3 + a^3c + ab^3 - a^3b$$

So, $\begin{vmatrix} 1 & 1 & 1 \\ a & b & c \\ a^3 & b^3 & c^3 \end{vmatrix} = (a - b)(b - c)(c - a)(a + b + c).$

13. $$\begin{vmatrix} x & 0 & 0 & d \\ -1 & x & 0 & c \\ 0 & -1 & x & b \\ 0 & 0 & -1 & a \end{vmatrix} = x\begin{vmatrix} x & 0 & c \\ -1 & x & b \\ 0 & -1 & a \end{vmatrix} - d\begin{vmatrix} -1 & x & 0 \\ 0 & -1 & x \\ 0 & 0 & -1 \end{vmatrix} = x\underbrace{(ax^2 + bx + c)}_{\text{From Exercise 12}} - d\left(-\begin{vmatrix} -1 & x \\ 0 & -1 \end{vmatrix}\right) = ax^3 + bx^2 + cx + d$$

15.
$$\begin{aligned} 4S + 4N &= 184 \\ S + 6F &= 146 \\ 2N + 4F &= 104 \end{aligned}$$

$$D = \begin{vmatrix} 4 & 4 & 0 \\ 1 & 0 & 6 \\ 0 & 2 & 4 \end{vmatrix} = -64$$

$$S = \frac{\begin{vmatrix} 184 & 4 & 0 \\ 146 & 0 & 6 \\ 104 & 2 & 4 \end{vmatrix}}{-64} = \frac{-2048}{-64} = 32$$

$$N = \frac{\begin{vmatrix} 4 & 184 & 0 \\ 1 & 146 & 6 \\ 0 & 104 & 4 \end{vmatrix}}{-64} = \frac{-896}{-64} = 14$$

$$F = \frac{\begin{vmatrix} 4 & 4 & 184 \\ 1 & 0 & 146 \\ 0 & 2 & 104 \end{vmatrix}}{-64} = \frac{-1216}{-64} = 19$$

Element	*Atomic mass*
Sulfur	32
Nitrogen	14
Fluoride	19

17. $A = \begin{bmatrix} 1 & -2 & 2 \\ 1 & 1 & -3 \\ 1 & -1 & 4 \end{bmatrix} \Rightarrow A^{-1} = \begin{bmatrix} \frac{1}{11} & \frac{6}{11} & \frac{4}{11} \\ -\frac{7}{11} & \frac{2}{11} & \frac{5}{11} \\ -\frac{2}{11} & -\frac{1}{11} & \frac{3}{11} \end{bmatrix}$

$$\begin{bmatrix} 23 & 13 & -34 \\ 31 & -34 & 63 \\ 25 & -17 & 61 \\ 24 & 14 & -37 \\ 41 & -17 & -8 \\ 20 & -29 & 40 \\ 38 & -56 & 116 \\ 13 & -11 & 1 \\ 22 & -3 & -6 \\ 41 & -53 & 85 \\ 28 & -32 & 16 \end{bmatrix} \begin{bmatrix} \frac{1}{11} & \frac{6}{11} & \frac{4}{11} \\ -\frac{7}{11} & \frac{2}{11} & \frac{5}{11} \\ -\frac{2}{11} & -\frac{1}{11} & \frac{3}{11} \end{bmatrix} = \begin{bmatrix} 0 & 18 & 5 \\ 13 & 5 & 13 \\ 2 & 5 & 18 \\ 0 & 19 & 5 \\ 16 & 20 & 5 \\ 13 & 2 & 5 \\ 18 & 0 & 20 \\ 8 & 5 & 0 \\ 5 & 12 & 5 \\ 22 & 5 & 14 \\ 20 & 8 & 0 \end{bmatrix}$$

0 18 5 13 5 13 2 5 18 0
_ R E M E M B E R _
19 5 16 20 5 13 2 5 18 0
S E P T E M B E R _
20 8 5 0 5 12 5 22 5 14 20 8 0
T H E _ E L E V E N T H _

Message: REMEMBER SEPTEMBER THE ELEVENTH

19. $A = \begin{bmatrix} 6 & 4 & 1 \\ 0 & 2 & 3 \\ 1 & 1 & 2 \end{bmatrix} \qquad A^{-1} = \begin{bmatrix} \frac{1}{16} & -\frac{7}{16} & \frac{5}{8} \\ \frac{3}{16} & \frac{11}{16} & -\frac{9}{8} \\ -\frac{1}{8} & -\frac{1}{8} & \frac{3}{4} \end{bmatrix}$

$|A| = 16$ and $\left|A^{-1}\right| = \dfrac{1}{16}$

Conjecture: $\left|A^{-1}\right| = \dfrac{1}{|A|}$

Practice Test for Chapter 7

1. Put the matrix in reduced row-echelon form.

$$\begin{bmatrix} 1 & -2 & 4 \\ 3 & -5 & 9 \end{bmatrix}$$

For Exercises 2–4, use matrices to solve the system of equations.

2. $\begin{cases} 3x + 5y = 3 \\ 2x - y = -11 \end{cases}$

3. $\begin{cases} 2x + 3y = -3 \\ 3x + 2y = 8 \\ x + y = 1 \end{cases}$

4. $\begin{cases} x + 3z = -5 \\ 2x + y = 0 \\ 3x + y - z = 3 \end{cases}$

5. Multiply $\begin{bmatrix} 1 & 4 & 5 \\ 2 & 0 & -3 \end{bmatrix}\begin{bmatrix} 1 & 6 \\ 0 & -7 \\ -1 & 2 \end{bmatrix}$.

6. Given $A = \begin{bmatrix} 9 & 1 \\ -4 & 8 \end{bmatrix}$ and $B = \begin{bmatrix} 6 & -2 \\ 3 & 5 \end{bmatrix}$, find $3A - 5B$.

7. Find $f(A)$.

$f(x) = x^2 - 7x + 8$, $A = \begin{bmatrix} 3 & 0 \\ 7 & 1 \end{bmatrix}$

8. True or false:

$(A + B)(A + 3B) = A^2 + 4AB + 3B^2$ where A and B are matrices.

(Assume that A^2, AB, and B^2 exist.)

For Exercises 9–10, find the inverse of the matrix, if it exists.

9. $\begin{bmatrix} 1 & 2 \\ 3 & 5 \end{bmatrix}$

10. $\begin{bmatrix} 1 & 1 & 1 \\ 3 & 6 & 5 \\ 6 & 10 & 8 \end{bmatrix}$

11. Use an inverse matrix to solve the systems.

(a) $\begin{cases} x + 2y = 4 \\ 3x + 5y = 1 \end{cases}$

(b) $\begin{cases} x + 2y = 3 \\ 3x + 5y = -2 \end{cases}$

For Exercises 12–14, find the determinant of the matrix.

12. $\begin{bmatrix} 6 & -1 \\ 3 & 4 \end{bmatrix}$

13. $\begin{bmatrix} 1 & 3 & -1 \\ 5 & 9 & 0 \\ 6 & 2 & -5 \end{bmatrix}$

14. $\begin{bmatrix} 1 & 4 & 2 & 3 \\ 0 & 1 & -2 & 0 \\ 3 & 5 & -1 & 1 \\ 2 & 0 & 6 & 1 \end{bmatrix}$

15. Evaluate $\begin{vmatrix} 6 & 4 & 3 & 0 & 6 \\ 0 & 5 & 1 & 4 & 8 \\ 0 & 0 & 2 & 7 & 3 \\ 0 & 0 & 0 & 9 & 2 \\ 0 & 0 & 0 & 0 & 1 \end{vmatrix}$.

16. Use a determinant to find the area of the triangle with vertices $(0, 7)$, $(5, 0)$, and $(3, 9)$.

17. Use a determinant to find the equation of the line passing through $(2, 7)$ and $(-1, 4)$.

For Exercises 18–20, use Cramer's Rule to find the indicated value.

18. Find x.

$$\begin{cases} 6x - 7y = 4 \\ 2x + 5y = 11 \end{cases}$$

19. Find z.

$$\begin{cases} 3x + z = 1 \\ y + 4z = 3 \\ x - y = 2 \end{cases}$$

20. Find y.

$$\begin{cases} 721.4x - 29.1y = 33.77 \\ 45.9x + 105.6y = 19.85 \end{cases}$$

CHAPTER 8
Sequences, Series, and Probability

CHAPTER 8
Sequences, Series, and Probability

Section 8.1 Sequences and Series

1. infinite sequence

3. recursively

5. index; upper; lower

7. $a_n = 4n - 7$

$a_1 = 4(1) - 7 = -3$

$a_2 = 4(2) - 7 = 1$

$a_3 = 4(3) - 7 = 5$

$a_4 = 4(4) - 7 = 9$

$a_5 = 4(5) - 7 = 13$

9. $a_n = (-1)^{n+1} + 4$

$a_1 = (-1)^{1+1} + 4 = 5$

$a_2 = (-1)^{2+1} + 4 = 3$

$a_3 = (-1)^{3+1} + 4 = 5$

$a_4 = (-1)^{4+1} + 4 = 3$

$a_5 = (-1)^{5+1} + 4 = 5$

11. $a_n = (-2)^n$

$a_1 = (-2)^1 = -2$

$a_2 = (-2)^2 = 4$

$a_3 = (-2)^3 = -8$

$a_4 = (-2)^4 = 16$

$a_5 = (-2)^5 = -32$

13. $a_n = \frac{2}{3}$

$a_1 = \frac{2}{3}$

$a_2 = \frac{2}{3}$

$a_3 = \frac{2}{3}$

$a_4 = \frac{2}{3}$

$a_5 = \frac{2}{3}$

15. $a_n = \frac{1}{3}n^3$

$a_1 = \frac{1}{3}(1)^3 = \frac{1}{3}$

$a_2 = \frac{1}{3}(2)^3 = \frac{8}{3}$

$a_3 = \frac{1}{3}(3)^3 = 9$

$a_4 = \frac{1}{3}(4)^3 = \frac{64}{3}$

$a_5 = \frac{1}{3}(5)^3 = \frac{125}{3}$

17. $a_n = \dfrac{n}{n + 2}$

$a_1 = \dfrac{1}{1 + 2} = \dfrac{1}{3}$

$a_2 = \dfrac{2}{2 + 2} = \dfrac{1}{2}$

$a_3 = \dfrac{3}{3 + 2} = \dfrac{3}{5}$

$a_4 = \dfrac{4}{4 + 2} = \dfrac{2}{3}$

$a_5 = \dfrac{5}{5 + 2} = \dfrac{5}{7}$

19. $a_n = n(n - 1)(n - 2)$

$a_1 = (1)(0)(-1) = 0$

$a_2 = (2)(1)(0) = 0$

$a_3 = (3)(2)(1) = 6$

$a_4 = (4)(3)(2) = 24$

$a_5 = (5)(4)(3) = 60$

21. $a_n = (-1)^n\left(\dfrac{n}{n + 1}\right)$

$a_1 = (-1)^1\dfrac{1}{1 + 1} = -\dfrac{1}{2}$

$a_2 = (-1)^2\dfrac{2}{2 + 1} = \dfrac{2}{3}$

$a_3 = (-1)^3\dfrac{3}{3 + 1} = -\dfrac{3}{4}$

$a_4 = (-1)^4\dfrac{4}{4 + 1} = \dfrac{4}{5}$

$a_5 = (-1)^5\dfrac{5}{5 + 1} = -\dfrac{5}{6}$

23. $a_{25} = (-1)^{25}(3(25) - 2) = -73$

25. $a_{11} = \dfrac{4(11)}{2(11)^2 - 3} = \dfrac{44}{239}$

27. $a_n = \frac{2}{3}n$

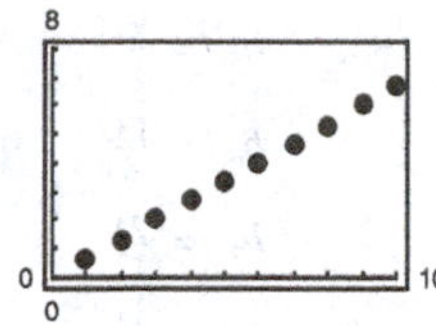

29. $a_n = 16(-0.5)^{n-1}$

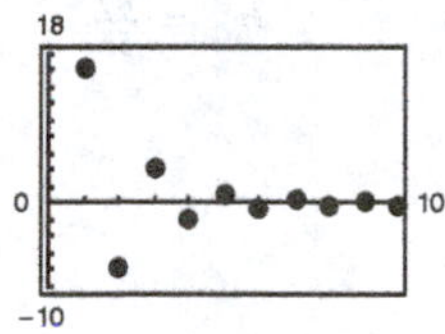

31. $a_n = \dfrac{2n}{n + 1}$

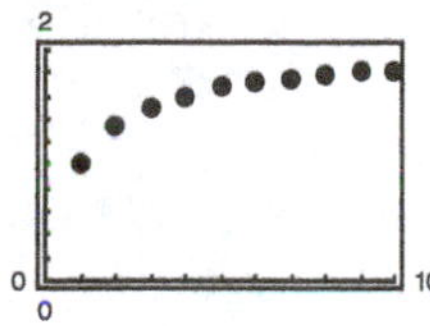

33. $a_n = \dfrac{8}{n + 1}$

$a_1 = 4,\ a_{10} = \dfrac{8}{11}$

The sequence decreases.

Matches graph (c).

34. $a_n = \dfrac{8n}{n + 1}$

$a_1 = 4,\ a_3 = \dfrac{24}{4} = 6$

The sequence increases.

Matches graph (b).

35. $a_n = 4(0.5)^{n-1}$

$a_1 = 4,\ a_{10} = \dfrac{1}{128}$

The sequence decreases.

Matches graph (d).

36. $a_n = n\left(2 - \dfrac{n}{10}\right)$

$a_1 = \dfrac{9}{10},\ a_{10} = 10$

The sequence increases.

Matches graph (a).

37. 3, 7, 11, 15, 19, …

n:	1	2	3	4	5	…	n
Terms:	3	7	11	15	19	…	a_n

Apparent pattern:

Each term is one less than four times n, which implies that $a_n = 4n - 1$.

39. 3, 10, 29, 66, 127, …

n:	1	2	3	4	5	…	n
Terms:	3	10	29	66	127	…	a_n

Apparent pattern:

Each term is more than n cubed, which implies that $a_n = n^3 + 2$.

41. 1, −1, 1, −1, 1, …

n:	1	2	3	4	5	…	n
Terms:	1	−1	1	−1	1	…	a_n

Apparent pattern:

Each term is either 1 or −1 which implies that $a_n = (-1)^{n+1}$.

43. $-\dfrac{2}{3}, \dfrac{3}{4}, -\dfrac{4}{5}, \dfrac{5}{6}, -\dfrac{6}{7}, \ldots$

$a_n = (-1)^n\left(\dfrac{n + 1}{n + 2}\right)$

45. $\dfrac{2}{1}, \dfrac{3}{3}, \dfrac{4}{5}, \dfrac{5}{7}, \dfrac{6}{9}, \ldots$

$a_n = \dfrac{n + 1}{2n - 1}$

47. $1, \dfrac{1}{2}, \dfrac{1}{6}, \dfrac{1}{24}, \dfrac{1}{120}, \ldots$

n:	1	2	3	4	5	…	n
Terms:	1	$\frac{1}{2}$	$\frac{1}{6}$	$\frac{1}{24}$	$\frac{1}{120}$	…	a_n

Apparent pattern:

Each term is the reciprocal of $n!$, which implies that $a_n = \dfrac{1}{n!}$.

49. $\frac{1}{1}, \frac{3}{1}, \frac{9}{2}, \frac{27}{6}, \frac{81}{24}, \ldots$

n: 1 2 3 4 ... n

Terms: $\frac{1}{1}$ $\frac{3}{1}$ $\frac{9}{2}$ $\frac{27}{6}$ ... a_n

Apparent pattern:

Each term is a power of three, 3^{n-1} divided by a factorial, $(n-1)!$. After trial and error, $a_n = \frac{3^{n-1}}{(n-1)!}$.

51. $a_1 = 28$ and $a_{k+1} = a_k - 4$

$a_1 = 28$

$a_2 = a_1 - 4 = 28 - 4 = 24$

$a_3 = a_2 - 4 = 24 - 4 = 20$

$a_4 = a_3 - 4 = 20 - 4 = 16$

$a_5 = a_4 - 4 = 16 - 4 = 12$

53. $a_1 = 81$ and $a_{k+1} = \frac{1}{3}a_k$

$a_1 = 81$

$a_2 = \frac{1}{3}a_1 = \frac{1}{3}(81) = 27$

$a_3 = \frac{1}{3}a_2 = \frac{1}{3}(27) = 9$

$a_4 = \frac{1}{3}a_3 = \frac{1}{3}(9) = 3$

$a_5 = \frac{1}{3}a_4 = \frac{1}{3}(3) = 1$

55. $a_0 = 1, a_1 = 2, a_k = a_{k-2} + \frac{1}{2}a_{k-1}$

$a_0 = 1$

$a_1 = 2$

$a_2 = a_0 + \frac{1}{2}a_1 = 1 + \frac{1}{2}(1) = 2$

$a_3 = a_1 + \frac{1}{2}a_2 = 2 + \frac{1}{2}(2) = 3$

$a_4 = a_2 + \frac{1}{2}a_3 = 2 + \frac{1}{2}(3) = \frac{7}{2}$

57. $a_1 = 1, a_2 = 1, a_k = a_{k-1} + a_{k-2}, k \geq 1$

$a_1 = 1$	$b_1 = \frac{1}{1} = 1$
$a_2 = 1$	$b_2 = \frac{2}{1} = 2$
$a_3 = 1 + 1 = 2$	$b_3 = \frac{3}{2}$
$a_4 = 2 + 1 = 3$	$b_4 = \frac{5}{3}$
$a_5 = 3 + 2 = 5$	$b_5 = \frac{8}{5}$
$a_6 = 5 + 3 = 8$	$b_6 = \frac{13}{8}$
$a_7 = 8 + 5 = 13$	$b_7 = \frac{21}{13}$
$a_8 = 13 + 8 = 21$	$b_8 = \frac{34}{21}$
$a_9 = 21 + 13 = 34$	$b_9 = \frac{55}{34}$
$a_{10} = 34 + 21 = 55$	$b_{10} = \frac{89}{55}$
$a_{11} = 55 + 34 = 89$	
$a_{12} = 89 + 55 = 144$	

59. $a_n = \frac{5}{n!}$

$a_0 = \frac{5}{0!} = \frac{5}{1} = 5$

$a_1 = \frac{5}{1!} = \frac{5}{1} = 5$

$a_2 = \frac{5}{2!} = \frac{5}{2}$

$a_3 = \frac{5}{3!} = \frac{5}{6}$

$a_4 = \frac{5}{4!} = \frac{5}{24}$

61. $a_n = \frac{(-1)^n(n+3)!}{n!}$

$a_0 = \frac{(-1)^0(0+3)!}{0!} = \frac{3!}{0!} = 6$

$a_1 = \frac{(-1)^1(1+3)!}{1!} = \frac{4!}{1!} = -24$

$a_2 = \frac{(-1)^2(2+3)!}{2!} = \frac{5!}{2!} = 60$

$a_3 = \frac{(-1)^3(3+3)!}{3!} = -\frac{6!}{3!} = -120$

$a_4 = \frac{(-1)^4(4+3)!}{4!} = \frac{7!}{4!} = 210$

63. $\dfrac{4!}{6!} = \dfrac{\cancel{1 \cdot 2 \cdot 3 \cdot 4}}{\cancel{1 \cdot 2 \cdot 3 \cdot 4} \cdot 5 \cdot 6} = \dfrac{1}{5 \cdot 6} = \dfrac{1}{30}$

65. $\dfrac{(n+1)!}{n!} = \dfrac{\cancel{1 \cdot 2 \cdot 3 \cdots n} \cdot (n+1)}{\cancel{1 \cdot 2 \cdot 3 \cdots n}} = \dfrac{n+1}{1}$

$= n + 1$

67. $\sum_{i=0}^{4} 3i^2 = 3 \cdot 0^2 + 3 \cdot 1^2 + 3 \cdot 2^2 + 3 \cdot 3^2 + 3 \cdot 4^2 = 90$

69. $\sum_{j=3}^{5} \dfrac{1}{j^2 - 3} = \dfrac{1}{3^2 - 3} + \dfrac{1}{4^2 - 3} + \dfrac{1}{5^2 - 3} = \dfrac{124}{429}$

71. $\sum_{k=2}^{5} (k+1)^2(k-3) = (3)^2(-1) + (4)^2(0) + (5)^2(1) + (6)^2(2) = 88$

73. $\sum_{i=1}^{4} \dfrac{i!}{2^i} = \dfrac{1!}{2^1} + \dfrac{2!}{2^2} + \dfrac{3!}{2^3} + \dfrac{4!}{2^4} = \dfrac{1}{2} + \dfrac{2}{4} + \dfrac{6}{8} + \dfrac{24}{16} = \dfrac{1}{2} + \dfrac{1}{2} + \dfrac{3}{4} + \dfrac{3}{2} = \dfrac{13}{4}$

75. $\sum_{k=0}^{4} \dfrac{(-1)^k}{k!} = \dfrac{3}{8}$

77. $\sum_{n=0}^{25} \dfrac{1}{4^n} \approx 1.33$

79. $\dfrac{1}{3(1)} + \dfrac{1}{3(2)} + \dfrac{1}{3(3)} + \cdots + \dfrac{1}{3(9)} = \sum_{i=1}^{9} \dfrac{1}{3i}$

81. $\left[2\left(\dfrac{1}{8}\right) + 3\right] + \left[2\left(\dfrac{2}{8}\right) + 3\right] + \left[2\left(\dfrac{3}{8}\right) + 3\right] + \cdots + \left[2\left(\dfrac{8}{8}\right) + 3\right] = \sum_{i=1}^{8}\left[2\left(\dfrac{i}{8}\right) + 3\right]$

83. $3 - 9 + 27 - 81 + 243 - 729 = \sum_{i=1}^{6} (-1)^{i+1} 3^i$

85. $\dfrac{1^2}{2} + \dfrac{2^2}{6} + \dfrac{3^2}{24} + \dfrac{4^2}{120} + \dfrac{5^2}{720} + \dfrac{6^2}{5040} + \dfrac{7^2}{40{,}320} = \sum_{i=1}^{7} \dfrac{i^2}{(i+1)!}$

87. $\dfrac{1}{4} + \dfrac{3}{8} + \dfrac{7}{16} + \dfrac{15}{32} + \dfrac{31}{64} = \sum_{i=1}^{5} \dfrac{2^i - 1}{2^{i+1}}$

89. (a) $\sum_{i=1}^{3} \left(\dfrac{1}{2}\right)^i = \left(\dfrac{1}{2}\right) + \left(\dfrac{1}{2}\right)^2 + \left(\dfrac{1}{2}\right)^3 = \dfrac{7}{8}$

(b) $\sum_{i=1}^{4} \left(\dfrac{1}{2}\right)^i = \left(\dfrac{1}{2}\right) + \left(\dfrac{1}{2}\right)^2 + \left(\dfrac{1}{2}\right)^3 + \left(\dfrac{1}{2}\right)^4 = \dfrac{15}{16}$

(c) $\sum_{i=1}^{5} \left(\dfrac{1}{2}\right)^i = \left(\dfrac{1}{2}\right) + \left(\dfrac{1}{2}\right)^2 + \left(\dfrac{1}{2}\right)^3 + \left(\dfrac{1}{2}\right)^4 + \left(\dfrac{1}{2}\right)^5 = \dfrac{31}{32}$

91. (a) $\sum_{n=1}^{3} 4\left(-\dfrac{1}{2}\right)^n = 4\left(-\dfrac{1}{2}\right) + 4\left(-\dfrac{1}{2}\right)^2 + 4\left(-\dfrac{1}{2}\right)^3 = -\dfrac{3}{2}$

(b) $\sum_{n=1}^{4} 4\left(-\dfrac{1}{2}\right)^n = 4\left(-\dfrac{1}{2}\right) + 4\left(-\dfrac{1}{2}\right)^2 + 4\left(-\dfrac{1}{2}\right)^3 + 4\left(-\dfrac{1}{2}\right)^4 = -\dfrac{5}{4}$

(c) $\sum_{n=1}^{5} 4\left(-\dfrac{1}{2}\right)^n = 4\left(-\dfrac{1}{2}\right) + 4\left(-\dfrac{1}{2}\right)^2 + 4\left(-\dfrac{1}{2}\right)^3 + 4\left(-\dfrac{1}{2}\right)^4 + 4\left(-\dfrac{1}{2}\right)^5 = -\dfrac{11}{8}$

93. $\sum_{i=1}^{\infty} 6\left(\tfrac{1}{10}\right)^i = 0.6 + 0.06 + 0.006 + 0.0006 + \cdots = \tfrac{2}{3}$

95. By using a calculator,

$$\sum_{k=1}^{10} 7\left(\frac{1}{10}\right)^k \approx 0.7777777777$$

$$\sum_{k=1}^{50} 7\left(\frac{1}{10}\right)^k \approx 0.7777777778$$

$$\sum_{k=1}^{100} 7\left(\frac{1}{10}\right)^k \approx \frac{7}{9}.$$

The terms approach zero as $n \to \infty$.

So, $\sum_{k=1}^{\infty} 7\left(\frac{1}{10}\right)^k = \frac{7}{9}$.

97. (a) $A_1 = \$10{,}087.50$
$A_2 \approx \$10{,}175.77$
$A_3 \approx \$10{,}264.80$
$A_4 \approx \$10{,}354.62$
$A_5 \approx \$10{,}445.22$
$A_6 \approx \$10{,}536.62$
$A_7 \approx \$10{,}628.81$
$A_8 \approx \$10{,}721.82$

(b) $A_{40} \approx \$14{,}169.09$

(c) No; The balance after 20 years, $A_{80} = \$20{,}076.31$ is not twice the balance after 10 years, $A_{40} \approx \$14{,}169.09$.

99. True, $\sum_{i=1}^{4}\left(i^2 + 2i\right) = \sum_{i=1}^{4} i^2 + 2\sum_{i=1}^{4} i$ by the Properties of Sums.

101. $\dfrac{327.15 + 785.69 + 433.04 + 265.38 + 604.12 + 590.30}{6} \approx \$500.95$

103.
$$\begin{aligned}\sum_{i=1}^{n}\left(x_i - \bar{x}\right)^2 &= \sum_{i=1}^{n}\left(x_i^2 - 2x_i\bar{x} + \bar{x}^2\right) = \sum_{i=1}^{n} x_i^2 - 2\bar{x}\sum_{i=1}^{n} x_i + n\bar{x}^2\\ &= \sum_{i=1}^{n} x_i^2 - 2\cdot\frac{1}{n}\sum_{i=1}^{n} x_i\sum_{i=1}^{n} x_i + n\cdot\frac{1}{n}\sum_{i=1}^{n} x_i\cdot\frac{1}{n}\sum_{i=1}^{n} x_i\\ &= \sum_{i=1}^{n} x_i^2 + \sum_{i=1}^{n} x_i\sum_{i=1}^{n} x_i\left(-\frac{2}{n} + \frac{1}{n}\right) = \sum_{i=1}^{n} x_i^2 - \frac{1}{n}\left(\sum_{i=1}^{n} x_i\right)^2\end{aligned}$$

105. The error was not summing the constant, $\sum_{k=1}^{4} 3 = 12$.

$$\begin{aligned}\sum_{k=1}^{4}\left(3 + 2k^2\right) &= \sum_{k=1}^{4} 3 + \sum_{k=1}^{4} 2k^2 = (3 + 3 + 3 + 3) + \left(2(1)^2 + 2(2)^2 + 2(2)^3 + 2(4)^2\right)\\ &= (12) + (2 + 8 + 18 + 32) = 75\end{aligned}$$

107. (a) and (b)

Number of blue cube faces	0	1	2	3
$3 \times 3 \times 3$	1	6	12	8
$4 \times 4 \times 4$	8	24	24	8
$5 \times 5 \times 5$	27	54	36	8
$6 \times 6 \times 6$	64	96	48	8

(c)

Number of blue cube faces	0	1	2	3
$n \times n \times n$	$(n-2)^3$	$6(n-2)^2$	$12(n-2)$	8

Section 8.2 Arithmetic Sequences and Partial Sums

1. arithmetic; common

3. recursion

5. 1, 2, 4, 8, 16,…

Not an arithmetic sequence

7. 10, 8, 6, 4, 2, …

Arithmetic sequence, $d = -2$

9. $\frac{5}{4}, \frac{3}{2}, \frac{7}{4}, 2, \frac{9}{4}, \ldots$

Arithmetic sequence, $d = \frac{1}{4}$

11. $1^2, 2^2, 3^2, 4^2, 5^2, \ldots$

Not an arithmetic sequence

13. $a_n = 5 + 3n$

8, 11, 14, 17, 20

Arithmetic sequence, $d = 3$

15. $a_n = 3 - 4(n - 2)$

7, 3, −1, −5, −9

Arithmetic sequence, $d = -4$

17. $a_n = (-1)^n$

−1, 1, −1, 1, −1

Not an arithmetic sequence

19. $a_n = (2^n)n$

2, 8, 24, 64, 160

Not an arithmetic sequence

21. $a_1 = 1, d = 3$

$a_n = a_1 + (n - 1)d = 1 + (n - 1)(3) = 3n - 2$

23. $a_1 = 100, d = -8$

$a_n = a_1 + (n - 1)d = 100 + (n - 1)(-8)$

$= -8n + 108$

25. $4, \frac{3}{2}, -1, -\frac{7}{2}, \ldots$

$d = -\frac{5}{2}$

$a_n = a_1 + (n - 1)d = 4 + (n - 1)\left(-\frac{5}{2}\right) = -\frac{5}{2}n + \frac{13}{2}$

27. $a_1 = 5, a_4 = 15$

$a_4 = a_1 + 3d \Rightarrow 15 = 5 + 3d \Rightarrow d = \frac{10}{3}$

$a_n = a_1 + (n - 1)d = 5 + (n - 1)\left(\frac{10}{3}\right) = \frac{10}{3}n + \frac{5}{3}$

29. $a_3 = 94, a_6 = 103$

$a_6 = a_3 + 3d \Rightarrow 103 = 94 + 3d \Rightarrow d = 3$

$a_1 = a_3 - 2d \Rightarrow a_1 = 94 - 2(3) = 88$

$a_n = a_1 + (n - 1)d = 88 + (n - 1)(3)$

$= 3n + 85$

31. $a_1 = 5, d = 6$

$a_1 = 5$

$a_2 = 5 + 6 = 11$

$a_3 = 11 + 6 = 17$

$a_4 = 17 + 6 = 23$

$a_5 = 23 + 6 = 29$

33. $a_1 = 2, a_{12} = -64$

$-64 = 2 + (12 + 1)d$

$-66 = 11d$

$-6 = d$

$a_1 = 2$

$a_2 = 2 - 6 = -4$

$a_3 = -4 - 6 = -10$

$a_4 = -10 - 6 = -16$

$a_5 = -16 - 6 = -22$

35. $a_8 = 26, a_{12} = 42$

$a_{12} = a_8 + 4d$

$42 = 26 + 4d \Rightarrow d = 4$

$a_8 = a_1 + 7d$

$26 = a_1 + 28 \Rightarrow a_1 = -2$

$a_1 = -2$

$a_2 = -2 + 4 = 2$

$a_3 = 2 + 4 = 6$

$a_4 = 6 + 4 = 10$

$a_5 = 10 + 4 = 14$

37. $a_1 = 15, a_{n+1} = a_n + 4$

$a_2 = 15 + 4 = 19$

$a_3 = 19 + 4 = 23$

$a_4 = 23 + 4 = 27$

$a_5 = 27 + 4 = 31$

39. $a_{n+1} = a_n - 2 \Rightarrow a_n = a_{n+1} + 2$

$a_5 = 7$

$a_4 = 7 + 2 = 9$

$a_3 = 9 + 2 = 11$

$a_2 = 11 + 2 = 13$

$a_1 = 13 + 2 = 15$

41. $a_1 = 5, a_2 = -1 \Rightarrow d = -1 - 5 = -6$

$a_n = a_1 + (n-1)d \Rightarrow a_{10} = 5 + 9(-6) = -49$

43. $a_1 = \dfrac{1}{8}, a_2 = \dfrac{3}{8} \Rightarrow d = \dfrac{3}{4} - \dfrac{1}{8} = \dfrac{5}{8}$

$a_n = a_1 + (n-1)d \Rightarrow a_7 = \dfrac{1}{8} + 6\left(\dfrac{5}{8}\right) = \dfrac{31}{8}$

45. $S_{10} = \frac{10}{2}(2 + 20) = 110$

47. $S_5 = \frac{5}{2}(-1 + (-9)) = -25$

49. $a_n = 2n - 1$

$a_1 = 1, a_{100} = 199$

$S_{100} = \frac{100}{2}(1 + 199) = 10{,}000$

51. 8, 20, 32, 44, …

$a_1 = 8, d = 12, n = 50$

$a_{50} = 8 + 49(12) = 596$

$S_{10} = \frac{50}{2}(8 + 596) = 15{,}100$

53. $0, -9, -18, -27, \ldots, n = 40$

$a_1 = 0, d = -9$

$a_{40} = 0 + (39)(-9) = -351$

$S_{100} = \frac{40}{2}(0 + (-351)) = -7020$

55. $a_1 = 1, a_{50} = 50, n = 50$

$\sum_{n=1}^{50} n = \frac{50}{2}(1 + 50) = 1275$

57. $a_1 = 9, a_{500} = 508, n = 500$

$\sum_{n=1}^{500} (n + 8) = \frac{500}{2}(9 + 508) = 129{,}250$

59. $a_1 = 14, a_{100} = -580, n = 100$

$\sum_{n=1}^{100} (-6n + 20) = \frac{100}{2}(14 + (-580)) = -28{,}300$

61. $a_n = -\frac{3}{4}n + 8$

$d = -\frac{3}{4}$ so the sequence is decreasing and $a_1 = 7\frac{1}{4}$.

Matches (b).

62. $a_n = 3n - 5$

$d = 3$ so the sequence is increasing and $a_1 = -2$.

Matches (d).

63. $a_n = 2 + \frac{3}{4}n$

$d = \frac{3}{4}$ so the sequence is increasing and $a_1 = 2\frac{3}{4}$.

Matches (c).

64. $a_n = 25 - 3n$

$d = -3$ so the sequence is decreasing and $a_1 = 22$.

Matches (a).

65. $a_n = 15 - \frac{3}{2}n$

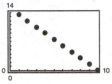

67. $a_n = 0.2n + 3$

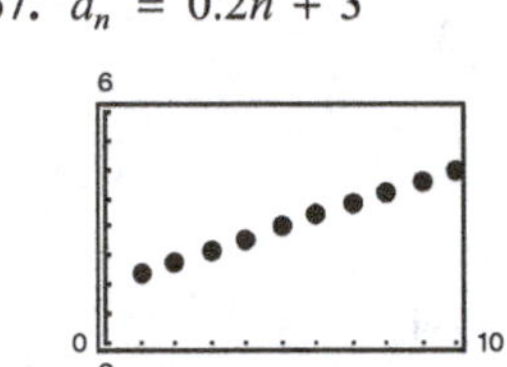

69. (a) $a_1 = 32{,}500, d = 1500$

$a_6 = a_1 + 5d = 32{,}500 + 5(1500) = \$40{,}000$

(b) $S_6 = \frac{6}{2}[32{,}500 + 40{,}000] = \$217{,}500$

71. $a_1 = 15, d = 3, n = 36$

$a_{36} = 15 + 35(3) = 120$

$S_{36} = \frac{36}{2}(15 + 120) = 2430$ seats

73. $a_1 = 16, a_2 = 48, a_3 = 80, a_4 = 112$

$d = 32$

$a_n = dn + c = 32n + c$

$c = a_n - dn$

$c = a_1 - d = 16 - 32 = -16$

$a_n = 32n - 16$

Distance $= \sum_{n=1}^{7} (32n - 16) = 784$ ft

75. $a_1 = 15{,}000$

$d = 5{,}000$

$n = 1, \ldots, 10$

$a_n = dn + c = 5000n + c$

$c = a_1 - d = 15{,}000 - 5000 = 10{,}000$

$a_n = 5000n + 10{,}000$

$$\text{Total sales} = \sum_{n=1}^{10} (5000n + 10{,}000)$$

$$= \tfrac{10}{2}(15{,}000 + 60{,}000)$$

$$= \$375{,}000$$

Answers will vary.

77. (a)

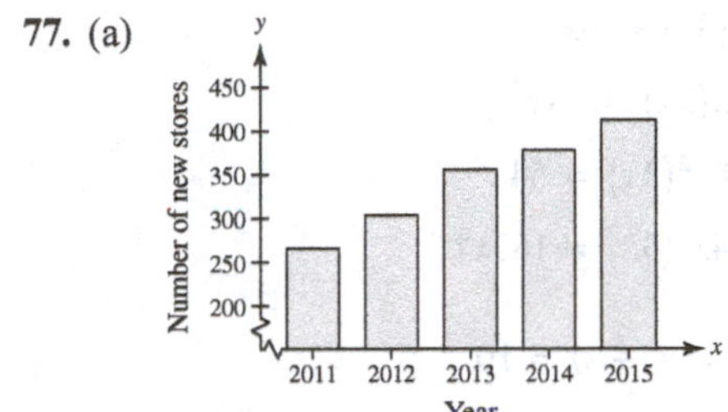

(b) The increase in net number of new stores $\dfrac{413 - 266}{5 - 1} = \dfrac{147}{4} = 36.75$ each year from 2011 to 2015. Because the common difference is $d = 36.75$ and the first term is $a_1 = 266$, the arithmetic sequence is

$a_n = 266 + 36.75(n - 1) = 229.25 + 36.75n$

(c)

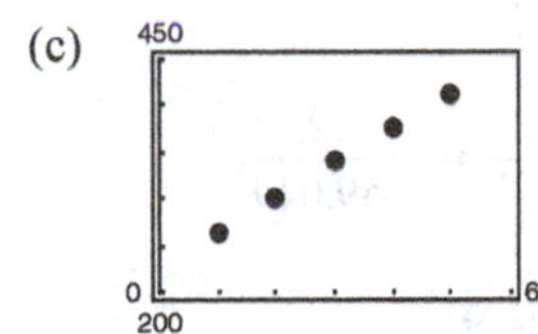

(d) $\displaystyle\sum_{n=1}^{5} (229.25 + 36.75n) = \tfrac{5}{2}(266 + 413)$

$= 1697.5 \approx 1698$ stores

79. True; given a_1 and a_2 then $d = a_2 - a_1$ and $a_n = a_1 + (n - 1)d$.

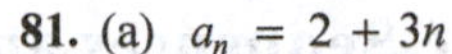

81. (a) $a_n = 2 + 3n$

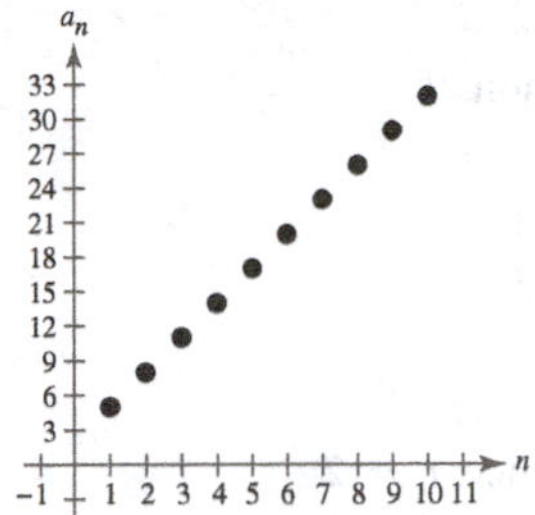

(b) $y = 3x + 2$

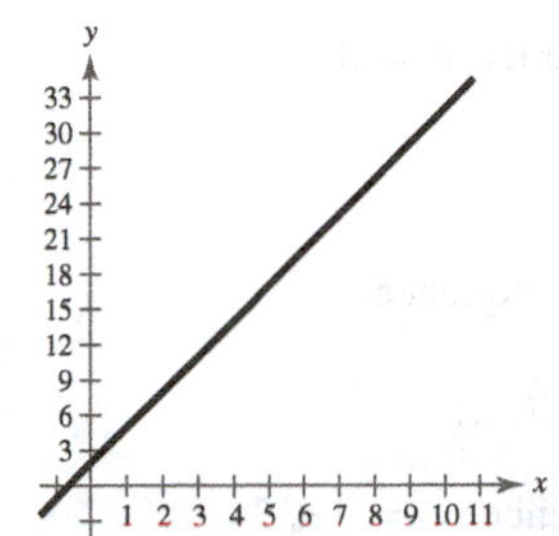

(c) The graph of $a_n = 2 + 3n$ contains only points at the positive integers. The graph of $y = 3x + 2$ is a solid line which contains these points.

(d) The slope $m = 3$ is equal to the common difference $d = 3$. In general, these should be equal.

83. $a_1 = x, d = 2x$

$a_n = x + (n - 1)2x$

$a_n = 2xn - x$

$a_1 = 2x(1) - x = x$	$a_6 = 2x(6) - x = 11x$
$a_2 = 2x(2) - x = 3x$	$a_7 = 2x(7) - x = 13x$
$a_3 = 2x(3) - x = 5x$	$a_8 = 2x(8) - x = 15x$
$a_4 = 2x(4) - x = 7x$	$a_9 = 2x(9) - x = 17x$
$a_5 = 2x(5) - x = 9x$	$a_{10} = 2x(10) - x = 19x$

85. The error was the value of $n = 50$, $a_{50} = 1 + (50 - 1)(2) = 99$. So $S_{50} = \tfrac{50}{2}(1 + 99) = 2500$.

87. (a) $1 + 3 = 4$

$1 + 3 + 5 = 9$

$1 + 3 + 5 + 7 = 16$

$1 + 3 + 5 + 7 + 9 = 36$

$1 + 3 + 5 + 7 + 9 + 11 = 36$

(b) $S_n = n^2$

$S_7 = 1 + 3 + 5 + 7 + 9 + 11 + 13 = 49 = 7^2$

(c) $S_n = \dfrac{n}{2}[1 + (2n - 1)] = \dfrac{n}{2}(2n) = n^2$

Section 8.3 Geometric Sequences and Series

1. geometric; common

3. $S_n = a_1\left(\frac{1 - r^n}{1 - r}\right)$

5. 3, 6, 12, 24, ...

Geometric sequence, $r = 2$

7. $\frac{1}{27}, \frac{1}{9}, \frac{1}{3}, 1, \ldots$

Geometric sequence, $r = 3$

9. $1, \frac{1}{2}, \frac{1}{3}, \frac{1}{4}, \ldots$

Not a geometric sequence

11. $1, -\sqrt{7}, 7, -7\sqrt{7}, \ldots$

Geometric sequence, $r = -\sqrt{7}$

13. $a_1 = 4, r = 3$

$a_1 = 4$

$a_2 = 4(3) = 12$

$a_3 = 12(3) = 36$

$a_4 = 36(3) = 108$

$a_5 = 108(3) = 324$

15. $a_1 = 1, r = \frac{1}{2}$

$a_1 = 1$

$a_2 = 1\left(\frac{1}{2}\right) = \frac{1}{2}$

$a_3 = \frac{1}{2}\left(\frac{1}{2}\right) = \frac{1}{4}$

$a_4 = \frac{1}{4}\left(\frac{1}{2}\right) = \frac{1}{8}$

$a_5 = \frac{1}{8}\left(\frac{1}{2}\right) = \frac{1}{16}$

17. $a_1 = 1, r = e$

$a_1 = 1$

$a_2 = 1(e) = e$

$a_3 = (e)(e) = e^2$

$a_4 = (e^2)(e) = e^3$

$a_5 = (e^3)(e) = e^4$

19. $a_1 = 3, r = \sqrt{5}$

$a_1 = 3$

$a_2 = 3\left(\sqrt{5}\right)^1 = 3\sqrt{5}$

$a_3 = 3\left(\sqrt{5}\right)^2 = 15$

$a_4 = 3\left(\sqrt{5}\right)^3 = 15\sqrt{5}$

$a_5 = 3\left(\sqrt{5}\right)^4 = 75$

21. $a_1 = 2, r = 3x$

$a_1 = 2$

$a_2 = 2(3x) = 6x$

$a_3 = 6x(3x) = 18x^2$

$a_4 = 18x^2(3x) = 54x^3$

$a_5 = 54x^3(3x) = 162x^4$

23. $a_1 = 4, r = \frac{1}{2}, n = 10$

$a_n = a_1r^{n-1} = 4\left(\frac{1}{2}\right)^{n-1}$

$a_{10} = 4\left(\frac{1}{2}\right)^9 = \left(\frac{1}{2}\right)^7 = \frac{1}{128}$

25. $a_1 = 6, r = -\frac{1}{3}, n = 12$

$a_n = a_1r^{n-1} = 6\left(-\frac{1}{3}\right)^{n-1}$

$a_{12} = 6\left(-\frac{1}{3}\right)^{11} = -\frac{2}{3^{10}} = -\frac{2}{59,049}$

27. $a_1 = 100, r = e^x, n = 9$

$a_n = a_1r^{n-1} = 100\left(e^x\right)^{n-1} = 100e^{x(n-1)}$

$a_9 = 100\left(e^x\right)^8 = 100e^{8x}$

29. $a_1 = 1, r = \sqrt{2}, n = 12$

$a_n = 1\left(\sqrt{2}\right)^{n-1} = \left(\sqrt{2}\right)^{n-1}$

$a_{12} = \left(\sqrt{2}\right)^{12-1} = 32\sqrt{2}$

31. $a_1 = 500, r = 1.02, n = 40$

$a_n = a_1r^{n-1} = 500(1.02)^{n-1}$

$a_{40} = 500(1.02)^{39} \approx 1082.372$

33. $64, 32, 16, \ldots$

$$r = \frac{32}{64} = \frac{1}{2}$$

$$a_n = 64\left(\tfrac{1}{2}\right)^{n-1}$$

35. $9, 18, 36, \ldots$

$$r = \frac{18}{9} = 2$$

$$a_n = 9(2)^{n-1}$$

37. $6, -9, \frac{27}{2}, \ldots$

$$r = \frac{-9}{6} = -\frac{3}{2}$$

$$a_n = 6\left(-\tfrac{3}{2}\right)^{n-1}$$

39. $6, 18, 54, \ldots$

$$r = \frac{18}{6} = 3$$

$$a_n = 6(3)^{n-1}$$

$$a_8 = 6(3)^{8-1}$$

$$= 6(3)^7$$

$$= 13{,}122$$

41. $\frac{1}{3}, \frac{-1}{6}, \frac{1}{12}, \ldots$

$$r = \frac{-\frac{1}{6}}{\frac{1}{3}} = -\frac{1}{2}$$

$$a_n = \frac{1}{3}\left(-\tfrac{1}{2}\right)^{n-1}$$

$$a_9 = \frac{1}{3}\left(-\tfrac{1}{2}\right)^{9-1}$$

$$= \frac{1}{3}\left(-\tfrac{1}{2}\right)^{8}$$

$$= \frac{1}{768}$$

43. $a_1 = 16, a_4 = \frac{27}{4}$

$$a_4 = a_1 r^3$$

$$\tfrac{27}{4} = 16r^3$$

$$\tfrac{27}{64} = r^3$$

$$\tfrac{3}{4} = r$$

$$a_n = 16\left(\tfrac{3}{4}\right)^{n-1}$$

$$a_3 = 16\left(\tfrac{3}{4}\right)^2 = 9$$

45. $a_4 = -18, a_7 = \frac{2}{3}$

$$a_7 = a_4 r^3$$

$$\frac{2}{3} = -18r^3$$

$$-\frac{1}{27} = r^3$$

$$-\frac{1}{3} = r$$

$$a_6 = \frac{a_7}{r} = \frac{2/3}{-1/3} = -2$$

47. $a_n = 18\left(\frac{2}{3}\right)^{n-1}$

$a_1 = 18$ and $r = \frac{2}{3}$

Because $0 < r < 1$, the sequence is decreasing.

Matches (a).

48. $a_n = 18\left(-\frac{2}{3}\right)^{n-1}$

Because $r = \left(-\frac{2}{3}\right) > -1$, the sequence alternates as it approaches 0.

Matches (c).

49. $a_n = 18\left(\frac{3}{2}\right)^{n-1}$

Because $a_1 = 18$ and $r = \frac{3}{2} > 1$, the sequence is increasing.

Matches (b).

50. $a_n = 18\left(-\frac{3}{2}\right)^{n-1}$

Because $r = \left(-\frac{3}{2}\right) < -1$, the sequence alternates as it approaches ∞.

Matches (d).

51. 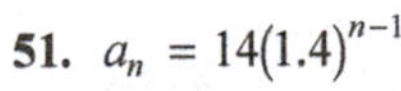 $a_n = 14(1.4)^{n-1}$

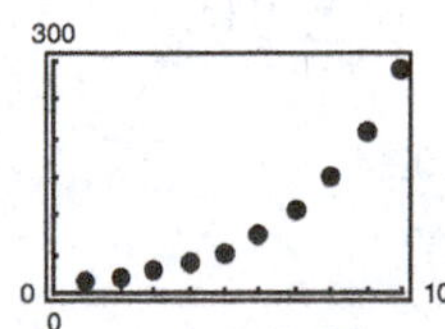

53. $a_n = 8(-0.3)^{n-1}$

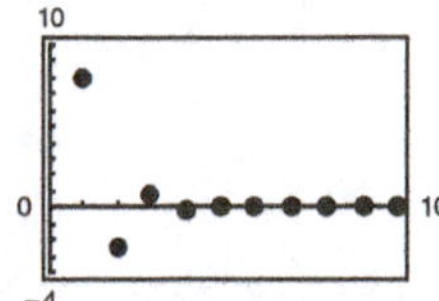

55. $\sum_{n=1}^{7} 4^{n-1} = 1 + 4^1 + 4^2 + 4^3 + 4^4 + 4^5 + 4^6 \Rightarrow a_1 = 1, r = 4$

$$S_7 = \frac{1(1 - 4^7)}{1 - 4} = 5461$$

57. $\sum_{n=1}^{6} (-7)^{n-1} = 1 + (-7) + (-7)^2 + \cdots + (-7)^5 \Rightarrow a_1 = 1, r = -7$

$$S_6 = \frac{1(1 - (-7)^6)}{1 - (-7)} = -14{,}706$$

59. $\sum_{n=0}^{20} 3\left(\frac{3}{2}\right)^n = \sum_{n=1}^{21} 3\left(\frac{3}{2}\right)^{n-1} = 3 + 3\left(\frac{3}{2}\right)^1 + 3\left(\frac{3}{2}\right)^2 + \cdots + 3\left(\frac{3}{2}\right)^{20} \Rightarrow a_1 = 3, r = \frac{3}{2}$

$$S_{21} = 3\left[\frac{1 - \left(\frac{3}{2}\right)^{21}}{1 - \frac{3}{2}}\right] = -6\left[1 - \left(\frac{3}{2}\right)^{21}\right] \approx 29{,}921.311$$

61. $\sum_{n=0}^{5} 200(1.05)^n = 200 + \sum_{n=1}^{5} 200(1.05)^5 = 200 + \left[200(1.05)^1 + 200(1.05)^2 + \cdots + 200(1.05)^5\right]$

$a_1 = 210, r = 1.05$

$$S_5 = 200 + 210\left[\frac{1 - (1.05)^5}{1 + (1.05)}\right] \approx 1360.383$$

63. $\sum_{n=0}^{40} 2\left(-\frac{1}{4}\right)^n = 2 + 2\left(-\frac{1}{4}\right) + 2\left(-\frac{1}{4}\right)^2 + \cdots + 2\left(-\frac{1}{4}\right)^{40} \Rightarrow a_1 = 2, r = -\frac{1}{4}, n = 41$

$$S_{41} = 2\left[\frac{1 - \left(-\frac{1}{4}\right)^{41}}{1 - \left(-\frac{1}{4}\right)}\right] = \frac{8}{5}\left[1 - \left(-\frac{1}{4}\right)^{41}\right] \approx 1.6 = \frac{8}{5}$$

65. $10 + 30 + 90 + \cdots + 7290$

$r = 3$ and $7290 = 10(3)^{n-1}$

$729 = 3^{n-1}$

$6 = n - 1 \Rightarrow n = 7$

So, the sum can be written as $\sum_{n=1}^{7} 10(3)^{n-1}$.

67. $0.1 + 0.4 + 1.6 + \cdots + 102.4$

$r = 4$ and $102.4 = 0.1(4)^{n-1}$

$1024 = 4^{n-1} \Rightarrow 5 = n - 1 \Rightarrow n = 6$

So, the sum can be written as $\sum_{n=1}^{6} 0.1(4)^{n-1}$.

69. $\sum_{n=0}^{\infty} \left(\frac{1}{2}\right)^n = 1 + \left(\frac{1}{2}\right)^1 + \left(\frac{1}{2}\right)^2 + \cdots$

$a_1 = 1, r = \frac{1}{2}$

$$\sum_{n=0}^{\infty} \left(\frac{1}{2}\right)^n = \frac{a_1}{1 - r} = \frac{1}{1 - \left(\frac{1}{2}\right)} = 2$$

71. $\sum_{n=0}^{\infty} \left(-\frac{1}{2}\right)^n = 1 + \left(-\frac{1}{2}\right)^1 + \left(-\frac{1}{2}\right)^2 + \cdots$

$a_1 = 1, r = -\frac{1}{2}$

$$\sum_{n=0}^{\infty} \left(-\frac{1}{2}\right)^n = \frac{a_1}{1 - r} = \frac{1}{1 - \left(-\frac{1}{2}\right)} = \frac{2}{3}$$

73. $\sum_{n=0}^{\infty}(0.8)^n = 1 + (0.8)^1 + (0.8)^2 + \cdots$

$a_1 = 1,\ r = 0.8$

$$\sum_{n=0}^{\infty}(0.8)^n = \frac{1}{1-0.8} = 5$$

75. $8 + 6 + \frac{9}{2} + \frac{27}{8} + \cdots = \sum_{n=0}^{\infty} 8\left(\frac{3}{4}\right)^n = \frac{8}{1-\frac{3}{4}} = 32$

77. $\frac{1}{9} - \frac{1}{3} + 1 - 3 + \cdots = \sum_{n=0}^{\infty}\frac{1}{9}(-3)^n$

The sum is undefined because

$|r| = |-3| = 3 > 1.$

79. $0.\overline{36} = \sum_{n=0}^{\infty} 0.36(0.01)^n = \frac{0.36}{1-0.01} = \frac{0.36}{0.99} = \frac{36}{99} = \frac{4}{11}$

81. $f(x) = 6\left[\frac{1-(0.5)^x}{1-(0.5)}\right],\ \sum_{n=0}^{\infty} 6\left(\frac{1}{2}\right)^n = \frac{6}{1-\frac{1}{2}} = 12$

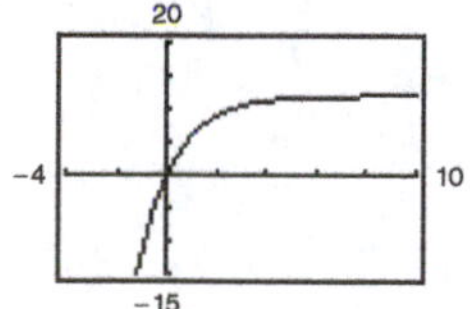

The horizontal asymptote of $f(x)$ is $y = 12$.

This corresponds to the sum of the series.

83. $V_5 = 175{,}000(0.70)^5 = \$29{,}412.25$

85. Let $N = 12t$ be the total number of deposits.

$$A = P\left(1+\frac{r}{12}\right) + P\left(1+\frac{r}{12}\right)^2 + \cdots + P\left(1+\frac{r}{12}\right)^N$$

$$= \left(1+\frac{r}{12}\right)\left[P + P\left(1+\frac{r}{12}\right) + \cdots + P\left(1+\frac{r}{12}\right)^{N-1}\right]$$

$$= P\left(1+\frac{r}{12}\right)\sum_{n=1}^{N}\left(1+\frac{r}{12}\right)^{n-1}$$

$$= P\left(1+\frac{r}{12}\right)\left[\frac{1-\left(1+\frac{r}{12}\right)^N}{1-\left(1+\frac{r}{12}\right)}\right]$$

$$= P\left(1+\frac{r}{12}\right)\left(-\frac{12}{r}\right)\left[1-\left(1+\frac{r}{12}\right)^N\right]$$

$$= P\left(\frac{12}{r}+1\right)\left[-1+\left(1+\frac{r}{12}\right)^N\right]$$

$$= P\left[\left(1+\frac{r}{12}\right)^N - 1\right]\left(1+\frac{12}{r}\right)$$

$$= P\left[\left(1+\frac{r}{12}\right)^{12t} - 1\right]\left(1+\frac{12}{r}\right)$$

87. $\sum_{n=0}^{\infty} 400(0.75)^n = \frac{400}{1-0.75} = \$1600$

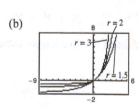

89. $27^2\left(\frac{1}{9}\right) + 27^2\left(\frac{1}{9}\right)\left(\frac{8}{9}\right) + 27^2\left(\frac{1}{9}\right)\left(\frac{8}{9}\right)^2 + 27^2\left(\frac{1}{9}\right)\left(\frac{8}{9}\right)^3 = \sum_{n=0}^{3} 27^2\left(\frac{1}{9}\right)\left(\frac{8}{9}\right)^n = \frac{2465}{9} = 273\frac{8}{9}$ square inches

91. $a_n = (45{,}000)(1.05)^{n-1}$

$$S_{40} = (45{,}000)\frac{1-(1.05)^{40}}{1-(1.05)} = \$5{,}435{,}989.84$$

93. False. A sequence is geometric if the ratios of consecutive terms are the same.

95. $y = \left(\frac{1-r^x}{1-r}\right)$

(a)

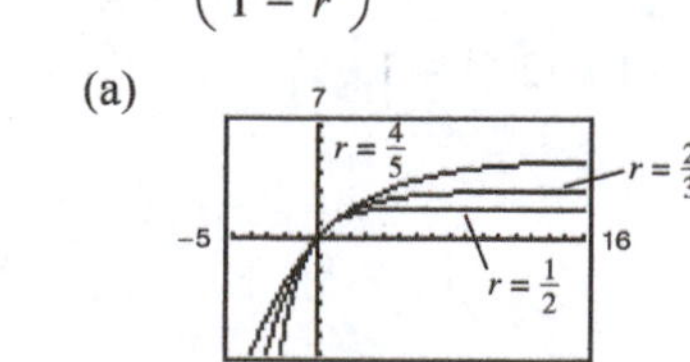

As $x \to \infty$, $y \to \frac{1}{1-r}$.

(b)

As $x \to \infty$, $y \to \infty$.

Section 8.4 Mathematical Induction

1. mathematical induction

3. arithmetic

5. $P_k = \dfrac{5}{k(k+1)}$

$P_{k+1} = \dfrac{5}{(k+1)[(k+1)+1]} = \dfrac{5}{(k+1)(k+2)}$

7. $P_k = k^2(k+3)^2$

$P_{k+1} = (k+1)^2[(k+1)+3]^2 = (k+1)^2(k+4)^2$

9. $P_k = \dfrac{3}{(k+2)(k+3)}$

$P_{k+1} = \dfrac{3}{[(k+1)+2][(k+1)+3]} = \dfrac{3}{(k+3)(k+4)}$

11. 1. When $n = 1$, $S_1 = 2 = 1(1+1)$.

2. Assume that

$S_k = 2 + 4 + 6 + 8 + \cdots + 2k = k(k+1).$

Then,

$$\begin{aligned} S_{k+1} &= 2 + 4 + 6 + 8 + \cdots + 2k + 2(k+1) \\ &= S_k + 2(k+1) = k(k+1) + 2(k+1) = (k+1)(k+2). \end{aligned}$$

So, we conclude that the formula is valid for all positive integer values of n.

13. 1. When $n = 1$, $S_1 = 2 = \frac{1}{2}(5(1) - 1)$.

2. Assume that

$S_k = 2 + 7 + 12 + 17 + \cdots + (5k - 3) = \frac{k}{2}(5k - 1).$

Then,

$$\begin{aligned} S_{k+1} &= 2 + 7 + 12 + 17 + \cdots + (5k-3) + [5(k+1) - 3] \\ &= S_k + (5k + 5 - 3) = \frac{k}{2}(5k - 1) + 5k + 2 \\ &= \frac{5k^2 - k + 10k + 4}{2} = \frac{5k^2 + 9k + 4}{2} \\ &= \frac{(k+1)(5k+4)}{2} = \frac{(k+1)}{2}[5(k+1) - 1]. \end{aligned}$$

So, we conclude that this formula is valid for all positive integer values of n.

15. 1. When $n = 1$, $S_1 = 1 = 2^1 - 1$.

2. Assume that

$S_k = 1 + 2 + 2^2 + 2^3 + \cdots + 2^{k-1} = 2^k - 1.$

Then,

$$S_{k+1} = 1 + 2 + 2^2 + 2^3 + \cdots + 2^{k-1} + 2^k = S_k + 2^k = 2^k - 1 + 2^k = 2(2^k) - 1 = 2^{k+1} - 1.$$

So, we conclude that this formula is valid for all positive integer values of n.

17. 1. When $n = 1$, $S_1 = 1 = \dfrac{1(1+1)}{2}$.

2. Assume that $S_k = 1 + 2 + 3 + 4 + \cdots + k = \dfrac{k(k+1)}{2}$.

Then,

$$S_{k+1} = 1 + 2 + 3 + 4 + \cdots + k + (k+1) = S_k + (k+1) = \frac{k(k+1)}{2} + \frac{2(k+1)}{2} = \frac{(k+1)(k+2)}{2}.$$

So, we conclude that this formula is valid for all positive integer values of n.

19. 1. When $n = 1$, $S_1 = 1^2 = \dfrac{1(2(1)-1)(2(1)+1)}{3}$

2. Assume that $S_k = 1^2 + 3^2 + \cdots + (2k-1)^2 = \dfrac{k(2k-1)(2k+1)}{3}$

Then,

$$\begin{aligned} S_{k+1} &= 1^2 + 3^2 + \cdots + (2k-1)^2 + (2k+1)^2 \\ &= S_k + (2k+1)^2 = \frac{k(2k-1)(2k+1)}{3} + (2k+1)^2 \\ &= (2k+1)\left[\frac{k(2k-1)}{3} + (2k+1)\right] = \frac{2k+1}{3}\left[2k^2 - k + 6k + 3\right] \\ &= \frac{2k+1}{3}(2k+3)(k+1) = \frac{(k+1)(2(k+1)-1)(2(k+1)+1)}{3} \end{aligned}$$

So, we conclude that this formula is valid for all positive integer values of n.

21. 1. When $n = 1$, $S_1 = 1 = \dfrac{(1)^2(1+1)^2\left(2(1)^2 + 2(1) - 1\right)}{12}$.

2. Assume that $S_k = \displaystyle\sum_{i=1}^{k} i^5 = \frac{k^2(k+1)^2\left(2k^2 + 2k - 1\right)}{12}$.

Then,

$$\begin{aligned} S_{k+1} &= \sum_{i=1}^{k+1} i^5 = \left(\sum_{i=1}^{k} i^5\right) + (k+1)^5 \\ &= \frac{k^2(k+1)^2\left(2k^2 + 2k - 1\right)}{12} + \frac{12(k+1)^5}{12} \\ &= \frac{(k+1)^2\left[k^2\left(2k^2 + 2k - 1\right) + 12(k+1)^3\right]}{12} \\ &= \frac{(k+1)^2\left[2k^4 + 2k^3 - k^2 + 12\left(k^3 + 3k^2 + 3k + 1\right)\right]}{12} \\ &= \frac{(k+1)^2\left[2k^4 + 14k^3 + 35k^2 + 36k + 12\right]}{12} \\ &= \frac{(k+1)^2\left(k^2 + 4k + 4\right)\left(2k^2 + 6k + 3\right)}{12} \\ &= \frac{(k+1)^2(k+2)^2\left[2(k+1)^2 + 2(k+1) - 1\right]}{12}. \end{aligned}$$

So, we conclude that this formula is valid for all positive integer values of n.

Note: The easiest way to complete the last two steps is to "work backwards." Start with the desired expression for S_{k+1} and multiply out to show that it is equal to the expression you found for $S_k + (k+1)^5$.

23. 1. When $n = 1, S_1 = 2 = \dfrac{1(2)(3)}{3}$.

2. Assume that

$$S_k = 1(2) + 2(3) + 3(4) + \cdots + k(k+1) = \frac{k(k+1)(k+2)}{3}.$$

Then,

$$S_{k+1} = 1(2) + 2(3) + 3(4) + \cdots + k(k+1) + (k+1)(k+2)$$

$$= S_k + (k+1)(k+2) = \frac{k(k+1)(k+2)}{3} + \frac{3(k+1)(k+2)}{3} = \frac{(k+1)(k+2)(k+3)}{3}.$$

So, we conclude that this formula is valid for all positive integer values of n.

25. 1. When $n = 4$, $4! = 24$ and $2^4 = 16$, thus $4! > 2^4$.

2. Assume

$k! > 2^k, k > 4$.

Then,

$(k+1)! = k!(k+1) > 2^k(2)$ since $k! > 2^k$ and $k + 1 > 2$.

Thus, $(k+1)! > 2^{k+1}$.

So, by extended mathematical induction, the inequality is valid for all integers n such that $n \geq 4$.

27. 1. When $n = 2, \dfrac{1}{\sqrt{1}} + \dfrac{1}{\sqrt{2}} \approx 1.707$ and $\sqrt{2} \approx 1.414$, thus $\dfrac{1}{\sqrt{1}} + \dfrac{1}{\sqrt{2}} > \sqrt{2}$.

2. Assume that

$$\frac{1}{\sqrt{1}} + \frac{1}{\sqrt{2}} + \frac{1}{\sqrt{3}} + \cdots + \frac{1}{\sqrt{k}} > \sqrt{k}, k > 2.$$

Then,

$$\frac{1}{\sqrt{1}} + \frac{1}{\sqrt{2}} + \frac{1}{\sqrt{3}} + \cdots + \frac{1}{\sqrt{k}} + \frac{1}{\sqrt{k+1}} > \sqrt{k} + \frac{1}{\sqrt{k+1}}.$$

Now it is sufficient to show that $\sqrt{k} + \dfrac{1}{\sqrt{k+1}} > \sqrt{k+1}, k > 2$, or equivalently $\left(\text{multiplying by } \sqrt{k+1}\right)$,

$\sqrt{k}\sqrt{k+1} + 1 > k + 1$.

This is true because $\sqrt{k}\sqrt{k+1} + 1 > \sqrt{k}\sqrt{k} + 1 = k + 1$.

Therefore, $\dfrac{1}{\sqrt{1}} + \dfrac{1}{\sqrt{2}} + \dfrac{1}{\sqrt{3}} + \cdots + \dfrac{1}{\sqrt{k}} + \dfrac{1}{\sqrt{k+1}} > \sqrt{k+1}$.

So, by extended mathematical induction, the inequality is valid for all integers n such that $n \geq 2$.

29. 1. When $n = 1, \left(\dfrac{x}{y}\right)^2 < \left(\dfrac{x}{y}\right)$ and $(0 < x < y)$.

2. Assume that

$$\left(\frac{x}{y}\right)^{k+1} < \left(\frac{x}{y}\right)^k$$

$$\left(\frac{x}{y}\right)^{k+1} < \left(\frac{x}{y}\right)^k \Rightarrow \left(\frac{x}{y}\right)\left(\frac{x}{y}\right)^{k+1} < \left(\frac{x}{y}\right)\left(\frac{x}{y}\right)^k \Rightarrow \left(\frac{x}{y}\right)^{k+2} < \left(\frac{x}{y}\right)^{k+1}.$$

So, $\left(\dfrac{x}{y}\right)^{n+1} < \left(\dfrac{x}{y}\right)^n$ for all integers $n \geq 1$.

31. 1. When $n = 1, \left[1^3 + 3(1)^2 + 2(1)\right] = 6$ and 3 is a factor.

2. Assume that 3 is a factor of $k^3 + 3k^2 + 2k$.

Then,
$$\begin{aligned}(k+1)^3 + 3(k+1)^2 + 2(k+1) &= k^3 + 3k^2 + 3k + 1 + 3k^2 + 6k + 3 + 2k + 2\\ &= \left(k^3 + 3k^2 + 2k\right) + \left(3k^2 + 9k + 6\right)\\ &= \left(k^3 + 3k^2 + 2k\right) + 3\left(k^2 + 3k + 2\right).\end{aligned}$$

Because 3 is a factor of each term, 3 is a factor of the sum.

So, 3 is a factor of $\left(n^3 + 3n^2 + 2n\right)$ for every positive integer n.

33. Prove 3 is a factor of $2^{2n+1} + 1$ for all positive integers n.

1. When $n = 1, 2^{2\cdot1+1} + 1 = 2^3 + 1 = 8 + 1 = 9$ and 3 is a factor.

2. Assume 3 is a factor of $2^{2k+1} + 1$.

Then,

$$2^{2(k+1)+1} + 1 = 2^{2k+2+1} + 1 = 2^{(2k+1)+2} + 1 = 2^{2k+1}\cdot 2^2 + 1 = 4\cdot 2^{2k+1} + 1 = 4\left(2^{2k+1} + 1\right) - 3.$$

Because 3 is a factor of each term, 3 is a factor of the sum.

So, 3 is a factor of $2^{2n+1} + 1$ for all positive integers n.

35. 1. When $n = 1, (ab)^1 = a^1b^1 = ab$.

2. Assume that $(ab)^k = a^kb^k$.

Then,
$$\begin{aligned}(ab)^{k+1} &= (ab)^k(ab)\\ &= a^kb^kab\\ &= a^{k+1}b^{k+1}.\end{aligned}$$

So, $(ab)^n = a^nb^n$.

37. 1. When $n = 2, (x_1x_2)^{-1} = \dfrac{1}{x_1x_2} = \dfrac{1}{x_1}\cdot\dfrac{1}{x_2} = x_1^{-1}x_2^{-1}$.

2. Assume that

$$(x_1x_2x_3\cdots x_k)^{-1} = x_1^{-1}x_2^{-1}x_3^{-1}\cdots x_k^{-1}.$$

Then,

$$\begin{aligned}(x_1x_2x_3\cdots x_kx_{k+1})^{-1} &= \left[(x_1x_2x_3\cdots x_k)x_{k+1}\right]^{-1}\\ &= (x_1x_2x_3\cdots x_k)^{-1}x_{k+1}^{-1}\\ &= x_1^{-1}x_2^{-1}x_3^{-1}\cdots x_k^{-1}x_{k+1}^{-1}.\end{aligned}$$

So, the formula is valid.

39. 1. When $n = 1, x(y_1) = xy_1$.

2. Assume that $x(y_1 + y_2 + \cdots + y_k) = xy_1 + xy_2 + \cdots + xy_k$.

Then,

$$\begin{aligned}xy_1 + xy_2 + \cdots + xy_k + xy_{k+1} &= x(y_1 + y_2 + \cdots + y_k) + xy_{k+1}\\ &= x\left[(y_1 + y_2 + \cdots + y_k) + y_{k+1}\right]\\ &= x(y_1 + y_2 + \cdots + y_k + y_{k+1}).\end{aligned}$$

So, the formula holds.

41. $S_n = 1 + 5 + 9 + 13 + \cdots + (4n - 3)$

$S_1 = 1 = 1 \cdot 1$

$S_2 = 1 + 5 = 6 = 2 \cdot 3$

$S_3 = 1 + 5 + 9 = 15 = 3 \cdot 5$

$S_4 = 1 + 5 + 9 + 13 = 28 = 4 \cdot 7$

From this sequence, it appears that $S_n = n(2n - 1)$.

This can be verified by mathematical induction.

1. The formula has already been verified for $n = 1$.
2. Assume that the formula is valid for $n = k$: $1 + 5 + 9 + 13 + \cdots + (4n - 3) = n(2n - 1)$.

   Then,

$$\begin{aligned} S_{k+1} &= \left[1 + 5 + 9 + 13 + \cdots + (4k - 3)\right] + \left[4(k + 1) - 3\right] \\ &= k(2k - 1) + (4k + 1) \\ &= 2k^2 + 3k + 1 \\ &= (k + 1)(2k + 1) \\ &= (k + 1)\left[2(k + 1) - 1\right] \end{aligned}$$

   So, the formula is valid.

43. $S_n = \dfrac{1}{4} + \dfrac{1}{12} + \dfrac{1}{24} + \dfrac{1}{40} + \cdots + \dfrac{1}{2n(n + 1)}$

$S_1 = \dfrac{1}{4} = \dfrac{1}{2(2)}$

$S_2 = \dfrac{1}{4} + \dfrac{1}{12} = \dfrac{4}{12} = \dfrac{2}{6} = \dfrac{2}{2(3)}$

$S_3 = \dfrac{1}{4} + \dfrac{1}{12} + \dfrac{1}{24} = \dfrac{9}{24} = \dfrac{3}{8} = \dfrac{3}{2(4)}$

$S_4 = \dfrac{1}{4} + \dfrac{1}{12} + \dfrac{1}{24} + \dfrac{1}{40} = \dfrac{16}{40} = \dfrac{4}{10} = \dfrac{4}{2(5)}$

From the sequence, it appears that $S_n = \dfrac{n}{2(n + 1)}$.

This can be verified by mathematical induction.

1. The formula has already by verified for $n = 1$.
2. Assume that the formula is valid for $n = k$: $\dfrac{1}{4} + \dfrac{1}{12} + \dfrac{1}{24} + \dfrac{1}{40} + \cdots + \dfrac{1}{2n(n + 1)} = \dfrac{n}{2(n + 1)}$

   Then, $S_{k+1} = \left[\dfrac{1}{4} + \dfrac{1}{12} + \dfrac{1}{40} + \cdots + \dfrac{1}{2k(k + 1)}\right] + \dfrac{1}{2(k + 1)(k + 2)} = \dfrac{k}{2(k + 1)} + \dfrac{1}{2(k + 1)(k + 2)}$

$$= \frac{k(k + 2) + 1}{2(k + 1)(k + 2)} = \frac{k^2 + 2k + 1}{2(k + 1)(k + 2)} = \frac{(k + 1)^2}{2(k + 1)(k + 2)} = \frac{k + 1}{2(k + 2)}.$$

   So, the formula is valid.

45. $\displaystyle\sum_{n=1}^{15} n = \frac{15(15 + 1)}{2} = 120$

47. $\displaystyle\sum_{n=1}^{6} n^2 = \frac{6(6 + 1)\left[2(6) + 1\right]}{6} = 91$

49. $\displaystyle\sum_{n=1}^{5} n^4 = \frac{5(5+1)\big[2(5)+1\big]\big[3(5)^2+3(5)-1\big]}{30} = 979$

51. $\displaystyle\sum_{n=1}^{6}(n^2 - n) = \sum_{n=1}^{6} n^2 - \sum_{n=1}^{6} n$

$\displaystyle= \frac{6(6+1)\big[2(6)+1\big]}{6} - \frac{6(6+1)}{2}$

$= 91 - 21 = 70$

53. $\displaystyle\sum_{i=1}^{6}(6i - 8i^3) = 6\sum_{i=1}^{6} i - 8\sum_{i=1}^{6} i^3 = 6\left[\frac{6(6+1)}{2}\right] - 8\left[\frac{(6)^2(6+1)^2}{4}\right] = 6(21) - 8(441) = -3402$

55. 5, 14, 23, 32, 41, 50, …

Linear

Note: This is an arithmetic sequence.

$a_1 = 5, d = 9$

$a_n = 5 + (n-1)(9)$

$a_n = 9n - 4$

57. 4, 10, 20, 34, 52, 74, …

Quadratic

$$\begin{cases} a + b + c = 4 \\ 4a + 2b + c = 10 \\ 9a + 3b + c = 20 \end{cases}$$

Solving this system yields $a = 2, b = 0$, and $c = 2$.

So, $a_n = 2n^2 + 2$.

59. −1, 11, 31, 59, 95, 139, …

Quadratic

$$\begin{cases} a + b + c = -1 \\ 4a + 2b + c = 11 \\ 9a + 3b + c = 31 \end{cases}$$

Solving this system yields $a = 4, b = 0$, and $c = -5$.

So, $a_n = 4n^2 - 5$.

61. $a_1 = 0, a_n = a_{n-1} + 3$

$a_1 = a_1 = 0$

$a_2 = a_1 + 3 = 0 + 3 = 3$

$a_3 = a_2 + 3 = 3 + 3 = 6$

$a_4 = a_3 + 3 = 6 + 3 = 9$

$a_5 = a_4 + 3 = 9 + 3 = 12$

$a_6 = a_5 + 3 = 12 + 3 = 15$

a_n: 0 3 6 9 12 15

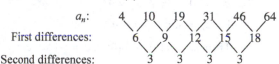

First differences: 3 3 3 3 3

Second differences: 0 0 0 0

Because the first differences are equal, the sequence has a linear model.

63. $a_1 = 4, a_n = a_{n-1} + 3n$

$a_1 = a_1 = 4$

$a_2 = a_1 + 3n = 4 + 3(2) = 10$

$a_3 = a_2 + 3n = 10 + 3(3) = 19$

$a_4 = a_3 + 3n = 19 + 3(4) = 31$

$a_5 = a_4 + 3n = 31 + 3(5) = 46$

$a_6 = a_5 + 3n = 46 + 3(6) = 64$

a_n: 4 10 19 31 46 64

First differences: 6 9 12 15 18

Second differences: 3 3 3 3

Because the second differences are all the same, the sequence has a quadratic model.

65. $a_1 = 3, a_n = a_{n-1} + n^2$

$a_1 = a_1 = 3$

$a_2 = a_1 + 2^2 = 3 + 4 = 7$

$a_3 = a_2 + 3^2 = 7 + 9 = 16$

$a_4 = a_3 + 4^2 = 16 + 16 = 32$

$a_5 = a_4 + 5^2 = 32 + 25 = 57$

$a_6 = a_5 + 6^2 = 57 + 36 = 93$

a_n: 3 7 16 32 57 93

First differences: 4 9 16 25 36

Second differences: 5 7 9 11

Neither linear nor quadratic.

67. $a_1 = 5, a_n = 4n - a_{n-1}$

$a_1 = a_1 = 5$

$a_2 = 4(2) - a_1 = 8 - 5 = 3$

$a_3 = 4(3) - a_2 = 12 - 3 = 9$

$a_4 = 4(4) - a_3 = 16 - 9 = 7$

$a_5 = 4(5) - a_4 = 20 - 7 = 13$

$a_6 = 4(6) - a_5 = 24 - 13 = 11$

a_n: 5 3 9 7 13 11

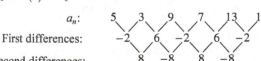

First differences: −2 6 −2 6 −2

Second differences: 8 −8 8 −8

Neither linear nor quadratic.

69. $a_0 = 3, a_1 = 3, a_4 = 15$

Let $a_n = an^2 + bn + c$.

Then:

$$a_0 = a(0)^2 + b(0) + c = 3 \Rightarrow \qquad c = 3$$

$$a_1 = a(1)^2 + b(1) + c = 3 \Rightarrow \quad a + b + c = 3$$

$$a + b = 0$$

$$a_4 = a(4)^2 + b(4) + c = 15 \Rightarrow 16a + 4b + c = 15$$

$$16a + 4b = 12$$

$$4a + b = 3$$

By elimination: $-a - b = 0$

$$\underline{4a + b = 3}$$

$$3a = 3$$

$$a = 1 \Rightarrow b = -1$$

So, $a_n = n^2 - n + 3$.

71. $a_0 = -1, a_2 = 5, a_4 = 15$

Let $a_n = an^2 + bn + c$.

Then: $a_0 = a(0)^2 + b(0) + c = -1 \Rightarrow c = -1$

$$a_2 = a(2)^2 + b(2) + c = 5 \Rightarrow \quad 4a + 2b + c = 5$$

$$4a + 2b = 6$$

$$2a + b = 3$$

$$a_4 = a(4)^2 + b(4) + c = 15 \Rightarrow 16a + 4b + c = 15$$

$$16a + 4b = 16$$

$$4a + b = 4$$

By elimination: $-2a - b = -3$

$$\underline{4a + b = 4}$$

$$2a = 1$$

$$a = \frac{1}{2}$$

$$4\left(\frac{1}{2}\right) + b = 4$$

$$b = 2$$

So, $a_n = \frac{1}{2}n^2 + 2n - 1$.

73. $a_1 = 0, a_2 = 7, a_4 = 27$

Let $a_n = an^2 + bn + c$.

Then: $a_1 = a(1)^2 + b(1) + c = 0 \Rightarrow a + b + c = 0$

$$a_2 = a(2)^2 + b(2) + c = 7 \Rightarrow 4a + 2b + c = 7$$

$$a_4 = a(4)^2 + b(4) + c = 27 \Rightarrow 16a + 4b + c = 27$$

$$\begin{cases} a + b + c = 0 \\ 4a + 2b + c = 7 \\ 16a + 4b + c = 27 \end{cases}$$

$$\begin{cases} a + b + c = 0 \\ -2b - 3c = 7 \\ -12b - 15c = 27 \end{cases}$$

$$\begin{cases} a + b + c = 0 \\ -2b - 3c = 7 \\ 3c = -15 \end{cases}$$

$$3c = -15 \rightarrow c = -5$$

$$-2b - 3(-5) = 7 \rightarrow b = 4$$

$$a + 4 + (-5) = 0 \rightarrow a = 1$$

So, $a_n = n^2 + 4n - 5$.

75. (a) n: 10 11 12 13 14 15

Terms: 4785 4801 4816 4831 4846 4859

First differences: 16 15 15 15 13

Sample Answer: Using common difference $d = 15$, $a_n = 15n + 4636$

(b) Using a graphing utility, a linear model for the data is $a_n \approx 14.9n + 4637$. So, the models are comparable.

(c) Using $a_n = 15n + 4636$, the number of residents in 2021 is $a_{16} = 15(21) + 4636 = 4{,}951{,}000$.

Using $a_n = 14.9n + 4637$, the number of residents in 2021 is $a_{16} = 14.9(21) + 4637 = 4{,}949{,}900$.

So, the predictions are similar.

77. False. P_1 must be proven to be true.

Section 8.5 The Binomial Theorem

1. expanding

3. Binomial Theorem; Pascal's Triangle

5. $_5C_3 = \dfrac{5!}{3! \cdot 2!} = \dfrac{5 \cdot 4}{2 \cdot 1} = 10$

7. $_{12}C_0 = \dfrac{12!}{0! \cdot 12!} = 1$

9. $\dbinom{10}{4} = \dfrac{10!}{6! \cdot 4!} = \dfrac{10 \cdot 9 \cdot 8 \cdot 7 \cdot 6!}{6!(24)} = 210$

11. $\dbinom{100}{98} = \dfrac{100!}{2! \cdot 98!} = \dfrac{100 \cdot 99}{2 \cdot 1} = 4950$

13.

1
1 1
1 2 1
1 3 3 1
1 4 6 4 1
1 5 10 10 5 1
1 6 15 (20) 15 6 1

$\binom{6}{3} = 20$, the 4th entry in the 6th row.

15.

1
1 1
1 2 1
1 3 3 1
1 4 6 4 1
1 (5) 10 10 5 1

$\binom{5}{1} = 5$, the 2nd entry in the 5th row.

17. $(x+1)^6 = {}_6C_0x^6 + {}_6C_1x^5(1) + {}_6C_2x^4(1)^2 + {}_6C_3x^3(1)^3 + {}_6C_4x^2(1)^4 + {}_6C_5x(1)^5 + {}_6C_6(1)^6$

$= x^6 + 6x^5 + 15x^4 + 20x^3 + 15x^2 + 6x + 1$

19. $(y-3)^3 = {}_3C_0y^3 - {}_3C_1y^2(3) + {}_3C_2y(3)^2 - {}_3C_3(3)^3$

$= 1y^3 - 3y^2(3) + 3y(3)^2 - 1(3)^3$

$= y^3 - 9y^2 + 27y - 27$

21. $(r+3s)^3 = {}_3C_0r^3 + {}_3C_1r^2(3s) + {}_3C_2r(3s)^2 + {}_3C_3(3s)^3$

$= 1r^3 + 3r^2(3s) + 3r(3s)^2 + 1(3s)^3$

$= r^3 + 9r^2s + 27rs^2 + 27s^3$

23. $(3a-4b)^5 = {}_5C_0(3a)^5 - {}_5C_1(3a)^4(4b) + {}_5C_2(3a)^3(4b)^2 - {}_5C_3(3a)^2(4b)^3 + {}_5C_4(3a)(4b)^4 - {}_5C_5(4b)^5$

$= (1)(243a^5) - 5(81a^4)(4b) + 10(27a^3)(16b^2) - 10(9a^2)(64b^3) + 5(3a)(256b^4) - (1)(1024b^5)$

$= 243a^5 - 1620a^4b + 4320a^3b^2 - 5760a^2b^3 + 3840ab^4 - 1024b^5$

25. $(a+6)^4 = {}_4C_0a^4 + {}_4C_1a^3(6) + {}_4C_2a^2(6)^2 + {}_4C_3a(6)^3 + {}_4C_4(6)^4$

$= 1a^4 + 4a^3(6) + 6a^2(6)^2 + 4a(6)^3 + 1(6)^4$

$= a^4 + 24a^3 + 216a^2 + 864a + 1296$

27. $(y-1)^6 = {}_6C_0y^6 - {}_6C_1y^5(1) + {}_6C_2y^4(1)^2 - {}_6C_3y^3(1)^3 + {}_6C_4y^2(1)^4 - {}_6C_5y(1)^5 + {}_6C_6(1)^6$

$= 1y^6 - 6y^5(1) + 15y^4(1)^2 - 20y^3(1)^3 + 15y^2(1)^4 - 6y(1)^5 + 1(1)^6$

$= y^6 - 6y^5 + 15y^4 - 20y^3 + 15y^2 - 6y + 1$

29. 4th Row of Pascal's Triangle: 1 4 6 4 1

$(3-2z)^4 = 3^4 - 4(3)^3(2z) + 6(3)^2(2z)^2 - 4(3)(2z)^3 + (2z)^4$

$= 81 - 216z + 216z^2 - 96z^3 + 16z^4$

31. 5th Row of Pascal's Triangle: 1 5 10 10 5 1

$(x+2y)^5 = 1x^5 + 5x^4(2y) + 10x^3(2y)^2 + 10x^2(2y)^3 + 5x(2y)^4 + 1(2y)^5$

$= x^5 + 10x^4y + 40x^3y^2 + 80x^2y^3 + 80xy^4 + 32y^5$

33. $(x^2+y^2)^4 = {}_4C_0(x^2)^4 + {}_4C_1(x^2)^3(y^2) + {}_4C_2(x^2)^2(y^2)^2 + {}_4C_3(x^2)(y^2)^3 + {}_4C_4(y^2)^4$

$= (1)(x^8) + (4)(x^6y^2) + (6)(x^4y^4) + (4)(x^2y^6) + (1)(y^8)$

$= x^8 + 4x^6y^2 + 6x^4y^4 + 4x^2y^6 + y^8$

35. $\left(\frac{1}{x} + y\right)^5 = {}_5C_0\left(\frac{1}{x}\right)^5 + {}_5C_1\left(\frac{1}{x}\right)^4 y + {}_5C_2\left(\frac{1}{x}\right)^3 y^2 + {}_5C_3\left(\frac{1}{x}\right)^2 y^3 + {}_5C_4\left(\frac{1}{x}\right)y^4 + {}_5C_5 y^5$

$= 1\left(\frac{1}{x}\right)^5 + 5\left(\frac{1}{x}\right)^4 y + 10\left(\frac{1}{x}\right)^3 y^2 + 10\left(\frac{1}{x}\right)^2 y^3 + 5\left(\frac{1}{x}\right)y^4 + 1y^5$

$= \frac{1}{x^5} + \frac{5y}{x^4} + \frac{10y^2}{x^3} + \frac{10y^3}{x^2} + \frac{5y^4}{x} + y^5$

37. $2(x-3)^4 + 5(x-3)^2 = 2\left[{}_4C_0x^4 - {}_4C_1x^3(3) + {}_4C_2x^2(3)^2 - {}_4C_3x(3)^3 + {}_4C_4(3)^4\right] + 5\left[{}_2C_0x^2 + {}_2C_1x(3) + {}_2C_2(3)^2\right]$

$= 2\left[x^4 - 4(x^3)(3) + 6(x^2)(3)^2 - 4(x)(3^3) + 3^4\right] + 5\left[x^2 - 2(x)(3) + 3^2\right]$

$= 2(x^4 - 12x^3 + 54x^2 - 108x + 81) + 5(x^2 - 6x + 9)$

$= 2x^4 - 24x^3 + 113x^2 - 246x + 207$

39. The 4th term in the expansion of $(x + y)^{10}$ is

${}_{10}C_3x^{10-3}y^3 = 120x^7y^3.$

41. The 3rd term in the expansion of $(x - 6y)^5$ is

${}_5C_2x^{5-2}(-6y)^2 = 10x^3(36y^2) = 360x^3y^2.$

43. The 8th term in the expansion of $(4x + 3y)^9$ is

${}_9C_7(4x)^{9-7}(3y)^7 = 36(16x^2)(2187y^7)$

$= 1{,}259{,}712x^2y^7.$

45. The 10th term in the expansion of $(10x - 3y)^{12}$ is

${}_{12}C_9(10x)^{12-9}(-3y)^9 = 220(1000x^3)(-19{,}683y^9)$

$= -4{,}330{,}260{,}000x^3y^9.$

47. The new term involving ax^3 in the expansion of $(x + 2)^6$ is

${}_6C_3x^3(2)^3 = \frac{6!}{3! \cdot 3!} \cdot 2^3x^3 = 160x^3.$

The coefficient is 160.

49. The term involving x^2y^8 in the expansion of $(4x - y)^{10}$

is ${}_{10}C_8(4x)^2(-y)^8 = \frac{10!}{(10-8)!8!} \cdot 16x^2y^8 = 720x^2y^8.$

The coefficient is 720.

51. The term involving x^4y^5 in the expansion of $(2x - 5x)^9$

is ${}_9C_5(2x)^4(-5y)^5 = 126(16x^4)(-3125y^5)$

$= -6{,}300{,}000x^4y^5.$

The coefficient is –6,300,000.

53. The term involving $x^8y^6 = (x^2)^4y^6$ in the expansion of

$(x^2 + y)^{10}$ is ${}_{10}C_6(x^2)^4y^6 = \frac{10!}{4!6!}(x^2)^4y^6 = 210x^8y^6.$

The coefficient is 210.

55. $(\sqrt{x} + 5)^3 = (\sqrt{x})^3 + 3(\sqrt{x})^2(5) + 3(\sqrt{x})(5^2) + 5^3$

$= x^{3/2} + 15x + 75x^{1/2} + 125$

57. $(x^{2/3} - y^{1/3})^3 = (x^{2/3})^3 - 3(x^{2/3})^2(y^{1/3}) + 3(x^{2/3})(y^{1/3})^2 - (y^{1/3})^3 = x^2 - 3x^{4/3}y^{1/3} + 3x^{2/3}y^{2/3} - y$

59. $(3\sqrt{t} + \sqrt[4]{t})^4 = (3\sqrt{t})^4 + 4(3\sqrt{t})^3(\sqrt[4]{t}) + 6(3\sqrt{t})^2(\sqrt[4]{t})^2 + 4(3\sqrt{t})(\sqrt[4]{t})^3 + (\sqrt[4]{t})^4$

$= 81t^2 + 108t^{7/4} + 54t^{3/2} + 12t^{5/4} + t$

61. $\frac{f(x+h) - f(x)}{h} = \frac{(x+h)^3 - x^3}{h} = \frac{x^3 + 3x^2h + 3xh^2 + h^3 - x^3}{h}$

$= \frac{h(3x^2 + 3xh + h^2)}{h} = 3x^2 + 3xh + h^2, h \neq 0$

63. $$\frac{f(x+h)-f(x)}{h} = \frac{(x+h)^6 - x^6}{h}$$
$$= \frac{x^6 + 6x^5h + 15x^4h^2 + 20x^3h^3 + 15x^2h^4 + 6xh^5 + h^6 - x^6}{h}$$
$$= \frac{h\left(6x^5 + 15x^4h + 20x^3h^2 + 15x^2h^3 + 6xh^4 + h^5\right)}{h}$$
$$= 6x^5 + 15x^4h + 20x^3h^2 + 15x^2h^3 + 6xh^4 + h^5,\ h \neq 0$$

65. $$\frac{f(x+h)-f(x)}{h} = \frac{\sqrt{x+h}-\sqrt{x}}{h} = \frac{\sqrt{x+h}-\sqrt{x}}{h}\cdot\frac{\sqrt{x+h}+\sqrt{x}}{\sqrt{x+h}+\sqrt{x}} = \frac{(x+h)-x}{h\left(\sqrt{x+h}+\sqrt{x}\right)} = \frac{1}{\sqrt{x+h}+\sqrt{x}},\ h \neq 0$$

67. $$(1+i)^4 = {}_4C_0(1)^4 + {}_4C_1(1)^3i + {}_4C_2(1)^2i^2 + {}_4C_3(1)i^3 + {}_4C_4i^4 = 1 + 4i - 6 - 4i + 1 = -4$$

69. $$(2-3i)^6 = {}_6C_0 2^6 - {}_6C_1 2^5(3i) + {}_6C_2 2^4(3i)^2 - {}_6C_3 2^3(3i)^3 + {}_6C_4 2^2(3i)^4 - {}_6C_5 2(3i)^5 + {}_6C_6(3i)^6$$
$$= (1)(64) - (6)(32)(3i) + 15(16)(-9) - 20(8)(-27i) + 15(4)(81) - 6(2)(243i) + (1)(-729)$$
$$= 64 - 576i - 2160 + 4320i + 4860 - 2916i - 729$$
$$= 2035 + 828i$$

71. $$\left(-\frac{1}{2}+\frac{\sqrt{3}}{2}i\right)^3 = \frac{1}{8}\left[(-1)^3 + 3(-1)^2\left(\sqrt{3}i\right) + 3(-1)\left(\sqrt{3}i\right)^2 + \left(\sqrt{3}i\right)^3\right] = \frac{1}{8}\left[-1 + 3\sqrt{3}i + 9 - 3\sqrt{3}i\right] = 1$$

73. $$(1.02)^8 = (1+0.02)^8$$
$$= 1 + 8(0.02) + 28(0.02)^2 + 56(0.02)^3 + 70(0.02)^4 + 56(0.02)^5 + 28(0.02)^6 + 8(0.02)^7 + (0.02)^8$$
$$= 1 + 0.16 + 0.0112 + 0.000448 + \cdots$$
$$\approx 1.172$$

75. $$(2.99)^{12} = (3-0.01)^{12}$$
$$= 3^{12} - 12(3)^{11}(0.01) + 66(3)^{10}(0.01)^2 - 220(3)^9(0.01)^3 + 495(3)^8(0.01)^4$$
$$-792(3)^7(0.01)^5 + 924(3)^6(0.01)^6 - 792(3)^5(0.01)^7 + 495(3)^4(0.01)^8$$
$$-220(3)^3(0.01)^9 + 66(3)^2(0.01)^{10} - 12(3)(0.01)^{11} + (0.01)^{12}$$
$$\approx 531{,}441 - 21{,}257.64 + 389.7234 - 4.3303 + 0.0325 - 0.0002 + \cdots \approx 510{,}568.785$$

77. $${}_7C_4\left(\frac{1}{2}\right)^4\left(\frac{1}{2}\right)^3 = \frac{7!}{3!4!}\left(\frac{1}{16}\right)\left(\frac{1}{8}\right) = 35\left(\frac{1}{16}\right)\left(\frac{1}{8}\right) \approx 0.273$$

79. $${}_8C_4\left(\frac{1}{3}\right)^4\left(\frac{2}{3}\right)^4 = \frac{8!}{4!4!}\left(\frac{1}{81}\right)\left(\frac{16}{81}\right) = 70\left(\frac{1}{81}\right)\left(\frac{16}{81}\right) \approx 0.171$$

81. $$f(x) = x^2 - 4x$$
$$g(x) = f(x+4)$$
$$= (x+4)^3 - 4(x+4)$$
$$= x^3 + 3x^2(4) + 3x(4)^2 + (4)^3 - 4x - 16$$
$$= x^3 + 12x^2 + 48x + 64 - 4x - 16$$
$$= x^3 + 12x^2 + 44x + 48$$

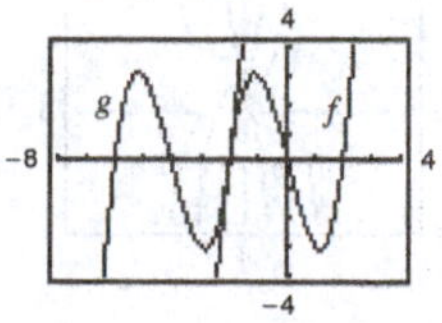

The graph of g is the same as the graph of f shifted four units to the left.

83.

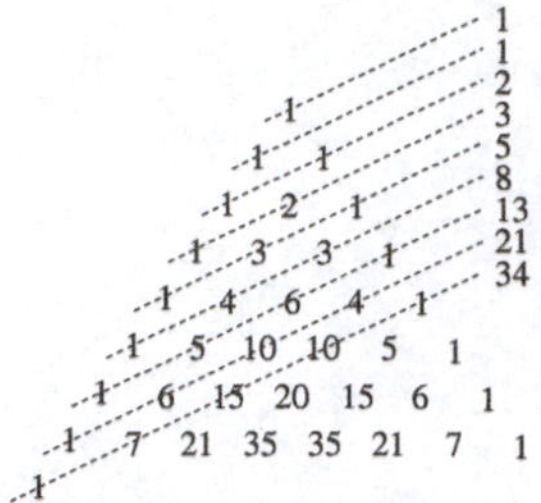

The first nine terms of the sequence are 1, 1, 2, 3, 5, 8, 13, 21, 34, …

After the first two terms, the next terms are formed by adding the previous two terms.

$a_1 = 1, a_2 = 1$

$a_3 = a_1 + a_2 = 1 + 1 = 2$

$a_4 = a_2 + a_3 = 1 + 2 = 3$

$a_5 = a_3 + a_4 = 2 + 3 = 5$

$a_6 = a_4 + a_5 = 3 + 5 = 8$

$a_7 = a_5 + a_6 = 5 + 8 = 13$

This is called the Fibonacci sequence.

85. (a) $f(t) = -0.056t^2 + 1.62t + 16.4$

$$\begin{aligned} g(t) &= f(t+5) \\ &= -0.056(t+5)^2 + 1.62(t+5) + 16.4 \\ &= -0.056(t^2 + 10t + 25) + 1.62(t+5) + 16.4 \\ &= -0.056t^2 - 0.56t - 1.4 + 1.62t + 8.1 + 16.4 \\ &= -0.056t^2 + 1.06t + 23.1 \end{aligned}$$

(b) and (c)

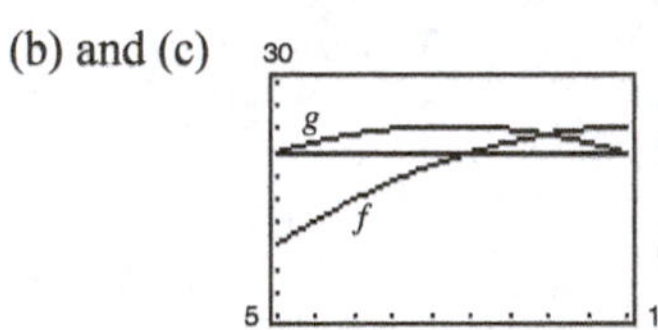

Graphically, child support collections exceeded $27 billion in about 2010.

87. True. The coefficients from the Binomial Theorem can be used to find the numbers in Pascal's Triangle.

89. The first and last numbers in each row are 1. Every other number in each row is formed by adding the two numbers immediately above the number.

91. The functions $f(x) = (1-x)^3$ and $k(x) = 1 - 3x + 3x^2 + x^3$

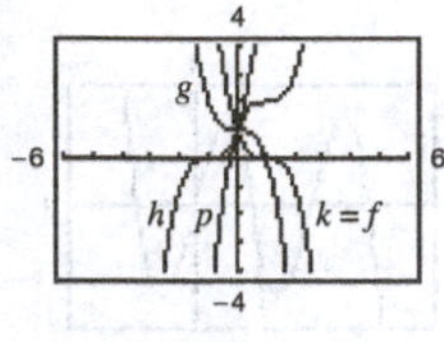

f and k have identical graphs, because $k(x)$ is the expansion of $f(x)$.

93. $${}_nC_{n-r} = \frac{n!}{(n-(n-r))!(n-r)!} = \frac{n!}{r!(n-r)!} = \frac{n!}{(n-r)!r!} = {}_nC_r$$

95.
$$\begin{aligned} {}_nC_r + {}_nC_{r-1} &= \frac{n!}{(n-r)!r!} + \frac{n!}{(n-r+1)!(r-1)!} \\ &= \frac{n!(n-r+1)!(r-1)! + n!(n-r)!r!}{(n-r)!r!(n-r+1)!(r-1)!} \\ &= \frac{n!\left[(n-r+1)!(r-1)! + r!(n-r)!\right]}{(n-r)!r!(n-r+1)!(r-1)!} \\ &= \frac{n!(r-1)!\left[(n-r+1)! + r(n-r)!\right]}{(n-r)!r!(n-r+1)!(r-1)!} \\ &= \frac{n!(n-r)!\left[(n-r+1) + r\right]}{(n-r)!r!(n-r+1)!} \\ &= \frac{n![n+1]}{r!(n-r+1)!} \\ &= \frac{(n+1)!}{\left[(n+1)-r\right]!r!} \\ &= {}_{n+1}C_r \end{aligned}$$

97.

n	r	${}_nC_r$	${}_nC_{n-r}$
9	5	126	126
7	1	7	7
12	4	495	495
6	0	1	1
10	7	120	120

${}_nC_r = {}_nC_{n-r}$

The table illustrates the symmetry of Pascal's Triangle.

Section 8.6 Counting Principles

1. Fundamental Counting Principle

3. $_nP_r = \dfrac{n!}{(n-r)!}$

5. combinations

7. Odd integers: 1, 3, 5, 7, 9, 11
6 ways

9. Prime integers: 2, 3, 5, 7, 11
5 ways

11. Divisible by 4: 4, 8, 12
3 ways

13. Sum is 9: $1+8, 2+7, 3+6, 4+5, 5+4,$
$6+3, 7+2, 8+1$
8 ways

15. Amplifiers: 3 choices
Compact disc players: 2 choices
Speakers: 5 choices
Total: $3 \cdot 2 \cdot 5 = 30$ ways

17. Math courses: 2
Science courses: 3
Social sciences and humanities courses: 5
Total: $2 \cdot 3 \cdot 5 = 30$ schedules

19. $2^6 = 64$

21. $26 \cdot 26 \cdot 26 \cdot 10 \cdot 10 \cdot 10 \cdot 10 = 175{,}760{,}000$
distinct license plate numbers

23. (a) $9 \cdot 10 \cdot 10 = 900$
(b) $9 \cdot 9 \cdot 8 = 648$
(c) $9 \cdot 10 \cdot 2 = 180$
(d) $6 \cdot 10 \cdot 10 = 600$

25. $40^3 = 64{,}000$

27. (a) $8 \cdot 7 \cdot 6 \cdot 5 \cdot 4 \cdot 3 \cdot 2 \cdot 1 = 40{,}320$
(b) $8 \cdot 1 \cdot 6 \cdot 1 \cdot 4 \cdot 1 \cdot 2 \cdot 1 = 384$

29. $5! = 120$ ways

31. $_5P_2 = \dfrac{5!}{(5-2)!} = \dfrac{5!}{3!} = 5 \cdot 4 = 20$

33. $_{12}P_2 = \dfrac{12!}{(12-2)!} = \dfrac{12!}{10!} = 12 \cdot 11 = 132$

35. $_{15}P_3 = 2730$

37. $_{50}P_4 = 5{,}527{,}200$

39. The number of permutations of 9 possible donors taken 3 at a time is
$_9P_3 = \dfrac{9!}{(9-3)!} = \dfrac{9!}{6!} = 9 \cdot 8 \cdot 7 = 504$ possible orders.

41. $_{15}P_9 = \dfrac{15!}{6!} = 1{,}816{,}214{,}400$
different batting orders

43. $\dfrac{7!}{2!1!3!1!} = \dfrac{7!}{2!3!} = 420$

45. $\dfrac{7!}{2!1!1!1!1!1!} = \dfrac{7!}{2!} = 7 \cdot 6 \cdot 5 \cdot 4 \cdot 3 = 2520$

47.

ABCD	BACD	CABD	DABC
ABDC	BADC	CADB	DACB
ACBD	BCAD	CBAD	DBAC
ACDB	BCDA	CBDA	DBCA
ADBC	BDAC	CDAB	DCAB
ADCB	BDCA	CDBA	DCBA

49. $_6C_4 = \dfrac{6!}{4!(6-4)!} = \dfrac{6!}{4!2!} = 15$

51. $_9C_9 = \dfrac{9!}{9!(9-9)!} = \dfrac{9!}{9!0!} = 1$

53. $_{16}C_2 = 120$

55. $_{20}C_6 = 38{,}760$

57. There are $_6C_2 = 15$ different combinations: AB, AC, AD, AE, AF, BC, BD, BE, BF, CD, CE, CF, DE, DF, EF

59. $_{40}C_{12} = \dfrac{40!}{28!12!} = 5{,}586{,}853{,}480$ ways

61. $_{35}C_5 = \dfrac{35!}{30!5!} = 324{,}632$ ways

63. There are 22 good units and 3 defective units.

(a) $_{22}C_4 = \dfrac{22!}{4!8!} = 7315$ ways

(b) $_{22}C_2 \cdot {_3C_2} = \dfrac{22!}{2!20!} \cdot \dfrac{3!}{2!1!} = 231 \cdot 3 = 693$ ways

(c)
$$\begin{aligned} {_{22}C_4} + {_{22}C_3} \cdot {_3C_1} + {_{22}C_2} \cdot {_3C_2} &= \frac{22!}{4!18!} + \frac{22!}{3!19!} \cdot \frac{3!}{1!2!} + \frac{22!}{2!20!} \cdot \frac{3!}{2!1!} \\ &= 7315 + 1540 \cdot 3 + 231 \cdot 3 \\ &= 12{,}628 \text{ ways} \end{aligned}$$

65. (a) Select type of card for three of a kind: $_{13}C_1$

Select three of four cards for three of a kind: $_4C_3$

Select type of card for pair: $_{12}C_1$

Select two of four cards for pair: $_4C_2$

$$_{13}C_1 \cdot {_4C_3} \cdot {_{12}C_1} \cdot {_4C_2} = \frac{13!}{(13-1)!1!} \cdot \frac{4!}{(4-3)!3!} \cdot \frac{12!}{(12-1)!1!} \cdot \frac{4!}{(4-2)!2!} = 3744$$

(b) Select two jacks: $_4C_2$

Select three aces: $_4C_3$

$$_4C_2 \cdot {_4C_3} = \frac{4!}{(4-2)!2!} \cdot \frac{4!}{(4-3)!3!} = 24$$

67. $_7C_1 \cdot {_{12}C_3} \cdot {_{20}C_2} = \dfrac{7!}{(7-1)!1!} \cdot \dfrac{12!}{(12-3)!3!} \cdot \dfrac{20!}{(20-2)!2!} = 292{,}600$

69. $_5C_2 - 5 = 10 - 5 = 5$ diagonals

71. $_8C_2 - 8 = 28 - 8 = 20$ diagonals

73. $_9C_2 = \dfrac{9!}{2!7!} = 36$ lines

75. $4 \cdot {_{n+1}P_2} = {_{n+2}P_3}$ **Note:** $n \geq 1$ for this to be defined.

$$4 \cdot \frac{(n+1)!}{(n-1)!} = \frac{(n+2)!}{(n-1)!}$$

$$4(n+1)(n) = (n+2)(n+1)n$$ (We can divide by $(n+1)n$ because $n \neq -1$ and $n \neq 0$.)

$$4 = n + 2$$

$$2 = n$$

77. $_{n+1}P_3 = 4 \cdot {_nP_2}$ **Note:** $n \geq 2$ for this to be defined.

$$\frac{(n+1)!}{(n-2)!} = 4 \cdot \frac{n!}{(n-2)!}$$

$$(n+1)(n)(n-1) = 4(n)(n-1)$$ (We can divide by $n(n-1)$ because $n \neq 0$, and $n \neq 1$.)

$$n + 1 = 4$$

$$n = 3$$

79. $14 \cdot {}_nP_3 = {}_{n+2}P_4$ **Note:** $n \geq 3$ for this to be defined.

$$14\left(\frac{n!}{(n-3)!}\right) = \frac{(n+2)!}{(n-2)!}$$

$$14n(n-1)(n-2) = (n+2)(n+1)n(n-1)$$ (We can divide here by $n(n-1)$ because $n \neq 0$, $n \neq 1$.)

$$14(n-2) = (n+2)(n+1)$$

$$14n - 28 = n^2 + 3n + 2$$

$$0 = n^2 - 11n + 30$$

$$0 = (n-5)(n-6)$$

$$n = 5 \quad \text{or} \quad n = 6$$

81. ${}_nP_4 = 10 \cdot {}_{n-1}P_3$ **Note:** $n \geq 4$ for this to be defined.

$$\frac{n!}{(n-4)!} = 10 \cdot \frac{(n-1)!}{(n-4)!}$$

$$n(n-1)(n-2)(n-3) = 10(n-1)(n-2)(n-3)$$ (We can divide by $(n-1)(n-2)(n-3)$ because $n \neq 1$, $n \neq 2$, and $n \neq 3$.)

$$n = 10$$

83. False.

It is an example of a combination.

85. ${}_{10}P_6 > {}_{10}C_6$

Changing the order of any of the six elements selected results in a different permutation but the same combination.

87. $${}_nP_{n-1} = \frac{n!}{(n-(n-1))!} = \frac{n!}{1!} = \frac{n!}{0!} = {}_nP_n$$

89. $${}_nC_{n-1} = \frac{n!}{(n-(n-1))!(n-1)!} = \frac{n!}{(1)!(n-1)!}$$

$$= \frac{n!}{(n-1)!1!} = {}_nC_1$$

91. ${}_{100}P_{80} \approx 3.836 \times 10^{139}$

This number is too large for some calculators to evaluate.

Section 8.7 Probability

1. experiment; outcomes

3. probability

5. mutually exclusive

7. complement

9. $\{(H, 1), (H, 2), (H, 3), (H, 4), (H, 5), (H, 6), (T, 1), (T, 2), (T, 3), (T, 4), (T, 5), (T, 6)\}$

11. {ABC, ACB, BAC, BCA, CAB, CBA}

13. {AB, AC, AD, AE, BC, BD, BE, CD, CE, DE}

15. $E = \{HHT, HTH, THH\}$

$$P(E) = \frac{n(E)}{n(S)} = \frac{3}{8}$$

17. $E = \{HHH, HHT, HTH, HTT\}$

$$P(E) = \frac{n(E)}{n(S)} = \frac{4}{8} = \frac{1}{2}$$

19. $E = \{HHH, HHT, HTH, HTT, THH, THT, TTH\}$

$$P(E) = \frac{n(E)}{n(S)} = \frac{7}{8}$$

21. $E = \{K♣, K♦, K♥, K♠, Q♣, Q♦, Q♥, Q♠, J♣, J♦, J♥, J♠\}$

$$P(E) = \frac{n(E)}{n(S)} = \frac{12}{52} = \frac{3}{13}$$

23. $E = \{K\blacklozenge, K\heartsuit, Q\blacklozenge, Q\heartsuit, J\blacklozenge, J\heartsuit\}$

$$P(E) = \frac{n(E)}{n(S)} = \frac{6}{52} = \frac{3}{26}$$

25. $E = \{(1, 5), (2, 4), (3, 3), (4, 2), (5, 1)\}$

$$P(E) = \frac{n(E)}{n(S)} = \frac{5}{36}$$

27. Use the complement.

$$E' = \{(5, 6), (6, 5), (6, 6)\}$$

$$P(E') = \frac{n(E')}{n(S)} = \frac{3}{36} = \frac{1}{12}$$

$$P(E) = 1 - P(E') = 1 - \frac{1}{12} = \frac{11}{12}$$

29. $E_3 = \{(1, 2), (2, 1)\}, n(E_3) = 2$

$E_5 = \{(1, 4), (2, 3), (3, 2), (4, 1)\}, n(E_5) = 4$

$E_7 = \{(1, 6), (2, 5), (3, 4), (4, 3), (5, 2), (6, 1)\}, n(E_7) = 6$

$E = E_3 \cup E_5 \cup E_7$

$n(E) = 2 + 4 + 6 = 12$

$$P(E) = \frac{n(E)}{n(S)} = \frac{12}{36} = \frac{1}{3}$$

31. $P(E) = \dfrac{{}_3C_2}{{}_6C_2} = \dfrac{3}{15} = \dfrac{1}{5}$

33. $P(E) = \dfrac{{}_4C_2}{{}_6C_2} = \dfrac{6}{15} = \dfrac{2}{5}$

35. (a) $0.12(8.3) \approx 0.996$ million $= 996{,}000$

(b) $18\% = \frac{9}{50}$

(c) $54\% = \frac{27}{50}$

(d) $12\% + 4\% = 16\% = \frac{4}{25}$

37. (a) $\dfrac{104}{128} = \dfrac{13}{16}$

(b) $\dfrac{24}{128} = \dfrac{3}{16}$

(c) $\dfrac{52 - 48}{128} = \dfrac{1}{32}$

39. $1 - 0.37 - 0.44 = 0.19 = 19\%$

41. (a) $\dfrac{{}_{15}C_{10}}{{}_{20}C_{10}} = \dfrac{3003}{184{,}756} = \dfrac{21}{1292} \approx 0.016$

(b) $\dfrac{{}_{15}C_8 \cdot {}_5C_2}{{}_{20}C_{10}} = \dfrac{64{,}350}{184{,}756} = \dfrac{225}{646} \approx 0.348$

(c) $\dfrac{{}_{15}C_9 \cdot {}_5C_1}{{}_{20}C_{10}} + \dfrac{{}_{15}C_{10}}{{}_{20}C_{10}} + \dfrac{25{,}025 + 3003}{184{,}756} = \dfrac{28{,}028}{184{,}756}$

$$= \frac{49}{323}$$

$$\approx 0.152$$

43. (a) $\dfrac{1}{{}_5P_5} = \dfrac{1}{120}$

(b) $\dfrac{1}{{}_4P_4} = \dfrac{1}{24}$

45. (a) $\frac{20}{52} = \frac{5}{13}$

(b) $\frac{26}{52} = \frac{1}{2}$

(c) $\frac{16}{52} = \frac{4}{13}$

47. (a) $\dfrac{{}_9C_4}{{}_{12}C_4} = \dfrac{126}{495} = \dfrac{14}{55}$ (4 good units)

(b) $\dfrac{{}_9C_2 \cdot {}_3C_2}{{}_{12}C_4} = \dfrac{108}{495} = \dfrac{12}{55}$ (2 good units)

(c) $\dfrac{{}_9C_3 \cdot {}_3C_1}{{}_{12}C_4} = \dfrac{252}{495} = \dfrac{28}{55}$ (3 good units)

At least 2 good units: $\dfrac{12}{55} + \dfrac{28}{55} + \dfrac{14}{55} = \dfrac{54}{55}$

49. (a) $P(EE) = \frac{20}{40} \cdot \frac{20}{40} = \frac{1}{4}$

(b) $P(EO \text{ or } OE) = 2\left(\frac{20}{40}\right)\left(\frac{20}{40}\right) = \frac{1}{2}$

(c) $P(N_1 < 30, N_2 < 30) = \frac{29}{40} \cdot \frac{29}{40} = \frac{841}{1600}$

(d) $P(N_1N_1) = \frac{30}{40} \cdot \frac{1}{40} = \frac{1}{40}$

51. $P(E') = 1 - P(E) = 1 - 0.73 = 0.27$

53. $P(E') = 1 - P(E) = 1 - \frac{1}{5} = \frac{4}{5}$

55. $P(E) = 1 - P(E') = 1 - 0.29 = 0.71$

57. $P(E) = 1 - P(E') = 1 - \frac{14}{25} = \frac{11}{25}$

59. (a) $P(SS) = (0.985)^2 \approx 0.9702$

(b) $P(FF) = (0.015)^2 \approx 0.0002$

(c) $P(S) = 1 - P(FF) = 1 - (0.015)^2 \approx 0.9998$

61. (a) $\frac{1}{38}$

(b) $\frac{18}{38} = \frac{9}{19}$

(c) $\frac{2}{38} + \frac{18}{38} = \frac{20}{38} = \frac{10}{19}$

(d) $\frac{1}{38} \cdot \frac{1}{38} = \frac{1}{1444}$

(e) $\frac{18}{38} \cdot \frac{18}{38} \cdot \frac{18}{38} = \frac{5832}{54{,}872} = \frac{729}{6859}$

63. $1 - \frac{(45)^2}{(60)^2} = 1 - \left(\frac{45}{60}\right)^2 = 1 - \left(\frac{3}{4}\right)^2 = 1 - \frac{9}{16} = \frac{7}{16}$

65. True. Two events are independent if the occurrence of one has no effect on the occurrence of the other.

67. (a) As you consider successive people with distinct birthdays, the probabilities must decrease to take into account the birth dates already used. Because the birth dates of people are independent events, multiply the respective probabilities of distinct birthdays.

(b) $\frac{365}{365} \cdot \frac{364}{365} \cdot \frac{363}{365} \cdot \frac{362}{365}$

(c) $P_1 = \frac{365}{365} = 1$

$P_2 = \frac{365}{365} \cdot \frac{364}{365} = \frac{364}{365}P_1 = \frac{365 - (2 - 1)}{365}P_1$

$P_3 = \frac{365}{365} \cdot \frac{364}{365} \cdot \frac{363}{365} = \frac{363}{365}P_2 = \frac{365 - (3 - 1)}{365}P_2$

$P_n = \frac{365}{365} \cdot \frac{364}{365} \cdot \frac{363}{365} \cdot \dots \cdot \frac{365 - (n - 1)}{365} = \frac{365 - (n - 1)}{365}P_{n-1}$

(d) Q_n is the probability that the birthdays are not distinct which is equivalent to at least two people having the same birthday.

(e)

n	10	15	20	23	30	40	50
P_n	0.88	0.75	0.59	0.49	0.29	0.11	0.03
Q_n	0.12	0.25	0.41	$\boxed{0.51}$	0.71	0.89	0.97

(f) 23; $Q_n > 0.5$ for $n \geq 23$.

Review Exercises for Chapter 8

1. $a_n = 3 + \frac{12}{n}$

$a_1 = 3 + \frac{12}{1} = 15$

$a_2 = 3 + \frac{12}{2} = 9$

$a_3 = 3 + \frac{12}{3} = 7$

$a_4 = 3 + \frac{12}{4} = 6$

$a_5 = 3 + \frac{12}{5} = \frac{27}{5}$

3. $a_n = \frac{120}{n!}$

$a_1 = \frac{120}{1!} = 120$

$a_2 = \frac{120}{2!} = 60$

$a_3 = \frac{120}{3!} = 20$

$a_4 = \frac{120}{4!} = 5$

$a_5 = \frac{120}{5!} = 1$

5. $-2, 2, -2, 2, -2, \ldots$

$a_n = 2(-1)^n$

7. $4, 2, \frac{4}{3}, 1, \frac{4}{5}, \ldots$

$a_n = \frac{4}{n}$

9. $\frac{3!}{5!} = \frac{3 \cdot 2 \cdot 1}{5 \cdot 4 \cdot 3 \cdot 2 \cdot 1} = \frac{1}{20}$

11. $\frac{(n-1)!}{(n+1)!} = \frac{(n-1)(n-2) \cdot \ldots \cdot 3 \cdot 2 \cdot 1}{(n+1)(n)(n-1)(n-2) \cdot \ldots \cdot 3 \cdot 2 \cdot 1}$

$= \frac{1}{n(n+1)}$

13. $\sum_{j=1}^{4} \frac{6}{j^2} = \frac{6}{1^2} + \frac{6}{2^2} + \frac{6}{3^2} + \frac{6}{4^2}$

$= 6 + \frac{3}{2} + \frac{2}{3} + \frac{3}{8}$

$= \frac{205}{24}$

15. $\frac{1}{2(1)} + \frac{1}{2(2)} + \frac{1}{2(3)} + \cdots + \frac{1}{2(20)} = \sum_{k=1}^{20} \frac{1}{2k}$

17. $\sum_{i=1}^{\infty} \frac{4}{10^i} = \sum_{i=1}^{\infty} 4\left(\frac{1}{10^i}\right) = \frac{\frac{4}{10}}{1 - \frac{1}{10}} = \frac{4}{9}$

19. (a) $A_1 = \$10{,}018.75$

$A_2 \approx \$10{,}037.54$

$A_3 \approx \$10{,}056.36$

$A_4 \approx \$10{,}075.21$

$A_5 \approx \$10{,}094.10$

$A_6 \approx \$10{,}113.03$

$A_7 \approx \$10{,}131.99$

$A_8 \approx \$10{,}150.99$

$A_9 \approx \$10{,}170.02$

$A_{10} \approx \$10{,}189.09$

(b) The balance in the account after 10 years is

$A_{120}\ 10{,}000\left(1 + \frac{0.0225}{12}\right)^{120} \approx \$12{,}520.59$

21. $5, -1, -7, -13, -19, \ldots$

Arithmetic sequence, $d = -6$

23. $\frac{1}{8}, \frac{1}{4}, \frac{1}{2}, 1, 2, \ldots$

Not an arithmetic sequence.

25. $a_1 = 7, d = 12$

$a_n = 7 + (n-1)12 = 7 + 12n - 12 = 12n - 5$

27. $a_3 = 96, a_7 = 24$

$a_7 = a_3 + 4d \Rightarrow 24 = 96 + 4d \Rightarrow -72 = 4d \Rightarrow d = -18$

$a_1 = a_3 - 2d \Rightarrow a_1 = 96 - 2(-18) = 132$

$a_n = 132 + (n-1)(-18)$

$= -18n + 150$

29. $a_1 = 4, d = 17$

$a_1 = 4$

$a_2 = 4 + 17 = 21$

$a_3 = 21 + 17 = 38$

$a_4 = 38 + 17 = 55$

$a_5 = 45 + 17 = 72$

31. $\sum_{k=1}^{100} 9k$ is arithmetic. Therefore, $a_1 = 9, a_{100} = 900,$

$S_{700} = \frac{100}{2}(9 + 900) = 45{,}450.$

33. $\sum_{j=1}^{10} (2j - 3)$ is arithmetic. Therefore,

$a_1 = -1, a_{10} = 17, S_{10} = \frac{10}{2}[-1 + 17] = 80.$

35. $\sum_{k=1}^{11} \left(\frac{2}{3}k + 4\right)$ is arithmetic. Therefore, $a_1 = \frac{14}{3}, a_{11} = \frac{34}{3}, S_{11} = \frac{11}{2}\left[\frac{14}{3} + \frac{34}{3}\right] = 88.$

37. $a_n = 43{,}800 + (n - 1)(1950)$

(a) $a_5 = 43{,}800 + 4(1950) = \$51{,}600$

(b) $S_5 = \frac{5}{2}(43{,}800 + 51{,}600) = \$238{,}500$

39. 2, 6, 18, 54, 162, …

$r = \dfrac{6}{2} = 3$

Geometric sequence, $r = 3$

41. $\frac{1}{5}, -\frac{3}{5}, \frac{9}{5}, -\frac{27}{5}, \ldots$

Geometric sequence, $r = -3$

43. $a_1 = 2, r = 15$

$a_1 = 2$

$a_2 = 2(15) = 30$

$a_3 = 30(15) = 450$

$a_4 = 450(15) = 6750$

$a_5 = 6750(15) = 101{,}250$

45. $a_1 = 9, a_3 = 4$

$a_3 = a_1r^2$

$4 = 9r^2$

$\frac{4}{9} = r^2 \Rightarrow r = \pm\frac{2}{3}$

$a_1 = 9$ | $a_1 = 9$

$a_2 = 9\left(\frac{2}{3}\right) = 6$ | $a_2 = 9\left(-\frac{2}{3}\right) = -6$

$a_3 = 6\left(\frac{2}{3}\right) = 4$ or $a_3 = -6\left(-\frac{2}{3}\right) = 4$

$a_4 = 4\left(\frac{2}{3}\right) = \frac{8}{3}$ | $a_4 = 4\left(-\frac{2}{3}\right) = -\frac{8}{3}$

$a_5 = \frac{8}{3}\left(\frac{2}{3}\right) = \frac{16}{9}$ | $a_5 = -\frac{8}{3}\left(-\frac{2}{3}\right) = \frac{16}{9}$

47. $a_1 = 100, r = 1.05$

$a_n = 100(1.05)^{n-1}$

$a_{10} = 100(1.05)^9 \approx 155.133$

49. $a_1 = 18, a_2 = -9$

$a_2 = a_1r$

$-9 = 18r$

$-\frac{1}{2} = r$

$a_n = 18\left(-\frac{1}{2}\right)^{n-1}$

$a_{10} = 18\left(-\frac{1}{2}\right)^9 = \frac{-9}{256}$

51. $\displaystyle\sum_{i=1}^{7} 2^{i-1} = \frac{1 - 2^7}{1 - 2} = 127$

53. $\displaystyle\sum_{i=1}^{4} \left(\frac{1}{2}\right)^i = \frac{1}{2} + \frac{1}{4} + \frac{1}{8} + \frac{1}{16} = \frac{15}{16}$

55. $\displaystyle\sum_{i=1}^{5} (2)^{i-1} = 1 + 2 + 4 + 8 + 16 = 31$

57. $\displaystyle\sum_{i=1}^{5} 10(0.6)^{i-1} = 23.056$

59. $\displaystyle\sum_{i=1}^{\infty} \left(\frac{7}{8}\right)^{i-1} = \frac{1}{1 - \frac{7}{8}} = 8$

61. $\displaystyle\sum_{k=1}^{\infty} 4\left(\frac{2}{3}\right)^{k-1} = \frac{4}{1 - \frac{2}{3}} = 12$

63. (a) $a_n = 120{,}000(0.7)^n$

(b) $a_5 = 120{,}000(0.7)^5$

$= \$20{,}168.40$

65. 1. When $n = 1, 3 = 1(1 + 2)$.

2. Assume that $S_k = 3 + 5 + 7 + \cdots + (2k + 1) = k(k + 2)$.

Then, $S_{k+1} = 3 + 5 + 7 + \cdots + (2k + 1) + [2(k + 1) + 1] = S_k + (2k + 3)$

$= k(k + 2) + 2k + 3$

$= k^2 + 4k + 3$

$= (k + 1)(k + 3)$

$= (k + 1)[(k + 1) + 2]$.

So, by mathematical induction, the formula is valid for all positive integer values of n.

67. 1. When $n = 1$, $a = a\left(\dfrac{1-r}{1-r}\right)$.

2. Assume that $S_k = \displaystyle\sum_{i=0}^{k-1} ar^i = \dfrac{a(1-r^k)}{1-r}$.

Then $S_{k+1} = \displaystyle\sum_{i=0}^{k} ar^i = \left(\sum_{i=0}^{k-1} ar^i\right) + ar^k = \dfrac{a(1-r^k)}{1-r} + ar^k$

$$= \frac{a(1 - r^k + r^k - r^{k+1})}{1-r} = \frac{a(1-r^{k+1})}{1-r}.$$

So, by mathematical induction, the formula is valid for all positive integer values of n.

69. $S_1 = 9 = 1(9) = 1[2(1) + 7]$

$S_2 = 9 + 13 = 22 = 2(11) = 2[2(2) + 7]$

$S_3 = 9 + 13 + 17 = 39 = 3(13) = 3[2(3) + 7]$

$S_4 = 9 + 13 + 17 + 21 = 60 = 4(15) = 4[2(4) + 7]$

$S_n = n(2n + 7)$

1. When $n = 1$, $S_1 = 9 = 1(2 + 7)$

2. Assume $S_k = 9 + 13 + \ldots + [4k + 5] = k(2k + 7)$

$$\begin{aligned} S_{k+1} &= S_k + a_{k+1} \\ &= k(2k+7) + [4(k+1) + 5] \\ &= 2k^2 + 7k + 4k + 9 \\ &= 2k^2 + 11k + 9 \\ &= (k+1)(2k+9) \\ &= (k+1)[2(k+1) + 7] \end{aligned}$$

So, the formula holds for all positive integers n.

71. $S_1 = 1$

$S_2 = 1 + \dfrac{3}{5} = \dfrac{8}{5}$

$S_3 = 1 + \dfrac{3}{5} + \dfrac{9}{25} = \dfrac{49}{25}$

$S_4 = 1 + \dfrac{3}{5} + \dfrac{9}{25} + \dfrac{27}{125} = \dfrac{272}{125}$

From these sums, there is no apparent pattern. Because the series is geometric, the formula for the sum is

$$S_n = \frac{5}{2}\left[1 - \left(\frac{3}{5}\right)^n\right].$$

1. When $n = 1$, $S_1 = 1 = \dfrac{5}{2}\left[1 - \left(\dfrac{3}{5}\right)^1\right]$

2. Assume $S_k = 1 + \dfrac{3}{5} + \dfrac{9}{25} + \cdots + \left(\dfrac{3}{5}\right)^{k-1}$

$$= \frac{5}{2}\left[1 - \left(\frac{3}{5}\right)^k\right]$$

$$= \frac{1 - \left(\frac{3}{5}\right)^k}{1 - \frac{3}{5}}$$

$$\begin{aligned} S_{k+1} = S_k + a_{k+1} &= \frac{1 - \left(\frac{3}{5}\right)^k}{1 - \frac{3}{5}} + \left(\frac{3}{5}\right)^{(k+1)-1} \\ &= \frac{1 - \left(\frac{3}{5}\right)^k}{1 - \frac{3}{5}} + \left(\frac{3}{5}\right)^k \\ &= \frac{1 - \left(\frac{3}{5}\right)^k + \left(1 - \frac{3}{5}\right)\left(\frac{3}{5}\right)^k}{1 - \frac{3}{5}} \\ &= \frac{1 - \left(\frac{3}{5}\right)^k + \left(\frac{3}{5}\right)^k - \left(\frac{3}{5}\right)^{k+1}}{1 - \frac{3}{5}} \\ &= \frac{1 - \left(\frac{3}{5}\right)^{k+1}}{1 - \frac{3}{5}} \\ &= \frac{5}{2}\left(1 - \left(\frac{3}{5}\right)^{k+1}\right) \end{aligned}$$

So, the formula holds for all positive integers n.

73. $\sum_{n=1}^{75} n = \frac{75(76)}{2} = 2850$

75. $a_1 = f(1) = 5,\ a_n = a_{n-1} + 5$

$a_1 = 5$

$a_2 = 5 + 5 = 10$

$a_3 = 10 + 5 = 15$

$a_4 = 15 + 5 = 20$

$a_5 = 20 + 5 = 25$

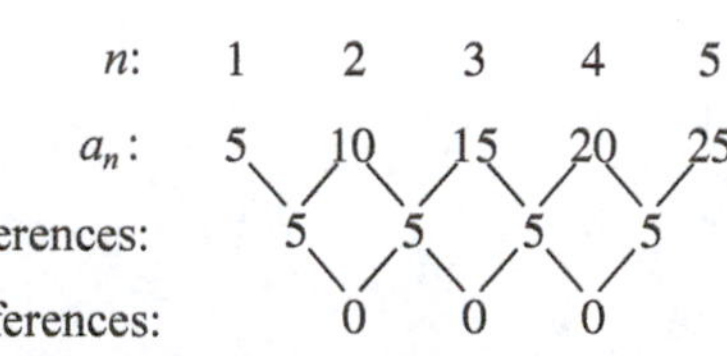

Because the first differences are all the same, the sequence has a linear model.

77. ${}_6C_4 = \frac{6!}{2!4!} = 15$

79.

1
1 1
1 2 1
1 3 3 1
1 4 6 4 1
1 5 10 10 5 1
1 6 15 20 15 6 1
1 7 (21) 35 35 21 7 1

$\binom{7}{2} = 21$, the 3rd entry in the 7th row.

81. $(x + 4)^4 = x^4 + 4x^3(4) + 6x^2(4)^2 + 4x(4)^3 + 4^4 = x^4 + 16x^3 + 96x^2 + 256x + 256$

83. $(4 - 5x)^3 = {}_3C_0(4^3) + {}_3C_1(4^2)(-5x) + {}_3C_2(4)(-5x)^2 + {}_3C_3(-5x)^3$

$= 4^3 - 3(4)^2(5x) + 3(4)(5x)^2 - (5x)^3$

$= 64 - 240x + 300x^2 + 125x^3$

85. First number: 1 2 3 4 5 6

Second number: 6 5 4 3 2 1

From this list, you can see that a sum of 7 occurs 6 different ways.

87. (10)(10)(10)(10) = 10,000 different telephone numbers

89. $5 \cdot 4 \cdot 3 \cdot 2 \cdot 1 = 120$

91. ${}_{32}C_{12} = \frac{32!}{20!12!} = 225{,}792{,}840$

93. (a) $P(E) = \frac{n(E)}{n(S)} = \frac{2}{10} = \frac{1}{5} = 0.2$

(b) $P(E) = \frac{n(E)}{n(S)} = \frac{6}{10} = \frac{3}{5} = 0.6$

95. (a) 25% + 18% = 43%

(b) 100% − 18% = 82%

97. $\left(\frac{1}{6}\right)\left(\frac{1}{6}\right)\left(\frac{1}{6}\right)\left(\frac{1}{6}\right) = \frac{1}{1296}$

99. $1 - \frac{13}{52} = 1 - \frac{1}{4} = \frac{3}{4}$

101. False. $\frac{(n+2)!}{n!} = \frac{(n+2)(n+1)\cancel{n!}}{\cancel{n!}}$

$= (n + 2)(n + 1)$

$\neq \frac{n+2}{n}$

103. True. $\sum_{k=1}^{8} 3k = 3\sum_{k=1}^{8} k$ by the Properties of Sums.

105. The domain of an infinite sequence is the set of natural numbers.

107. Each term of the sequence is defined in terms of preceding terms.

Problem Solving for Chapter 8

1. $a_n = \dfrac{n+1}{n^2+1}$

(a)

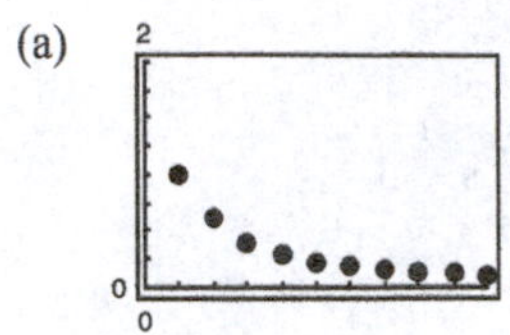

(c)

n	1	10	100	1000	10,000
a_n	1	0.1089	0.0101	0.0010	0.0001

(b) $a_n \to 0$ as $n \to \infty$

(d) $a_n \to 0$ as $n \to \infty$

3. Distance: $\displaystyle\sum_{n=1}^{\infty} 20\left(\frac{1}{2}\right)^{n-1} = \frac{20}{1-\frac{1}{2}} = 40$

Time: $\displaystyle\sum_{n=1}^{\infty} \left(\frac{1}{2}\right)^{n-1} = \frac{1}{1-\frac{1}{2}} = 2$

In two seconds, both Achilles and the tortoise will be 40 feet away from Achilles' starting point.

5. Let $a_n = dn + c$, an arithmetic sequence with a common difference of d.

(a) If C is added to each term, then the resulting sequence, $b_n = a_n + C = dn + c + C$, is still arithmetic with a common difference of d.

(b) If each term is multiplied by a nonzero constant C, then the resulting sequence, $b_n = C(dn + c) = Cdn + Cc$, is still arithmetic. The common difference is Cd.

(c) If each term is squared, the resulting sequence, $b_n = a_n^2 = (dn + c)^2$, is not arithmetic.

7. $a_n = \begin{cases} \dfrac{a_{n-1}}{2}, & \text{if } a_{n-1} \text{ is even} \\ 3a_{n-1} + 1 & \text{if } a_{n-1} \text{ is odd} \end{cases}$

(a)

$a_1 = 7$
$a_2 = 3(7) + 1 = 22$
$a_3 = \frac{22}{3} = 11$
$a_4 = 3(11) + 1 = 34$
$a_5 = \frac{34}{2} = 17$
$a_6 = 3(17) + 1 = 52$
$a_7 = \frac{52}{2} = 26$
$a_8 = \frac{26}{2} = 13$
$a_9 = 3(13) + 1 = 40$
$a_{10} = \frac{40}{2} = 20$

$a_{11} = \frac{20}{2} = 10$
$a_{12} = \frac{10}{2} = 5$
$a_{13} = 3(5) + 1 = 16$
$a_{14} = \frac{16}{2} = 8$
$a_{15} = \frac{8}{2} = 4$
$a_{16} = \frac{4}{2} = 2$
$a_{17} = \frac{2}{2} = 1$
$a_{18} = 3(1) + 1 = 4$
$a_{19} = \frac{4}{2} = 2$
$a_{20} = \frac{2}{2} = 1$

(b)

$a_1 = 4$
$a_2 = 2$
$a_3 = 1$
$a_4 = 4$
$a_5 = 2$
$a_6 = 1$
$a_7 = 4$
$a_8 = 2$
$a_9 = 1$
$a_{10} = 4$

$a_1 = 5$
$a_2 = 16$
$a_3 = 8$
$a_4 = 4$
$a_5 = 2$
$a_6 = 1$
$a_7 = 4$
$a_8 = 2$
$a_9 = 1$
$a_{10} = 4$

$a_1 = 12$
$a_2 = 6$
$a_3 = 3$
$a_4 = 10$
$a_5 = 5$
$a_6 = 16$
$a_7 = 8$
$a_8 = 4$
$a_9 = 2$
$a_{10} = 1$

Eventually the terms repeat: 4, 2, 1

9. The numbers 1, 5, 12, 22, 35, 51, … can be written recursively as $P_n = P_{n-1} + (3n - 2)$. Show that $P_n = n(3n - 1)/2$.

1. For $n = 1$: $1 = \dfrac{1(3 - 1)}{2}$

2. Assume $P_k = \dfrac{k(3k - 1)}{2}$.

Then,
$$\begin{aligned} P_{k+1} &= P_k + \left[3(k + 1) - 2\right] \\ &= \frac{k(3k - 1)}{2} + (3k + 1) = \frac{k(3k - 1) + 2(3k + 1)}{2} \\ &= \frac{3k^2 + 5k + 2}{2} = \frac{(k + 1)(3k + 2)}{2} \\ &= \frac{(k + 1)\left[3(k + 1) - 1\right]}{2}. \end{aligned}$$

So, by mathematical induction, the formula is valid for all integers $n \ge 1$.

11. Side lengths: $1, \dfrac{1}{2}, , \dfrac{1}{8}, \ldots$

$$S_n = \left(\frac{1}{2}\right)^{n-1} \text{ for } n \ge 1$$

Areas: $\dfrac{\sqrt{3}}{4}, \dfrac{\sqrt{3}}{4}\left(\dfrac{1}{2}\right)^2, \dfrac{\sqrt{3}}{4}\left(\dfrac{1}{4}\right)^2, \dfrac{\sqrt{3}}{4}\left(\dfrac{1}{8}\right)^2, \ldots$

$$A_n = \frac{\sqrt{3}}{4}\left[\left(\frac{1}{2}\right)^{n-1}\right]^2 = \frac{\sqrt{3}}{4}\left(\frac{1}{2}\right)^{2n-2} = \frac{\sqrt{3}}{4}S_n^2$$

13. $\frac{1}{3}$

15. (a) Odds in favor of choosing a blue marble $= \dfrac{\text{number of blue marbles}}{\text{number of yellow marbles}} = \dfrac{3}{7}$

Odds against choosing a blue marble $= \dfrac{\text{number of yellow marbles}}{\text{number of blue marbles}} = \dfrac{7}{3}$

(b) Odds against choosing a red marble $= \dfrac{\text{number of non-red marbles}}{\text{number of red marbles}}$

$$\frac{4}{1} = \frac{x}{6}$$

$$24 = x \quad \text{(number of non-red marbles)}$$

$$\text{Total marbles} = 6 + 24 = 30$$

(c) $P(E) = \dfrac{n(E)}{n(S)} = \dfrac{n(E)}{n(E) + n(E')} = \dfrac{n(E)/n(E')}{n(E)/n(E') + n(E')/n(E')}$

$P(E) = \dfrac{\text{odds in favor of } E}{\text{odds in favor of } E + 1}$

(d) $P(E) = \dfrac{n(E)}{n(S)}$ $\qquad P(E') = \dfrac{n(E')}{n(S)}$

$n(S)P(E) = n(E)$ $\qquad n(S)P(E') = n(E')$

Odds in favor of event $E = \dfrac{n(E)}{n(E')} = \dfrac{n(S)P(E)}{n(S)P(E')} = \dfrac{P(E)}{P(E')}$

Practice Test for Chapter 8

1. Write out the first five terms of the sequence $a_n = \dfrac{2n}{(n+2)!}$.

2. Write an expression for the nth term of the sequence $\frac{4}{3}, \frac{5}{9}, \frac{6}{27}, \frac{7}{81}, \frac{8}{243}, \ldots$.

3. Find the sum $\sum_{i=1}^{6} (2i - 1)$.

4. Write out the first five terms of the arithmetic sequence where $a_1 = 23$ and $d = -2$.

5. Find a_n for the arithmetic sequence with $a_1 = 12$, $d = 3$, and $n = 50$.

6. Find the sum of the first 200 positive integers.

7. Write out the first five terms of the geometric sequence with $a_1 = 7$ and $r = 2$.

8. Evaluate $\sum_{n=1}^{10} 6\left(\frac{2}{3}\right)^{n-1}$.

9. Evaluate $\sum_{n=0}^{\infty} (0.03)^n$.

10. Use mathematical induction to prove that $1 + 2 + 3 + 4 + \cdots + n = \dfrac{n(n+1)}{2}$.

11. Use mathematical induction to prove that $n! > 2^n$, $n \geq 4$.

12. Evaluate ${}_{13}C_4$.

13. Expand $(x + 3)^5$.

14. Find the term involving x^7 in $(x - 2)^{12}$.

15. Evaluate ${}_{30}P_4$.

16. How many ways can six people sit at a table with six chairs?

17. Twelve cars run in a race. How many different ways can they come in first, second, and third place? (Assume that there are no ties.)

18. Two six-sided dice are tossed. Find the probability that the total of the two dice is less than 5.

19. Two cards are selected at random from a deck of 52 playing cards without replacement. Find the probability that the first card is a King and the second card is a black ten.

20. A manufacturer has determined that for every 1000 units it produces, 3 will be faulty. What is the probability that an order of 50 units will have one or more faulty units?

Appendix A Errors and the Algebra of Calculus

1. numerator

3. The middle term needs to be included.

$(x + 3)^2 = x^2 + 6x + 9$

5. $\sqrt{x + 9} \neq \sqrt{x} + 3$

Do not apply the radical to the terms.

$\sqrt{x + 9}$ does not simplify.

7. $\dfrac{2x^2 + 1}{5x} \neq \dfrac{2x + 1}{5}$

Divide out common factors not common terms.

$\dfrac{2x^2 + 1}{5x}$ cannot be simplified.

9. $(4x)^2 \neq 4x^2$

The exponent applies to the coefficient also.

$(4x)^2 = 16x^2$

11. $\dfrac{3}{x} + \dfrac{4}{y} = \dfrac{3}{x} \cdot \dfrac{y}{y} + \dfrac{4}{y} \cdot \dfrac{x}{x} = \dfrac{3y + 4x}{xy}$

To add fractions, they must have a common denominator.

13.
$$\begin{aligned} 2x(x + 2)^{-1/2} + (x + 2)^{1/2} &= (x + 2)^{-1/2}\left[2x + (x + 2)\right] \\ &= (x + 2)^{-1/2}(3x + 2) \\ &= \frac{3x + 2}{(x + 2)^{1/2}} \end{aligned}$$

15.
$$\begin{aligned} 4x^3(2x - 1)^{3/2} - 2x(2x - 1)^{-1/2} &= 2x(2x - 1)^{-1/2}\left[2x^2(2x - 1)^2 - 1\right] \\ &= 2x(2x - 1)^{-1/2}\left[2x^2(4x^2 - 4x + 1) - 1\right] \\ &= 2x(2x - 1)^{-1/2}(8x^4 - 8x^3 + 2x^2 - 1) \\ &= \frac{2x(8x^4 - 8x^3 + 2x^2 - 1)}{(2x - 1)^{1/2}} \end{aligned}$$

17. $\dfrac{5x + 3}{4} = \dfrac{1}{4}(5x + 3)$

The required factor is $5x + 3$.

19. $\frac{2}{3}x^2 + \frac{1}{3}x + 5 = \frac{2}{3}x^2 + \frac{1}{3}x + \frac{15}{3} = \frac{1}{3}(2x^2 + x + 15)$

The required factor is $2x^2 + x + 15$.

21. $x^{1/3} - 5x^{4/3} = x^{1/3}(1 - 5x^{3/3}) = x^{1/3}(1 - 5x)$

The required factor is $1 - 5x$.

23.
$$\begin{aligned} \tfrac{1}{10}(2x + 1)^{5/2} - \tfrac{1}{6}(2x + 1)^{3/2} &= \tfrac{3}{30}(2x + 1)^{3/2}(2x + 1)^1 - \tfrac{5}{30}(2x + 1)^{3/2} \\ &= \tfrac{1}{30}(2x + 1)^{3/2}\left[3(2x + 1) - 5\right] \\ &= \tfrac{1}{30}(2x + 1)^{3/2}(6x - 2) \\ &= \tfrac{1}{30}(2x + 1)^{3/2}2(3x - 1) \\ &= \tfrac{1}{15}(2x + 1)^{3/2}(3x - 1) \end{aligned}$$

The required factor is $3x - 1$.

25. $x^2(x^3 - 1)^4 = \frac{1}{3}(x^3 - 1)^4(3x^2)$

The required factor is $\frac{1}{3}$.

27. $\dfrac{4x + 6}{(x^2 + 3x + 7)^3} = \dfrac{2(2x + 3)}{(x^2 + 3x + 7)^3} = \dfrac{2}{1} \cdot \dfrac{(2x + 3)}{1} \cdot \dfrac{1}{(x^2 + 3x + 7)^3} = (2)\dfrac{1}{(x^2 + 3x + 7)^3}(2x + 3)$

The required factor is 2.

29. $4x^2 + \dfrac{6y^2}{10} = 4x^2\dfrac{(1/4)}{(1/4)} + \dfrac{6y^2}{10}\dfrac{(1/2)}{(1/2)}$

$= \dfrac{x^2}{1/4} + \dfrac{3y^2}{5}$

31. $\dfrac{25x^2}{36} + \dfrac{4y^2}{9} = \dfrac{25x^2}{36}\dfrac{(1/25)}{(1/25)} + \dfrac{4y^2}{9}\dfrac{(1/4)}{(1/4)}$

$= \dfrac{x^2}{36/25} + \dfrac{y^2}{9/4}$

33. $\dfrac{x^2}{\frac{3}{10}}\dfrac{(10)}{(10)} - \dfrac{y^2}{\frac{4}{5}}\dfrac{(5)}{(5)} = \dfrac{10x^2}{3} - \dfrac{5y^2}{4}$

35. $\dfrac{7}{(x+3)^5} = 7(x+3)^{-5}$

37. $\dfrac{2x^5}{(3x+5)^4} = 2x^5(3x+5)^{-4}$

39. $\dfrac{4}{3x} + \dfrac{4}{x^4} - \dfrac{7x}{\sqrt[3]{2x}} = \dfrac{4}{3}3x^{-1} + 4x^{-4} - 7x(2x)^{-1/3}$

41. $\dfrac{x^2 + 6x + 12}{3x} = \dfrac{x^2}{3x} + \dfrac{6x}{3x} + \dfrac{12}{3x}$

$= \dfrac{x}{3} + 2 + \dfrac{4}{x}$

43. $\dfrac{4x^3 - 7x^2 + 1}{x^{1/3}} = \dfrac{4x^3}{x^{1/3}} - \dfrac{7x^2}{x^{1/3}} + \dfrac{1}{x^{1/3}}$

$= 4x^{3-1/3} - 7x^{2-1/3} + \dfrac{1}{x^{1/3}}$

$= 4x^{8/3} - 7x^{5/3} + \dfrac{1}{x^{1/3}}$

45. $\dfrac{3 - 5x^2 - x^4}{\sqrt{x}} = \dfrac{3}{\sqrt{x}} - \dfrac{5x^2}{\sqrt{x}} - \dfrac{x^4}{\sqrt{x}}$

$= \dfrac{3}{\sqrt{x}} - 5x^{2-1/2} - x^{4-1/2}$

$= \dfrac{3}{x^{1/2}} - 5x^{3/2} - x^{7/2}$

47. $\dfrac{-2(x^2-3)^{-3}(2x)(x+1)^3 - 3(x+1)^2(x^2-3)^{-2}}{\left[(x+1)^3\right]^2} = \dfrac{(x^2-3)^{-3}(x+1)^2\left[-4x(x+1) - 3(x^2-3)\right]}{(x+1)^6}$

$= \dfrac{-4x^2 - 4x - 3x^2 + 9}{(x^2-3)^3(x+1)^4}$

$= \dfrac{-7x^2 - 4x + 9}{(x^2-3)^3(x+1)^4}$

49. $\dfrac{(6x+1)^3(27x^2+2) - (9x^3+2x)(3)(6x+1)^2(6)}{\left[(6x+1)^3\right]^2} = \dfrac{(6x+1)^2\left[(6x+1)(27x^2+2) - 18(9x^3+2x)\right]}{(6x+1)^6}$

$= \dfrac{162x^3 + 12x + 27x^2 + 2 - 162x^3 - 36x}{(6x+1)^4}$

$= \dfrac{27x^2 - 24x + 2}{(6x+1)^4}$

51. $\dfrac{(x+2)^{3/4}(x+3)^{-2/3} - (x+3)^{1/3}(x+2)^{-1/4}}{\left[(x+2)^{3/4}\right]^2} = \dfrac{(x+2)^{-1/4}(x+3)^{-2/3}\left[(x+2) - (x+3)\right]}{(x+2)^{6/4}}$

$= \dfrac{x + 2 - x - 3}{(x+2)^{1/4}(x+3)^{2/3}(x+2)^{6/4}}$

$= -\dfrac{1}{(x+3)^{2/3}(x+2)^{7/4}}$

53. $$\frac{2(3x-1)^{1/3}-(2x+1)(1/3)(3x-1)^{-2/3}(3)}{(3x-1)^{2/3}} = \frac{(3x-1)^{-2/3}\left[2(3x-1)-(2x+1)\right]}{(3x-1)^{2/3}}$$
$$= \frac{6x-2-2x-1}{(3x-1)^{2/3}(3x-1)^{2/3}}$$
$$= \frac{4x-3}{(3x-1)^{4/3}}$$

55. $$\frac{1}{\left(x^2+4\right)^{1/2}}\cdot\frac{1}{2}\left(x^2+4\right)^{-1/2}(2x) = \frac{1}{\left(x^2+4\right)^{1/2}}\cdot\frac{1}{\left(x^2+4\right)^{1/2}}\cdot\frac{1}{2}(2x)$$
$$= \frac{1}{\left(x^2+4\right)^{1}}(x)$$
$$= \frac{x}{x^2+4}$$

57. $$\left(x^2+5\right)^{1/2}\left(\frac{3}{2}\right)(3x-2)^{1/2}(3)+(3x-2)^{3/2}\left(\frac{1}{2}\right)\left(x^2+5\right)^{-1/2}(2x) = \frac{9}{2}\left(x^2+5\right)^{1/2}(3x-2)^{1/2}+x\left(x^2+5\right)^{-1/2}(3x-2)^{3/2}$$
$$= \frac{9}{2}\left(x^2+5\right)^{1/2}(3x-2)^{1/2}+\frac{2}{2}x\left(x^2+5\right)^{-1/2}(3x-2)^{3/2}$$
$$= \frac{1}{2}\left(x^2+5\right)^{-1/2}(3x-2)^{1/2}\left[9\left(x^2+5\right)^{1}+2x(3x-2)^{1}\right]$$
$$= \frac{1}{2}\left(x^2+5\right)^{-1/2}(3x-2)^{1/2}\left(9x^2+45+6x^2-4x\right)$$
$$= \frac{(3x-2)^{1/2}\left(15x^2-4x+45\right)}{2\left(x^2+5\right)^{1/2}}$$

59. (a) $$y_1 = x^2\left(\frac{1}{3}\right)\left(x^2+1\right)^{-2/3}(2x)+\left(x^2+1\right)^{1/3}(2x)$$
$$= 2x\left(x^2+1\right)^{-2/3}\left[\frac{x^2}{3}+\left(x^2+1\right)\right]$$
$$= 2x\left(x^2+1\right)^{-2/3}\left[\frac{x^2}{3}+\frac{3\left(x^2+1\right)}{3}\right]$$
$$= \frac{2x}{\left(x^2+1\right)^{2/3}}\cdot\frac{4x^2+3}{3}$$
$$= \frac{2x\left(4x^2+3\right)}{3\left(x^2+1\right)^{2/3}}$$
$$= y_2$$

(b)

x	-2	-1	$-\frac{1}{2}$	0	1	2	$\frac{5}{2}$
y_1	−8.7	−2.9	−1.1	0	2.9	8.7	12.5
y_2	−8.7	−2.9	−1.1	0	2.9	8.7	12.5

(c) The graphs appear to coincide, so they are equal.

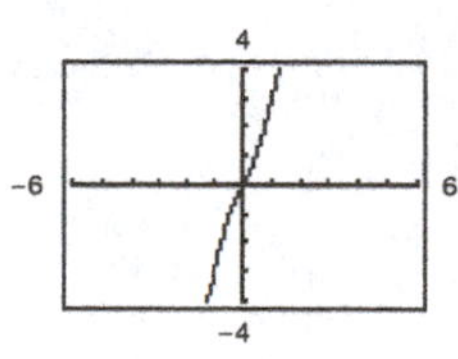

61. You cannot move term-by-term from the denominator to the numerator.

CHECKPOINTS

CHECKPOINTS
Chapter P

Checkpoints for Section P.1

1. (a) Natural numbers: $\left\{\frac{6}{3}, 8\right\}$

(b) Whole numbers: $\left\{\frac{6}{3}, 8\right\}$

(c) Integers: $\left\{-22, -1, \frac{6}{3}, 8\right\}$

(d) Rational numbers: $\left\{-22, -7.5, -1 - \frac{1}{4}, \frac{6}{3}, 8\right\}$

(e) Irrational numbers: $\left\{-\pi, \frac{1}{2}\sqrt{2}\right\}$

2.

(a) The point representing the real number $\frac{5}{2} = 2.5$ lies halfway between 2 and 3, on the real number line.

(b) The point representing the real number -1.6 lies between -2 and -1 but closer to -2, on the real number line.

(c) The point representing the real number $-\frac{3}{4}$ lies between -1 and 0 but closer to -1, on the real number line.

(d) The point representing the real number 0.7 lies between 0 and 1 but closer to 1, on the real number line.

3. (a) Because -5 lies to the left of 1 on the real number line, you can say that -5 is *less than* 1, and write $-5 < 1$.

(b) Because $\frac{3}{2}$ lies to the left of 7 on the real number line, you can say that $\frac{3}{2}$ is *less than* 7, and write $\frac{3}{2} < 7$.

(c) Because $-\frac{2}{3}$ lies to the right of $-\frac{3}{4}$ on the real number line, you can say that $-\frac{2}{3}$ is *greater than* and $-\frac{3}{4}$, write $-\frac{2}{3} > -\frac{3}{4}$.

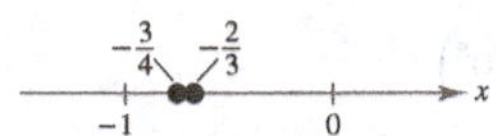

4. (a) The inequality $x > -3$ denotes all real numbers greater than -3.

(b) The inequality $0 < x \le 4$ means that $x > 0$ and $x \le 4$. This double inequality denotes all real numbers between 0 and 4, including 4 but not including 0.

5. The interval consists of real numbers greater than or equal to -2 and less than 5.

6. The inequality $-2 \le x < 4$ can represent the statement "x is less than 4 and at least -2."

7. (a) $|1| = 1$

(b) $-\left|\frac{3}{4}\right| = -\left(\frac{3}{4}\right) = -\frac{3}{4}$

(c) $\frac{2}{|-3|} = \frac{2}{3}$

(d) $-|0.7| = -(0.7) = -0.7$

8. (a) If $x > -3$, then $\frac{|x+3|}{x+3} = \frac{x+3}{x+3} = 1$.

(b) If $x < -3$, then $\frac{|x+3|}{x+3} = \frac{-(x+3)}{x+3} = -1$.

9. (a) $|-3| < |4|$ because $|-3| = 3$ and $|4| = 4$, and 3 is less than 4.

(b) $-|-4| = -|-4|$ because $-|-4| = -4$ and $-|4| = -4$.

(c) $|-3| > -|-3|$ because $|-3| = 3$ and $-|-3| = -3$, and 3 is greater than -3.

10. (a) The distance between 35 and -23 is $|35 - (-23)| = |58| = 58$.

(b) The distance between -35 and -23 is $|-35 - (-23)| = |-12| = 12$.

(c) The distance between 35 and 23 is $|35 - 23| = |12| = 12$.

11.

Algebraic Expression	Terms	Coefficients
$-2x + 4$	$-2x$, 4	-2, 4

12.

Expression	Value of Variable	Substitute	Value of Expression
$4x - 5$	$x - 0$	$4(0) - 5$	$0 - 5 = -5$

13. (a) $x + 9 = 9 + x$: This statement illustrates the Commutative Property of Addition. In other words, you obtain the same result whether you add x and 9, or 9 and x.

(b) $5(x^3 \cdot 2) = (5x^3)2$: This statement illustrates the Associative Property of Multiplication. In other words, to form the product $5 \cdot x^3 \cdot 2$, it does not matter whether 5 and $(x^3 \cdot 2)$, or $5x^3$ and 2 are multiplied first.

(c) $(2 + 5x^2)y^2 = 2y^2 + 5x^2 \cdot y^2$: This statement illustrates the Distributive Property. In other words, the terms 2 and $5x^2$ are multiplied by y^2.

14. (a) $\frac{3}{5} \cdot \frac{x}{6} = \frac{3x}{30} = \frac{3x \div 3}{30 \div 3} = \frac{x}{10}$

(b) $\frac{x}{10} + \frac{2x}{5} = \frac{x}{10} + \frac{2x}{5} \cdot \frac{2}{2}$

$= \frac{x}{10} + \frac{2x}{5} = \frac{x}{10} + \frac{2x}{5} \cdot \frac{2}{2}$

$= \frac{x}{10} + \frac{4x}{10}$

$= \frac{x + 4x}{10}$

$= \frac{5x \div 5}{10 \div 5}$

$= \frac{x}{2}$

Checkpoints for Section P.2

1. (a) $-3^4 = -(3)(3)(3)(3) = -81$

(b) $(-3)^4 = (-3)(-3)(-3)(-3) = 81$

(c) $3^2 \cdot 3 = 3^{2+1} = 3^3 = (3)(3)(3) = 27$

(d) $\frac{3^5}{3^8} = 3^{5-8} = 3^{-3} = \frac{1}{3^3} = \frac{1}{27}$

2. (a) When $x = 4$, the expression $-x^{-2}$ has a value of

$-x^{-2} = -(4)^{-2} = -\frac{1}{4^2} = -\frac{1}{16}.$

(b) When $x = 4$, the expression $\frac{1}{4}(-x)^4$ has a value of

$\frac{1}{4}(-x)^4 = \frac{1}{4}(-4)^4 = \frac{1}{4}(256) = 64.$

3. (a) $(2x^{-2}y^3)(-x^4y) = (2)(-1)(x^{-2})(x^4)(y^3)(y) = -2x^2y^4$

(b) $(4a^2b^3)^0 = 1, a \neq 0, b \neq 0$

(c) $(-5z)^3(z^2) = (-5)^3(z)^3z^2$

$= -125z^5$

(d) $\left(\frac{3x^4}{x^2y^2}\right)^2 = \left(\frac{3x^2}{4^2}\right)^2 = \frac{3^2(x^2)^2}{(y^2)^2}$

$= \frac{9x^4}{y^4}, x \neq 0$

4. (a) $2a^{-2} = \frac{2}{a^2}$ Property 3

(b) $\frac{3a^{-3}b^4}{15ab^{-1}} = \frac{3b^4 \cdot b}{15a \cdot a^3}$ Property 3

$= \frac{b^5}{5a^4}$ Property 1

(c) $\left(\frac{x}{10}\right)^{-1} = \frac{x^{-1}}{10^{-1}}$ Property 7

$= \frac{10}{x}$ Property 3

(d) $(-2x^2)^3(4x^3)^{-1} = (-2)^3(x^2)^3 \cdot 4^{-1} \cdot (x^3)^{-1}$ Property 5

$= \frac{-8x^6}{4x^3}$ Properties 3 and 6

$= -2x^3$ Property 2

5. $45{,}850 = 4.585 \times 10^4$

6. $-2.718 \times 10^{-3} = -0.002718$

7. $(24{,}000{,}000{,}000)(0.00000012)(300{,}000)$

$= (2.4 \times 10^{10})(1.2 \times 10^{-7})(3.0 \times 10^5)$

$= (2.4)(1.2)(3.0)(10^8)$

$= 8.64 \times 10^8$

$= 864{,}000{,}000$

8. (a) $-\sqrt{144} = -12$ because

$-\left(\sqrt{144}\right) = -\left(\sqrt{12^2}\right) = -(12) = -12.$

(b) $\sqrt{-144}$ is not a real number because no real number raised to the second power produces -144.

(c) $\sqrt{\frac{25}{64}} = \frac{5}{8}$ because $\left(\frac{5}{8}\right)^2 = \frac{5^2}{8^2} = \frac{25}{64}.$

(d) $-\sqrt[3]{\frac{8}{27}} = -\frac{2}{3}$ because

$-\left(\sqrt[3]{\frac{8}{27}}\right) = -\left(\frac{\sqrt[3]{8}}{\sqrt[3]{27}}\right) = -\left(\frac{2}{3}\right).$

9. (a) $\frac{\sqrt{125}}{\sqrt{5}} = \sqrt{\frac{125}{5}}$ Property 3

$= \sqrt{25}$ Simplify.

$= 5$ Simplify.

(b) $\sqrt[3]{125^2} = \left(\sqrt[3]{125}\right)^2$ Property 1

$= (5)^2$ Simplify.

$= 25$ Simplify.

(c) $\sqrt[3]{x^2} \cdot \sqrt[3]{x} = \sqrt[3]{x^2 \cdot x}$ Property 2

$= \sqrt[3]{x^3}$ Simplify.

$= x$ Property

(d) $\sqrt{\sqrt{x}} = \sqrt[2\cdot 2]{x}$ Property 4

$= \sqrt[4]{x}$ Simplify.

10. (a) $\sqrt{32} = \sqrt{16 \cdot 2} = \sqrt{4^2 \cdot 2} = 4\sqrt{2}$

(b) $\sqrt[3]{250} = \sqrt[3]{125 \cdot 2} = \sqrt[3]{5^3 \cdot 2} = 5\sqrt[3]{2}$

(c) $\sqrt{24a^5} = \sqrt{4 \cdot 6 \cdot a^4 \cdot a} = \sqrt{4a^4 \cdot 6a}$

$= \sqrt{\left(2a^2\right)^2 \cdot 6a}$

$= 2a^2\sqrt{6a}$

(d) $\sqrt[3]{-135x^3} = \sqrt[3]{(-27) \cdot 5 \cdot x^3}$

$= \sqrt[3]{(-3x)^3 \cdot 5}$

$= -3x\sqrt[3]{5}$

11. (a) $3\sqrt{8} + \sqrt{18} = 3\sqrt{4 \cdot 2} + \sqrt{9 \cdot 2}$ Find square factors.

$= 3 \cdot 2\sqrt{2} + 3\sqrt{2}$ Find square roots.

$= 6\sqrt{2} + 3\sqrt{2}$ Multiply.

$= (6 + 3)\sqrt{2}$ Combine like radicals.

$= 9\sqrt{2}$ Simplify.

(b) $\sqrt[3]{81x^5} - \sqrt[3]{24x^2} = \sqrt[3]{27x^3 \cdot 3x^2} - \sqrt[3]{8 \cdot 3x^2}$ Find cube factors.

$= 3x\sqrt[3]{3x^2} - 2\sqrt[3]{3x^2}$ Find cube roots.

$= (3x - 2)\sqrt[3]{3x^2}$ Combine like radicals.

12. (a) $\frac{5}{3\sqrt{2}} = \frac{5}{3\sqrt{2}} \cdot \frac{\sqrt{2}}{\sqrt{2}}$ $\sqrt{2}$ is rationalizing factor.

$= \frac{5\sqrt{2}}{3(2)}$ Multiply.

$= \frac{5\sqrt{2}}{6}$ Simplify.

(b) $\frac{1}{\sqrt[3]{25}} = \frac{1}{\sqrt[3]{25}} \cdot \frac{\sqrt[3]{5}}{\sqrt[3]{5}}$ $\sqrt[3]{5}$ is rationalizing factor.

$= \frac{\sqrt[3]{5}}{\sqrt[3]{125}}$ Multiply.

$= \frac{\sqrt[3]{5}}{5}$ Simplify.

13. $\dfrac{8}{\sqrt{6}-\sqrt{2}} = \dfrac{8}{\sqrt{6}-\sqrt{2}} \cdot \dfrac{\sqrt{6}+\sqrt{2}}{\sqrt{6}+\sqrt{2}}$ Multiply numerator and denominator by conjugate of denominator.

$= \dfrac{8(\sqrt{6}+\sqrt{2})}{6+\sqrt{12}-\sqrt{12}-2}$ Use Distributive Property.

$= \dfrac{8(\sqrt{6}+\sqrt{2})}{4}$ Simplify.

$= 2(\sqrt{6}+\sqrt{2})$ Simplify.

14. $\dfrac{2-\sqrt{2}}{3} = \dfrac{2-\sqrt{2}}{3} \cdot \dfrac{2+\sqrt{2}}{2+\sqrt{2}}$ Multiply numerator and denominator by conjugate of numerator.

$= \dfrac{4+2\sqrt{2}-2\sqrt{2}-2}{3(2+\sqrt{2})}$ Multiply.

$= \dfrac{2}{3(2+\sqrt{2})}$ Simplify.

15. (a) $\sqrt[3]{27} = 27^{1/3}$

(b) $\sqrt{x^3y^5z} = (x^3y^5z)^{1/2}$

$= x^{3\cdot 1/2}y^{5\cdot 1/2}z^{1/2}$

$= x^{3/2}y^{5/2}z^{1/2}$

(c) $3x\sqrt[3]{x^2} = 3x(x^2)^{1/3}$

$= 3x \cdot x^{2/3}$

$= 3x^{1+2/3}$

$= 3x^{5/3}$

16. (a) $(x^2-7)^{-1/2} = \dfrac{1}{(x^2-7)^{1/2}} = \dfrac{1}{\sqrt{x^2-7}}$

(b) $-3b^{1/3}c^{2/3} = -3(bc^2)^{1/3} = -3\sqrt[3]{bc^2}$

(c) $a^{0.75} = a^{3/4} = \sqrt[4]{a^3}$

(d) $(x^2)^{2/5} = x^{4/5} = \sqrt[5]{x^4}$

17. (a) $(-125)^{-2/3} = (\sqrt[3]{-125})^{-2} = (-5)^{-2} = \dfrac{1}{(-5)^2} = \dfrac{1}{25}$

(b) $(4x^2y^{3/2})(-3x^{-1/3})(y^{-3/5}) = -12x^{(2)-(1/3)}y^{(3/2)-(3/5)} = -12x^{5/3}y^{9/10},\ x \neq 0, y \neq 0$

(c) $\sqrt[3]{\sqrt[4]{27}} = \sqrt[12]{27} = \sqrt[12]{(3)^3} = 3^{3/12} = 3^{1/4} = \sqrt[4]{3}$

(d) $(3x+2)^{5/2}(3x+2)^{-1/2} = (3x+2)^{(5/2)-(1/2)} = (3x+2)^2,\ x \neq -2/3$

Checkpoints for Section P.3

1.

Polynomial	Standard Form	Degree	Leading Coefficient
$6 - 7x^3 + 2x$	$-7x^3 + 2x + 6$	3	-7

2. $(2x^3 - x + 3) - (x^2 - 2x - 3)$

$= 2x^3 - x + 3 - x^2 + 2x + 3$

$= 2x^3 - x^2 + (-x + 2x) + (3 + 3)$

$= 2x^3 - x^2 + x + 6$

3. F O I L

$(3x-1)(x-5) = 3x^2 - 15x - x + 5$

$= 3x^2 - 16x + 5$

4.

$$\begin{array}{r}
x^2 + 2x + 3 \\
\times\ x^2 - 2x + 3 \\
\hline
x^4 + 2x^3 + 3x^2 \quad \leftarrow x^2(x^2+2x+3) \\
-2x^3 - 4x^2 - 6x \quad \leftarrow -2x(x^2+2x+3) \\
3x^2 + 6x + 9 \quad \leftarrow 3(x^2+2x+3) \\
\hline
x^4 + 0x^3 + 2x^2 + 0x + 9
\end{array}$$

So, $(x^2 + 2x + 3)(x^2 - 2x - 3) = x^4 + 2x^2 + 9.$

5. This product has the form $(u + v)(u - v) = u^2 - v^2$.

$$(3x - 2)(3x + 2) = (3x) - (2)^2 = 9x^2 - 4$$

6. This product has the form $(u + v) = u^2 + 2uv + v^2$.

$$(x + 10)^2 = (x)^2 + 2(x)(10) + (10)^2 = x^2 + 20x + 100$$

7. This product has the form $(u - v)^3 = u^3 - 3u^2v + 3uv^2 - v^3$.

$$\begin{aligned}(4x - 1)^3 &= (4x)^3 - 3(4x)^2(1) + 3(4x)(1)^2 - (1)^3\\ &= 64x^3 - 48x^2 + 12x - 1\end{aligned}$$

8. This product has the form $(u + v)(u - v) = u^2 - v^2$.

$$\begin{aligned}(x - 2 + 3y)(x - 2 - 3y) &= \left[(x - 2) + 3y\right]\left[(x - 2) - 3y\right]\\ &= (x - 2)^2 - (3y)^2\\ &= x^2 - 4x + 4 - 9y^2\\ &= x^2 - 9y^2 - 4x + 4\end{aligned}$$

9. The volume of a rectangular box is equal to the product of its length, width, and height.

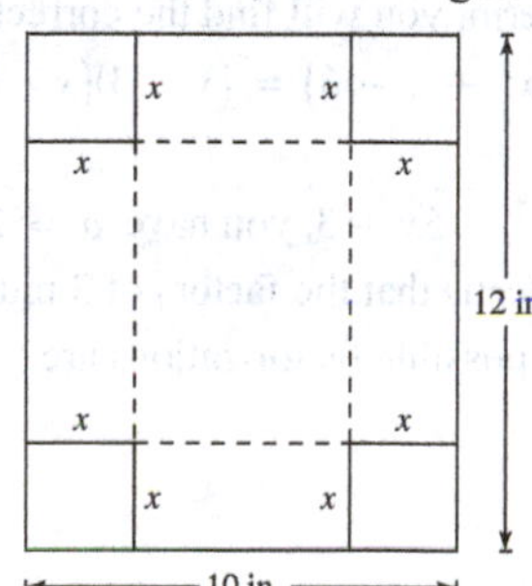

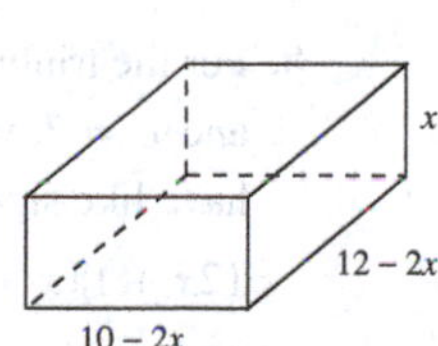

The length is $10 - 2x$, the width is $12 - 2x$, and the height is x. The volume of the box is

$$\begin{aligned}\text{Volume} &= (10 - 2x)(12 - 2x)(x)\\ &= \left(120 - 44x + 4x^2\right)(x)\\ &= 120x - 44x^2 + 4x^3\end{aligned}$$

When $x = 2$ inches the volume of the box is

$$\begin{aligned}\text{Volume} &= 120(2) - 44(2)^2 + 4(2)^3\\ &= 96 \text{ cubic inches.}\end{aligned}$$

When $x = 3$ inches, the volume of the box is

$$\begin{aligned}\text{Volume} &= 120(3) - 44(3)^2 + 4(3)^3\\ &= 72 \text{ cubic inches.}\end{aligned}$$

Checkpoints for Section P.4

1. (a) $5x^3 - 15x^2 = 5x^2(x) - 5x^2(3)$ $\quad 5x^2$ is a common factor.

$$= 5x^2(x - 3)$$

(b) $-3 + 6x - 12x^3 = -12x^3 + 6x - 3$

$$\begin{aligned}&= -3\left(4x^3\right) + (-3)(-2x) + (-3)(1) \quad -3 \text{ is a common factor.}\\ &= -3\left(4x^3 - 2x + 1\right)\end{aligned}$$

(c) $(x + 1)\left(x^2\right) - (x + 1)(2) = (x + 1)\left(x^2 - 2\right)$ $\quad (x + 1)$ is a common factor.

2. $100 - 4y^2 = 4(25 - y^2)$ 4 is a common factor.

$= 4\left[(5)^2 - (y)^2\right]$

$= 4(5 + y)(5 - y)$ Difference of two squares.

3. $(x-1)^2 - 9y^4 = (x-1)^2 - (3y^2)^2$

$= \left[(x-1) + 3y^2\right]\left[(x-1) - 3y^2\right]$

$= (x - 1 + 3y^2)(x - 1 - 3y^2)$

4. $9x^2 - 30x + 25 = (3x)^2 - 2(3x)(5) + 5^2$

$= (3x - 5)^2$

5. $64x^3 - 1 = (4x)^3 - (1)^3$

$= (4x - 1)(16x^2 - 4x + 1)$

6. (a) $x^3 + 216 = (x)^3 + (6)^3$

$= (x + 6)(x^2 - 6x + 36)$

(b) $5y^3 + 135 = 5(y^3 + 27)$

$= 5\left[(y)^3 + (3)^3\right]$

$= 5(y + 3)(y^2 - 3y + 9)$

7. For the trinomial $x^2 + x - 6$, you have $a = 1$, $b = 1$, and $c = -6$. Because b is positive and c is negative, one factor of -6 is positive and one is negative. So, the possible factorizations of $x^2 + x - 6$ are

$(x - 3)(x + 2)$,

$(x + 3)(x - 2)$,

$(x + 6)(x - 1)$, and

$(x - 6)(x + 1)$.

Testing the middle term, you will find the correct factorization to be $(x^2 + x - 6) = (x + 3)(x - 2)$.

8. For the trinomial $2x^2 - 5x + 3$, you have $a = 2$ and $c = 3$, which means that the factors of 3 must have like signs. The possible factorizations are

$(2x + 1)(x + 3)$,

$(2x - 1)(x - 3)$,

$(2x + 3)(x + 1)$, and

$(2x - 3)(x - 1)$.

Testing the middle term, you will find the correct factorization to be $2x^2 - 5x + 3 = (2x - 3)(x - 1)$.

9. $x^3 + x^2 - 5x - 5 = (x^3 + x^2) - (5x + 5)$ Group terms.

$= x^2(x + 1) - 5(x + 1)$ Factor each group.

$= (x + 1)(x^2 - 5)$ Distributive Property

10. $2x^2 + 5x - 12 = 2x^2 + 8x - 3x - 12$ Rewrite middle term.

$= (2x^2 + 8x) - (3x + 12)$ Group terms.

$= 2x(x + 4) - 3(x + 4)$ Factor groups.

$= (x + 4)(2x - 3)$ Distributive Property

Checkpoints for Section P.5

1. (a) The domain of the polynomial $4x^2 + 3,\ x \geq 0$ is the set of all real numbers that are greater than or equal to 0. The domain is specifically restricted.

(b) The domain of the radical expression $\sqrt{x + 7}$ is the set of all real numbers greater than or equal to -7, because the square root of a negative number is not a real number.

(c) The domain of the rational expression $\dfrac{1 - x}{x}$ is the set of all real number except $x = 0$, which would result in division by zero, which is undefined.

2. $\dfrac{4x + 12}{x^2 - 3x - 18} = \dfrac{4(x + 3)}{(x - 6)(x + 3)}$ Factor completely.

$= \dfrac{4}{x - 6}, x \neq -3$ Divide out common factor.

3. $\dfrac{3x^2 - x - 2}{5 - 4x - x^2} = \dfrac{3x^2 - x - 2}{-x^2 - 4x + 5} = \dfrac{(3x + 2)(x - 1)}{-(x + 5)(x - 1)}$ Write in standard form.

$= -\dfrac{3x + 2}{x + 5}, x \neq 1$ Divide out common factor.

4. $\dfrac{15x^2 + 5x}{x^3 - 3x^2 - 18x} \cdot \dfrac{x^2 - 2x - 15}{3x^2 - 8x - 3} = \dfrac{5x(3x + 1)}{x(x - 6)(x + 3)} \cdot \dfrac{(x - 5)(x + 3)}{(3x + 1)(x - 3)}$

$= \dfrac{5(x - 5)}{(x - 6)(x - 3)}, x \neq -3, x \neq -\dfrac{1}{3}, x \neq 0$

5. $\dfrac{x^3 - 1}{x^2 - 1} \div \dfrac{x^2 + x + 1}{x^2 + 2x + 1} = \dfrac{x^3 - 1}{x^2 - 1} \cdot \dfrac{x^2 + 2x + 1}{x^2 + x + 1}$ Invert and multiply.

$= \dfrac{(x - 1)(x^2 + x + 1)}{(x + 1)(x - 1)} \cdot \dfrac{(x + 1)(x + 1)}{x^2 + x + 1}$ Factor completely.

$= x + 1, x \neq \pm 1$ Divide out common factors.

6. $\dfrac{x}{2x - 1} - \dfrac{1}{x + 2} = \dfrac{x(x + 2) - (2x - 1)}{(2x - 1)(x + 2)}$ Basic definition

$= \dfrac{x^2 + 2x - 2x + 1}{(2x - 1)(x + 2)}$ Distributive Property

$= \dfrac{x^2 + 1}{(2x - 1)(x + 2)}$ Combine like terms.

7. The LCD of the ration expression $\dfrac{4}{x} - \dfrac{x + 5}{x^2 - 4} + \dfrac{4}{x + 2}$ is $x(x + 2)(x - 2)$.

$\dfrac{4}{x} - \dfrac{x + 5}{(x + 2)(x - 2)} + \dfrac{4}{x + 2} = \dfrac{4(x + 2)(x - 2)}{x(x + 2)(x - 2)} - \dfrac{x(x + 5)}{x(x + 2)(x - 2)} + \dfrac{4x(x - 2)}{x(x + 2)(x - 2)}$ Rewrite using the LCD.

$= \dfrac{4(x + 2)(x - 2) - x(x + 5) + 4x(x - 2)}{x(x + 2)(x - 2)}$ Distributive Property

$= \dfrac{4x^2 - 16 - x^2 - 5x + 4x^2 - 8x}{x(x + 2)(x - 2)}$

$= \dfrac{7x^2 - 13x - 16}{x(x + 2)(x - 2)}$

8. $\dfrac{\left(\dfrac{1}{x+2}+1\right)}{\left(\dfrac{x}{3}-1\right)} = \dfrac{\left(\dfrac{1+1(x+2)}{x+2}\right)}{\left(\dfrac{x-1(3)}{3}\right)}$ Combine fractions.

$= \dfrac{\left(\dfrac{x+3}{x+2}\right)}{\left(\dfrac{x-3}{3}\right)}$ Simplify.

$= \dfrac{x+3}{x+2} \cdot \dfrac{3}{x-3}$ Invert and multiply.

$= \dfrac{3(x+3)}{(x+2)(x-3)}$

9. $(x-1)^{-1/3} - x(x-1)^{-4/3} = (x-1)^{-4/3}\left[(x-1)^{(-1/3)-(-4/3)} - x\right]$

$= (x-1)^{-4/3}\left[(x-1)^1 - x\right]$

$= -\dfrac{1}{(x-1)^{4/3}}$

10. $\dfrac{\dfrac{x^2}{(x^2-2)^{1/2}} + \dfrac{1}{(x^2-2)^{-1/2}}}{x^2-2} = \dfrac{x^2(x^2-2)^{-1/2} + (x^2-2)^{1/2}}{x^2-2}$

$= \dfrac{x^2(x^2-2)^{-1/2} + (x^2-2)^{1/2}}{x^2-2} \cdot \dfrac{(x^2-2)^{1/2}}{(x^2-2)^{1/2}}$

$= \dfrac{x^2(x^2-2)^0 + (x^2-2)^1}{(x^2-2)^{3/2}}$

$= \dfrac{x^2 + x^2 - 2}{(x^2-2)^{3/2}}$

$= \dfrac{2x^2 - 2}{(x^2-2)^{3/2}}$

$= \dfrac{2(x+1)(x-1)}{(x^2-2)^{3/2}}$

11. $\dfrac{\sqrt{9+h}-3}{h} = \dfrac{\sqrt{9+h}-3}{h} \cdot \dfrac{\sqrt{9+h}+3}{\sqrt{9+h}+3}$

$= \dfrac{\left(\sqrt{9+h}\right)^2 - (3)^2}{h\left(\sqrt{9+h}+3\right)}$

$= \dfrac{(9+h) - 9}{h\left(\sqrt{9+h}+3\right)}$

$= \dfrac{h}{h\left(\sqrt{9+h}+3\right)}$

$= \dfrac{1}{\sqrt{9+h}+3}, h \neq 0$

Checkpoints for Section P.6

1.

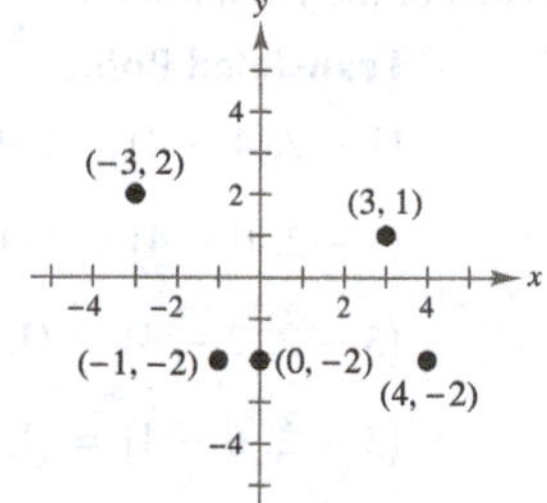

2. To sketch a scatter plot of the data shown in the table, first draw a vertical axis to represent the number of employees E (in thousands) and a horizontal axis to represent the year. Then plot the resulting points. Note that the break in the t-axis indicates that the numbers through 2004 have been omitted.

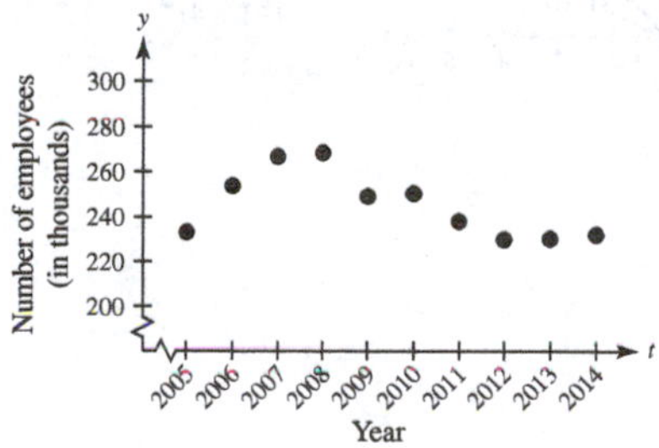

3. Let $(x_1, y_1) = (3, 1)$ and $(x_2, y_2) = (-3, 0)$.

Then apply the Distance Formula.

$$\begin{aligned} d &= \sqrt{(x_2 - x_1)^2 + (y_2 - y_1)^2} \\ &= \sqrt{(-3 - 3)^2 + (0 - 1)^2} \\ &= \sqrt{(-6)^2 + (-1)^2} \\ &= \sqrt{36 + 1} \\ &= \sqrt{37} \\ &\approx 6.08 \end{aligned}$$

So, the distance between the points is about 6.08 units.

4. The three points are plotted in the figure.

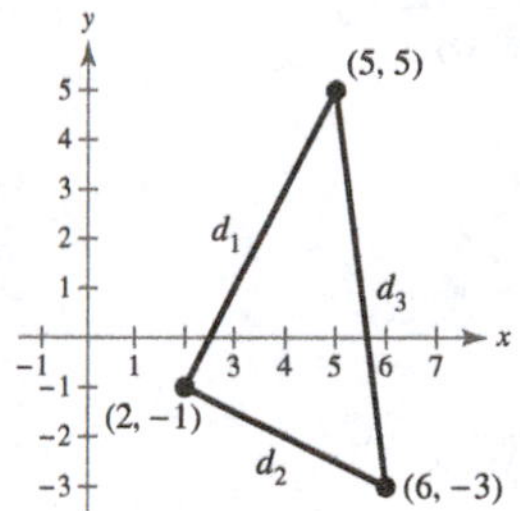

Using the Distance Formula, the lengths of the three sides are as follows.

$$\begin{aligned} d_1 &= \sqrt{(5 - 2)^2 + (5 - (-1))^2} \\ &= \sqrt{3^2 + 6^2} \\ &= \sqrt{9 + 36} \\ &= \sqrt{45} \end{aligned}$$

$$\begin{aligned} d_2 &= \sqrt{(6 - 2)^2 + (-3 - (-1))^2} \\ &= \sqrt{4^2 + (-2)^2} \\ &= \sqrt{16 + 4} \\ &= \sqrt{20} \end{aligned}$$

$$\begin{aligned} d_3 &= \sqrt{(6 - 5)^2 + (-3 - 5)^2} \\ &= \sqrt{(1)^2 + (-8)^2} \\ &= \sqrt{1 + 64} \\ &= \sqrt{65} \end{aligned}$$

Because $(d_1)^2 + (d_2)^2 = 45 + 20 = \left(\sqrt{65}\right)^2 = (d_3)^2$ you can conclude by the Pythagorean Theorem that the triangle must be a right triangle.

5. Let $(x_1, y_1) = (-2, 8)$ and $(x_2, y_2) = (-4, -$`0$)$.

$$\begin{aligned} \text{Midpoint} &= \left(\frac{x_1 + x_2}{2}, \frac{y_1 + y_2}{2}\right) \\ &= \left(\frac{-2 + 4}{2}, \frac{8 + (-10)}{2}\right) \\ &= \left(\frac{2}{2}, -\frac{2}{2}\right) \\ &= 1, -1 \end{aligned}$$

The midpoint of the line segment is $(1, -1)$.

6. You can find the length of the pass by finding the distance between the points $(10, 10)$ and $(25, 32)$.

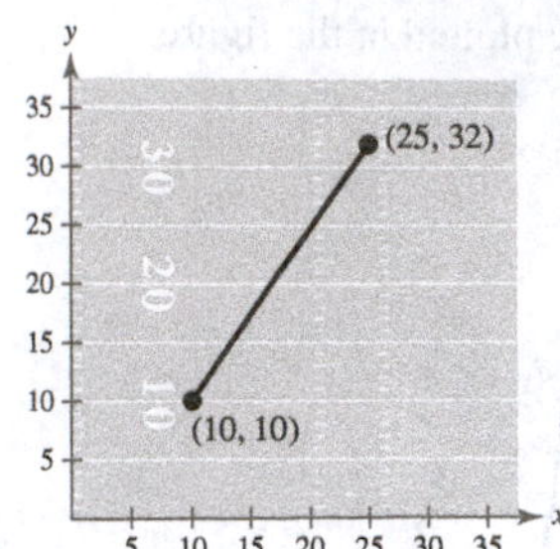

$$\begin{aligned} d &= \sqrt{(x_2 - x_1)^2 + (y_2 - y_1)^2} \\ &= \sqrt{(25 - 10)^2 + (32 - 10)^2} \\ &= \sqrt{15^2 + 22^2} \\ &= \sqrt{225 + 484} \\ &= \sqrt{709} \\ &\approx 26.6 \text{ years} \end{aligned}$$

So, the pass is about 26.6 yard long.

7. Assuming that the annual revenue from Yahoo! Inc. followed a linear pattern, you can estimate the 2013 annual revenue by finding the midpoint of the line segment connecting the points $(2012, 5.0)$ and $(2014, 4.6)$.

$$\begin{aligned} \text{Midpoint} &= \left(\frac{x_1 + x_2}{2}, \frac{y_1 + y_2}{2}\right) \\ &= \left(\frac{2012 + 2014}{2}, \frac{5.0 + 4.6}{2}\right) \\ &= (2013, 4.8) \end{aligned}$$

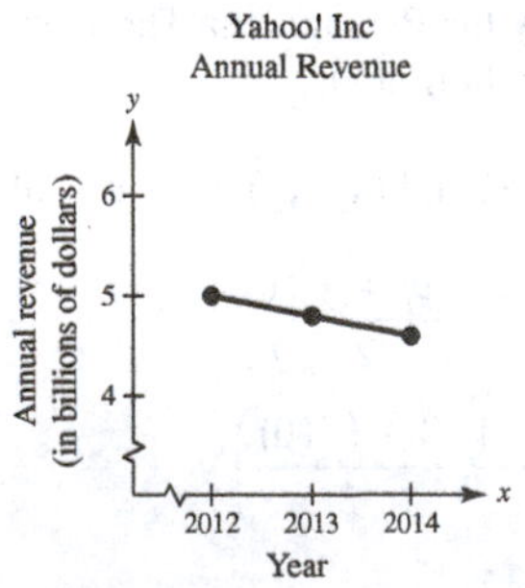

So, you can estimate the annual revenue for Yahoo! Inc. was $4.8 billion in 2013.

8. To shift the vertices two units to the left, subtract 2 from each of the x-coordinates. To shift the vertices four units down, subtract 4 from each of the y-coordinates.

Original point	**Translated Point**
(1, 4)	$(1 - 2, 4 - 4) = (-1, 0)$
(1, 0)	$(1 - 2, 0 - 4) = (-1, -4)$
(3, 2)	$(3 - 2, 2 - 4) = (1, -2)$
(3, 6)	$(3 - 2, 6 - 4) = (1, 2)$

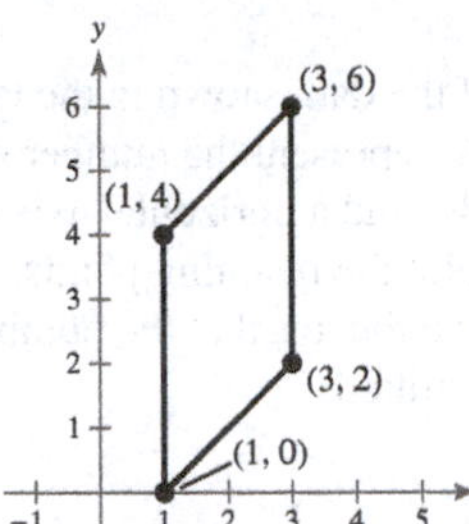

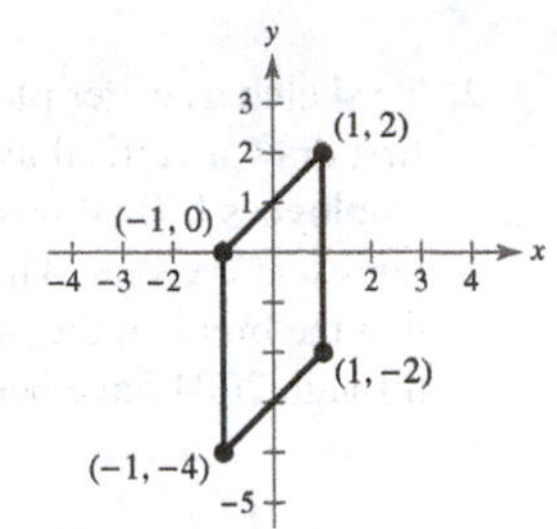

Chapter 1

Checkpoints for Section 1.1

1. (a)

$y = 14 - 6x$	Write original equation.
$-5 \stackrel{?}{=} 14 - 6(3)$	Substitute 3 for x and -5 for y.
$-5 \stackrel{?}{=} 14 - 18$	
$-5 \neq -4$	$(3, -5)$ is not a solution.

(b)

$y = 14 - 6x$	Write original equation.
$26 \stackrel{?}{=} 14 - 6(-2)$	Substitute -2 for x and 26 for y.
$26 \stackrel{?}{=} 14 + 12$	
$26 = 26$	$(-2, 26)$ is a solution. ✓

2. (a) To graph $3x + y = 2$, first isolate the variable y.

$y = -3x + 2$

Next construct a table of values that consists of several solution points. Then plot the points and connect them.

x	$y = -3x + 2$	(x, y)
-2	$y = -3(-2) + 2 = 8$	$(-2, 8)$
-1	$y = -3(-1) + 2 = 5$	$(-1, 5)$
0	$y = -3(0) + 2 = 2$	$(0, 2)$
1	$y = -3(1) + 2 = -1$	$(1, -1)$
2	$y = -3(2) + 2 = -4$	$(2, -4)$

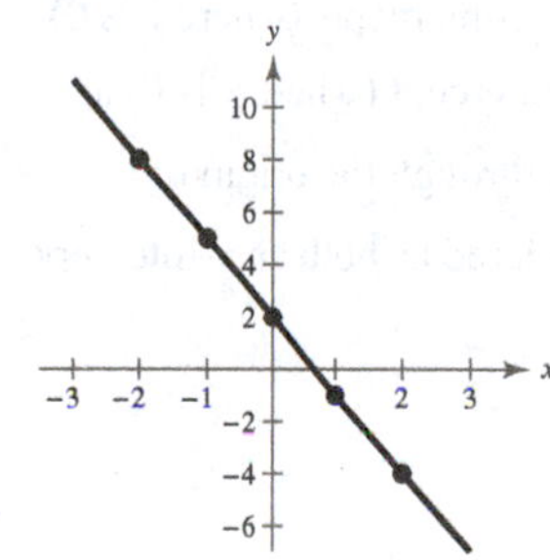

(b) To graph $-2x + y = 1$, first isolate the variable y.

$y = 2x + 1$

Next construct a table of values that consists of several solution points. Then plot the points and connect them.

x	$y = x^2 + 3$	(x, y)
-2	$y = 2(-2) + 1 = -3$	$(-2, -3)$
-1	$y = 2(-1) + 1 = -1$	$(-1, -1)$
0	$y = 2(0) + 1 = 1$	$(0, 1)$
1	$y = 2(1) + 1 = 3$	$(1, 3)$
2	$y = 2(2) + 1 = 5$	$(2, 5)$

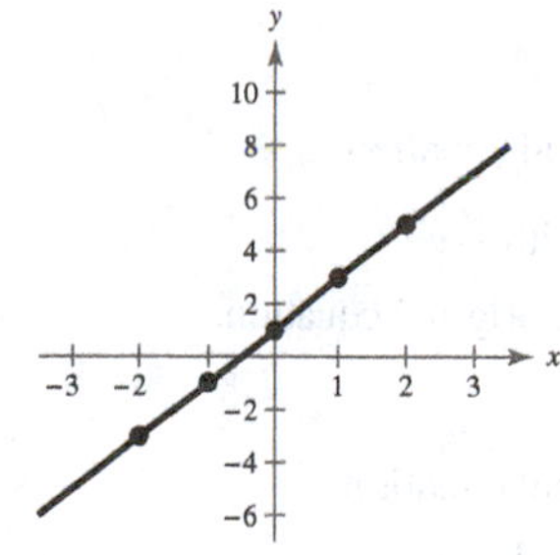

3. (a) To graph $y = x^2 + 3$, construct a table of values that consists of several solution points. Then plot the points and connect them with a smooth curve.

x	$y = x^2 + 3$	(x, y)
-2	$y = (-2)^2 + 3 = 7$	$(-2, 7)$
-1	$y = (-1)^2 + 3 = 4$	$(-1, 4)$
0	$y = (0)^2 + 3 = 3$	$(0, 3)$
1	$y = (1)^2 + 3 = 4$	$(1, 4)$
2	$y = 2(2) + 3 = 7$	$(2, 7)$

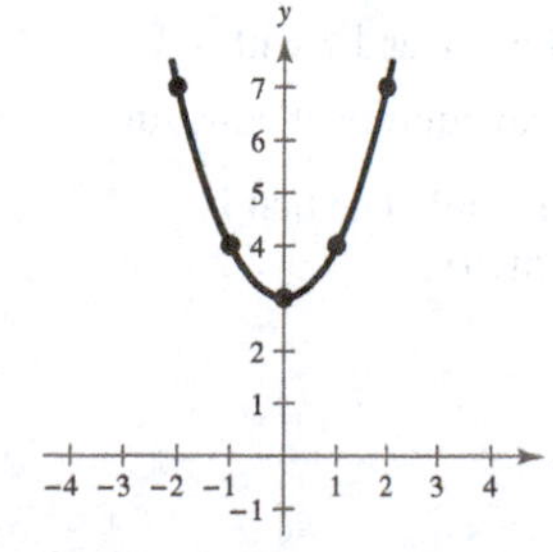

(b) To graph $y = 1 - x^2$, construct a table of values that consists of several solution points. Then plot the points and connect them with a smooth curve.

x	$y = 1 - x^2$	(x, y)
-2	$y = 1 - (-2)^2 = -3$	$(-2, -3)$
-1	$y = 1 - (-1)^2 = 0$	$(-1, 0)$
0	$y = 1 - (0)^2 = 1$	$(0, 1)$
1	$y = 1 - (1)^2 = 0$	$(1, 0)$
2	$y = 1 - (2)^2 = -3$	$(2, -3)$

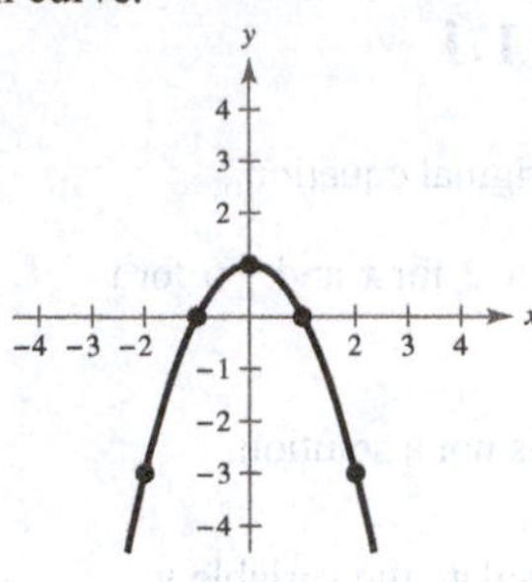

4. From the figure, you can see that the graph of the equation $y = -x^2 - 5x$ has x-intercepts (where y is 0) at $(0, 0)$ and $(-5, 0)$ and a y-intercept (where x is 0) at $(0, 0)$. Since the graph passes through the origin or $(0, 0)$, that point can be considered as both an x-intercept and a y-intercept.

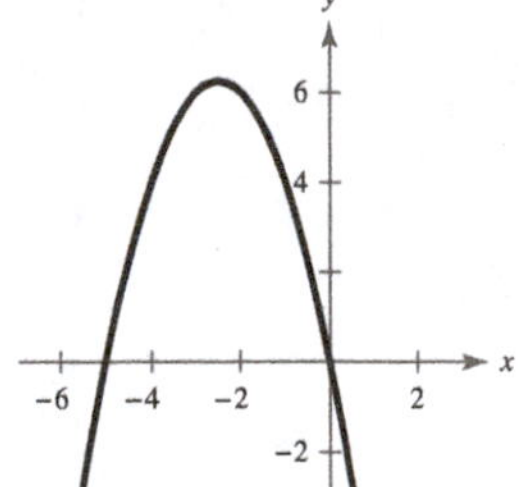

5. x-Axis:

$y^2 = 6 - x$ Write original equation.

$(-y)^2 = 6 - x$ Replace y with $-y$.

$y^2 = 6 - x$ Result is the original equation.

y-Axis:

$y^2 = 6 - x$ Write original equation.

$y^2 = 6 - (-x)$ Replace x with $-x$.

$y^2 = 6 + x$ Result is *not* an equivalent equation.

Origin:

$y^2 = 6 - x$ Write original equation.

$(-y)^2 = 6 - (-x)$ Replace y with $-y$ and x with $-x$.

$y^2 = 6 + x$ Result is *not* an equivalent equation.

Of the three tests for symmetry, the only one that is satisfied is the test for x-axis symmetry.

6. Of the three test of symmetry, the only one that is satisfied is the test for y-axis symmetry because $y = (-x)^2 - 4$ is equivalent to $y = x^2 - 4$. Using symmetry, you only need to find solution points to the right of the y-axis and then reflect them about the y-axis to obtain the graph.

7. The equation $y = |x - 2|$ fails all three tests for symmetry and consequently its graph is not symmetric with respect to either axis or to the origin. So, construct a table of values. Then plot and connect the points.

x	$y = \|x - 2\|$	(x, y)
-2	$y = \|(-2) - 2\| = 4$	$(-2, 4)$
-1	$y = \|(-1) - 2\| = 3$	$(-1, 3)$
0	$y = \|(0) - 2\| = 2$	$(0, 2)$
1	$y = \|(1) - 2\| = 1$	$(1, 1)$
2	$y = \|(2) - 2\| = 0$	$(3, 0)$
3	$y = \|(2) - 2\| = 0$	$(3, 1)$
4	$y = \|(2) - 2\| = 0$	$(4, 2)$

From the table, you can see that the x-intercept is $(2, 0)$ and the y-intercept is $(0, 2)$.

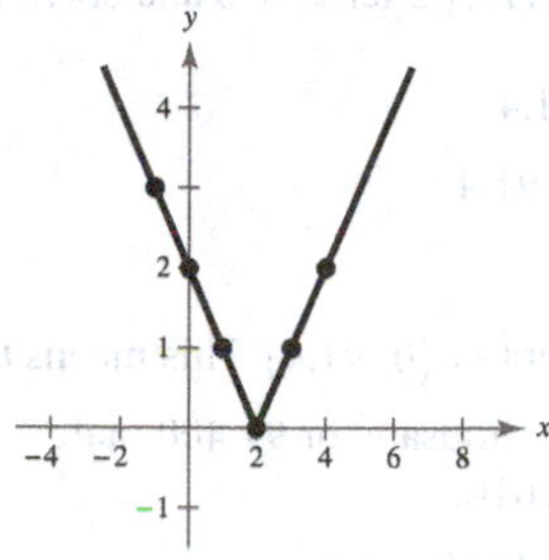

8. The radius of the circle is the distance between $(1, -2)$ and $(-3, -5)$.

$$\begin{aligned} r &= \sqrt{(x-h)^2 + (y-k)^2} \\ &= \sqrt{[1-(-3)]^2 + [-2-(-5)]^2} \\ &= \sqrt{4^2 + 3^2} \\ &= \sqrt{16+9} \\ &= \sqrt{25} \\ &= 5 \end{aligned}$$

Using $(h, k) = (-3, -5)$ and $r = 5$, the equation of the circle is

$$(x-h)^2 + (y-k)^2 = r^2$$
$$[x-(-3)]^2 + [y-(-5)]^2 = (5)^2$$
$$(x+3)^2 + (y+5)^2 = 25.$$

9. From the graph, you can estimate that a height of 75 inches corresponds to a maximum weight of about 221 pounds.

Let $x = 75$: $y = 0.040x^2 - 0.11x + 3.9$
$$= 0.040(75)^2 - 0.11(75) + 3.9$$
$$= 220.65$$

Algebraically, you can conclude that a height of 75 inches corresponds to a maximum weight of 220.65 pounds. So, the graphical estimate of 221 is fairly good.

Checkpoints for Section 1.2

1. (a)

$7 - 2x = 15$	Write original equation.
$-2x = 8$	Subtract 7 from each side.
$x = -4$	Divide each side by -2.

Check: $7 - 2x = 15$
$$7 - 2(-4) \stackrel{?}{=} 15$$
$$7 + 8 \stackrel{?}{=} 15$$
$$15 = 15$$

(b)

$7x - 9 = 5x + 7$	Write original equation.
$2x - 9 = 7$	Subtract $5x$ from each side.
$2x = 16$	Add 9 from each side.
$x = 8$	Divide each side by 2.

Check: $7x - 9 = 5x + 7$
$$7(8) - 9 = 5(8 + 7)$$
$$56 - 9 = 40 + 7$$
$$47 = 47 \checkmark$$

2.

$4(x+2) - 12 = 5(x-6)$	Write original equation.
$4x + 8 - 12 = 5x - 30$	Distributive Property
$4x - 4 = 5x - 30$	Simplify.
$-x = -26$	Simplify.
$x = 26$	Divide each side by -1.

Check: $4(x+2) - 12 = 5(x-6)$
$$4(26+2) - 12 \stackrel{?}{=} 5(26-6)$$
$$4(28) - 12 \stackrel{?}{=} 5(20)$$
$$112 - 12 \stackrel{?}{=} 100$$
$$100 = 100$$

3. $\frac{4x}{9} - \frac{1}{3} = x + \frac{5}{3}$ Write original equation.

$(9)\left(\frac{4x}{9}\right) - (9)\left(\frac{1}{3}\right) = (9)x + 9\left(\frac{5}{3}\right)$ Multiply each term by the LCD.

$4x - 3 = 9x + 15$ Simplify.

$-5x = 18$ Combine like terms.

$x = -\frac{18}{5}$ Divide each side by -5.

4. $\frac{3x}{x-4} = 5 + \frac{12}{x-4}$ Write original equation.

$(x-4)\left(\frac{3x}{x-4}\right) = (x-4)5 + (x-4)\left(\frac{12}{x-4}\right)$ Multiply each term by LCD.

$3x = 5x - 20 + 12, x \neq 4$ Simplify.

$-2x = -8$ Divide each side by -2.

$x = 4$ Extraneous solution

In the original equation, $x = 4$ yields a denominator of zero. So, $x = 4$ is an extraneous solution, and the original equation has no solution.

5. To find the *x*-intercept, set *y* equal to zero, and solve for *x*.

(a) $y = -3x - 2$ Write original equation.

$0 = -3x - 2$ Substitute 0 for *y*.

$2 = -3x$ Add 2 to each side.

$-\frac{2}{3} = x$ Divide each side by -3.

So, the *x*-intercept is $\left(-\frac{2}{3}, 0\right)$.

To find the *y*-intercept, set *x* equal to zero, and solve for *y*.

$y = -3x - 2$ Write original equation.

$y = -3(0) - 2$ Substitute 0 for *x*.

$y = -2$ Simplify.

So, the *y*-intercept is $(0, -2)$.

(b) To find the *x*-intercept, set *y* equal to zero, and solve for *x*.

$5x + 3y = 15$ Write original equation.

$5x + 3(0) = 15$ Substitute 0 for *y*.

$5x = 15$ Simplify.

$x = 3$ Divide each side by 5.

So, the *x*-intercept is $(3, 0)$.

To find the *y*-intercept, set *x* equal to zero, and solve for *y*.

$5x + 3y = 15$ Write original equation.

$5(0) + 3y = 15$ Substitute 0 for *x*.

$3y = 15$ Simplify.

$y = 5$ Divide each side by 3.

So, the *y*-intercept is $(0, 5)$.

6. (a) To find the *y*-intercept, let $t = 0$ and solve for *y*.

$y = 3.66t + 91.4$

$y = 3.66(0) + 91.4$

$y = 91.4$

So, the *y*-intercept is $(0, 91.4)$. This means there were about 91.4 thousand or 91,400 male participants in 2010.

(b) Let $y = 128$ and solve for t.

$y = 3.66t + 91.4$

$128 = 3.66t + 91.4$

$36.6 = 3.66t$

$\frac{36.6}{3.66} = t$

$10 = t$

Because $t = 0$ represents 2010, $t = 10$ must represent 2020. So, from this model, there will be 128,000 male participants in 2020.

Checkpoints for Section 1.3

1. Because there are 52 weeks in a year and you will be paid weekly, it follows that you will receive 52 paychecks during the year.

Verbal Model: [Income for year] = [52 paychecks] · [Amount of each paycheck] + [Bonus]

Labels:
Income for year = 58,400 (dollars)
Amount each paycheck = x (dollars)
Bonus = 1200 (dollars)

Equation:

$58{,}400 = 52x + 1200$	Write equation.
$57{,}200 = 52x$	Subtract 1200 from each side.
$1100 = x$	Divide each side by 52.

So, your gross pay for each paycheck is \$1100.

2. *Verbal Model:* [Increase in price] = [Percent] · [Original price]

Labels:
Original price = 15 (dollars per share)
Increase in price = 18 − 15 = 3 (dollars per share)
Percent = r (in decimal form)

Equation:

$3 = r \cdot 15$	Write equation.
$\frac{1}{5} = r$	Divide each side by 15.
$0.2 = r$	Rewrite the fraction as a decimal.

The stock's value increased by 0.2 or 20%.

3. The total amount of your family's loan payments is \$17,920. The loan payments equal 28% of the annual income.

Verbal Model: [Loan payments] = [Percent] · [Annual income]

Labels:
Loan payments = 17,920 (dollars)
Annual income = x (dollars)
Percent = 28% = 0.28 (in decimal form)

Equation:

$17{,}920 = 0.28 \cdot x$	Write equation.
$64{,}000 = x$	Divide each side by 0.28.

Your family's annual income is \$64,000.

4. Draw a diagram,

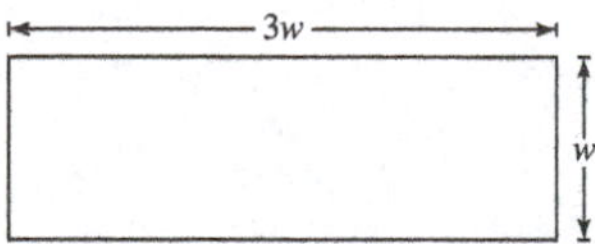

Verbal Model: 2 · [Length] + 2 · [Width] = [Perimeter]

Labels:
Perimeter = 112 (feet)
Width = w (feet)
Length = l = $3w$ (feet)

Equation:

$2(3w) + 2(w) = 112$	Write equation.
$6w + 2w = 112$	Simplify.
$8w = 12$	Combine like terms.
$w = 14$	Divide each side by 8.

Because the length is three times the width, $l = 3w = 3(14) = 42$.

So, the dimensions of the family room are 14 feet by 42 feet.

5. *Verbal Model:* $\boxed{\text{Distance}} = \boxed{\text{Rate}} \cdot \boxed{\text{Time}}$

Labels: Distance $= 14$ (miles)

Rate $= \dfrac{\text{first five miles}}{\text{time to travel first five miles}} = \dfrac{5}{0.5} = 10$ (miles per hour)

Time $= t$ (hours)

Equation: $14 = 10 \cdot t$ Write equation.

$\dfrac{14}{10} = t$ Divide each side by 10.

$1.4 = t$ Simplify.

The entire trip will take 1.4 hours, or 1 hour and 24 minutes.

6. To solve this problem, use the result from geometry that the ratios of the corresponding sides of similar triangles are equal.

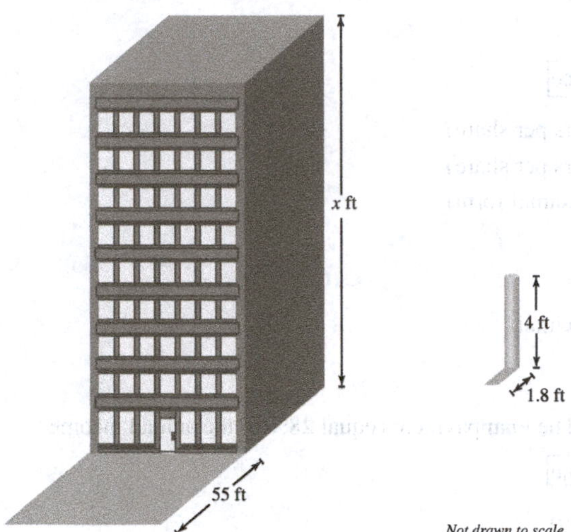

Not drawn to scale

Verbal Model: $\dfrac{\text{Height of building}}{\text{Length of building's shadow}} = \dfrac{\text{Height of post}}{\text{Length of post's shadow}}$

Labels: Height of building $= x$ (feet)

Length of building's shadow $= 55$ (feet)

Height of post $= 4$ (feet)

Length of post's shadow $= 1.8$ (feet)

Equation: $\dfrac{x}{55} = \dfrac{4}{1.8}$

$x = \dfrac{4}{1.8} \cdot 55$

$x \approx 122.2$

So, the building is about 122.2 feet high.

7. Let x represent the amount invested at $2\frac{1}{2}\%$. Then the amount invested at $3\frac{1}{2}\%$ is $5000 - x$.

Verbal Model: $\boxed{\text{Interest from } 2\frac{1}{2}\%} + \boxed{\text{Interest from } 3\frac{1}{2}\%} = \boxed{\text{Total interest}}$

Labels: Interest from $2\frac{1}{2}\% = Prt = (x)(0.025)(1)$ (dollars)

Interest from $3\frac{1}{2}\% = Prt = (5000 - x)(0.035)(1)$ (dollars)

Total interest $= 151.25$ (dollars)

Equation:

$$\begin{aligned} 0.025x + 0.035(5000 - x) &= 151.25 \\ 0.025x + 175 - 0.035x &= 151.25 \\ -0.01x + 175 &= 151.25 \\ -0.01x &= -23.75 \\ x &= 2375 \end{aligned}$$

So, \$2375 was invested at $2\frac{1}{2}\%$ and $5000 - 2375 = \$2625$ was invested at $3\frac{1}{2}\%$.

8. *Verbal Model:* $\boxed{\text{Profit from 24-inch televisions}} + \boxed{\text{Profit from 50-inch televisions}} = \boxed{\text{Total profit}}$

Labels: Inventory of 24-inch televisions $= x$ (dollars)

Inventory of 50-inch televisions $= 30{,}000 - x$ (dollars)

Profit from 24-inch televisions $= 0.24x$ (dollars)

Profit from 50-inch televisions $= 0.42(30{,}000 - x)$ (dollars)

Total profit $= 0.36(30{,}000) = 10{,}800$ (dollars)

Equation:

$$\begin{aligned} 0.24x + 0.42(30{,}000 - x) &= 10{,}800 \\ 0.24x + 12{,}600 - 0.42x &= 10{,}800 \\ 12{,}600 - 0.42x &= -1800 \\ -0.18x &= -1800 \\ x &= 10{,}000 \end{aligned}$$

So, \$10,000 was invested in 24-inch televisions and $30{,}000 - 10{,}000 = \$20{,}000$ was invested in 50-inch televisions.

9. The formula for the volume of a cylindrical container is $V = \pi r^2 h$. To find the height of the container, solve for h.

$$h = \frac{V}{\pi r^2}$$

Then, using $V = 84$ and $r = 3$, find the height.

$$h = \frac{84}{\pi(3)^2}$$

$$h = \frac{84}{9\pi}$$

$$h \approx 2.97$$

So, the height of the container is about 2.97 inches. You can use unit analysis to check that your answer is reasonable.

$$\frac{84 \text{ in.}^3}{9\pi \text{ in.}^2} = \frac{84 \text{ in.} \cdot \cancel{\text{in.}} \cdot \cancel{\text{in.}}}{9\pi \cancel{\text{in.}} \cdot \cancel{\text{in.}}} = \frac{84}{9\pi} \text{ in.} \approx 2.97 \text{ in.}$$

Checkpoints for Section 1.4

1. $2x^2 - 3x + 1 = 6$ Write original equation.

$2x^2 - 3x - 5 = 0$ Write in general form.

$(2x - 5)(x + 1) = 0$ Factor.

$2x - 5 = 0 \Rightarrow x = \frac{5}{2}$ Set 1st factor equal to 0.

$x + 1 = 0 \Rightarrow x = -1$ Set 2nd factor equal to 0.

The solutions are $x = -1$ and $x = \frac{5}{2}$.

Check: $x = -1$

$$2x^2 - 3x + 1 = 6$$
$$2(-1)^2 - 3(-1) + 1 \stackrel{?}{=} 6$$
$$2(1) + 3 + 1 \stackrel{?}{=} 6$$
$$6 = 6 \checkmark$$

$x = \frac{5}{2}$

$$2x^2 - 3x + 1 = 6$$
$$2\left(\tfrac{5}{2}\right)^2 - 3\left(\tfrac{5}{2}\right) + 1 \stackrel{?}{=} 6$$
$$2\left(\tfrac{25}{4}\right) - \tfrac{15}{2} + 1 \stackrel{?}{=} 6$$
$$6 = 6 \checkmark$$

2. (a) $3x^2 = 36$ Write original equation.

$x^2 = 12$ Divide each side by 3.

$x = \pm\sqrt{12}$ Extract square roots.

$x = \pm 2\sqrt{3}$

The solutions are $x = \pm 2\sqrt{3}$.

Check: $x = -2\sqrt{3}$

$$3x^2 = 36$$
$$3\left(-2\sqrt{3}\right)^2 \stackrel{?}{=} 36$$
$$3(12) \stackrel{?}{=} 36$$
$$36 = 36 \checkmark$$

$x = 2\sqrt{3}$

$$3x^2 = 36$$
$$3\left(2\sqrt{3}\right)^2 \stackrel{?}{=} 36$$
$$3(12) \stackrel{?}{=} 36$$
$$36 = 36 \checkmark$$

(b) $(x - 1)^2 = 10$

$$x - 1 = \pm\sqrt{10}$$
$$x = 1 \pm \sqrt{10}$$

The solutions are $x = 1 \pm \sqrt{10}$.

Check: $x = 1 - \sqrt{10}$

$$(x - 1)^2 = 10$$
$$\left[\left(1 - \sqrt{10}\right) - 1\right]^2 \stackrel{?}{=} 10$$
$$\left(-\sqrt{10}\right)^2 \stackrel{?}{=} 10$$
$$10 = 10 \checkmark$$

$x = 1 + \sqrt{10}$

$$(x - 1)^2 = 10$$
$$\left[\left(1 + \sqrt{10}\right) - 1\right] \stackrel{?}{=} 10$$
$$\left(\sqrt{10}\right)^2 \stackrel{?}{=} 10$$
$$10 = 10 \checkmark$$

3.

$x^2 - 4x - 1 = 0$	Write original equation.
$x^2 - 4x = 1$	Add 1 to each side.
$x^2 - 4x + (2)^2 = 1 + (2)^2$ (half of 4)^2	Add 2^2 to each side.
$(x - 2)^2 = 5$	Simplify.
$x - 2 = \pm\sqrt{5}$	Extract square roots.
$x = 2 \pm \sqrt{5}$	Add 2 to each side.

The solutions are $x = 2 \pm \sqrt{5}$.

Check: $x = 2 - \sqrt{5}$

$$x^2 - 4x - 1 = 0$$
$$\left(2 - \sqrt{5}\right)^2 - 4\left(2 - \sqrt{5}\right) - 1 \stackrel{?}{=} 0$$
$$\left(4 - 4\sqrt{5} + 5\right) - 8 + 4\sqrt{5} - 1 \stackrel{?}{=} 0$$
$$4 + 5 - 8 - 1 \stackrel{?}{=} 0$$
$$0 = 0 \checkmark$$

$x = 2 + \sqrt{5}$ also checks. ✓

4.

$2x^2 - 4x + 1 = 0$	Write original equation.
$2x^2 - 4x = -1$	Subtract 1 from each side.
$x^2 - 2x = -\frac{1}{2}$	Divide each side by 2.
$x^2 - 2x + (1)^2 = -\frac{1}{2} + (1)^2$	Add 1^2 to each side.
$(x - 1)^2 = \frac{1}{2}$	Simplify.
$x - 1 = \pm\sqrt{\frac{1}{2}}$	Extract square roots.
$x - 1 = \pm\frac{\sqrt{2}}{2}$	Rationalize the denominator.
$x = 1 \pm\frac{\sqrt{2}}{2}$	Add 1 to each side.

Check: $x = 1 \pm \frac{\sqrt{2}}{2}$

$$2x^2 - 4x + 1 = 0$$
$$2\left(1 - \frac{\sqrt{2}}{2}\right)^2 - 4\left(1 - \frac{\sqrt{2}}{2}\right) + 1 \stackrel{?}{=} 0$$
$$2\left(1 - \sqrt{2} + \frac{1}{2}\right) - 4 + 2\sqrt{2} + 1 \stackrel{?}{=} 0$$
$$2 - 2\sqrt{2} + 1 - 4 + 2\sqrt{2} + 1 \stackrel{?}{=} 0$$
$$2 + 1 - 4 + 1 \stackrel{?}{=} 0$$
$$0 = 0 \checkmark$$

The solution $x = 1 + \frac{\sqrt{2}}{2}$ also checks. ✓

5.

$3x^2 - 10x - 2 = 0$	Original equation
$3x^2 - 10x = 2$	Add 2 to each side.
$x^2 - \frac{10}{3}x = \frac{2}{3}$	Divide each side by 3.
$x^2 - \frac{10}{3}x + \left(\frac{5}{3}\right)^2 = \frac{2}{3} + \left(\frac{5}{3}\right)^2$	Add $\left(\frac{5}{3}\right)^2$ to each side.
$\left(x - \frac{5}{3}\right)^2 = \frac{31}{9}$	Simplify.
$x - \frac{5}{3} = \pm\frac{\sqrt{31}}{3}$	Extract square roots.
$x = \frac{5}{3} \pm\frac{\sqrt{31}}{3}$	Add $\frac{5}{3}$ to each side.

The solutions are $\frac{5}{3} \pm \frac{\sqrt{31}}{3}$.

6. $3x^2 + 2x = 10$ Write original equation.

$3x^2 + 2x - 10 = 0$ Write equation in general form, set equal to 0.

$x = \dfrac{-b \pm \sqrt{b^2 - 4ac}}{2a}$ Quadratic Formula

$x = \dfrac{-2 \pm \sqrt{(2)^2 - 4(3)(-10)}}{2(3)}$ Substitute $a = 3$, $b = 2$ and $c = -10$.

$x = \dfrac{-2 \pm \sqrt{4 + 120}}{6}$ Simplify.

$x = \dfrac{-2 \pm \sqrt{124}}{6}$ Simplify.

$x = \dfrac{-2 \pm 2\sqrt{31}}{6}$ Simplify.

$x = \dfrac{2(-1 \pm \sqrt{31})}{6}$ Factor our common factor.

$x = \dfrac{-1 \pm \sqrt{31}}{3}$ Simplify.

The solutions are $\dfrac{-1 \pm \sqrt{31}}{3}$.

Check: $x = \dfrac{-1 \pm \sqrt{31}}{3}$

$$3x^2 + 2x - 10 = 0$$

$$3\left(\frac{-1 + \sqrt{31}}{3}\right)^2 + 2\left(\frac{-1 + \sqrt{31}}{3}\right) - 10 \stackrel{?}{=} 0$$

$$\frac{1}{3} - \frac{2\sqrt{31}}{3} + \frac{31}{3} - \frac{2}{3} + \frac{2\sqrt{31}}{3} - 10 \stackrel{?}{=} 0$$

$$10 - 10 \stackrel{?}{=} 0$$

$$0 = 0 \checkmark$$

The solution $x = \dfrac{-1 - \sqrt{31}}{3}$ also checks. ✓

7. *Verbal Model:* $\boxed{\text{Width of room}} \cdot \boxed{\text{Length of room}} = \boxed{\text{Area of room}}$

Labels: Width $= w$ (feet)

Length $= l = w + 6$ (feet)

Area $= 112$ (square feet)

Equation:

$$w(w + 6) = 112$$

$$w^2 + 6w = 112$$

$$w^2 + 6w - 112 = 0$$

$$(w - 8)(w + 14) = 0$$

$$w - 8 = 0 \Rightarrow w = 8$$

$$w + 14 = 0 \Rightarrow w = -14$$

Choosing the positive value, you can conclude the width is 8 feet and the length is $w + 6 = 14$ feet. You can check this solution by observing that the length is 6 feet longer than the width and the product of the length and width is 112 square feet.

8. To set up a mathematical model for the height of the rock, use the position equation

$s = -16t^2 + V_0 t + s_0.$

Because the object is dropped, the initial velocity is $V_0 = 0$ feet per second. Because the initial height is $s_0 = 196$ feet, you have $s = -16t^2 + 196$.

To find the time the rock hits the ground, let the height s be zero and solve the equation for t.

$s = -16t^2 + 196$	Write positive equation.
$0 = -16t^2 + 196$	Substitute 0 for height.
$16t^2 = 196$	Add 16^2 to each side.
$t^2 = \dfrac{196}{16}$	Divide each side by 16.
$t = \sqrt{\dfrac{196}{16}}$	Extract positive square root.
$t = \dfrac{14}{4}$	Simplify.
$t = 3.5$	Simplify.

The rock will take 3.5 seconds to hit the ground.

9. To find the year in which the number of active Facebook users reached 1.2 billion or 1200 million, you can solve the equation

$F = -10.024t^2 + 448.79t - 2990.5, \; 9 \le t \le 15.$

Set $F = 1200$.

$-10.024\,t^2 + 448.79t - 2990.5 = 1200$

To begin, write the equation in general form.

$-10.024t^2 + 448.79t - 4190.5 = 0$

Then apply the Quadratic Formula.

$$t = \frac{-b \pm \sqrt{b^2 - 4ac}}{2a}$$

$$t = \frac{-448.79 \pm \sqrt{(448.59)^2 - 4(-10.024)(-4190.5)}}{2(-10.024)}$$

$$t = \frac{-448.79 \pm \sqrt{33{,}390.1761}}{-20.048}$$

$t \approx 13.3, 31.5$

Choose $t \approx 13.2$ because it is in the domain of t. Because $t = 9$ corresponds to 2009, it follows that $t = 13.3$ corresponds to 2013. So, the number of active Facebook users reached 1.2 billion during the year 2013. From the table, you can also see that during the year 2013 the number of active Facebook users reached 1.2 billion.

10. 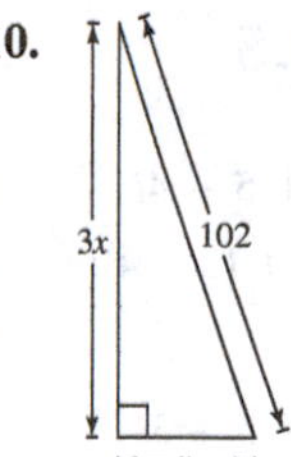

Use the Pythagorean Theorem.

$a^2 + b^2 = c^2$	Pythagorean Theorem
$(3x)^2 + (x)^2 = (102)^2$	Substitute for a, b, and c.
$9x^2 + x^2 = 10{,}404$	Simplify.
$10x^2 = 10{,}404$	Simplify.
$x^2 = 1040.4$	Divide each side by 10.
$x = \sqrt{1040.4}$	Extract positive square root.
$x \approx 32.3$ feet	

The total distance covered by walking on the L-shaped sidewalk is

$x + 3x = 4x$

$= 4\sqrt{1040.4}$

≈ 129.02 feet.

Walking on the diagonal saves a person about $129.02 - 102 = 27.02$ feet.

Checkpoints for Section 1.5

1. (a) $(7 + 3i) + (5 - 4i) = 7 + 3i + 5 - 4i$ Remove parentheses.
$= (7 + 5) + (3 - 4)i$ Group like terms.
$= 12 - i$ Write in standard form.

(b) $(3 + 4i) - (5 - 3i) = 3 + 4i - 5 + 3i$ Remove parentheses.
$= (3 - 5) + (4 + 3)i$ Group like terms.
$= -2 + 7i$ Write in standard form.

(c) $2i + (-3 - 4i) - (-3 - 3i) = 2i - 3 - 4i + 3 + 3i$ Remove parentheses.
$= (-3 + 3) + (2 - 4 + 3)i$ Group like terms.
$= i$ Write in standard form.

(d) $(5 - 3i) + (3 + 5i) - (8 + 2i) = 5 - 3i + 3 + 5i - 8 - 2i$ Remove parentheses.
$= (5 + 3 - 8) + (-3 + 5 - 2)i$ Group like terms.
$= 0 + 0i$ Simplify.
$= 0$ Write in standard form.

2. (a) $-5(3 - 2i) = -5(3) - (-5)(2i)$ Distributive Property
$= -15 + 10i$ Simplify

(b) $(2 - 4i)(3 + 3i) = 6 + 6i - 12i - 12i^2$ FOIL Method
$= 6 + 6i - 12i - 12(-1)$ $i^2 = -1$
$= 6 - 6i + 12$ Simplify.
$= 18 - 6i$ Write in standard form.

(c) $(4 + 5i)(4 - 5i) = 4(4 - 5i) + 5i(4 - 5i)$ Distributive Property
$= 16 - 20i + 20i - 25i^2$ Distributive Property
$= 16 - 20i + 20i - 25(-1)$ $i^2 = -1$
$= 16 + 25$ Simplify.
$= 41$ Write in standard form.

(d) $(4 + 2i)^2 = (4 + 2i)(4 + 2i)$ Square of a binomial
$= 4(4 + 2i) + 2i(4 + 2i)$ Distributive Property
$= 16 + 8i + 8i + 4i^2$ Distributive Property
$= 16 + 8i + 8i + 4(-1)$ $i^2 = -1$
$= (16 - 4) + (8i + 8i)$ Group like terms.
$= 12 + 16i$ Write in standard form.

3. (a) The complex conjugate of $3 + 6i$ is $3 - 6i$.

$$\begin{aligned}(3 + 6i)(3 - 6i) &= (3)^2 - (6i)^2\\ &= 9 - 36i^2\\ &= 9 - 36(-1)\\ &= 45\end{aligned}$$

(b) The complex conjugate of $2 - 5i$ is $2 + 5i$.

$$\begin{aligned}(2 - 5i)(2 + 5i) &= (2)^2 - (5i)^2\\ &= 4 - 25i^2\\ &= 4 - 25(-1)\\ &= 29\end{aligned}$$

4. $\dfrac{2+i}{2-i} = \dfrac{2+i}{2-i} \cdot \dfrac{2+i}{2+i}$ Multiply numerator and denominator by complex conjugate of the denominator.

$= \dfrac{4+2i+2i+i^2}{4-i^2}$ Expand.

$= \dfrac{4-1+4i}{4-(-1)}$ $i^2 = -1$

$= \dfrac{3+4i}{5}$ Simplify.

$= \dfrac{3}{5} + \dfrac{4}{5}i$ Write in standard form.

5. $\sqrt{-14}\sqrt{-2} = \sqrt{14}i\sqrt{2}i = \sqrt{28}i^2 = 2\sqrt{7}(-1) = -2\sqrt{7}$

6. To solve $8x^2 + 14x + 9 = 0$, use the Quadratic formula

$x = \dfrac{-b \pm \sqrt{b^2 - 4ac}}{2a}$.

$x = \dfrac{-14 \pm \sqrt{14^2 - 4(8)(9)}}{2(8)}$ Substitute $a = 8$, $b = 14$, and $c = 9$.

$= \dfrac{-14 \pm \sqrt{-92}}{16}$ Simplify.

$= \dfrac{-14 \pm 2\sqrt{23}i}{16}$ Write $\sqrt{-92}$ in standard form.

$= \dfrac{-14}{16} + \dfrac{2\sqrt{23}i}{16}$ Write in standard form.

$= \dfrac{-7}{8} \pm \dfrac{\sqrt{23}i}{8}$ Simplify.

Checkpoints for Section 1.6

1.

$$9x^4 - 12x^2 = 0 \quad \text{Write original equation.}$$

$$3x^2(3x^2 - 4) = 0 \quad \text{Factor out common factor.}$$

$$3x^2(\sqrt{3}x - 2)(\sqrt{3}x + 2) = 0 \quad \text{Factor completely.}$$

$$3x^2 = 0 \Rightarrow \quad x = 0 \quad \text{Set 1st factor equal to 0.}$$

$$\sqrt{3}x - 2 = 0 \Rightarrow \quad x = \frac{2}{\sqrt{3}} = \frac{2\sqrt{3}}{3} \quad \text{Set 2nd factor equal to 0.}$$

$$\sqrt{3}x + 2 = 0 \Rightarrow \quad x = \frac{-2}{\sqrt{3}} = \frac{-2\sqrt{3}}{3} \quad \text{Set 3rd factor equal to 0.}$$

Check these solutions by substituting in the original equation.

Check: $x = 0$

$$9x^4 - 12x^2 = 0$$

$$9(0)^4 - 12(0)^2 \stackrel{?}{=} 0$$

$$0 = 0 \checkmark$$

$$x = \frac{2\sqrt{3}}{3}$$

$$9x^4 - 12x^2 = 0$$

$$9\left(\frac{2\sqrt{3}}{3}\right)^4 - 12\left(\frac{2\sqrt{3}}{3}\right)^2 \stackrel{?}{=} 0$$

$$9\left(\frac{16}{9}\right) - 12\left(\frac{4}{3}\right) \stackrel{?}{=} 0$$

$$16 - 16 \stackrel{?}{=} 0$$

$$0 = 0 \checkmark$$

The solution $x = \dfrac{-2\sqrt{3}}{3}$ also checks. ✓

2. (a)

$x^3 - 5x^2 - 2x + 10 = 0$	Write original equation.
$x^2(x-5) - 2(x-5) = 0$	Factor by grouping.
$(x-5)(x^2-2) = 0$	Distributive Property
$x - 5 = 0 \Rightarrow x = 5$	Set 1st factor equal to 0.
$x^2 - 2 = 0 \Rightarrow x^2 = 2$	Set 2nd factor equal to 0.
$= \pm\sqrt{2}$	

Check: $x = 5$

$$x^3 - 5x^2 - 2x + 10 = 0$$
$$(5)^3 - 5(5)^2 - 2(5) + 10 \stackrel{?}{=} 0$$
$$125 - 125 - 10 + 10 \stackrel{?}{=} 0$$
$$0 = 0 \checkmark$$

$x = \sqrt{2}$

$$x^3 - 5x^2 - 2x + 10 = 0$$
$$\left(\sqrt{2}\right)^3 - 5\left(\sqrt{2}\right)^2 - 2\left(\sqrt{2}\right) + 10 \stackrel{?}{=} 0$$
$$2\sqrt{2} - 10 - 2\sqrt{2} + 10 \stackrel{?}{=} 0$$
$$0 = 0 \checkmark$$

The solution $x = -\sqrt{2}$ also checks. ✓

(b) $6x^3 - 27x^2 - 54x = 0$

$3x(2x^2 - 9x - 18) = 0$	Factor out common factor.
$3x(2x+3)(x-6) = 0$	Factor quadratic factor.
$3x = 0 \Rightarrow x = 0$	Set 1st factor equal to 0.
$2x + 3 = 0 \Rightarrow x = -\frac{3}{2}$	Set 2nd factor equal to 0.
$x - 6 = 0 \Rightarrow x = 6$	Set 3rd factor equal to 0.

Check: $x = 0$

$$6x^3 - 27x^2 - 54x = 0$$
$$6(0)^3 - 27(0)^2 - 54(0) \stackrel{?}{=} 0$$
$$0 = 0 \checkmark$$

$x = -\frac{3}{2}$

$$6x^3 - 27x^2 - 54x = 0$$
$$6\left(-\tfrac{3}{2}\right)^3 - 27\left(-\tfrac{3}{2}\right)^2 - 54\left(-\tfrac{3}{2}\right) \stackrel{?}{=} 0$$
$$6\left(-\tfrac{27}{8}\right)^3 - 27\left(\tfrac{9}{4}\right)^2 + 27(3) \stackrel{?}{=} 0$$
$$-\tfrac{81}{4} - \tfrac{243}{4} + 81 \stackrel{?}{=} 0$$
$$0 = 0 \checkmark$$

$x = 6$

$$6x^3 - 27x^2 - 54x = 0$$
$$6(6)^3 - 27(6)^2 - 54(6) \stackrel{?}{=} 0$$
$$6(216) - 27(36) - 324 \stackrel{?}{=} 0$$
$$0 = 0 \checkmark$$

3. (a) This equation is of quadratic type with $u = x^2$.

$x^4 - 7x^2 + 12 = 0$	Write original equation.
$(x^2)^2 - 7(x^2) + 12 = 0$	Quadratic form
$u^2 - 7u + 12 = 0$	$u = x^2$
$(u - 3)(u - 4) = 0$	Factor.
$u - 3 = 0 \Rightarrow u = 3$	Set 1st factor equal to 0.
$u - 4 = 0 \Rightarrow u = 4$	Set 2nd factor equal to 0.

Replace u with x^2 and solve for x in each equation.

$u = 3$	$u = 4$
$x^2 = 3$	$x^2 = 4$
$x = \pm\sqrt{3}$	$x = \pm 2$

The solutions are $x = -\sqrt{3}$, $x = \sqrt{3}$, $x = -2$, and $x = 2$.

Check: $x = -\sqrt{3}$

$$x^4 - 7x^2 + 12 = 0$$
$$\left(-\sqrt{3}\right)^4 - 7\left(-\sqrt{3}\right)^2 + 12 \stackrel{?}{=} 0$$
$$9 - 7(3) + 12 \stackrel{?}{=} 0$$
$$-21 + 21 \stackrel{?}{=} 0$$
$$0 = 0 \checkmark$$

The solution $x = \sqrt{3}$ also checks. ✓

$x = -2$

$$x^4 - 7x^2 + 12 = 0$$
$$(-2)^4 - 7(-2)^2 + 12 \stackrel{?}{=} 0$$
$$16 - 28 + 12 \stackrel{?}{=} 0$$
$$0 = 0 \checkmark$$

The solution $x = 2$ also checks. ✓

(b) This equation is of quadratic type with $u = x^2$.

$9x^4 - 37x^2 + 4 = 0$	Write original equation.
$9(x^2)^2 - 37(x^2) + 4 = 0$	Quadratic form
$9u^2 - 37u + 4 = 0$	$u = x^2$
$(9u - 1)(u - 4) = 0$	Factor.
$9u - 1 = 0 \Rightarrow u = \frac{1}{9}$	Set 1st factor equal to 0.
$u - 4 = 0 \Rightarrow u = 4$	Set 2nd factor equal to 0.

Replace u with x^2 and solve for x in each equation.

$u = \frac{1}{9}$	$u = 4$
$x^2 = \frac{1}{9}$	$x^2 = 4$
$x = \pm\frac{1}{3}$	$x = \pm 2$

The solutions are $-\frac{1}{3}$, $\frac{1}{3}$, -2, and 2.

Check: $x = -\frac{1}{3}$

$$9x^4 - 37x^2 + 4 = 0$$
$$9\left(-\tfrac{1}{3}\right)^4 - 37\left(-\tfrac{1}{3}\right)^2 + 4 \stackrel{?}{=} 0$$
$$9\left(\tfrac{1}{81}\right) - 37\left(\tfrac{1}{9}\right) + 4 \stackrel{?}{=} 0$$
$$\tfrac{1}{9} - \tfrac{37}{9} + 4 \stackrel{?}{=} 0$$
$$-4 + 4 = 0 \checkmark$$

The solution $x = \frac{1}{3}$ also checks. ✓

$x = -2$

$$9x^4 - 37x^2 + 4 = 0$$
$$9(-2)^4 - 37(-2)^2 + 4 \stackrel{?}{=} 0$$
$$9(16) - 37(4) + 4 \stackrel{?}{=} 0$$
$$144 - 148 + 4 = 0 \checkmark$$

The solution $x = 2$ also checks. ✓

4. $-\sqrt{40 - 9x} + 2 = x$ Write original equation.

$-\sqrt{40 - 9x} = x - 2$ Isolated radical.

$\left(-\sqrt{40 - 9x}\right)^2 = (x - 2)^2$ Square each side.

$40 - 9x = x^2 - 4x + 4$ Simplify.

$0 = x^2 + 5x - 36$ Write in general form.

$0 = (x - 4)(x + 9)$ Factor.

$x - 4 = 0 \Rightarrow x = 4$ Set 1st factor equal to 0.

$x + 9 = 0 \Rightarrow x = -9$ Set 2nd factor equal to 0.

Check: $x = 4$

$$-\sqrt{40 - 9x} + 2 = x$$

$$-\sqrt{40 - 9(4)} + 2 \stackrel{?}{=} 4$$

$$-\sqrt{4} + 2 \stackrel{?}{=} 4$$

$$-2 + 2 \stackrel{?}{=} 4$$

$$0 \neq 4 \;\times$$

$x = 4$ is an extraneous solution.

$x = -9$

$$-\sqrt{40 - 9(-9)} + 2 \stackrel{?}{=} -9$$

$$-\sqrt{121} + 2 \stackrel{?}{=} -9$$

$$-11 + 2 \stackrel{?}{=} -9$$

$$-9 \stackrel{?}{=} -9 \;\checkmark$$

The only solution is $x = -9$.

5. $(x - 5)^{2/3} = 16$ Write original equation.

$\sqrt[3]{(x - 5)^2} = 16$ Rewrite in radical form.

$(x - 5)^2 = 4096$ Cube each side.

$x - 5 = \pm 64$ Extract square roots.

$x = 5 \pm 64$ Add 5 to each side.

$x = -59, x = 69$

Check: $x = -59$

$$(x - 5)^{2/3} = 16$$

$$(-59 - 5)^{2/3} \stackrel{?}{=} 16$$

$$(-64)^{2/3} \stackrel{?}{=} 16$$

$$(-4)^2 \stackrel{?}{=} 16$$

$$16 = 16 \;\checkmark$$

$x = 69$

$$(x - 5)^{2/3} = 16$$

$$(69 - 5)^{2/3} \stackrel{?}{=} 16$$

$$(64)^{2/3} \stackrel{?}{=} 16$$

$$(4)^2 \stackrel{?}{=} 16$$

$$16 = 16 \;\checkmark$$

The solutions are $x = -59$ and $x = 64$.

6. The LCD of the three terms is $x(x + 3)$.

$\frac{4}{x} + \frac{2}{x+3} = -3$	Write original equation.
$x(x+3)\frac{4}{x} + x(x+3)\frac{2}{x+3} = x(x+3)(-3)$	Multiply each term by LCD.
$4(x+3) + 2x = -3x(x+3),\ x \neq -3, 0$	Simplify.
$6x + 12 = -3x^2 - 9x$	Simplify.
$3x^2 + 15x + 12 = 0$	Write in general form.
$3(x^2 + 5x + 4) = 0$	Factor.
$x^2 + 5x + 4 = 0$	Divide each side by 3.
$(x+1)(x+4) = 0$	Factor.
$x + 1 = 0 \Rightarrow x = -1$	Set 1st factor equal to 0.
$x + 4 = 0 \Rightarrow x = -4$	Set 2nd factor equal to 0.

Both $x = -4$ and $x = -1$ are possible solutions.

Check: $x = -4$

$$\frac{4}{x} + \frac{2}{x+3} = -3$$
$$\frac{4}{(-4)} + \frac{2}{(-4)+3} \stackrel{?}{=} -3$$
$$-1 + (-2) \stackrel{?}{=} -3$$
$$-3 = -3 \checkmark$$

$x = -1$

$$\frac{4}{x} + \frac{2}{x+3} = -3$$
$$\frac{4}{(-1)} + \frac{2}{(-1)+3} \stackrel{?}{=} -3$$
$$-4 + (1) \stackrel{?}{=} -3$$
$$-3 = -3 \checkmark$$

So, the solutions are $x = -4$ and $x = -1$.

7. $|x^2 + 4x| = 7x + 18$

First Equation

$x^2 + 4x = 7x + 18$	Use positive expression.
$x^2 - 3x - 18 = 0$	Write in general form.
$(x + 3)(x - 6) = 0$	Factor.
$x + 3 = 0 \Rightarrow x = -3$	Set 1st factor equal to 0.
$x - 6 = 0 \Rightarrow x = 6$	Set 2nd factor equal to 0.

Second Equation

$-(x^2 + 4x) = 7x + 18$	Use negative expression.
$0 = x^2 + 11x + 18$	Write in general form.
$(x + 9)(x + 2) = 0$	Factor.
$x + 9 = 0 \Rightarrow x = -9$	Set 1st factor equal to 0.
$x + 2 = 0 \Rightarrow x = -2$	Set 2nd factor equal to 0.

The possible solutions are $x = -3$, $x = 6$, and $x = -9$ and $x = -2$.

Check: $x = -3$

$$|x^2 + 4x| = 7x + 18$$

$$|(-3)^2 + 4(-3)| \stackrel{?}{=} 7(-3) + 18$$

$$|-3| \stackrel{?}{=} -3$$

$$3 \neq -3$$

$x = -3$ does not check.

$x = 6$

$$|(6)^2 + 4(6)| \stackrel{?}{=} 7(6) + 18$$

$$|60| \stackrel{?}{=} 60$$

$$60 = 60 \checkmark$$

$x = 4$ checks.

Check: $x = -9$

$$|x^2 + 4x| = 7x + 18$$

$$|(-9)^2 + 4(-9)| \stackrel{?}{=} 7(-9) + 18$$

$$|45| \stackrel{?}{=} -45$$

$$45 \neq -45$$

$x = -9$ does not check.

$x = -2$

$$|(-2)^2 + 4(-2)| \stackrel{?}{=} 7(-2) + 18$$

$$|-4| \stackrel{?}{=} 4$$

$$4 = 4 \checkmark$$

$x = -2$ checks.

8. *Verbal Model:* $\boxed{\text{Cost per student}} \cdot \boxed{\text{Number of students}} = \boxed{\text{Cost of trip}}$

Labels:

Cost of trip	$= 560$	(dollars)
Original number of students	$= x$	(people)
New number of students	$= x + 8$	(people)
Original cost per student	$= \dfrac{560}{x}$	(dollars per person)
New cost per students	$= \dfrac{560}{x} - 3.50$	(dollars per person)

Equation:

$$\left(\frac{560}{x} - 3.50\right)(x + 8) = 560$$
$$\left(\frac{560 - 3.50x}{x}\right)(x + 8) = 560$$
$$(560 - 3.50x)(x + 8) = 560x,\ x \neq 0$$
$$560x + 4480 - 3.5x^2 - 28x = 560x$$
$$-3.5x^2 - 28x + 4480 = 0$$
$$x^2 + 8x - 1280 = 0$$
$$(x - 32)(x + 40) = 0$$
$$x - 32 = 0 \Rightarrow w = 32$$
$$x + 40 = 0 \Rightarrow w = -40$$

Check:

$$\left(\frac{560}{x} - 3.50\right)(x + 8) \stackrel{?}{=} 560$$
$$\left(\frac{560}{32} - 3.50\right)(32 + 8) \stackrel{?}{=} 560$$
$$(14)(40) \stackrel{?}{=} 560$$
$$560 \stackrel{?}{=} 560 \checkmark$$

Only the positive value of x makes sense in the context of the problem, so you can conclude that $x = 32$ students were in the original group.

9. *Formula:* $A = P\left(1 + \dfrac{r}{n}\right)^{nt}$

Labels:

Balance	$= A = 3544.06$	(dollars)
Principal	$= P = 2500$	(dollars)
Time	$= t = 5$	(years)
Compounding periods per year	$= n = 12$	(compoundings per year)
Annual interest rate	$= r$	(percent in decimal form)

Equation:

$$3544.06 = 2500\left(1 + \frac{r}{12}\right)^{(12)(5)}$$
$$\frac{3544.06}{2500} = \left(1 + \frac{r}{12}\right)^{60}$$
$$1.4176 \approx \left(1 + \frac{r}{12}\right)^{60}$$
$$(1.4176)^{1/60} = 1 + \frac{r}{12}$$
$$1.0058 \approx 1 + \frac{r}{12}$$
$$0.0058 \approx \frac{r}{12}$$
$$0.07 \approx r$$

Check:

$$A = P\left(1 + \frac{r}{n}\right)^{nt}$$
$$3544.06 \stackrel{?}{=} 2500\left(1 + \frac{0.07}{12}\right)^{(12)(5)}$$
$$3544.06 \stackrel{?}{=} 3544.06 \checkmark$$

The annual interest rate is about 0.07 or 7%.

Checkpoints for Section 1.7

1. (a) $[-1, 3]$ corresponds to $-1 \le x \le 3$. The interval is bounded.

(b) $(-1, 6)$ corresponds to $-1 < x < 6$. The interval is bounded.

(c) $(-\infty, 4)$ corresponds to $x < 4$. The interval is unbounded.

(d) $[0, \infty)$ corresponds to $x \ge 0$. The interval is unbounded.

2.

$7x - 3 \le 2x + 7$	Write original inequality.
$5x \le 10$	Subtract $2x$ and add 3 to each side.
$x \le 2$	Divide each side by 5.

The solution set is all real numbers less than or equal to 2.

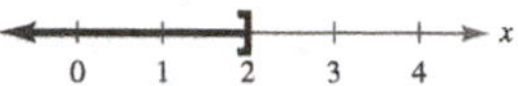

3. (a) **Algebraic solution**

$2 - \frac{5}{3}x > x - 6$	Write original inequality.
$6 - 5x > 3x - 18$	Multiply each side by 3.
$-8x > -24$	Subtract $3x$ and subtract 6 from each side.
$x < 3$	Divide each side by -8 reverse the inequality symbol.

The solution set is all real numbers that are less than 3.

(b) **Graphical solution**

Use a graphing utility to graph $y_1 = 2 - \frac{5}{3}x$ and $y_2 = x - 6$ in the same viewing window. Use the *intersect* feature to determine that the graphs intersect at $(3, -3)$. The graph of y_1 lies above the graph of y_2 to the left of their point of intersection, which implies that $y_1 > y_2$ for all $x < 3$.

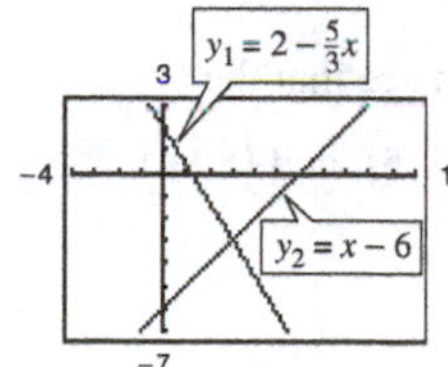

4.

$1 < 2x + 7 < 11$	Write original inequality.
$1 - 7 < 2x + 7 - 7 < 11 - 7$	Subtract 7 from each part.
$-6 < 2x < 4$	Simplify.
$-\frac{6}{2} < \frac{2x}{2} < \frac{4}{2}$	Divide each part by 2.
$-3 < x < 2$	Simplify.

The solution set is all real numbers greater than -3 and less than 2, which is denoted by $(-3, 2)$.

$-4 \quad -3 \quad -2 \quad -1 \quad 0 \quad 1 \quad 2 \quad 3$

5. $|x - 20| \leq 4$ Write original inequality.

$-4 \leq x - 20 \leq 4$ Write equivalent inequalities.

$-4 + 20 \leq x - 20 + 20 \leq 4 = 20$ Add 20 to each part.

$16 \leq x \leq 24$ Simplify.

The solution set is all real numbers that are greater than or equal to 16 and less than or equal to 24, which is denoted by [16, 24].

12 14 16 18 20 22 24 26 28 x

6. Let h represent your additional hours in one month. Write and solve an inequality.

Plan B: $C = 10.25h + 8$ and new Plan A: $C = 10h + 25$

$10.25h + 8 > 10h + 25$

$0.25h > 17$

$h > 68$

Plan B costs more when you use more than 68 additional hours in one month.

7. Let x represent the actual weight of your bag. The difference of the actual weight and the weight on the scale is at most $\frac{1}{64}$ pound. That is, $\left|x - \frac{1}{2}\right| \leq -\frac{1}{64}$.

You can solve the inequality as follows.

$-\frac{1}{64} \leq x - \frac{1}{2} \leq \frac{1}{64}$

$\frac{31}{64} \leq x \leq \frac{33}{64}$

The least your bag can weigh is $\frac{31}{64}$ pound, which would have cost $\left(\frac{31}{64} \text{ pound}\right) \times (\$9.89 \text{ per pound}) = \$4.79$.

The most your bag can weigh is $\frac{33}{64}$ pound, which would have cost $\left(\frac{33}{64} \text{ pound}\right) \times (\$9.89 \text{ per pound}) = \$5.10$.

So, you might have been under charged by as much as $\$5.10 - \$4.95 = \$0.15$ or over charged as much as $\$4.95 - \$4.79 = \$0.16$.

Checkpoints for Section 1.8

1. By factoring the polynomial $x^2 - x - 20 < 0$ as $x^2 - x - 20 = (x + 4)(x - 5)$ you can see that the key numbers are $x = -4$ and $x = 5$. So, the polynomial's test intervals are $(-\infty, -4)$, $(-4, 5)$, and $(5, \infty)$.

In each test interval, choose a representative x-value and evaluate the polynomial.

Test interval	x-value	Polynomial value	Conclusion
$(-\infty, -4)$	$x = -5$	$(-5)^2 - (-5) - 20 = 10$	Positive
$(-4, 5)$	$x = 0$	$(0)^2 - (0) - 20 = -20$	Negative
$(5, \infty)$	$x = 6$	$(6)^2 - (6) - 20 = 10$	Positive

From this, you can conclude that the inequality is satisfied for all x-values in $(-4, 5)$.

This implies that the solution of the inequality $x^2 - x - 20 < 0$ is the interval $(-4, 5)$.

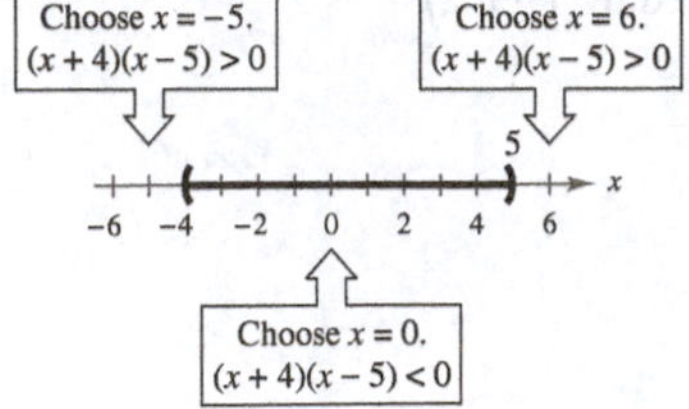

2. (a) **Algebraic solution**

$2x^2 + 3x < 5$	Write original inequality.
$2x^2 + 3x - 5 < 0$	Write in general form.
$(2x + 5)(x - 1) < 0$	Factor.

Key numbers: $x = -\frac{5}{2}$ and $x = 1$

Test intervals: $\left(-\infty, -\frac{5}{2}\right), \left(-\frac{5}{2}, 1\right), (1, \infty)$

Test: Is $(2x + 5)(x - 1) < 0$?

After testing the intervals, you can see that the polynomial $2x^2 + 3x - 5$ is negative on the open interval $\left(-\frac{5}{2}, 1\right)$.

So, the solution set of the inequality is $\left(-\frac{5}{2}, 1\right)$.

(b) **Graphical solution**

First write the polynomial inequality $2x^2 + 3x < 5$ as $2x^2 + 3x - 5 > 0$. Then use a graphing utility to graph $y = 2x^2 + 3x - 5$. You can see that the graph is below the x-axis when x is greater than $-\frac{5}{2}$ and when x is less than 1. So, the solution set is $\left(-\frac{5}{2}, 1\right)$.

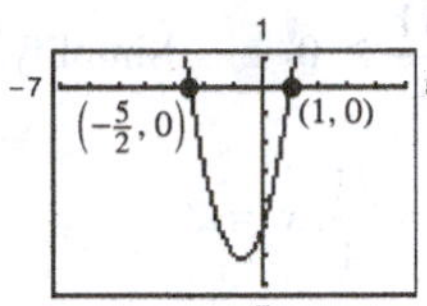

3.

$3x^3 - x^2 - 12x > -4$	Write original inequality.
$3x^2 - x^2 - 12x + 4 > 0$	Write in general form.
$x^2(3x - 1) - 4(3x - 1) > 0$	Factor.
$(3x - 1)(x^2 - 4) > 0$	Factor.
$(3x - 1)(x + 2)(x - 2) > 0$	Factor.

The key numbers are $x = -2$, $x = \frac{1}{3}$, and $x = 2$, and the test intervals are $(-\infty, -2)$, $\left(-2, \frac{1}{3}\right)$, $\left(\frac{1}{3}, 2\right)$, and $(2, \infty)$.

Test interval	x-value	Polynomial value	Conclusion
$(-\infty, -2)$	$x = -3$	$3(-3)^3 - (-3)^2 - 12(-3) + 4 = -50$	Negative
$\left(-2, \frac{1}{3}\right)$	$x = 0$	$3(0)^3 - (0)^2 - 12(0) + 4 = 4$	Positive
$\left(\frac{1}{3}, 2\right)$	$x = 1$	$3(1)^3 - (1)^2 - 12(1) + 4 = -6$	Negative
$(2, \infty)$	$x = 3$	$3(3)^3 - (3)^2 - 12(3) + 4 = 40$	Positive

From this, you can conclude that the inequality is satisfied on the open intervals $\left(-2, \frac{1}{3}\right)$ and $(2, \infty)$.

So, the solution set is $\left(-2, \frac{1}{3}\right) \cup (2, \infty)$.

$$\frac{1}{3}$$

−4 −3 −2 −1 0 1 2 3 x

4. (a) The solution set of $x^2 + 6x + 9 < 0$ is empty. In other words, the quadratic $x^2 + 6x + 9$ is not less than 0 for any value of x.

(b) The solution set of $x^2 + 4x + 4 \le 0$ consists of the single real number $\{-2\}$, because the quadratic $x^2 + 4x + 4$ has only one key number, $x = -2$, and it is the only value that satisfies the inequality.

(c) The solution set of $x^2 - 6x + 970$ consists of all real numbers except $x = 3$. In interval notation, the solution set can be written as $(-\infty, 3) \cup (3, \infty)$.

(d) The solution set of $x^2 - 2x + 1 \ge 0$ consists of the entire set of real numbers $(-\infty, \infty)$. In other words, the value of the quadratic $x^2 - 2x + 1$ is non-negative for every real value of x.

5. (a)

$$\frac{x-2}{x-3} \geq -3 \quad \text{Write original inequality.}$$

$$\frac{x-2}{x-3} + 3 \geq 0 \quad \text{Write in general form.}$$

$$\frac{x-2}{x-3} + \frac{3(x-3)}{x-3} \geq 0 \quad \text{Rewrite fraction using LCD.}$$

$$\frac{x-2+3x-9}{x-3} \geq 0 \quad \text{Add fractions.}$$

$$\frac{4x-11}{x-3} \geq 0 \quad \text{Simplify.}$$

Key numbers: $x = \frac{11}{4}, x = 3$

Test intervals: $\left(-\infty, \frac{11}{4}\right), \left(\frac{11}{4}, 3\right), (3, \infty)$

Test: Is $\frac{4x-11}{x-3} \geq 0$?

Test interval	x-value	Polynomial value	Conclusion
$\left(-\infty, \frac{11}{4}\right)$	$x = 0$	$\frac{4(0)-11}{0-3} = \frac{11}{3}$	Positive
$\left(\frac{11}{4}, 3\right)$	$x = 2.9$	$\frac{4(2.9)-11}{2.9-3} = \frac{0.6}{-0.1} = -6$	Negative
$(3, \infty)$	$x = 4$	$\frac{4(4)-11}{4-3} = \frac{5}{1} = 5$	Positive

After testing these intervals, you can see that the inequality is satisfied on the open intervals $\left(-\infty, \frac{11}{4}\right)$ and $(3, \infty)$. Moreover, because $\frac{4x-11}{x-3} = 0$ when $x = \frac{11}{4}$, you can conclude that the solution set consists of all real numbers in the intervals $\left(-\infty, \frac{11}{4}\right] \cup (3, \infty)$.

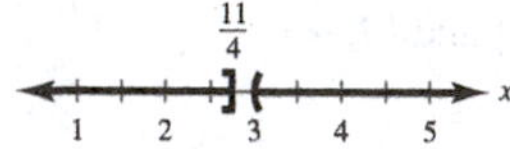

(b)

$$\frac{4x-1}{x-6} > 3 \quad \text{Write original inequality.}$$

$$\frac{4x-1}{x-6} - 3 > 0 \quad \text{Write in general form.}$$

$$\frac{4x-1-3(x-6)}{x-6} > 0 \quad \text{Combine fractions with LCD.}$$

$$\frac{4x-1-3x+18}{x-6} > 0 \quad \text{Simplify.}$$

$$\frac{x+17}{x-6} > 0 \quad \text{Simplify.}$$

Key numbers: $x = -17, x = 6$

Test intervals: $(-\infty, -17)$, $(-17, 6)$, and $(6, \infty)$

Test: Is $\dfrac{x+17}{x-6} > 0$?

Test interval	x-value	Polynomial value	Conclusion
$(-\infty, -17)$	-20	$\dfrac{(-20)+17}{(-20)-6} = \dfrac{-3}{-26} = \dfrac{3}{26}$	Positive
$(-17, 6)$	0	$\dfrac{(0)+17}{(0)-6} = -\dfrac{17}{6}$	Negative
$(6, \infty)$	8	$\dfrac{(8)+17}{(8)-6} = \dfrac{25}{2}$	Positive

After testing these intervals, you can see that the inequality is satisfied on the open intervals $(-\infty, -17)$ and $(6, \infty)$.

So, you can conclude that the solution set consists of all real numbers in the intervals $(-\infty, -17) \cup (6, \infty)$.

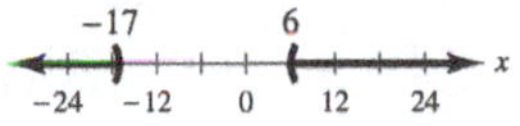

6. *Verbal Model:* [Profit] = [Review] − [Cost]

Equation:

$$P = R - C$$

$$P = x(60 - 0.0001x) - (12x + 1{,}800{,}000)$$

$$P = -0.0001x^2 + 48x = 1{,}800{,}000$$

To answer the question, solve the inequality as follows.

$$P \geq 3{,}600{,}000$$

$$-0.0001x^2 + 48x - 1{,}800{,}000 \geq 3{,}600{,}000$$

$$-0.0001x^2 + 48x - 5{,}400{,}000 \geq 0$$

$$0.0001x^2 - 48x + 5{,}400{,}000 \leq 0$$

$$x^2 - 480{,}000x + 54{,}000{,}000{,}000 \leq 0$$

$$(x - 180{,}000)(x - 300{,}000) \leq 0$$

After finding the key point and testing the intervals, you can find the solution set is $[180{,}000,\ 300{,}000]$.

So, by selling at least 180,000 units but not more than 300,000 units, the profit is at least $3,600,000.

7. Algebraic solution

Recall that the domain of an expression is the set of all x-values for which the expression is defined.

Because $\sqrt{x^2 - 7x + 10}$ is defined only of $x^2 - 7x + 10$ is non-negative, the domain is given by $x^2 - 7x + 10 \geq 0$.

$x^2 - 7x + 10 \geq 0$ Write in general form.

$(x - 2)(x - 5) \geq 0$ Factor.

So, the inequality has two key numbers: $x = 2$ and $x = 5$.

Key numbers: $x = 2, x = 5$

Test intervals: $(-\infty, 2), (2, 5), (5, \infty)$

Test: Is $(x - 2)(x - 5) \geq 0$?

A test shows that the inequality is satisfied in the unbounded half-closed intervals $(-\infty, 2]$ or $[5, \infty)$. So, the domain of the expression $\sqrt{x^2 - 7x + 10}$ is $(-\infty, 2] \cup [5, \infty)$.

Graphical solution

Begin by sketching the graph of the equation $y = \sqrt{x^2 - 7x + 10}$. From the graph, you can determine that the x-values extend up to 2 (including 2) and from 5 and beyond (including 5). So, the domain of the expression $\sqrt{x^2 - 7x + 10}$ is $(-\infty, 2] \cup [5, \infty)$.

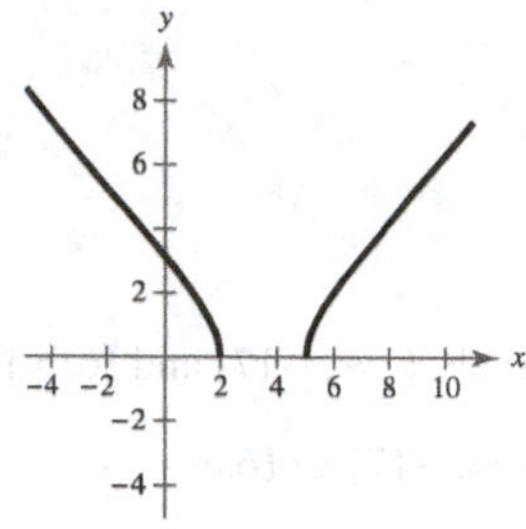

Chapter 2

Checkpoints for Section 2.1

1. (a) $y = 3x + 2$: Because $b = 2$, the y-intercept is $(0, 2)$. Because the slope is $m = 3$, the line rises three units for each unit the line moves to the right.

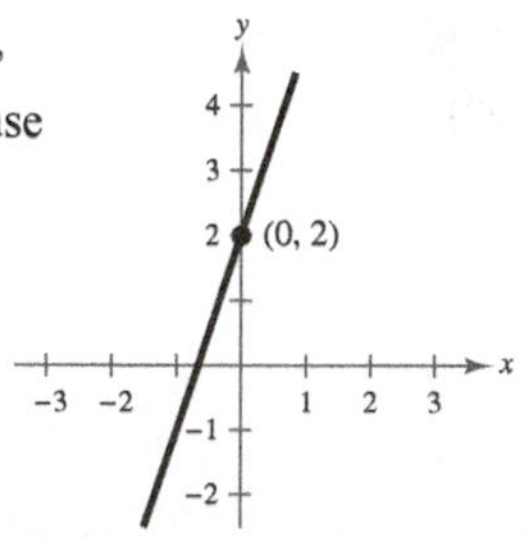

(b) By writing $y = -3$: this equation in the form $y = (0)x - 3$, you can see that the y-intercept is $(0, -3)$ and the slope is $m = 0$. A zero slope implies that the line is horizontal.

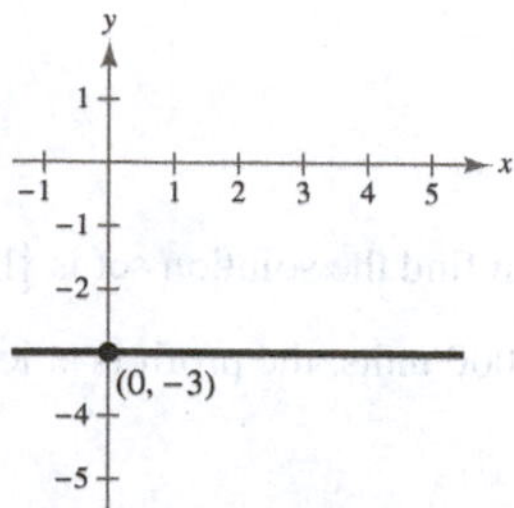

(c) $4x + y = 5$: By writing this equation in slope-intercept form

$4x + y = 5$

$y = -4x + 5$

you can see that the y-intercept is $(0, 5)$. Because the slope is $m = -4$, the line falls four units for each unit the line moves to the right.

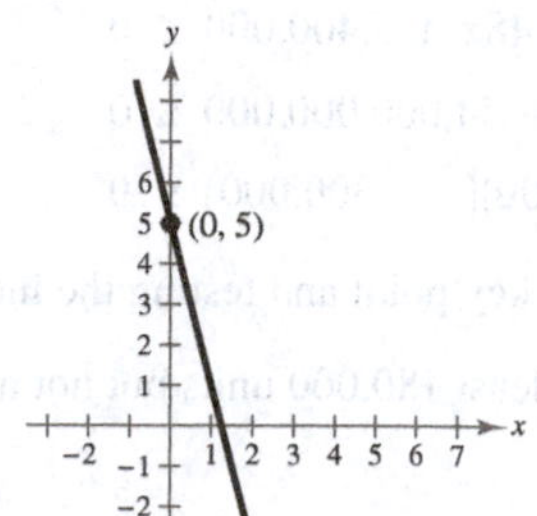

2. (a) The slope of the line passing through $(-5, -6)$ and $(2, 8)$ is $m = \dfrac{8-(-6)}{2-(-5)} = \dfrac{14}{7} = 2$.

(b) The slope of the line passing through $(4, 2)$ and $(2, 5)$ is $m = \dfrac{5-2}{2-4} = \dfrac{3}{-2} = -\dfrac{3}{2}$.

(c) The slope of the line passing through $(0, 0)$ and $(0, -6)$ is $m = \dfrac{-6-0}{0-0} = \dfrac{-6}{0}$. Because division by 0 is undefined, the slope is undefined and the line is vertical.

(d) The slope of the line passing through $(0, -1)$ and $(3, -1)$ is $m = \dfrac{-1-(-1)}{3-0} = \dfrac{0}{3} = 0$.

3. (a) Use the point-slope form with $m = 2$ and $(x_1, y_1) = (3, -7)$.

$$y - y_1 = m(x - x_1)$$
$$y - (-7) = 2(x - 3)$$
$$y + 7 = 2x - 6$$
$$y = 2x - 13$$

The slope-intercept form of this equation is $y = 2x - 13$.

(b) Use the point-slope form with $m = \dfrac{-2}{3}$ and $(x_1, y_1) = (1, 1)$

$$y - y_1 = m(x - x_1)$$
$$y - 1 = \frac{-2}{3}(x - 1)$$
$$y - 1 = \frac{-2}{3}x + \frac{2}{3}$$
$$y = \frac{-2}{3}x + \frac{5}{3}$$

The slope-intercept form of this equation is $y = -\dfrac{2}{3}x + \dfrac{5}{3}$.

(c) Use the point-slope form with $m = 0$ and $(x_1, y_1) = (1, 1)$.

$$y - y_1 = m(x - x_1)$$
$$y - 1 = 0(x - 1)$$
$$y - 1 = 0$$
$$y = 1$$

The slope-intercept of the equation is the line $y = 1$.

4. By writing the equation of the given line in slope-intercept form

$$5x - 3y = 8$$
$$-3y = -5x + 8$$
$$y = \frac{5}{3}x - \frac{8}{3}$$

You can see that it has a slope of $m = \dfrac{5}{3}$.

(a) Any line parallel to the given line must also have a slope of $m = \dfrac{5}{3}$. So, the line through $(-4, 1)$ that is parallel to the given line has the following equation.

$$y - y_1 = m(x - x_1)$$
$$y - 1 = \frac{5}{3}(x - (-4))$$
$$y - 1 = \frac{5}{3}(x + 4)$$
$$y - 1 = \frac{5}{3}x + \frac{20}{3}$$
$$y = \frac{5}{3}x + \frac{23}{3}$$

(b) Any line perpendicular to the given line must also have a slope of $m = -\dfrac{3}{5}$ because $-\dfrac{3}{5}$ is the negative reciprocal of $\dfrac{5}{3}$. So, the line through $(-4, 1)$ that is perpendicular to the given line has the following equation.

$$y - y_1 = m(x - x_1)$$
$$y - 1 = -\frac{3}{5}(x - (-4))$$
$$y - 1 = -\frac{3}{5}(x + 4)$$
$$y - 1 = -\frac{3}{5}x - \frac{12}{5}$$
$$y = -\frac{3}{5}x - \frac{7}{5}$$

5. The horizontal length of the ramp is 32 feet or $12(32) = 384$ inches.

So, the slope of the ramp is

$$\text{Slope} = \frac{\text{vertical change}}{\text{horizontal change}} = \frac{36 \text{ in.}}{384 \text{ in.}} \approx 0.094.$$

Because $\dfrac{1}{12} \approx 0.083$, the slope of the ramp is steeper than recommended.

6. The y-intercept $(0, 1500)$ tells you that the value of the copier when it was purchased $(t = 0)$ was \$1500. The slope of $m = -300$ tells you that the value of the copier decreases \$300 each year after it was purchased.

7. Let V represent the value of the machine at the end of year t. The initial value of the machine can be represented by the data point $(0, 24{,}750)$ and the salvage value of the machine can be represented by the data point $(6, 0)$. The slope of the line is

$$m = \frac{0 - 24{,}750}{6 - 0}$$
$$m = -\$4125$$

The slope represents the annual depreciation in dollars per year. Using the point-slope form, you can write the equation of the line as follows

$$V - 24{,}750 = -4125(t - 0)$$
$$V - 24{,}750 = -4125t$$
$$V = -4125t + 24{,}750$$

The equation $V = -4125t + 24{,}750$ represents the book value of the machine each year.

8. Let $t = 3$ represent 2013. Then the two given values are represented by the data points $(3, 6.5)$ and $(4, 7.2)$.

The slope of the line through these points is

$$m = \frac{7.2 - 6.5}{4 - 3} = -0.7.$$

You can find the equation that relates the sales y and the year t to be,

$$y - 6.5 = -0.7(t - 3)$$
$$y - 6.5 = -0.7t + 2.1$$
$$y = -0.7t + 4.4$$

According to this equation, the sales for Foot Locker in 2017 will be

$$y = -0.7(7) + 4.4$$
$$= 4.9 + 4.4$$
$$= \$9.3 \text{ billion.}$$

Checkpoints for Section 2.2

1. (a) This mapping *does not* describe y as a function of x. The input value of -1 is assigned or matched to two different y-values.

(b) The table *does* describe y as a function of x. Each input value is matched with exactly one output value.

2. (a) The given value of $x = -1$ corresponds to two values of y. So y is not a function of x.

(b) To each value of x there corresponds exactly one value of y. So, y is a function of x.

3. (a) Replacing x with 2 in $f(x) = 10 - 3x^2$ yields the following.

$$f(2) = 10 - 3(2)^2$$
$$= 10 - 12$$
$$= -2$$

(b) Replacing x with -4 yields the following.

$$f(-4) = 10 - 3(-4)^2$$
$$= 10 - 48$$
$$= -38$$

(c) Replacing x with $x - 1$ yields the following.

$$f(x - 1) = 10 - 3(x - 1)^2$$
$$= 10 - 3(x^2 - 2x + 1)$$
$$= 10 - 3x^2 + 6x - 3$$
$$= -3x^2 + 6x + 7$$

4. Because $x = -2$ is less than 0, use $f(x) = x^2 + 1$ to obtain $f(-2) = (-2)^2 + 1 = 4 + 1 = 5$.

Because $x = 2$ is greater than or equal to 0, use $f(x) = x - 1$ to obtain $f(2) = 2 - 1 = 1$.

For $x = 3$, use $f(x) = x - 1$ to obtain $f(3) = 3 - 1 = 2$.

5. Set $f(x) = 0$ and solve for x.

$$f(x) = 0$$
$$x^2 - 16 = 0$$
$$(x + 4)(x - 4) = 0$$
$$x + 4 = 0 \Rightarrow x = -4$$
$$x - 4 = 0 \Rightarrow x = 4$$

So, $f(x) = 0$ when $x = -4$ or $x = 4$.

6. $x^2 + 6x - 24 = 4x - x^2$ Set $f(x)$ equal to $g(x)$.

$2x^2 + 2x - 24 = 0$ Write in general form.

$2(x^2 + x - 12) = 0$ Factor out common factor.

$x^2 + x - 12 = 0$ Divide each side by 2.

$(x + 4)(x - 3) = 0$ Factor.

$x + 4 = 0 \Rightarrow x = -4$ Set 1st factor equal to 0.

$x - 3 = 0 \Rightarrow x = 3$ Set 2nd factor equal to 0.

So, $f(x) = g(x)$, when $x = -4$ or $x = 3$.

7. (a) The domain of f consists of all first coordinates in the set of ordered pairs.

Domain $= \{-2, -1, 0, 1, 2\}$

(b) Excluding x-values that yield zero in the denominator, the domain of g is the set of all real numbers x except $x = 3$.

(c) Because the function represents the circumference of a circle, the values of the radius r must be positive. So, the domain is the set of real numbers r such that $r > 0$.

(d) This function is defined only for x-values for which $x - 16 \geq 0$. You can conclude that $x \geq 16$. So, the domain is the interval $[16, \infty)$.

8. Use the formula for surface area of a cylinder, $s = 2\pi r^2 + 2\pi rh$.

(a) $$s(r) = 2\pi r^2 + 2\pi r(4r) = 2\pi r^2 + 8\pi r^2 = 10\pi r^2$$

(b) $$s(h) = 2\pi\left(\frac{h}{4}\right)^2 + 2\pi\left(\frac{h}{4}\right)h = 2\pi\left(\frac{h^2}{16}\right) + \frac{\pi h^2}{2} = \frac{1}{8}\pi h^2 + \frac{1}{2}\pi h^2 = \frac{5}{8}\pi h^2$$

9. When $x = 60$, you can find the height of the baseball as follows

$f(x) = -0.004x^2 + 0.3x + 6$ Write original function.

$f(60) = -0.004(60)^2 + 0.3(60) + 6$ Substitute 60 for x.

$= 9.6$ Simplify.

When $x = 60$, the height of the ball thrown from the second baseman is 9.6 feet. So, the first baseman cannot catch the baseball without jumping.

10. From 2009 through 2011, use $S(t) = 69t + 151$.

2009: $S(t) = 69(9) + 151 = 772$ fuel stations.

2010: $S(t) = 69(10) + 151 = 841$ fuel stations.

2011: $S(t) = 69(11) + 151 = 910$ fuel stations.

From 2012 through 2015, use $S(t) = 160t - 803$.

2012: $S(12) = 160(12) - 803 = 1117$ fuel stations.

2013: $S(13) = 160(13) - 803 = 1277$ fuel stations.

2014: $S(14) = 160(14) - 803 = 1437$ fuel stations.

2015: $S(15) = 160(15) - 803 = 1597$ fuel stations.

11. $\dfrac{f(x+h)-f(x)}{h} = \dfrac{\left[(x+h)^2 + 2(x+h) - 3\right] - (x^2 + 2x - 3)}{h}$

$= \dfrac{x^2 + 2xh + h^2 + 2x + 2h - 3 - x^2 - 2x + 3}{h}$

$= \dfrac{2xh + h^2 + 2h}{h}$

$= \dfrac{2(2x + h + 2)}{h}$

$= 2x + h + 2, \quad h \neq 0$

Checkpoints for Section 2.3

1. (a) The open dot at $(-3, -6)$ indicates that $x = -3$ is not in the domain of *f*. So, the domain of *f* is all real numbers, except $x \neq -3$, or $(-\infty, -3) \cup (-3, \infty)$.

(b) Because $(0, 3)$ is a point on the graph of f, it follows that $f(0) = 3$. Similarly, because the point $(3, -6)$ is a point on the graph of *f*, it follows that $f(3) = -6$.

(c) Because the graph of *f* does not extend above $f(0) = 3$, the range of *f* is the interval $(-\infty, 3]$.

2.

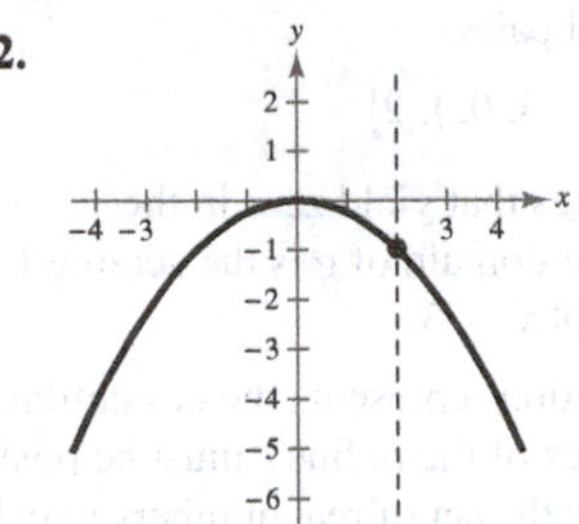

This *is* a graph of *y* as a function of *x*, because every vertical line intersects the graph at most once. That is, for a particular input *x*, there is at most one output *y*.

3. To find the zeros of a function, set the function equal to zero, and solve for the independent variable.

(a) $2x^2 + 13x - 24 = 0$ Set $f(x)$ equal to 0.

$(2x - 3)(x + 8) = 0$ Factor.

$2x - 3 = 0 \Rightarrow x = \dfrac{3}{2}$ Set 1st factor equal to 0.

$x + 8 = 0 \Rightarrow x = -8$ Set 2nd factor equal to 0.

The zeros of *f* are $x = \dfrac{3}{2}$ and $x = -8$. The graph of *f* has $\left(\dfrac{3}{2}, 0\right)$ and $(-8, 0)$ as its *x*-intercepts.

(b) $\sqrt{t - 25} = 0$ Set $g(t)$ equal to 0.

$\left(\sqrt{t - 25}\right)^2 = (0)^2$ Square each side.

$t - 25 = 0$ Simplify.

$t = 25$ Add 25 to each side.

The zero of *g* is $t = 25$.

The graph of *g* has $(25, 0)$ as its *t*-intercept.

(c) $\dfrac{x^2 - 2}{x - 1} = 0$ Set $h(x)$ equal to zero.

$(x - 1)\left(\dfrac{x^2 - 2}{x - 1}\right) = (x - 1)(0)$ Multiply each side by $x - 1$.

$x^2 - 2 = 0$ Simplify.

$x^2 = 2$ Add 2 to each side.

$x = \pm\sqrt{2}$ Extract square roots.

The zeros of *h* are $x = \pm\sqrt{2}$. The graph of *h* has $\left(\sqrt{2}, 0\right)$ and $\left(-\sqrt{2}, 0\right)$ as its *x*-intercepts.

4.

This function is increasing on the interval $(-\infty, -2)$, decreasing on the interval $(-2, 0)$, and increasing on the interval $(0, \infty)$.

5.

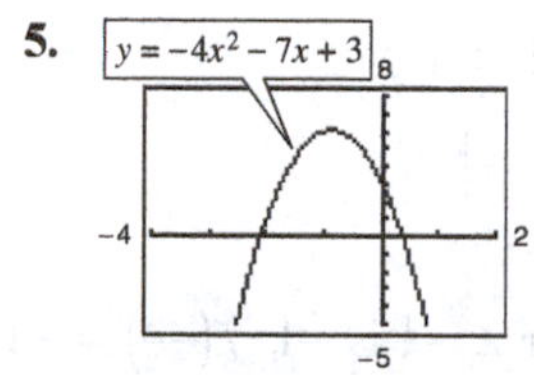

By using the zoom and the trace features or the maximum feature of a graphing utility, you can determine that the function has a relative maximum at the point $\left(-\frac{7}{8}, \frac{97}{16}\right)$ or $(-0.875, 6.0625)$.

6. (a) The average rate of change of f from $x_1 = -3$ to $x_2 = -2$ is

$$\frac{f(x_2) - f(x_1)}{x_2 - x_1} = \frac{f(-2) - f(-3)}{-2 - (-3)}$$

$$= \frac{0 - 3}{1} = -3.$$

(b) The average rate of change of f from $x_1 = -2$ to $x_2 = 0$ is

$$\frac{f(x_2) - f(x_1)}{x_2 - x_1} = \frac{f(0) - f(-2)}{0 - (-2)}$$

$$= \frac{0 - 0}{2} = 0.$$

7. (a) The average speed of the car from $t_1 = 0$ to $t_2 = 1$ second is

$$\frac{s(t_2) - s(t_1)}{t_2 - t_1} = \frac{20 - 0}{1 - 0} = 20 \text{ feet per second.}$$

(b) The average speed of the car from $t_1 = 1$ to $t_2 = 4$ seconds is

$$\frac{s(t_2) - s(t_1)}{t_2 - t_1} = \frac{160 - 20}{4 - 1}$$

$$= \frac{140}{3}$$

$$\approx 46.7 \text{ feet per second.}$$

8. (a) The function $f(x) = 5 - 3x$ is neither odd nor even because $f(-x) \neq -f(x)$ and $f(-x) \neq f(x)$ as follows.

$$f(-x) = 5 - 3(-x)$$

$$= 5 + 3x \neq -f(x) \quad \text{not odd}$$

$$\neq f(x) \quad \text{not even}$$

So, the graph of f is not symmetric to the origin nor the y-axis.

(b) The function $g(x) = x^4 - x^2 - 1$ is even because $g(-x) = g(x)$ as follows.

$$g(-x) = (-x)^4 - (-x)^2 - 1$$

$$= x^4 - x^2 - 1$$

$$= g(x)$$

So, the graph of g is symmetric to the y-axis.

(c) The function $h(x) = 2x^3 + 3x$ is odd because $h(-x) = -h(x)$,

$$h(-x) = 2(-x)^3 + 3(-x)$$

$$= -2x^3 - 3x$$

$$= -(2x^3 + 3x)$$

$$= -h(x)$$

So, the graph of h is symmetric to the origin.

Checkpoints for Section 2.4

1. To find the equation of the line that passes through the points $(x_1, y_1) = (-2, 6)$ and $(x_2, y_2) = (4, -4)$, first find the slope of the line.

$$m = \frac{y_2 - y_1}{x_2 - x_1} = \frac{-9 - 6}{4 - (-2)} = \frac{-15}{6} = \frac{-5}{2}$$

Next, use the point-slope form of the equation of the line.

$y - y_1 = m(x - x_1)$ Point-slope form

$y - 6 = -\frac{5}{2}[x - (-2)]$ Substitute x_1, y_1 and m.

$y - 6 = -\frac{5}{2}(x + 2)$ Simplify.

$y - 6 = -\frac{5}{2}x - 5$ Simplify.

$y = -\frac{5}{2}x + 1$ Simplify.

$f(x) = -\frac{5}{2}x + 1$ Function notation

2. For $x = -\frac{3}{2}$, $f\left(-\frac{3}{2}\right) = \left[\!\left[-\frac{3}{2} + 2\right]\!\right]$

$= \left[\!\left[\frac{1}{2}\right]\!\right]$

$= 0$

Since the greatest integer $\le \frac{1}{2}$ is 0, $f\left(-\frac{3}{2}\right) = 0$.

For $x = 1$, $f(1) = [\![1 + 2]\!]$

$= [\![3]\!]$

$= 3$

Since the greatest integer ≤ 3 is 3, $f(1) = 3$.

For $x = -\frac{5}{2}$, $f\left(-\frac{5}{2}\right) = \left[\!\left[-\frac{5}{2} + 2\right]\!\right]$

$= \left[\!\left[-\frac{1}{2}\right]\!\right]$

$= -1$

Since the greatest integer $\le -\frac{1}{2}$ is -1, $f\left(-\frac{5}{2}\right) = -1$.

3. This piecewise-defined function consists of two linear functions. At $x = -4$ and to the left of $x = -4$, the graph is the line $y = -\frac{1}{2}x - 6$, and to the right of $x = -4$ the graph is the line $y = x + 5$. Notice that the point $(-4, -2)$ is a solid dot and $(-4, 1)$ is an open dot. This is because $f(-4) = -2$.

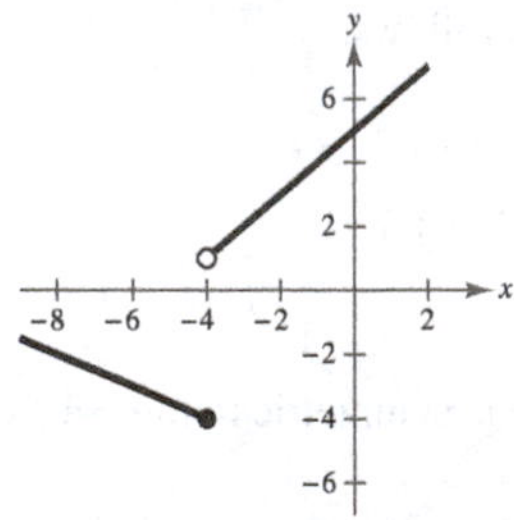

Checkpoints for Section 2.5

1. (a) Relative to the graph of $f(x) = x^3$, the graph of $h(x) = x^3 + 5$ is an upward shift of five units.

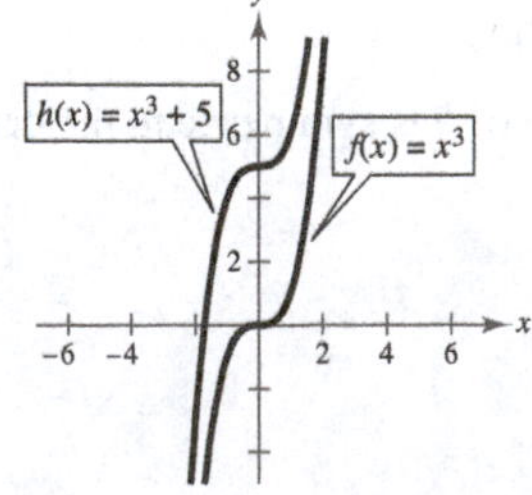

(b) Relative to the graph of $f(x) = x^3$, the graph of $g(x) = (x - 3)^3 + 2$ involves a right shift of three units and an upward shift of two units.

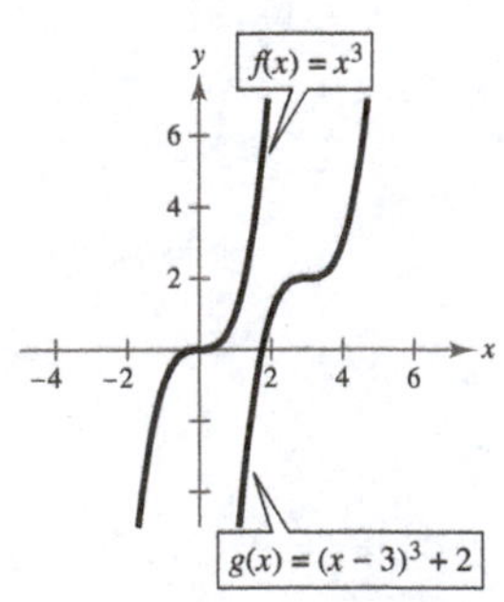

2. The graph of j is a horizontal shift of three units to the left *followed by* a reflection in the x-axis of the graph of $f(x) = x^4$. So, the equation for j is $j(x) = -(x + 3)^4$.

3. (a) **Algebraic Solution:**

The graph of g is a reflection of the graph of f in the x-axis because

$$g(x) = -\sqrt{x - 1}$$
$$= -f(x).$$

Graphical Solution:

Graph f and g on the same set of coordinate axes. From the graph, you can see that the graph of g is a reflection of the graph of f in the x-axis.

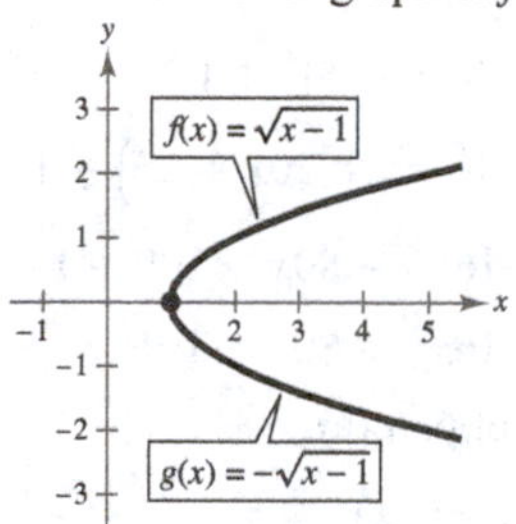

(b) **Algebraic Solution:**

The graph of h is a reflection of the graph of f in the y-axis because

$$h(x) = \sqrt{-x - 1}$$
$$= f(-x).$$

Graphical Solution:

Graph f and h on the same set of coordinate axes. From the graph, you can see that the graph h is a reflection of the graph of f, in the y-axis.

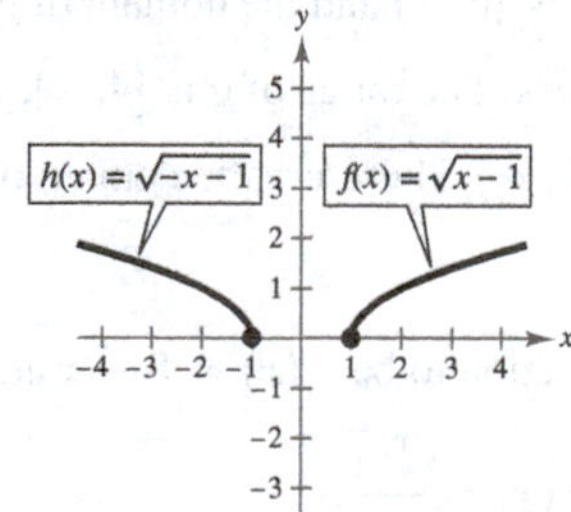

4. (a) Relative to the graph of $f(x) = x^2$, the graph of $g(x) = 4x^2 = 4f(x)$ is a vertical stretch (each y-value is multiplied by 4) of the graph of f.

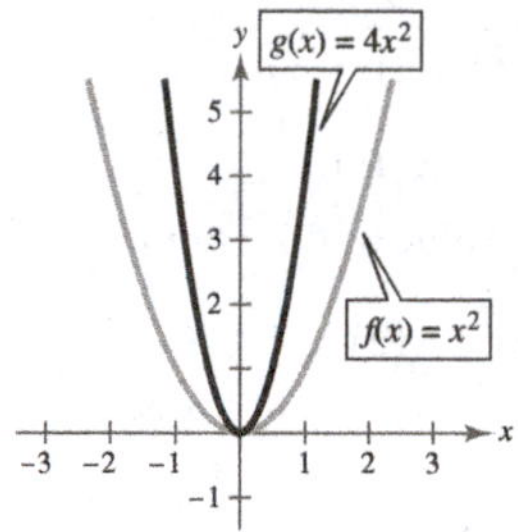

(b) Relative to the graph of $f(x) = x^2$, the graph of $h(x) = \frac{1}{4}x^2 = \frac{1}{4}f(x)$ is a vertical shrink (each y-value is multiplied by $\frac{1}{4}$) of the graph of f.

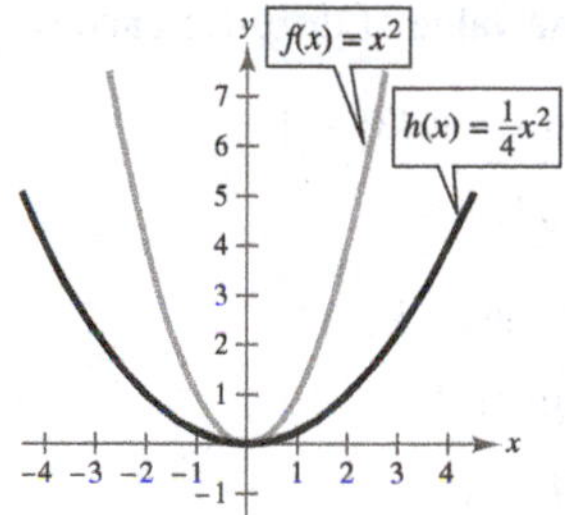

5. (a) Relative to the graph of $f(x) = x^2 + 3$, the graph of $g(x) = f(2x) = (2x)^2 + 3 = 4x^2 + 3$ is a horizontal shrink $(c > 1)$ of the graph of f.

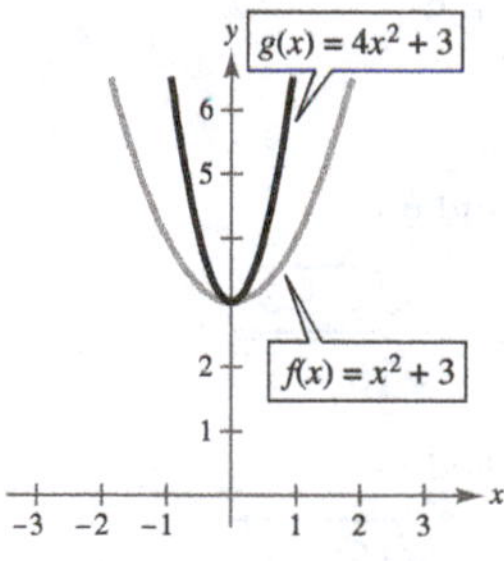

(b) Relative to the graph of $f(x) = x^2 + 3$, the graph of $h(x) = f\left(\frac{1}{2}x\right) = \left(\frac{1}{2}x\right)^2 + 3 = \frac{1}{4}x^2 + 3$ is a horizontal stretch $(0 < c < 1)$ of the graph of f.

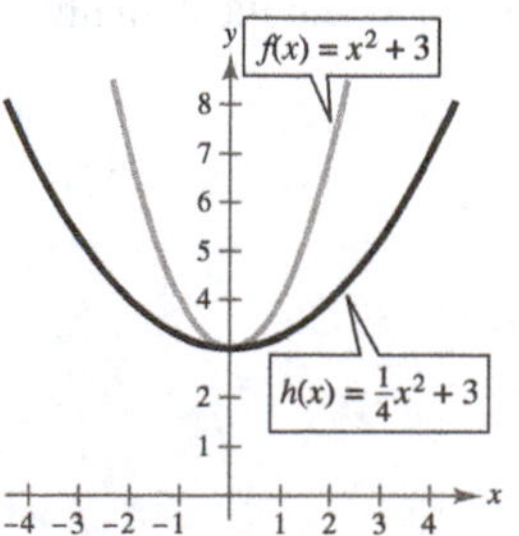

Checkpoints for Section 2.6

1. The sum of f and g is

$$(f + g)(x) = f(x) + g(x)$$
$$= (x^2) + (1 - x)$$
$$= x^2 - x + 1.$$

When $x = 2$, the value of this sum is

$$(f + g)(2) = (2)^2 - (2) + 1$$
$$= 3.$$

2. The difference of f and g is

$$(f - g)(x) = f(x) - g(x)$$
$$= (x^2) - (1 - x)$$
$$= x^2 + x - 1.$$

When $x = 3$, the value of the difference is

$$(f - g)(3) = (3)^2 + (3) - 1$$
$$= 11.$$

3. The product of f and g is

$$(fg) = f(x)g(x)$$
$$= (x^2)(1 - x)$$
$$= x^2 - x^3$$
$$= -x^3 + x^2.$$

When $x = 3$, the value of the product is

$$(fg)(3) = -(3)^3 - (3)^2$$
$$= -27 + 9$$
$$= -18.$$

4. The quotient of f and g is

$$\left(\frac{f}{g}\right)(x) = \frac{f(x)}{g(x)} = \frac{\sqrt{x - 3}}{\sqrt{16 - x^2}}.$$

The quotient of g and f is

$$\left(\frac{g}{f}\right)(x) = \frac{g(x)}{f(x)} = \frac{\sqrt{16 - x^2}}{\sqrt{x - 3}}.$$

The domain of f is $[3, \infty)$ and the domain of g is $[-4, 4]$. The intersection of these two domains is $[3, 4]$. So, the domain of f/g is $[3, 4)$ and the domain of g/f is $(3, 4]$.

5. (a) The composition of f with g is as follows.

$$(f \circ g)(x) = f(g(x))$$
$$= f(4x^2 + 1)$$
$$= 2(4x^2 + 1) + 5$$
$$= 8x^2 + 2 + 5$$
$$= 8x^2 + 7$$

(b) The composition of g with f is as follows.

$$(g \circ f)(x) = g(f(x))$$
$$= g(2x + 5)$$
$$= 4(2x + 5)^2 + 1$$
$$= 4(4x^2 + 20x + 25) + 1$$
$$= 16x^2 + 80x + 100 + 1$$
$$= 16x^2 + 80x + 101$$

(c) Use the result of part (a).

$$(f \circ g)\left(-\tfrac{1}{2}\right) = 8\left(-\tfrac{1}{2}\right)^2 + 7$$
$$= 8\left(\tfrac{1}{4}\right) + 7$$
$$= 2 + 7$$
$$= 9$$

6. The composition of f with g is as follows.

$$(f \circ g)(x) = f(g(x))$$
$$= f(x^2 + 4)$$
$$= \sqrt{x^2 + 4}$$

The domain of f is $[0, \infty)$ and the domain of g is the set of all real numbers. The range of g is $[4, \infty)$, which is in the range of f, $[0, \infty)$. Therefore the domain of $f \circ g$ is all real numbers.

7. Let the inner function to be $g(x) = 8 - x$ and the outer function to be $f(x) = \dfrac{\sqrt[3]{x}}{5}$.

$$h(x) = \frac{\sqrt[3]{8 - x}}{5}$$
$$= f(8 - x)$$
$$= f(g(x))$$

8. (a) $(N \circ T)(t) = N(T(t))$

$$= 8(2t + 2)^2 - 14(2t + 2) + 200$$
$$= 8(4t^2 + 8t + 4) - 28t - 28 + 200$$
$$= 32t^2 + 64t + 32 - 28t - 28 + 200$$
$$= 32t^2 + 36t + 204$$

The composite function $(N \circ T)(t)$ represents the number of bacteria in the food as a function of the amount of time the food has been out of refrigeration.

(b) Let $(N \circ T)(t) = 1000$ and solve for t.

$$32t^2 + 36t + 204 = 1000$$
$$32t^2 + 36t - 796 = 0$$
$$4(8t^2 + 9t - 199) = 0$$
$$8t^2 + 9t - 199 = 0$$

Use the quadratic formula:

$$t = \frac{-9 \pm \sqrt{(9)^2 - 4(8)(-199)}}{2(8)}$$
$$= \frac{-9 \pm \sqrt{6449}}{16}$$

$t \approx 4.5$ and $t \approx -5.6$.

Using $t \approx 4.5$ hours, the bacteria count reaches approximately 1000 about 4.5 hours after the food is removed from the refrigerator.

Checkpoints for Section 2.7

1. The function f multiplies each input by $\frac{1}{5}$. To "undo" this function, you need to multiply each input by 5. So, the inverse function of $f(x) = \frac{1}{5}x$ is $f^{-1}(x) = 5x$.

To verify this, show that

$f(f^{-1}(x)) = x$ and $f^{-1}(f(x)) = x$.

$$f(f^{-1}(x)) = f(5x) = \tfrac{1}{5}(5x) = x$$
$$f^{-1}(f(x)) = f^{-1}\left(\tfrac{1}{5}x\right) = 5\left(\tfrac{1}{5}x\right) = x$$

So, the inverse function of $f(x) = \frac{1}{5}x$ is $f^{-1}(x) = 5x$.

2. By forming the composition of f and g, you have

$$f(g(x)) = f(7x + 4) = \frac{(7x + 4) - 4}{7} = \frac{7x}{7} = x.$$

So, it appears that g is the inverse function of f. To confirm this, form the composition of g and f.

$$g(f(x)) = g\left(\frac{x - 4}{7}\right) = 7\left(\frac{x - 4}{7}\right) + 4 = x - 4 + 4 = x$$

By forming the composition of f and h, you can see that h is *not* the inverse function of f, since the result is not the identity function x.

$$f(h(x)) = f\left(\frac{7}{x - 4}\right) = \frac{\left(\frac{7}{x - 4}\right) - 4}{7} = \frac{23 - 4x}{7(x - 4)} \neq x$$

So, g is the inverse function of f.

3. First, sketch the graphs of $f(x) = 4x - 1$ and $g(x) = \frac{1}{4}(x + 1)$ as shown. You can see that they appear to be reflections of each other in the line $y = x$. This reflective property can be tested using a few points and the fact that if the point (a, b) is on the graph of f then the point (b, a) is on the graph of g.

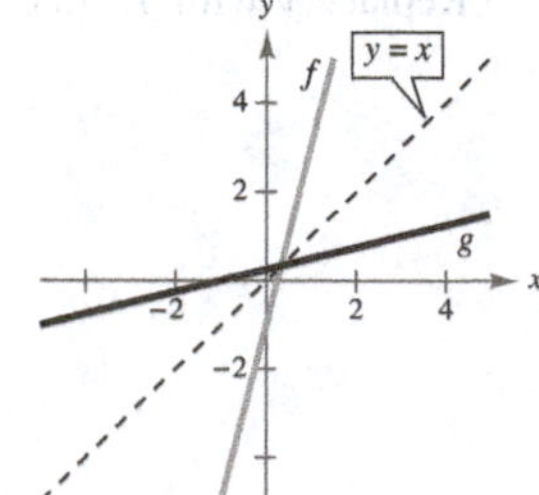

Graph of $f(x) = 4x - 1$	Graph of $g(x) = \frac{1}{4}(x + 1)$
$(-1, -5)$	$(-5, -1)$
$(0, -1)$	$(-1, 0)$
$(1, 3)$	$(3, 1)$
$(2, 7)$	$(7, 2)$

4. First, sketch the graphs of $f(x) = x^2 + 1, x \geq 0$ and $g(x) = \sqrt{x - 1}$ as shown. You can see that they appear to be reflections of each other in the line $y = x$. This reflective property can be tested using a few points and the fact that if the point (a, b) is on the graph of f then the point (b, a) is on the graph of g.

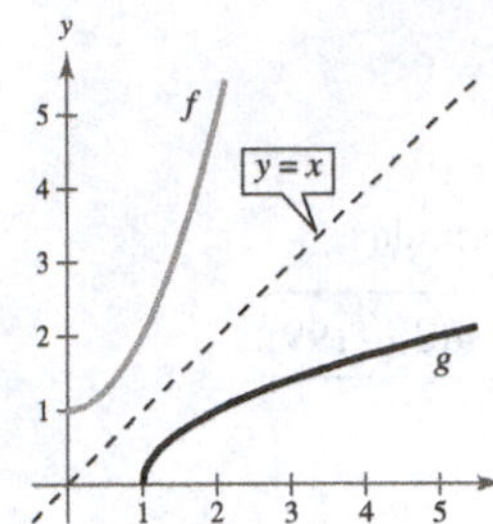

Graph of $f(x) = x^2 + 1, x \geq 0$	Graph of $g(x) = \sqrt{x - 1}$
(0, 1)	(1, 0)
(1, 2)	(2, 1)
(2, 5)	(5, 2)
(3, 10)	(10, 3)

5. (a) The graph of $f(x) = \frac{1}{2}(3 - x)$ is shown.

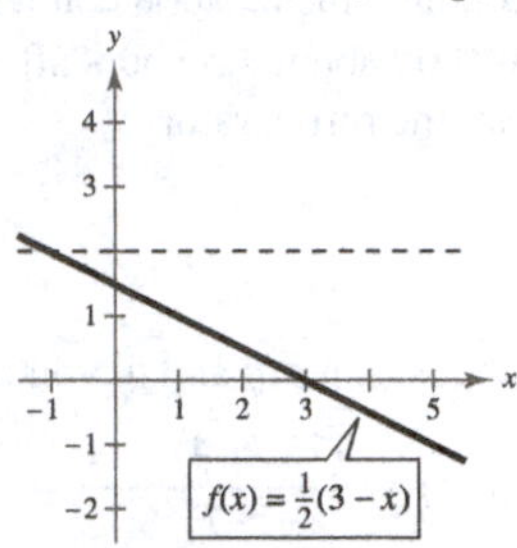

Because no horizontal line intersects the graph of f at more than one point, f is a one-to-one function and *does* have an inverse function.

(b) The graph of $f(x) = |x|$ is shown.

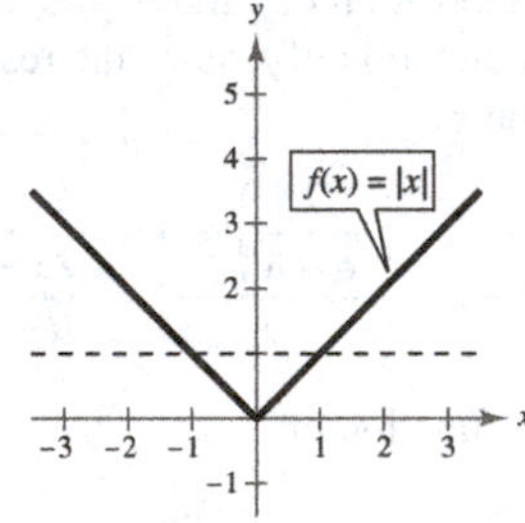

Because it is possible to find a horizontal line that intersects the graph of f at more than one point, f is *not* a one-to-one function and *does not* have an inverse function.

6. The graph of $f(x) = \dfrac{5 - 3x}{x + 2}$ is shown.

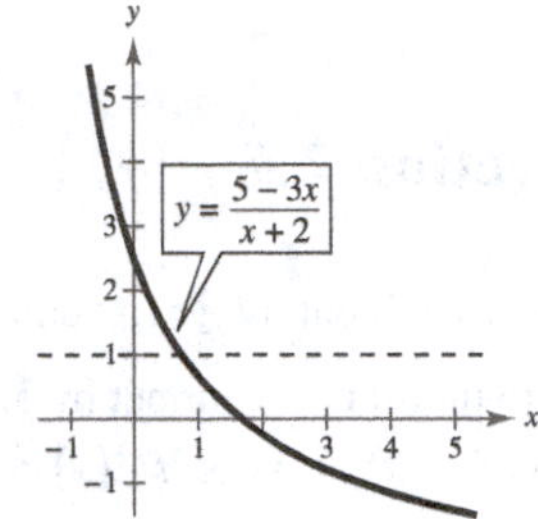

This graph passes the Horizontal Line Test. So, you know f is one-to-one and has an inverse function.

$f(x) = \dfrac{5 - 3x}{x + 2}$	Write original function.
$y = \dfrac{5 - 3x}{x + 2}$	Replace $f(x)$ with y.
$x = \dfrac{5 - 3y}{y + 2}$	Interchange x and y.
$x(y + 2) = 5 - 3y$	Multiply each side by $y + 2$.
$xy + 2x = 5 - 3y$	Distribute Property
$xy + 3y = 5 - 2x$	Collect like terms with y.
$y(x + 3) = 5 - 2x$	Factor.
$y = \dfrac{5 - 2x}{x + 3}$	Solve for y.
$f^{-1}(x) = \dfrac{5 - 2x}{x + 3}$	Replace y with $f^{-1}(x)$.

7. The graph of $f(x) = \sqrt[3]{10 + x}$ is shown.

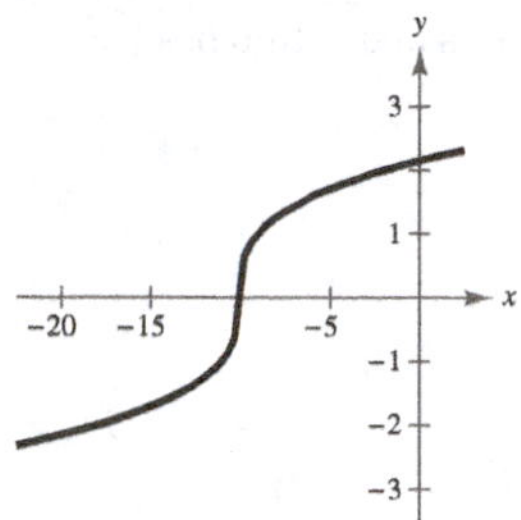

Because this graph passes the Horizontal Line Test, you know that f is one-to-one and has an inverse function.

$$\begin{aligned} f(x) &= \sqrt[3]{10 + x} \\ y &= \sqrt[3]{10 + x} \\ x &= \sqrt[3]{10 + y} \\ x^3 &= 10 + y \\ y &= x^3 - 10 \\ f^{-1}(x) &= x^3 - 10 \end{aligned}$$

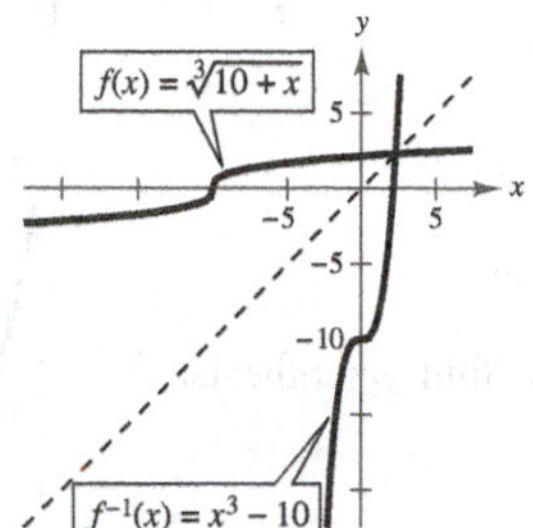

The graphs of f and f^{-1} are reflections of each other in the line $y = x$. So, the inverse of $f(x) = \sqrt[3]{10 + x}$ is $f^{-1}(x) = x^3 - 10$.

To verify, check that $f(f^{-1}(x)) = x$ and $f^{-1}(f(x)) = x$.

$$\begin{aligned} f(f^{-1}(x)) &= f(x^3 - 10) \\ &= \sqrt[3]{10 + (x^3 - 10)} \\ &= \sqrt[3]{x^3} \\ &= x \end{aligned} \qquad \begin{aligned} f^{-1}(f(x)) &= f^{-1}\left(\sqrt[3]{10 + x}\right) \\ &= \left(\sqrt[3]{10 + x}\right)^3 - 10 \\ &= 10 + x - 10 \\ &= x \end{aligned}$$

Chapter 3

Checkpoints for Section 3.1

1. (a) Compared with the graph of $y = x^2$, each output of $f(x) = \frac{1}{4}x^2$ "shrinks" by a factor of $\frac{1}{4}$, creating a broader parabola.

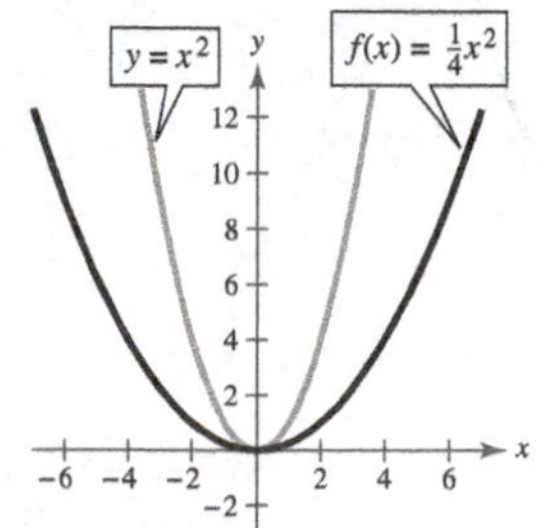

(b) Compared with the graph of $y = x^2$, each output of $f(x) = -\frac{1}{6}x^2$ is reflected in the x-axis and "shrinks" by a factor of $\frac{1}{6}$, creating a broader parabola.

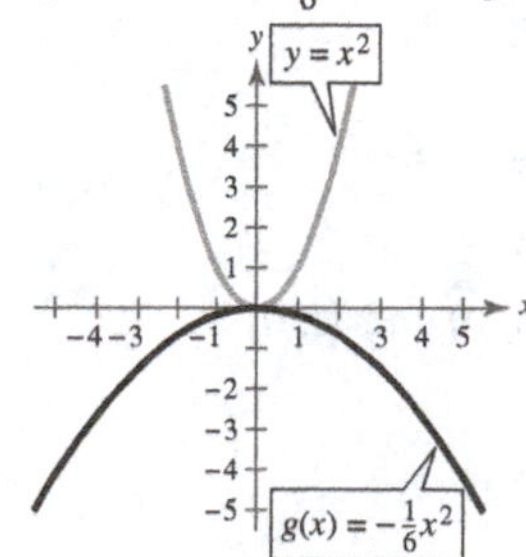

(c) Compared with $y = x^2$, each output of of $h(x) = \frac{5}{2}x^2$ "stretches" by a factor of $\frac{5}{2}$, creating a narrower parabola.

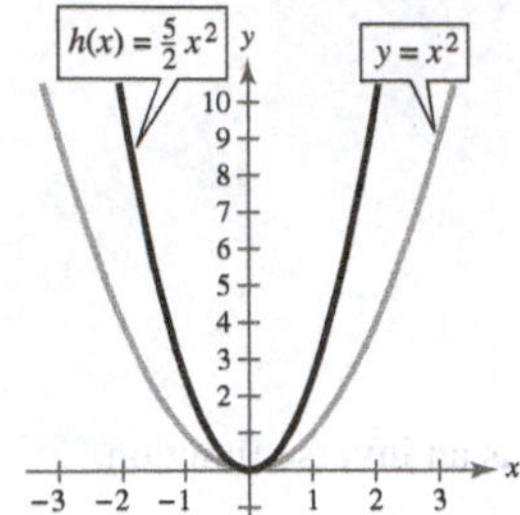

(d) Compared with $y = -4x^2$, each output of $k(x)$ is reflected in the x-axis and "stretches" by a factor of 4.

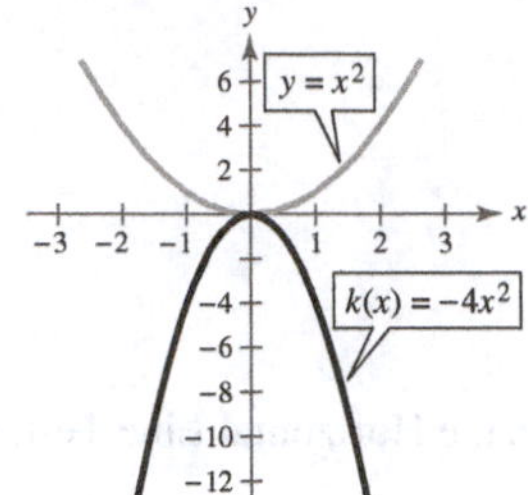

2.

$f(x) = 3x^2 - 6x + 4$	Write original function.
$= 3(x^2 - 2x) + 4$	Factor 3 out of x-terms.
$= 3(x^2 - 2x + 1 - 1) + 4$	Add and subtract 1 within parenthesis.
$= 3(x^2 - 2x + 1) - 3(1) + 4$	Regroup terms.
$= 3(x^2 - 2x + 1) - 3 + 4$	Simplify.
$= 3(x - 1)^2 + 1$	Write in standard form.

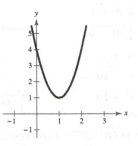

You can see that the graph of f is a parabola that opens upward and has its vertex at $(1, 1)$. This corresponds to a right shift of one unit and an upward shift of one unit relative to the graph of $y = 3x^2$, which is a "stretch" of $y = x^2$.

The axis of the parabola is the vertical line through the vertex, $x = 1$.

3.

$f(x) = x^2 - 4x + 3$	Write original function.
$= (x^2 - 4x + 4 - 4) + 3$	Add and subtract 4 within parenthesis.
$= (x^2 - 4x + 4) - 4 + 3$	Regroup terms.
$= (x^2 - 4x + 4) - 1$	Simplify.
$= (x^2 - 2)^2 - 1$	Write in standard form.

In standard form, you can see that f is a parabola that opens upward with vertex $(2, -1)$.

The x-intercepts of the graph are determined as follows.

$$x^2 - 4x + 3 = 0$$

$$(x - 3)(x - 1) = 0$$

$$x - 3 = 0 \Rightarrow x = 3$$

$$x - 1 = 0 \Rightarrow x = 1$$

So, the x-intercepts are $(3, 0)$ and $(1, 0)$.

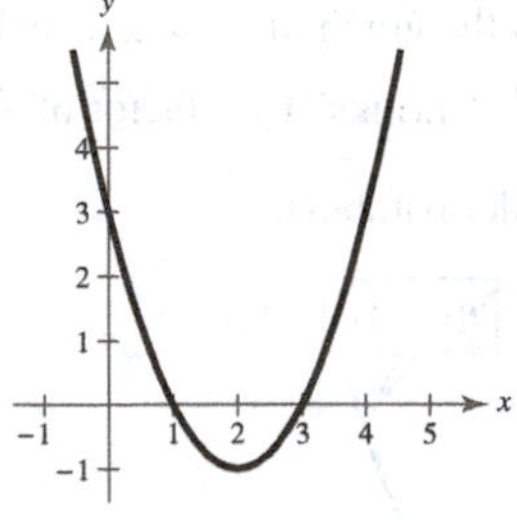

4. The vertex is $(h, k) = (-4, 11)$ so the equation has the form

$f(x) = a(x + 4)^2 + 11.$

The parabola passes through the point $(-6, 15)$ so it follows that $f(-6) = 15$.

$f(x) = a(x + 4)^2 + 11$	Write standard form.
$15 = a(-6 + 4)^2 + 11$	Substitute -6 for x and 15 for $f(x)$.
$15 = a(-2)^2 + 11$	Simplify.
$4 = 4a$	Subtract 11 from each side.
$1 = a$	Divide each side by 4.

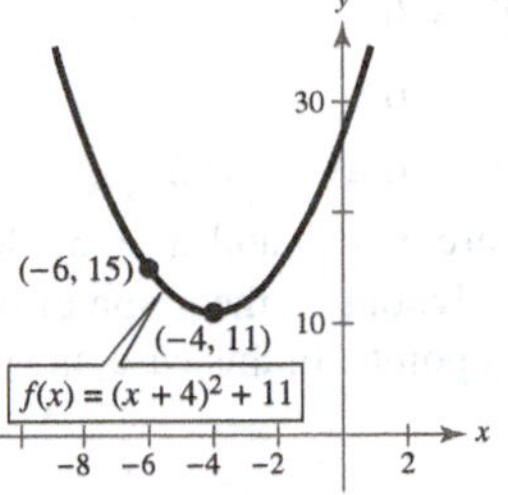

The equation in standard form is $f(x) = (x + 4)^2 + 11$.

5. For this quadratic function,

$f(x) = ax^2 + bx + c = -0.007x^2 + x + 4$

which implies that $a = -0.007$ and $b = 1$.

Because $a < 0$, the function has a maximum at $x = -\dfrac{b}{2a}$. So, the baseball reaches its maximum height when it is

$$x = -\frac{b}{2a} = -\frac{1}{2(-0.007)} = \frac{1}{0.014} \approx 71.4 \text{ feet from home plate.}$$

At this distance, the maximum height is $f(71.4) = -0.007(71.4)^2 + (71.4) + 4 \approx 39.7$ feet.

Checkpoints for Section 3.2

1. (a) The graph of $f(x) = (x + 5)^4$ is a left shift by five units of the graph of $y = x^4$.

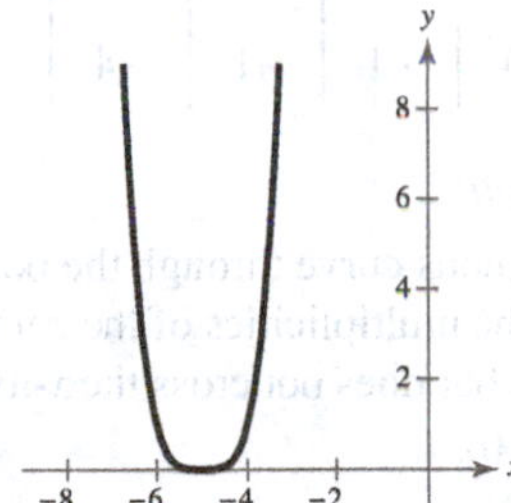

(b) The graph of $g(x) = x^4 - 7$ is a downward shift of seven units of the graph of $y = x^4$.

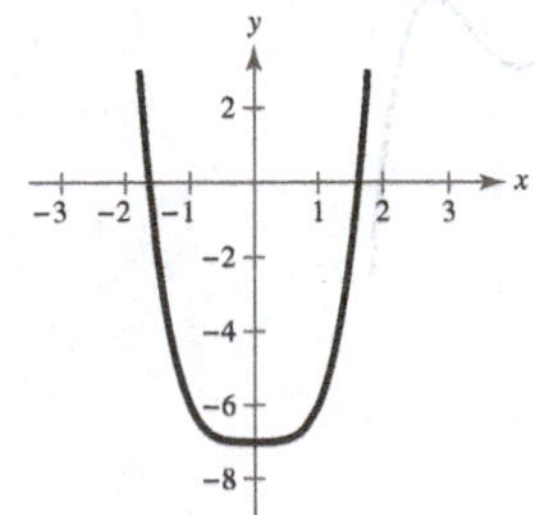

(c) The graph of $h(x) = 7 - x^4 = -x^4 + 7$ is a reflection in the x-axis then an upward shift of seven units of the graph of $y = x^4$.

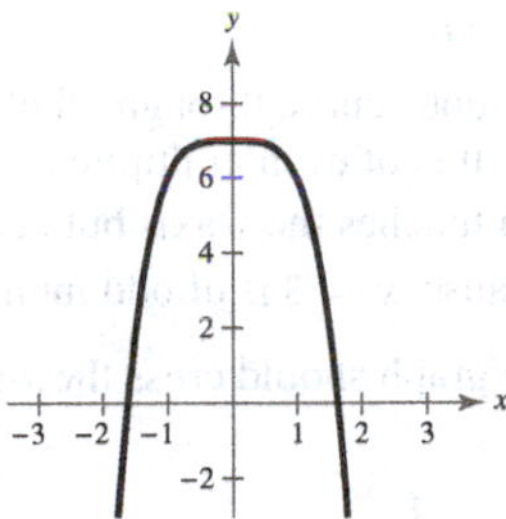

(d) The graph of $k(x) = \frac{1}{4}(x - 3)^4$ is a right shift by three units and a vertical "shrink" by a factor of $\frac{1}{4}$ of the graph of $y = x^4$.

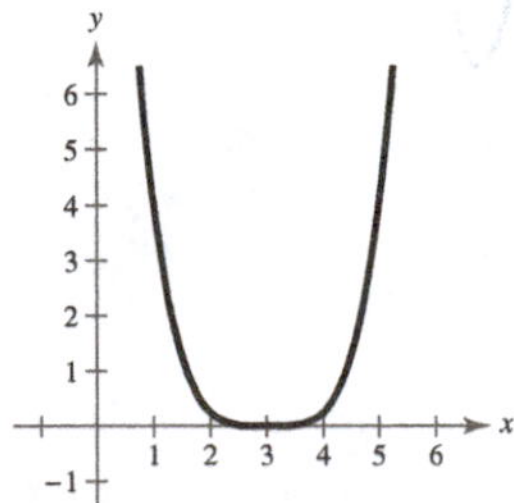

2. (a) Because the degree is odd and the leading coefficient is positive, the graph falls to the left and rises to the right.

(b) Because the degree is odd and the leading coefficient is negative, the graph rises to the left and falls to the right.

3. To find the real zeros of $f(x) = x^3 - 12x^2 + 36x$, set $f(x)$ equal to zero, and solve for x.

$$x^3 - 12x^2 + 36x = 0$$
$$x(x^2 - 12x + 36) = 0$$
$$x(x-6)^2 = 0$$
$$x = 0$$
$$x - 6 = 0 \Rightarrow x = 6$$

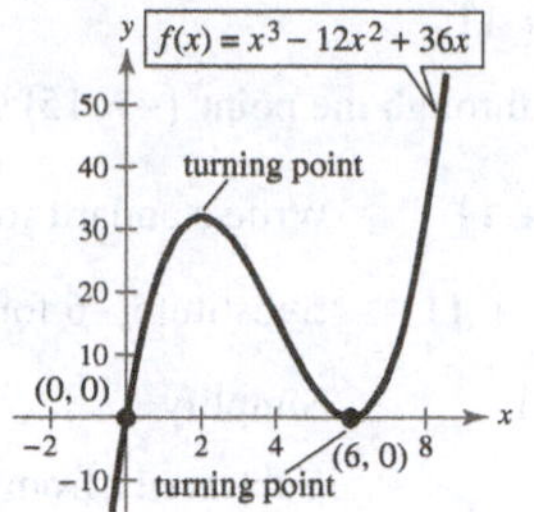

So, the real zeros are $x = 0$ and $x = 6$. Because the function is a third-degree polynomial, the graph of f can have at most $3 - 1 = 2$ turning points. In this case, the graph of f has two turning points.

4. 1. *Apply the Leading Coefficient Test.*

Because the leading coefficient is positive and the degree is odd, you know that the graph eventually falls to the left and rises to the right.

2. *Find the Real Zeros of the Polynomial.*

By factoring $f(x) = 2x^3 - 6x^2$

$$= 2x^2(x - 3)$$

you can see that the real zeros of f are $x = 0$ (even multiplicity) and $x = 3$ (odd multiplicity). So, the x-intercepts occur at $(0, 0)$ and $(3, 0)$

3. *Plot a Few Additional Points.*

x	-1	1	2	4
$f(x)$	-8	-4	-8	32

4. *Draw the Graph.*

Draw a continuous curve through all of the points. Because $x = 0$ is of even multiplicity, you know that the graph touches the x-axis but does not cross it at $(0, 0)$. Because $x = 3$ is of odd multiplicity, you know that the graph should cross the x-axis at $(3, 0)$.

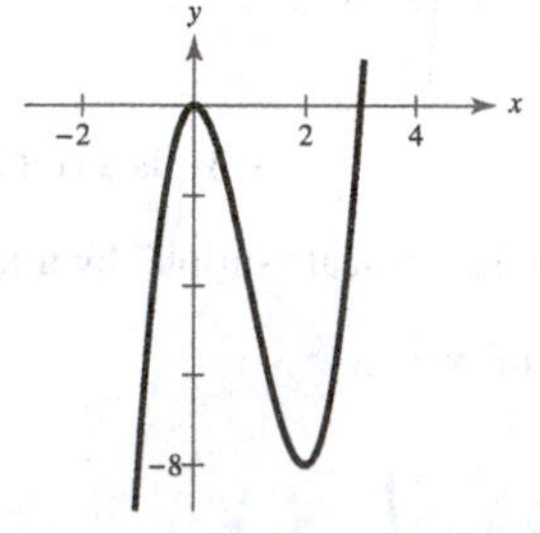

5. 1. *Apply the Leading Coefficient Test.*

Because the leading coefficient is negative and the degree is even, you know that the graph eventually falls to the left and falls to the right.

2. *Find the Real Zeros of the Polynomial.*

By factoring $f(x) = -\frac{1}{4}x^4 + \frac{3}{2}x^3 - \frac{9}{4}x^2$

$$= -\tfrac{1}{4}x^2(x^2 - 6x + 9)$$
$$= -\tfrac{1}{4}x^2(x - 3)^2$$

you can see that the real zeros of f are $x = 0$ (even multiplicity) and $x = 3$ (even multiplicity). So, the x-intercepts occur at $(0, 0)$ and $(3, 0)$.

3. *Plot a Few Additional Points.*

x	-1	1	2	4
$f(x)$	-4	-1	-1	-4

4. *Draw the graph.*

Draw a continuous curve through the points. As indicated by the multiplicities of the zeros, the graph touches but does not cross the x-axis at $(0, 0)$ and $(3, 0)$.

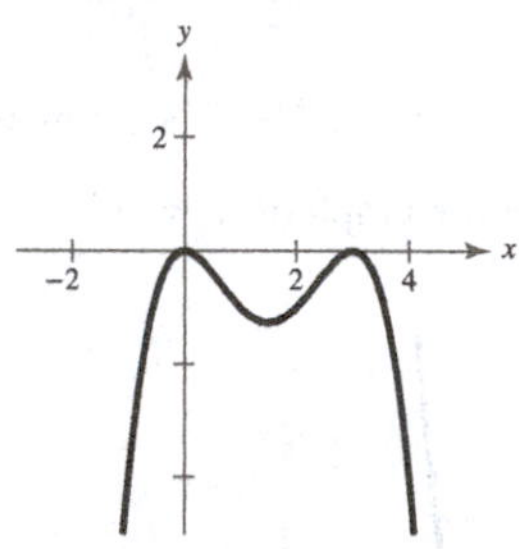

6. Begin by computing a few function values of $f(x) = x^3 - 3x^2 - 2$.

x	-1	0	1	2	3	4
$f(x)$	-6	-2	-4	-6	-2	14

Because $f(3)$ is negative and $f(4)$ is positive, you can apply the Intermediate Value Theorem to conclude that the function has a real zero between $x = 3$ and $x = 4$. To find this real zero more closely, divide the interval $[3, 4]$ into tenths and evaluate the function at each point.

x	3.1	3.2	3.3	3.4	3.5	3.6	3.7	3.8	3.9
$f(x)$	-1.039	0.048	1.267	2.624	4.125	5.776	7.583	9.552	11.689

So, f must have a real zero between 3.1 and 3.2.
To find a more accurate approximation, you can compute the function value between $f(3.1)$ and $f(3.2)$ and apply the Intermediate Value Theorem again to verify that $x \approx 3.196$.

Checkpoints for Section 3.3

1. To divide $9x^3 + 36x^2 - 49x - 196$ by $x + 4$ using long division, you can set up the operation as shown.

$$\begin{array}{r l}
9x^2 - 49 & \\
x + 4 \overline{)9x^3 + 36x^2 - 49x - 196} & \\
\underline{9x^3 + 36x^2 } & \text{Multiply by: } 9x^2(x + 4) \\
-49x - 196 & \text{Subtract.} \\
\underline{-49x - 196} & \text{Multiply by: } -49(x + 4). \\
0 & \text{Subtract.}
\end{array}$$

From this division, you can conclude that

$$\begin{aligned}
9x^3 + 36x^2 - 49x - 196 &= (x + 4)(9x^2 - 49) \\
&= (x + 4)(3x + 7)(3x - 7).
\end{aligned}$$

2. To divide $x^3 - 2x^2 - 9$ by $x - 3$ using long division, you can set up the operation as shown. Because there is no x-term in the dividend, rewrite the dividend as $x^3 - 2x^2 + 0x - 9$ before you apply the Division Algorithm.

$$\begin{array}{r l}
x^2 + x + 3 & \\
x - 3 \overline{)x^3 - 2x^2 + 0x - 9} & \\
\underline{x^3 - 3x^2 } & \text{Multiply } x^2 \text{ by } x - 3. \\
x^2 + 0x - 9 & \text{Subtract.} \\
\underline{x^2 - 3x } & \text{Multiply } x \text{ by } x - 3. \\
3x - 9 & \text{Subtract.} \\
\underline{3x - 9} & \text{Multiply 3 by } x - 3. \\
0 & \text{Subtract.}
\end{array}$$

So, $x - 3$ divides evenly into $x^3 - 2x^2 - 9$, and you can write $\dfrac{x^3 - 2x^2 - 9}{x - 3} = x^2 + x + 3, x \neq 3$.

Check the result by multiplying

$$\begin{aligned}
(x + 3)(x^2 + x + 3) &= x^3 - 3x^2 + x^2 - 3x + 3x - 9 \\
&= x^3 - 2x^2 - 9.
\end{aligned}$$

3. To divide $-x^3 + 9x + 6x^4 - x^2 - 3$ by $1 + 3x$ using long division, begin by rewriting the dividend and divisor in descending powers of x.

$$\begin{array}{rll}
 & 2x^3 - x^2 \qquad + 3 & \\
3x + 1 \,\big)\!\!\! & \overline{6x^4 - x^3 - x^2 + 9x - 3} & \\
 & \underline{6x^4 + 2x^3} & \text{Multiply } 2x^3 \text{ by } 3x + 1. \\
 & -3x^3 - x^2 + 9x - 3 & \text{Subtract.} \\
 & \underline{-3x^3 - x^2} & \text{Multiply } -x^2 \text{ by } 3x + 1. \\
 & 9x - 3 & \text{Subtract.} \\
 & \underline{9x + 3} & \text{Multiply 3 by } 3x + 1. \\
 & -6 & \text{Subtract.}
\end{array}$$

So, you have $\dfrac{6x^4 - x^3 - x^2 + 9x - 3}{3x + 1} = 2x^3 - x^2 + 3 - \dfrac{6}{3x + 1}$.

Check the result by multiplying $(3x + 1)\left(2x^3 - x^2 + 3 - \dfrac{6}{3x + 1}\right) = 6x^4 - 3x^3 + 9x + 2x^3 - x^2 + 3 - 6$

$$= 6x^4 - x^3 - x^2 + 9x - 3$$

4. To divide $5x^3 + 8x^2 - x + 6$ by $x + 2$ using synthetic division, you can set up the array as shown.

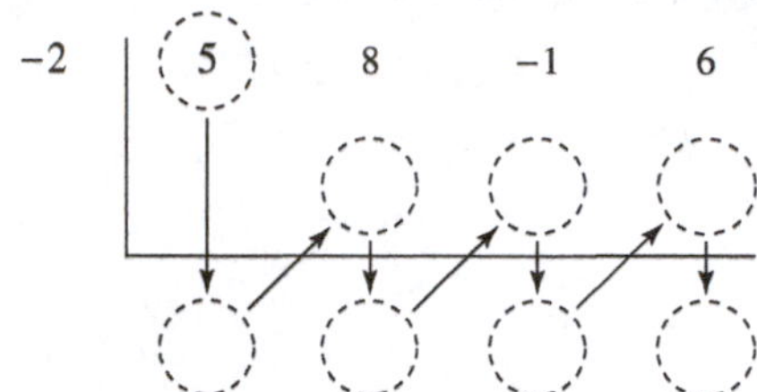

Then, use the synthetic division pattern by adding terms in columns and multiplying the results by -2.

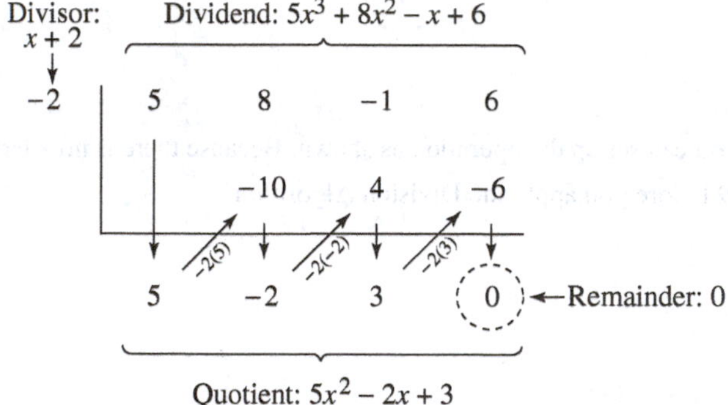

So, you have, $\dfrac{5x^3 + 8x^2 - x + 6}{x + 2} = 5x^2 - 2x + 3$.

5. (a) $f(-1)$

$$\begin{array}{r|rrrr} -1 & 4 & 10 & -3 & -8 \\ & & -4 & -6 & 9 \\ \hline & 4 & 6 & -9 & 1 \end{array}$$

Because the remainder is $r = 1$, $f(-1) = 1$.

Check: $f(-1) = 4(-1)^3 + 10(-1)^2 - 3(-1) - 8$
$= 4(-1) + 10(1) + 3 - 8$
$= 1$

(b) $f(4)$

$$\begin{array}{r|rrrr} 4 & 4 & 10 & -3 & -8 \\ & & 16 & 104 & 404 \\ \hline & 4 & 26 & 101 & 396 \end{array}$$

Because the remainder is $r = 396$, $f(4) = 396$.

Check: $f(4) = 4(4)^3 + 10(4)^2 - 3(4) - 8$
$= 4(64) + 10(16) - 12 - 8$
$= 396$

(c) $f\left(\frac{1}{2}\right)$

$$\begin{array}{r|rrrr} \frac{1}{2} & 4 & 10 & -3 & -8 \\ & & 2 & 6 & \frac{3}{2} \\ \hline & 4 & 12 & 3 & -\frac{13}{2} \end{array}$$

Because the remainder is $r = -\frac{13}{2}$, $f\left(\frac{1}{2}\right) = -\frac{13}{2}$.

Check: $f\left(\frac{1}{2}\right) = 4\left(\frac{1}{2}\right)^3 + 10\left(\frac{1}{2}\right)^2 - 3\left(\frac{1}{2}\right) - 8$
$= 4\left(\frac{1}{8}\right) + 10\left(\frac{1}{4}\right) - \frac{3}{2} - 8$
$= \frac{1}{2} + \frac{5}{2} - \frac{3}{2} - 8$
$= -\frac{13}{2}$

(d) $f(-3)$

$$\begin{array}{r|rrrr} -3 & 4 & 10 & -3 & -8 \\ & & -12 & 6 & -9 \\ \hline & 4 & -2 & 3 & -17 \end{array}$$

Because the remainder is $r = -17$, $f(-3) = -17$.

Check: $f(-3) = 4(-3)^3 + 10(-3)^2 - 3(-3) - 8$
$= 4(-27) + 10(9) + 9 - 8$
$= -17$

6. Algebraic Solution:

Using synthetic division with the factor $(x + 3)$, you obtain the following.

$$\begin{array}{r|rrrr} -3 & 1 & 0 & -19 & -30 \\ & & -3 & 9 & 30 \\ \hline & 1 & -3 & -10 & 0 \end{array} \rightarrow \text{ 0 remainder, so } f(-3) = 0 \text{ and } (x+3) \text{ is a factor.}$$

Because the resulting quadratic expression factors as $x^2 - 3x - 10 = (x - 5)(x + 2)$, the complete factorization of $f(x)$ is $f(x) = x^3 - 19x - 30 = (x + 3)(x - 5)(x + 2)$.

Graphical Solution:

From the graph of $f(x) = x^3 - 19x - 30$, you can see there are three x-intercepts. These occur at $x = -3$, $x = -2$, and $x = 5$. This implies that $(x + 3)$, $x + 2$, and $(x - 5)$ are factors of $f(x)$.

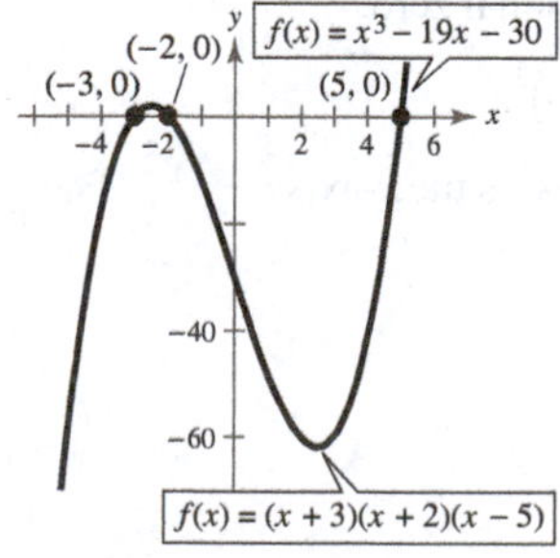

Checkpoints for Section 3.4

1. $f(x) = x^4 - 1 = (x^2 + 1)(x^2 - 1)$

$= (x^2 + 1)(x + 1)(x - 1)$

$x^2 + 1 \Rightarrow x = \pm i$

$x + 1 \Rightarrow x = -1$

$x - 1 \Rightarrow x = 1$

The fourth-degree polynomial function $f(x) = x^4 - 1$ has exactly four zeros, $x = \pm i$ and $x = \pm 1$

2. $f(x) = x^3 + 2x^2 + 6x - 4$

Possible rational zeros: ± 1, ± 2, and ± 4

By testing these possible zeros,

$f(-1) = (-1)^3 + 2(-1)^2 + 6(-1) - 4 = -9$

$f(1) = (1)^3 + 2(1)^2 + 6(1) - 4 = 5$

$f(-2) = (-2)^3 + 2(-2)^2 + 6(-2) - 4 = -16$

$f(2) = (2)^3 + 2(2)^2 + 6(2) - 4 = 24$

$f(-4) = (-4)^3 + 2(-4)^2 + 6(-4) - 4 = -60$

$f(4) = (4)^3 + 2(4)^2 + 6(4) - 4 = 116$

you can conclude that the polynomial $f(x) = x^3 + 2x^2 + 6x - 4$ has no *rational* zeros.

3. Because the leading coefficient is 1, the possible rational zeros are the factors of the constant term.

Possible rational zeros: ± 1, ± 5, ± 25, ± 125

$$\begin{array}{r|rrrr} 5 & 1 & -15 & 75 & -125 \\ & & 5 & -50 & 125 \\ \hline & 1 & -10 & 25 & 0 \end{array} \rightarrow \text{0 remainder, so } x = 5 \text{ is a factor.}$$

$$\begin{array}{r|rrr} 5 & 1 & -10 & 25 \\ & & 5 & -25 \\ \hline & 1 & -5 & 0 \end{array} \rightarrow \text{0 remainder, so } x = 5 \text{ is a factor.}$$

By applying synthetic division successively, you can determine that $x = 5$ is the only rational zero.

So, $f(x) = x^3 - 15x^2 + 75x - 125$ factors as $f(x) = (x - 5)(x - 5)(x - 5) = (x - 5)^3$.

Because the rational zero $x = 5$ has multiplicity of three, which is odd, the graph of f crosses the x-axis at the x-intercept, $(5, 0)$.

4. The leading coefficient is 2 and the constant term is 6.

Possible rational zeros: $\dfrac{\text{Factors of } 6}{\text{Factors of } 2} = \dfrac{\pm1, \pm2, \pm3, \pm6}{\pm1, \pm2} = \pm1, \pm2, \pm3, \pm6, \pm\dfrac{1}{2}, \pm\dfrac{3}{2}$

Choose a value of x and use synthetic division.

$x = -3$

$$\begin{array}{r|rrrr} -3 & 2 & 1 & -13 & 6 \\ & & -6 & 15 & -6 \\ \hline & 2 & -5 & 2 & 0 \end{array} \quad \rightarrow \quad \text{0 remainder, so } x + 3 \text{ is a factor.}$$

So, $f(x) = 2x^3 + x^2 - 13x + 6$ factors as $f(x) = (x + 3)(2x^2 - 5x + 2) = (x + 3)(2x - 1)(x - 2)$ and you can conclude that the rational zeros of f are $x = -3$, $x = \dfrac{1}{2}$, and $x = 2$.

5. The leading coefficient is -2 and the constant term is 18.

Possible rational zeros: $\dfrac{\text{Factor of } 18}{\text{Factors of } -2} = \dfrac{\pm1, \pm2, \pm3, \pm6, \pm9, \pm18}{\pm1, \pm2} = \pm1, \pm2, \pm3, \pm6, \pm9, \pm18, \pm\dfrac{1}{2}, \pm\dfrac{3}{2}, \pm\dfrac{9}{2}$

A graph can assist you to narrow the list to reasonable possibilities.

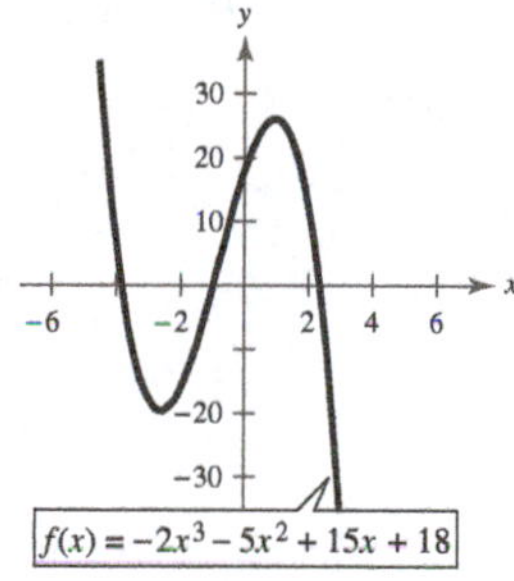

Start by testing $x = -1$.

$$\begin{array}{r|rrrr} -1 & -2 & -5 & 15 & 18 \\ & & 2 & 3 & -18 \\ \hline & -2 & -3 & 18 & 0 \end{array} \quad \rightarrow \quad \text{0 remainder, so } x + 1 \text{ is a factor.}$$

So, $f(x) = -2x^3 - 5x^2 + 15x + 18$ factors as $f(x) = (x + 1)(-2x^2 - 3x + 18)$

$-2x^2 - 3x + 18 = 0$

$x = \dfrac{-b \pm \sqrt{b^2 - 4ac}}{2a}$

$x = \dfrac{-(-3) \pm \sqrt{(-3)^2 - 4(-2)(18)}}{2(-2)}$

$x = \dfrac{3 \pm \sqrt{153}}{-4}$

$x = \dfrac{-3 \pm 3\sqrt{17}}{4}$

$x = \dfrac{-3(1 \pm \sqrt{17})}{4} \approx -3.8423, 2.3423$

And you can conclude that the rational zeros of f are $x = -1$ and $x = \dfrac{-3(1 \pm \sqrt{37})}{4} \approx -3.8423, 2.3423.$

6. Because $-7i$ is a zero, you know that the conjugate $7i$ must also be a zero.

So, the four zeros are $2, -2, 7i,$ and $-7i$.

Then, using the Linear Factorization Theorem, $f(x)$ can be written as

$f(x) = a(x - 2)(x + 2)(x - 7i)(x + 7i).$

For simplicity, let $a = 1$. Then multiply the factors with real coefficients to obtain $(x + 2)(x - 2) = x^2 - 4$ and multiply the complex conjugates to obtain $(x - 7i)(x + 7i) = x^2 + 49$.

So, you obtain the following fourth-degree polynomial function.

$$\begin{aligned} f(x) &= (x^2 - 4)(x^2 + 49) = x^4 + 49x^2 - 4x^2 - 196 \\ &= x^4 + 45x^2 - 196 \end{aligned}$$

7. Because $2i$ is a zero, you know that the conjugate $-2i$ must also be a zero.

This means that both $(x - 2i)$ and $(x + 2i)$ are factors of f.

$(x - 2i)(x + 2i) = x^2 - 4i^2 = x^2 + 4$. The other zeros are $x = -2$ and $x = 1$.

$$\begin{aligned} f(x) &= a(x + 2)(x - 1)(x^2 + 4) \\ &= a(x^2 + x - 2)(x^2 + 4) \\ &= a(x^4 + x^3 + 2x^2 + 4x - 8) \end{aligned}$$

Because $f(-1) = 10$

$$\begin{aligned} 10 &= a\left[(-1)^4 + (-1)^3 + 2(-1)^2 + 4(-1) - 8\right] \\ 10 &= a(-10) \\ -1 &= a \end{aligned}$$

So, $f(x) = (-1)(x^4 + x^3 + 2x^2 + 4x - 8)$

$$= -x^4 - x^3 - 2x^2 - 4x + 8$$

8. Because complex zeros occur in conjugate pairs you know that if $4i$ is a zero of f, so is $-4i$.

This means that both $(x - 4i)$ and $(x + 4i)$ are factors of f.

$(x - 4i)(x + 4i) = x^2 - 16i^2 = x^2 + 16$

Using long division, you can divide $x^2 + 16$ into $f(x)$ to obtain the following.

$$\begin{array}{r} 3x - 12 \\ x^2 + 16 \overline{\big) 3x^3 - 2x^2 + 48x - 32} \\ \underline{3x^3 \qquad\quad + 48x \qquad} \\ -2x^2 \qquad\quad - 32 \\ \underline{-2x^2 \qquad\quad - 32} \\ 0 \end{array}$$

So, you have $f(x) = (x^2 + 16)(3x - 2)$ and you can conclude that the real zeros of f are $x = -4i$, $x = 4i$, and $x = \frac{2}{3}$.

9. $f(x) = x^4 + 8x^2 - 9$

Because the leading coefficient is 1, the possible rational zeros are the factors of the constant term.
Possible rational zeros: $\pm1, \pm3$, and ±9

Synthetic division produces the following.

$$\begin{array}{r|rrrrr} 1 & 1 & 0 & 8 & 0 & -9 \\ & & 1 & 1 & 9 & 9 \\ \hline & 1 & 1 & 9 & 9 & 0 \end{array} \rightarrow \text{ 1 is a zero, so } x - 1 \text{ is a factor.}$$

$$\begin{array}{r|rrrr} -1 & 1 & 1 & 9 & 9 \\ & & -1 & 0 & -9 \\ \hline & 1 & 0 & 9 & 0 \end{array} \rightarrow \text{ } -1 \text{ is a zero, so } x + 1 \text{ is a factor.}$$

So, you have $f(x) = x^4 + 8x^2 - 9 = (x - 1)(x + 1)(x^2 + 9)$.
You can factor $x^2 + 9$ as $x^2 - (-9) = \left(x + \sqrt{-9}\right)\left(x - \sqrt{-9}\right) = (x + 3i)(x - 3i)$.

So, you have $f(x) = (x - 1)(x + 1)(x + 3i)(x - 3i)$ and you can conclude that the zeros of f are $x = 1, x = -1, x = 3i$, and $x = -3i$.

10. Determine the possible numbers of positive and negative real zeros of $f(x) = 2x^3 + 5x^2 + x + 8$.

The original polynomial has *zero* variations in sign.

$f(x) = 2x^3 + 5x^2 + x + 8$

The polynomial $f(-x) = 2(-x)^3 + 5(-x)^2 + (-x) + 8 = -2x^3 + 5x^2 - x + 8$ has *three* variations in sign.

$$\begin{array}{c} -\text{ to }+ \qquad - \text{ to } + \\ \downarrow \quad \downarrow \qquad \downarrow \quad \downarrow \\ f(x) = -2x^3 + 5x^2 - x + 8 \\ \uparrow \qquad \uparrow \\ + \text{ to } - \end{array}$$

So, from Descarte's Rule of Signs, the polynomial $f(x) = 2x^3 + 5x^2 + x + 8$ has either three negative real zeros or one negative real zero and no positive real zeros.
From the graph, you can see that the function has only one negative real zero.

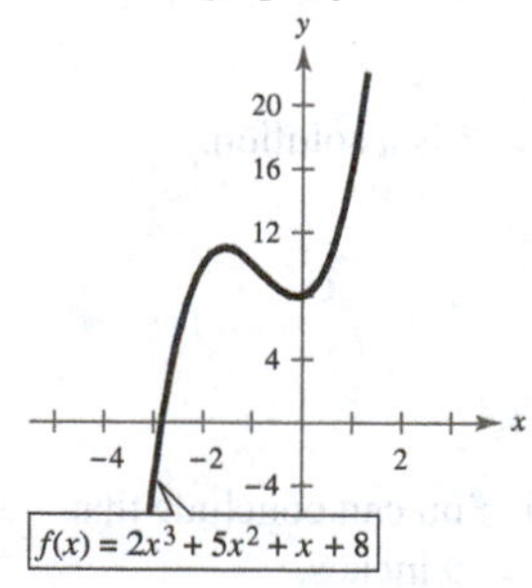

11. The possible real zeros are as follows.

$$\frac{\text{Factors of } -3}{\text{Factors of } 8} = \frac{\pm1 \pm3}{\pm1 \pm2 \pm4 \pm8} = \pm\frac{1}{8}, \pm\frac{1}{4}, \pm\frac{3}{8}, \pm\frac{1}{2}, \pm\frac{3}{4}, \pm1, \pm\frac{3}{2}, \pm3$$

The original polynomial $f(x)$ has three variations in sign. The polynomial

$$f(-x) = 8(-x)^3 - 4(-x)^2 + 6(-x) - 3 = -8x^3 - 4x^2 - 6x - 3$$

has no variations in sign. So, you can apply Descarte's Rule of Signs to conclude that there are either three positive real zeros or one positive real zero, and no negative real zeros.

Using $x = 1$, synthetic division produces the following.

$$\begin{array}{r|rrrr} 1 & 8 & -4 & 6 & -3 \\ & & 8 & 4 & 10 \\ \hline & 8 & 4 & 10 & 7 \end{array} \rightarrow \quad 1 \text{ is not a zero.}$$

So, $x = 1$ is not a zero, but because the last row has all positive entries, you know that $x = 1$ is an upper bound for the real zeros. So, you can restrict the search to real zeros between 0 and 1. Using $x = \frac{1}{2}$, synthetic division produces the following.

$$\begin{array}{r|rrrr} \frac{1}{2} & 8 & -4 & 6 & -3 \\ & & 4 & 0 & 3 \\ \hline & 8 & 0 & 6 & 0 \end{array} \rightarrow \quad \tfrac{1}{2} \text{ is a real zero.}$$

$$\begin{aligned} f(x) &= 8x^3 - 4x^2 + 6x - 3 \\ &= \left(x - \frac{1}{2}\right)(8x^2 + 6) \end{aligned}$$

Because $8x^2 + 6$ has no real zeros, it follows that $x = \frac{1}{2}$ is the only real zero.

12. The volume of a pyramid is $V = \frac{1}{3}Bh$, where B is the area of the base and h is the height. The area at the base is x^2 and the height is $x + 2$. So, the volume of the pyramid is $V = \frac{1}{3}x^2(x + 2)$. Substituting 147 for the volume yields the following.

$$\begin{aligned} 147 &= \tfrac{1}{3}x^2(x + 2) \\ 441 &= x^3 + 2x^2 \\ 0 &= x^3 + 2x^2 - 441 \end{aligned}$$

The possible rational zeros are $x = \pm1, \pm3, \pm7, \pm9, \pm21, \pm49, \pm63, \pm147$, and ±441.

Use synthetic division to test some of the possible solutions. So, you can determine that $x = 7$ is a solution.

$$\begin{array}{r|rrrr} 7 & 1 & 2 & 0 & -441 \\ & & 7 & 63 & 441 \\ \hline & 1 & 9 & 63 & 0 \end{array}$$

The other two solutions that satisfy $x^2 + 9x + 63 = 0$ are imaginary and can be discarded. You can conclude that the base of the candle mold should be 7 inches by 7 inches and the height should be $7 + 2 = 9$ inches.

Checkpoints for Section 3.5

1. 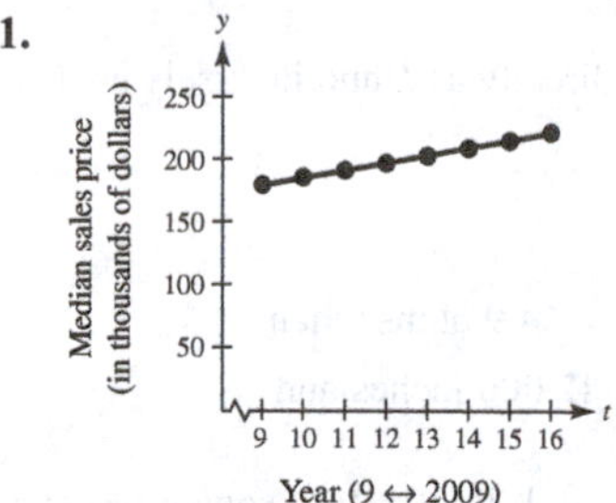

The actual data are plotted, along with the graph of the linear model. From the graph, it appears that the model is a "good fit" for the actual data. You can see how well the model fits by comparing the actual values of y with the values of y given by the model. The values given by the model are labeled y^* in the table below.

t	9	10	11	12	13	14	15	16
y	179.4	185.4	191.0	196.7	202.6	208.7	214.9	221.4
y^*	179.1	185.1	191.1	197.0	203.0	208.9	214.9	220.9

2. 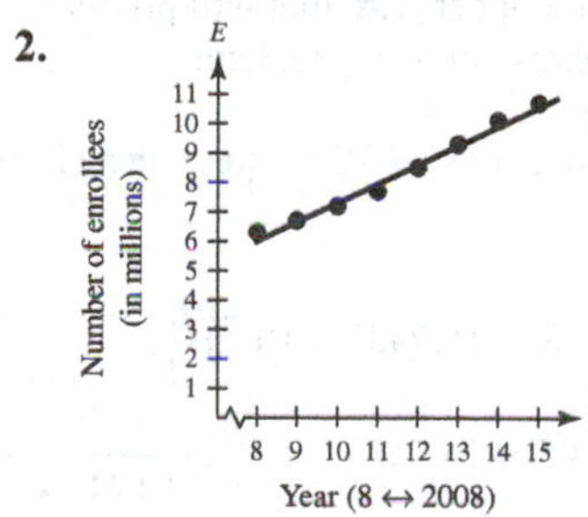

Let $t = 8$ represent 2008. The scatter plot for the data is shown. Using the *regression* feature of a graphing utility, you can determine that the equation of the least squares regression line is

$E = 0.65t + 0.8$

To check this model, compare the actual E values with the E^* values given by the model, which are labeled E^* in the table. The correlation coefficient for this model is $r \approx 0.991$, which implies that the model is a good fit.

t	8	9	10	11	12	13	14	15
E	6.3	6.7	7.2	7.7	8.5	9.3	10.1	10.7
E^*	6.0	6.7	7.3	8.0	8.6	9.3	9.9	10.6

3. *Verbal Model*: [Simple interest] = [r] [Amount of investment]

Labels:
Simple interest $= I$ (dollars)
Amount of investment $= P$ (dollars)
Interest rate $= r$ (percent in decimal form)

Equation: $I = rP$

To solve for r, substitute the given information into the equation $I = rP$, and then solve for r.

$I = rP$ Write direct variation model.

$187.50 = r(2500)$ Substitute 187.50 for t and 2500 for P.

$\dfrac{187.50}{2500} = r$ Divide each side by 2500.

$0.075 = r$ Simplify.

So, the mathematical model is $I = 0.075P$.

4. Letting s be the distance (in feet) the object falls and letting t be the time (in seconds) that the object falls, you have $s = Kt^2$.

Because $s = 144$ feet when $t = 3$ seconds, you can see that $K = \frac{144}{9}$ as follows.

$s = Kt^2$ Write direct variation model.

$144 = K(3)^2$ Substitute 144 for s and 3 for t.

$144 = 9K$ Simplify.

$\frac{144}{9} = K$ Divide each side by 9.

$16 = K$

So, the equation relating distance to time is $s = 16t^2$.

To find the distance the object falls in 6 seconds , let $t = 6$.

$s = 16t^2$ Write direct variation model.

$s = 16(6)^2$ Substitute 6 for t.

$s = 16(36)$ Simplify.

$s = 576$ Simplify.

So, the object falls 576 feet in 6 seconds.

5. Let p be the price and let x be the demand. Because x varies inversely as p, you have

$$x = \frac{k}{p}$$

Now because $x = 600$ then $p = 2.75$ you have

$x = \frac{k}{p}$ Write inverse variation model.

$600 = \frac{k}{2.75}$ Substitute 600 for x and 2.75 for p.

$(600)(2.75) = k$ Multiply each side by 2.75.

$1650 = k.$ Simplify.

So, the equation relating price and demand is

$$x = \frac{1650}{p}.$$

When $p = 3.25$ the demand is

$x = \frac{1650}{p}$ Write inverse variation model.

$= \frac{1650}{3.25}$ Substitute 3.25 for p.

≈ 508 units. Simplify.

So, the demand for the product is 508 units when the price of the product is \$3.25.

6. Let R be the resistance (in ohms), let L be the length (in inches), and let A be the cross-sectional area (in square inches).

Because R varies directly as L and inversely as A, you have

$$R = \frac{kL}{A}.$$

Now, because $R = 64.9$ ohms when $L = 1000$ feet $= 12{,}000$ inches and

$A = \pi\left(\frac{0.0126}{2}\right)^2 \approx 1.2469 \times 10^{-4}$ square inches,

you have

$$64.9 = \frac{k(12,000)}{1.2469 \times 10^{-4}}$$

$6.7437 \times 10^{-7} \approx k.$

So, the equation relating resistance, length, and the cross-sectional area is $R = 6.7437 \times 10^{-7}\frac{L}{A}$.

To find the length of copper wire that will produce a resistance of 33.5 ohms, let $R = 33.5$ and

$A = \pi\left(\frac{0.0201}{2}\right)^2 \approx 3.1731 \times 10^{-4}$ square inches, and

solve for L.

$$R = \left(6.7437 \times 10^{-7}\right)\frac{L}{A}$$

$$33.5 = \left(6.7437 \times 10^{-7}\right)\frac{L}{\left(3.1731 \times 10^{-4}\right)}$$

$$(33.5)\left(3.1731 \times 10^{-4}\right) = \left(6.7437 \times 10^{-7}\right)L$$

$$\frac{(33.5)\left(3.1731 \times 10^{-4}\right)}{\left(6.7437 \times 10^{-7}\right)} = L$$

$15{,}762.7$ inches $\approx L$

So, the length of the wire is approximately 15,762.7 inches or about 1,314 feet.

7. E = kinetic energy, m = mass, and V = velocity.

Because E varies jointly with the object's mass, m and the square of the object's velocity, V you have

$E = kmV^2$

For $E = 6400$ joules, $m = 50$ kg, and $V = 16$m/sec, you have

$$E = kmV^2$$
$$6400 = k(50)(16)^2$$
$$6400 = k(12800)$$
$$\frac{1}{2} = k$$

So, the equation relating kinetic energy, mass, and velocity is $E = \frac{1}{2}mV$.

When $m = 70$ and $V = 20$, the kinetic energy is

$$E = \frac{1}{2}mV^2 = \frac{1}{2}(70)(20)^2 = \frac{1}{2}(70)(400)$$
$$= 14{,}000 \text{ joules.}$$

Chapter 4

Checkpoints for Section 4.1

1. Because the denominator is zero when $x = 1$, the domain of f is all real numbers except $x = 1$.

To determine the behavior of f, this excluded value, evaluate $f(x)$ to the left and to the right of $x = 1$.

x	0	0.5	0.9	0.99	0.999	$\to 1$
$f(x)$	0	−3	−27	−297	−2997	$\to -\infty$

x	$1 \leftarrow$	1.001	1.01	1.1	1.5	2
$f(x)$	$\infty \leftarrow$	3003	303	33	9	6

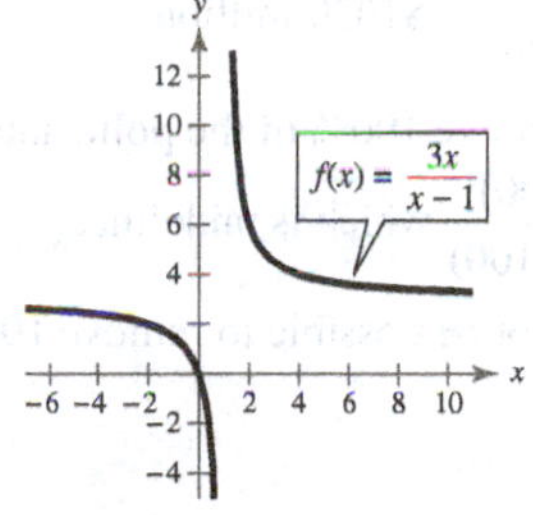

As x approaches 1 from the left, $f(x)$ decreases without bound.

As x approaches 1 from the right, $f(x)$ increases without bound.

2. For this rational function the degree of the numerator is equal to the degree of the denominator. The leading coefficient of the numerator is 5 and the leading coefficient of the denominator is 1, so the graph has the line $y = \frac{5}{1} = 5$ as a horizontal asymptote. To find any vertical asymptotes set the denominator equal to zero and solve the resulting equation for x.

$$x^2 - 1 = 0$$
$$(x + 1)(x - 1) = 0$$
$$x + 1 = 0 \to x = -1$$
$$x - 1 = 0 \to x = 1$$

The equation has two real solutions, $x = 1$ and $x = -1$, so the graph has the lines $x = 1$ and $x = -1$ as vertical asymptotes.

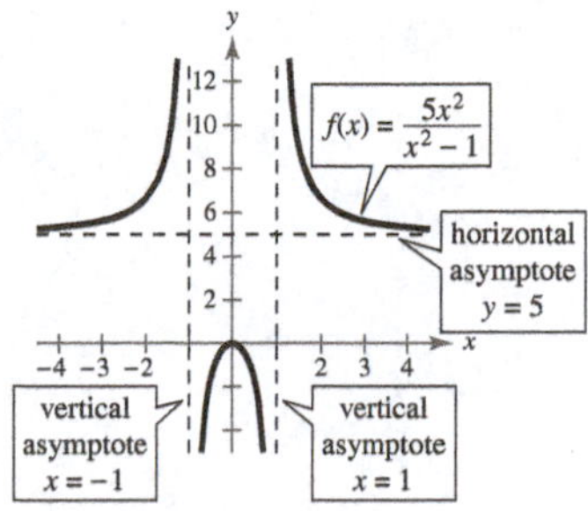

3. For this rational function, the degree of the numerator is equal to the degree of the denominator. The leading coefficient of the numerator is 3 and the leading coefficient of the denominator is 1, so the graph of the function has the line $y = \frac{3}{1} = 3$ as a horizontal asymptote. To find any vertical asymptotes, first factor the numerator and denominator as follows.

$$f(x) = \frac{3x^2 + 7x - 6}{x^2 + 4x + 3} = \frac{(3x - 2)\cancel{(x + 3)}}{(x + 1)\cancel{(x + 3)}}$$

$$= \frac{3x - 2}{x + 1},\ x \neq -3$$

By setting the denominator $x + 1$ (of the simplified function) equal to zero, you can determine that the graph has the line $x = -1$ as a vertical asymptote.

4. (a) The cost to remove 20% of the pollutants is

$$C = \frac{255(20)}{100 - (20)} = \$63.75 \text{ million.}$$

The cost to remove 45% of the pollutants is

$$C = \frac{255(45)}{100 - 45} \approx \$208.64 \text{ million.}$$

The cost to remove 80% of the pollutants is

$$C = \frac{255(80)}{100 - 80} = \$1020 \text{ million.}$$

(b) The cost to remove 100% of the pollutants is

$$C = \frac{255(100)}{100 - (100)} \text{ which is undefined.}$$

So, it would not be possible to remove 100% of the pollutants.

5. (a) When $x = 1000$,

$$\overline{C} = \frac{0.4(1000) + 8000}{1000} = \$8.40 \text{ per unit.}$$

When $x = 8000$,

$$\overline{C} = \frac{0.4(8000) + 8000}{8000} = \$1.40 \text{ per unit.}$$

When $x = 20{,}000$,

$$\overline{C} = \frac{0.4(20{,}000) + 8000}{20{,}000} = \$0.80 \text{ per unit.}$$

When $x = 100{,}000$,

$$\overline{C} = \frac{0.4(100{,}000) + 8000}{100{,}000} = \$0.48 \text{ per unit.}$$

(b) The graph has the line $\overline{C} = \frac{0.4}{1} = 0.4$ as a horizontal asymptote.

So, as x approaches infinity, $\overline{C}$ approaches 0.4. This means that as the number of units increases without bound the average cost per unit approaches \$0.40 per unit.

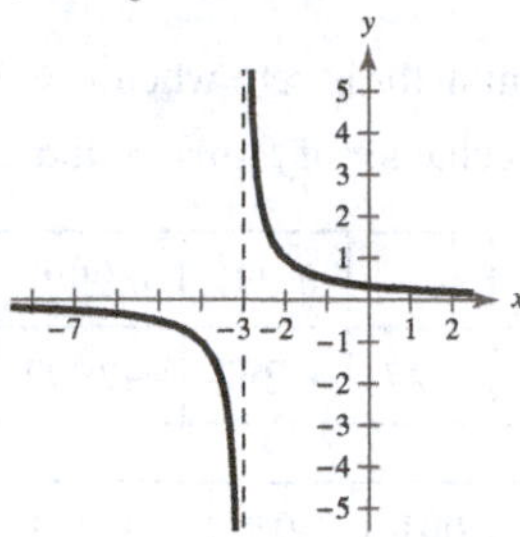

Checkpoints for Section 4.2

1. $f(x) = \dfrac{1}{x + 3}$

y-intercept: $\left(0, \frac{1}{3}\right)$, because $f(0) = \frac{1}{3}$

x-intercept: none, because $1 \neq 0$

Vertical asymptote: $x = -3$, zero of denominator

Horizontal asymptote: $y = 0$, because degree of $N(x) <$ degree of $D(x)$

Additional points:

x	-5	-4	-2	-1	1	2
$f(x)$	$-\frac{1}{2}$	-1	1	$\frac{1}{2}$	$\frac{1}{4}$	$\frac{1}{5}$

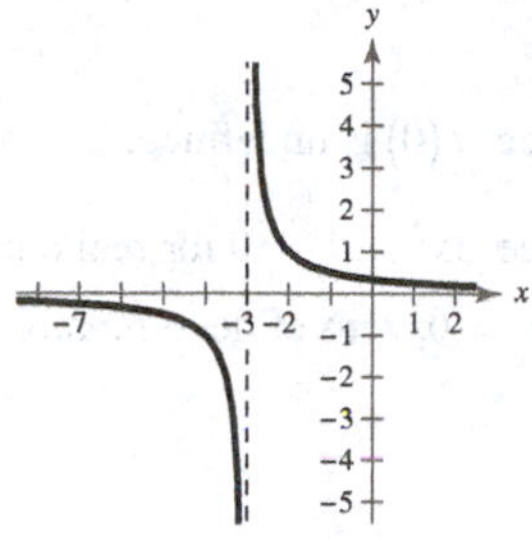

The domain of f is all real numbers except $x = -3$.

2. $g(x) = \dfrac{3 + 2x}{1 + x}$

y-intercept: $(0, 3)$, because $g(0) = 3$

x-intercept: $\left(-\frac{3}{2}, 0\right)$, because $g\left(-\frac{3}{2}\right) = 0$

Vertical asymptote: $x = -1$, zero of denominator

Horizontal asymptote: $y = 2$, because degree of $N(x) =$ degree of $D(x)$

Additional points:

x	-3	-2	1	3
$g(x)$	$\frac{3}{2}$	1	$\frac{5}{2}$	$\frac{9}{4}$

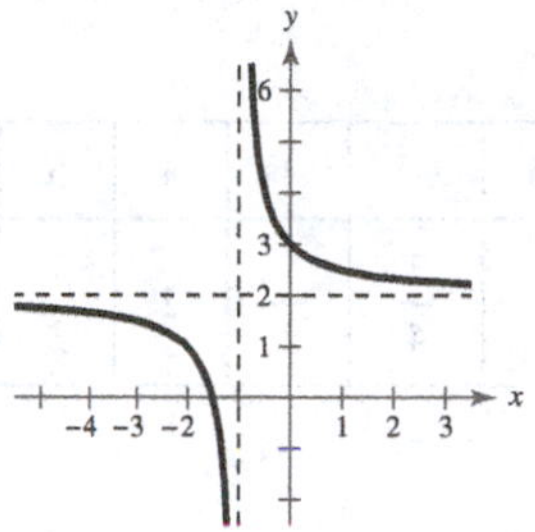

The domain of f is all real number except $x = -1$.

3. $f(x) = \dfrac{3x}{x^2 + x - 2} = \dfrac{3x}{(x + 2)(x - 1)}$

y-intercept: $(0, 0)$, because $f(0) = 0$

x-intercept: $(0, 0)$, because $f(0) = 0$

Vertical asymptotes: $x = -2$, $x = 1$, zeros of denominator

Horizontal asymptote: $y = 0$, because degree of $N(x) <$ degree of $D(x)$

Additional points:

x	-3	-1	2	3
$f(x)$	$-\dfrac{9}{4}$	$\dfrac{3}{2}$	$\dfrac{3}{2}$	$\dfrac{9}{10}$

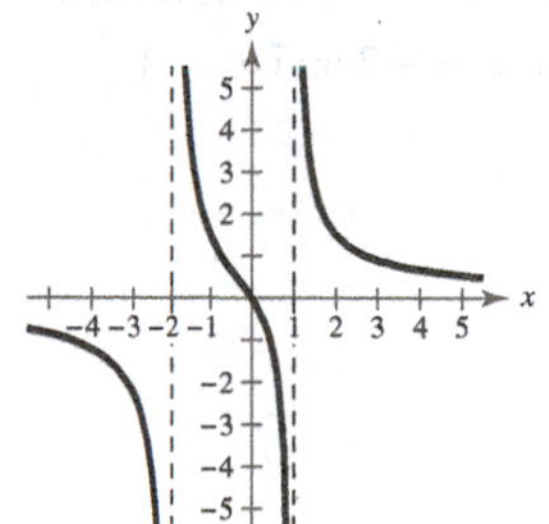

The domain of f is all real numbers except $x = -2$ and $x = 1$.

4. $f(x) = \dfrac{x^2 - 4}{x^2 - x - 6}$

$= \dfrac{(x + 2)(x - 2)}{(x - 3)(x + 2)}$

$= \dfrac{x - 2}{x - 3}, x \neq -2$

y-intercept: $\left(0, \frac{2}{3}\right)$, because $f(0) = \frac{2}{3}$

x-intercept: (2, 0), because $f(2) = 0$

Vertical asymptote: $x = 3$, zero of (simplified) denominator

Horizontal asymptote: $y = 1$, because degree of $N(x) =$ degree of $D(x)$

Additional points:

x	-7	-5	-1	1	4	5
$f(x)$	$\frac{9}{10}$	$\frac{7}{8}$	$\frac{3}{4}$	$\frac{1}{2}$	2	$\frac{3}{2}$

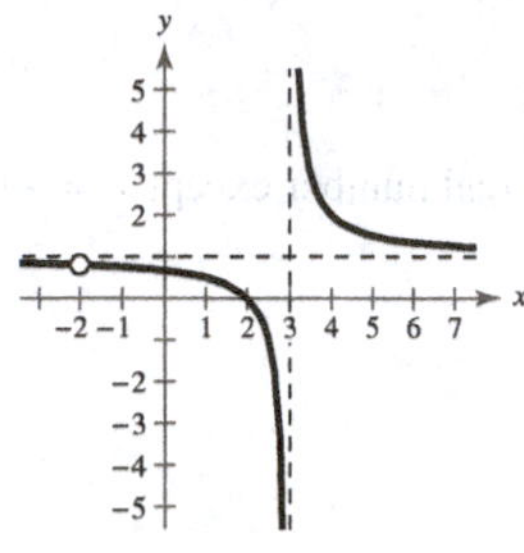

Notice that there is a hole at $x = -2$ because the function is not defined when $x = -2$, the domain is all real number except $x = -2$ and $x = 3$.

5. $f(x) = \dfrac{3x^2 + 1}{x}$

First divide $3x^2 + 1$ by x, either by long division:

$$\begin{array}{r} 3x \\ x\overline{)3x^2 + 0x + 1} \\ \underline{3x^2} \\ 1 \end{array}$$

So $\dfrac{3x^2 + 1}{x} = 3x + \dfrac{1}{x}$

or by separating, the numerator and simplifying:

$\dfrac{3x^2 + 1}{x} = \dfrac{3x^2}{x} + \dfrac{1}{x} = 3x + \dfrac{1}{x}$

So, the start asymptote is $y = 3x$, since $\dfrac{3x^2 + 1}{x} = 3x + \dfrac{1}{x}$.

y-intercept: none, since $f(0)$ is undefined.

x-intercept: none, since $3x^2 + 1 \neq 0$ for real numbers.

Vertical asymptote: $x = 0$, zero of denominator

Start asymptote: $y = 3x$

Additional points:

x	-2	-1	-0.5	0.5	1	2
$f(x)$	$-\frac{13}{2}$	-4	$-\frac{7}{2}$	$\frac{7}{2}$	4	$\frac{13}{2}$

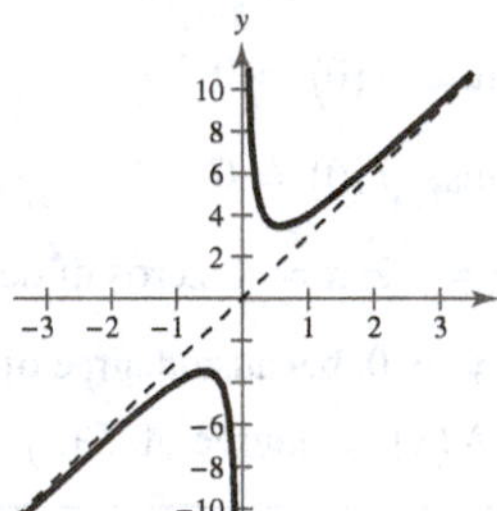

The domain of f is all real numbers except $x = 0$.

6. Graphical Solution

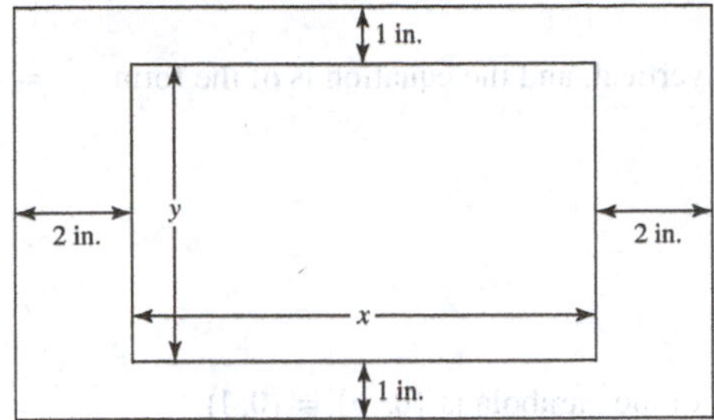

Let A be the area to be minimized.

$A = (x + 4)(y + 2)$

The printed area inside the margins is modeled by

$40 = xy$ or $y = \frac{40}{x}$.

To find the minimum area, rewrite the equation for A in terms of just one variable by substituting $\frac{40}{x}$ for y.

$$\begin{aligned} A &= (x + 4)\left(\frac{40}{x} + 2\right) \\ &= (x + 4)\left(\frac{40 + 2x}{x}\right) \\ &= \frac{(x + 4)(40 + 2x)}{x}, \ x > 0 \end{aligned}$$

The graph of this rational function is shown below. Because x represents the width of the printed area, you need to consider only the portion of the graph for which x is positive. Using a graphing utility, you can approximate the minimum value of A to occur when $x \approx 8.9$ inches. The corresponding value of y is

$\frac{40}{8.9} \approx 4.5$ inches.

So, the dimensions should be $8.9 + 4 = 129$ inches by $4.5 + 2 = 6.5$ inches.

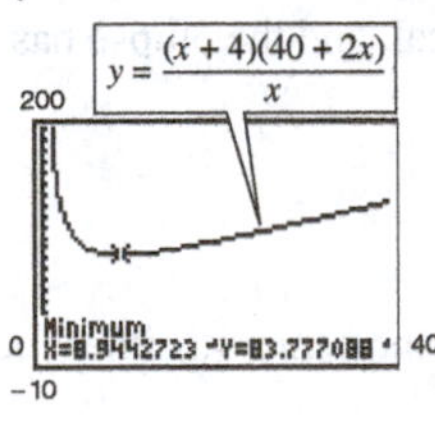

Numerical Solution

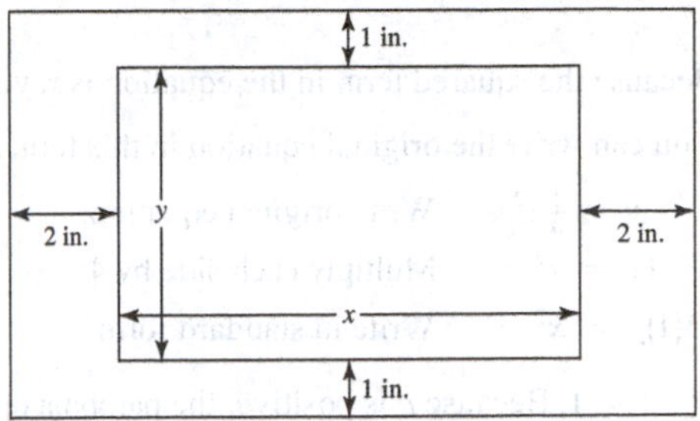

Let A be the area to be minimized.

$A = (x + 4)(y + z)$

The printed area inside the margins is modeled by

$40 = xy$ or $y = \frac{40}{x}$.

To find the minimum area, rewrite the equation for A in terms of just one variable by substituting $\frac{40}{x}$ for y.

$$\begin{aligned} A &= (x + 4)\left(\frac{40}{x} + 2\right) \\ &= (x + 4)\left(\frac{40 + 2x}{x}\right) \\ &= \frac{(x + 4)(40 + 2x)}{x}, \ x > 0 \end{aligned}$$

Use the *table* feature of a graphing utility to create a table of values for the function

$$y_1 = \frac{(x + 4)(40 + 2x)}{x}, \ x > 0$$

beginning at $x = 6$. From the table, you can see that the minimum value of y_1 occurs when x is somewhere between 8 and 9, as shown.

To approximate the minimum value of y_1 to one decimal place, change the table so that it starts at $x = 8$ and increases by 0.1. The minimum value of y_1 occurs when $x \approx 8.9$ as shown.

The corresponding value of y is $\frac{40}{8.9} \approx 4.5$ inches.

So, the dimensions should be $8.9 + 4 = 12.9$ inches by $4.5 + 2 = 6.5$ inches.

x	y_1
6	86.667
7	84.857
8	84
9	83.778
10	84
11	84.545

x	y_1
8.8	83.782
8.9	83.778
9.0	83.778
9.1	83.782

Checkpoints for Section 4.3

1. Because the squared term in the equation is x, you know that the axis is vertical, and the equation is of the form $x^2 = 4py$. You can write the original equation in this form as follows.

$y = \frac{1}{4}x^2$ Write original equation.

$4y = x^2$ Multiply each side by 4.

$4(1)y = x^2$ Write in standard form.

So, $p = 1$. Because p is positive, the parabola opens upward. The focus of the parabola is $(0, p) = (0, 1)$ and the directrix of the parabola is $y = -p = -1$.

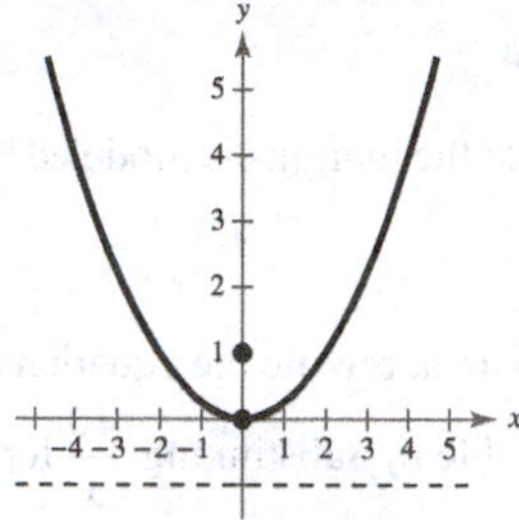

2. The axis of the parabola is vertical, with vertex $(0, 0)$ and focus $\left(0, \frac{3}{8}\right)$. So, the standard form is $x^2 = 4py$.

Because the focus is $p = \frac{3}{8}$ units from the vertex, the equation is as follows.

$$x^2 = 4\left(\frac{3}{8}\right)y$$

$$x^2 = \frac{3}{2}y$$

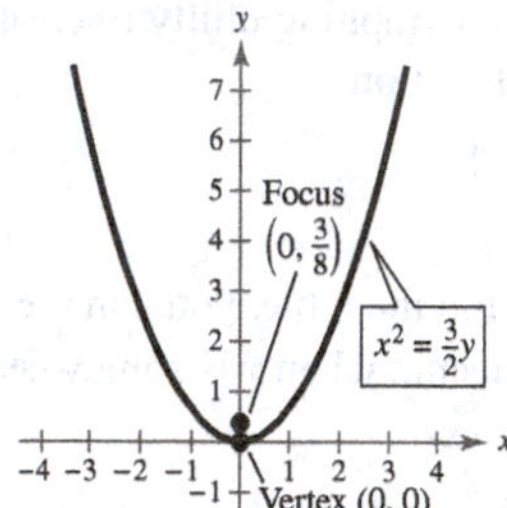

3. The foci occur at $(0, -3)$ and $(0, 3)$. So, the center of the ellipse is $(0, 0)$, the major axis is vertical, and the ellipse has the equation of the form

$$\frac{y^2}{a^2} + \frac{x^2}{b^2} = 1.$$

The length of the major axis is $2a = 10$. This implies that $a = 5$. Moreover, the distance from the center to either focus is $c = 3$.

Finally, $b^2 = a^2 - c^2 = (5)^2 - (3)^2 = 25 - 9 = 16$.

Substituting $a^2 = 3^2$ and $b^2 = 4^2$ yields the following equation in standard form.

$$\frac{y^2}{5^2} + \frac{x^2}{4^2} = 1$$

$$\frac{y^2}{25} + \frac{x^2}{16} = 1$$

4. Algebraic Solution

$x^2 + 9y^2 = 81$ Write original equation.

$\frac{x^2}{81} + \frac{9y^2}{81} = \frac{81}{81}$ Divide each side by 81.

$\frac{x^2}{81} + \frac{y^2}{9} = 1$ Simplify.

$\frac{x^2}{9^2} + \frac{y^2}{3^2} = 1$ Write in standard form.

Because the denominator of the x^2-term is larger than the denominator of the y^2-term, you can conclude that the major axis is horizontal. Moreover, because

$a^2 = 9^2$

the endpoints of the major axis lie nine units to the left and to the right of center $(0, 0)$. So, the vertices of the ellipse are $(9, 0)$ and $(-9, 0)$.

Similarly, because the denominator of the y^2-term is $b^2 = 3^2$ the endpoints of the minor axis (or co-vertices) lie three units up and down from the center $(0, 0)$.

So, the endpoints are $(0, 3)$ and $(0, -3)$

The ellipse is shown below.

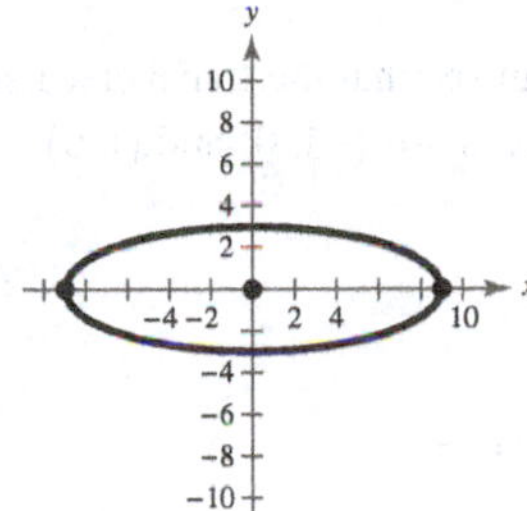

Graphical Solution

Solve the equation of the ellipse for y.

$$x^2 + 9y^2 = 81$$

$$9y^2 = 81 - x^2$$

$$y^2 = \frac{81 - x^2}{9}$$

$$y = \pm\sqrt{\frac{81 - x^2}{9}}$$

$$y = \pm\frac{\sqrt{81 - x^2}}{\sqrt{9}}$$

$$y = \pm\frac{1}{3}\sqrt{81 - x^2}$$

Then graph both equations,

$y_1 = \frac{1}{3}\sqrt{81 - x^2}$ and $y_2 = \frac{1}{3}\sqrt{81 - x^2}$

in the same viewing window. Be sure to use a square setting.

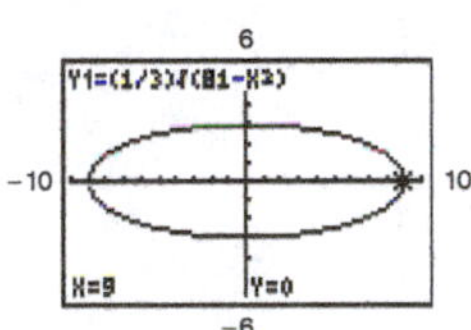

The center of the ellipse is $(0, 0)$ and the major axis is horizontal. The vertices are $(9, 0)$ and $(-9, 0)$.

5. You can determine that $c = 6$, because the foci are at $(0, \pm 6)$. Also, $a = 3$, because the vertices are at $(0, \pm 3)$.

$$\begin{aligned} b^2 &= c^2 - a^2 \\ &= 6^2 - 3^2 \\ &= 36 - 9 \\ &= 27 \end{aligned}$$

Because the transverse axis is vertical, the standard form of the equation is

$$\frac{y^2}{a^2} + \frac{x^2}{b^2} = 1.$$

Substitute $a^2 = 3^2$ and $b^2 = 27 = \left(3\sqrt{3}\right)^2$ to obtain the equation as follows.

$\frac{y^2}{3^2} - \frac{x^2}{\left(3\sqrt{3}\right)^2} = 1$ Write in standard form.

$\frac{y^2}{9} - \frac{x^2}{27} = 1$ Simplify.

6. Algebraic Solution

$2x^2 - \frac{y^2}{2} = 2$ Write original equation.

$\frac{x^2}{1} - \frac{y^2}{4} = 1$ Simplify

$\frac{x^2}{1^2} - \frac{y^2}{2^2} = 1$ Write in standard form.

Because the x^2-term is positive, you can conclude that the transverse axis is horizontal. Also since the center is $(0, 0)$ and $a = 1$, the vertices occur at $(1, 0)$ and $(-1, 0)$. Because $b = 2$, the endpoints of the conjugate axis occur at $(0, 2)$ and $(0, -2)$, and you can sketch the rectangle $2a$ wide by $2b$ tall or 2 units by 4 units, as shown. Finally, by drawing the asymptotes through the corners of this rectangle, you can complete the sketch, as shown. Note that the asymptotes are $y = k \pm \frac{b}{a}(x - h)$, or $y = 2x$ and $y = -2x$.

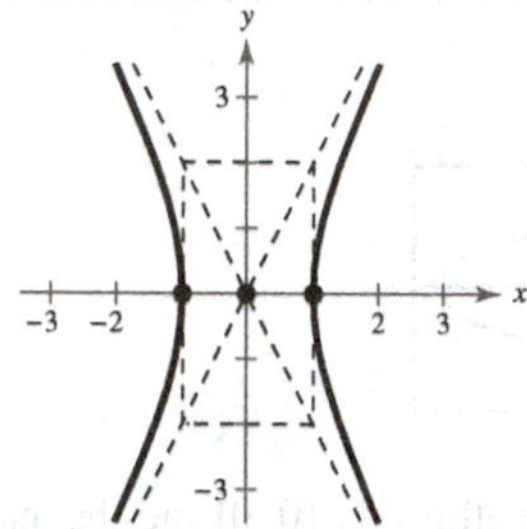

Graphical Solution

Solve the equation of the hyperbola for y as follows.

$$2x^2 - \frac{y^2}{2} = 2$$

$$x^2 - \frac{y^2}{4} = 1$$

$$4x^2 - y^2 = 4$$

$$4x^2 - 4 = y^2$$

$$\pm\sqrt{4x^2 - 4} = y$$

$$y = \pm\sqrt{4x^2 - 4}$$

or

$$y = \pm 2\sqrt{x^2 - 1}$$

Then use a graphing utility to graph $y_1 = \sqrt{4x^2 - 4}$ and $y_2 = -\sqrt{4x^2 - 4}$ in the same viewing window. Be sure to use a square setting.

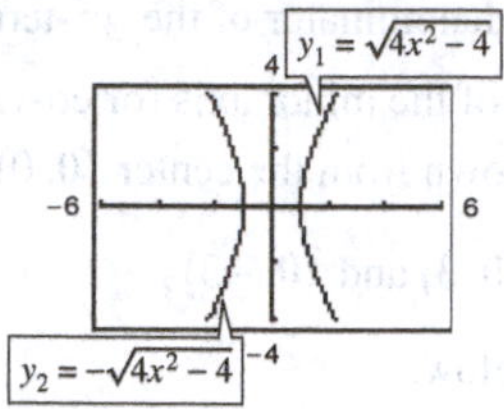

From the graph, you can see that the transverse axis is horizontal and the vertices are $(-1, 0)$ and $(1, 0)$.

7. Because the transverse axis is horizontal, the asymptotes are of the forms $y = \frac{b}{a}x$ and $y = -\frac{b}{a}x$.

Using the fact that the asymptotes are $y = 4x$ and $y = -4x$, you can determine that $\frac{b}{a} = 4$.

Because $a = 1$, you can determine that $b = 4$. Finally, you can conclude that the hyperbola has the following equation.

$\frac{x^2}{1^2} - \frac{y^2}{4^2} = 1$ Write in standard form.

$x^2 - \frac{y^2}{16} = 1$ Simplify.

Checkpoints for Section 4.4

1. (a) The graph of $\frac{(x+1)^2}{3^2} + \frac{(y-2)^2}{5^2} = 1$ is an *ellipse* whose center is the point $(-1, 2)$. The major axis of the ellipse is vertical and of length $2(5) = 10$, and the minor axis of ellipse is horizontal and of length $2(3) = 6$. The graph has been shifted one unit to the left and two units up from standard position.

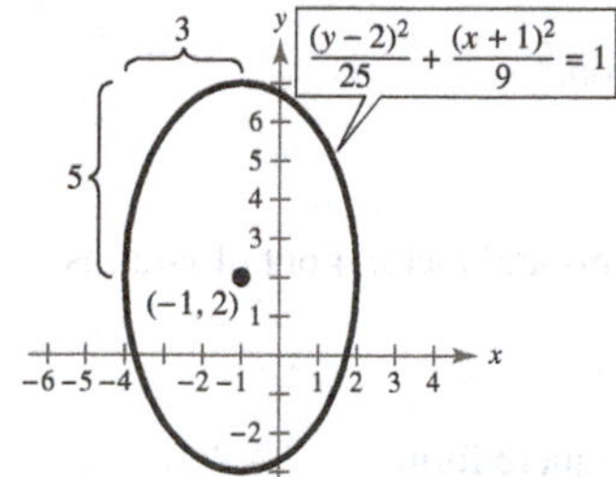

(b) The graph of $(x+1)^2 + (y-1)^2 = 2^2$ is a *circle* whose center is the point $(-1, 1)$ and whose radius is 2. The graph of the circle has been shifted one unit to the left and one unit up from the standard position.

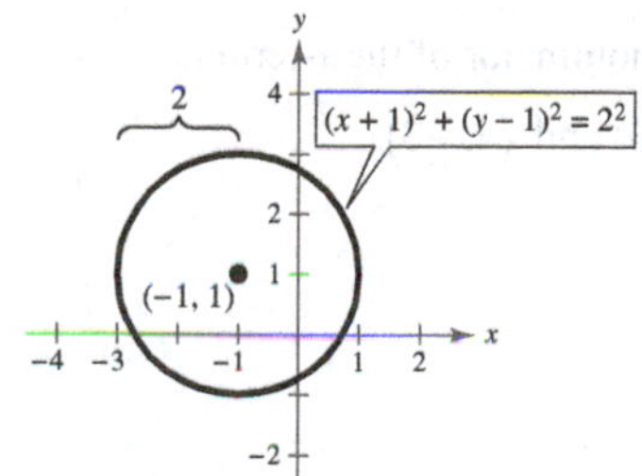

(c) The graph of $(x+4)^2 = 4(2)(y-3)$ is a *parabola* whose vertex is the point $(-4, 3)$. The axis of the parabola is vertical. Because $p = 2$, the focus lies two units above the vertex as $(-4, 5)$ and the directrix is the horizontal line $y = k - p = 3 - 2 = 1$. The graph of the parabola has been shifted four units to the left and three units up from standard position.

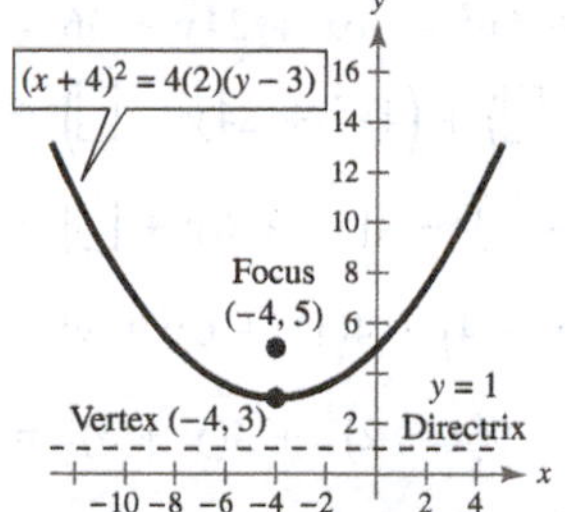

(d) The graph of $\frac{(x-3)^2}{2^2} - \frac{(y-1)^2}{1^2} = 1$ is a *hyperbola* whose center is the point $(3, 1)$. The transverse axis is horizontal and of length $2(2) = 4$, and the conjugate axis is vertical and of length $2(1) = 2$. The graph has been shifted three units to the right and one unit up from standard position.

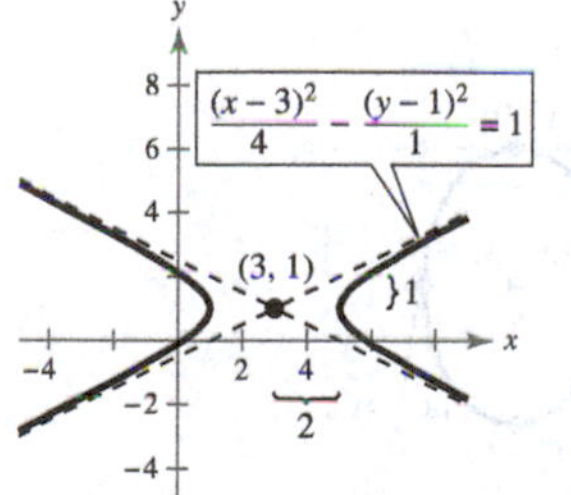

2. Complete the square to right the equation in standard form.

$y^2 - 6y + 4x + 17 = 0$	Write original equation.
$y^2 - 6y = -4x - 17$	Collect y terms on one side of the equaton.
$y^2 - 6y + 9 = -4x - 17 + 9$	Complete the square.
$(y-3)^2 = -4x - 8$	Write in completed square form.
$(y-3)^2 = 4(-1)(x+2)$	Write in standard form.

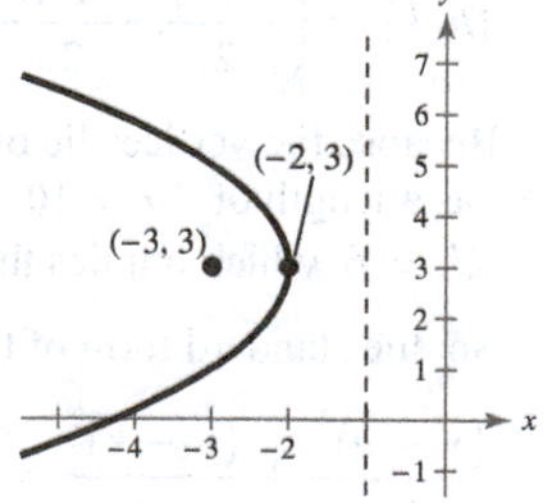

In this standard form, it follows that $h = -2$, $k = 3$, and $p = -1$.

Because the axis is horizontal and p is negative, the parabola opens to the left.

The vertex is $(h, k) = (-2, 3)$, the focus is $(h + p, k) = (-3, 3)$, and the directrix is $x = h - p = -1$.

3. Because the directrix is $y = 0$, which is a horizontal line, the axis is vertical.

The standard form is $(x - h)^2 = 4p(y - k)$.

Because the vertex lies one unit above the directrix, it follows that $p = 1$.

So, the standard form of the parabola is as follows.

$$(x - h)^2 = 4p(y - k)$$

$$\left[x - (-1)\right]^2 = 4(1)(y - 1)$$

$$(x + 1)^2 = 4(y - 1)$$

4. Complete the square to write the equation in standard form.

$9x^2 + 4y^2 - 36x + 24y + 36 = 0$	Write original equation.
$(9x^2 - 36x + \square) + (4y^2 + 24y + \square) = -36$	Group terms.
$9(x^2 - 4x + \square) + 4(y^2 + 6y + \square) = -36$	Factor 9 out of x-terms and factor 4 out of y-terms.
$9(x^2 - 4x + 4) + 4(y^2 + 6y + 9) = -36 + 9(4) + 4(9)$	Complete the square.
$9(x - 2)^2 + 4(y + 3)^2 = 36$	Write in completed square form.
$\dfrac{(x - 2)^2}{4} + \dfrac{(y + 3)^2}{9} = 1$	Divide each side by 36.
$\dfrac{(x - 2)^2}{2^2} + \dfrac{(y + 3)^2}{3^2} = 1$	Write in standard form.

So, the center is $(h, k) = (2, -3)$. Because the denominator of the y-term is $a^2 = 3^2$, the endpoints of the major axis lie three units above and below the center at $(2, 0)$ and $(2, -6)$. Because the denominator of the x-term is $b^2 = 2^2$, the endpoints of the minor axis lie two units left and right of the center at $(0, -3)$ and $(4, -3)$.

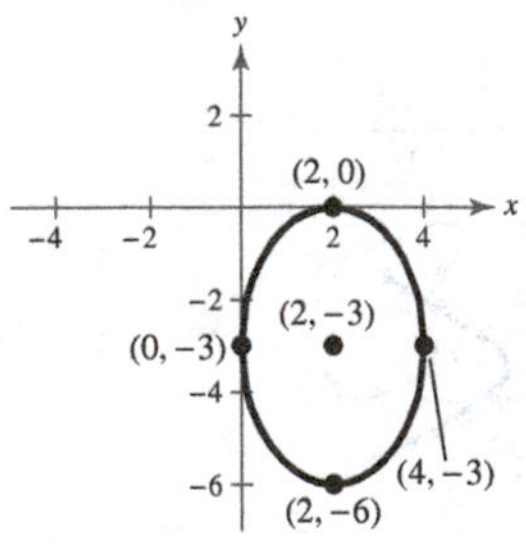

5. The center of the ellipse lies at the midpoint of its vertices. So, the center is

$$(h, k) = \left(\frac{3 + 3}{2}, \frac{0 + 10}{2}\right) = (3, 5).$$

Because the vertices lie on a vertical line and are 10 units apart, it follows that the major axis is vertical and has a length of $2a = 10$. So, $a = 5$. Moreover, because the minor axis has a length of 6 it follows that $2b = 6$, which implies that $b = 3$.

So, the standard form of the ellipse is as follows.

$\dfrac{(x - h)^2}{b^2} + \dfrac{(y - k)^2}{a^2} = 1$	Major axis is vertical.
$\dfrac{(x - 3)^2}{3^2} + \dfrac{(y - 5)^2}{5^2} = 1$	Write in standard form.
$\dfrac{(x - 3)^2}{9} + \dfrac{(y - 5)^2}{25} = 1$	Simplify.

6. Complete the square to write the equation in standard form.

$9x^2 - y^2 - 18x - 6y - 9 = 0$	Write original equation.
$(9x^2 - 18x + \square) - (y^2 - 6y + \square) = 9$	Group terms.
$9(x^2 - 2x + \square) - (y^2 + 6y + \square) = 9$	Factor 9 out of x-terms.
$9(x^2 - 2x + 1) - (y^2 + 6y + 9) = 9 + 9(1) - 9$	Complete the squares.
$9(x - 1)^2 - (y + 3)^2 = 9$	Write in completed square form.
$(x - 1)^2 - \dfrac{(y + 3)^2}{9} = 1$	Divide each side by 9.
$\dfrac{(x - 1)^2}{1^2} - \dfrac{(y + 3)^2}{3^2} = 1$	Write in standard form.

From this standard form, you can see that the transverse axis is horizontal and the center lies at $(h, k) = (1, -3)$.

Because the denominator of the x-term is $a^2 = 1$, you know that the vertices occur one unit to the left and right of the center at $(0, -3)$ and $(2, -3)$.

To sketch the hyperbola, draw a rectangle whose sides pass through the vertices. Because the denominator of the y-term is $b^2 = 3^2$, locate the top and bottom of the rectangle 3 units up and down from the center. Finally, sketch the asymptotes by drawing lines through the opposite corners of the rectangle. Using these asymptotes, you can complete the graph of the hyperbola.

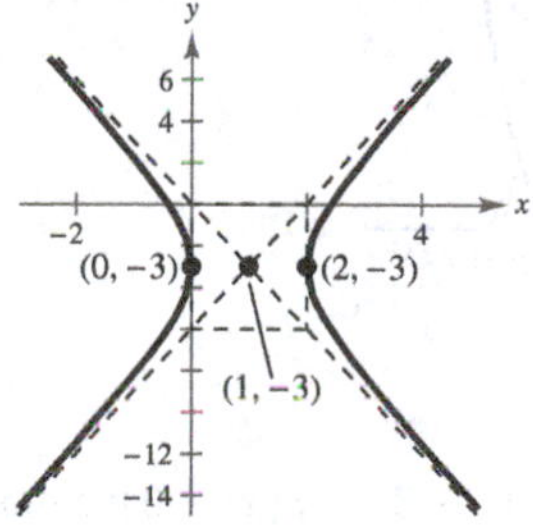

7. The center of the hyperbola lies at the midpoint of its vertices. So, the center is

$$(h, k) = \left(\frac{3 + 5}{2}, \frac{-1 + (-1)}{2}\right)$$
$$= (4, -1).$$

Because the vertices lie on a horizontal line and are two units apart, it follows that the transverse axis is horizontal and has a length of $2a = 2$. So, $a = 1$. Because the foci are three units from center, it follows that $c = 3$. So, $b^2 = c^2 - a^2 = 3^2 - 1^2 = 9 - 1 = 8$.

Because the transverse axis is horizontal, the standard form of the equation is

$\dfrac{(x - h)^2}{a^2} - \dfrac{(y - k)^2}{b^2} = 1$	Standard form, horizontal transverse axis.
$\dfrac{(x - 4)^2}{1^2} - \dfrac{(y - (-1))^2}{(2\sqrt{2})^2} = 1$	Write in standard form.
$(x - 4)^2 - \dfrac{(y + 1)^2}{8} = 1$	Simplify.

Chapter 5

Checkpoints for Section 5.1

1. Function Value

$f(\sqrt{2}) = 8^{-\sqrt{2}}$

Graphing Calculator Keystrokes

8 [∧] [(] [(−)] [√] 2 [)] [Enter]

Display

0.052824803759

2. The table lists some values for each function, and the graph shows a sketch of the two functions. Note that both graphs are increasing and the graph of $g(x) = 9^x$ is increasing more rapidly than the graph of $f(x) = 3^x$.

x	-3	-2	-1	0	1	2
3^x	$\frac{1}{27}$	$\frac{1}{9}$	$\frac{1}{3}$	1	3	9
9^x	$\frac{1}{729}$	$\frac{1}{81}$	$\frac{1}{9}$	1	9	81

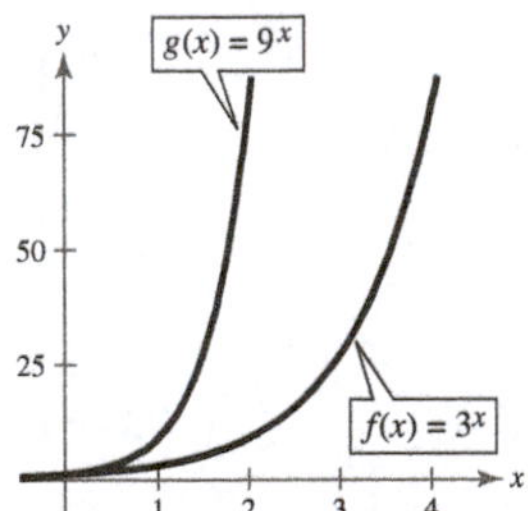

3. The table lists some values for each function and, the graph shows a sketch for each function. Note that both graphs are decreasing and the graph of $g(x) = 9^{-x}$ is decreasing more rapidly than the graph of $f(x) = 3^{-x}$

x	-2	-1	0	1	2	3
9^{-x}	64	8	1	$\frac{1}{8}$	$\frac{1}{64}$	$\frac{1}{512}$
3^{-x} $g(x)$	9	3	1	$\frac{1}{3}$	$\frac{1}{9}$	$\frac{1}{27}$

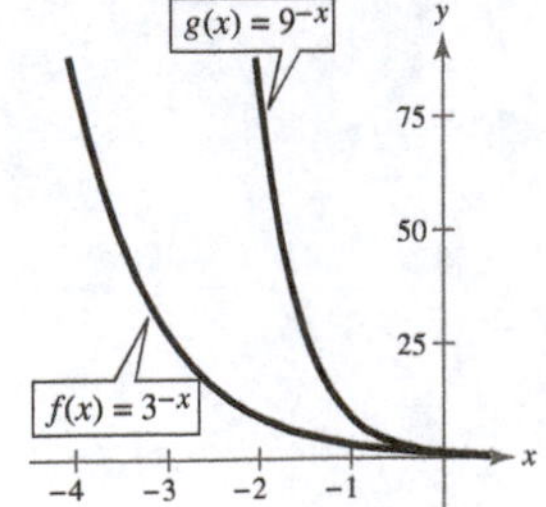

4. (a)

$8 = 2^{2x-1}$	Write Original equation.
$2^3 = 2^{2x-1}$	$8 = 2^3$
$3 = 2x - 1$	One-to-One Property
$4 = 2x$	
$2 = x$	Solve for x.

(b)

$\left(\frac{1}{3}\right)^{-x} = 27$	Write Original equation.
$3^x = 27$	$\left(\frac{1}{3}\right)^{-x} = 3^x$
$3^x = 3^3$	$27 = 3^3$
$x = 3$	One-to-One Property

5. (a) Because $g(x) = 4^{x-2} = f(x-2)$, the graph of g can be obtained by shifting the graph of f two units to the right.

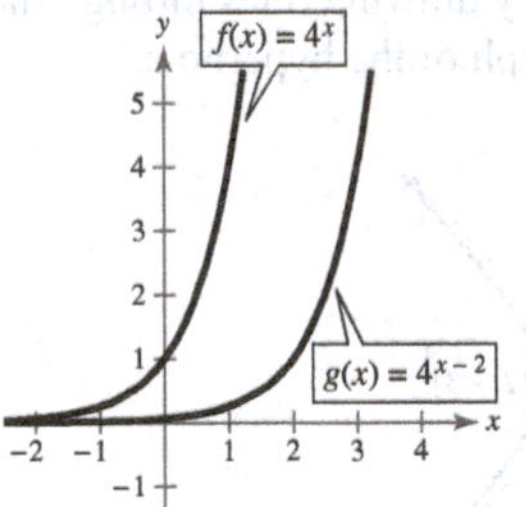

(b) Because $h(x) = 4^x + 3 = f(x) + 3$ the graph of h can be obtained by shifting the graph of f up three units.

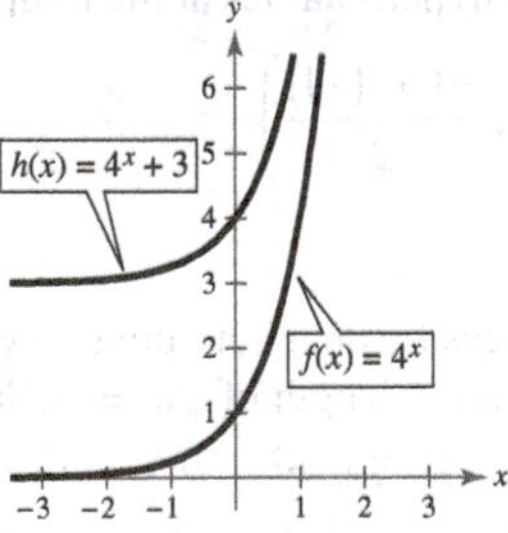

(c) Because $k(x) = 4^{-x} - 3 = f(-x) - 3$, the graph of k can be obtained by reflecting the graph of f in the y-axis and shifting the graph of f down three units.

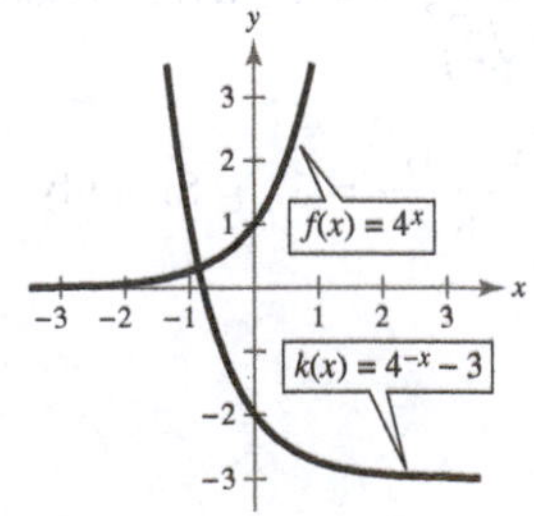

6.

Function Value	Graphing Calculator Keystrokes	Display
(a) $f(0.3) = e^{0.3}$	[e^x] 0.3 [Enter]	1.3498588
(b) $f(-1.2) = e^{-1.2}$	[e^x] [(−)] 1.2 [Enter]	0.3011942
(c) $f(6, 2) = e^{6.2}$	[e^x] 6.2 [Enter]	492.7490411

7. To sketch the graph of $f(x) = 5e^{0.17x}$, use a graphing utility to construct a table of values. After constructing the table, plot the points and draw a smooth curve.

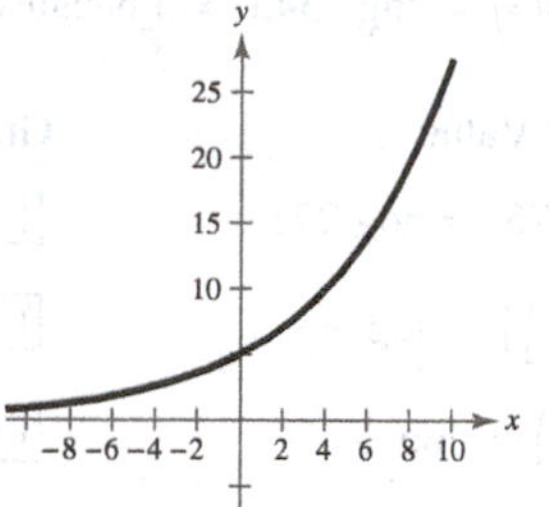

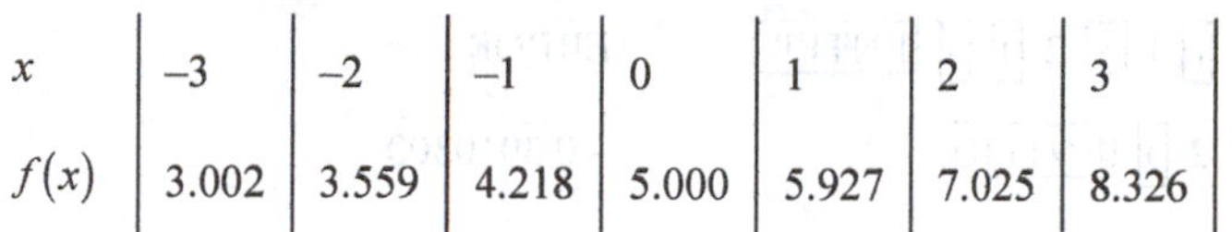

x	−3	−2	−1	0	1	2	3
$f(x)$	3.002	3.559	4.218	5.000	5.927	7.025	8.326

8. (a) For quarterly compounding, you have $n = 4$. So, in 7 years at 4%, the balance is as follows.

$A = P\left(1 + \frac{r}{n}\right)^{nt}$ Formula for compound interest.

$= 6000\left(1 + \frac{0.04}{4}\right)^{4(7)}$ Substitute P, r, n, and t.

$\approx \$7927.75$ Use a calculator.

(b) For monthly compounding, you have $n = 12$. So in 7 years at 4%, the balance is as follows.

$A = P\left(1 + \frac{r}{n}\right)^{nt}$ Formula for compound interest.

$= 6000\left(1 + \frac{0.04}{12}\right)^{12(7)}$ Substitute P, r, n, and t.

$\approx \$7935.08$ Use a calculator.

(c) For continuous compounding, the balance is as follows.

$A = Pe^{rt}$ Formula for continuous compounding.

$= 6000e^{0.04(7)}$ Substitute P, r, and t.

$\approx \$7938.78$ Use a calculator.

9. Use the model for the amount of Plutonium that remains from an initial amount of 10 pounds after t years, where $t = 0$ represents the year 1986.

$P = 10\left(\frac{1}{2}\right)^{t/24,100}$

To find the amount that remains in the year 2089, let $t = 103$.

$P = 10\left(\frac{1}{2}\right)^{t/24,100}$ Write original model.

$P = 10\left(\frac{1}{2}\right)^{103/24,100}$ Substitute 103 for t.

$P \approx 9.970$ Use a calculator.

In the year 2089, 9.970 pounds of plutonium will remain.

To find the amount that remains after 125,000 years, let $t = 125,000$.

$P = 10\left(\frac{1}{2}\right)^{t/24,100}$ Write original model.

$P = 10\left(\frac{1}{2}\right)^{125,000/24,100}$ Substitute 125,000 for t.

$P \approx 0.275$ Use a calculator.

After 125,000 years 0.275 pound of plutonium will remain.

Checkpoints for Section 5.2

1. (a) $f(1) = \log_6 1 = 0$ because $6^0 = 1$.

(b) $f\left(\frac{1}{125}\right) = \log_5 \frac{1}{125} = -3$ because $5^{-3} = \frac{1}{125}$.

(c) $f(343) = \log_7 343 = 3$ because $7^3 = 343$.

2.

Function Value	Graphing Calculator Keystrokes	Display
(a) $f(275) = \log 275$	[LOG] 275 [ENTER]	2.4393327
(b) $f\left(-\frac{1}{2}\right) = \log -\frac{1}{2}$	[LOG] [(] [(−)] [(] 1 [÷] 2 [)] [)] [ENTER]	ERROR
(c) $f\left(\frac{1}{2}\right) = \log \frac{1}{2}$	[LOG] [(] 1 [÷] 2 [)] [ENTER]	−0.3010300

3. (a) Using Property 2, $\log_9 9 = 1$.

(b) Using Property 3, $20^{\log_{20} 3} = 3$.

(c) Using Property 1, $\log_{\sqrt{3}} 1 = 0$.

4.
$$\log_5 (x^2 + 3) = \log_5 12$$
$$x^2 + 3 = 12$$
$$x^2 = 9$$
$$x = \pm 3$$

5. (a) For $f(x) = 8^x$, construct a table of values. Then plot the points and draw a smooth curve.

x	-2	-1	0	1	2
$f(x) = 8^x$	$\frac{1}{64}$	$\frac{1}{8}$	1	8	64

(b) Because $g(x) = \log_8 x$ is the inverse function of $f(x) = 8^x$, the graph of g is obtained by plotting the points $(f(x), x)$ and connecting them with a smooth curve. The graph of g is a reflection of the graph of f in the line $y = x$.

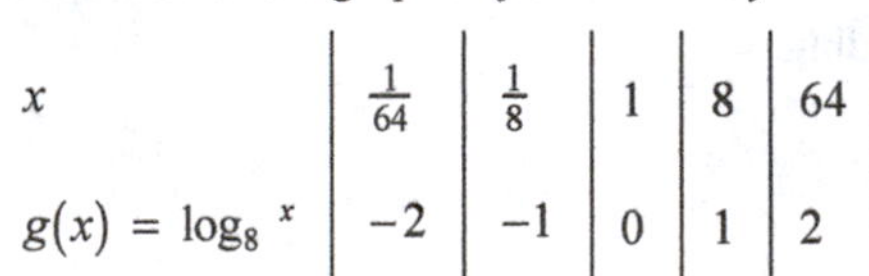

x	$\frac{1}{64}$	$\frac{1}{8}$	1	8	64
$g(x) = \log_8 x$	-2	-1	0	1	2

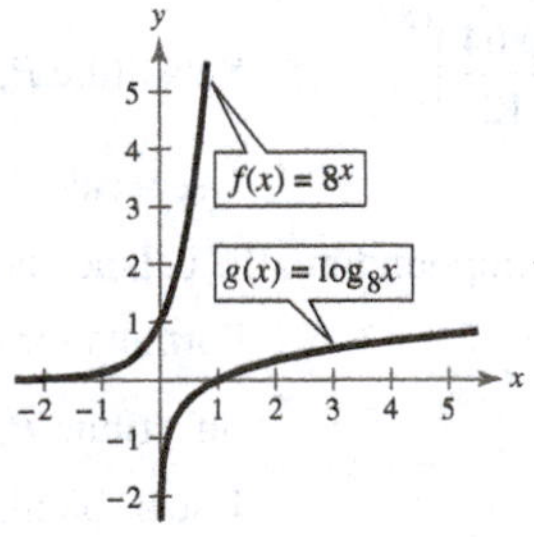

6. Begin by constructing a table of values. Note that some of the values can be obtained without a calculator by using the properties of logarithms. Then plot the points and draw a smooth curve.

x	$\frac{1}{9}$	$\frac{1}{3}$	1	3	9
$f(x) = \log_3 x$	-2	-1	0	1	2

The vertical asymptote is $x = 0$, the y-axis.

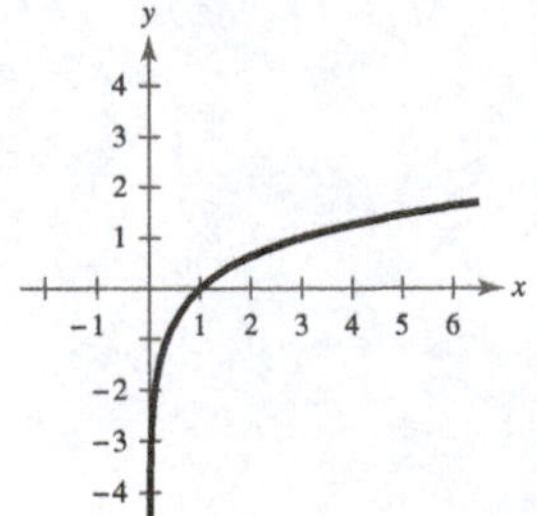

7. (a) Because $g(x) = -1 + \log_3 x = f(x) - 1$, the graph of g can be obtained by shifting the graph of f one unit down.

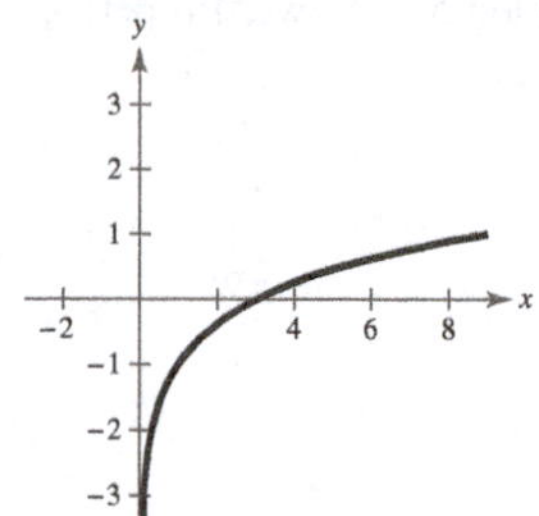

(b) Because $h(x) = \log_3 (x + 3) = f(x + 3)$, the graph of h can be obtained by shifting the graph of f three units to the left.

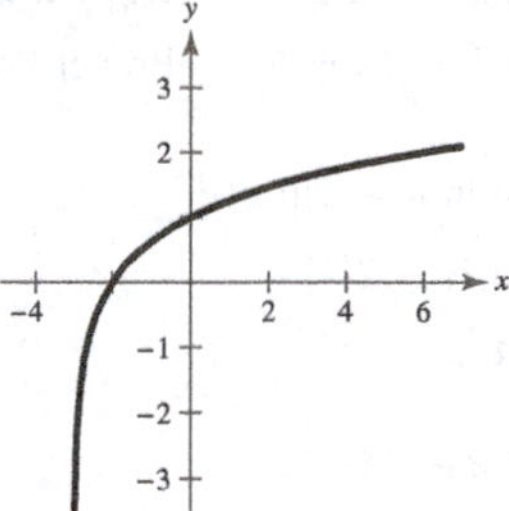

8.

Function Value	Graphing Calculator Keystrokes	Display
$f(0.01) = \ln 0.01$	LN 0.01 ENTER	−4.6051702
$f(4) = \ln 4$	LN 4 ENTER	1.3862944
$f(\sqrt{3} + 2) = \ln(\sqrt{3} + 2)$	LN ((√ 3) + 2) ENTER	1.3169579
$f(\sqrt{3} - 2) = \ln(\sqrt{3} - 2)$	LN ((√ 3) − 2) ENTER	ERROR

9. (a) $\ln e^{1/3} = \frac{1}{3}$ Inverse Property

(b) $5 \ln 1 = 5(0) = 0$ Property 1

(c) $\frac{3}{4} \ln e = \frac{3}{4}(1) = \frac{3}{4}$ Property 2

(d) $e^{\ln 7} = 7$ Inverse Property

10. Because $\ln(x + 3)$ is defined only when $x + 3 > 0$, it follows that the domain of f is $(-3, \infty)$. The graph of f is shown.

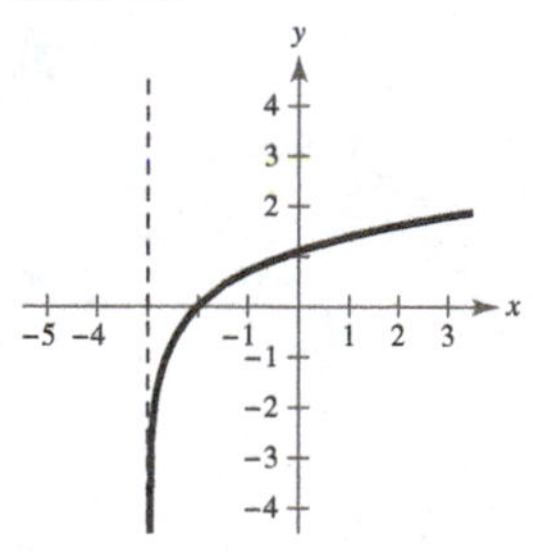

11. (a) After 1 month, the average score was the following.

$f(1) = 75 - 6 \ln(1 + 1)$ Substitute 1 for t.

$= 75 - 6 \ln 2$ Simplify.

$\approx 75 - 6(0.6931)$ Use a calculator.

≈ 70.84 Solution

(b) After 9 months, the average score was the following.

$f(9) = 75 - 6 \ln(9 + 1)$ Substitute 9 for t.

$= 75 - 6 \ln 10$ Simplify.

$\approx 75 - 6(2.3026)$ Use a calculator.

≈ 61.18 Solution

(c) After 12 months, the average score was the following.

$f(12) = 75 - 6 \ln(12 + 1)$ Substitute 9 for t.

$= 75 - 6 \ln 13$ Simplify.

$\approx 75 - 6(2.5649)$ Use a calculator.

≈ 59.61 Solution

Checkpoints for Section 5.3

1. $\log_2 12 = \dfrac{\log 12}{\log 2}$ $\log_a x = \dfrac{\log x}{\log a}$

$\approx \dfrac{1.07918}{0.30103}$ Use a calculator.

≈ 3.5850 Simplify.

2. $\log_2 12 = \dfrac{\ln 12}{\ln 2}$ $\log_a x = \dfrac{\ln x}{\ln a}$

$\approx \dfrac{2.48491}{0.69315}$ Use a calculator.

≈ 3.5850 Simplify.

3. (a) $\log 75 = \log(3 \cdot 25)$ Rewrite 75 as $3 \cdot 25$.

$= \log 3 + \log 25$ Product Property

$= \log 3 + \log 5^2$ Rewrite 25 as 5^2.

$= \log 3 + 2\log 5$ Power Property

(b) $\log \frac{9}{125} = \log 9 - \log 125$ Quotient Property

$= \log 3^2 - \log 5^3$ Rewrite 9 as 3^2 and 125 as 5^2.

$= 2\log 3 - 3\log 5$ Power Property

4. $\ln e^6 - \ln e^2 = 6\ln e - 2\ln e$

$= 6(1) - 2(1)$

$= 4$

5. $\log_3 \dfrac{4x^2}{\sqrt{y}} = \log_3 \dfrac{4x^2}{y^{1/2}}$ Rewrite using rational exponent.

$= \log_3 4x^2 - \log_3 y^{1/2}$ Quotient Property

$= \log_3 4 + \log_3 x^2 - \log_3 y^{1/2}$ Product Property

$= \log_3 4 + 2\log_3 x - \dfrac{1}{2}\log_3 y$ Power Property

6. $2\left[\log(x+3) - 2\log(x-2)\right] = 2\left[\log(x+3) - \log(x-2)^2\right]$ Power Property

$= 2\left[\log\left(\dfrac{x+3}{(x-2)^2}\right)\right]$ Quotient Property

$= \log\left(\dfrac{x+3}{(x-2)^2}\right)^2$ Power Property

$= \log\dfrac{(x+3)^2}{(x-2)^4}$ Simplify.

7. To solve this problem, take the natural logarithm of each of the x- and y-values of the ordered pairs.

$(\ln x, \ln y)$: $(-0.994, -0.673)$, $(0.000, 0.000)$, $(1.001, 0.668)$, $(2.000, 1.332)$, $(3.000, 2.000)$

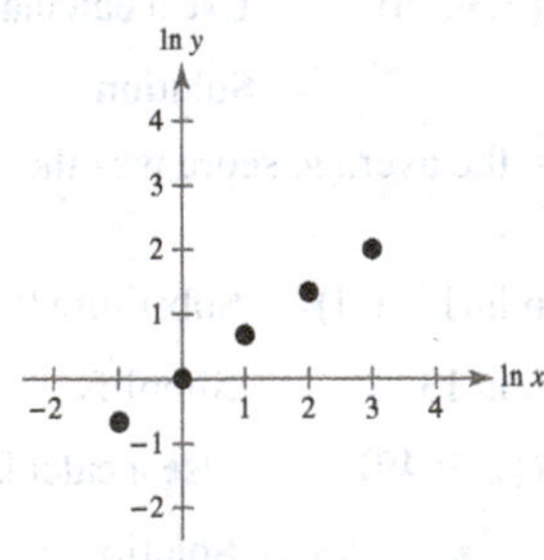

By plotting the ordered pairs, you can see that all five points appear to lie in a line. Choose any two points to determine the slope of the line. Using the points $(0, 0)$ and $(1.001, 0.668)$, the slope of the line is

$m = \dfrac{0.668 - 0}{1 - 0} = 0.668 \approx \dfrac{2}{3}$.

By the point-slope form, the equation of the line is $y = \dfrac{2}{3}x$, where $y = \ln y$ and $x = \ln x$. So, the logarithmic equation is $\ln y = \dfrac{2}{3}\ln x$.

Checkpoints for Section 5.4

1.

Original Equation	Rewritten Equation	Solution	Property
(a) $2^x = 512$	$2^x = x^9$	$x = 9$	One-to-One
(b) $\log_6 x = 3$	$6^{\log_6 x} = 6^3$	$x = 216$	Inverse
(c) $5 - e^x = 0$ $5 = e^x$	$\ln 5 = \ln e^x$	$\ln 5 = x$	Inverse
(d) $9^x = \frac{1}{3}$	$3^{2x} = 3^{-1}$	$2x = -1$ $x = -\frac{1}{2}$	One-to-One

2. (a)

$e^{2x} = e^{x^2-8}$ Write original equation.

$2x = x^2 - 8$ One-to-One Property

$0 = x^2 - 2x - 8$ Write in general form.

$0 = (x - 4)(x + 2)$ Factor.

$x - 4 = 0 \Rightarrow x = 4$ Set 1st factor equal to 0.

$x + 2 = 0 \Rightarrow x = -2$ Set 2nd factor equal to 0.

The solutions are $x = 4$ and $x = -2$

Check $x = -2$:

$$e^{2x} = e^{x^2-8}$$
$$e^{2(-2)} \stackrel{?}{=} e^{(-2)^2-8}$$
$$e^{-4} \stackrel{?}{=} e^{4-8}$$
$$e^{-4} = e^{-4} \checkmark$$

$x = 4$:

$$e^{2(4)} \stackrel{?}{=} e^{(4)^2-8}$$
$$e^8 \stackrel{?}{=} e^{16-8}$$
$$e^8 = e^8 \checkmark$$

(b)

$2(5^x) = 32$ Write original equation.

$5^x = 16$ Divide each side by 2.

$\log_5 5^x = \log_5 16$ Take log(base 5) of each side.

$x = \log_5 16$ Inverse Property

$x = \dfrac{\ln 16}{\ln 5} \approx 1.723$ Change of base formula

The solution is $x = \log_5 16 \approx 1.723$.

Check $x = \log_5 16$:

$$2(5^x) = 32$$
$$2\left[5^{(\log_5 16)}\right] \stackrel{?}{=} 32$$
$$2(16) \stackrel{?}{=} 32$$
$$32 = 32 \checkmark$$

3. $e^x - 7 = 23$ Write original equation.

$e^x = 30$ Add 7 to each side.

$\ln e^x = \ln 30$ Take natural log of each side.

$x = \ln 30 \approx 3.401$ Inverse Property

Check $x = \ln 30$:

$e^x - 7 = 23$

$e^{(\ln 30) - 7} \stackrel{?}{=} 23$

$30 - 7 \stackrel{?}{=} 23$

$23 = 23$ ✓

4. $6(2^{t+5}) + 4 = 11$ Write original equation.

$6(2^{t+5}) = 7$ Subtract 4 from each side.

$2^{t+5} = \dfrac{7}{6}$ Divide each side by 6.

$\log_2 2^{t+5} = \log_2 \left(\dfrac{7}{6}\right)$ Take log (base 2) of each side.

$t + 5 = \log_2 \left(\dfrac{7}{6}\right)$ Inverse Property

$t = \log_2 \left(\dfrac{7}{6}\right) - 5$ Subtract 5 from each side.

$t = \dfrac{\ln\left(\dfrac{1}{6}\right)}{\ln 2} - 5$ Change of base formula.

$t \approx -4.778$ Use a calculator.

The solution is $t = \log_2 \left(\dfrac{7}{6}\right) - 5 \approx -4.778$.

Check $t \approx -4.778$:

$6(2^{t+5}) + 4 = 11$

$6\left[2^{(-4.778+5)}\right] + 4 \stackrel{?}{=} 11$

$6(1.166) + 4 \stackrel{?}{=} 11$

$10.998 \approx 11$ ✓

5. Algebraic Solution

$$e^{2x} - 7e^x + 12 = 0 \quad \text{Write original equation.}$$

$$\left(e^x\right)^2 - 7e^x + 12 = 0 \quad \text{Write in quadratic form.}$$

$$\left(e^x - 3\right)\left(e^x - 4\right) = 0 \quad \text{Factor.}$$

$$e^x - 3 = 0 \Rightarrow e^x = 3 \quad \text{Set 1st factor equal to 0.}$$

$$x = \ln 3 \quad \text{Solution}$$

$$e^x - 4 = 0 \Rightarrow e^x = 4 \quad \text{Set 2nd factor equal to 0.}$$

$$x = \ln 4 \quad \text{Solution}$$

The solutions are $x = \ln 3 \approx 1.099$ and $x = \ln 4 \approx 1.386$.

Check $x = \ln 3$:

$$e^{2x} - 7e^x + 12 = 0$$

$$e^{2(\ln 3)} - 7e^{(\ln 3)} + 12 \stackrel{?}{=} 0$$

$$e^{\ln\left(3^2\right)} - 7e^{\ln 3} + 12 \stackrel{?}{=} 0$$

$$3^2 - 7(3) + 12 \stackrel{?}{=} 0$$

$$0 = 0 \checkmark$$

$x = \ln 4$:

$$e^{2(\ln 4)} - 7e^{(\ln 4)} + 12 = 0$$

$$e^{\ln 4^2} - 7e^{\ln 4} + 12 \stackrel{?}{=} 0$$

$$4^2 - 7(4) + 12 \stackrel{?}{=} 0$$

$$0 = 0 \checkmark$$

Graphical Solution

Use a graphing utility to graph $y = e^{2x} - 7e^x + 12$ and then find the zeros.

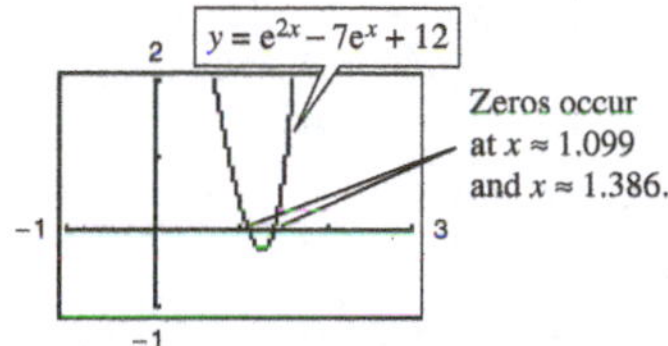

So, you can conclude that the solutions are $x \approx 1.099$ and $x \approx 1.386$.

6. (a) $\ln x = \dfrac{2}{3}$ Write original equation.

$$e^{\ln x} = e^{2/3} \quad \text{Exponentiate each side.}$$

$$x = e^{2/3} \quad \text{Inverse Property}$$

(b) $\log_2(2x - 3) = \log_2(x + 4)$ Write original equation.

$$2x - 3 = x + 4 \quad \text{One-to-One Property}$$

$$x = 7 \quad \text{Solution}$$

(c) $\log 4x - \log(12 + x) = \log 2$ Write Original equation.

$$\log\left(\frac{4x}{12 + x}\right) = \log 2 \quad \text{Quotient Property of Logarithms}$$

$$\frac{4x}{12 + x} = 2 \quad \text{One-to-One Property}$$

$$4x = 2(12 + x) \quad \text{Multiply each side by } (12 + x).$$

$$4x = 24 + 2x \quad \text{Distribute.}$$

$$2x = 24 \quad \text{Subtract } 2x \text{ from each side.}$$

$$x = 12 \quad \text{Solution}$$

7. Algebraic Solution

$7 + 3 \ln x = 5$	Write original equation.
$3 \ln x = -2$	Subtract 7 from each side.
$\ln x = -\frac{2}{3}$	Divide each side by 3.
$e^{\ln x} = e^{-2/3}$	Exponentiate each side.
$x = e^{-2/3}$	Inverse Property
$x \approx 0.513$	Use a calculator.

Graphical Solution

Use a graphing utility to graph $y_1 = 7 + 3 \ln x$ and $y_2 = 5$. Then find the intersection point.

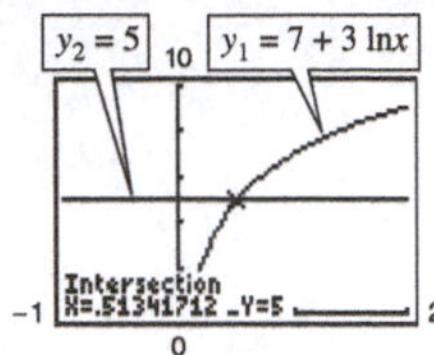

The point of intersection is about $(0.513, 5)$. So, the solution is $x \approx 0.513$.

8.

$3 \log_4 6x = 9$	Write original equation.
$\log_4 6x = 3$	Divide each side by 3.
$4^{\log_4 6x} = 4^3$	Exponentiate each side (base 4).
$6x = 64$	Inverse Property
$x = \frac{32}{3}$	Divide each side by 6 and simplify.

Check $x = \frac{32}{3}$:

$$3 \log_4 6x = 9$$

$$3 \log_4 6\left(\frac{32}{3}\right) \stackrel{?}{=} 9$$

$$3 \log_4 64 \stackrel{?}{=} 9$$

$$3 \log_4 4^3 \stackrel{?}{=} 9$$

$$3 \cdot 3 \stackrel{?}{=} 9$$

$$9 = 9 \checkmark$$

9. Algebraic Solution

$\log x + \log(x-9) = 1$	Write original equation.
$\log[x(x-9)] = 1$	Product Property of Logarithms
$10^{\log[x(x-9)]} = 10^1$	Exponentiate each side (base 10).
$x(x-9) = 10$	Inverse Property
$x^2 - 9x - 10 = 0$	Write in general form.
$(x-10)(x+1) = 0$	Factor.
$x - 10 = 0 \Rightarrow x = 10$	Set 1st factor equal to 0.
$x + 1 = 0 \Rightarrow x = -1$	Set 2nd factor equal to 0.

Check $x = 10$:

$$\log x + \log(x-9) = 1$$
$$\log(10) + \log(10-9) \stackrel{?}{=} 1$$
$$\log 10 + \log 1 \stackrel{?}{=} 1$$
$$1 + 0 \stackrel{?}{=} 1$$
$$1 = 1 \checkmark$$

$x = -1$:

$$\log x + \log(x-9) = 1$$
$$\log(-1) + \log(-1-9) \stackrel{?}{=} 1$$
$$\log(-1) + \log(-10) \stackrel{?}{=} 1$$

-1 and -10 are not in the domain of $\log x$. So, it does not check.

The solutions appear to be $x = 10$ and $x = -1$. But when you check these in the original equation, you can see that $x = 10$ is the only solution.

Graphical Solution

First, rewrite the original solution as

$\log x + \log(x-9) - 1 = 0.$

Then use a graphing utility to graph the equation $y = \log x + \log(x-9) - 1$ and find the zeros.

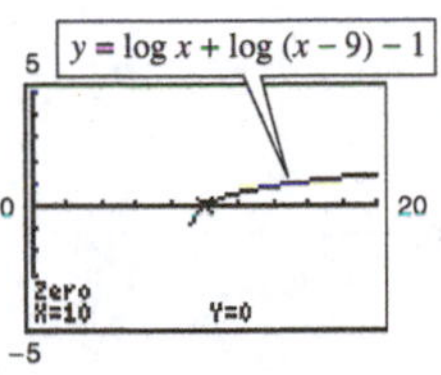

10. Using the formula for continuous compounding, the balance is

$A = Pe^{rt}$

$A = 500e^{0.0525t}.$

To find the time required for the balance to double, let $A = 1000$ and solve the resulting equation for t

$500e^{0.0525t} = 1000$	Let $A = 1000$.
$e^{0.0525t} = 2$	Divide each side by 500.
$\ln e^{0.0525t} = \ln 2$	Take natural log of each side.
$0.0525t = \ln 2$	Inverse Property
$t = \dfrac{\ln 2}{0.0525}$	Divide each side by 0.0525.
$t \approx 13.20$	Use a calculator.

The balance in the account will double after approximately 13.20 years.

Because the interest rate is lower than the interest rate in Example 2, it will take more time for the account balance to double.

11. To find when sales reached \$180 billion, let $y = 80$ and solve for t.

$-614 + 342.2 \ln t = y$	Write original equation
$-614 + 342.2 \ln t = 180$	Substitute 180 for y.
$342.2 \ln t = 794$	Add 614 to each side.
$\ln t = \dfrac{794}{342.2}$	Divide each side by 342.2.
$e^{\ln t} = e^{794/342.2}$	Exponentiate each side (base e).
$t = e^{794/342.2}$	Inverse Property
$t \approx 10.2$	Use a calculator.

The solution is $t \approx 10.2$. Because $t = 9$ represents 2009, it follows that $t = 10$ represents 2010. So, sales reached \$180 billion in 2010.

Checkpoints for Section 5.5

1. Algebraic Solution

To find when the amount of U.S. online advertising spending will reach \$100 billion, let $s = 100$ and solve for t.

$0.00036e^{0.7563t} = S$	Write original model.
$0.00036e^{0.7563t} = 300$	Substitute 300 for s.
$e^{0.7563t} \approx 833{,}333.33$	Divide each side by 0.0036.
$\ln e^{0.7563t} \approx \ln 833{,}333.33$	Take natural log of each side.
$0.7563t \approx 13.6332$	Inverse Property
$t \approx 18.03$	Divide each side by 0.7563.

According to the model, the amount of U.S. online advertising spending will reach \$300 million in 2018.

Graphical Solution

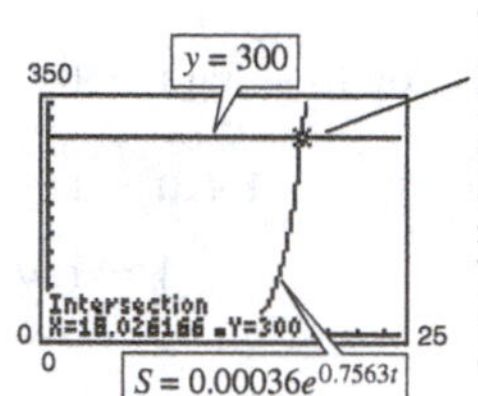

The intersection point of the model and the line $y = 300$ is about (18.03, 300). So, according to the model, the amount of U.S. online advertising spending will reach \$300 billion in 2018.

2. Let y be the number of bacteria at time t. From the given information you know that $y = 100$ when $t = 1$ and $y = 200$ when $t = 2$. Substituting this information into the model $y = ae^{bt}$ produces $100 = ae^{(1)b}$ and $200 = ae^{(2)b}$. To solve for b, solve for a in the first equation.

$100 = ae^b$	Write first equation.
$\dfrac{100}{e^b} = a$	Solve for a.

Then substitute the result into the second equation.

$200 = ae^{2b}$	Write second equation
$200 = \left(\dfrac{100}{e^b}\right)e^{2b}$	Substitute $\dfrac{100}{e^b}$ for a.
$\dfrac{200}{100} = e^b$	Simplify and divide each side by 100.
$2 = e^b$	Simplify.
$\ln 2 = \ln e^b$	Take natural log of each side
$\ln 2 = b$	Inverse Property

Use $b = \ln 2$ and the equation you found for a.

$a = \dfrac{100}{e^{\ln 2}}$	Substitute ln 2 for b.
$= \dfrac{100}{2}$	Inverse Property
$= 50$	Simplify.

So, with $a = 50$ and $b = \ln 2$, the exponential growth model is $y = 50e^{(\ln 2)t}$.

After 3 hours, the number of bacteria will be $y = 50e^{\ln 2(3)} = 400$ bacteria.

3. Algebraic Solution

In the carbon dating model, substitute the given value of R to obtain the following.

$$\frac{1}{10^{12}}e^{-t/8223} = R \qquad \text{Write original model.}$$

$$\frac{e^{-t/8223}}{10^{12}} = \frac{1}{10^{14}} \qquad \text{Substitute } \frac{1}{10^{14}} \text{ for } R.$$

$$e^{-t/8223} = \frac{1}{10^{2}} \qquad \text{Multiply each side by } 10^{12}.$$

$$e^{-t/8223} = \frac{1}{100} \qquad \text{Simplify.}$$

$$\ln e^{-t/8223} = \ln\frac{1}{100} \qquad \text{Take natural log of each side.}$$

$$-\frac{t}{8223} \approx -4.6052 \qquad \text{Inverse Property}$$

$$t \approx 37{,}869 \qquad \text{Multiply each side by } -8223.$$

So, to the nearest thousand years, the age of the fossil is about 38,000 years.

Graphical Solution

Use a graphing utility to graph the formula for the ratio of carbon 14 to carbon 12 at any time t as

$$y_1 = \frac{1}{10^{12}}e^{-x/8223}.$$

In the same viewing window, graph $y_2 = \dfrac{1}{10^{14}}$

Use the *intersect* feature to estimate that $x \approx 18{,}934$ when $y = 1/10^{13}$.

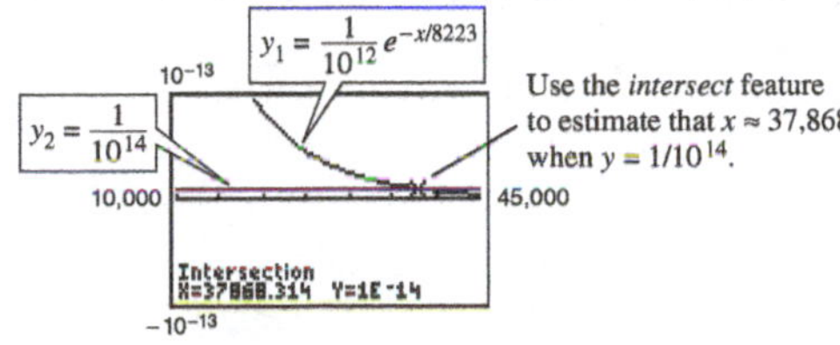

So, to the nearest thousand years, the age of fossil is about 38,000 years.

4. The graph of the function is shown below. On this bell-shaped curve, the maximum value of the curve represents the average score. From the graph, you can estimate that the average reading score for college-bound seniors in the United States in 2015 was 495.

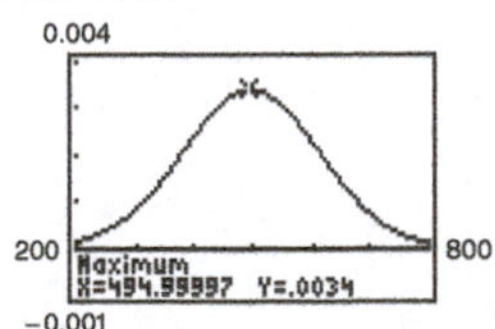

5. To find the number of days that 250 students are infected, let $y = 250$ and solve for t.

$\dfrac{5000}{1 + 4999e^{-0.8t}} = y$ Write original model.

$\dfrac{5000}{1 + 4999e^{-0.8t}} = 250$ Substitute 250 for y.

$\dfrac{5000}{250} = 1 + 4999e^{-0.8t}$ Divide each side by 250 and multiply each side by $1 + 4999e^{-0.8t}$.

$20 = 1 + 4999e^{-0.8t}$ Simplify.

$19 = 4999e^{-0.8t}$ Subtract 1 from each side.

$\dfrac{19}{4999} = e^{-0.8t}$ Divide each side by 4999.

$\ln\left(\dfrac{19}{4999}\right) = \ln e^{-0.8t}$ Take natural log of each side.

$\ln\left(\dfrac{19}{4999}\right) = -0.8t$ Inverse Property

$-5.5726 \approx -0.8t$ Use a calculator.

$t \approx 6.97$ Divide each side by -0.8.

So, after about 7 days, 250 students will be infected.

Graphical Solution

To find the number of days that 250 students are infected, use a graphing utility to graph.

$$y_1 = \frac{5000}{1 + 4999e^{-0.8x}} \text{ and } y_2 = 250$$

in the same viewing window. Use the *intersect* feature of the graphing utility to find the point of intersection of the graphs.

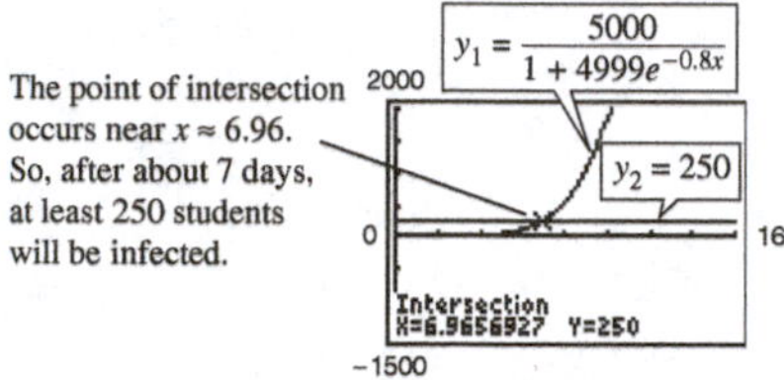

6. (a) Because $I_0 = 1$ and $R = 6.0$, you have the following.

$R = \log \dfrac{I}{I_0}$

$6.0 = \log \dfrac{I}{1}$ Substitute 1 for I_0 and 6.0 for R.

$10^{6.0} = 10^{\log I}$ Exponentiate each side (base 10).

$10^{6.0} = I$ Inverse Property

$1{,}000{,}000 = I$ Simplify.

(b) Because $I_0 = 1$ and $R = 7.9$, you have the following.

$7.9 = \log \dfrac{I}{1}$ Substitute 1 for I_0 and 7.9 for R.

$10^{7.9} = 10^{\log I}$ Exponentiate each side (base 10).

$10^{7.9} = I$ Inverse Property

$79{,}432{,}823 \approx I$ Simplify.

Chapter 6

Checkpoints for Section 6.1

1. $\begin{cases} x - y = 0 & \text{Equation 1} \\ 5x - 3y = 6 & \text{Equation 2} \end{cases}$

Begin by solving for y in Equation 1.

$x - y = 0$

$y = x$

Next substitute this expression for y into Equation 2 and solve the resulting single-variable equation for x.

$5x - 3y = 6$ Write Equation 2.

$5x - 3(x) = 6$ Substitute x for y.

$2x = 6$ Collect like terms.

$x = 3$ Divide each side by 2.

Finally, solve for y by back-substituting $x = 3$ into equation $y = x$, to obtain the corresponding value for y.

$y = x$ Write revised Equation 1.

$y = 3$ Substitute 3 for x.

The solution is the ordered pair $(3, 3)$.

Check

Substitute $(3, 3)$ into Equation 1:

$x - y = 0$ Write Equation 1.

$3 - 3 \stackrel{?}{=} 0$ Substitute for x and y

$0 = 0$ Solution checks in Equation 1.

Substitute $(3, 3)$ into Equation 2:

$5x - 3y = 6$ Write Equation 2.

$5(3) - 3(3) \stackrel{?}{=} 6$ Substitute for x and y.

$15 - 9 \stackrel{?}{=} 6$

$6 = 6$ Solution checks in Equation 2.

Because $(3, 3)$ satisfies both equations in the system, it is a solution of the system of equations.

2. *Verbal Model:* [Amount in 6.5% fund] + [Amount in 8.5% fund] = [Total investment]

[Interest for 6.5% fund] + [Interest for 8.5% fund] = [Total interest]

Labels:

Amount in 6.5% fund $= x$	(dollars)
Interest for 6.5% fund $= 0.065x$	(dollars)
Amount in 8.5% fund $= y$	(dollars)
Interest for 8.5% fund $= 0.085y$	(dollars)
Total investment $= 25{,}000$	(dollars)
Total interest $= 2600$	(dollars)

System: $\begin{cases} x + y = 25{,}000 & \text{Equation 1} \\ 0.065x + 0.085y = 2000 & \text{Equation 2} \end{cases}$

To begin, it is convenient to multiply each side of Equation 2 by 1000. This eliminates the need to work with decimals.

$1000(.065x + 0.085y) = 1000(2000)$ Multiply each side of Equation 2 by 1000.

$65x + 85y = 2{,}000{,}000$ Revised Equation 2

To solve this system, you can solve for x in Equation 1.

$x = 25{,}000 - y$ Revised Equation 1

Then, substitute this expression for x into revised Equation 2 and solve the resulting equation for y.

$65x + 85y = 2{,}000{,}000$ Write revised Equation 2.

$65(25{,}000 - y) + 85y = 2{,}000{,}000$ Substitute 1 $25000 - y$ for x.

$1{,}625{,}000 - 65y + 85y = 2{,}000{,}000$ Distributive Property

$20y = 375{,}000$ Combine like terms.

$y = 18{,}750$ Divide each side by 20.

Next, back-substitute $y = 18{,}750$ to solve for x.

$x = 25000 - y$ Write revised Equation 1.

$x = 25000 - (18750)$ Substitute 18750 for y.

$x = 6250$ Subtract.

The solution is $(6250, 18{,}750)$. So, \$6250 is invested at 6.5% and \$18,750 is invested at 8.5%.

3. $\begin{cases} -2x + y = 5 & \text{Equation 1} \\ x^2 - y + 3x = 1 & \text{Equation 2} \end{cases}$

Begin by solving for y in Equation 1 to obtain $y = 2x + 5$. Next, substitute this expression for y into Equation 2 and solve for x.

$x^2 - y + 3x = 1$ Write Equation 2.

$x^2 - (2x + 5) + 3x = 1$ Substitute $2x + 5$ for y into Equation 2.

$x^2 - 2x - 5 + 3x = 1$ Simplify.

$x^2 + x - 6 = 0$ Write in standard form.

$(x + 3)(x - 2) = 0$ Factor.

$x + 3 = 0 \Rightarrow x = -3$ Solve for x.

$x - 2 = 0 \Rightarrow x = 2$

Back-substituting these values of x to solve for the corresponding values of y produces the following solutions.

$y = 2x + 5$

$y = 2(-3) + 5 = -1$

$y = 2(2) + 5 = 9$

So, the solutions of the system are $(-3, -1)$ and $(2, 9)$.

4. $\begin{cases} 2x - y = -3 & \text{Equation 1} \\ 2x^2 + 4x - y^2 = 0 & \text{Equation 2} \end{cases}$

Begin by solving for y in Equation 1 to obtain $y = 2x + 3$. Next, substitute this expression for y into Equation 2 and solve for x.

$2x^2 + 4x - y^2 = 0$ Write Equation 2.

$2x^2 + 4x - (2x + 3)^2 = 0$ Substitute $2x + 3$ for y into Equation 2.

$2x^2 + 4x - (4x^2 + 12x + 9) = 0$ Simplify.

$-2x^2 - 8x - 9 = 0$ Combine like terms.

$2x^2 + 8x + 9 = 0$ Write in standard form

$x = \dfrac{-(8) \pm \sqrt{(8)^2 - 4(2)(9)}}{2(2)}$ Use the Quadratic Formula.

$x = \dfrac{-8 \pm \sqrt{-8}}{4}$ Simplify.

Because the discriminant is negative, the equation $2x^2 + 8x + 9 = 0$ has no (real) solution. So, the original system of equations has no (real) solution.

5.

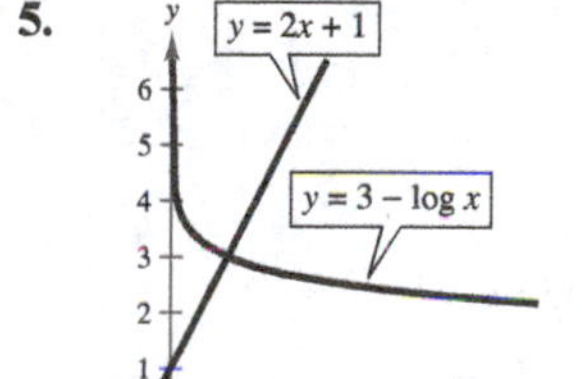

There is only one point of intersection of the graphs of the two equations, and (1, 3) is the solution point.

Check (1, 3) in Equation 1:

$y = 3 - \log x$ Write Equation 1.

$3 \stackrel{?}{=} 3 - \log 1$ Substitute for x and y.

$3 \stackrel{?}{=} 3 - 0$

$3 = 3$ Solution checks in Equation 1.

Check (1, 3) in Equation 2:

$-2x + y = 1$ Write Equation 2.

$-2(1) + 3 \stackrel{?}{=} 1$ Substitute for x and y.

$-2 + 3 \stackrel{?}{=} 1$

$1 = 1$ Solution checks in Equation 2.

6. Algebraic Solution

The total cost of producing x units is

$\boxed{\text{Totalcost}} = \boxed{\text{Cost per unit}} \cdot \boxed{\text{Number of units}} + \boxed{\text{Initial cost}}$

$C = 12x + 300{,}000.$ Equation 1

The revenue obtained by selling x units is

$\boxed{\text{Total revenue}} = \boxed{\text{Price per unit}} \cdot \boxed{\text{Number of units}}$

$R = 70x.$ Equation 2

Because the break-even point occurs when $R = C$, you have $C = 70x$, and the system of equations to solve is

$$\begin{cases} C = 12x + 300{,}000 \\ C = 70x \end{cases}.$$

Solve by substitution.

$70x = 12x + 300{,}000$ Substitute $70x$ for C in Equation 1.

$58x = 300{,}000$ Subtract $12x$ from each side.

$x \approx 5172$ Divide each side by 58

So, the company must sell 5172 pairs of shoes to break even.

Graphical Solution

The system of equations to solve is

$$\begin{cases} C = 12x + 300{,}000 \\ C = 70x \end{cases}.$$

Use a graphing utility to graph $y_1 = 12x + 300{,}000$ and $y_2 = 70x$ in the same viewing window.

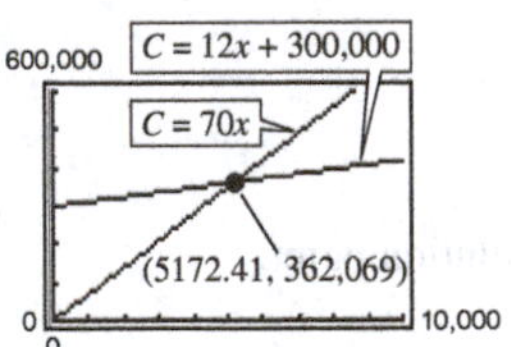

So, the company must sell about 5172 pairs of shoes to break even.

7. Algebraic Solution

Because both equations are already solved for S in terms of x, substitute either expression for S into the other equation and solve for x.

$$\begin{cases} S = 108 - 9.4x & \text{Animated} \\ S = 16 + 9x & \text{Horror} \end{cases}$$

$16 + 9x = 108 - 9.4x$ Substitute for S in Equation 1.

$9.4x + 9x = 108 - 16$ Add $9.4x$ and -16 to each side.

$18.4x = 92$

$x = 5$ Divide each side by 18.4.

So, since $x = 1$ corresponds to week one, the weekly ticket sales in millions of dollars for the two movies will be equal after 5 weeks.

Numerical Solution

You can create a table of values for each model to determine when ticket sales for the two movies will be equal.

Number of weeks x	1	2	3	4	5	6	7
Sales S (Animated)	98.6	89.2	79.8	70.4	61	51.6	70
Sales S (Horror)	25	34	43	52	61	42.2	79

So, from the table, the weekly ticket sales in millions of dollars for the two movies will be equal after 5 weeks.

Checkpoints for Section 6.2

1. Because the coefficients of y differ only in sign, eliminate the y-terms by adding the two equations.

$2x + y = 4$	Write Equation 1.
$2x - y = -1$	Write Equation 2.
$4x \quad = 3$	Add equations.
$x \quad = \frac{3}{4}$	Solve for x.

Solve for y by back-substituting $x = \frac{3}{4}$ into Equation 1.

$$2\left(\tfrac{3}{4}\right) + y = 4$$
$$\tfrac{3}{2} + y = 4$$
$$y = \tfrac{5}{2}$$

The solution is $\left(\frac{3}{4}, \frac{5}{2}\right)$.

Check this in the original system.

$2\left(\frac{3}{4}\right) + \left(\frac{5}{2}\right) \stackrel{?}{=} 4$	Write Equation 1.
$\frac{3}{2} + \frac{5}{2} = 4$	Solution checks in Equation 1. ✓
$2\left(\frac{3}{4}\right) - \left(\frac{5}{2}\right) \stackrel{?}{=} -1$	Write Equation 2.
$\frac{3}{2} - \frac{5}{2} = -1$	Solution checks in Equation 2. ✓

2. To obtain coefficients that differ only in sign, multiply Equation 2 by 3.

$2x + 3y = 17 \Rightarrow$	$2x + 3y = 17$	Write Equation 1.
$5x - y = 17 \Rightarrow$	$15x - 3y = 51$	Multiply Equation 2 by 3.
	$17x \quad = 68$	Add Equations.
	$x \quad = 4$	Solve for x.

Solve for y by back-substituting $x = 4$ into Equation 2.

$5x - y = 17$	Write Equation 2.
$5(4) - y = 17$	Substitute 4 for x.
$20 - y = 17$	Simplify.
$y = 3$	Solve for y.

The solution is $(4, 3)$.

Check this in the original system.

$2(4) + 3(3) \stackrel{?}{=} 17$	Write Equation 1.
$8 + 9 = 17$	Solution Checks in Equation 1. ✓
$5(4) - (3) \stackrel{?}{=} 17$	Write Equation 2.
$20 - 3 = 17$	Solution Checks in Equation 2. ✓

3. Algebraic Solution

You can obtain coefficients that differ only in sign by multiplying Equation 1 by 2 and multiplying Equation 2 by -3.

$$\begin{aligned} 3x + 2y = 7 &\Rightarrow \quad 6x + \ \ 4y = 14 && \text{Multiply Equation 1 by 2.} \\ 2x + 5y = 1 &\Rightarrow \underline{-6x - 15y = -3} && \text{Multiply Equation 2 by } -3. \\ & \qquad\quad\ -11y = 11 && \text{Add Equations.} \\ & \qquad\qquad\ \ y = -1 && \text{Solve for } y. \end{aligned}$$

Solve for x by back-substituting $y = 1$ into Equation 1.

$$\begin{aligned} 3x + 2y &= 7 && \text{Write Equation 1.} \\ 3x + 2(-1) &= 7 && \text{Substitute } -1 \text{ for } y \\ 3x - 2 &= 7 \\ 3x &= 9 \\ x &= 3 \end{aligned}$$

The solution is $(3, -1)$.

Graphical Solution

Solve each equation for y and use a graphing utility to graph the equations in the same viewing window.

From the graph, the solution is $(3, -1)$.

Check this in the original system.

$$\begin{aligned} 3(3) + 2(-1) &\stackrel{?}{=} 7 && \text{Write Equation 1.} \\ 9 - 2 &= 7 && \text{Solution checks in Equation 1. } \checkmark \\ 2(3) + 5(-1) &\stackrel{?}{=} 1 && \text{Write Equation 2.} \\ 6 - 5 &= 1 && \text{Solution checks in Equation 2. } \checkmark \end{aligned}$$

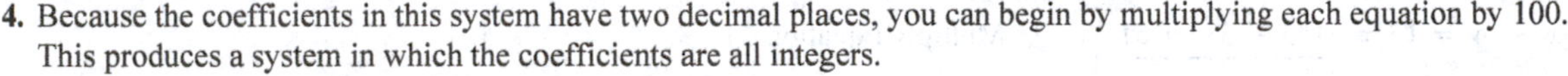

4. Because the coefficients in this system have two decimal places, you can begin by multiplying each equation by 100. This produces a system in which the coefficients are all integers.

$$\begin{aligned} 0.03x + 0.04y = 0.75 &\Rightarrow 3x + 4y = 75 \\ 0.02x + 0.06y = 0.90 &\Rightarrow 2x + 6y = 90 \end{aligned}$$

Now, to obtain coefficients that differ only in sign, multiply Equation 1 by 2 and Equation 2 by –3.

$$\begin{aligned} 3x + 4y = 75 &\Rightarrow \quad 6x + \ \ 8y = \ \ 150 && \text{Multiply Equation 1 by 2.} \\ 2x + 6y = 90 &\Rightarrow \underline{-6x - 18y = -270} && \text{Multiply Equation 2 by } -3. \\ & \qquad\quad\ -10y = -120 && \text{Add Equations.} \\ & \qquad\qquad\quad y = \ \ 12 && \text{Solve for } y. \end{aligned}$$

Back-substitute $y = 12$ into revised Equation 1 to solve for x.

$$\begin{aligned} 3x + 4y &= 75 \\ 3x + 4(12) &= 75 \\ 3x + 48 &= 75 \\ 3x &= 27 \\ x &= 9 \end{aligned}$$

The solution is $(9, 12)$.

Check this in the original system, as follows.

$$\begin{aligned} 0.03(9) + 0.04(12) &\stackrel{?}{=} 0.75 && \text{Write Equation 1.} \\ 0.27 + 0.48 &= 0.75 && \text{Solution Checks in Equation 1. } \checkmark \\ 0.02(9) + 0.06(12) &\stackrel{?}{=} 0.90 && \text{Write Equation 2.} \\ 0.18 + 0.72 &= 0.90 && \text{Solution Checks in Equation 2. } \checkmark \end{aligned}$$

5. First, write each equation in slope-intercept form.

$$\begin{cases} 2x + 3y = 6 & \Rightarrow y = \frac{2}{3}x + 2 \\ 4x - 6y = -9 & \Rightarrow y = \frac{2}{3}x + \frac{3}{2} \end{cases}$$

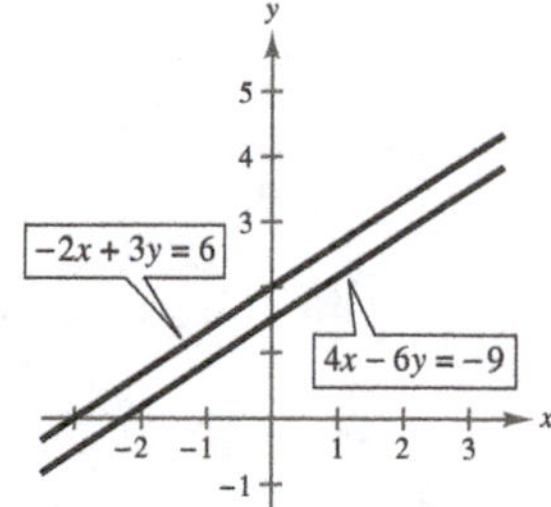

The graph of the system is a pair of parallel lines. The lines have no point of intersection, so the system has no solution. The system is inconsistent.

6. To obtain coefficients that differ only in sign, multiply Equation 1 by 2.

$$6x - 5y = 3 \Rightarrow \quad 12x - 10y = 6 \qquad \text{Multiply Equation by 2.}$$
$$-12x + 10y = 5 \Rightarrow \underline{-12x + 10y = 5} \qquad \text{Write Equation 2.}$$
$$0 = 11 \qquad \text{Add equations.}$$

Because there are no values of x and y for which $0 = 11$, you can conclude that the system is inconsistent and has no solution. The graph shows the lines corresponding to the two equations in this system. Note that the two lines are parallel, so they have no point of intersection.

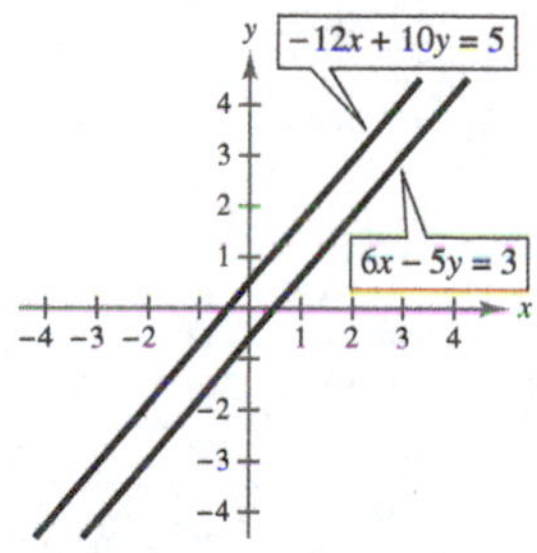

7. To obtain coefficients that differ only in sign, multiply Equation 1 by 8.

$$\frac{1}{2}x - \frac{1}{8}y = -\frac{3}{8} \Rightarrow \quad 4x - y = -3 \qquad \text{Multiply Equation by 8}$$
$$-4x + y = 3 \Rightarrow \underline{-4x + y = 3} \qquad \text{Write Equation 2.}$$
$$0 = 0 \qquad \text{Add equations.}$$

Because the two equations are equivalent (have the same solution set), the system has infinitely many solutions. The solution set consists of all points (x, y) lying on the line $-4x + y = 3$ as shown. Letting $x = a$, where a is any real number, the solutions of the system are $(a, 4a + 3)$

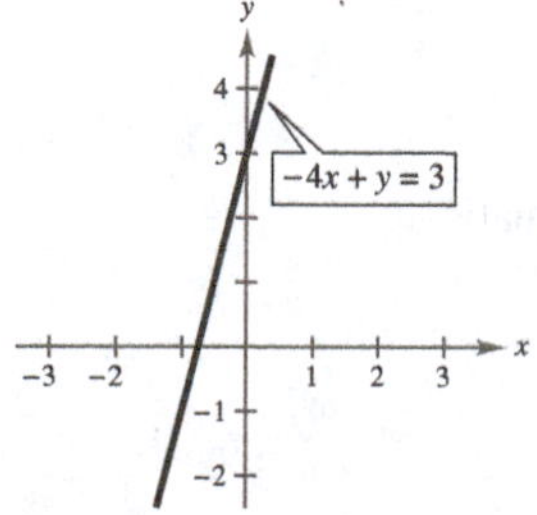

8. The two unknown quantities are the speeds of the wind and of the plane. If r_1 is the speed of the plane and r_2 is the speed of the wind, then

$r_1 - r_2 =$ speed of the plane against the wind

$r_1 + r_2 =$ speed of the plane with the wind.

Using the formula

distance = (rate)(time)

for these two speeds, you obtain the following equations.

$$2000 = (r_1 - r_2)\left(4 + \frac{24}{60}\right)$$

$$2000 = (r_1 - r_2)\left(4 + \frac{6}{60}\right)$$

These two equations simplify as follows.

$$\begin{cases} 5000 = 11r_1 - 11r_2 & \text{Equation 1} \\ 20{,}000 = 41r_1 + 41r_2 & \text{Equation 2} \end{cases}$$

To solve this system by elimination, multiply Equation 1 by 41 and Equation 2 by 11.

$$\begin{aligned} 205{,}000 &= 451r_1 - 451r_2 && \text{Multiply Equation 1 by 41.} \\ \underline{220{,}000} &\underline{= 451r_1 + 451r_2} && \text{Multiply Equation 2 by 11.} \\ 425{,}000 &= 902r_1 && \text{Add equations.} \end{aligned}$$

So, $r_1 = \dfrac{425{,}000}{902} \approx 471.18$ miles per hour

and $r_2 = \frac{1}{11}(11r_1 - 5000)$

$$r_2 = \frac{1}{11}\left(11 \cdot \frac{425{,}000}{902} - 5000\right) \approx 16.63 \text{ miles per hour.}$$

Check this solution in the original system of equations.

$$2000 \approx (471.18 - 16.63)\left(4 + \frac{24}{60}\right) \checkmark$$

$$2000 \approx (471.18 + 16.63)\left(4 + \frac{6}{60}\right) \checkmark$$

9. Because p is written in terms of x, begin by substituting the value of p given in the supply equation into the demand equation.

$$\begin{aligned} p &= 567 - 0.00002x && \text{Write demand equation.} \\ 492 + 0.00003x &= 567 - 0.00002x && \text{Substitute } 492 + 0.00003x \text{ for } p. \\ 0.00005x &= 75 && \text{Combine like terms.} \\ x &= 1{,}500{,}000 && \text{Solve for } x. \end{aligned}$$

So, the equilibrium point occurs when the demand and supply are each 1.5 million units. Obtain the price that corresponds to this x-value by back-substituting $x = 1{,}500{,}000$ into either of the original equations. For instance, back-substituting into the demand equation produces

$$p = 567 - 0.00002(1{,}500{,}000) = 567 - 30 = \$537.$$

The solution is (1,500,000, 537). Check this by substituting into the demand and supply equations.

$$p = 567 - 0.00002x$$

$$537 = 567 - 0.00002(1{,}500{,}000) \checkmark$$

$$p = 492 + 0.00003x$$

$$537 = 492 + 0.00003(1{,}500{,}000) \checkmark$$

Checkpoints for Section 6.3

1. $\begin{cases} x - y + 5z = 22 \\ y + 3z = 6 \\ z = 3 \end{cases}$

From Equation 3, you know the value of z. To solve for y, back-substitute $z = 3$ into Equation 2 to obtain the following.

$y + 3z = 6$ Write Equation 2.

$y + 3(3) = 6$ Substitute 3 for z.

$y = -3$ Solve for y.

Then back-substitute $y = -3$ and $z = 3$ into Equation 1 to obtain the following.

$x - y + 5z = 22$ Write Equation 1.

$x - (-3) + 5(3) = 22$ Substitute -3 for y and 3 for z.

$x + 18 = 4$ Combine like terms.

$x = 4$ Solve for x.

The solution is $x = 4$, $y = -3$, and $z = 3$, which can be written as the ordered triple $(4, -3, 3)$. Check this in the original system of equations.

Check

Equation 1: $x - y + 5z = 22$

$4 - (-3) + 5(3) = 22$

$4 + 3 + 15 = 22$ ✓

Equation 2: $y + 3z = 6$

$(-3) + 3(3) \stackrel{?}{=} 6$

$-3 + 9 = 6$ ✓

Equation 3: $z = 3$

$(3) = 3$ ✓

2. $\begin{cases} 2x + y = 3 \\ x + 2y = 3 \end{cases}$ Write Equation 1. Write Equation 2.

$\begin{cases} x + 2y = 3 \\ 2x + y = 3 \end{cases}$ Interchange the two equations in the system.

$\begin{cases} -2x - 4y = -6 \\ 2x + y = 3 \end{cases}$ Multiply the first equation by -2.

$$\begin{array}{r} -2x - 4y = -6 \\ 2x + y = 3 \\ \hline -3y = -3 \\ y = 1 \end{array}$$

Add the multiple of the first equation to the second equation to obtain a new second equation.

$\begin{cases} x + 2y = 3 \\ y = 1 \end{cases}$ New system in row-echelon form.

Now back-substitute $y = 1$ into the first equation in row-echelon form and solve for x.

$x + 2(1) = 3$ Substitute 1 for y.

$x = 1$ Solve for x.

The solution is $x = 1$ and $y = 1$, which can be written as the ordered pair $(1, 1)$.

3. Because the leading coefficient of the first equation is 1, begin by keeping the x in the upper left position and eliminating the other x-terms from the first column.

$$\begin{array}{rl} -2x - 2y - 2z = -12 & \text{Multiply Equation 1 by } -2. \\ \underline{2x - y + z = 3} & \text{Write Equation 2.} \\ -3y - z = -9 & \text{Add revised Equation 1 to Equation 2.} \end{array}$$

$$\begin{cases} x + y + z = 6 \\ -3y - z = -9 \\ 3x + y - z = 2 \end{cases}$$ Adding -2 times the first equation to the second equation produces a new second equation.

$$\begin{array}{rl} -3x - 3y - 3z = -18 & \text{Multiply Equation 1 by } -3. \\ \underline{3x + y - z = 2} & \text{Write Equation 3.} \\ -2y - 4z = -16 & \text{Add revised Equation 1 to Equation 3.} \end{array}$$

$$\begin{cases} x + y + z = 6 \\ -3y - z = -9 \\ -2y - 4z = -16 \end{cases}$$ Adding -3 times the first equation to the third equation produces a new third equation.

Now that you have eliminated all but the x in the upper position of the first column, work on the second column.

$$\begin{cases} x + y + z = 6 \\ -3y - z = -9 \\ -y - 2z = -8 \end{cases}$$ Multiplying the third equation by 2, produces a new third equation.

$$\begin{array}{rl} -3y - z = -9 & \text{Write Equation 2.} \\ \underline{3y + 6z = 24} & \text{Multiply Equation 3 by } -3. \\ 5z = 15 & \text{Add equations.} \end{array}$$

$$\begin{cases} x + y + z = 6 \\ -3y - z = -9 \\ 5z = 15 \end{cases}$$ Adding the second equation to -3 times the third equation produces a new third equation.

$$\begin{cases} x + y + z = 6 \\ y + \frac{1}{3}z = 3 \\ 5z = 15 \end{cases}$$ Multiplying the second equation by $-\frac{1}{3}$ produces a new second equation.

$$\begin{cases} x + y + z = 6 \\ y + \frac{1}{3}z = 3 \\ z = 3 \end{cases}$$ Multiplying the third equation by $\frac{1}{5}$ produces a new third equation.

To solve for y, back-substitute $z = 3$ into Equation 2 to obtain the following.

$$y + \tfrac{1}{3}(3) = 3$$
$$y = 2.$$

Then back-substitute $y = 2$ and $z = 3$ into Equation 1 to obtain the following.

$$x + (2) + (3) = 6$$
$$x = 1$$

The solution is $x = 1$, $y = 2$, and $z = 3$, which can be written as $(1, 2, 3)$.

4. $\begin{cases} x + y - 2z = 3 \\ 3x - 2y + 4z = 1 \\ 2x - 3y + 6z = 8 \end{cases}$

$\begin{cases} x + y - 2z = 3 \\ -5y + 10z = -8 \\ 2x - 3y + 6z = 8 \end{cases}$ Adding -3 times the first equation to the second equation produces a new second equation.

$\begin{cases} x + y - 2z = 3 \\ -5y + 10z = -8 \\ -5y + 10z = 2 \end{cases}$ Adding -2 times the first equation to the third equation produces a new third equation.

$\begin{cases} x + y - 2z = 3 \\ -5y + 10z = -8 \\ 0 = 10 \end{cases}$ Adding -1 times the second equation to the third equation produces a new third equation.

Because $0 = 10$ is a false statement, this is an inconsistent system and has no solution. Moreover, because this system is equivalent to the original system, the original system has no solution.

5. $\begin{cases} x + 2y - 7z = -4 \\ 2x + 3y + z = 5 \\ 3x + 7y - 36z = -25 \end{cases}$

$\begin{cases} x + 2y - 7z = -4 \\ -y + 15z = 13 \\ 3x + 7y - 36z = -25 \end{cases}$ Adding -2 times the first equation to the second equation produces a new second equation.

$\begin{cases} x + 2y - 7z = -4 \\ -y + 15z = 13 \\ y - 15z = -13 \end{cases}$ Adding -3 times the first equation to the third equation produces a new third equation.

$\begin{cases} x + 2y - 7z = -4 \\ -y + 15z = 13 \\ 0 = 0 \end{cases}$ Adding the second equation to the third equation to produces a new third equation.

This result means that Equation 3 depends on Equations 1 and 2 in the sense that it gives no additional information about the variables. Because $0 = 0$ is a true statement, this system has infinitely many solutions. However, it is incorrect to say that the solution is "infinite." You must also specify the correct form of the solution. So, the original system is equivalent to the system.

$\begin{cases} x + 2y - 7z = -4 \\ -y + 15z = 13. \end{cases}$

In the second equation, solve for y in terms of z to obtain the following.

$$\begin{aligned} -y + 15z &= 13 \\ -y &= -15z + 13 \\ y &= 15z - 13 \end{aligned}$$

Back-substituting in the first equation produces the following.

$$\begin{aligned} x + 2y - 7z &= -4 \\ x + 2(15z - 13) - 7z &= -4 \\ x + 30z - 26 - 7z &= -4 \\ x &= -23z + 22 \end{aligned}$$

Finally, letting $z = a$ where a is a real number, the solutions of the given system are all of the form $x = -23a + 22$, $y = 15a - 13$, and $z = a$. So, every ordered triple of the form $(-23a + 22, 15a - 13, a)$ is a solution of the system.

6. $\begin{cases} x - y + 4z = 3 \\ 4x \qquad\quad - z = 0 \end{cases}$

$\begin{cases} x - y + \quad z = \quad 3 \\ \qquad 4y - 17z = -12 \end{cases}$ Adding -4 times the first equation to the second equation produces a new second equation.

$\begin{cases} x - y + \quad z = 3 \\ \qquad y - \frac{17}{4}z = -3 \end{cases}$ Multiplying the second equation by $\frac{1}{4}$ produces a new second equation.

Solve for y in terms of z to obtain the following.

$$y - \tfrac{17}{4}z = -3$$
$$y = \tfrac{17}{4}z - 3$$

Solve for x by back-substituting $y = \frac{17}{4}z - 3$ into Equation 1.

$$x - y + 4z = 3$$
$$x - \left(\tfrac{17}{4}z - 3\right) + 4z = 3$$
$$x - \tfrac{17}{4}z + 3 + 4z = 3$$
$$x = \tfrac{1}{4}z$$

Finally, by letting $z = a$, where a is a real number, you have the solution

$x = \frac{1}{4}a$, $y = \frac{17}{4}a - 3$, and $z = a$.

So, every ordered triple of the form $\left(\frac{1}{4}a, \frac{17}{4}a - 3, a\right)$ is a solution of the system. Because the original system had three variables and only two equations, the system cannot have a unique solution and has infinitely many solutions.

7. By substituting the three values of t and s into the position equation, you can obtain three linear equations in a, v_0 and s_0.

When $t = 1$: $\frac{1}{2}a(1)^2 + v_0(1) + s_0 = 104 \Rightarrow a + 2v_0 + 2s_0 = 208$

When $t = 2$: $\frac{1}{2}a(2)^2 + v_0(2) + s_0 = 76 \Rightarrow 2a + 2v_0 + s_0 - 76$

When $t = 3$: $\frac{1}{2}a(3)^2 + v_0(3) + s_0 = 16 \Rightarrow 9a + 6v_0 + 2s_0 = 32$

This produces the following system of linear equation.

$$\begin{cases} a + 2v_0 + 2s_0 = 208 \\ 2a + 2v_0 + s_0 = 76 \\ 9a + 6v_0 + 2s_0 = 32 \end{cases}$$

Now solve the system using Gaussian Elimination.

$$\begin{cases} a + 2v_0 + 2s_0 = 208 \\ -2v_0 - 3s_0 = -340 \\ 9a + 6v_0 + 2s_0 = 32 \end{cases}$$

Adding -2 times the first equation to the second equation produces a new second equation.

$$\begin{cases} a + 2v_0 + 2s_0 = 208 \\ -2v_0 - 3s_0 = -340 \\ -12v_0 - 16s_0 = -1840 \end{cases}$$

Adding -9 times the first equation to the third equation produces a new third equation.

$$\begin{cases} a + 2v_0 + 2s_0 = 208 \\ -2v_0 - 3s_0 = -340 \\ 2s_0 = 200 \end{cases}$$

Adding -6 times the second equation to the third equation produces a new third equation.

$$\begin{cases} a + 2v_0 + 2s_0 = 208 \\ v_0 + \frac{3}{2}s_0 = 170 \\ s_0 = 100 \end{cases}$$

Multiplying the second equation by $-\frac{1}{2}$ produces a new second equation and multiplying the third equation by $\frac{1}{2}$ produces a new third equation.

So, $s_0 = 100$.

Find v_0 by back-substituting $s_0 = 100$ into Equation 2.

$$v_0 + \frac{3}{2}(100) = 170$$
$$v_0 = 20$$

Find a by back-substituting $s_0 = 100$ and $v_0 = 20$ into Equation 1.

$$a + 2(20) + 2(100) = 208$$
$$a = -32$$

So, the solution of this system is $a = -32$, $v_0 = 20$, and $s_0 = 100$, which can be written as $(-32, 20, 100)$.

This results in a position equation of $s = \frac{1}{2}(-32)t^2 + 20t + 100$

$$= -16t^2 + 20t + 100$$

and implies that the object was thrown upward at a velocity of 20 feet per second from a height of 100 feet.

8. Because the graph of $y = ax^2 + bx + c$ passes through the points $(0, 0)$, $(3, -3)$, and $(6, 0)$, you can write the following.

When $x = 0$, $y = 0$: $a(0)^2 + b(0) + c = 0$

When $x = 3$, $y = -3$: $a(3)^2 + b(3) + c = -3$

When $x = 6$, $y = 0$: $a(6)^2 + b(6) + c = 0$

This produces the following system of linear equations.

$$\begin{cases} c = 0 & \text{Equation 1} \\ 9a + 3b + c = -3 & \text{Equation 2} \\ 36a + 6b + c = 0 & \text{Equation 3} \end{cases}$$

You can reorder these equations as shown.

$$\begin{cases} 36a + 6b + c = 0 \\ 9a + 3b + c = -3 \\ c = 0 \end{cases}$$

$$\begin{cases} 36a + 6b + c = 0 \\ -6b - 3c = 12 \\ c = 0 \end{cases}$$

Adding -4 times the second equation to the first equation produces a new second equation.

$$\begin{cases} a + \frac{1}{6}b + \frac{1}{36}c = 0 \\ b + \frac{1}{2}c = -2 \\ c = 0 \end{cases}$$

Multiplying the first equation by $\frac{1}{36}$ produces a new first equation and multiplying the second equation by $-\frac{1}{6}$ produces a new second equation.

So, $c = 0$,

$$b + \tfrac{1}{2}(0) = -2$$
$$b = -2,$$

and $a + \frac{1}{6}(-2) + \frac{1}{36}(0) = 0$

$$a = \tfrac{1}{3}.$$

The solution of this system is $a = \frac{1}{3}$, $b = -2$, and $c = 0$.

So, the equation of the parabola is $y = \frac{1}{3}x^2 - 2x$.

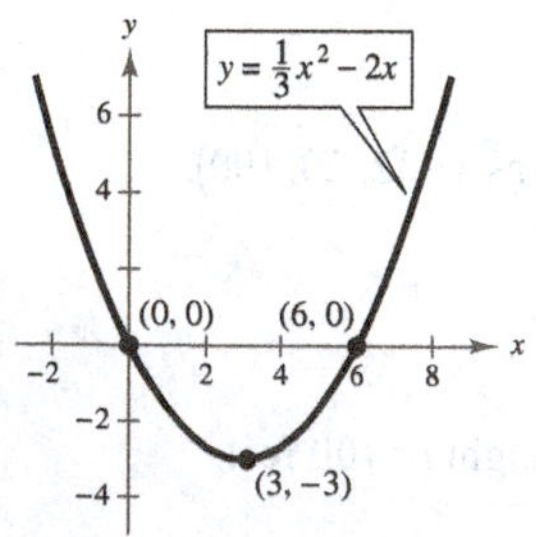

Checkpoints for Section 6.4

1. The expression is proper, so you should begin by factoring the denominator. Because

$2x^2 - x - 1 = (2x + 1)(x - 1)$

you should include one partial fraction with a constant numerator for each linear factor of the denominator. Write the form of the decomposition as follows.

$$\frac{x + 5}{2x^2 - x - 1} = \frac{A}{2x + 1} + \frac{B}{x - 1}$$

Multiplying each side of this equation by the least common denominator, $(2x + 1)(x - 1)$, leads to the **basic equation**

$x + 5 = A(x - 1) + B(2x + 1).$

Because this equation is true for all x, substitute any *convenient* values of x that will help determine the constants A and B. Values of x that are especially convenient are those that make the factors $x - 1$ and $2x + 1$ equal to zero. For instance, to solve for B, let $x = 1$. Then

$$\begin{aligned} 1 + 5 &= A(1 - 1) + B[2(1) + 1] && \text{Substitute 1 for } x \\ 6 &= A(0) + B(3) \\ 6 &= 3B \\ 2 &= B. \end{aligned}$$

To solve for A, let $x = -\frac{1}{2}$ and then

$$\begin{aligned} -\frac{1}{2} + 5 &= A\left(\frac{1}{2} - 1\right) + B\left[2\left(\frac{1}{2}\right) + 1\right] && \text{Substitute } \frac{1}{2} \text{ for } x \\ \frac{9}{2} &= A\left(-\frac{3}{2}\right) + B(0) \\ \frac{9}{2} &= -\frac{3}{2}A \\ -3 &= A. \end{aligned}$$

So, the partial fraction decomposition is

$$\frac{x + 5}{2x^2 - x - 1} = \frac{-3}{2x + 1} + \frac{2}{x - 1}.$$

Check this result by combining the two partial fractions on the right side of the equation, or by using your graphing utility.

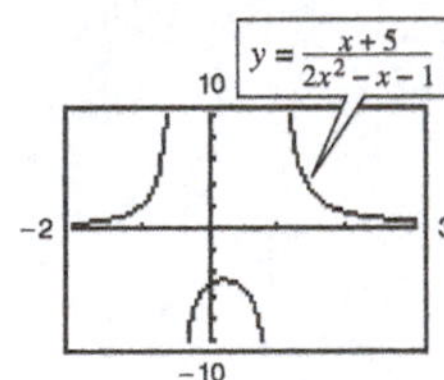

2. This rational expression is improper, so you should begin by dividing the numerator by the denominator.

$$\frac{x^4 + x^3 + x + 4}{x^3 + x^2} \Rightarrow x^3 + x^2 \overline{\smash{\big)} x^4 + x^3 + 0x^2 + x + 4}$$

$$\begin{array}{r} x \\ x^3 + x^2 \overline{\smash{\big)} x^4 + x^3 + 0x^2 + x + 4} \\ \underline{x^4 + x^3 } \\ x + 4 \end{array}$$

So, $\dfrac{x^4 + x^3 + x + 4}{x^3 + x^2} = x + \dfrac{x + 4}{x^3 + x^2}.$

Because the denominator of the remainder factors as $x^3 + x^2 = x^2(x + 1)$, you should include one partial fraction with a constant numerator for each power of x and $x + 1$, and write the form of the decomposition as follows.

$$\frac{x + 4}{x^3 + x^2} = \frac{A}{x} + \frac{B}{x^2} + \frac{C}{x + 1}$$

Multiplying each side by the LCD, $x^2(x + 1)$, leads to the basic equation

$$x + 4 - Ax(x + 1) + B(x + 1) + Cx^2.$$

Letting $x = -1$ eliminates the A- and B-terms and yields the following.

$$\begin{aligned} -1 + 4 &= A(-1)(-1 + 1) + B(-1 + 1) + C(-1)^2 \\ 3 &= 0 + 0 + C \\ 3 &= C \end{aligned}$$

Letting $x = 0$, eliminates the A- and C-terms.

$$\begin{aligned} 0 + 4 &= A(0)(0 + 1) + B(0 + 1) + C(0)^2 \\ 4 &= 0 + B + 0 \\ 4 &= B \end{aligned}$$

At this point, you have exhausted the most convenient values of x, so to find the value of A, use any other value of x along with the known values of B and C.

So, using $x = 1$, $B = 4$ and $C = 3$,

$$\begin{aligned} 1 + 4 &= A(1)(1 + 1) + 4(1 + 1) + 3(1)^2 \\ 5 &= 2A + 8 + 3 \\ -6 &= 2A \\ -3 &= A. \end{aligned}$$

So, the partial fraction decomposition is

$$\frac{x^4 + x^3 + x + 4}{x^3 + x^2} = x + \frac{-3}{x} + \frac{4}{x^2} + \frac{3}{x + 1}.$$

3. This expression is proper, so begin by factoring the denominator. Because the denominator factors as

$x^3 + x = x(x^2 + 1)$

you should include one partial fraction with a constant numerator and one partial fraction with a linear numerator, and write the form of the decomposition as follows.

$$\frac{2x^2 - 5}{x^3 + x} = \frac{A}{x} + \frac{Bx + C}{x^2 + 1}$$

Multiplying each side by the LCD, $x(x^2 + 1)$ yields the basic equation

$$2x^2 - 5 = A(x^2 + 1) + (Bx + C)x$$

Expanding this basic equation and collecting like terms produces

$$\begin{aligned} 2x^2 - 5 &= Ax^2 + A + Bx^2 + Cx \\ &= (A + B)x^2 + Cx + A. \qquad \text{Polynomial form} \end{aligned}$$

Finally, because two polynomials are equal if and only if the coefficients of like terms are equal, equate the coefficients of like terms on opposite sides of the equation.

$$2x^2 + 0x - 5 = (A + B)x^2 + Cx + A$$

Now write the following system of linear equations.

$$\begin{cases} A + B \quad = 2 & \text{Equation 1} \\ \quad\quad C = 0 & \text{Equation 2} \\ A \quad\quad = -5 & \text{Equation 3} \end{cases}$$

From this system, you can see that $A = -5$ and $C = 0$.

Moreover, back-substituting $A = -5$ into Equation 1 yields $-5 + B = 2 \Rightarrow B = 7$.

So, the partial fraction decomposition is $\dfrac{2x^2 - 5}{x^3 + x} = \dfrac{-5}{x} + \dfrac{7x}{x^2 + 1}$.

4. Include one partial fraction with a linear numerator for each power of $(x^2 + 4)$.

$$\frac{x^3 + 3x^2 - 2x + 7}{(x^2 + 4)^2} = \frac{Ax + B}{x^2 + 4} + \frac{Cx + D}{(x^2 + 4)^2} \qquad \text{Write form of decomposition.}$$

Multiplying each side by the LCD, $(x^2 + 4)^2$, yields the basic equation

$$\begin{aligned} x^3 + 3x^2 - 2x + 7 &= (Ax + B)(x^2 + 4) + Cx + D \qquad \text{Basic equation} \\ &= Ax^3 + 4Ax + Bx^2 + 4B + Cx + D \\ &= Ax^3 + Bx^2 + (4A + C)x + (4B + D). \qquad \text{Polynomial form} \end{aligned}$$

Equating coefficients of like terms on opposite sides of the equation

$$x^3 + 3x^2 - 2x + 7 = Ax^3 + Bx^2 + (4A + C)x + (4B + D)$$

produces the following system of linear equations.

$$\begin{cases} A \quad\quad\quad\quad = 1 & \text{Equation 1} \\ \quad B \quad\quad\quad = 3 & \text{Equation 2} \\ 4A + \quad C \quad = -2 & \text{Equation 3} \\ \quad 4B + \quad D = 7 & \text{Equation 4} \end{cases}$$

Use the values $A = 1$ and $B = 3$ to obtain the following.

$4(1) + C = -2$ Substitute 1 for A in Equation 3.

$C = -6$

$4(3) + D = 7$ Substitute 3 for B in Equation 4.

$D = -5$

So, using $A = 1$, $B = 3$, $C = -6$, and $D = -5$

The partial fraction decomposition is $\dfrac{x^3 + 3x^2 - 2x + 7}{(x^2 + 4)^2} = \dfrac{x + 3}{x^2 + 4} + \dfrac{-6x - 5}{(x^2 + 4)^2}$.

Check this result by combining the two partial fractions on the right side of the equation, or by using your graphing utility.

5. Include one partial fraction with a constant numerator for each power of x and one partial fraction with a linear numerator for each power of $(x^2 + 2)$.

$$\frac{4x - 8}{x^2(x^2 + 2)^2} = \frac{A}{x} + \frac{B}{x^2} + \frac{Cx + D}{x^2 + 2} + \frac{Ex + F}{(x^2 + 2)^2} \quad \text{Write form of decomposition.}$$

Multiplying each side by the LCD, $x^2(x^2 + 2)^2$, yields the basic equation

$$\begin{aligned}4x - 8 &= Ax(x^2 + 2)^2 B(x^2 + 2)^2 + (Cx + D)x^2(x^2 + 2) + (Ex + F)x^2\\ &= Ax(x^4 + 4x^2 + 4) + B(x^4 + 4x^2 + 4) + (Cx + D)(x^4 + 2x^2) + (Ex + F)x^2\\ &= Ax^5 + 4Ax^3 + 4Ax + Bx^4 + 4Bx^2 + 4B + Cx^5 + 2Cx^3 + Dx^4 + 2Dx^2 + Ex^3 + Fx^2\\ &= (A + C)x^5 + (B + D)x^4 + (4A + 2C + E)x^3 + (4B + 2D + F)x^2 + (4A)x + 4B\end{aligned}$$

Equating coefficients yields this system of linear equations.

$$\begin{cases} A + C = 0 & \text{Equation 1}\\ B + D = 0 & \text{Equation 2}\\ 4A + 2C + E = 0 & \text{Equation 3}\\ 4B + 2D + F = 0 & \text{Equation 4}\\ 4A = 4 & \text{Equation 5}\\ 4B = -8 & \text{Equation 6}\end{cases}$$

So, from Equations 5 and 6, $A = 1$ and $B = -2$.

Then back-substituting into Equations 1 and 2, $1 + C = 0 \Rightarrow C = -1$ and $-2 + D = 0 \Rightarrow D = 2$.

Using these values and Equations 3 and 4, you have

$4(1) + 2(-1) + E = 0 \Rightarrow E = -2$ and $4(-2) + 2(2) + F = 0 \Rightarrow F = 4$.

So, $A = 1$, $B = -2$, $C = -1$, $D = 2$, $E = -2$, and $F = 4$.

The partial fraction decomposition is

$$\frac{4x - 8}{x^2(x^2 + 2)^2} = \frac{1}{x} + \frac{-2}{x^2} + \frac{-x + 2}{x^2 + 2} + \frac{-2x + 4}{(x^2 + 2)^2}.$$

Checkpoints for Section 6.5

1. Begin by graphing the corresponding equation $(x + 2)^2 + (y - 2)^2 = 16$, which is a circle, with center $(-2, 2)$ and a radius of 4 units as shown.

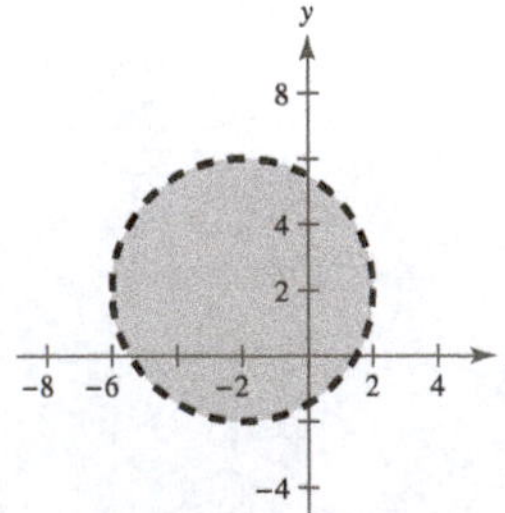

Test a point inside the circle such as $(-2, 2)$ and a point outside the circle such as $(4, 2)$.

The points that satisfy the inequality are those lying inside the circle but not on the circle.

$(-2, 2)$: $(x + 2)^2 + (y - 2)^2 \overset{?}{<} 16$

$(-2 + 2)^2 + (2 - 2)^2 \overset{?}{<} 16$

$0 < 16$

$(-2, 2)$ is a solution.

$(4, 2)$: $(x + 2)^2 + (y - 2)^2 < 16$

$(4 + 2)^2 + (2 - 2)^2 \overset{?}{<} 16$

$36 \not< 16$

$(4, 2)$ is not a solution.

2. The graph of the corresponding equation $x = 3$ is a vertical line. The points that satisfy the inequality $x \geq 3$ are those lying to the right of (or on) this line.

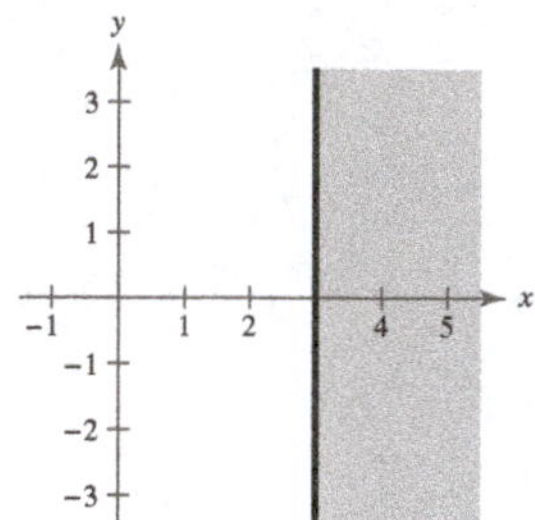

3. The graph of the corresponding equation $x + y = -2$ is a line as shown. Because the origin $(0, 0)$ satisfies the inequality, the graph consists of the half-plane lying above the line.

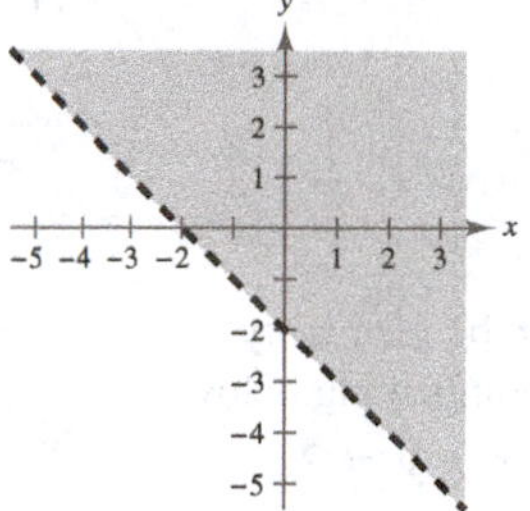

4. The graphs of each of these inequalities are shown independently.

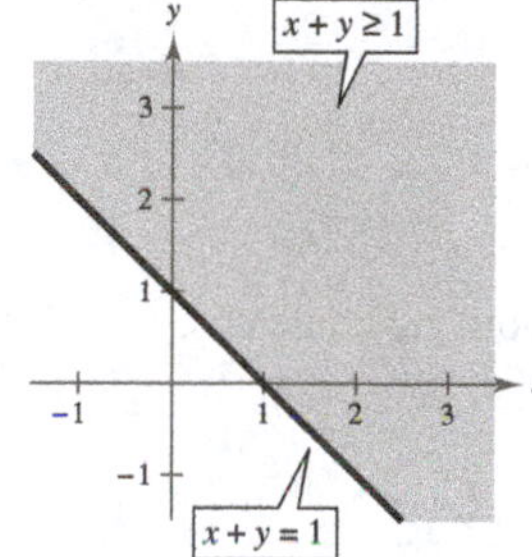

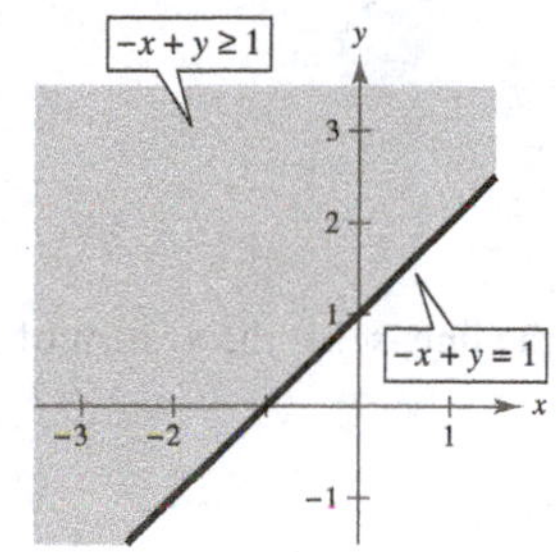

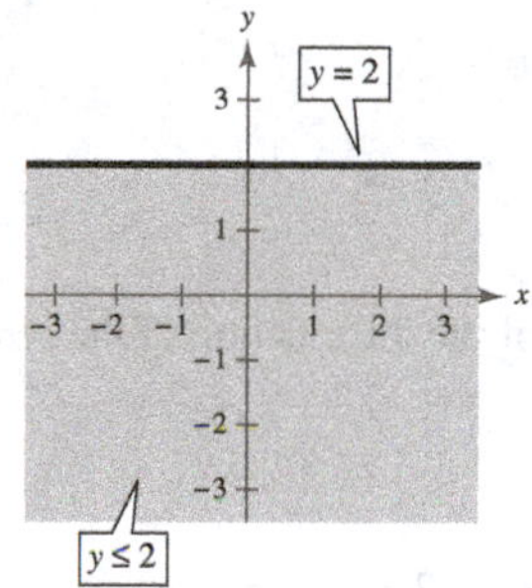

By superimposing the graphs on the same coordinate system, the region common to all three graphs can be found. To find the vertices of the region, solve the three systems of corresponding equations by taking pairs of equations representing the boundaries of the individual regions.

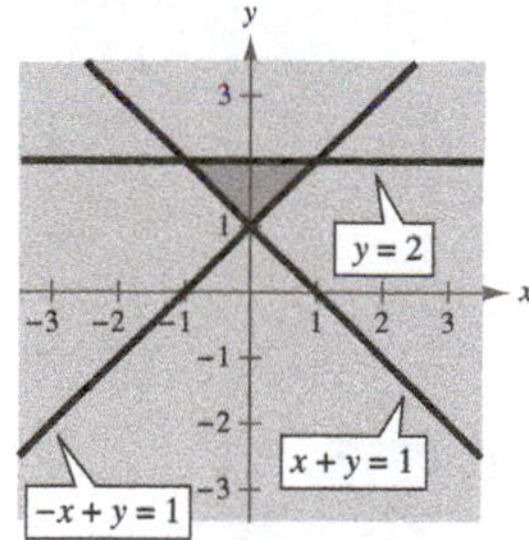

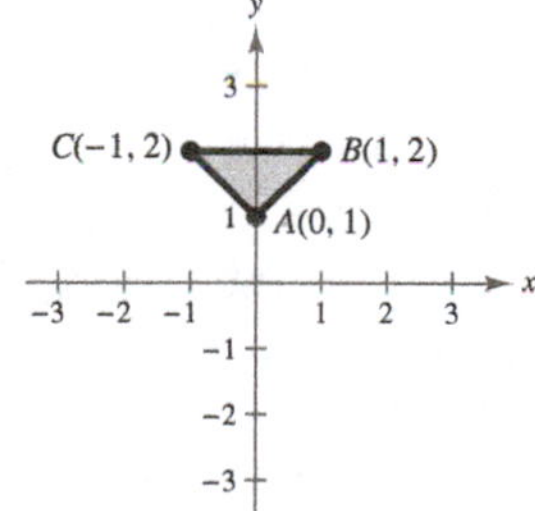

Vertex A: $(0, 1)$

$$\begin{cases} x + y = 1 \\ -x + y = 1 \end{cases}$$

Vertex B: $(1, 2)$

$$\begin{cases} -x + y = 1 \\ y = 2 \end{cases}$$

Vertex C: $(-1, 2)$

$$\begin{cases} x + y = 1 \\ y = 2 \end{cases}$$

Note that the vertices of the region are represented by solid dots. This means that the vertices *are* solutions of the system of equations as well as all of the points that lie on the lines.

5. The points that satisfy the inequality $x - y^2 > 0$ are the points inside the parabola (but not on) the parabola $x = y^2$.

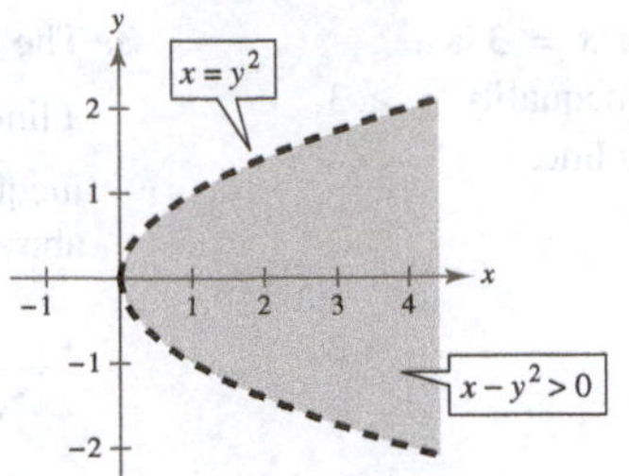

The points satisfying the inequality $x + y < 2$ are the points lying below (but not on) the line $x + y = 2$.

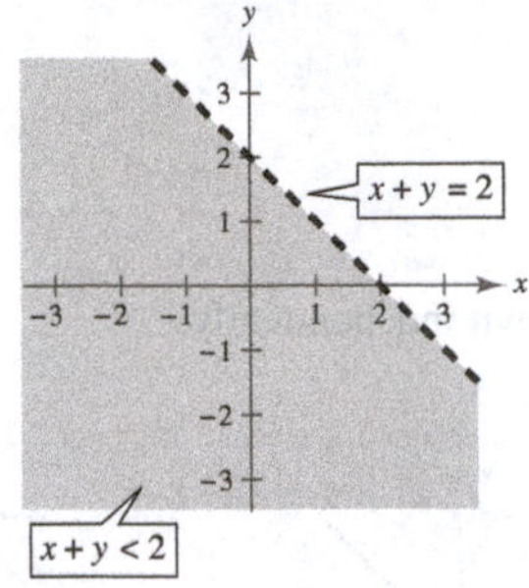

To find the points of the intersection of the parabola and the line, solve the system of corresponding equations.

$$\begin{cases} x - y^2 = 0 \\ x + y = 2 \end{cases}$$

$$x + y = 2 \Rightarrow y = 2 - x$$

$$\begin{aligned} x - y^2 &= 0 \\ x - (2 - x)^2 &= 0 \\ x - \left(4 - 4x + x^2\right) &= 0 \\ x - 4 + 4x - x^2 &= 0 \\ -x^2 + 5x - 4 &= 0 \\ -\left(x^2 - 5x + 4\right) &= 0 \\ (x - 4)(x - 1) &= 0 \end{aligned}$$

$$x - 4 = 0 \qquad x - 1 = 0$$

$$x = 4 \qquad x = 1$$

When $x = 4$, $y = 2 - 4 = -2$.

When $x = 1$, $y = 2 - 1 = 1$.

Using the method of substitution, you can find the solutions to be $(4, -2)$ and $(1, 1)$.

So, the region containing all points that satisfy the system is indicated by the shaded region.

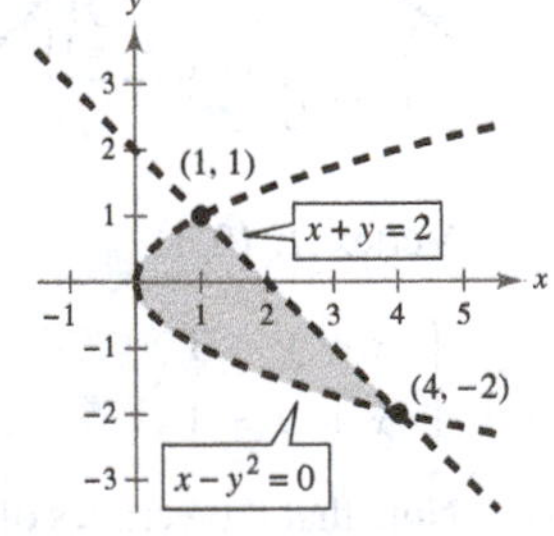

6. From the way the system is written, it should be clear that the system has no solution because the quantity $2x - y$ cannot be both less than -3 and greater than 1. The graph of the inequality $2x - y < -3$ is the half-plane lying above the line $2x - y = -3$ and the graph of the inequality $2x - y > 1$ is the half-plane lying below the line $2x - y = 1$ as shown. These two half-planes have no points in common. So, the system of inequalities has no solution.

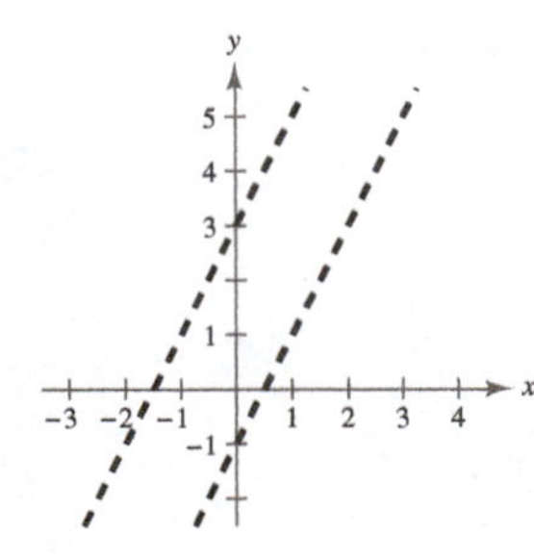

7. The graph of the inequality $x^2 - y < 0$ is the region inside the parabola $x^2 - y = 0$. The graph of the inequality $x - y < -2$ is the half-plane that lies above the line $x - y = -2$. The intersection of these regions is an infinite region having points of intersection at $(-1, 1)$ and $(2, 4)$, as shown below. So, the solution set of the system of inequalities is unbounded.

Points of intersection:

$$\begin{cases} x^2 - y = 0 \Rightarrow x^2 = y \\ x - y = -2 \end{cases}$$

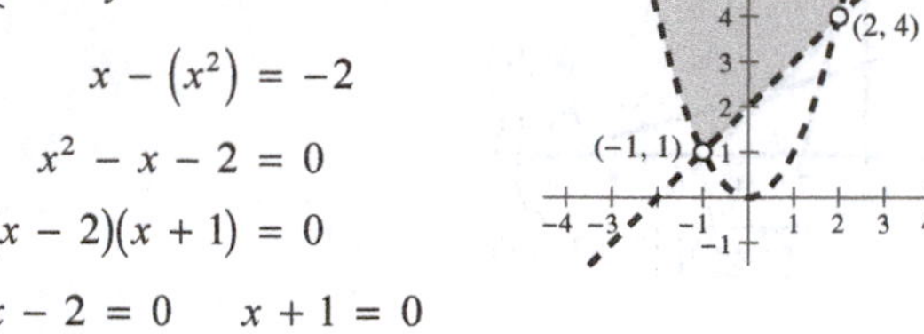

$$x - (x^2) = -2$$

$$x^2 - x - 2 = 0$$

$$(x - 2)(x + 1) = 0$$

$$x - 2 = 0 \quad x + 1 = 0$$

$$x = 2 \quad x = -1$$

When $x = 2$, $y = (2)^2 = 4$.

When $x = -1$ $y = (-1)^2 = 1$.

8. Begin by finding the equilibrium point (when supply and demand are equal) by solving the equation

$492 + 0.00003x = 567 - 0.00002x$.

In checkpoint 9 in Section 9.2, you saw that the solution is $x = 1{,}500{,}000$ units, which corresponds to an equilibrium price of $p = 537$. So, the consumer surplus and producer surplus are the areas of the following triangular regions.

Consumer Surplus

$$\begin{cases} p \le 567 - 0.00002x \\ p \ge 537 \\ x \ge 0 \end{cases}$$

Producer Surplus

$$\begin{cases} p \ge 492 + 0.00003x \\ p \le 537 \\ x \ge 0 \end{cases}$$

The consumer and producer surpluses are the areas of the shaded triangles shown.

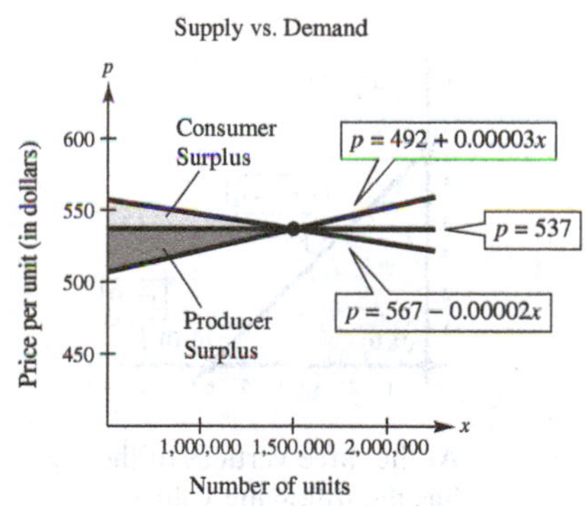

Consumer surplus $= \frac{1}{2}(\text{base})(\text{height})$

$= \frac{1}{2}(1{,}500{,}000)(30)$

$= \$22{,}500{,}000$

Producer surplus $= \frac{1}{2}(\text{base})(\text{height})$

$= \frac{1}{2}(1{,}500{,}000)(45)$

$= \$33{,}750{,}000$

9. Begin by letting x represent the number of bottles of brand X coral nutrients and y represent the number of bottles of brand Y coral nutrients.

To meet the minimum required amounts of nutrients, the following inequalities must be satisfied.

$$\begin{cases} 8x + 2y \geq 16 & \text{Nutrient A} \\ x + y \geq 5 & \text{Nutrient B} \\ 2x + 7y \geq 20 & \text{Nutrient C} \\ x \geq 0 \\ y \geq 0 \end{cases}$$

The graph of this system of inequalities is shown.

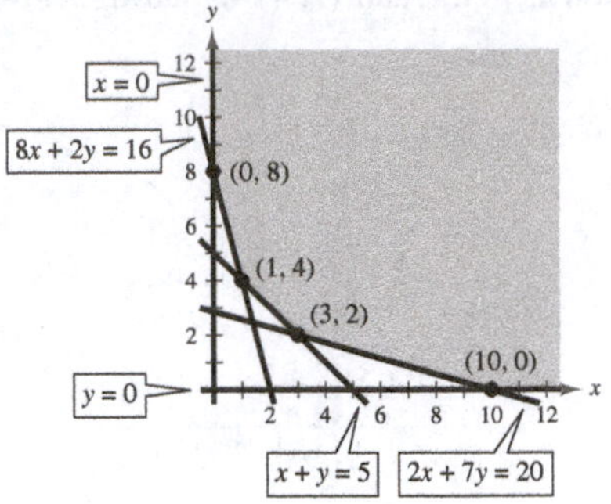

Nutrients A and B

$$\begin{cases} 8x + 2y = 16 \\ x + y = 5 \end{cases}$$

$$\begin{cases} 8x + 2y = 16 \\ -2x - 2y = -10 \end{cases}$$

$$\begin{cases} 6x = 6 \rightarrow x = 1 \\ 1 + y = 5 \rightarrow y = 4 \end{cases}$$

Nutrients B and C

$$\begin{cases} x + y = 5 \\ 2x + 7y = 20 \end{cases}$$

$$\begin{cases} 2x + 2y = 10 \\ -2x - 7y = -20 \end{cases}$$

$$\begin{cases} -5y = -10 \rightarrow y = 2 \\ x + 2 = 5 \rightarrow x = 3 \end{cases}$$

Checkpoints for Section 6.6

1. The constraints form the region shown.

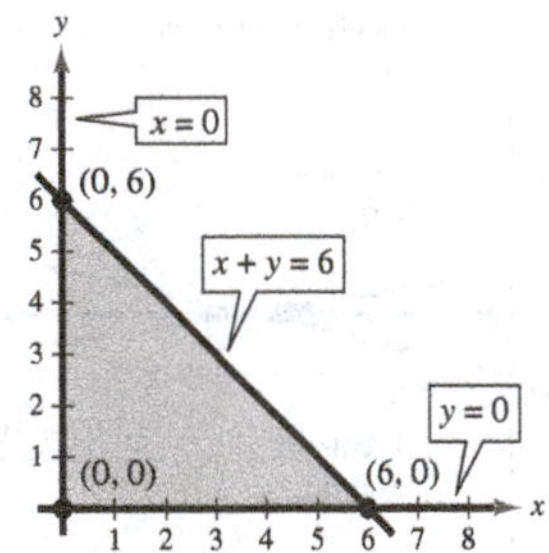

At the three vertices of the region, the objective function has the following values,

At $(0, 0)$: $z = 4(0) + 5(0) = 0$

At $(0, 6)$: $z = 4(0) + 5(6) = 30$

At $(6, 0)$: $z = 4(6) + 5(0) = 24$

So the maximum value of z is 30, and this occurs when $x = 0$ and $y = 6$.

2. The constraints form the region shown.

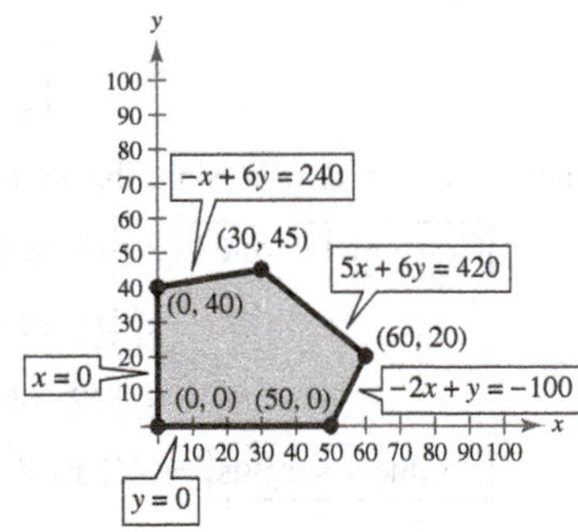

By testing the objective function at each vertex, you obtain the following.

At $(0, 0)$: $z = 12(0) + 8(0) = 0$

At $(0, 40)$: $z = 12(0) + 8(40) = 320$

At $(30, 45)$: $z = 12(30) + 8(45) = 720$

At $(60, 20)$: $z = 12(60) + 8(20) = 880$

At $(50, 0)$: $z = 12(50) + 8(0) = 600$

So, the minimum value of z is 0, which occurs when $x = 0$ and $y = 0$.

3. Using the values of z at the vertices shown in Checkpoint Example 2, the maximum value of z is

$$z = 12(60) + 8(20)$$
$$= 880$$

and occurs when $x = 60$ and $y = 20$.

4. The constraints form the region shown.

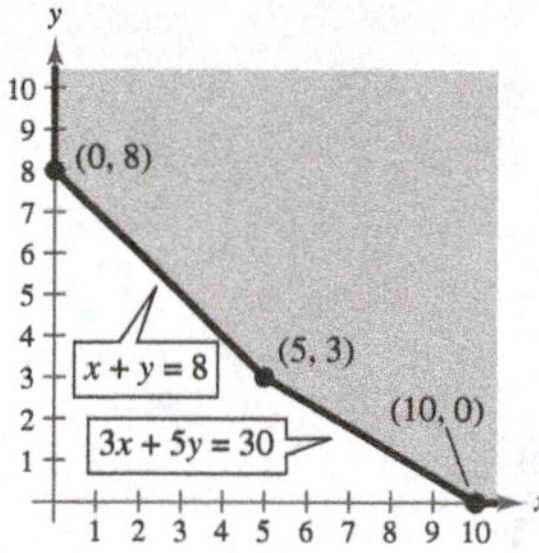

By testing the objective function at each vertex, you obtain the following.

At $(0, 8)$: $z = 3(0) + 7(8) = 56$

At $(5, 3)$: $z = 3(5) + 7(3) = 36$

At $(10, 0)$: $z = 3(10) + 7(0) = 30$

So, the minimum of z is 30, which occurs when $x = 10$ and $y = 0$.

5. Let x be the number of boxes of chocolate-covered creams and let y be the number of boxes of chocolate-covered nuts. So, the objective function (for the combined profit) is

$P = 2.5x + 2y$ Objective function

To find the maximum monthly profit, test the values of P at the vertices of the region.

At $(0, 0)$: $P = 2.5(0) + 2(0) = 0$

At $(800, 400)$: $P = 2.5(800) + 2(400) = 2800$

At $(1050, 150)$: $P = 2.5(1050) + 2(150) = 2925$ Maximum Profit

At $(600, 0)$: $P = 2.5(600) + 2(0) = 1500$

So, the maximum monthly profit is \$2925 and it occurs when the monthly production consists of 1050 boxes of chocolate-covered creams and 150 boxes of chocolate-covered nuts.

6. Begin by letting *x* represent the number of bottles of brand X coral nutrients and *y* represent the number of bottles of brand Y coral nutrients.

To meet the minimum required amounts of nutrients, the following inequalities must be satisfied.

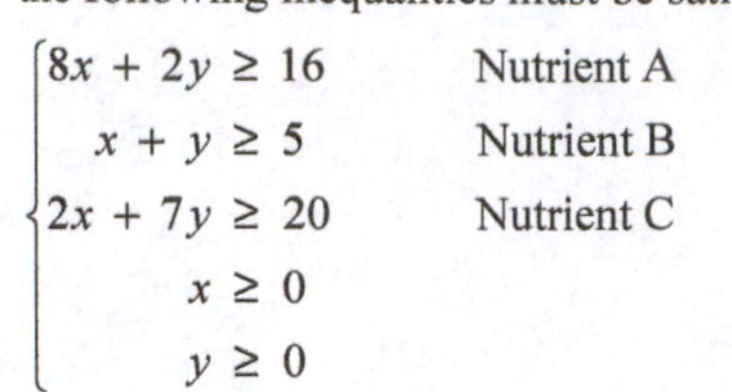

$$\begin{cases} 8x + 2y \geq 16 & \text{Nutrient A} \\ x + y \geq 5 & \text{Nutrient B} \\ 2x + 7y \geq 20 & \text{Nutrient C} \\ x \geq 0 & \\ y \geq 0 & \end{cases}$$

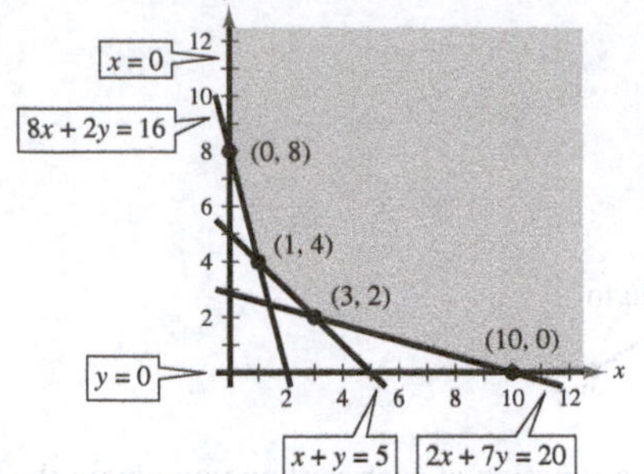

The graph of this system of inequalities is shown.

Nutrients A and B

$$\begin{cases} 8x + 2y = 16 \\ x + y = 5 \end{cases}$$

$$\begin{cases} 8x + 2y = 16 \\ -2x - 2y = -10 \end{cases}$$

$$\begin{cases} 6x = 6 \rightarrow x = 1 \\ 1 + y = 5 \rightarrow y = 4 \end{cases}$$

Nutrients B and C

$$\begin{cases} x + y = 5 \\ 2x + 7y = 20 \end{cases}$$

$$\begin{cases} 2x + 2y = 10 \\ -2x - 7y = -20 \end{cases}$$

$$\begin{cases} -5y = -10 \rightarrow y = 2 \\ x + 2 = 5 \rightarrow x = 3 \end{cases}$$

The figure shows the graph of the region corresponding to the constraints.

The cost function is given by $C = 15x + 30y$.

Because you want to incur as little cost as possible, you want to determine the minimum cost.

At $(0, 8)$: $C = 15(0) + 30(8) = 240$

At $(1, 4)$: $C = 15(1) + 30(4) = 135$

At $(3, 2)$: $C = 15(3) + 30(2) = 105$ Minimum Cost

At $(10, 0)$: $C = 15(10) + 30(0) = 150$

So, the minimum cost is \$7.20 and occurs when 3 bottles of Brand X and 2 bottles of Brand Y are added.

Chapter 7

Checkpoints for Section 7.1

1. The matrix has *two* rows and *three* columns. The order of the matrix is 2×3.

2. $\begin{cases} x + y + z = 2 \\ 2x - y + 3z = -1 \\ -x + 2y - z = 4 \end{cases}$

All of the variables are aligned in the system. Next, use the coefficients and constant terms as the matrix entries.

$$\begin{matrix} R_1 \\ R_2 \\ R_3 \end{matrix} \begin{bmatrix} 1 & 1 & 1 & \vdots & 2 \\ 2 & -1 & 3 & \vdots & -1 \\ -1 & 2 & -1 & \vdots & 4 \end{bmatrix}$$

The augmented matrix has three rows and four columns, so it is a 3×4 matrix.

3. Add -3 times the first row of the original matrix to the second row.

Original Matrix **New Row-Equivalent Matrix**

$$\begin{bmatrix} 1 & 0 & 2 \\ 3 & 1 & 7 \\ 2 & -6 & 14 \end{bmatrix} \quad -3R_1 + R_2 \rightarrow \begin{bmatrix} 1 & 0 & 2 \\ 0 & 1 & 1 \\ 2 & -6 & 14 \end{bmatrix}$$

4. Linear System | **Associated Augmented Matrix**

$$\begin{cases} 2x + y - z = -3 \\ 4x - 2y + 2z = -2 \\ -6x + 5y + 4z = 10 \end{cases} \qquad \begin{bmatrix} 2 & 1 & -1 & : & -3 \\ 4 & -2 & 2 & : & -2 \\ -6 & 5 & 4 & : & 10 \end{bmatrix}$$

Multiply the first equation by $\frac{1}{2}$.

$$\begin{cases} x + \frac{1}{2}y - \frac{1}{2}z = -\frac{3}{2} \\ 4x - 2y + 2z = -2 \\ -6x + 5y + 4z = 10 \end{cases} \qquad \frac{1}{2}R_1 \rightarrow \begin{bmatrix} 1 & \frac{1}{2} & -\frac{1}{2} & : & -\frac{3}{2} \\ 4 & -2 & 2 & : & -2 \\ -6 & 5 & 4 & : & 10 \end{bmatrix}$$

Add -4 times the first equation to the second equation.

$$\begin{cases} x + \frac{1}{2}y - \frac{1}{2}z = -\frac{3}{2} \\ -4y + 4z = 4 \\ -6x + 5y + 4z = 10 \end{cases} \qquad -4R_1 + R_2 \rightarrow \begin{bmatrix} 1 & \frac{1}{2} & -\frac{1}{2} & : & -\frac{3}{2} \\ 0 & -4 & 4 & : & 4 \\ -6 & 5 & 4 & : & 10 \end{bmatrix}$$

Multiply the second equation by $-\frac{1}{4}$.

$$\begin{cases} x + \frac{1}{2}y - \frac{1}{2}z = -\frac{3}{2} \\ y - z = -1 \\ -6x + 5y + 4z = 10 \end{cases} \qquad -\frac{1}{4}R_2 \rightarrow \begin{bmatrix} 1 & \frac{1}{2} & -\frac{1}{2} & : & -\frac{3}{2} \\ 0 & 1 & -1 & : & -1 \\ -6 & 5 & 4 & : & 10 \end{bmatrix}$$

Add 6 times the first equation to the third equation.

$$\begin{cases} x + \frac{1}{2}y - \frac{1}{2}z = -\frac{3}{2} \\ y - z = -1 \\ 8y + z = 1 \end{cases} \qquad 6R_1 + R_3 \rightarrow \begin{bmatrix} 1 & \frac{1}{2} & -\frac{1}{2} & : & -\frac{3}{2} \\ 0 & 1 & -1 & : & -1 \\ 0 & 8 & 1 & : & 1 \end{bmatrix}$$

Add -8 times the second equation to the third equation.

$$\begin{cases} x + \frac{1}{2}y - \frac{1}{2}z = -\frac{3}{2} \\ y - z = -1 \\ 9z = 9 \end{cases} \qquad -8R_2 + R_3 \rightarrow \begin{bmatrix} 1 & \frac{1}{2} & -\frac{1}{2} & : & -\frac{3}{2} \\ 0 & 1 & -1 & : & -1 \\ 0 & 0 & 9 & : & 9 \end{bmatrix}$$

Multiply the third equation by $\frac{1}{9}$.

$$\begin{cases} x + \frac{1}{2}y - \frac{1}{2}z = -\frac{3}{2} \\ y - z = -1 \\ z = 1 \end{cases} \qquad \frac{1}{9}R_3 \rightarrow \begin{bmatrix} 1 & \frac{1}{2} & -\frac{1}{2} & : & -\frac{3}{2} \\ 0 & 1 & -1 & : & -1 \\ 0 & 0 & 1 & : & 1 \end{bmatrix}$$

At this point, you can use back-substitution to find x and y.

$$y - z = -1$$
$$y - (1) = -1$$
$$y = 0$$
$$x + \frac{1}{2}y - \frac{1}{2}z = -\frac{3}{2}$$
$$x + \frac{1}{2}(0) - \frac{1}{2}(1) = -\frac{3}{2}$$
$$x = -1$$

The solution is $x = -1$, $y = 0$, and $z = 1$.

5. The Matrix is in row-echelon form, because the row consisting entirely of zeros occurs at the bottom of the matrix, and for each row that does not consist entirely of zeros, the first nonzero entry is 1. Furthermore the matrix is in reduced row-echelon form, since every column that has a leading 1 has zeros in every position above and below its leading 1.

6.
$$\begin{bmatrix} -3 & 5 & 3 & \vdots & -19 \\ 3 & 4 & 4 & \vdots & 8 \\ 4 & -8 & -6 & \vdots & 26 \end{bmatrix}$$ Write augmented matrix.

$$\begin{matrix} R_3 + R_1 \rightarrow \\ \\ \\ \end{matrix} \begin{bmatrix} 1 & -3 & -3 & \vdots & 7 \\ 3 & 4 & 4 & \vdots & 8 \\ 4 & -8 & -6 & \vdots & 26 \end{bmatrix}$$ Add R_3 to R_1 so first column has leading 1 in upper left corner.

$$\begin{matrix} \\ -3R_1 + R_2 \rightarrow \\ -4R_1 + R_3 \rightarrow \end{matrix} \begin{bmatrix} 1 & -3 & -3 & \vdots & 7 \\ 0 & 13 & 13 & \vdots & -13 \\ 0 & 4 & 6 & \vdots & -2 \end{bmatrix}$$ Perform operations on R_2 and R_3 so first column has zeros below its leading 1.

$$\begin{matrix} \\ R_2 + (-3)R_3 \rightarrow \\ \\ \end{matrix} \begin{bmatrix} 1 & -3 & -3 & \vdots & 5 \\ 0 & 1 & -5 & \vdots & -7 \\ 0 & 4 & 6 & \vdots & -2 \end{bmatrix}$$ Perform operations on R_2 so second column has a leading 1.

$$\begin{matrix} \\ \\ -4R_2 + R_3 \rightarrow \end{matrix} \begin{bmatrix} 1 & -3 & -3 & \vdots & 7 \\ 0 & 1 & -5 & \vdots & -7 \\ 0 & 0 & 26 & \vdots & 26 \end{bmatrix}$$ Perform operations on R_3 so second column has a zero below its leading 1.

$$\begin{matrix} \\ \\ \frac{1}{26}R_3 \rightarrow \end{matrix} \begin{bmatrix} 1 & -3 & -3 & \vdots & 7 \\ 0 & 1 & -5 & \vdots & -7 \\ 0 & 0 & 1 & \vdots & 1 \end{bmatrix}$$ Perform operations on R_3 so third column has a leading 1.

The matrix is now in row-echelon form, and the corresponding system is

$$\begin{cases} x - 3y - 3z = 7 \\ y - 5z = -7 \\ z = 1 \end{cases}$$

Using back-substitution, you can determine that the solution is $x = 4$, $y = -2$ and $z = 1$.

7.
$$\begin{bmatrix} 1 & 1 & 1 & \vdots & 1 \\ 1 & 2 & 2 & \vdots & 2 \\ 1 & -1 & -1 & \vdots & 1 \end{bmatrix}$$ Write augmented matrix.

$$\begin{matrix} \\ -R_1 + R_2 \rightarrow \\ -R_1 + R_3 \rightarrow \end{matrix} \begin{bmatrix} 1 & 1 & 1 & \vdots & 1 \\ 0 & 1 & 1 & \vdots & 1 \\ 0 & -2 & -2 & \vdots & 0 \end{bmatrix}$$ Perform row operations.

$$\begin{matrix} \\ \\ 2R_2 + R_3 \rightarrow \end{matrix} \begin{bmatrix} 1 & 1 & 1 & \vdots & 1 \\ 0 & 1 & 1 & \vdots & 1 \\ 0 & 0 & 0 & \vdots & 2 \end{bmatrix}$$ Perform row operations.

Note that the third row of this matrix consists entirely of zeros except for the last entry. This means that the original system of linear equations is inconsistent. You can see why this is true by converting back to a system of linear equations.

$$\begin{cases} x + y + z = 1 \\ y + z = 1 \\ 0 = 2 \end{cases}$$

Because the third equation is not possible, the system has no solution.

8. $\begin{bmatrix} -3 & 7 & 2 & \vdots & 1 \\ -5 & 3 & -5 & \vdots & -8 \\ 2 & -2 & -3 & \vdots & 15 \end{bmatrix}$

$R_2 + R_1 \rightarrow \begin{bmatrix} -1 & 5 & -1 & \vdots & 16 \\ -5 & 3 & -5 & \vdots & -8 \\ 2 & -2 & -3 & \vdots & 15 \end{bmatrix}$

$-R_1 \rightarrow \begin{bmatrix} 1 & -5 & 1 & \vdots & -16 \\ -5 & 3 & -5 & \vdots & -8 \\ 2 & -2 & -3 & \vdots & 15 \end{bmatrix}$

$5R_1 + R_2 \rightarrow \begin{bmatrix} 1 & -5 & 1 & \vdots & -16 \\ 0 & -22 & 0 & \vdots & -88 \\ 2 & -2 & -3 & \vdots & 15 \end{bmatrix}$

$-2R_1 + R_3 \rightarrow \begin{bmatrix} 1 & -5 & 1 & \vdots & -16 \\ 0 & -22 & 0 & \vdots & -88 \\ 0 & 8 & -5 & \vdots & 47 \end{bmatrix}$

$-\frac{1}{22}R_2 \rightarrow \begin{bmatrix} 1 & -5 & 1 & \vdots & -16 \\ 0 & 1 & 0 & \vdots & 4 \\ 0 & 8 & -5 & \vdots & 47 \end{bmatrix}$

$-8R_2 + R_3 \rightarrow \begin{bmatrix} 1 & -5 & 1 & \vdots & -16 \\ 0 & 1 & 0 & \vdots & 4 \\ 0 & 0 & -5 & \vdots & 15 \end{bmatrix}$

$-\frac{1}{5}R_3 \rightarrow \begin{bmatrix} 1 & -5 & 1 & \vdots & -16 \\ 0 & 1 & 0 & \vdots & 4 \\ 0 & 0 & 1 & \vdots & -3 \end{bmatrix}$

At this point, the matrix is in row-echelon form. Now, apply elementary row operations until you obtain zeros above each of the leading is, as follows.

$5R_2 + R_1 \rightarrow \begin{bmatrix} 1 & 0 & 1 & \vdots & 4 \\ 0 & 1 & 0 & \vdots & 4 \\ 0 & 0 & 1 & \vdots & -3 \end{bmatrix}$

$-R_3 + R_1 \rightarrow \begin{bmatrix} 1 & 0 & 0 & \vdots & 7 \\ 0 & 1 & 0 & \vdots & 4 \\ 0 & 0 & 1 & \vdots & -3 \end{bmatrix}$

The matrix is now in reduced row-echelon form. Converting back to a system of linear equations, you have

$$\begin{cases} x = 7 \\ y = 4. \\ z = -3 \end{cases}$$

So, the solution is $x = 7$, $y = 4$, and $z = -3$, which can be written as the ordered triple $(7, 4, -3)$.

9. $\begin{bmatrix} 2 & -6 & 6 & \vdots & 46 \\ 2 & -3 & 0 & \vdots & 31 \end{bmatrix}$

$\begin{bmatrix} 1 & -3 & 3 & \vdots & 23 \\ 2 & -3 & 0 & \vdots & 31 \end{bmatrix}$

$-2R_1 + R_2 \rightarrow \begin{bmatrix} 1 & -3 & 3 & \vdots & 23 \\ 0 & 3 & -6 & \vdots & -15 \end{bmatrix}$

$\frac{1}{3}R_2 \rightarrow \begin{bmatrix} 1 & -3 & 3 & \vdots & 23 \\ 0 & 1 & -2 & \vdots & -5 \end{bmatrix}$

$R_1 + 3R_2 \rightarrow \begin{bmatrix} 1 & 0 & -3 & \vdots & 8 \\ 0 & 1 & -2 & \vdots & -5 \end{bmatrix}$

The corresponding system of equations is

$$\begin{cases} x - 3z = 8 \\ y - 2z = -5. \end{cases}$$

Solving for x and y in terms of z, you have $x = 3z + 8$ and $y = 2z - 5$.

To write a solution of the system that does not use any of the three variables of the system, let a represent any real number and let $z = a$. Substituting a for z in the equations for x and y, you have

$x = 3z + 8 = 3a + 8$ and $y = 2z - 5 = 2a - 5$.

So, the solution set can be written as an ordered triple of the form

$(3a + 8, 2a - 5, a)$

where a is any real number. Remember that a solution set of this form represents an infinite number of solutions. Try substituting values for a to obtain a few solutions. Then check each solution in the original system of equations.

Checkpoints for Section 7.2

1. $\begin{bmatrix} a_{11} & a_{12} \\ a_{21} & a_{22} \end{bmatrix} = \begin{bmatrix} 6 & 3 \\ -2 & 4 \end{bmatrix}$

Because two matrices are equal when their corresponding entries are equal you can conclude that $a_{11} = 6$, $a_{12} = 3$, $a_{21} = -2$, and $a_{22} = 4$.

2. (a) $\begin{bmatrix} 4 & -1 \\ 2 & -3 \end{bmatrix} + \begin{bmatrix} 2 & -1 \\ 0 & 6 \end{bmatrix} = \begin{bmatrix} 4+2 & -1+(-1) \\ 2+0 & -3+6 \end{bmatrix} = \begin{bmatrix} 6 & -2 \\ 2 & 3 \end{bmatrix}$

(b) $\begin{bmatrix} 2 & -1 \\ 3 & 4 \\ 0 & -2 \end{bmatrix} + \begin{bmatrix} -2 & 1 \\ -3 & -4 \\ 0 & 2 \end{bmatrix} = \begin{bmatrix} 0 & 0 \\ 0 & 0 \\ 0 & 0 \end{bmatrix}$

(c) $\begin{bmatrix} 3 & 9 & 6 \\ 0 & 4 & -2 \\ 1 & -1 & 0 \end{bmatrix} + \begin{bmatrix} 3 & 9 & 6 \\ 0 & 2 & -4 \end{bmatrix}$, not possible. The matrices are not of the same dimension.

(d) $\begin{bmatrix} 1 \\ -1 \\ 1 \end{bmatrix} + \begin{bmatrix} -1 \\ 1 \\ 1 \end{bmatrix} = \begin{bmatrix} 0 \\ 0 \\ 2 \end{bmatrix}$

3. $A = \begin{bmatrix} 4 & -1 \\ 0 & 4 \\ -3 & 8 \end{bmatrix}$ and $B = \begin{bmatrix} 0 & 4 \\ -1 & 3 \\ 1 & 7 \end{bmatrix}$

(a) $A - B = \begin{bmatrix} 4 & -1 \\ 0 & 4 \\ -3 & 8 \end{bmatrix} - \begin{bmatrix} 0 & 4 \\ -1 & 3 \\ 1 & 7 \end{bmatrix} = \begin{bmatrix} 4 & -5 \\ 1 & 1 \\ -4 & 1 \end{bmatrix}$

(b) $3A = 3\begin{bmatrix} 4 & -1 \\ 0 & 4 \\ -3 & 8 \end{bmatrix} = \begin{bmatrix} 12 & -3 \\ 0 & 12 \\ -9 & 24 \end{bmatrix}$

(c) $3A - 2B = 3\begin{bmatrix} 4 & -1 \\ 0 & 4 \\ -3 & 8 \end{bmatrix} - 2\begin{bmatrix} 0 & 4 \\ -1 & 3 \\ 1 & 7 \end{bmatrix} = \begin{bmatrix} 12 & -3 \\ 0 & 12 \\ -9 & 24 \end{bmatrix} - \begin{bmatrix} 0 & 8 \\ -2 & 6 \\ 2 & 14 \end{bmatrix} = \begin{bmatrix} 12 & -11 \\ 2 & 6 \\ -11 & 10 \end{bmatrix}$

4. $\begin{bmatrix} 3 & -8 \\ 0 & 2 \end{bmatrix} + \begin{bmatrix} -2 & 3 \\ 6 & -5 \end{bmatrix} + \begin{bmatrix} 0 & 7 \\ 4 & -1 \end{bmatrix} = \begin{bmatrix} 3+(-2)+0 & -8+3+7 \\ 0+6+4 & 2+(-5)+(-1) \end{bmatrix} = \begin{bmatrix} 1 & 2 \\ 10 & -4 \end{bmatrix}$

5. $2\left(\begin{bmatrix} 1 & 3 \\ -2 & 2 \end{bmatrix} + \begin{bmatrix} -4 & 0 \\ -3 & 1 \end{bmatrix}\right) = 2\begin{bmatrix} 1 & 3 \\ -2 & 2 \end{bmatrix} + 2\begin{bmatrix} -4 & 0 \\ -3 & 1 \end{bmatrix}$

$= \begin{bmatrix} 2 & 6 \\ -4 & 4 \end{bmatrix} + \begin{bmatrix} -8 & 0 \\ -6 & 2 \end{bmatrix}$

$= \begin{bmatrix} -6 & 6 \\ -10 & 6 \end{bmatrix}$

6. Begin by solving the matrix equation for X to obtain

$2X - A = B$

$2X = B + A$

$X = \frac{1}{2}(B + A).$

Now, using the matrices A and B you have the following

$X = \frac{1}{2}\left(\begin{bmatrix} 4 & -1 \\ -2 & 5 \end{bmatrix} + \begin{bmatrix} 6 & 1 \\ 0 & 3 \end{bmatrix}\right) = \frac{1}{2}\begin{bmatrix} 10 & 0 \\ -2 & 8 \end{bmatrix} = \begin{bmatrix} 5 & 0 \\ -1 & 4 \end{bmatrix}$

7. $AB = \begin{bmatrix} -1 & 4 \\ 2 & 0 \\ 1 & 2 \end{bmatrix}\begin{bmatrix} 1 & -2 \\ 0 & 7 \end{bmatrix}$

$$= \begin{bmatrix} (-1)(1) + (4)(0) & (-1)(-2) + (4)(7) \\ (2)(1) + (0)(0) & (2)(-2) + (0)(7) \\ (1)(1) + (2)(0) & (1)(-2) + (2)(7) \end{bmatrix}$$

$$= \begin{bmatrix} -1 & 30 \\ 2 & -4 \\ 1 & 12 \end{bmatrix}$$

8. $AB = \begin{bmatrix} 0 & 4 & -3 \\ 2 & 1 & 7 \\ 3 & -2 & 1 \end{bmatrix}\begin{bmatrix} -2 & 0 \\ 0 & -4 \\ 1 & 2 \end{bmatrix} = \begin{bmatrix} (0)(-2) + (4)(0) + (-3)(1) & (0)(0) + (4)(-4) + (-3)(2) \\ (2)(-2) + (1)(0) + (7)(1) & (2)(0) + (1)(-4) + (7)(2) \\ (3)(-2) + (-2)(0) + (1)(1) & (3)(0) + (-2)(-4) + (1)(2) \end{bmatrix} = \begin{bmatrix} -3 & -22 \\ 3 & 10 \\ -5 & 10 \end{bmatrix}$

9. (a) $BA = \begin{bmatrix} 1 \\ -3 \end{bmatrix}[3 \;\; -1] = \begin{bmatrix} (1)(3) & (1)(-1) \\ (-3)(3) & (-3)(-1) \end{bmatrix} = \begin{bmatrix} 3 & -1 \\ -9 & 3 \end{bmatrix}$

(b) $[3 \;\; -1]\begin{bmatrix} 1 \\ -3 \end{bmatrix} = [(3)(1) + (-1)(-3)] = [6]$

(c) $\begin{bmatrix} 3 & 1 & 2 \\ 7 & 0 & -2 \end{bmatrix}\begin{bmatrix} 6 & 4 \\ 2 & -1 \end{bmatrix}$ is not defined, since the first matrix has dimensions 2×3 and the second matrix has dimensions 2×2. The number of columns of the first matrix is not equal to the number of rows of the second matrix.

10. $A^2 = AA = \begin{bmatrix} 2 & 1 \\ 3 & -2 \end{bmatrix}\begin{bmatrix} 2 & 1 \\ 3 & -2 \end{bmatrix} = \begin{bmatrix} (2)(2) + (1)(3) & (2)(1) + (1)(-2) \\ (3)(2) + (-2)(3) & (3)(1) + (-2)(-2) \end{bmatrix} = \begin{bmatrix} 7 & 0 \\ 0 & 7 \end{bmatrix}$

11. $\begin{cases} -2x_1 - 3x_2 = -4 \\ 6x_1 + x_2 = -36 \end{cases}$

(a) In matrix form $Ax = B$, the system can be written as follows.

$$\begin{bmatrix} -2 & -3 \\ 6 & 1 \end{bmatrix}\begin{bmatrix} x_1 \\ x_2 \end{bmatrix} = \begin{bmatrix} -4 \\ -36 \end{bmatrix}$$

(b) The augmented matrix is formed by adjoining matrix B to matrix A.

$$[A \vdots B] = \begin{bmatrix} -2 & -3 & \vdots & -4 \\ 6 & 1 & \vdots & -36 \end{bmatrix}$$

Use Gauss-Jordan elimination to rewrite the matrix.

$$-\tfrac{1}{2}R_1 \rightarrow \begin{bmatrix} 1 & \frac{3}{2} & \vdots & 2 \\ 6 & 1 & \vdots & -36 \end{bmatrix}$$

$$-6R_1 + R_2 \rightarrow = \begin{bmatrix} 1 & \frac{3}{2} & \vdots & 2 \\ 0 & -8 & \vdots & -48 \end{bmatrix}$$

$$-\tfrac{1}{8}R_2 \rightarrow = \begin{bmatrix} 1 & \frac{3}{2} & \vdots & -2 \\ 0 & 1 & \vdots & 6 \end{bmatrix}$$

$$-\tfrac{3}{2}R_2 + R_1 \rightarrow = \begin{bmatrix} 1 & 0 & \vdots & 7 \\ 0 & 1 & \vdots & 6 \end{bmatrix}$$

$$= [I \vdots X]$$

So, the solution of the matrix equation is $X = \begin{bmatrix} x_1 \\ x_2 \end{bmatrix} = \begin{bmatrix} -7 \\ 6 \end{bmatrix}$.

12. The equipment lists E and the costs per item C can be written in matrix form as

$$E = \begin{bmatrix} 12 & 15 \\ 45 & 38 \\ 15 & 17 \end{bmatrix} \text{ and } C = [100 \quad 3 \quad 65].$$

The total cost of equipment for each team is given by the following product.

$$CE = [100 \quad 3 \quad 65]\begin{bmatrix} 12 & 15 \\ 45 & 38 \\ 15 & 17 \end{bmatrix} = [(100)(12) + (3)(45) + (65)(15) \quad (100)(15) + (3)(38) + (65)(17)] = [2310 \quad 2719]$$

So, the total cost of equipment for the women's team is \$2310 and the total cost of equipment for the men's team is \$2719.

Checkpoints for Section 7.3

1. To show that B is the inverse of A, show that $AB = I = BA$, as follows

$$AB = \begin{bmatrix} 2 & -1 \\ -3 & 1 \end{bmatrix}\begin{bmatrix} -1 & -1 \\ -3 & -2 \end{bmatrix} = \begin{bmatrix} -2+3 & -2+2 \\ 3-3 & 3-2 \end{bmatrix} = \begin{bmatrix} 1 & 0 \\ 0 & 1 \end{bmatrix}$$

$$BA = \begin{bmatrix} -1 & -1 \\ -3 & -2 \end{bmatrix}\begin{bmatrix} 2 & -1 \\ -3 & 1 \end{bmatrix} = \begin{bmatrix} -2+3 & 1-1 \\ -6+6 & 3-2 \end{bmatrix} = \begin{bmatrix} 1 & 0 \\ 0 & 1 \end{bmatrix}$$

Because $AB = I = BA$, B is the inverse of A.

2. To find the inverse of A, solve the matrix equation $AX = I$ for X.

$$\overset{A}{\begin{bmatrix} 1 & -2 \\ -1 & 3 \end{bmatrix}}\overset{X}{\begin{bmatrix} X_{11} & X_{12} \\ X_{21} & X_{22} \end{bmatrix}} \overset{=}{=} \overset{I}{\begin{bmatrix} 1 & 0 \\ 0 & 1 \end{bmatrix}}$$

$$\begin{bmatrix} X_{11} - 2X_{21} & X_{12} - 2X_{22} \\ -X_{11} + 3X_{21} & -X_{12} + 3X_{22} \end{bmatrix} = \begin{bmatrix} 1 & 0 \\ 0 & 1 \end{bmatrix}$$

Equating corresponding entries, you obtain two system of linear equations.

$$\begin{cases} X_{11} - 2X_{21} = 1 \\ -X_{11} + 3X_{21} = 0 \end{cases} \qquad \begin{cases} X_{12} - 2X_{22} = 0 \\ -X_{12} + 3X_{22} = 1 \end{cases}$$

Solving the first system yields

$X_{11} = 3$ and $X_{21} = 1$.

Solving the second system yields.

$X_{12} = 2$ and $X_{22} = 1$.

So, the inverse of A is $X = A^{-1} = \begin{bmatrix} 3 & 2 \\ 1 & 1 \end{bmatrix}$.

You can check this by finding AA^{-1} and $A^{-1}A$.

Check:

$$AA^{-1} = \begin{bmatrix} 1 & -2 \\ -1 & 3 \end{bmatrix}\begin{bmatrix} 3 & 2 \\ 1 & 1 \end{bmatrix} = \begin{bmatrix} 1 & 0 \\ 0 & 1 \end{bmatrix} = I \checkmark$$

$$A^{-1}A = \begin{bmatrix} 3 & 2 \\ 1 & 1 \end{bmatrix}\begin{bmatrix} 1 & -2 \\ -1 & 3 \end{bmatrix} = \begin{bmatrix} 1 & 0 \\ 0 & 1 \end{bmatrix} = I \checkmark$$

3. Begin by adjoining the identity matrix to A to form the matrix

$$[A \,\vdots\, I] = \left[\begin{array}{ccc|ccc} 1 & -2 & -1 & 1 & 0 & 0 \\ 0 & -1 & 2 & 0 & 1 & 0 \\ 1 & -2 & 0 & 0 & 0 & 1 \end{array}\right].$$

Use elementary row operations to obtain the form $[I \,\vdots\, A^{-1}]$

$$\begin{array}{r} \\ -R_2 \to \\ -R_1 + R_3 \to \end{array}\left[\begin{array}{ccc|ccc} 1 & -2 & -1 & 1 & 0 & 0 \\ 0 & 1 & -2 & 0 & -1 & 0 \\ 0 & 0 & 1 & -1 & 0 & 1 \end{array}\right]$$

$$\begin{array}{r} 2R_2 + R_1 \to \\ 2R_3 - R_2 \to \\ \\ \end{array}\left[\begin{array}{ccc|ccc} 1 & 0 & -5 & 1 & -2 & 0 \\ 0 & 1 & 0 & -2 & -1 & 2 \\ 0 & 0 & 1 & -1 & 0 & 1 \end{array}\right]$$

$$\begin{array}{r} 5R_3 + R_1 \to \\ \\ \\ \end{array}\left[\begin{array}{ccc|ccc} 1 & 0 & 0 & -4 & -2 & 5 \\ 0 & 1 & 0 & -2 & -1 & 2 \\ 0 & 0 & 1 & -1 & 0 & 1 \end{array}\right] = [I \,\vdots\, A^{-1}]$$

So, the matrix A is invertible and its inverse is

$$A^{-1} = \begin{bmatrix} -4 & -2 & 5 \\ -2 & -1 & 2 \\ -1 & 0 & 1 \end{bmatrix}.$$

Confirm this result by multiplying AA^{-1} to obtain I.

Check:

$$AA^{-1} = \begin{bmatrix} 1 & -2 & -1 \\ 0 & -1 & 2 \\ 1 & -2 & 0 \end{bmatrix}\begin{bmatrix} -4 & -2 & 5 \\ -2 & -1 & 2 \\ -1 & 0 & 1 \end{bmatrix} = \begin{bmatrix} 1 & 0 & 0 \\ 0 & 1 & 0 \\ 0 & 0 & 1 \end{bmatrix} = I$$

4. For the matrix A, apply the formula for the inverse of a 2×2 matrix to obtain

$ad - bc = (5)(4) - (-1)(3) = 23$

Because this quantity is not zero, the matrix is invertible. The inverse is formed by interchanging the entries on the main diagonal, changing the signs of the other two entries, and multiplying by the scalar $\frac{1}{23}$, as follows.

$$A^{-1} = \frac{1}{ad - bc}\begin{bmatrix} d & -b \\ -c & a \end{bmatrix} \qquad \text{Formula for the inverse of a } 2 \times 2 \text{ matrix}$$

$$= \frac{1}{23}\begin{bmatrix} 4 & 1 \\ -3 & 5 \end{bmatrix} \qquad \text{Substitute for } a, b, c, d, \text{ and the determinant}$$

$$= \begin{bmatrix} \frac{4}{23} & \frac{1}{23} \\ \frac{-3}{23} & \frac{5}{23} \end{bmatrix} \qquad \text{Multiply by the scalar } \frac{1}{23}.$$

5. Begin by writing the system in the matrix form $AX = B$.

$$\begin{bmatrix} 2 & 3 & 1 \\ 3 & 3 & 1 \\ 2 & 4 & 1 \end{bmatrix}\begin{bmatrix} x \\ y \\ z \end{bmatrix} = \begin{bmatrix} -1 \\ 1 \\ -2 \end{bmatrix}$$

Then, use Gauss-Jordan elimination to find A^{-1}.

$$[A \vdots I] = \begin{bmatrix} 2 & 3 & 1 & \vdots & 1 & 0 & 0 \\ 3 & 3 & 1 & \vdots & 0 & 1 & 0 \\ 2 & 4 & 1 & \vdots & 0 & 0 & 1 \end{bmatrix} \quad \tfrac{1}{2}R_1 \rightarrow \begin{bmatrix} 1 & \frac{3}{2} & \frac{1}{2} & \vdots & \frac{1}{2} & 0 & 0 \\ 3 & 3 & 1 & \vdots & 0 & 1 & 0 \\ 2 & 4 & 1 & \vdots & 0 & 0 & 1 \end{bmatrix}$$

$$\begin{matrix} \\ -3R_1 + R_2 \rightarrow \\ -2R_1 + R_3 \rightarrow \end{matrix}\begin{bmatrix} 1 & \frac{3}{2} & \frac{1}{2} & \vdots & \frac{1}{2} & 0 & 0 \\ 0 & -\frac{3}{2} & -\frac{1}{2} & \vdots & -\frac{3}{2} & 1 & 0 \\ 0 & 1 & 0 & \vdots & -1 & 0 & 1 \end{bmatrix}$$

$$\begin{matrix} \\ R_2 \rightarrow \\ R_3 \rightarrow \end{matrix}\begin{bmatrix} 1 & \frac{3}{2} & \frac{1}{2} & \vdots & \frac{1}{2} & 0 & 0 \\ 0 & 1 & 0 & \vdots & -1 & 0 & 1 \\ 0 & -\frac{3}{2} & -\frac{1}{2} & \vdots & -\frac{3}{2} & 1 & 0 \end{bmatrix}$$

$$\begin{matrix} \\ \\ \frac{3}{2}R_2 + R_3 \rightarrow \end{matrix}\begin{bmatrix} 1 & \frac{3}{2} & \frac{1}{2} & \vdots & \frac{1}{2} & 0 & 0 \\ 0 & 1 & 0 & \vdots & -1 & 0 & 1 \\ 0 & 0 & -\frac{1}{2} & \vdots & -3 & 1 & \frac{3}{2} \end{bmatrix}$$

$$\begin{matrix} \\ \\ -2R_3 \rightarrow \end{matrix}\begin{bmatrix} 1 & \frac{3}{2} & \frac{1}{2} & \vdots & \frac{1}{2} & 0 & 0 \\ 0 & 1 & 0 & \vdots & -1 & 0 & 1 \\ 0 & 0 & 1 & \vdots & 6 & -2 & -3 \end{bmatrix}$$

$$\begin{matrix} -\frac{3}{2}R_2 + R_1 \rightarrow \\ \\ \\ \end{matrix}\begin{bmatrix} 1 & 0 & \frac{1}{2} & \vdots & 2 & 0 & \frac{3}{2} \\ 0 & 1 & 0 & \vdots & -1 & 0 & 1 \\ 0 & 0 & 1 & \vdots & 6 & -2 & -3 \end{bmatrix}$$

$$\begin{matrix} -\frac{1}{2}R_3 + R_1 \rightarrow \\ \\ \\ \end{matrix}\begin{bmatrix} 1 & 0 & 0 & \vdots & -1 & 1 & 0 \\ 0 & 1 & 0 & \vdots & -1 & 0 & 1 \\ 0 & 0 & 1 & \vdots & 6 & -2 & -3 \end{bmatrix} = \begin{bmatrix} I & \vdots & A^{-1} \end{bmatrix}$$

$$A^{-1} = \begin{bmatrix} -1 & 1 & 0 \\ -1 & 0 & 1 \\ 6 & -2 & -3 \end{bmatrix}$$

Finally, multiply B by A^{-1} on the left to obtain the solution.

$$X = A^{-1}B = \begin{bmatrix} -1 & 1 & 0 \\ -1 & 0 & 1 \\ 6 & -2 & -3 \end{bmatrix}\begin{bmatrix} -1 \\ 1 \\ -2 \end{bmatrix} = \begin{bmatrix} 2 \\ -1 \\ -2 \end{bmatrix}$$

The solution of the system is $x = 2$, $y = -1$, and $z = -2$

Checkpoints for Section 7.4

1. (a) $\det(A) = \begin{vmatrix} 1 & 2 \\ 3 & -1 \end{vmatrix}$
$= 1(-1) - 3(2)$
$= -1 - 6$
$= -7$

(b) $\det(B) = \begin{vmatrix} 5 & 0 \\ -4 & 2 \end{vmatrix}$
$= 5(2) - (-4)(0)$
$= 10 + 0$
$= 10$

(c) $\det(C) = \begin{vmatrix} 3 & 6 \\ 2 & 4 \end{vmatrix}$
$= 3(4) - (2)(6)$
$= 12 - 12$
$= 0$

2. To find the minor M_{11}, delete the first row and first column of A and evaluate the determinant of the resulting matrix.

$$\begin{bmatrix} 1 & 2 & 3 \\ 0 & -1 & 5 \\ 2 & 1 & 4 \end{bmatrix}, \quad M_{11} = \begin{vmatrix} -1 & 5 \\ 1 & 4 \end{vmatrix} = -1(4) - 1(5) = -9$$

Continuing this pattern, you obtain the minors.

$$M_{12} = \begin{vmatrix} 0 & 5 \\ 2 & 4 \end{vmatrix} = 0(4) - 2(5) = -10$$

$$M_{13} = \begin{vmatrix} 0 & -1 \\ 2 & 1 \end{vmatrix} = 0(1) - 2(-1) = 2$$

$$M_{21} = \begin{vmatrix} 2 & 3 \\ 1 & 4 \end{vmatrix} = 2(4) - 1(3) = 5$$

$$M_{22} = \begin{vmatrix} 1 & 3 \\ 2 & 4 \end{vmatrix} = 1(4) - 2(3) = -2$$

$$M_{23} = \begin{vmatrix} 1 & 2 \\ 2 & 1 \end{vmatrix} = 1(1) - 2(2) = -3$$

$$M_{31} = \begin{vmatrix} 2 & 3 \\ -1 & 5 \end{vmatrix} = 2(5) - (-1)(3) = 13$$

$$M_{32} = \begin{vmatrix} 1 & 3 \\ 0 & 5 \end{vmatrix} = 1(5) - 0(3) = 5$$

$$M_{33} = \begin{vmatrix} 1 & 2 \\ 0 & -1 \end{vmatrix} = 1(-1) - 0(2) = -1$$

Now, to find the cofactors, combine these minors with the checker board pattern of signs for a 3×3 matrix,

$$\begin{bmatrix} + & - & + \\ - & + & - \\ + & - & + \end{bmatrix}.$$

$C_{11} = -9$ $\quad C_{12} = 10$ $\quad C_{13} = 2$
$C_{21} = -5$ $\quad C_{22} = -2$ $\quad C_{23} = 3$
$C_{31} = 13$ $\quad C_{32} = -5$ $\quad C_{33} = -1$

3. The cofactors of the entries in the first row are as follows.

$$C_{11} = +M_{11} = \begin{vmatrix} 5 & 0 \\ 4 & 1 \end{vmatrix} = 5(1) - 4(0) = 5$$

$$C_{12} = -M_{12} = \begin{vmatrix} 3 & 0 \\ -1 & 1 \end{vmatrix} = -(3(1) - (-1)(0)) = -3$$

$$C_{13} = +M_{13} = \begin{vmatrix} 3 & 5 \\ -1 & 4 \end{vmatrix} = 3(4) - (-1)(5) = 17$$

So, by the definition of a determinant, you have the following.

$|A| = a_{11}C_{11} + a_{12}C_{12} + a_{13}C_{13}$
$= 3(5) + 4(-3) + (-2)(17)$
$= -31$

4. Notice that these are two zeros in the third column. So, you can eliminate some of the work in the expansion by using the third column.

$$|A| = a_{13}C_{13} + a_{23}C_{23} + a_{33}C_{33} + a_{43}C_{43}$$
$$= -4C_{13} + 3C_{23} + 0C_{33} + 0C_{43}$$

Because C_{33} and C_{43} have zero coefficients, you need only to find the cofactors of C_{13} and C_{23}.

$$C_{13} = (-1)^{1+3}\begin{vmatrix} 2 & -2 & 6 \\ 1 & 5 & 1 \\ 3 & 1 & -5 \end{vmatrix} = \begin{vmatrix} 2 & -2 & 6 \\ 1 & 5 & 1 \\ 3 & 1 & -5 \end{vmatrix}$$

Expanding by cofactors along the first row yields the following.

$$C_{13} = (2)(-1)^{1+1}\begin{vmatrix} 5 & 1 \\ 1 & -5 \end{vmatrix} + (-2)(-1)^{1+2}\begin{vmatrix} 1 & 1 \\ 3 & -5 \end{vmatrix} + (6)(-1)^{1+3}\begin{vmatrix} 1 & 5 \\ 3 & 1 \end{vmatrix}$$
$$= (2)(1)(-26) + (-2)(-1)(-8) + (6)(1)(-14)$$
$$= -152$$

$$C_{23} = (-1)^{2+3}\begin{vmatrix} 2 & 6 & 2 \\ 1 & 5 & 1 \\ 3 & 1 & -5 \end{vmatrix} = -\begin{vmatrix} 2 & 6 & 2 \\ 1 & 5 & 1 \\ 3 & 1 & -5 \end{vmatrix}$$

Expanding by cofactors along the first row yields the following.

$$C_{23} = -\left((2)(-1)^{1+1}\begin{vmatrix} 5 & 1 \\ 1 & -5 \end{vmatrix} + (6)(-1)^{1+2}\begin{vmatrix} 1 & 1 \\ 3 & -5 \end{vmatrix} + (2)(-1)^{1+3}\begin{vmatrix} 1 & 5 \\ 3 & 1 \end{vmatrix}\right)$$
$$= -((2)(1)(-26) + (6)(-1)(-8) + (2)(1)(-14))$$
$$= 32$$

So, $|A| = -4C_{13} + 3C_{23} + 0C_{33} + 0C_{43}$
$$= -4(-152) + 3(32) + 0 + 0$$
$$= 704.$$

Checkpoints for Section 7.5

1. To begin, find the determinant of the coefficient matrix.

$$D = \begin{vmatrix} 3 & 4 \\ 5 & 3 \end{vmatrix} = 9 - 20 = -11$$

Because this determinant is not zero, you can apply Cramer's Rule.

$$x = \frac{D_x}{D} = \frac{\begin{vmatrix} 1 & 4 \\ 9 & 3 \end{vmatrix}}{-11} = \frac{3 - 36}{-11} = \frac{-33}{-11} = 3$$

$$y = \frac{D_y}{D} = \frac{\begin{vmatrix} 3 & 1 \\ 5 & 9 \end{vmatrix}}{-11} = \frac{27 - 5}{-11} = \frac{22}{-11} = -2$$

So, the solution is $x = 3$ and $y = -2$.

2. To find the determinant of the coefficient matrix, expand along the first row, as follows.

$$\begin{bmatrix} 4 & -1 & 1 \\ 2 & 2 & 3 \\ 5 & -2 & 6 \end{bmatrix}$$

$$D = 4(-1)^2\begin{vmatrix} 2 & 3 \\ -2 & 6 \end{vmatrix} + (-1)(-1)^3\begin{vmatrix} 2 & 3 \\ 5 & 6 \end{vmatrix} + (1)(-1)^4\begin{vmatrix} 2 & 2 \\ 5 & -2 \end{vmatrix} = 4(18) + (1)(-3) + (1)(-14) = 55$$

Because this determinant is not zero, you can apply Cramer's Rule. Next, find D_x, D_y, and D_z.

$$\begin{aligned} D_x &= \begin{vmatrix} 12 & -1 & 1 \\ 1 & 2 & 3 \\ 22 & -2 & 6 \end{vmatrix} \\ &= (12)(-1)^2\begin{vmatrix} 2 & 3 \\ -2 & 6 \end{vmatrix} + (-1)(-1)^3\begin{vmatrix} 1 & 3 \\ 22 & 6 \end{vmatrix} + (1)(-1)^4\begin{vmatrix} 1 & 2 \\ 22 & -2 \end{vmatrix} \\ &= (12)(18) + (1)(-60) + (1)(-46) \\ &= 110 \end{aligned}$$

$$\begin{aligned} D_y &= \begin{vmatrix} 4 & 12 & 1 \\ 2 & 1 & 3 \\ 5 & 22 & 6 \end{vmatrix} \\ &= (4)(-1)^2\begin{vmatrix} 1 & 3 \\ 22 & 6 \end{vmatrix} + (12)(-1)^3\begin{vmatrix} 2 & 3 \\ 5 & 6 \end{vmatrix} + (1)(-1)^4\begin{vmatrix} 2 & 1 \\ 5 & 22 \end{vmatrix} \\ &= (4)(-60) + (-12)(-3) + (1)(39) \\ &= -165 \end{aligned}$$

$$\begin{aligned} D_z &= \begin{vmatrix} 4 & -1 & 12 \\ 2 & 2 & 1 \\ 5 & -2 & 22 \end{vmatrix} \\ &= (4)(-1)^2\begin{vmatrix} 2 & 1 \\ -2 & 22 \end{vmatrix} + (-1)(-1)^3\begin{vmatrix} 2 & 1 \\ 5 & 22 \end{vmatrix} + (12)(-1)^4\begin{vmatrix} 2 & 2 \\ 5 & -2 \end{vmatrix} \\ &= (4)(46) + (1)(39) + (12)(-14) \\ &= 55 \end{aligned}$$

Finally, you can determine the values of x, y, and z as follows.

$$x = \frac{D_x}{D} = \frac{110}{55} = 2$$

$$y = \frac{D_y}{D} = \frac{-165}{55} = -3$$

$$z = \frac{D_z}{D} = \frac{55}{55} = 1$$

So, the solution is $x = 2$, $y = -3$, and $z = 1$.

3.

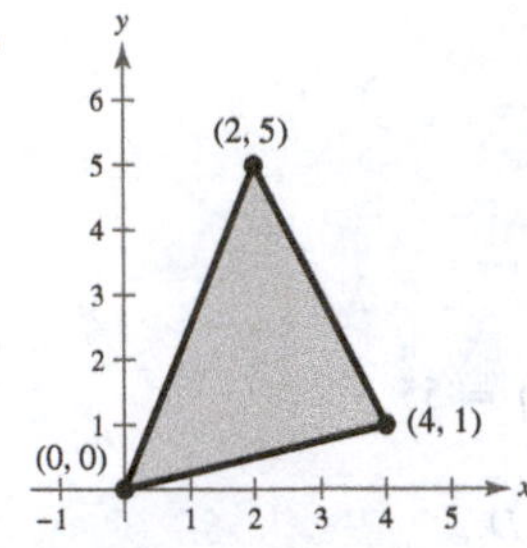

Let $(x_1, y_1) = (0, 0)$, $(x_2, y_2) = (4, 1)$, and $(x_3, y_3) = (2, 5)$. Then, to find the area of the triangle, evaluate the determinant.

$$\begin{vmatrix} x_1 & y_1 & 1 \\ x_2 & y_2 & 1 \\ x_3 & y_3 & 1 \end{vmatrix} = \begin{vmatrix} 0 & 0 & 1 \\ 4 & 1 & 1 \\ 2 & 5 & 1 \end{vmatrix}$$

$$= (0)(-1)^2\begin{vmatrix} 1 & 1 \\ 5 & 1 \end{vmatrix} + (0)(-1)^3\begin{vmatrix} 4 & 1 \\ 2 & 1 \end{vmatrix} + (1)(-1)^4\begin{vmatrix} 4 & 1 \\ 2 & 5 \end{vmatrix}$$

$$= 0 + 0 + (1)(18)$$

$$= 18$$

Using this value, you can conclude that the area of the triangle is

$$\text{Area} = \frac{1}{2}\begin{vmatrix} 0 & 0 & 1 \\ 4 & 1 & 1 \\ 2 & 5 & 1 \end{vmatrix} = \frac{1}{2}(18) = 9 \text{ square units.}$$

4.

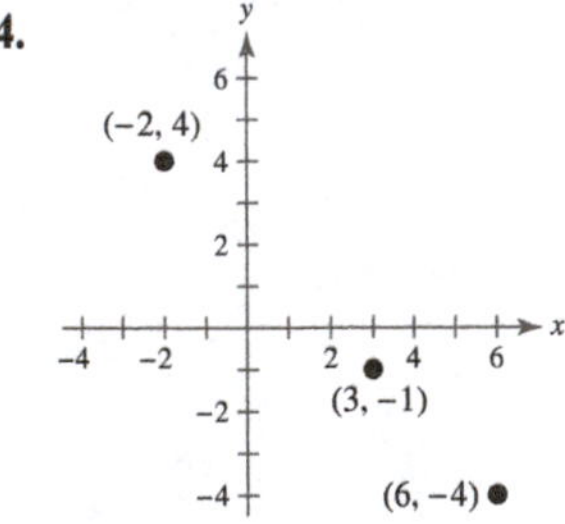

To determine if the points are collinear, let $(x_1, y_1) = (-2, 4)$, $(x_2, y_2) = (3, -1)$, and $(x_3, y_3) = (6, -4)$. Then, evaluate the determinant as follows.

$$\begin{vmatrix} x_1 & y_1 & 1 \\ x_2 & y_2 & 1 \\ x_3 & y_3 & 1 \end{vmatrix} = \begin{vmatrix} -2 & 4 & 1 \\ 3 & -1 & 1 \\ 6 & -4 & 1 \end{vmatrix}$$

$$= (-2)(-1)^2\begin{vmatrix} -1 & 1 \\ -4 & 1 \end{vmatrix} + (4)(-1)^3\begin{vmatrix} 3 & 1 \\ 6 & 1 \end{vmatrix} + (1)(-1)^4\begin{vmatrix} 3 & -1 \\ 6 & -4 \end{vmatrix}$$

$$= (-2)(3) + (-4)(-3) + (1)(-6)$$

$$= 0$$

Because the value of this determinant is equal to zero, you can conclude that the three points are collinear.

5. 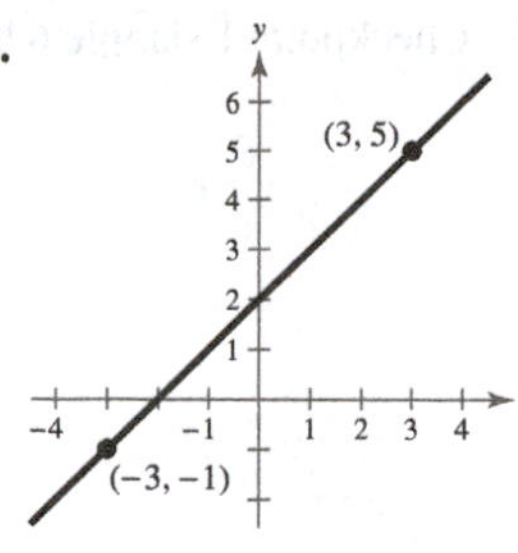

To find an equation of the line, let $(x_1, y_1) = (-3, -1)$ and $(x_2, y_2) = (3, 5)$. Applying the determinant formula for the equation of a line produces the following.

$$\begin{vmatrix} x & y & 1 \\ -3 & -1 & 1 \\ 3 & 5 & 1 \end{vmatrix} = 0$$

To evaluate this determinant, expand by cofactors along the first row.

$$x(-1)^2\begin{vmatrix} -1 & 1 \\ 5 & 1 \end{vmatrix} + y(-1)^3\begin{vmatrix} -3 & 1 \\ 3 & 1 \end{vmatrix} + (1)(-1)^4\begin{vmatrix} -3 & -1 \\ 3 & 5 \end{vmatrix} = 0$$

$$(x)(-6) + (-y)(-6) + (1)(-12) = 0$$

$$-6x + 6y - 12 = 0$$

$$x - y + 2 = 0$$

So, an equation passing through the two points is $x - y + 2 = 0$.

6. To find the image of the square with vertices $(0, 0)$, $(2, 0)$, $(0, 2)$ and $(2, 2)$ after a vertical stretch by a factor of $k = 2$, multiply the vertices matrix by $\begin{bmatrix} 1 & 0 \\ 0 & 2 \end{bmatrix}$

$\begin{bmatrix} 1 & 0 \\ 0 & 2 \end{bmatrix}\begin{bmatrix} 0 \\ 0 \end{bmatrix} = \begin{bmatrix} 0 \\ 0 \end{bmatrix}, \begin{bmatrix} 1 & 0 \\ 0 & 2 \end{bmatrix}\begin{bmatrix} 2 \\ 0 \end{bmatrix} = \begin{bmatrix} 2 \\ 0 \end{bmatrix}, \begin{bmatrix} 1 & 0 \\ 0 & 2 \end{bmatrix}\begin{bmatrix} 0 \\ 2 \end{bmatrix} = \begin{bmatrix} 0 \\ 4 \end{bmatrix}$ and $\begin{bmatrix} 1 & 0 \\ 0 & 2 \end{bmatrix}\begin{bmatrix} 2 \\ 2 \end{bmatrix} = \begin{bmatrix} 2 \\ 4 \end{bmatrix}$. The vertices of the image are $(0, 0)$, $(2, 0)$, $(0, 4)$ and $(2, 4)$.

7. To find the area of the parallelogram with vertices $(0, 0)$, (a, b), (c, d) and $(a + c, b + d)$, you can use the formula

$\text{Area} = |\det(A)|$ where $A = \begin{bmatrix} a & b \\ c & d \end{bmatrix}$. If the parallelogram has vertices $(0, 0)$, $(5, 5)$, $(2, 4)$ and $(7, 9)$ then $a = 2$, $b = 4$, $c = 5$, and $d = 5$.

$$\begin{aligned} \text{Area} = |\det(A)| &= \begin{vmatrix} 2 & 4 \\ 5 & 5 \end{vmatrix} \\ &= |(2)(5) - (4)(5)| \\ &= |-10| = 10 \text{ square units.} \end{aligned}$$

8. Partioning the message (including blank spaces, but ignoring any punctuation) into groups of three produces the following uncoded 1×3 row matrices.

[15	23	12]	[19	0	1]	[18	5	0]	[14	15	3]	[20	21	18]	[14	1	12]
O	W	L	S		A	R	E		N	O	C	T	U	R	N	A	L

9. The coded row matrices are obtained by multiplying each of the uncoded row matrices found in Checkpoint Example 6 by the matrix A, as follows.

Uncoded Matrix	Encoding Matrix A		Coded Matrix
$\begin{bmatrix}15 & 23 & 12\end{bmatrix}$	$\begin{bmatrix}1 & -1 & 0\\ 1 & 0 & -1\\ 6 & -2 & -3\end{bmatrix}$	=	$\begin{bmatrix}110 & -39 & -59\end{bmatrix}$
$\begin{bmatrix}19 & 0 & 1\end{bmatrix}$	$\begin{bmatrix}1 & -1 & 0\\ 1 & 0 & -1\\ 6 & -2 & -3\end{bmatrix}$	=	$\begin{bmatrix}25 & -21 & -3\end{bmatrix}$
$\begin{bmatrix}18 & 5 & 0\end{bmatrix}$	$\begin{bmatrix}1 & -1 & 0\\ 1 & 0 & -1\\ 6 & -2 & -3\end{bmatrix}$	=	$\begin{bmatrix}23 & -18 & -5\end{bmatrix}$
$\begin{bmatrix}14 & 15 & 3\end{bmatrix}$	$\begin{bmatrix}1 & -1 & 0\\ 1 & 0 & -1\\ 6 & -2 & -3\end{bmatrix}$	=	$\begin{bmatrix}47 & -20 & -24\end{bmatrix}$
$\begin{bmatrix}20 & 21 & 18\end{bmatrix}$	$\begin{bmatrix}1 & -1 & 0\\ 1 & 0 & -1\\ 6 & -2 & -3\end{bmatrix}$	=	$\begin{bmatrix}149 & -56 & -75\end{bmatrix}$
$\begin{bmatrix}14 & 1 & 12\end{bmatrix}$	$\begin{bmatrix}1 & -1 & 0\\ 1 & 0 & -1\\ 6 & -2 & -3\end{bmatrix}$	=	$\begin{bmatrix}87 & -38 & -37\end{bmatrix}$

So, the cryptogram is $110\ \ -39\ \ -59\ \ 25\ \ -21\ \ -3\ \ 23\ \ -18\ \ -5\ \ 47\ \ -20\ \ -24\ \ 149\ \ -56\ \ -75\ \ 87\ \ -38\ \ -37$.

10. First find the decoding matrix A^{-1} using matrix A from the Checkpoint Example 7.

$$[A \ \vdots \ I] = \begin{bmatrix} 1 & -1 & 0 & \vdots & 1 & 0 & 0 \\ 1 & 0 & -1 & \vdots & 0 & 1 & 0 \\ 6 & -2 & -3 & \vdots & 0 & 0 & 1 \end{bmatrix}$$

$$\begin{matrix} R_2 \rightarrow \\ R_1 \rightarrow \\ -6R_1 + R_3 \rightarrow \end{matrix} \begin{bmatrix} 1 & 0 & -1 & \vdots & 0 & 1 & 0 \\ 1 & -1 & 0 & \vdots & 1 & 0 & 0 \\ 0 & 4 & -3 & \vdots & -6 & 0 & 1 \end{bmatrix}$$

$$\begin{matrix} \\ -R_1 + R_2 \rightarrow \\ \\ \end{matrix} \begin{bmatrix} 1 & 0 & -1 & \vdots & 0 & 1 & 0 \\ 0 & -1 & 1 & \vdots & 1 & -1 & 0 \\ 0 & 4 & -3 & \vdots & -6 & 0 & 1 \end{bmatrix}$$

$$\begin{matrix} \\ \\ 4R_2 + R_3 \rightarrow \end{matrix} \begin{bmatrix} 1 & 0 & -1 & \vdots & 0 & 1 & 0 \\ 0 & -1 & 1 & \vdots & 1 & -1 & 0 \\ 0 & 0 & 1 & \vdots & -2 & -4 & 1 \end{bmatrix}$$

$$\begin{matrix} \\ -R_2 \rightarrow \\ \\ \end{matrix} \begin{bmatrix} 1 & 0 & -1 & \vdots & 0 & 1 & 0 \\ 0 & 1 & -1 & \vdots & -1 & 1 & 0 \\ 0 & 0 & 1 & \vdots & -2 & -4 & 1 \end{bmatrix}$$

$$\begin{matrix} R_1 + R_3 \rightarrow \\ R_2 + R_3 \rightarrow \\ \\ \end{matrix} \begin{bmatrix} 1 & 0 & 0 & \vdots & -2 & -3 & 1 \\ 0 & 1 & 0 & \vdots & -3 & -3 & 1 \\ 0 & 0 & 1 & \vdots & -2 & -4 & 1 \end{bmatrix} = \left[I \ \vdots \ A^{-1}\right]$$

Partition the message into groups of three to form the coded row matrices. Finally, multiply each coded row matrix by A^{-1} (on the right).

Coded Matrix	**Decoding Matrix A^{-1}**		**Decoded Matrix**
$[110 \ \ -39 \ \ -59]$	$\begin{bmatrix} -2 & -3 & 1 \\ -3 & -3 & 1 \\ -2 & -4 & 1 \end{bmatrix}$	=	$[15 \ \ 23 \ \ 12]$
$[25 \ \ -21 \ \ -3]$	$\begin{bmatrix} -2 & -3 & 1 \\ -3 & -3 & 1 \\ -2 & -4 & 1 \end{bmatrix}$	=	$[19 \ \ 0 \ \ 1]$
$[23 \ \ -18 \ \ -5]$	$\begin{bmatrix} -2 & -3 & 1 \\ -3 & -3 & 1 \\ -2 & -4 & 1 \end{bmatrix}$	=	$[18 \ \ 5 \ \ 0]$
$[47 \ \ -20 \ \ -24]$	$\begin{bmatrix} -2 & -3 & 1 \\ -3 & -3 & 1 \\ -2 & -4 & 1 \end{bmatrix}$	=	$[14 \ \ 15 \ \ 3]$
$[149 \ \ -56 \ \ -75]$	$\begin{bmatrix} -2 & -3 & 1 \\ -3 & -3 & 1 \\ -2 & -4 & 1 \end{bmatrix}$	=	$[20 \ \ 21 \ \ 18]$
$[87 \ \ -38 \ \ -37]$	$\begin{bmatrix} -2 & -3 & 1 \\ -3 & -3 & 1 \\ -2 & -4 & 1 \end{bmatrix}$	=	$[14 \ \ 1 \ \ 12]$

So, the message is as follows.

[15	23	12]	[19	0	1]	[18	5	0]	[14	15	3]	[20	21	18]	[14	1	12]
O	W	L	S		A	R	E		N	O	C	T	U	R	N	A	L

Chapter 8

Checkpoints for Section 8.1

1. The first four terms of the sequence given by $a_n = 2n + 1$ are as follows.

$a_1 = 2(1) + 1 = 3$ 1st term

$a_2 = 2(2) + 1 = 5$ 2nd term

$a_3 = 2(3) + 1 = 7$ 3rd term

$a_4 = 2(4) + 1 = 9$ 4th term

2. The first four terms of the sequence given by $a_n = \frac{2 + (-1)^n}{n}$ are as follows.

$$a_1 = \frac{2 + (-1)^1}{1} = \frac{2 - 1}{1} = 1$$

$$a_2 = \frac{2 + (-1)^2}{2} = \frac{2 + 1}{2} = \frac{3}{2}$$

$$a_3 = \frac{2 + (-1)^3}{3} = \frac{2 - 1}{3} = \frac{1}{3}$$

$$a_4 = \frac{2 + (-1)^4}{4} = \frac{2 + 1}{4} = \frac{3}{4}$$

3. (a) n: 1 2 3 4 ... n

Terms: 1 5 9 13 ... a_n

Apparent pattern: Each term is 3 less than 4 times n, which implies that $a_n = 4n - 3$.

(b) n: 1 2 3 4 ... n

Terms: 2 −4 6 −8 ... a_n

Apparent pattern: The absolute value of each term is 2 times n, and the terms have alternating signs, with those in the even positions being negative. This implies that $a_n = (-1)^{n+1}2n$.

4. The first five terms of the sequence are as follows.

$a_1 = 6$ and $a_{k+1} = a_k + 1$

$a_1 = 6$ 1st term is given.

$a_2 = a_{1+1} = a_1 + 1 = 6 + 1 = 7$ Use recursion formula.

$a_3 = a_{2+1} = a_2 + 1 = 7 + 1 = 8$ Use recursion formula.

$a_4 = a_{3+1} = a_3 + 1 = 8 + 1 = 9$ Use recursion formula.

$a_5 = a_{4+1} = a_4 + 1 = 9 + 1 = 10$ Use recursion formula.

5. The first five terms of the sequence are as follows,

$a_0 = 1$ 0th term is given.

$a_1 = 3$ 1st term is given.

$a_2 = a_{2-2} + a_{2-1} = a_0 + a_1 = 1 + 3 = 4$ Use recursion formula.

$a_3 = a_{3-2} + a_{3-1} = a_1 + a_2 = 3 + 4 = 7$ Use recursion formula.

$a_4 = a_{4-2} + a_{4-1} = a_2 + a_3 = 4 + 7 = 11$ Use recursion formula.

6. Algebraic Solution

$$a_0 = \frac{3^0+1}{0!} = \frac{1+1}{1} = 2$$

$$a_1 = \frac{3^1+1}{1!} = \frac{3+1}{1} = 4$$

$$a_2 = \frac{3^2+1}{2!} = \frac{9+1}{2} = \frac{10}{2} = 5$$

$$a_3 = \frac{3^3+1}{3!} = \frac{27+1}{6} = \frac{28}{6} = \frac{14}{3}$$

$$a_4 = \frac{3^4+1}{4!} = \frac{81+1}{24} = \frac{82}{24} = \frac{41}{12}$$

Graphical Solution

Using a graphing utility set to *dot* and *sequence* modes, enter the sequence. Next, graph the sequence.

You can estimate the first five terms of the sequence as follows.

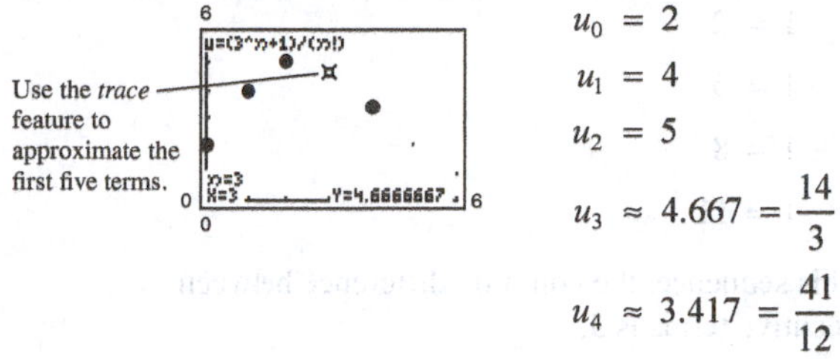

$u_0 = 2$

$u_1 = 4$

$u_2 = 5$

$u_3 \approx 4.667 = \frac{14}{3}$

$u_4 \approx 3.417 = \frac{41}{12}$

7. $\dfrac{4!(n+1)!}{3!n!} = \dfrac{(\cancel{1\cdot 2\cdot 3}\cdot 4)[\cancel{1\cdot 2\cdot 3 \ldots n}\cdot(n+1)]}{(\cancel{1\cdot 2\cdot 3})(\cancel{1\cdot 2\cdot 3\ldots n})}$

$= 4(n+1)$

8. $\sum_{i=1}^{4}(4i+1) = [4(1)+1] + [4(2)+1] + [4(3)+1] + [4(4)+1]$

$= 5 + 9 + 13 + 17$

$= 44$

9. (a) The fourth partial sum is as follows.

$$\sum_{i=1}^{4}\frac{5}{10^i} = \frac{5}{10^1} + \frac{5}{10^2} + \frac{5}{10^3} + \frac{5}{10^4}$$

$$= 0.5 + 0.05 + 0.005 + 0.0005$$

$$= 0.5555$$

(b) The sum of the series is as follows.

$$\sum_{i=1}^{\infty}\frac{5}{10^i} = \frac{5}{10^1} + \frac{5}{10^2} + \frac{5}{10^3} + \frac{5}{10^4} + \frac{5}{10^5} + \cdots$$

$$= 0.5 + 0.05 + 0.005 + 0.0005 + 0.00005 + \cdots$$

$$= 0.55555\ldots$$

$$= \frac{5}{9}$$

10. (a) The first three terms of the sequence are as follows.

$A_0 = 1000\left(1 + \frac{0.03}{12}\right)^0 = \$1000$ Original deposit

$A_1 = 1000\left(1 + \frac{0.03}{12}\right)^1 = \$1002.50$ First-month balance

$A_2 = 1000\left(1 + \frac{0.03}{12}\right)^2 \approx \$1005.01$ Second-month balance.

(b) The 48th term of the sequence is

$A_{48} = 1000\left(1 + \frac{0.03}{12}\right)^{48} \approx \$1127.33$ Four-year balance

Checkpoints for Section 8.2

1. The sequence whose nth term is $3n - 1$ is arithmetic. The first four terms are as follows.

$3(1) - 1 = 2$

$3(2) - 1 = 5$

$3(3) - 1 = 8$

$3(4) - 1 = 11$

For this sequence, the common difference between consecutive terms is 3.

$\underbrace{2, 5}_{5-2=3}, 8, 11, \ldots$

2. You know that the formula for the nth term is of the form $a_n = a_1 + (n - 1)d$. Because the common difference is $d = 5$ and the first term is $a_1 = -1$, the formula must have the form

$a_n = a_1 + (n - 1)d = -1 + 5(n - 1).$

So, the formula for the nth term is $a_n = 5n - 6$.

The sequence therefore has the following form.

$-1, 4, 9, 14, \ldots, 5n - 6, \ldots$

The figure below shows a graph of the first 15 terms of the sequence. Notice that the points lie on a line.

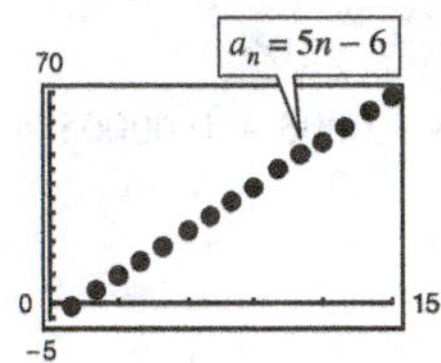

3. You know that $a_8 = 25$ and $a_{12} = 41$. So, you must add the common difference d four times to the eighth term to obtain the 12th term. Therefore, the eighth term and the 12th terms of the sequence are related by

$a_{12} = a_8 + 4d.$

Using $a_8 = 25$ and $a_{12} = 41$, solve for d.

$a_{12} = a_8 + 4d$

$41 = 25 + 4d$

$16 = 4d$

$4 = d$

Use the formula for the nth term of an arithmetic sequence to find a_1.

$a_n = a_1 + (n - 1)d$

$a_8 = a_1 + (8 - 1)(4)$

$25 = a_1 + (7)(4)$

$-3 = a_1$

So, the formula for the nth term of the sequence is

$a_n = -3 + (n - 1)(4) = -3 + 4n - 4 = 4n - 7.$

The sequence is as follows.

a_1	a_2	a_3	a_4	a_5	a_6	a_7	a_8	a_9	a_{10}	a_{11}	$\ldots$
−3	1	5	9	13	17	21	25	29	33	37	$\ldots$

4. For this arithmetic sequence, the common difference is $d = 15 - 7 = 8$.

There are two ways to find the tenth term. One way is to write the first ten terms (by repeatedly adding 8).

7, 15, 23, 31, 39, 47, 55, 63, 71, 79

So, the tenth term is 79.

Another way to find the tenth term is to first find a formula for the nth term. Because the common difference is $d = 8$ and the first term is $a_1 = 7$, the formula must have the form

$a_n = a_1 + (n - 1)d = 7 + (n - 1)(8).$

Therefore, a formula for the nth term is $a_n = 8n - 1$ which implies that the tenth term is

$a_{10} = 8(10) - 1 = 79.$

5. To begin, notice that the sequence is arithmetic (with a common difference of $d = 37 - 40 = -3$). Moreover, the sequence has 7 terms. So, the sum of the sequence is

$S_n = \frac{n}{2}(a_1 + a_n)$ Sum of a finite arithmetic sequence

$= \frac{7}{2}(40 + 22)$ Substitute 7 for n, 40 for a_1, and 22 for a_n.

$= 217.$ Simplify.

6. (a) The integers from 1 to 35 form an arithmetic sequence that has 35 terms. So, you can use the formula for the sum of a finite arithmetic sequence, as follows.

$S_n = 1 + 2 + 3 + \ldots + 34 + 35$

$= \frac{n}{2}(a_1 + a_n)$ Sum of a finite arithmetic sequence

$= \frac{35}{2}(1 + 35)$ Substitute 35 for n, 1 for a_1, and 35 for a_n.

$= 630$ Simplify.

(b) The sum of the integers from 1 to $2N$ form an arithmetic sequence that has $2N$ terms.

$S_n = 1 + 2 + 3 + \ldots + (2N - 1) + 2N$

$= \frac{n}{2}(a_1 + a_n)$ Sum of a finite arithmetic sequence.

$= \frac{2N}{2}(1 + 2N)$ Substitute $2N$ for n, 1 for a_1 and $2N$ for a_n.

$= N(1 + 2N)$ Simplify.

7. For this arithmetic sequence, $a_1 = 6$ and $d = 12 - 6 = 6$.

So, $a_n = a_1 + (n - 1)d = 6 + 6(n - 1)$ and the nth term is $a_n = 6n$.

Therefore, $a_{120} = 6(120) = 720$, and the sum of the first 120 terms is

$$S_{120} = \frac{n}{2}(a_1 + a_{120})$$
$$= \frac{120}{2}(6 + 720)$$
$$= 60(726)$$
$$= 43,560.$$

8. For this arithmetic sequence, $a_1 = 78$ and $d = 76 - 78 = -2$.

So, $a_n = 78 + (-2)(n - 1)$ and the nth term is $a_n = -2n + 80$.

Therefore, $a_{30} = -2(30) + 80 = 20$, and the sum of the first 30 terms is

$$S_{30} = \frac{n}{2}(a_1 + a_{30})$$
$$= \frac{30}{2}(78 + 20)$$
$$= 15(98)$$
$$= 1470.$$

9. The annual sales form an arithmetic sequence in which $a_1 = 160,000$ and $d = 20,000$.

So, $a_n = 160,000 + 20,000(n - 1)$ and the nth term of the sequence is $a_n = 20,000n + 140,000$.

Therefore, the 10th term of the sequence is

$$a_{10} = 20,000(10) + 140,000$$
$$= 340,000.$$

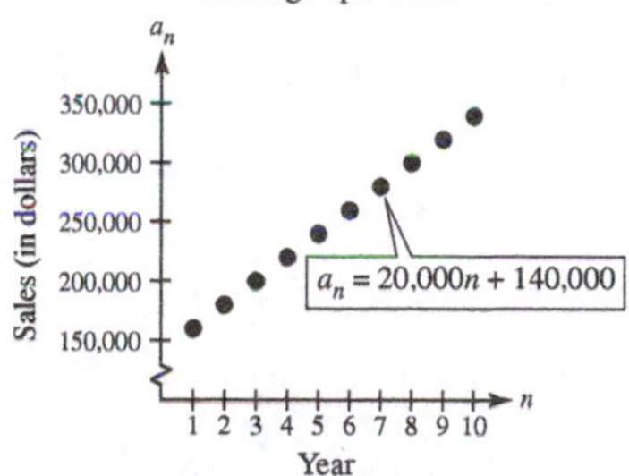

The sum of the first 10 terms of the sequence is

$$S_{10} = \frac{n}{2}(a_1 + a_{10})$$
$$= \frac{10}{2}(160,000 + 340,000)$$
$$= 5(500,000)$$
$$= 2,500,000.$$

So, the total sales for the first 10 years will be \$2,500,000.

Checkpoints for Section 8.3

1. The sequence whose *n*th term is $6(-2)^n$ is geometric.

For this sequence, the common ratio of consecutive terms is -2.

The first four terms, beginning with $n = 1$ are as follows.

$a_1 = 6(-2)^1 = 6(-2) = -12$

$a_2 = 6(-2)^2 = 6(4) = 24$

$a_3 = 6(-2)^3 = 6(-8) = -48$

$a_4 = 6(-2)^4 = 6(16) = 96$

So the sequence of terms are

$\underbrace{-12, 24,}_{\frac{24}{-12}=-2} -48, 96, \ldots, 6(-2)^n, \ldots$

2. Starting with $a_1 = 2$, repeatedly multiply by 4 to obtain the following.

$a_1 = 2$	1st term
$a_2 = 2(4^1) = 8$	2nd term
$a_3 = 2(4^2) = 32$	3rd term
$a_4 = 2(4^3) = 128$	4th term
$a_5 = 2(4)^4 = 512$	5th term

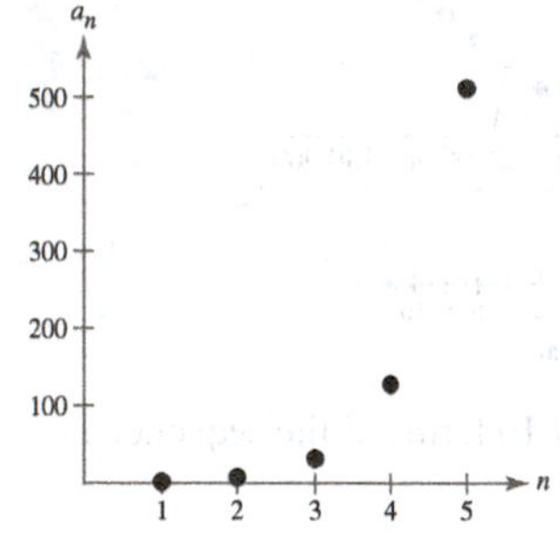

3. Algebraic Solution

Use the formula for the *n*th term of a geometric sequence.

$a_n = a_1 r^{n-1}$

$a_{12} = 14(1.2)^{12-1}$

$= 14(1.2)^{11}$

≈ 104.02

Numerical Solution

For this sequence, $r = 1.2$ and $a_1 = 14$. So $a_n = 14(1.2)^{n-1}$. Use a graphing utility to create a table that shows the terms of the sequence.

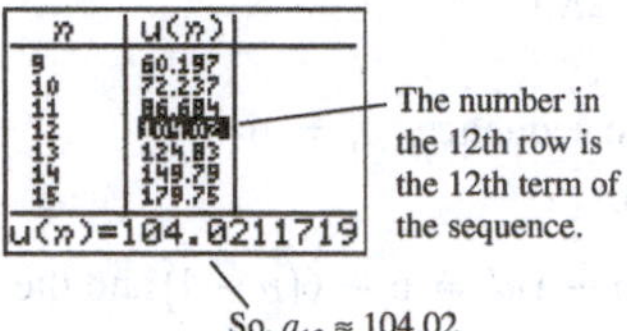

4. The first few terms of the geometric sequence are 4, 20, 100, ... You can find the common ratio of this geometric sequence by dividing any term by the previous term so $r = \frac{20}{4} = 5$. Because the first term is $a_1 = 4$, the formula for the *n*th term of a geometric sequence is as follows.

$a_n = a_1 r^{n-1}$

$a_n = 4(5)^{n-1}$

The 12th term of the sequence is as follows.

$a_{12} = 4(5)^{11} = 195{,}312{,}500$

5. The fifth term is related to the second term by the equation $a_5 = a_2 r^3$.

Because $a_5 = \frac{81}{4}$ and $a_2 = 6$, you can solve for *r* as follows.

$a_5 = a_2 r^3$	Multiply the second term by r^{5-3}.
$\frac{81}{4} = 6r^3$	Substitute $\frac{81}{4}$ for a_5 and 6 for a_2.
$\frac{27}{8} = r^3$	Divide each side by 6.
$\frac{3}{2} = r$	Take the cube root of each side.

You can obtain the eighth term by multiplying the fifth term by r^3.

$a_8 = a_5 r^3$	Multiply the fifth term by r^{8-5}.
$= \frac{81}{4}\left(\frac{3}{2}\right)^3$	Substitute $\frac{81}{3}$ for a_5 and $\frac{3}{2}$ for r
$= \frac{81}{4}\left(\frac{27}{8}\right)$	Evalutate power.
$= \frac{2187}{32}$	Multiply fractions.

6. You have $\sum_{i=1}^{10} 2(0.25)^{i-1} = 2(0.25)^0 + 2(0.25)^1 + 2(0.25)^2 + \ldots + 2(0.25)^9$.

Now, $a_1 = 2$, $r = 0.25$, and $n = 10$, so applying the formula for the sum of a finite geometric sequence, you obtain the following.

$$S_n = a_1\left(\frac{1 - r^n}{1 - r}\right) \qquad \text{Sum of a finite geometric series}$$

$$\sum_{i=1}^{10} 2(0.25)^{i-1} = 2\left[\frac{1 - (0.25)^{10}}{1 - 0.25}\right] \qquad \text{Substitute 2 for } a_1\text{, 0.25 for } r\text{, and 10 for } n.$$

$$\approx 2.667 \qquad \text{Use a calculator.}$$

7. (a) $\sum_{n=0}^{\infty} 5(0.5)^n = 5 + 5(0.5) + 5(0.5)^2 + \ldots + 5(0.5)^n + \ldots$

$$= \frac{5}{1 - 0.5} \qquad \text{Use } \frac{a_1}{1 - r} \text{ and Subsitute 5 for } a_1 \text{ and 0.5 for } r.$$

$$= \frac{5}{0.5}$$

$$= 10$$

(b) To find the common ratio, divide any term by the preceding term.

So, $r = \frac{1}{5} = 0.2$.

The sum of the infinite geometric series is as follows.

$$5 + 1 + 0.2 + 0.04 + \ldots = 5(0.2)^0 + 5(0.2)^1 + 5(0.2)^2 + 5(0.2)^3 + \ldots$$

$$= \frac{a_1}{1 - r}$$

$$= \frac{5}{1 - 0.2}$$

$$= 6.25$$

8. To find the balance in the account after 48 months, consider each of the 48 deposits separately. The first deposit will gain interest for 48 months, and its balance will be

$$A_{48} = 70\left(1 + \frac{0.02}{12}\right)^{48}.$$

The second deposit will gain interest for 47 months, and its balance will be

$$A_{47} = 70\left(1 + \frac{0.02}{12}\right)^{47}.$$

The last deposit will gain interest for only 1 month, and its balance will be

$$A_1 = 70\left(1 + \frac{0.02}{12}\right)^1.$$

The total balance in the annuity will be the sum of the balances of the 48 deposits.

Using the formula for the sum of a finite geometric sequence, with $A_1 = 70(1.0017)$ $r = 1.0017$, and $n = 48$ you have

$$S_n = A_1\left(\frac{1 - r^n}{1 - r}\right)$$

$$S_{48} = 70\left(1 + \frac{0.02}{12}\right)\left[\frac{1 - \left(1 + \frac{0.02}{12}\right)^{48}}{1 - \left(1 + \frac{0.02}{12}\right)}\right] \approx \$3500.85.$$

Checkpoints for Section 8.4

1. (a) $P_{k+1} : S_{k+1} = \dfrac{6}{(k+1)(k+1+3)}$

$= \dfrac{6}{(k+1)(k+4)}$

(b) $P_{k+1} : k+1+2 \le 3(k+1-1)^2$

$k + 3 \le 3k^2$

(c) $P_{k+1} : 2^{4(k+1)-2} + 1 > 5(k+1)$

$2^{4k+4-2} + 1 > 5k + 5$

$2^{4k+2} + 1 > 5k + 5$

2. Mathematical induction consists of two distinct parts.

1. First, you must show that the formula is true when $n = 1$. When $n = 1$, the formula is valid, because $S_1 = 1(1+4) = 5$.
2. The second part of mathematical induction has two steps. The first step is to *assume* that the formula is valid for some integer k. The second step is to use this assumption to prove that the formula is valid for the *next* integer, $k + 1$. Assuming that the formula

$S_k = 5 + 7 + 9 + 11 + \ldots + (2k+3) = k(k+4)$

is true, you must show that the formula

$S_{k+1} = (k+1)(k+1+4)$
$= (k+1)(k+5)$ is true.

$$\begin{aligned}
S_{k+1} &= 5 + 7 + 9 + 11 + \ldots + (2k+3) + [2(k+1)+3] \\
&= [5 + 7 + 9 + 11 + \ldots + (2k+3)] + (2k+2+3) \\
&= S_k + 2(k+5) \\
&= k(k+4) + 2k + 5 \\
&= k^2 + 4k + 2k + 5 \\
&= k^2 + 6k + 5 \\
&= (k+1)(k+5)
\end{aligned}$$

Combining the results of parts (1) and (2), you can conclude by mathematical induction that the formula is valid for all integers $n \ge 1$.

3. 1. When $n = 1$, the formula is valid, because $S_1 = (1)(1-1) = \dfrac{(1)(1-1)(1+1)}{3} = 0.$

2. Assuming that $S_k = 1(1-1) + 2(2-1) + 3(3-1) + \ldots + k(k-1) = \dfrac{k(k-1)(k+1)}{3}$

You must show that $S_{k+1} = \dfrac{(k+1)(k+1-1)(k+1+1)}{3} = \dfrac{k(k+1)(k+2)}{3}.$

To do this, write the following.

$$\begin{aligned}
S_{k+1} &= S_k + a_{k+1} \\
&= [1(1-1) + 2(2-1) + 3(3-1) + \ldots + k(k-1)] + (k+1)(k+1-1) && \text{Substitution} \\
&= \frac{k(k-1)(k+1)}{3} + k(k+1) && \text{By assumption} \\
&= \frac{k(k-1)(k+1) + 3k(k+1)}{3} && \text{Combine fractions.} \\
&= \frac{k(k+1)[(k-1)+3]}{3} && \text{Factor.} \\
&= \frac{k(k+1)(k+2)}{3} && \text{Simplify.}
\end{aligned}$$

So, S_k implies S_{k+1}.

Combining the results of parts (1) and (2), you can conclude by mathematical induction that the formula is valid for all positive integers n.

4. 1. For $n = 1$ and $n = 2$, the statement is true because

$1! \geq 1$ and $2! \geq 2$.

2. Assuming that

$k! \geq k$

You need to show that $(k + 1)! \geq k + 1$. For $n = k$, you have $(k + 1)! = (k + 1)k! \geq (k + 1)k$.

Because $(k + 1)! > k + 1$ for all $k > 1$, it follows that $(k + 1)! \geq k + 1$.

Combining the results of parts (1) and (2), you can conclude by mathematical induction that $n! \geq n$ for all integers $n \geq 1$.

5. 1. For $n = 1$, the statement is true because

$3^1 + 1 = 4$.

So, 2 is a factor.

2. Assuming that 2 is a factor of $3^k + 1$, you must show that 2 is a factor of $3^{k+1} + 1$.

To do this, write the following.

$3^{k+1} + 1 = 3^{k+1} - 3^k + 3^k + 1$ Subtract and add 3^k.

$= 3^k(3 - 1) + 3^k + 1$ Regroup terms.

$= 3^k \cdot 2 + 3^k + 1$ Simplify.

Because 2 is a factor of $3^k \cdot 2$ and 2 is also a factor of $3^k + 1$, it follows that 2 is a factor of $3^{k+1} + 1$. Combining the results of parts (1) and (2), you can conclude by mathematical induction that 2 is a factor of $3^n + 1$ for all positive integers n.

6. Begin by writing the first few sums.

$S_1 = 3 = 1(3)$

$S_2 = 3 + 7 = 10 = 2(5)$

$S_3 = 3 + 7 + 11 = 21 = 3(7)$

$S_4 = 3 + 7 + 11 + 15 = 36 = 4(9)$

$S_5 = 3 + 7 + 11 + 15 + 19 = 55 = 5(11)$

From this sequence, it appears that the formula for the kth sum is

$S_k = 3 + 7 + 11 + 15 + 19 + \ldots + 4k - 1 = k(2k + 1)$.

To prove the validity of this hypothesis, use mathematical induction. Note that you have already verified the formula for $n = 1$, so begin by assuming that the formula is valid for $n = k$ and trying to show that it is valid for $n = k + 1$.

$$\begin{aligned}
S_{k+1} &= [3 + 7 + 11 + 15 + \ldots + 4k - 1] + 4(k + 1) - 1 \\
&= k(2k + 1) + 4k + 3 \\
&= 2k^2 + k + 4k + 3 \\
&= 2k^2 + 5k + 3 \\
&= (k + 1)(2k + 3) \\
&= (k + 1)[2(k + 1) + 1]
\end{aligned}$$

So, by mathematical induction, the hypothesis is valid.

7. (a) Using the formula for the sum of the first n positive integers, you obtain

$$\sum_{i=1}^{20} i = 1 + 2 + 3 + \ldots + 20 = \frac{20(20+1)}{2} = \frac{20(21)}{2} = 210.$$

(b)
$$\begin{aligned}\sum_{i=1}^{5} 2i^2 + 3i^3 &= \sum_{i=1}^{5} 2i^2 + \sum_{i=1}^{5} 3i^3 \\ &= 2\sum_{i=1}^{5} i^2 + 3\sum_{i=1}^{5} i^3 \\ &= 2\left[\frac{5(5+1)[2(5)+1]}{6}\right] + 3\left[\frac{(5)^2(5+1)^2}{4}\right] \\ &= 2\left[\frac{(5)(6)(11)}{6}\right] + 3\left[\frac{(25)(36)}{4}\right] \\ &= 2(55) + 3(225) \\ &= 785\end{aligned}$$

8. Begin by finding the first and second differences.

1		2		3		4		5		6	
−2		0		4		10		18		28	
	2		4		6		8		10		
		2		2		2		2			

You know from the second differences that the model is quadratic and has the form $a_n = an^2 + bn + c$. By substituting 1, 2, and 3 for n, you can obtain a system of three linear equations in three variables.

$$a_1 = a(1)^2 + b(1) + c = -2$$

$$a_2 = a(2)^2 + b(2) + c = 0$$

$$a_3 = a(3)^2 + b(3) + c = 4$$

$$\begin{cases} a + b + c = -2 \\ 4a + 2b + c = 0 \\ 9a + 3b + c = 4 \end{cases}$$

Solving this system using techniques from Chapter 9, you can find the solution to be $a = 1$, $b = -1$, and $c = -2$.

So, the quadratic model is $a_n = n^2 - n - 2$.

Checkpoints for Section 8.5

1. (a) $\binom{11}{5} = \frac{11!}{6!5!} = \frac{(11 \cdot 10 \cdot 9 \cdot 8 \cdot 7) \cdot \cancel{6!}}{\cancel{6!}\,5!} = \frac{11 \cdot 10 \cdot 9 \cdot 8 \cdot 7}{5 \cdot 4 \cdot 3 \cdot 2 \cdot 1} = 462$

(b) ${}_9C_2 = \frac{9!}{7!2!} = \frac{(9 \cdot 8) \cdot \cancel{7!}}{\cancel{7!}\,2!} = \frac{9 \cdot 8}{2 \cdot 1} = 36$

(c) $\binom{5}{0} = \frac{\cancel{5!}}{\cancel{5!}\,0!} = 1$

(d) ${}_{15}C_{15} = \frac{\cancel{15!}}{0!\,\cancel{15!}} = 1$

2. (a) $_7C_5 = \dfrac{7!}{2!5!} = \dfrac{7\cdot 6\cdot \cancel{5}\cdot \cancel{4}\cdot \cancel{3}}{\cancel{3}\cdot \cancel{4}\cdot \cancel{5}\cdot 2\cdot 1} = 21$

(b) $\dbinom{7}{2} = \dfrac{7!}{5!2!} = \dfrac{7.6}{2.1} = 21$

(c) $_{14}C_{13} = \dfrac{14!}{1!13!} = \dfrac{14}{1} = 14$

(d) $\dbinom{14}{1} = \dfrac{14!}{13!1!} = \dfrac{14}{1} = 14$

3.

1
1 1
1 2 1
1 3 3 1
1 4 6 4 1
1 5 10 10 5 1
1 6 15 20 15 6 1
1 7 21 35 35 21 7 1
1 8 28 56 70 56 28 8 1 8th row
1 9 36 84 126 126 84 36 9 1
↓ ↓ ↓ ↓ ↓ ↓ ↓ ↓ ↓ ↓
$_9C_0$ $_9C_1$ $_9C_2$ $_9C_3$ $_9C_4$ $_9C_5$ $_9C_6$ $_9C_7$ $_9C_8$ $_9C_9$

4. The binomial coefficients from the fourth row of Pascal's Triangle are 1, 4, 6, 4, 1.
So, the expansion is as follows.

$$\begin{aligned}(x+2)^4 &= (1)x^4 + (4)x^3(2) + (6)x^2(2^2) + (4)x(2^3) + (1)(2^4)\\ &= x^4 + 8x^3 + 24x^2 + 32x + 16\end{aligned}$$

5. (a) $$\begin{aligned}(y-2)^4 &= (1)y^4 - (4)y^3(2) + (6)y^2(2^2) - (4)y(2^3) + (1)(2)^4\\ &= y^4 - 8y^3 + 24y^2 - 32y + 16\end{aligned}$$

(b) $$\begin{aligned}(2x-y)^5 &= (1)(2x)^5 - (5)(2x)^4y + (10)(2x)^3y^2 - (10)(2x)^2y^3 + (5)(2x)y^4 - (1)y^5\\ &= 32x^5 - 80x^4y + 80x^3y^2 - 40x^2y^3 + 10xy^4 - y^5\end{aligned}$$

6. Use the third row of Pascal's Triangle to write the expansion of $(5+y^2)^3 = (y^2+5)^3$, as follows.

$$\begin{aligned}(y^2+5)^3 &= (1)(y^2)^3 + (3)(y^2)^2(5) + (3)(y^2)(5^2) + (1)(5^3)\\ &= y^6 + 15y^4 + 75y^2 + 125\end{aligned}$$

7. (a) Remember that the formula is for the $(r+1)$th term, so r is one less than the number of the term you need. So, to find the fifth term in this binomial expansion, use $r = 4$, $n = 8$, $x = a$, and $y = 2b$, as shown.

$$\begin{aligned}{}_nC_rx^{n-r}y^r &= {}_8C_4a^{8-4}(2b)^4 = (70)(a^4)(2b)^4\\ &= 70(2^4)a^4b^4\\ &= 1120a^4b^4\end{aligned}$$

(b) In this case, $n = 11$, $r = 7$, $x = 3a$, and $y = -2b$. Substitute these values to obtain the following.

$$\begin{aligned}{}_nC_rx^{n-r}y^r &= {}_{11}C_7(3a)^4(-2b)^7\\ &= (330)(81a^4)(-128b^7)\\ &= -3{,}421{,}440a^4b^7\end{aligned}$$

So, the coefficient is $-3{,}421{,}440$.

Checkpoints for Section 8.6

1. To solve this problem, count the different ways to obtain a sum of 14 using two numbers from 1 to 8.

First number:	6	7	8
Second number:	8	7	6

So, a sum of 14 can occur in three different ways.

2. To solve this problem, count the different ways to obtain a sum of 14 *using two different numbers* from 1 to 8.

First number:	6	8
Second number:	8	6

So, a sum of 14 can occur in 2 ways.

3. There are three events in this situation. The first event is the choice of the first number, the second event is the choice of the second number, and the third event is the choice of the third number. Because there is a choice of 30 numbers for each event, it follows that the number of different lock combinations is

$30 \cdot 30 \cdot 30 = 27{,}000.$

4. Because the product's catalog number is made up of one letter from the English alphabet followed by a five-digit number, there are 26 choices for the first digit and 10 choices for each of the other 5 digits.

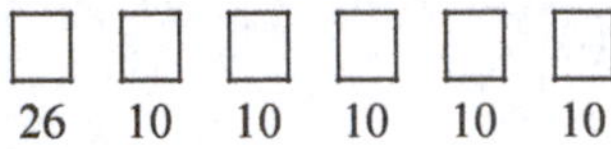

26 10 10 10 10 10

So, the number of possible catalog numbers is $26 \cdot 10 \cdot 10 \cdot 10 \cdot 10 \cdot 10 = 2{,}600{,}000.$

5. *First position*: Any of the *four* letters

Second position: Any of the remaining *three* letters

Third position: Either of the remaining *two* letters

Fourth position: The *one* remaining letter

So, the numbers of choices for the four positions are as follows.

Permutations of four letters

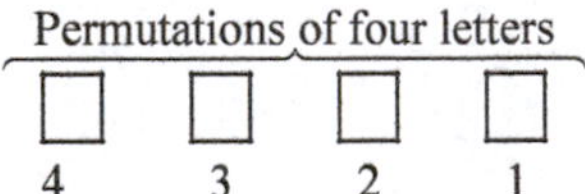

4 3 2 1

The total number of permutations of the four letters is

$$4! = 4 \cdot 3 \cdot 2 \cdot 1 = 24.$$

6. Here are the different possibilities.

President (first position): *Five* choices

Vice-President (second position): *Four* choices

Using the Fundamental Counting Principle, multiply these two numbers to obtain the following.

Different orders of offices

☐ ☐

President Vice-President

5 4

So, there are $5 \cdot 4 = 20$ different ways there can be a President and Vice-President.

7. The word M I T O S I S has seven letters, of which there are two I's, two S's, and one M, T, and O. So, the number of distinguishable ways the letters can be written is

$$\frac{n!}{n_1!\,n_2!} = \frac{7!}{2!\,2!} = \frac{7 \cdot 6 \cdot 5 \cdot 4 \cdot 3 \cdot \cancel{2!}}{2!\,\cancel{2!}} = 1260.$$

8. The following subsets represent the different combinations of two letters that can be chosen from the seven letters.

{A, B} {B, C} {C, D} {D, E} {E, F} {F, G}

{A, C} {B, D} {C, E} {D, F} {E, G}

{A, D} {B, E} {C, F} {D, G}

{A, E} {B, F} {C, G}

{A, F} {B, G}

{A, G}

From this list, you can conclude that there are 21 different ways that two letters can be chosen from seven letters.

9. To find the number of three card poker hands, use the formula for the number of combinations of 52 elements taken three at a time, as follows.

$$\begin{aligned} {}_{52}C_3 &= \frac{52!}{(52-3)!\,3!} \\ &= \frac{52!}{49!\,3!} \\ &= \frac{52 \cdot 51 \cdot 50 \cdot \cancel{49!}}{\cancel{49!}\,3!} \\ &= \frac{52 \cdot 51 \cdot 50}{3 \cdot 2 \cdot 1} \\ &= 22{,}100 \end{aligned}$$

10. There are ${}_{10}C_6$ ways of choosing six girls from a group of ten girls and ${}_{15}C_6$ ways of choosing boys from a group of fifteen boys. By the Fundamental Counting Principle, there are ${}_{10}C_6 \cdot {}_{15}C_6$ ways of choosing six girls and six boys.

$$\begin{aligned}{}_{10}C_6 \cdot {}_{15}C_6 &= \frac{10!}{4! \cdot 6!} \cdot \frac{15!}{9! \cdot 6!}\\ &= 210 \cdot 5005\\ &= 1{,}051{,}050\end{aligned}$$

So, there are 1,051,050 12-member swim teams possible.

Checkpoints for Section 8.7

1. Because either coin can land heads up or tails up, and the six-sided die can land with a 1 through 6 up. So, the sample space is

$$\begin{aligned}S = \{&HH1, HH2, HH3, HH4, HH5, HH6\\ &HT1, HT2, HT3, HT4, HT5, HT6,\\ &TH1, TH2, TH3, TH4, TH5, TH6,\\ &TT1, TT2, TT3, TT4, TT5, TT6\}\end{aligned}$$

2. (a) Let $E = \{TTT\}$ and $S = \{HHH, HHT, HTH, HTT, TTT, THT, TTH, THH\}$.

The probability of getting three tails is

$$P(E) = \frac{n(E)}{n(S)} = \frac{1}{8}.$$

(b) Because there are 52 cards in a standard deck of playing cards and there are 13 diamonds, the probability of drawing a diamond is

$$\begin{aligned}P(E) &= \frac{n(E)}{n(S)}\\ &= \frac{13}{52}\\ &= \frac{1}{4}.\end{aligned}$$

3. Because there are six possible outcomes on each die, use the Fundamental Counting Principle to conclude that there are $6 \cdot 6$ or 36 different outcomes when you toss two dice. To find the probability of rolling a total of 5, you must first count the number of ways in which this can occur.

First Die	Second Die
1	4
2	3
3	2
4	1

So, a total of 5 can be rolled in four ways, which means that the probability of rolling a 5 is

$$P(E) = \frac{n(E)}{n(S)} = \frac{4}{36} = \frac{1}{9}.$$

4. For a standard deck of 52 playing cards, there are 13 clubs. So, the probability of drawing a club is

$$P(E) = \frac{n(E)}{n(S)} = \frac{13}{52} = \frac{1}{4}.$$

For a set consisting of the aces, the sample space is 4 cards. So, the probability of drawing the ace at hearts is

$$P(E) = \frac{n(E)}{n(S)} = \frac{1}{4}.$$

So, the probability of drawing a club from a standard deck of cards is the same as drawing the ace of hearts from the set of aces.

5. The total number of colleges and universities is 4622. Because there are 640 colleges and universities in the Pacific region, the probability that the institution is in that region is

$$P(E) = \frac{n(E)}{n(S)} = \frac{640}{4622} \approx 0.138.$$

6. To find the number of elements in the sample space, use the formula for the number of combinations of 43 elements taken five at a time.

$$n(S) = {}_{43}C_5$$
$$= \frac{43 \cdot 42 \cdot 41 \cdot 40 \cdot 39}{5 \cdot 4 \cdot 3 \cdot 2 \cdot 1}$$
$$= 962{,}598$$

When a player buys one ticket, the probability of winning is

$$P(E) = \frac{1}{962{,}598} \approx 0.000001.$$

7. Because the deck has 4 aces, the probability of drawing an ace (event A) is

$$P(A) = \frac{4}{52}.$$

Similarly, because the deck has 13 spades, the probability of drawing a spade (event B) is

$$P(B) = \frac{13}{52}.$$

Because one of the cards is an ace *and* a spade, the ace of spades, it follows that

$$P(A \cap B) = \frac{1}{52}.$$

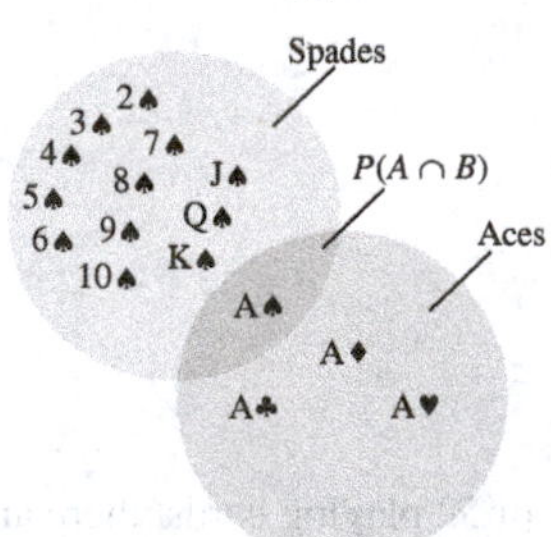

Finally, applying the formula for the probability of the union of two events, the probability of drawing an ace or spade is as follows.

$$P(A \cup B) = P(A) + P(B) - P(A \cap B)$$
$$= \frac{4}{52} + \frac{13}{52} - \frac{1}{52}$$
$$= \frac{16}{52} = \frac{4}{13} \approx 0.308$$

8. To begin, add the number of employees to find that the total is 529. Next, let event A represent choosing an employee with 30-34 years of service, event B with 35-39 years of service, event C with 40-44 years of service, and event D with 45 or more years of service.

Because events A, B, C, and D have no outcomes in common, these four events are mutually exclusive and

$$P(A \cup B \cup C \cup D) = P(A) + P(B) + P(C) + +P(D)$$
$$= \frac{35}{529} + \frac{21}{529} + \frac{8}{529} + \frac{2}{529}$$
$$= \frac{66}{529}$$
$$\approx 0.125.$$

So, the probability of choosing an employee who has 30 or more years of service is about 0.125.

9. The probability of selecting a number from 1 to 11 from a set of numbers from 1 to 30 is

$$P(A) = \frac{11}{30}.$$

So, the probability that both numbers are less than 12 is

$$P(A) \cdot P(A) = \frac{11}{30} \cdot \frac{11}{30} = \frac{121}{900} \approx 0.134.$$

10. Let event A represent selecting a person who expected much of the workforce to be automated within 50 years. The probability of event A is 0.65. Each of the 5 occurrence of event A is an independent event, so the probability that all 5 people expected much of the workforce to be automated within 50 years is

$$\left[P(A)\right]^5 = (0.65)^5 \approx 0.116.$$

11. To solve this problem as stated, you would need to find the probabilities of having exactly one faulty unit, exactly two faulty units, exactly three faulty units, and so on. However, using complements, you can find the probability that all units are perfect and then subtract this value from 1. Because the probability that any given unit is perfect is 499/500, the probability that all 300 units are perfect is

$$P(A) = \left(\frac{499}{500}\right)^{300} \approx 0.548.$$

So, the probability that at least one unit is faulty is

$$P(A^1) = 1 - P(A) \approx 1 - 0.548 = 0.452.$$

Appendix

Checkpoints for Appendix A

1. Do not apply radicals term-by-term when adding terms.

Leave as $\sqrt{x^2 + 4}$.

2. $x(x-2)^{-1/2} + 6(x-2)^{1/2} = (x-2)^{-1/2}\left[x(x-2)^0 + 6(x-2)^1\right]$

$$= (x-2)^{-1/2}[x + 6x - 12]$$

$$= (x-2)^{-1/2}(7x - 12)$$

3. The expression on the left side of the equation is three times the expression on the right side. To make both sides equal, insert a factor of 3.

$$\frac{6x - 3}{\left(x^2 - x + 4\right)^2} = (3)\frac{1}{\left(x^2 - x + 4\right)^2}(2x - 1)$$

4. To write the expression on the left side of the equation in the form given on the right side, first multiply the numerator and denominator of the first term by $\frac{1}{9}$. Then multiply the numerator and denominator of the second term by $\frac{1}{25}$.

$$\frac{9x^2}{16} + 25y^2 = \frac{9x^2}{16}\left(\frac{1/9}{1/9}\right) + \frac{25y^2}{1}\left(\frac{1/25}{1/25}\right) = \frac{x^2}{16/25} + \frac{y^2}{1/25}$$

5. $\dfrac{-6x}{\left(1 - 3x^2\right)^2} + \dfrac{1}{\sqrt[3]{x}} = -6x\left(1 - 3x^2\right)^{-2} + x^{-1/3}$

6. (a) $\dfrac{x^4 - 2x^3 + 5}{x^3} = \dfrac{x^4}{x^3} - \dfrac{2x^3}{x^3} + \dfrac{5}{x^3} = x - 2 + \dfrac{5}{x^3}$

(b) $\dfrac{x^2 - x + 5}{\sqrt{x}} = \dfrac{x^2}{x^{1/2}} - \dfrac{x}{x^{1/2}} + \dfrac{5}{x^{1/2}} = x^{3/2} - x^{1/2} + 5x^{-1/2}$

Chapter P Practice Test Solutions

1. $\dfrac{|-42| - 20}{15 - |-4|} = \dfrac{42 - 20}{15 - 4} = \dfrac{22}{11} = 2$

2. $\dfrac{x}{z} - \dfrac{z}{y} = \dfrac{x}{z} \cdot \dfrac{y}{y} - \dfrac{z}{y} \cdot \dfrac{z}{z} = \dfrac{xy - z^2}{yz}$

3. $|x - 7| \le 4$

4. $10(-5)^3 = 10(-125) = -1250$

5. $(-4x^3)(-2x^{-5})\left(\dfrac{1}{16}x\right) = (-4)(-2)\left(\dfrac{1}{16}\right)x^{3+(-5)+1}$
$= \dfrac{8}{16}x^{-1}$
$= \dfrac{1}{2x}$

6. $0.0000412 = 4.12 \times 10^{-5}$

7. $125^{2/3} = \left(\sqrt[3]{125}\right)^2 = (5)^2 = 25$

8. $\sqrt[4]{64x^7y^9} = \sqrt[4]{16 \cdot 4x^4x^3y^8y} = 2xy^2\sqrt[4]{4x^3y}$

9. $\dfrac{6}{\sqrt{12}} = \dfrac{6}{2\sqrt{3}} \cdot \dfrac{\sqrt{3}}{\sqrt{3}} = \dfrac{6\sqrt{3}}{6} = \sqrt{3}$

10. $3\sqrt{80} - 7\sqrt{500} = 3(4\sqrt{5}) - 7(10\sqrt{5})$
$= 12\sqrt{5} - 70\sqrt{5}$
$= -58\sqrt{5}$

11. $(8x^4 - 9x^2 + 2x - 1) - (3x^3 + 5x + 4) = 8x^4 - 3x^3 - 9x^2 - 3x - 5$

12. $(x - 3)(x^2 + x - 7) = x^3 + x^2 - 7x - 3x^2 - 3x + 21 = x^3 - 2x^2 - 10x + 21$

13. $[(x - 2) - y]^2 = (x - 2)^2 - 2y(x - 2) + y^2 = x^2 - 4x + 4 - 2xy + 4y + y^2 = x^2 + y^2 - 2xy - 4x + 4y + 4$

14. $16x^4 - 1 = (4x^2 + 1)(4x^2 - 1)$
$= (4x^2 + 1)(2x + 1)(2x - 1)$

15. $6x^2 + 5x - 4 = (2x - 1)(3x + 4)$

16. $x^3 - 64 = x^3 - 4^3 = (x - 4)(x^2 + 4x + 16)$

17. $-\dfrac{3}{x} + \dfrac{x}{x^2 + 2} = \dfrac{-3(x^2 + 2) + x^2}{x(x^2 + 2)} = \dfrac{-2x^2 - 6}{x(x^2 + 2)} = -\dfrac{2(x^2 + 3)}{x(x^2 + 2)}$

18. $\dfrac{x - 3}{4x} \div \dfrac{x^2 - 9}{x^2} = \dfrac{x - 3}{4x} \cdot \dfrac{x^2}{(x + 3)(x - 3)} = \dfrac{x}{4(x + 3)}, \ x \ne 0, 3$

19. $\dfrac{1 - \dfrac{1}{x}}{1 - \dfrac{1}{1 - (1/x)}} = \dfrac{\dfrac{x - 1}{x}}{1 - \dfrac{1}{(x - 1)/x}} = \dfrac{\dfrac{x - 1}{x}}{1 - \dfrac{x}{x - 1}} = \dfrac{\dfrac{x - 1}{x}}{\dfrac{-1}{x - 1}} = \dfrac{x - 1}{x} \cdot \dfrac{x - 1}{-1} = -\dfrac{(x - 1)^2}{x}, \ x \ne 1$

20. (a)

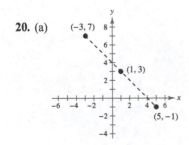

(b) $d = \sqrt{[5 - (-3)]^2 + (-1 - 7)^2}$
$= \sqrt{(8)^2 + (-8)^2}$
$= \sqrt{64 + 64}$
$= \sqrt{128}$
$= 8\sqrt{2}$

(c) $\left(\dfrac{-3 + 5}{2}, \dfrac{7 + (-1)}{2}\right) = (1, 3)$

Chapter 1 Practice Test Solutions

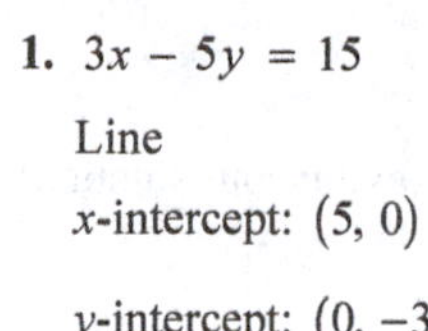

1. $3x - 5y = 15$

   Line

   x-intercept: $(5, 0)$

   y-intercept: $(0, -3)$

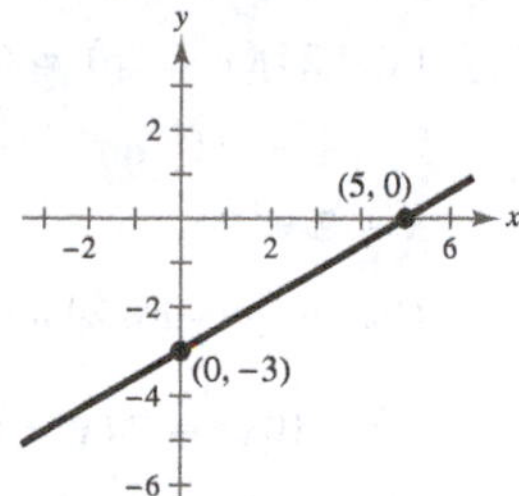

2. $y = \sqrt{9 - x}$

   Domain: $(-\infty, 9]$

   x-intercept: $(9, 0)$

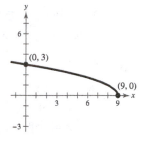

   y-intercept: $(0, 3)$

3. $$\begin{aligned} 5x + 4 &= 7x - 8 \\ 4 + 8 &= 7x - 5x \\ 12 &= 2x \\ x &= 6 \end{aligned}$$

4. $$\begin{aligned} \frac{x}{3} - 5 &= \frac{x}{5} + 1 \\ 15\left(\frac{x}{3} - 5\right) &= 15\left(\frac{x}{5} + 1\right) \\ 5x - 75 &= 3x + 15 \\ 2x &= 90 \\ x &= 45 \end{aligned}$$

5. $$\begin{aligned} \frac{3x + 1}{6x - 7} &= \frac{2}{5} \\ 5(3x + 1) &= 2(6x - 7) \\ 15x + 5 &= 12x - 14 \\ 3x &= -19 \\ x &= -\frac{19}{3} \end{aligned}$$

6. $$\begin{aligned} (x - 3)^2 + 4 &= (x + 1)^2 \\ x^2 - 6x + 9 + 4 &= x^2 + 2x + 1 \\ -8x &= -12 \\ x &= \frac{-12}{-8} \\ x &= \frac{3}{2} \end{aligned}$$

7. $$\begin{aligned} A &= \frac{1}{2}(a + b)h \\ 2A &= ah + bh \\ 2A - bh &= ah \\ \frac{2A - bh}{h} &= a \end{aligned}$$

8. Percent $= \frac{301}{4300} = 0.07 = 7\%$

9. Let x = number of quarters.

   Then $53 - x$ = number of nickels.

   $$\begin{aligned} 25x + 5(53 - x) &= 605 \\ 20x + 265 &= 605 \\ 20x &= 340 \\ x &= 17 \text{ quarters} \\ 53 - x &= 36 \text{ nickels} \end{aligned}$$

10. Let x = amount in the $9\frac{1}{2}\%$ fund.

    Then $15{,}000 - x$ = amount in 11% fund.

    $$\begin{aligned} 0.095x + 0.11(15{,}000 - x) &= 1582.50 \\ -0.015x + 1650 &= 1582.50 \\ -0.015x &= -67.5 \\ x &= \$4500 \text{ at } 9\tfrac{1}{2}\% \\ 15{,}000 - x &= \$10{,}500 \text{ at } 11\% \end{aligned}$$

11. $$\begin{aligned} 28 + 5x - 3x^2 &= 0 \\ (4 - x)(7 + 3x) &= 0 \\ 4 - x = 0 &\Rightarrow x = 4 \\ 7 + 3x = 0 &\Rightarrow x = -\tfrac{7}{3} \end{aligned}$$

12. $$\begin{aligned} (x - 2)^2 &= 24 \\ x - 2 &= \pm\sqrt{24} \\ x - 2 &= \pm 2\sqrt{6} \\ x &= 2 \pm 2\sqrt{6} \end{aligned}$$

13. $$\begin{aligned} x^2 - 4x - 9 &= 0 \\ x^2 - 4x + 2^2 &= 9 + 2^2 \\ (x - 2)^2 &= 13 \\ x - 2 &= \pm\sqrt{13} \\ x &= 2 \pm \sqrt{13} \end{aligned}$$

14. $x^2 + 5x - 1 = 0$

$a = 1, b = 5, c = -1$

$$x = \frac{-5 \pm \sqrt{(5)^2 - 4(1)(-1)}}{2(1)}$$

$$= \frac{-5 \pm \sqrt{25 + 4}}{2}$$

$$= \frac{-5 \pm \sqrt{29}}{2}$$

15. $3x^2 - 2x + 4 = 0$

$a = 3, b = -2, c = 4$

$$x = \frac{-(-2) \pm \sqrt{(-2)^2 - 4(3)(4)}}{2(3)}$$

$$= \frac{2 \pm \sqrt{4 - 48}}{6}$$

$$= \frac{2 \pm \sqrt{-44}}{6}$$

$$= \frac{2 \pm 2i\sqrt{11}}{6}$$

$$= \frac{1 \pm i\sqrt{11}}{3} = \frac{1}{3} \pm \frac{\sqrt{11}}{3}i$$

16. $60{,}000 = xy$

$$y = \frac{60{,}000}{x}$$

$$2x + 2y = 1100$$

$$2x + 2\left(\frac{60{,}000}{x}\right) = 1100$$

$$x + \frac{60{,}000}{x} = 550$$

$$x^2 + 60{,}000 = 550x$$

$$x^2 - 550x + 60{,}000 = 0$$

$$(x - 150)(x - 400) = 0$$

$x = 150$ or $x = 400$

$y = 400$ $\quad y = 150$

Length: 400 feet; width: 150 feet

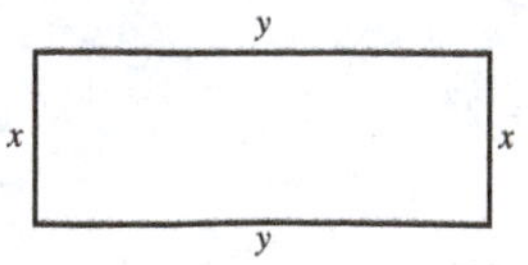

17. $x(x + 2) = 624$

$$x^2 + 2x - 624 = 0$$

$$(x - 24)(x + 26) = 0$$

$x = 24$ or $x = -26$, (extraneous solution)

$x + 2 = 26$

The integers are 24 and 26.

18. $x^2 - 10x^2 + 24x = 0$

$$x(x^2 - 10x + 24) = 0$$

$$x(x - 4)(x - 6) = 0$$

$x = 0, x = 4, x = 6$

19. $\sqrt[3]{6 - x} = 4$

$$6 - x = 64$$

$$-x = 58$$

$$x = -58$$

20. $(x^2 - 8)^{2/5} = 4$

$$x^2 - 8 = \pm 4^{5/2}$$

$x^2 - 8 = 32$ or $x^2 - 8 = -32$

$x^2 = 40$ $\quad x^2 = -24$

$x = \pm\sqrt{40}$ $\quad x = \pm\sqrt{-24}$

$x = \pm 2\sqrt{10}$ $\quad x = \pm 2\sqrt{6}i$

21. $x^4 - x^2 - 12 = 0$

$$(x^2 - 4)(x^2 + 3) = 0$$

$x^2 = 4$ or $x^2 = -3$

$x^2 = \pm 2$ $\quad x = \pm\sqrt{3}i$

22. $4 - 3x > 16$

$$-3x > 12$$

$$x < -4$$

23. $\left|\frac{x - 3}{2}\right| < 5$

$$-5 < \frac{x - 3}{2} < 5$$

$$-10 < x - 3 < 10$$

$$-7 < x < 13$$

24.
$$\frac{x+1}{x-3} < 2$$
$$\frac{x+1}{x-3} - 2 < 0$$
$$\frac{x+1-2(x-3)}{x-3} < 0$$
$$\frac{7-x}{x-3} < 0$$

Critical numbers: $x = 7$ and $x = 3$

Test intervals: $(-\infty, 3), (3, 7), (7, \infty)$

Test: Is $\frac{7-x}{x-3} < 0$?

Solution intervals: $(-\infty, 3) \cup (7, \infty)$

25. $|3x - 4| \geq 9$

$3x - 4 \leq -9$ or $3x - 4 \geq 9$

$3x \leq -5$ $\qquad 3x \geq 13$

$x \leq -\frac{5}{3}$ $\qquad x \geq \frac{13}{3}$

Chapter 2 Practice Test Solutions

1.
$$m = \frac{-1-4}{3-2} = -5$$
$$y - 4 = -5(x - 2)$$
$$y - 4 = -5x + 10$$
$$y = -5x + 14$$

2. $y = \frac{4}{3}x - 3$

3. $2x + 3y = 0$
$$y = -\frac{2}{3}x$$
$$m_1 = -\frac{2}{3}$$
$$\perp m_2 = \frac{3}{2} \text{ through } (4, 1)$$
$$y - 1 = \frac{3}{2}(x - 4)$$
$$y - 1 = \frac{3}{2}x - 6$$
$$y = \frac{3}{2}x - 5$$

4. $(5, 32)$ and $(9, 44)$
$$m = \frac{44-32}{9-5} = \frac{12}{4} = 3$$
$$y - 32 = 3(x - 5)$$
$$y - 32 = 3x - 15$$
$$y = 3x + 17$$

When $x = 20$, $y = 3(20) + 17$

$y = \$77.$

5.
$$f(x-3) = (x-3)^2 - 2(x-3) + 1$$
$$= x^2 - 6x + 9 - 2x + 6 + 1$$
$$= x^2 - 8x + 16$$

6.
$$f(3) = 12 - 11 = 1$$
$$\frac{f(x) - f(3)}{x - 3} = \frac{(4x - 11) - 1}{x - 3}$$
$$= \frac{4x - 12}{x - 3}$$
$$= \frac{4(x-3)}{x-3} = 4, x \neq 3$$

7. $f(x) = \sqrt{36 - x^2} = \sqrt{(6 + x)(6 - x)}$

Domain: $[-6, 6]$

Range: $[0, 6]$, because

$(6 + x)(6 - x) \geq 0$ on this interval.

8. (a) $6x - 5y + 4 = 0$

$y = \frac{6x + 4}{5}$ is a function of x.

(b) $x^2 + y^2 = 9$

$y = \pm\sqrt{9 - x^2}$ is not a function of x.

(c) $y^3 = x^2 + 6$

$y = \sqrt[3]{x^2 + 6}$ is a function of x.

9. Parabola

Vertex: $(0, -5)$

Intercepts:

$(0, -5), (\pm\sqrt{5}, 0)$

y-axis symmetry

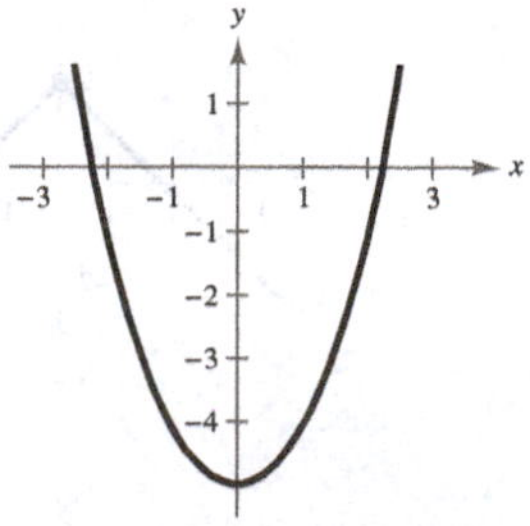

10. Intercepts: $(0, 3)$, $(-3, 0)$

x	-4	-3	-2	-1	0	1	2
y	1	0	1	2	3	4	5

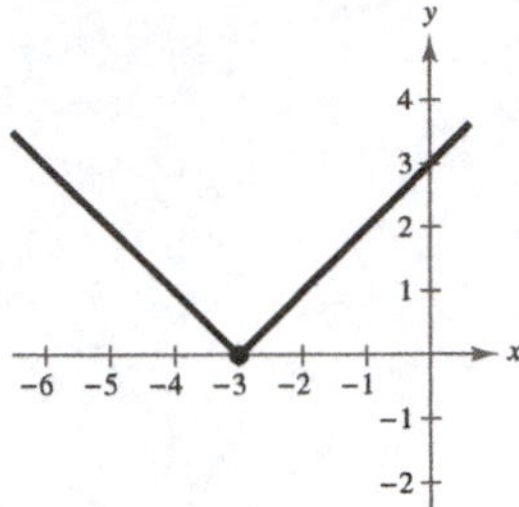

11.

x	0	1	2	3
y	1	3	5	7

x	-1	-2	-3
y	2	6	12

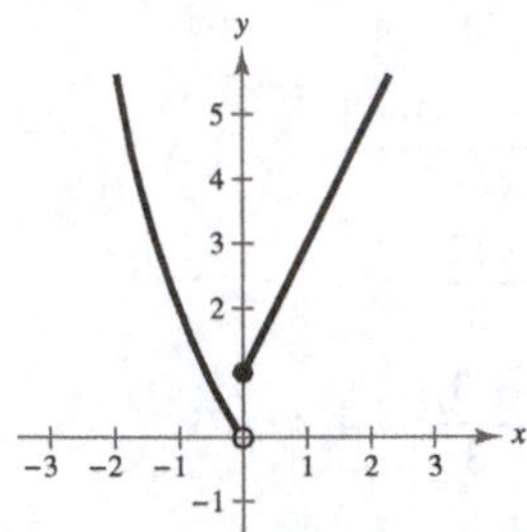

12. (a) $f(x + 2)$

Horizontal shift two units to the left

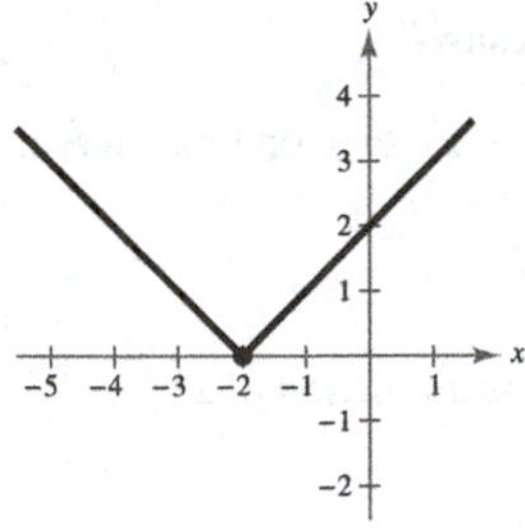

(b) $-f(x) + 2$

Reflection in the x-axis and a vertical shift two units upward

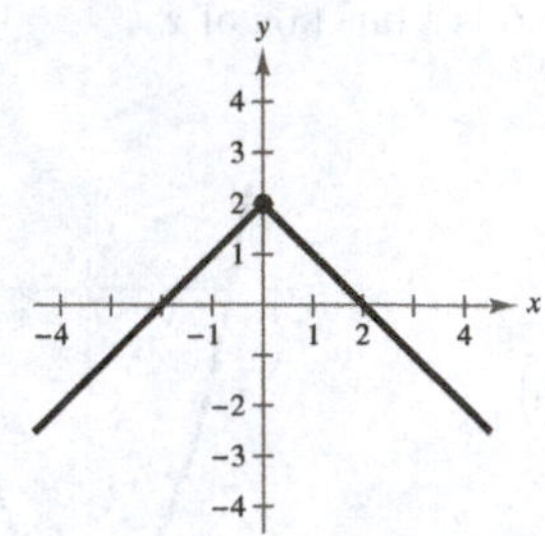

13. (a) $$\begin{aligned}(g - f)(x) &= g(x) - f(x)\\ &= (2x^2 - 5) - (3x + 7)\\ &= 2x^2 - 3x - 12\end{aligned}$$

(b) $$\begin{aligned}(fg)(x) &= f(x)g(x)\\ &= (3x + 7)(2x^2 - 5)\\ &= 6x^3 + 14x^2 - 15x - 35\end{aligned}$$

14. $$\begin{aligned}f(g(x)) &= f(2x + 3)\\ &= (2x + 3)^2 - 2(2x + 3) + 16\\ &= 4x^2 + 12x + 9 - 4x - 6 + 16\\ &= 4x^2 + 8x + 19\end{aligned}$$

15. $$\begin{aligned}f(x) &= x^3 + 7\\ y &= x^3 + 7\\ x &= y^3 + 7\\ x - 7 &= y^3\\ \sqrt[3]{x - 7} &= y\\ f^{-1}(x) &= \sqrt[3]{x - 7}\end{aligned}$$

16. (a) $f(x) = |x - 6|$ does not have an inverse. Its graph does not pass the Horizontal Line Test.

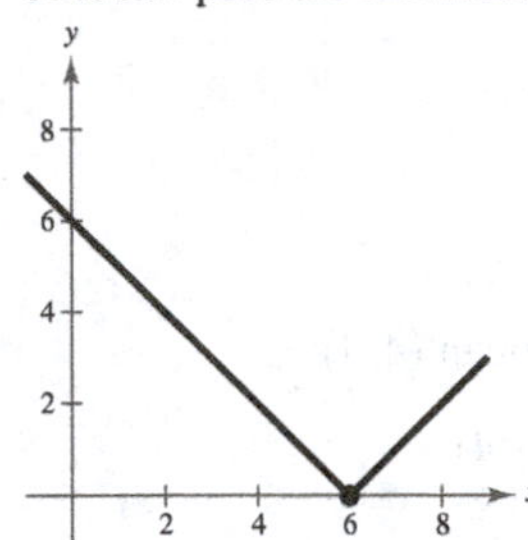

(b) $f(x) = ax + b,\ a \neq 0$ does have an inverse.

$$\begin{aligned}y &= ax + b\\ x &= ay + b\\ \frac{x - b}{a} &= y\\ f^{-1}(x) &= \frac{x - b}{a}\end{aligned}$$

(c) $f(x) = x^3 - 19$ does have an inverse.

$$\begin{aligned}y &= x^3 - 19\\ x &= y^3 - 19\\ x + 19 &= y^3\\ \sqrt[3]{x + 19} &= y\\ f^{-1}(x) &= \sqrt[3]{x + 19}\end{aligned}$$

17. $f(x) = \sqrt{\dfrac{3-x}{x}},\ 0 < x \le 3,\ y \ge 0$

$$y = \sqrt{\frac{3-x}{x}}$$
$$x = \sqrt{\frac{3-y}{y}},\ 0 < y \le 3,\ x \ge 0$$
$$x^2 = \frac{3-y}{y}$$
$$x^2y = 3 - y$$
$$x^2y + y = 3$$
$$y(x^2 + 1) = 3$$
$$y = \frac{3}{x^2 + 1}$$
$$f^{-1}(x) = \frac{3}{x^2 + 1},\ x \ge 0$$

18. False. The slopes of 3 and $\frac{1}{3}$ are not **negative** reciprocals.

19. True. Let $y = (f \circ g)(x)$. Then $x = (f \circ g)^{-1}(y)$.

Also,

$$(f \circ g)(x) = y$$
$$f(g(x)) = y$$
$$g(x) = f^{-1}(y)$$
$$x = g^{-1}(f^{-1}(y))$$
$$x = (g^{-1} \circ f^{-1})(y).$$

Since $x = x$, we have $(f \circ g)^{-1}(y) = (g^{-1} \circ f^{-1})(y)$.

20. True. It must pass the Vertical Line Test to be a function and it must pass the Horizontal Line Test to have an inverse.

Chapter 3 Practice Test Solutions

1. x-intercepts: $(1, 0)$, $(5, 0)$

y-intercept: $(0, 5)$

Vertex: $(3, -4)$

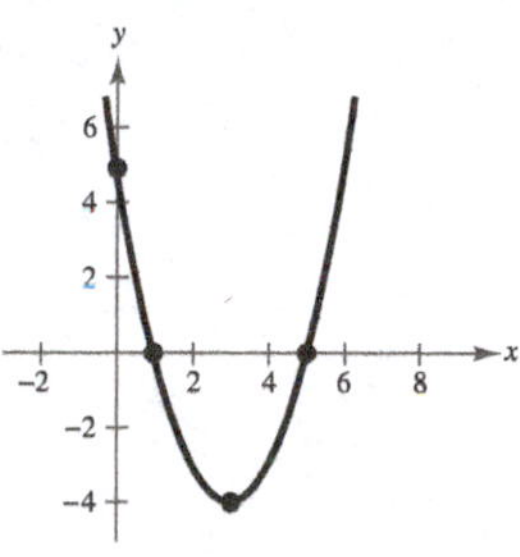

2. $a = 0.01$, $b = -90$

$$\frac{-b}{2a} = \frac{90}{2(0.01)} = 4500 \text{ units}$$

3. Vertex $(1, 7)$ opening downward through $(2, 5)$

$$y = a(x-1)^2 + 7 \quad \text{Standard form}$$
$$5 = a(2-1)^2 + 7$$
$$5 = a + 7$$
$$a = -2$$
$$y = -2(x-1)^2 + 7$$
$$= -2(x^2 - 2x + 1) + 7$$
$$= -2x^2 + 4x + 5$$

4. $y = \pm a(x-2)(3x-4)$ where a is any real nonzero number.

$$y = \pm(3x^2 - 10x + 8)$$

5. Leading coefficient: -3

Degree: 5 (odd)

Falls to the right.

Rises to the left.

6. $0 = x^5 - 5x^3 + 4x$

$$= x(x^4 - 5x^2 + 4)$$
$$= x(x^2 - 1)(x^2 - 4)$$
$$= x(x+1)(x-1)(x+2)(x-2)$$
$$x = 0,\ x = \pm 1,\ x = \pm 2$$

7. $f(x) = x(x-3)(x+2)$

$$= x(x^2 - x - 6)$$
$$= x^3 - x^2 - 6x$$

8. $f(x) = x^3 - 12x$

Intercepts: $(0, 0)$, $(\pm 2\sqrt{3}, 0)$

Rises to the right.

Falls to the left.

x	-2	-1	0	1	2
y	16	11	0	-11	-16

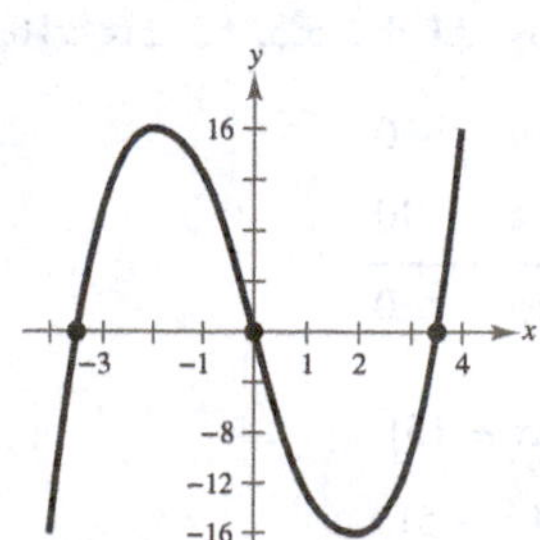

9.
$$\begin{array}{r} 3x^3 + 9x^2 + 20x + 62 \\ x-3\overline{)3x^4 + 0x^3 - 7x^2 + 2x - 10} \\ \underline{3x^4 - 9x^3} \qquad\qquad\qquad\qquad \\ 9x^3 - 7x^2 \qquad\qquad\quad \\ \underline{9x^3 - 27x^2} \qquad\qquad\quad \\ 20x^2 + 2x \qquad\quad \\ \underline{20x^2 - 60x} \qquad\quad \\ 62x - 10 \\ \underline{62x - 186} \\ 176 \end{array}$$

$$\frac{3x^4 - 7x^2 + 2x - 10}{x - 3} = 3x^3 + 9x^2 + 20x + 6 + \frac{176}{x - 3}$$

10.
$$\begin{array}{r} x - 2 \\ x^2 + 2x - 1\overline{)x^3 + 0x^2 + 0x - 11} \\ \underline{x^3 + 2x^2 - x} \qquad \\ -2x^2 + x - 11 \\ \underline{-2x^2 - 4x + 2} \\ 5x - 13 \end{array}$$

$$\frac{x^3 - 11}{x^2 + 2x - 1} = x - 2 + \frac{5x - 13}{x^2 + 2x - 1}$$

11.
$$\begin{array}{r|rrrrrr} -5 & 3 & 13 & 0 & 0 & 12 & -1 \\ & & -15 & 10 & -50 & 250 & -1310 \\ \hline & 3 & -2 & 10 & -50 & 262 & -1311 \end{array}$$

$$\frac{3x^5 + 13x^4 + 12x - 1}{x + 5} = 3x^4 - 2x^3 + 10x^2 - 50x + 262 - \frac{1311}{x + 5}$$

12.
$$\begin{array}{r|rrrr} -6 & 7 & 40 & -12 & 15 \\ & & -42 & 12 & 0 \\ \hline & 7 & -2 & 0 & 15 \end{array}$$

$f(-6) = 15$

13. $0 = x^3 - 19x - 30$

Possible rational zeros: $\pm1, \pm2, \pm3, \pm5, \pm6, \pm10, \pm15, \pm30$

$$\begin{array}{r|rrrr} -2 & 1 & 0 & -19 & -30 \\ & & -2 & 4 & 30 \\ \hline & 1 & -2 & -15 & 0 \end{array}$$

$0 = (x + 2)(x^2 - 2x - 15)$

$0 = (x + 2)(x + 3)(x - 5)$

Zeros: $x = -2, x = -3, x = 5$

14. $0 = x^4 + x^3 - 8x^2 - 9x - 9$

Possible rational zeros: $\pm 1, \pm 3, \pm 9$

$$\begin{array}{r|rrrrr} 3 & 1 & 1 & -8 & -9 & -9 \\ & & 3 & 12 & 12 & 9 \\ \hline & 1 & 4 & 4 & 3 & 0 \end{array}$$

$0 = (x - 3)(x^3 + 4x^2 + 4x + 3)$

Possible rational zeros of $x^3 + 4x^2 + 4x + 3$: $\pm 1, \pm 3$

$$\begin{array}{r|rrrr} -3 & 1 & 4 & 4 & 3 \\ & & -3 & -3 & -3 \\ \hline & 1 & 1 & 1 & 0 \end{array}$$

$0 = (x - 3)(x + 3)(x^2 + x + 1)$

Use the Quadratic Formula. The zeros of $x^2 + x + 1$ are $x = \dfrac{-1 \pm \sqrt{3}i}{2}$.

Zeros:

$x = 3, x = -3, x = -\dfrac{1}{2} + \dfrac{\sqrt{3}}{2}i, x = -\dfrac{1}{2} - \dfrac{\sqrt{3}}{2}i$

The real zeros are $x = 3$ and $x = -3$.

15. $0 = 6x^3 - 5x^2 + 4x - 15$

Possible rational zeros:

$\pm 1, \pm 3, \pm 5, \pm 15, \pm\frac{1}{2}, \pm\frac{3}{2}, \pm\frac{5}{2}, \pm\frac{15}{2}, \pm\frac{1}{3}, \pm\frac{5}{3}, \pm\frac{1}{6}, \pm\frac{5}{6}$

16. $0 = x^3 - \frac{20}{3}x^2 + 9x - \frac{10}{3}$

$0 = 3x^3 - 20x^2 + 27x - 10$

Possible rational zeros:

$\pm 1, \pm 2, \pm 5, \pm 10, \pm\frac{1}{3}, \pm\frac{2}{3}, \pm\frac{5}{3}, \pm\frac{10}{3}$

$$\begin{array}{r|rrrr} 1 & 3 & -20 & 27 & -10 \\ & & 3 & -17 & 10 \\ \hline & 3 & -17 & 10 & 0 \end{array}$$

$0 = (x - 1)(3x^2 - 17x + 10)$

$0 = (x - 1)(3x - 2)(x - 5)$

Zeros: $x = 1, x = \frac{2}{3}, x = 5$

17. Possible rational zeros: $\pm 1, \pm 2, \pm 5, \pm 10$

$$\begin{array}{r|rrrrr} 1 & 1 & 1 & 3 & 5 & -10 \\ & & 1 & 2 & 5 & 10 \\ \hline & 1 & 2 & 5 & 10 & 0 \end{array}$$

$$\begin{array}{r|rrrr} -2 & 1 & 2 & 5 & 10 \\ & & -2 & 0 & -10 \\ \hline & 1 & 0 & 5 & 0 \end{array}$$

$$\begin{aligned} f(x) &= (x - 1)(x + 2)(x^2 + 5) \\ &= (x - 1)(x + 2)(x + \sqrt{5}i)(x - \sqrt{5}i) \end{aligned}$$

18.
$$\begin{aligned} f(x) &= (x - 2)[x - (3 + i)][x - (3 - i)] \\ &= (x - 2)[(x - 3) - i][(x - 3) + i] \\ &= (x - 2)\left[(x - 3)^2 - (i)^2\right] \\ &= (x - 2)[x^2 - 6x + 10] \\ &= x^3 - 8x^2 + 22x - 20 \end{aligned}$$

19.
$$\begin{array}{r|rrrr} 3i & 1 & 4 & 9 & 36 \\ & & 3i & 12i - 9 & -36 \\ \hline & 1 & 4 + 3i & 12i & 0 \end{array}$$

Thus, $f(3i) = 0$.

20. $z = \dfrac{kx^2}{\sqrt{y}}$

Chapter 4 Practice Test Solutions

1. Vertical asymptote: $x = 0$

Horizontal asymptote: $y = \frac{1}{2}$

x-intercept: $(1, 0)$

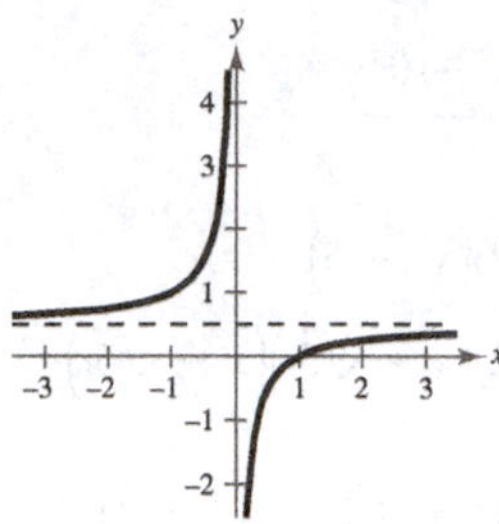

2. Vertical asymptote: $x = 0$

Slant asymptote: $y = 3x$

x-intercepts: $\left(\pm\frac{2}{\sqrt{3}}, 0\right)$

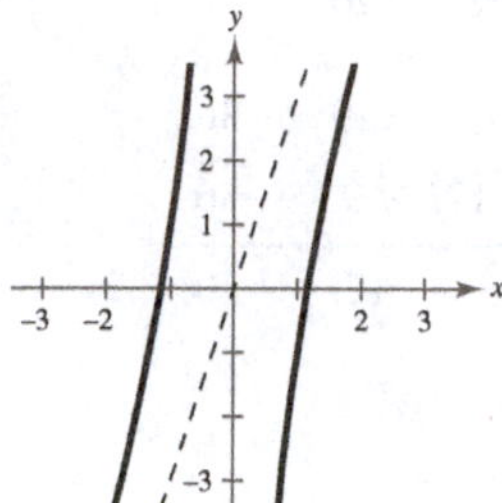

3. $y = 8$ is a horizontal asymptote since the degree of the numerator equals the degree of the denominator. There are no vertical asymptotes.

4. $x = 1$ is a vertical asymptote.

$$\frac{4x^2 - 2x + 7}{x - 1} = 4x + 2 + \frac{9}{x - 1}$$

so $y = 4x + 2$ is a slant asymptote.

5. $f(x) = \dfrac{x - 5}{(x - 5)^2} = \dfrac{1}{x - 5}$

Vertical asymptote: $x = 5$

Horizontal asymptote: $y = 0$

y-intercept: $\left(0, -\frac{1}{5}\right)$

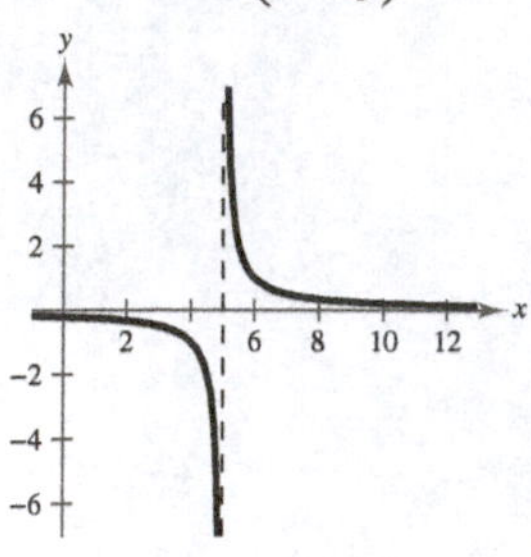

6. $(x - 0)^2 = 4(5)(y - 0)$

Vertex: $(0, 0)$

Focus: $(0, 5)$

Directrix: $y = -5$

7. $(y - 0)^2 = 4(7)(x - 0)$

$y^2 = 28x$

8. $a = 12, b = 5, h = k = 0,$

$c = \sqrt{144 - 25} = \sqrt{119}$

Center: $(0, 0)$

Foci: $\left(\pm\sqrt{119}, 0\right)$

Vertices: $(\pm 12, 0)$

9. Center: $(0, 0)$

$c = 4, 2b = 6 \Rightarrow b = 3,$

$a = \sqrt{16 + 9} = 5$

$$\frac{x^2}{25} + \frac{y^2}{9} = 1$$

10. $a = 12, b = 13, c = \sqrt{144 + 169} = \sqrt{313}$

Center: $(0, 0)$

Foci: $\left(0, \pm\sqrt{313}\right)$

Vertices: $(0, \pm 12)$

Asymptotes: $y = \pm\frac{12}{13}x$

11. Center: $(0, 0)$

$a = 4, \pm\frac{1}{2} = \pm\frac{b}{4} \Rightarrow b = 2$

$$\frac{x^2}{16} - \frac{y^2}{4} = 1$$

12. $p = 4$

$(x - 6)^2 = 4(4)(y + 1)$

$(x - 6)^2 = 16(y + 1)$

13.
$$16x^2 - 96x + 9y^2 + 36y = -36$$
$$16(x^2 - 6x + 9) + 9(y^2 + 4y + 4) = -36 + 144 + 36$$
$$16(x - 3)^2 + 9(y + 2)^2 = 144$$
$$\frac{(x-3)^2}{9} + \frac{(y+2)^2}{16} = 1$$

$a = 4, b = 3, c = \sqrt{16 - 9} = \sqrt{7}$

Center: $(3, -2)$

Foci: $(3, -2 \pm \sqrt{7})$

Vertices: $(3, -2 \pm 4)$ OR $(3, 2)$ and $(3, -6)$

14. Center: $(3, 1)$

$a = 4, 2b = 2 \Rightarrow b = 1$

$$\frac{(x-3)^2}{16} + \frac{(y-1)^2}{1} = 1$$

15. $$\frac{(x+3)^2}{1/4} - \frac{(y-1)^2}{1/9} = 1$$

$a = \frac{1}{2}, b = \frac{1}{3}, c = \sqrt{\frac{1}{4} + \frac{1}{9}} = \frac{\sqrt{13}}{6}$

Center: $(-3, 1)$

Vertices: $\left(-3 \pm \frac{1}{2}, 1\right)$ OR $\left(-\frac{5}{2}, 1\right)$ and $\left(-\frac{7}{2}, 1\right)$

Foci: $\left(-3 \pm \frac{\sqrt{13}}{6}, 1\right)$

Asymptotes: $y = \pm\frac{1/3}{1/2}(x + 3) + 1 = \pm\frac{2}{3}(x + 3) + 1$

16. Center: $(3, 0)$

$a = 4, c = 7, b = \sqrt{49 - 16} = \sqrt{33}$

$$\frac{y^2}{16} - \frac{(x-3)^2}{33} = 1$$

Chapter 5 Practice Test Solutions

1. $x^{3/5} = 8$

$$x = 8^{5/3} = \left(\sqrt[3]{8}\right)^5 = 2^5 = 32$$

2. $3^{x-1} = \frac{1}{81}$

$3^{x-1} = 3^{-4}$

$x - 1 = -4$

$x = -3$

3. $f(x) = 2^{-x} = \left(\frac{1}{2}\right)^x$

x	-2	-1	0	1	2
$f(x)$	4	2	1	$\frac{1}{2}$	$\frac{1}{4}$

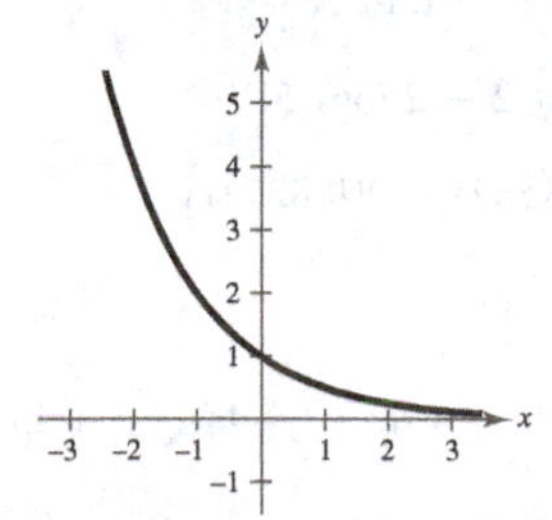

4. $g(x) = e^x + 1$

x	-2	-1	0	1	2
$g(x)$	1.14	1.37	2	3.72	8.39

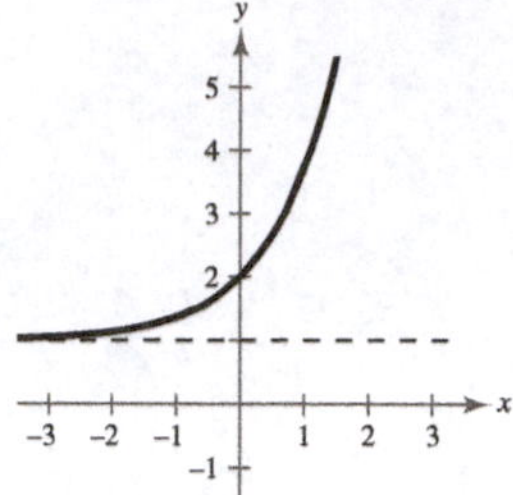

5. (a) $A = P\left(1 + \frac{r}{n}\right)^{nt}$

$$A = 5000\left(1 + \frac{0.09}{12}\right)^{12(3)} \approx \$6543.23$$

(b) $A = P\left(1 + \frac{r}{n}\right)^{nt}$

$$A = 5000\left(1 + \frac{0.09}{4}\right)^{4(3)} \approx \$6530.25$$

(c) $A = Pe^{rt}$

$$A = 5000e^{(0.09)(3)} \approx \$6549.82$$

6. $7^{-2} = \frac{1}{49}$

$$\log_7 \frac{1}{49} = -2$$

7. $x - 4 = \log_2 \frac{1}{64}$

$$2^{x-4} = \frac{1}{64}$$

$$2^{x-4} = 2^{-6}$$

$$x - 4 = -6$$

$$x = -2$$

8. $\log_b \sqrt[4]{\frac{8}{25}} = \frac{1}{4} \log_b \frac{8}{25}$

$$= \frac{1}{4}[\log_b 8 - \log_b 25]$$

$$= \frac{1}{4}[\log_b 2^3 - \log_b 5^2]$$

$$= \frac{1}{4}[3 \log_b 2 - 2 \log_b 5]$$

$$= \frac{1}{4}[3(0.3562) - 2(0.8271)]$$

$$= -0.1464$$

9. $5 \ln x - \frac{1}{2} \ln y + 6 \ln z = \ln x^5 - \ln \sqrt{y} + \ln z^6$

$$= \ln\left(\frac{x^5 z^6}{\sqrt{y}}\right), z > 0$$

10. $\log_9 28 = \frac{\log 28}{\log 9} \approx 1.5166$

11. $\log N = 0.6646$

$$N = 10^{0.6646} \approx 4.62$$

12.

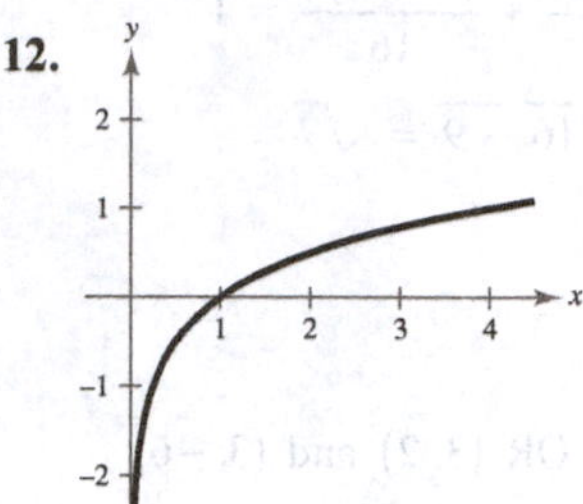

13. Domain:

$$x^2 - 9 > 0$$

$$(x + 3)(x - 3) > 0$$

$$x < -3 \text{ or } x > 3$$

14.

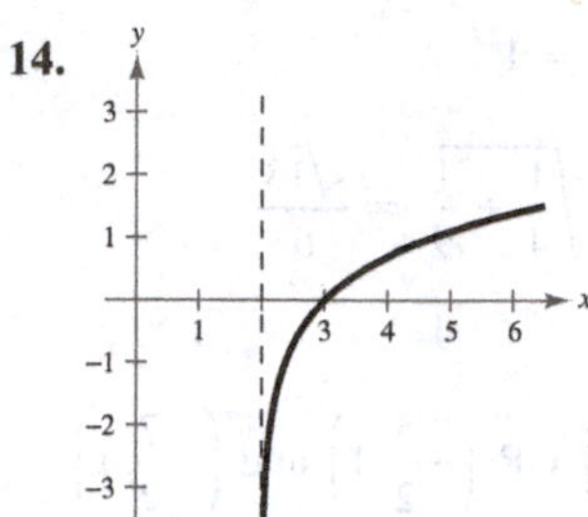

15. False. $\frac{\ln x}{\ln y} \neq \ln(x - y)$ because $\frac{\ln x}{\ln y} = \log_y x$.

16. $5^3 = 41$

$$x = \log_5 41 = \frac{\ln 41}{\ln 5} \approx 2.3074$$

17. $x - x^2 = \log_5 \frac{1}{25}$

$$5^{x-x^2} = \frac{1}{25}$$

$$5^{x-x^2} = 5^{-2}$$

$$x - x^2 = -2$$

$$0 = x^2 - x - 2$$

$$0 = (x + 1)(x - 2)$$

$$x = -1 \text{ or } x = 2$$

18. $\log_2 x + \log_2(x - 3) = 2$

$$\begin{aligned}\log_2[x(x-3)] &= 2\\ x(x-3) &= 2^2\\ x^2 - 3x &= 4\\ x^2 - 3x - 4 &= 0\\ (x+1)(x-4) &= 0\\ x &= 4\\ x &= -1 \text{ (extraneous)}\end{aligned}$$

$x = 4$ is the only solution.

19. $$\frac{e^x + e^{-x}}{3} = 4$$

$$\begin{aligned}e^x(e^x + e^{-x}) &= 12e^x\\ e^{2x} + 1 &= 12e^x\\ e^{2x} - 12e^x + 1 &= 0\\ e^x &= \frac{12 \pm \sqrt{144 - 4}}{2}\end{aligned}$$

$e^x \approx 11.9161$ or $e^x \approx 0.0839$

$x = \ln 11.9161$ $\quad x = \ln 0.0839$

$x \approx 2.478$ $\quad x \approx -2.478$

20. $$\begin{aligned}A &= Pe^{rt}\\ 12{,}000 &= 6000e^{0.13t}\\ 2 &= e^{0.13t}\\ 0.13t &= \ln 2\\ t &= \frac{\ln 2}{0.13}\\ t &\approx 5.3319 \text{ years or 5 years 4 months}\end{aligned}$$

Chapter 6 Practice Test Solutions

1. $\begin{cases} x + y = 1 \\ 3x - y = 15 \Rightarrow y = 3x - 15 \end{cases}$

$$\begin{aligned}x + (3x - 15) &= 1\\ 4x &= 16\\ x &= 4\\ y &= -3\end{aligned}$$

Solution: $(4, -3)$

2. $\begin{cases} x - 3y = -3 \Rightarrow x = 3y - 3 \\ x^2 + 6y = 5 \end{cases}$

$$\begin{aligned}(3y-3)^2 + 6y &= 5\\ 9y^2 - 18y + 9 + 6y &= 5\\ 9y^2 - 12y + 4 &= 0\\ (3y-2)^2 &= 0\\ y &= \tfrac{2}{3}\\ x &= -1\end{aligned}$$

Solution: $\left(-1, \frac{2}{3}\right)$

3. $\begin{cases} x + y + z = 6 \Rightarrow z = 6 - x - y \\ 2x - y + 3z = 0 \Rightarrow 2x - y + 3(6 - x - y) = 0 \Rightarrow -x - 4y = -18 \Rightarrow x = 18 - 4y \\ 5x + 2y - z = -3 \Rightarrow 5x + 2y - (6 - x - y) = -3 \Rightarrow 6x + 3y = 3 \end{cases}$

$$\begin{aligned}6(18 - 4y) + 3y &= 3\\ -21y &= -105\\ y &= 5\\ x &= 18 - 4y = -2\\ z &= 6 - x - y = 3\end{aligned}$$

Solution: $(-2, 5, 3)$

4. $x + y = 110 \Rightarrow y = 110 - x$

$xy = 2800$

$$x(110 - x) = 2800$$
$$0 = x^2 - 110x + 2800$$
$$0 = (x - 40)(x - 70)$$

$x = 40$ or $x = 70$

$y = 70$ $\quad$ $y = 40$

Solution: The two numbers are 40 and 70.

5. $2x + 2y = 170 \Rightarrow y = \dfrac{170 - 2x}{2} = 85 - x$

$$xy = 1500$$
$$x(85 - x) = 1500$$
$$0 = x^2 - 85x + 1500$$
$$0 = (x - 25)(x - 60)$$

$x = 25$ or $x = 60$

$y = 60$ $\quad$ $y = 25$

Dimensions: 60 ft × 25 ft

6. $\begin{cases} 2x + 15y = 4 \Rightarrow 2x + 15y = 4 \\ x - 3y = 23 \Rightarrow 5x - 15y = 115 \end{cases}$

$$7x = 119$$
$$x = 17$$
$$y = \frac{x - 23}{3} = -2$$

Solution: $(17, -2)$

7. $\begin{cases} x + y = 2 \Rightarrow 19x + 19y = 38 \\ 38x - 19y = 7 \Rightarrow 38x - 19y = 7 \end{cases}$

$$57x = 45$$
$$x = \tfrac{15}{19}$$
$$y = 2 - x = \tfrac{38}{19} - \tfrac{15}{19} = \tfrac{23}{19}$$

Solution: $\left(\frac{15}{19}, \frac{23}{19}\right)$

8. $\begin{cases} 0.4x + 0.5y = 0.112 \Rightarrow 0.28x + 0.35y = 0.0784 \\ 0.3x - 0.7y = -0.131 \Rightarrow 0.15x - 0.35y = -0.0655 \end{cases}$

$$0.43x = 0.0129$$

$$x = \frac{0.0129}{0.43} = 0.03$$
$$y = \frac{0.112 - 0.4x}{0.5} = 0.20$$

Solution: $(0.03, 0.20)$

9. Let x = amount in 11% fund and y = amount in 13% fund.

$x + y = 17{,}000 \Rightarrow y = 17{,}000 - x$

$$0.11x + 0.13y = 2080$$
$$0.11x + 0.13(17{,}000 - x) = 2080$$
$$-0.02x = -130$$
$$x = \$6500 \text{ at } 11\%$$
$$y = \$10{,}500 \text{ at } 13\%$$

10. $(4, 3), (1, 1), (-1, -2), (-2, -1)$

Use a calculator.

$y = ax + b = \frac{11}{14}x - \frac{1}{7}$

11. $\begin{cases} x + y = -2 \\ 2x - y + z = 11 \\ 4y - 3z = -20 \end{cases}$

$\begin{cases} x + y = -2 \\ -3y + z = 15 \\ 4y - 3z = -20 \end{cases}$ $\quad -2\text{Eq.}1 + \text{Eq.}2$

$\begin{cases} x + y = -2 \\ y - 2z = -5 \\ 4y - 3z = -20 \end{cases}$ $\quad \text{Eq.}3 + \text{Eq.}2$

$\begin{cases} x + y = -2 \\ y - 2z = -5 \\ 5z = 0 \end{cases}$ $\quad -4\text{Eq.}2 + \text{Eq.}3$

$\begin{cases} x + y = -2 \\ y - 2z = -5 \\ z = 0 \end{cases}$

$y - 2(0) = -5 \Rightarrow y = -5$

$x + (-5) = -2 \Rightarrow x = 3$

Solution: $(3, -5, 0)$

12. $\begin{cases} 4x - y + 5z = 4 \\ 2x + y - z = 0 \\ 2x + 4y + 8z = 0 \end{cases}$

$\begin{cases} 2x + 4y + 8z = 0 \\ 2x + y - z = 0 \\ 4x - y + 5z = 4 \end{cases}$ $\quad$ Interchange equations.

$\begin{cases} 2x + 4y + 8z = 0 \\ -3y - 9z = 0 \\ -9y - 11z = 4 \end{cases}$ $\quad \begin{matrix} -\text{Eq.}1 + \text{Eq.}2 \\ -2\text{Eq.}1 + \text{Eq.}3 \end{matrix}$

$\begin{cases} 2x + 4y + 8z = 0 \\ -3y - 9z = 0 \\ 16z = 4 \end{cases}$ $\quad -3\text{Eq.}2 + \text{Eq.}3$

$\begin{cases} x + 2y + 4z = 0 \\ y + 3z = 0 \\ z = \frac{1}{4} \end{cases}$ $\quad \begin{matrix} \frac{1}{2}\text{Eq.}1 \\ -\frac{1}{3}\text{Eq.}2 \\ \frac{1}{16}\text{Eq.}3 \end{matrix}$

$y + 3\left(\frac{1}{4}\right) = 0 \Rightarrow y = -\frac{3}{4}$

$x + 2\left(-\frac{3}{4}\right) + 4\left(\frac{1}{4}\right) = 0 \Rightarrow x = \frac{1}{2}$

Solution: $\left(-\frac{1}{2}, -\frac{3}{4}, \frac{1}{4}\right)$

13. $\begin{cases} 3x + 2y - z = 5 \\ 6x - y + 5z = 2 \end{cases}$

$\begin{cases} 3x + 2y - z = 5 \\ -5y + 7z = -8 \end{cases}$ $\quad -2\text{Eq.}1 + \text{Eq.}2$

$\begin{cases} x + \frac{2}{3}y - \frac{1}{3}z = \frac{5}{3} \\ y - \frac{7}{5}z = \frac{8}{5} \end{cases}$ $\quad \begin{matrix} \frac{1}{3}\text{Eq.}1 \\ -\frac{1}{5}\text{Eq.}2 \end{matrix}$

Let $a = z$.

Then $y = \frac{7}{5}a + \frac{8}{5}$, and $x + \frac{2}{3}\left(\frac{7}{5}a + \frac{8}{5}\right) - \frac{1}{3}a = \frac{5}{3}$

$x + \frac{3}{5}a = \frac{3}{5}$

$x = -\frac{3}{5}a + \frac{3}{5}.$

Solution: $\left(-\frac{3}{5}a + \frac{3}{5}, \frac{7}{5}a + \frac{8}{5}, a\right)$ where a is any real number

14. $y = ax^2 + bx + c$ passes through $(0, -1)$, $(1, 4)$, and $(2, 13)$.

At $(0, -1)$: $-1 = a(0)^2 + b(0) + c \Rightarrow c = -1$

At $(1, 4)$: $4 = a(1)^2 + b(1) - 1 \Rightarrow 5 = a + b \Rightarrow 5 = a + b$

At $(2, 13)$: $13 = a(2)^2 + b(2) - 1 \Rightarrow 14 = 4a + 2b \Rightarrow -7 = -2a - b$

$-2 = -a$

$a = 2$

$b = 3$

So, the equation of the parabola is $y = 2x^2 + 3x - 1$.

15. $s = \frac{1}{2}at^2 + v_0t + s_0$ passes through (1, 12), (2, 5), and (3, 4).

At (1, 12): $12 = \frac{1}{2}a + v_0 + s_0$
At (2, 5): $5 = 2a + 2v_0 + s_0$
At (3, 4): $4 = \frac{9}{2}a + 3v_0 + s_0$

$$\begin{cases} a + 2v_0 + 2s_0 = 24 \\ 2a + 2v_0 + s_0 = 5 \\ 9a + 6v_0 + 2s_0 = 8 \end{cases}$$

$$\begin{cases} a + 2v_0 + 2s_0 = 24 \\ -2v_0 - 3s_0 = -43 & -2\text{Eq.1} + \text{Eq.2} \\ -12v_0 - 16s_0 = -208 & -9\text{Eq.1} + \text{Eq.3} \end{cases}$$

$$\begin{cases} a + 2v_0 + 2s_0 = 24 \\ -2v_0 - 3s_0 = -43 \\ 2s_0 = 50 & -6\text{Eq.2} + \text{Eq.3} \end{cases}$$

$$\begin{cases} a + 2v_0 + 2s_0 = 24 \\ v_0 + \frac{3}{2}s_0 = \frac{43}{2} & -\frac{1}{2}\text{Eq.2} \\ s_0 = 25 & \frac{1}{2}\text{Eq.3} \end{cases}$$

$$s_0 = 25$$
$$v_0 + \frac{3}{2}(25) = \frac{43}{2} \Rightarrow v_0 = 16$$
$$a + 2(-16) + 2(25) = 24 \Rightarrow a = 6$$

So, $s = \frac{1}{2}(6)t^2 - 16t + 25 = 3t^2 - 16t + 25$.

16. $x^2 + y^2 \geq 9$

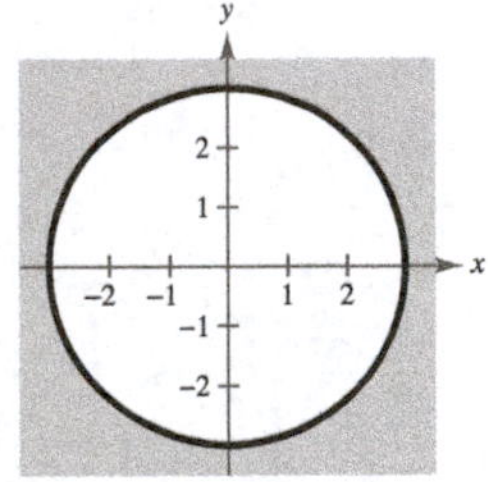

17. $\begin{cases} x + y \leq 6 \\ x \geq 2 \\ y \geq 0 \end{cases}$

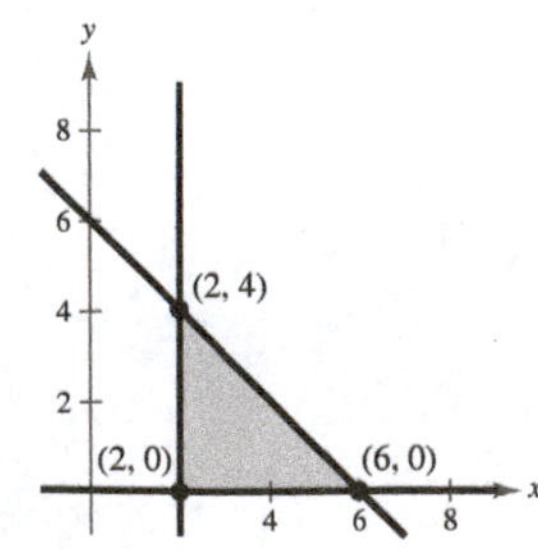

18. Line through (0, 0) and (0, 7): $x = 0$

Line through (0, 0) and (2, 3):

$y = \frac{3}{2}x$ or $3x - 2y = 0$

Line through (0, 7) and (2, 3):

$y = -2x + 7$ or $2x + y = 7$

Inequalities: $\begin{cases} x \geq 0 \\ 3x - 2y \leq 0 \\ 2x + y \leq 7 \end{cases}$

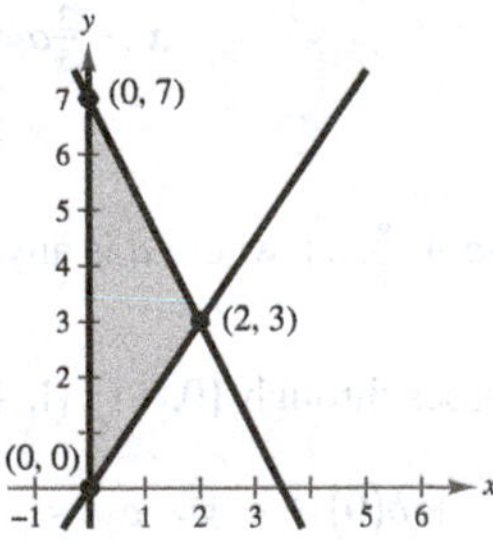

19. Vertices $(0, 0), (0, 7), (6, 0), (3, 5)$

$z = 30x + 26y$

At $(0, 0)$: $z = 0$

At $(0, 7)$: $z = 182$

At $(6, 0)$: $z = 180$

At $(3, 5)$: $z = 220$

The maximum value of z occurs at $(3, 5)$ and is 220.

20.

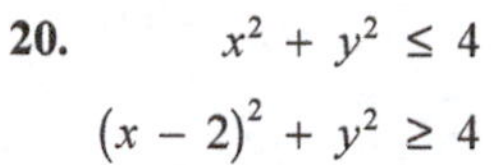

$$x^2 + y^2 \le 4$$
$$(x - 2)^2 + y^2 \ge 4$$

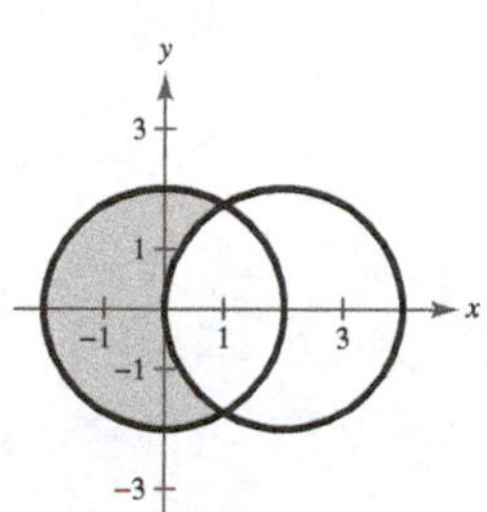

21. $\dfrac{1 - 2x}{x^2 + x} = \dfrac{1 - 2x}{x(x + 1)} = \dfrac{A}{x} + \dfrac{B}{x + 1}$

$1 - 2x = A(x + 1) + Bx$

When $x = 0, 1 = A$.

When $x = -1, 3 = -B \Rightarrow B = -3$.

$\dfrac{1 - 2x}{x^2 + x} = \dfrac{1}{x} - \dfrac{3}{x + 1}$

22. $\dfrac{6x - 17}{(x - 3)^2} = \dfrac{A}{x - 3} + \dfrac{B}{(x - 3)^2}$

$6x - 17 = A(x - 3) + B$

When $x = 3, 1 = B$.

When $x = 0, -17 = -3A + B \Rightarrow A = 6$.

$\dfrac{6x - 17}{(x - 3)^2} = \dfrac{6}{x - 3} + \dfrac{1}{(x - 3)^2}$

Chapter 7 Practice Test Solutions

1. $\begin{bmatrix} 1 & -2 & 4 \\ 3 & -5 & 9 \end{bmatrix}$

$-3R_1 + R_2 \rightarrow \begin{bmatrix} 1 & -2 & 4 \\ 0 & 1 & -3 \end{bmatrix}$

$2R_2 + R_1 \rightarrow \begin{bmatrix} 1 & 0 & -2 \\ 0 & 1 & -3 \end{bmatrix}$

2. $\begin{cases} 3x + 5y = 3 \\ 2x - y = -11 \end{cases}$

$\begin{bmatrix} 3 & 5 & \vdots & 3 \\ 2 & -1 & \vdots & -11 \end{bmatrix}$

$-R_2 + R_1 \rightarrow \begin{bmatrix} 1 & 6 & \vdots & 14 \\ 2 & -1 & \vdots & -11 \end{bmatrix}$

$-2R_1 + R_2 \rightarrow \begin{bmatrix} 1 & 6 & \vdots & 14 \\ 0 & -13 & \vdots & -39 \end{bmatrix}$

$-\frac{1}{13}R_2 \rightarrow \begin{bmatrix} 1 & 6 & \vdots & 14 \\ 0 & 1 & \vdots & 3 \end{bmatrix}$

$-6R_2 + R_1 \rightarrow \begin{bmatrix} 1 & 0 & \vdots & -4 \\ 0 & 1 & \vdots & 3 \end{bmatrix}$

$x = -4, y = 3$

Solution: $(-4, 3)$

3. $\begin{cases} 2x + 3y = -3 \\ 3x - 2y = 8 \\ x + y = 1 \end{cases}$

$\begin{bmatrix} 2 & 3 & \vdots & -3 \\ 3 & 2 & \vdots & 8 \\ 1 & 1 & \vdots & 1 \end{bmatrix}$

$R_3 \leftrightarrow R_1$: $\begin{bmatrix} 1 & 1 & \vdots & 1 \\ 3 & 2 & \vdots & 8 \\ 2 & 3 & \vdots & -3 \end{bmatrix}$

$\begin{matrix} \\ -3R_1 + R_2 \rightarrow \\ -2R_1 + R_3 \rightarrow \end{matrix} \begin{bmatrix} 1 & 1 & \vdots & 1 \\ 0 & -1 & \vdots & 5 \\ 0 & 1 & \vdots & -5 \end{bmatrix}$

$-R_2 \rightarrow \begin{bmatrix} 1 & 1 & \vdots & 1 \\ 0 & 1 & \vdots & -5 \\ 0 & 1 & \vdots & -5 \end{bmatrix}$

$\begin{matrix} -R_2 + R_1 \rightarrow \\ \\ -R_2 + R_3 \rightarrow \end{matrix} \begin{bmatrix} 1 & 0 & \vdots & 6 \\ 0 & 1 & \vdots & -5 \\ 0 & 0 & \vdots & 0 \end{bmatrix}$

$x = 6, y = -5$

Solution: $(6, -5)$

4. $\begin{cases} x + 3z = -5 \\ 2x + y = 0 \\ 3x + y - z = -3 \end{cases}$

$$\begin{bmatrix} 1 & 0 & 3 & \vdots & -5 \\ 2 & 1 & 0 & \vdots & 0 \\ 3 & 1 & -1 & \vdots & 3 \end{bmatrix}$$

$$\begin{matrix} \\ -2R_1 + R_2 \rightarrow \\ -3R_1 + R_3 \rightarrow \end{matrix}\begin{bmatrix} 1 & 0 & 3 & \vdots & -5 \\ 0 & 1 & -6 & \vdots & 10 \\ 0 & 1 & -10 & \vdots & 18 \end{bmatrix}$$

$$\begin{matrix} \\ \\ -R_2 + R_3 \rightarrow \end{matrix}\begin{bmatrix} 1 & 0 & 3 & \vdots & -5 \\ 0 & 1 & -6 & \vdots & 10 \\ 0 & 0 & -4 & \vdots & 8 \end{bmatrix}$$

$$\begin{matrix} \\ \\ -\frac{1}{4}R_3 \rightarrow \end{matrix}\begin{bmatrix} 1 & 0 & 3 & \vdots & -5 \\ 0 & 1 & -6 & \vdots & 10 \\ 0 & 0 & 1 & \vdots & -2 \end{bmatrix}$$

$$\begin{matrix} -3R_3 + R_1 \rightarrow \\ 6R_3 + R_2 \rightarrow \\ \\ \end{matrix}\begin{bmatrix} 1 & 0 & 0 & \vdots & 1 \\ 0 & 1 & 0 & \vdots & -2 \\ 0 & 0 & 1 & \vdots & -2 \end{bmatrix}$$

$x = 1, y = -2, z = -2$

Solution: $(1, -2, -2)$

5. $$\begin{bmatrix} 1 & 4 & 5 \\ 2 & 0 & -3 \end{bmatrix}\begin{bmatrix} 1 & 6 \\ 0 & -7 \\ -1 & 2 \end{bmatrix} = \begin{bmatrix} (1)(1) + (4)(0) + (5)(-1) & (1)(6) + (4)(-7) + (5)(2) \\ (2)(1) + (0)(0) + (-3)(-1) & (2)(6) + (0)(-7) + (-3)(2) \end{bmatrix} = \begin{bmatrix} -4 & -12 \\ 5 & 6 \end{bmatrix}$$

6. $$3A - 5B = 3\begin{bmatrix} 9 & 1 \\ -4 & 8 \end{bmatrix} - 5\begin{bmatrix} 6 & -2 \\ 3 & 5 \end{bmatrix} = \begin{bmatrix} 27 & 3 \\ -12 & 24 \end{bmatrix} - \begin{bmatrix} 30 & -10 \\ 15 & 25 \end{bmatrix} = \begin{bmatrix} -3 & 13 \\ -27 & -1 \end{bmatrix}$$

7. $$f(A) = \begin{bmatrix} 3 & 0 \\ 7 & 1 \end{bmatrix}^2 - 7\begin{bmatrix} 3 & 0 \\ 7 & 1 \end{bmatrix} + 8\begin{bmatrix} 1 & 0 \\ 0 & 1 \end{bmatrix}$$
$$= \begin{bmatrix} 3 & 0 \\ 7 & 1 \end{bmatrix}\begin{bmatrix} 3 & 0 \\ 7 & 1 \end{bmatrix} - \begin{bmatrix} 21 & 0 \\ 49 & 7 \end{bmatrix} + \begin{bmatrix} 8 & 0 \\ 0 & 8 \end{bmatrix}$$
$$= \begin{bmatrix} 9 & 0 \\ 28 & 1 \end{bmatrix} - \begin{bmatrix} 21 & 0 \\ 49 & 7 \end{bmatrix} + \begin{bmatrix} 8 & 0 \\ 0 & 8 \end{bmatrix}$$
$$= \begin{bmatrix} -4 & 0 \\ -21 & 2 \end{bmatrix}$$

8. False.

$$(A + B)(A + 3B) = A(A + 3B) + B(A + 3B)$$
$$= A^2 + 3AB + BA + 3B^2$$ and, in general, $AB \neq BA$.

9.

$$\left[\begin{array}{cc:cc} 1 & 2 & 1 & 0 \\ 3 & 5 & 0 & 1 \end{array}\right]$$

$$\begin{array}{r} \\ -3R_1 + R_2 \rightarrow \end{array}\left[\begin{array}{cc:cc} 1 & 2 & 1 & 0 \\ 0 & -1 & -3 & 1 \end{array}\right]$$

$$\begin{array}{r} 2R_2 + R_1 \rightarrow \\ \end{array}\left[\begin{array}{cc:cc} 1 & 0 & -5 & 2 \\ 0 & -1 & -3 & 1 \end{array}\right]$$

$$\begin{array}{r} \\ -R_2 \rightarrow \end{array}\left[\begin{array}{cc:cc} 1 & 0 & -5 & 2 \\ 0 & 1 & 3 & -1 \end{array}\right]$$

$$A^{-1} = \begin{bmatrix} -5 & 2 \\ 3 & -1 \end{bmatrix}$$

10.

$$\left[\begin{array}{ccc:ccc} 1 & 1 & 1 & 1 & 0 & 0 \\ 3 & 6 & 5 & 0 & 1 & 0 \\ 6 & 10 & 8 & 0 & 0 & 1 \end{array}\right]$$

$$\begin{array}{r} \\ -3R_1 + R_2 \rightarrow \\ -6R_1 + R_3 \rightarrow \end{array}\left[\begin{array}{ccc:ccc} 1 & 1 & 1 & 1 & 0 & 0 \\ 0 & 3 & 2 & -3 & 1 & 0 \\ 0 & 4 & 2 & -6 & 0 & 1 \end{array}\right]$$

$$\begin{array}{r} \\ -R_3 + R_2 \rightarrow \\ \\ \end{array}\left[\begin{array}{ccc:ccc} 1 & 1 & 1 & 1 & 0 & 0 \\ 0 & -1 & 0 & 3 & 1 & -1 \\ 0 & 4 & 2 & -6 & 0 & 1 \end{array}\right]$$

$$\begin{array}{r} R_2 + R_1 \rightarrow \\ \\ 4R_2 + R_3 \rightarrow \end{array}\left[\begin{array}{ccc:ccc} 1 & 0 & 1 & 4 & 1 & -1 \\ 0 & -1 & 0 & 3 & 1 & -1 \\ 0 & 0 & 2 & 6 & 4 & -3 \end{array}\right]$$

$$\begin{array}{r} \\ -R_2 \rightarrow \\ \frac{1}{2}R_3 \rightarrow \end{array}\left[\begin{array}{ccc:ccc} 1 & 0 & 1 & 4 & 1 & -1 \\ 0 & 1 & 0 & -3 & -1 & 1 \\ 0 & 0 & 1 & 3 & 2 & -\frac{3}{2} \end{array}\right]$$

$$\begin{array}{r} -R_3 + R_1 \rightarrow \\ \\ \\ \end{array}\left[\begin{array}{ccc:ccc} 1 & 0 & 0 & 1 & -1 & \frac{1}{2} \\ 0 & 1 & 0 & -3 & -1 & 1 \\ 0 & 0 & 1 & 3 & 2 & -\frac{3}{2} \end{array}\right]$$

$$A^{-1} = \begin{bmatrix} 1 & -1 & \frac{1}{2} \\ -3 & -1 & 1 \\ 3 & 2 & -\frac{3}{2} \end{bmatrix}$$

11. (a) $\begin{cases} x + 2y = 4 \\ 3x + 5y = 1 \end{cases}$

$$A = \begin{bmatrix} 1 & 2 \\ 3 & 5 \end{bmatrix}$$

$$A^{-1} = \frac{1}{5-6}\begin{bmatrix} 5 & -2 \\ -3 & 1 \end{bmatrix} = \begin{bmatrix} -5 & 2 \\ 3 & -1 \end{bmatrix}$$

$$\begin{bmatrix} x \\ y \end{bmatrix} = A^{-1}B = \begin{bmatrix} -5 & 2 \\ 3 & -1 \end{bmatrix}\begin{bmatrix} 4 \\ 1 \end{bmatrix} = \begin{bmatrix} -18 \\ 11 \end{bmatrix}$$

$x = -18,\ y = 11$

Solution: $(-18, 11)$

(b) $\begin{cases} x + 2y = 3 \\ 3x + 5y = -2 \end{cases}$

Again, $A^{-1} = \begin{bmatrix} -5 & 2 \\ 3 & -1 \end{bmatrix}$.

$$\begin{bmatrix} x \\ y \end{bmatrix} = A^{-1}B = \begin{bmatrix} -5 & 2 \\ 3 & -1 \end{bmatrix}\begin{bmatrix} 3 \\ -2 \end{bmatrix} = \begin{bmatrix} -19 \\ 11 \end{bmatrix}$$

$x = -19,\ y = 11$

Solution: $(-19, 11)$

12. $\begin{vmatrix} 6 & -1 \\ 3 & 4 \end{vmatrix} = 24 - (-3) = 27$

13. $\begin{vmatrix} 1 & 3 & -1 \\ 5 & 9 & 0 \\ 6 & 2 & -5 \end{vmatrix} = -1\begin{vmatrix} 5 & 9 \\ 6 & 2 \end{vmatrix} - 5\begin{vmatrix} 1 & 3 \\ 5 & 9 \end{vmatrix}$

$= -(-44) - 5(-6) = 74$

14. Expand along Row 2.

$$\begin{vmatrix} 1 & 4 & 2 & 3 \\ 0 & 1 & -2 & 0 \\ 3 & 5 & -1 & 1 \\ 2 & 0 & 6 & 1 \end{vmatrix} = \begin{vmatrix} 1 & 2 & 3 \\ 3 & -1 & 1 \\ 2 & 6 & 1 \end{vmatrix} + 2\begin{vmatrix} 1 & 4 & 3 \\ 3 & 5 & 1 \\ 2 & 0 & 1 \end{vmatrix}$$

$= 51 + 2(-29) = -7$

15. $$\begin{vmatrix} 6 & 4 & 3 & 0 & 6 \\ 0 & 5 & 1 & 4 & 8 \\ 0 & 0 & 2 & 7 & 3 \\ 0 & 0 & 0 & 9 & 2 \\ 0 & 0 & 0 & 0 & 1 \end{vmatrix} = 6\begin{vmatrix} 5 & 1 & 4 & 8 \\ 0 & 2 & 7 & 3 \\ 0 & 0 & 9 & 2 \\ 0 & 0 & 0 & 1 \end{vmatrix} = 6(5)\begin{vmatrix} 2 & 7 & 3 \\ 0 & 9 & 2 \\ 0 & 0 & 1 \end{vmatrix} = 6(5)(2)\begin{vmatrix} 9 & 2 \\ 0 & 1 \end{vmatrix} = 6(5)(2)(9) = 540$$

16. Area $= \dfrac{1}{2}\begin{vmatrix} 0 & 7 & 1 \\ 5 & 0 & 1 \\ 3 & 9 & 1 \end{vmatrix} = \dfrac{1}{2}(31) = \dfrac{31}{2}$

17. $\begin{vmatrix} x & y & 1 \\ 2 & 7 & 1 \\ -1 & 4 & 1 \end{vmatrix} = 3x - 3y + 15 = 0$ or $x - y + 5 = 0$

18. $x = \dfrac{\begin{vmatrix} 4 & -7 \\ 11 & 5 \end{vmatrix}}{\begin{vmatrix} 6 & -7 \\ 2 & 5 \end{vmatrix}} = \dfrac{97}{44}$

19. $z = \dfrac{\begin{vmatrix} 3 & 0 & 1 \\ 0 & 1 & 3 \\ 1 & -1 & 2 \end{vmatrix}}{\begin{vmatrix} 3 & 0 & 1 \\ 0 & 1 & 4 \\ 1 & -1 & 0 \end{vmatrix}} = \dfrac{14}{11}$

20. $y = \dfrac{\begin{vmatrix} 721.4 & 33.77 \\ 45.9 & 19.85 \end{vmatrix}}{\begin{vmatrix} 721.4 & -29.1 \\ 45.9 & 105.6 \end{vmatrix}} = \dfrac{12{,}769.747}{77{,}515.530} \approx 0.1647$

Chapter 8 Practice Test Solutions

1. $a_n = \dfrac{2n}{(n+2)!}$

$a_1 = \dfrac{2(1)}{3!} = \dfrac{2}{6} = \dfrac{1}{3}$

$a_2 = \dfrac{2(2)}{4!} = \dfrac{4}{24} = \dfrac{1}{6}$

$a_3 = \dfrac{2(3)}{5!} = \dfrac{6}{120} = \dfrac{1}{20}$

$a_4 = \dfrac{2(4)}{6!} = \dfrac{8}{720} = \dfrac{1}{90}$

$a_5 = \dfrac{2(5)}{7!} = \dfrac{10}{5040} = \dfrac{1}{504}$

Terms: $\dfrac{1}{3}, \dfrac{1}{6}, \dfrac{1}{20}, \dfrac{1}{90}, \dfrac{1}{504}$

2. $a_n = \dfrac{n+3}{3^n}$

3. $\displaystyle\sum_{i=1}^{6} (2i - 1) = 1 + 3 + 5 + 7 + 9 + 11 = 36$

4. $a_1 = 23, d = -2$

$a_2 = 23 + (-2) = 21$

$a_3 = 21 + (-2) = 19$

$a_4 = 19 + (-2) = 17$

$a_5 = 17 + (-2) = 15$

Terms: 23, 21, 19, 17, 15

5. $a_1 = 12, d = 3, n = 50$

$a_n = a_1 + (n-1)d$

$a_{50} = 12 + (50 - 1)3 = 159$

6. $a_1 = 1$

$a_{200} = 200$

$S_n = \dfrac{n}{2}(a_1 + a_n)$

$S_{200} = \dfrac{200}{2}(1 + 200) = 20{,}100$

7. $a_1 = 7, r = 2$

$a_2 = 7(2) = 14$

$a_3 = 7(2)^2 = 28$

$a_4 = 7(2)^3 = 56$

$a_5 = 7(2)^4 = 112$

Terms: 7, 14, 28, 56, 112

8. $\displaystyle\sum_{n=1}^{10} 6\left(\frac{2}{3}\right)^{n-1}, a_1 = 6, r = \frac{2}{3}, n = 10$

$$S_n = \frac{a_1(1 - r^n)}{1 - r} = \frac{6\left[1 - \left(\frac{2}{3}\right)^{10}\right]}{1 - \frac{2}{3}} = 18\left(1 - \frac{1024}{59{,}049}\right) = \frac{116{,}050}{6561} \approx 17.6879$$

9. $\sum_{n=0}^{\infty} (0.03)^n = \sum_{n=1}^{\infty} (0.03)^{n-1}, a_1 = 1, r = 0.03$

$$S = \frac{a_1}{1-r} = \frac{1}{1-0.03} = \frac{1}{0.97} = \frac{100}{97} \approx 1.0309$$

10. For $n = 1, 1 = \frac{1(1+1)}{2}$.

Assume that $S_k = 1 + 2 + 3 + 4 + \cdots + k = \frac{k(k+1)}{2}$.

Then $S_{k+1} = 1 + 2 + 3 + 4 + \cdots + k + (k+1) = \frac{k(k+1)}{2} + k + 1$

$$= \frac{k(k+1)}{2} + \frac{2(k+1)}{2}$$

$$= \frac{(k+1)(k+2)}{2}.$$

Thus, by the principle of mathematical induction, $1 + 2 + 3 + 4 + \cdots + n = \frac{n(n+1)}{2}$ for all integers $n \ge 1$.

11. For $n = 4$, $4! > 2^4$. Assume that $k! > 2^k$.

Then $(k+1)! = (k+1)(k!) > (k+1)2^k > 2 \cdot 2^k = 2^{k+1}$.

Thus, by the extended principle of mathematical induction, $n! > 2^n$ for all integers $n \ge 4$.

12. ${}_{13}C_4 = \frac{13!}{(13-4)!4!} = 715$

13. $(x+3)^5 = x^5 + 5x^4(3) + 10x^3(3)^2 + 10x^2(3)^3 + 5x(3)^4 + (3)^5$

$$= x^5 + 15x^4 + 90x^3 + 270x^2 + 405x + 243$$

14. $-{}_{12}C_5x^7(2)^5 = -25{,}344x^7$

15. ${}_{30}P_4 = \frac{30!}{(30-4)!} = 657{,}720$

16. $6! = 720$ ways

17. ${}_{12}P_3 = 1320$

18. $P(2) + P(3) + P(4) = \frac{1}{36} + \frac{2}{36} + \frac{3}{36}$

$$= \frac{6}{36} = \frac{1}{6}$$

19. $P(\text{K}, \text{B10}) = \frac{4}{52} \cdot \frac{2}{51} = \frac{2}{663}$

20. Let A = probability of no faulty units.

$$P(A) = \left(\frac{997}{1000}\right)^{50} \approx 0.8605$$

$$P(A') = 1 - P(A) \approx 0.1395$$

PART II

Chapter Test Solutions for Chapter P

1. $-\frac{10}{3} < -\frac{5}{3}$

2. $d\left(-\frac{7}{4}, \frac{5}{4}\right) = \left|\frac{5}{4} - \left(-\frac{7}{4}\right)\right| = \left|\frac{12}{4}\right| = 3$

3. $(5 - x) + 0 = 5 - x$

   Additive Identity Property

4. (a) $\left(-\frac{3}{5}\right)^3 = -\frac{27}{125}$

   (b) $\left(\frac{3^2}{2}\right)^{-3} = \left(\frac{2}{9}\right)^3 = \frac{8}{729}$

   (c) $\frac{5^3 \cdot 7^{-1}}{5^2 \cdot 7} = \frac{5^{3-2}}{7^{1+1}} = \frac{5}{7^2} = \frac{5}{49}$

   (d) $\left(2^3\right)^{-2} = 2^{-6} = \frac{1}{2^6} = \frac{1}{64}$

5. (a) $\sqrt{5} \cdot \sqrt{125} = \sqrt{625} = 25$

   (b) $\frac{\sqrt{27}}{\sqrt{2}} = \frac{\sqrt{9 \cdot 3}}{\sqrt{2}} = \frac{3\sqrt{3}}{\sqrt{2}} = \frac{3\sqrt{3}}{\sqrt{2}} \cdot \frac{\sqrt{2}}{\sqrt{2}} = \frac{3\sqrt{6}}{2}$

   (c) $\frac{5.4 \times 10^8}{3 \times 10^3} = \frac{5.4}{3} \cdot \frac{10^8}{10^3} = 1.8 \times 10^5$

   (d) $\left(4.0 \times 10^8\right)\left(2.4 \times 10^{-3}\right) = (4.0)(2.4) \times \left(10^{8-3}\right)$
   $= 9.6 \times 10^5$

6. (a) $3z^2\left(2z^3\right)^2 = 3z^2\left(4z^6\right) = 12z^8$

   (b) $(u - 2)^{-4}(u - 2)^{-3} = (u - 2)^{-7} = \frac{1}{(u - 2)^7}$

   (c) $\left(\frac{x^{-2}y^2}{3}\right)^{-1} = \frac{x^2y^{-2}}{3^{-1}} = \frac{3x^2}{y^2}$

7. (a) $9z\sqrt{8z} - 3\sqrt{2z^3} = 18z\sqrt{2z} - 3z\sqrt{2z}$
   $= 15z\sqrt{2z}$

   Since $\sqrt{8z}$ appears in the same expression, we may assume that $z \geq 0$. It is not necessary to use an absolute value when simplifying $\sqrt{2z^3}$.

   (b) $\left(4x^{3/5}\right)\left(x^{1/3}\right) = 4x^{(3/5)+(1/3)} = 4x^{14/15}$

   (c) $\sqrt[3]{\frac{16}{v^5}} = \sqrt[3]{\frac{8}{v^6} \cdot 2v} = \frac{2}{v^2}\sqrt[3]{2v}$

8. Standard form: $-2x^5 - x^4 + 3x^3 + 3$

   Degree: 5

   Leading coefficient: -2

9. $\left(x^2 + 3\right) - \left[3x + \left(8 - x^2\right)\right] = x^2 + 3 - 3x - 8 + x^2$
   $= 2x^2 - 3x - 5$

10. $\left(x + \sqrt{5}\right)\left(x - \sqrt{5}\right) = x^2 - \left(\sqrt{5}\right)^2 = x^2 - 5$

11. $\frac{5x}{x - 4} + \frac{20}{4 - x} = \frac{5x}{x - 4} - \frac{20}{x - 4}$
    $= \frac{5x - 20}{x - 4}$
    $= \frac{5(x - 4)}{x - 4}$
    $= 5, x \neq 4$

12. $\left[\frac{\frac{x^3}{(x - 1)^3}}{\frac{x}{(x - 1)^4}}\right] = \frac{x^3}{(x - 1)^3} \cdot \frac{(x - 1)^4}{x} = x^2(x - 1), x \neq 0, 1$

13. (a) $2x^4 - 3x^3 - 2x^2 = x^2\left(2x^2 - 3x - 2\right)$
    $= x^2(2x + 1)(x - 2)$

    (b) $x^3 + 2x^2 - 4x - 8 = x^2(x + 2) - 4(x + 2)$
    $= (x + 2)\left(x^2 - 4\right)$
    $= (x + 2)(x + 2)(x - 2)$
    $= (x + 2)^2(x - 2)$

14. (a) $\frac{16}{\sqrt[3]{16}} = \frac{16}{\sqrt[3]{16}} \cdot \frac{\sqrt[3]{4}}{\sqrt[3]{4}} = \frac{16\sqrt[3]{4}}{\sqrt[3]{64}} = \frac{16\sqrt[3]{4}}{4} = 4\sqrt[3]{4}$

    (b) $\frac{4}{1 - \sqrt{2}} = \frac{4}{1 - \sqrt{2}} \cdot \frac{1 + \sqrt{2}}{1 + \sqrt{2}}$
    $= \frac{4\left(1 + \sqrt{2}\right)}{1 - 2}$
    $= -4\left(1 + \sqrt{2}\right)$

15. The domain of $\frac{6 - x}{1 - x}$ is all real numbers x except $x = 1$.

16. $\dfrac{y^2 + 8y + 16}{2y - 4} \cdot \dfrac{8y - 16}{(y + 4)^3} = \dfrac{(y + 4)^2}{2(y - 2)} \cdot \dfrac{8(y - 2)}{(y + 4)^3} = \dfrac{4}{y + 4},\ y \neq 2$

17. $P = R - C$

$= 15x - (1480 + 6x)$

$= 9x - 1480$

When $x = 225$,

$P = 9(225) - 1480$

$= \$545.$

18.

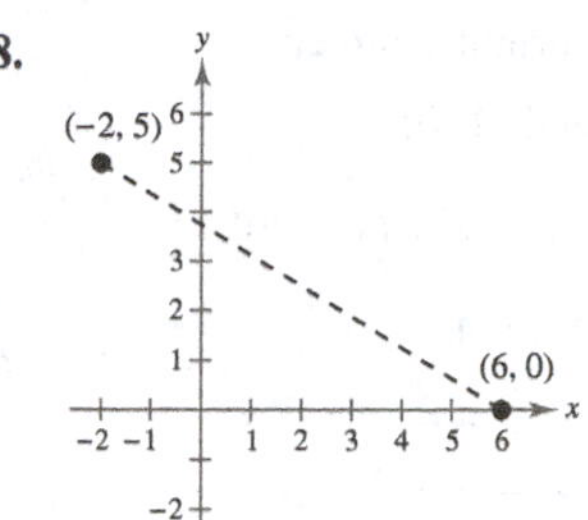

Midpoint: $\left(\dfrac{-2 + 6}{2}, \dfrac{5 + 0}{2}\right) = \left(2, \dfrac{5}{2}\right)$

Distance: $d = \sqrt{(-2 - 6)^2 + (5 - 0)^2}$

$= \sqrt{64 + 25}$

$= \sqrt{89}$

19. Shaded area = Total Area − Unshaded Area

Shaded Area $= \dfrac{1}{2}(3x + 1)(3x + 1) - \dfrac{1}{2}(2x)(2x)$

$= \dfrac{1}{2}\left[\left(9x^2 + 6x + 1\right) - 4x^2\right]$

$= \dfrac{1}{2}\left(5x^2 + 6x + 1\right)$

$= \dfrac{5}{2}x^2 + 3x + \dfrac{1}{2}$ square units

Chapter Test Solutions for Chapter 1

1. $y = 4 - \frac{3}{4}x$

No symmetry

x-intercept: $\left(\frac{16}{3}, 0\right)$

y-intercept: $(0, 4)$

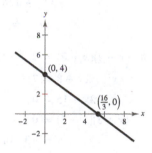

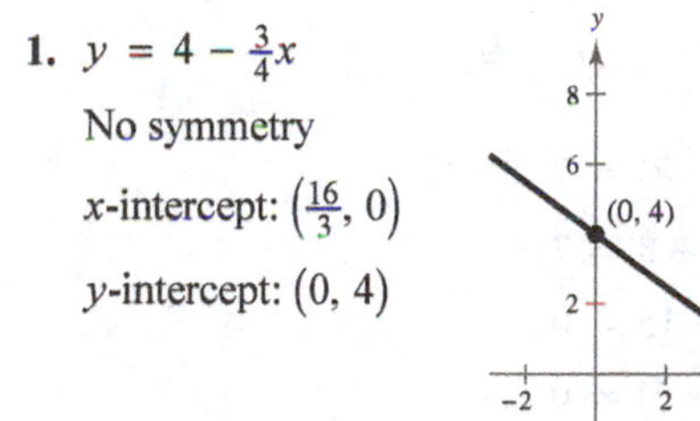

2. $y = 4 - \frac{3}{4}|x|$

y-axis symmetry

x-intercepts: $\left(\pm\frac{16}{3}, 0\right)$

y-intercept: $(0, 4)$

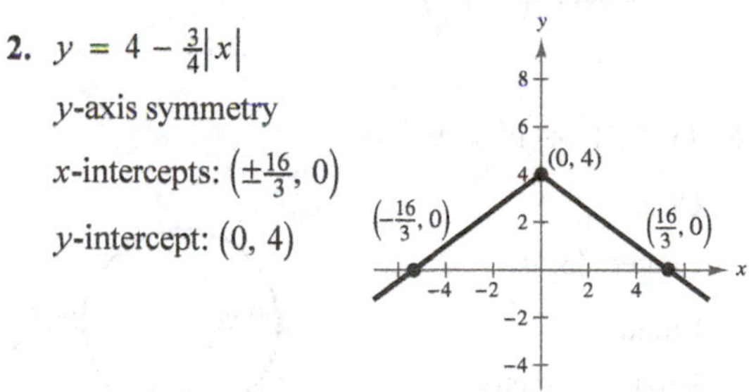

3. $y = 4 - (x - 2)^2$

Parabola; vertex: $(2, 4)$

No x-axis, y-axis, or origin symmetry

x-intercepts: $(0, 0)$ and $(4, 0)$

$$\begin{aligned} 0 &= 4 - (x - 2)^2 \\ (x - 2)^2 &= 4 \\ x - 2 &= \pm 2 \\ x &= 2 \pm 2 \\ x &= 4 \quad \text{or} \quad x = 0 \end{aligned}$$

y-intercept: $(0, 0)$

4. $y = x - x^3$

Origin symmetry

x-intercepts: $(0, 0)$, $(1, 0)$, $(-1, 0)$

$$\begin{aligned} 0 &= x - x^3 \\ 0 &= x(1 + x)(1 - x) \\ x &= 0, x = \pm 1 \end{aligned}$$

y-intercept: $(0, 0)$

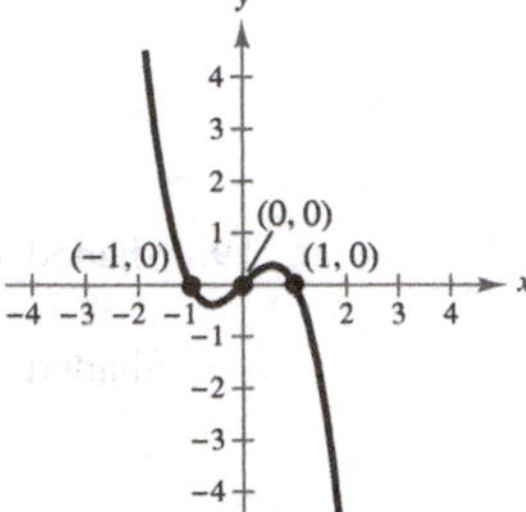

5. $y = \sqrt{5 - x}$

Domain: $x \le 5$

No symmetry

x-intercept: $(5, 0)$

y-intercept: $\left(0, \sqrt{5}\right)$

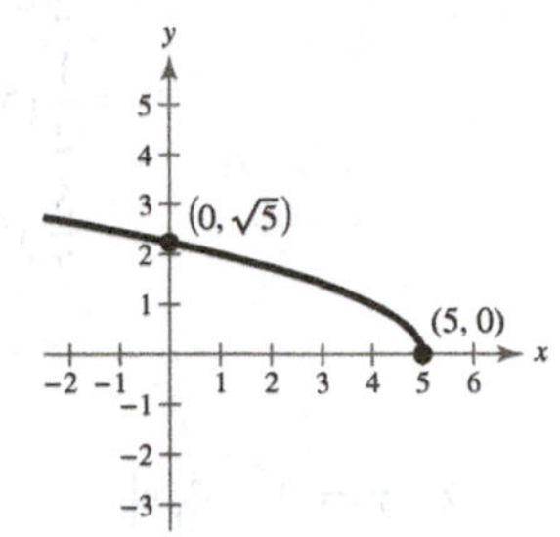

6. $(x - 3)^2 + y^2 = 9$

Circle

Center: $(3, 0)$

Radius: 3

x-axis symmetry

x-intercepts: $(0, 0)$, $(6, 0)$

y-intercept: $(0, 0)$

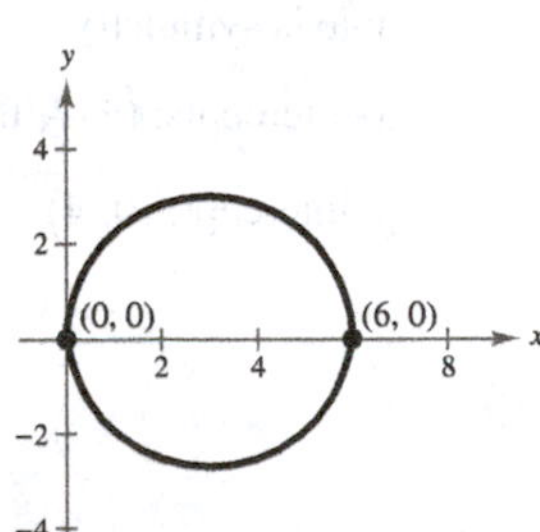

7.
$$\begin{aligned} \tfrac{2}{3}(x - 1) + \tfrac{1}{4}x &= 10 \\ 12\left[\tfrac{2}{3}(x - 1) + \tfrac{1}{4}x\right] &= 12(10) \\ 8(x - 1) + 3x &= 120 \\ 8x - 8 + 3x &= 120 \\ 11x &= 128 \\ x &= \tfrac{128}{11} \end{aligned}$$

8.
$$\begin{aligned} (x - 4)(x + 2) &= 7 \\ x^2 - 2x - 8 &= 7 \\ x^2 - 2x - 15 &= 0 \\ (x + 3)(x - 5) &= 0 \\ x = -3 \quad &\text{or} \quad x = 5 \end{aligned}$$

9. $\dfrac{x - 2}{x + 2} + \dfrac{4}{x + 2} + 4 = 0, x \ne -2$

$$\frac{x + 2}{x + 2} = -4$$

$1 \ne -4 \Rightarrow$ No solution because the variable is divided out.

10. $x^4 + x^2 - 6 = 0$

$$(x^2 - 2)(x^2 + 3) = 0$$

$$x^2 = 2 \Rightarrow x = \pm\sqrt{2}$$

$$x^2 = -3 \Rightarrow x = \pm\sqrt{3}i$$

11. $2\sqrt{x} - \sqrt{2x+1} = 1$

$$-\sqrt{2x+1} = 1 - 2\sqrt{x}$$

$$\left(-\sqrt{2x+1}\right)^2 = \left(1 - 2\sqrt{x}\right)^2$$

$$2x + 1 = 1 - 4\sqrt{x} + 4x$$

$$-2x = -4\sqrt{x}$$

$$x = 2\sqrt{x}$$

$$x^2 = 4x$$

$$x^2 - 4x = 0$$

$$x(x - 4) = 0$$

$$x = 0 \quad \text{or} \quad x = 4$$

Only $x = 4$ is a solution to the original equation. $x = 0$ is extraneous.

12. $|3x - 1| = 7$

$3x - 1 = 7$ or $3x - 1 = -7$

$3x = 8$ $\qquad$ $3x = -6$

$x = \frac{8}{3}$ $\qquad$ $x = -2$

13. $-3 \le 2(x + 4) < 14$

$$-3 \le 2x + 8 < 14$$

$$-11 \le 2x < 6$$

$$-\tfrac{11}{2} \le x < 3$$

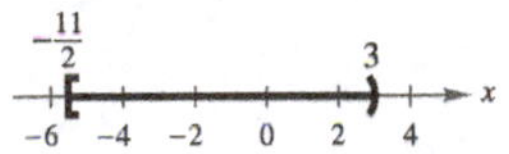

14. $\dfrac{2}{x} > \dfrac{5}{x + 6}$

$$\frac{2}{x} - \frac{5}{x+6} > 0$$

$$\frac{2(x+6) - 5x}{x(x+6)} > 0$$

$$\frac{-3x + 12}{x(x+6)} > 0$$

$$\frac{-3(x - 4)}{x(x+6)} > 0$$

Critical numbers: $x = 4$, $x = 0$, $x = -6$

Test intervals: $(-\infty, -6)$, $(-6, 0)$, $(0, 4)$, $(4, \infty)$

Test: Is $\dfrac{-3(x-4)}{x(x+6)} > 0$?

Solution set: $(-\infty, -6) \cup (0, 4)$

In inequality notation: $x < -6$ or $0 < x < 4$

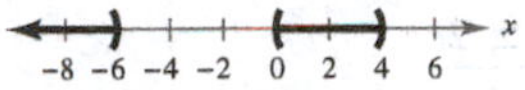

15. $2x^2 + 5x > 12$

$$2x^2 + 5x - 12 > 0$$

$$(2x - 3)(x + 4) > 0$$

Critical numbers: $x = \frac{3}{2}$, $x = -4$

Test intervals: $(-\infty, -4)$, $\left(-4, \frac{3}{2}\right)$, $\left(\frac{3}{2}, \infty\right)$

Test: Is $(2x - 3)(x + 4) > 0$?

Solution set: $(-\infty, -4) \cup \left(\frac{3}{2}, \infty\right)$

In inequality notation: $x < -4$ or $x > \frac{3}{2}$

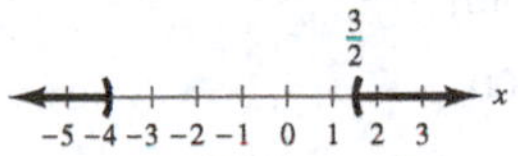

16. $|3x + 5| \ge 10$

$3x + 5 \le -10$ or $3x + 5 \ge 10$

$3x \le -15$ $\qquad$ $3x \ge 5$

$x \le -5$ $\qquad$ $x \ge \frac{5}{3}$

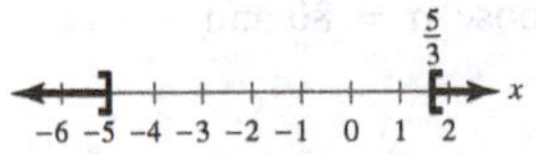

17. (a) $\sqrt{-16} - 2(7 + 2i) = 4i - 14 - 4i$

$$= -14$$

(b) $(5 - i)(3 + 4i) = 15 + 20i - 3i - 4i^2$

$$= 15 + 17i - 4(-1)$$

$$= 19 + 17i$$

18. $\dfrac{8}{1+2i} = \dfrac{8}{1+2i} \cdot \dfrac{1-2i}{1-2i}$

$= \dfrac{8-16i}{1-4i^2}$

$= \dfrac{8-16i}{1-4(-1)}$

$= \dfrac{8-16i}{5}$

$= \dfrac{8}{5} - \dfrac{16}{5}i$

19. $2x^2 - 6x + 5 = 0$

$$x = \frac{-(-6) \pm \sqrt{(-6)^2 - 4(2)(5)}}{2(2)}$$

$$x = \frac{6 \pm \sqrt{36 - 40}}{4}$$

$$x = \frac{6 \pm \sqrt{-4}}{4}$$

$$x = \frac{6 \pm 2i}{4}$$

$$x = \frac{3}{2} \pm \frac{1}{2}i$$

20. (a)

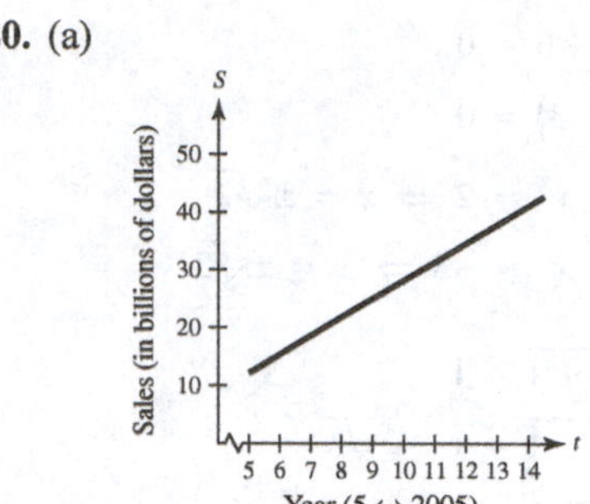

(b) Using trace and zoom features the sale in 2017 will be about \$50.8 billion.

(c) $S = 3.2205(17) - 3.908$

≈ 50.8

The sales in 2017 will be about \$50.8 billion.

21. $V = \dfrac{4}{3}\pi r^3$

$\dfrac{4}{3}\pi r^3 = 455.9$

$r = \sqrt[3]{\dfrac{455.9(3)}{4\pi}} \approx 4.774$ inches

22. $(100 \text{ km/hr})(2\frac{1}{4} \text{ hr}) + (x \text{ km/hr})(1\frac{1}{3} \text{ hr}) = 350 \text{ km}$

$$225 + \tfrac{4}{3}x = 350$$

$$\tfrac{4}{3}x = 125$$

$$x = \tfrac{375}{4} = 93\tfrac{3}{4} \text{ km/hr}$$

23. $a + b = 100 \Rightarrow b = 100 - a$

Area of ellipse = Area of circle

$\pi ab = \pi(40)^2$

$a(100 - a) = 1600$

$0 = a^2 - 100a + 1600$

$0 = (a - 80)(a - 20)$

$a = 80 \Rightarrow b = 20$

or

$a = 20 \Rightarrow b = 80$

Because $a > b$, choose $a = 80$ and $b = 20$.

Chapter Test Solutions for Chapter 2

1. $(-2, 5), (1, -7)$

$$m = \frac{-7 - 5}{1 - (-2)} = \frac{-12}{3} = -4$$

$$y - 5 = -4(x - (-2))$$
$$y - 5 = -4(x + 2)$$
$$y - 5 = -4x - 8$$
$$y = -4x - 3$$

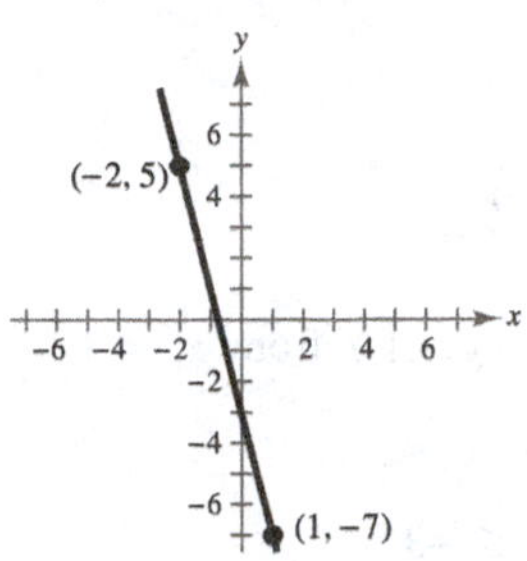

2. $(-4, -7), (1, \frac{4}{3})$

$$m = \frac{\frac{4}{3} - (-7)}{1 - (-4)} = \frac{\frac{25}{3}}{5} = \frac{5}{3}$$

$$y - (-7) = \frac{5}{3}(x - (-4))$$
$$y + 7 = \frac{5}{3}(x + 4)$$
$$y + 7 = \frac{5}{3}x + \frac{20}{3}$$
$$y = \frac{5}{3}x - \frac{1}{3}$$

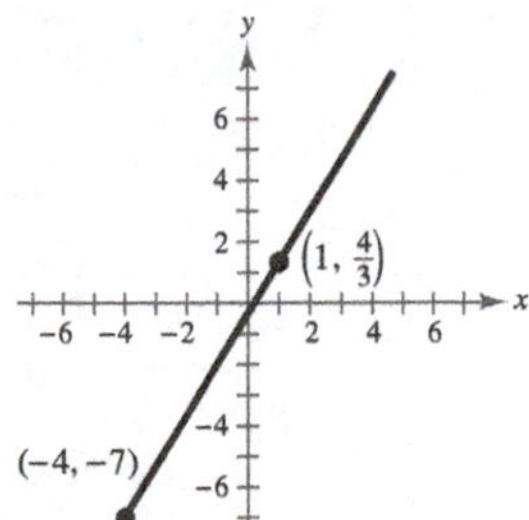

3. $5x + 2y = 3$

$$2y = -5x + 3$$
$$y = -\tfrac{5}{2}x + \tfrac{3}{2}$$

(a) Parallel line:

$$m = -\tfrac{5}{2}$$
$$y - 4 = -\tfrac{5}{2}(x - 0)$$
$$y - 4 = -\tfrac{5}{2}x$$
$$y = -\tfrac{5}{2}x + 4$$

(b) Perpendicular line:

$$m = \tfrac{2}{5}$$
$$y - 4 = \tfrac{2}{5}(x - 0)$$
$$y - 4 = \tfrac{2}{5}x$$
$$y = \tfrac{2}{5}x + 4$$

4. $f(x) = |x + 2| - 15$

(a) $f(-8) = -9$

(b) $f(14) = 1$

(c) $f(x - 6) = |x - 4| - 15$

5. $f(x) = \dfrac{\sqrt{x + 9}}{x^2 - 81}$

(a) $f(7) = \dfrac{4}{-32} = -\dfrac{1}{8}$

(b) $f(-5) = \dfrac{2}{-56} = -\dfrac{1}{28}$

(c) $f(x - 9) = \dfrac{\sqrt{x}}{(x - 9)^2 - 81} = \dfrac{\sqrt{x}}{x^2 - 18x}$

6. (a) $f(x) = \dfrac{x - 5}{2x^2 - x} = \dfrac{x - 5}{x(2x - 1)}$, the domain of f is all real numbers x such that $x \neq 0, \dfrac{1}{2}$.

(b) $f(x) = \dfrac{x - 5}{2x^2 - x} = 0 \Rightarrow x = 5$

7. (a) $f(x) = 10 - \sqrt{3 - x}$

$3 - x \geq 0$

$3 \geq x$

$x \leq 3$

Domain: All real numbers x such that $x \leq 3$

(b) $10 - \sqrt{3 - x} = 0 \Rightarrow 10 = \sqrt{3 - x}$

$100 = 3 - x$

$x = -97$

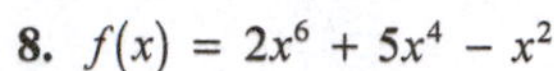

8. $f(x) = 2x^6 + 5x^4 - x^2$

(a)

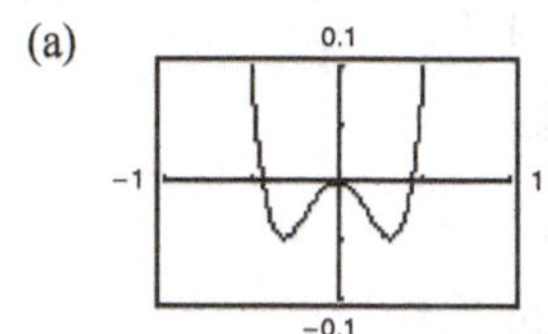

(b) Increasing on $(-0.31, 0), (0.31, \infty)$

Decreasing on $(-\infty, -0.31), (0, 0.31)$

(c) y-axis symmetry $\Rightarrow$ the function is even.

9. $f(x) = 4x\sqrt{3 - x}$

(a)

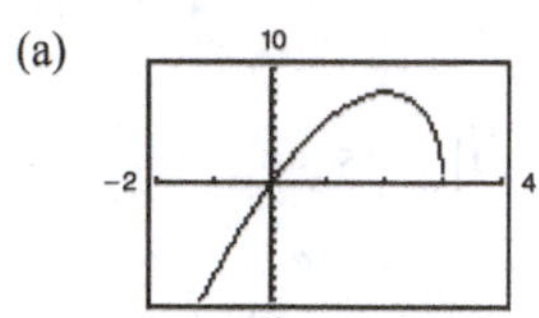

(b) Increasing on $(-\infty, 2)$

Decreasing on $(2, 3)$

(c) The function is neither odd nor even.

10. $f(x) = |x + 5|$

(a)

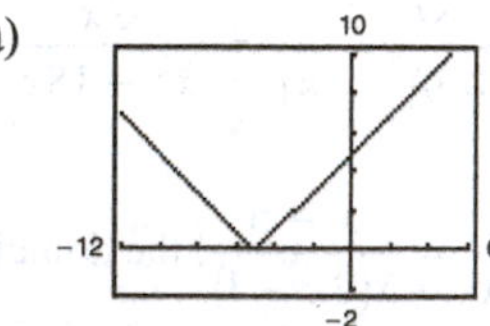

(b) Increasing on $(-5, \infty)$

Decreasing on $(-\infty, -5)$

(c) The function is neither odd nor even.

11. $f(x) = -x^3 + 2x - 1$

Relative minimum: $(-0.82, -2.09)$

Relative maximum: $(0.82, 0.09)$

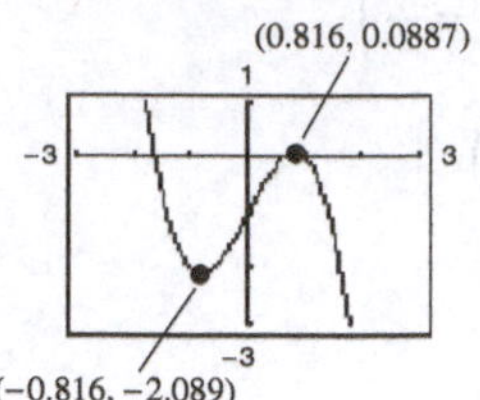

12. $f(x) = -2x^2 + 5x - 3$

$$\frac{f(3) - f(1)}{3 - 1} = \frac{-6 - 0}{2} = -3$$

The average rate of change of f from $x_1 = 1$ to $x_2 = 3$ is -3.

13. $f(x) = \begin{cases} 3x + 7, & x \leq -3 \\ 4x^2 - 1, & x > -3 \end{cases}$

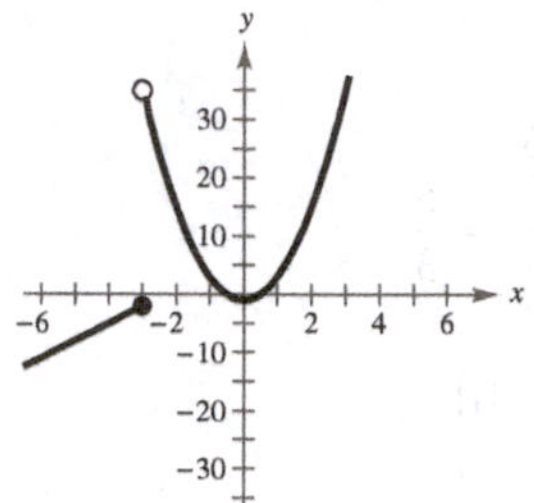

14. $h(x) = 4[\![x]\!]$

(a) The parent function is $f(x) = [\![x]\!]$

(b) The graph of h is a vertical stretch of the graph of f.

(c)

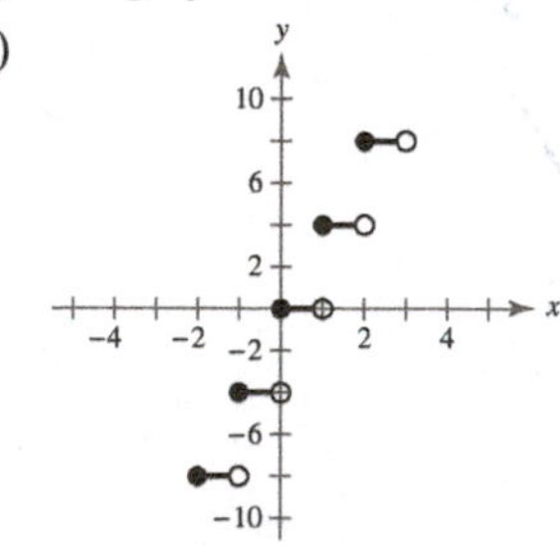

15. $h(x) = -\sqrt{x+5} + 8$

(a) Parent function: $f(x) = \sqrt{x}$

(b) Transformation: Reflection in the x-axis, a horizontal shift 5 units to the left, and a vertical shift 8 units upward

(c)

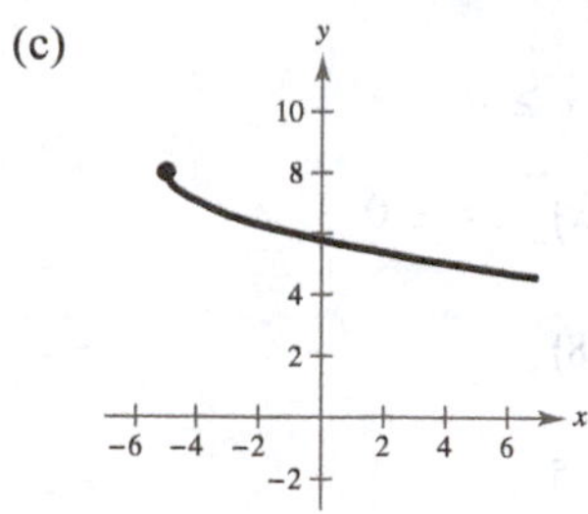

16. $h(x) = -2(x-5)^3 + 3$

(a) Parent function: $f(x) = x^3$

(b) Transformation:

Vertical stretch, reflection in x-axis, horizontal shift 5 units to the right, vertical shift 3 units upward

(c)

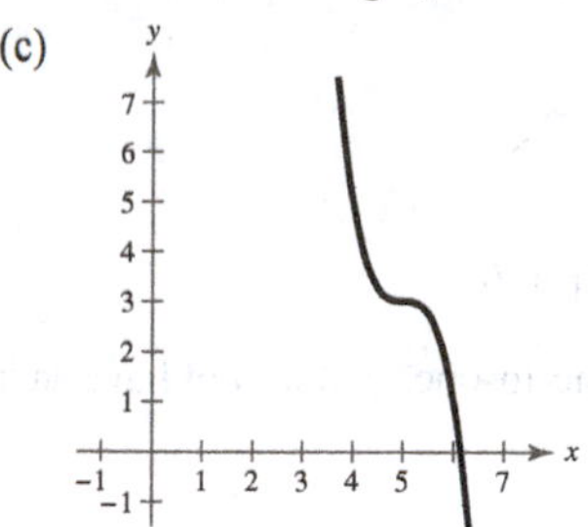

17. $f(x) = 3x^2 - 7,\ g(x) = -x^2 - 4x + 5$

(a) $(f+g)(x) = (3x^2 - 7) + (-x^2 - 4x + 5) = 2x^2 - 4x - 2$

(b) $(f-g)(x) = (3x^2 - 7) - (-x^2 - 4x + 5) = 4x^2 + 4x - 12$

(c) $(fg)(x) = (3x^2 - 7)(-x^2 - 4x + 5) = -3x^4 - 12x^3 + 22x^2 + 28x - 35$

(d) $\left(\dfrac{f}{g}\right)(x) = \dfrac{3x^2 - 7}{-x^2 - 4x + 5},\ x \neq -5, 1$

(e) $(f \circ g)(x) = f(g(x)) = f(-x^2 - 4x + 5) = 3(-x^2 - 4x + 5)^2 - 7 = 3x^4 + 24x^3 + 18x^2 - 120x + 68$

(f) $(g \circ f)(x) = g(f(x)) = g(3x^2 - 7) = -(3x^2 - 7)^2 - 4(3x^2 - 7) + 5 = -9x^4 + 30x^2 - 16$

18. $f(x) = \dfrac{1}{x},\ g(x) = 2\sqrt{x}$

(a) $(f+g)(x) = \dfrac{1}{x} + 2\sqrt{x} = \dfrac{1 + 2x^{3/2}}{x},\ x > 0$

(b) $(f-g)(x) = \dfrac{1}{x} - 2\sqrt{x} = \dfrac{1 - 2x^{3/2}}{x},\ x > 0$

(c) $(fg)(x) = \left(\dfrac{1}{x}\right)(2\sqrt{x}) = \dfrac{2\sqrt{x}}{x},\ x > 0$

(d) $\left(\dfrac{f}{g}\right)(x) = \dfrac{1/x}{2\sqrt{x}} = \dfrac{1}{2x\sqrt{x}} = \dfrac{1}{2x^{3/2}},\ x > 0$

(e) $(f \circ g)(x) = f(g(x)) = f(2\sqrt{x}) = \dfrac{1}{2\sqrt{x}} = \dfrac{\sqrt{x}}{2x},\ x > 0$

(f) $(g \circ f)(x) = g(f(x)) = g\left(\dfrac{1}{x}\right) = 2\sqrt{\dfrac{1}{x}} = \dfrac{2}{\sqrt{x}} = \dfrac{2\sqrt{x}}{x},\ x > 0$

19. $f(x) = x^3 + 8$

Since f is one-to-one, f has an inverse.

$$y = x^3 + 8$$
$$x = y^3 + 8$$
$$x - 8 = y^3$$
$$\sqrt[3]{x - 8} = y$$
$$f^{-1}(x) = \sqrt[3]{x - 8}$$

20. $f(x) = |x^2 - 3| + 6$

Since f is not one-to-one, f does not have an inverse.

21. $f(x) = 3x\sqrt{x} = 3x^{3/2}, x \geq 0$

Because f is one-to-one, f has an inverse.

$$y = 3x^{3/2}$$
$$x = 3y^{3/2}$$
$$\tfrac{1}{3}x = y^{3/2}$$
$$\left(\tfrac{1}{3}x\right)^{2/3} = y, x \geq 0$$
$$f^{-1}(x) = \left(\tfrac{1}{3}x\right)^{2/3}, x \geq 0$$

22. (6, 58) and (10, 78)

$$m = \frac{78 - 58}{10 - 6} = 5$$
$$C - 58 = 5(x - 6)$$
$$C = 5x + 28$$

When $x = 25$: $C = 5(25) + 28 = \$153$

Cumulative Test Solutions for Chapters P–2

1. $\dfrac{8x^2y^{-3}}{30x^{-1}y^2} = \dfrac{8x^2x}{30y^2y^3} = \dfrac{4x^3}{15y^5}, x \neq 0$

2. $\sqrt{20x^2y^3} = \sqrt{4x^2y^2 5y} = 2|x|y\sqrt{5y}$

3. $4x - [2x + 3(2 - x)] = 4x - [2x + 6 - 3x]$
$= 4x - [-x + 6]$
$= 5x - 6$

4. $(x - 2)(x^2 + x - 3) = x^3 + x^2 - 3x - 2x^2 - 2x + 6$
$= x^3 - x^2 - 5x + 6$

5. $\dfrac{3}{x + 5} - \dfrac{2}{s - 3} = \dfrac{3(s - 3) - 2(s + 5)}{(s + 5)(s - 3)}$
$= \dfrac{3s - 9 - 2s - 10}{(s + 5)(s - 3)}$
$= \dfrac{s - 19}{(s + 5)(s - 3)}$

6. $36 - (x + 1)^2 = [6 + (x + 1)][6 - (x + 1)]$
$= (x + 7)(5 - x)$

7. $x - 5x^2 - 6x^3 = x(1 - 5x - 6x^2)$
$= x(1 + x)(1 - 6x)$

8. $54x^3 + 16 = 2(27x^3 + 8)$
$= 2\left((3x)^3 + 2^3\right)$
$= 2(3x + 2)(9x^2 - 6x + 4)$

9. Area = area of rectangle + area of triangle

$A = x(2(x + 1)) + \frac{1}{2}(2(x + 1))(2x + 1)$
$= x(2x + 2) + \frac{1}{2}(2x + 2)(2x + 1)$
$= 2x^2 + 2x + \frac{1}{2}(4x^2 + 6x + 2)$
$= 2x^2 + 2x + 2x^2 + 3x + 1$
$= 4x^2 + 5x + 1$

10. Area = area of rectangle + area of triangle

$A = x(2x + 1) + \frac{1}{2}(x + 1)(x + 1)$
$= x(2x + 1) + \frac{1}{2}(x + 1)(x + 1)$
$= 2x^2 + x + \frac{1}{2}(x^2 + 2x + 1)$
$= 2x^2 + x + \frac{1}{2}x^2 + x + \frac{1}{2}$
$= \frac{5}{2}x^2 + 2x + \frac{1}{2}$

11. $x - 3y + 12 = 0$

Line

x-intercept: $(-12, 0)$

y-intercept: $(0, 4)$

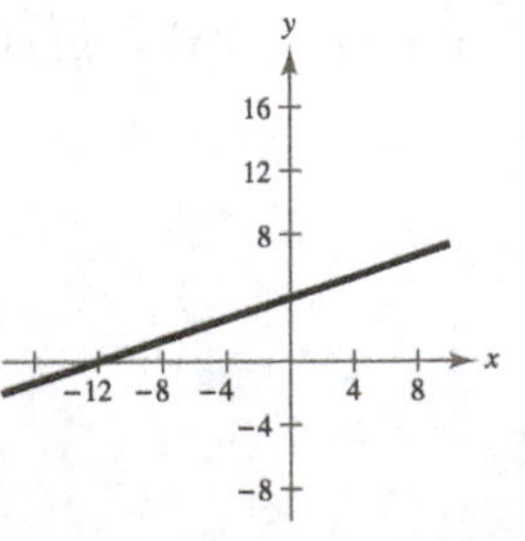

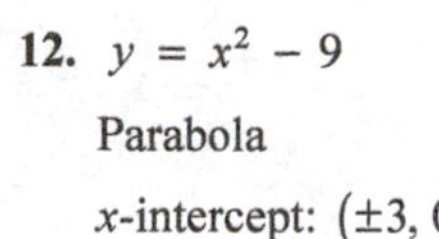

12. $y = x^2 - 9$

Parabola

x-intercept: $(\pm 3, 0)$

y-intercept: $(0, -9)$

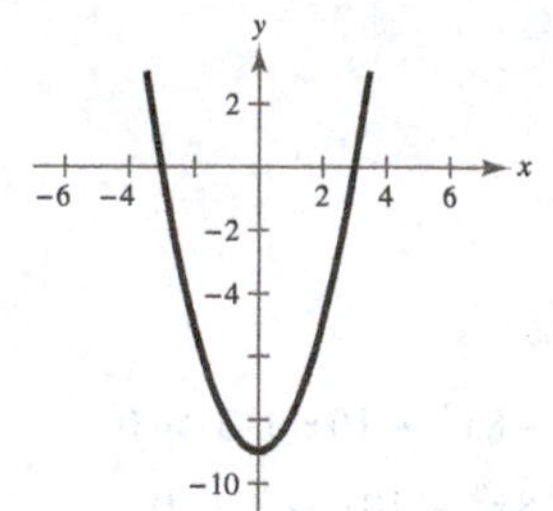

13. $y = \sqrt{4 - x}$

Domain: $x \le 4$

x-intercept: $(4, 0)$

y-intercept: $(0, 2)$

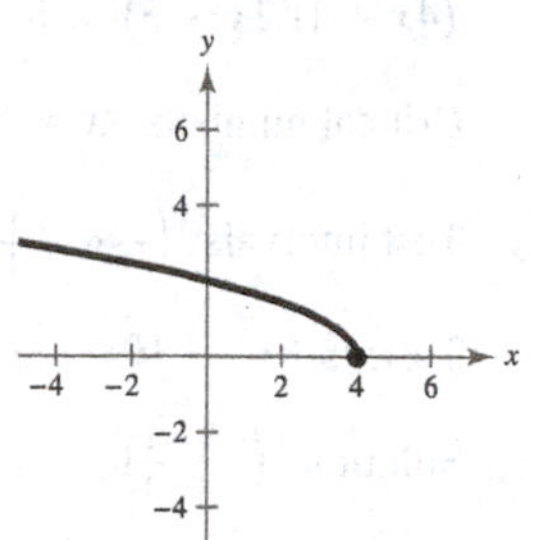

14. $3x - 5 = 6x + 8$

$$-3x = 13$$

$$x = -\tfrac{13}{3}$$

15. $-(x + 3) = 14(x - 6)$

$$-x - 3 = 14x - 84$$

$$-15x = -81$$

$$x = \frac{-81}{-15} = \frac{27}{5}$$

16. $\dfrac{1}{x - 2} = \dfrac{10}{4x + 3}$

$$4x + 3 = 10(x - 2)$$

$$4x + 3 = 10x - 20$$

$$-6x = -23$$

$$x = \frac{23}{6}$$

17. $x^2 - 4x + 3 = 0$

$$(x - 1)(x - 3) = 0$$

$$x - 1 = 0 \Rightarrow x = 1$$

$$x - 3 = 0 \Rightarrow x = 3$$

18. $-2x^2 + 4x + 6 = 0$

$$-2(x^2 - 2x - 3) = 0$$

$$x^2 - 2x - 3 = 0$$

$$x^2 - 2x = 3$$

$$x^2 - 2x + 1 = 3 + 1$$

$$(x - 1)^2 = 4$$

$$x - 1 = \pm\sqrt{4}$$

$$x = 1 \pm 2 \Rightarrow x = 3, -1$$

19. $3x^2 + 9x + 1 = 0$

$a = 3, b = 9, c = 1$

$$x = \frac{-9 \pm \sqrt{9^2 - 4(3)(1)}}{2(3)}$$

$$= \frac{-9 \pm \sqrt{81 - 12}}{6}$$

$$= \frac{-9 \pm \sqrt{69}}{6}$$

$$= -\frac{3}{2} \pm \frac{\sqrt{69}}{6}$$

20. $3x^2 + 5x - 6 = 0$

$a = 3, b = 5, c = -6$

$$x = \frac{-5 \pm \sqrt{5^2 - 4(3)(-6)}}{2(3)}$$

$$= \frac{-5 \pm \sqrt{25 + 72}}{6}$$

$$= \frac{-5 \pm \sqrt{97}}{6}$$

21. $\tfrac{2}{3}x^2 = 24$

$$x^2 = 36$$

$$x = \pm 6$$

22. $\dfrac{1}{2}x^2 - 7 = 25$

$$\frac{1}{2}x^2 = 32$$

$$x^2 = 64$$

$$x = \pm\sqrt{64}$$

$$x = \pm 8$$

23. $x^4 + 12x^3 + 4x^2 + 48x = 0$

$$x^3(x + 12) + 4x(x + 12) = 0$$

$$(x^3 + 4x)(x + 12) = 0$$

$$x(x^2 + 4)(x + 12) = 0$$

$$x = 0$$

$$x^2 + 4 = 0 \Rightarrow x = \pm 2i$$

$$x + 12 = 0 \Rightarrow x = -12$$

24. $8x^3 - 48x^2 + 72x = 0$

$$8x(x^2 - 6x + 9) = 0$$

$$8x(x - 3)^2 = 0$$

$$8x = 0 \Rightarrow x = 0$$

$$x - 3 = 0 \Rightarrow x = 3$$

25. $x^{3/2} + 21 = 13$

$$x^{3/2} = -8$$

$$x = \sqrt[3]{(-8)^2}$$

$x = 4$ is an extraneous solution, so there are no real solutions

26. $\sqrt{x + 10} = x - 2$

$$x + 10 = x^2 - 4x + 4$$

$$0 = x^2 - 5x - 6$$

$$0 = (x - 6)(x + 1)$$

$$x = 6 \text{ or } x = -1$$

Only $x = 6$ is a solution to the original equation. $x = -1$ is extraneous.

27. $|2(x - 1)| = 8$

$2(x - 1) = -8$ or $2(x - 1) = 8$

$x - 1 = -4$ $\quad$ $x - 1 = 4$

$x = -$ $\quad$ $x = 5$

28. $|x - 12| = -2$

No solution. The absolute value of a number cannot be negative.

29. $|x + 1| \le 6$

$$-6 \le x + 1 \le 6$$

$$-7 \le x \le 5$$

30. $|5 + 6x| > 3$

$5 + 6x < -3$ or $5 + 6x > 3$

$6x < -8$ $\quad$ $6x > -2$

$x < -\frac{4}{3}$ $\quad$ $x > -\frac{1}{3}$

31. $5x^2 + 12x + 7 \ge 0$

$$(5x + 7)(x + 1) \ge 0$$

Critical numbers: $x = -\frac{7}{5}, -1$

Test intervals: $\left(-\infty, -\frac{7}{5}\right), \left(-\frac{7}{5}, -1\right), (-1, \infty)$

Test: Is $5x^2 + 12x + 7 \ge 0$?

Solution: $x \le -\frac{7}{5}, x \ge -1$

32. $-8x^2 + 10x + 3 > 0$

$$8x^2 - 10x - 3 < 0$$

$$(4x + 1)(2x - 3) < 0$$

Critical numbers: $x = -\frac{1}{4}, \frac{3}{2}$

Test intervals: $\left(-\infty, -\frac{1}{4}\right), \left(-\frac{1}{4}, \frac{3}{2}\right), \left(\frac{3}{2}, \infty\right)$

Test: Is $8x^2 - 10x - 3 < 0$?

Solution: $\left(-\frac{1}{4}, \frac{3}{2}\right)$

33. $\left(-\frac{1}{2}, 1\right)$ and $(3, 8)$

$$m = \frac{8 - 1}{3 - (-1/2)} = \frac{7}{7/2} = 2$$

$$y - 8 = 2(x - 3)$$

$$y - 8 = 2x - 6$$

$$y = 2x + 2$$

34. It fails the Vertical Line Test. For some values of x there correspond two values of y.

35. $f(x) = \dfrac{x}{x - 2}$

(a) $f(6) = \dfrac{6}{4} = \dfrac{3}{2}$

(b) $f(2)$ is undefined because division by zero is undefined.

(c) $f(s + 2) = \dfrac{s + 2}{(s + 2) - 2} = \dfrac{s + 2}{s}$

36. $f(x) = 5 + \sqrt{4 - x}$

$$f(-x) = 5 + \sqrt{4 - (-x)} = 5 + \sqrt{4 + x}$$

$$-f(x) = -5 - \sqrt{4 - x}$$

$$f(-x) \ne f(x) \text{ and } f(-x) \ne -f(x)$$

The function is neither even nor odd.

37. $f(x) = 2x^3 - 4x$

$f(-x) = 2(-x)^3 - 4(-x)$
$= -2x^3 + 4x$
$= -(2x^3 - 4x) = -f(x)$

$f(-x) = -f(x)$

The function is odd.

38. $f(x) = x^4 + 1$

$f(-x) = (-x)^4 + 1$
$= x^4 + 1$

$f(-x) = f(x)$

The function is even.

39. $y = \sqrt[3]{x}$

(a) $r(x) = \frac{1}{2}\sqrt[3]{x}$ is a vertical shrink by a factor of $\frac{1}{2}$.

(b) $h(x) = \sqrt[3]{x} + 2$ is a vertical shift two units upward.

(c) $g(x) = \sqrt[3]{x + 2}$ is a horizontal shift two units to the left.

40. $f(x) = x - 4$, $g(x) = 3x + 1$

(a) $(f + g)(x) = f(x) + g(x)$
$= (x - 4) + (3x + 1)$
$= 4x - 3$

(b) $(f - g)(x) = f(x) - g(x)$
$= (x - 4) - (3x + 1)$
$= -2x - 5$

(c) $(fg)(x) = f(x)g(x)$
$= (x - 4)(3x + 1)$
$= 3x^2 - 11x - 4$

(d) $\left(\frac{f}{g}\right)(x) = \frac{f(x)}{g(x)} = \frac{x - 4}{3x + 1}$

Domain: All real numbers x except $x = -\frac{1}{3}$

41. $f(x) = \sqrt{x - 1}$, $g(x) = x^2 + 1$

(a) $(f + g)(x) = f(x) + g(x)$
$= \sqrt{x - 1} + x^2 + 1$

(b) $(f - g)(x) = f(x) - g(x)$
$= \sqrt{x - 1} - x^2 - 1$

(c) $(fg)(x) = f(x)g(x)$
$= \sqrt{x - 1}(x^2 + 1)$
$= x^2\sqrt{x - 1} + \sqrt{x - 1}$

(d) $\left(\frac{f}{g}\right)(x) = \frac{f(x)}{g(x)} = \frac{\sqrt{x - 1}}{x^2 + 1}$

Domain: all real numbers x such that $x \geq 1$

42. $f(x) = 2x^2$, $g(x) = \sqrt{x + 6}$

(a) $(f \circ g)(x) = f(g(x))$
$= f(\sqrt{x + 6})$
$= 2(\sqrt{x + 6})^2$
$= 2(x + 6)$
$= 2x + 12$

Domain: all real numbers x such that $x \geq -6$

(b) $(g \circ f)(x) = g(f(x))$
$= g(2x^2)$
$= \sqrt{2x^2 + 6}$

Domain: all real numbers x

43. $f(x) = x - 2$, $g(x) = |x|$

(a) $(f \circ g)(x) = f(g(x))$
$= f(|x|)$
$= |x| - 2$

Domain: all real numbers x

(b) $(g \circ f)(x) = g(f(x))$
$= g(x - 2)$
$= |x - 2|$

Domain: all real numbers x

44. $h(x) = 3x - 4$

Because h is one-to-one, h has an inverse.

$y = 3x - 4$
$x = 3y - 4$
$x + 4 = 3y$
$\frac{1}{3}(x + 4) = y$
$h^{-1}(x) = \frac{1}{3}(x + 4)$

45. Cost per person: $\dfrac{36{,}000}{n}$

If three additional people join the group, the cost per person is $\dfrac{36{,}000}{n+3}$.

$$\frac{36{,}000}{n} = \frac{36{,}000}{n+3} + 1000$$
$$36{,}000(n+3) = 36{,}000n + 1000n(n+3)$$
$$36(n+3) = 36n + n(n+3)$$
$$36n + 108 = 36n + n^2 + 3n$$
$$0 = n^2 + 3n - 108$$
$$0 = (n+12)(n-9)$$

Choosing the positive value, the group has $n = 9$ people.

46. Rate $= 10 - 0.05(n - 60),\ n \ge 60$

(a) Revenue $=$ (number of people)(rate per person)

$$\begin{aligned} R(n) &= n[10 - 0.05(n - 60)] \\ &= 10n - 0.05n(n - 60) \\ &= 10n - 0.05n^2 + 3n \\ &= -0.05n^2 + 13n,\ n \ge 60 \end{aligned}$$

(b)

900

60 280

0

The revenue is maximum when $n = 130$ passengers.

47. $s(t) = -16t^2 + 36t + 8$

$$\frac{s(2) - s(0)}{2 - 0} = \frac{16 - 8}{2} = 4$$

The average rate of change in the height of the object from $t_1 = 0$ to $t_2 = 2$ seconds is 4 feet per second.

Chapter Test Solutions for Chapter 3

1. (a)

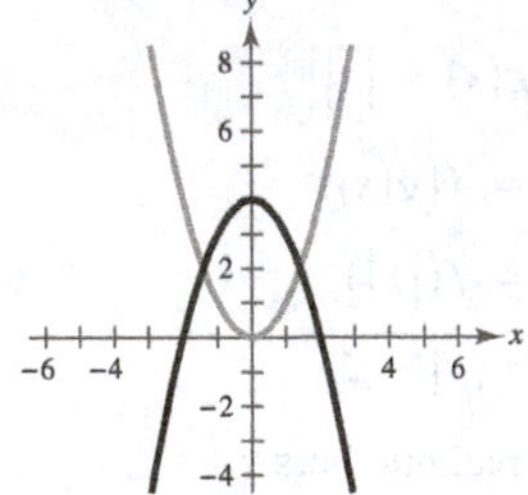

The graph of g is a reflection in the x-axis and a vertical shift up of four units of the graph of $y = x^2$.

(b)

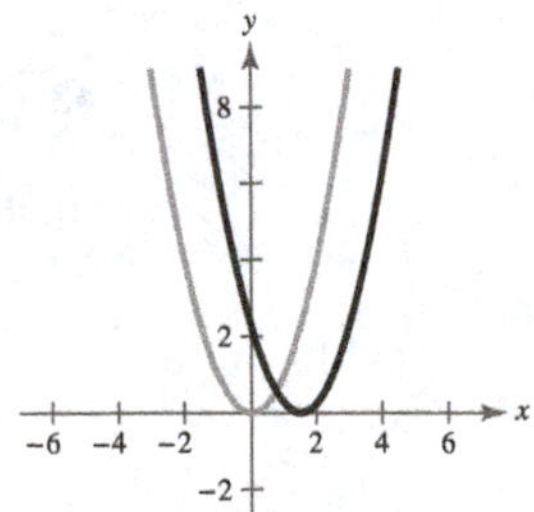

The graph of g is a horizontal shift right $\frac{3}{2}$ units of the graph of $y = x^2$.

2.
$$\begin{aligned} f(x) &= x^2 - 2x - 3 \\ &= (x^2 - 2x + 1) - 3 - 1 \\ &= (x - 1)^2 - 4 \end{aligned}$$

Vertex: $(1, -4)$

Find the x-intercepts: $x^2 - 2x - 3 = 0$

$$(x - 3)(x + 1) = 0$$
$$x - 3 = 0 \Rightarrow x = 3$$
$$x + 1 = 0 \Rightarrow x = -1$$

The x-intercepts are $(3, 0)$ and $(-1, 0)$

Find the y-intercept: $f(0) = 0^2 - 2(0) - 3 = -3$

The y-intercept is $(0, -3)$

3. Vertex: $(3, -6)$

$$y = a(x - 3)^2 - 6$$

Point on the graph: $(0, 3)$

$$3 = a(0 - 3)^2 - 6$$
$$9 = 9a \Rightarrow a = 1$$

So, $y = (x - 3)^2 - 6$.

4. (a) $y = -\frac{1}{20}x^2 + 3x + 5$

$$= -\tfrac{1}{20}(x^2 - 60x + 900 - 900) + 5$$
$$= -\tfrac{1}{20}\left[(x - 30)^2 - 900\right] + 5$$
$$= -\tfrac{1}{20}(x - 30)^2 + 50$$

Vertex: $(30, 50)$

The maximum height is 50 feet.

(b) The constant term, $c = 5$, determines the height at which the ball was thrown. Changing this constant results in a vertical translation of the graph, and therefore, changes the maximum height.

5. $h(t) = -\frac{3}{4}t^5 + 2t^2$

The degree is odd and the leading coefficient is negative. The graph rises to the left and falls to the right.

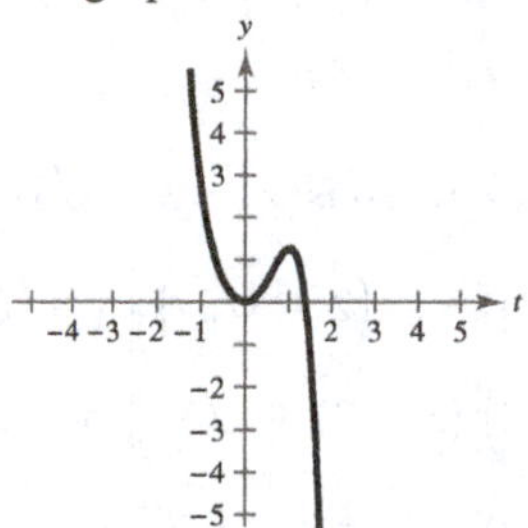

6.
$$\begin{array}{r}
3x + \dfrac{x-1}{x^2+1} \\
x^2 + 0x + 1 \overline{)\,3x^3 + 0x^2 + 4x - 1} \\
\underline{3x^3 + 0x^2 + 3x} \\
x - 1
\end{array}$$

Thus, $\dfrac{3x^3 + 4x - 1}{x^2 + 1} = 3x + \dfrac{x - 1}{x^2 + 1}$.

7.
$$\begin{array}{r|rrrrr}
-2 & 2 & 0 & -3 & 4 & -1 \\
 & & -4 & 8 & -10 & 12 \\
\hline
 & 2 & -4 & 5 & -6 & 11
\end{array}$$

So, $\dfrac{2x^4 - 3x^2 + 4x - 1}{x + 2} = 2x^3 - 4x^2 + 5x - 6 + \dfrac{11}{x + 2}$.

8.
$$\begin{array}{r|rrrr}
\sqrt{3} & 2 & -5 & -6 & 15 \\
 & & 2\sqrt{3} & 6 - 5\sqrt{3} & -15 \\
\hline
 & 2 & 2\sqrt{3} - 5 & -5\sqrt{3} & 0
\end{array}$$

$$\begin{array}{r|rrr}
-\sqrt{3} & 2 & 2\sqrt{3} - 5 & -5\sqrt{3} \\
 & & -2\sqrt{3} & 5\sqrt{3} \\
\hline
 & 2 & -5 & 0
\end{array}$$

$$2x^3 - 5x^2 - 6x + 15 = (x - \sqrt{3})(x + \sqrt{3})(2x - 5)$$

The real zeros of $f(x)$ are $x = \pm\sqrt{3}$ and $x = \frac{5}{2}$.

9. $g(t) = 2t^4 - 3t^3 + 16t - 24$

Possible rational zeros:

$\pm1, \pm2, \pm3, \pm4, \pm6, \pm8, \pm12, \pm24, \pm\frac{1}{2}, \pm\frac{3}{2}$

From the graph, we have $t = -2$ and $t = \frac{3}{2}$.

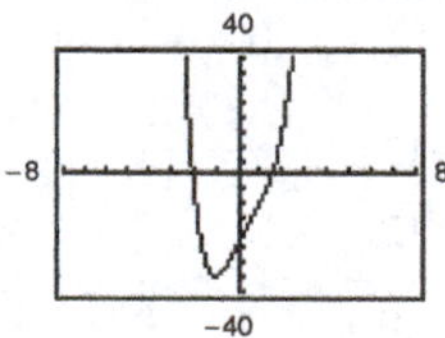

10. $h(x) = 3x^5 + 2x^4 - 3x - 2$

Possible rational zeros: $\pm1, \pm2, \pm\frac{1}{3}, \pm\frac{2}{3}$

From the graph, we have $x = \pm1$ and $x = -\frac{2}{3}$.

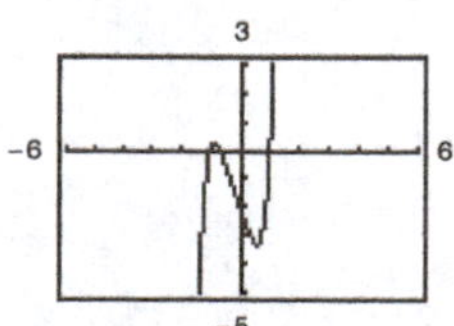

11. If $x = 3i$ is a zero, so is $x = -3i$

$$\begin{aligned} f(x) &= x(x-2)(x-3i)(x+3i) \\ &= (x^2-2x)(x^2+9) \\ &= x^4 - 2x^3 + 9x^2 - 18x \end{aligned}$$

12. If $x = 2 + \sqrt{3}\,i$ is a zero, so is $x = 2 - \sqrt{3}\,i$

$$\begin{aligned} f(x) &= (x-1)(x-1)\left[x - \left(2 - \sqrt{3}i\right)\right]\left[x - \left(2 + \sqrt{3}i\right)\right] \\ &= \left(x^2 - 2x + 1\right)\left[(x-2) + \sqrt{3}i\right]\left[(x-2) - \sqrt{3}i\right] \\ &= \left(x^2 - 2x + 1\right)\left[(x-2)^2 - \left(\sqrt{3}i\right)^2\right] \\ &= \left(x^2 - 2x + 1\right)\left[\left(x^2 - 4x + 4\right) + 3\right] \\ &= \left(x^2 - 2x + 1\right)\left(x^2 - 4x + 7\right) \\ &= x^4 - 6x^3 + 16x^2 - 18x + 7 \end{aligned}$$

13. $f(x) = 3x^3 + 14x^2 - 7x - 10$

Possible rational zeros: $\pm1, \pm2, \pm5, \pm10, \pm\frac{1}{3}, \pm\frac{2}{3}, \pm\frac{5}{3}, \pm\frac{10}{3}$

$$\begin{array}{r|rrrr} 1 & 3 & 14 & -7 & -10 \\ & & 3 & 17 & 10 \\ \hline & 3 & 17 & 10 & 0 \end{array}$$

$$\begin{aligned} f(x) &= (x-1)\left(3x^2 + 17x + 10\right) \\ &= (x-1)(3x+2)(x+5) \end{aligned}$$

The zeros of $f(x)$ are $x = -5$, $x = -\frac{2}{3}$, and $x = 1$.

14. $f(x) = x^4 - 9x^2 - 22x - 24$

Possible rational zeros: $\pm1, \pm2, \pm3, \pm4, \pm6, \pm8, \pm12, \pm24$

$$\begin{array}{r|rrrrr} -2 & 1 & 0 & -9 & -22 & -24 \\ & & -2 & 4 & 10 & 24 \\ \hline & 1 & -2 & -5 & -12 & 0 \end{array}$$

$$\begin{array}{r|rrrr} 4 & 1 & -2 & -5 & -12 \\ & & 4 & 8 & 12 \\ \hline & 1 & 2 & 3 & 0 \end{array}$$

$$f(x) = (x+2)(x-4)\left(x^2 + 2x + 3\right)$$

By the Quadratic Formula the zeros of $x^2 + 2x + 3$ are $x = -1 \pm \sqrt{2}i$. The zeros of f are: $x = -2, 4, -1 \pm \sqrt{2}i$.

15. $v = k\sqrt{s}$

$24 = k\sqrt{16}$

$6 = k$

$v = 6\sqrt{s}$

16. $A = kxy$

$500 = k(15)(8)$

$500 = k(120)$

$\frac{25}{6} = k$

$A = \frac{25}{6}xy$

17. $b = \dfrac{k}{a}$

$32 = \dfrac{k}{1.5}$

$48 = k$

$b = \dfrac{48}{a}$

18. $y = -232.8t + 8890$; The model represents the data well.

Chapter Test Solutions for Chapter 4

1. $y = \dfrac{3x}{x+1}$

Domain: all real numbers x except $x = -1$

Vertical asymptote: $x = -1$

Horizontal asymptote: $y = 3$

2. $f(x) = \dfrac{3 - x^2}{3 + x^2} = \dfrac{-x^2 + 3}{x^2 + 3}$

Domain: all real numbers x

Vertical asymptote: None

Horizontal asymptote: $y = \dfrac{-1}{1} = -1$

3. $g(x) = \dfrac{x - 4}{x^2 - 9x + 20}$

$= \dfrac{x - 4}{(x - 4)(x - 5)}$

$= \dfrac{1}{x - 5}, x \neq 4$

Domain: all real numbers $x \neq 4, 5$

Vertical asymptote: $x = 5$

Horizontal asymptote: $y = 0$

4. $h(x) = \dfrac{3}{x^2} - 1 = \dfrac{3 - x^2}{x^2}$

x-intercepts: $(\pm\sqrt{3}, 0)$

Vertical asymptote: $x = 0$

Horizontal asymptote: $y = -1$

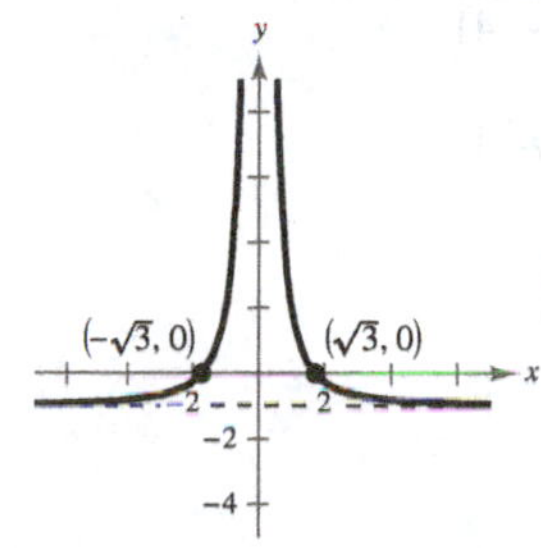

5. $g(x) = \dfrac{x^2 + 2}{x - 1} = x + 1 + \dfrac{3}{x - 1}$

y-intercept: $(0, -2)$

Vertical asymptote: $x = 1$

Slant asymptote: $y = x + 1$

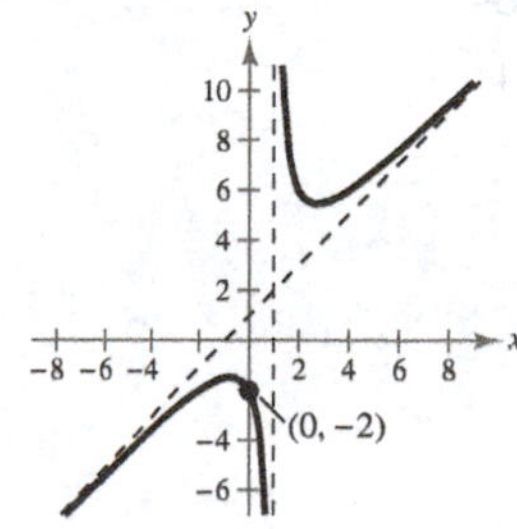

6. $f(x) = \dfrac{x+1}{x^2+x-12} = \dfrac{x+1}{(x+4)(x-3)}$

x-intercept: $(-1, 0)$

y-intercept: $\left(0, -\dfrac{1}{12}\right)$

Vertical asymptotes: $x = -4$, $x = 3$

Horizontal asymptote: $y = 0$

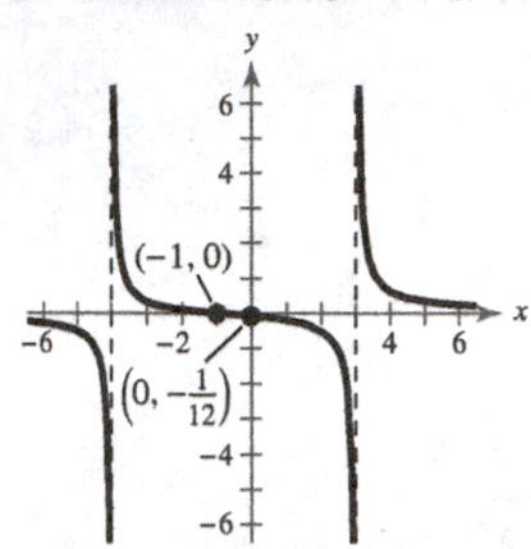

7. $f(x) = \dfrac{2x^2-5x-12}{x^2-16}$

$= \dfrac{(2x+3)(x-4)}{(x+4)(x-4)}$

$= \dfrac{2x+3}{x+4}, x \neq 4$

x-intercept: $\left(-\dfrac{3}{2}, 0\right)$

y-intercept: $\left(0, \dfrac{3}{4}\right)$

Vertical asymptote: $x = -4$

Horizontal asymptote: $y = 2$

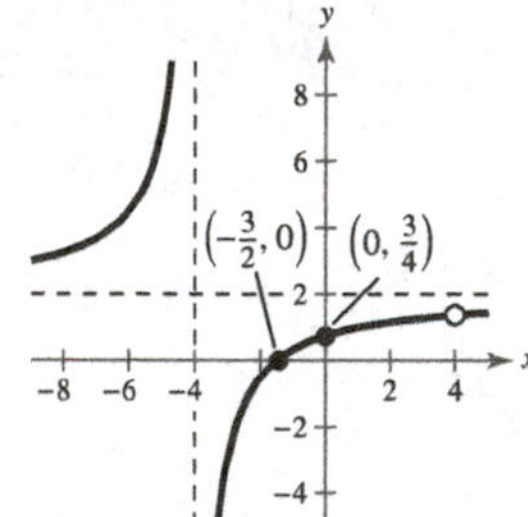

8. $f(x) = \dfrac{2x^2+9}{5x^2+9}$

y-intercept: $(0, 1)$

Horizontal asymptote: $y = \dfrac{2}{5}$

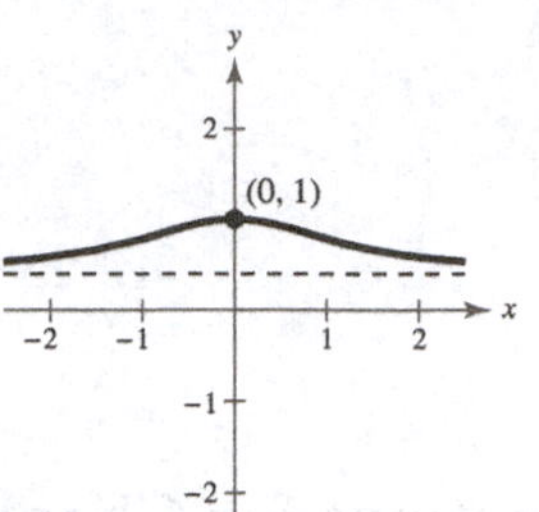

9. $g(x) = \dfrac{2x^3+3x^2-8x-12}{x^2-x-2}$

$= 2x + 5 + \dfrac{x-2}{(x-2)(x+1)}$

$= 2x + 5 + \dfrac{1}{x+1}$

$= \dfrac{2x^2+7x+6}{x+1}$

$= \dfrac{(2x+3)(x+2)}{x+1}, x \neq 2$

x-intercept: $\left(-\dfrac{3}{2}, 0\right), (-2, 0)$

y-intercept: $(0, 6)$

Vertical asymptote: $x = -1$

Slant asymptote: $y = 2x + 5$

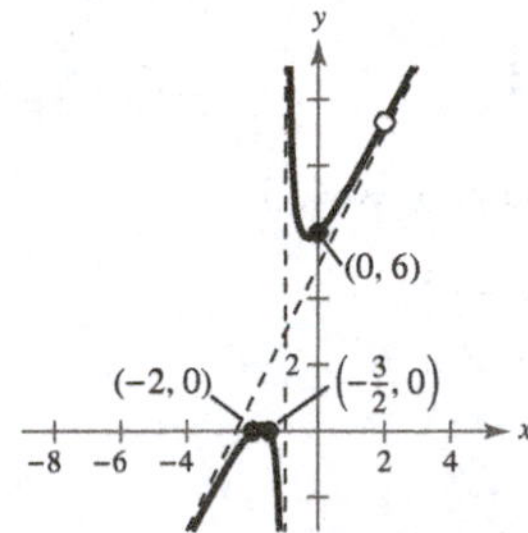

10.

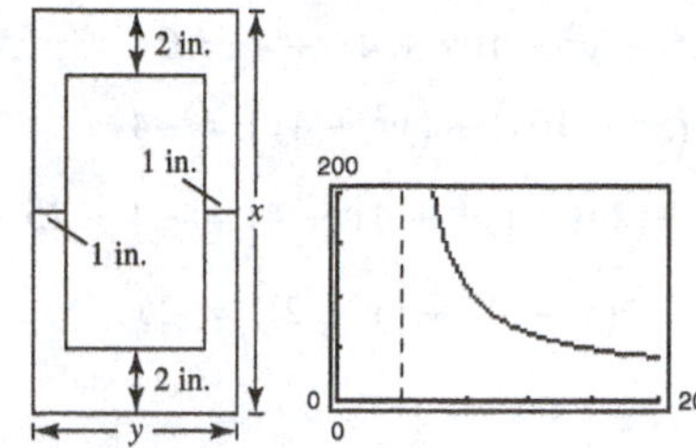

Minimize $A = xy$.

Given: $(x - 4)(y - 2) = 36$

$$y = \frac{36}{x - 4} + 2$$

$$A = x\left(\frac{36}{x - 4} + 2\right) = x\left(\frac{2x + 28}{x - 4}\right) = \frac{2x(x + 14)}{x - 4}$$

Domain: $x > 4$

From the graph of A we see that the minimum occurs when $x \approx 12.49$ inches and

$y = \frac{36}{x - 4} + 2 \approx 6.24$ inches. The dimensions are 6.24 inches by 12.49 inches.

Note: The exact values are $x = 4 + 6\sqrt{2}$ and $y = 2 + 3\sqrt{2}$.

11. (a) $A = \frac{1}{2}xy$

$$= \frac{1}{2}x\left[1 + \frac{2}{x - 2}\right]$$

$$= \frac{x}{2} + \frac{x}{x - 2}$$

$$= \frac{x^2}{2(x - 2)}$$

In context, we have $x > 2$ for the domain.

(b)

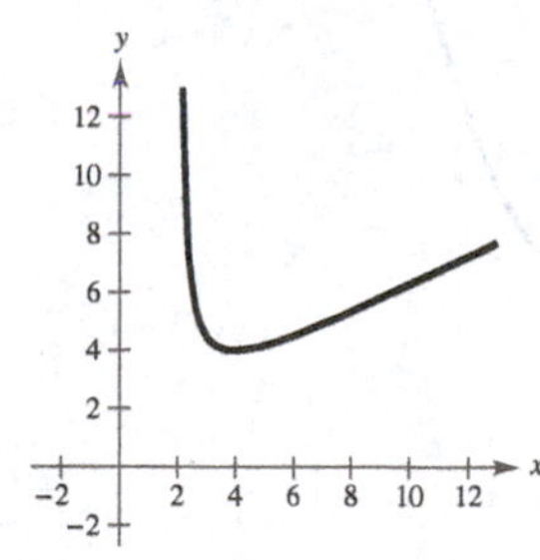

The minimum area occurs at $x = 4$ and is $A = 4$.

12. $\frac{x^2}{1} - \frac{y^2}{4} = 1$

Hyperbola

Center: $(0, 0)$

$a = 1, b = 2, c = \sqrt{5}$

Horizontal transverse axis

Vertices: $(\pm 1, 0)$

Foci: $(\pm\sqrt{5}, 0)$

Asymptotes: $y = \pm 2x$

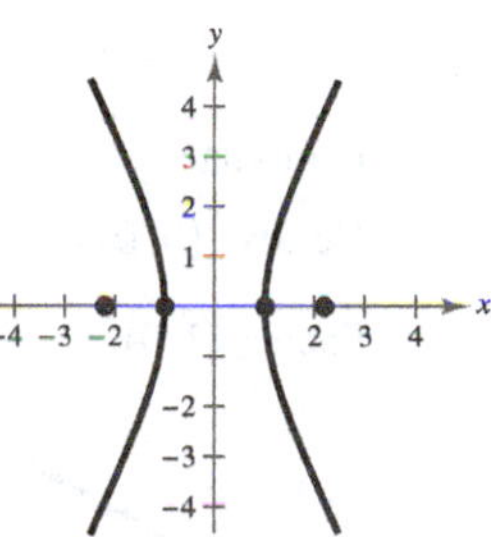

13. $4y^2 - 5x^2 = 80$

$$\frac{y^2}{20} - \frac{x^2}{16} = 1$$

Hyperbola

Center: $(0, 0)$

$a = 2\sqrt{5}, b = 4, c = 6$

Vertical transverse axis

Vertices: $\left(0, \pm 2\sqrt{5}\right)$

Foci: $(0, \pm 6)$

Asymptotes: $y = \pm\frac{\sqrt{5}}{2}x$

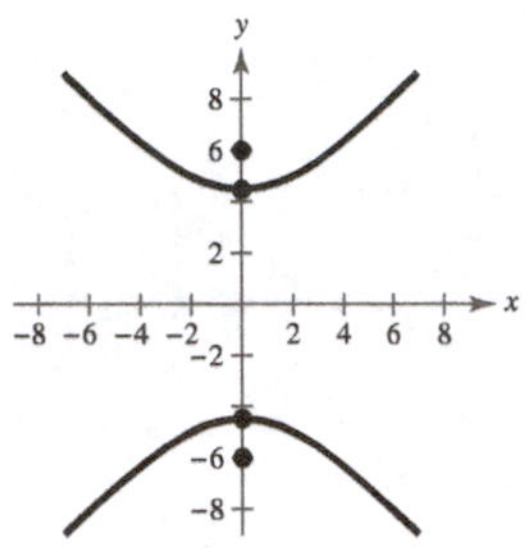

14. $y^2 - 4x = 0$

$$y^2 = 4x$$

$$y^2 = 4(1)x \Rightarrow p = 1$$

Parabola

Vertex: $(0, 0)$

Focus: $(1, 0)$

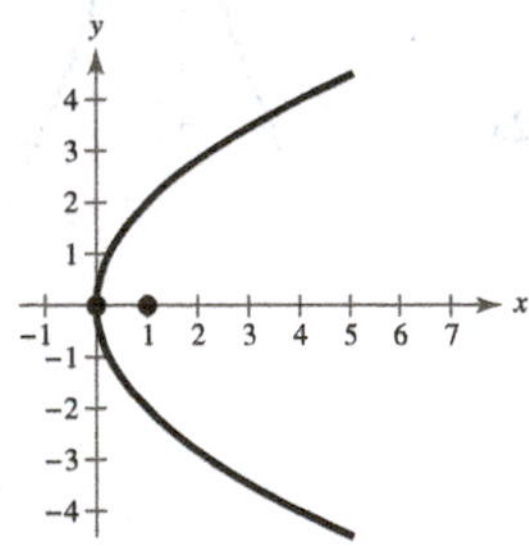

15.

$$x^2 + y^2 - 10x + 4y + 4 = 0$$

$$\left(x^2 - 10x\right) + \left(y^2 + 4y\right) = -4$$

$$\left(x^2 - 10x + 25\right) + \left(y^2 + 4y + 4\right) = -4 + 25 + 4$$

$$(x - 5)^2 + (y + 2)^2 = 25$$

Circle

Center: $(5, -2)$

Radius: 5

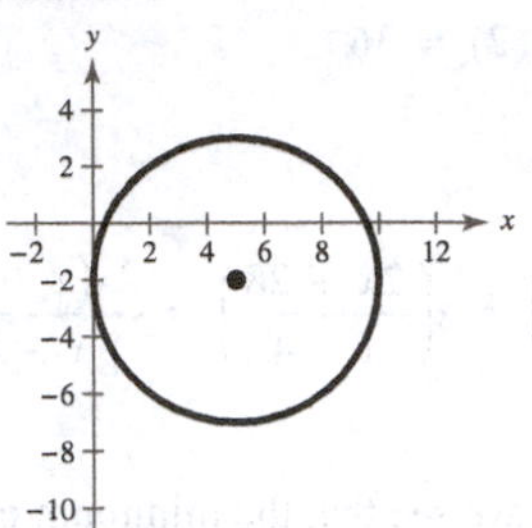

16. $x^2 - 10x - 2y + 19 = 0$

$$x^2 - 10x = 2y - 19$$

$$x^2 - 10x + 25 = 2y - 19 + 25$$

$$(x - 5)^2 = 2(y + 3)$$

Parabola

Vertex: $(5, -3)$

Focus: $\left(5, -\frac{5}{2}\right)$

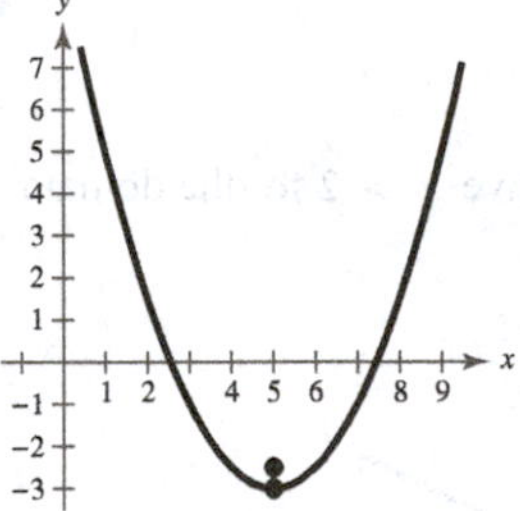

17. $$x^2 + 3y^2 - 2x + 36y + 100 = 0$$
$$(x^2 - 2x) + 3(y^2 + 12y) = -100$$
$$(x^2 - 2x + 1) + 3(y^2 + 12y + 36) = -100 + 1 + 108$$
$$(x - 1)^2 + 3(y + 6)^2 = 9$$
$$\frac{(x - 1)^2}{9} + \frac{(y + 6)^2}{3} = 1$$

Ellipse

Center: $(1, -6)$

$a = 3, b = \sqrt{3}, c = \sqrt{6}$

Horizontal major axis

Vertices: $(-2, -6), (4, -6)$

Foci: $\left(1 \pm \sqrt{6}, -6\right)$

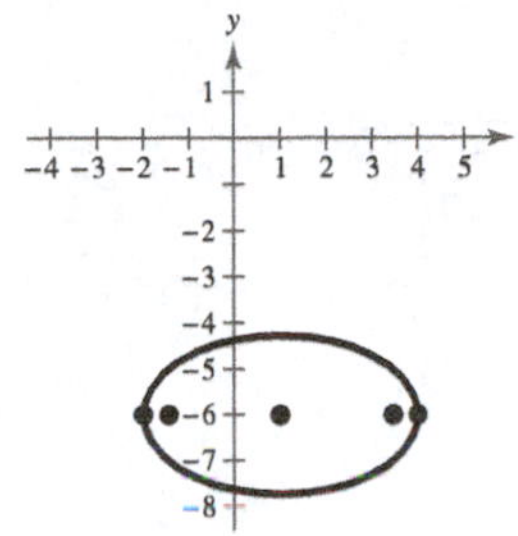

18. Vertices: $(0, 2)$ and $(8, 2)$

Center: $(4, 2)$

Horizontal major axis: $a = 4$

Minor axis of length 4: $2b = 4 \Rightarrow b = 2$

$$\frac{(x - 4)^2}{16} + \frac{(y - 2)^2}{4} = 1$$

$$a^2 = b^2 + c^2$$
$$c^2 = a^2 - b^2 = 16 - 4 = 12$$
$$c = \sqrt{12} = 2\sqrt{3}$$

Eccentricity: $e = \frac{c}{a} = \frac{2\sqrt{3}}{4} = \frac{\sqrt{3}}{2}$

19. Hyperbola

Vertices: $(0, \pm 3)$

Center: $(0, 0)$

Vertical transverse axis: $a = 3$

Asymptotes: $y = \pm\frac{3}{2}x$

$$\pm\frac{a}{b} = \pm\frac{3}{2} \Rightarrow b = 2$$

$$\frac{(y - k)^2}{a^2} - \frac{(x - h)^2}{b^2} = 1$$

$$\frac{y^2}{9} - \frac{x^2}{4} = 1$$

20.

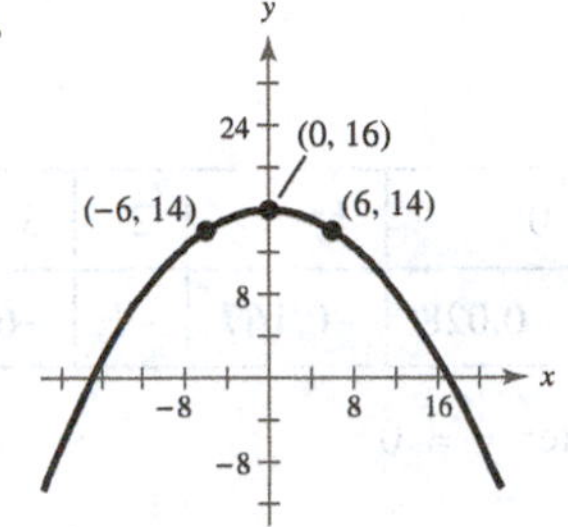

$$x^2 = 4p(y - 16)$$
$$36 = 4p(14 - 16)$$
$$-\tfrac{9}{2} = p$$
$$x^2 = -18(y - 16)$$

When $y = 0$: $x^2 = -18(-16) \Rightarrow x \approx 17 \Rightarrow 2x \approx 34$

At ground level, the archway is approximately 34 meters.

21. $a = \frac{1}{2}(768{,}900) = 384{,}450$

$b = \frac{1}{2}(767{,}746) = 383{,}873$

$c = \sqrt{384{,}450^2 - 383{,}873^2} \approx 21{,}055.2$

Least distance (perigee): $a - c \approx 363{,}395$ km

Greatest distance (apogee): $a + c \approx 405{,}505$ km

Chapter Test Solutions for Chapter 5

1. $0.7^{2.5} \approx 0.410$

2. $3^{-\pi} \approx 0.032$

3. $e^{-7/10} \approx 0.497$

4. $e^{3.1} \approx 22.198$

5. $f(x) = 10^{-x}$

x	-1	$-\frac{1}{2}$	0	$\frac{1}{2}$	1
$f(x)$	10	3.162	1	0.316	0.1

Horizontal asymptote: $y = 0$

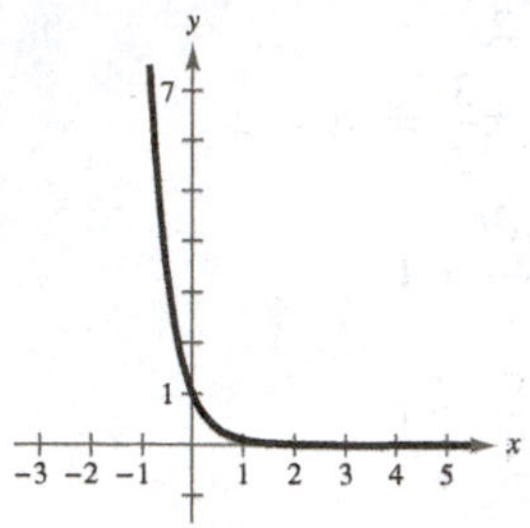

6. $f(x) = -6^{x-2}$

x	-1	0	1	2	3
$f(x)$	-0.005	-0.028	-0.167	-1	-6

Horizontal asymptote: $y = 0$

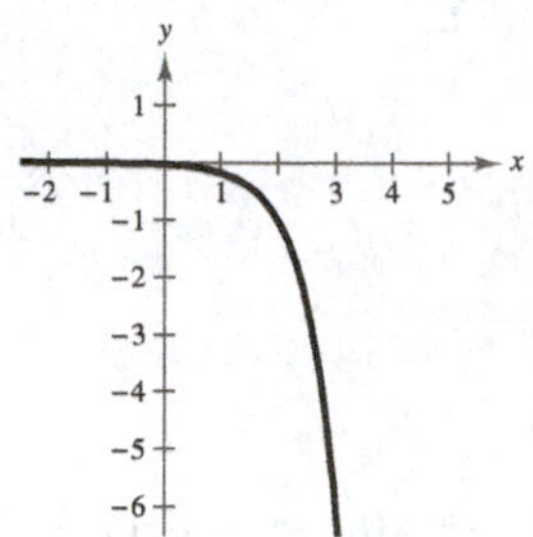

7. $f(x) = 1 - e^{2x}$

x	-1	$-\frac{1}{2}$	0	$\frac{1}{2}$	1
$f(x)$	0.865	0.632	0	-1.718	-6.389

Horizontal asymptote: $y = 1$

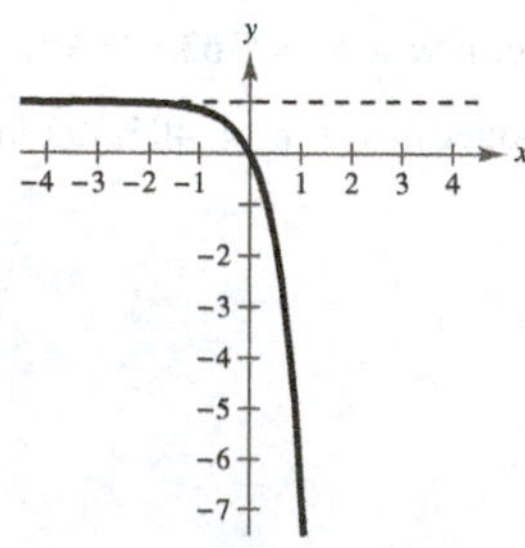

8. (a) $\log_7 7^{-0.89} = -0.89$

(b) $4.6 \ln e^2 = 4.6(2) = 9.2$

9. $f(x) = 4 + \log x$

Domain: $(0, \infty)$

x-intercept:

$$4 + \log x = 0$$
$$\log x = -4$$
$$10^{\log x} = 10^{-4}$$
$$x = 10^{-4}$$
$$(10^{-4}, 0) = (0.0001, 0)$$

Vertical asymptote: $x = 0$

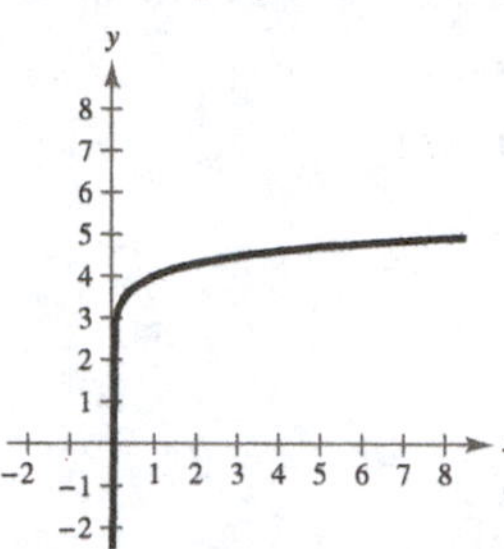

10. $f(x) = \ln(x - 4)$

Domain: $(4, \infty)$

$$\ln(x - 4) = 0$$

x-intercept: $e^{\ln(x-4)} = e^0$

$$x - 4 = 1$$
$$x = 5$$
$$(5, 0)$$

Vertical asymptote: $x = 4$

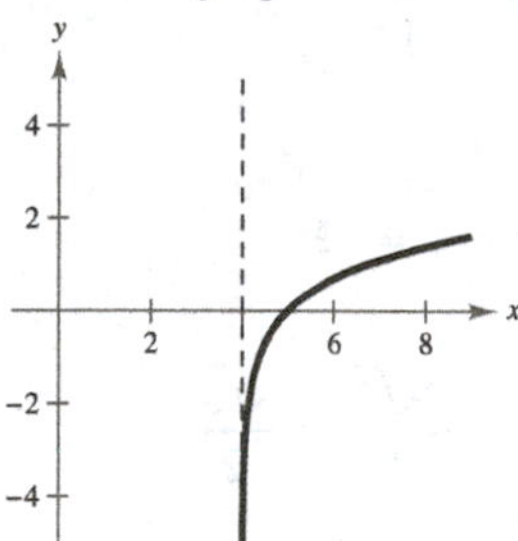

11. $f(x) = 1 + \ln(x + 6)$

Domain: $(-6, \infty)$

x-intercept:

$$\begin{aligned} 1 + \ln(x + 6) &= 0 \\ \ln(x + 6) &= -1 \\ e^{\ln(x+6)} &= e^{-1} \\ x + 6 &= e^{-1} \\ x &= -6 + e^{-1} \end{aligned}$$

$(-6 + e^{-1}, 0) \approx (-5.632, 0)$

Vertical asymptote: $x = -6$

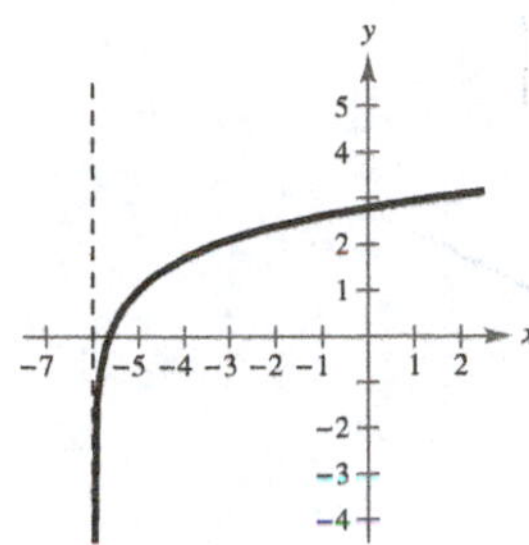

12. $\log_5 35 = \dfrac{\ln 35}{\ln 5} = \dfrac{\log 35}{\log 5} \approx 2.209$

13. $\log_{16} 0.63 = \dfrac{\log 0.63}{\log 16} \approx -0.167$

14. $\log_{3/4} 24 = \dfrac{\log 24}{\log(3/4)} \approx -11.047$

15. $\log_2 3a^4 = \log_2 3 + \log_2 a^4 = \log_2 3 + 4\log_2|a|$

16.
$$\begin{aligned} \ln\frac{\sqrt{x}}{7} &= \ln\left(\sqrt{x}\right) - \ln 7 \\ &= \ln\sqrt{x} - \ln 7 \\ &= \ln x^{1/2} - \ln 7 \\ &= \frac{1}{2}\ln x - \ln 7 \end{aligned}$$

17.
$$\begin{aligned} \ln\frac{10x^2}{y^3} &= \ln\left(10x^2\right) - \ln y^3 \\ &= \ln 10 + \ln x^2 - \ln y^3 \\ &= \ln 10 + 2\ln x - 3\ln y \end{aligned}$$

18. $\log_3 13 + \log_3 y = \log_3 13y$

19. $4\ln x - 4\ln y = \ln x^4 - \ln y^4 = \ln\dfrac{x^4}{y^4}$

20. $3\ln x - \ln(x + 3) + 2\ln y = \ln x^3 - \ln(x + 3) + \ln y^2 = \ln\dfrac{x^3y^2}{x + 3}$

21.
$$\begin{aligned} 5^x &= \tfrac{1}{25} \\ 5^x &= 5^{-2} \\ x &= -2 \end{aligned}$$

22.
$$\begin{aligned} 3e^{-5x} &= 132 \\ e^{-5x} &= 44 \\ -5x &= \ln 44 \\ x &= \frac{\ln 44}{-5} \approx -0.757 \end{aligned}$$

23.
$$\begin{aligned} \frac{1025}{8 + e^{4x}} &= 5 \\ 1025 &= 5\left(8 + e^{4x}\right) \\ 205 &= 8 + e^{4x} \\ 197 &= e^{4x} \\ \ln 197 &= 4x \\ x &= \frac{\ln 197}{4} \approx 1.321 \end{aligned}$$

24.
$$\begin{aligned} \ln x &= \tfrac{1}{2} \\ x &= e^{1/2} \approx 1.649 \end{aligned}$$

25.
$$\begin{aligned} 18 + 4\ln x &= 7 \\ 4\ln x &= -11 \\ \ln x &= -\tfrac{11}{4} \\ x &= e^{-11/4} \approx 0.0639 \end{aligned}$$

26.
$$\begin{aligned} \log x + \log(x - 15) &= 2 \\ \log\left[x(x - 15)\right] &= 2 \\ x(x - 15) &= 10^2 \\ x^2 - 15x - 100 &= 0 \\ (x - 20)(x + 5) &= 0 \end{aligned}$$

$x - 20 = 0$ or $x + 5 = 0$

$x = 20$ $\qquad$ $x = -5$

The value $x = -5$ is extraneous. The only solution is $x = 20$.

27. $y = ae^{bt}$

$(0, 2745)$: $2745 = ae^{b(0)} \Rightarrow a = 2745$

$$y = 2745e^{bt}$$

$(9, 11{,}277)$: $11{,}277 = 2745e^{b(9)}$

$$\frac{11{,}277}{2745} = e^{9b}$$

$$\ln\left(\frac{11{,}277}{2745}\right) = 9b$$

$$\frac{1}{9}\ln\left(\frac{11{,}277}{2745}\right) = b \Rightarrow b \approx 0.1570$$

So, $y = 2745e^{0.1570t}$.

28. $y = ae^{bt}$

$$\frac{1}{2}a = ae^{b(21.77)}$$

$$\frac{1}{2} = e^{21.77b}$$

$$\ln\left(\frac{1}{2}\right) = 21.77b$$

$$b = \frac{\ln(1/2)}{21.77} \approx -0.0318$$

$$y = ae^{-0.0318t}$$

When $t = 19$: $y = ae^{-0.0318(19)} \approx 0.55a$

So, 55% will remain after 19 years.

29. $H = 70.228 + 5.104x + 9.222 \ln x, \frac{1}{4} \le x \le 6$

(a)

x	H (cm)
$\frac{1}{4}$	58.720
$\frac{1}{2}$	66.388
1	75.332
2	86.828
3	95.671
4	103.43
5	110.59
6	117.38

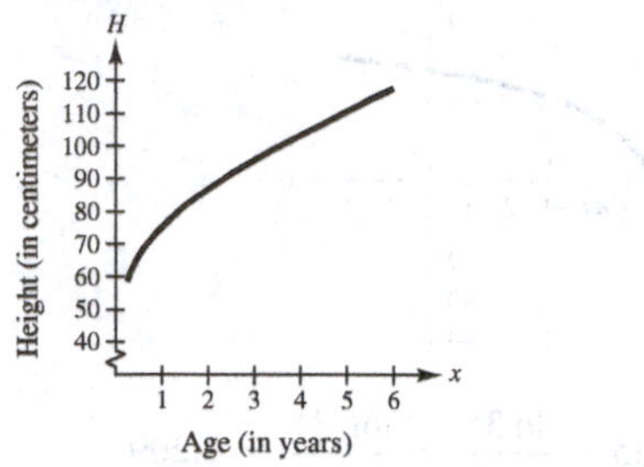

(b) Estimate: 103

When $x = 4$, $H \approx 103.43$ cm.

Cumulative Test Solutions for Chapters 3–5

1. Vertex: $(-8, 5)$

Point: $(-4, -7)$

$$y - k = a(x - h)^2$$

$$y - 5 = a(x + 8)^2$$

$$-7 - 5 = a(-4 + 8)^2$$

$$-12 = 16a$$

$$-\tfrac{3}{4} = a$$

$$y = -\tfrac{3}{4}(x + 8)^2 + 5$$

2. $h(x) = -x^2 + 10x - 21$

$$= -(x^2 - 10x) - 21$$

$$= -(x^2 - 10x + 25 - 25) - 21$$

$$= -(x^2 - 10x + 25) + 25 - 21$$

$$= -(x - 5)^2 + 4$$

Parabola

Vertex: $(5, 4)$

$$-(x - 5)^2 + 4 = 0$$

$$4 = (x - 5)^2$$

$$\pm 2 = x - 5$$

$$x = 7, 3$$

Intercepts: $(3, 0)$, $(7, 0)$

3. $f(t) = -\frac{1}{2}(t-1)^2(t+2)^2$

4th degree polynomial,

Falls to the left

Falls to the right

$$-\frac{1}{2}(t-1)^2(t+2)^2 = 0$$

$$(t-1)^2 = 0 \rightarrow t = 1$$

$$(t+2)^2 = 0 \rightarrow t = -2$$

zeros are of even multiplicity

x-intercepts: $(1, 0), (-2, 0)$

$$f(0) = -\frac{1}{2}(0-1)^2(0+2)^2$$
$$= -\frac{1}{2}(-1)^2(2)^2 = -2$$

y-intercept: $(0, -2)$

4. $g(s) = s^3 - 3s^2$

Cubic

Falls to the left,

Rises to the right

$$s^3 - 3x^2 = 0$$
$$s^2(s-3) = 0$$
$$s = 0 \text{ (even multiplicity)}$$
$$s - 3 = 0 \rightarrow s = 3 \text{ (odd multiplicity)}$$

Intercepts: $(0, 0), (3, 0)$

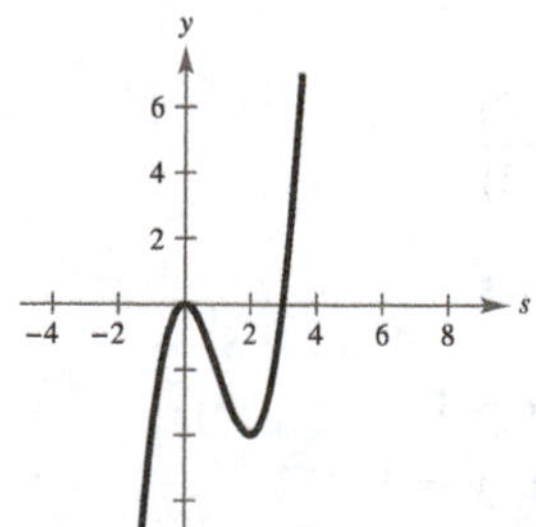

5. $f(x) = x^3 + 2x^2 + 4x + 8$

$$= x^2(x+2) + 4(x+2)$$
$$= (x+2)(x^2+4)$$

$$x + 2 = 0 \Rightarrow x = -2$$
$$x^2 + 4 = 0 \Rightarrow x = \pm 2i$$

The zeros of $f(x)$ are -2 and $\pm 2i$.

6. $f(x) = x^4 + 4x^3 - 21x^2$

$$= x^2(x^2 + 4x - 21)$$
$$= x^2(x+7)(x-3)$$

The zeros of $f(x)$ are $0, -7,$ and 3.

7.
$$\begin{array}{r}
3x - 2 + \dfrac{-3x+2}{2x^2+1} \\
2x^2 + 0x + 1\overline{\big)\,6x^3 - 4x^2 + 0x + 0} \\
\underline{6x^3 + 0x^2 + 3x} \\
-4x^2 - 3x + 0 \\
\underline{-4x^2 + 0x - 2} \\
-3x + 2
\end{array}$$

Thus, $\dfrac{6x^3 - 4x^2}{2x^2 + 1} = 3x - 2 - \dfrac{3x - 2}{2x^2 + 1}$.

8.

2	3	0	2	–5	3
		6	12	28	46
	3	6	14	23	49

Thus, $\dfrac{3x^4 + 2x^2 - 5x + 3}{x - 2} = 3x^3 + 6x^2 + 14x + 23 + \dfrac{49}{x - 2}$.

9. $g(x) = x^3 + 3x^2 - 6$

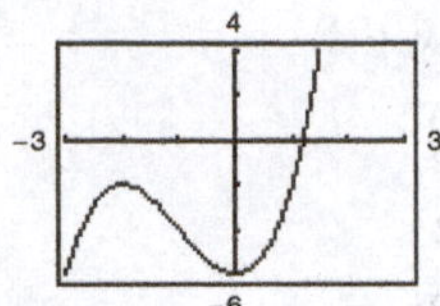

$x \approx 1.196$

10. Because $2 + \sqrt{3}i$ is a zero, so is $2 - \sqrt{3}i$.

$$\begin{aligned} f(x) &= (x + 5)(x + 2)\left[x - \left(2 + \sqrt{3}i\right)\right]\left[x - \left(2 - \sqrt{3}i\right)\right] \\ &= \left(x^2 + 7x + 10\right)\left[(x - 2) - \sqrt{3}i\right]\left[(x - 2) + \sqrt{3}i\right] \\ &= \left(x^2 + 7x + 10\right)\left[(x - 2)^2 + 3\right] \\ &= \left(x^2 + 7x + 10\right)\left(x^2 - 4x + 7\right) \\ &= x^4 + 3x^3 - 11x^2 + 9x + 70 \end{aligned}$$

11. $f(x) = \dfrac{2x}{x - 3}$

Domain: all real numbers x except $x = 3$

Vertical asymptote: $x = 3$

Horizontal asymptote: $y = 2$

Intercept: $(0, 0)$

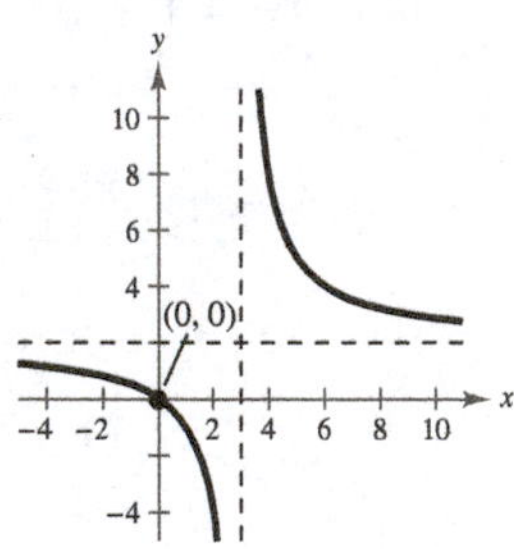

12. $f(x) = \dfrac{4x^2}{x - 5} = 4x + 20 + \dfrac{100}{x - 5}$

Domain: all real numbers x except $x = 5$

Vertical asymptote: $x = 5$

Slant asymptote: $y = 4x + 20$

Intercept: $(0, 0)$

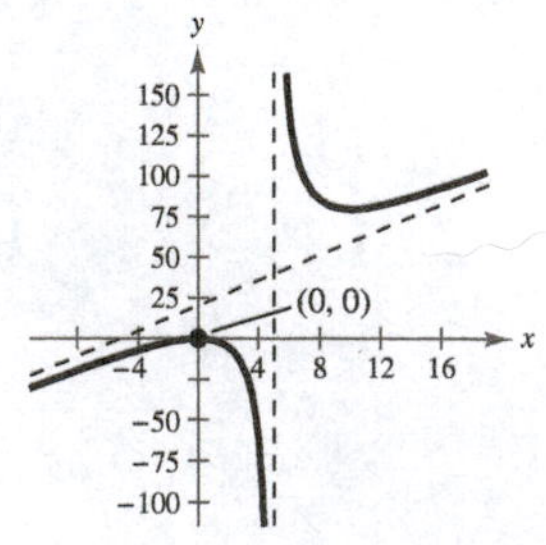

13. $f(x) = \dfrac{2x}{x^2 + 2x - 3}$

$= \dfrac{2x}{(x + 3)(x - 1)}$

Intercept: $(0, 0)$

Vertical asymptotes: $x = -3, x = 1$

Horizontal asymptote: $y = 0$

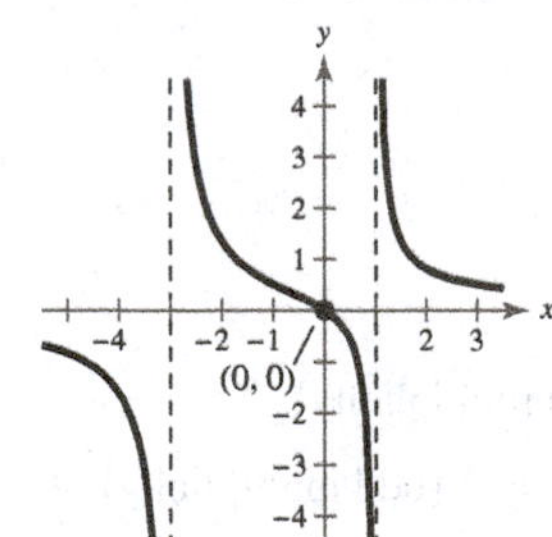

14. $f(x) = \dfrac{x^2 - 4}{x^2 + x - 2}$

$= \dfrac{(x + 2)(x - 2)}{(x + 2)(x - 1)}$

$= \dfrac{x - 2}{x - 1}, x \neq -2$

Vertical asymptote: $x = 1$

Horizontal asymptote: $y = 1$

x-intercept: $(2, 0)$

y-intercept: $(0, 2)$

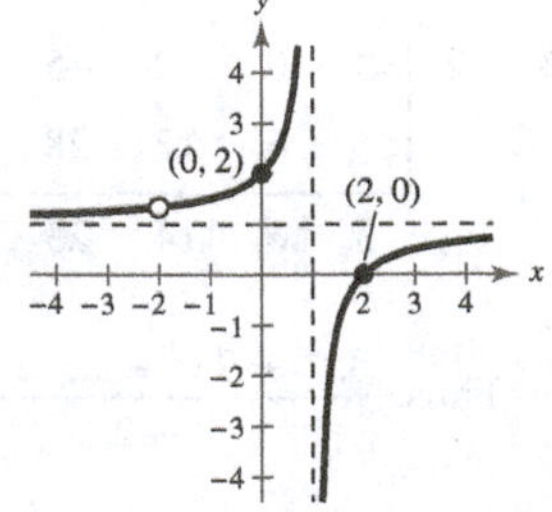

15. $f(x) = \dfrac{x^3 - 2x^2 - 9x + 18}{x^2 + 4x + 3}$

$= \dfrac{x^2(x-2) - 9(x-2)}{(x+1)(x+3)}$

$= \dfrac{(x-2)(x^2-9)}{(x+1)(x+3)}$

$= \dfrac{(x-2)(x+3)(x-3)}{(x+1)(x+3)}$

$= \dfrac{(x-2)(x-3)}{x+1}$

$= \dfrac{x^2 - 5x + 6}{x+1}$

$= x - 6 + \dfrac{12}{x+1}, x \neq -3$

Vertical asymptote: $x = -1$

Slant asymptote: $y = x - 6$

x-intercepts: $(2, 0), (3, 0)$

y-intercept: $(0, 6)$

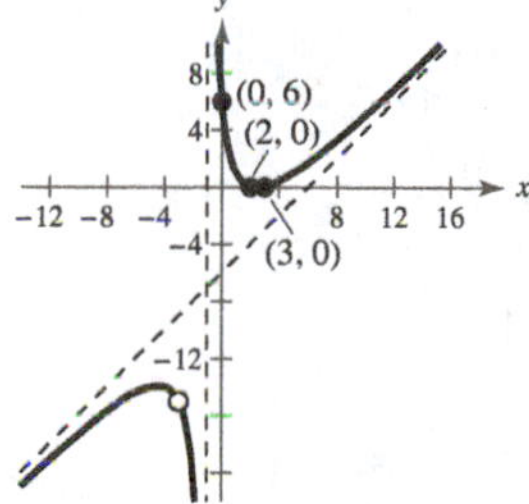

16. $\dfrac{(x+3)^2}{16} - \dfrac{(y+4)^2}{25} = 1$

Hyperbola

Center: $(-3, -4)$

Vertices: $(-7, -4), (1, -4)$

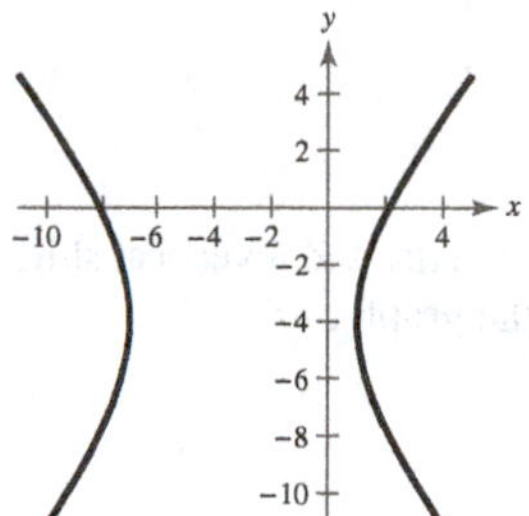

17. $\dfrac{(x-2)^2}{4} + \dfrac{(y+1)^2}{9} = 1$

Ellipse

Center: $(2, -1)$

Vertices: $(2, -4), (2, 2)$

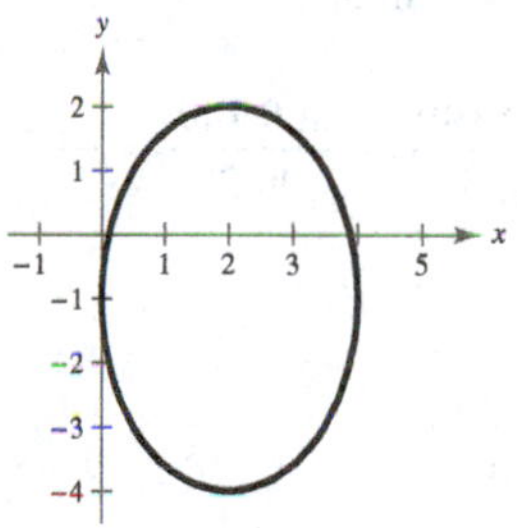

18. Parabola

Vertex: $(3, -2) \Rightarrow y = a(x-3)^2 - 2$

Point: $(0, 4) \Rightarrow 4 = a(0-3)^2 - 2$

$6 = 9a \Rightarrow a = \frac{2}{3}$

Equation: $y = \frac{2}{3}(x-3)^2 - 2$

$y + 2 = \frac{2}{3}(x-3)^2$

$\frac{3}{2}(y+2) = (x-3)^2$

$(x-3)^2 = \frac{3}{2}(y+2)$

19. Hyperbola

Vertices: $(-1, -5)$ and $(-1, 1) \Rightarrow$ Center: $(-1, -2) \Rightarrow$ vertical transverse axis and $a = 3$

Foci: $(-1, -7)$ and $(-1, 3) \Rightarrow c = 5$

$c^2 = a^2 + b^2 \Rightarrow 25 = 9 + b^2 \Rightarrow b^2 = 16$

Equation: $\dfrac{(y+2)^2}{9} - \dfrac{(x+1)^2}{16} = 1$

20. $f(x) = \left(\frac{2}{5}\right)^x$

$g(x) = -\left(\frac{2}{5}\right)^{-x+3}$

g is a reflection in the x-axis, a reflection in the y-axis, and a horizontal shift three units to the right of the graph of f.

21. $f(x) = 2.2^x$

$g(x) = -2.2^x + 4$

g is a reflection in the x-axis, and a vertical shift four units upward of the graph of f.

22. $\log 98 \approx 1.991$

23. $\log\left(\frac{6}{7}\right) \approx -0.067$

24. $\ln\sqrt{31} \approx 1.717$

25. $\ln\left(\sqrt{30} - 4\right) \approx 0.390$

26. $\log_5 4.3 = \dfrac{\log_{10} 4.3}{\log_{10} 5} = \dfrac{\ln 4.3}{\ln 5} \approx 0.906$

27. $\log_3 0.149 = \dfrac{\log_{10} 0.149}{\log_{10} 3} = \dfrac{\ln 0.149}{\ln 3} \approx -1.733$

28. $\log_{1/2} 17 = \dfrac{\log_{10} 17}{\log_{10}(1/2)} = \dfrac{\ln 17}{\ln(1/2)} \approx -4.087$

29.
$$\begin{aligned}\ln\left(\frac{x^2 - 25}{x^4}\right) &= \ln\left(x^2 - 25\right) - \ln x^4 \\ &= \ln\left[(x + 5)(x - 5)\right] - \ln x^4 \\ &= \ln(x + 5) + \ln(x - 5) - 4\ln x, x > 5\end{aligned}$$

30.
$$\begin{aligned}2\ln x - \frac{1}{2}\ln(x + 5) &= \ln x^2 - \ln\sqrt{x + 5} \\ &= \ln\frac{x^2}{\sqrt{x + 5}}, x > 0\end{aligned}$$

31.
$$\begin{aligned}6e^{2x} &= 72 \\ e^{2x} &= 12 \\ 2x &= \ln 12 \\ x &= \frac{\ln 12}{2} \approx 1.242\end{aligned}$$

32.
$$\begin{aligned}4^{x-5} + 21 &= 30 \\ 4^{x-5} &= 9 \\ x - 5 &= \log_4 9 \\ x &= 5 + \log_4 9 \\ x &= 5 + \frac{\ln 9}{\ln 4} \\ x &\approx 6.585\end{aligned}$$

33.
$$\begin{aligned}e^{2x} - 13e^x + 42 &= 0 \\ \left(e^x - 6\right)\left(e^x - 7\right) &= 0\end{aligned}$$

$$e^x - 6 = 0 \Rightarrow e^x = 6 \Rightarrow x = \ln 6 \approx 1.792$$

$$e^x - 7 = 0 \Rightarrow e^x = 7 \Rightarrow x = \ln 7 \approx 1.946$$

34.
$$\begin{aligned}\log_2 x + \log_2 5 &= 6 \\ \log_2 5x &= 6 \\ 5x &= 2^6 \\ x &= \tfrac{64}{5} = 12.8\end{aligned}$$

35.
$$\begin{aligned}\ln 4x - \ln 2 &= 8 \\ \ln\frac{4x}{2} &= 8 \\ \ln 2x &= 8 \\ 2x &= e^8 \\ x &= \frac{e^8}{2} \approx 1490.479\end{aligned}$$

36.
$$\begin{aligned}\ln\sqrt{x + 2} &= 3 \\ \tfrac{1}{2}\ln(x + 2) &= 3 \\ \ln(x + 2) &= 6 \\ x + 2 &= e^6 \\ x &= e^6 - 2 \approx 401.429\end{aligned}$$

37. $A = 2500e^{(0.075)(25)} \approx \$16,302.05$

38.
$$\begin{aligned} N &= 175e^{kt} \\ 420 &= 175e^{k(8)} \\ 2.4 &= e^{8k} \\ \ln 2.4 &= 8k \\ \frac{\ln 2.4}{8} &= k \\ k &\approx 0.1094 \\ N &= 175e^{0.1094t} \\ 350 &= 175e^{0.1094t} \\ 2 &= e^{0.1094t} \\ \ln 2 &= 0.1094t \\ t &= \frac{\ln 2}{0.1094} \approx 6.3 \text{ hours to double} \end{aligned}$$

39. Let $P = 32$ and solve for t.

$$\begin{aligned} 32 &= 20.913e^{0.0184t} \\ \frac{32}{20.913} &= e^{0.0184t} \\ 1.530 &\approx e^{0.0184t} \\ \ln 1.530 &\approx \ln e^{0.0184t} \\ 0.4253 &\approx 0.0184t \\ 23.1 &\approx t \end{aligned}$$

According to the model, the population of Texas will reach 32 million during 2023.

40. $p = \dfrac{1200}{1 + 3e^{-t/5}}$

(a) $p(0) = \dfrac{1200}{1 + 3e^{0}} = \dfrac{1200}{4} = 300$ birds

(b) $p(5) = \dfrac{1200}{1 + 3e^{-1}} \approx 570$ birds

(c)
$$\begin{aligned} 800 &= \frac{1200}{1 + 3e^{-t/5}} \\ 800\left(1 + 3e^{-t/5}\right) &= 1200 \\ 1 + 3e^{-t/5} &= 1.5 \\ 3e^{-t/5} &= 0.5 \\ e^{-t/5} &= \frac{1}{6} \\ -\frac{t}{5} &= \ln\left(\frac{1}{6}\right) \\ t &= -5\ln\left(\frac{1}{6}\right) \approx 9 \text{ years} \end{aligned}$$

Chapter Test Solutions for Chapter 6

1. $\begin{cases} x + y = -9 \Rightarrow x = -y - 9 \\ 5x - 8y = 20 \end{cases}$

$$\begin{aligned} 5(-y - 9) - 8y &= 20 \\ -13y &= 65 \\ y &= -5 \end{aligned}$$

$x - 5 = -9 \Rightarrow x = -4$

Solution: $(-4, -5)$

2. $\begin{cases} y = x - 1 \\ y = (x - 1)^3 \end{cases}$

$$\begin{aligned} x - 1 &= (x - 1)^3 \\ x - 1 &= x^3 - 3x^2 + 3x - 1 \\ 0 &= x^3 - 3x^2 + 2x \\ 0 &= x(x - 1)(x - 2) \end{aligned}$$

$x = 0$ or $x = 1$ or $x = 2$

$y = -1 \quad y = 0 \quad y = 1$

Solutions: $(0, -1), (1, 0), (2, 1)$

3. $\begin{cases} x - y = 4 \Rightarrow x = y + 4 \\ 2x - y^2 = 0 \Rightarrow 2(y + 4) - y^2 = 0 \end{cases}$

$$\begin{aligned} 0 &= y^2 - 2y - 8 \\ 0 &= (y + 2)(y - 4) \end{aligned}$$

$y = -2$ or $y = 4$

$x = 2 \quad x = 8$

Solutions: $(2, -2), (8, 4)$

4. $\begin{cases} 3x - 6y = 0 \Rightarrow y = \frac{1}{2}x \\ 2x + 5y = 18 \Rightarrow y = -\frac{2}{5}x + \frac{18}{5} \end{cases}$

Solution: $(4, 2)$

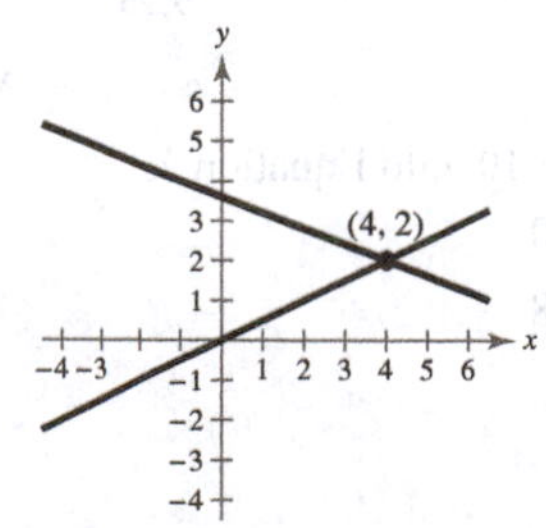

5. 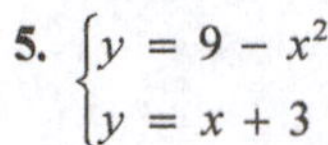

$$\begin{cases} y = 9 - x^2 \\ y = x + 3 \end{cases}$$

Solutions: $(-3, 0), (2, 5)$

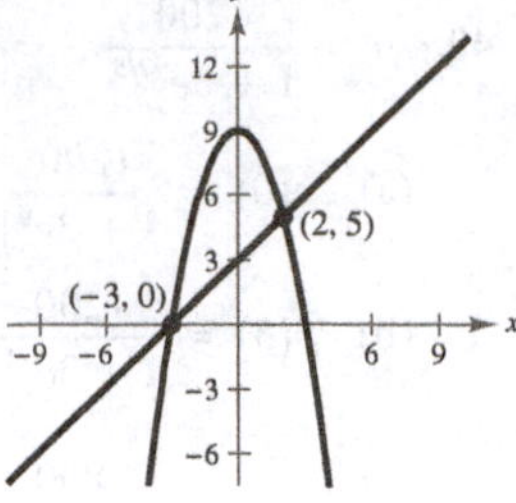

6. $$\begin{cases} y - \ln x = 4 \Rightarrow y = \ln x + 4 \\ 7x - 2y - 5 = -6 \Rightarrow y = \dfrac{7}{2}x + \dfrac{1}{2} \end{cases}$$

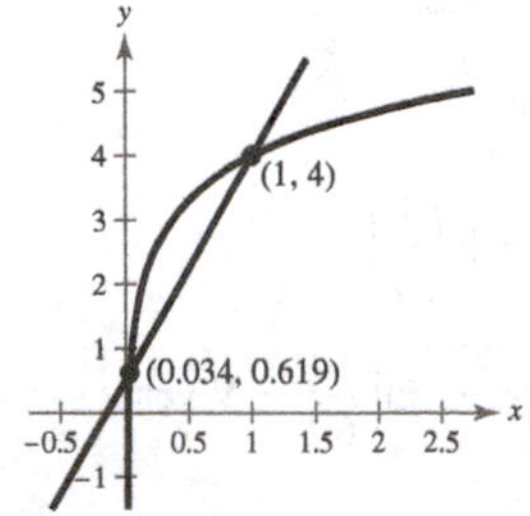

Solutions: $(1, 4)$ and $\approx (0.034, 0.619)$

7. $$\begin{cases} 3x + 4y = -26 & \text{Equation 1} \\ 7x - 5y = 11 & \text{Equation 2} \end{cases}$$

Multiply Equation 1 by 5: $15x + 20y = -130$

Multiply Equation 2 by 4: $28x - 20y = 44$

Add the equations to eliminate y:
$$\begin{array}{rcr} 15x + 20y & = & -130 \\ 28x - 20y & = & 44 \\ \hline 43x & = & -86 \\ x & = & -2 \end{array}$$

Back-substitute $x = -2$ into Equation 1:

$3(-2) + 4y = -26$

$y = -5$

Solution: $(-2, -5)$

8. $$\begin{cases} 1.4x - y = 17 & \text{Equation 1} \\ 0.8x + 6y = -10 & \text{Equation 2} \end{cases}$$

Multiply Equation 1 by 6: $8.4x - 6y = 102$

Add this to Equation 2 to eliminate y:
$$\begin{array}{rcr} 8.4x - 6y & = & 102 \\ 0.8x + 6y & = & -10 \\ \hline 9.2x & = & 92 \\ x & = & 10 \end{array}$$

Back-substitute $x = 10$ into Equation 2:

$0.8(10) + 6y = -10$

$6y = -18$

$y = -3$

Solution: $(10, -3)$

9. $$\begin{cases} x - 2y + 3z = 11 \\ 2x - z = 3 \\ 3y + z = -8 \end{cases}$$

$$\begin{cases} x - 2y + 3z = 11 \\ 4y - 7z = -19 & -2\text{Eq.}1 + \text{Eq.}2 \\ 3y + z = -8 \end{cases}$$

$$\begin{cases} x - 2y + 3z = 11 \\ y - 8z = -11 & -\text{Eq.}3 + \text{Eq.}2 \\ 3y + z = -8 \end{cases}$$

$$\begin{cases} x - 2y + 3z = 11 \\ y - 8z = -11 \\ 25z = 25 & -3\text{Eq.}2 + \text{Eq.}3 \end{cases}$$

$$\begin{cases} x - 2y + 3z = 11 \\ y - 8z = -11 \\ z = 1 & \frac{1}{25}\text{Eq.}3 \end{cases}$$

$y - 8(1) = -11 \Rightarrow y = -3$

$x - 2(-3) + 3(1) = 11 \Rightarrow x = 2$

Solution: $(2, -3, 1)$

10. $$\begin{cases} 3x + 2y + z = 17 & \text{Equation 1} \\ -x + y + z = 4 & \text{Equation 2} \\ x - y - z = 3 & \text{Equation 3} \end{cases}$$

Interchange Equations 1 and 3.

$$\begin{cases} x - y - z = 3 \\ -x + y + z = 4 \\ 3x + 2y + z = 17 \end{cases}$$

$$\begin{cases} x - y - z = 3 \\ 0 \neq 7 & \text{Eq. } 1 + \text{Eq. } 2 \\ 3x + 2y + z = 17 \end{cases}$$

Inconsistent

No solution

11. $$\frac{2x + 5}{x^2 - x - 2} = \frac{2x + 5}{(x - 2)(x + 1)} = \frac{A}{x - 2} + \frac{B}{x + 1}$$

$2x + 5 = A(x + 1) + B(x - 2)$

Let $x = 2$: $9 = 3A \Rightarrow A = 3$

Let $x = -1$: $3 = -3B \Rightarrow B = -1$

$$\frac{2x + 5}{x^2 - x - 2} = \frac{3}{x - 2} - \frac{1}{x + 1}$$

12. $\dfrac{3x^2 - 2x + 4}{x^2(2 - x)} = \dfrac{A}{x} + \dfrac{B}{x^2} + \dfrac{C}{2 - x}$

$3x^2 - 2x + 4 = Ax(2 - x) + B(2 - x) + Cx^2$

Let $x = 0$: $4 = 2B \Rightarrow B = 2$

Let $x = 2$: $12 = 4C \Rightarrow C = 3$

Let $x = 1$: $5 = A + B + C = A + 2 + 3 \Rightarrow A = 0$

$\dfrac{3x^2 - 2x + 4}{x^2(2 - x)} = \dfrac{2}{x^2} + \dfrac{3}{2 - x}$

13. $\dfrac{x^4 + 5}{x^3 - x} = x + \dfrac{x^2 + 5}{x^3 - x}$, use long division first to create a proper fraction.

$\dfrac{x^2 + 5}{x^3 - x} = \dfrac{x^2 + 5}{x(x + 1)(x - 1)} = \dfrac{A}{x} + \dfrac{B}{x - 1} + \dfrac{c}{x - 1}$

$x^2 + 5 = A(x + 1)(x - 1) + Bx(x - 1) + Cx(x + 1)$

Let $x = 0$: $5 = -A \Rightarrow A = -5$

Let $x = -1$: $6 = 2B \Rightarrow B = 3$

Let $x = 1$: $6 = 2C \Rightarrow C = 3$

$\dfrac{x^4 + 5}{x^3 - x} = x - \dfrac{5}{x} + \dfrac{3}{x + 1} + \dfrac{3}{x - 1}$

14. $\dfrac{x^2 - 4}{x^3 + 2x} = \dfrac{x^2 - 4}{x(x^2 + 2)} = \dfrac{A}{x} + \dfrac{Bx + C}{x^2 + 2}$

$x^2 - 4 = A(x^2 + 2) + (Bx + C)x$

$= Ax^2 + 2A + Bx^2 + Cx$

$= (A + B)x^2 + Cx + 2A$

Equate the coefficients of like terms:

$1 = A + B, 0 = C, -4 = 2A$

So, $A = -2, B = 3, C = 0$.

$\dfrac{x^2 - 4}{x^3 + 2x} = -\dfrac{2}{x} + \dfrac{3x}{x^2 + 2}$

15. $\begin{cases} 2x + y \le 4 \\ 2x - y \ge 0 \\ x \ge 0 \end{cases}$

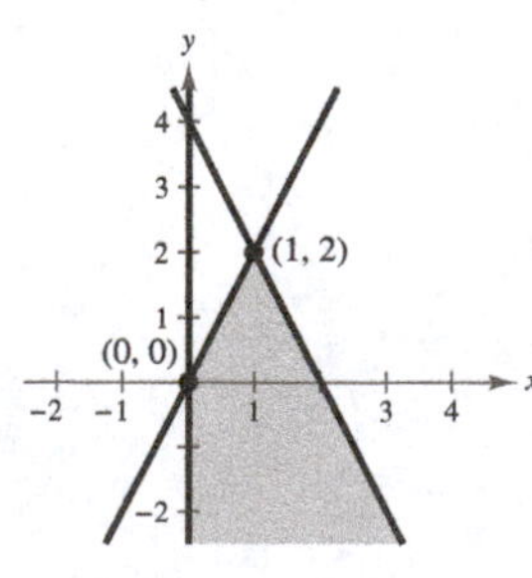

16. $\begin{cases} y < -x^2 + x + 4 \\ y > 4x \end{cases}$

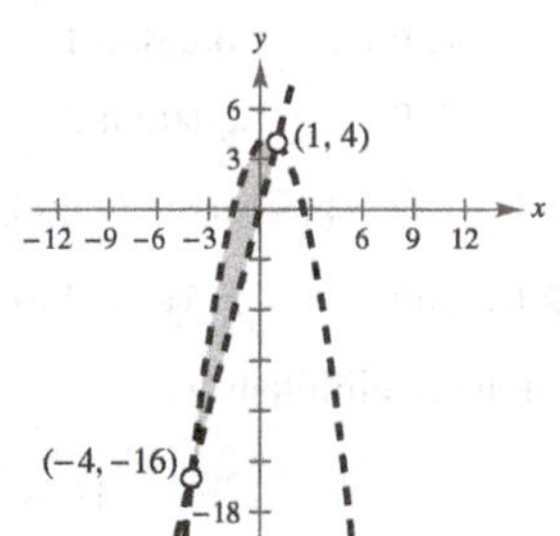

17. $\begin{cases} x^2 + y^2 \le 36 \\ x \ge 2 \\ y \ge -4 \end{cases}$

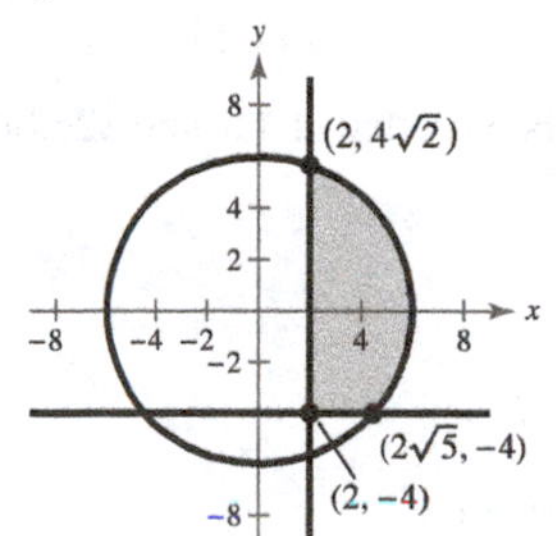

18. Maximize $z = 20x + 12y$ subject to:

$\begin{cases} x \ge 0, y \ge 0 \\ x + 4y \le 32 \\ 3x + 2y \le 36 \end{cases}$

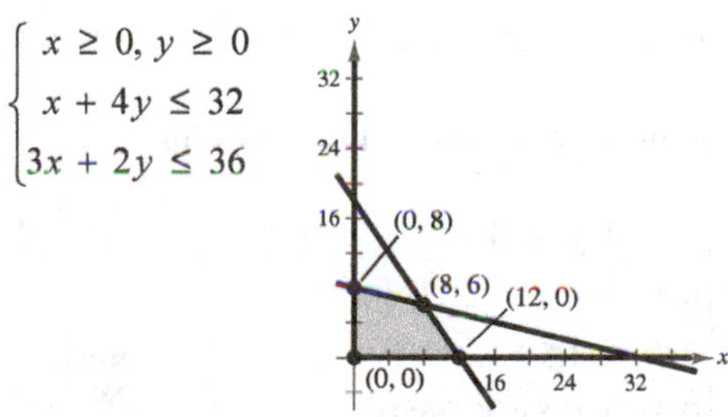

At $(0, 0)$ we have $z = 0$.

At $(0, 8)$ we have $z = 96$.

At $(8, 6)$ we have $z = 232$.

At $(12, 0)$ we have $z = 240$.

The maximum value, $z = 240$, occurs at $(12, 0)$.

The minimum value, $z = 0$ occurs at $(0, 0)$.

19. Let x = amount of money invested at 4%.

Let y = amount of money invested at 5.5%.

$$\begin{cases} x + y = 50{,}000 & \text{Equation 1} \\ 0.04x + 0.055y = 2390 & \text{Equation 2} \end{cases}$$

Multiply Equation 1 by -4: $-4x - 4y = -200{,}000$

Multiply Equation 2 by 100: $4x + 5.5y = 239{,}000$

Add these two equations to eliminate x:

$$\begin{aligned} -4x - 4y &= -200{,}000 \\ 4x + 5.5y &= 239{,}000 \\ \hline 1.5y &= 39{,}000 \\ y &= 26{,}000 \end{aligned}$$

Back-substitute $y = 26{,}000$ into Equation 1:

$$\begin{aligned} x + 26{,}000 &= 50{,}000 \\ x &= 24{,}000 \end{aligned}$$

So, \$24,000 should be invested at 4% and \$26,000 should be invested at 5.5%.

20. $y = ax^2 + bx + c$

$(0, 6)$: $6 = c$

$(-2, 2)$: $2 = 4a - 2b + c$

$\left(3, \frac{9}{2}\right)$: $\frac{9}{2} = 9a + 3b + c$

Solving this system yields: $a = -\frac{1}{2}$, $b = 1$, and $c = 6$.

So, $y = -\frac{1}{2}x^2 + x + 6$.

21. Optimize $P = 30x + 40y$ subject to:

$$\begin{cases} x \ge 0, y \ge 0 \\ 0.5x + 0.75y \le 3750 \\ 2.0x + 1.5y \le 8950 \\ 0.5x + 0.5y \le 2650 \end{cases}$$

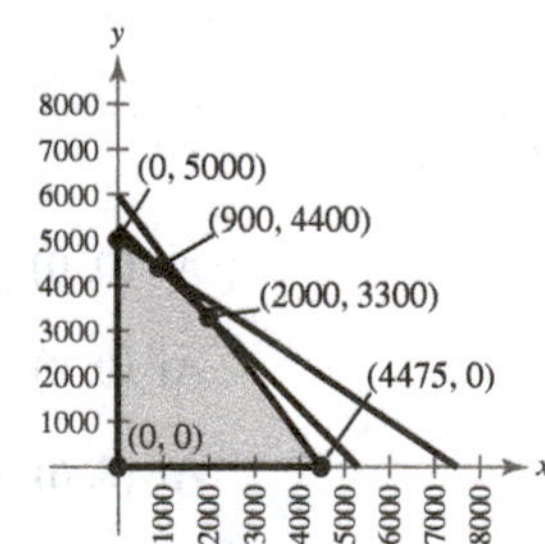

At $(0, 0)$: $P = 0$

At $(0, 5000)$: $P = \$200{,}000$

At $(900, 4400)$: $P = \$203{,}000$

At $(2000, 3300)$: $P = \$192{,}000$

At $(4475, 0)$: $P = \$134{,}250$

The manufacturer should produce 900 units of Model I and produce any of 4400 units of Model II to realize an optimal profit of \$203,000.

Chapter Test Solutions for Chapter 7

1. $\begin{bmatrix} 1 & -1 & 5 \\ 6 & 2 & 3 \\ 5 & 3 & -3 \end{bmatrix}$

$\begin{matrix} \\ -6R_1 + R_2 \rightarrow \\ -5R_1 + R_3 \rightarrow \end{matrix} \begin{bmatrix} 1 & -1 & 5 \\ 0 & 8 & -27 \\ 0 & 8 & -28 \end{bmatrix}$

$\begin{matrix} \\ \\ -R_2 + R_3 \rightarrow \end{matrix} \begin{bmatrix} 1 & -1 & 5 \\ 0 & 8 & -27 \\ 0 & 0 & -1 \end{bmatrix}$

$\begin{matrix} \\ \frac{1}{8}R_2 \rightarrow \\ -R_3 \rightarrow \end{matrix} \begin{bmatrix} 1 & -1 & 5 \\ 0 & 1 & -\frac{27}{8} \\ 0 & 0 & 1 \end{bmatrix}$

$\begin{matrix} R_2 + R_1 \rightarrow \\ \\ \\ \end{matrix} \begin{bmatrix} 1 & 0 & \frac{13}{8} \\ 0 & 1 & -\frac{27}{8} \\ 0 & 0 & 1 \end{bmatrix}$

$\begin{matrix} -\frac{13}{8}R_3 + R_1 \rightarrow \\ \frac{27}{8}R_3 + R_2 \rightarrow \\ \\ \end{matrix} \begin{bmatrix} 1 & 0 & 0 \\ 0 & 1 & 0 \\ 0 & 0 & 1 \end{bmatrix}$

2. $\begin{bmatrix} 1 & 0 & -1 & 2 \\ -1 & 1 & 1 & -3 \\ 1 & 1 & -1 & 1 \\ 3 & 2 & -3 & 4 \end{bmatrix}$

$\begin{matrix} \\ R_1 + R_2 \rightarrow \\ -R_1 + R_3 \rightarrow \\ -3R_1 + R_4 \rightarrow \end{matrix} \begin{bmatrix} 1 & 0 & -1 & 2 \\ 0 & 1 & 0 & -1 \\ 0 & 1 & 0 & -1 \\ 0 & 2 & 0 & -2 \end{bmatrix}$

$\begin{matrix} \\ \\ -R_2 + R_3 \rightarrow \\ -2R_2 + R_4 \rightarrow \end{matrix} \begin{bmatrix} 1 & 0 & -1 & 2 \\ 0 & 1 & 0 & -1 \\ 0 & 0 & 0 & 0 \\ 0 & 0 & 0 & 0 \end{bmatrix}$

3. $\left[\begin{array}{ccc:c} 4 & 3 & -2 & 14 \\ -1 & -1 & 2 & -5 \\ 3 & 1 & -4 & 8 \end{array}\right]$

$\begin{matrix} 3R_2 + R_1 \rightarrow \\ \\ \\ \end{matrix} \left[\begin{array}{ccc:c} 1 & 0 & 4 & -1 \\ -1 & -1 & 2 & -5 \\ 3 & 1 & -4 & 8 \end{array}\right]$

$\begin{matrix} \\ R_1 + R_2 \rightarrow \\ -3R_1 + R_3 \rightarrow \end{matrix} \left[\begin{array}{ccc:c} 1 & 0 & 4 & -1 \\ 0 & -1 & 6 & -6 \\ 0 & 1 & -16 & 11 \end{array}\right]$

$\begin{matrix} \\ \\ R_2 + R_3 \rightarrow \end{matrix} \left[\begin{array}{ccc:c} 1 & 0 & 4 & -1 \\ 0 & -1 & 6 & -6 \\ 0 & 0 & -10 & 5 \end{array}\right]$

$\begin{matrix} \\ -R_2 \rightarrow \\ -\frac{1}{10}R_3 \rightarrow \end{matrix} \left[\begin{array}{ccc:c} 1 & 0 & 4 & -1 \\ 0 & 1 & -6 & 6 \\ 0 & 0 & 1 & -\frac{1}{2} \end{array}\right]$

$\begin{matrix} -4R_3 + R_1 \rightarrow \\ 6R_3 + R_2 \rightarrow \\ \\ \end{matrix} \left[\begin{array}{ccc:c} 1 & 0 & 0 & 1 \\ 0 & 1 & 0 & 3 \\ 0 & 0 & 1 & -\frac{1}{2} \end{array}\right]$

Solution: $\left(1, 3, -\frac{1}{2}\right)$

4. $A = \begin{bmatrix} 6 & 5 \\ -5 & -5 \end{bmatrix}$, $B = \begin{bmatrix} 5 & 0 \\ -5 & -1 \end{bmatrix}$, $C = \begin{bmatrix} 2 & -1 & 4 \\ 0 & 6 & -3 \end{bmatrix}$

(a) $A - B = \begin{bmatrix} 6 & 5 \\ -5 & -5 \end{bmatrix} - \begin{bmatrix} 5 & 0 \\ -5 & -1 \end{bmatrix} = \begin{bmatrix} 6-5 & 5-0 \\ -5-(-5) & -5-(-1) \end{bmatrix} = \begin{bmatrix} 1 & 5 \\ 0 & -4 \end{bmatrix}$

(b) $3C = 3\begin{bmatrix} 2 & -1 & 4 \\ 0 & 6 & -3 \end{bmatrix} = \begin{bmatrix} 3(2) & 3(-1) & 3(4) \\ 3(0) & 3(6) & 3(-3) \end{bmatrix} = \begin{bmatrix} 6 & -3 & 12 \\ 0 & 18 & -9 \end{bmatrix}$

(c) $3A - 2B = 3\begin{bmatrix} 6 & 5 \\ -5 & -5 \end{bmatrix} - 2\begin{bmatrix} 5 & 0 \\ -5 & -1 \end{bmatrix} = \begin{bmatrix} 3(6)-2(5) & 3(5)-2(0) \\ 3(-5)-2(-5) & 3(-5)-2(-1) \end{bmatrix} = \begin{bmatrix} 8 & 15 \\ -5 & -13 \end{bmatrix}$

(d) $BC = \begin{bmatrix} 5 & 0 \\ -5 & -1 \end{bmatrix}\begin{bmatrix} 2 & -1 & 4 \\ 0 & 6 & -3 \end{bmatrix}$

$= \begin{bmatrix} (5)(2)+(0)(0) & (5)(-1)+(0)(6) & (5)(4)+(0)(-3) \\ (-5)(2)+(-1)(0) & (-5)(-1)+(-1)(6) & (-5)(4)+(-1)(-3) \end{bmatrix}$

$= \begin{bmatrix} 10 & -5 & 20 \\ -10 & -1 & -17 \end{bmatrix}$

(e) C^2 is not possible.

5. $A = \begin{bmatrix} a & b \\ c & d \end{bmatrix}$, $A^{-1} = \dfrac{1}{ad - bc}\begin{bmatrix} d & -b \\ -c & a \end{bmatrix}$

$A = \begin{bmatrix} -4 & 3 \\ 5 & -2 \end{bmatrix}$

$ad - bc = (-4)(-2) - (3)(5) = -7$

$A^{-1} = -\dfrac{1}{7}\begin{bmatrix} -2 & -3 \\ -5 & -4 \end{bmatrix} = \begin{bmatrix} \frac{2}{7} & \frac{3}{7} \\ \frac{5}{7} & \frac{4}{7} \end{bmatrix}$

6. $\left[\begin{array}{rrr|rrr} -2 & 4 & -6 & 1 & 0 & 0 \\ 2 & 1 & 0 & 0 & 1 & 0 \\ 4 & -2 & 5 & 0 & 0 & 1 \end{array}\right]$

$\begin{array}{r} \\ R_1 + R_2 \rightarrow \\ 2R_1 + R_3 \rightarrow \end{array}\left[\begin{array}{rrr|rrr} -2 & 4 & -6 & 1 & 0 & 0 \\ 0 & 5 & -6 & 1 & 1 & 0 \\ 0 & 6 & -7 & 2 & 0 & 1 \end{array}\right]$

$\begin{array}{r} -\frac{1}{2}R_1 \rightarrow \\ -R_3 + R_2 \rightarrow \\ \\ \end{array}\left[\begin{array}{rrr|rrr} 1 & -2 & 3 & -\frac{1}{2} & 0 & 0 \\ 0 & -1 & 1 & -1 & 1 & -1 \\ 0 & 6 & -7 & 2 & 0 & 1 \end{array}\right]$

$\begin{array}{r} -2R_2 + R_1 \rightarrow \\ \\ 6R_2 + R_3 \rightarrow \end{array}\left[\begin{array}{rrr|rrr} 1 & 0 & 1 & \frac{3}{2} & -2 & 2 \\ 0 & -1 & 1 & -1 & 1 & -1 \\ 0 & 0 & -1 & -4 & 6 & -5 \end{array}\right]$

$\begin{array}{r} \\ -R_2 \rightarrow \\ -R_3 \rightarrow \end{array}\left[\begin{array}{rrr|rrr} 1 & 0 & 1 & \frac{3}{2} & -2 & 2 \\ 0 & 1 & -1 & 1 & -1 & 1 \\ 0 & 0 & 1 & 4 & -6 & 5 \end{array}\right]$

$\begin{array}{r} -R_3 + R_1 \rightarrow \\ R_3 + R_2 \rightarrow \\ \\ \end{array}\left[\begin{array}{rrr|rrr} 1 & 0 & 0 & -\frac{5}{2} & 4 & -3 \\ 0 & 1 & 0 & 5 & -7 & 6 \\ 0 & 0 & 1 & 4 & -6 & 5 \end{array}\right]$

$A^{-1} = \begin{bmatrix} -\frac{5}{2} & 4 & -3 \\ 5 & -7 & 6 \\ 4 & -6 & 5 \end{bmatrix}$

7. $\begin{cases} -4x + 3y = 6 \\ 5x - 2y = 24 \end{cases}$

$$\begin{bmatrix} -4 & 3 \\ 5 & -2 \end{bmatrix}\begin{bmatrix} x \\ y \end{bmatrix} = \begin{bmatrix} 6 \\ 24 \end{bmatrix}$$

$$\begin{bmatrix} x \\ y \end{bmatrix} = \begin{bmatrix} -4 & 3 \\ 5 & -2 \end{bmatrix}^{-1}\begin{bmatrix} 6 \\ 24 \end{bmatrix} = \begin{bmatrix} \frac{2}{7} & \frac{3}{7} \\ \frac{5}{7} & \frac{4}{7} \end{bmatrix}\begin{bmatrix} 6 \\ 24 \end{bmatrix} = \begin{bmatrix} 12 \\ 18 \end{bmatrix}$$

Solution: $(12, 18)$

8. $\begin{vmatrix} -6 & 4 \\ 10 & 12 \end{vmatrix} = (-6)(12) - (4)(10) = -112$

9. $\begin{vmatrix} \frac{5}{2} & -\frac{3}{8} \\ -8 & \frac{6}{5} \end{vmatrix} = \left(\frac{5}{2}\right)\left(\frac{6}{5}\right) - \left(-\frac{3}{8}\right)(-8) = 3 - 3 = 0$

10. Expand along Column 3.

$$\begin{vmatrix} 6 & -7 & 2 \\ 3 & -2 & 0 \\ 1 & 5 & 1 \end{vmatrix} = 2\begin{vmatrix} 3 & -2 \\ 1 & 5 \end{vmatrix} + \begin{vmatrix} 6 & -7 \\ 3 & -2 \end{vmatrix} = 2(17) + 9 = 43$$

11. $\begin{cases} 7x + 6y = 9 \\ -2x - 11y = -49 \end{cases}$ $\quad D = \begin{vmatrix} 7 & 6 \\ -2 & -11 \end{vmatrix} = -65$

$$x = \frac{\begin{vmatrix} 9 & 6 \\ -49 & -11 \end{vmatrix}}{-65} = \frac{195}{-65} = -3$$

$$y = \frac{\begin{vmatrix} 7 & 9 \\ -2 & -49 \end{vmatrix}}{-65} = \frac{-325}{-65} = 5$$

Solution: $(-3, 5)$

12. $\begin{cases} 6x - y + 2z = -4 \\ -2x + 3y - z = 10 \\ 4x - 4y + z = -18 \end{cases}$ $\quad D = \begin{vmatrix} 6 & -1 & 2 \\ -2 & 3 & -1 \\ 4 & -4 & 1 \end{vmatrix} = -12$

$$x = \frac{\begin{vmatrix} -4 & -1 & 2 \\ 10 & 3 & -1 \\ -18 & -4 & 1 \end{vmatrix}}{-12} = \frac{24}{-12} = -2$$

$$y = \frac{\begin{vmatrix} 6 & -4 & 2 \\ -2 & 10 & -1 \\ 4 & -18 & 1 \end{vmatrix}}{-12} = \frac{-48}{-12} = 4$$

$$z = \frac{\begin{vmatrix} 6 & -1 & -4 \\ -2 & 3 & 10 \\ 4 & -4 & -18 \end{vmatrix}}{-12} = \frac{-72}{-12} = 6$$

Solution: $(-2, 4, 6)$

13. $A = -\frac{1}{2}\begin{vmatrix} -5 & 0 & 1 \\ 4 & 4 & 1 \\ 3 & 2 & 1 \end{vmatrix} = -\frac{1}{2}(-14) = 7$

14.
$$\begin{matrix} \text{K} & \text{N} & \text{O} \\ \text{C} & \text{K} & - \\ \text{O} & \text{N} & - \\ \text{W} & \text{O} & \text{O} \\ \text{D} & - & - \end{matrix}\begin{bmatrix} 11 & 14 & 15 \\ 3 & 11 & 0 \\ 15 & 14 & 0 \\ 23 & 15 & 15 \\ 4 & 0 & 0 \end{bmatrix}\begin{bmatrix} 1 & -1 & 0 \\ 1 & 0 & -1 \\ 6 & -2 & -3 \end{bmatrix} = \begin{bmatrix} 115 & -41 & -59 \\ 14 & -3 & -11 \\ 29 & -15 & -14 \\ 128 & -53 & -60 \\ 4 & -4 & 0 \end{bmatrix}$$

Message: $[11\ \ 14\ \ 15], [3\ \ 11\ \ 0], [15\ \ 14\ \ 0], [23\ \ 15\ \ 15], [4\ \ 0\ \ 0]$

Encoded Message: 115 −41 −59 14 −3 −11 29 −15 −14 128 −53 −60 4 −4 0

15. Let x = amount of 60% solution and y = amount of 20% solution.

$$\begin{cases} x + y = 100 \Rightarrow y = 100 - x \\ 0.60x + 0.20y = 0.50(100) \Rightarrow 6x + 2y = 500 \end{cases}$$

By substitution,

$$\begin{aligned} 6x + 2(100 - x) &= 500 \\ 6x + 200 - 2x &= 500 \\ 4x &= 300 \\ x &= 75 \\ y &= 100 - x = 25. \end{aligned}$$

75 liters of 60% solution and 25 liters of 20% solution

Chapter Test Solutions for Chapter 8

1. $a_n = \dfrac{(-1)^n}{3n + 2}$

$a_1 = -\dfrac{1}{5}$

$a_2 = \dfrac{1}{8}$

$a_3 = -\dfrac{1}{11}$

$a_4 = \dfrac{1}{14}$

$a_5 = -\dfrac{1}{17}$

2. $\dfrac{3}{1!}, \dfrac{4}{2!}, \dfrac{5}{3!}, \dfrac{6}{4!}, \dfrac{7}{5!}, \ldots$

$a_n = \dfrac{n + 2}{n!}$

3. $8 + 21 + 34 + 47 + \ldots$

$a_5 = 60, a_6 = 73, a_7 = 86$

$S_7 = 8 + 21 + 34 + 47 + 60 + 73 + 86 = 329$

4. $a_5 = 45, a_{12} = 24$

$$\begin{aligned} a_{12} &= a_5 + 7d \\ 24 &= 45 + 7d \\ -21 &= 7d \\ -3 &= d \\ a_1 &= a_5 - 4d \\ a_1 &= 45 - 4(-3) \\ &= 57 \\ a_n &= a_1 + (n - 1)d \\ &= 57 + (n - 1)(-3) \\ &= -3n + 60 \end{aligned}$$

5. $a_2 = 14, a_6 = 224$

$$\begin{aligned} a_6 &= a_2 r^4 \\ 224 &= 14r^4 \\ 16 &= r^4 \\ 2 &= r \\ a_2 &= a_1 r \\ 14 &= a_1(2) \\ 7 &= a_1 \\ a_n &= 7(2)^{n-1} \end{aligned}$$

or

$$\begin{aligned} &= 7 \cdot 2^n 2^{-1} \\ &= \frac{7}{2}(2)^n \end{aligned}$$

6. $$\begin{aligned} \sum_{i=1}^{50} (2i^2 + 5) &= 2\sum_{i=1}^{50} i^2 + \sum_{i=1}^{50} 5 \\ &= 2\left[\frac{50(51)(101)}{6}\right] + 50(5) \\ &= 86{,}100 \end{aligned}$$

7. $\sum_{n=1}^{9}(12n-7) = 12\sum_{n=1}^{9} n - \sum_{n=1}^{9} 7$

$= 12\left[\frac{9(10)}{2}\right] - 9(7)$

$= 477$

8. $\sum_{i=1}^{\infty} 4\left(\frac{1}{2}\right)^i = \frac{2}{1-\frac{1}{2}} = 4$

9. $\sum_{n=1}^{\infty}\left(-\frac{1}{3}\right)^n = \frac{-\frac{1}{3}}{1-\left(-\frac{1}{3}\right)} = \frac{-\frac{1}{3}}{\frac{4}{3}} = -\frac{1}{4}$

10. $5 + 10 + 15 + \cdots + 5n = \frac{5n(n+1)}{2}$

When $n = 1$, $S_1 = 5 = \frac{5(1)(2)}{2}$, so the formula is valid.

Assume that

$S_k = 5 + 10 + 15 + \cdots + 5k = \frac{5k(k+1)}{2}$, then

$S_{k+1} = S_k + a_{k+1}$

$= \frac{5k(k+1)}{2} + 5(k+1)$

$= \frac{5k(k+1)}{2} + \frac{10(k+1)}{2}$

$= \frac{5k(k+1) + 10(k+1)}{2}$

$= \frac{5(k+1)(k+2)}{2}$

$= \frac{5(k+1)[(k+1)+1]}{2}$.

So, the formula is valid for all integers $n \geq 1$.

11. $(x+6y)^4 = x^4 + {}_4C_1x^3(6y) + {}_4C_2x^2(6y)^2 + {}_4C_3x(6y)^3 + {}_4C_4(6y)^4$

$= x^4 + 24x^3y + 216x^2y^2 + 864xy^3 + 1296y^4$

12. 5th Row of Pascal's Triangle: 1 5 10 10 5 1 and 3rd Row of Pascal's Triangle: 1 3 3 1

$3(x-2)^5 + 4(x-2)^3 = 3\left[(1)x^5 + (5)x^4(-2) + (10)x^3(-2)^2 + (10)x^2(-2)^3 + (5)x(-2)^4 + (1)(-2)^5\right]$

$+ 4\left[(1)x^3 + (3)x^2(-2) + (3)x(-2)^2 + (1)(-2)^3\right]$

$= 3(x^5 - 10x^4 + 40x^3 - 80x^2 + 80x - 32) + 4(x^3 - 6x^2 + 12x - 8)$

$= 3x^5 - 30x^4 + 124x^3 - 264x^2 + 288x - 128$

13. ${}_nC_rx^{n-r}y^r = {}_7C_3(3a)^4(-2b)^3$

$= 35(81a^4)(-8b^3)$

$= -22{,}680a^4b^3$

So, the coefficient of a^4b^3 is $-22{,}680$.

14. (a) ${}_9P_2 = \frac{9!}{7!} = 72$

(b) ${}_{70}P_3 = \frac{70!}{67!} = 328{,}440$

15. (a) ${}_{11}C_4 = \frac{11!}{7!4!} = 330$

(b) ${}_{66}C_4 = \frac{66!}{62!4!} = 720{,}720$

16. $(26)(10)(10)(10) = 26{,}000$ distinct license plates

17. $\underbrace{(1)}_{\text{owner}} \cdot \underbrace{(3)(2)}_{\text{bow seats}} \cdot \underbrace{(5)(4)(3)(2)(1)}_{\text{remaining seats}} = 720$ seating arrangements

18. $\frac{20}{300} = \frac{1}{15} \approx 0.0667$

19. $\frac{1}{{}_{30}C_4} = \frac{1}{27{,}405}$

20. $P(E') = 1 - P(E)$

$= 1 - 0.90$

$= 0.10$ or 10%

Cumulative Test Solutions for Chapters 6–8

1. $\begin{cases} y = 3 - x^2 \\ 2(y - 2) = x - 1 \Rightarrow 2(3 - x^2 - 2) = x - 1 \end{cases}$

$$2(1 - x^2) = x - 1$$
$$2 - 2x^2 = x - 1$$
$$0 = 2x^2 + x - 3$$
$$0 = (2x + 3)(x - 1)$$
$$x = -\tfrac{3}{2} \text{ or } x = 1$$
$$y = \tfrac{3}{4} \qquad y = 2$$

Solutions: $\left(-\frac{3}{2}, \frac{3}{4}\right)$, $(1, 2)$

2. $\begin{cases} x + 3y = -6 \Rightarrow 4x + 12y = -24 \\ 2x + 4y = -10 \Rightarrow -6x - 12y = 30 \end{cases}$

$$-2x = 6$$
$$x = -3 \Rightarrow y = -1$$

Solution: $(-3, -1)$

3. $\begin{cases} -2x + 4y - z = -16 \\ x - 2y + 2z = 5 \\ x - 3y - z = 13 \end{cases}$

Interchange equations.

$\begin{cases} x - 2y + 2z = 5 & \text{Eq.1} \\ -2x + 4y - z = -16 & \text{Eq.2} \\ x - 3y - z = 13 & \text{Eq.3} \end{cases}$

$\begin{cases} x - 2y + 2z = 5 & \\ 3z = -6 & 2\text{Eq.1} + \text{Eq.2} \\ -y - 3z = 8 & -\text{Eq.1} + \text{Eq.3} \end{cases}$

From Equation 2, $z = -2$. Substituting this into Equation 3 yields $y = -2$. Using these in Equation 1 yields $x = 5$.

Solution: $(5, -2, -2)$

4. $\begin{cases} x + 3y - 2z = -7 \\ -2x + y - z = -5 \\ 4x + y + z = 3 \end{cases}$

$\begin{cases} x + 3y - 2z = -7 & \\ 7y - 5z = -19 & 2\text{Eq. }1 + \text{Eq.}2 \\ -11y + 9z = 31 & -4\text{Eq. }1 + \text{Eq.}3 \end{cases}$

$\begin{cases} x + 3y - 2z = -7 & \\ y - \frac{5}{7}z = -\frac{19}{7} & \frac{1}{7}\text{Eq.}2 \\ -11y + 9z = 31 & \end{cases}$

$\begin{cases} x + \frac{1}{7}z = \frac{8}{7} & -3\text{Eq.}2 + \text{Eq.}1 \\ y - \frac{5}{7}z = -\frac{19}{7} & \\ \frac{8}{7}z = \frac{8}{7} & 11\text{Eq.}2 + \text{Eq.}3 \end{cases}$

$\begin{cases} x + \frac{1}{7}z = \frac{8}{7} & \\ y - \frac{5}{7}z = -\frac{19}{7} & \\ z = 1 & \frac{7}{8}\text{Eq.}3 \end{cases}$

$\begin{cases} x = 1 & -\frac{1}{7}\text{Eq.}3 + \text{Eq.}1 \\ y = -2 & \frac{5}{7}\text{Eq.}3 + \text{Eq.}2 \\ z = 1 & \end{cases}$

Solution: $(1, -2, 1)$

5. $\begin{cases} x + y = 200 \Rightarrow y = 200 - x \\ 0.75x + 1.25y = 0.95(200) \end{cases}$

$$0.75x + 1.25(200 - x) = 190$$
$$0.75x + 250 - 1.25x = 190$$
$$-0.50x = -60$$
$$x = 120$$
$$y = 200 - x = 80$$

120 pounds of \$0.75 seed and 80 pounds of \$1.25 seed.

6. $y = ax^2 + bx + c$

$(0, 6)$: $6 = a(0)^2 + b(0) + c \Rightarrow c = 6$

$(2, 3)$: $3 = a(2)^2 + b(2) + 6 \Rightarrow 4a + 2b = -3$

$$2a + b = -\tfrac{3}{2}$$

$(4, 2)$: $2 = a(4)^2 + b(4) + 6 \Rightarrow 16a + 4b = -4$

$$4a + b = -1$$

Solving the system:

$\begin{cases} 2a + b = -\frac{3}{2} \\ 4a + b = -1 \end{cases}$ yields $a = \frac{1}{4}$ and $b = -2$.

So, the equation of the parabola is $y = \frac{1}{4}x^2 - 2x + 6$.

7. $\dfrac{2x^2 - x - 6}{x(x^2 + 2)} = \dfrac{A}{x} + \dfrac{Bx + C}{x^2 + 2}$

$$\begin{aligned} 2x^2 - x - 6 &= A(x^2 + 2) + x(Bx + C) \\ &= Ax^2 + 2A + Bx^2 + Cx \\ &= (A + B)x^2 + Cx + 2A \\ 2A &= -6 \rightarrow A = -3 \\ C &= -1 \\ A + B &= 2 \rightarrow B = 5 \end{aligned}$$

$$\frac{2x^2 - x - 6}{x(x^2 + 2)} = -\frac{3}{x} + \frac{5x - 1}{x^2 + 2}$$

8. $\begin{cases} 2x + \ \ y \geq -3 \\ \ \ x - 3y \leq \ \ 2 \end{cases}$

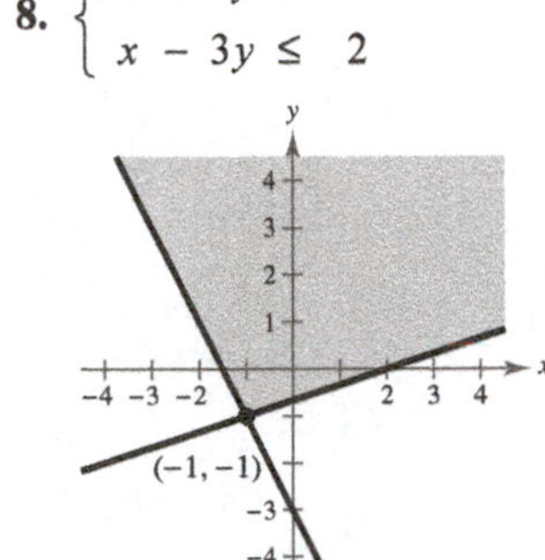

9. $\begin{cases} \ \ x - \ \ y > 6 \\ 5x + 2y < 10 \end{cases}$

$\left(\frac{22}{7}, -\frac{20}{7}\right)$

10. Objective function: $z = 3x + 2y$

Subject to: $x + 4y \leq 20$

$2x + y \leq 12$

$x \geq 0, y \geq 0$

At $(0, 0)$: $z = 0$

At $(0, 5)$: $z = 10$

At $(4, 4)$: $z = 30$

At $(6, 0)$: $z = 18$

Minimum of $z = 0$ at $(0, 0)$

Maximum of $z = 20$ at $(4, 4)$

11. $\begin{cases} -x + 2y - \ \ z = \ \ 9 \\ 2x - \ \ y + 2z = -9 \\ 3x + 3y - 4z = \ \ 7 \end{cases}$

$$\left[\begin{array}{ccc:c} -1 & 2 & -1 & 9 \\ 2 & -1 & 2 & -9 \\ 3 & 3 & -4 & 7 \end{array}\right]$$

12.

$$\left[\begin{array}{ccc:c} -1 & 2 & -1 & 9 \\ 2 & -1 & 2 & -9 \\ 3 & 3 & -4 & 7 \end{array}\right]$$

$$\begin{array}{r} \\ 2R_1 + R_2 \rightarrow \\ 3R_1 + R_3 \rightarrow \end{array} \left[\begin{array}{ccc:c} -1 & 2 & -1 & 9 \\ 0 & 3 & 0 & 9 \\ 0 & 9 & -7 & 34 \end{array}\right]$$

$$\begin{array}{r} -R_1 \rightarrow \\ \\ -3R_2 + R_3 \rightarrow \end{array} \left[\begin{array}{ccc:c} 1 & -2 & 1 & -9 \\ 0 & 3 & 0 & 3 \\ 0 & 0 & -7 & 7 \end{array}\right]$$

$$\begin{array}{r} \\ \frac{1}{3}R_2 \rightarrow \\ -\frac{1}{7}R_2 \rightarrow \end{array} \left[\begin{array}{ccc:c} 1 & -2 & 1 & -9 \\ 0 & 1 & 0 & 3 \\ 0 & 0 & 1 & -1 \end{array}\right]$$

$$\begin{array}{r} 2R_2 + R_1 \rightarrow \\ \\ \\ \end{array} \left[\begin{array}{ccc:c} 1 & 0 & 1 & -3 \\ 0 & 1 & 0 & 3 \\ 0 & 0 & 1 & -1 \end{array}\right]$$

$$\begin{array}{r} -R_3 + R_1 \rightarrow \\ \\ \\ \end{array} \left[\begin{array}{ccc:c} 1 & 0 & 0 & -2 \\ 0 & 1 & 0 & 3 \\ 0 & 0 & 1 & -1 \end{array}\right]$$

Solution: $(-2, 3, -1)$

In Exercises 13-18,

$$A = \begin{bmatrix} -1 & 3 \\ 6 & 2 \end{bmatrix}, B = \begin{bmatrix} -2 & 5 \\ 0 & -1 \end{bmatrix} \text{ and } C = \begin{bmatrix} 4 & 0 & 1 \\ -3 & 2 & -1 \end{bmatrix}$$

13. $A + B = \begin{bmatrix} -1 & 3 \\ 6 & 2 \end{bmatrix} + \begin{bmatrix} -2 & 5 \\ 0 & -1 \end{bmatrix} = \begin{bmatrix} -3 & 8 \\ 6 & 1 \end{bmatrix}$

14.
$$\begin{aligned} 2A - 5B &= 2A + (-5)B \\ &= 2\begin{bmatrix} -1 & 3 \\ 6 & 2 \end{bmatrix} + (-5)\begin{bmatrix} -2 & 5 \\ 0 & -1 \end{bmatrix} \\ &= \begin{bmatrix} -2 & 6 \\ 12 & 4 \end{bmatrix} + \begin{bmatrix} 10 & -25 \\ 0 & 5 \end{bmatrix} \\ &= \begin{bmatrix} 8 & -19 \\ 12 & 9 \end{bmatrix} \end{aligned}$$

15. $AC = \begin{bmatrix} -1 & 3 \\ 6 & 2 \end{bmatrix}\begin{bmatrix} 4 & 0 & 1 \\ -3 & 2 & -1 \end{bmatrix} = \begin{bmatrix} (-1)(4) + (3)(-3) & (-1)(0) + (3)(2) & (-1)(1) + (3)(-1) \\ (6)(4) + (2)(-3) & (6)(0) + (2)(2) & (6)(1) + (2)(-1) \end{bmatrix} = \begin{bmatrix} -13 & 6 & -4 \\ 18 & 4 & 4 \end{bmatrix}$

16. CB not possible. The number of columns of C is not equal to the number of rows of B.

17. $A^2 = \begin{bmatrix} -1 & 3 \\ 6 & 2 \end{bmatrix}\begin{bmatrix} -1 & 3 \\ 6 & 2 \end{bmatrix} = \begin{bmatrix} (-1)(-1) + (3)(6) & (-1)(3) + (3)(2) \\ (6)(-1) + (2)(6) & (6)(3) + (2)(2) \end{bmatrix} = \begin{bmatrix} 19 & 3 \\ 6 & 22 \end{bmatrix}$

18.
$$\begin{aligned} BA - B^2 &= \begin{bmatrix} -2 & 5 \\ 0 & -1 \end{bmatrix}\begin{bmatrix} -1 & 3 \\ 6 & 2 \end{bmatrix} - \begin{bmatrix} -2 & 5 \\ 0 & -1 \end{bmatrix}\begin{bmatrix} -2 & 5 \\ 0 & -1 \end{bmatrix} \\ &= \begin{bmatrix} -2(-1) + 5(6) & -2(3) + 5(2) \\ 0(-1) + (-1)(6) & 0(3) + (-1)(2) \end{bmatrix} - \begin{bmatrix} -2(-2) + 5(0) & -2(5) + 5(-1) \\ 0(-2) + (-1)(0) & 0(5) + (-1)(-1) \end{bmatrix} \\ &= \begin{bmatrix} 32 & 4 \\ -6 & -2 \end{bmatrix} - \begin{bmatrix} 4 & -15 \\ 0 & 1 \end{bmatrix} \\ &= \begin{bmatrix} 28 & 19 \\ -6 & -3 \end{bmatrix} \end{aligned}$$

19.
$$\left[\begin{array}{ccc:ccc} 1 & 2 & -1 & 1 & 0 & 0 \\ 3 & 7 & -10 & 0 & 1 & 0 \\ -5 & -7 & -15 & 0 & 0 & 1 \end{array}\right]$$

$$\begin{array}{r} \\ -3R_1 + R_2 \rightarrow \\ 5R_1 + R_3 \rightarrow \end{array}\left[\begin{array}{ccc:ccc} 1 & 2 & -1 & 1 & 0 & 0 \\ 0 & 1 & -7 & -3 & 1 & 0 \\ 0 & 3 & -20 & 5 & 0 & 1 \end{array}\right]$$

$$\begin{array}{r} -2R_2 + R_1 \rightarrow \\ \\ -3R_2 + R_3 \rightarrow \end{array}\left[\begin{array}{ccc:ccc} 1 & 0 & 13 & 7 & -2 & 0 \\ 0 & 1 & -7 & -3 & 1 & 0 \\ 0 & 0 & 1 & 14 & -3 & 1 \end{array}\right]$$

$$\begin{array}{r} -13R_3 + R_1 \rightarrow \\ 7R_3 + R_2 \rightarrow \\ \\ \end{array}\left[\begin{array}{ccc:ccc} 1 & 0 & 0 & -175 & 37 & -13 \\ 0 & 1 & 0 & 95 & -20 & 7 \\ 0 & 0 & 1 & 14 & -3 & 1 \end{array}\right]$$

$$\begin{bmatrix} 1 & 2 & -1 \\ 3 & 7 & -10 \\ -5 & -7 & -15 \end{bmatrix}^{-1} = \begin{bmatrix} -175 & 37 & -13 \\ 95 & -20 & 7 \\ 14 & -3 & 1 \end{bmatrix}$$

20. Expand along Row 1.

$$\begin{vmatrix} 7 & 1 & 0 \\ -2 & 4 & -1 \\ 3 & 8 & 5 \end{vmatrix} = 7\begin{vmatrix} 4 & -1 \\ 8 & 5 \end{vmatrix} - 1\begin{vmatrix} -2 & -1 \\ 3 & 5 \end{vmatrix} = 7(28) - 1(-7) = 203$$

21. To produce a reflection in the x-axis, use the matrix $\begin{bmatrix} 1 & 0 \\ 0 & -1 \end{bmatrix}$ and multiply by each vertex matrix $\begin{bmatrix} 0 \\ 2 \end{bmatrix}, \begin{bmatrix} 0 \\ 5 \end{bmatrix}, \begin{bmatrix} 3 \\ 2 \end{bmatrix}, \begin{bmatrix} 3 \\ 5 \end{bmatrix}$.

$$\begin{bmatrix} 1 & 0 \\ 0 & -1 \end{bmatrix}\begin{bmatrix} 0 \\ 2 \end{bmatrix} = \begin{bmatrix} (1)(0) + (0)(2) \\ (0)(0) + (-1)(2) \end{bmatrix} = \begin{bmatrix} 0 \\ -2 \end{bmatrix}$$

$$\begin{bmatrix} 1 & 0 \\ 0 & -1 \end{bmatrix}\begin{bmatrix} 0 \\ 5 \end{bmatrix} = \begin{bmatrix} (1)(0) + (0)(5) \\ (0)(0) + (-1)(5) \end{bmatrix} = \begin{bmatrix} 0 \\ -5 \end{bmatrix}$$

$$\begin{bmatrix} 1 & 0 \\ 0 & -1 \end{bmatrix}\begin{bmatrix} 3 \\ 2 \end{bmatrix} = \begin{bmatrix} (1)(3) + (0)(2) \\ (0)(3) + (-1)(2) \end{bmatrix} = \begin{bmatrix} 3 \\ -2 \end{bmatrix}$$

$$\begin{bmatrix} 1 & 0 \\ 0 & -1 \end{bmatrix}\begin{bmatrix} 3 \\ 5 \end{bmatrix} = \begin{bmatrix} (1)(3) + (0)(5) \\ (0)(3) + (-1)(5) \end{bmatrix} = \begin{bmatrix} 3 \\ -5 \end{bmatrix}$$

22. Let x = total sales of gym shoes (in millions),
y = total sales of jogging shoes (in millions),
z = total sales of walking shoes (in millions).

$$\begin{bmatrix} 0.079 & 0.064 & 0.029 \\ 0.050 & 0.060 & 0.020 \\ 0.103 & 0.159 & 0.085 \end{bmatrix}\begin{bmatrix} x \\ y \\ z \end{bmatrix} = \begin{bmatrix} 479.88 \\ 365.88 \\ 1248.89 \end{bmatrix}$$

$$\begin{bmatrix} x \\ y \\ z \end{bmatrix} = \begin{bmatrix} 0.079 & 0.064 & 0.029 \\ 0.050 & 0.060 & 0.020 \\ 0.103 & 0.159 & 0.085 \end{bmatrix}^{-1}\begin{bmatrix} 479.88 \\ 365.88 \\ 1248.89 \end{bmatrix} \approx \begin{bmatrix} 2539 \\ 2362 \\ 4418 \end{bmatrix}$$

So, sales for each type of shoe amounted to:
Gym shoes: \$2539 million
Jogging shoes: \$2362 million
Walking shoes: \$4418 million

23. $\begin{cases} 8x - 3y = -52 \\ 3x + 5y = 5 \end{cases}$, $\quad D = \begin{vmatrix} 8 & -3 \\ 3 & 5 \end{vmatrix} = 49$

$$x = \frac{\begin{vmatrix} -52 & -3 \\ 5 & 5 \end{vmatrix}}{49} = \frac{-245}{49} = -5$$

$$y = \frac{\begin{vmatrix} 8 & -52 \\ 3 & 5 \end{vmatrix}}{49} = \frac{196}{49} = 4$$

Solution: $(-5, 4)$

24. $\begin{cases} 5x + 4y + 3z = 7 \\ -3x - 8y + 7z = -9, \\ 7x - 5y - 6z = -53 \end{cases}$ $\quad D = \begin{vmatrix} 5 & 4 & 3 \\ -3 & -8 & 7 \\ 7 & -5 & -6 \end{vmatrix} = 752$

$$x = \frac{\begin{vmatrix} 7 & 4 & 3 \\ -9 & -8 & 7 \\ -53 & -5 & -6 \end{vmatrix}}{752} = \frac{-2256}{752} = -3$$

$$y = \frac{\begin{vmatrix} 5 & 7 & 3 \\ -3 & -9 & 7 \\ 7 & -53 & -6 \end{vmatrix}}{752} = \frac{3008}{752} = 4$$

$$z = \frac{\begin{vmatrix} 5 & 4 & 7 \\ -3 & -8 & -9 \\ 7 & -5 & -53 \end{vmatrix}}{752} = \frac{1504}{752} = 2$$

Solution: $(-3, 4, 2)$

25. $A = \pm\frac{1}{2}\begin{vmatrix} -2 & 3 & 1 \\ 1 & 5 & 1 \\ 4 & 1 & 1 \end{vmatrix} = -\frac{1}{2}(-18) = 9$

26. $a_n = \frac{(-1)^{n+1}}{2n+3}$

$a_1 = \frac{1}{5}$

$a_2 = -\frac{1}{7}$

$a_3 = \frac{1}{9}$

$a_4 = -\frac{1}{11}$

$a_5 = \frac{1}{13}$

27. $\frac{2!}{4}, \frac{3!}{5}, \frac{4!}{6}, \frac{5!}{7}, \frac{6!}{8}, \ldots$

$a_n = \frac{(n+1)!}{n+3}$

28. 6, 18, 30, 42, …

$a_n = 12n - 6$

$a_1 = 6, a_{16} = 186$

$S_{16} = \frac{16}{2}(6 + 186) = 1536$

29. (a)

$$\begin{aligned} a_6 &= 20.6 \\ a_9 &= 30.2 \\ a_9 &= a_6 + 3d \\ 30.2 &= 20.6 + 3d \\ 9.6 &= 3d \\ 3.2 &= d \\ a_{20} &= a_9 + 11d = 30.2 + 11(3.2) = 65.4 \end{aligned}$$

(b)

$$\begin{aligned} a_1 &= a_6 - 5d \\ a_1 &= 20.6 - 5(3.2) \\ &= 4.6 \\ a_n &= a_1 + (n-1)d \\ &= 4.6 + (n-1)(3.2) \\ &= 3.2n + 1.4 \end{aligned}$$

30. $a_n = 3(2)^{n-1}$

$a_1 = 3$

$a_2 = 6$

$a_3 = 12$

$a_4 = 24$

$a_5 = 48$

31.

$$\begin{aligned} \sum_{i=0}^{\infty} 1.9\left(\frac{1}{10}\right)^{i-1} &= \sum_{i=0}^{\infty} 1.9\left(\frac{1}{10}\right)^{i}\left(\frac{1}{10}\right)^{-1} \\ &= \sum_{i=0}^{\infty} 19\left(\frac{1}{10}\right)^{i} \\ &= \frac{19}{1 - \frac{1}{10}} = 19\left(\frac{10}{9}\right) = \frac{190}{9} \end{aligned}$$

32. 1. When $n = 2$, $3! = 6$ and $2^2 = 4$, thus $3! > 2^2$.

2. Assume

$(k+1)! > 2^k, k > 2.$

Then, we need to show that $(k+2)! > 2^{k+1}$.

$(k+2)! = (k+1)!(k+2) > 2^k(2)$ since $k + 2 > 2$.

Thus, $(k+2)! > 2^{k+1}$.

Therefore, by mathematical induction, the formula is valid for all integers n such that $n \geq 2$.

33.

$$\begin{aligned} (w-9)^4 &= w^4 + {}_4C_1w^3(-9) + {}_4C_2w^2(-9)^2 + {}_4C_3w(-9)^3 + (-9)^4 \\ &= w^4 - 36w^3 + 486w^2 - 2916w + 6561 \end{aligned}$$

34. ${}_{14}P_3 = \frac{14!}{(14-3)!} = \frac{14!}{11!} = 2184$

35. ${}_{25}P_2 = \frac{25!}{(25-2)!} = \frac{25!}{23!} = 600$

36. $\binom{8}{4} = {}_8C_4 = \frac{8!}{(8-4)!4!} = \frac{8!}{4!4!} = 70$

37. ${}_{11}C_6 = \frac{11!}{(11-6)!6!} = \frac{11!}{5!6!} = 462$

38. B A S K E T B A L L

$$\frac{10!}{2!2!2!1!1!1!1!} = 453{,}600 \text{ distinguishable permutations}$$

39. A N T A R C T I C A

$$\frac{10!}{3!2!2!1!1!1!} = 151{,}200 \text{ distinguishable permutations}$$

40. $_{10}P_3 = \dfrac{10!}{(10 - 3)!} = \dfrac{10!}{7!} = 720$

41. The first digit is 4 or 5, so the probability of picking it correctly is $\frac{1}{2}$. Then there are two numbers left for the second digit so its probability is also $\frac{1}{2}$. If these two are correct, then the third digit must be the remaining number. The probability of winning is

$\left(\frac{1}{2}\right)\left(\frac{1}{2}\right)(1) = \frac{1}{4}$.